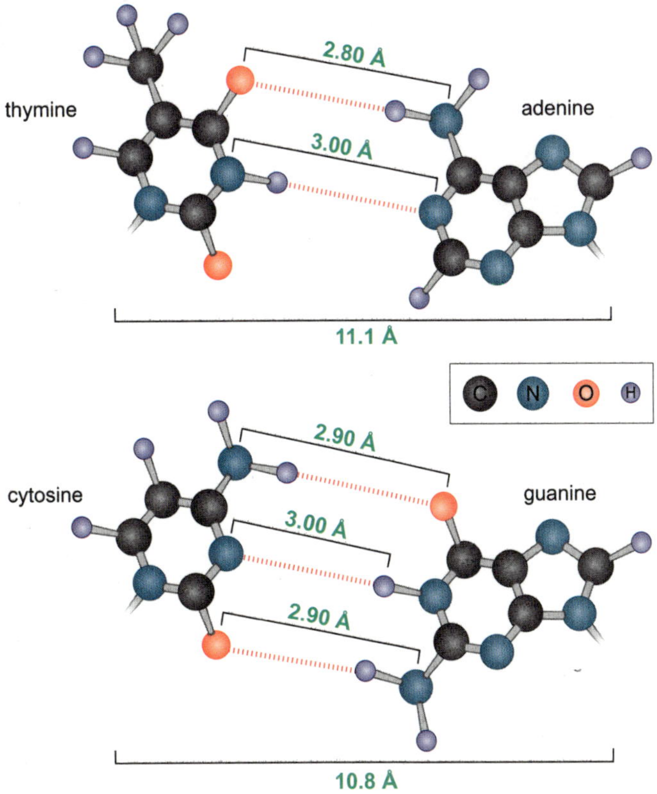

The position and length of the hydrogen bonds between the base pairs.

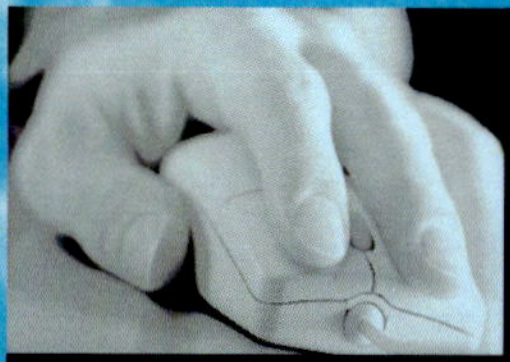

Watson, Baker, Bell, Gann, Levine, Losick
Molecular Biology of the Gene, 6e

Log on.

Explore.

Succeed.

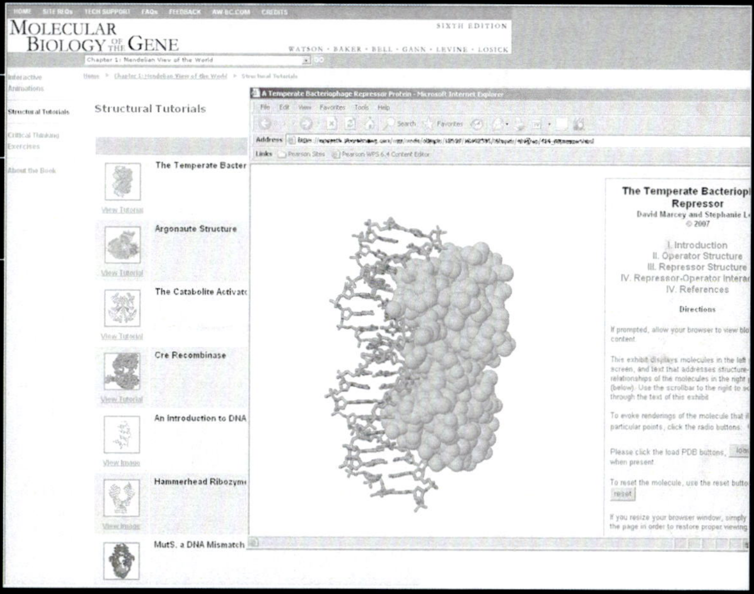

To help you succeed in molecular biology, your professor has arranged for you to enjoy access to a great media resource, http://www.aw-bc.com/watson. You'll find that this website, which accompanies your textbook, will enhance your understanding of course concepts.

What your system needs to use these media resources:

WINDOWS
Windows 2000/XP
1024x768 screen resolution
Thousands of colors
Browsers: Internet Explorer 6.x; Firefox 7.x
Plug-ins: Latest versions of Flash, Shockwave, QuickTime, and Acrobat Reader

MACINTOSH
Mac OS 10.2.4, 10.3.2
1024x768 screen resolution
Thousands of colors.
Browsers: Firefox 1.x; Safari 1.3
Plug-ins: Latest versions of Flash, Shockwave, QuickTime, and Acrobat Reader

How to log on to http://www.aw-bc.com/watson

1. Go to http://www.aw-bc.com/watson.
2. Click *Molecular Biology of the Gene, 6e*
3. Click "Register."
4. Scratch off the silver foil coating below to reveal your pre-assigned access code.
5. Enter your pre-assigned access code exactly as it appears below.
6. Complete the online registration form to create your own personal user Login Name and Password.
7. Once your personal Login Name and Password are confirmed by email, go back to http://www.aw-bc.com/watson, type in your new Login Name and Password, and click "Log In."

Your Access Code is:

*

If there is no silver foil covering the access code above, the code may no longer be valid. In that case, you need to purchase access online using a major credit card.
Go to http://www.aw-bc.com/watson, and click **Buy Now**.

Technical Questions?

For technical support, please visit http://247.aw.com. Email technical support is 24/7.

Important: Please read the License Agreement, located on the launch screen, before using http://www.aw-bc.com/watson. By using the website, you indicate that you have read, understood, and accepted the terms of this agreement.

MOLECULAR BIOLOGY OF THE GENE

SIXTH EDITION

MOLECULAR BIOLOGY OF THE GENE

SIXTH EDITION

James D. Watson
Cold Spring Harbor Laboratory

Tania A. Baker
Massachusetts Institute of Technology

Stephen P. Bell
Massachusetts Institute of Technology

Alexander Gann
Cold Spring Harbor Laboratory

Michael Levine
University of California, Berkeley

Richard Losick
Harvard University

PEARSON
Benjamin Cummings

COLD SPRING HARBOR LABORATORY PRESS
Cold Spring Harbor, New York

BENJAMIN CUMMINGS

Editor-in-Chief: Beth Wilbur
Executive Editor: Gary Carlson
Managing Editor: Michael Early
Production Supervisor: Lori Newman
Illustrators: Dragonfly Media Group
Manufacturing Buyer: Dorothy Cox
Executive Marketing Manager: Lauren Harp
Text Printer: RR Donnelley and Sons, Willard
Cover Printer: Phoenix Color Corp.

COLD SPRING HARBOR LABORATORY PRESS

Publisher and Sponsoring Editor: John Inglis
Editorial Director: Alexander Gann
Director, Book Development, Marketing & Sales: Jan Argentine
Project Manager and Developmental Editor: Kaaren Janssen
Project Coordinator: Inez Sialiano
Production Manager: Denise Weiss
Production Editor: Kathleen Bubbeo
Desktop Editor: Susan Schaefer
Permissions Coordinator: Carol Brown
Crystal Structure Images: Leemor Joshua-Tor
Cover Designer: Ed Atkeson

Front cover image: See Preface.

Back cover images: First edition, Bill Prokos; Second edition, courtesy of O.L. Miller, Jr., Oak Ridge National Laboratory, and Barbara A. Hamkalo and C.A. Thomas, Jr., Harvard Medical School; Third edition, courtesy of O. Croissant, C. Cremisi, P. Pignatti, and M. Yaniv, Institut Pasteur; Fourth edition, image prepared by the Graphic Systems Research Group at the IBM U.K. Scientific Centre, using coordinates courtesy of Brian W. Matthews, University of Oregon; Fifth edition, Tomo Narashima.

Part opener images: Part 2 image is reprinted, with permission, from Willard H.F. *Nature* **423:** 810–813 (Fig. 1), © Macmillan. Part 5 image is adapted, with permission, from Collins S.R. et al. 2007. *Mol. Cell. Proteomics* **6:** 439–450 (Fig. 3D), © American Society for Biochemistry & Molecular Biology.

Library of Congress Cataloging-in-Publication Data

Molecular biology of the gene / James D. Watson ... [et al.]. -- 6th ed.
p. ; cm.
Includes bibliographical references and index.
ISBN 978-0-8053-9592-1 (cloth : alk. paper)
1. Molecular biology. 2. Molecular genetics. I. Watson, James D., 1928-
[DNLM: 1. Molecular Biology. 2. Cytogenetics. 3. Gene Expression Regulation. 4. Gene Expression. 5. Genetic Techniques. QH 506 M7191 2008]

QH506.M6627 2008
572.8--dc22

2007038009

ISBN 0-8053-9592-X/978-0-8053-9592-1

6 7 8 9 10—DOW—12

PEARSON
Benjamin Cummings

www.aw-bc.com

CSHL PRESS

www.cshlpress.com

Preface

THREE SIGNIFICANT CHANGES HAVE TAKEN PLACE in molecular biology since the fifth edition of *Molecular Biology of the Gene* appeared four years ago. The first is in the cost and influence of genome sequencing; the second is the explosion in our appreciation of RNA as a regulatory molecule; and the third is the widespread touting of systems biology as the future of our field in the postgenomic era. The influence of all three developments is evident in this, the sixth edition.

Numerous species—together with one of the authors—have had their genomes sequenced in the past four years. And currently each genome can be done at a cost almost 500-fold lower than was possible in 2003. In that year we knew the sequences of the model systems—yeast, worm, *Arabidopsis*, fly, mouse, human—and perhaps two or three other animals. Now we have a much wider representation from the tree of life: we have the chimp, rat, and honeybee genomes; the cat and dog; several more insects; rice; the sea anemone; the opossum; and many others besides. And we have many more representatives of each group: as an example, 13 *Drosophila* species have by now been sequenced. And finally, as noted above, we have entered the era of the individual human genome—as well as James Watson's, Craig Venter's genome has been completely sequenced.

Along with this mushrooming in the amount of sequencing, the *influence* of sequencing is becoming ever more apparent and routine. There are several cases in this new edition where evolutionary and mechanistic arguments are based on the ability to compare genomes—in some cases between closely related species, in others more distantly.

Genome sequences have contributed to the second big development—the growing understanding of all that RNA does. The extent and importance of alternative splicing and the widespread use of regulatory RNAs represent areas where the text has been extensively revised.

Systems biology remains a rather ill-defined term, evoking somewhat different things for different people. For some it reflects high-throughput methodology, for example; for others it reflects mathematical modeling of biological systems. In this edition, we examine one aspect of this emerging field—the one with most obvious and immediate significance to topics covered in this book—namely, the representation and modeling of gene regulatory networks.

Organization of the Book

Although much has changed, many core values and organizing principles of the book have not. Thus, the new edition retains the structure of the last, being divided into five parts.

Part 1 lays the groundwork for what follows—it summarizes the history of genetics and molecular biology and explains the chemical principles that determine the structure and function of macromolecules.

Part 2 covers the organization and maintenance of the genetic material. Beginning with a description of the structures of DNA and RNA, it also covers the organization of genomes in chromosomes and explains DNA replication, recombination, and repair.

Part 3 deals with expression of the genome through transcription, splicing, and translation.

Part 4 is about how the expression of genes within the genome is regulated. This includes mechanisms of transcriptional regulation, the role of regulatory RNAs in controlling gene expression, gene regulation in the control of development, and how changes in regulation underlie much of evolution. Finally, we consider genome-wide studies of gene expression and how gene regulatory networks are modeled and analyzed in the emerging fields of systems and synthetic biology.

Part 5 comprises two chapters: one on the techniques of molecular biology, genomics, proteomics, and bioinformatics and the other on the model organisms whose study has revealed many of the underlying principles of molecular biology.

As well as revision of each chapter, there are significant additions to this edition. Most notably, there are two new chapters and a new range of boxed features.

New Chapters and Boxes

Chapter 18: Regulatory RNAs One of the striking developments since the fifth edition is the extent to which RNAs are found to be regulators of gene expression. In this chapter, we cover everything from riboswitches and small RNAs in bacteria, through RNA interference and microRNAs in eukaryotes, to the role of RNA regulators in X-chromosome inactivation. As well as mechanistic details of how these regulators work, we discuss how they were discovered and how they have afforded us new ways to manipulate gene expression artificially.

Chapter 20: Genome Analysis and Systems Biology The genomes of many animals have been sequenced and found to be strikingly similar. This finding emphasizes that it is largely differences in the regulation, rather than the identity, of the genes within a genome that determine phenotype. Techniques now exist to study genome-wide gene expression patterns; we discuss how these techniques work and what they have revealed. In the second half of the chapter, we learn about the representation of gene regulatory networks as so-called wiring diagrams (an example is shown on the cover of this book as is discussed below). These diagrams emphasize the information flow and types of decision-making steps within such pathways.

Boxes The sixth edition includes an array of new boxes, each assigned to one of four color-coded categories:

- *KEY EXPERIMENTS* describe critical experiments in each field, ranging from the historical to the very recent. Without disrupting the narrative flow of the main text, they give students insight into how scientific problems are solved.
- *TECHNIQUES* describe techniques of particular relevance to the chapter in question, complementing the more general techniques described in the last two chapters of the book.
- *MEDICAL CONNECTIONS* highlight links between the basic molecular biology described and a variety of human diseases. These boxes emphasize how basic research informs current medical understanding.
- *ADVANCED CONCEPTS* explore selected concepts in more depth to allow students a glimpse of cutting-edge ideas.

As well as being included in the detailed table of contents, there is a separate listing of all the Boxes by topic on page xxxi.

Media Icons New web references and icons within the text direct students to explore the animations and structural tutorials on the companion website.

Supplements

Companion Website www.aw-bc.com/watson
The Companion Website combines the contents from the previous edition's "Student Media" CD-ROM into one fully integrated resource that helps students better understand complex structures and molecular interactions. Two new structural tutorials on Argonaute and RecBCD have been added in the sixth edition in addition to the popular structural media tutorials of the previous edition.

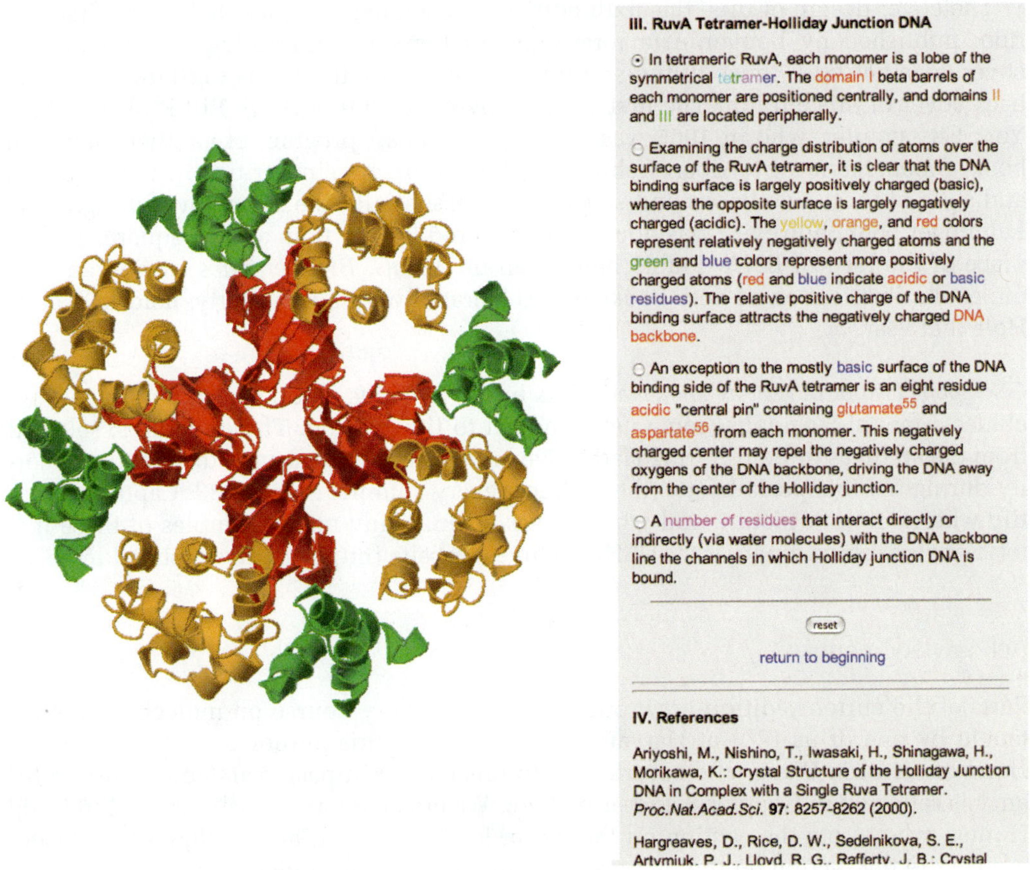
III. RuvA Tetramer-Holliday Junction DNA
In tetrameric RuvA, each monomer is a lobe of the symmetrical tetramer. The domain I beta barrels of each monomer are positioned centrally, and domains II and III are located peripherally.
Examining the charge distribution of atoms over the surface of the RuvA tetramer, it is clear that the DNA binding surface is largely positively charged (basic), whereas the opposite surface is largely negatively charged (acidic). The yellow, orange, and red colors represent relatively negatively charged atoms and the green and blue colors represent more positively charged atoms (red and blue indicate acidic or basic residues). The relative positive charge of the DNA binding surface attracts the negatively charged DNA backbone.
An exception to the mostly basic surface of the DNA binding side of the RuvA tetramer is an eight residue acidic "central pin" containing glutamate55 and aspartate56 from each monomer. This negatively charged center may repel the negatively charged oxygens of the DNA backbone, driving the DNA away from the center of the Holliday junction.
A number of residues that interact directly or indirectly (via water molecules) with the DNA backbone line the channels in which Holliday junction DNA is bound.
reset
return to beginning
IV. References
Ariyoshi, M., Nishino, T., Iwasaki, H., Shinagawa, H., Morikawa, K.: Crystal Structure of the Holliday Junction DNA in Complex with a Single Ruva Tetramer. Proc.Nat.Acad.Sci. 97: 8257-8262 (2000).
Hargreaves, D., Rice, D. W., Sedelnikova, S. E.,

Glossary
An online glossary defines terms that appear in bold in the text.

Instructor Resource CD-ROM 978-0-3215-2766-0/0-3215-2766-6
This dual-platform CD-ROM contains all art and tables from the book in jpeg and PowerPoint format in high-resolution (150-dpi) files. This CD-ROM also contains the answers to the end-of-chapter Critical Thinking questions posed to students on the Companion Website.

Transparency Acetates 0-3215-3639-8/978-0-3215-3639-6
The transparency package features approximately 300 four-color illustrations from the text selected by the authors.

Cover Design and Part-Opener Photographs

Cover image The cover is a representation from systems biology of a frequently observed motif in the circuitry governing gene expression (see Chapter 20). The motif, known as an incoherent feed-forward loop, consists of two regulatory genes, A and B, and a target gene C whose expression is indicated as the output. In response to an input, gene A is turned on. The gene A product is an activator both of the target gene C and of the second regulatory gene B. The gene B product is a repressor of gene C. The representation "AND" in the figure indicates that the circuit operates with the logic of a so-called AND gate, a term borrowed from electrical engineering, which means that the output requires (in the case shown) the presence of the gene A activator AND the absence of the gene B repressor. Incoherent feed-forward loops have the property of causing target genes to be expressed in a pulse. That is, gene C is turned on by the activator in response to the input and then switched off by the delayed accumulation of the repressor.

The cover design of this, the sixth edition, consciously mimics that of the first edition, published by Benjamin (a forerunner of Benjamin Cummings) in 1965. That cover, together with those of the second through fifth editions, is reproduced on the back cover of this edition. The first edition cover was designed by Bill Prokos, a New York City painter, who in 1965 was working on the art program of another Benjamin book, *Bioenergetics*, by Albert Lehninger. Benjamin saw similarities in the ambition and style of these two books and so gave both essentially the same cover design. The Lehninger book, however, was dominated by a red (rather than a blue) square, and, in place of the turn of DNA double helix seen on *MBoG, Bioenergetics* carried an ATP molecule. Publication of both books was celebrated with a joint party, held in Woods Hole that year.

Cold Spring Harbor Laboratory Photographs As in the previous edition, each part opener includes a few photographs, some newly added to this edition. These pictures, selected from the archives of Cold Spring Harbor Laboratory, were all taken at the lab, the majority during the Symposia hosted there almost every summer since 1933. Captions identify who is in each picture and when it was taken. Many more examples of these historic photos can be found at the CSHL archives website (http://archives.cshl.edu/).

Acknowledgments

Parts of the current edition grew out of an introductory course on molecular biology taught by one of us (RL) at Harvard University, and this author is grateful to Steve Harrison and Jim Wang who contributed to this course in past years and whose influence is reflected in Chapter 6 and elsewhere. We are also particularly grateful to Craig Hunter, who wrote the section on the worm for Chapter 21, and to Rob Martienssen, who wrote the section on the plant *Arabidopsis* for that same chapter.

We have shown sections of the manuscript to various colleagues and their comments have been most useful in ensuring the text and figures reflect current ideas and information. Specifically, we thank Katsura Asano, Jamie Cate, Amy Caudy, Richard Ebright, Mike Eisen, Chris Fromme, Brenton Graveley, Ann Hochschild, Jim Hu, Richard Jorgensen, David Jeruzalmi, Leemor Joshua-Tor, Sandy Johnson, Adrian Krainer, Karoline Luger, Julian Lewis, Sue Lovett, Rob Martienssen, Bill McGinnis, Matt Michael, Lily Mirels, Nipam Patel, Mark Ptashne, Danny Reinberg, and Bruce Stillman.

We also thank those who provided us with figures, or the wherewithal to create them: Sean Carroll, Seth Darst, Richard Ebright, Brenton Graveley, Ann Hochschild, Julian Lewis, Bill McGinnis, Phoebe Rice, Dan Rokhsar, Nori Satoh, Matt Scott, Peter Sorger, Tom Steitz, Andrzej Stasiak, Dan Voytas, and Steve West.

We are indebted to Leemor Joshua-Tor who rendered all the structure figures so beautifully. Her skill and patience were much appreciated. We are also grateful to those who provided their software[1]: Per Kraulis, Robert Esnouf, Ethan Merrit, Barry Honig, and Warren Delano. Coordinates were obtained from the Protein Data Bank (www.rcsb.org/pdb/); and citations to those who solved each structure are included in the figure legends.

Our art program was again executed by a talented team from the Dragonfly Media Group, led by Craig Durant and assisted by Helen Wortham. Denise Weiss and Ed Atkeson produced the cover design. We thank Clare Clark and the CSHL Archive for providing the photos for the part openers and for much help tracking them down.

We thank Susan Winslow at Benjamin Cummings who did much to ensure this new edition happened and even more to make things as easy as possible once it began in earnest. Gary Carlson took over late in the game, and we are grateful to him for letting things continue seamlessly. In overseeing the book's development, Jan Argentine managed endlessly to accommodate our demands, reorganizing our tasks to provide the time and support we needed, while still protecting the sacred schedule. Kaaren Janssen edited our text and redrew our messy sketches, generating momentum with her inexhaustible desire to help, and Inez Sialiano kept organized the resulting drafts and corrections, and Carol Brown the permissions. In production, the patience of Kathleen Bubbeo and Susan Schaefer was as tireless as it needed to be in accommodating the changes we continued to generate even after we were supposed to have finished. And we are grateful to Denise Weiss, who not only oversaw production, but did much herself to make the book look so good from page layout to design. John Inglis kept a watchful eye on proceedings, providing useful advice whenever needed.

And finally we thank our families and friends for again providing strong support, despite having to put up with the demands of the project just as much as we did.

James D. Watson
Tania A. Baker
Stephen P. Bell
Alexander Gann
Michael Levine
Richard Losick

[1]Per Kraulis granted permission to use MolScript (Kraulis P.J. 1991. MOLSCRIPT: A program to produce both detailed and schematic plots of protein structures. *J. Appl. Crystallog.* **24:** 946–950). Robert Esnouf gave permission to use BobScript (Esnouf R.M. 1997. An extensively modified version of MolScript that includes greatly enhanced coloring capabilities. *J. Mol. Graph. Model.* **15:** 132–134). In addition, Ethan Merritt gave us use of Raster 3D (Merritt E.A. and Bacon D.J. 1997. Raster3D: Photorealistic molecular graphics. *Methods Enzymol.* **277:** 505–524), and Barry Honig granted permission to use GRASP (Nicholls A., Sharp K.A., and Honig B. 1991. Protein folding and association: Insights from the interfacial and thermodynamic properties of hydrocarbons. *Proteins* **11:** 281–296). Warren DeLano agreed to the use of PyMOL (DeLano W.L. 2002. *The PyMOL Molecular Graphics System*. DeLano Scientific, Palo Alto, California).

About the Authors

JAMES D. WATSON was Director of Cold Spring Harbor Laboratory from 1968 to 1993, was its President from 1994 to 2003, and is now its Chancellor. He spent his undergraduate years at the University of Chicago and received his Ph.D. in 1950 from Indiana University. Between 1950 and 1953, he did postdoctoral research in Copenhagen and Cambridge, England. While at Cambridge, he began the collaboration that resulted in the elucidation of the double-helical structure of DNA in 1953. (For this discovery, Watson, Francis Crick, and Maurice Wilkins were awarded the Nobel Prize in 1962.) Later in 1953, he went to the California Institute of Technology. He moved to Harvard in 1955, where he taught and did research on RNA synthesis and protein synthesis until 1976. He was the first Director of the National Center for Genome Research of the National Institutes of Health from 1989 to 1992, and his own genome was sequenced in 2007. Dr. Watson was sole author of the first, second, and third editions of *Molecular Biology of the Gene*, and a co-author of the fourth and fifth editions. These were published in 1965, 1970, 1976, 1987, and 2003, respectively. He is also a co-author of two other textbooks: *Molecular Biology of the Cell* and *Recombinant DNA*.

TANIA A. BAKER is the Whitehead Professor of Biology at the Massachusetts Institute of Technology and an Investigator of the Howard Hughes Medical Institute. She received a B.S. in biochemistry from the University of Wisconsin, Madison, and a Ph.D. in biochemistry from Stanford University in 1988. Her graduate research was carried out in the laboratory of Professor Arthur Kornberg and focused on mechanisms of initiation of DNA replication. She did postdoctoral research in the laboratory of Dr. Kiyoshi Mizuuchi at the National Institutes of Health, studying the mechanism and regulation of DNA transposition. Her current research explores mechanisms and regulation of genetic recombination, enzyme-catalyzed protein unfolding, and ATP-dependent protein degradation. Professor Baker received the 2001 Eli Lilly Research Award from the American Society of Microbiology and the 2000 MIT School of Science Teaching Prize for Undergraduate Education, is a fellow of the American Academy of Arts and Sciences since 2004, and was elected to the National Academy of Sciences in 2007. She is co-author (with Arthur Kornberg) of the book *DNA Replication*, Second Edition.

STEPHEN P. BELL is a Professor of Biology at the Massachusetts Institute of Technology and an Investigator of the Howard Hughes Medical Institute. He received B.A. degrees from the Department of Biochemistry, Molecular Biology, and Cell Biology and the Integrated Sciences Program at Northwestern University and a Ph.D. in biochemistry at the University of California, Berkeley in 1991. His graduate research was carried out in the laboratory of Dr. Robert Tjian and focused on eukaryotic transcription. He did postdoctoral research in the laboratory of Dr. Bruce Stillman at Cold Spring Harbor Laboratory, working on the initiation of eukaryotic DNA replication. His current research focuses on the mechanisms controlling the duplication of eukaryotic chromosomes. Professor Bell received the 2001 ASBMB–Schering-Plough Scientific Achievement Award, the 1998 Everett Moore Baker Memorial Award for Excellence in Undergraduate Teaching at MIT, and the 2006 MIT School of Science Teaching Award.

ALEXANDER GANN is Editorial Director of Cold Spring Harbor Laboratory Press and a faculty member of the Watson School of Biological Sciences at Cold Spring Harbor Laboratory. He received his B.Sc. in microbiology from University College London and a Ph.D. in molecular biology from The University of Edinburgh in 1989. His graduate research was carried out in the laboratory of Noreen Murray and focused on DNA recognition by restriction enzymes. He did postdoctoral research in the laboratory of Mark Ptashne at Harvard, working on transcriptional regulation, and that of Jeremy Brockes at the Ludwig Institute for Cancer Research at University College London, where he worked on newt limb regeneration. He was a Lecturer at Lancaster University, United Kingdom, from 1996 to 1999, before moving to Cold Spring Harbor Laboratory. He is co-author (with Mark Ptashne) of the book *Genes & Signals* (2002).

MICHAEL LEVINE is a Professor of Molecular and Cell Biology at the University of California, Berkeley, and is also Co-Director of the Center for Integrative Genomics. He received his B.A. from the Department of Genetics at University of California, Berkeley, and his Ph.D. with Alan Garen in the Department of Molecular Biophysics and Biochemistry from Yale University in 1981. As a postdoctoral fellow with Walter Gehring and Gerry Rubin from 1982 to 1984, he studied the molecular genetics of *Drosophila* development. Professor Levine's research group currently studies the gene networks responsible for the gastrulation of the *Drosophila* and *Ciona* (sea squirt) embryos. He holds the F. Williams Chair in Genetics and Development at the University of California, Berkeley. He was awarded the Monsanto Prize in Molecular Biology from the National Academy of Sciences in 1996 and was elected to the American Academy of Arts and Sciences in 1996 and the National Academy of Sciences in 1998.

RICHARD LOSICK is the Maria Moors Cabot Professor of Biology, a Harvard College Professor, and a Howard Hughes Medical Institute Professor in the Faculty of Arts and Sciences at Harvard University. He received his A.B. in chemistry at Princeton University and his Ph.D. in biochemistry at the Massachusetts Institute of Technology. Upon completion of his graduate work, Professor Losick was named a Junior Fellow of the Harvard Society of Fellows when he began his studies on RNA polymerase and the regulation of gene transcription in bacteria. Professor Losick is a past Chairman of the Departments of Cellular and Developmental Biology and Molecular and Cellular Biology at Harvard University. He received the Camille and Henry Dreyfuss Teacher-Scholar Award, is a member of the National Academy of Sciences, a Fellow of the American Academy of Arts and Sciences, a Fellow of the American Association for the Advancement of Science, a Fellow of the American Academy of Microbiology, a member of the American Philosophical Society, and a former Visiting Scholar of the Phi Beta Kappa Society. Professor Losick is the 2007 winner of the Selman A. Waksman Award in Microbiology of the National Academy of Sciences.

Class Testers and Reviewers

WE WISH TO THANK ALL OF THE INSTRUCTORS for their thoughtful suggestions and comments on versions of many chapters in this book.

Chapter Reviewers

Ann Aguanno, *Marymount Manhattan College*
Charles F. Austerberry, *Creighton University*
David G. Bear, *University of New Mexico Health Sciences Center*
Margaret E. Beard, *Holy Cross*
Gail S. Begley, *Northeastern University*
Sanford Bernstein, *San Diego State University*
Michael Blaber, *Florida State University*
Nicole Bournias, *California State University, San Bernardino*
John Boyle, *Mississippi State University*
Suzanne Bradshaw, *University of Cincinnati*
John G. Burr, *University of Texas at Dallas*
Michael A. Campbell, *Pennsylvania State University, Erie, The Behrend College*
Shirley Coomber, *King's College, University of London*
Anne Cordon, *University of Toronto*
Sumana Datta, *Texas A&M University*
Jeff DeJong, *University of Texas at Dallas*
Jurgen Denecke, *University of Leeds*
Susan M. DiBartolomeis, *Millersville University*
Santosh R. D'Mello, *University of Texas at Dallas*
Robert J. Duronio, *University of North Carolina, Chapel Hill*
Steven W. Edwards, *University of Liverpool*
Allen Gathman, *Southeast Missouri State University*
Anthony D.M. Glass, *University of British Columbia*
Elliott S. Goldstein, *Arizona State University*
Ann Grens, *Indiana University, South Bend*
Gregory B. Hecht, *Rowan University*
Robert B. Helling, *University of Michigan*
David C. Higgs, *University of Wisconsin, Parkside*
Mark Kainz, *Colgate University*
Gregory M. Kelly, *University of Western Ontario*
Ann Kleinschmidt, *Allegheny College*
Dan Krane, *Wright State University*
Mark Levinthal, *Purdue University*
Gary J. Lindquester, *Rhodes College*
Curtis Loer, *University of San Diego*
Virginia McDonough, *Hope College*
Michael J. McPherson, *University of Leeds*
Victoria Meller, *Tufts University*
William L. Miller, *North Carolina State University*
Dragana Miskovic, *University of Waterloo*
David Mullin, *Tulane University*
Jeffrey D. Newman, *Lycoming College*
James B. Olesen, *Ball State University*
Anthony J. Otsuka, *Illinois State University*
Karen Palter, *Temple University*
James G. Patton, *Vanderbilt University*
Ian R. Phillips, *Queen Mary, University of London*
Steve Picksley, *University of Bradford*
Todd P. Primm, *University of Texas at El Paso*
Eva Sapi, *University of New Haven*
Jon B. Scales, *Midwestern State University*

Michael Schultze, *University of York*
Venkat Sharma, *University of West Florida*
Erica L. Shelley, *University of Toronto at Mississauga*
Elizabeth A. Shephard, *University College, London*
Margaret E. Stevens, *Ripon College*
Akif Uzman, *University of Houston, Downtown*
Quinn Vega, *Montclair State University*
Jeffrey M. Voight, *Albany College of Pharmacy*
Robert Wiggers, *Stephen F. Austin State University*
Bruce C. Wightman, *Muhlenberg College*
Bob Zimmermann, *University of Massachusetts*

Class Testers

Charles F. Austerberry, *Creighton University*
Christine E. Bezotté, *Elmira College*
Astrid Helfant, *Hamilton College*
Gerald Joyce, *The Scripps Research Institute*
Jocelyn Krebs, *University of Alaska, Anchorage*
Cran Lucas, *Louisiana State University in Shreveport*
Anthony J. Otsuka, *Illinois State University*
Charles Polson, *Florida Institute of Technology*
Ming-Che Shih, *University of Iowa*

Brief Contents

Detailed Contents

PART 1
CHEMISTRY AND GENETICS, 1

CHAPTER 1 • The Mendelian View of the World, 5

CHAPTER 2 • Nucleic Acids Convey Genetic Information, 19

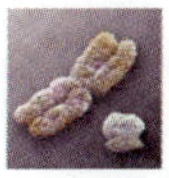

CHAPTER 9 • The Mutability and Repair of DNA, 257

CHAPTER 10 • Homologous Recombination at the Molecular Level, 283

CHAPTER 11 • Site-Specific Recombination and Transposition of DNA, 319

PART 3
EXPRESSION OF THE GENOME, 371

CHAPTER 12 • Mechanisms of Transcription, 377

CHAPTER 15 • The Genetic Code, 521

PART 4
REGULATION, 541

CHAPTER 16 • Transcriptional Regulation in Prokaryotes, 547

CHAPTER 17 • Transcriptional Regulation in Eukaryotes, 589

CHAPTER 18 • Regulatory RNAs, 633

CHAPTER 20 • Genome Analysis and Systems Biology, 703

PART 5

METHODS, 733

CHAPTER 21 • Techniques of Molecular Biology, 739

Box Contents

Advanced Concepts

Key Experiments

Medical Connections

Techniques

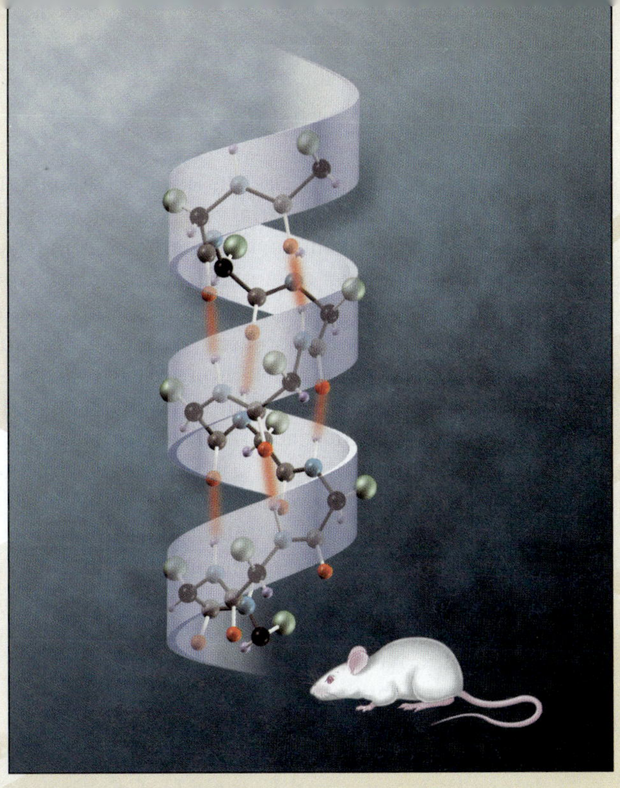

PART

1

CHEMISTRY AND GENETICS

OUTLINE

UNLIKE THE REST OF THIS BOOK, THE FIVE CHAPTERS that make up Part 1 contain material largely unchanged from earlier editions. This is because the material remains as important as ever—even in these days of genome sequencing. Specifically, Chapters 1 and 2 provide an historical account of how the field of genetics and the molecular basis of genetics was established. Key ideas and experiments are described. Chapters 3–5 present the chemistry that lies at the heart of molecular biology. We discuss the fundamental chemical principles that underlie the structures of the macromolecules that figure so prominently throughout the rest of the book—DNA, RNA, and protein—and the interactions between those molecules. Although the bulk of the material is retained from earlier editions, some of it has been reorganized and more recent examples have been included.

Chapter 1 addresses the founding events in the history of genetics. We discuss everything from Mendel's famous experiments on peas, which uncovered the basic laws of heredity, to the one gene encodes one enzyme hypothesis of Garrod. Chapter 2 describes the revolutionary development of molecular biology that was started with Avery's discovery that DNA was the genetic material, and continued with James D. Watson and Francis H. Crick's proposal that the structure of DNA is a double helix, and the elucidation of the genetic code and the "central dogma" (DNA "makes" RNA which "makes" protein). Chapter 2 concludes with a discussion of recent developments stemming from the complete sequencing of the genomes of many organisms and the impact this sequencing has on modern biology.

The basic chemistry presented in Chapters 3–5 focuses on the nature of chemical bonds—both weak and strong—and describes their roles in biology.

Our discussion opens, in Chapter 3, with weak chemical interactions, namely hydrogen bonds, and van der Waals and hydrophobic interactions. These forces mediate most interactions between macromolecules—between proteins or between proteins and DNA, for example. These weak bonds are critical for the activity and regulation of the majority of cellular processes. Thus, enzymes bind their substrates using weak chemical interactions; and transcriptional regulators bind sites on DNA to switch genes on and off using the same class of bonds.

Individual weak interactions are very weak indeed and thus dissociate quickly after forming. This reversibility is important for their roles in biology. Inside cells, molecules must interact dynamically (reversibly) or the whole system would seize up. At the same time, certain interactions must, at least in the short term, be stable. To accommodate these apparently conflicting demands, multiple weak interactions tend to be used together.

Strong bonds hold together the components that make up each macromolecule. Thus, proteins are made up of amino acids linked in a specific order by strong bonds, and DNA is made up of similarly linked nucleotides. (The atoms that make up the amino acids and nucleotides are also joined together by strong bonds.) These bonds are described in Chapter 4.

In Chapter 5, we see how the strong and weak bonds together give macromolecules distinctive three-dimensional shapes (and thus bestow upon them specific functions). Thus, just as weak bonds mediate interactions between macromolecules, so too they act between, for example, nonadjacent amino acids within a given protein. In so doing, they determine how the primary chain of amino acids folds into a three-dimensional shape. Likewise, it is weak bonds that hold together the two chains of the DNA molecule.

We also consider, in Chapter 5, how the function of a protein can be regulated. One way is by changing the shape of the protein, a mechanism called allosteric regulation. Thus, in one conformation, a given protein may perform a specific enzymatic function or bind a specific target molecule. In another conformation, however, it may lose that ability. Such a change in shape can be triggered by the binding of another protein or a small molecule such as a sugar. In other cases, an allosteric effect can be induced by a covalent modification. For example, attaching one or more phosphate groups to a protein can trigger a change in the shape of that protein. Another way a protein can be controlled is by regulating when it is brought into contact with a target molecule. In this way a given protein can be recruited to work on different target proteins in response to different signals.

PHOTOS FROM THE COLD SPRING HARBOR LABORATORY ARCHIVES

Vernon Ingram, Marshall W. Nirenberg, and Matthias Staehelin, 1963 Symposium on Synthesis and Structure of Macromolecules. Ingram demonstrated that genes control the amino acid sequence of proteins; the mutation causing sickle-cell anemia produces a single amino acid change in the hemoglobin protein (Chapter 2). Nirenberg was key in unraveling the genetic code, using protein synthesis directed by artificial RNA templates in vitro (Chapters 2 and 14). For this achievement, he shared in the 1968 Nobel Prize in Physiology or Medicine. Staehelin worked on the small RNA molecules, tRNAs, which translate the genetic code into amino acid sequences of proteins (Chapters 2 and 14).

Raymond Appleyard, George Bowen, and Martha Chase, 1953 Symposium on Viruses. Appleyard and Bowen, both phage geneticists, are here shown with Chase, who, in 1952, together with Alfred Hershey, did the simple experiment that finally convinced most people that the genetic material is DNA (Chapter 2).

Melvin Calvin, Francis Crick, George Gamow, and James Watson, 1963 Symposium on Synthesis and Structure of Macromolecules. Calvin won the 1961 Nobel Prize in Chemistry for his work on CO_2 assimilation by plants. For their proposed structure of DNA, Crick and Watson shared in the 1962 Nobel Prize in Physiology or Medicine (Chapters 2 and 6). Gamow, a physicist attracted to the problem of the genetic code (Chapters 2 and 15), founded an informal group of like-minded scientists called the RNA Tie Club. (He is wearing the club tie, which he designed, in this picture.)

Calvin Bridges, 1934 Symposium on Aspects of Growth. Bridges (shown reading the newspaper) was part of T.H. Morgan's famous "fly group" that pioneered the development of the fruit fly *Drosophila* as a model genetic organism (Chapters 1 and 22). With him is Dr. T. Buckholtz.

Joan Steitz and Fritz Lipmann, 1969 Symposium on The Mechanism of Protein Synthesis. Steitz's research focuses on the structure and function of RNA molecules, particularly those involved in RNA splicing (Chapter 13), and she was an author of the fourth edition of this book. Lipmann showed that the high-energy phosphate group in ATP is the source of energy that drives many biological processes (Chapter 4). For this he shared, with Hans Krebs, the 1953 Nobel Prize in Physiology or Medicine.

Max Perutz, 1971 Symposium on Structure and Function of Proteins at the Three-Dimensional Level. Perutz shared, with John Kendrew, the 1962 Nobel Prize for Chemistry; using X-ray crystallography, and after 25 years of effort, they were the first to solve the atomic structures of proteins—hemoglobin and myoglobin, respectively (Chapter 5).

CHAPTER

1

The Mendelian View of the World

IT IS EASY TO CONSIDER HUMAN BEINGS UNIQUE among living organisms. We alone have developed complicated languages that allow meaningful and complex interplay of ideas and emotions. Great civilizations have developed and changed our world's environment in ways inconceivable for any other form of life. There has always been a tendency, therefore, to think that something special differentiates humans from every other species. This belief has found expression in the many forms of religion through which we seek the origin of and explore the reasons for our existence and, in so doing, try to create workable rules for conducting our lives. Little more than a century ago, it seemed natural to think that, just as every human life begins and ends at a fixed time, the human species and all other forms of life must also have been created at a fixed moment.

This belief was first seriously questioned almost 150 years ago, when Charles Darwin and Alfred R. Wallace proposed their theories of evolution, based on the selection of the most fit. They stated that the various forms of life are not constant but continually give rise to slightly different animals and plants, some of which adapt to survive and multiply more effectively. At the time of this theory, they did not know the origin of this continuous variation, but they did correctly realize that these new characteristics must persist in the progeny if such variations are to form the basis of evolution.

At first, there was a great furor against Darwin, most of it coming from people who did not like to believe that humans and the rather obscene-looking apes could have a common ancestor, even if this ancestor had lived some 10 million years ago. There was also initial opposition from many biologists who failed to find Darwin's evidence convincing. Among these was the famous naturalist Jean L. Agassiz, then at Harvard, who spent many years writing against Darwin and Darwin's champion, Thomas H. Huxley, the most successful of the popularizers of evolution. But by the end of the 19th century, the scientific argument was almost complete; both the current geographic distribution of plants and animals and their selective occurrence in the fossil records of the geologic past were explicable only by postulating that continuously evolving groups of organisms had descended from a common ancestor. Today, evolution is an accepted fact for everyone except a fundamentalist minority, whose objections are based not on reasoning but on doctrinaire adherence to religious principles.

An immediate consequence of Darwinian theory is the realization that life first existed on our Earth more than 4 billion years ago in a simple form, possibly resembling the bacteria—the simplest variety

OUTLINE

of life known today. The existence of such small bacteria tells us that the essence of the living state is found in very small organisms. Evolutionary theory further suggests that the basic principles of life apply to all living forms.

MENDEL'S DISCOVERIES

Gregor Mendel's experiments traced the results of breeding experiments (genetic crosses) between strains of peas differing in well-defined characteristics, like seed shape (round or wrinkled), seed color (yellow or green), pod shape (inflated or wrinkled), and stem length (long or short). His concentration on well-defined differences was of great importance; many breeders had previously tried to follow the inheritance of more gross qualities, like body weight, and were unable to discover any simple rules about their transmission from parents to offspring (see Box 1-1, Mendelian Laws).

The Principle of Independent Segregation

After ascertaining that each type of parental strain bred true—that is, produced progeny with particular qualities identical to those of the

ADVANCED CONCEPTS

Box 1-1 Mendelian Laws

The most striking attribute of a living cell is its ability to transmit hereditary properties from one cell generation to another. The existence of heredity must have been noticed by early humans, who witnessed the passing of characteristics, like eye or hair color, from parents to offspring. Its physical basis, however, was not understood until the first years of the 20th century, when, during a remarkable period of creative activity, the chromosomal theory of heredity was established.

Hereditary transmission through the sperm and egg became known by 1860, and in 1868 Ernst Haeckel, noting that sperm consists largely of nuclear material, postulated that the nucleus is responsible for heredity. Almost 20 years passed before the chromosomes were singled out as the active factors, because the details of mitosis, meiosis, and fertilization had to be worked out first. When this was accomplished, it could be seen that, unlike other cellular constituents, the chromosomes are equally divided between daughter cells. Moreover, the complicated chromosomal changes that reduce the sperm and egg chromosome number to the haploid number during meiosis became understandable as necessary for keeping the chromosome number constant. These facts, however, merely suggested that chromosomes carry heredity.

Proof came at the turn of the century with the discovery of the basic rules of heredity. The concepts were first proposed by Gregor Mendel in 1865 in a paper entitled "Experiments in Plant Hybridization" given to the Natural Science Society at Brno. In his presentation, Mendel described in great detail the patterns of transmission of traits in pea plants, his conclusions of the principles of heredity, and their relevance to the controversial theories of evolution. The climate of scientific opinion, however, was not favorable, and these ideas were completely ignored, despite some early efforts on Mendel's part to interest the prominent biologists of his time. In 1900, 16 years after Mendel's death, three plant breeders working independently on different systems confirmed the significance of Mendel's forgotten work. Hugo de Vries, Karl Correns, and Erich von Tschermak-Seysenegg, all doing experiments related to Mendel's, reached similar conclusions before they knew of Mendel's work.

parents—Mendel performed a number of crosses between parents (P) differing in single characteristics (such as seed shape or seed color). All the progeny (F_1 = first filial generation) had the appearance of *one* parent only. For example, in a cross between peas having yellow seeds and peas having green seeds, all the progeny had yellow seeds. The trait that appears in the F_1 progeny is called **dominant,** whereas the trait that does not appear in F_1 is called **recessive.**

The meaning of these results became clear when Mendel set up genetic crosses between F_1 offspring. These crosses gave the important result that the recessive trait reappeared in approximately 25% of the F_2 progeny, whereas the dominant trait appeared in 75% of these offspring. For each of the seven traits he followed, the ratio in F_2 of dominant to recessive traits was always approximately 3:1. When these experiments were carried to a third (F_3) progeny generation, all the F_2 peas with recessive traits bred true (produced progeny with the recessive traits). Those with dominant traits fell into two groups: one-third bred true (produced only progeny with the dominant trait); the remaining two-thirds again produced mixed progeny in a 3:1 ratio of dominant to recessive.

Mendel correctly interpreted his results as follows (Fig. 1-1): the various traits are controlled by pairs of factors (which we now call **genes**), one factor derived from the male parent, the other from the female. For example, pure-breeding strains of round peas contain two versions (or **alleles**) of the roundness gene (*RR*), whereas pure-breeding wrinkled strains have two copies of the wrinkledness (*rr*) allele. The round-strain gametes each have one gene for roundness (*R*); the wrinkled-strain gametes each have one gene for wrinkledness (*r*). In a cross between *RR* and *rr*, fertilization produces an F_1 plant with both alleles (*Rr*). The seeds look round because *R* is dominant over *r*. We refer to the appearance or physical structure of an individual as its **phenotype,** and to its genetic composition as its **genotype.** Individuals with identical phenotypes may possess different genotypes; thus, to determine the genotype of an organism, it is frequently necessary to perform genetic crosses for several generations. The term **homozygous** refers to a gene pair in which both the maternal and paternal genes are identical (e.g., *RR* or *rr*). In contrast, those gene pairs in which paternal and maternal genes are different (e.g., *Rr*) are called **heterozygous.**

One or several letters or symbols may be used to represent a particular gene. The dominant allele of the gene may be indicated by a capital letter (*R*), by a superscript + (r^+), or by a + standing alone. In our discussions here, we use the first convention in which the dominant allele is represented by a capital letter and the recessive allele by the lowercase letter.

It is important to notice that a given gamete contains only one of the two copies (one allele) of the genes present in the organism it comes from (e.g., either *R* or *r*, but never both) and that the two types of gametes are produced in equal numbers. Thus, there is a 50:50 chance that a given gamete from an F_1 pea will contain a particular gene (*R* or *r*). This choice is purely random. We do not expect to find *exact* 3:1 ratios when we examine a limited number of F_2 progeny. The ratio will sometimes be slightly higher and other times slightly lower. But as we look at increasingly larger samples, we expect that the ratio of peas with the dominant trait to peas with the recessive trait will approximate the 3:1 ratio more and more closely.

The reappearance of the recessive characteristic in the F_2 generation indicates that recessive alleles are neither modified nor lost in the F_1 (*Rr*) generation, but that the dominant and recessive genes are

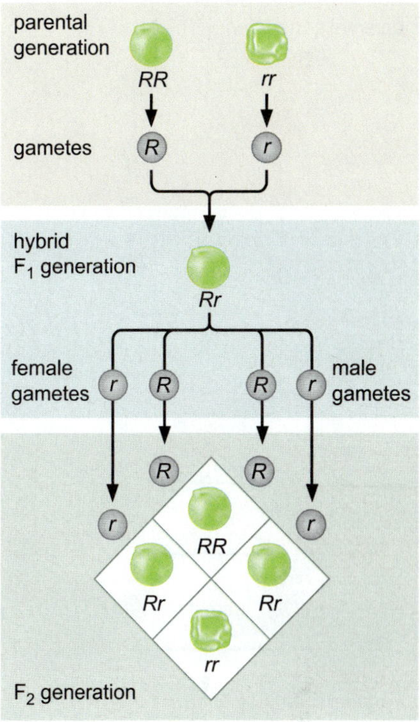

FIGURE 1-1 How Mendel's first law (independent segregation) explains the 3:1 ratio of dominant to recessive phenotypes among the F_2 progeny. *R* represents the dominant gene and *r* the recessive gene. The round seed represents the dominant phenotype, the wrinkled seed the recessive phenotype.

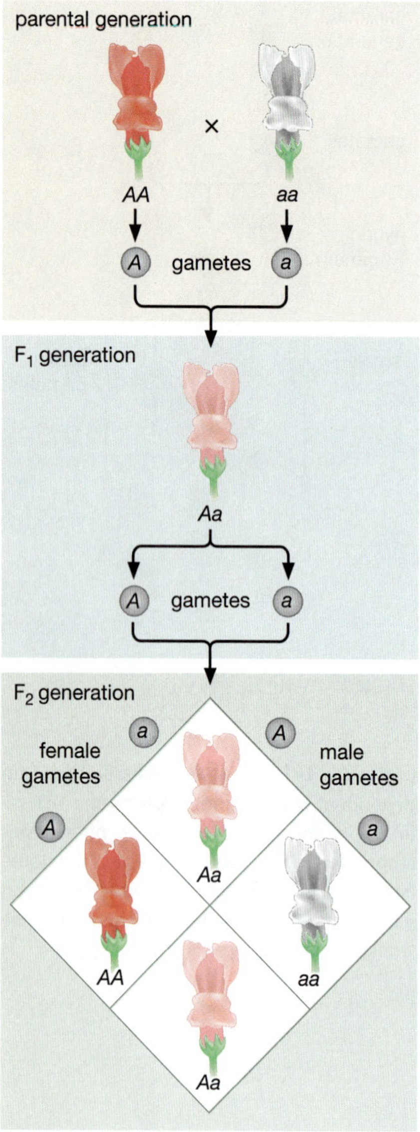

FIGURE 1-2 The inheritance of flower color in the snapdragon. One parent is homozygous for red flowers (*AA*) and the other homozygous for white flowers (*aa*). No dominance is present, and the heterozygous F_1 flowers are pink. The 1:2:1 ratio of red, pink, and white flowers in the F_2 progeny is shown by appropriate coloring.

independently transmitted and so are able to segregate independently during the formation of sex cells. This **principle of independent segregation** is frequently referred to as Mendel's first law.

Some Alleles Are neither Dominant nor Recessive

In the crosses reported by Mendel, one member of each gene pair was clearly dominant to the other. Such behavior, however, is not universal. Sometimes the heterozygous phenotype is intermediate between the two homozygous phenotypes. For example, the cross between a pure-breeding red snapdragon (*Antirrhinum*) and a pure-breeding white variety gives F_1 progeny of the intermediate pink color. If these F_1 progeny are crossed among themselves, the resulting F_2 progeny contain red, pink, and white flowers in the proportion of 1:2:1 (Fig. 1-2). Thus, it is possible here to distinguish heterozygotes from homozygotes by their phenotype. We also see that Mendel's laws do not depend on whether one allele of a gene pair is dominant over the other.

Principle of Independent Assortment

Mendel extended his breeding experiments to peas differing by more than one characteristic. As before, he started with two strains of peas, each of which bred pure when mated with itself. One of the strains had round yellow seeds; the other, wrinkled green seeds. Since round and yellow are dominant over wrinkled and green, the entire F_1 generation produced round yellow seeds. The F_1 generation was then crossed within itself to produce a number of F_2 progeny, which were examined for seed appearance (phenotype). In addition to the two original phenotypes (round yellow; wrinkled green), two new types (**recombinants**) emerged: wrinkled yellow and round green.

Again Mendel found he could interpret the results by the postulate of genes, if he assumed that each gene pair was independently transmitted to the gamete during sex-cell formation. This interpretation is shown in Figure 1-3. Any one gamete contains only one type of allele from each gene pair. Thus, the gametes produced by an F_1 (*RrYy*) will have the composition *RY, Ry, rY,* or *ry,* but never *Rr, Yy, YY,* or *RR.* Furthermore, in this example, all four possible gametes are produced with equal frequency. There is no tendency of genes arising from one parent to stay together. As a result, the F_2 progeny phenotypes appear in the ratio nine round yellow, three round green, three wrinkled yellow, and one wrinkled green as depicted in the Punnett square, named after the British mathematician who introduced it (in the lower part of Fig. 1-3). This **principle of independent assortment** is frequently called Mendel's second law.

CHROMOSOMAL THEORY OF HEREDITY

A principal reason for the original failure to appreciate Mendel's discovery was the absence of firm facts about the behavior of chromosomes during meiosis and mitosis. This knowledge was available, however, when Mendel's laws were confirmed in 1900 and was seized upon in 1903 by American biologist Walter S. Sutton. In his classic paper "The Chromosomes in Heredity," Sutton emphasized the importance of the fact that the diploid chromosome group consists of two morphologically similar sets and that, during meiosis, every gamete

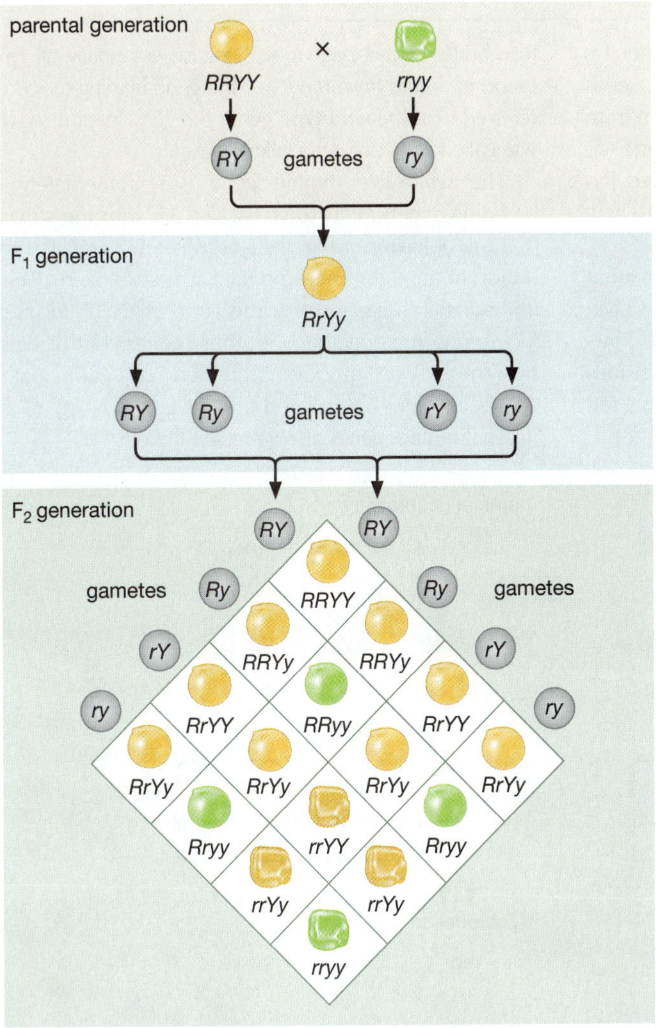

FIGURE 1-3 How Mendel's second law (independent assortment) operates. In this example, the inheritance of yellow (*Y*) and green (*y*) seed color is followed together with the inheritance of round (*R*) and wrinkled (*r*) seed shapes. The *R* and *Y* alleles are dominant over *r* and *y*. The genotypes of the various parents and progeny are indicated by letter combinations, and four different phenotypes are distinguished by appropriate shading.

receives only one chromosome of each homologous pair. He then used this fact to explain Mendel's results by assuming that genes are parts of the chromosome. He postulated that the yellow- and green-seed genes are carried on a certain pair of chromosomes and that the round- and wrinkled-seed genes are carried on a different pair. This hypothesis immediately explains the experimentally observed 9:3:3:1 segregation ratios. Although Sutton's paper did not prove the chromosomal theory of heredity, it was immensely important, for it brought together for the first time the independent disciplines of genetics (the study of breeding experiments) and cytology (the study of cell structure).

GENE LINKAGE AND CROSSING OVER

Mendel's principle of independent assortment is based on the fact that genes located on different chromosomes behave independently during meiosis. Often, however, two genes do not assort independently because they are located on the same chromosome (**linked genes;** see Box 1-2, Genes Are Linked to Chromosomes). Many examples of nonrandom assortment were found as soon as a large

Box 1-2 Genes Are Linked to Chromosomes

Initially, all breeding experiments used genetic differences already existing in nature. For example, Mendel used seeds obtained from seed dealers, who must have obtained them from farmers. The existence of alternative forms of the same gene (alleles) raises the question of how they arose. One obvious hypothesis states that genes can change (mutate) to give rise to new genes (**mutant genes**). This hypothesis was first seriously tested, beginning in 1908, by the great American biologist Thomas Hunt Morgan and his young collaborators, geneticists Calvin B. Bridges, Hermann J. Muller, and Alfred H. Sturtevant. They worked with the tiny fly *Drosophila melanogaster.* The first mutant found was a male with white eyes instead of the normal red eyes. The white-eyed variant appeared spontaneously in a culture bottle of red-eyed flies. Because essentially all *Drosophila* found in nature have red eyes, the gene leading to red eyes was referred to as the **wild-type gene;** the gene leading to white eyes was called a mutant gene (allele).

The white-eye mutant gene was immediately used in breeding experiments (Box 1-2 Fig. 1), with the striking result that the behavior of the allele completely paralleled the distribution of an *X* chromosome (i.e., was sex-linked). This finding immediately suggested that this gene might be located on the *X* chromosome, together with those genes controlling sex. This hypothesis was quickly confirmed by additional genetic crosses using newly isolated mutant genes. Many of these additional mutant genes also were sex-linked.

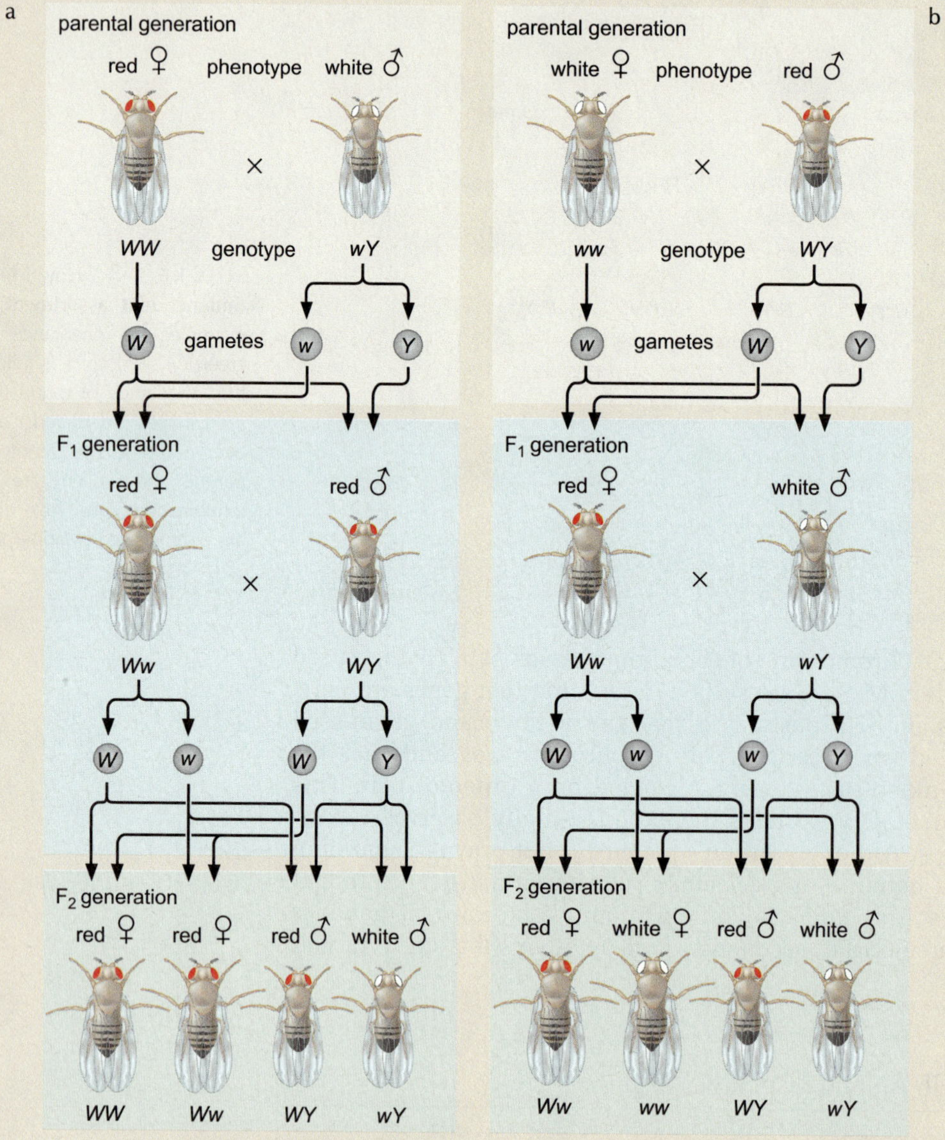

BOX 1-2 FIGURE 1 The inheritance of a sex-linked gene in *Drosophila*. Genes located on sex chromosomes can express themselves differently in male and female progeny, because if there is only one *X* chromosome present, recessive genes on this chromosome are always expressed. Here are two crosses, both involving a recessive gene (*w*, for white eye) located on the *X* chromosome. (a) The male parent is a white-eyed (*wY*) fly, and the female is homozygous for red eye (*WW*). (b) The male has red eyes (*WY*) and the female white eyes (*ww*). The letter *Y* stands here not for an allele, but for the *Y* chromosome, present in male *Drosophila* in place of a homologous *X* chromosome. There is no gene on the *Y* chromosome corresponding to the *w* or *W* gene on the *X* chromosome.

number of mutant genes became available for breeding analysis. In every well-studied case, the number of linked groups was identical to the haploid chromosome number. For example, there are four groups of linked genes in *Drosophila* and four morphologically distinct chromosomes in a haploid cell.

Linkage, however, is in effect never complete. The probability that two genes on the same chromosome will remain together during meiosis ranges from just less than 100% to nearly 50%. This variation in linkage suggests that there must be a mechanism for exchanging genes on homologous chromosomes. This mechanism is called **crossing over.** Its cytological basis was first described by Belgian cytologist F.A. Janssens. At the start of meiosis, through the process of **synapsis,** the homologous chromosomes form pairs with their long axes parallel. At this stage, each chromosome has duplicated to form two chromatids. Thus, synapsis brings together four chromatids (a tetrad), which coil about one another. Janssens postulated that, possibly because of tension resulting from this coiling, two of the chromatids might sometimes break at a corresponding place on each. These events could create four broken ends, which might rejoin crossways, so that a section of each of the two chromatids would be joined to a section of the other (Fig. 1-4). In this manner, recombinant chromatids might be produced that contain a segment derived from each of the original homologous chromosomes. Formal proof of Janssens's hypothesis that chromosomes physically interchange material during synapsis came more than 20 years later, when in 1931, Barbara McClintock and Harriet B. Creighton, working at Cornell University with the corn plant *Zea mays,* devised an elegant cytological demonstration of chromosome breakage and rejoining (Fig. 1-5).

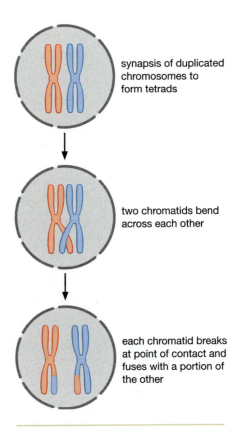

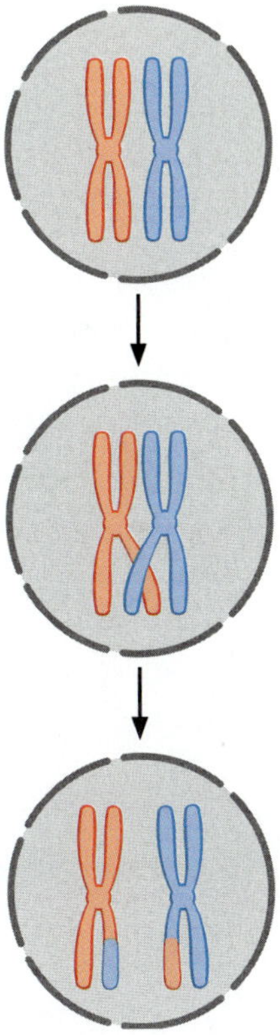

FIGURE 1-4 Janssens's hypothesis of crossing over.

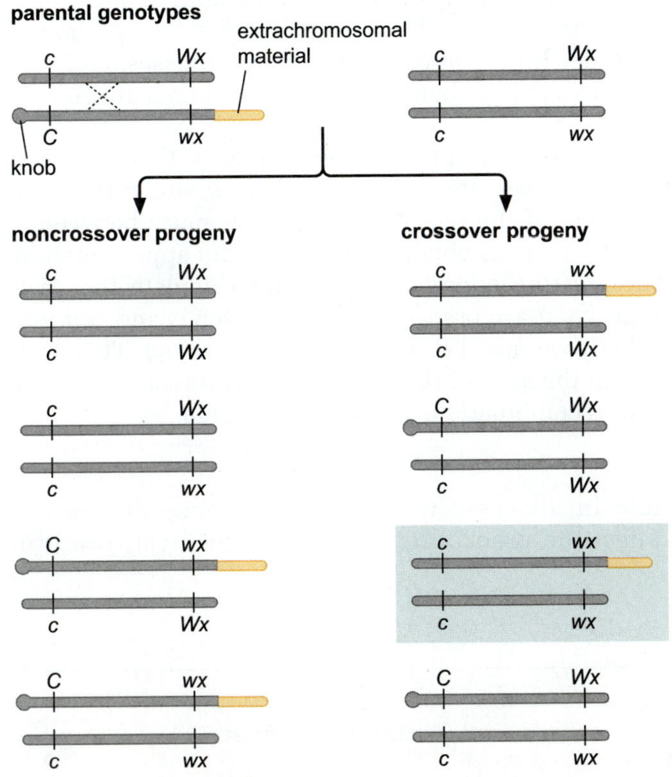

FIGURE 1-5 Demonstration of physical exchanges between homologous chromosomes. In most organisms, pairs of homologous chromosomes have identical shapes. Occasionally, however, the two members of a pair are not identical; one is marked by the presence of extrachromosomal material or compacted regions that reproducibly form knob-like structures. McClintock and Creighton found one such pair and used it to show that crossing over involves actual physical exchanges between the paired chromosomes. In the experiment shown here, the homozygous *c, wx* progeny had to arise by crossing over between the *C* and *wx* loci. When such *c, wx* offspring were cytologically examined, knob chromosomes were seen, showing that a knobless *Wx* region had been physically replaced by a knobbed *wx* region. The colored box in the figure identifies the chromosomes of the homozygous *c, wx* offspring.

CHROMOSOME MAPPING

Thomas Hunt Morgan and his students, however, did not await formal cytological proof of crossing over before exploiting the implication of Janssens's hypothesis. They reasoned that genes located close together on a chromosome would assort with one another much more regularly (close linkage) than genes located far apart on a chromosome. They immediately saw this as a way to locate (map) the relative positions of genes on chromosomes and thus to produce a **genetic map.** The way they used the frequencies of the various recombinant classes is very straightforward. Consider the segregation of three genes all located on the same chromosome. The arrangement of the genes can be determined by means of three crosses, in each of which two genes are followed (two-factor crosses). A cross between *AB* and *ab* yields four progeny types: the two parental genotypes (*AB* and *ab*) and two recombinant genotypes (*Ab* and *aB*). A cross between *AC* and *ac* similarly gives two parental combinations as well as the *Ac* and *aC* recombinants, whereas a cross between *BC* and *bc* produces the parental types and the recombinants *Bc* and *bC*. Each cross will produce a specific ratio of parental to recombinant progeny. Consider, for example, the fact that the first cross gives 30% recombinants, the second cross 10%, and the third cross 25%. This tells us that genes *a* and *c* are closer together than *a* and *b* or *b* and *c* and that the genetic distances between *a* and *b* and *b* and *c* are more similar. The gene arrangement that best fits these data is *a-c-b* (Fig. 1-6).

The correctness of gene order suggested by crosses of two gene factors can usually be unambiguously confirmed by three-factor crosses. When the three genes used in the preceding example are followed in the cross *ABC* x *abc*, six recombinant genotypes are found (Fig. 1-7). They fall into three groups of reciprocal pairs. The rarest of these groups arises from a double crossover. By looking for the least frequent class, it is often possible to instantly confirm (or deny) a postulated arrangement. The results in Figure 1-7 immediately confirm the order hinted at by the two-factor crosses. Only if the order is *a-c-b* does the fact that the rare recombinants are *AcB* and *aCb* make sense.

The existence of multiple crossovers means that the amount of recombination between the outside markers *a* and *b* (*ab*) is usually less than the sum of the recombination frequencies between *a* and *c* (*ac*) and *c* and *b* (*cb*). To obtain a more accurate approximation of the distance between the outside markers, we calculate the probability (*ac* x *cb*) that when a crossover occurs between *c* and *b*, a crossover also occurs between *a* and *c*, and vice versa (*cb* x *ac*). This probability subtracted from the sum of the frequencies expresses more accurately the amount of recombination. The simple formula

$$ab = ac + cb - 2(ac)(cb)$$

is applicable in all cases where the occurrence of one crossover does not affect the probability of another crossover. Unfortunately,

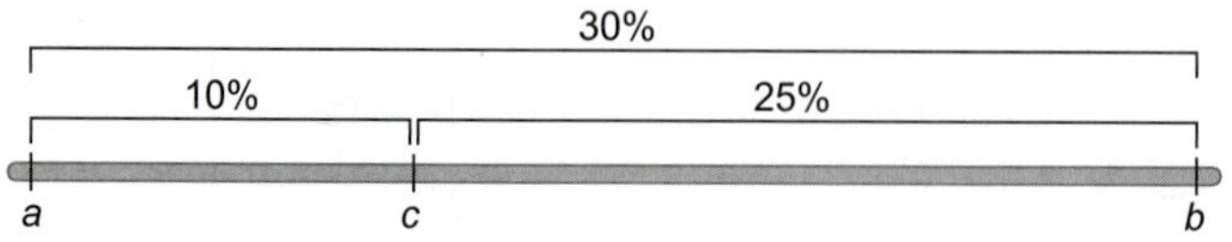

FIGURE 1-6 **Assignment of the tentative order of three genes on the basis of three two-factor crosses.**

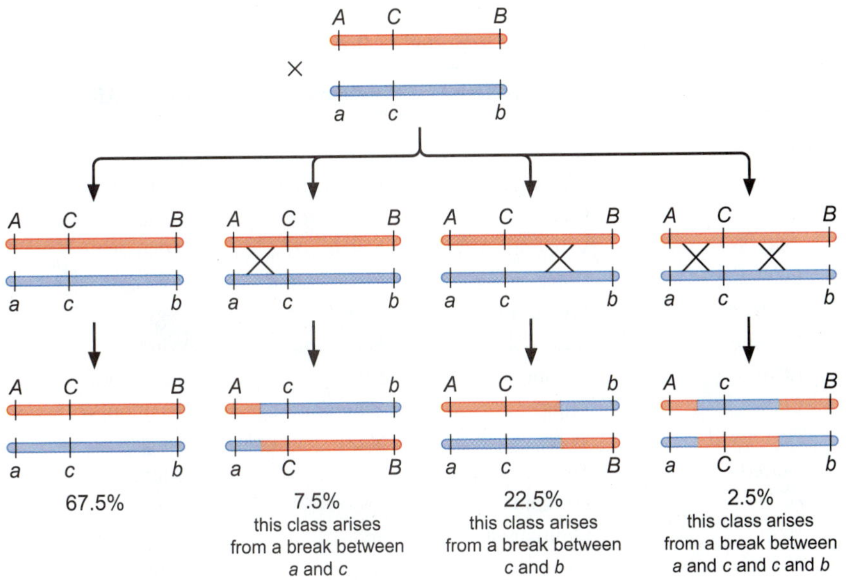

FIGURE 1-7 The use of three-factor crosses to assign gene order. The least frequent pair of reciprocal recombinants must arise from a double crossover. The percentages listed for the various classes are the theoretical values expected for an infinitely large sample. When finite numbers of progeny are recorded, the exact values will be subject to random statistical fluctuations.

accurate mapping is often disturbed by *interference* phenomena, which can either increase or decrease the probability of correlated crossovers.

Using such reasoning, the Columbia University group headed by Morgan had by 1915 assigned locations to more than 85 mutant genes in *Drosophila* (Table 1-1), placing each of them at distinct spots on one of the four linkage groups, or chromosomes. Most importantly, all the genes on a given chromosome were located on a line. The gene arrangement was strictly linear and never branched. The genetic map of one of the chromosomes of *Drosophila* is shown in Figure 1-8. Distances between genes on such a map are measured in **map units,** which are related to the frequency of recombination between the genes. Thus, if the frequency of recombination between two genes is found to be 5%, the genes are said to be separated by five map units. Because of the high probability of double crossovers between widely spaced genes, such assignments of map units can be considered accurate only if recombination between closely spaced genes is followed.

Even when two genes are at the far ends of a very long chromosome, they assort together at least 50% of the time because of multiple crossovers. The two genes will be separated if an odd number of crossovers occurs between them, but they will end up together if an even number occurs between them. Thus, in the beginning of the genetic analysis of *Drosophila,* it was often impossible to determine whether two genes were on different chromosomes or at the opposite ends of one long chromosome. Only after large numbers of genes had been mapped was it possible to demonstrate convincingly that the number of linkage groups equalled the number of cytologically visible chromosomes. In 1915, Morgan, with his students Alfred H. Sturtevant, Hermann J. Muller, and Calvin B. Bridges, published their definitive book *The Mechanism of Mendelian Heredity,* which first announced the general validity of the chromosomal basis of heredity. We now rank this concept, along with the theories of evolution and the cell, as a major achievement in our quest to understand the nature of the living world.

TABLE 1-1 The 85 Mutant Genes Reported in *Drosophila melanogaster* in 1915

Name	Region Affected	Name	Region Affected
Group 1			
Abnormal	Abdomen	*Lethal, 13*	Body, death
Bar	Eye	*Miniature*	Wing
Bifid	Venation	*Notch*	Venation
Bow	Wing	*Reduplicated*	Eye color
Cherry	Eye color	*Ruby*	Leg
Chrome	Body color	*Rudimentary*	Wing
Cleft	Venation	*Sable*	Body color
Club	Wing	*Shifted*	Venation
Depressed	Wing	*Short*	Wing
Dotted	Thorax	*Skee*	Wing
Eosin	Eye color	*Spoon*	Wing
Facet	Ommatidia	*Spot*	Body color
Forked	Spine	*Tan*	Antenna
Furrowed	Eye	*Truncate*	Wing
Fused	Venation	*Vermilion*	Eye color
Green	Body color	*White*	Eye color
Jaunty	Wing	*Yellow*	Body color
Lemon	Body color		
Group 2			
Antlered	Wing	*Jaunty*	Wing
Apterous	Wing	*Limited*	Abdominal band
Arc	Wing	*Little crossover*	Chromosome 2
Balloon	Venation	*Morula*	Ommatidia
Black	Body color	*Olive*	Body color
Blistered	Wing	*Plexus*	Venation
Comma	Thorax mark	*Purple*	Eye color
Confluent	Venation	*Speck*	Thorax mark
Cream II	Eye color	*Strap*	Wing
Curved	Wing	*Streak*	Pattern
Dachs	Leg	*Trefoil*	Pattern
Extra vein	Venation	*Truncate*	Wing
Fringed	Wing	*Vestigial*	Wing
Group 3			
Band	Pattern	*Pink*	Eye color
Beaded	Wing	*Rough*	Eye
Cream III	Eye color	*Safranin*	Eye color
Deformed	Eye	*Sepia*	Eye color
Dwarf	Size of body	*Sooty*	Body color
Ebony	Body color	*Spineless*	Spine
Giant	Size of body	*Spread*	Wing
Kidney	Eye	*Trident*	Pattern
Low crossing over	Chromosome 3	*Truncate*	Wing
Maroon	Eye color	*Whitehead*	Pattern
Peach	Eye color	*White ocelli*	Simple eye
Group 4			
Bent	Wing	*Eyeless*	Eye

The mutations fall into four linkage groups. Because four chromosomes were cytologically observed, this indicated that the genes are situated on the chromosomes. Notice that mutations in various genes can act to alter a single character, such as body color, in different ways.

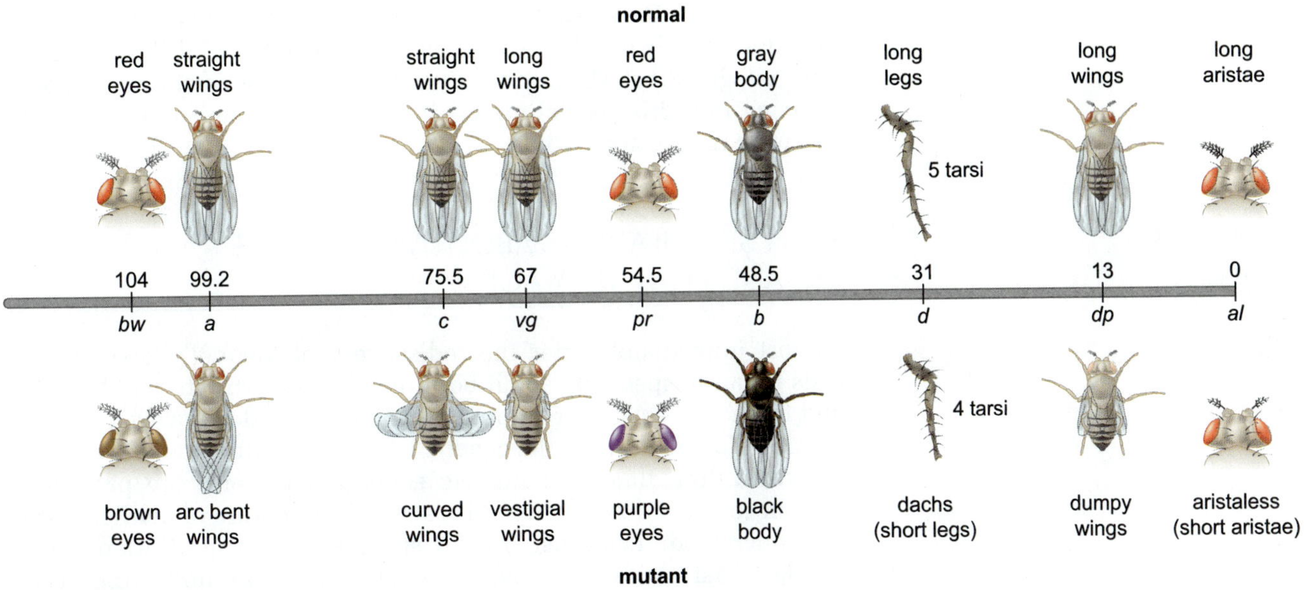

FIGURE 1-8 The genetic map of chromosome 2 of *Drosophila melanogaster*.

THE ORIGIN OF GENETIC VARIABILITY THROUGH MUTATIONS

It now became possible to understand the hereditary variation that is found throughout the biological world and that forms the basis of the theory of evolution. Genes are normally copied exactly during chromosome duplication. Rarely, however, changes (**mutations**) occur in genes to give rise to altered forms, most—*but not all*—of which function less well than the wild-type alleles. This process is necessarily rare; otherwise, many genes would be changed during every cell cycle, and offspring would not ordinarily resemble their parents. There is, instead, a strong advantage in there being a small but finite mutation rate; it provides a constant source of new variability, necessary to allow plants and animals to adapt to a constantly changing physical and biological environment.

Surprisingly, however, the results of the Mendelian geneticists were not avidly seized upon by the classical biologists, then the authorities on the evolutionary relations between the various forms of life. Doubts were raised about whether genetic changes of the type studied by Morgan and his students were sufficient to permit the evolution of radically new structures, like wings or eyes. Instead, these biologists believed that there must also occur more powerful "macromutations," and that it was these events that allowed great evolutionary advances.

Gradually, however, doubts vanished, largely as a result of the efforts of the mathematical geneticists Sewall Wright, Ronald A. Fisher, and John Burden Sanderson Haldane. They showed that, considering the great age of Earth, the relatively low mutation rates found for *Drosophila* genes, together with only mild selective advantages, would be sufficient to allow the gradual accumulation of new favorable attributes. By the 1930s, biologists began to reevaluate their knowledge on the origin of species and to understand the work of the mathematical geneticists. Among these new Darwinians were biologist Julian Huxley (a grandson of Darwin's original publicist, Thomas Huxley), geneticist Theodosius Dobzhansky, paleontologist George Gaylord Simpson, and ornithologist Ernst Mayr. In the 1940s all four wrote major works, each showing from his special viewpoint how Mendelianism and Darwinism were indeed compatible.

EARLY SPECULATIONS ABOUT WHAT GENES ARE AND HOW THEY ACT

Almost immediately after the rediscovery of Mendel's laws, geneticists began to speculate about both the chemical structure of the gene and the way it acts. No real progress could be made, however, because the chemical identity of the genetic material remained unknown. Even the realization that both nucleic acids and proteins are present in chromosomes did not really help, since the structure of neither was at all understood. The most fruitful speculations focused attention on the fact that genes must be, in some sense, self-duplicating. Their structure must be exactly copied every time one chromosome becomes two. This fact immediately raised the profound chemical question of how a complicated molecule could be precisely copied to yield exact replicas.

Some physicists also became intrigued with the gene, and when quantum mechanics burst on the scene in the late 1920s, the possibility arose that in order to understand the gene, it would first be necessary to master the subtleties of the most advanced theoretical physics. Such thoughts, however, never really took root, since it was obvious that even the best physicists or theoretical chemists would not concern themselves with a substance whose structure still awaited elucidation. There was only one fact that they might ponder: Muller's and L.J. Stadler's independent 1927 discoveries that X-rays induce mutations. Because there is a greater possibility that an X-ray will hit a larger gene than a smaller gene, the frequency of mutations induced in a given gene by a given X-ray dose yields an estimate of the size of this gene. But even here, so many special assumptions were required that virtually no one, not even Muller and Stadler themselves, took the estimates very seriously.

PRELIMINARY ATTEMPTS TO FIND A GENE–PROTEIN RELATIONSHIP

The most fruitful early endeavors to find a relationship between genes and proteins examined the ways in which gene changes affect which proteins are present in the cell. At first these studies were difficult, because no one knew anything about the proteins that were present in structures such as the eye or the wing. It soon became clear that genes with simple metabolic functions would be easier to study than genes affecting gross structures. One of the first useful examples came from a study of a hereditary disease affecting amino acid metabolism. Spontaneous mutations occur in humans affecting the ability to metabolize the amino acid phenylalanine. When individuals homozygous for the mutant trait eat food containing phenylalanine, their inability to convert the amino acid to tyrosine causes a toxic level of phenylpyruvic acid to build up in the bloodstream. Such diseases, examples of "inborn errors of metabolism," suggested to English physician Archibald E. Garrod, as early as 1909, that the wild-type gene is responsible for the presence of a particular enzyme, and that in a homozygous mutant, the enzyme is congenitally absent.

Garrod's general hypothesis of a gene–enzyme relationship was extended in the 1930s by work on flower pigments by Haldane and Rose Scott-Moncrieff in England, studies on the hair pigment of the guinea pig by Wright in the United States, and research on the pigments of insect eyes by A. Kuhn in Germany and by Boris Ephrussi and George W. Beadle, working first in France and then in California. In all cases, the evidence revealed that a particular gene affected a particular step in the formation of the respective pigment whose absence changed, say, the color of a fly's eyes from red to ruby. However, the lack of fundamental knowledge about the structures of the relevant enzymes ruled out deeper examination of the gene–enzyme relationship, and no assurance could be given either that most genes control the synthesis of proteins (by then it was suspected that all enzymes were proteins) or that all proteins are under gene control.

As early as 1936, it became apparent to the Mendelian geneticists that future experiments of the sort successful in elucidating the basic features of Mendelian genetics were unlikely to yield productive evidence about how genes act. Instead, it would be necessary to find biological objects more suitable for chemical analysis. They were aware, moreover, that contemporary knowledge of nucleic acid and protein chemistry was completely inadequate for a fundamental chemical attack on even the most suitable biological systems. Fortunately, however, the limitations in chemistry did not deter them from learning how to do genetic experiments with chemically simple molds, bacteria, and viruses. As we shall see, the necessary chemical facts became available almost as soon as the geneticists were ready to use them.

SUMMARY

Heredity is controlled by chromosomes, which are the cellular carriers of genes. Hereditary factors were first discovered and described by Mendel in 1865, but their importance was not realized until the start of the 20th century. Each gene can exist in a variety of different forms called alleles. Mendel proposed that a hereditary factor (now known to be a gene) for each hereditary trait is given by each parent to each of its offspring. The physical basis for this behavior is the distribution of homologous chromosomes during meiosis: one (randomly chosen) of each pair of homologous chromosomes is distributed to each haploid cell. When two genes are on the same chromosome, they tend to be inherited together (linked). Genes affecting different characteristics are sometimes inherited independently of each other, because they are located on different chromosomes. In any case, linkage is seldom complete because homologous chromosomes attach to each other during meiosis and often break at identical spots and rejoin crossways (crossing over). Crossing over transfers genes initially located on a paternally derived chromosome onto gene groups originating from the maternal parent.

Different alleles from the same gene arise by inheritable changes (mutations) in the gene itself. Normally, genes are extremely stable and are copied exactly during chromosome duplication; mutation occurs only rarely and usually has harmful consequences. Mutation does, however, play a positive role, because the accumulation of rare favorable mutations provides the basis for genetic variability that is presupposed by the theory of evolution.

For many years, the structure of genes and the chemical ways in which they control cellular characteristics were a mystery. As soon as large numbers of spontaneous mutations had been described, it became obvious that a one gene–one characteristic relationship does not exist and that all complex characteristics are under the control of many genes. The most sensible idea, postulated by Garrod in 1909, was that genes affect the synthesis of enzymes. However, the tools of Mendelian geneticists—organisms such as the corn plant, the mouse, and even the fruit fly *Drosophila*—were not suitable for detailed chemical investigations of gene–protein relations. For this type of analysis, work with much simpler organisms was to become indispensable.

BIBLIOGRAPHY

Ayala F.J. and Kiger J.A., Jr. 1984. *Modern genetics,* 2nd ed. Benjamin Cummings, Menlo Park, California.

Beadle G.W. and Ephrussi B. 1937. Development of eye color in *Drosophila:* Diffusible substances and their interrelations. *Genetics* **22:** 76–86.

Carlson E.A. 1966. *The gene: A critical history.* Saunders, Philadelphia.

——— 1981. *Genes, radiation, and society: The life and work of H.J. Muller.* Cornell University Press, Ithaca, New York.

Caspari E. 1948. Cytoplasmic inheritance. *Adv. Genet.* **2:** 1–66.

Correns C. 1937. *Nicht Mendelnde vererbung* (ed. F. von Wettstein). Borntraeger, Berlin.

Dobzhansky T. 1941. *Genetics and the origin of species,* 2nd ed. Columbia University Press, New York.

Fisher R.A. 1930. *The genetical theory of natural selection.* Clarendon Press, Oxford.

Garrod A.E. 1908. Inborn errors of metabolism. *Lancet* **2:** 1–7, 73–79, 142–148, 214–220.

Haldane J.B.S. 1932. *The courses of evolution.* Harper & Row, New York.

Huxley J. 1943. *Evolution: The modern synthesis.* Harper & Row, New York.

Lea D.E. 1947. *Actions of radiations on living cells.* Macmillan, New York.

Mayr E. 1942. *Systematics and the origin of species.* Columbia University Press, New York.

——— 1982. *The growth of biological thought: Diversity, evolution, and inheritance.* Harvard University Press, Cambridge, Massachusetts.

McClintock B. 1951. Chromosome organization and gene expression. *Cold Spring Harbor Symp. Quant. Biol.* **16:** 13–57.

——— 1984. The significance of responses of genome to challenge. *Science* **226:** 792–800.

McClintock B. and Creighton H.B. 1931. A correlation of cytological and genetical crossing over in *Zea mays. Proc. Natl. Acad. Sci.* **17:** 492–497.

Moore J. 1972a. *Heredity and development,* 2nd ed. Oxford University Press, Oxford.

——— 1972b. *Readings in heredity and development.* Oxford University Press, Oxford.

Morgan T.H. 1910. Sex-linked inheritance in *Drosophila. Science* **32:** 120–122.

Morgan T.H., Sturtevant A.H., Muller H.J., and Bridges C.B. 1915. *The mechanism of Mendelian heredity.* Holt, Rinehart & Winston, New York.

Muller H.J. 1927. Artificial transmutation of the gene. *Science* **46:** 84–87.

Olby R.C. 1966. *Origins of Mendelism.* Constable and Company Ltd., London.

Peters J.A. 1959. *Classic papers in genetics.* Prentice-Hall, Englewood Cliffs, New Jersey.

Rhoades M.M. 1946. Plastid mutations. *Cold Spring Harbor Symp. Quant. Biol.* **11:** 202–207.

Sager R. 1972. *Cytoplasmic genes and organelles.* Academic Press, New York.

Scott-Moncrieff R. 1936. A biochemical survey of some Mendelian factors for flower color. *J. Genetics* **32:** 117–170.

Simpson G.G. 1944. *Tempo and mode in evolution.* Columbia University Press, New York.

Sonneborn T.M. 1950. The cytoplasm in heredity. *Heredity* **4:** 11–36.

Stadler L.J. 1928. Mutations in barley induced by X-rays and radium. *Science* **110:** 543–548.

Sturtevant A.H. 1913. The linear arrangement of six sex-linked factors in *Drosophila* as shown by mode of association. *J. Exp. Zool.* **14:** 39–45.

Sturtevant A.H. and Beadle G.W. 1962. *An introduction to genetics.* Dover, New York.

Sutton W.S. 1903. The chromosome in heredity. *Biol. Bull.* **4:** 231–251.

Wilson E.B. 1925. *The cell in development and heredity,* 3rd ed. Macmillan, New York.

Wright S. 1931. Evolution in Mendelian populations. *Genetics* **16:** 97–159.

——— 1941. The physiology of the gene. *Physiol. Rev.* **21:** 487–527.

CHAPTER

2 Nucleic Acids Convey Genetic Information

OUTLINE

THAT SPECIAL MOLECULES MIGHT CARRY genetic information was appreciated by geneticists long before the problem claimed the attention of chemists. By the 1930s, geneticists began speculating as to what sort of molecules could have the kind of stability that the gene demanded, yet be capable of permanent, sudden change to the mutant forms that must provide the basis of evolution. Until the mid-1940s, there appeared to be no direct way to attack the chemical essence of the gene. It was known that chromosomes possessed a unique molecular constituent, deoxyribonucleic acid (DNA). Despite this, there was no way to show that DNA carried genetic information, as opposed to serving merely as a molecular scaffold for a still undiscovered class of proteins especially tailored to carry genetic information. It was generally assumed that genes would be composed of amino acids because, at that time, they appeared to be the only biomolecules with sufficient complexity to convey genetic information.

It therefore made sense to approach the nature of the gene by asking how genes function within cells. In the early 1940s, research on the mold *Neurospora*, spearheaded by George W. Beadle and Edward Tatum, was generating increasingly strong evidence supporting the 30-year-old hypothesis of Archibald E. Garrod that genes work by controlling the synthesis of specific enzymes (the one gene–one enzyme hypothesis). Thus, given that all known enzymes had, by this time, been shown to be proteins, the key problem was the way genes participate in the synthesis of proteins. From the very start of serious speculation, the simplest hypothesis was that genetic information within genes determines the order of the 20 different amino acids within the polypeptide chains of proteins.

In attempting to test this proposal, intuition was of little help even to the best biochemists, because there is no logical way to use enzymes as tools to determine the order of each amino acid added to a polypeptide chain. Such schemes would require, for the synthesis of a single type of protein, as many ordering enzymes as there are amino acids in the respective protein. But because all enzymes known at that time were themselves proteins (we now know that RNA can also act as an enzyme), still additional ordering enzymes would be necessary to synthesize the ordering enzymes. This situation clearly poses a paradox, unless we assume a fantastically interrelated series of syntheses in which a given protein has many different enzymatic specificities. With such an assumption, it might be possible (and then only with great difficulty) to visualize a workable cell. It did not seem likely, however, that most proteins would be found to carry out multiple tasks. In fact, all the current knowledge pointed to the opposite conclusion of one protein, one function.

AVERY'S BOMBSHELL: DNA CAN CARRY GENETIC SPECIFICITY

The idea that DNA might be the key genetic molecule emerged most unexpectedly from studies on pneumonia-causing bacteria. In 1928 English microbiologist Frederick Griffith made the startling observation that nonvirulent strains of the bacteria became virulent when mixed with their heat-killed pathogenic counterparts. That such **transformations** from nonvirulence to virulence represented hereditary changes was shown by using descendants of the newly pathogenic strains to transform still other nonpathogenic bacteria. This raised the possibility that, when pathogenic cells are killed by heat, their genetic components remain undamaged. Moreover, once liberated from the heat-killed cells, these components can pass through the cell wall of the living recipient cells and undergo subsequent genetic recombination with the recipient's genetic apparatus (Fig. 2-1). Subsequent research has confirmed this genetic interpretation. Pathogenicity reflects the action of the capsule gene, which codes for a key enzyme involved in the synthesis of the carbohydrate-containing capsule that surrounds most pneumonia-causing bacteria. When the *S* (smooth) allele of the capsule gene is present, a capsule is formed around the cell that is necessary for pathogenesis (the formation of a capsule also gives a smooth appearance to the colonies formed from these cells). When the *R* (rough) allele of this gene is present, no capsule is formed, the respective cells are not pathogenic, and the colonies these cells are round around the edges.

Within several years after Griffith's original observation, extracts of the killed bacteria were found capable of inducing hereditary transformations, and a search began for the chemical identity of the transforming agent. At that time, the vast majority of biochemists still believed that genes were proteins. It therefore came as a great surprise

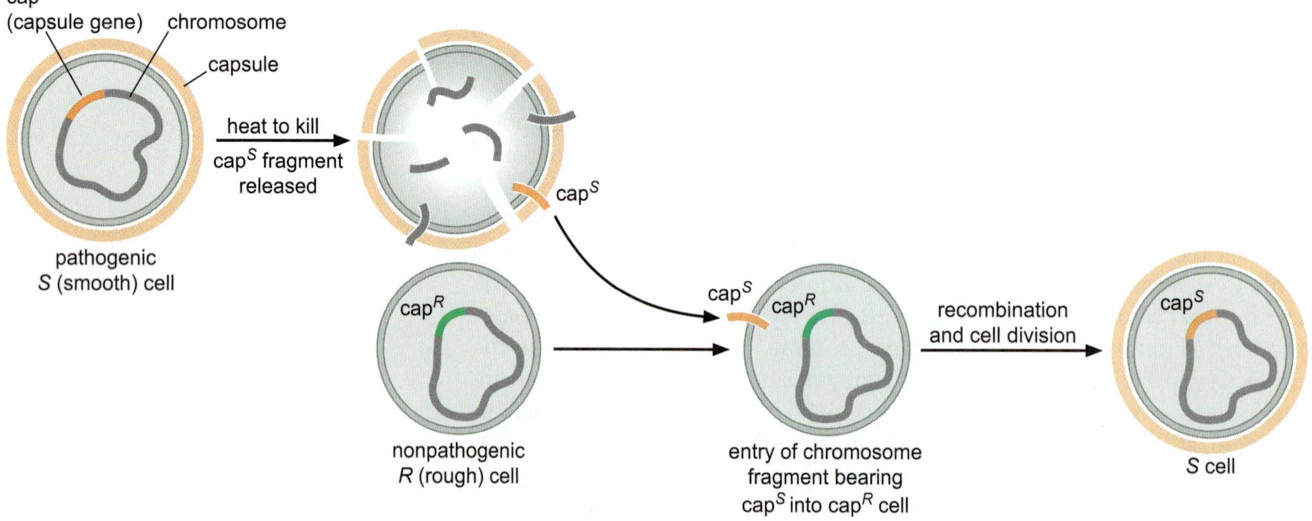

FIGURE 2-1 Transformation of a genetic characteristic of a bacterial cell (*Streptococcus pneumoniae*) by addition of heat-killed cells of a genetically different strain. Here we show an *R* cell receiving a chromosomal fragment containing the capsule gene from a heat-treated *S* cell. Since most *R* cells receive other chromosomal fragments, the efficiency of transformation for a given gene is usually less than 1%.

when in 1944, after some 10 years of research, U.S. microbiologist Oswald T. Avery and his colleagues at the Rockefeller Institute in New York, Colin M. MacLeod and Maclyn McCarty, made the momentous announcement that the active genetic principle was DNA (Fig. 2-2). Supporting their conclusion were key experiments showing that the transforming activity of their highly purified active fractions was destroyed by deoxyribonuclease, a recently purified enzyme that specifically degrades DNA molecules to their nucleotide building blocks but has no effect on the integrity of protein molecules or RNA. In contrast, the addition of either ribonuclease (which degrades RNA) or various proteolytic enzymes (which degrade proteins) had no influence on the transforming activity.

Viral Genes Are Also Nucleic Acids

Equally important confirmatory evidence came from chemical studies with viruses and virus-infected cells. By 1950 it was possible to obtain a number of essentially pure viruses and to determine which types of molecules were present in them. This work led to the very important generalization that all viruses contain nucleic acid. Because there was at that time a growing realization that viruses contain genetic material, the question immediately arose as to whether the nucleic acid component was the carrier of viral genes. A crucial test of the question came from isotopic study of the multiplication of T2, a bacterial virus (typically called a **bacteriophage,** or **phage**) composed of a DNA core and a protective shell built up by the aggregation of a number of different protein molecules. In these experiments, performed in 1952 by Alfred D. Hershey and Martha Chase working at Cold Spring Harbor Laboratory in Long Island, New York, the protein coat was labeled with the radioactive isotope ^{35}S and the DNA core with the radioactive isotope ^{32}P. The labeled virus was then used to follow the fates of the phage protein and nucleic acid as phage multiplication proceeded, particularly to see which labeled atoms from the parental phage entered the host cell and later appeared in the progeny phage.

Clear-cut results emerged from these experiments; much of the parental nucleic acid and none of the parental protein was detected in the progeny phage (Fig. 2-3). Moreover, it was possible to show that little of the parental protein even enters the bacteria; instead, it stays attached to the outside of the bacterial cell, performing no function after the DNA component has passed inside. This point was neatly shown by violently agitating infected bacteria after the entrance of the DNA; the protein coats were shaken off without affecting the ability of the bacteria to form new phage particles.

With some viruses it is now possible to do an even more convincing experiment. For example, purified DNA from the mouse polyoma virus can enter mouse cells and initiate a cycle of viral multiplication producing many thousands of new polyoma particles. The primary function of viral protein is thus to protect and transport its genetic/nucleic acid component in its movement from one cell to another.

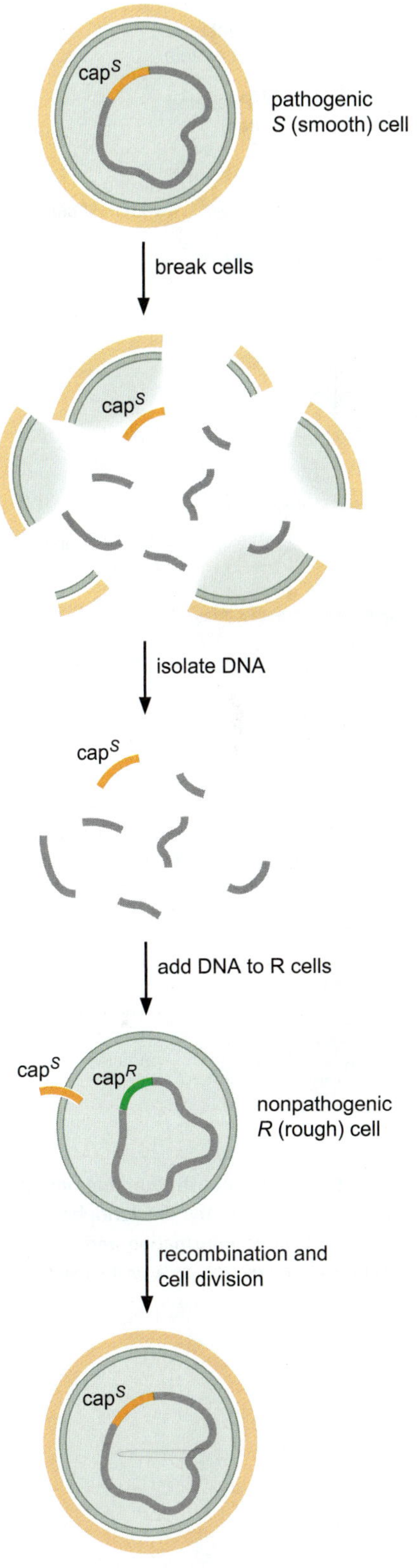

FIGURE 2-2 Isolation of a chemically pure transforming agent. (Adapted, with permission, from Stahl F.W. 1964. *The mechanics of inheritance*, Fig. 2.3. © Pearson Education, Inc.)

THE DOUBLE HELIX

While work was proceeding on the X-ray analysis of protein structure, a smaller number of scientists were trying to solve the X-ray diffraction pattern of DNA. The first diffraction patterns were taken in 1938

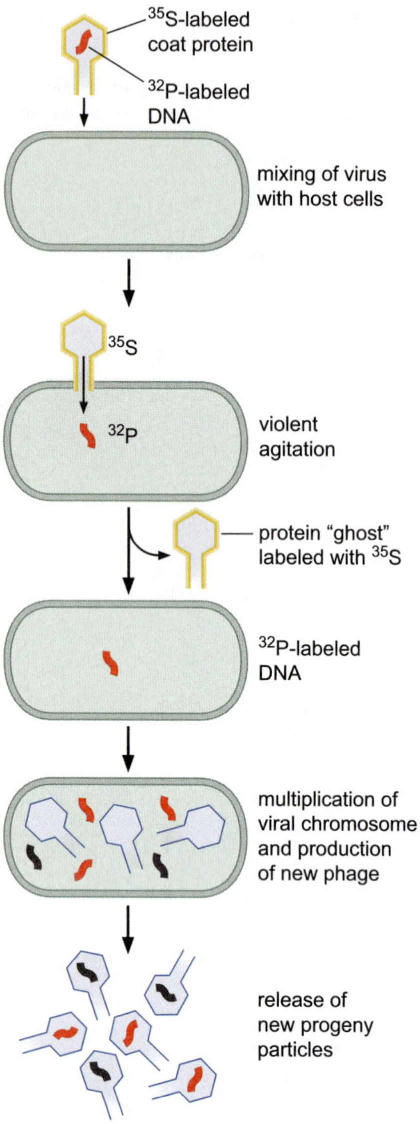

FIGURE 2-3 Demonstration that only the DNA component of the bacteriophage T2 carries the genetic information and that the protein coat serves only as a protective shell.

by William Astbury using DNA supplied by Ola Hammarsten and Torbjörn Caspersson. It was not until the early 1950s that high-quality X-ray diffraction photographs were taken by Maurice Wilkins and Rosalind Franklin (Fig. 2-4). These photographs suggested not only that the underlying DNA structure was helical but that it was composed of more than one polynucleotide chain—either two or three. At the same time, the covalent bonds of DNA were being unambiguously established. In 1952 a group of organic chemists working in the laboratory of Alexander Todd showed that 3′–5′ phosphodiester bonds regularly link together the nucleotides of DNA (Fig. 2-5).

In 1951, because of interest in Linus Pauling's α helix protein motif (which we shall consider in Chapter 5), an elegant theory of diffraction of helical molecules was developed by William Cochran, Francis H. Crick, and Vladimir Vand. This theory made it easy to test possible DNA structures on a trial-and-error basis. The correct solution, a complementary double helix (see Chapter 6), was found in 1953 by Crick and James D. Watson, then working in the laboratory of Max Perutz and John Kendrew in Cambridge, United Kingdom. Their arrival at the correct answer depended largely on finding the stereochemically most favorable configuration compatible with the X-ray diffraction data of Wilkins and Franklin.

In the double helix, the two DNA chains are held together by hydrogen bonds (a weak noncovalent chemical bond; see Chapter 3) between pairs of bases on the opposing strands (Fig. 2-6). This base pairing is very specific: the purine adenine only base-pairs to the pyrimidine thymine, whereas the purine guanine only base-pairs to the pyrimidine cytosine. In double-helical DNA, the number of A residues must be equal to the number of T residues, whereas the number of G and C residues must likewise be equal (see Box 2-1, Chargaff's Rules). As a result, the sequence of the bases of the two chains of a given double helix have a complementary relationship, and the sequence of any DNA strand exactly defines that of its partner strand.

The discovery of the double helix initiated a profound revolution in the way many geneticists analyzed their data. The gene was no longer

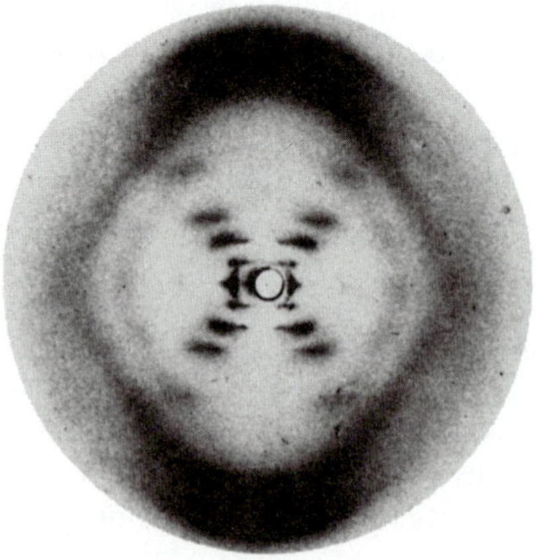

FIGURE 2-4 The key X-ray photograph involved in the elucidation of the DNA structure. This photograph, taken by Rosalind Franklin at King's College, London, in the winter of 1952–1953, confirmed the guess that DNA was helical. The helical form is indicated by the crossways pattern of X-ray reflections (photographically measured by darkening of the X-ray film) in the center of the photograph. The very heavy black regions at the top and bottom reveal that the 3.4-Å-thick purine and pyrimidine bases are regularly stacked next to each other, perpendicular to the helical axis. (Printed, with permission, from Franklin R.E. and Gosling R.G. 1953. *Nature* 171: 740–741. © Macmillan.)

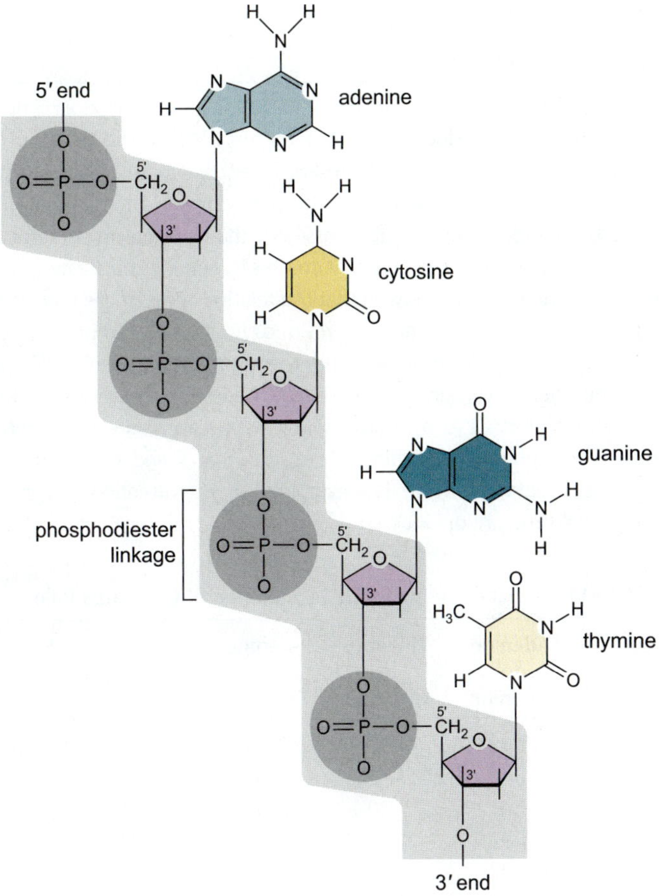

FIGURE 2-5 A portion of a DNA polynucleotide chain, showing the 3′ → 5′ phosphodiester linkages that connect the nucleotides. Phosphate groups connect the 3′ carbon of one nucleotide with the 5′ carbon of the next.

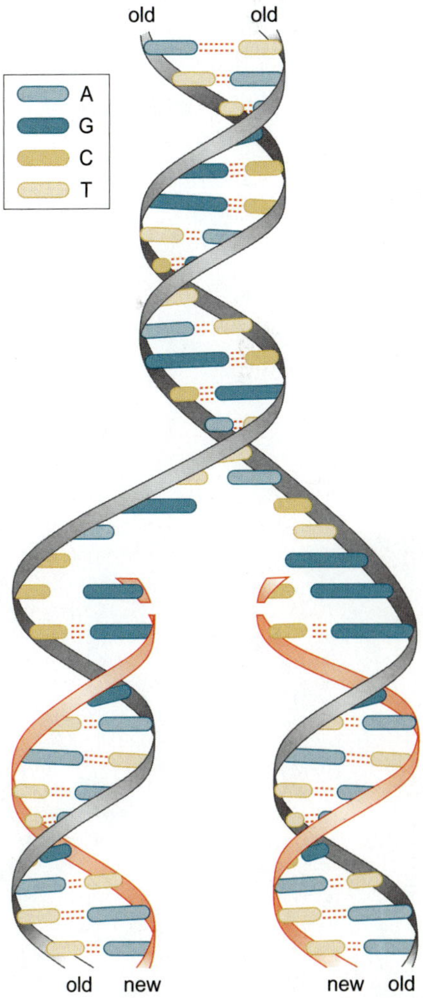

FIGURE 2-6 The replication of DNA. The newly synthesized strands are shown in orange.

a mysterious entity, the behavior of which could be investigated only by genetic experiments. Instead, it quickly became a real molecular object about which chemists could think objectively, as they did about smaller molecules such as pyruvate and ATP. Most of the excitement, however, came not merely from the fact that the structure was solved, but also from the nature of the structure. Before the answer was known, there had always been the worry that it would turn out to be dull, revealing nothing about how genes replicate and function. Fortunately, the answer was immensely exciting. The two intertwined strands of complementary structures suggested that one strand serves as the specific surface (template) upon which the other strand is made (Fig. 2-6). If this hypothesis were true, then the fundamental problem of gene replication, about which geneticists had puzzled for so many years, was, in fact, conceptually solved.

Finding the Polymerases That Make DNA

Rigorous proof that a single DNA chain is the template that directs the synthesis of a complementary DNA chain had to await the development of test-tube (in vitro) systems for DNA synthesis. These came much faster than anticipated by molecular geneticists, whose world until then had been far removed from that of the biochemist

KEY EXPERIMENTS

Box 2-1 Chargaff's Rules

Biochemist Erwin Chargaff used a technique called "paper chromatography" to analyze the nucleotide composition of DNA. By 1949 his data showed not only that the four different nucleotides are not present in equal amounts, but also that the exact ratios of the four nucleotides vary from one species to another (Box 2-1 Table 1). These findings opened up the possibility that it is the precise arrangement of nucleotides within a DNA molecule that confers its genetic specificity.

Chargaff's experiments also showed that the relative ratios of the four bases were not random. The number of adenine (A) residues in all DNA samples was equal to the number of thymine (T) residues, and the number of guanine (G) residues equaled the number of cytosine (C) residues. In addition, regardless of the DNA source, the ratio of purines to pyrimidines was always approximately 1 (purines = pyrimidines). The fundamental significance of the A = T and G = C relationships (Chargaff's rules) could not emerge, however, until serious attention was given to the three-dimensional structure of DNA.

BOX 2-1 TABLE 1 Data Leading to the Formulation of Chargaff's Rules

Source	Adenine to Guanine	Thymine to Cytosine	Adenine to Thymine	Guanine to Cytosine	Purines to Pyrimidines
Ox	1.29	1.43	1.04	1.00	1.1
Human	1.56	1.75	1.00	1.00	1.0
Hen	1.45	1.29	1.06	0.91	0.99
Salmon	1.43	1.43	1.02	1.02	1.02
Wheat	1.22	1.18	1.00	0.97	0.99
Yeast	1.67	1.92	1.03	1.20	1.0
Hemophilus influenzae	1.74	1.54	1.07	0.91	1.0
Escherichia coli K2	1.05	0.95	1.09	0.99	1.0
Avian tubercle bacillus	0.4	0.4	1.09	1.08	1.1
Serratia marcescens	0.7	0.7	0.95	0.86	0.9
Bacillus schatz	0.7	0.6	1.12	0.89	1.0

After Chargaff E. et al. 1949. *J. Biol. Chem.* 177: 405.

well versed in the procedures needed for enzyme isolation. Leading this biochemical assault on DNA replication was U.S. biochemist Arthur Kornberg, who by 1956 had demonstrated DNA synthesis in cell-free extracts of bacteria. Over the next several years, Kornberg went on to show that a specific polymerizing enzyme was needed to catalyze the linking together of the building-block precursors of DNA. Kornberg's studies revealed that the nucleotide building blocks for DNA are energy-rich precursors (dATP, dGTP, dCTP, and dTTP; Fig. 2-7). Further studies identified a single polypeptide, DNA polymerase I (DNA Pol I), that was capable of catalyzing the synthesis of new DNA strands. It links the nucleotide precursors by 3′–5′ phosphodiester bonds (Fig. 2-8). Furthermore, it works only in the presence of DNA, which is needed to order the four nucleotides in the polynucleotide product.

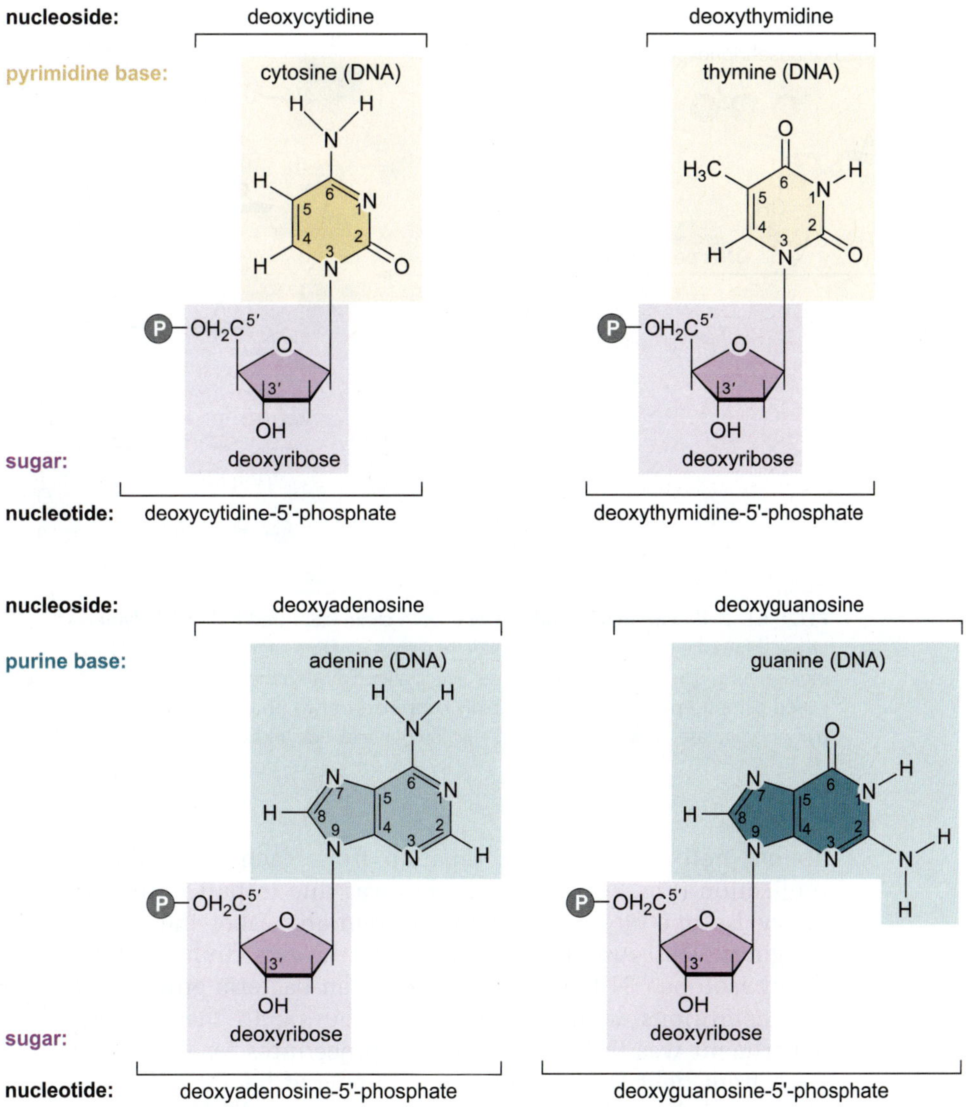

FIGURE 2-7 **The nucleotides of DNA.** The structures of the different components of each of the four nucleotides are shown.

DNA Pol I depends on a DNA template to determine the sequence of the DNA it is synthesizing. This was first demonstrated by allowing the enzyme to work in the presence of DNA molecules that contained varying amounts of A:T and G:C base pairs. In every case, the enzymatically synthesized product had the base ratios of the template DNA (Table 2-1). During this cell-free synthesis, no synthesis of proteins or any other molecular class occurs, unambiguously eliminating any non-DNA compounds as intermediate carriers of genetic specificity. Thus, there is no doubt that DNA is the direct template for its own formation.

Experimental Evidence Favors Strand Separation during DNA Replication

Simultaneously with Kornberg's research, in 1958 Matthew Meselson and Franklin W. Stahl, then at the California Institute of Technology, carried out an elegant experiment in which they separated daughter DNA molecules and, in so doing, showed that the two strands of the

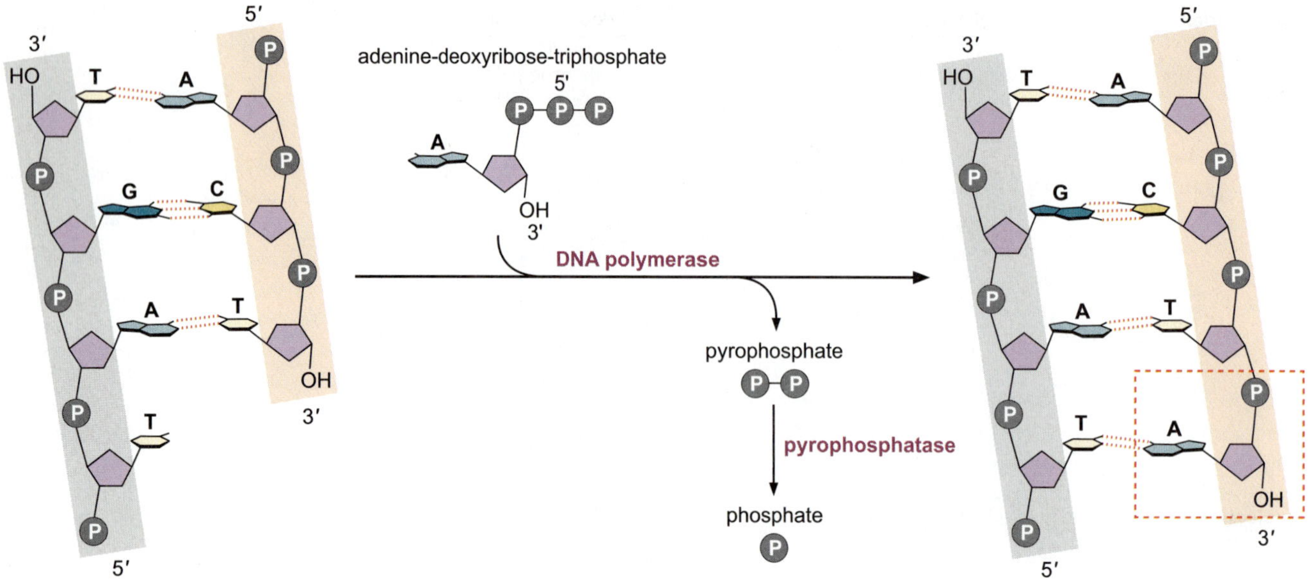

FIGURE 2-8 Enzymatic synthesis of a DNA chain catalyzed by DNA polymerase I. This image shows the addition of a nucleotide to a growing DNA strand as catalyzed by DNA polymerase. Although the DNA polymerase can catalyze DNA synthesis by itself, in the cell the released pyrophosphate molecule is rapidly converted to two phosphates by an enzyme called pyrophosphatase, making the forward reaction of nucleotide addition even more favorable.

double helix permanently separate from each other during DNA replication (Fig. 2-9). Their success was due in part to the use of the heavy isotope ^{15}N as a tag to differentially label the parental and daughter DNA strands. Bacteria grown in a medium containing the heavy isotope ^{15}N have denser DNA than bacteria grown under normal conditions with ^{14}N. Also contributing to the success of the experiment was the development of procedures for separating heavy DNA from light DNA in density gradients of heavy salts like cesium chloride. When high centrifugal forces are applied, the solution becomes more dense at the bottom of the centrifuge tube (which, when spinning, is the farthest from the axis of rotation). When the correct initial solution density is chosen, the individual DNA molecules will move to the central region of the centrifuge tube, where their density

TABLE 2-1 A Comparison of the Base Composition of Enzymatically Synthesized DNAs and Their DNA Templates

	Base Composition of the Enzymatic Product				$\frac{A+T}{G+C}$	$\frac{A+T}{G+C}$
Source of DNA Template	Adenine	Thymine	Guanine	Cytosine	In Product	In Template
Micrococcus lysodeikticus (a bacterium)	0.15	0.15	0.35	0.35	0.41	0.39
Aerobacter aerogenes (a bacterium)	0.22	0.22	0.28	0.28	0.80	0.82
Escherichia coli	0.25	0.25	0.25	0.25	1.00	0.97
Calf thymus	0.29	0.28	0.21	0.22	1.32	1.35
Phage T2	0.32	0.32	0.18	0.18	1.78	1.84

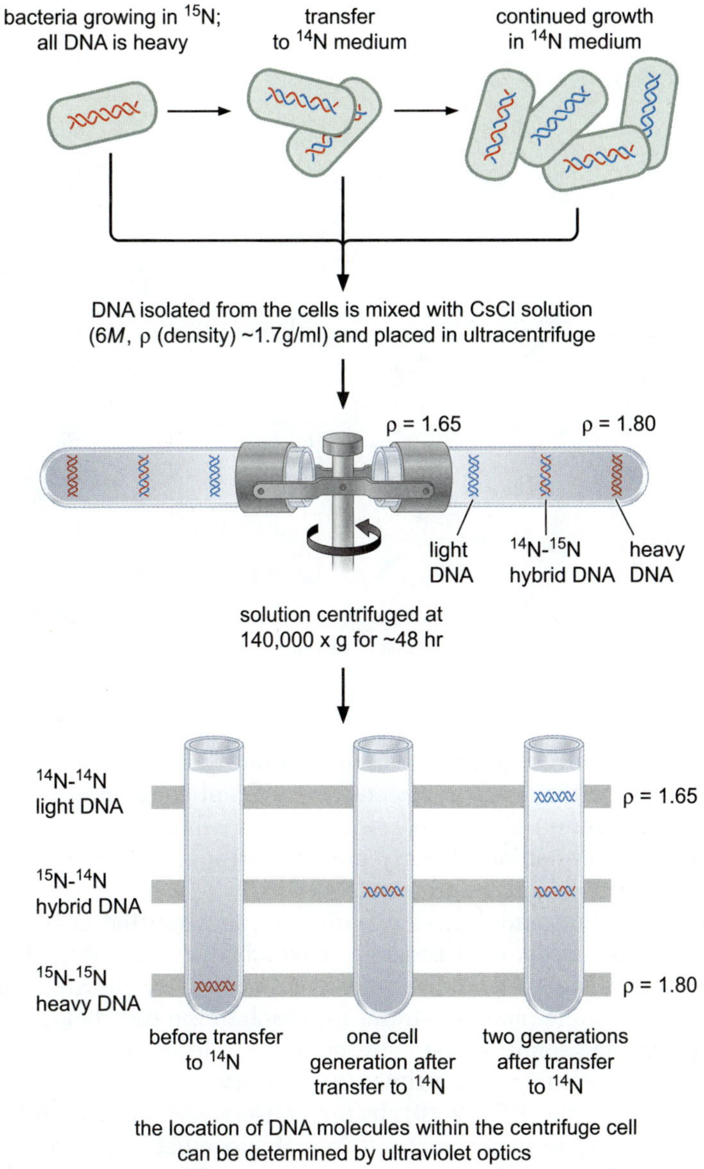

bacteria growing in 15N; all DNA is heavy
transfer to 14N medium
continued growth in 14N medium
DNA isolated from the cells is mixed with CsCl solution (6M, ρ (density) ~1.7g/ml) and placed in ultracentrifuge
ρ = 1.65
ρ = 1.80
light DNA
14N-15N hybrid DNA
heavy DNA
solution centrifuged at 140,000 x g for ~48 hr
14N-14N light DNA
15N-14N hybrid DNA
15N-15N heavy DNA
ρ = 1.65
ρ = 1.80
before transfer to 14N
one cell generation after transfer to 14N
two generations after transfer to 14N
the location of DNA molecules within the centrifuge cell can be determined by ultraviolet optics

FIGURE 2-9 Use of a cesium chloride (CsCl) density gradient to demonstrate the separation of complementary strands during DNA replication.

equals that of the salt solution. In this situation, DNA molecules in which both strands are composed of entirely ^{15}N precursors (heavy–heavy or HH DNA) will form a band at a higher density (closer to the bottom of the tube) than DNA molecules in which both strands are composed entirely of ^{14}N precursors (light–light or LL DNA). If bacteria containing heavy DNA are transferred to a light medium (containing ^{14}N) and allowed to grow, the precursor nucleotides available for use in DNA synthesis will be light; hence, DNA synthesized after transfer will be distinguishable from DNA made before transfer.

If DNA replication involves strand separation, definite predictions can be made about the density of the DNA molecules found after various growth intervals in a light medium. After one generation of growth, all the DNA molecules should contain one heavy strand and one light strand and thus be of intermediate density (heavy–light or HL DNA). This result is exactly what Meselson and Stahl observed. Likewise, after two generations of growth, half the DNA molecules were light and half hybrid, just as strand separation predicts. It is important to note that during isolation from the bacteria the DNA was

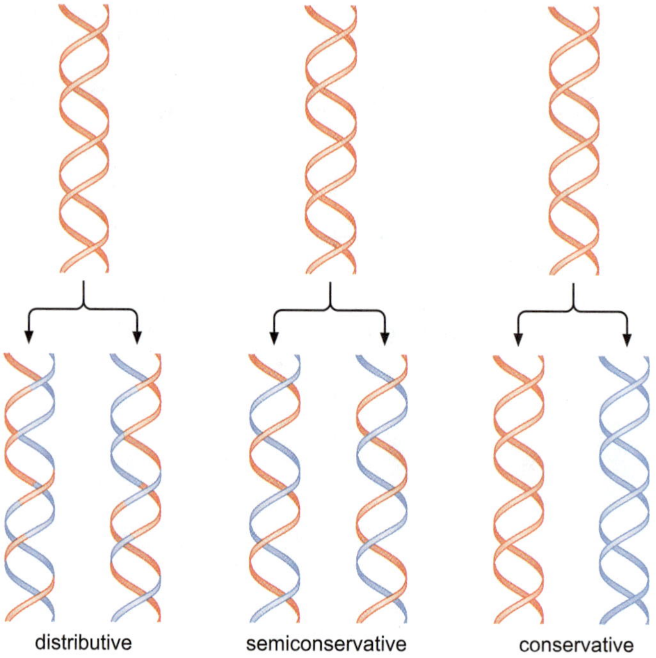

FIGURE 2-10 Three possible mechanisms for DNA replication. When the structure of DNA was discovered, several models were proposed to explain how it was replicated; three are illustrated here. The experiments proposed by Meselson and Stahl clearly distinguished among these models, demonstrating that DNA was replicated semiconservatively.

broken into small fragments, which ensured that the vast majority of the DNA was either fully replicated or not replicated at all. If the entire bacterial genome was maintained intact, then there would have been many intermediate-density molecules (neither HH, HL, nor LL) that were only partially replicated.

Thus, Meselson and Stahl's experiments showed that DNA **replication** is a semiconservative process in which the single strands of the double helix remain intact (are conserved) during a replication process that distributes one parental strand into each of the two daughter molecules (thus the "semi" in semiconservative). These experiments ruled out two other models at the time: the conservative and the dispersive replication schemes (Fig. 2-10). In the conservative model, both of the parental strands were proposed to remain together and the two new strands of DNA would form an entirely new DNA molecule. In this model, fully light DNA would be formed after one cell generation. In the dispersive model, which was favored by many at the time, the DNA strands were proposed to be broken as frequently as every ten base pairs and used to prime the synthesis of similarly short regions of DNA. These short DNA fragments would subsequently be joined to form complete DNA strands. In this complex model, all DNA strands would be composed of both old and new DNA (thus nonconservative) and fully light DNA would only be observed after many generations of growth.

THE GENETIC INFORMATION WITHIN DNA IS CONVEYED BY THE SEQUENCE OF ITS FOUR NUCLEOTIDE BUILDING BLOCKS

The finding of the double helix had effectively ended any controversy about whether DNA was the primary genetic substance. Even before strand separation during DNA replication was experimentally verified, the main concern of molecular geneticists had turned to how the genetic information of DNA functions to order amino acids during protein synthesis (see Box 2-2, Evidence that Genes Control Amino Acid Sequences

KEY EXPERIMENTS

Box 2-2 Evidence That Genes Control Amino Acid Sequences in Proteins

The first experimental evidence that genes (DNA) control amino acid sequences arose from the study of the hemoglobin present in humans suffering from the genetic disease sickle-cell anemia. If an individual has the *S* allele of the β-globin gene (which encodes one of the two polypeptides that together form hemoglobin) present in both homologous chromosomes (*SS*), a severe anemia results, characterized by the red blood cells having a sickle-cell shape. If only one of the two alleles of the β-globin gene are of the *S* form (+*S*), the anemia is less severe and the red blood cells appear almost normal in shape. The type of hemoglobin in red blood cells correlates with the genetic pattern. In the *SS* case, the hemoglobin is abnormal, characterized by a solubility different from that of normal hemoglobin, whereas in the +*S* condition, half the hemoglobin is normal and half abnormal.

Wild-type hemoglobin molecules are constructed from two kinds of polypeptide chains: α chains and β chains (see Box 2-2 Fig. 1). Each chain has a molecular weight of about 16,100 daltons (D). Two α chains and two β chains are present in each molecule, giving hemoglobin a molecular weight of about 64,400 D. The α chains and β chains are controlled by distinct genes so that a single mutation will affect either the α chain or the β chain, but not both. In 1957, Vernon M. Ingram at Cambridge University showed that sickle hemoglobin differs from normal hemoglobin by the change of one amino acid in the β chain: at position 6, the glutamic acid residue found in wild-type hemoglobin is replaced by valine. Except for this one change, the entire amino acid sequence is identical in normal and mutant hemoglobin. Because this change in amino acid sequence was observed only in patients with the *S* allele of the β-globin gene, the simplest hypothesis is that the *S* allele of the gene encodes the change in the β-globin gene. Subsequent studies of amino acid sequences in hemoglobin isolated from other forms of anemia completely supported this proposal; sequence analysis showed that each specific anemia is characterized by a single amino acid replacement at a unique site along the polypeptide chain (Box 2-2 Fig. 2).

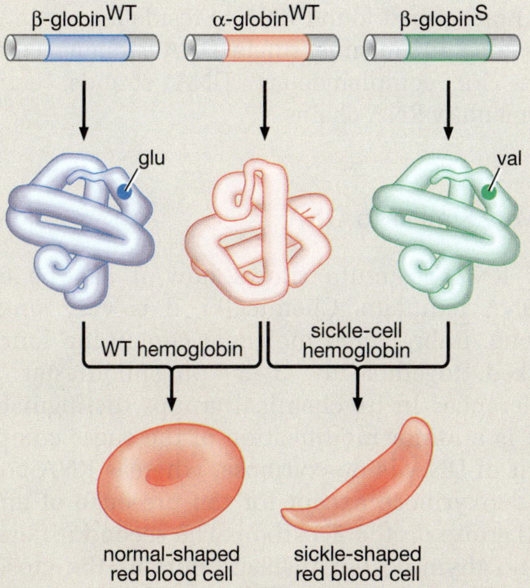

BOX 2-2 FIGURE 1 Formation of wild-type and sickle-cell hemoglobin. (Source of hemoglobin structures: Illustration, Irving Geis. Rights owned by Howard Hughes Medical Institute. Not to be reproduced without permission.)

α chain									
	position	1	2	16	30	57	58	68	141
	amino acid	Val	Leu	Lys^+	Glu^-	Gly	His^+	AspN	Arg
Hb variant	Hb I			Asp^-					
	Hb G Honolulu				GluN				
	Hb Norfolk					Asp^-			
	Hb M Boston						Tyr		
	Hb G Philadelphia							Lys^+	

β chain											
	position	1	2	3	6	7	26	63	67	125	150
	amino acid	Val	His^+	Leu	Glu^-	Glu^-	Glu^-	His^+	Val	Glu	His^+
Hb variant	Hb S				Val						
	Hb C				Lys^+						
	Hb G San José					Gly					
	Hb E						Lys^+				
	Hb M Saskatoon							Tyr			
	Hb Zürich							Arg^+			
	Hb M Milwaukee-1								Glu^-		
	Hb D β Punjab									GluN	

BOX 2-2 FIGURE 2 A summary of some established amino acid substitutions in human hemoglobin variants.

in Proteins). With all DNA chains capable of forming double helices, the essence of their genetic specificity had to reside in the linear sequences of their four nucleotide building blocks. Thus, as information-containing entities, DNA molecules were by then properly regarded as very long words (as we shall see later, they are now best considered very long sentences) built up from a four-letter alphabet (A, G, C, and T). Even with only four letters, the number of potential DNA sequences (4^N, where N is the number of letters in the sequence) is very, very large for even the smallest of DNA molecules; a virtually infinite number of different genetic messages can exist. Now we know that a typical bacterial gene is made up of approximately 1000 base pairs. The number of potential genes of this size is 4^{1000}, a number that is orders of magnitude larger than the number of known genes in any organism.

DNA Cannot Be the Template That Directly Orders Amino Acids during Protein Synthesis

Although DNA must carry the information for ordering amino acids, it was quite clear that the double helix itself could not be the template for protein synthesis. Experiments showing that protein synthesis occurs at sites where DNA is absent ruled out a direct role for DNA. Protein synthesis in all eukaryotic cells occurs in the cytoplasm, which is separated by the nuclear membrane from the chromosomal DNA.

Therefore, at least for eukaryotic cells, a second information-containing molecule had to exist that obtains its genetic specificity from DNA. This molecule would then move to the cytoplasm to function as the template for protein synthesis. Attention from the start focused on the still functionally obscure second class of nucleic acids, RNA. Torbjörn Caspersson and Jean Brachet had found RNA to reside largely in the cytoplasm; and it was easy to imagine single DNA strands, when not serving as templates for complementary DNA strands, acting as templates for complementary RNA chains.

RNA Is Chemically Very Similar to DNA

Mere inspection of RNA structure shows how it can be exactly synthesized on a DNA template. Chemically, it is very similar to DNA. It, too, is a long, unbranched molecule containing four types of nucleotides linked together by 3′–5′ phosphodiester bonds (Fig. 2-11). Two differences in its chemical groups distinguish RNA from DNA. The first is a minor modification of the sugar component (Fig. 2-12). The sugar of DNA is deoxyribose, whereas RNA contains ribose, identical to deoxyribose except for the presence of an additional OH (hydroxyl) group on the 2′ carbon. The second difference is that RNA contains no thymine but instead contains the closely related pyrimidine uracil. Despite these differences, however, polyribonucleotides have the potential for forming complementary helices of the DNA type. Neither the additional hydroxyl group nor the absence of the methyl group found in thymine but not in uridine affects RNA's ability to form double-helical structures held together by base pairing. Unlike DNA, however, RNA is typically found in the cell as a single-stranded molecule. If double-stranded RNA helices are formed, they most often are composed of two parts of the same single-stranded RNA molecule.

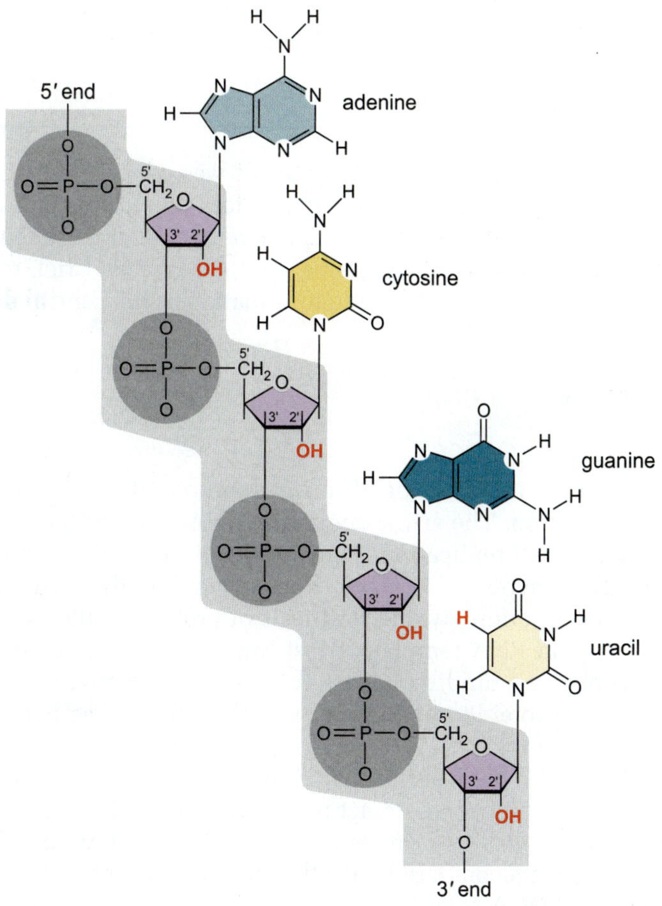

FIGURE 2-11 **A portion of a polyribonucleotide (RNA) chain.** Elements in red are distinct from DNA.

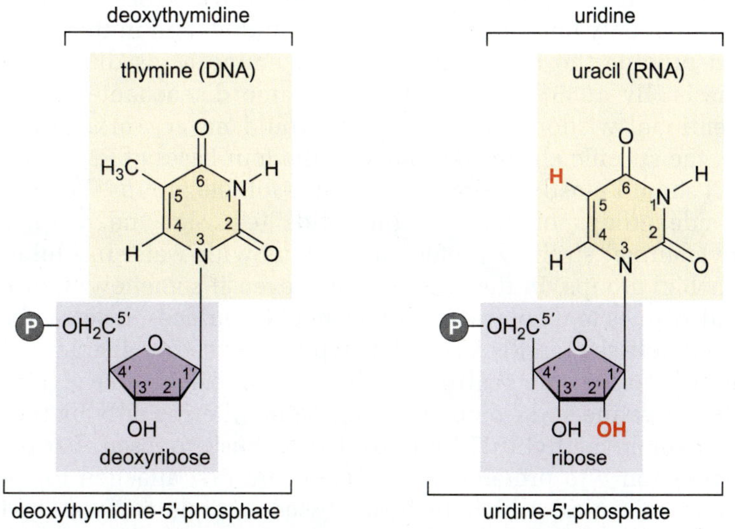

FIGURE 2-12 **Distinctions between the nucleotides of RNA and DNA.** A nucleotide of DNA is shown next to a nucleotide of RNA. All RNA nucleotides have the sugar ribose (instead of deoxyribose for DNA), which has a hydroxyl group on the 2′ carbon (shown in red). In addition, RNA has the pyrimidine base uracil instead of thymine. Uracil has a hydrogen at the 5 position of the pyrimidine ring (shown in red) rather than the methyl group found in that position for thymine. The three other bases that occur in DNA and RNA are identical.

THE CENTRAL DOGMA

By the fall of 1953, the working hypothesis was adopted that chromosomal DNA functions as the template for RNA molecules, which subsequently move to the cytoplasm, where they determine the arrangement of amino acids within proteins. In 1956 Francis Crick referred to this pathway for the flow of genetic information as the **central dogma**:

$$\text{Duplication} \quad \circlearrowright \text{DNA} \xrightarrow{\text{Transcription}} \text{RNA} \xrightarrow{\text{Translation}} \text{Protein.}$$

Here the arrows indicate the directions proposed for the transfer of genetic information. The arrow encircling DNA signifies that DNA is the template for its self-replication. The arrow between DNA and RNA indicates that RNA synthesis (called **transcription**) is directed by a DNA template. Correspondingly, the synthesis of proteins (called **translation**) is directed by an RNA template. Most importantly, the last two arrows were presented as unidirectional; that is, RNA sequences are never determined by protein templates nor was DNA then imagined ever to be made on RNA templates. The idea that proteins never serve as templates for RNA has stood the test of time. However, as we will see in Chapter 11, RNA chains sometimes do act as templates for DNA chains of complementary sequence. Such reversals of the normal flow of information are very rare events compared with the enormous number of RNA molecules made on DNA templates. Thus, the central dogma as originally proclaimed more than 50 years ago still remains essentially valid.

The Adaptor Hypothesis of Crick

At first it seemed simplest to believe that the RNA templates for protein synthesis were folded up to create cavities on their outer surfaces specific for the 20 different amino acids. The cavities would be so shaped that only one given amino acid would fit, and in this way RNA would provide the information to order amino acids during protein synthesis. By 1955, however, Crick became disenchanted with this conventional wisdom, arguing that it would never work. In the first place, the specific chemical groups on the four bases of RNA (A, U, G, and C) should mostly interact with water-soluble groups. Yet, the specific side groups of many amino acids (e.g., leucine, valine, and phenylalanine) strongly prefer interactions with water-insoluble (hydrophobic) groups. In the second place, even if somehow RNA could be folded so as to display some hydrophobic surfaces, it seemed at the time unlikely that an RNA template would be used to discriminate accurately between chemically very similar amino acids like glycine and alanine or valine and isoleucine, both pairs differing only by the presence of single methyl (CH_3) groups. Crick thus proposed that prior to incorporation into proteins, amino acids are first attached to specific adaptor molecules, which in turn possess unique surfaces that can bind specifically to bases on the RNA templates.

Discovery of Transfer RNA

The discovery of how proteins are synthesized required the development of cell-free extracts capable of making proteins from amino acid precursors as directed by added RNA molecules. These were first effectively developed beginning in 1953 by Paul C. Zamecnik and his collab-

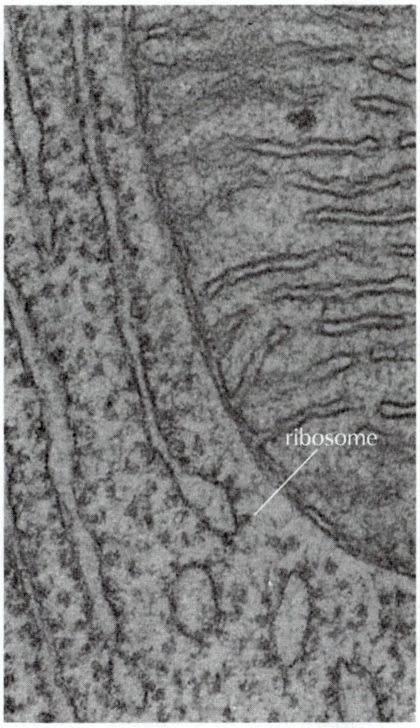

FIGURE 2-13 Electron micrograph of ribosomes attached to the endoplasmic reticulum. This electron micrograph (105,000x) shows a portion of a pancreatic cell. The upper right portion shows a portion of the mitochondrion and the lower left shows a large number of ribosomes (small circles of electron density) attached to the endoplasmic reticulum. Some ribosomes exist free in the cytoplasm; others are attached to the membranous endoplasmic reticulum. (Courtesy of K.R. Porter.)

orators. Key to their success were the recently available radioactively tagged amino acids, which they used to mark the trace amounts of newly made proteins, as well as high-quality, easy-to-use, preparative ultracentrifuges for fractionation of their cellular extracts. Early on, the cellular site of protein synthesis was pinpointed to be the ribosomes, small RNA-containing particles in the cytoplasm of all cells engaged in protein synthesis (Fig. 2-13).

Several years later, Zamecnik, by then collaborating with Mahlon B. Hoagland, went on to make the seminal discovery that prior to their incorporation into proteins, amino acids are first attached to what we now call **transfer RNA (tRNA)** molecules. Transfer RNA accounts for some 10% of all cellular RNA (Fig. 2-14).

To nearly everyone except Crick, this discovery was totally unexpected. He had, of course, previously speculated that his proposed "adaptors" might be short RNA chains, because their bases would be able to base-pair and "read" the appropriate groups on the RNA molecules that served as the templates for protein synthesis. As we shall relate later in greater detail (Chapter 14), the transfer RNA molecules of Zamecnik and Hoagland are in fact the adaptor molecules postulated by Crick. Each transfer RNA contains a sequence of adjacent bases (the anticodon) that bind specifically during protein synthesis to successive groups of bases (codons) along the RNA template.

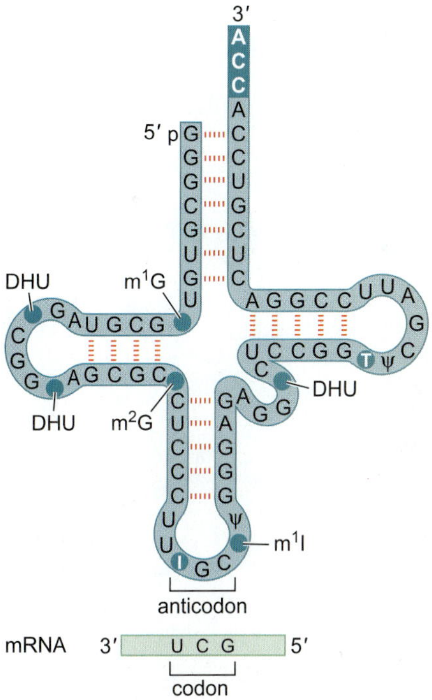

FIGURE 2-14 Yeast alanine tRNA structure, as determined by Robert W. Holley and his associates. The anticodon in this tRNA recognizes the codon for alanine in the mRNA. Several modified nucleosides exist in the structure: ψ = pseudouridine, T = ribothymidine, DHU = 5,6-dihydrouridine, I = inosine, m^1G = 1-methylguanosine, m^1I = 1-methylinosine, and m^2G = *N*,*N*-dimethylguanosine.

The Paradox of the Nonspecific-Appearing Ribosomes

About 85% of cellular RNA is found in ribosomes, and because its absolute amount is greatly increased in cells engaged in large-scale protein synthesis (e.g., pancreas cells and rapidly growing bacteria), **ribosomal RNA (rRNA)** was initially thought to be the template for ordering amino acids. But once the ribosomes of *Escherichia coli* were carefully analyzed, several disquieting features emerged. First, all *E. coli* ribosomes, as well as those from all other organisms, are composed of two unequally sized subunits, each containing RNA, that either stick together or fall apart in a reversible manner, depending on the surrounding ion concentration. Second, all the rRNA chains within the small subunits are of similar chain lengths (about 1500 bases in *E. coli*), as are the rRNA chains of the large subunits (about 3000 bases). Third, the base composition of both the small and large rRNA chains is approximately the same (high in G and C) in all known bacteria, plants, and animals, despite wide variations in the AT/GC ratios of their respective DNA. This was not to be expected if the rRNA chains were in fact a large collection of different RNA templates derived from a large number of different genes. Thus, neither the small nor large class of rRNA had the feel of template RNA.

Discovery of Messenger RNA (mRNA)

Cells infected with phage T4 provided the ideal system to find the true template. Following infection by this virus, cells stop synthesizing *E. coli* RNA; the only RNA synthesized is transcribed off the T4 DNA. Most strikingly, not only does T4 RNA have a base composition very similar to T4 DNA, but it does not bind to the ribosomal proteins that normally associate with rRNA to form ribosomes. Instead, after first attaching to previously existing ribosomes, T4 RNA moves across their surface to bring its bases into positions where they can bind to the appropriate tRNA–amino acid precursors for protein synthesis (Fig. 2-15). In so acting, T4 RNA orders the amino acids and is thus the long-sought-for RNA template for protein synthesis. Because it carries

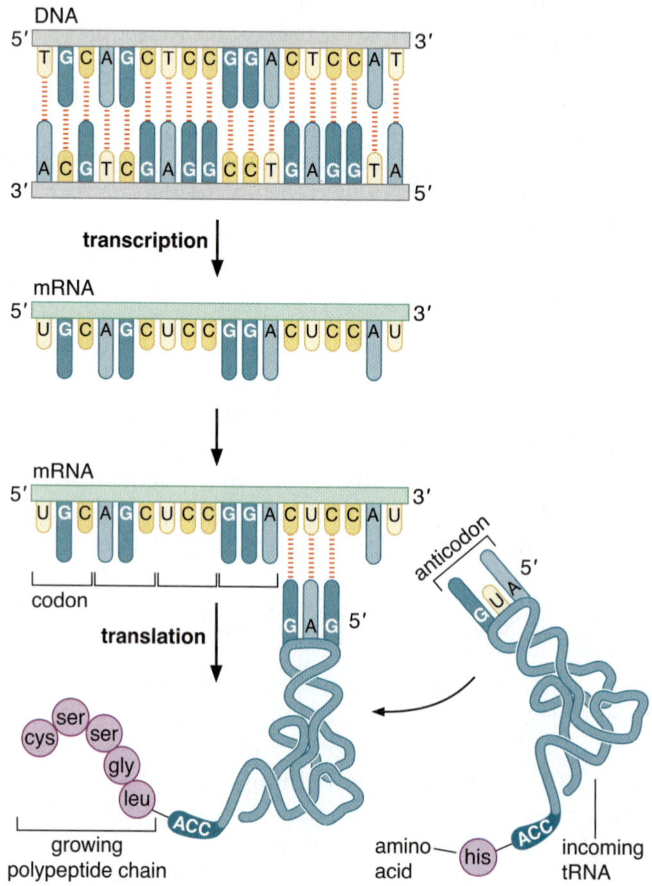

FIGURE 2-15 Transcription and translation. The nucleotides of mRNA are assembled to form a complementary copy of one strand of DNA. Each group of three is a codon that is complementary to a group of three nucleotides in the anticodon region of a specific tRNA molecule. When base pairing occurs, an amino acid carried at the other end of the tRNA molecule is added to the growing protein chain.

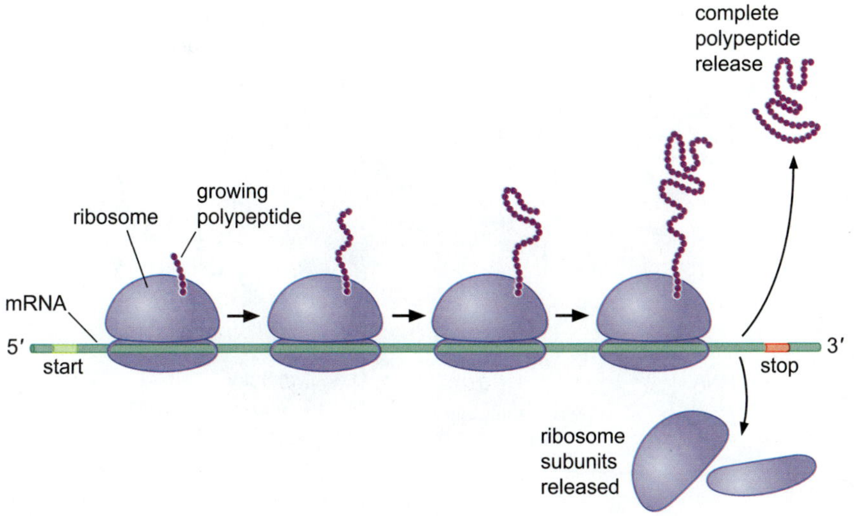

FIGURE 2-16 Diagram of a polyribosome. Each ribosome attaches at a start signal at the 5′ end of an mRNA chain and synthesizes a polypeptide as it proceeds along the molecule. Several ribosomes may be attached to one mRNA molecule at one time; the entire assembly is called a polyribosome.

the information from DNA to the ribosomal sites of protein synthesis, it is called **messenger RNA (mRNA)**. The observation of T4 RNA binding to *E. coli* ribosomes, first made in the spring of 1960, was soon followed with evidence for a separate messenger class of RNA within uninfected *E. coli* cells, thereby definitively ruling out a template role for any rRNA. Instead, in ways that we discuss more extensively in Chapter 14, the rRNA components of ribosomes, together with some 50 different ribosomal proteins that bind to them, serve as the factories for protein synthesis, functioning to bring the tRNA–amino acid precursors into positions where they can read off the information provided by the mRNA templates.

Only a few percent of total cellular RNA is mRNA. This RNA shows the expected large variations in length and nucleotide composition required to encode the many different proteins found in a given cell. Hence, it is easy to understand why mRNA was first overlooked. Because only a small segment of mRNA is attached at a given moment to a ribosome, a single mRNA molecule can simultaneously be read by several ribosomes. Most ribosomes are found as parts of **polyribosomes** (groups of ribosomes translating the same mRNA), which can include more than 50 members (Fig. 2-16).

Enzymatic Synthesis of RNA upon DNA Templates

As mRNA was being discovered, the first of the enzymes that synthesize (or transcribe) RNA using DNA templates was being independently isolated in the labs of biochemists Jerard Hurwitz and Samuel B. Weiss. Called **RNA polymerases,** these enzymes function only in the presence of DNA, which serves as the template upon which single-stranded RNA chains are made, and use the nucleotides ATP, GTP, CTP, and UTP as precursors (Fig. 2-17). These enzymes make RNA using appropriate segments of chromosomal DNA as their templates. Direct evidence that DNA lines up the correct ribonucleotide precursors came from seeing how the RNA base composition varied with the addition of DNA molecules of different AT/GC ratios. In every enzymatic synthesis, the RNA AU/GC ratio was roughly similar to the DNA AT/GC ratio (Table 2-2).

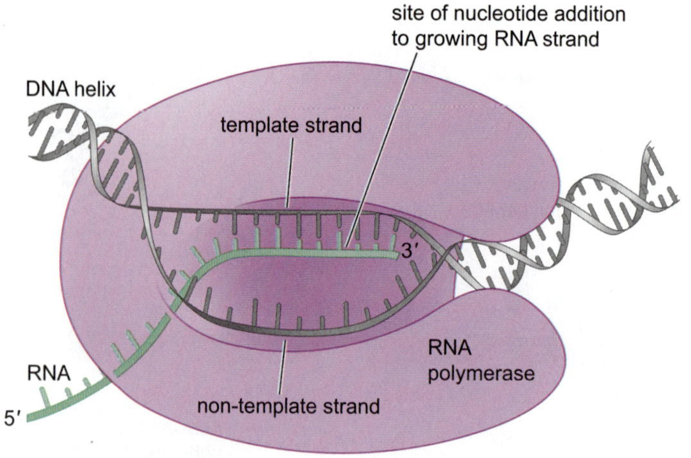

FIGURE 2-17 Enzymatic synthesis of RNA upon a DNA template, catalyzed by RNA polymerase.

During transcription, only one of the two strands of DNA is used as a template to make RNA. This makes sense, because the messages carried by the two strands, being complementary but not identical, are expected to code for completely different polypeptides. The synthesis of RNA always proceeds in a fixed direction, beginning at the 5′ end and concluding with the 3′-end nucleotide (see Fig. 2-17).

By this time, there was firm evidence for the postulated movement of RNA from the DNA-containing nucleus to the ribosome-containing cytoplasm of eukaryotic cells. By briefly exposing cells to radioactively labeled precursors, then adding a large excess of unlabeled ribonucleotides (a "pulse chase" experiment), mRNA synthesized during a short time window was labeled. These studies showed that mRNA is synthesized in the nucleus. Within an hour, most of this RNA had left the nucleus and was observed in the cytoplasm (Fig. 2-18).

Establishing the Genetic Code

Given the existence of 20 amino acids but only four bases, groups of several nucleotides must somehow specify a given amino acid. Groups of two, however, would specify only 16 (4 x 4) amino acids. So from 1954, the start of serious thinking about what the genetic code might be like, most attention was given to how triplets (groups of three)

TABLE 2-2 Comparison of the Base Composition of Enzymatically Synthesized RNAs with the Base Composition of Their Double-Helical DNA Templates

	Composition of the RNA Bases				$\frac{A+U}{G+C}$	$\frac{A+T}{G+C}$
Source of DNA Template	Adenine	Uracil	Guanine	Cytosine	Observed	In DNA
T2	0.31	0.34	0.18	0.17	1.86	1.84
Calf thymus	0.31	0.29	0.19	0.21	1.50	1.35
Escherichia coli	0.24	0.24	0.26	0.26	0.92	0.97
Micrococcus lysodeikticus (a bacterium)	0.17	0.16	0.33	0.34	0.49	0.39

might work, even though they obviously would provide more permutations (4 x 4 x 4) than needed if each amino acid was specified by only a single triplet. The assumption of colinearity was then very important. It held that successive groups of nucleotides along a DNA chain code for successive amino acids along a given polypeptide chain. An elegant mutational analysis on bacterial proteins, carried out in the early 1960s by Charles Yanofsky and Sydney Brenner, showed that colinearity does in fact exist. Equally important were the genetic analyses by Brenner and Crick, which in 1961 first established that groups of three nucleotides are used to specify individual amino acids.

But which specific groups of three bases (codons) determine which specific amino acids could only be learned by biochemical analysis. The major breakthrough came when Marshall Nirenberg and Heinrich Matthaei, then working together, observed in 1961 that the addition of the synthetic polynucleotide poly U (UUUUU . . .) to a cell-free system capable of making proteins leads to the synthesis of polypeptide chains containing only the amino acid phenylalanine. The nucleotide groups UUU thus must specify phenylalanine. Use of increasingly more complex polynucleotides as synthetic messenger RNAs rapidly led to the identification of more and more codons. Particularly important in completing the code was the use of polynucleotides like AGUAGU, put together by organic chemist Har Gobind Khorana. These further defined polynucleotides were critical to test more specific sets of codons. Completion of the code in 1966 revealed that 61 out of the 64 possible permuted groups corresponded to amino acids, with most amino acids being encoded by more than one nucleotide triplet (Table 2-3).

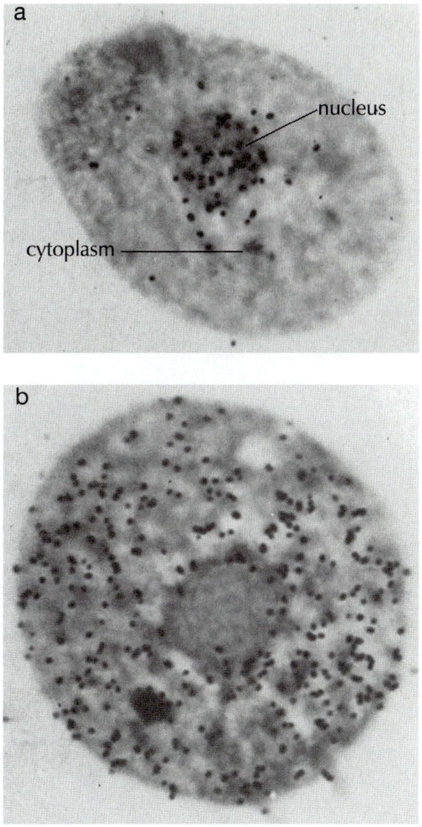

FIGURE 2-18 Demonstration that RNA is synthesized in the nucleus and moves to the cytoplasm. (a) Autoradiograph of a cell (*Tetrahymena*) exposed to radioactive cytidine for 15 min. Superimposed on a photograph of a thin section of the cell is a photograph of an exposed silver emulsion. Each dark spot represents the origin of an electron emitted from a ^{3}H (tritium) atom that has been incorporated into RNA. Almost all the newly made RNA is found within the nucleus. (b) Autoradiograph of a similar cell exposed to radioactive cytidine for 12 min and then allowed to grow for 88 min in the presence of nonradioactive cytidine. Practically all the label incorporated into RNA in the first 12 min has left the nucleus and moved into the cytoplasm. (Courtesy of D.M. Prescott, University of Colorado Medical School; reproduced, with permission, from Prescott D.M. 1964. *Progr. Nucleic Acid Res. Mol. Biol.* 3: 35. © Elsevier.)

TABLE 2-3 The Genetic Code

first position	second position: U		C		A		G		third position
U	UUU	Phe	UCU	Ser	UAU	Tyr	UGU	Cys	U
	UUC		UCC		UAC		UGC		C
	UUA	Leu	UCA		UAA	stop	UGA	stop	A
	UUG		UCG		UAG	stop	UGG	Trp	G
C	CUU	Leu	CCU	Pro	CAU	His	CGU	Arg	U
	CUC		CCC		CAC		CGC		C
	CUA		CCA		CAA	Gln	CGA		A
	CUG		CCG		CAG		CGG		G
A	AUU	Ile	ACU	Thr	AAU	Asn	AGU	Ser	U
	AUC		ACC		AAC		AGC		C
	AUA		ACA		AAA	Lys	AGA	Arg	A
	AUG	Met	ACG		AAG		AGG		G
G	GUU	Val	GCU	Ala	GAU	Asp	GGU	Gly	U
	GUC		GCC		GAC		GGC		C
	GUA		GCA		GAA	Glu	GGA		A
	GUG		GCG		GAG		GGG		G

ESTABLISHING THE DIRECTION OF PROTEIN SYNTHESIS

The nature of the genetic code, once determined, led to further questions about how a polynucleotide chain directs the synthesis of a polypeptide. As we have seen here and shall discuss in detail in Chapter 8, polynucleotide chains (both DNA and RNA) are synthesized by adding to the 3′ end of the growing strand (growth in the 5′ → 3′ direction). But what about the growing polypeptide chain? Is it assembled in an amino-terminal to carboxy-terminal direction, or the opposite?

This question was answered using a cell-free system for protein synthesis similar to the one used to identify the amino acid codons. Instead of providing synthetic mRNAs, however, the investigators provided β-globin mRNA to direct protein synthesis. A few minutes after initiation of protein synthesis, the cell-free system was treated with a radioactive amino acid for a few seconds (less than the time required to synthesize a complete globin chain) after which protein synthesis was immediately stopped. A brief radioactive labeling regime of this kind is known as **pulse-labeling.** Next, the β-globin chains that had *completed* their growth during the period of the pulse-labeling were separated from incomplete chains by gel electrophoresis (Chapter 20). Thus, all proteins analyzed would have completed their synthesis in the presence of radiolabeled precursors. The full-length polypeptides were then treated with an enzyme, the protease trypsin, that cleaves proteins at particular sites in the polypetide chain, thereby generating a series of peptide fragments. In the final step of the experiment, the amount of radioactivity that had been incorporated into each peptide fragment was measured (Fig. 2-19).

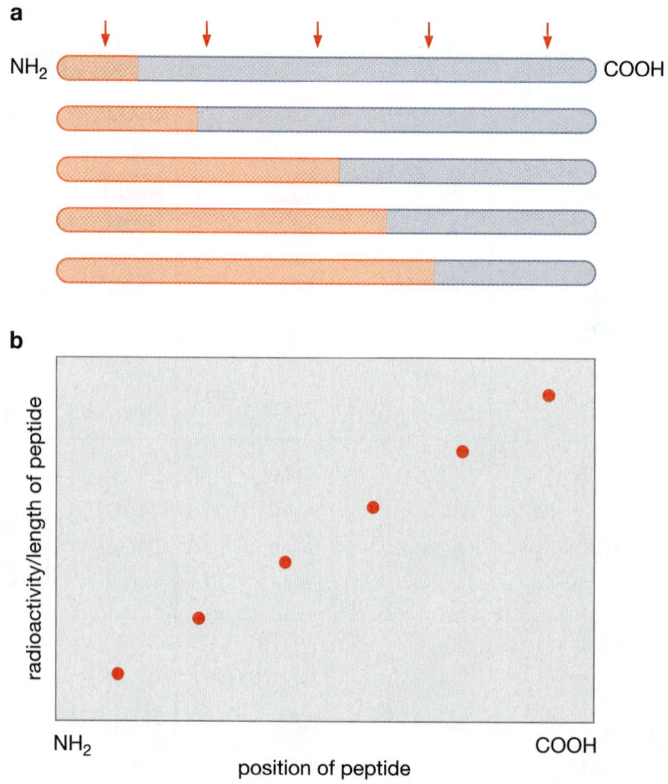

FIGURE 2-19 Incorporation of radioactively labeled amino acids into a growing polypeptide chain. (a) Distribution of radioactivity (shown in blue) among completed chains after a short period of labeling. The sites of trypsin cleavage of the β-globin protein are indicated by the red arrows. (b) Incorporation of label normalized to the length of each peptide is plotted as a function of position of the peptide within the completed chain.

Because in this experiment all proteins finished their synthesis in the presence of radioactive precursors, the peptides last to be synthesized will have the highest density of radiolabeled precursors (Fig. 2-19a). Conversely, peptides with the least amount of radioactive amino acid (normalized to the size of the peptide) would be derived from regions of the β-globin protein that were the first to be synthesized. The investigators observed that radioactive labeling was lowest for peptides from the amino-terminal region of globin and greatest for peptides from the carboxy-terminal region (Fig. 2-19b). This led to the conclusion that the direction of protein synthesis is from the amino terminus to the carboxyl terminus. In other words, new amino acids are added to the carboxyl terminus of the growing polypeptide chain.

Start and Stop Signals Are Also Encoded within DNA

Initially, it was guessed that translation of an mRNA molecule would commence at one end and finish when the entire mRNA message had been read into amino acids. But, in fact, translation both starts and stops at internal positions. Thus, signals must be present within DNA (and its mRNA products) to initiate and terminate translation. The stop signals were the first to be worked out. Three separate codons (UAA, UAG, and UGA), first known as **nonsense codons,** do not direct the addition of a particular amino acid. Instead, these codons serve as translational stop signals (sometimes called stop codons). The way translational start signals are encoded is more complicated. The amino acid methionine initiates all polypeptide chains, but the triplet (AUG) that codes for these initiating methionines also codes for methionine residues that are found at internal protein positions. In prokaryotes, the AUG codons that start new polypeptide chains are preceded by specific purine-rich blocks of nucleotides that serve to attach mRNA to ribosomes (see Chapter 14). In eukaryotes, the position of the AUG relative to the beginning of the mRNA is the critical determinant, with the first AUG always being selected as the start site of translation.

THE ERA OF GENOMICS

With the elucidation of the central dogma, it became clear by the mid-1960s how the genetic blueprint contained in the nucleotide sequence could determine phenotype. This meant that profound insights into the nature of living things and their evolution would be revealed from DNA sequences. In recent years the advent of rapid, automated DNA sequencing methods has led to the determination of complete genome sequences for hundreds of organisms. Even the human genome, a single copy of which is composed of more than 3 billion base pairs, has been elucidated and shown to contain more than 20,000 genes. The sequencing of the genomes of many organisms has made the comparative analysis of genome sequences very useful. By comparing the predicted amino acid sequences encoded by similar genes from different organisms one can frequently identify important regions of a protein. For example, the amino acids in DNA polymerases that are critical for binding the incoming nucleotide or that directly catalyze nucleotide addition are well conserved in the DNA polymerases from many different organisms. Similarly, amino acids that are important to DNA polymerase function in bacteria but not in eukaryotic cells will be conserved only in the amino acid sequences predicted by the genome sequences from bacteria.

Comparison of different genomes can also offer insights into DNA sequences that do not encode proteins. The identification of sequences that direct gene expression, DNA replication, chromosome segregation, and recombination can all be facilitated by comparing genome sequences. Because these regulatory sequences tend to diverge more rapidly, these comparisons are often made between closely related species (such as between different bacteria or between humans and other primates). The value of comparisons between closely related species has led to efforts to sequence the genomes of organisms closely related to well-studied model organisms such as the fruit fly *Drosophila melanogaster*, the yeast *Saccharomyces cerevisiae*, or multiple primates.

Comparative genomics between different individuals of the same organism has the potential to identify mutations that lead to disease. For example, recent efforts have developed methods to rapidly compare the sequences of a small subset of the human genome between many different individuals in an effort to identify disease genes. Finally, it is possible to envision a day when comparative genome analysis will reveal basic insights into the origins of complex behavior in humans, such as the acquisition of language, as well as the mechanisms underlying the evolutionary diversification of animal body plans.

The purpose of the forthcoming chapters is to provide a firm foundation for understanding how DNA functions as the template for biological complexity. The remaining chapters in Part 1 review the basic chemistry and biology relevant to the main themes of this book. Part 2, Maintenance of the Genome, describes the structure of the genetic material and its faithful duplication. Part 3, Expression of the Genome, shows how the genetic instructions contained in DNA are converted into proteins. Part 4, Regulation, describes strategies for differential gene activity that are used to generate complexity within organisms (e.g., embryogenesis) and diversity among organisms (e.g., evolution). Finally, Part 5, Methods, describes various laboratory techniques, bioinformatics approaches, and model systems that are commonly used to investigate biological problems.

SUMMARY

The discovery that DNA is the genetic material can be traced to experiments performed by Griffith, who showed that nonvirulent strains of bacteria could be genetically transformed with a substance derived from a heat-killed pathogenic strain. Avery, McCarty, and MacLeod subsequently demonstrated that the transforming substance was DNA. Further evidence that DNA is the genetic material was obtained by Hershey and Chase in experiments with radiolabeled bacteriophage. Building on Chargaff's rules and Franklin and Wilkins' X-ray diffraction studies, Watson and Crick proposed a double-helical structure of DNA. In this model, two polynucleotide chains are twisted around each other to form a regular double helix. The two chains within the double helix are held together by hydrogen bonds between pairs of bases. Adenine is always joined to thymine, and guanine is always bonded to cytosine. The existence of the base pairs means that the sequence of nucleotides along the two chains are not identical, but complementary. The finding of this relationship suggested a mechanism for the replication of DNA in which each strand serves as a template for its complement. Proof for this hypothesis came from (a) the observation of Meselson and Stahl that the two strands of each double helix separate during each round of DNA replication, and (b) Kornberg's discovery of an enzyme that uses single-stranded DNA as a template for the synthesis of a complementary strand.

As we have seen, according to the "central dogma" information flows from DNA to RNA to protein. This transformation is achieved in two steps. First, DNA is transcribed into an RNA intermediate (messenger RNA), and second, the mRNA is translated into protein. Translation of the mRNA requires RNA adaptor molecules called tRNAs. The key characteristic of the genetic code is that each triplet codon is recognized by a tRNA, which is associated with a cognate amino acid. Out of 64 (4 x 4 x 4) potential codons, 61 are used to specify the 20 amino acid building blocks of proteins, whereas 3 are used to provide chain-terminating signals. Knowledge of the genetic code allows us to predict protein-coding sequences from DNA sequences. The advent of rapid DNA sequencing methods has ushered in a new era of genomics, in which complete genome sequences are being determined for a wide variety of organisms, including humans. Comparing genome sequences offers a powerful method to identify critical regions of the genome that encode not only important elements of proteins but also regulatory regions that control the expression of genes and the duplication of the genome.

BIBLIOGRAPHY

Brenner S., Jacob F., and Meselson M. 1961. An unstable intermediate carrying information from genes to ribosomes for protein synthesis. *Nature* **190:** 576–581.

Brenner S., Stretton A.O.W., and Kaplan S. 1965. Genetic code: The nonsense triplets for chain termination and their suppression. *Nature* **206:** 994–998.

Cairns J., Stent G.S., and Watson J.D., eds. 1966. *Phage and the origins of molecular biology.* Cold Spring Harbor Laboratory, Cold Spring Harbor, New York.

Chargaff E. 1951. Structure and function of nucleic acids as cell constituents. *Fed. Proc.* **10:** 654–659.

Cold Spring Harbor Symposia on Quantitative Biology. 1966. Vol. 31: *The genetic code.* Cold Spring Harbor Laboratory, Cold Spring Harbor, New York.

Crick F.H.C. 1955. On degenerate template and the adaptor hypothesis. A note for the RNA Tie Club, unpublished. Mentioned in Crick's 1957 discussion, pp. 25–26, in The structure of nucleic acids and their role in protein synthesis. *Biochem. Soc. Symp.* no. 14, Cambridge University Press, Cambridge, England.

———. 1958. On protein synthesis. *Symp. Soc. Exp. Biol.* **12:** 548–555.

———. 1963. The recent excitement in the coding problem. *Prog. Nucleic Acid Res.* **1:** 164–217.

———. 1988. *What mad pursuit: A personal view of scientific discovery.* Basic Books, New York.

Crick F.H.C. and Watson J.D. 1954. The complementary structure of deoxyribonucleic acid. *Proc. Roy. Soc. A* **223:** 80–96.

Echols H. and Gross C.A., eds. 2001. *Operators and promoters: The story of molecular biology and its creators.* University of California Press, Berkeley, California.

Franklin R.E. and Gosling R.G. 1953. Molecular configuration in sodium thymonuclease. *Nature* **171:** 740–741.

Hershey A.D. and Chase M. 1952. Independent function of viral protein and nucleic acid on growth of bacteriophage. *J. Gen. Physiol.* **36:** 39–56.

Hoagland M.B., Stephenson M.L., Scott J.F., Hecht L.I., and Zamecnik P.C. 1958. A soluble ribonucleic acid intermediate in protein synthesis. *J. Biol. Chem.* **231:** 241–257.

Holley R.W., Apgar J., Everett G.A., Madison J.T., Marquisse M., Merrill S.H., Penswick J.R., and Zamir A. 1965. Structure of a ribonucleic acid. *Science* **147:** 1462–1465.

Ingram V.M. 1957. Gene mutations in human hemoglobin: The chemical difference between normal and sickle cell hemoglobin. *Nature* **180:** 326–328.

Jacob F. and Monod J. 1961. Genetic regulatory mechanisms in the synthesis of proteins. *J. Mol. Biol.* **3:** 318–356.

Judson H.F. 1996. *The eighth day of creation,* expanded edition. Cold Spring Harbor Laboratory Press, Cold Spring Harbor, New York.

Kornberg A. 1960. Biological synthesis of deoxyribonucleic acid. *Science* **131:** 1503–1508.

Kornberg A. and Baker T.A. 1992. *DNA replication.* W.H. Freeman, New York.

McCarty M. 1985. *The transforming principle: Discovering that genes are made of DNA.* Norton, New York.

Meselson M. and Stahl F.W. 1958. The replication of DNA in *Escherichia coli. Proc. Natl. Acad. Sci.* **44:** 671–682.

Nirenberg M.W. and Matthaei J.H. 1961. The dependence of cell-free protein synthesis in *E. coli* upon naturally occurring or synthetic polyribonucleotides. *Proc. Natl. Acad. Sci.* **47:** 1588–1602.

Olby R. 1975. *The path to the double helix.* University of Washington Press, Seattle.

Portugal F.H. and Cohen J.S. 1980. *A century of DNA: A history of the discovery of the structure and function of the genetic substance.* MIT Press, Cambridge, Massachusetts.

Sarabhai A.S., Stretton A.O.W., Brenner S., and Bolte A. 1964. Co-linearity of the gene with the polypeptide chain. *Nature* **201:** 13–17.

Stent G.S. and Calendar R. 1978. *Molecular genetics: An introductory narrative,* 2nd ed. Freeman, San Francisco.

Volkin E. and Astrachan L. 1956. Phosphorus incorporation in *E. coli* ribonucleic acid after infection with bacteriophage T2. *Virology* **2:** 146–161.

Watson J.D. 1963. Involvement of RNA in synthesis of proteins. *Science* **140:** 17–26.

———. 1968. *The double helix.* Atheneum, New York.

———. 1980. *The double helix: A Norton critical edition* (ed. G.S. Stent). Norton, New York.

———. 2000. *A passion for DNA: Genes, genomes and society.* Cold Spring Harbor Laboratory Press, Cold Spring Harbor, New York.

———. 2002. *Genes, girls, and Gamow: After the double helix.* Knopf, New York.

Watson J.D. and Crick F.H.C. 1953a. Genetical implications of the structure of deoxyribonucleic acid. *Nature* **171:** 964–967.

——— 1953b. Molecular structure of nucleic acids; a structure for deoxyribose nucleic acid. *Nature* **171:** 737–738.

Wilkins M.H.F., Stokes A.R., and Wilson H.R. 1953. Molecular structure of deoxypentose nucleic acid. *Nature* **171:** 738–740.

Yanofsky C., Carlton B.C., Guest J.R., Helinski D.R., and Henning U. 1964. On the colinearity of gene structure and protein structure. *Proc. Natl. Acad. Sci.* **51:** 266–272.

CHAPTER

3

The Importance of Weak Chemical Interactions

OUTLINE

THE MACROMOLECULES THAT WILL PREOCCUPY us throughout this book—and those of most concern to molecular biologists—are proteins and nucleic acids. These are made of amino acids and nucleotides, respectively, and in both cases the constituents are joined by covalent bonds to make polypeptide (protein) and polynucleotide (nucleic acid) chains. Covalent bonds are strong, stable bonds, and essentially never break spontaneously within biological systems. But weaker bonds also exist, and indeed are vital for life, partly because they can form and break under the physiological conditions present within cells. Weak bonds mediate the interactions between enzymes and their substrates, and between macromolecules—most strikingly, as we shall see in later chapters, between proteins and DNA or RNA, or proteins and other proteins. But equally important, weak bonds also mediate interactions between different parts of individual macromolecules, determining the shape of those molecules and hence their biological function. Thus, although a protein is a linear chain of covalently linked amino acids, its shape and function are determined by the stable three-dimensional structure it adopts. That shape is determined by a large collection of individually weak interactions that form between amino acids that do not need to be adjacent in the primary sequence. Likewise, it is the weak, noncovalent bonds that hold the two chains of a DNA double helix together.

In this chapter, we consider the nature of chemical bonds, concentrating in large part on the weak bonds so vital to the proper function of all biological macromolecules. In particular, we describe what it is that gives weak bonds their weak character. These bonds include van der Waals bonds, hydrophobic bonds, hydrogen bonds, and ionic bonds.

CHARACTERISTICS OF CHEMICAL BONDS

A **chemical bond** is an attractive force that holds atoms together. Aggregates of finite size are called molecules. Originally, it was thought that only covalent bonds hold atoms together in molecules; now, weaker attractive forces are known to be important in holding together many macromolecules. For example, the four polypeptide chains of hemoglobin are held together by the combined action of several weak bonds. It is thus now customary also to call weak positive interactions chemical bonds, even though they are not strong enough, when present singly, to effectively bind two atoms together.

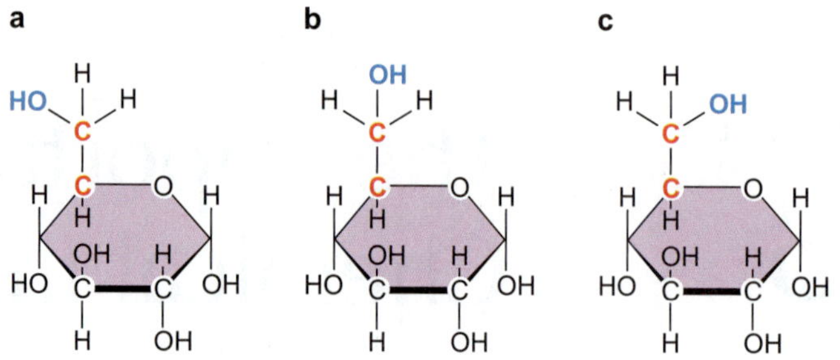

FIGURE 3-1 Rotation about the C_5–C_6 bond in glucose. This carbon–carbon bond is a single bond, and so any of the three configurations, a, b, or c, may occur.

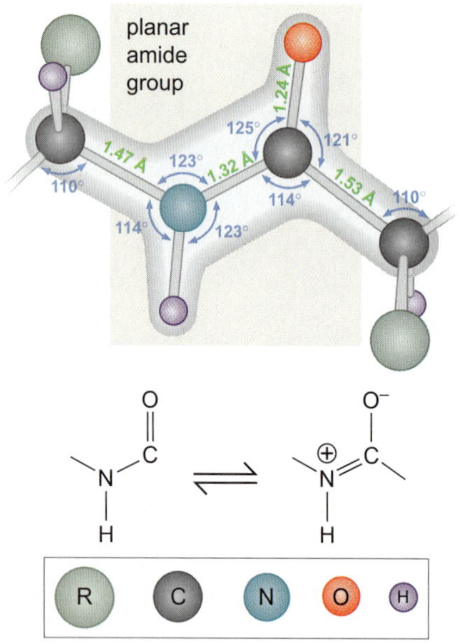

FIGURE 3-2 The planar shape of the peptide bond. Shown here is a portion of an extended polypeptide chain. Almost no rotation is possible about the peptide bond because of its partial double-bond character (see middle panel). All the atoms in the shaded area must lie in the same plane. Rotation is possible, however, around the remaining two bonds, which make up the polypeptide configurations. (Adapted, with permission, from Pauling L. 1960. *The nature of the chemical bond and the structure of molecules and crystals: An introduction to modern structural chemistry,* 3rd ed., p. 495. © 1960 Cornell University.)

Chemical bonds are characterized in several ways. An obvious characteristic of a bond is its strength. Strong bonds almost never fall apart at physiological temperatures. This is why atoms united by covalent bonds always belong to the same molecule. Weak bonds are easily broken, and when they exist singly, they exist fleetingly. Only when present in ordered groups do weak bonds last a long time. The strength of a bond is correlated with its length, so that two atoms connected by a strong bond are always closer together than the same two atoms held together by a weak bond. For example, two hydrogen atoms bound covalently to form a hydrogen molecule (H:H) are 0.74 Å apart, whereas the same two atoms held together by van der Waals forces are 1.2 Å apart.

Another important characteristic is the maximum number of bonds that a given atom can make. The number of covalent bonds that an atom can form is called its **valence**. Oxygen, for example, has a valence of two: it can never form more than two covalent bonds. There is more variability in the case of van der Waals bonds, in which the limiting factor is purely steric. The number of possible bonds is limited only by the number of atoms that can touch each other simultaneously. The formation of hydrogen bonds is subject to more restrictions. A covalently bonded hydrogen atom usually participates in only one hydrogen bond, whereas an oxygen atom seldom participates in more than two hydrogen bonds.

The angle between two bonds originating from a single atom is called the **bond angle**. The angle between two specific covalent bonds is always approximately the same. For example, when a carbon atom has four single covalent bonds, they are directed tetrahedrally (bond angle = 109°). In contrast, the angles between weak bonds are much more variable.

Bonds differ also in the **freedom of rotation** they allow. Single covalent bonds permit free rotation of bound atoms (Fig. 3-1), whereas double and triple bonds are quite rigid. Bonds with partial double-bond character, such as the peptide bond, are also quite rigid. For that reason, the carbonyl (C=O) and imino (N=C) groups bound together by the peptide bond must lie in the same plane (Fig. 3-2). Much weaker, ionic bonds, on the other hand, impose no restrictions on the relative orientations of bonded atoms.

Chemical Bonds Are Explainable in Quantum-Mechanical Terms

The nature of the forces, both strong and weak, that give rise to chemical bonds remained a mystery to chemists until the quantum theory of the atom (quantum mechanics) was developed in the 1920s. Then, for the first time, the various empirical laws about how chemical bonds are formed were put on a firm theoretical basis. It was realized that all chemical bonds, weak as well as strong, are based on electrostatic forces. Quantum mechanics provided explanations for covalent bonding by the sharing of electrons and also for the formation of weaker bonds.

Chemical-Bond Formation Involves a Change in the Form of Energy

The spontaneous formation of a bond between two atoms always involves the release of some of the internal energy of the unbonded atoms and its conversion to another energy form. The stronger the bond, the greater the amount of energy released upon its formation. The bonding reaction between two atoms A and B is thus described by

$$A + B \rightarrow AB + \text{energy}, \qquad \textbf{[Equation 3-1]}$$

where AB represents the bonded aggregate. The rate of the reaction is directly proportional to the frequency of collision between A and B. The unit most often used to measure energy is the calorie, the amount of energy required to raise the temperature of 1 g of water from 14.5°C to 15.5°C. Since thousands of calories are usually involved in the breaking of a mole of chemical bonds, most energy changes within chemical reactions are expressed in kilocalories per mole.

However, atoms joined by chemical bonds do not remain together forever, because there also exist forces that break chemical bonds. By far the most important of these forces arises from heat energy. Collisions with fast-moving molecules or atoms can break chemical bonds. During a collision, some of the kinetic energy of a moving molecule is given up as it pushes apart two bonded atoms. The faster a molecule is moving (the higher the temperature), the greater the probability that, upon collision, it will break a bond. Hence, as the temperature of a collection of molecules is increased, the stability of their bonds decreases. The breaking of a bond is thus always indicated by the formula

$$AB + \text{energy} \rightarrow A + B. \qquad \textbf{[Equation 3-2]}$$

The amount of energy that must be added to break a bond is exactly equal to the amount that was released upon formation of the bond. This equivalence follows from the first law of thermodynamics, which states that energy (except as it is interconvertible with mass) can be neither made nor destroyed.

Equilibrium between Bond Making and Breaking

Every bond is thus a result of the combined actions of bond-making and bond-breaking forces. When an equilibrium is reached in a closed system, the number of bonds forming per unit time will equal the number of bonds breaking. Then the proportion of bonded atoms is described by the mass action formula

$$K_{eq} = \frac{\text{conc}^{AB}}{\text{conc}^{A} \times \text{conc}^{B}}, \qquad \textbf{[Equation 3-3]}$$

where K_{eq} is the **equilibrium constant**, and conc^{A}, conc^{B}, and conc^{AB} are the concentrations of A, B, and AB, respectively, in moles per liter. Whether we start with only free A and B, with only the molecule AB, or with a combination of AB and free A and B, at equilibrium the proportions of A, B, and AB will reach the concentrations given by K_{eq}.

THE CONCEPT OF FREE ENERGY

There is always a change in the form of energy as the proportion of bonded atoms moves toward the equilibrium concentration. Biologically, the most useful way to express this energy change is through the physical chemist's concept of **free energy**, denoted by the symbol G, which honors the great 19th-century physicist Josiah Gibbs. We shall not give a rigorous description of free energy in this text nor show how it differs from the other forms of energy. For this, the reader must refer to a chemistry text that discusses the second law of thermodynamics. It must suffice to say here that *free energy is energy that has the ability to do work.*

The second law of thermodynamics tells us that a decrease in free energy (ΔG is negative) always occurs in spontaneous reactions. When equilibrium is reached, however, there is no further change in the amount of free energy ($\Delta G = 0$). The equilibrium state for a closed collection of atoms is thus the state that contains the least amount of free energy.

The free energy lost as equilibrium is approached is either transformed into heat or used to increase the amount of entropy. We shall not attempt to define entropy here except to say that the amount of entropy is a measure of the amount of disorder. The greater the disorder, the greater the amount of entropy. The existence of entropy means that many spontaneous chemical reactions (those with a net decrease in free energy) need not proceed with an evolution of heat. For example, when sodium chloride (NaCl) is dissolved in water, heat is absorbed rather than released. There is, nonetheless, a net decrease in free energy because of the increase in disorder of the sodium and chlorine ions as they move from a solid to a dissolved state.

K_{eq} Is Exponentially Related to ΔG

Clearly, the stronger the bond, and hence the greater the change in free energy (ΔG) that accompanies its formation, the greater the proportion of atoms that must exist in the bonded form. This commonsense idea is quantitatively expressed by the physicochemical formula

$$\Delta G = -RT \ln K_{eq} \quad \text{or} \quad K_{eq} = e^{-\Delta G/RT}, \qquad \textbf{[Equation 3-4]}$$

where R is the universal gas constant, T is the absolute temperature, ln is the logarithm (of K_{eq}) to the base e, K_{eq} is the equilibrium constant, and e = 2.718.

Insertion of the appropriate values of R (1.987 cal/deg-mole) and T (298 at 25°C) tells us that ΔG values as low as 2 kcal/mole can drive a bond-forming reaction to virtual completion if all reactants are present at molar concentrations (Table 3-1).

TABLE 3-1 The Numerical Relationship between the Equilibrium Constant and ΔG at 25°C

K_{eq}	ΔG (kcal/mole)
0.001	4.089
0.01	2.726
0.1	1.363
1.0	0
10.0	–1.363
100.0	–2.726
1000.0	–4.089

Covalent Bonds Are Very Strong

The ΔG values accompanying the formation of covalent bonds from free atoms, such as hydrogen or oxygen, are very large and negative in sign, usually –50 to –110 kcal/mole. Equation 3-4 tells us that K_{eq} of the bonding reaction will be correspondingly large, and so the concentration of hydrogen or oxygen atoms existing unbound will be very small. For example, with a ΔG value of –100 kcal/mole, if we start with 1 mole/liter of the reacting atoms, only 1 in 10^{40} atoms will remain unbound when equilibrium is reached.

WEAK BONDS IN BIOLOGICAL SYSTEMS

The main types of weak bonds important in biological systems are the van der Waals bonds, hydrophobic bonds, hydrogen bonds, and ionic bonds. Sometimes, as we shall soon see, the distinction between a hydrogen bond and an ionic bond is arbitrary.

Weak Bonds Have Energies between 1 and 7 kcal/mole

The weakest bonds are the van der Waals bonds. These have energies (1–2 kcal/mole) only slightly greater than the kinetic energy of heat motion. The energies of hydrogen and ionic bonds range between 3 and 7 kcal/mole.

In liquid solutions, almost all molecules form a number of weak bonds to nearby atoms. All molecules are able to form van der Waals bonds, whereas hydrogen and ionic bonds can form only between molecules that have a net charge (ions) or in which the charge is unequally distributed. Some molecules thus have the capacity to form several types of weak bonds. Energy considerations, however, tell us that molecules always have a greater tendency to form the stronger bond.

Weak Bonds Are Constantly Made and Broken at Physiological Temperatures

Weak bonds, at their weakest, have energies only slightly higher than the average energy of kinetic motion (heat) at 25°C (0.6 kcal/mole), but even the strongest of these bonds have only about ten times that energy. Nevertheless, because there is a significant spread in the energies of kinetic motion, many molecules with sufficient kinetic energy to break the strongest weak bonds still exist at physiological temperatures.

The Distinction between Polar and Nonpolar Molecules

All forms of weak interactions are based on attractions between electric charges. The separation of electric charges can be permanent or temporary, depending on the atoms involved. For example, the oxygen molecule (O:O) has a symmetric distribution of electrons between its two oxygen atoms, so each of its two atoms is uncharged. In contrast, there is a nonuniform distribution of charge in water (H:O:H), in which the bond electrons are unevenly shared (Fig. 3-3). They are held more strongly by the oxygen atom, which thus carries a considerable negative charge, whereas the two hydrogen atoms together have an equal amount of positive charge. The center of the positive charge is on one side of the center of the negative charge. A combination of separated positive and negative charges is called an electric **dipole moment**. Unequal electron sharing reflects dissimilar affinities of the bonding atoms for electrons. Atoms that have a tendency to gain electrons are called **electronegative** atoms. **Electropositive** atoms have a tendency to give up electrons.

Molecules (such as H_2O) that have a dipole moment are called **polar molecules. Nonpolar molecules** are those with no effective dipole moments. In methane (CH_4), for example, the carbon and hydrogen atoms have similar affinities for their shared electron pairs, so neither the carbon nor the hydrogen atom is noticeably charged.

The distribution of charge in a molecule can also be affected by the presence of nearby molecules, particularly if the affected molecule is

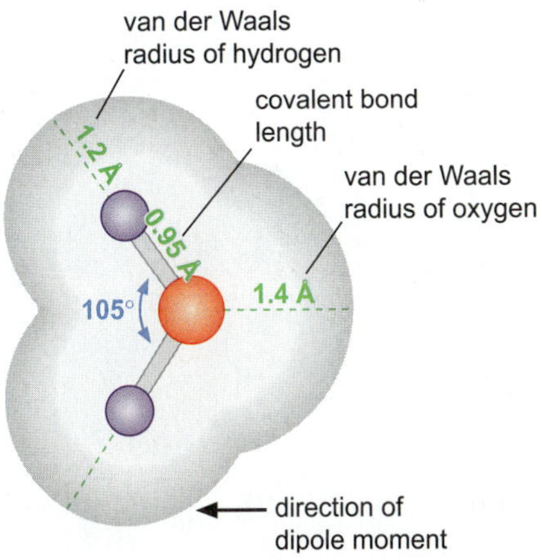

FIGURE 3-3 The structure of a water molecule. van der Waals radii are discussed below (see Fig. 3-5).

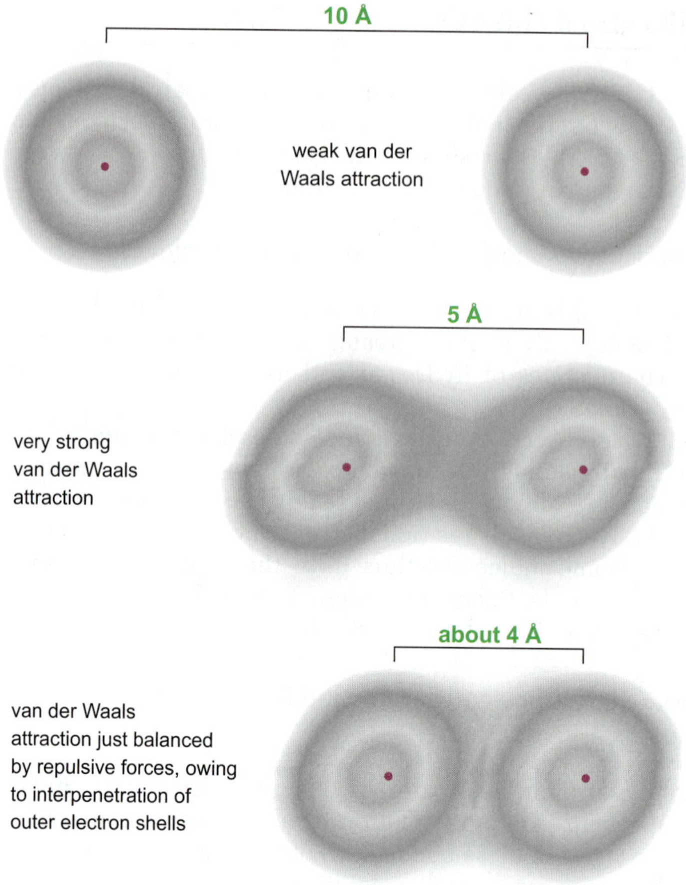

FIGURE 3-4 **Variation of van der Waals forces with distance.** The atoms shown in this diagram are atoms of the inert rare gas argon. (Adapted, with permission, from Pauling L. 1953. *General chemistry*, 2nd ed., p. 322. © W.H. Freeman.)

polar. The effect may cause a nonpolar molecule to acquire a slightly polar character. If the second molecule is not polar, its presence will still alter the nonpolar molecule, establishing a fluctuating charge distribution. Such induced effects, however, give rise to a much smaller separation of charge than is found in polar molecules, resulting in smaller interaction energies and correspondingly weaker chemical bonds.

van der Waals Forces

van der Waals bonding arises from a nonspecific attractive force originating when two atoms come close to each other. It is based not on the existence of permanent charge separations but rather on the induced fluctuating charges caused by the nearness of molecules. It therefore operates between all types of molecules, nonpolar as well as polar. It depends heavily on the distance between the interacting groups, because the bond energy is inversely proportional to the sixth power of distance (Fig. 3-4).

There also exists a more powerful van der Waals *repulsive* force, which comes into play at even shorter distances. This repulsion is caused by the overlapping of the outer electron shells of the atoms involved. The van der Waals attractive and repulsive forces balance at a certain distance specific for each type of atom. This distance is the so-called **van der Waals radius** (Fig. 3-5 and Table 3-2). The van der Waals bonding energy between two atoms separated by the sum of their van der Waals radii increases with the size of the respective

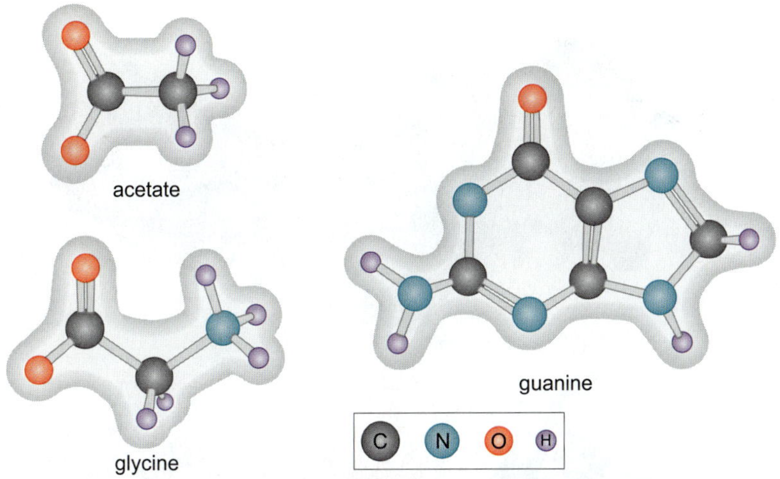

FIGURE 3-5 Drawings of several molecules with the van der Waals radii of the atoms shown as shaded clouds.

TABLE 3-2 van der Waals Radii of the Atoms in Biological Molecules

Atom	van der Waals Radius (Å)
H	1.2
N	1.5
O	1.4
P	1.9
S	1.85
CH_3 group	2.0
Half thickness of aromatic molecule	1.7

atoms. For two average atoms, it is only about 1 kcal/mole, which is just slightly more than the average thermal energy of molecules at room temperature (0.6 kcal/mole).

This means that van der Waals forces are an effective binding force at physiological temperatures only when several atoms in a given molecule are bound to several atoms in another molecule or another part of the same molecule. Then the energy of interaction is much greater than the dissociating tendency resulting from random thermal movements. For several atoms to interact effectively, the molecular fit must be precise, because the distance separating any two interacting atoms must not be much greater than the sum of their van der Waals radii (Fig. 3-6). The strength of interaction rapidly approaches zero when this distance is only slightly exceeded. Thus, the strongest type of van der Waals contact arises when a molecule contains a cavity exactly complementary in shape to a protruding group of another molecule, as is the case with an antigen and its specific antibody (Fig. 3-7). In this instance, the binding energies sometimes can be as large as 20–30 kcal/mole, so that antigen–antibody complexes seldom fall apart. The bonding pattern of polar molecules is rarely dominated by van der Waals interactions, because such molecules can acquire a lower energy state (lose more free energy) by forming other types of bonds.

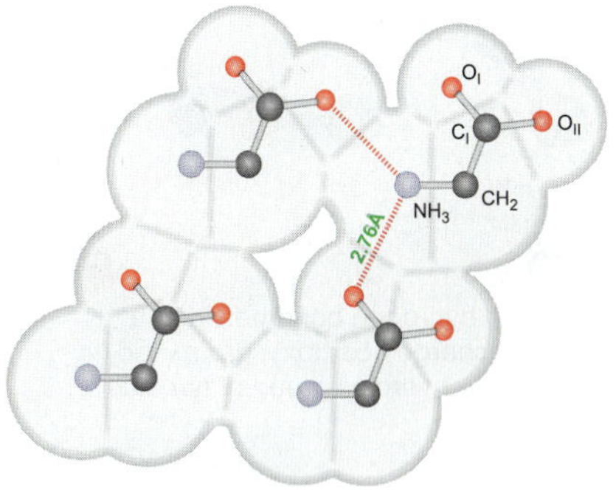

FIGURE 3-6 The arrangement of molecules in a layer of a crystal formed by the amino acid glycine. The packing of the molecules is determined by the van der Waals radii of the groups, except for the N—H·········O contacts, which are shortened by the formation of hydrogen bonds. (Adapted, with permission, from Pauling L. 1960. *The nature of the chemical bond and the structure of molecules and crystals: An introduction to modern structural chemistry,* 3rd ed., p. 262. © Cornell University.)

Hydrogen Bonds

A hydrogen bond is formed between a donor hydrogen atom with some positive charge and a negatively charged acceptor atom (Fig. 3-8). For example, the hydrogen atoms of the amino ($-NH_2$) group are attracted by the negatively charged keto (–C=O) oxygen atoms. Sometimes, the hydrogen-bonded atoms belong to groups with a unit of charge (such as NH_3^+ or COO^-). In other cases, both the donor hydrogen atoms and the negative acceptor atoms have less than a unit of charge.

The biologically most important hydrogen bonds involve hydrogen atoms covalently bound to oxygen atoms (O—H) or nitrogen atoms (N—H). Likewise, the negative acceptor atoms are usually nitrogen or oxygen. Table 3-3 lists some of the most important hydrogen bonds. In the absence of surrounding water molecules, bond energies range be-

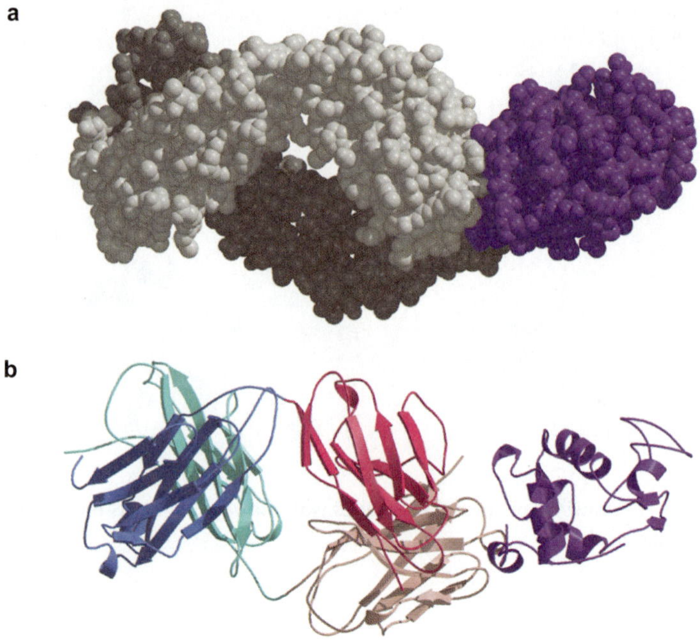

FIGURE 3-7 Antibody–antigen interaction. The structures, depicted as space filling (a) and as ribbons (b), show the complex between Fab D 1.3 and lysozyme (in purple). (Fischmann T.O. et al. 1991. *J. Biol. Chem.* 266: 12915.) Images prepared with MolScript, BobScript, and Raster 3D.

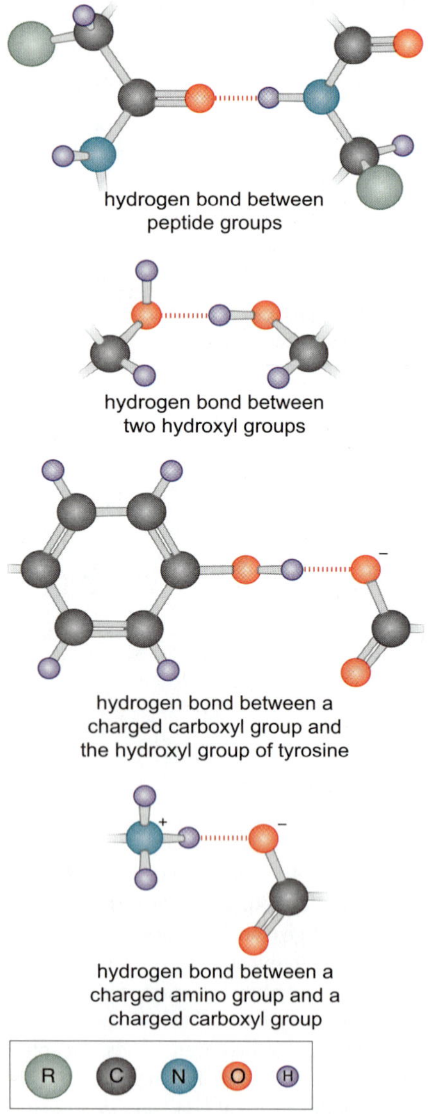

FIGURE 3-8 Examples of hydrogen bonds in biological molecules.

tween 3 and 7 kcal/mole, the stronger bonds involving the greater charge differences between donor and acceptor atoms. Hydrogen bonds are thus weaker than covalent bonds, yet considerably stronger than van der Waals bonds. A hydrogen bond, therefore, will hold two atoms closer together than the sum of their van der Waals radii, but not so close together as a covalent bond would hold them.

Hydrogen bonds, unlike van der Waals bonds, are highly directional. In the strongest hydrogen bonds, the hydrogen atom points directly at the acceptor atom (Fig. 3-9). If it points more than 30° away, the bond energy is much less. Hydrogen bonds are also much more specific than van der Waals bonds, because they demand the existence of molecules with complementary donor hydrogen and acceptor groups.

Some Ionic Bonds Are Hydrogen Bonds

Many organic molecules possess ionic groups that contain one or more units of net positive or negative charge. The negatively charged mononucleotides, for example, contain phosphate groups, which are negatively charged, whereas each amino acid (except proline) has a negative carboxyl group (COO^-) and a positive amino group (NH_3^+), both of which carry a unit of charge. These charged groups are usually neutralized by nearby, oppositely charged groups. The electrostatic forces acting between the oppositely charged groups are called ionic bonds. Their average bond energy in an aqueous solution is about 5 kcal/mole.

In many cases, either an inorganic cation like Na^+, K^+, or Mg^+ or an inorganic anion like Cl^- or SO_4^{2-} neutralizes the charge of ionized organic molecules. When this happens in aqueous solution, the neutralizing cations and anions do not carry fixed positions because inorganic ions are usually surrounded by shells of water molecules and so do not directly bind to oppositely charged groups. Thus, in water solutions, electrostatic bonds to surrounding inorganic cations or anions are usually not of primary importance in determining the molecular shapes of organic molecules.

On the other hand, highly directional bonds result if the oppositely charged groups can form hydrogen bonds to each other. For example, COO^- and NH_3^+ groups are often held together by hydrogen bonds. Because these bonds are stronger than those that involve groups with less than a unit of charge, they are correspondingly shorter. A strong hydrogen bond can also form between a group with a unit charge and a group having less than a unit charge. For example, a hydrogen atom belonging to an amino group (NH_2) bonds strongly to an oxygen atom of a carboxyl group (COO^-).

TABLE 3-3 Approximate Bond Lengths of Biologically Important Hydrogen Bonds

Bond	Approximate H Bond Length (Å)
O—H ⋯ O	2.70 ± 0.10
O—H ⋯ O^-	2.63 ± 0.10
O—H ⋯ N	2.88 ± 0.13
N—H ⋯ O	3.04 ± 0.13
N^+—H ⋯ O	2.93 ± 0.10
N—H ⋯ N	3.10 ± 0.13

Weak Interactions Demand Complementary Molecular Surfaces

Weak binding forces are effective only when the interacting surfaces are close. This proximity is possible only when the molecular surfaces have **complementary structures**, so that a protruding group (or positive charge) on one surface is matched by a cavity (or negative charge) on another. That is, the interacting molecules must have a lock-and-key relationship. In cells, this requirement often means that some molecules hardly ever bond to other molecules of the same kind, because such molecules do not have the properties of symmetry necessary for self-interaction. For example, some polar molecules contain donor hydrogen atoms and no suitable acceptor atoms, whereas other molecules can accept hydrogen bonds but have no hydrogen atoms to donate. On the other hand, there are many molecules with the necessary symmetry to permit strong self-interaction in cells. Water is the most important example of this.

Water Molecules Form Hydrogen Bonds

Under physiological conditions, water molecules rarely ionize to form H^+ and OH^- ions. Instead, they exist as polar H-O-H molecules with both the hydrogen and oxygen atoms forming strong hydrogen bonds. In each water molecule, the oxygen atom can bind to two external hydrogen atoms, whereas each hydrogen atom can bind to one adjacent oxygen atom. These bonds are directed tetrahedrally (Fig. 3-10), so in its solid and liquid forms, each water molecule tends to have four nearest neighbors, one in each of the four directions of a tetrahedron. In ice, the bonds to these neighbors are very rigid and the arrangement of molecules fixed. Above the melting temperature (0°C), the energy of thermal motion is sufficient to break the hydrogen bonds and to allow the water molecules to change their nearest neighbors continually. Even in the liquid form, however, at any given instant most water molecules are bound by four strong hydrogen bonds.

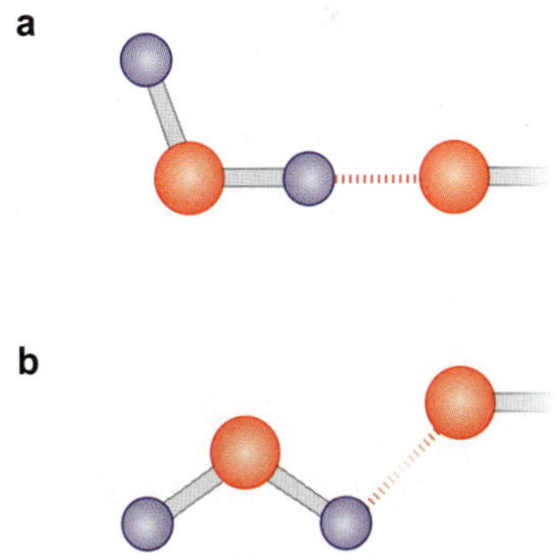

FIGURE 3-9 Directional properties of hydrogen bonds. (a) The vector along the covalent O-H bond points directly at the acceptor oxygen, thereby forming a strong bond. (b) The vector points away from the oxygen atom, resulting in a much weaker bond.

Weak Bonds between Molecules in Aqueous Solutions

The average energy of a secondary, weak, bond, although small compared to that of a covalent bond, is nonetheless strong enough compared to heat energy to ensure that most molecules in aqueous solution will form secondary bonds to other molecules. The proportion of

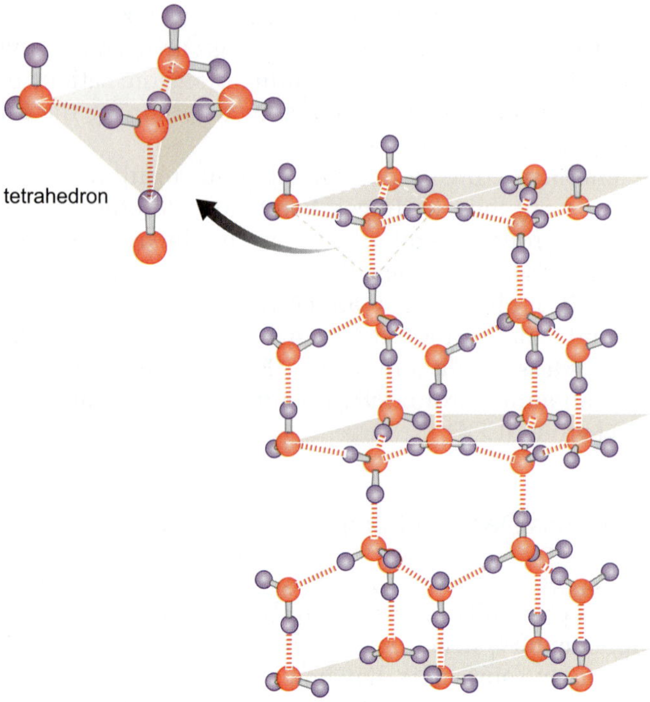

FIGURE 3-10 Diagram of a lattice formed by water molecules. The energy gained by forming specific hydrogen bonds between water molecules favors the arrangement of the molecules in adjacent tetrahedrons. Oxygen atoms are indicated by red spheres, hydrogen atoms by purple spheres. Although the rigidity of the arrangement depends on the temperature of the molecules, the pictured structure is nevertheless predominant in water as well as in ice. (Adapted, with permission, from Pauling L. 1960. *The nature of the chemical bond and the structure of molecules and crystals: An introduction to modern structural chemistry,* 3rd ed., p. 262. © Cornell University.)

bonded to nonbonded arrangements is given by Equation 3-4, corrected to take into account the high concentration of molecules in a liquid. It tells us that interaction energies as low as 2–3 kcal/mole are sufficient at physiological temperatures to force most molecules to form the maximum number of strong secondary bonds.

The specific structure of a solution at a given instant is markedly influenced by which solute molecules are present, not only because molecules have specific shapes, but also because molecules differ in which types of secondary bonds they can form. Thus, a molecule will tend to move until it is next to a molecule with which it can form the strongest possible bond.

Solutions, of course, are not static. Because of the disruptive influence of heat, the specific configuration of a solution is constantly changing from one arrangement to another of approximately the same energy content. Equally important in biological systems is the fact that metabolism is continually transforming one molecule into another and so automatically changing the nature of the secondary bonds that can be formed. The solution structure of cells is thus constantly disrupted not only by heat motion, but also by the metabolic transformations of the cell's solute molecules.

Organic Molecules That Tend to Form Hydrogen Bonds Are Water Soluble

The energy of hydrogen bonds per atomic group is much greater than that of van der Waals contacts; thus, molecules will form hydrogen bonds in preference to van der Waals contacts. For example, if we try to mix water with a compound that cannot form hydrogen bonds, such as benzene, the water and benzene molecules rapidly separate from each other, the water molecules forming hydrogen bonds among themselves while the benzene molecules attach to one another by van der Waals bonds. It is therefore impossible to insert a non-hydrogen-bonding organic molecule into water.

On the other hand, polar molecules such as glucose and pyruvate, which contain a large number of groups that form excellent hydrogen bonds (such as =O or OH), are soluble in water (i.e., they are hydrophilic as opposed to hydrophobic). While the insertion of such groups into a water lattice breaks water–water hydrogen bonds, it results simultaneously in the formation of hydrogen bonds between the polar organic molecule and water. These alternative arrangements, however, are not usually as energetically satisfactory as the water–water arrangements, so that even the most polar molecules ordinarily have only limited solubility (see Box 3-1).

Thus, almost all the molecules that cells acquire, either through food intake or through biosynthesis, are somewhat insoluble in water. These molecules, by their thermal movements, randomly collide with other

ADVANCED CONCEPTS

Box 3-1 The Uniqueness of Molecular Shapes and the Concept of Selective Stickiness

Even though most cellular molecules are built up from only a small number of chemical groups, such as OH, NH_2, and CH_3, there is great specificity as to which molecules tend to lie next to each other. This is because each molecule has unique bonding properties. One very clear demonstration comes from the specificity of stereoisomers. For example, proteins are always constructed from L-amino acids, never from their mirror images, the D-amino acids (Box 3-1 Fig. 1). Although the D- and L-amino acids have identical covalent bonds, their binding properties to asymmetric molecules are often very different. Thus, most enzymes are specific for L-amino acids. If an L-amino acid is able to attach to a specific enzyme, the D-amino acid is unable to bind.

Most molecules in cells can make good "weak" bonds with only a small number of other molecules, partly because most molecules in biological systems exist in an aqueous environment. The formation of a bond in a cell therefore depends not only on whether two molecules bind well to each other, but also on whether bond formation is overall more favorable than the alternative bonds that can form with solvent water molecules.

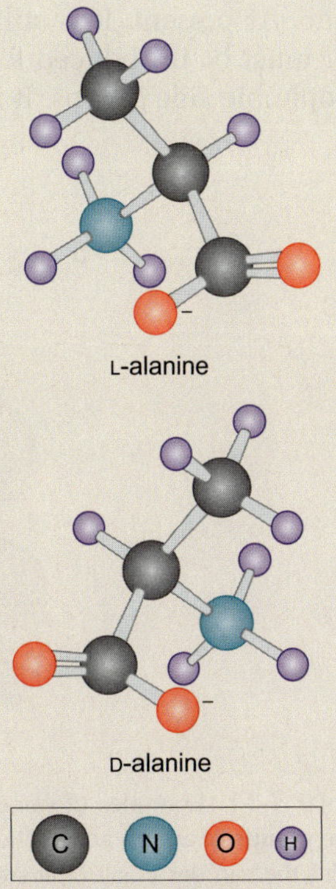

Box 3-1 Figure 1 The two stereoisomers of the amino acid alanine. (Adapted, with permission, from Pauling L. 1960. *The nature of the chemical bond and the structure of molecules and crystals: An introduction to modern structural chemistry*, 3rd ed., p. 465. © Cornell University; Pauling L. 1953. *General chemistry*, 2nd ed., p. 498. © W.H. Freeman.)

molecules until they find complementary molecular surfaces on which to attach and thereby release water molecules for water–water interactions.

Hydrophobic "Bonds" Stabilize Macromolecules

The strong tendency of water to exclude nonpolar groups is frequently referred to as **hydrophobic bonding**. Some chemists like to call all the bonds between nonpolar groups *in a water solution* hydrophobic bonds (Fig. 3-11). In a sense this term is a misnomer, for the phenomenon that it seeks to emphasize is the absence, not the presence, of bonds. (The bonds that tend to form between the nonpolar groups are due to van der Waals attractive forces.) On the other hand, the term *hydrophobic bond* is often useful, because it emphasizes the fact that nonpolar groups will try to arrange themselves so that they are not in contact with water molecules. Hydrophobic bonds are important both in the stabilization of proteins and complexes of proteins with other molecules and in the partitioning of proteins into membranes. They may account for as much as one-half the total free energy of protein folding.

Consider, for example, the different amounts of energy generated when the amino acids alanine and glycine are bound, in water, to a third molecule that has a surface complementary to alanine. A methyl group is present in alanine but not in glycine. When alanine is bound to the third molecule, the van der Waals contacts around the methyl group yield 1 kcal/mole of energy, which is not released when glycine is bound instead. From Equation 3-4, we know that this small energy difference alone would give only a factor of 6 between the binding of alanine and glycine. However, this calculation does not take into consideration the fact that water is trying to exclude alanine much more than glycine. The presence of alanine's CH_3 group upsets the water lattice much more seriously than does the hydrogen atom side group of glycine. At present, it is still difficult to predict how large a correction factor must be introduced for this disruption of the water lattice by the hydrophobic side groups. It is likely that the water tends to exclude ala-

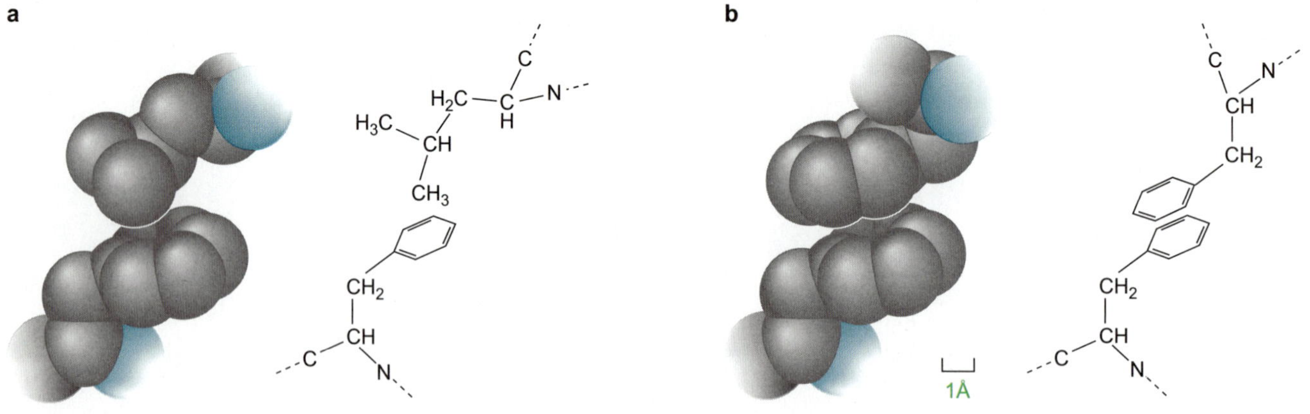

FIGURE 3-11 Examples of van der Waals (hydrophobic) bonds between the nonpolar side groups of amino acids. The hydrogens are not indicated individually. For the sake of clarity, the van der Waals radii are reduced by 20%. The structural formulas adjacent to each space-filling drawing indicate the arrangement of the atoms. (a) Phenylalanine–leucine bond. (b) Phenylalanine–phenylalanine bond. (Adapted, with permission, from Scheraga H.A. 1963. *The proteins*, 2nd ed. (ed. H. Neurath), p. 527. Academic Press. © Harold Scheraga.)

nine, thrusting it toward a third molecule, with a hydrophobic force of approximately 2–3 kcal/mole larger than the forces excluding glycine.

We thus arrive at the important conclusion that the energy difference between the binding of even the most similar molecules to a third molecule (when the difference between the similar molecules involves a nonpolar group) is at least 2–3 kcal/mole greater in the aqueous interior of cells than under nonaqueous conditions. Frequently, the energy difference is 3–4 kcal/mole, because the molecules involved often contain polar groups that can form hydrogen bonds.

The Advantage of ΔG between 2 and 5 kcal/mole

We have seen that the energy of just one secondary bond (2–5 kcal/mole) is often sufficient to ensure that a molecule preferentially binds to a selected group of molecules. Moreover, these energy differences are not so large that rigid lattice arrangements develop within a cell; that is, the interior of a cell never crystallizes, as it would if the energy of secondary bonds were several times greater. Larger energy differences would mean that the secondary bonds seldom break, resulting in low diffusion rates incompatible with cellular existence.

Weak Bonds Attach Enzymes to Substrates

Weak bonds are necessarily the basis by which enzymes and their substrates initially combine with each other. Enzymes do not indiscriminately bind all molecules, having noticeable affinity only for their own substrates.

Because enzymes catalyze both directions of a chemical reaction, they must have specific affinities for both sets of reacting molecules. In some cases, it is possible to measure an equilibrium constant for the binding of an enzyme to one of its substrates (Equation 3-4), which consequently enables us to calculate the ΔG upon binding. This calculation in turn hints at which types of bonds may be involved. For ΔG values between 5 and 10 kcal/mole, several strong secondary bonds are the basis of specific enzyme–substrate interactions. Also worth noting is that the ΔG of binding is never exceptionally high; thus, enzyme–substrate complexes can be both made and broken apart rapidly as a result of random thermal movement. This explains why enzymes can function quickly, sometimes as often as 10^6 times per second. If enzymes were bound to their substrates, or more importantly to their products, by more powerful bonds, they would act much more slowly.

Weak Bonds Mediate Most Protein–DNA and Protein–Protein Interactions

As we shall see throughout the book, interactions between proteins and DNA, and between proteins and other proteins, lie at the heart of how cells detect and respond to signals, express genes, replicate, repair, and recombine their DNA, and so on—as well as how those processes are regulated. Again, these interactions are mediated by weak chemical bonds of the sort we have described in this chapter. Despite the low energy of each individual bond, affinity in these interactions, and specificity as well, results from the combined effects of many such bonds between any two interacting molecules.

In Chapter 5, we return to these matters with a detailed look at how proteins are built, how they adopt particular structures, and how they bind DNA, RNA, and each other.

SUMMARY

Many important chemical events in cells do not involve the making or breaking of covalent bonds. The cellular location of most molecules depends on weak, or secondary, attractive or repulsive forces. In addition, weak bonds are important in determining the shape of many molecules, especially very large ones. The most important of these weak forces are hydrogen bonds, van der Waals interactions, hydrophobic bonds, and ionic bonds. Even though these forces are relatively weak, they are still large enough to ensure that the right molecules (or atomic groups) interact with each other. For example, the surface of an enzyme is uniquely shaped to allow specific attraction of its substrates.

The formation of all chemical bonds, weak interactions as well as strong covalent bonds, proceeds according to the laws of thermodynamics. A bond tends to form when the result would be a release of free energy (negative ΔG). For the bond to be broken, this same amount of free energy must be supplied. Because the formation of covalent bonds between atoms usually involves a very large negative ΔG, covalently bound atoms almost never separate spontaneously. In contrast, the ΔG values accompanying the formation of weak bonds are only several times larger than the average thermal energy of molecules at physiological temperatures. Single weak bonds are thus frequently being made and broken in living cells.

Molecules having polar (charged) groups interact quite differently from nonpolar molecules (in which the charge is symmetrically distributed). Polar molecules can form good hydrogen bonds, whereas nonpolar molecules can form only van der Waals bonds. The most important polar molecule is water. Each water molecule can form four hydrogen bonds to other water molecules. Although polar molecules tend to be soluble in water (to various degrees), nonpolar molecules are insoluble because they cannot form hydrogen bonds with water molecules.

Every distinct molecule has a unique molecular shape that restricts the number of molecules with which it can form strong secondary bonds. Strong secondary interactions demand both a complementary (lock-and-key) relationship between the two bonding surfaces and the involvement of many atoms. Although molecules bound together by only one or two secondary bonds frequently fall apart, a collection of these weak bonds can result in a stable aggregate. The fact that double-helical DNA never falls apart spontaneously demonstrates the extreme stability possible in such an aggregate.

BIBLIOGRAPHY

Branden C. and Tooze J. 1999. *Introduction to protein structure*, 2nd ed. Garland Publishing, New York.

Creighton T.E. 1992. *Proteins: Structure and molecular properties*, 2nd ed. W.H. Freeman, New York.

——— 1983. *Proteins*. W.H. Freeman, San Francisco.

Donohue J. 1968. Selected topics in hydrogen bonding. In *Structural chemistry and molecular biology* (ed. A. Rich and N. Davidson), pp. 443–465. W.H. Freeman, San Francisco.

Fersht A. 1999. *Structure and mechanism in protein science: A guide to enzyme catalysis and protein folding*. W.H. Freeman, New York.

Gray H.B. 1964. *Electrons and chemical bonding*. Benjamin Cummings, Menlo Park, California.

Klotz I.M. 1967. *Energy changes in biochemical reactions*. Academic Press, New York.

Kyte J. 1995. *Mechanism in protein chemistry*. Garland Publishing, New York.

——— 1995. *Structure in protein chemistry*. Garland Publishing, New York.

Lehninger A.L. 1971. *Bioenergetics*, 3rd ed. Benjamin Cummings, Menlo Park, California.

Lesk A. 2000. *Introduction to protein architecture: The structural biology of proteins*. Oxford University Press, New York.

Marsh R.E. 1968. Some comments on hydrogen bonding in purine and pyrimidine bases. In *Structural chemistry and molecular biology* (ed. A. Rich and N. Davidson), pp. 485–489. W.H. Freeman, San Francisco.

Morowitz H.J. 1970. *Entropy for biologists*. Academic Press, New York.

Pauling L. 1960. *The nature of the chemical bond*, 3rd ed. Cornell University Press, Ithaca, New York.

Tinoco I. (ed.), Sauer K., Wang J.C., and Puglisi J.D. 2001. *Physical chemistry: Principles and applications in life sciences*, 4th ed. Prentice Hall College Division, Upper Saddle River, New Jersey.

CHAPTER

4

The Importance of High-Energy Bonds

OUTLINE

IN THE PREVIOUS CHAPTER WE LOOKED AT THE formation of weak bonds from the thermodynamic viewpoint. Each time a potential weak bond was considered, the question was posed, Does its formation involve a gain or a loss of free energy? Only when ΔG is negative does the thermodynamic equilibrium favor a reaction. This same approach is equally valid for covalent bonds. The fact that enzymes are usually involved in the making or breaking of a covalent bond does not in any sense alter the requirement of a negative ΔG.

On superficial examination, however, many of the important covalent bonds in cells appear to be formed in violation of the laws of thermodynamics, particularly those bonds joining small molecules together to form large polymeric molecules. The formation of such bonds involves an increase in free energy. Originally, this fact suggested to some people that cells had the unique ability to work in violation of thermodynamics and that this property was, in fact, the real "secret of life."

Now, however, it is clear that these biosynthetic processes do not violate thermodynamics but rather are based on different reactions from those originally postulated. Nucleic acids, for example, do not form by the condensation of nucleoside phosphates; glycogen is not formed directly from glucose residues; proteins are not formed by the union of amino acids. Instead, the monomeric precursors, using energy present in ATP, are first converted to high-energy "activated" precursors, which then spontaneously (with the help of specific enzymes) unite to form larger molecules. In this chapter, we shall illustrate these ideas by concentrating on the thermodynamics of peptide (protein) and phosphodiester (nucleic acid) bonds. First, however, we must briefly look at some general thermodynamic properties of covalent bonds.

MOLECULES THAT DONATE ENERGY ARE THERMODYNAMICALLY UNSTABLE

There is great variation in the amount of free energy possessed by specific molecules. This is because covalent bonds do not all have the same bond energy. As an example, the covalent bond between oxygen and hydrogen is considerably stronger than the bond between hydrogen and hydrogen, or oxygen and oxygen. The formation of an O—H bond at the expense of O—O or H—H will thus release energy. Energy considerations, therefore, tell us that a sufficiently concentrated mixture of oxygen and hydrogen will be transformed into water.

A molecule thus possesses a larger amount of free energy if linked together by weak covalent bonds than if it is linked together by strong bonds. This idea seems almost paradoxical at first glance because it means that the stronger the bond, the less energy it can give off. But the notion automatically makes sense when we realize that an atom that has formed a very strong bond has already lost a large amount of free energy in this process. Therefore, the best food molecules (molecules that donate energy) are those molecules that contain weak covalent bonds and are therefore thermodynamically unstable.

For example, glucose is an excellent food molecule because there is a great decrease in free energy when it is oxidized by oxygen to yield carbon dioxide and water. On the other hand, carbon dioxide, composed of strong covalent double bonds between carbon and oxygen, known as **carbonyl bonds**, is not a food molecule in animals. In the absence of the energy donor ATP, carbon dioxide cannot be transformed spontaneously into more complex organic molecules, even with the help of specific enzymes. Carbon dioxide can be used as a primary source of carbon in plants only because the energy supplied by light quanta during photosynthesis results in the formation of ATP.

The chemical reactions by which molecules are transformed into other molecules containing less free energy do not occur at significant rates at physiological temperatures in the absence of a catalyst. This is because even a weak covalent bond is, in reality, very strong and is only rarely broken by thermal motion within a cell. For a covalent bond to be broken in the absence of a catalyst, energy must be supplied to push apart the bonded atoms. When the atoms are partially apart, they can recombine with new partners to form stronger bonds. In the process of recombination, the energy released is the sum of the free energy supplied to break the old bond plus the difference in free energy between the old and the new bond (Fig. 4-1).

The energy that must be supplied to break the old covalent bond in a molecular transformation is called the **activation energy**. The activation energy is usually less than the energy of the original bond because molecular rearrangements generally do not involve the production of completely free atoms. Instead, a collision between the two reacting molecules is required, followed by the temporary formation of a molecular complex called the **activated state**. In the activated state, the close proximity of the two molecules makes each other's bonds more labile, so that less energy is needed to break a bond than when the bond is present in a free molecule.

Most reactions of covalent bonds in cells are therefore described by

$$(A—B) + (C—D) \rightarrow (A—D) + (C—B). \qquad \textbf{[Equation 4-1]}$$

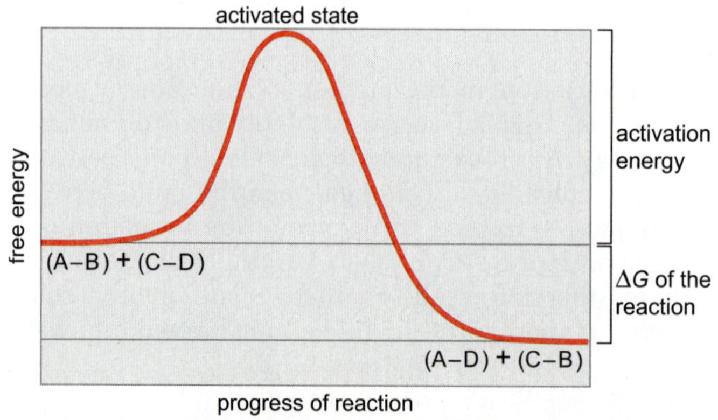

FIGURE **4-1** **The energy of activation of a chemical reaction: (A—B) + (C—D) → (A—D) + (C—B).** This reaction is accompanied by a decrease in free energy.

The mass action expression for such a reaction is

$$K_{eq} = \frac{\text{conc}^{A-D} \times \text{conc}^{C-B}}{\text{conc}^{A-B} \times \text{conc}^{C-D}}, \qquad \textbf{[Equation 4-2]}$$

where conc^{A-B}, conc^{C-D}, and so on are the concentrations of the several reactants in moles per liter. Here, also, the value of K_{eq} is related to ΔG by Equation 4-3 (see also Table 4-1):

$$\Delta G = -RT \ln K_{eq} \quad \text{or} \quad K_{eq} = e^{-\Delta G/RT}. \qquad \textbf{[Equation 4-3]}$$

Because energies of activation are generally between 20 and 30 kcal/mole, activated states practically never occur at physiological temperatures. High activation energies are thus barriers preventing spontaneous rearrangements of cellular-covalent bonds.

These barriers are enormously important. Life would be impossible if they did not exist, for all atoms would be in the state of least possible energy. There would be no way to temporarily store energy for future work. On the other hand, life would also be impossible if means were not found to selectively lower the activation energies of certain reactions. This also must happen if cell growth is to occur at a rate sufficiently fast so as not to be seriously impeded by random destructive forces, such as ionization or ultraviolet radiation.

TABLE 4-1 The Relationship between K_{eq} and ΔG ($\Delta G = -RT$ in K_{eq})

K_{eq}	ΔG (kcal/mole)
10^{-6}	8.2
10^{-5}	6.8
10^{-4}	5.1
10^{-3}	4.1
10^{-2}	2.7
10^{-1}	1.4
10^{0}	0.0
10^{1}	–1.4
10^{2}	–2.7
10^{3}	–4.1

ENZYMES LOWER ACTIVATION ENERGIES IN BIOCHEMICAL REACTIONS

Enzymes are absolutely necessary for life. The function of enzymes is to speed up the rate of the chemical reactions requisite to cellular existence by lowering the activation energies of molecular rearrangements to values that can be supplied by the heat of motion (Fig. 4-2). When a specific enzyme is present, there is no longer an effective barrier preventing the rapid formation of the reactants possessing the lowest amounts of free energy. Enzymes never affect the nature of an equilibrium: they merely speed up the rate at which it is reached. Thus, if the thermodynamic equilibrium is unfavorable for the formation of a molecule, the presence of an enzyme can in no way bring about thc molecule's accumulation.

Because enzymes must catalyze essentially every cellular molecular rearrangement, knowing the free energy of various molecules cannot by itself tell us whether an energetically feasible rearrangement will, in fact, occur. The rate of the reactions must always be considered. Only if a cell possesses a suitable enzyme will the reaction be important.

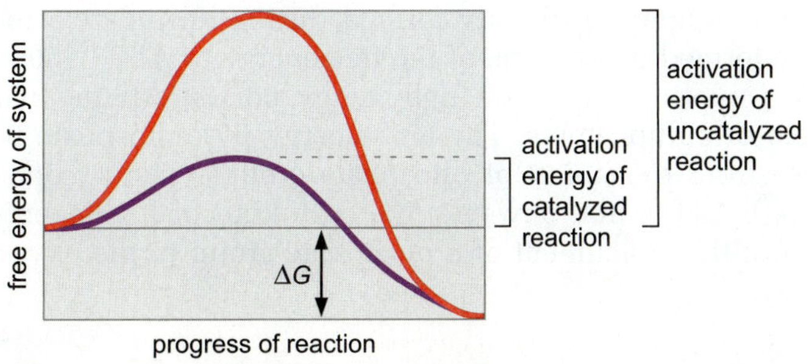

FIGURE 4-2 Enzymes lower activation energies and thus speed up the rate of the reaction. The enzyme-catalyzed reaction is shown by the purple curve. Note that ΔG remains the same because the equilibrium position remains unaltered.

FREE ENERGY IN BIOMOLECULES

Thermodynamics tells us that all biochemical pathways must be characterized by a decrease in free energy. This is clearly the case for degradative pathways, in which thermodynamically unstable food molecules are converted to more stable compounds, such as carbon dioxide and water, with the evolution of heat. All degradative pathways have two primary purposes: (1) to produce the small organic fragments necessary as building blocks for larger organic molecules and (2) to conserve a significant fraction of the free energy of the original food molecule in a form that can do work. This latter purpose is accomplished by coupling some of the steps in degradative pathways with the simultaneous formation of high-energy molecules such as ATP, which can store free energy.

Not all the free energy of a food molecule is converted into the free energy of high-energy molecules. If this were the case, a degradative pathway would not be characterized by a decrease in free energy, and there would be no driving force to favor the breakdown of food molecules. Instead, we find that all degradative pathways are characterized by a conversion of at least one-half the free energy of the food molecule into heat and/or entropy. For example, it is estimated that in cells, approximately 40% of the free energy of glucose is used to make new high-energy compounds, the remainder being dissipated into heat energy and entropy.

High-Energy Bonds Hydrolyze with Large Negative ΔG

A high-energy molecule contains one or more bonds whose breakdown by water, called **hydrolysis**, is accompanied by a large decrease in free energy. The specific bonds whose hydrolysis yields these large negative ΔG values are called **high-energy bonds**, a somewhat misleading term, since it is not the bond energy but the free energy of hydrolysis that is high. Nonetheless, the term *high-energy bond* is generally employed, and for convenience, we shall continue this usage by designating high-energy bonds with the symbol ~.

The energy of hydrolysis of the average high-energy bond (≈7 kcal/mole) is very much smaller than the amount of energy that would be released if a glucose molecule were to be completely degraded in one step (688 kcal/mole). A one-step breakdown of glucose would be inefficient in making high-energy bonds. This is undoubtedly the reason why biological glucose degradation requires so many steps. In this way, the amount of energy released per degradative step is of the same order of magnitude as the free energy of hydrolysis of a high-energy bond.

The most important high-energy compound is ATP. It is formed from inorganic phosphate (Ⓟ) and ADP, using energy obtained either from degradative reactions or from the sun, a process known as photosynthesis. There are, however, many other important high-energy compounds. Some are directly formed during degradative reactions; others are formed using some of the free energy of ATP. Table 4-2 lists the most important types of high-energy bonds. All involve either phosphate or sulfur atoms. The high-energy pyrophosphate bonds of ATP arise from the union of phosphate groups. The pyrophosphate linkage (Ⓟ~Ⓟ) is not, however, the only kind of high-energy phosphate bond: the attachment of a phosphate group to the oxygen atom

TABLE 4-2 Important Classes of High-Energy Bonds

Class	Molecular Example	Reaction	ΔG of Reaction (kcal/mole)
Pyrophosphate	P ~ P pyrophosphate	P ~ P ⇌ P + P	ΔG = –6
Nucleoside diphosphates	adenosine—P ~ P (ADP)	ADP ⇌ AMP + P	ΔG = –6
Nucleoside triphosphates	adenosine—P ~ P ~ P (ATP)	ATP ⇌ ADP + P ATP ⇌ AMP + P ~ P	ΔG = –7
Enol phosphates	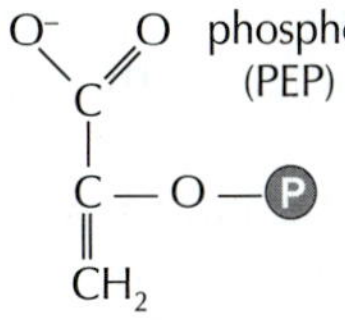 phosphoenolpyruvate (PEP)	PEP ⇌ pyruvate + P	ΔG = –12
Aminoacyl adenylates	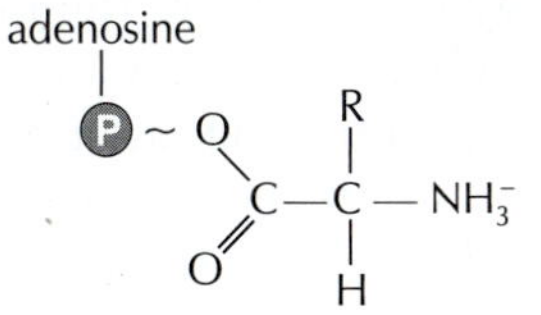	AMP ~ AA ⇌ AMP + AA	ΔG = –7
Guanidinium phosphates	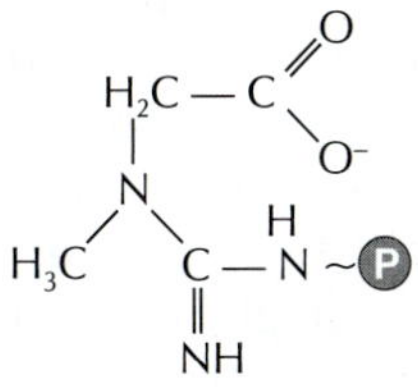 creatine phosphate	creatine ~ P ⇌ creatine + P	ΔG = –8
Thioesters	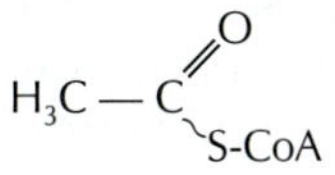 acetyl-CoA	acetyl CoA ⇌ CoA-SH + acetate	ΔG = –8

of a carboxyl group creates a high-energy acyl bond. It is now clear that high-energy bonds involving sulfur atoms play almost as important a role in energy metabolism as those involving phosphorus. The most important molecule containing a high-energy sulfur bond is acetyl-CoA. This bond is the main source of energy for fatty acid biosynthesis.

The wide range of ΔG values of high-energy bonds (see Table 4-2) means that calling a bond "high-energy" is sometimes arbitrary. The usual criterion is whether its hydrolysis can be coupled with another reaction to effect an important biosynthesis. For example, the negative ΔG accompanying the hydrolysis of glucose-6-phosphate is 3–4 kcal/mole. But this ΔG is not sufficient for efficient synthesis of peptide bonds, so this phosphate ester bond is not included among high-energy bonds.

HIGH-ENERGY BONDS IN BIOSYNTHETIC REACTIONS

The construction of a large molecule from smaller building blocks often requires the input of free energy. Yet, a biosynthetic pathway, like a degradative pathway, would not exist if it were not characterized by a net decrease in free energy. This means that many biosynthetic pathways demand an external source of free energy. These free-energy sources are the high-energy compounds. The making of many biosynthetic bonds is coupled with the breakdown of a high-energy bond, so that the net change of free energy is always negative. Thus, high-energy bonds in cells generally have a very short life. Almost as soon as they are formed during a degradative reaction, they are enzymatically broken down to yield the energy needed to drive another reaction to completion.

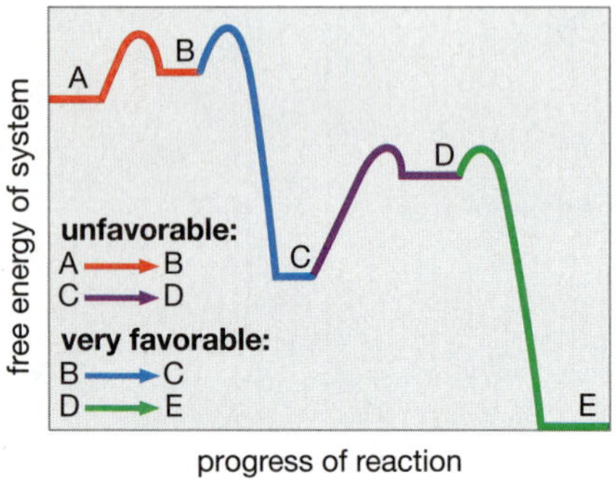

FIGURE 4-3 Free-energy changes in a multistep metabolic pathway, A → B → C → D → E. Two steps (A → B and C → D) do not favor the A → E direction of the reaction, because they have small positive ΔG values. However, they are insignificant owing to the very large negative ΔG values provided in steps B → C and D → E. Therefore, the overall reaction favors the A → E conversion.

Not all the steps in a biosynthetic pathway require the breakdown of a high-energy bond. Often, only one or two steps involve such a bond. Sometimes this is because the ΔG, even in the absence of an externally added high-energy bond, favors the biosynthetic direction. In other cases, ΔG is effectively zero or may even be slightly positive. These small positive ΔG values, however, are not significant so long as they are followed by a reaction characterized by the hydrolysis of a high-energy bond. Rather, it is the *sum* of all the free-energy changes in a pathway that is significant, as shown in Figure 4-3. It does not really matter that the K_{eq} of a specific biosynthetic step is slightly (80:20) in favor of degradation if the K_{eq} of the succeeding step is 100:1 in favor of the forward biosynthetic direction.

Likewise, not all the steps in a degradative pathway generate high-energy bonds. For example, only two steps in the lengthy glycolytic (Embden–Meyerhof) breakdown of glucose generate ATP. Moreover, there are many degradative pathways that have one or more steps requiring the breakdown of a high-energy bond. The glycolytic breakdown of glucose is again an example. It uses up two molecules of ATP for every four that it generates. Here, of course, as in every energy-yielding degradative process, more high-energy bonds must be made than consumed.

Peptide Bonds Hydrolyze Spontaneously

The formation of a dipeptide and a water molecule from two amino acids requires a ΔG of 1–4 kcal/mole, depending on which amino acids are being joined. These positive ΔG values by themselves tell us that polypeptide chains cannot form from free amino acids. In addition, we must take into account the fact that water molecules have a much, much higher concentration than any other cellular molecules (generally more than 100 times higher). All equilibrium reactions in which water participates are thus strongly pushed in the direction that consumes water molecules. This is easily seen in the definition of equilibrium constants. For example, the reaction forming a dipeptide,

$$\text{amino acid(A)} + \text{amino acid(B)} \rightarrow \text{dipeptide(A—B)} + H_2O, \quad \textbf{[Equation 4-4]}$$

has the equilibrium constant

$$K_{eq} = \frac{\text{conc}^{A—B} \times \text{conc}^{H_2O}}{\text{conc}^{A} \times \text{conc}^{B}}, \quad \textbf{[Equation 4-5]}$$

where concentrations are given in moles per liter. Thus, for a given K_{eq} value (related to ΔG by the formula $\Delta G = -RT \ln K_{eq}$), a much greater concentration of water means a correspondingly smaller concentration of the dipeptide. The relative concentrations are, therefore, very important. In fact, a simple calculation shows that hydrolysis may often proceed spontaneously even when the ΔG for the nonhydrolytic reaction is –3 kcal/mole.

Thus, in theory, proteins are unstable and, given sufficient time, will spontaneously degrade to free amino acids. On the other hand, in the absence of specific enzymes, these spontaneous rates are too slow to have a significant effect on cellular metabolism. That is, once a protein is made, it remains stable unless its degradation is catalyzed by a specific enzyme.

Coupling of Negative with Positive ΔG

Free energy must be added to amino acids before they can be united to form proteins. How this happens became clear with the discovery of the fundamental role of ATP as an energy donor. ATP contains three phosphate groups attached to an adenosine molecule (adenosine—O—Ⓟ~Ⓟ~Ⓟ). When one or two of the terminal ~Ⓟ groups are broken off by hydrolysis, there is a significant decrease of free energy:

Adenosine—O—Ⓟ~Ⓟ~Ⓟ + H_2O → Adenosine—O—Ⓟ~Ⓟ + Ⓟ
($\Delta G = -7$ kcal/mole), **[Equation 4-6]**

Adenosine—O—Ⓟ~Ⓟ~Ⓟ + H_2O → Adenosine—O—Ⓟ + Ⓟ~Ⓟ
($\Delta G = -8$ kcal/mole), **[Equation 4-7]**

Adenosine—O—Ⓟ~Ⓟ + H_2O → Adenosine—O—Ⓟ + Ⓟ
($\Delta G = -6$ kcal/mole). **[Equation 4-8]**

All these breakdown reactions have negative ΔG values considerably greater in absolute value (numerical value without regard to sign) than the positive ΔG values accompanying the formation of polymeric molecules from their monomeric building blocks. The essential trick underlying these biosynthetic reactions, which by themselves have a positive ΔG, is that they are coupled with the breakage of high-energy bonds, characterized by negative ΔG of greater absolute value. Thus, during protein synthesis, the formation of each peptide bond ($\Delta G = +0.5$ kcal/mole) is coupled with the breakdown of ATP to AMP and pyrophosphate, which has a ΔG of –8 kcal/mole (see Equation 4-7). This results in a net ΔG of –7.5 kcal/mole, more than sufficient to ensure that the equilibrium favors protein synthesis rather than breakdown.

ACTIVATION OF PRECURSORS IN GROUP TRANSFER REACTIONS

When ATP is hydrolyzed to ADP and phosphate, most of the free energy is liberated as heat. Because heat energy cannot be used to make covalent bonds, a coupled reaction cannot be the result of two completely separate reactions, one with a positive ΔG, the other with a negative ΔG. Instead, a coupled reaction is achieved by two or more successive reactions. These are always **group-transfer** reactions: reactions, not involving oxidations or reductions, in which molecules exchange functional groups. The enzymes that catalyze these reactions are called **transferases**. Consider the reaction

$$(\mathrm{A{-}X}) + (\mathrm{B{-}Y}) \rightarrow (\mathrm{A{-}B}) + (\mathrm{X{-}Y}). \qquad \textbf{[Equation 4-9]}$$

In this example, group X is exchanged with component B. Group-transfer reactions are arbitrarily defined to exclude water as a participant. When water is involved,

$$(\mathrm{A{-}B}) + (\mathrm{H{-}OH}) \rightarrow (\mathrm{A{-}OH}) + (\mathrm{B{-}H}). \qquad \textbf{[Equation 4-10]}$$

This reaction is called a hydrolysis, and the enzymes involved are called hydrolases.

The group-transfer reactions that interest us here are those involving groups attached by high-energy bonds. When such a high-energy group is transferred to an appropriate acceptor molecule, it becomes attached to the acceptor by a high-energy bond. Group transfer thus allows the transfer of high-energy bonds from one molecule to another. For example, Equations 4-11 and 4-12 show how energy present in ATP is transferred to form GTP, one of the precursors used in RNA synthesis:

$$\text{Adenosine}{-}\textcircled{P}{\sim}\textcircled{P}{\sim}\textcircled{P} + \text{Guanosine}{-}\textcircled{P} \rightarrow \text{Adenosine}{-}\textcircled{P}{\sim}\textcircled{P} + \text{Guanosine}{-}\textcircled{P}{\sim}\textcircled{P}, \qquad \textbf{[Equation 4-11]}$$

$$\text{Adenosine}{-}\textcircled{P}{\sim}\textcircled{P}{\sim}\textcircled{P} + \text{Guanosine}{-}\textcircled{P}{\sim}\textcircled{P} \rightarrow \text{Adenosine}{-}\textcircled{P}{\sim}\textcircled{P} + \text{Guanosine}{-}\textcircled{P}{\sim}\textcircled{P}{\sim}\textcircled{P}. \qquad \textbf{[Equation 4-12]}$$

The high-energy $\textcircled{P}{\sim}\textcircled{P}$ group on GTP allows it to unite spontaneously with another molecule. GTP is thus an example of what is called an activated molecule; correspondingly, the process of transferring a high-energy group is called group activation.

ATP Versatility in Group Transfer

ATP synthesis has a key role in the controlled trapping of the energy of molecules that serve as energy donors. In both oxidative and photosynthetic phosphorylations, energy is used to synthesize ATP from ADP and phosphate:

$$\text{Adenosine}{-}\textcircled{P}{\sim}\textcircled{P} + \textcircled{P} + \text{energy} \rightarrow \text{Adenosine}{-}\textcircled{P}{\sim}\textcircled{P}{\sim}\textcircled{P}. \qquad \textbf{[Equation 4-13]}$$

Because ATP is the original biological recipient of high-energy groups, it must be the starting point of a variety of reactions in which high-energy groups are transferred to low-energy molecules to give them the potential to react spontaneously. ATP's central role utilizes the fact that it contains two high-energy bonds whose splitting releases specific groups. This is seen in Figure 4-4, which shows three important groups arising from ATP: $\textcircled{P}{\sim}\textcircled{P}$, a pyrophosphate group; ~AMP, an adenosyl monophosphate group; and ${\sim}\textcircled{P}$, a phosphate group. It is important to notice that these high-energy groups retain their high-energy quality only when transferred to an appropriate acceptor molecule. For example, although the transfer of a ${\sim}\textcircled{P}$ group to a $\mathrm{COO^-}$ group yields a high-energy $\mathrm{COO}{\sim}\textcircled{P}$ acylphosphate group, the transfer of the same group to a sugar hydroxyl group (–C–OH), as in the formation of glucose-6-phosphate, gives rise to a low-energy bond (<5 kcal/mole decrease in ΔG upon hydrolysis).

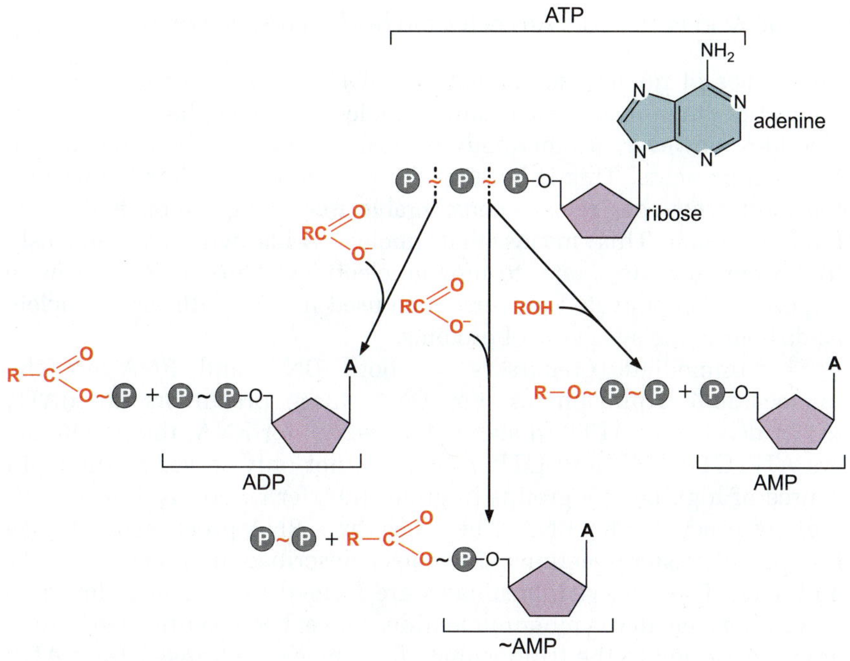

FIGURE 4-4 **Important group transfers involving ATP.**

Activation of Amino Acids by Attachment of AMP

The activation of an amino acid is achieved by transfer of an AMP group from ATP to the COO^- group of the amino acid, as shown by

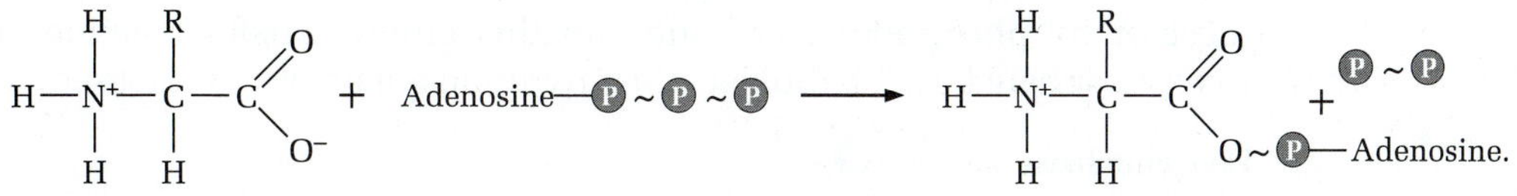

[Equation 4-14]

(In the equation, R represents the specific side group of the amino acid.) The enzymes that catalyze this type of reaction are called **aminoacyl synthetases**. Upon activation, an amino acid (AA) is thermodynamically capable of being efficiently used for protein synthesis. Nonetheless, the AA~AMP complexes are not the direct precursors of proteins. Instead, for a reason we shall explain in Chapter 14, a second group transfer must occur to transfer the amino acid, still activated at its carboxyl group, to the end of a tRNA molecule:

$$\text{AA\sim AMP + tRNA} \rightarrow \text{AA\sim tRNA + AMP.} \qquad \textbf{[Equation 4-15]}$$

A peptide bond then forms by the condensation of the AA~tRNA molecule onto the end of a growing polypeptide chain:

$$\text{AA\sim tRNA + growing polypeptide chain (of } n \text{ amino acids)} \rightarrow$$
$$\text{tRNA + growing polypeptide chain (of } n+1 \text{ amino acids).}$$

[Equation 4-16]

Thus, the final step of this "coupled reaction," like that of all other coupled reactions, necessarily involves the removal of the activating group and the conversion of a high-energy bond into one with a lower free energy of hydrolysis. This is the source of the negative ΔG that drives the reaction in the direction of protein synthesis.

Nucleic Acid Precursors Are Activated by the Presence of Ⓟ~Ⓟ

Both types of nucleic acid, DNA and RNA, are built up from mononucleotide monomers, also called nucleoside phosphates. Mononucleotides, however, are thermodynamically even less likely to combine than amino acids. This is because the phosphodiester bonds that link the former together release considerable free energy upon hydrolysis (–6 kcal/mole). This means that nucleic acids will spontaneously hydrolyze, at a slow rate, to mononucleotides. Thus, it is even more important that activated precursors be used in the synthesis of nucleic acids than in the synthesis of proteins.

The immediate precursors for both DNA and RNA are the nucleoside-5′-triphosphates. For DNA, these precursors are dATP, dGTP, dCTP, and dTTP (d stands for deoxy); for RNA, the precursors are ATP, GTP, CTP, and UTP. ATP, thus, not only serves as the main source of high-energy groups in group-transfer reactions, but is itself a direct precursor for RNA. The other three RNA precursors all arise by group-transfer reactions like those described in Equations 4-11 and 4-12. The deoxytriphosphates are formed in basically the same way: after the deoxymononucleotides have been synthesized, they are transformed to the triphosphate form by group transfer from ATP:

Deoxynucleoside—Ⓟ + ATP → Deoxynucleoside—Ⓟ~Ⓟ + ADP,
[Equation 4-17]

Deoxynucleoside—Ⓟ~Ⓟ + ATP →
Deoxynucleoside—Ⓟ~Ⓟ~Ⓟ + ADP. **[Equation 4-18]**

These triphosphates can then unite to form polynucleotides held together by phosphodiester bonds. In this group-transfer reaction, a pyrophosphate bond is broken and a pyrophosphate group released:

Deoxynucleoside—Ⓟ~Ⓟ~Ⓟ
+ growing polynucleotide chain (of *n* nucleotides),
Ⓟ~Ⓟ + growing polynucleotide chain (*n* + 1 nucleotides).
[Equation 4-19]

This reaction, unlike that which forms peptide bonds, does not have a negative ΔG. In fact, the ΔG is slightly positive (~0.5 kcal/mole). This situation immediately poses the question, as polynucleotides obviously form: what is the source of the necessary free energy?

The Value of Ⓟ~Ⓟ Release in Nucleic Acid Synthesis

The needed free energy comes from the splitting of the high-energy pyrophosphate group that is formed simultaneously with the high-energy phosphodiester bond. All cells contain a powerful enzyme, pyrophosphatase, which breaks down pyrophosphate molecules almost as soon as they are formed:

Ⓟ~Ⓟ → 2 Ⓟ ($\Delta G = -7$ kcal/mole). **[Equation 4-20]**

The large negative ΔG means that the reaction is effectively irreversible. This means that once Ⓟ~Ⓟ is broken down, it never re-forms.

The union of the nucleoside monophosphate group (Equation 4-17), coupled with the splitting of the pyrophosphate groups (Equation 4-20), has an equilibrium constant determined by the combined ΔG values of the two reactions: (0.5 kcal/mole) + (–7 kcal/mole). The resulting value ($\Delta G = -6.5$ kcal/mole) tells us that nucleic acids almost never break down to re-form their nucleoside triphosphate precursors.

Here we see a powerful example of the fact that often it is the free-energy change accompanying a *group of reactions* that determines whether a reaction in the group will take place. Reactions with small, positive ΔG values, which by themselves would never take place, are often part of important metabolic pathways in which they are followed by reactions with large negative ΔG values. At all times we must remember that a single reaction (or even a single pathway) never occurs in isolation; rather, the nature of the equilibrium is constantly being changed through the addition and removal of metabolites.

Ⓟ~Ⓟ Splits Characterize Most Biosynthetic Reactions

The synthesis of nucleic acids is not the only reaction where direction is determined by the release and splitting of Ⓟ~Ⓟ. In fact, essentially all biosynthetic reactions are characterized by one or more steps that release pyrophosphate groups. Consider, for example, the activation of an amino acid by the attachment of AMP. By itself, the transfer of a high-energy bond from ATP to the AA~AMP complex has a slightly positive ΔG. Therefore, it is the release and splitting of ATP's terminal pyrophosphate group that provides the negative ΔG that is necessary to drive the reaction.

The great utility of the pyrophosphate split is neatly demonstrated when we consider the problems that would arise if a cell attempted to synthesize nucleic acid from nucleoside diphosphates rather than triphosphates (Fig. 4-5). Phosphate, rather than pyrophosphate, would be liberated as the backbone phosphodiester linkages were made. The phosphodiester linkages, however, are not stable in the presence of significant quantities of phosphate, because they are formed without a significant release of free energy. Thus, the biosynthetic reaction would be easily reversible; if phosphate were to accumulate, the reaction would begin to move in the direction of nucleic acid breakdown according to the law of mass action. Moreover, it is not feasible for a cell to remove the phosphate groups as soon as they are generated (thereby preventing this reverse reaction), as all cells require a significant internal level of phosphate to grow. In contrast, a sequence of reactions that liberate pyrophosphate and then rapidly break it down into two phosphates disconnects the liberation of phosphate from the nucleic acid biosynthesis reaction, and thereby prevents the possibility of reversing the biosynthetic reaction (see Fig. 4-5). In consequence, it would be very difficult to accumulate enough phosphate in the cell to drive both reactions in the reverse, or breakdown, direction. It is clear that the use of nucleoside triphosphates as precursors of nucleic acids is not a matter of chance.

This same type of argument tells us why ATP, and not ADP, is the key donor of high-energy groups in all cells. At first this preference seemed arbitrary to biochemists. Now, however, we see that many reactions using ADP as an energy donor would occur equally well in both directions.

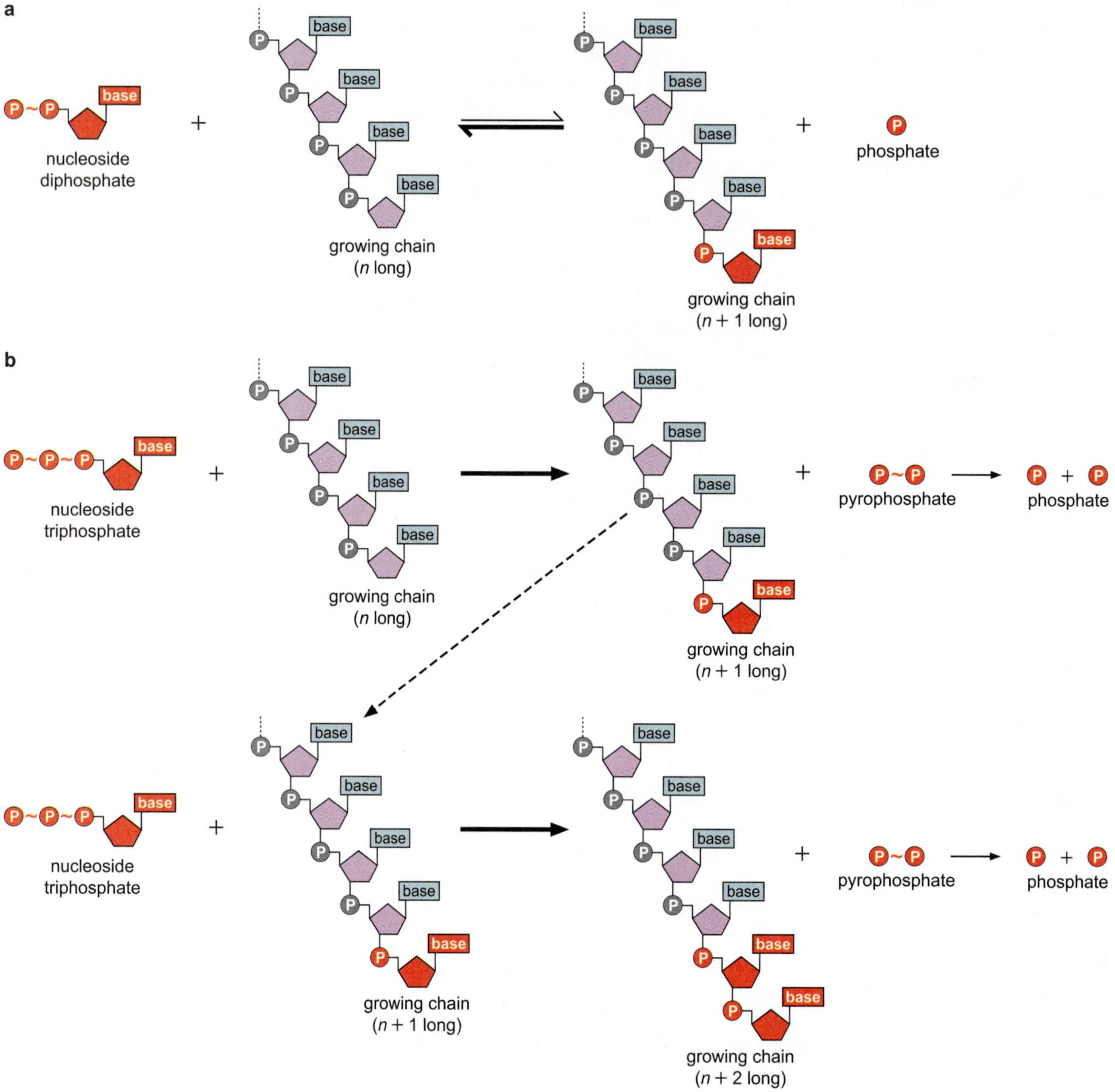

FIGURE 4-5 Two scenarios for nucleic acid biosynthesis. (a) Synthesis of nucleic acids using nucleoside diphosphates. (b) Synthesis of nucleic acids using nucleoside triphosphates.

SUMMARY

The biosynthesis of many molecules appears, at a superficial glance, to violate the thermodynamic law that spontaneous reactions always involve a decrease in free energy (ΔG is negative). For example, the formation of proteins from amino acids has a positive ΔG. This paradox is removed when we realize that the biosynthetic reactions do not proceed as initially postulated. Proteins, for example, are not formed from free amino acids. Instead, the precursors are first enzymatically converted to high-energy activated molecules, which, in the presence of a specific enzyme, spontaneously unite to form the desired biosynthetic product.

Many biosynthetic processes are thus the result of "coupled" reactions, the first of which supplies the energy that allows the spontaneous occurrence of the second reaction. The primary energy source in cells is ATP. It is formed from ADP and inorganic phosphate, either during degradative reactions (such as fermentation or respiration) or during photosynthesis. ATP contains several high-energy bonds whose hydrolysis has a large negative ΔG. Groups linked by high-energy bonds are called high-energy groups. High-energy groups can be transferred to other molecules by group-transfer reactions, thereby creating new high-energy compounds. These derivative high-energy molecules are then the immediate precursors for many biosynthetic steps.

Amino acids are activated by the addition of an AMP group, originating from ATP, to form an AA~AMP molecule. The energy of the high-energy bond in the AA~AMP molecule is similar to that of a high-energy bond of ATP. Nonetheless, the group-transfer reaction proceeds to completion because the high-energy Ⓟ~Ⓟ molecule, created when the AA~AMP molecule is formed, is broken down by the enzyme pyrophosphatase to low-energy groups. Thus, the reverse reaction, Ⓟ~Ⓟ + AA~AMP → ATP + AA, cannot occur.

Almost all biosynthetic reactions result in the release of Ⓟ~Ⓟ. Almost as soon as it is made, it is enzymatically broken down to two phosphate molecules, thereby making a reversal of the biosynthetic reaction impossible. The great utility of the Ⓟ~Ⓟ split provides an explanation for why ATP, not ADP, is the primary energy donor. ADP cannot transfer a high-energy group and at the same time produce Ⓟ~Ⓟ groups as a by-product.

BIBLIOGRAPHY

Berg J., Tymoczko J.L., and Stryer L. 2006. *Biochemistry*, 6th ed. Freeman, New York.

Kornberg A. 1962. On the metabolic significance of phosphorolytic and pyrophosphorolytic reactions. In *Horizons in biochemistry* (ed. M. Kasha and B. Pullman), pp. 251–264. Academic Press, New York.

Krebs H.A. and Kornberg H.L. 1957. A survey of the energy transformation in living material. *Ergeb. Physiol. Biol. Chem. Exp. Pharmakol.* **49:** 212.

Nelson D.L. and Cox M.M. 2000. *Lehninger principles of biochemistry*, 3rd ed. Worth Publishing, New York.

Nicholls D.G. and Ferguson S.J. 2002. *Bioenergetics 3*. Academic Press, San Diego.

Purich D.L. (ed.) 2002. Methods in enzymology: Enzyme kinetics and mechanism: Detection and characterization of enzyme reaction intermediates. *Methods in enzymology*, vol. 354. Academic Press, San Diego.

Silverman R.B. 2002. *The organic chemistry of enzyme-catalyzed reactions*. Academic Press, San Diego.

Tinoco I. (ed.), Sauer K., Wang J.C. and Puglisi J.D. 2001. *Physical chemistry: Principles and applications in life sciences*, 4th ed. Prentice Hall College Division, Upper Saddle River, New Jersey.

Voet D., Voet J.G., and Pratt C. 2002. *Fundamentals of biochemistry*. John Wiley & Sons, New York.

CHAPTER

5

Weak and Strong Bonds Determine Macromolecular Structure

DNA, RNA, AND PROTEIN ARE ALL POLYMERS of simple building blocks. As we learned in Chapter 4, synthesis of these polymers depends on the controlled, catalyzed linkage of activated building blocks. For DNA and RNA, these building blocks are nucleotides (see Fig. 2-12). For proteins, the building blocks are the 20 amino acids donated from their activated intermediates, the donor tRNAs. Assembly of these chains requires breakage of multiple high-energy bonds for the addition of each building block. For all these molecules, the order of the constituent building blocks determines their genetic and biochemical function.

Weak bonds play a critical role in determining the structure and function of these polymers. The primary information of RNA, DNA, and proteins is the order of their covalently linked building blocks. Nevertheless, it is only after they have formed extensive additional weak bonds between their different parts that these polymers adopt characteristic shapes that allow them to carry out their functions. The hydrogen bonds and ionic, hydrophobic, and van der Waals interactions described in Chapter 3 direct proteins to form critical binding sites and DNA to assume its double-helical structure. Indeed, the disruption of these interactions (e.g., by heat or detergent) without disruption of covalent bonds completely destroys the activity of all but a few biological polymers. In this chapter we briefly describe the structure of biological macromolecules and the forces that control their shape. DNA and RNA are discussed briefly here and more thoroughly in Chapter 6. We then focus on the diverse structures of proteins. The final sections of the chapter focus on the interactions between proteins and nucleic acids—an activity central to many of the processes we will encounter in this book—and the control of protein function.

HIGHER-ORDER STRUCTURES ARE DETERMINED BY INTRA- AND INTERMOLECULAR INTERACTIONS

DNA Can Form a Regular Helix

DNA molecules usually have regular helical configurations. This is because most DNA molecules contain two antiparallel polynucleotide strands that have complementary structures (see Chapter 6 for more details). Both internal and external noncovalent bonds stabilize the structure. The two strands are held together by hydrogen bonds between pairs of complementary purines and pyrimidines (Fig. 5-1). Adenine is always hydrogen-bonded to thymine, whereas guanine is

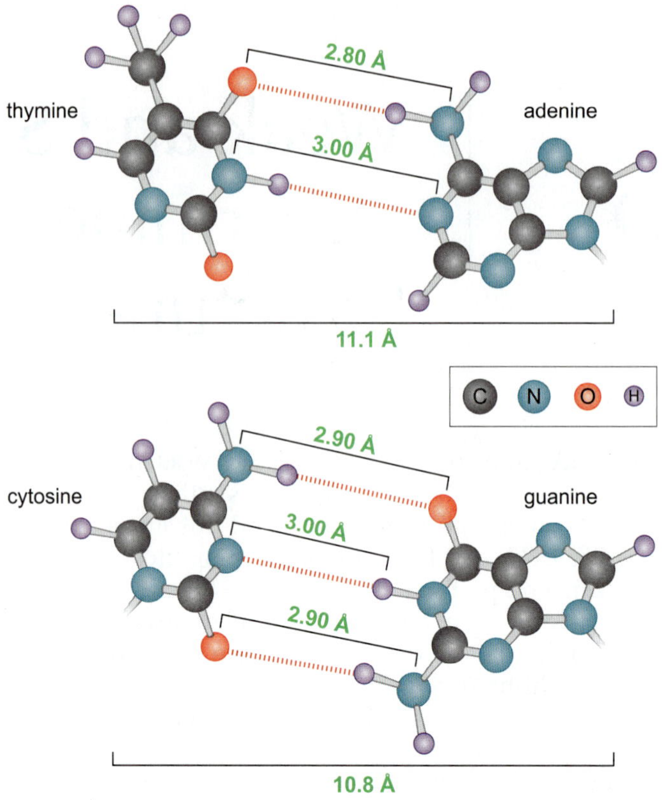

FIGURE 5-1 **The hydrogen-bonded base pairs of DNA.** The figure shows the position and length of the hydrogen bonds between the base pairs. The covalent bonds between the atoms within each base are shown, but double and single bonds are not distinguished (see Fig. 6-6).

hydrogen-bonded to cytosine. In addition, virtually all the surface atoms in the sugar and phosphate groups form bonds to water molecules.

The purine–pyrimidine base pairs are found in the center of the DNA molecule. This arrangement allows their flat surfaces to stack on top of each other, creating shared (π–π) electrons between the bases and limiting their contact with water. This arrangement, known as base stacking, would be much less satisfactory if only one polynucleotide chain were present. Because pyrimidines are smaller than the purines, single-stranded DNA would result in the unfavorable exposure of hydrophobic surface between adjacent bases. The presence of complementary base pairs in double-helical DNA makes a regular structure possible, since each base pair is of the same size.

The double-helical DNA molecule is very stable for two reasons. First, disruption of the double helix would bring the hydrophobic purines and pyrimidines into greater contact with water, which is very unfavorable. Second, double-stranded DNA molecules contain a *very large number of weak bonds,* arranged so that most of them cannot break without simultaneously breaking many others. Thus, for example, even though thermal motion is constantly breaking apart the purine–pyrimidine pairs at the ends of each molecule, the two chains do not usually fall apart because other hydrogen bonds in the molecule are still intact (Fig. 5-2). Once a given bond is broken, the most likely next event is the reforming of the same hydrogen bonds to restore the original molecular configuration, rather than the breaking of additional bonds. Sometimes, of course, the first breakage is followed by a second, and so forth. Such multiple breaks, however, are quite rare, so that double helices held together by more than ten base pairs are very stable at room temperature. When DNA strands do come apart without reforming, this typically starts at one end of the molecule and proceeds inward. This is because the interactions between the bases at the end of the DNA are the least supported by adjacent interactions. That is, they have only one neighboring base pair to help secure the interaction. As described in more detail below, the same principle—the use of multiple weak bonds—governs the stability of proteins.

Ordered collections of secondary bonds become less and less stable as the temperature is raised above physiological temperatures. At elevated temperatures, the simultaneous breakage of several weak bonds is more frequent. After a significant number have broken, a molecule usually loses its original form (the process of denaturation) and assumes an inactive, or denatured, configuration. Thus, as the temperature rises, more interactions are required to maintain the double-stranded nature of DNA.

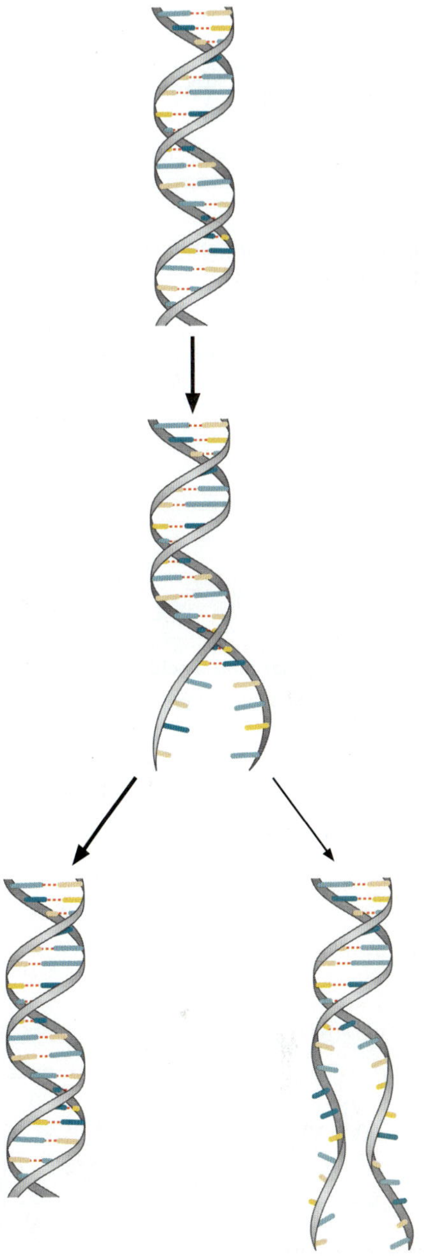

FIGURE 5-2 The breaking of terminal base pairs in DNA by random thermal motion. The figure shows that, once some bonds have broken at the termini, they can re-form (lower left) or additional bonds can break. At physiological temperature, the former is more likely.

RNA Forms a Wide Variety of Structures

In contrast to the highly regular structure of the DNA double helix, RNA is usually found as a single-stranded molecule. Some RNA molecules (such as messenger RNAs) function as transient carriers of genetic information and are constantly associated with proteins and thus do not have an independent, stable, tertiary fold. Other RNA molecules fold into unique tertiary structures. For these RNAs, intramolecular interactions between distinct regions lead to the formation of specific elements of secondary structure. These interactions are principally between the bases of the RNA and include traditional Watson–Crick base pairing, unusual base pairing found only in RNA, and hydrophobic base stacking. RNA differs from DNA in that the ribose sugar of the backbone carries a 2′-hydroxyl group. In the folded structure of RNA molecules, these 2′-hydroxyl groups often participate in interactions that stabilize the structure. The binding of divalent metal ions (such as Mg^{2+}, Mn^{2+}, and Ca^{2+}) to the RNA is often critical to the formation of a stable, folded conformation because these ions can shield the negative charge of the RNA backbone, allowing regions of the molecule to pack more closely together.

The precisely folded, compact nature of RNA tertiary structure is illustrated by the high-resolution structures of some important RNA molecules, for example, tRNA—a molecule that participates in protein synthesis (see Fig. 14-5). These structures reveal that base stacking plays a major role in RNA conformation; for example, 72 out of the 76 bases in tRNA are involved in stacking interactions. As in the DNA double-helix structure, stacking of RNA bases on top of one another is energetically favorable. For this reason, short, base-paired helical regions of RNA stack on top of one another to form longer, discontinuous helical regions. These regions of stacked helices then pack against each other via additional tertiary interactions.

We have only briefly discussed the features of DNA and RNA structure here. In Chapter 6, we will describe in much more detail the interactions that govern the structures of these critical cellular molecules. For the remainder of this chapter we focus on the forces influencing the structure of proteins.

Chemical Features of Protein Building Blocks

In contrast to the four nucleotide building blocks used for RNA or DNA, the 20 amino acid building blocks used for protein synthesis are

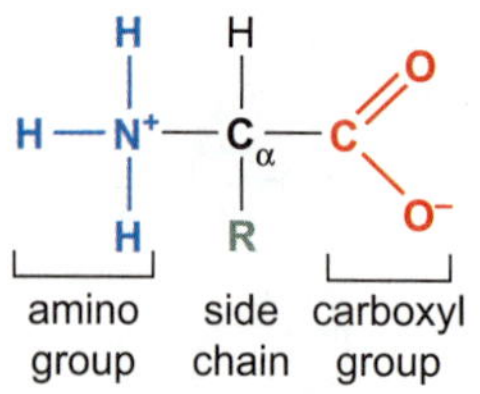

FIGURE 5-3 The common structural features of amino acids.

highly diverse. The common structural features of the amino acids are the central carbon (C_α) linked to a hydrogen, a primary amino group, and a carboxylic acid group (Fig. 5-3). The fourth linkage is to a variable side chain called the **R group.** The R groups of the 20 amino acids can be categorized by their size, shape, and chemical composition (Fig. 5-4). By this approach, the R groups fall into four categories: neutral-nonpolar, neutral-polar, acidic, and basic. The neutral-nonpolar side chains are composed of simple carbon chains or aromatic rings and make principally hydrophobic contacts. The neutral-polar side chains include hydroxyl, sulfhydryl, amide, and imidazole moities and make primarily hydrogen-bonding interactions. The charged (acidic and basic) side chains include primary and secondary amines and carboxylates and make ionic and hydrogen-bonding interactions. All four types of side chain participate in van der Waals contacts, as these associations are only dependent on the proximity of atoms rather than their specific chemical makeup (Chapter 3).

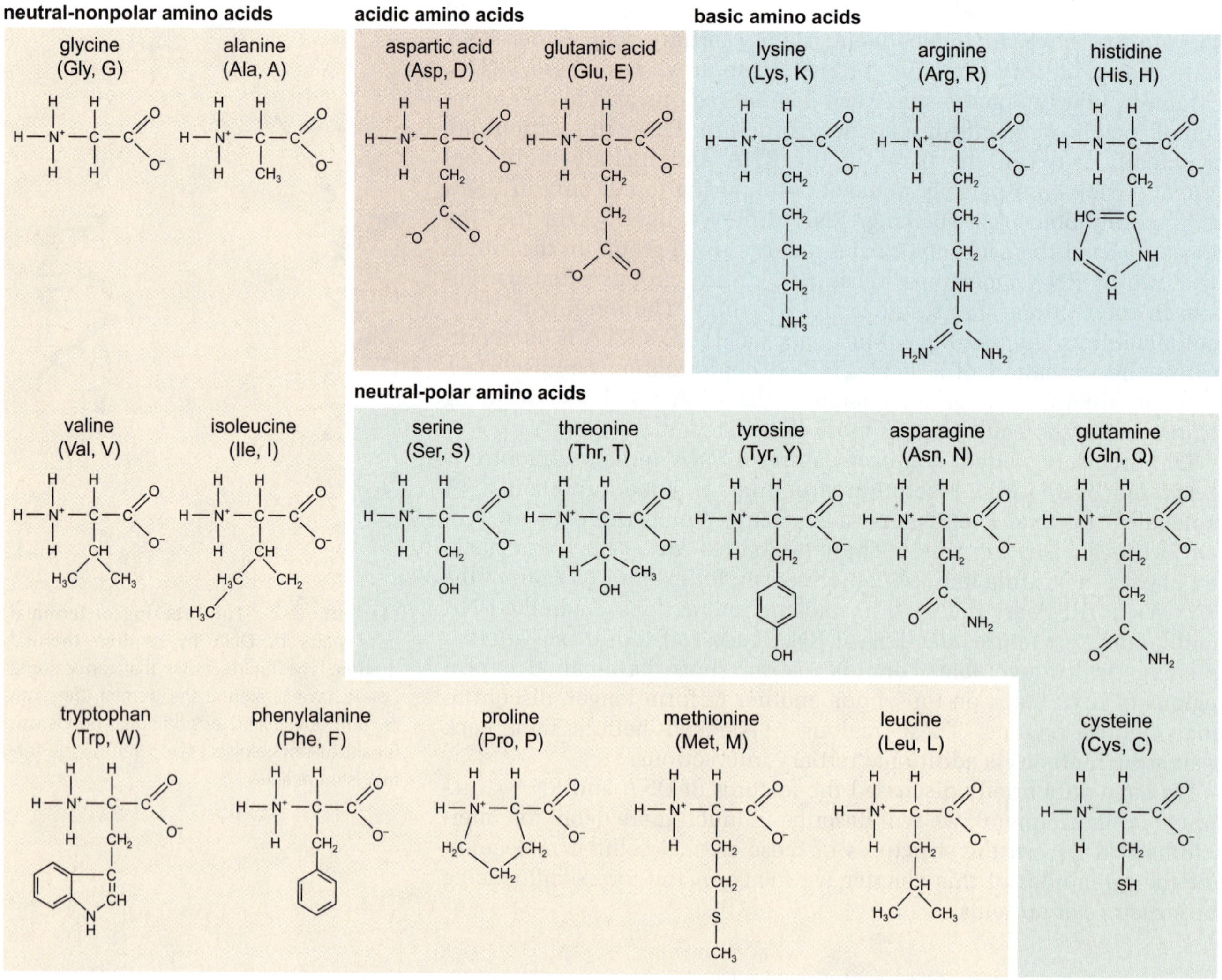

FIGURE 5-4 The 20 different amino acids that occur in proteins. Commonly used abbreviations for amino acids, including the single-letter code, are shown in parentheses.

The Peptide Bond

As we discussed in Chapter 3 (Fig. 3-2), the primary covalent linkage between amino acids in proteins is the **peptide bond** (Fig. 5-5). This bond is made when the primary amine group of one amino acid is covalently joined to the carboxylic acid group of a second amino acid. This linkage has a partially double-bonded character. Because this type of bond involves more than one pair of electrons, rotation around this linkage is limited; completely free rotation about a bond is possible only when atoms are attached by single bonds. (For example, the methyl groups of ethane, $H_3C–CH_3$, rotate about the carbon–carbon bond.) In contrast to the peptide bond, all of the other linkages in the peptide backbone are single bonds and thus rotate freely. Theoretically, these bonds could exist in an infinite number of conformations; however, in the context of a protein, steric interference between adjacent peptide groups limits their rotation. The orientation of adjacent planar peptide bonds can be described by two bond angles: ϕ and ψ (Fig. 5-6). Within proteins, these angles are constrained by the need to maximize formation of secondary bonds among functional groups within the peptide backbone while minimizing steric interference.

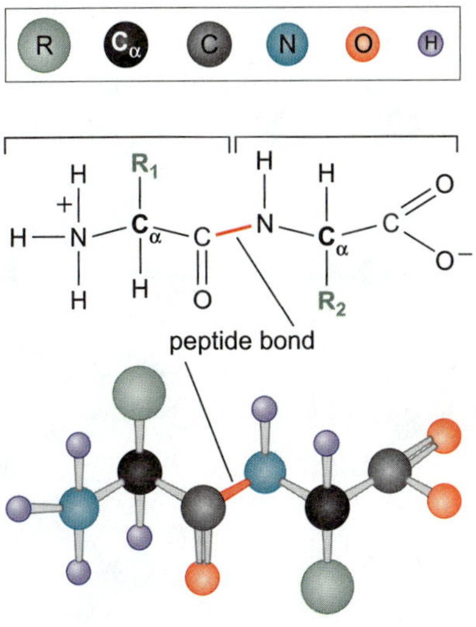

FIGURE 5-5 Peptide bond. The brackets indicate the two amino acid residues that are joined by a peptide bond.

There Are Four Levels of Protein Structure

The final three-dimensional structure or shape of a protein is formed through the sequential association of increasingly distant amino acids. The types of interactions observed within a protein can be divided into four classes (Fig. 5-7). The linear sequence of amino acids in the polypeptide chain is the **primary structure.** Nearby amino acids associate with one another to form regions of **secondary structure.** The elements of secondary structure are usually formed through interactions between those parts of the amino acids that make up the polypeptide backbone rather than the side chains. As we will see below, α helices and β sheets are the major elements of secondary structure. These elements then pack together in a defined manner to generate a given polypeptide's **tertiary structure,** which is the overall conformation of a single polypeptide chain. Many proteins are composed of multiple polypeptide chains known as **protein subunits.** The manner in which these subunits associate with one another is referred to as the protein's **quarternary structure.**

The information contained within the primary structure is nearly always sufficient to determine the eventual tertiary structure of

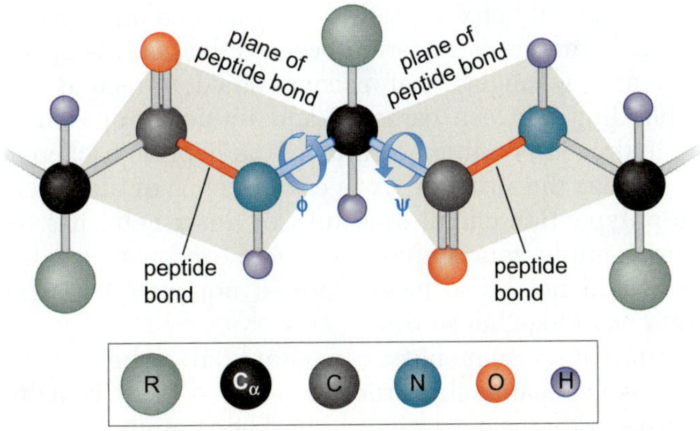

FIGURE 5-6 ϕ and ψ angles of rotation about the C_α-N and C_α-C bonds. The shaded areas represent the planes of the peptide bonds. (Illustration, Irving Geis. Rights owned by Howard Hughes Medical Institute. Not to be reproduced without permission.)

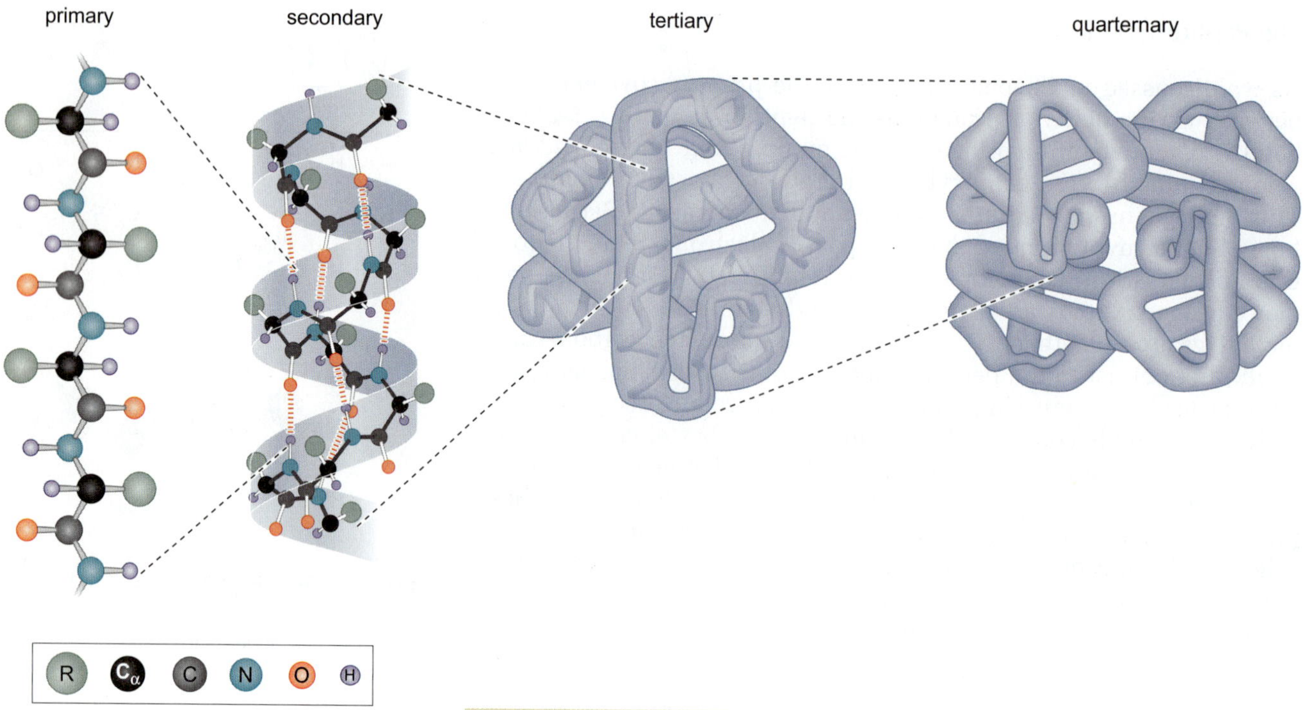

FIGURE 5-7 **Four levels of protein structure.** (Illustration, Irving Geis. Rights owned by Howard Hughes Medical Institute. Not to be reproduced without permission.)

a polypeptide. This was demonstrated in a classic experiment in which the single-polypeptide enzyme ribonuclease was subjected to harsh conditions that interfere with hydrogen bonding and other weak chemical interactions (but not covalent bonds) leading to the complete denaturation (or unfolding) of the polypeptide. When the denatured ribonuclease was restored to conditions that allow the formation of weak chemical bonds, the enzyme rapidly regained both its normal three-dimensional structure and RNA cleaving activity. For a description of how protein structures are worked out experimentally, see Box 5-1, Determination of Protein Structure.

α Helices and β Sheets Are the Common Forms of Secondary Structure

The most stable arrangement of a polypeptide backbone is the α helix. This is a right-handed helix, repeating every 5.4 Å along the helical axis (Fig. 5-8). This structure is preferred because the peptide backbone has favorable ϕ and ψ angles that accommodate a regular pattern of hydrogen bonding between carbonyl and imino groups on the same chain. The hydrogen-bonding potential of the peptide backbone is fully utilized to stablize the structure. As a consequence of the precise geometry of the polypeptide chain, each turn of the α helix has 3.6 amino acids. If, for example, four amino acids were used per turn, the hydrogen bonds would not be so neatly formed nor would the individual backbone atoms fit together so well.

Many amino acid sequences can adopt an α helical secondary structure. This is because the structure of the α helix is stabilized by contacts between the nearly universal backbone atoms of the carbonyl

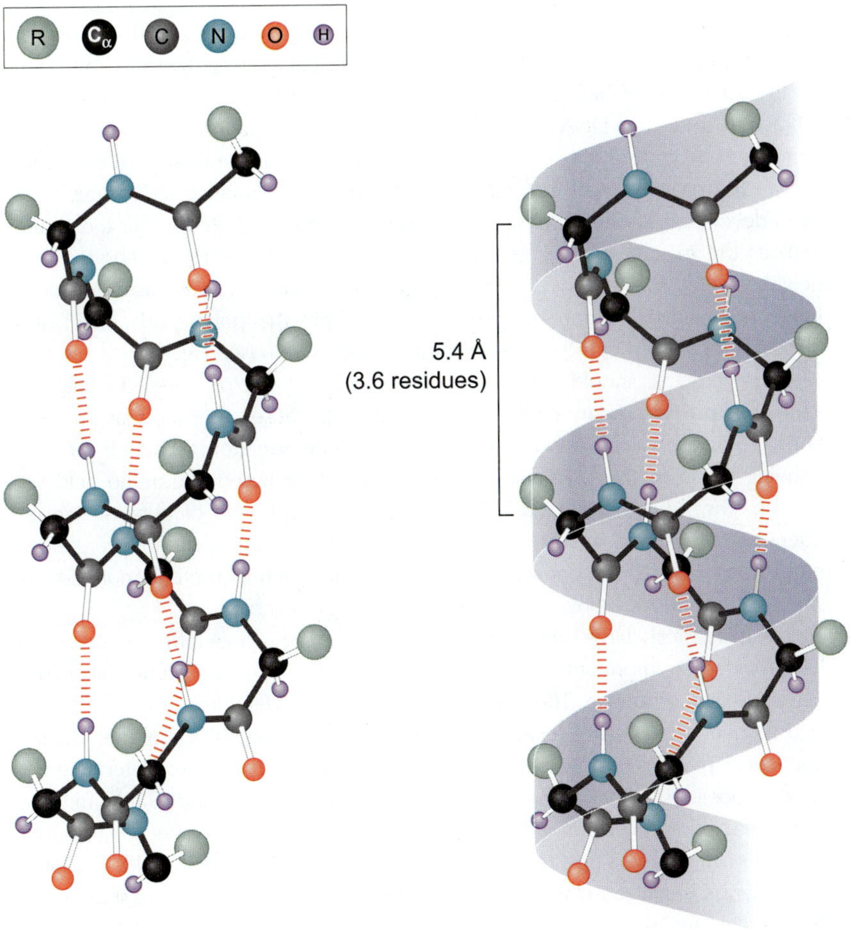

FIGURE 5-8 A polypeptide chain folded into a helical configuration called the α helix. (Illustration, Irving Geis. Rights owned by Howard Hughes Medical Institute. Not to be reproduced without permission.)

and imino groups in the polypeptide chain. The only amino acid that lacks these atoms is proline, which cannot participate as a donor in the hydrogen bonding that stabilizes the helix because of its cyclic chemical structure. Thus, proline is a **helix-breaking residue.** Although their structures do not prevent it, glycine, tyrosine, and serine are also rarely found in α helices. Another consequence of the fact that α helices are constructed through exclusively backbone contacts is that the side chains project away from the helix. This puts these side chains in an ideal position to interact with another region of the protein or another macromolecule, such as DNA.

The second common secondary structural element is the β sheet (Fig. 5-9). In contrast to the α helix, the β sheet is a highly extended form of the polypeptide backbone. Stablization of the β-sheet structure comes from alignment of regions of polypeptide in this extended conformation such that hydrogen bonds can form between carbonyl groups of one β strand and NH groups on the adjacent strand. Typically, a region of β sheet is composed of four to six separate stretches of polypeptide (each forming an individual β strand), each eight to ten amino acids in length. In the β sheet, adjacent amino acids are related by a rotation of 180° and thus their respective side groups emerge from opposite sides of the β sheet (see Fig. 5-9b).

TECHNIQUES

Box 5-1 Determination of Protein Structure

There are two principal methods to determine the three-dimensional structure of proteins. The first to be developed was X-ray crystallography. This method relies on the formation of highly ordered crystals of pure protein. As with the original diffraction studies of DNA fibers (see Box 6-2), the irradiation of protein crystals with high-energy X-rays results in diffraction patterns that are related to the structure of the protein. More recently, nuclear magnetic resonance techniques have been developed to elucidate the conformation of smaller proteins. This technique exploits the magnetic properties of certain atoms (such as ^{1}H) to monitor how neighboring atoms influence each other. This information can be used to determine the relative location of specific atoms within the polypeptide chain and these distances predict the overall structure of the protein (see Fig. 5-7).

In principle it should be possible to predict a protein's three-dimensional structure from its primary amino acid sequence, because, after all, that information is sufficient for a protein to adopt a unique conformation. Although progress is being made in the prediction of protein structure based on amino acid sequence, the full determination of the energetic constraints of a particular sequence is still beyond the most powerful computational approaches. Nevertheless, prediction of certain secondary structural elements (such as the common α-helix structure introduced below) is becoming increasingly reliable.

The increasingly large number of available experimentally determined structures has provided an important resource for making protein structure predictions based on amino acid sequence. These atomic structures have identified families of amino acid sequences that share related three-dimensional shapes. By comparing the sequences of proteins of unknown structure with those that have been determined, it is often possible to make structural predictions based on the identified similarity. Combining this information with computer algorithms that predict secondary structures is proving to be a powerful method for predicting how proteins fold. The long-term outlook is that these approaches will allow at least an approximate structure to be predicted for any protein from its primary sequence alone.

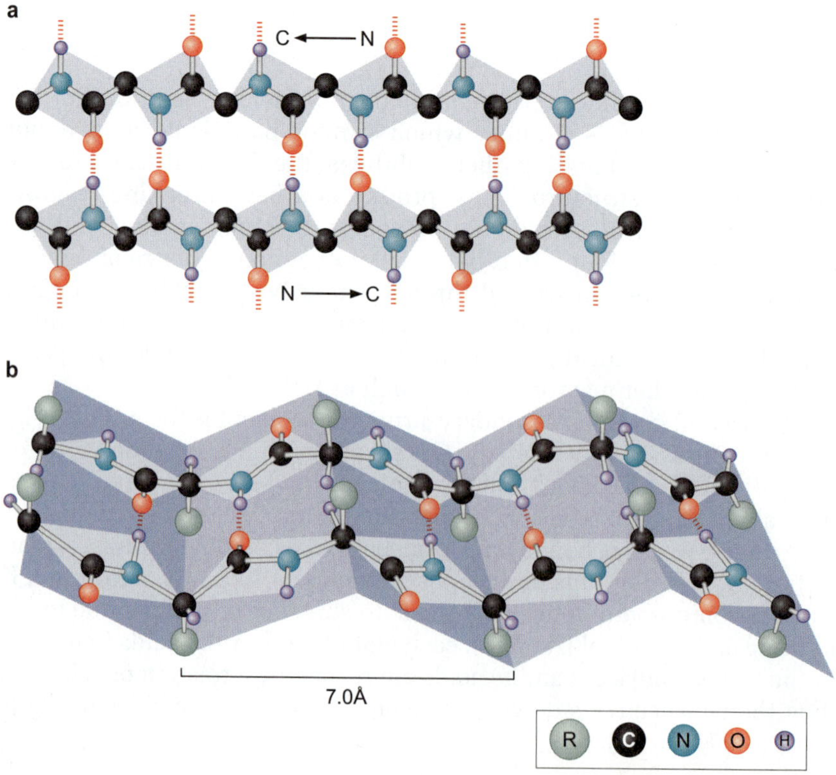

FIGURE 5-9 β sheets are held together by hydrogen bonds. (a) A β sheet is shown from above. Note that the oxygens and nitrogens of the backbone are fully hydrogen-bonded. (b) A β sheet shown from a side view. This illustrates the location of the side groups, which alternate between emerging from above or below the plane of the β sheet. This representation emphasizes the pleated shape of the β sheet. The example given here is an antiparallel β sheet (see Fig. 5-10). (Illustration, Irving Geis. Rights owned by Howard Hughes Medical Institute. Not to be reproduced without permission.)

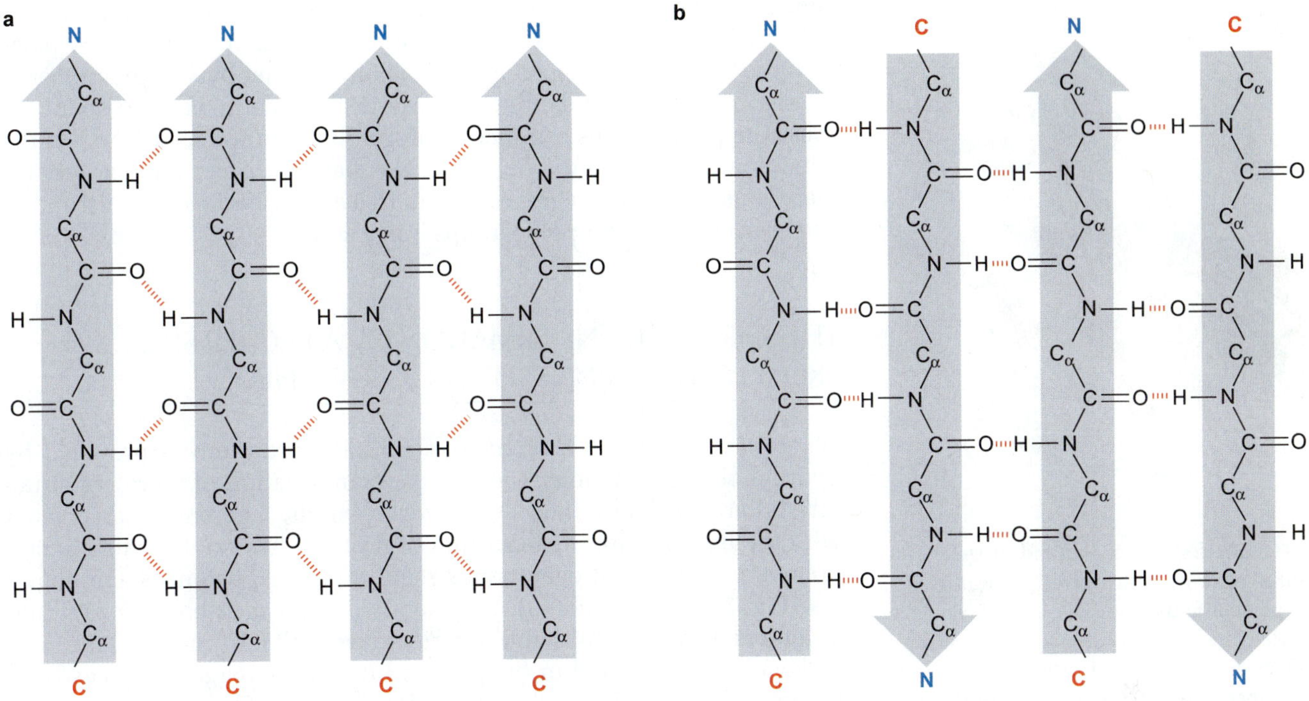

FIGURE 5-10 **Two types of β sheets.** (a) Parallel β sheet: schematic diagram showing hydrogen bond pattern. Note that the chains run in the same amino to carboxyl direction. (b) Antiparallel β sheet: schematic diagram showing the hydrogen bonding pattern. Note that the main NH and O atoms within a β sheet are hydrogen-bonded to each other. (Adapted, with permission, from Branden C. and Tooze J. 1999. *Introduction to protein structure,* 2nd ed., p. 19, Fig 2.6a and p. 18, Fig 2.5b. © Garland Science/Taylor & Francis LLC.)

β sheets come in predominantly one of two forms. These differ in the relative orientations of their chains (Fig. 5-10). In one, the adjacent chains run in the same amino-to-carboxyl direction to produce a **parallel** β sheet. In the other, the adjacent chains run in opposite directions to yield an **antiparallel** β sheet. Although less common, there are also β sheets that have both parallel and antiparallel components. In both parallel and antiparallel β sheets, all the peptide groups lie approximately in the plane of the sheet. Structural studies have revealed that in most cases the individual strands of β sheets tend to be twisted along their length in a right-handed manner (Fig. 5-11). Thus, instead of flat sheets of protein, regions of β sheet tend to curve to generate a compact protein module.

For a protein to fold properly, both the backbone and the side chains must adopt conformations that maximize favorable interactions. The α helix and β sheet are both very stable conformations of the polypeptide backbone. But for each side chain to make the maximum number of weak bonds, proteins have to adopt more varied shapes. The three-dimensional structures of the polypeptide chains of proteins are thus compromises between the tendency of the backbones to form either α helices or β sheets and the tendency of the side groups to twist the backbone into less regular configurations that maximize the strength of the secondary bonds formed by those side groups (Fig. 5-12).

As we discuss in more detail below, one of the strongest influences on protein folding can be attributed to the burial of hydrophobic (nonpolar) amino acid side groups into the core of the protein's structure. This leads to the prediction that in aqueous solutions, proteins containing very

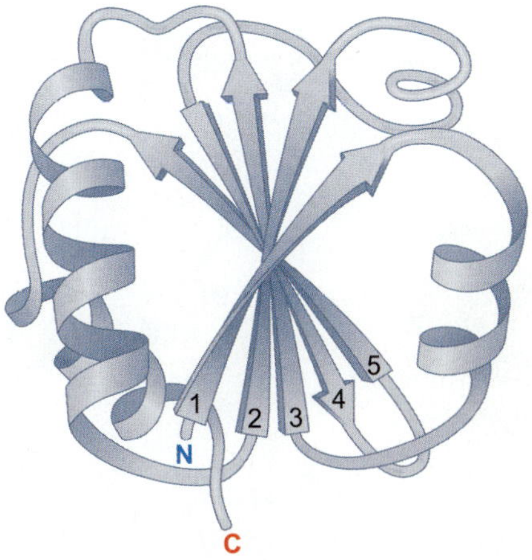

FIGURE 5-11 **β sheets twist in a right-handed manner along their length.** The schematic shows the mixed structure of β sheets and α helices of the *Escherichia coli* protein thioredoxin. β strands are drawn as arrows from the amino to the carboxyl end of the protein. (Adapted, with permission, from Branden C. and Tooze J. 1999. *Introduction to protein structure,* 2nd ed., p. 20, Fig 2.7a. Garland Science/Taylor & Francis LLC. © B. Furugren.)

large numbers of nonpolar side groups will tend to internalize the nonpolar residues and be more stable than proteins containing mostly polar groups. If we disrupt a polar molecule held together by a large number of internal hydrogen bonds, the decrease in free energy is often small because the polar groups can then hydrogen-bond to water instead. On the other hand, when we disrupt molecules having many nonpolar groups, there is usually a much greater loss in free energy because the disruption necessarily inserts nonpolar groups into water.

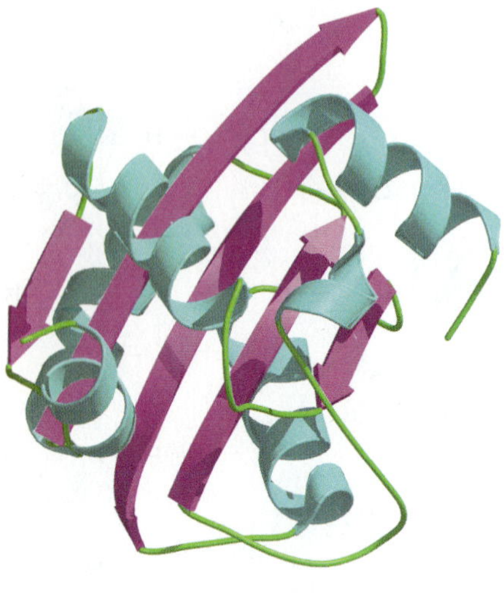

FIGURE 5-12 Regular and irregular features of protein structures. Irregular configurations in the backbone (green) allow the maximum formation of secondary structures (β sheet in purple and α helix in turquoise) by other regions of the protein. The structure shown is that of the DNA-binding domain of the E1 protein of papillomavirus. (Enemark E.J. et al. 2000. *Mol. Cell* 6: 149.) Image prepared with MolScript, BobScript, and Raster 3D.

THE SPECIFIC CONFORMATION OF A PROTEIN RESULTS FROM ITS PATTERN OF HYDROGEN BONDS

Whereas a portion of the energy stabilizing a protein is provided by hydrophobic interactions, the specific conformation of a protein structure is largely determined by hydrogen bonds. The energy associated with the hydrophobic stabilization of proteins has no directional component, whereas hydrogen bonds require precise distances and angles (see Fig. 3-9 and Table 3-3). In general, all hydrogen-bond donors and acceptors within a protein's interior have suitable mates. Failure to make a hydrogen bond in the protein interior is energetically costly, at the rate of a few kilocalories per hydrogen bond. The vitally important role of hydrogen bonds in proteins is to destabilize incorrect structures as much as to stabilize the correct one.

The necessity of satisfying all the hydrogen-bond donors and acceptors on the polypeptide backbone (two per residue) drives formation of the large sections of α helices and β sheets found in most proteins. The only way that a polypeptide can traverse the nonaqueous interior of a protein, as it must, and satisfy the hydrogen-bonding necessity is through formation of regular secondary structures. Side chains do not have enough donors and acceptors to do the job. Thus, all large proteins contain significant regions of β sheets, α helices, or both. Despite the small number of secondary-structure building blocks, the variety of protein structures that can be built from these is vast. Even proteins that are composed entirely of β sheets or α helices adopt structures spanning a wide range (Fig. 5-13).

Of course, some polypeptide sections must be less regular to allow their chains to turn at the ends of α helices and individual strands of β sheets (β strands). **Turns** are loops of amino acids that link α helices and β strands but do not exhibit a defined secondary structure themselves. Turns can vary in length from only a few amino acids to extended segments that are substantially longer. They are, however, generally relatively short so as to minimize the number of unfulfilled hydrogen bonds that accompany their formation (e.g., see Fig. 5-14).

In addition, the less regular structures of these loops are critical for the formation of binding sites for small molecules, the active sites of enzymes, and the surfaces involved in protein–protein interactions. This will become apparent in the three-dimensional protein structures we discuss in the rest of this chapter and the remainder of the text.

α Helices Come Together to Form Coiled-Coils

Many polypeptides interact with one another through the supercoiling of α helices around each other. Typically, this can occur only when the nonpolar side chains along each α helix are arranged so that their side groups contact the other helix. The twisting of the helices around each

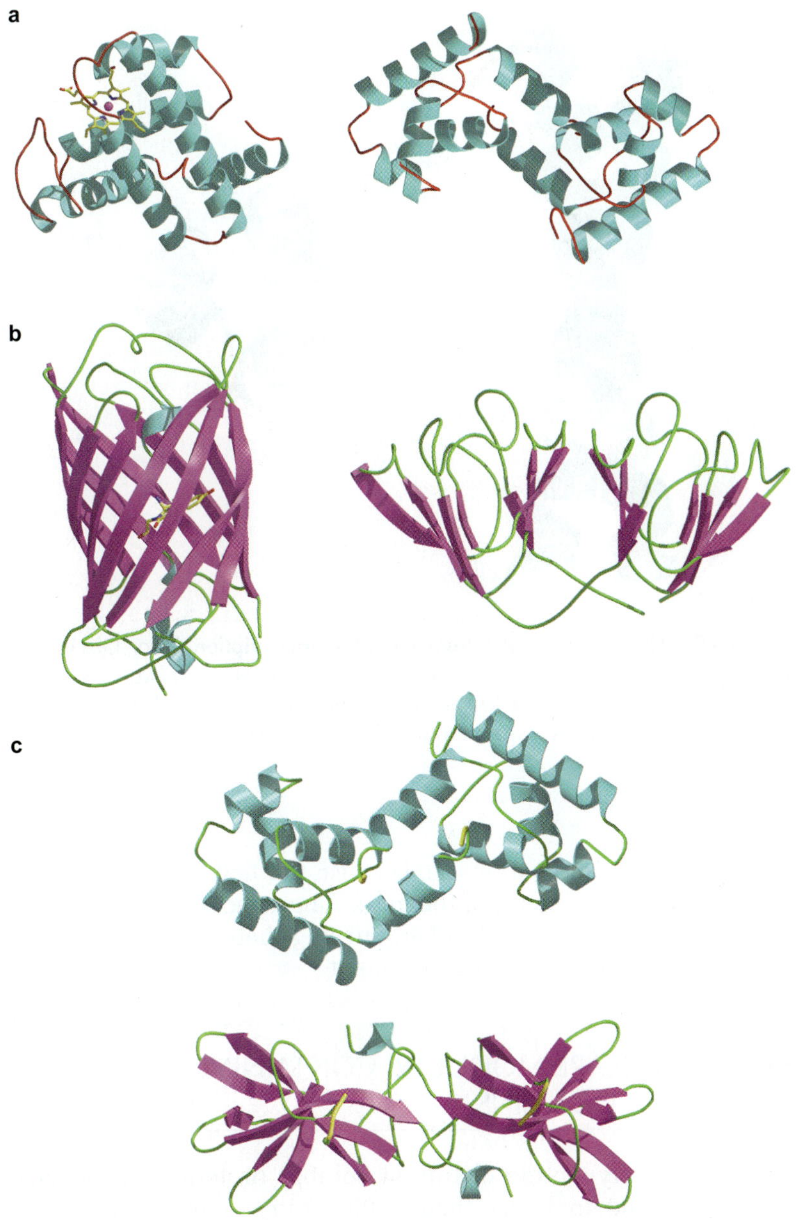

FIGURE 5-13 **Polypeptide chain folding.** (a) Proteins composed of α helices: myoglobin and the amino-terminal domain of λ repressor. (b) Proteins composed of β sheets: the green fluorescent protein (GFP) and gamma crystalline. (c) Comparison of the amino-terminal domain of λ repressor, composed of α helices, with the carboxy-terminal domain of λ repressor, composed of β sheets. (a, Vojtechovsky J. et al. 1999. *Biophys. J.* 77: 2153; and Beamer L.J. and Pabo C.O. 1992. *J. Mol. Biol.* 227: 177. b, Ormo M. et al. 1996. *Science* 273: 1392 and Chirgadze Y.N. et al. 1996. *Acta Crystallogr. D. Biol. Crystallogr.* 52: 712. c, Beamer L.J. and Pabo C.O. 1992. *J. Mol. Biol.* 227: 177; and Bell C.E. et al. 2000. *Cell* 101: 801.) All images prepared with MolScript, BobScript, and Raster 3D.

other reflects the nonintegral (3.6 residues per turn) nature of the α helix, which allows the side groups to pack neatly together only when the α helices interact at an angle of 18° from parallel. If the α helices remained perfectly rigid, they could stay in contact for only a few residues. But by supercoiling in a left-handed direction, neatly packed, highly stable, **coiled-coils** are created (Fig. 5-15).

One example of a coiled-coil is found in the leucine zipper family of DNA-binding proteins. These DNA-binding factors have two subunits that come together to form a dimer through the use of a coiled-coil region. This coiled-coil region is called a leucine zipper because of the repeating appearance of leucine or other amino acids with an aliphatic side group, such as valine or isoleucine. These leucines appear in a regular pattern as follows. Two turns of an α helix will represent a segment of approximately seven amino acids. The aliphatic amino acids are located within each seven-amino-acid stretch at the first and fourth positions. This positioning ensures that

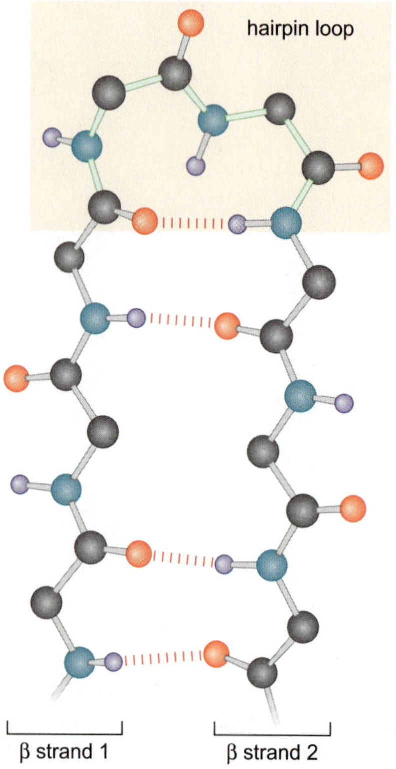

FIGURE 5-14 Adjacent antiparallel β strands are joined by hairpin loops. Schematic showing an example of a two-residue hairpin loop. The bonds within the hairpin loop (in shaded area at top of structure) are green.

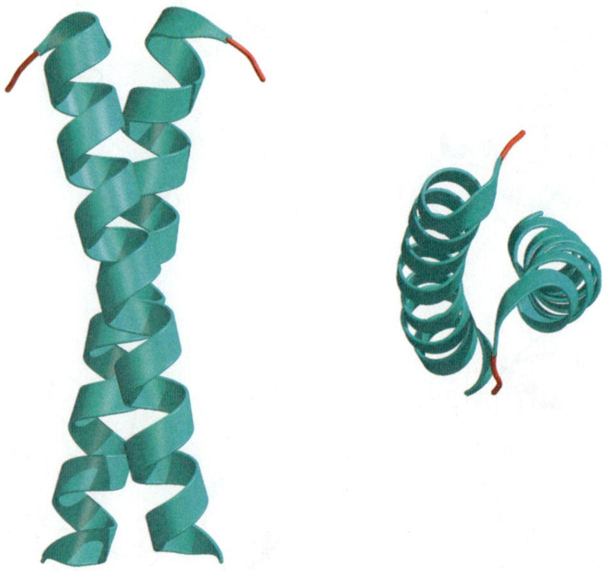

FIGURE 5-15 The leucine zipper from the yeast transcription factor Gcn4. The leucine zipper is an example of a coiled-coil (see text). Here we show two views of the leucine zipper: from the side (on the left) and from above (on the right). (Ellenberger T.E. et al. 1992. *Cell* 71: 1223.) Images prepared with MolScript, BobScript, and Raster 3D.

one side of the α helix is aliphatic, because the first and fourth positions will be on the same face of the helix. These faces in two adjacent helices are packed against each other, burying their hydrophobic side chains away from the aqueous environment.

MOST PROTEINS ARE MODULAR, CONTAINING TWO OR THREE DOMAINS

The individual polypeptide chains of soluble proteins vary in size from less than 100 to larger than 2000 amino acid residues. The smallest polypeptides that form folded proteins have molecular weights of about 11,000 D (~100 residues), but most are between 20,000 and 70,000 D for a single subunit.

Single proteins larger than about 20,000 D often fold into two or more **domains** (Fig. 5-16). The term "domain" is used to describe a part of the structure that appears independent from the rest, as if it would be stable in solution on its own, which is often the case. Typically, a single domain is formed from a continuous amino acid sequence and not portions of sequence scattered throughout the polypeptide. This is an important point when considering how multidomain proteins have evolved (see Box 5-2, Large Proteins Are Often Constructed of Several Smaller Polypeptide Chains, and Fig. 13-28).

Proteins Are Composed of a Surprisingly Small Number of Domain Structural Motifs

Determination of the first half-dozen protein structures showed a bewildering variety of protein folding motifs, implying the existence of

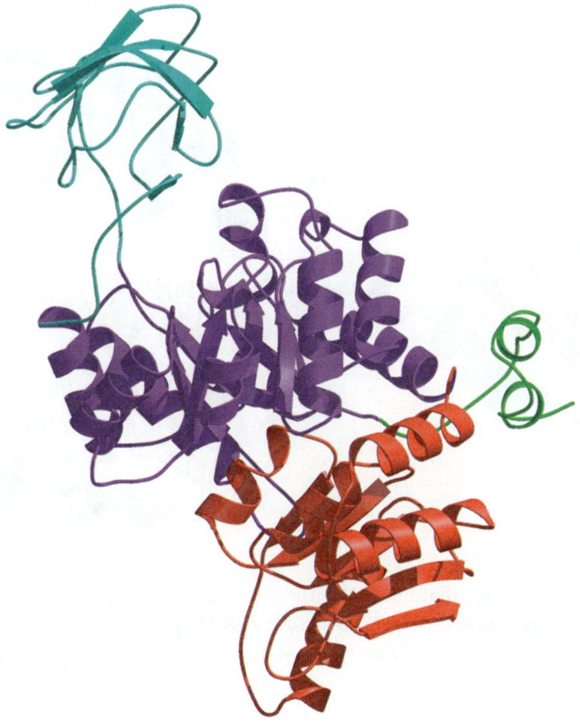

FIGURE 5-16 **Pyruvate kinase is composed of distinct domains.** The distinct domains of the enzyme are shown in turquoise, purple, and red. (Allen S.C. and Muirhead H. 1996. *Acta Crystallogr. D. Biol. Crystallgr.* 52: 499.) Image prepared with MolScript, BobScript, and Raster 3D.

ADVANCED CONCEPTS

Box 5-2 Large Proteins Are Often Constructed of Several Smaller Polypeptide Chains

Most large proteins are regular aggregates of several smaller polypeptide chains. The relationship among the polypeptide chains making up such a protein is termed its quarternary structure (see Fig. 5.7). For example, the macromolecular complexes responsible for the synthesis of RNA (RNA polymerase) and protein (ribosome) are each assemblies of multiple subunits. The complexes are about 500,000 and 2,500,000 D, respectively, but do not include any individual subunits greater than 200,000 D. The ribosome is composed of both protein and RNA subunits. This type of factor is called a ribonuclear protein (RNP).

Why are large protein complexes composed of multiple subunits rather than a single large subunit? The use of multiple subunits to build large protein complexes reflects a building principle applicable to all complex structures, nonliving as well as living. This principle states that it is much easier to reduce the impact of construction mistakes if faulty subunits can be discarded before they are incorporated into the final product. For example, let us consider two alternative ways of constructing a molecule with a million atoms. In scheme 1, we build the structure atom by atom; in scheme 2, we first build a thousand smaller units, each with a thousand atoms, but subsequently put the subunits together into the million-atom product. Now consider that our building process randomly makes mistakes, inserting the wrong atom once every 100,000 times. Let us assume that each mistake results in a nonfunctional product.

Under scheme 1, each molecule will contain, on the average, ten wrong atoms, and so almost no good products will be synthesized. Under scheme 2, however, mistakes will occur in only 1% of the subunits. If there is a device to reject the bad subunits, then good products can be made easily, and the cell will hardly be bothered by the occurrence of the occasional nonfunctional subunit. This is the same construction strategy that forms the basis of the assembly line, in which complicated industrial products, such as radios and automobiles, are constructed. At each stage of assembly, there are devices to throw away bad subunits. In industrial assembly lines, mistakes were initially removed by human hands; now, automation often replaces manual control. In cells, mistakes are sometimes controlled by the specificity of enzymes. If a monomeric subunit is wrongly put together, it usually will not be recognized by the polymer-making enzyme and hence will not be incorporated into a macromolecule. In other cases, faulty substances are rejected because they are unable to spontaneously become part of stable molecular aggregates.

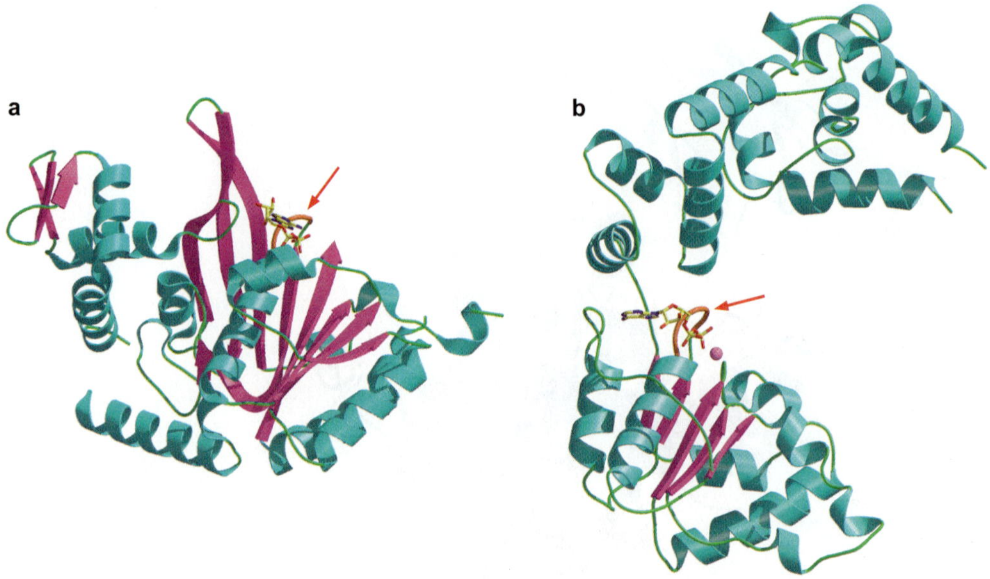

FIGURE 5-17 **Enzymes that bind ATP.** The red arrows point to the ATP molecules bound within each structure. (a) RecA; (b) DnaA. (a, Story R.M. and Steitz T.A. 1992. *Nature* 355: 374. b, Erzberger J.P. et al. 2002. *EMBO J.* 21: 4763–4773.) Images prepared with MolScript, BobScript, and Raster 3D.

an infinite number of protein structures. Now that we know the three-dimensional structures of thousands of proteins, however, it appears that a relatively small number of different domains account for most of the large variety of protein structures. Although an accurate estimate is not possible, the number of truly unique domain motifs will be orders of magnitude smaller than the number of unique proteins.

Specific kinds of domain motifs are often associated with particular kinds of activities. One frequently observed motif has been termed the **dinucleotide fold** because it is frequently found in enzymes that bind ATP (Fig. 5-17). This domain binds ATP through a central, parallel β sheet with α helices on both sides. The nucleotide binding site is on the carboxyl end of the β strands. What varies is the number and detailed arrangement of the α helices and, to a lesser extent, the order of the β strands. Related domains of similar structure serve the same function in many different proteins.

Different Protein Functions Arise from Different Domain Combinations

The various functional properties of proteins appear to arise from their modular construction in much the same way as computers with different specifications can be assembled from the appropriate modular components. Numerous examples can be given. There are, for example, many dehydrogenase enzymes, each working on a specific substrate. Each enzyme consists of two domains, one a common dinucleotide binding domain that binds the coenzyme NAD^+, the other a domain that binds substrate and has the catalytic site. The structure of the latter domain varies among different dehydrogenases.

Repressor and activator proteins that regulate gene expression (Chapter 16) provide another example of modular construction. The

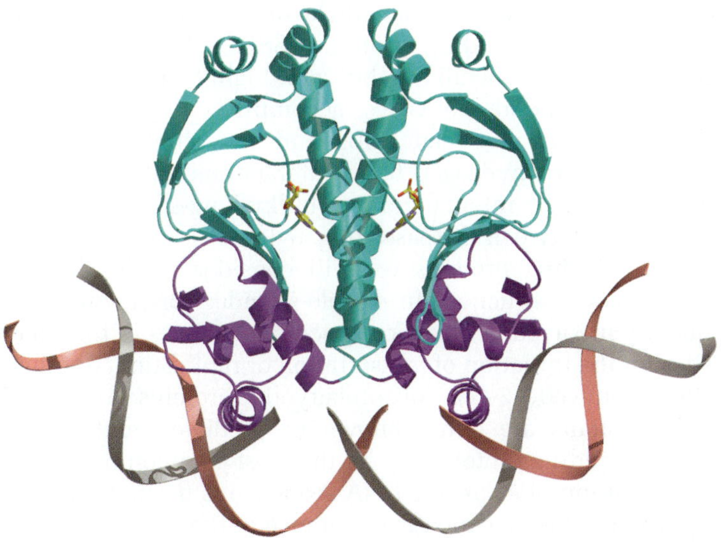

FIGURE 5-18 CAP complex with cAMP interacting with bent DNA. The larger domain of CAP (shown in turquoise) binds cyclic AMP (shown in red and yellow in the center of that domain). The smaller, DNA-binding domain (shown in purple) recognizes specific DNA sequences (the double helix is shown in red and gray). (Schultz S.C. et al. 1991. *Science* 253: 1001.) Image prepared with MolScript, BobScript, and Raster 3D.

Lac repressor and the catabolite gene activator protein (CAP) of *E. coli*, for example, each contain two domains. The crystal structure of CAP shows that the larger domain of that protein binds a molecule of cyclic AMP (cAMP) in its interior, whereas the smaller domain recognizes specific DNA sequences (Fig. 5-18). There are significant amino acid sequence similarities between the cAMP-binding domain of CAP and the regulatory subunit of cAMP-dependent protein kinase, suggesting that the cAMP-binding domain of both proteins evolved from the same precursor. In CAP, this cAMP-binding domain is attached to the DNA-binding domain, so that changes in cAMP levels control transcription levels. In the kinase, the cAMP-binding domain regulates the activity of the first enzyme in a cascade of enzymes that result in the breakdown of stored glycogen.

In the case of Lac repressor, the protein also has a DNA-binding domain and in addition a domain that binds the sugar lactose. The DNA-binding domains of CAP and Lac repressor are again very similar; however, the domains to which those are attached—the lactose-binding domain of Lac repressor and the cAMP-binding domain of CAP—are quite different.

WEAK BONDS CORRECTLY POSITION PROTEINS ALONG DNA AND RNA MOLECULES

DNA-binding proteins mediate many of the central processes in biology. The bonds that hold these proteins onto DNA are the same collection of weak bonds that give proteins, DNA, and RNA their own specific three-dimensional configurations. The most abundant DNA-binding proteins have a structural role in packaging and compacting the huge amount of DNA that must be fitted into the cell. For example, the nucleus of a human cell is only 10 μm (10^{-5} m) across but contains roughly 2 m of double-stranded DNA (see Chapter 7).

There are many ways that proteins can recognize DNA. Some protein–DNA interactions are specific for particular sequences of DNA, whereas others are more specific for DNA in specific conformations. For example, when DNA is unwound in the cell during DNA

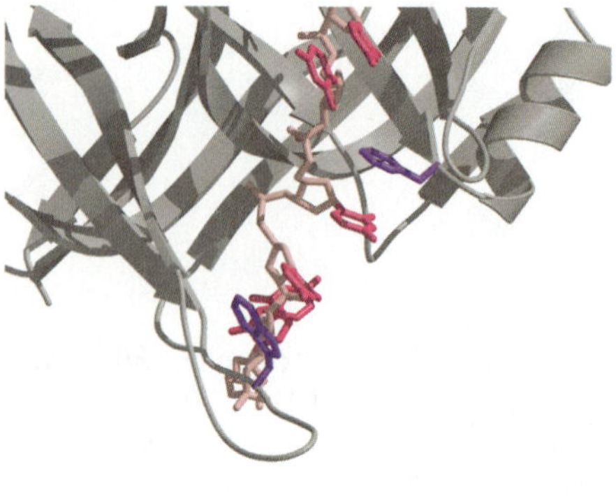

FIGURE 5-19 **Single-strand DNA-binding protein (SSB) interaction with single strand of DNA.** Part of SSB is shown (in gray) with two aromatic side chains in purple. Single-stranded DNA is shown in red. (Raghumathan S. et al. 2000. *Nat. Struct. Biol.* 8: 648.) Image prepared with MolScript, BobScript, and Raster 3D.

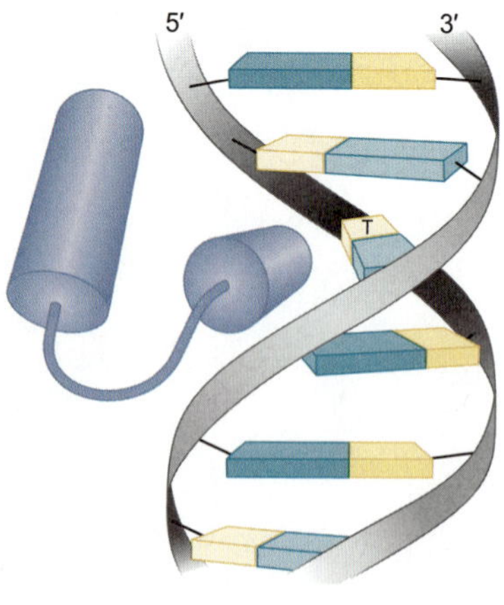

FIGURE 5-20 **Schematic of interaction between the recognition helix of λ repressor monomer and the major groove of its specific DNA-binding site.** (Adapted, with permission, from Jordan S.R. and Pabo C.O. 1988. *Science* 242: 893–899. © AAAS.)

replication or recombination, the single strands are rapidly bound by single-stranded DNA-binding proteins (SSBs). These proteins bind with little sequence specificity but are highly specific for single- versus double-stranded DNA. To accomplish this specificity, the primary interactions between SSBs and the single-stranded DNA are through ionic or hydrogen bond interactions with the phosphate backbone or through intercalation of bulky ring-shaped side chains (e.g., Tyr or Trp) between the bases (Fig. 5-19).

Most DNA-binding proteins we will consider in this book recognize specific DNA sequences in double-stranded DNA. Such proteins are frequently involved in choosing specific sequences in the genome to act as sites for the initiation of transcription or replication, or other DNA transactions. Indeed, 2–3% of prokaryotic proteins and 6–7% of eukaryotic proteins are either known or predicted to be sequence-specific DNA-binding proteins. By far the most common mechanism for protein recognition of a specific DNA sequence is through the insertion of an α helix into the so-called major groove of the DNA (see Fig. 5-20). As was evident in Figure 5-2 and is shown explicitly in Figure 6-1, the double helix has a wide groove known as the major groove and a narrow, or minor, groove. Recognition using an α helix that inserts in the major groove is advantageous for several reasons.

1. The width and depth of the major groove is a very good match to the dimensions of an α helix. This match allows weak interactions to occur between the DNA and approximately half of the surface of the α helix.
2. The major groove is rich in hydrogen bond acceptors and donors located on the edges of the bases (see Fig. 6-10). More importantly, the pattern of hydrogen bonding elements is distinct for each of the base pairs. This allows the pattern of hydrogen bond donors and acceptors to act as a code for the sequence of the DNA, in the same way that hydrogen bonding between the base pairs ensures the appropriate recognition of complementary DNA sequences during DNA hybridization. A diagram of the pattern of hydrogen-bonding donor and acceptor residues in the major groove for each base pair illustrates the distinct pattern for each base pair (see Fig. 6-10). Note that not only can a G:C base pair be easily distinguished from an A:T base pair, but A:T and T:A, and G:C and C:G base pairs can also be distinguished. In contrast, the pattern of base pairs in the minor groove has significantly less information and generally only allows the distinction of A:T and G:C.
3. α helices have a dipole moment that leads to their amino-terminal end being positively charged. This positively charged end frequently makes weak interactions with the phosphate backbone adjacent to the major groove.

The **helix-turn-helix motif** was the first protein motif involved in sequence-specific DNA binding to be identified. This motif is composed of two adjacent α helices that are separated by a short turn (Figs. 5-20 and 5-21). One α helix, called the **recognition helix,** is responsible for DNA sequence recognition. The second α helix is located approximately perpendicular to the first α helix. Although these two helices form the core of the DNA recognition motif, other nearby regions of helix-turn-helix DNA-binding proteins frequently stabilize the arrangement of these two α helices and contact the DNA. Other DNA-binding motifs also insert α helices into the major groove, such as the zinc finger and leucine zipper DNA-binding motifs (as we shall discuss in Chapter 17).

Whereas the use of an α helix is the predominant form of specific DNA recognition, some proteins do use different strategies. An extreme example of this is seen with the TATA-binding protein (TBP), which determines the site of transcriptional initiation at many eukaryotic promoters (see Chapter 12). TBP uses an extensive region of β sheet to recognize the minor groove of the so-called TATA box (Fig. 5-22). So, in this case, we see the use of β sheet instead of α helix and interactions with the minor groove rather than the major groove (for a detailed discussion of this matter, see Chapter 12).

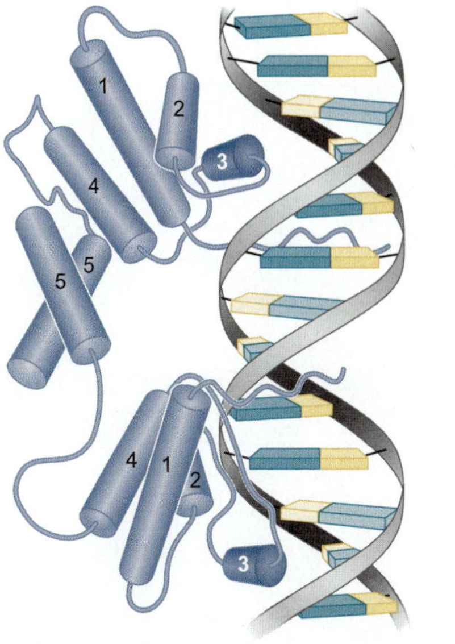

FIGURE 5-21 Geometry of λ repressor–operator complex. The schematic shows two monomers of λ repressor bound to the operator. The helices in each monomer are labeled 1–5. It is helix 3 that inserts into the major groove as shown in Fig. 5-20. (Adapted, with permission, from Jordan S.R. and Pabo C.O. 1988. *Science* 242: 893–899, Fig. 2b. © AAAS.)

Proteins Scan along DNA to Locate a Specific DNA-Binding Site

Many DNA-binding proteins make substantial contacts with the DNA backbone as well as with the specific base pairs of their recognition sites. Mediating these backbone contacts are patches of positively charged amino acids located at sites very close to those that bind to the base pairs. These associations rely primarily on electrostatic attraction between these positive patches and the negatively charged phosphate backbone of the DNA. Because the backbone has a similar negatively charged surface regardless of the sequence, these protein–DNA backbone contacts contribute substantially to both the specific and nonspecific affinity of a protein for DNA. Thus, even a highly specific DNA-binding protein will have a substantial affinity for nonspecific DNA sites as well.

For example, in the case of some well-characterized regulators of gene expression (such as the lactose repressor we mentioned earlier) the affinity for their recognition sequences is about 10^5-fold greater than their affinity for nonspecific DNA. As a consequence, if you looked inside a cell, you would see that molecules of these proteins are typically bound at a number of nonspecific sites as well as at their specific target sequence. This is due to the much larger number of nonspecific sites compared to the specific sites. Indeed, every nucleotide in the genome can be considered the beginning of a potential (and almost always nonspecific) binding site. Thus in *E. coli*, which has ~5×10^6 bp in its circular genome, there would be ~5×10^6 nonspecific binding sites. So, although the ratio of specific to nonspecific

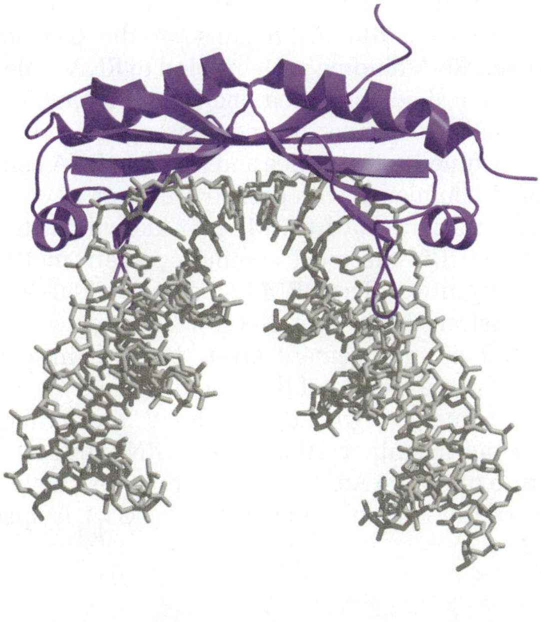

FIGURE 5-22 Structure of the TBP–TATA box complex. The backbone of TBP is shown in purple at the top of the figure; the DNA helix below is shown in gray. (Nikolov D.B. et al. 1995. *Nature* 377: 119.) Image prepared with MolScript, BobScript, and Raster 3D. Extended DNA on either side of image modeled by Leemor Joshua-Tor.

DNA-binding affinity is high (10^5-fold), the ratio of nonspecific-to-specific sites is even higher (5×10^6-fold). This comparison explains why the cell would have to contain multiple copies of the repressor protein to ensure continued occupancy of its specific DNA-binding site. Under these conditions, most of the repressor protein molecules will be bound to nonspecific sites.

Nonspecific protein–DNA interactions are not just an unavoidable consequence of proteins using the charge of the DNA backbone in DNA recognition. These interactions are believed to speed up the rate at which a given regulatory protein finds its appropriate target. Nonspecifically bound proteins are constrained, by their charge interaction, to diffuse linearly along DNA, rather than simply floating freely in the cell. This restricted diffusion allows a DNA-binding protein to sample sites at random in their "search" for a specific binding site. In this way, proteins will reach their targets faster than if they were free to diffuse throughout the cell.

A small subset of DNA-binding proteins does not merely diffuse on DNA, but actively tracks along the DNA. These proteins use directional movement on DNA to perform key functions during DNA replication, repair, and recombination (see Chapters 8–10). Because this movement is directional, it requires energy. Thus, these proteins hydrolyze ATP to direct changes in their binding to DNA.

Diverse Strategies for Protein Recognition of RNA

As introduced above, RNA is structurally more diverse than DNA. RNA-binding proteins have various roles in RNA function, from stabilizing the RNA to enzymatically processing the RNA. The structures of several RNA-binding proteins bound to their target molecules reveal various strategies for protein–RNA recognition.

Some RNA-binding proteins interact specifically with double-stranded RNA. In these cases, the proteins recognize features that distinguish the RNA from the DNA double helix. For example, the presence of the 2′-hydroxyl group is clearly a distinguishing feature of RNA, as is the fact that RNA forms predominantly an A-form helix (see Chapter 6), which has both deeper and narrower grooves than the B-form helix characteristic of DNA. In contrast to the DNA-binding proteins discussed above, the RNA-binding proteins do not engage the nucleic acid by inserting α-helical regions into the RNA grooves.

Many important RNA-binding proteins bind to RNA molecules that are not in a regular helical conformation. Included are proteins that interact with messenger RNA molecules during transcription and RNA processing. Likewise, machineries that splice and translate RNA contain subunits consisting of RNA complexed with protein (Chapters 13 and 14).

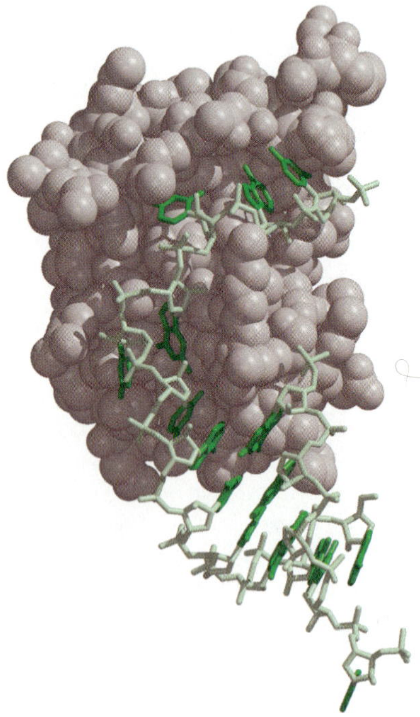

FIGURE 5-23 Structure of spliceosomal protein–RNA complex: U1A binds hairpin II of U1 snRNA. The protein is shown in gray; the U1 snRNA is shown in green. (Oubridge C. et al. 1994. *Nature* 372: 432.) Image prepared with MolScript, BobScript, and Raster 3D.

The RNA-recognition motif (RRM; also known as the ribonuclear protein [RNP] motif) is the most common protein motif involved in specific RNA recognition. The RRM is made up of 80–90 amino acids that form a four-stranded antiparallel β sheet and two α helices that pack against it. This arrangement gives the domain a characteristic split αβ topology. An example of this common domain is found in the U1A protein that interacts with the U1 small nuclear (sn) RNA, both components of the machinery that splices RNA transcripts (Chapter 13); the structure of the U1A:U1snRNA complex is shown in Figure 5-23. The shape of the RNA-binding surface of U1A is specific for this particular RNA.

Many structures of RRMs are known by now, and in most cases RNA recognition is mediated by the surface of the β sheet. Specifically, three conserved residues within the motif make specific contacts with the RNA: an Arg or Lys residue makes a salt bridge to the phosphodiester backbone, and two aromatic residues make stacking interactions with the ribonucleotides.

A single RRM can recognize a relatively short region of RNA—typically as short as a four-, and at most an eight-, nucleotide stretch. Thus, in most proteins that recognize RNA, more than one RNA-recognition domain is used. In U1A, the structure shown in the figure is just one of two RRM domains in the protein. In some proteins there can be three or four domains that act together to increase affinity and specificity of RNA binding. And there are other RNA-recognition domains—in addition to RRM—that can be used. Thus, Figure 5-24 shows a range of proteins and the combination of RNA-recognition domains they use.

As we have already noted, RNA recognition is important in splicing and translation; but it is also central to other aspects of gene expression and RNA processing including, as we shall see in detail in Chapter 18, regulation of gene expression by RNA molecules themselves—in particular, through RNA interference (RNAi) by microRNAs and siRNAs. There we will discuss RNA recognition by the proteins Dicer and Argonaute using PAZ domains (the example of Dicer is also shown in Fig. 5-24).

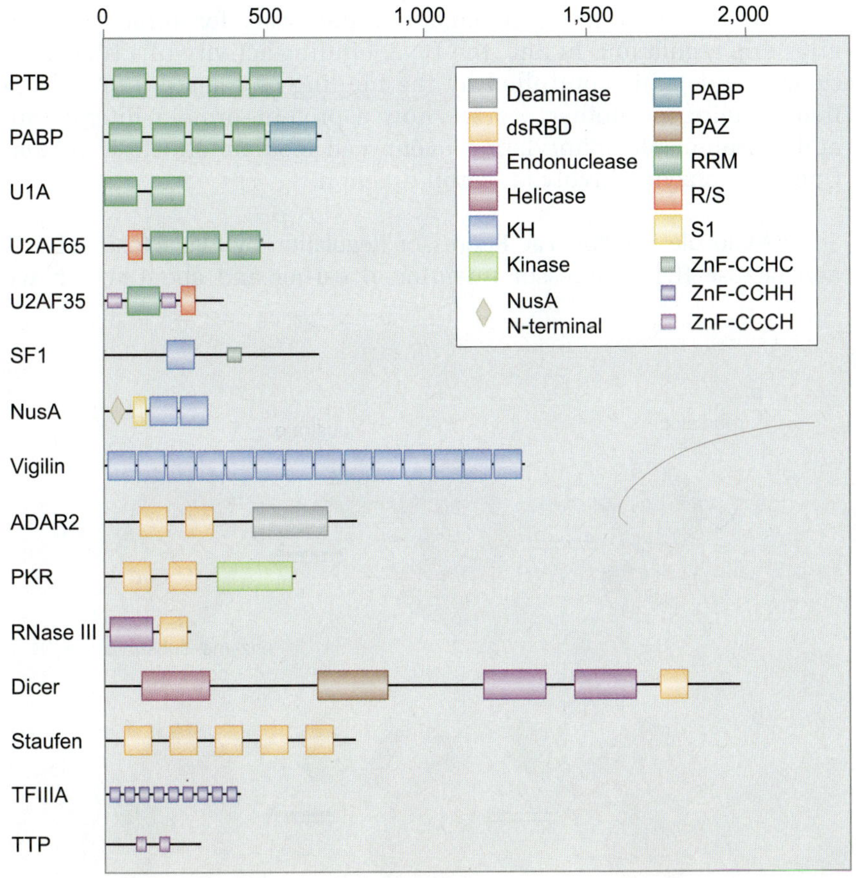

FIGURE 5-24 RNA-binding domains within proteins. Shown are examples from some RNA-binding protein families. The type, number, and arrangements of domains vary in different proteins, as shown. Other functional domains are also indicated. A number of these proteins will turn up in later chapters when we discuss transcription (e.g., TFIIIA, PABP, and NusA), splicing (e.g., U1A and U2AF), and regulatory RNAs (e.g., Dicer). Different domains are represented as colored boxes: the RNA-recognition motif (RRM; by far the most common RNA-binding protein module), the K-homology (KH) domain (which can bind both single-stranded RNA and DNA), the dsRBD (a sequence-independent dsRNA-binding module), and RNA-binding zinc finger (ZnF) domains. Enzymatic domains and less common functional modules are also shown. PABP, poly(A)-binding protein; PTB, polypyrimidine-tract binding; R/S, Arg/Ser-rich domain; SF1, splicing factor-1; TTP, tristetraprolin; U2AF, U2 auxiliary factor. (Redrawn, with permission, from Lunde B.M. et al. 2007. *Nat. Rev. Mol. Cell Biol.* 8: 479, Fig. 1. © Macmillan.)

ALLOSTERY: REGULATION OF A PROTEIN'S FUNCTION BY CHANGING ITS SHAPE

The binding of either small or large molecules (ligands) to a protein can cause a substantial change in the conformation of that protein. Such ligand-induced conformational changes can have a variety of effects, from increasing the affinity of the protein for a second ligand, to switching the enzymatic activity of a protein on or off. This is known as **allosteric regulation** and is a prevalent control mechanism in biological systems. "Allostery" means "other shape," and the basic mechanism is as follows: a ligand binding at one site on a protein changes the shape of that protein; as a result of that change, an active site, or another binding site, elsewhere on the protein is altered in a way that increases or decreases its activity (Fig. 5-25). Examples of proteins controlled in this way range from metabolic enzymes to transcriptional regulatory proteins.

The ligand (the **allosteric effector**) is very often a small molecule—a sugar or an amino acid. But allosteric regulation of a given protein can also be mediated by the binding of another protein, and a very similar effect can, in some cases, be triggered by enzymatic modification of a single amino acid residue within the regulated protein. We will see examples of allosteric regulation by all three mechanisms in this section.

The Structural Basis of Allosteric Regulation Is Known for Examples Involving Small Ligands, Protein–Protein Interactions, and Protein Modification

Here we consider the detailed structural basis for three cases of allosteric regulation. In one, the DNA-binding activity of a transcriptional regulator is controlled by the binding of a small molecule to that protein. In another, we see how a protein–protein interaction, and a protein phosphorylation event, can mediate allosteric regulation of an enzyme involved in cell division.

Small Molecule Effector: Lac Repressor Regulation by Allolactose The *E. coli* lactose (Lac) repressor (mentioned earlier and about which we

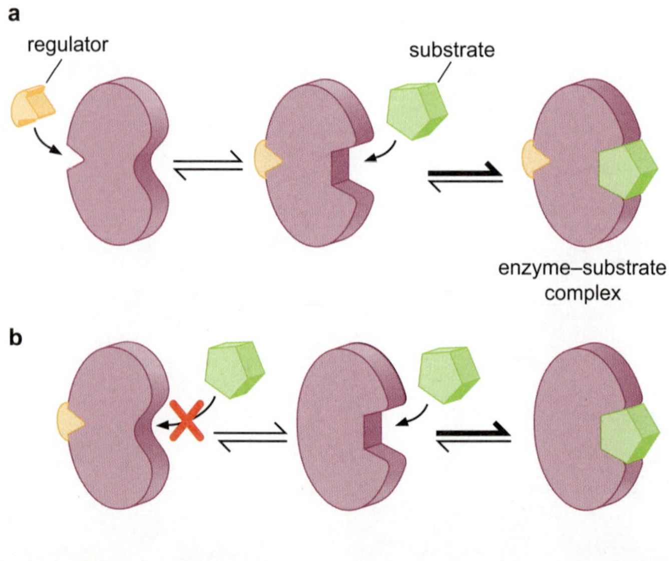

FIGURE 5-25 Allosteric regulation of enzyme activity. (a) The enzyme is controlled positively. Binding of regulator causes a change in the shape of the enzyme to a form that can bind substrate. (b) The enzyme is controlled negatively. In this case, the binding of the regulator locks the enzyme in a shape that cannot bind substrate.

shall learn more in Chapter 16) is controlled allosterically—indeed, it was one of the earliest characterized examples of an allosterically controlled protein. The protein is involved in gene regulation, and, when bound to DNA, it prevents transcription of the genes required for the cell to use the sugar lactose as a carbon source. However, when lactose is present in the environment, a specific form of this sugar (β-1-6-allolactose) *induces* expression of the lactose genes. The allolactose inducer functions by directly binding to the Lac repressor protein and destabilizing its interaction with DNA.

Structural analysis reveals that the Lac repressor changes shape upon inducer binding. (Those structural studies used the artificial inducer molecule isopropyl-β-D-thiogalactoside [IPTG].) This change in shape, in turn, explains how the DNA-binding activity of the protein is weakened. Lac repressor is a large protein (a tetramer of 155 kD) and contains distinct domains involved in DNA binding, protein multimerization, and inducer binding. The very amino-terminal region of the protein (amino acids 1–49) is a helix-turn-helix motif that specifically binds the DNA major groove within the control region of the promoter, as we saw above in the case of λ repressor. Adjacent to this region is an additional helix, known as the hinge helix, that makes minor groove contacts. The inducer-binding pocket, in contrast, is in the middle of the large core domain (composed of residues 62–333).

Comparing the DNA-bound structure of Lac repressor with that of the protein free from DNA (and bound to inducer) provides a picture of why these two states are essentially mutually exclusive. Binding of inducer causes a distortion in the disposition of the amino-terminal half of the large core domain. This conformational change, in turn, disrupts the structure of the hinge helix, which weakens DNA binding; the structure of the adjacent helix-turn-helix domain is rendered more flexible as well, a change likely to lower the protein's affinity for its specific DNA site (Fig. 5-26).

The allosteric modification of the enzyme aspartate transcarbamoylase by its ligand, CTP, provides another example of a small molecule effector (Fig. 5-27). In that case the ligand induces a well-characterized change in protein tertiary structure.

Protein Effector: Cdk Activation by Cyclin We now turn to a case of allosteric regulation of an enzyme by the interaction between that enzyme and a regulatory protein. The enzyme (called Cdk2) is a member of a family of kinases known as cyclin-dependent kinases (Cdks) that regulate progression through the cell cycle. It is inactive until

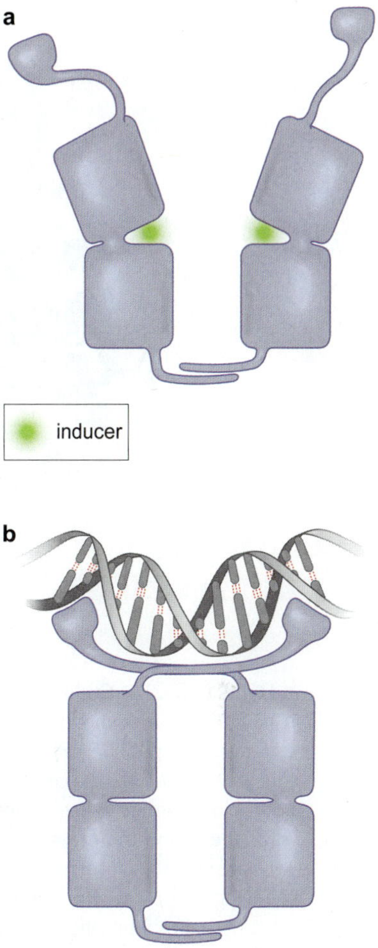

FIGURE 5-26 Allosteric changes of Lac repressor. Each part of the figure shows a dimer of Lac repressor. (a) The top of the figure shows the dimer of the inducer–Lac repressor complex. Binding of inducer causes a change in the structure that reduces affinity of repressor for the operator. (b) The bottom of the figure shows the dimer in the absence of inducer. In this case, the hinge helices form and the amino-terminal domain makes contact with the operator sequence. (Adapted, with permission, from Lewis M. et al. 1996. *Science* 271: 1247–1254, Fig. 12. © AAAS.)

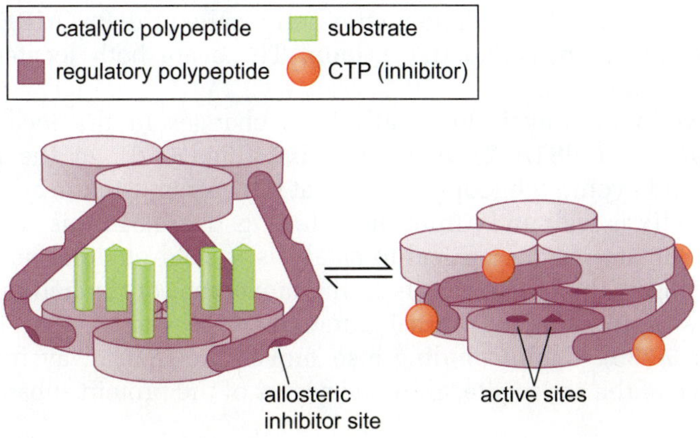

FIGURE 5-27 The allosteric modification of aspartate transcarbamoylase (ATCase) by CTP.

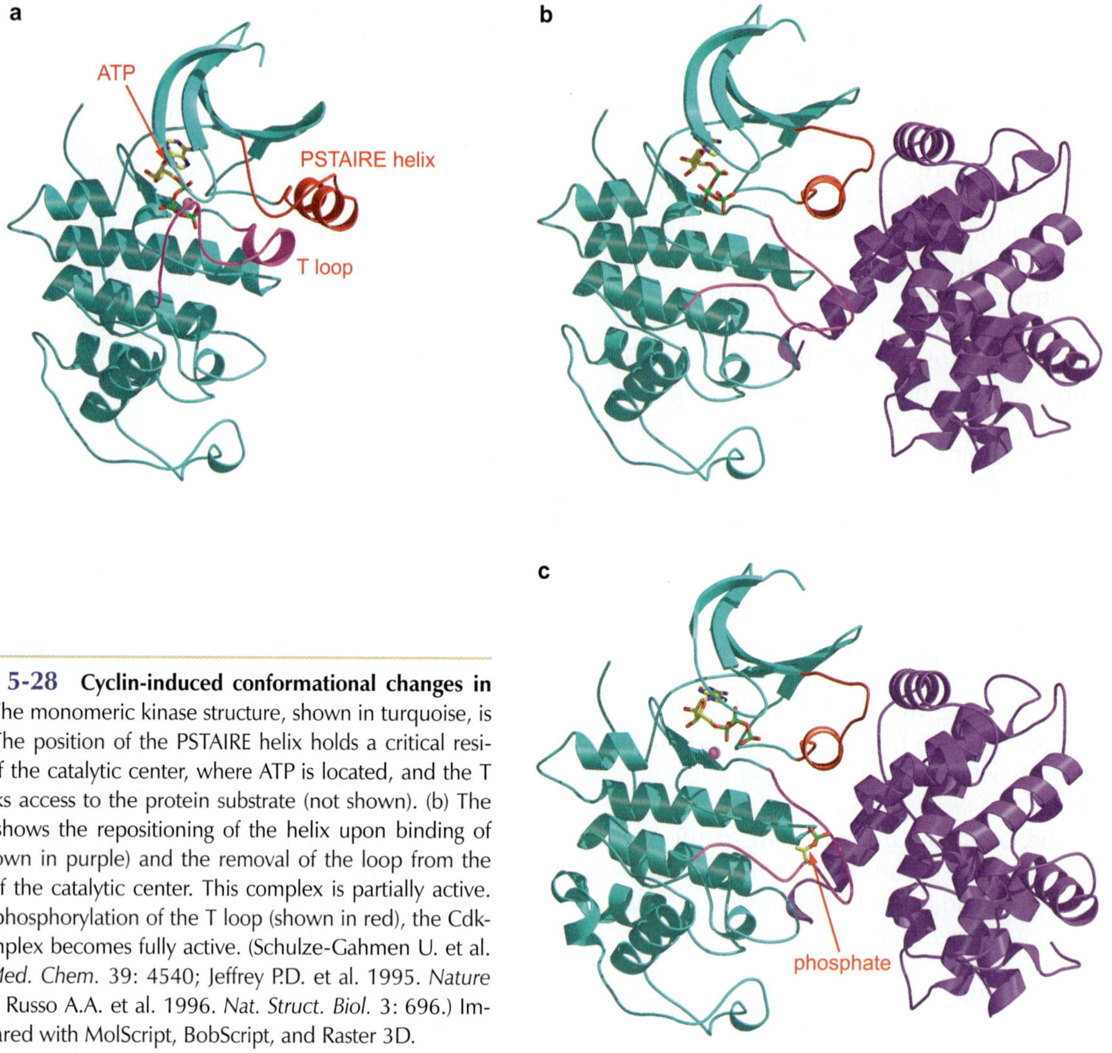

FIGURE 5-28 Cyclin-induced conformational changes in Cdk. (a) The monomeric kinase structure, shown in turquoise, is inactive. The position of the PSTAIRE helix holds a critical residue out of the catalytic center, where ATP is located, and the T loop blocks access to the protein substrate (not shown). (b) The structure shows the repositioning of the helix upon binding of cyclin (shown in purple) and the removal of the loop from the opening of the catalytic center. This complex is partially active. (c) Upon phosphorylation of the T loop (shown in red), the Cdk-cyclin complex becomes fully active. (Schulze-Gahmen U. et al. 1996. *J. Med. Chem.* 39: 4540; Jeffrey P.D. et al. 1995. *Nature* 376: 313; Russo A.A. et al. 1996. *Nat. Struct. Biol.* 3: 696.) Images prepared with MolScript, BobScript, and Raster 3D.

complexed with a regulatory protein called a cyclin. Binding of that second protein induces a conformational change that alters the structure of Cdk2 around its active site, partially activating its function. Further conformational changes induced by phosphorylation of a specific threonine residue nearby activate the enzyme further (see below).

The structural details of the allosteric event mediated by cyclin binding have been established. The structure of Cdk2, free from cyclin, looks very like that of other kinases. Two elements of Cdk2 structure are critical for its regulation: an α helix, called the **PSTAIRE helix,** and a flexible loop, called the **T loop.** These are both located near the kinase active site.

Cyclin binding induces allosteric changes in the location of the T loop and PSTAIRE helix of Cdk2 (Fig. 5-28). In the absence of a bound cyclin, the loop is located at the entrance to the active site and the helix is well away from that site (Fig. 5-28a). In this conformation, a glutamate residue critical to catalysis is held outside the active site. Binding of the cyclin results in the movement of the helix into the active site, allowing the critical glutamate residue to take part in catalysis (Fig. 5-28b). Cyclin binding also moves the loop away from the entrance of the active site, allowing access of the protein substrate.

Phosphorylation as Effector: Cdk Activation by CAK As we have just seen, Cdks are activated by binding cyclins. Full activation of Cdk requires a second allosteric change in that enzyme, mediated by phosphorylation. This phosphorylation takes place on a threonine residue within the T loop mentioned above. This modification leads to further reorganization of the active site of the Cdk. Once added, the phosphate group is bound by three arginines, each from a different region around the catalytic cleft. These interactions fix the catalytic cleft in a conformation favorable for high activity.

The phosphorylation is performed by another kinase (called CAK). Many kinases are activated by a similar phosphorylation event. The two events that together activate Cdks—binding of cyclin and phosphorylation—occur in that order. This is because cyclin binding not only increases the activity of the enzyme, but also makes the T loop accessible for phosphorylation by CAK.

Not All Regulation of Proteins Is Mediated by Allosteric Events

Some proteins are controlled in ways that do not involve allostery. For example, one protein can recruit another to particular locations or substrates and in that way control what that protein acts on. When we discuss regulation of RNA polymerase (the enzyme that transcribes genes into mRNA), we will see that what (in that case) is usually meant by regulation is the choice of which gene is transcribed at any given time. This is done by regulatory proteins that bind the DNA with one surface and the RNA polymerase with another. These interactions bring the enzyme to the gene (or genes) that bear appropriate binding sites for that particular regulator. This is an example of **cooperative binding** of proteins to DNA (see Chapter 16).

SUMMARY

DNA, RNA, and proteins are all polymers, each composed of a defined set of subunits joined by covalent bonds. For example, DNA is made up of chains of nucleotides, and proteins are chains of amino acids. The three-dimensional shape of each such polymer is further determined by multiple weak, or secondary, interactions between those subunits. Thus, in the case of DNA, hydrogen bonds and stacking interactions between the bases of nucleotides account for the double-helical character of that molecule. Likewise, the stable three-dimensional structure of a given protein requires multiple weak interactions between (nonadjacent) amino acids within the polypeptide chain. We discussed the nature of these weak bonds in Chapter 3; in this chapter we looked at how those weak interactions determine the shapes of molecules and the interactions between and among them, particularly proteins. (We shall consider the structures of DNA and RNA in more detail in Chapter 6.)

There are multiple levels to the structural organization of a protein. The initial covalent linkage of the amino acids is the primary structure. Each amino acid is linked to the next by a peptide bond. Secondary structure is formed by interactions between amino acids typically found rather near each other in the primary structure of the protein. The α helix and β sheet are examples of secondary structural elements. The tertiary structure of a protein is the final three-dimensional shape of a polypeptide chain and is determined by the arrangement of the various elements of secondary structure in an energetically favorable way. For many proteins there is another level of structural organization—the quarternary structure. This refers to multimerization of individual polypeptide chains into dimer or higher-order structures. Many proteins work as multimers—hemoglobin is a tetramer, for example, and many DNA-binding proteins work as dimers.

Many native proteins contain several discrete folded sections (domains) that are stable by themselves and that arise from a continuous amino acid sequence. Combinations of such domains account for a large variety of all known proteins. The number of truly unique domains is probably only a few hundred. Each domain is often associated with a specific functional activity—for example, DNA binding.

The specific shape of each macromolecule restricts the number of other molecules with which it can interact. Strong secondary interactions *between* molecules demand both a complementary (lock-and-key) relationship between the two bonding surfaces and the involvement of many atoms. Although molecules bound together by only one or two secondary bonds frequently fall apart, a collection of these weak bonds can result in a quite stable complex. The fact that double-helical DNA does not fall apart spontaneously into single strands shows just how stable such complexes can be. Although complexes held together by multiple weak bonds are not observed to fall apart spontaneously, their assembly can occur spontaneously, with the correct bonds forming in a step-by-step manner (the principle of self-assembly).

The binding of specific proteins to specific sequences along DNA molecules also involves the formation of weak bonds, usually hydrogen bonds between groups on DNA bases and appropriate acceptor or donor groups on proteins. Most regulatory proteins use an α helix to recognize specific DNA sequences. That "recognition helix" fits into the major groove of DNA, and the amino acids in the helix contact the edges of bases in a sequence-specific manner. These contacts are stabilized by the binding energy of the specific interactions. DNA-binding proteins also contain regions that allow nonspecific bonding to the DNA backbone. These nonspecific backbone interactions permit linear diffusion along DNA, allowing proteins to reach their specific target sequences more quickly. A few proteins use β sheets (rather than α helices) to recognize specific DNA sequences, and interactions with the minor groove, but these are much less common. RNA-binding proteins are also vital to many aspects of gene expression and RNA processing. We described the case of one specific RNA-recognition domain—this and other similar domains often work together in concert to increase both affinity and specificity.

Proteins perform many functions, such as catalysis or DNA binding. These activities are commonly regulated by the binding of small ligands or other proteins to the protein in question, or through enzymatic modifications of residues within that protein. These ligands, or modifications, often regulate protein function through allostery. That is, the ligand binds (or the modification targets) a site on the protein separate from the region of that protein that mediates its main function (the active site of an enzyme, DNA-binding domain, etc). This binding or modification triggers a change in the shape of the protein that increases or decreases the activity of the active site, or DNA-binding domain, essentially switching the activity on or off.

In other cases, a protein may be controlled by modification or binding of a second protein, in ways that do not involve allostery. For example, modification can create a site on a protein that is recognized by a second protein. Such protein–protein interactions can recruit proteins to particular locations or substrates and, in that way, control what they do.

BIBLIOGRAPHY

Books

Branden C. and Tooze J. 1999. *Introduction to protein structure,* 2nd ed. Garland Publishing, New York.

Pauling L. 1960. *The nature of the chemical bond*, 3rd ed. Cornell University Press, Ithaca, New York.

Petsko G.A. and Ringe D. 2003. *Protein structure and function (primers in biology).* New Science Press, Waltham, Massachusetts.

The Specific Conformation of a Protein Results from Its Pattern of Hydrogen Bonds

Chothia C. 1984. Principles that determine the structures of proteins. *Annu. Rev. Biochem.* **53:** 537–572.

Most Proteins Are Modular, Containing Two or Three Domains

Rose G.E. 1979. Hiererarchical organization of domains in globular proteins. *J. Mol. Biol.* **134:** 447–470.

Steitz T.A., Weber I.T., and Matthew J.B. 1982. Catabolite gene activator protein: Structure, homology, with other proteins, and cyclic AMP and DNA binding. *Cold Spring Harbor Symp. Quant. Biol.* **47:** 419–426.

Weak Bonds Correctly Position Proteins along DNA and RNA Molecules

Auweter S.D., Oberstrass F.C., and Allain F.H.-T. 2006. Sequence-specific binding of single-stranded RNA: Is there a code for recognition? *Nucleic Acids Res.* **34:** 4943–4959.

De Guzman R.N., Turner R.B., and Summers M.F. 1999. Protein–RNA recognition. *Biopolymers* **48:** 181–195.

Lunde B.M., Moore C., and Varani G. 2007. RNA-binding proteins: Modular design for efficient function. *Nat. Rev. Mol. Cell Biol.* **8:** 479–490.

Sperling R. and Wachtel E.J. 1981. The histones. *Adv. Prot. Chem.* **34:** 1–52.

Allostery: Regulation of a Protein's Function by Changing Its Shape

Bell C.E. and Lewis M. 2001. The Lac repressor: A second generation of structural and functional studies. *Curr. Opin. Struct. Biol.* **11:** 19–25.

Pace H.C., Kercher M.A., Lu P., Markiewicz P., Miller J.H., Chang G., and Lewis M. 1997. Lac repressor genetic map in real space. *Trends Biochem. Sci.* **22:** 334–338.

Pavletich N.P. 1999. Mechanisms of cyclin-dependent kinase regulation: Structures of cdks, their cyclin activators, and cip and INK4 inhibitors. *J. Mol. Biol.* **287:** 821–828.

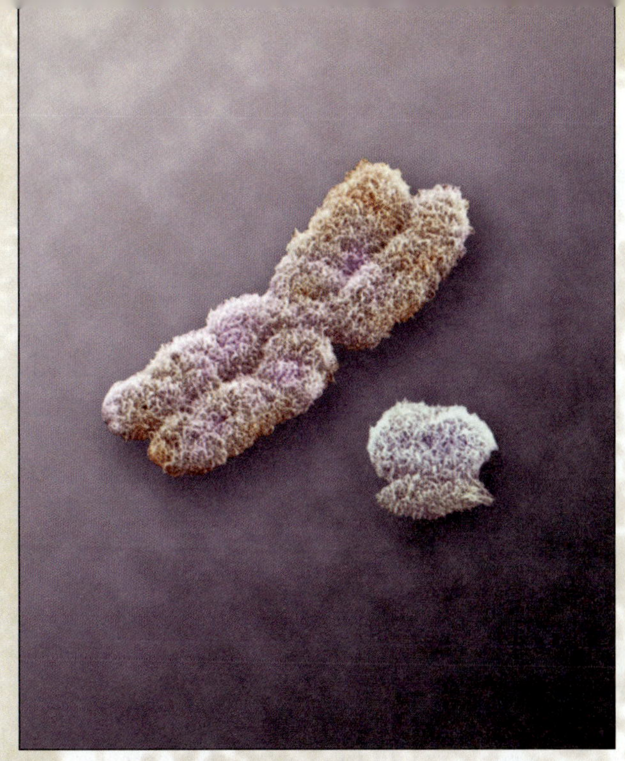

PART

2

MAINTENANCE OF THE GENOME

PART 2 IS DEDICATED TO THE STRUCTURE OF DNA and the processes that propagate, maintain, and alter it from one cell generation to the next. In Chapters 6–11, we shall examine DNA and its close relative, RNA, and address the following questions:

- How do the structures of DNA and RNA account for their functions?
- How are DNA molecules that make up the chromosomes of eukaryotic organisms, which are extraordinarily long compared to the size of the cell, packaged within the nucleus?
- How is DNA replicated completely during the cell cycle, and how is this achieved with high fidelity?
- How is DNA protected from spontaneous and environmental damage, and how is damage, once inflicted, reversed?
- How are DNA sequences exchanged between chromosomes in processes known as recombination and transposition, and what are the biological roles of these processes?

In answering these questions, we shall see that the DNA molecule is subject both to conservative processes that act to maintain it unaltered from generation to generation and to other processes that bring about profound changes in the genetic material that help drive organism diversity and evolution. In the cell, DNA is subjected to forces that peel apart its strands, twist it into topologically constrained structures, wrap it around and through protein assemblies, and break and reseal its backbone. These manipulations are mediated by numerous enzymes and molecular machines that propagate, maintain, and alter the genetic material.

Chapter 6 explores the structure of DNA in atomic detail, from the chemistry of its bases and backbone, to the base-pairing interactions and other forces that hold the two strands together. DNA is often topologically constrained, and Chapter 6 considers the biological effects of such constraints, together with enzymes that alter topology. This chapter also explores the structure of RNA. Despite its similarity to DNA, RNA has its own distinctive structural features and properties, including the remarkable capacity to catalyze several cellular processes.

Chapter 7 describes how the very large DNA molecules that make up chromosomes vary in their organization and size between different organisms. The large size of chromosomal DNA requires that it is not naked but packaged into a more compact form to fit inside the cell. The packaged form of DNA is called chromatin. Being packaged in this way not only reduces the length of the chromosomes but also alters the accessibility and behavior of the DNA. In addition, chromatin can be modified to increase or decrease that accessibility. These changes contribute to ensuring it is replicated, recombined, and transcribed at the right time and in the right place. Chapter 7 introduces us to the histone and nonhistone components of chromatin, to the structure of chromatin, and to the enzymes that modulate the accessibility of the chromosomal DNA.

The structure of DNA offered a likely mechanism for how genetic material is duplicated. Chapter 8 describes this copying mechanism in detail. We describe the enzymes that synthesize DNA and the complex molecular machines that allow both strands of the DNA to be replicated simultaneously. We also discuss how the process of DNA replication is initiated and how this event is carefully regulated by cells to ensure the appropriate chromosome number is maintained.

But the replication machinery is not infallible. Each round of replication results in errors, which, if left uncorrected, would become mutations in daughter DNA molecules. In addition, DNA is a fragile molecule that undergoes damage spontaneously and from chemicals and radiation. Such damage must be detected and mended if the genetic material is to avoid rapidly accumulating an unacceptable load of mutations. Chapter 9 is devoted to the mechanisms that detect and repair damage in DNA. Organisms from bacteria to humans rely on similar, and often highly conserved, mechanisms for preserving the integrity of their DNA. Failure of these systems has catastrophic consequences, such as cancer.

The final two chapters of Part 2 reveal a complementary aspect of DNA metabolism. In contrast to the conservative processes of replication and repair, which seek to preserve the genetic material with minimal alteration, the processes considered in these chapters are designed to bring about new arrangements of DNA sequences. Chapter 10 covers the topic of homologous recombination—the process of breakage and reunion by which very similar chromosomes (homologs) exchange equivalent segments of DNA. Homologous recombination, which allows both the generation of genetic diversity and the replacement of missing or damaged sequences, is a major mechanism for repairing broken DNA molecules. Models for pathways of homologous recombination are described, as well as the fascinating set of molecular "machines" that search for homologous sequences between DNA molecules and then create and resolve the intermediates predicted by the pathway models.

Finally, Chapter 11 brings us to two specialized kinds of recombination known as site-specific recombination and transposition. These processes lead to the vast accumulation of some sequences within the genomes of many organisms, including humans. We will discuss the molecular mechanisms and biological consequences of these forms of genetic exchange.

PHOTOS FROM THE COLD SPRING HARBOR LABORATORY ARCHIVES

Barbara McClintock and Robin Holliday, 1984 Symposium on Recombination at the DNA Level. McClintock proposed the existence of transposons to account for the results of her genetic studies with maize, carried out through the 1940s (Chapter 11); the Nobel Prize in Physiology or Medicine in recognition of this work came more than 30 years later, in 1983. Holliday proposed the fundamental model of homologous recombination that bears his name (Chapter 10).

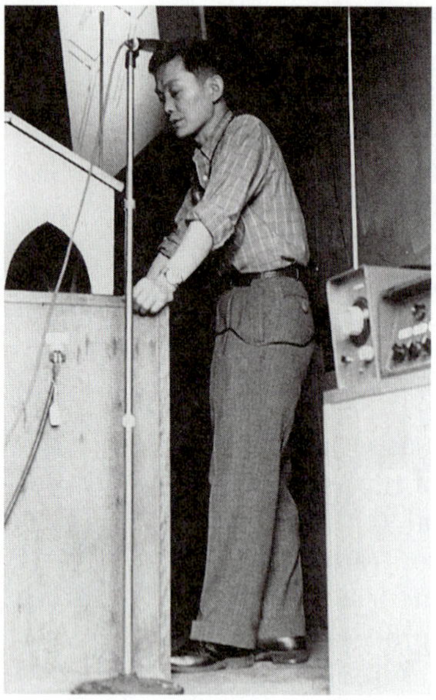

Reiji Okazaki, 1968 Symposium on Replication of DNA in Microorganisms. Okazaki had at this time just shown how, during DNA replication, one of the new strands is synthesized in short fragments that are only later joined together. The existence of these "Okazaki fragments" explained how an enzyme that synthesizes DNA in only one direction can nevertheless make two strands of opposite polarity simultaneously (Chapter 8).

Matthew Meselson, 1968 Symposium on Replication of DNA in Microorganisms. Meselson was Stahl's partner in the experiment showing that DNA replication is semiconservative (see photo below, and Chapter 2). Meselson later made major contributions to a number of fields, including purification of the first restriction enzyme, published the year this photo was taken. Furthermore, he is widely known for his work toward preventing the production and use of chemical and biological weapons.

Arthur Kornberg, 1978 Symposium on DNA: Replication and Recombination. Kornberg's extensive contributions to the study of DNA replication (Chapter 8) began with purifying the first enzyme that could synthesize DNA, a DNA polymerase from *Escherichia coli.* His experiments showed that a DNA template was required for new DNA synthesis, confirming a prediction of the model for DNA replication proposed by James Watson and Francis Crick. For this work Kornberg shared, with Severo Ochoa, the 1959 Nobel Prize in Physiology or Medicine.

Franklin Stahl and Max Delbrück, 1958 Symposium on Exchange of Genetic Material: Mechanism and Consequences. Stahl, together with Matt Meselson (see above), demonstrated that DNA is replicated by a semiconservative mechanism. This was once famously called "the most beautiful experiment in biology" (Chapter 2). Stahl subsequently contributed much to our understanding of homologous recombination (Chapter 10). Delbrück was the influential cofounder of the so-called "Phage Group"—a group of scientists who spent their summers at Cold Spring Harbor Laboratory and developed bacteriophage as the first model system of molecular biology (Chapter 22).

Carol Greider, Titia de Lange, and Elizabeth Blackburn, 2001 Telomeres Meetings. Blackburn discovered the repeated sequences characteristic of telomeres at the ends of chromosomes. Later, while a graduate student in Blackburn's lab, Greider discovered telomerase, the enzyme that maintains the telomeres (Chapter 8). Shown between them here is de Lange, whose work focuses on proteins that bind to and protect telomeres within the cell.

Jack Szostak (right), 1981 Yeast Meeting. Szostak, together with Elizabeth Blackburn (above), showed that the telomere sequences she had identified in *Tetrahymena*, when attached to the ends of linear DNA molecules in yeast, protect that DNA. They proposed that an enzyme—later identified by Carol Greider (also shown above)—made telomeres (Chapter 8). Szostak is talking to an unidentified person; in the background is Jim Hicks, who was at the time working at Cold Spring Harbor Laboratory with Jeff Strathern and Amar Klar on the mechanism of yeast mating-type switching (Chapters 11 and 17).

Paul Modrich, 1993 Symposium on DNA and Chromosomes. A pioneer in the DNA repair field (Chapter 9), Modrich worked out much of the mechanistic basis of mismatch repair.

CHAPTER

6 The Structures of DNA and RNA

THE DISCOVERY THAT DNA IS THE PRIME GENETIC molecule, carrying all the hereditary information within chromosomes, immediately focused attention on its structure. It was hoped that knowledge of the structure would reveal how DNA carries the genetic messages that are replicated when chromosomes divide to produce two identical copies of themselves. During the late 1940s and early 1950s, several research groups in the United States and in Europe engaged in serious efforts—both cooperative and rival—to understand how the atoms of DNA are linked together by covalent bonds and how the resulting molecules are arranged in three-dimensional space. Not surprisingly, there initially were fears that DNA might have very complicated and perhaps bizarre structures that differed radically from one gene to another. Great relief, if not general elation, was thus expressed when the fundamental DNA structure was found to be the double helix. It told us that all genes have roughly the same three-dimensional form and that the differences between two genes reside in the order and number of their four nucleotide building blocks along the complementary strands.

Now, some 50 years after the discovery of the double helix, this simple description of the genetic material remains true and has not had to be appreciably altered to accommodate new findings. Nevertheless, we have come to realize that the structure of DNA is not quite as uniform as was first thought. For example, the chromosomes of some small viruses have single-stranded, not double-stranded, molecules. Moreover, the precise orientation of the base pairs varies slightly from base pair to base pair in a manner that is influenced by the local DNA sequence. Some DNA sequences even permit the double helix to twist in the left-handed sense, as opposed to the right-handed sense originally formulated for DNA's general structure. And some DNA molecules are linear, whereas others are circular. Still additional complexity comes from the supercoiling (further twisting) of the double helix, often around cores of DNA-binding proteins.

Likewise, we now realize that RNA, which at first glance appears to be very similar to DNA, has its own distinctive structural features. It is principally found as a single-stranded molecule. Yet by means of intrastrand base pairing, RNA exhibits extensive double-helical character and is capable of folding into a wealth of diverse tertiary structures. These structures are full of surprises, such as nonclassical base pairs, base–backbone interactions, and knot-like configurations. Most remarkable of all, and of profound evolutionary significance, some RNA molecules are enzymes that carry out reactions that are at the core of information transfer from nucleic acid to protein.

OUTLINE

Visit Web Content for Structural Tutorials and Interactive Animations

Clearly, the structures of DNA and RNA are richer and more intricate than was at first appreciated. Indeed, there is no one generic structure for DNA and RNA. As we see in this chapter, there are in fact variations on common themes of structure that arise from the unique physical, chemical, and topological properties of the polynucleotide chain.

DNA STRUCTURE

DNA Is Composed of Polynucleotide Chains

WEB STRUCTURAL TUTORIAL

The most important feature of DNA is that it is usually composed of two **polynucleotide chains** twisted around each other in the form of a double helix (Fig. 6-1; see Structural Tutorial 6-1). Figure 6-1a presents the structure of the double helix in a schematic form. Note that if inverted 180° (e.g., by turning this book upside down), the double helix looks superficially the same, because of the complementary nature of the two DNA strands. The space-filling model of the double helix, in Figure 6-1b, shows the components of the DNA molecule and their relative positions in the helical structure. The backbone of each strand of the helix is composed of alternating sugar and phosphate residues; the bases project inward but are accessible through the major and minor grooves.

Let us begin by considering the nature of the nucleotide, the fundamental building block of DNA. The nucleotide consists of a phosphate joined to a sugar, known as **2′-deoxyribose,** to which a base is attached. The phosphate and the sugar have the structures shown in Figure 6-2. The sugar is called 2′-deoxyribose because there is no hydroxyl at

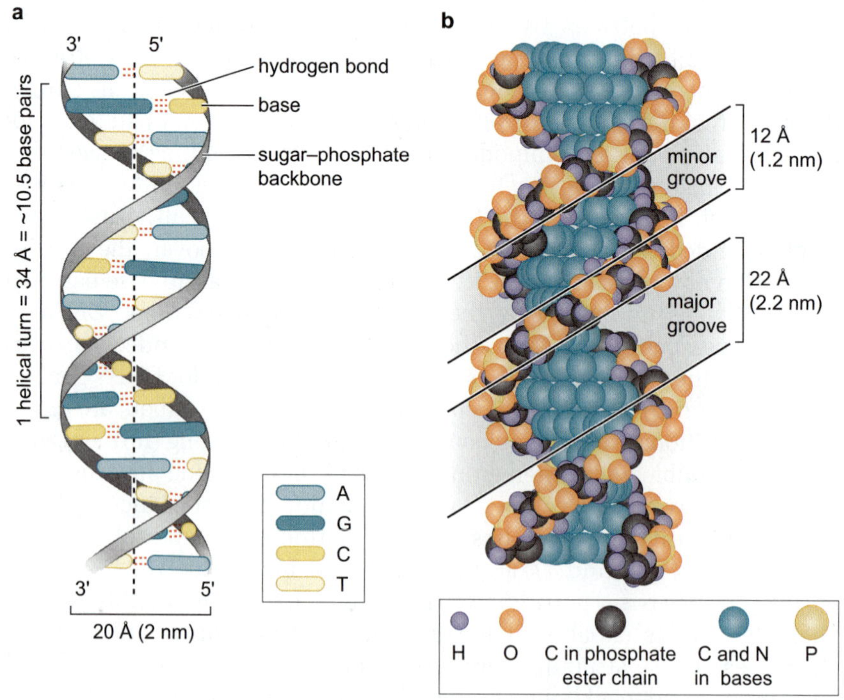

FIGURE 6-1 The helical structure of DNA. (a) Schematic model of the double helix. One turn of the helix (34Å or 3.4 nm) spans approximately 10.5 base pairs. (b) Space-filling model of the double helix. The sugar and phosphate residues in each strand form the backbone, which is traced by the yellow, gray, and red circles, showing the helical twist of the overall molecule. The bases project inward but are accessible through major and minor grooves.

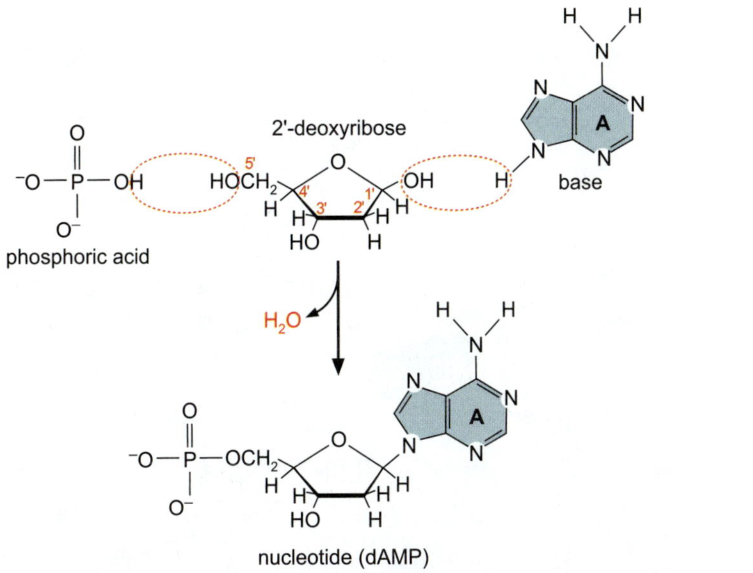

FIGURE 6-2 **Formation of nucleotide by removal of water.** The numbers of the carbon atoms in 2′-deoxyribose are labeled in red.

position 2′ (just two hydrogens). Note that the positions on the sugar are designated with primes to distinguish them from positions on the bases (see the discussion below).

We can think of how the base is joined to 2′-deoxyribose by imagining the removal of a molecule of water between the hydroxyl on the 1′ carbon of the sugar and the base to form a glycosidic bond (Fig. 6-2). The sugar and base alone are called a **nucleoside.** Likewise, we can imagine linking the phosphate to 2′-deoxyribose by removing a water molecule from between the phosphate and the hydroxyl on the 5′ carbon to make a 5′ phosphomonoester. Adding a phosphate (or more than one phosphate) to a **nucleoside** creates a **nucleotide.** Thus, by making a glycosidic bond between the base and the sugar, and by making a phosphoester bond between the sugar and the phosphoric acid, we have created a nucleotide (Table 6-1).

TABLE 6-1 **Adenine and Related Compounds**

	Base Adenine	**Nucleoside 2'-deoxyadenosine**	**Nucleotide 2'-deoxyadenosine 5'-phosphate**
Structure	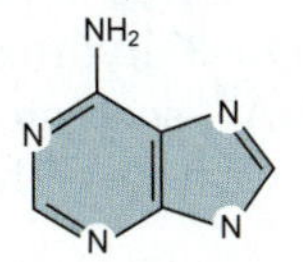	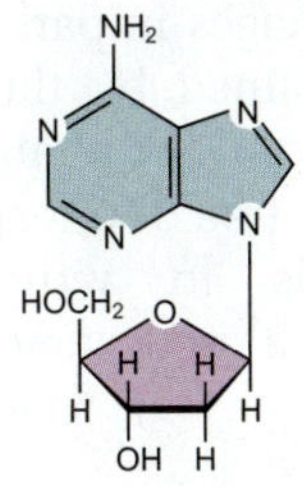	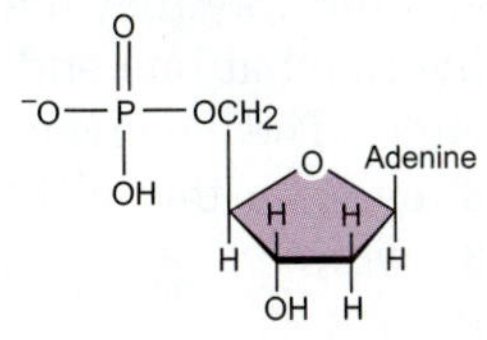
Molecular weight	135.1	251.2	331.2

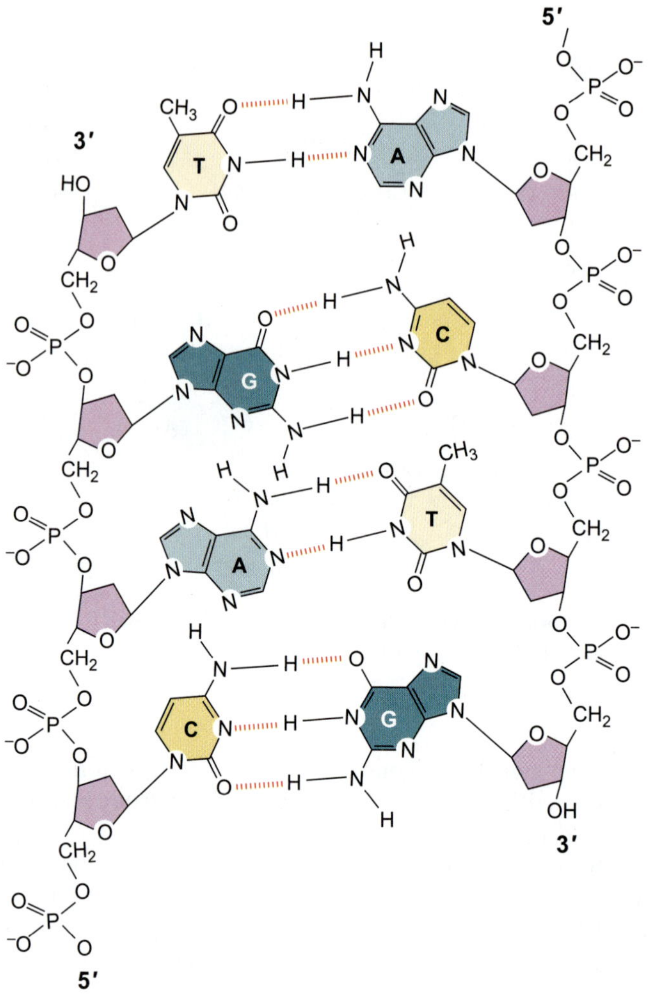

FIGURE 6-3 Detailed structure of polynucleotide polymer. The structure shows base pairing between purines (in blues) and pyrimidines (in yellows), and the phosphodiester linkages of the backbone. (Adapted from Dickerson R.E. 1983. *Sci. Am.* 249: 94. Illustration, Irving Geis. Image from Irving Geis Collection/Howard Hughes Medical Institution. Not to be reproduced without permission.)

Nucleotides are, in turn, joined to each other in polynucleotide chains through the 3′-hydroxyl of 2′-deoxyribose of one nucleotide and the phosphate attached to the 5′-hydroxyl of another nucleotide (Fig. 6-3). This is a **phosphodiester linkage** in which the phosphoryl group between the two nucleotides has one sugar esterified to it through a 3′-hydroxyl and a second sugar esterified to it through a 5′-hydroxyl. Phosphodiester linkages create the repeating, sugar–phosphate backbone of the polynucleotide chain, which is a regular feature of DNA. In contrast, the order of the bases along the polynucleotide chain is irregular. This irregularity as well as the long length is, as we shall see, the basis for the enormous information content of DNA.

The phosphodiester linkages impart an inherent polarity to the DNA chain. This polarity is defined by the asymmetry of the nucleotides and the way they are joined. DNA chains have a free 5′-phosphate or 5′-hydroxyl at one end and a free 3′-phosphate or 3′-hydroxyl at the other end. The convention is to write DNA sequences from the 5′ end (on the left) to the 3′ end, generally with a 5′-phosphate and a 3′-hydroxyl.

Each Base Has Its Preferred Tautomeric Form

The bases in DNA fall into two classes, **purines** and **pyrimidines.** The purines are **adenine** and **guanine,** and the pyrimidines are **cytosine** and

thymine. The purines are derived from the double-ringed structure shown in Figure 6-4. Adenine and guanine share this essential structure but with different groups attached. Likewise, cytosine and thymine are variations on the single-ringed structure shown in Figure 6-4. The figure also shows the numbering of the positions in the purine and pyrimidine rings. The bases are attached to the deoxyribose by glycosidic linkages at N1 of the pyrimidines or at N9 of the purines.

Each of the bases exists in two alternative **tautomeric states,** which are in equilibrium with each other. The equilibrium lies far to the side of the conventional structures shown in Figure 6-4, which are the predominant states and the ones important for base pairing. The nitrogen atoms attached to the purine and pyrimidine rings are in the amino form in the predominant state and only rarely assume the imino configuration. Likewise, the oxygen atoms attached to the guanine and thymine normally have the keto form and only rarely take on the enol configuration. As examples, Figure 6-5 shows tautomerization of cytosine into the imino form (Fig. 6-5a) and guanine into the enol form (Fig. 6-5b). As we shall see, the capacity to form an alternative tautomer is a frequent source of errors during DNA synthesis.

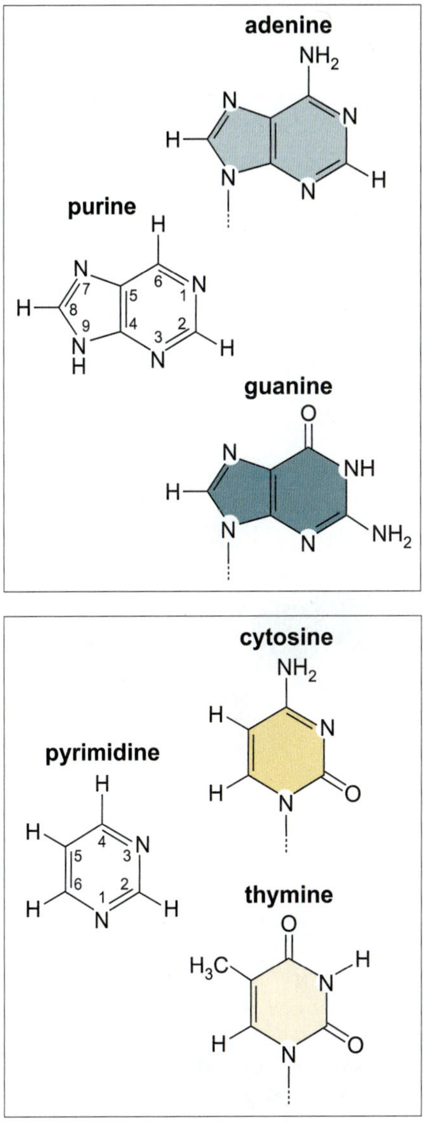

FIGURE 6-4 **Purines and pyrimidines.** The dotted lines indicate the sites of attachment of the bases to the sugars. For simplicity, hydrogens are omitted from the sugars and bases in subsequent figures, except where pertinent to the illustration.

The Two Strands of the Double Helix Are Held Together by Base Pairing in an Antiparallel Orientation

The double helix is composed of two polynucleotide chains that are held together by weak, noncovalent bonds between pairs of bases, as shown in Figure 6-3. Adenine on one chain is always paired with thymine on the other chain and, likewise, guanine is always paired with cytosine. The two strands have the same helical geometry but base pairing holds them together with the opposite polarity. That is,

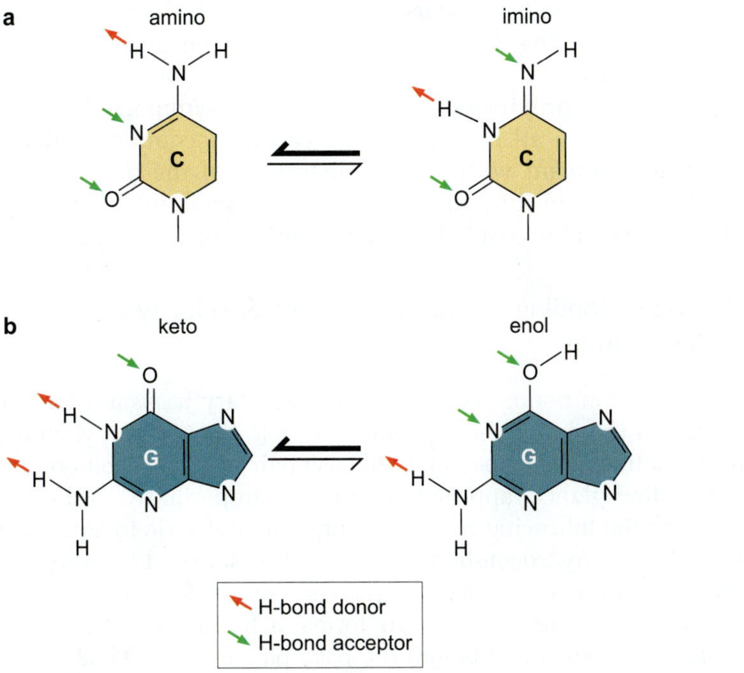

FIGURE 6-5 **Base tautomers.** Amino ⇌ imino and keto ⇌ enol tautomerism. (a) Cytosine is usually in the amino form but rarely forms the imino configuration. (b) Guanine is usually in the keto form but is rarely found in the enol configuration.

the base at the 5′ end of one strand is paired with the base at the 3′ end of the other strand. The strands are said to have an antiparallel orientation. This antiparallel orientation is a stereochemical consequence of the way that adenine and thymine, and guanine and cytosine, pair with each together.

The Two Chains of the Double Helix Have Complementary Sequences

The pairing between adenine and thymine, and between guanine and cytosine, results in a complementary relationship between the sequence of bases on the two intertwined chains and gives DNA its self-encoding character. For example, if we have the sequence 5′-ATGTC-3′ on one chain, the opposite chain must have the complementary sequence 3′-TACAG-5′.

The strictness of the rules for this "Watson–Crick" pairing derives from the complementarity both of shape and of hydrogen-bonding properties between adenine and thymine and between guanine and cytosine (Fig. 6-6). Adenine and thymine match up so that a hydrogen bond can form between the exocyclic amino group at C6 on adenine and the carbonyl at C4 in thymine; and likewise, a hydrogen bond can form between N1 of adenine and N3 of thymine. A corresponding arrangement can be drawn between a guanine and a cytosine, so that there is both hydrogen bonding and shape complementarity in this base pair as well. A G:C base pair has three hydrogen bonds, because the exocyclic NH_2 at C2 on guanine lies opposite to, and can hydrogen bond with, a carbonyl at C2 on cytosine. Likewise, a hydrogen bond can form between N1 of guanine and N3 of cytosine and between the carbonyl at C6 of guanine and the exocyclic NH_2 at C4 of cytosine. Watson–Crick base pairing requires that the bases be in their preferred tautomeric states.

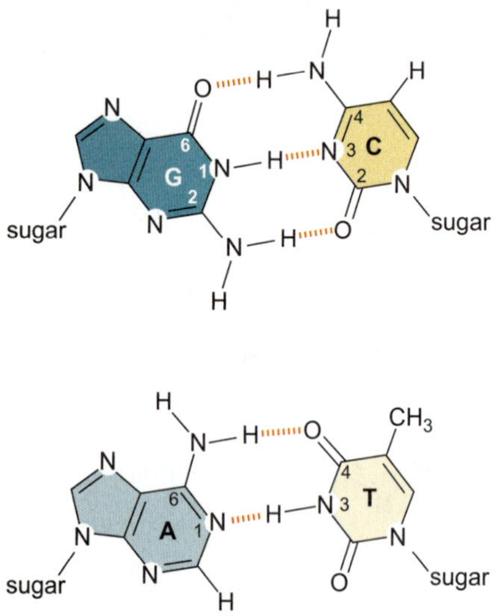

FIGURE 6-6 A:T and G:C base pairs. The figure shows hydrogen bonding between the bases.

An important feature of the double helix is that the two base pairs have exactly the same geometry; having an A:T base pair or a G:C base pair between the two sugars does not perturb the arrangement of the sugars because the distance between the sugar attachment points are the same for both base pairs. Neither does T:A or C:G. In other words, there is an approximately twofold axis of symmetry that relates the two sugars, and all four base pairs can be accommodated within the same arrangement without any distortion of the overall structure of the DNA. In addition, the base pairs can stack neatly on top of each other between the two helical sugar–phosphate backbones.

Hydrogen Bonding Is Important for the Specificity of Base Pairing

The hydrogen bonds between complementary bases are a fundamental feature of the double helix, contributing to the thermodynamic stability of the helix and the specificity of base pairing. Hydrogen bonding might not, at first glance, appear to contribute importantly to the stability of DNA for the following reason. An organic molecule in aqueous solution has all of its hydrogen-bonding properties satisfied by water molecules that come on and off very rapidly. As a result, for every hydrogen bond that is made when a base pair forms, a hydrogen bond with water is broken that was there before the base pair formed. Thus, the net energetic contribution of hydrogen bonds to the stability of the double helix would appear to be modest. However, when polynucleotide strands are separate, water molecules are lined up on the bases. When strands come together in the double helix, the water molecules are displaced from the bases. This creates disorder and increases entropy, thereby stabilizing the double helix. Hydrogen bonds are not the only force that stabilizes the double helix. A second important contribution comes from stacking interactions between the bases. The bases are flat, relatively water-insoluble molecules, and they tend to stack above each other roughly perpendicular to the direction of the helical axis. Electron cloud interactions (π–π) between bases in the helical stacks contribute significantly to the stability of the double helix.

Hydrogen bonding is particularly important for the specificity of base pairing. Suppose we tried to pair an adenine with a cytosine. Then we would have a hydrogen-bond acceptor (N1 of adenine) lying opposite a hydrogen-bond acceptor (N3 of cytosine) with no room to put a water molecule in between to satisfy the two acceptors (Fig. 6-7). Likewise, two hydrogen-bond donors, the NH_2 groups at C6 of adenine and C4 of cytosine, would lie opposite each other. Thus, an A:C base pair would be unstable because water would have to be stripped off the donor and acceptor groups without restoring the hydrogen bond formed within the base pair.

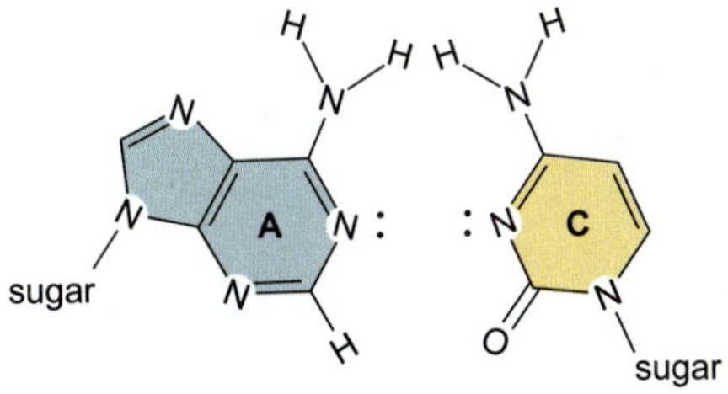

FIGURE 6-7 **A:C incompatibility.** The structure shows the inability of adenine to form the proper hydrogen bonds with cytosine. The base pair is therefore unstable.

Bases Can Flip Out from the Double Helix

As we have seen, the energetics of the double helix favor the pairing of each base on one polynucleotide strand with the complementary base on the other strand. Sometimes, however, individual bases can protrude from the double helix in a remarkable phenomenon known as **base flipping** (Fig. 6-8). As we shall see in Chapter 9, certain enzymes that methylate bases or remove damaged bases do so with the base in an extrahelical configuration in which it is flipped out from the double helix, enabling the base to sit in the catalytic cavity of the enzyme. Furthermore, enzymes involved in homologous recombination and DNA repair are believed to scan DNA for homology or lesions by flipping out one base after another. This is not energetically expensive because only one base is flipped out at a time. Clearly, DNA is more flexible than might be assumed at first glance.

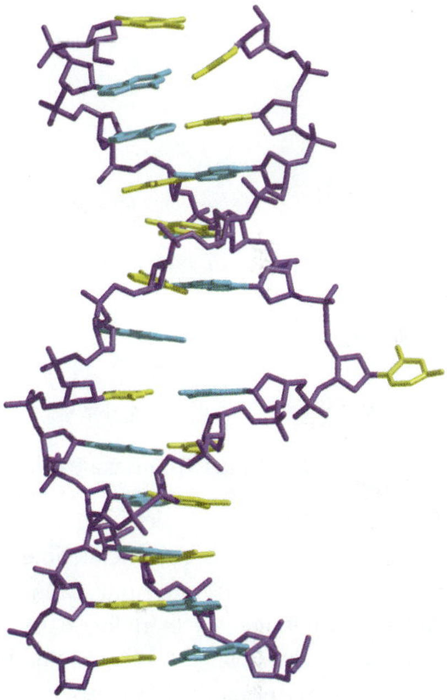

FIGURE 6-8 **Base flipping.** Structure of isolated DNA from the methylase structure, showing the flipped cytosine residue and the small distortions to the adjacent base pairs. (Klimasauskas S. et al. 1994. *Cell* 76: 357.) Image prepared with BobScript, MolScript, and Raster 3D.

DNA Is Usually a Right-Handed Double Helix

Applying the handedness rule from physics, we can see that each of the polynucleotide chains in the double helix is right-handed. In your mind's eye, hold your right hand up to the DNA molecule in Figure 6-9 with your thumb pointing up and along the long axis of the helix and your fingers following the grooves in the helix. Trace along one

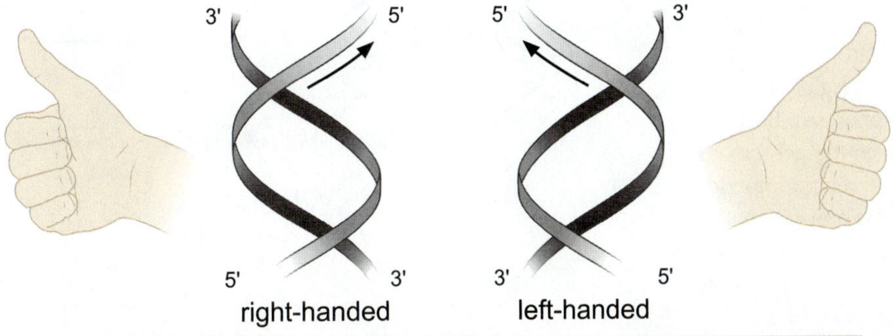

FIGURE 6-9 **Left- and right-handed helices.** The two polynucleotide chains in the double helix wrap around one another in a right-handed manner.

strand of the helix in the direction in which your thumb is pointing. Notice that you go around the helix in the same direction as your fingers are pointing. This does not work if you use your left hand. Try it!

A consequence of the helical nature of DNA is its periodicity. Each base pair is displaced (twisted) from the previous one by about 36°. Thus, in the X-ray crystal structure of DNA it takes a stack of about ten base pairs to go completely around the helix (360°) (Fig. 6-1a). That is, the helical periodicity is generally ten base pairs per turn of the helix. For further discussion, see Box 6-1, DNA Has 10.5 Base Pairs per Turn of the Helix in Solution: The Mica Experiment.

The Double Helix Has Minor and Major Grooves

As a result of the double-helical structure of the two chains, the DNA molecule is a long extended polymer with two grooves that are not equal in size to each other. Why are there a minor groove and a major groove? It is a simple consequence of the geometry of the base pair. The angle at which the two sugars protrude from the base pairs (i.e., the angle between the glycosidic bonds) is about 120° (for the narrow angle) or 240° (for the wide angle) (see Figs. 6-1b and 6-6). As a result, as more and more base pairs stack on top of each other, the narrow angle between the sugars on one edge of the base pairs generates a **minor**

KEY EXPERIMENTS

Box 6-1 DNA Has 10.5 Base Pairs per Turn of the Helix in Solution: The Mica Experiment

This value of ten base pairs per turn varies somewhat under different conditions. A classic experiment that was carried out in the 1970s demonstrated that DNA absorbed on a surface has somewhat greater than ten base pairs per turn. Short segments of DNA were allowed to bind to a mica surface. The presence of 5′-terminal phosphates on the DNAs held them in a fixed orientation on the mica. The mica-bound DNAs were then exposed to DNase I, an enzyme (a deoxyribonuclease) that cleaves the phosphodiester bonds in the DNA backbone. Because the enzyme is bulky, it is able to cleave phosphodiester bonds only on the DNA surface furthest from the mica (think of the DNA as a cylinder lying down on a flat surface) because of the steric difficulty of reaching the sides or bottom surface of the DNA. As a result, the length of the resulting fragments should reflect the periodicity of the DNA, the number of base pairs per turn.

After the mica-bound DNA was exposed to DNase, the resulting fragments were separated by electrophoresis in a polyacrylamide gel, a jelly-like matrix (Box 6-1 Fig. 1; see also Chapter 20 for an explanation of gel electrophoresis). Because DNA is negatively charged, it migrates through the gel toward the positive pole of the electric field. The gel matrix impedes movement of the fragments in a manner that is proportional to their length such that larger fragments migrate more slowly than smaller fragments. When the experiment is carried out, we see clusters of DNA fragments of average sizes 10 and 11, 21, 31, and 32 base pairs and so forth, that is, in multiples of 10.5, which is the number of base pairs per turn. This value of 10.5 base pairs per turn is close to that of DNA in solution as inferred by other methods (see the section titled The Double Helix Exists in Multiple Conformations). The strategy of using DNase to probe the structure of DNA is now used to analyze the interaction of DNA with proteins (see Chapter 17).

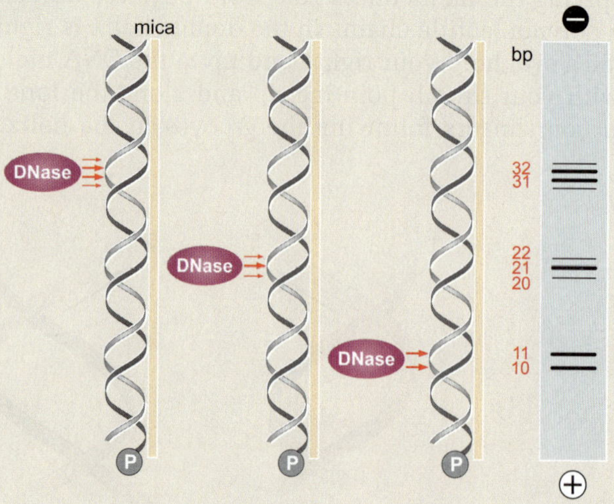

BOX 6-1 FIGURE 1 **The mica experiment.**

groove and the large angle on the other edge generates a **major groove.** (If the sugars pointed away from each other in a straight line, i.e., at an angle of 180°, then the two grooves would be of equal dimensions and there would be no minor and major grooves.)

The Major Groove Is Rich in Chemical Information

The edges of each base pair are exposed in the major and minor grooves, creating a pattern of hydrogen-bond donors and acceptors and of van der Waals surfaces that identifies the base pair (see Fig. 6-10). The edge of an A:T base pair displays the following chemical groups in the following order in the major groove: a hydrogen-bond acceptor (the N7 of adenine), a hydrogen-bond donor (the exocyclic amino group on C6 of adenine), a hydrogen-bond acceptor (the carbonyl group on C4 of thymine) and a bulky hydrophobic surface (the methyl group on C5 of thymine). Similarly, the edge of a G:C base pair displays the following groups in the major groove: a hydrogen-bond acceptor (at N7 of guanine), a hydrogen-bond acceptor (the carbonyl on C6 of guanine), a hydrogen-bond donor (the exocyclic amino group on C4 of cytosine), and a small nonpolar hydrogen (the hydrogen at C5 of cytosine).

Thus, there are characteristic patterns of hydrogen bonding and of overall shape that are exposed in the major groove that distinguish an

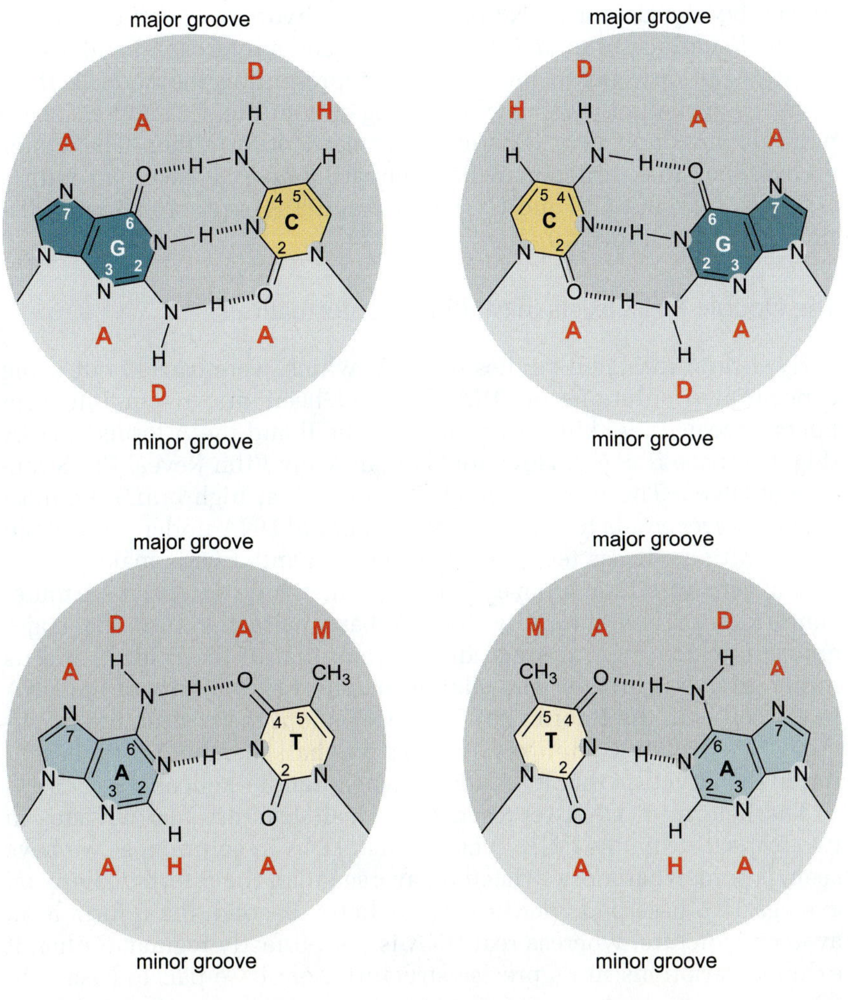

FIGURE 6-10 Chemical groups exposed in the major and minor grooves from the edges of the base pairs. The letters in red identify hydrogen bond acceptors (A), hydrogen bond donors (D), nonpolar hydrogens (H), and methyl groups (M).

A:T base pair from a G:C base pair, and, for that matter, A:T from T:A, and G:C from C:G. We can think of these features as a code in which **A** represents a **hydrogen-bond acceptor, D** a **hydrogen-bond donor, M** a **methyl group,** and **H** a **nonpolar hydrogen.** In such a code, **A D A M** in the major groove signifies an A:T base pair, and **A A D H** stands for a G:C base pair. Likewise, **M A D A** stands for a T:A base pair and **H D A A** is characteristic of a C:G base pair. In all cases, this code of chemical groups in the major groove specifies the identity of the base pair. These patterns are important because they allow proteins to unambiguously recognize DNA sequences without having to open and thereby disrupt the double helix. Indeed, as we shall see, a principal decoding mechanism relies upon the ability of amino acid side chains to protrude into the major groove and to recognize and bind to specific DNA sequences.

The minor groove is not as rich in chemical information, and what information is available is less useful for distinguishing between base pairs. The small size of the minor groove is less able to accommodate amino acid side chains. Also, A:T and T:A base pairs and G:C and C:G base pairs look similar to one another in the minor groove. An A:T base pair has a hydrogen-bond acceptor (at N3 of adenine), a nonpolar hydrogen (at N2 of adenine), and a hydrogen-bond acceptor (the carbonyl on C2 of thymine). Thus, its code is **A H A**. But this code is the same if read in the opposite direction, and hence an A:T base pair does not look very different from a T:A base pair from the point of view of the hydrogen-bonding properties of a protein poking its side chains into the minor groove. Likewise, a G:C base pair exhibits a hydrogen-bond acceptor (at N3 of guanine), a hydrogen-bond donor (the exocyclic amino group on C2 of guanine), and a hydrogen-bond acceptor (the carbonyl on C2 of cytosine), representing the code **A D A**. Thus, from the point of view of hydrogen bonding, C:G and G:C base pairs do not look very different from each other either. The minor groove does look different when comparing an A:T base pair with a G:C base pair, but G:C and C:G, or A:T and T:A, cannot be easily distinguished (see Fig. 6-10).

The Double Helix Exists in Multiple Conformations

Early X-ray diffraction studies of DNA, which were carried out using concentrated solutions of DNA that had been drawn out into thin fibers, revealed two kinds of structures, the B and the A forms of DNA (Fig. 6-11; see Box 6-2, How Spots on an X-ray Film Reveal the Structure of DNA). The B form, which is observed at high humidity, most closely corresponds to the average structure of DNA under physiological conditions. It has ten base pairs per turn and a wide major groove and a narrow minor groove. The A form, which is observed under conditions of low humidity, has 11 base pairs per turn. Its major groove is narrower and much deeper than that of the B form, and its minor groove is broader and shallower. The vast majority of the DNA in the cell is in the B form, but DNA does adopt the A structure in certain DNA–protein complexes. Also, as we shall see, the A form is similar to the structure that RNA adopts when double-helical.

The B form of DNA represents an ideal structure that deviates in two respects from the DNA in cells. First, DNA in solution, as we have seen, is somewhat more twisted on average than the B form, having on average 10.5 base pairs per turn of the helix. Second, the B form is an average structure, whereas real DNA is not perfectly regular. Rather, it exhibits variations in its precise structure from base pair to base pair.

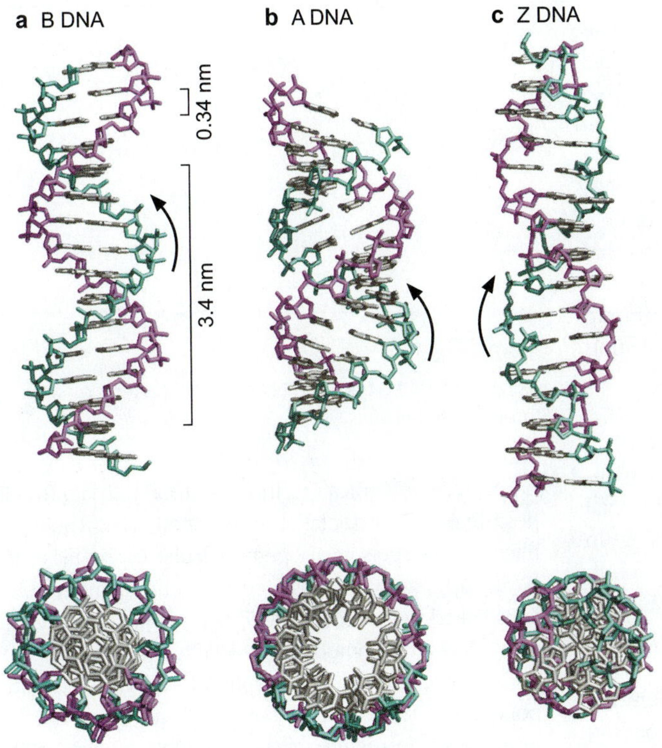

FIGURE 6-11 Models of the B, A, and Z forms of DNA. The sugar–phosphate backbone of each chain is on the outside in all structures (one purple and one green) with the bases (silver) oriented inward. Side views are shown at the top, and views along the helical axis at the bottom. (a) The B form of DNA, the usual form found in cells, is characterized by a helical turn every ten base pairs (3.4 nm); adjacent stacked base pairs are 0.34 nm apart. The major and minor grooves are also visible. (b) The more compact A form of DNA has 11 base pairs per turn and exhibits a large tilt of the base pairs with respect to the helix axis. In addition, the A form has a central hole (bottom). This helical form is adopted by RNA–DNA and RNA–RNA helices. (c) Z DNA is a left-handed helix and has a zigzag (hence "Z") appearance. (Courtesy of C. Kielkopf and P.B. Dervan.)

This was revealed by comparison of the crystal structures of individual DNAs of different sequences. For example, the two members of each base pair do not always lie exactly in the same plane. Rather, they can display a "propeller twist" arrangement in which the two flat bases counterrotate relative to each other along the long axis of the base pair, giving the base pair a propeller-like character (Fig. 6-12). Moreover, the precise rotation per base pair is not a constant. As a result, the width of the major and minor grooves varies locally. Thus, DNA molecules are never perfectly regular double helices. Instead, their exact conformation depends on which base pair (A:T, T:A, G:C, or C:G) is present at each position along the double helix and on the identity of neighboring base pairs. Still, the B form is for many purposes a good first approximation of the structure of DNA in cells.

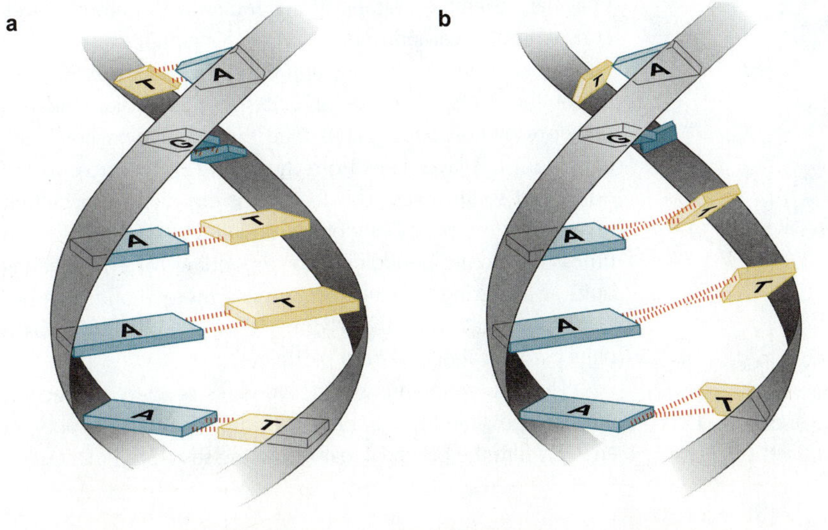

FIGURE 6-12 The propeller twist between the purine and pyrimidine base pairs of a right-handed helix. (a) The structure shows a sequence of three consecutive A:T base pairs with normal Watson–Crick bonding. (b) Propeller twist causes rotation of the bases about their long axis. (Adapted, with permission, from Aggarwal A.K. et al. 1988. *Science* 242: 899–907, Fig. 5b. © AAAS.)

Box 6-2 How Spots on an X-ray Film Reveal the Structure of DNA

One of the most enduring images in the history of molecular biology is the famous photograph taken by Rosalind Franklin of the X-ray diffraction pattern of an oriented fiber of DNA molecules. Franklin's image is of great historic significance because it provided critical evidence in support of the Watson–Crick model for B-form DNA. Also, Francis Crick, who had helped develop the theory of the diffraction of helical molecules, was able to infer from the pattern of spots that the strands of DNA are twisted around each other. At first glance, Franklin's image shows no recognizable relationship to a double helix. How then did this mysterious pattern of spots help unravel the atomic structure of the genetic material?

As seen in the figure, Franklin's image consists of a central "Maltese" cross (highlighted in red in Box 6-2 Fig. 1), which is composed of broad spots (the breadth of the spots reflecting disorder in the fiber). The spots are evenly spaced along horizontal "layer" lines (numbered in the figure). Notice that counting up and down from the center of the cross, the spots at the fourth layer line are missing. Notice also that the Maltese cross and the intensely dark regions at the top and bottom of the image create a series of four diamond-shaped areas (two examples of which are highlighted in blue). As we now explain, it can be understood in qualitative terms from a few simple considerations about the nature of wave diffraction that this seemingly arcane pattern of spots corresponds to the structure of the double helix.

The principle underlying X-ray diffraction is that when waves pass through a periodic array, interference occurs between the waves if the wavelength of the waves is similar to the repeat distance of the array. (Hence, X-rays, which have a very short wavelength [0.15 nm] are used for revealing atomic structure.) If the oscillations of the waves are aligned, the waves reinforce each other (constructive interference), but if the troughs of one set of waves are aligned with the peaks of another set of waves, the waves cancel each other out (destructive interference). Thus, a beam of waves passing through an array consisting of a horizontal set of lines would generate a row of spots

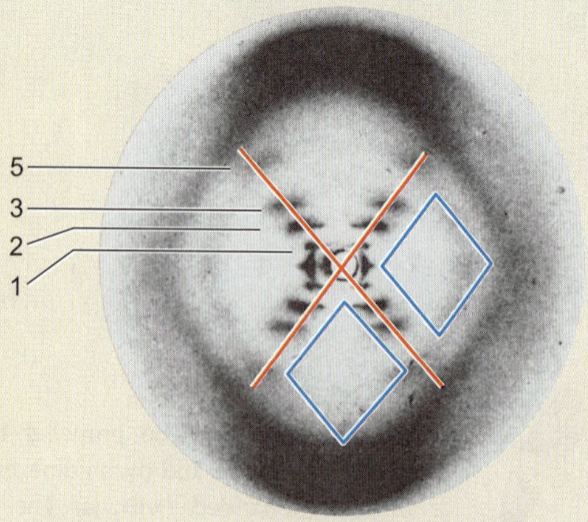

BOX 6-2 FIGURE 1 Rosalind Franklin's X-ray diffraction image of DNA revealing the Maltese cross. (Modified, with permission, from Franklin R.E. and Gosling R.G. 1953. *Nature* 171: 740–741. © Macmillan.)

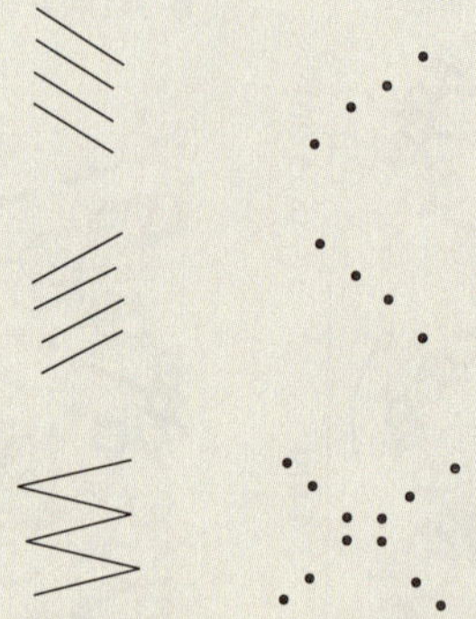

BOX 6-2 FIGURE 2 Diffraction pattern of waves passing through parallel lines.

perpendicular (vertical) to the lines (Box 6-2 Fig. 2). Now suppose that the horizontal lines are tilted. This would result in a tilted row of spots (again perpendicular to the tilt of the lines). Next, suppose that waves are passing through two sets of tilted lines linked to each other in zigzag fashion as in the figure: this results in a cross composed of two tilted rows of spots.

Now let us turn our attention to DNA. Imagine the backbone of one strand of the double helix projected onto a flat surface. Loosely speaking, this would create a linked series of zigs and zags (or, more properly, a sinusoidal curve). If we think of the zigs as generating one set of tilted lines and the zags as generating another set, then waves passing through the zigs and zags will generate two rows of spots that cross each other as in the example above. This is the basis for the Maltese cross in the Franklin photograph, and hence the cross reveals that DNA is helical. Knowledge of the wavelength of X-rays and measurements of the spacing between the layer lines further reveals that the helix has a periodicity of 3.4 nm. Of course, DNA consists of two helical backbones, not one. This, too, is revealed in the Franklin photograph. The helices of DNA are out of register with each other by three-eighths of a helical repeat. It turns out that this offset between the helices creates an additional destructive interference that obliterates the fourth layer line. Thus, the missing fourth layer line shows that DNA is a double helix and tells us how the two helices are aligned relative to each other.

Finally, the DNA backbone is not a smooth line as in our imaginary example. Rather, it is granular at the atomic level, consisting of sugar–phosphate units. This granularity results in additional intensities, particularly north and south of the center of the cross, to create a pattern of four diamonds. In higher-resolution photographs than the one shown here, one can count ten layer lines from the center of the cross to the north and south poles. This feature of the diffraction pattern reveals that the periodicity of the double helix (3.4 nm) is ten times the atomic periodicity, corresponding to ten repeating units at a spacing of 0.34 nm. Because there is one base per sugar–phosphate unit, the B form of DNA consists of ten base pairs per helical period (turn of the helix).

Thus, a rudimentary understanding of wave diffraction makes it possible to coax out of a simple pattern of spots on an X-ray film the principal features of the structure of DNA.

DNA Can Sometimes Form a Left-Handed Helix

DNA containing alternative purine and pyrimidine residues can fold into left-handed as well as right-handed helices. To understand how DNA can form a left-handed helix, we need to consider the glycosidic bond that connects the base to the 1′ position of 2′-deoxyribose. This bond can be in one of two conformations called ***syn*** and ***anti*** (Fig. 6-13). In right-handed DNA, the glycosidic bond is always in the *anti* conformation. In the left-handed helix, the fundamental repeating unit usually is a purine–pyrimidine dinucleotide, with the glycosidic bond in the *anti* conformation at pyrimidine residues and in the *syn* conformation at purine residues. It is this *syn* conformation at the purine nucleotides that is responsible for the left-handedness of the helix. The change to the *syn* position in the purine residues to alternating *anti–syn* conformations gives the backbone of left-handed DNA a zigzag look (hence its designation of **Z DNA;** see Fig. 6-11), which distinguishes it from right-handed forms. The rotation that effects the change from *anti* to *syn* also causes the sugar group to undergo a change in its pucker. Note, as shown in Figure 6-13, that C3′ and C2′ can switch locations. In solution alternating purine–pyrimidine residues assume the left-handed conformation only in the presence of high concentrations of positively charged ions (e.g., Na^+) that shield the negatively charged phosphate groups. At lower salt concentrations, they form typical right-handed conformations. The physiological significance of Z DNA is uncertain and left-handed helices probably account at most for only a small proportion of a cell's DNA. Further details of the A, B, and Z forms of DNA are presented in Table 6-2.

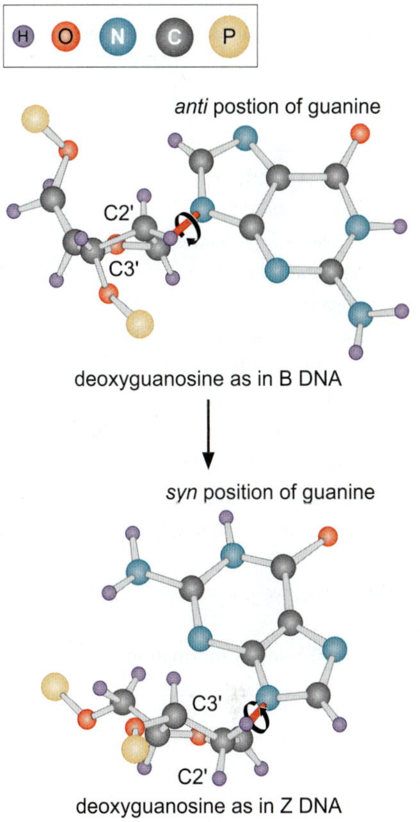

FIGURE 6-13 ***Syn*** **and** ***anti*** **positions of guanine in B and Z DNA.** In right-handed B DNA, the glycosyl bond (colored red) connecting the base to the deoxyribose group is always in the *anti* position, whereas in left-handed Z DNA it rotates in the direction of the arrow, forming the *syn* conformation at the purine (here guanine) residues but remains in the regular *anti* position (no rotation) in the pyrimidine residues. (Adapted, with permission, from Wang A.J.H. et al. 1982. *Cold Spring Harbor Symp. Quant. Biol.* **47:** 41. © Cold Spring Harbor Laboratory Press.)

DNA Strands Can Separate (Denature) and Reassociate

Because the two strands of the double helix are held together by relatively weak (noncovalent) forces, you might expect that the two strands could come apart easily. Indeed, the original structure for the double helix suggested that DNA replication would occur in just this manner. The complementary strands of the double helix can also be made to come apart when a solution of DNA is heated above physiological temperatures (to near 100°C) or under conditions of high pH, a process known as **denaturation.** However, this complete separation of DNA strands by denaturation is reversible. When heated solutions of denatured DNA are slowly cooled, single strands often meet their complementary strands and re-form regular double helices (Fig. 6-14). The capacity to renature denatured DNA molecules permits artificial hybrid DNA molecules to be formed by slowly cooling mixtures of denatured DNA from two different sources. Likewise, hybrids can be formed between complementary strands of DNA and RNA. As we shall see in Chapter 20, the ability to form hybrids between two single-stranded nucleic acids, called **hybridization,** is the basis for several indispensable techniques in molecular biology, such as Southern blot hybridization (see Chapter 21) and DNA microarray analysis (see Chapter 19, Box 19-1).

Important insights into the properties of the double helix were obtained from classic experiments carried out in the 1950s in which the denaturation of DNA was studied under a variety of conditions. In these experiments, DNA denaturation was monitored by measuring the absorbance of ultraviolet light passed through a solution of DNA. DNA maximally absorbs ultraviolet light at a wavelength of about 260 nm. It is the bases that are principally responsible for this absorption. When the temperature of a solution of DNA is raised to near the boiling point

TABLE 6-2 **A Comparison of the Structural Properties of A, B, and Z DNAs as Derived from Single-Crystal X-ray Analysis**

	Helix Type		
	A	B	Z
Overall proportions	Short and broad	Longer and thinner	Elongated and slim
Rise per base pair	2.3 Å	3.32 Å	3.8 Å
Helix-packing diameter	25.5 Å	23.7 Å	18.4 Å
Helix rotation sense	Right-handed	Right-handed	Left-handed
Base pairs per helix repeat	1	1	2
Base pairs per turn of helix	~11	~10	12
Rotation per base pair	33.6°	35.9°	–60° per 2 bp
Pitch per turn of helix	24.6 Å	33.2 Å	45.6 Å
Tilt of base normals to helix axis	+19°	–1.2°	–9°
Base-pair mean propeller twist	+18°	+16°	~0°
Helix axis location	Major groove	Through base pairs	Minor groove
Major-groove proportions	Extremely narrow but very deep	Wide and of intermediate depth	Flattened out on helix surface
Minor-groove proportions	Very broad but shallow	Narrow and of intermediate depth	Extremely narrow but very deep
Glycosyl-bond conformation	*anti*	*anti*	*anti* at C, *syn* at G

Adapted, with permission, from Dickerson R.E. et al. 1982. *Cold Spring Harbor Symp. Quant. Biol.* 47: 14.

of water, the optical density, called **absorbance,** at 260 nm markedly increases, a phenomenon known as **hyperchromicity.** The explanation for this increase is that duplex DNA absorbs less ultraviolet light by about 40% than do individual DNA chains. This hypochromicity is due to base stacking, which diminishes the capacity of the bases in duplex DNA to absorb ultraviolet light.

If we plot the optical density of DNA as a function of temperature, we observe that the increase in absorption occurs abruptly over a relatively narrow temperature range. The midpoint of this transition is the **melting point** or T_m (Fig. 6-15). Like ice, DNA melts: it undergoes a transition from a highly ordered double-helical structure to a much less ordered structure of individual strands. The sharpness of the increase in absorbance at the melting temperature tells us that the denaturation and renaturation of complementary DNA strands is a highly cooperative, zippering-like process. Renaturation, for example, probably occurs by means of a slow nucleation process in which a relatively small stretch of bases on one strand find and pair with their complement on the complementary strand (middle panel of Fig. 6-14). The remainder of the two strands then rapidly zipper up from the nucleation site to re-form an extended double helix (lower panel of Fig. 6-14).

The melting temperature of DNA is a characteristic of each DNA that is largely determined by the G:C content of the DNA and the ionic strength of the solution. The higher the percent of G:C base pairs in the DNA (and hence the lower the content of A:T base pairs), the higher the melting point (Fig. 6-16). Likewise, the higher the salt concentration of the solution, the greater the temperature at which the DNA denatures. How do we explain this behavior? G:C base pairs contribute more to the stability of DNA than do A:T base pairs because of the greater number of hydrogen bonds for the former (three in a G:C base pair versus two for A:T) but also, importantly, because the stacking interactions of G:C base pairs with adjacent base pairs are more favorable than the corresponding interactions of A:T base pairs with their neighboring base pairs. The

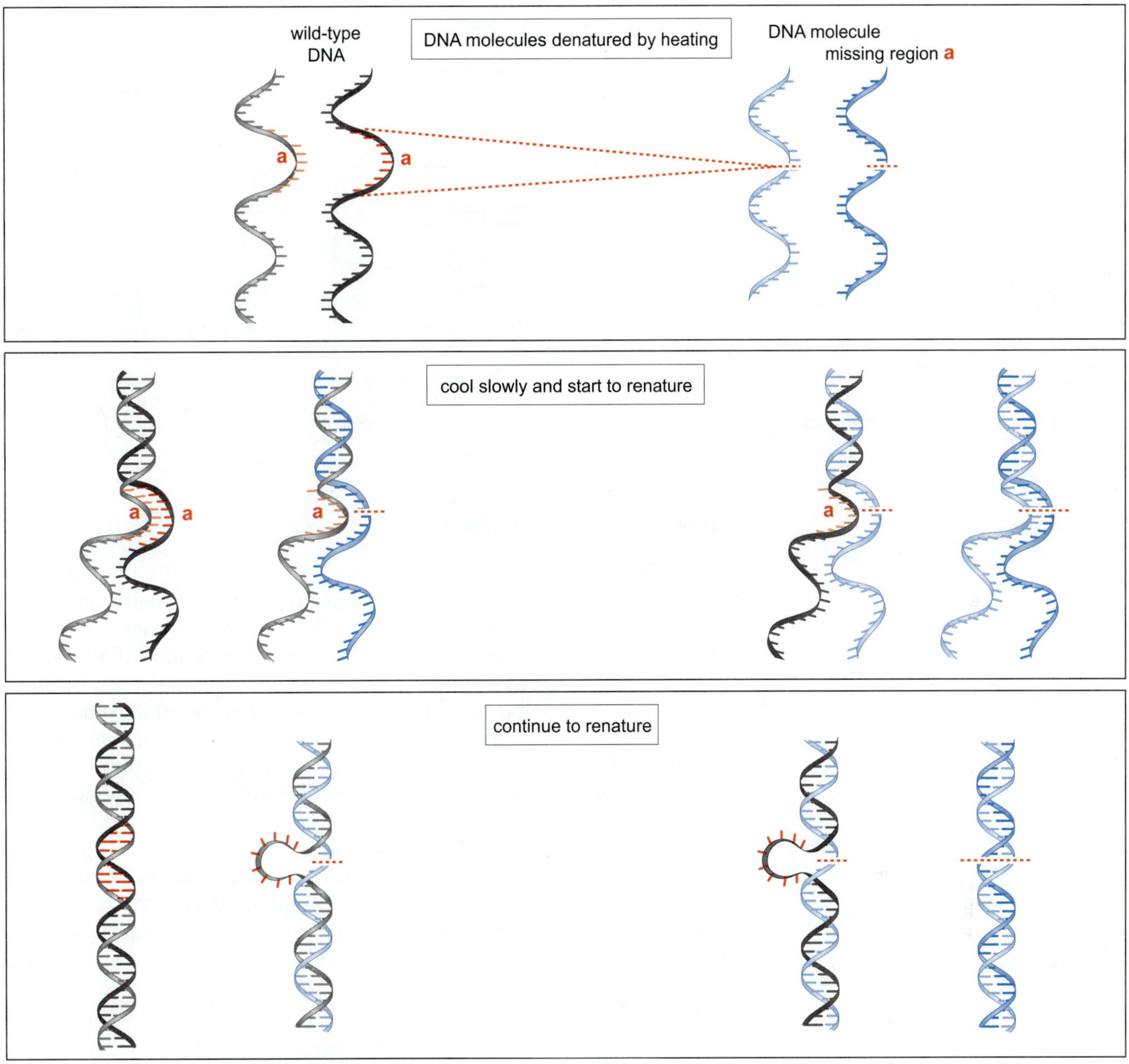

FIGURE 6-14 Reannealing and hybridization. A mixture of two otherwise identical double-stranded DNA molecules, one normal wild-type DNA and the other a mutant missing a short stretch of nucleotides (marked as region **a** in red), are denatured by heating. The denatured DNA molecules are allowed to renature by incubation just below the melting temperature. This treatment results in two types of renatured molecules. One type is composed of completely renatured molecules in which two complementary wild-type strands re-form a helix and two complementary mutant strands re-form a helix. The other type are hybrid molecules, composed of a wild-type and a mutant strand, exhibiting a short unpaired loop of DNA (region **a**).

effect of ionic strength reflects another fundamental feature of the double helix. The backbones of the two DNA strands contain phosphoryl groups that carry a negative charge. These negative charges are close enough across the two strands that, if not shielded, they tend to cause the strands to repel each other, facilitating their separation. At high ionic strength, the negative charges are shielded by cations, thereby stabilizing the helix. Conversely, at low ionic strength the unshielded negative charges render the helix less stable.

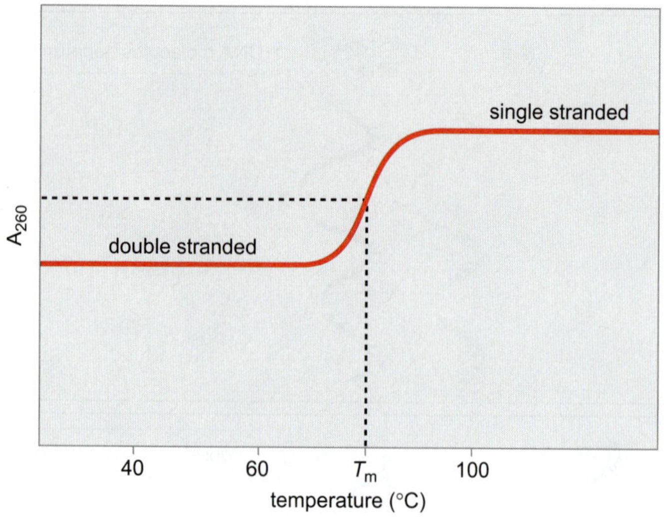

FIGURE 6-15 DNA denaturation curve.

Some DNA Molecules Are Circles

It was initially believed that all DNA molecules are linear and have two free ends. Indeed, the chromosomes of eukaryotic cells each contain a single (extremely long) DNA molecule. But now we know that some DNAs are circles. For example, the chromosome of the small monkey DNA virus SV40 is a circular, double-helical DNA molecule of about 5000 base pairs. Also, most (but not all) bacterial chromosomes are circular; *Escherichia coli* has a circular chromosome of about 5 million base pairs. Additionally, many bacteria have small autonomously replicating genetic elements known as **plasmids,** which are generally circular DNA molecules.

Interestingly, some DNA molecules are sometimes linear and sometimes circular. The most well-known example is that of the bacteriophage λ, a DNA virus of *E. coli*. The phage λ genome is a linear double-stranded molecule in the virion particle. However, when the λ genome is injected into an *E. coli* cell during infection, the DNA

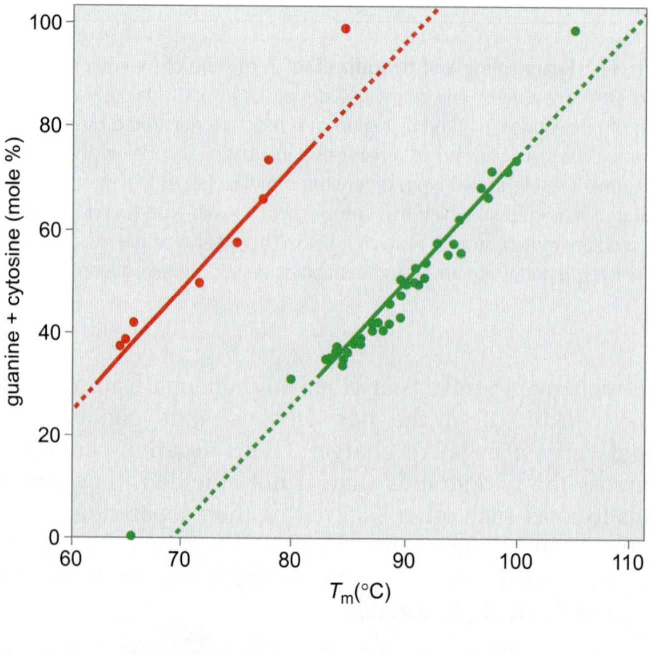

FIGURE 6-16 Dependence of DNA denaturation on G + C content and on salt concentration. The greater the G + C content, the higher the temperature must be to denature the DNA strand. DNA from different sources was dissolved in solutions of low (red line) and high (green line) concentrations of salt at pH 7.0. The points represent the temperature at which the DNA denatured graphed against the G + C content. (Data from Marmur J. and Doty P. 1962. *J. Mol. Biol.* 5: 120. © Elsevier.)

circularizes. This occurs by base pairing between single-stranded regions that protrude from the ends of the DNA and that have complementary sequences, also known as "sticky ends."

DNA TOPOLOGY

As DNA is a flexible structure, its exact molecular parameters are a function of both the surrounding ionic environment and the nature of the DNA-binding proteins with which it is complexed. Because their ends are free, linear DNA molecules can freely rotate to accommodate changes in the number of times the two chains of the double helix twist about each other. But if the two ends are covalently linked to form a circular DNA molecule and if there are no interruptions in the sugar–phosphate backbones of the two strands, then the absolute number of times the chains can twist about each other cannot change. Such a covalently closed, circular DNA is said to be topologically constrained. Even the linear DNA molecules of eukaryotic chromosomes are subject to topological constraints because of their extreme length, entrainment in chromatin, and interaction with other cellular components (see Chapter 7). Despite these constraints, DNA participates in numerous dynamic processes in the cell. For example, the two strands of the double helix, which are twisted around each other, must rapidly separate in order for DNA to be duplicated and to be transcribed into RNA. Thus, understanding the topology of DNA and how the cell both accommodates and exploits topological constraints during DNA replication, transcription, and other chromosomal transactions is of fundamental importance in molecular biology.

Linking Number Is an Invariant Topological Property of Covalently Closed, Circular DNA

Let us consider the topological properties of **covalently closed, circular DNA**, which is referred to as **cccDNA**. Because there are no interruptions in either polynucleotide chain, the two strands of cccDNA cannot be separated from each other without the breaking of a covalent bond. If we wished to separate the two circular strands without permanently breaking any bonds in the sugar–phosphate backbones, we would have to pass one strand through the other strand repeatedly (we will encounter an enzyme that can perform just this feat!). The number of times one strand would have to be passed through the other strand in order for the two strands to be entirely separated from each other is called the **linking number** (Fig. 6-17). The linking number, which is always an integer, is an invariant topological property of cccDNA, no matter how much the shape of the DNA molecule is distorted.

Linking Number Is Composed of Twist and Writhe

WEB ANIMATION

The linking number is the sum of two geometric components called the **twist** and the **writhe** (see Interactive Animation 6-1). Let us consider twist first. Twist is simply the number of helical turns of one strand about the other, that is, the number of times one strand completely wraps around the other strand. Consider a cccDNA that is lying flat on a plane. In this flat conformation, the linking number is fully composed of twist. Indeed, the twist can be easily determined by counting the num-

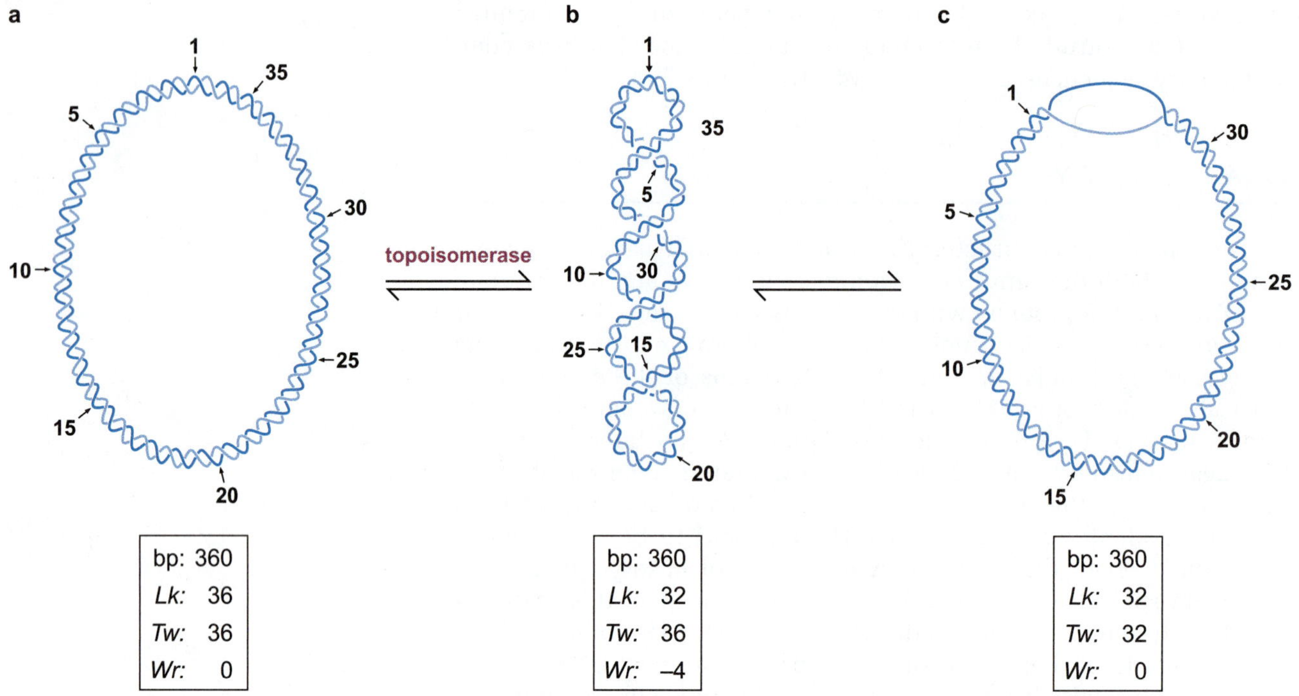

FIGURE 6-17 Topological states of covalently closed, circular (ccc) DNA. The figure shows conversion of the relaxed (a) to the negatively supercoiled (b) form of DNA. The strain in the supercoiled form may be taken up by supertwisting (b) or by local disruption of base pairing (c). (Adapted from a diagram provided by Dr. M. Gellert.) (Modified, with permission, from Kornberg A. and Baker T.A. 1992. *DNA replication,* I 1–21, p. 32. © W.H. Freeman.)

ber of times the two strands cross each other (see Fig. 6-17a). The helical crossovers (twist) in a right-handed helix are defined as positive such that the linking number of DNA will have a positive value.

But cccDNA is generally not lying flat on a plane. Rather, it is usually torsionally stressed such that the long axis of the double helix crosses over itself, often repeatedly, in three-dimensional space (Fig. 6-17b). This is called *writhe.* To visualize the distortions caused by torsional stress, think of the coiling of a telephone cord that has been overtwisted.

Writhe can take two forms. One form is the **interwound** or **plectonemic writhe,** in which the long axis is twisted around itself, as depicted in Figures 6-17b and 6-18a. The other form of writhe is a **toroid** or **spiral** in which the long axis is wound in a cylindrical manner, as often occurs when DNA wraps around protein (Fig. 6-18b). The **writhing number** (***Wr***) is the total number of interwound and/or spiral writhes in cccDNA. For example, the molecule shown in Figure 6-17b has a writhing number of 4.

Interwound writhe and spiral writhe are topologically equivalent to each other and are readily interconvertible geometric properties of cccDNA. Also, twist and writhe are interconvertible. A molecule of cccDNA can readily undergo distortions that convert some of its twist to writhe or some of its writhe to twist without the breakage of any covalent bonds. The only constraint is that the sum of the **twist number** (***Tw***) and the writhing number (*Wr*) must remain equal to the **linking number** (***Lk***). This constraint is described by the equation

$$Lk = Tw + Wr.$$

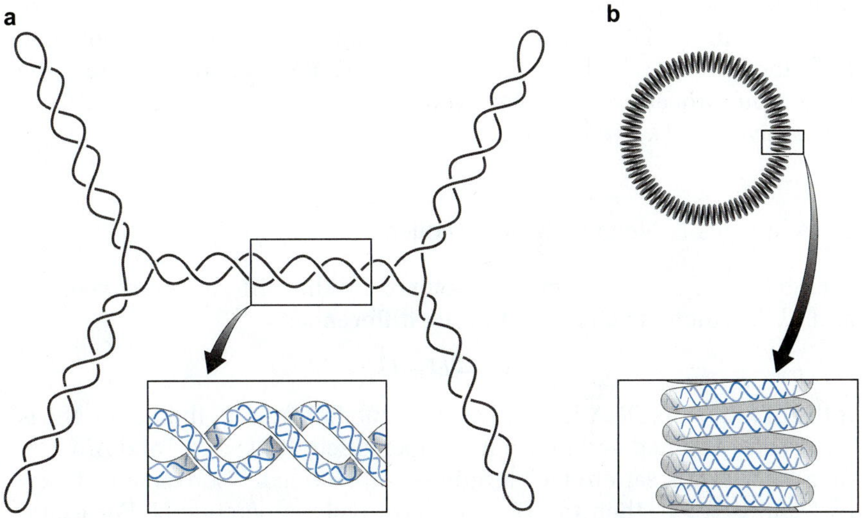

FIGURE 6-18 Two forms of writhe of supercoiled DNA. The figure shows interwound (a) and toroidal (b) writhe of cccDNA of the same length. (a) The interwound or plectonemic writhe is formed by twisting of the double-helical DNA molecule over itself as depicted in the example of a branched molecule. (b) Toroidal or spiral writhe is depicted in this example by cylindrical coils. (Modified, with permission, from Kornberg A. and Baker T.A. 1992. *DNA replication*, I 1–22, p. 33. © W.H. Freeman. Used by permission of Dr. Nicholas Cozzarelli.)

Lk^O Is the Linking Number of Fully Relaxed cccDNA under Physiological Conditions

Consider cccDNA that is free of **supercoiling** (i.e., it is said to be **relaxed**) and whose twist corresponds to that of the B form of DNA in solution under physiological conditions (about 10.5 base pairs per turn of the helix). The linking number (Lk) of such cccDNA under physiological conditions is assigned the symbol $\boldsymbol{Lk^O}$. Lk^O for such a molecule is the number of base pairs divided by 10.5. For a cccDNA of 10,500 base pairs, $Lk = +1000$. (The sign is positive because the twists of DNA are right-handed.) One way to see this is to imagine pulling one strand of the 10,500-base-pair cccDNA out into a flat circle. If we did this, then the other strand would cross the flat circular strand 1000 times.

How can we remove supercoils from cccDNA if it is not already relaxed? One procedure is to treat the DNA mildly with the enzyme DNase I, so as to break on average one phosphodiester bond (or a small number of bonds) in each DNA molecule. Once the DNA has been "nicked" in this manner, it is no longer topologically constrained and the strands can rotate freely, allowing writhe to dissipate (Fig. 6-19). If the nick is then repaired, the resulting cccDNA molecules will be relaxed and will have on average an Lk that is equal to Lk^O. (Be-

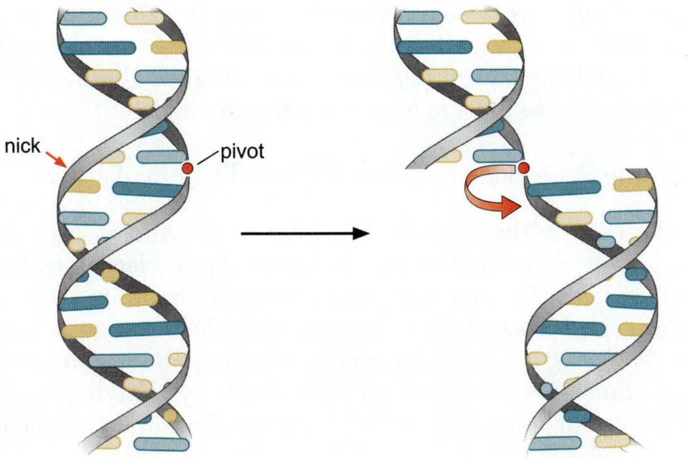

FIGURE 6-19 Relaxing DNA with DNase I.

cause of rotational fluctuation at the time the nick is repaired, some of the resulting cccDNAs will have an *Lk* that is somewhat higher than Lk^O and others will have an *Lk* that is somewhat lower. Thus, the relaxation procedure will generate a narrow spectrum of cccDNAs whose average *Lk* is equal to Lk^O.)

DNA in Cells Is Negatively Supercoiled

The extent of supercoiling is measured by the difference between *Lk* and Lk^O, which is called the **linking difference:**

$$\Delta Lk = Lk - Lk^O.$$

If the ΔLk of a cccDNA is significantly different from 0, then the DNA is torsionally strained and hence it is supercoiled. If $Lk < Lk^O$ and $\Delta Lk < 0$, then the DNA is said to be "negatively supercoiled." Conversely, if $Lk > Lk^O$ and $\Delta Lk > 0$, then the DNA is "positively supercoiled." For example, the molecule shown in Figure 6-17b is negatively supercoiled and has a linking difference of –4 because its *Lk* (32) is 4 less than that (36) for the relaxed form of the molecule shown in Figure 6-17a.

Because ΔLk and Lk^O are dependent on the length of the DNA molecule, it is more convenient to refer to a normalized measure of supercoiling. This is the **superhelical density,** which is assigned the symbol σ and is defined as

$$\sigma = \Delta Lk / Lk^O.$$

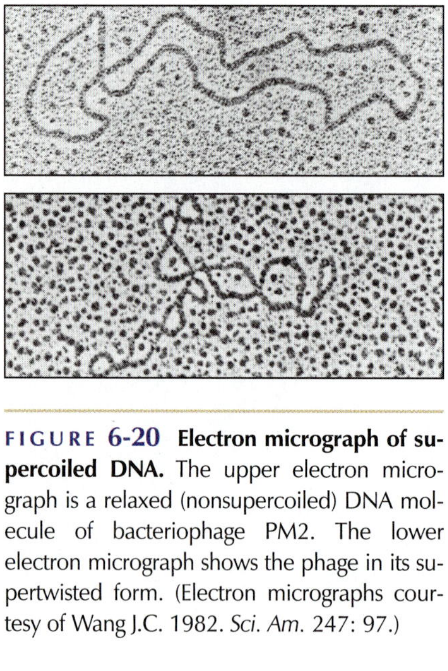

FIGURE 6-20 Electron micrograph of supercoiled DNA. The upper electron micrograph is a relaxed (nonsupercoiled) DNA molecule of bacteriophage PM2. The lower electron micrograph shows the phage in its supertwisted form. (Electron micrographs courtesy of Wang J.C. 1982. *Sci. Am.* 247: 97.)

Circular DNA molecules purified from both bacteria and eukaryotes are usually negatively supercoiled, having values of σ of about –0.06. The electron micrograph shown in Figure 6-20 compares the structures of bacteriophage DNA in its relaxed form with its supercoiled form.

What does superhelical density mean biologically? Negative supercoils can be thought of as a store of free energy that aids in processes that require strand separation, such as DNA replication and transcription. Because $Lk = Tw + Wr$, negative supercoils can be converted into untwisting of the double helix (compare Fig. 6-17a with 6-17b). Regions of negatively supercoiled DNA, therefore, have a tendency to partially unwind. Thus, strand separation can be accomplished more easily in negatively supercoiled DNA than in relaxed DNA.

The only organisms that have been found to have positively supercoiled DNA are certain thermophiles, microorganisms that live under conditions of extreme high temperatures, such as in hot springs. In this case, the positive supercoils can be thought of as a store of free energy that helps keep the DNA from denaturing at the elevated temperatures. Insofar as positive supercoils can be converted into more twist (positively supercoiled DNA can be thought of as being overwound), strand separation requires more energy in thermophiles than in organisms whose DNA is negatively supercoiled.

Nucleosomes Introduce Negative Supercoiling in Eukaryotes

As we shall see in the next chapter, DNA in the nucleus of eukaryotic cells is packaged in small particles known as **nucleosomes** in which the double helix is wrapped almost two times around the outside circumference of a protein core. You will be able to recognize this wrapping as the toroid or spiral form of writhe. Importantly, it occurs in a left-handed manner. (Convince yourself of this by applying the handedness rule in your mind's eye to DNA wrapped around the nucleosome in Chapter 7, Fig. 7-18.) It turns out that writhe in the form of left-handed spirals is equivalent to negative supercoils. Thus, the packaging of DNA into nucleosomes introduces negative superhelical density.

Topoisomerases Can Relax Supercoiled DNA

As we have seen, the linking number is an invariant property of DNA that is topologically constrained. It can be changed only by introducing interruptions into the sugar–phosphate backbone. A remarkable class of enzymes known as **topoisomerases** are able to do just this by introducing transient single-strand or double-strand breaks into the DNA (see Interactive Animation 6-2).

WEB ANIMATION

Topoisomerases are of two general types. Type II topoisomerases change the linking number in steps of two. They make transient double-strand breaks in the DNA through which they pass a segment of uncut duplex DNA before resealing the break. This type of reaction is shown schematically in Figure 6-21. Type II topoisomerases require the energy of ATP hydrolysis for their action. Type I topoisomerases, in contrast, change the linking number of DNA in steps of one. They make transient single-strand breaks in the DNA, allowing the uncut strand to pass through the break before resealing the nick (Fig. 6-22). In contrast to the type II topoisomerases, type I topoisomerases do not require ATP. How topoisomerases relax DNA and promote other related reactions in a controlled and concerted manner is explained below.

Prokaryotes Have a Special Topoisomerase That Introduces Supercoils into DNA

Both prokaryotes and eukaryotes have type I and type II topoisomerases that are capable of removing supercoils from DNA. In addition, however, prokaryotes have a special type II topoisomerase known as DNA gyrase that introduces, rather than removes, negative supercoils. DNA gyrase is responsible for the negative supercoiling of chromosomes in prokaryotes. This negative supercoiling facilitates the unwinding of the DNA duplex, which stimulates many reactions of DNA including initiation of both transcription and DNA replication.

Topoisomerases Also Unknot and Disentangle DNA Molecules

In addition to relaxing supercoiled DNA, topoisomerases promote several other reactions important to maintaining the proper DNA structure within cells. The enzymes use the same transient DNA break and strand

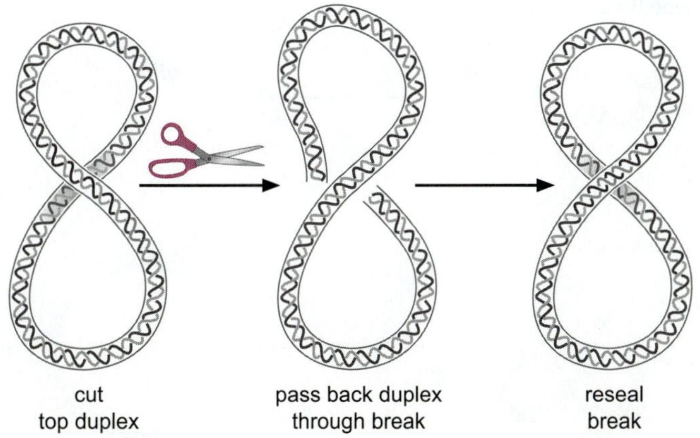

FIGURE 6-21 Schematic for changing the linking number in DNA with topoisomerase II. Topoisomerase II binds to DNA, creates a double-strand break, passes uncut DNA through the gap, and then reseals the break.

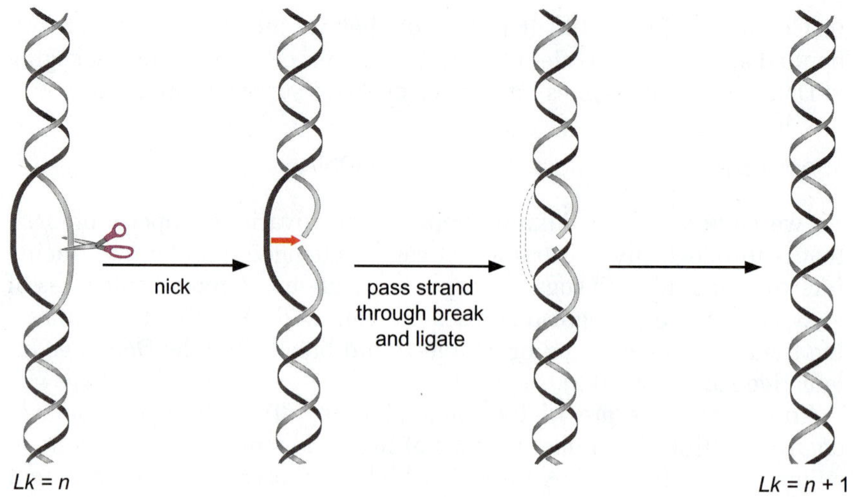

FIGURE 6-22 Schematic mechanism of action for topoisomerase I. The enzyme cuts a single strand of the DNA duplex, passes the uncut strand through the break, and then reseals the break. The process increases the linking number by +1.

passage reaction that they use to relax DNA to carry out these reactions.

Topoisomerases can both **catenate** and **decatenate** circular DNA molecules. Circular DNA molecules are said to be catenated if they are linked together like two rings of a chain (Fig. 6-23a). Of these two activities, the ability of topoisomerases to decatenate DNA is of clear biological importance. As we will see in Chapter 8, catenated DNA molecules are commonly produced as a round of DNA replication is finished (see Fig. 8-34). Topoisomerases play the essential role of unlinking these DNA molecules to allow them to separate into the two daughter cells for cell division. De-

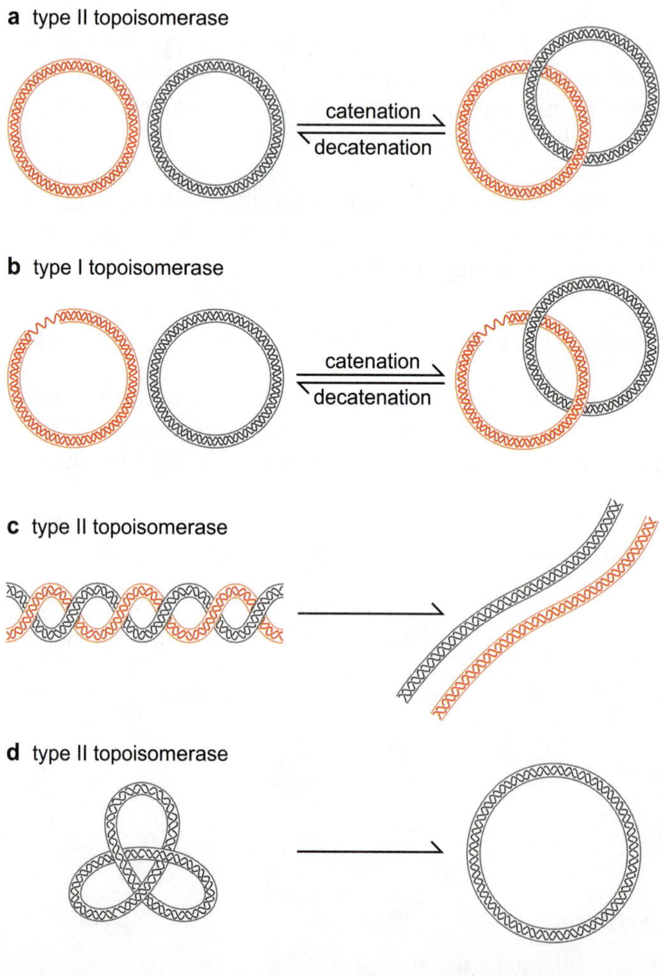

FIGURE 6-23 Topoisomerases decatenate, disentangle, and unknot DNA. (a) Type II topoisomerases can catenate and decatenate covalently closed, circular DNA molecules by introducing a double-strand break in one DNA and passing the other DNA molecule through the break. (b) Type I topoisomerases can catenate and decatenate molecules only if one DNA strand has a nick or a gap. This is because these enzymes cleave only one DNA strand at a time. (c) Entangled long linear DNA molecules, generated, e.g., during the replication of eukaryotic chromosomes, can be disentangled by a topoisomerase. (d) DNA knots can also be unknotted by topoisomerase action.

catenation of two covalently closed circular DNA molecules requires passage of the two DNA strands of one molecule through a double-strand break in the second DNA molecule. This reaction therefore depends on a type II topoisomerase. The requirement for decatenation explains why type II topoisomerases are essential cellular proteins. However, if at least one of the two catenated DNA molecules carries a nick or a gap, then a type I enzyme may also unlink the two molecules (Fig. 6-23b).

Although we often focus on circular DNA molecules when considering topological issues, the long linear chromosomes of eukaryotic organisms also experience topological problems. For example, during a round of DNA replication, the two double-stranded daughter DNA molecules will often become entangled (Fig. 6-23c). These sites of entanglement, just like the links between catenated DNA molecules, block the separation of the daughter chromosomes during mitosis. Therefore, DNA disentanglement, generally catalyzed by a type II topoisomerase, is also required for a successful round of DNA replication and cell division in eukaryotes.

On occasion, a DNA molecule becomes knotted (Fig. 6-23d). For example, some site-specific recombination reactions, which we shall discuss in detail in Chapter 11, give rise to knotted DNA products. Once again, a type II topoisomerase can "untie" a knot in duplex DNA. If the DNA molecule is nicked or gapped, then a type I enzyme also can do this job.

Topoisomerases Use a Covalent Protein–DNA Linkage to Cleave and Rejoin DNA Strands

To perform their functions, topoisomerases must cleave a DNA strand (or two strands) and then rejoin the cleaved strand (or strands). Topoisomerases are able to promote both DNA cleavage and rejoining without the assistance of other proteins or high-energy co-factors (e.g., ATP; also see below) because they use a covalent-intermediate mechanism. DNA cleavage occurs when a tyrosine residue in the active site of the topoisomerase attacks a phosphodiester bond in the backbone of the target DNA (Fig. 6-24). This attack generates a break in the DNA, whereby the topoisomerase is covalently joined to one of the broken ends via a phosphotyrosine linkage. The other end of the DNA terminates with a free OH group. This end is also held tightly by the enzyme, as we will see below.

The phospho-tyrosine linkage conserves the energy of the phosphodiester bond that was cleaved. Therefore, the DNA can be resealed simply by reversing the original reaction: the OH group from one broken DNA end attacks the phospho-tyrosine bond re-forming the DNA phosphodiester bond. This reaction rejoins the DNA strand and releases the topoisomerase, which can then go on to catalyze another reaction cycle. Although as noted above, type II topoisomerases require ATP hydrolysis for activity, the energy released by this hydrolysis is used to promote conformational changes in the topoisomerase–DNA complex rather than to cleave or rejoin DNA.

Topoisomerases Form an Enzyme Bridge and Pass DNA Segments through Each Other

Between the steps of DNA cleavage and DNA rejoining, the topoisomerase promotes passage of a second segment of DNA through the break. Topoisomerase function thus requires that DNA cleavage, strand passage, and DNA rejoining all occur in a highly coordinated manner. Structures of several different topoisomerases have provided insight into how the reaction cycle occurs. Here we will explain a model for how a type I topoisomerase relaxes DNA.

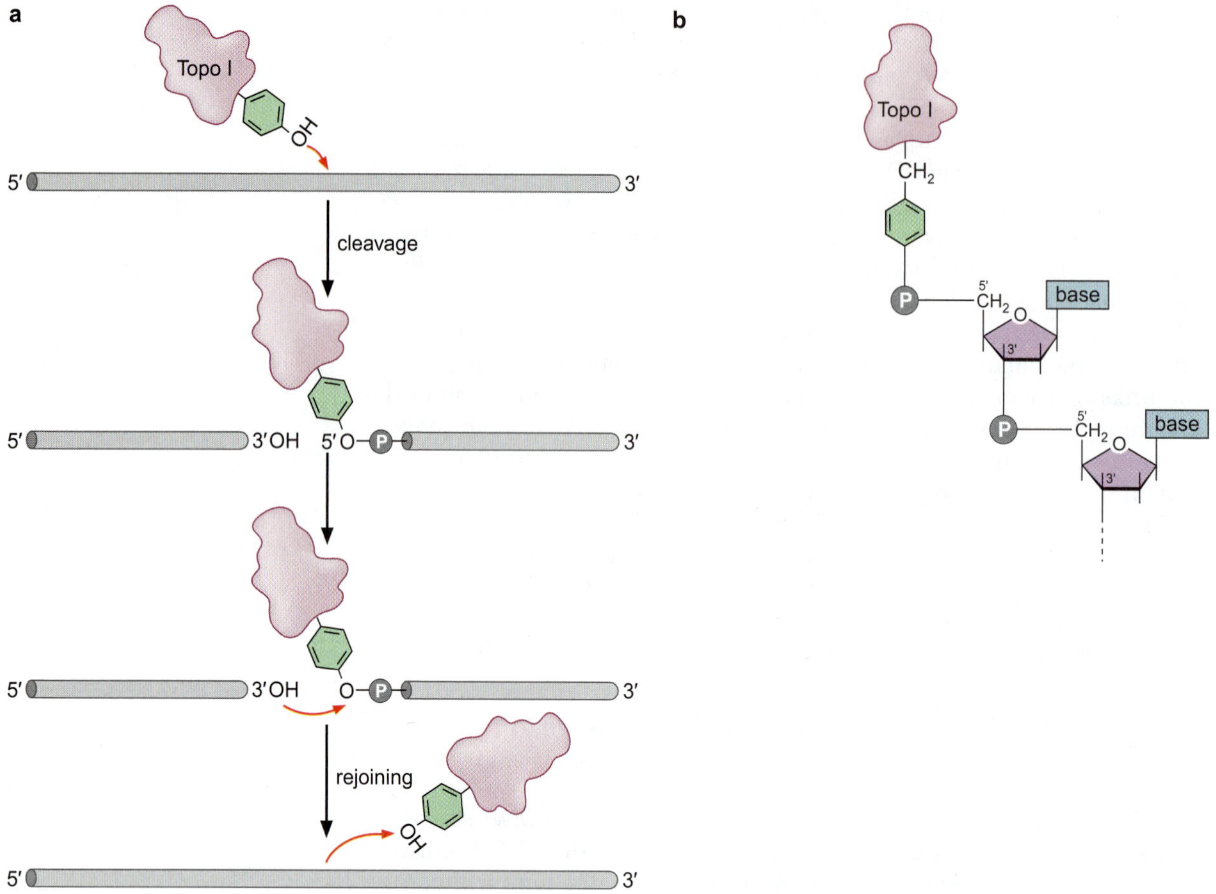

FIGURE 6-24 Topoisomerases cleave DNA using a covalent tyrosine–DNA intermediate. (a) Schematic of the cleavage and rejoining reaction. For simplicity, only a single strand of DNA is shown. See Figure 6-25 for a more realistic picture. The same mechanism is used by type II topoisomerases, although two enzyme subunits are required, one to cleave each of the two DNA strands. Topoisomerases sometimes cut to the 5′ side and sometimes to the 3′ side. (b) Close-up view of the phosphotyrosine covalent intermediate.

To initiate a relaxation cycle, the topoisomerase binds to a segment of duplex DNA in which the two strands are melted (Fig. 6-25a). Melting of the DNA strands is favored in highly negatively supercoiled DNA (see above), making this DNA an excellent substrate for relaxation. One of the DNA strands binds in a cleft in the enzyme that places it near the active-site tyrosine. This strand is cleaved to generate the covalent DNA-tyrosine intermediate (Fig. 6-25b). The success of the reaction requires that the other end of the newly cleaved DNA also be tightly bound by the enzyme. After cleavage, the topoisomerase undergoes a large conformational change to open up a gap in the cleaved strand, with the enzyme bridging the gap. The second (uncleaved) DNA strand then passes though the gap, and binds to a DNA-binding site in an internal "donut-shaped" hole in the protein (Fig. 6-25c). After strand passage occurs, a second conformational change in the topoisomerase–DNA complex brings the cleaved DNA ends back together (Fig. 6-25d); rejoining of the DNA strand occurs by attack of the OH end on the phosopho-tyrosine bond (see above). After rejoining, the enzyme must open up one final time to release the DNA (Fig. 6-25e). This product DNA is identical to the starting DNA molecule, except that the linking number has been increased by 1.

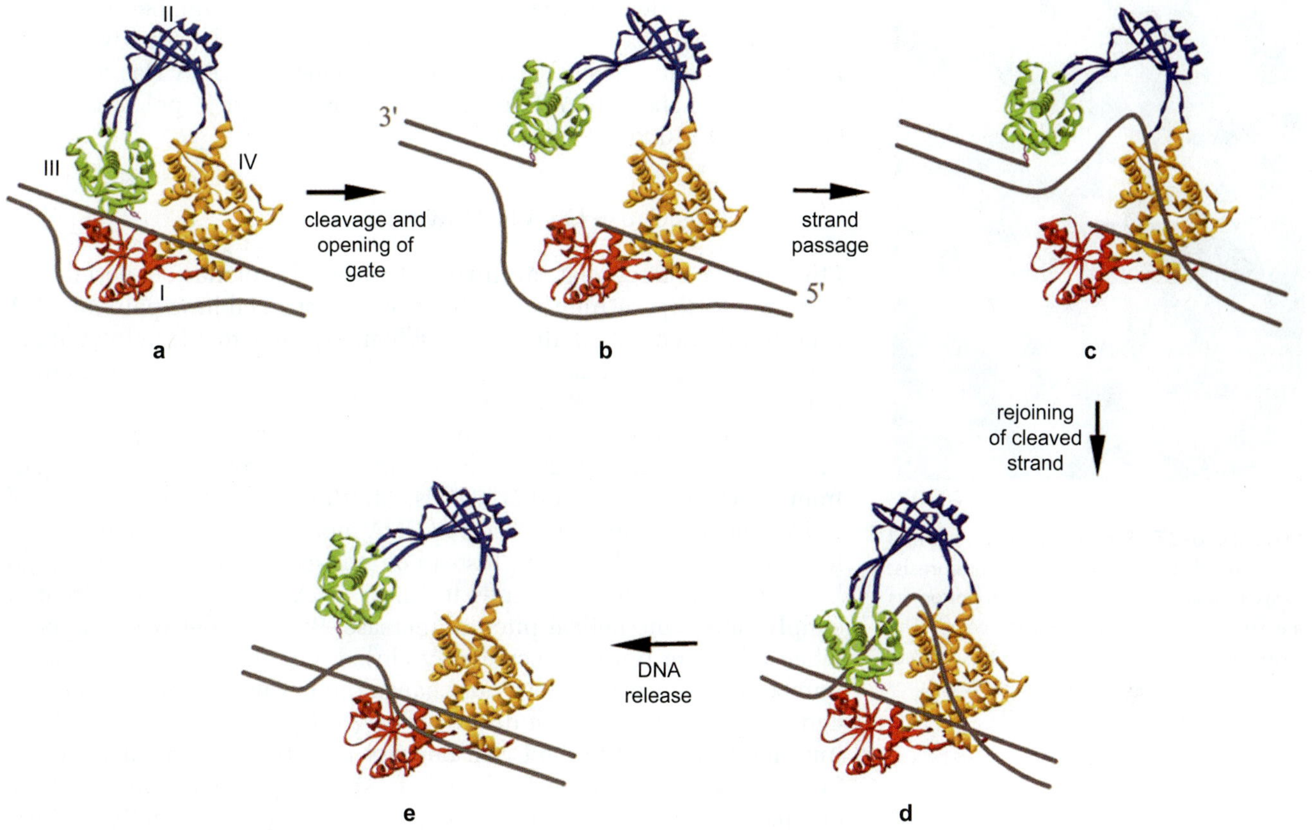

FIGURE 6-25 Model for the reaction cycle catalyzed by a type I topoisomerase. The figure shows a series of proposed steps for the relaxation of one turn of a negatively supercoiled plasmid DNA. The two strands of DNA are shown as dark gray (and not drawn to scale). The four domains of the protein are labeled in panel a. Domain I is shown in red, II is blue, III is green, and IV is orange. (Adapted, with permission, from Champoux J. 2001. *Annu. Rev. Biochem.* 70: 369–413. © Annual Reviews.)

This general mechanism, in which the enzyme provides a "protein bridge" during the strand passage reaction, can also be applied to the type II topoisomerases. The type II enzymes, however, are dimeric (or in some cases tetrameric). Two topoisomerase subunits, with their active site tyrosine residues, are required to cleave the two DNA strands and make the double-stranded DNA break that is an essential feature of the type II topoisomerase mechanism.

DNA Topoisomers Can Be Separated by Electrophoresis

Covalently closed, circular DNA molecules of the same length but of different linking numbers are called **DNA topoisomers.** Even though topoisomers have the same molecular weight, they can be separated from each other by electrophoresis through a gel of agarose (see Chapter 20 for an explanation of **gel electrophoresis**). The basis for this separation is that the greater the writhe, the more compact the shape of a cccDNA. Once again, think of how supercoiling a telephone cord causes it to become more compact. The more compact the DNA, the more easily (up to a point) it is able to migrate through the gel matrix (Fig. 6-26). Thus, a fully relaxed cccDNA migrates more slowly than a highly supercoiled topoisomer of the same circular DNA. Figure 6-27 shows a ladder of DNA topoisomers resolved by gel electrophoresis. Molecules in adjacent

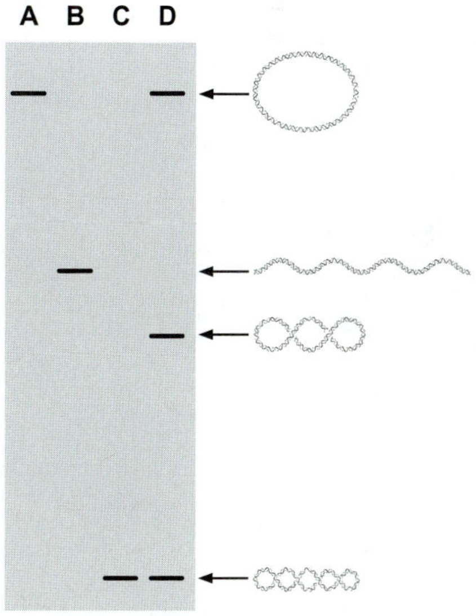

FIGURE 6-26 Schematic of electrophoretic separation of DNA topoisomers. Lane A represents relaxed or nicked circular DNA; lane B, linear DNA; lane C, highly supercoiled cccDNA; and lane D, a ladder of topoisomers.

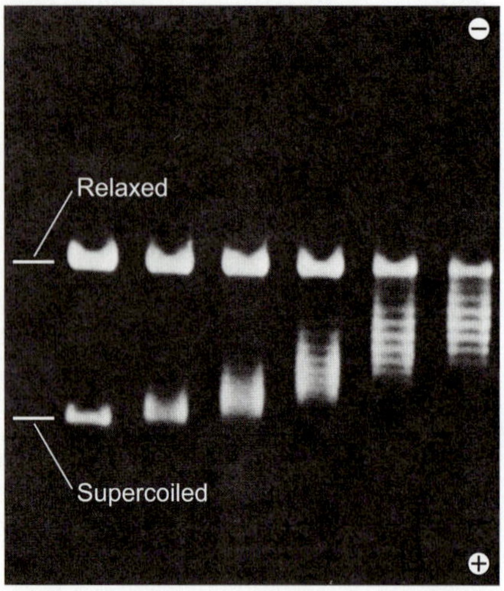

FIGURE 6-27 Separation of relaxed and supercoiled DNA by gel electrophoresis. Relaxed and supercoiled DNA topoisomers are resolved by gel electrophoresis. The speed with which the DNA molecules migrate increases as the number of superhelical turns increases. (Courtesy of J.C. Wang.)

rungs of the ladder differ from each other by a linking number difference of just 1. Obviously, electrophoretic mobility is highly sensitive to the topological state of DNA (see Box 6-3, Proving that DNA Has a Helical Periodicity of about 10.5 Base Pairs per Turn from the Topological Properties of DNA Rings).

Ethidium Ions Cause DNA to Unwind

Ethidium is a large, flat, multiringed cation. Its planar shape enables ethidium to slip, or intercalate, between the stacked base pairs of DNA (Fig. 6-28). Because it fluoresces when exposed to ultraviolet light, and because its fluorescence increases dramatically after intercalation, ethidium is used as a stain to visualize DNA.

When an ethidium ion intercalates between two base pairs, it causes the DNA to unwind by 26°, reducing the normal rotation per base pair from ~36° to ~10°. In other words, ethidium decreases the twist of DNA. Imagine the extreme case of a DNA molecule that has an ethidium ion between every base pair. Instead of ten base pairs per turn it would have 36! When ethidium binds to linear DNA or to a nicked circle, it simply causes the helical pitch to increase. But consider what happens when ethidium binds to covalently closed, circular DNA. The linking number of the cccDNA does not change (no covalent bonds are broken and resealed), but the twist decreases by 26° for each molecule of ethidium that has bound to the DNA. Because $Lk = Tw + Wr$, this decrease in Tw must be compensated for by a corresponding increase in Wr. If the circular DNA is initially negatively supercoiled (as is normally the case for circular DNAs isolated from cells), then the addition of ethidium will increase Wr. In other words, the addition of ethidium will relax the DNA. If enough ethidium is added, the negative supercoiling will be brought to 0, and if even more ethidium is added, Wr will increase above 0 and the DNA will become positively supercoiled.

Because the binding of ethidium increases Wr, its presence greatly affects the migration of cccDNA during gel electrophoresis. In the presence of nonsaturating amounts of ethidium, negatively supercoiled circular DNAs are more relaxed and migrate more slowly, whereas relaxed cccDNAs become positively supercoiled and migrate more rapidly.

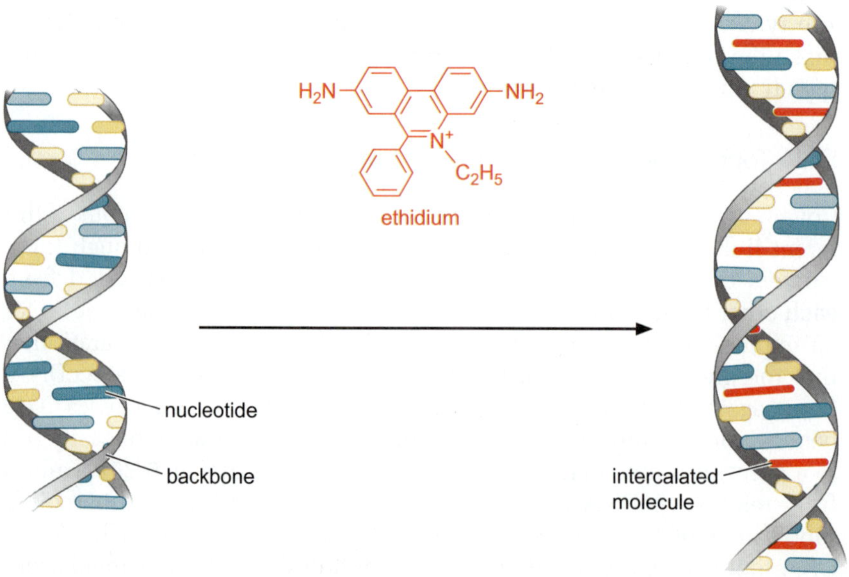

FIGURE 6-28 Intercalation of ethidium into DNA. Ethidium increases the spacing of successive base pairs, distorts the regular sugar–phosphate backbone, and decreases the twist of the helix.

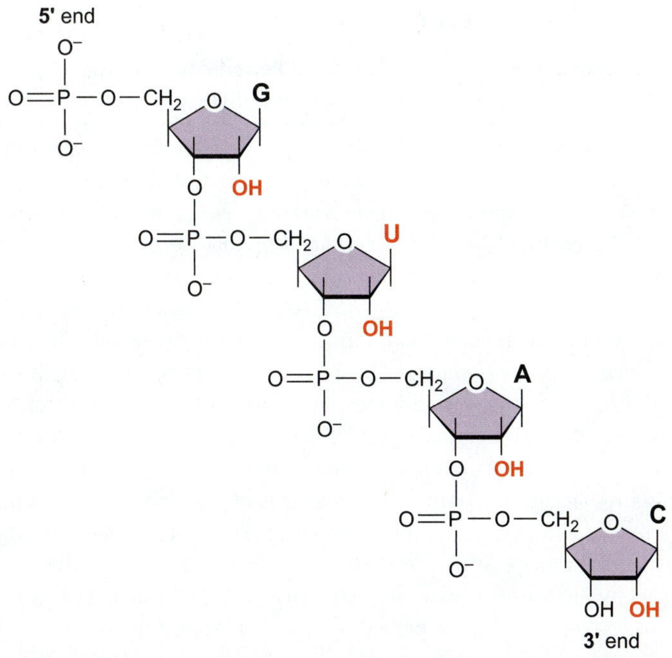

FIGURE 6-29 Structural features of RNA. The figure shows the structure of the backbone of RNA, composed of alternating phosphate and ribose moieties. The features of RNA that distinguish it from DNA are highlighted in red.

RNA STRUCTURE

RNA Contains Ribose and Uracil and Is Usually Single-Stranded

We now turn our attention to RNA, which differs from DNA in three respects (Fig. 6-29). First, the backbone of RNA contains ribose rather than 2′-deoxyribose. That is, ribose has a hydroxyl group at the 2′ position. Second, RNA contains **uracil** in place of thymine. Uracil has the same single-ringed structure as thymine, except that it lacks the **5 methyl** group. Thymine is in effect **5 methyl-uracil**. Third, RNA is usually found as a single polynucleotide chain. Except for the case of certain viruses, RNA is not the genetic material and does not need to be capable of serving as a template for its own replication. Rather, RNA functions as the intermediate, the mRNA, between the gene and the protein-synthesizing machinery. Another function of RNA is as an adaptor, the tRNA, between the codons in the mRNA and amino acids. RNA can also play a structural role, as in the case of the RNA components of the ribosome. Yet another role for RNA is as a regulatory molecule, which through sequence complementarity binds to, and interferes with the translation of, certain mRNAs. Finally, some RNAs (including one of the structural RNAs of the ribosome) are enzymes that catalyze essential reactions in the cell. In all of these cases, the RNA is copied as a single strand off only one of the two strands of the DNA template, and its complementary strand does not exist. RNA is capable of forming long double helices, but these are unusual in nature.

RNA Chains Fold Back on Themselves to Form Local Regions of Double Helix Similar to A-Form DNA

Despite being single-stranded, RNA molecules often exhibit a great deal of double-helical character (Fig. 6-30). This is because RNA chains

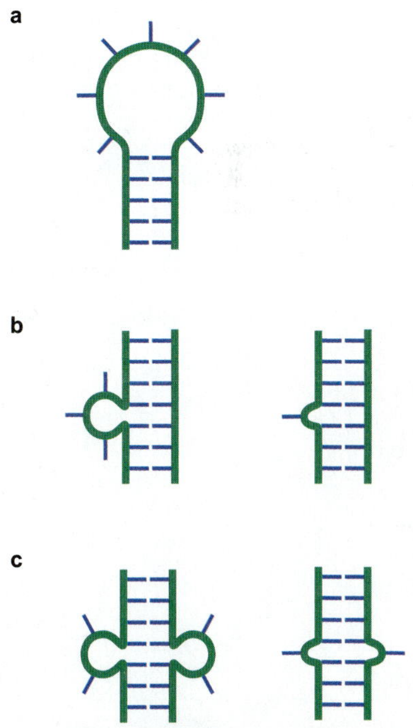

FIGURE 6-30 Double-helical characteristics of RNA. In an RNA molecule having regions of complementary sequences, the intervening (noncomplementary) stretches of RNA may become "looped out" to form one of the structures illustrated in the figure: (a) hairpin, (b) bulge, or (c) loop.

KEY EXPERIMENTS

Box 6-3 Proving that DNA Has a Helical Periodicity of about 10.5 Base Pairs per Turn from the Topological Properties of DNA Rings

The observation that DNA topoisomers can be separated from each other electrophoretically is the basis for a simple experiment that proves that DNA has a helical periodicity of about 10.5 base pairs (bp) per turn in solution. Consider three cccDNAs of sizes 3990, 3995, and 4011 bp that were relaxed to completion by treatment with type I topoisomerase. When subjected to electrophoresis through agarose, the 3990- and 4011-bp DNAs exhibit essentially identical mobilities. Because of thermal fluctuation, topoisomerase treatment actually generates a narrow spectrum of topoisomers, but for simplicity let us consider the mobility of only the most abundant topoisomer (that corresponding to the cccDNA in its most relaxed state). The mobilities of the most abundant topoisomers for the 3990- and 4011-bp DNAs are indistinguishable because the 21-bp difference between them is negligible compared to the sizes of the rings. The most abundant topoisomer for the 3995-bp ring, however, is found to migrate slightly more rapidly than the other two rings even though it is only 5 bp larger than the 3990-bp ring. How are we to explain this anomaly? The 3990- and 4011-bp rings in their most relaxed states are expected to have linking numbers equal to Lk^O, that is, 380 in the case of the 3990-bp ring (dividing the size by 10.5 bp) and 382 in the case of the 4011-bp ring. Because Lk is equal to Lk^O, the linking difference ($\Delta Lk = Lk - Lk^O$) in both cases is 0 and there is no writhe. But because the linking number must be an integer, the most relaxed state for the 3995-bp ring would be either of two topoisomers having linking numbers of 380 or 381. However, Lk^O for the 3995-bp ring is 380.5. Thus, even in its most relaxed state, a covalently closed circle of 3995 bp would necessarily have about half a unit of writhe (its linking difference would be 0.5), and hence it would migrate more rapidly than the 3990- and 4011-bp circles. In other words, to explain how rings that differ in length by 21 bp (two turns of the helix) have the same mobility, whereas a ring that differs in length by only 5 bp (about half a helical turn) exhibits a different mobility, we must conclude that DNA in solution has a helical periodicity of about 10.5 bp per turn.

frequently fold back on themselves to form base-paired segments between short stretches of complementary sequences. If the two stretches of complementary sequence are near each other, the RNA may adopt one of various **stem-loop structures** in which the intervening RNA is looped out from the end of the double-helical segment as in a hairpin, a bulge, or a simple loop.

The stability of such stem-loop structures is in some instances enhanced by the special properties of the loop. For example, a stem-loop with the "tetraloop" sequence UUCG is unexpectedly stable because of special base-stacking interactions in the loop (Fig. 6-31). Base pairing can also take place between sequences that are not contiguous to form complex structures aptly named **pseudoknots** (Fig. 6-32). The regions of base pairing in RNA can be a regular double helix or they can contain discontinuities, such as noncomplementary nucleotides that bulge out from the helix.

A feature of RNA that adds to its propensity to form double-helical structures is an additional, non-Watson–Crick base pair. This is the G:U base pair, which has hydrogen bonds between N3 of uracil and the car-

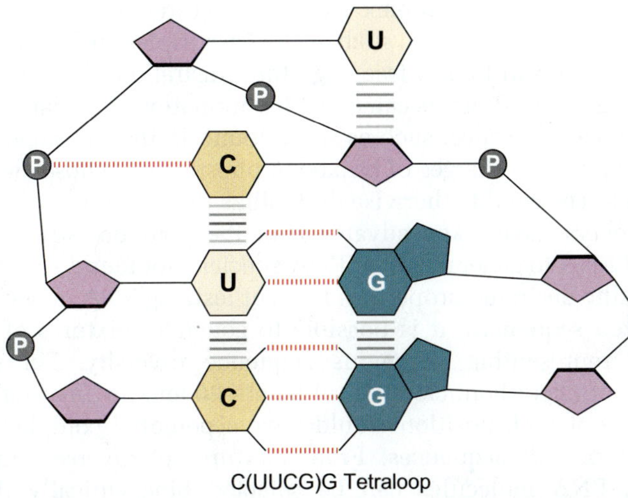

FIGURE 6-31 **Tetraloop.** Base-stacking interactions promote and stabilize the tetraloop structure. The gray circles between the riboses shown in purple represent the phosphate moieties of the RNA backbone. Horizontal lines represent base-stacking interactions.

bonyl on C6 of guanine and between the carbonyl on C2 of uracil and N1 of guanine (Fig. 6-33). Because G:U base pairs can occur as well as the four conventional, Watson–Crick base pairs, RNA chains have an enhanced capacity for self-complementarity. Thus, RNA frequently exhibits local regions of base pairing but not the long-range, regular helicity of DNA.

The presence of 2′-hydroxyls in the RNA backbone prevents RNA from adopting a B-form helix. Rather, double-helical RNA resembles the A-form structure of DNA. As such, the minor groove is wide and shallow, and hence accessible, but recall that the minor groove offers little sequence-specific information. Meanwhile, the major groove is so narrow and deep that it is not very accessible to amino acid side chains from interacting proteins. Thus, the RNA double helix is quite distinct from the DNA double helix in its detailed atomic structure and less well suited for sequence-specific interactions with proteins (although some proteins do bind to RNA in a sequence-specific manner).

RNA Can Fold Up into Complex Tertiary Structures

Freed of the constraint of forming long-range regular helices, RNA can adopt a wealth of tertiary structures. This is because RNA has enormous rotational freedom in the backbone of its non-base-paired regions. Thus,

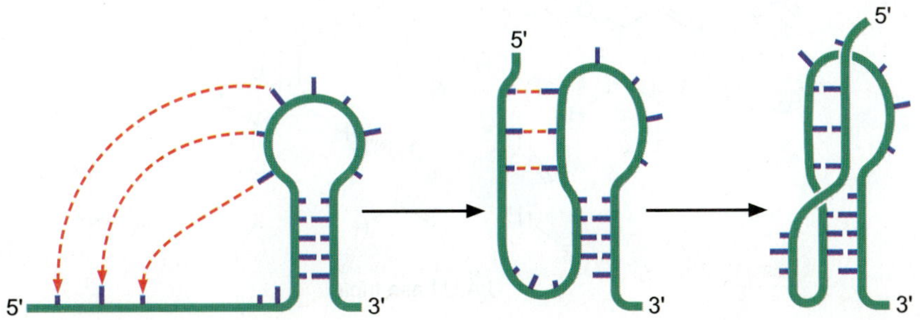

FIGURE 6-32 **Pseudoknot.** The pseudoknot structure is formed by base pairing between noncontiguous complementary sequences.

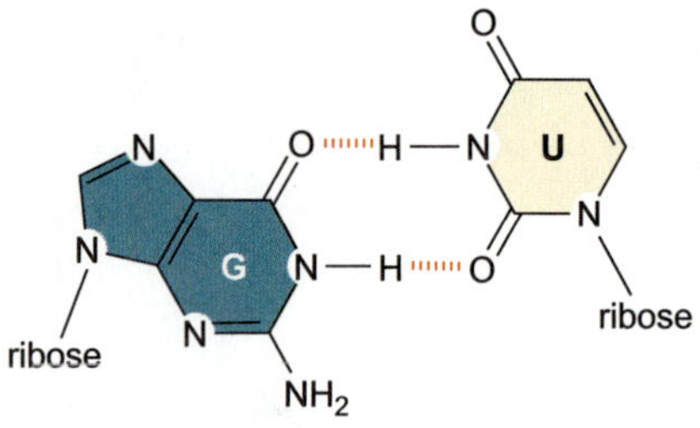

FIGURE 6-33 G:U base pair. The structure shows hydrogen bonds that allow base pairing to occur between guanine and uracil.

RNA can fold up into complex tertiary structures frequently involving unconventional base pairing, such as the base triples and base–backbone interactions seen in tRNAs (see, e.g., the illustration of the U:A:U base triple in Fig. 6-34). Proteins can assist the formation of tertiary structures by large RNA molecules, such as those found in the ribosome. Proteins shield the negative charges of backbone phosphates, whose electrostatic repulsive forces would otherwise destabilize the structure.

Researchers have taken advantage of the potential structural complexity of RNA to generate novel RNA species (not found in nature) that have specific desirable properties. By synthesizing RNA molecules with randomized sequences, it is possible to generate mixtures of oligonucleotides representing enormous sequence diversity. For example, a mixture of oligoribonucleotides of length 20 and having four possible nucleotides at each position would have a potential complexity of 4^{20} sequences or 10^{12} sequences! From mixtures of diverse oligoribonucleotides, RNA molecules can be selected biochemically that have particular properties, such as an affinity for a specific small molecule.

Some RNAs Are Enzymes

It was widely believed for many years that only proteins could be enzymes. An enzyme must be able to bind a substrate, carry out a chemical reaction, release the product, and repeat this sequence of events many times. Proteins are well-suited to this task because they are composed of many different kinds of amino acids (20) and they can fold into complex tertiary structures with binding pockets for the substrate and small molecule co-factors and an active site for catalysis. Now we know that RNAs, which as we have seen can similarly adopt complex tertiary structures, can also be biological catalysts (see Interactive Animation 6-3). Such RNA enzymes are known as **ribozymes,** and they exhibit many of the features of a classical enzyme, such as an active site, a binding site for a substrate, and a binding site for a co-factor, such as a metal ion.

WEB ANIMATION

One of the first ribozymes to be discovered was **RNase P,** a ribonuclease that is involved in generating tRNA molecules from larger, precursor RNAs. RNase P is composed of both RNA and protein; however, the RNA moiety alone is the catalyst. The protein moiety of RNase P facilitates the reaction by shielding the negative charges on the RNA so that it can bind effectively to its negatively charged substrate. The RNA moiety is able to catalyze cleavage of the tRNA precursor in the absence of the protein if a small, positively charged counter ion, such as the peptide spermidine, is used to shield the repulsive, negative charges. Other ribozymes carry out transesterification reactions involved in the removal of intervening sequences known as **introns** from precursors to certain mRNAs, tRNAs, and ribosomal RNAs in a process known as **RNA splicing** (see Chapter 13).

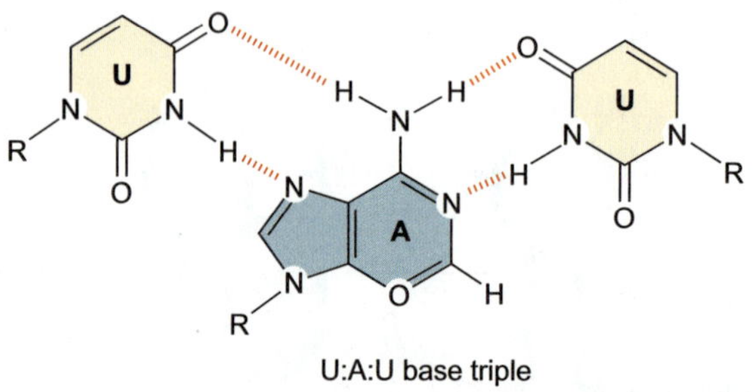

FIGURE 6-34 U:A:U base triple. The structure shows one example of hydrogen bonding that allows unusual triple base pairing.

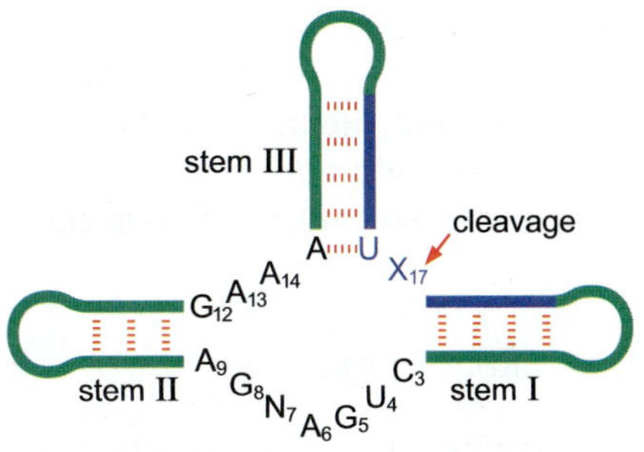

FIGURE 6-35 Secondary structure of the hammerhead ribozyme. The molecule is shown with the two halves of each stem connected with a loop, but none of the three stems need be a loop: in fact, in the viroid, the two halves of stem III are not joined with a loop. Watson–Crick base-pair interactions are shown in red; the scissile bond is shown by a red arrow; approximate minimal substrate strands are labeled in blue; (U) uracil; (A) adenine; (C) cytosine; (G) guanine. (Redrawn, with permission, from McKay D.B. and Wedekind J.E. 1999. *The RNA world,* 2nd ed. [ed. R.F. Gesteland et al.], p. 267, Fig. 1A. © Cold Spring Harbor Laboratory Press.)

The Hammerhead Ribozyme Cleaves RNA by the Formation of a 2′, 3′ Cyclic Phosphate

WEB STRUCTURAL TUTORIAL

Before concluding our discussion of RNA, let us look in more detail at the structure and function of one particular ribozyme, the **hammerhead** (see Structural Tutorial 6-2). The hammerhead is a sequence-specific ribonuclease that is found in certain infectious RNA agents of plants known as **viroids,** which depend on self-cleavage to propagate. When the viroid replicates, it produces multiple copies of itself in one continuous RNA chain. Single viroids arise by cleavage, and this cleavage reaction is carried out by the RNA sequence around the junction. One such self-cleaving sequence is called the hammerhead because of the shape of its secondary structure, which consists of three base-paired stems (I, II, and III) surrounding a core of noncomplementary nucleotides required for catalysis (Fig. 6-35). The tertiary structure of the hammerhead, however, looks more like a wishbone (Fig. 6-36).

To understand how the hammerhead works, let us first look at how RNA undergoes hydrolysis under alkaline conditions. At high pH, the 2′-hydroxyl of the ribose in the RNA backbone can become deprotonated, and the resulting negatively charged oxygen can attack the scissile phosphate at the 3′ position of the same ribose. This reaction breaks the RNA chain, producing a 2′, 3′ cyclic phosphate and a free 5′-hydroxyl. Each ribose in an RNA chain can undergo this reaction, completely cleaving the parent molecule into nucleotides. (Why is DNA not similarly susceptible to alkaline hydrolysis?) Many protein ribonucleases also cleave their RNA substrates via the formation of a 2′, 3′ cyclic phosphate. Working at normal cellular pH, these protein enzymes use a metal ion, bound at their active site, to activate the 2′-hydroxyl of the RNA. The hammerhead also cleaves RNA via the formation of a 2′, 3′ cyclic phosphate, but the mechanism of the cleavage reaction is not yet understood.

Because the normal reaction of the hammerhead is self-cleavage, it is not really a catalyst; each molecule normally promotes a reaction one time only, thus having a turnover number of 1. But the hammerhead can be engineered to function as a true ribozyme by dividing the

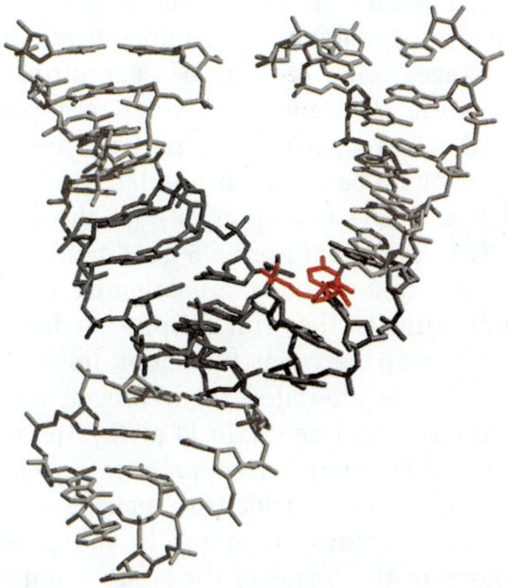

FIGURE 6-36 Tertiary structure of the hammerhead ribozyme. This view of the refined hammerhead ribozyme structure shows stem I (top right), stem II (top left), and stem loop III (bottom). The site of cleavage at cytosine 17 (see Fig. 6-35) is shown in red. (Scott W.G. et al. 1995. *Cell* 81: 991.) Image prepared with MolScript, BobScript, and Raster 3D.

molecule into two portions—one, the ribozyme, that contains the catalytic core and the other, the substrate, that contains the cleavage site. The substrate binds to the ribozyme at stems I and III (Fig. 6-35). After cleavage, the substrate is released and replaced by a fresh uncut substrate, thereby allowing repeated rounds of cleavage.

Did Life Evolve from an RNA World?

The discovery of ribozymes has profoundly altered our view of how life might have evolved. We can now imagine that there was a primitive form of life based entirely on RNA. In this world, RNA would have functioned as the genetic material and as the enzymatic machinery. This RNA world would have preceded life as we know it today, in which information transfer is based on DNA, RNA, and protein. A hint that the protein world might have arisen from an RNA world is the discovery that the component in the ribosome that is responsible for the formation of the peptide bond, the peptidyl transferase, is an RNA molecule (see Chapter 14). Unlike RNase P, the hammerhead, and other previously known ribozymes that act on phosphorous centers, the peptidyl transferase acts on a carbon center to create the peptide bond. It thus links RNA chemistry to the most fundamental reaction in the protein world, peptide bond formation. Perhaps then the ribosome ribozyme is a relic of an earlier form of life in which all enzymes were RNAs.

SUMMARY

DNA is usually in the form of a right-handed double helix. The helix consists of two polydeoxynucleotide chains. Each chain is an alternating polymer of deoxyribose sugars and phosphates that are joined together via phosphodiester linkages. One of four bases protrudes from each sugar: adenine and guanine, which are purines, and thymine and cytosine, which are pyrimidines. Although the sugar–phosphate backbone is regular, the order of bases is irregular, and this is responsible for the information content of DNA. Each chain has a 5′ to 3′ polarity, and the two chains of the double helix are oriented in an antiparallel manner—that is, they run in opposite directions.

Pairing between the bases holds the chains together. Pairing is mediated by hydrogen bonds and is specific: Adenine on one chain is always paired with thymine on the other chain, whereas guanine is always paired with cytosine. This strict base pairing reflects the fixed locations of hydrogen atoms in the purine and pyrimidine bases in the forms of those bases found in DNA. Adenine and cytosine almost always exist in the amino as opposed to the imino tautomeric forms, whereas guanine and thymine almost always exist in the keto as opposed to enol forms. The complementarity between the bases on the two strands gives DNA its self-coding character.

The two strands of the double helix fall apart (denature) upon exposure to high temperature, extremes of pH, or any agent that causes the breakage of hydrogen bonds. Upon slow return to normal cellular conditions, the denatured single strands can specifically reassociate to biologically active double helices (renature or anneal).

DNA in solution has a helical periodicity of about 10.5 base pairs per turn of the helix. The stacking of base pairs upon each other creates a helix with two grooves. Because the sugars protrude from the bases at an angle of about 120°, the grooves are unequal in size. The edges of each base pair are exposed in the grooves, creating a pattern of hydrogen-bond donors and acceptors and of van der Waals surfaces that identifies the base pair. The wider—or *major*—groove is richer in chemical information than the narrow—or *minor*—groove and is more important for recognition by nucleotide sequence-specific binding proteins.

Almost all cellular DNAs are extremely long molecules, with only one DNA molecule within a given chromosome. Eukaryotic cells accommodate this extreme length in part by wrapping the DNA around protein particles known as nucleosomes. Most DNA molecules are linear but some DNAs are circles, as is often the case for the chromosomes of prokaryotes and for certain viruses.

DNA is flexible. Unless the molecule is topologically constrained, it can freely rotate to accommodate changes in the number of times the two strands twist about each other. DNA is topologically constrained when it is in the form of a covalently closed circle, or when it is entrained in chromatin. The linking number is an invariant topological property of covalently closed, circular DNA. It is the number of times one strand would have to be passed through the other strand in order to separate the two circular strands. The linking number is the sum of two interconvertible geometric properties: twist, which is the number of times the two strands are wrapped around each other; and the writhing number, which is the number of times the long axis of the DNA crosses over itself in space. DNA is relaxed under physiological conditions when it has about 10.5 base pairs per turn and is free of writhe. If the linking number is decreased, then the DNA becomes torsionally stressed, and it is said to be negatively supercoiled. DNA in cells is usually negatively supercoiled by about 6%.

The left-handed wrapping of DNA around nucleosomes introduces negative supercoiling in eukaryotes. In prokaryotes, which lack histones, the enzyme DNA gyrase is responsible for generating negative supercoils. DNA gyrase is a member of the type II family of topoisomerases. These enzymes change the linking number of DNA in steps of two by making a transient break in the double helix and passing a region of duplex DNA through the break. Some type II topoisomerases relax supercoiled DNA, whereas DNA gyrase generates negative supercoils. Type I topoisomerases also relax supercoiled DNAs but do so in steps of one in which one DNA strand is passed through a transient nick in the other strand.

RNA differs from DNA in the following ways: Its backbone contains ribose rather than 2′-deoxyribose; it contains the pyrimidine uracil in place of thymine; and it usually exists as a single polynucleotide chain, without a complementary chain. As a consequence of being a single strand, RNA can fold back on itself to form short stretches of double helix between regions that are complementary to each other. RNA allows a greater range of base pairing than does DNA. Thus, as well as A:U and C:G pairing, U can also pair with G. This capacity to form a non-Watson–Crick base pair adds to the propensity of RNA to form double-helical segments. Freed of the constraint of forming long-range regular helices, RNA can form complex tertiary structures, which are often based on unconventional interactions between bases and the sugar–phosphate backbone.

Some RNAs act as enzymes—they catalyze chemical reactions in the cell and in vitro. These RNA enzymes are known as ribozymes. Most ribozymes act on phosphorous centers, as in the case of the ribonuclease RNase P. RNase P is composed of protein and RNA, but it is the RNA moiety that is the catalyst. The hammerhead is a self-cleaving RNA, which cuts the RNA backbone via the formation of a 2′, 3′ cyclic phosphate. Peptidyl transferase is an example of a ribozyme that acts on a carbon center. This ribozyme, which is responsible for the formation of the peptide bond, is one of the RNA components of the ribosome. The discovery of RNA enzymes that can act on phosphorus or carbon centers suggests that life might have evolved from a primitive form in which RNA functioned both as the genetic material and as the enzymatic machinery.

BIBLIOGRAPHY

Books

Bloomfield V.A., Crothers D.M., Tinoco I., Jr., and Heast J.E. 2000. *Nucleic acids: Structures, properties, and functions.* University Science Books, Sausalito, California.

Cold Spring Harbor Symposium on Quantitative Biology. 1982. Volume 47: Structures of DNA. Cold Spring Harbor Laboratory Press, Cold Spring Harbor, New York.

Gesteland R.F., Cech T.R., and Atkin J.F., eds. 1999. *The RNA world,* 3rd ed. Cold Spring Harbor Laboratory Press, Cold Spring Harbor, New York.

DNA Structure

Chambers D.A., ed. 1995. DNA: The double-helix—Perspective and prospective at forty years. *Ann. N.Y. Acad. Sci.* **758:** 1–472.

Dickerson R.E. 1983. The DNA helix and how it is read. *Sci. Am.* **249:** 94–111.

Franklin R.E. and Gosling R.G. 1953. Molecular configuration in sodium thymonucleate. *Nature* **171:** 740–741.

Roberts R.J. 1995. On base flipping. *Cell* **82:** 9–12.

Watson J.D. and Crick F.H.C. 1953. Molecular structure of nucleic acids: A structure for deoxyribonucleic acids. *Nature* **171:** 737–738.

———. 1953. Genetical implications of the structure of deoxyribonucleic acids. *Nature* **171:** 964–967.

Wilkins M.H.F., Stokes A.R., and Wilson H.R. 1953. Molecular structure of deoxypentose nucleic acids. *Nature* **171:** 738–740.

DNA Topology

Dröge P. and Cozzarelli N.R. 1992. Topological structure of DNA knots and catenanes. *Methods Enzymol.* **212:** 120–130.

Wang J.C. 2002. Cellular roles of DNA topoisomerases: A molecular perspective. *Nat. Rev. Mol. Cell Biol.* **3:** 430–440.

RNA Structure

Darnell J.E., Jr. 1985. RNA. *Sci. Am.* **253:** 68–78.

Doudna J.A. and Cech T.R. 2002. The chemical repertoire of natural ribozymes. *Nature* **418:** 222–228.

Doudna J.A. and Lorsch. J.R. 2005. Ribozyme catalysis: Not different, just worse. *Nat. Struct. Mol. Biol.* **15:** 394–402.

Nelson J.A. and Uhlenbeck O.C. 2006. When to believe what you see. *Mol. Cell* **23:** 447–450.

CHAPTER

7

Genome Structure, Chromatin, and the Nucleosome

In Chapter 6, we considered the structure of DNA in isolation. Within the cell, however, DNA is associated with proteins, and each DNA molecule and its associated protein is called a **chromosome.** This organization holds true for prokaryotic and eukaryotic cells and even for viruses. Packaging of the DNA into chromosomes serves several important functions. First, the chromosome is a compact form of the DNA that readily fits inside the cell. Second, packaging the DNA into chromosomes serves to protect the DNA from damage. Completely naked DNA molecules are relatively unstable in cells. In contrast, chromosomal DNA is extremely stable. Third, only DNA packaged into a chromosome can be transmitted efficiently to both daughter cells when a cell divides. Finally, the chromosome confers an overall organization to each molecule of DNA. This organization regulates gene expression as well as the recombination between parental chromosomes that generates the diversity observed among different individuals of any organism.

Half of the molecular mass of a eukaryotic chromosome is protein. In eukaryotic cells, a given region of DNA with its associated proteins is called **chromatin**, and the majority of the associated proteins are small, basic proteins called **histones**. Although not nearly as abundant, other proteins, frequently referred to as the **nonhistone proteins**, are also associated with eukaryotic chromosomes. These proteins include the numerous DNA-binding proteins that regulate the transcription, replication, repair, and recombination of cellular DNA. Each of these topics is discussed in more detail in the next five chapters.

The protein component of chromatin performs another essential function: compacting the DNA. The following calculation makes the importance of this function clear. A human cell contains 3×10^9 bp per haploid set of chromosomes. As we learned in Chapter 6, the average thickness of each base pair (the "rise") is 3.4 Å. Therefore, if the DNA molecules in a haploid set of chromosomes were laid out end to end, the total length of DNA would be $\sim 10^{10}$ Å, or 1 m! For a diploid cell (as human cells typically are), this length is doubled to 2 m. Because the diameter of a typical human cell nucleus is only 10–15 μm, it is obvious that the DNA must be compacted by many orders of magnitude to fit in such a small space. How is this achieved?

Most compaction in human cells (and all other eukaryotic cells) is the result of the regular association of DNA with histones to form structures called **nucleosomes**. The formation of nucleosomes is the first step in a process that allows the DNA to be folded into much more compact structures that reduce the linear length by as much as

10,000-fold. Compacting the DNA does not come without cost. Association of the DNA with histones and other packaging proteins limits the accessibility of the DNA. This reduced accessibility can interfere with the proteins that direct the replication, repair, recombination, and—perhaps most significantly—transcription of the DNA. Eukaryotic cells exploit the inhibitory properties of chromatin to regulate gene expression and many other events involving DNA. Alterations to individual nucleosomes allow specific regions of the chromosomal DNA to interact with other proteins. These alterations are mediated by enzymes that modify and remodel nucleosomes. These processes are both dynamic and local, allowing enzymes and regulatory proteins access to different regions of the chromosome at different times. Therefore, understanding the structure of nucleosomes and the regulation of their association with DNA is critical to understanding the regulation of most events involving DNA in eukaryotic cells.

Although prokaryotic cells typically have smaller genomes, the need to compact their DNA is still substantial. *Escherichia coli* must pack its ~1-mm chromosome into a cell that is only 1 μm in length. It is less clear how prokaryotic DNA is compacted. Bacteria have no histones or nucleosomes, for example, but they do have other small basic proteins that may serve similar functions. In this chapter, we focus on the better understood chromosomes and chromatin of eukaryotic cells. We first consider the underlying DNA sequences of chromosomes from different organisms, focusing in particular on the change in protein-coding content. We then discuss the overall mechanisms that ensure that chromosomes are accurately transmitted as cells divide. The remainder of the chapter focuses on the structure and regulation of eukaryotic chromatin and its fundamental building block, the nucleosome.

GENOME SEQUENCE AND CHROMOSOME DIVERSITY

Before we discuss the structure of chromosomes in detail, it is important to understand the features of the DNA molecules that are their foundation. The sequencing of the genomes of thousands of organisms has provided a wealth of information concerning the makeup of chromosomal DNAs and how their characteristics have changed as organisms have increased in complexity.

Chromosomes Can Be Circular or Linear

The traditional view is that prokaryotic cells have a single circular chromosome and eukaryotic cells have multiple linear chromosomes (Table 7-1). As more prokaryotic organisms have been studied, this view has been challenged. Although the most studied prokaryotes (such as *E. coli* and *Bacillus subtilis*) do indeed have single circular chromosomes, there are now numerous examples of prokaryotic cells that have multiple chromosomes, linear chromosomes, or even both. In contrast, all eukaryotic cells have multiple linear chromosomes. Depending on the eukaryotic organism, the number of chromosomes typically varies from 2 to less than 50, but in rare instances can reach thousands (e.g., in the macronucleus of the protozoa *Tetrahymena,* Table 7-1).

Circular and linear chromosomes each pose specific challenges that must be overcome for maintenance and replication of the

TABLE 7-1 Variation in Chromosome Makeup in Different Organisms

Species		Number of Chromosomes	Chromosome Copy Number	Form of Chromosome(s)	Genome Size (Mb)
Prokaryotes					
Mycoplasma genitalium		1	1	Circular	0.58
Escherichia coli K-12		1	1	Circular	4.6
Agrobacterium tumefaciens		4	1	3 circular 1 linear	5.67
Sinorhizobium meliloti		3	1	Circular	6.7
Eukaryotes					
Saccharomyces cerevisiae (budding yeast)		16	1 or 2	Linear	12.1
Schizosaccharomyces pombe (fission yeast)		3	1 or 2	Linear	12.5
Caenorhabditis elegans (roundworm)		6	2	Linear	97
Arabidopsis thaliana (weed)		5	2	Linear	125
Drosophila melanogaster (fruit fly)		4	2	Linear	180
Tetrahymena thermophilus (protozoa)	Micronucleus	5	2	Linear	125
	Macronucleus	225	10–10,000	Linear	
Fugu rubripes (fish)		22	2	Linear	393
Mus musculus (mouse)		19 + X and Y	2	Linear	2,600
Homo sapiens		22 + X and Y	2	Linear	3,200

genome. Circular chromosomes require topoisomerases to separate the daughter molecules after they are replicated. Without these enzymes, the two daughter molecules would remain interlocked, or catenated, with each other after replication (see Fig. 6-23). In contrast, the ends of the linear eukaryotic chromosomes have to be protected from enzymes that normally degrade DNA ends and present a different set of difficulties during DNA replication, as we shall see in Chapter 8.

Every Cell Maintains a Characteristic Number of Chromosomes

Prokaryotic cells typically have only one *complete* copy of their chromosome(s) that is packaged into a structure called the **nucleoid** (Fig. 7-1b). When prokaryotic cells are dividing rapidly, however, portions of the chromosome in the process of replicating are present in two and sometimes even four copies. Prokaryotes also frequently carry one or more smaller independent circular DNAs, called **plasmids.** Unlike the larger chromosomal DNA, plasmids typically are not essential for bacterial growth. Instead, they carry genes that confer desirable traits to the bacteria, such as antibiotic resistance. Also, unlike chromosomal DNA, plasmids are often present in many complete copies per cell.

The majority of eukaryotic cells are **diploid**; that is, they contain two copies of each chromosome (see Fig. 7-1c). The two copies of a given chromosome are called **homologs**; one is derived from each parent. But, not all cells in a eukaryotic organism are diploid; a subset of

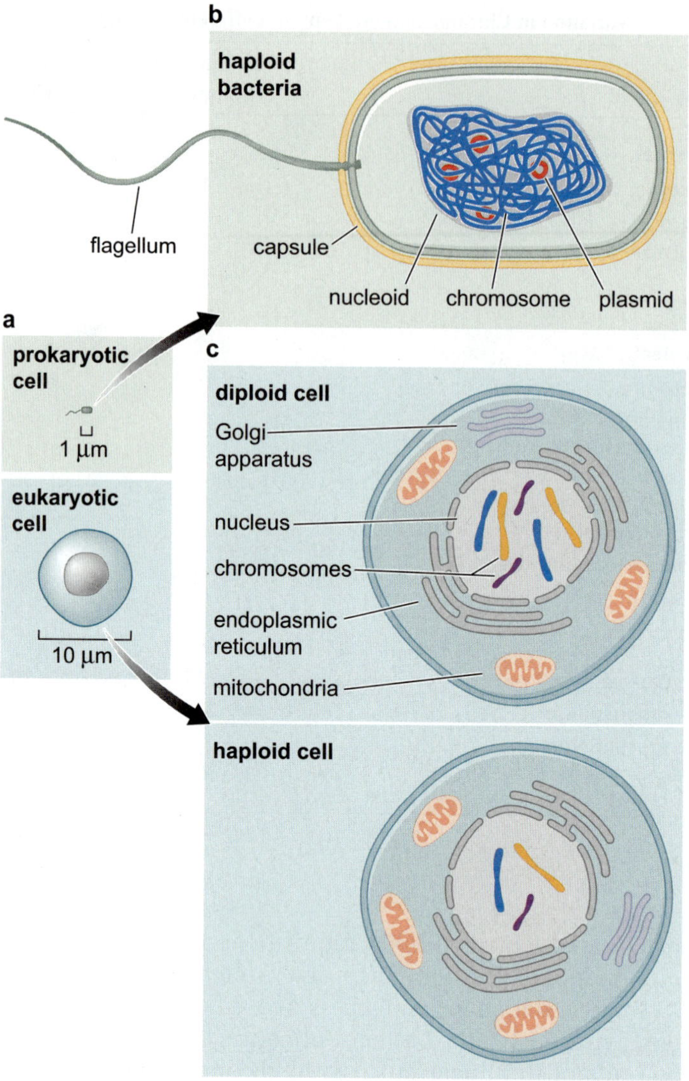

FIGURE 7-1 Comparison of typical prokaryotic and eukaryotic cells. (a) The diameter of a typical eukaryotic cell is ~10 μm. The typical prokaryotic cell is ~1 μm long. (b) Prokaryotic chromosomal DNA is located in the nucleoid and occupies a substantial portion of the internal region of the cell. Unlike the eukaryotic nucleus, the nucleoid is not separated from the remainder of the cell by a membrane. Plasmid DNA is shown in red. (c) Eukaryotic chromosomes are located in the membrane-bound nucleus. Haploid (one copy) and diploid (two copies) cells are distinguished by the number of copies of each chromosome present in the nucleus. (Adapted, with permission, from Brown T.A. 2002. *Genomes,* 2nd ed., p. 32, Fig. 2.1. © BIOS Scientific Publishers by permission of Taylor & Francis.)

eukaryotic cells are either haploid or polyploid. **Haploid** cells contain a single copy of each chromosome and are involved in sexual reproduction (e.g., sperm and eggs are haploid cells). **Polyploid** cells have more than two copies of each chromosome. Indeed, some organisms maintain the majority of their adult cells in a polyploid state. In extreme cases, there can be hundreds or even thousands of copies of each chromosome. This type of global genome amplification allows a cell to generate larger amounts of RNA and, in turn, protein. For example, megakaryocytes are specialized polyploid cells (~28 copies of each chromosome) that produce thousands of platelets that lack chromosomes but are an essential component of human blood (there are ~200,000 platelets per milliliter of blood). By becoming polyploid, megakaryocytes can maintain the very high levels of metabolism necessary to produce large numbers of platelets. The segregation of such a large number of chromosomes is difficult; therefore, polyploid cells have almost always stopped dividing. No matter the number, eukaryotic chromosomes are always contained within a membrane-bound organelle called the **nucleus** (see Fig. 7-1c).

Genome Size Is Related to the Complexity of the Organism

Genome size (the length of DNA associated with one haploid complement of chromosomes) varies substantially between different organisms (Table 7-2). Because more genes are required to direct the formation of more complex organisms (at least when comparing bacteria, single-cell eukaryotes, and multicellular eukaryotes; see Chapter 19), it is not surprising that genome size is roughly correlated with an organism's apparent complexity. Thus, prokaryotic cells typically have genomes smaller than 10 Mb. The genomes of single-cell eukaryotes are typically less than 50 Mb, although some complex protozoans can have genomes greater than 200 Mb. Multicellular organisms have even larger genomes that can reach sizes greater than 100,000 Mb.

Although there is a rough correlation between genome size and organism complexity, it is far from perfect. Many organisms of apparently similar complexities have very different genome sizes: a fruit fly has a genome ~25 times smaller than a locust, and the rice genome is ~40 times smaller than wheat (see Table 7-2). These examples point out that the number of genes rather than genome size is more closely related to organism complexity. This becomes clear when we examine the relative gene densities of different genomes.

TABLE 7-2 Comparison of the Gene Density in Different Organisms' Genomes

Species	Genome Size (Mb)	Approximate Number of Genes	Gene Density (Genes/Mb)
Prokaryotes (bacteria)			
Mycoplasma genitalium	0.58	500	860
Streptococcus pneumoniae	2.2	2,300	1,060
Escherichia coli K-12	4.6	4,400	950
Agrobacterium tumefaciens	5.7	5,400	960
Sinorhizobium meliloti	6.7	6,200	930
Eukaryotes (animals)			
Fungi			
Saccharomyces cerevisiae	12	5,800	480
Schizosaccharomyces pombe	12	4,900	410
Protozoa			
Tetrahymena thermophila	125	27,000	220
Invertebrates			
Caenorhabditis elegans	103	20,000	190
Drosophila melanogaster	180	14,700	82
Ciona intestinalis	160	16,000	100
Locusta migratoria	5,000	nd	nd
Vertebrates			
Fugu rubripes (pufferfish)	393	22,000	56
Homo sapiens	3,200	20,000	6.25
Mus musculus (mouse)	2,600	22,000	8.5
Plants			
Arabidopsis thaliana	120	26,500	220
Oryza sativa (rice)	430	~45,000	~100
Zea mays (corn)	2,200	>45,000	>20
Triticum aestivum (wheat)	16,000	nd	nd
Fritillaria assyriaca (tulip)	~120,000	nd	nd

nd, not determined.

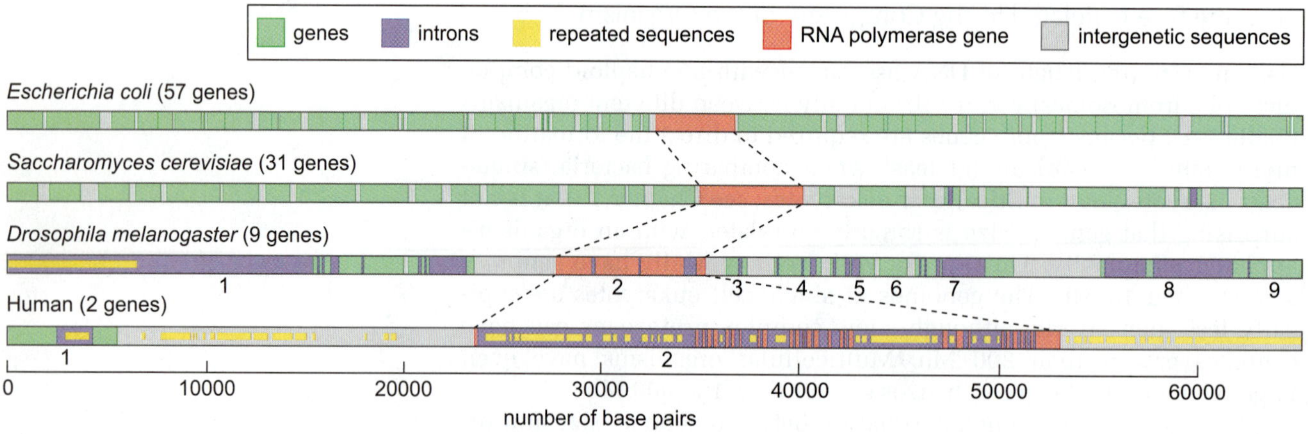

FIGURE 7-2 Comparison of chromosomal gene density for different organisms. A 65-kb region of DNA including the gene for the largest subunit of RNA polymerase (RNA polymerase II for the eukaryotic cells) is illustrated for each organism. In each case, the RNA polymerase encoding DNA is indicated in red. Coding DNA for other genes is indicated in green, intron DNA in purple, repeated DNA in yellow, and unique intergenic DNA in gray. Note how the number of genes included in the 65-kb region decreases as organism complexity increases.

The *E. coli* Genome Is Composed Almost Entirely of Genes

The great majority of the single chromosome of the bacteria *E. coli* encodes proteins or structural RNAs (Fig. 7-2). The majority of the noncoding sequences are dedicated to regulating gene transcription (as we shall see in Chapter 16). Because a single site of transcription initiation is often used to control the expression of several genes, even these regulatory regions are kept to a minimum in the genome. One critical element of the *E. coli* genome is not a gene or a sequence that regulates gene expression. Instead, the *E. coli* origin of replication is dedicated to directing the assembly of the replication machinery (as we shall discuss in Chapter 8). Despite its important role, this region is still very small, occupying only a few hundred base pairs of the 4.6-Mb *E. coli* genome.

More Complex Organisms Have Decreased Gene Density

What explains the dramatically different genome sizes of organisms of apparently similar complexity (such as the fruit fly and locust)? The differences are largely related to gene density. One simple measure of gene density is the average number of genes per megabase of genomic DNA. For example, if an organism has 5000 genes and a genome size of 50 Mb, then the gene density for that organism is 100 genes/Mb. When the gene densities of different organisms are compared, it becomes clear that different organisms use the gene-encoding potential of DNA with varying efficiencies. There is a roughly inverse correlation between organism complexity and gene density; the less complex the organism, the higher the gene density. For example, the highest gene densities are found for viruses that in some instances use both strands of the DNA to encode overlapping genes.

Although overlapping genes are rare, bacterial gene density is consistently near 1000 genes/Mb.

Gene density in eukaryotic organisms is consistently lower and more variable than in their prokaryotic counterparts (see Table 7-2). Among eukaryotes, there is a general trend for gene density to decrease with increasing organism complexity. The simple unicellular eukaryote *Saccharomyces cerevisiae* has a gene density close to that of prokaryotes (~500 genes/Mb). In contrast, the human genome is estimated to have a 50-fold lower gene density. In Figure 7-2, the amount of DNA sequence devoted to the expression of a related gene conserved across all organisms (the large subunit of RNA polymerase) is compared, illustrating the vast differences in gene size. Organisms with much larger genomes than humans are likely to have much lower gene densities. What is responsible for this reduction in gene density?

Genes Make Up Only a Small Proportion of the Eukaryotic Chromosomal DNA

Two factors contribute to the decreased gene density observed in eukaryotic cells: increases in gene size and increases in the DNA between genes, called **intergenic sequences**. The major reason that gene size is larger in more complex organisms is not that the average protein is bigger or that more DNA is required to encode the same protein. Instead, protein-encoding genes in eukaryotes frequently have discontinuous protein-coding regions. These interspersed non-protein-coding regions, called **introns**, are removed from the RNA after transcription in a process called **RNA splicing** (Fig. 7-3); we shall consider RNA splicing in detail in Chapter 13. The presence of introns can increase dramatically the length of DNA required to encode a gene (Table 7-3). For example, the average transcribed region of a human gene is ~27 kb (this should not be confused with the gene density), whereas the average protein-coding region of a human gene is 1.3 kb. A simple calculation reveals that only 5% of the average human protein-encoding gene directly encodes the desired protein. The remaining 95% is made up of introns. Consistent with their higher gene density, simpler eukaryotes have far fewer introns. For example, in the yeast *S. cerevisiae*, only 3.5% of genes have introns, none of which is greater than 1 kb (see Table 7-3).

An explosion in the amount of intergenic sequences in more complex organisms is responsible for the remaining decreases in gene

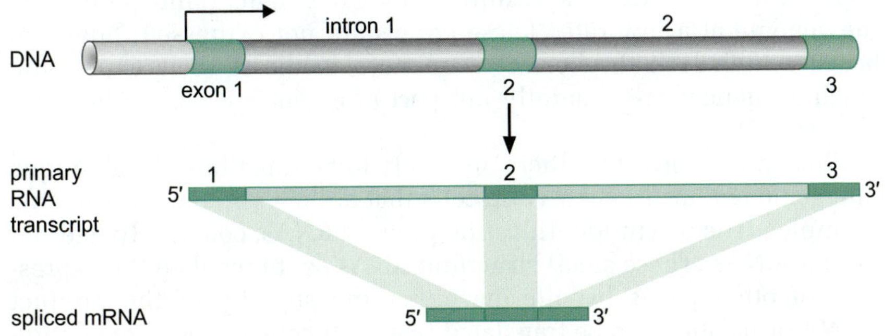

FIGURE 7-3 **Schematic of RNA splicing.** Transcription of pre-mRNA is initiated at the arrow shown above exon 1. This primary transcript is then processed (by splicing) to remove noncoding introns to produce messenger RNA.

TABLE 7-3 Contribution of Introns and Repeated Sequences to Different Genomes

Species	Gene Density (Genes/Mb)	Average Number of Introns per Gene	% of Repetitive DNA
Prokaryotes (bacteria)			
Escherichia coli K-12	950	0	<1
Eukaryotes (animals)			
Fungi			
Saccharomyces cerevisiae	480	0.04	3.4
Invertebrates			
Caenorhabditis elegans	190	5	6.3
Drosophila melanogaster	82	3	12
Vertebrates			
Fugu rubripes	56	5	2.7
Homo sapiens	6.25	6	46
Plants			
Arabidopsis thaliana	220	3	nd
Oryza sativa (rice)	~100	nd	42

nd, not determined.

density. Intergenic DNA is the portion of a genome that does not encode proteins or structural RNAs. More than 60% of the human genome is composed of intergenic sequences, and much of this DNA has no known function (Fig. 7-4). There are two kinds of intergenic DNAs: unique and repeated. About one-quarter of the intergenic DNA is unique. One contributor to an increase in unique intergenic sequences is an increase in regions of the DNA that are required to direct and regulate transcription, called **regulatory sequences**. As organisms become more complex and encode for more genes, the regulatory sequences required to coordinate gene expression also grow in complexity and size. The unique regions of the human intergenic DNA also include many apparently nonfunctional relics, including nonfunctional mutant genes, gene fragments, and pseudogenes. The mutant genes and gene fragments arise from simple random mutagenesis or mistakes in DNA recombination. Pseudogenes arise from the action of an enzyme called **reverse transcriptase** (Fig. 7-5 and Chapter 11). This enzyme copies RNA into double-stranded DNA (referred to as copy DNA or cDNA). Reverse transcriptase is only expressed by certain types of viruses that require this enzyme to reproduce. But, as a side effect of infection by such a virus, cellular mRNAs can be copied into DNA, and the resulting DNA fragments reintegrated into the genome at a low rate. These copies are not expressed, however, because they lack the correct sequences to direct their expression (such sequences are generally not part of a gene's RNA product; see Chapter 12).

Finally, it is clear that there are likely to be functions of the unique intergenic regions in eukaryotic cells that are not yet understood. One example is the recent identification of microRNAs, commonly referred to as **miRNAs**. These small structural RNAs act to regulate the expression of other genes by altering either the stability of the product mRNA or its ability to be translated (we shall consider gene regulation

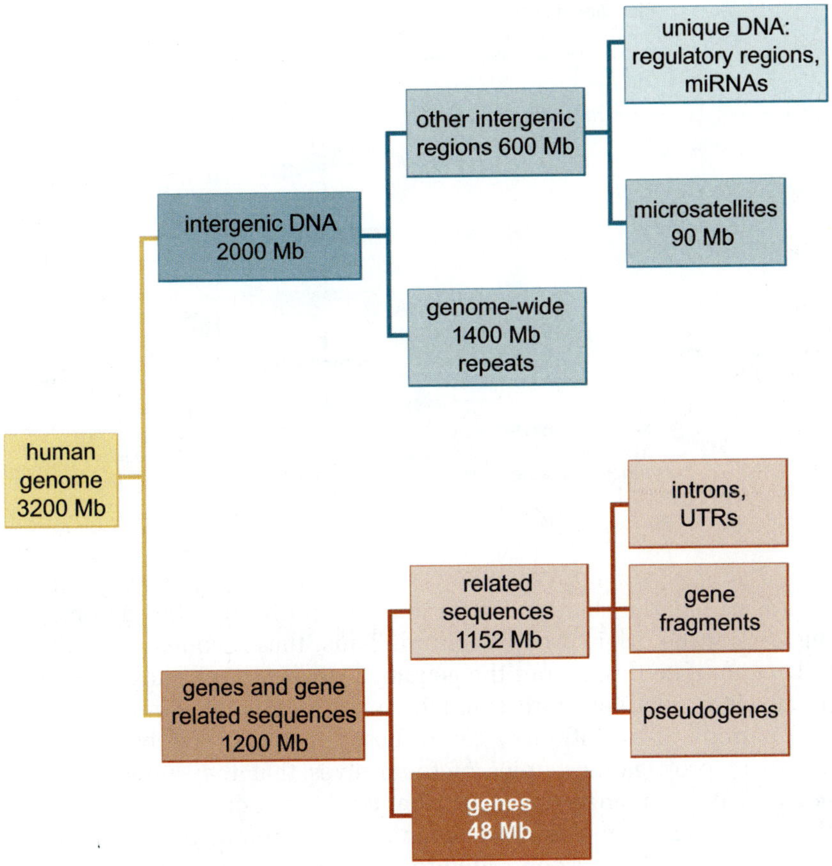

FIGURE 7-4 **Organization and content of the human genome.** The human genome is composed of many different types of DNA sequences, the majority of which do not encode proteins. Shown are the distribution and amount of each of the various types of sequences. (Adapted from Brown T.A. 2002. *Genomes,* 2nd ed., p. 23, Box 1.4. © BIOS Scientific Publishers by permission of Taylor & Francis.)

by small RNAs in Chapter 18). Because these sequences have been discovered only recently, they are not included as genes in Table 7-2; however, it has been estimated that human cells may have more than 400 miRNAs. Another function likely to be encoded in the unique intergenic regions are origins of replication, which have yet to be identified in most eukaryotic organisms.

The Majority of Human Intergenic Sequences Are Composed of Repetitive DNA

Almost half of the human genome is composed of DNA sequences that are repeated many times in the genome. There are two general classes of repeated DNA: microsatellite DNA and genome-wide repeats. **Microsatellite DNA** is composed of very short (<13 bp), tandemly repeated sequences. The most common microsatellite sequences are dinucleotide repeats (e.g., CACACACACACACACA). These repeats arise from difficulties in accurately duplicating the DNA and represent nearly 3% of the human genome.

Genome-wide repeats are much larger than their microsatellite counterparts. Each genome-wide repeat unit is greater than 100 bp in length and many are greater than 1 kb. These sequences can be found either as single copies dispersed throughout the genome or as closely spaced clusters. Although there are numerous classes of such repeats, their common feature is that all are forms of **transposable elements**.

Transposable elements are sequences that can "move" from one place in the genome to another. During **transposition**, as this move-

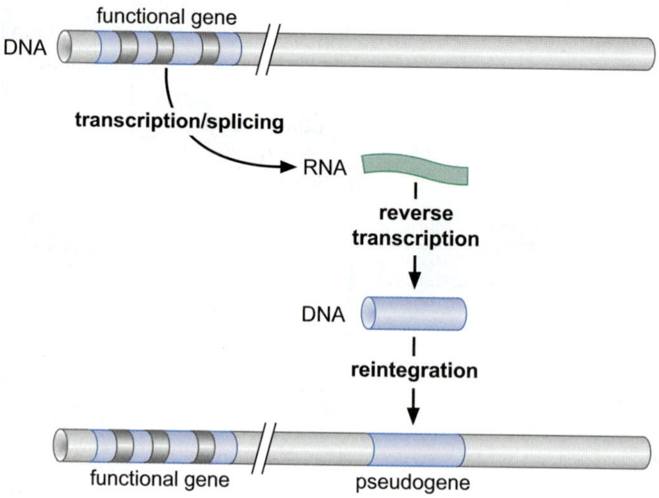

FIGURE 7-5 Processed pseudogenes arise from integration of reverse-transcribed messenger RNAs. When reverse transcriptase is present in a cell, messenger RNA (mRNA) molecules can be copied into double-stranded DNA. In rare instances, these DNA molecules can integrate into the genome creating pseudogenes. Because introns are rapidly removed from newly transcribed RNAs, these pseudogenes have the common characteristic of lacking introns. This distinguishes the pseudogene from the copy of the gene from which it was derived. In addition, pseudogenes lack the appropriate promoter sequences to direct their transcription as these are not part of the mRNA from which they are derived.

ment is called, the element moves to a new position in the genome, often leaving the original copy behind. Thus, these sequences multiply and accumulate throughout the genome. Movement of transposable elements is a relatively rare event in human cells. Nevertheless, over long periods of evolutionary time, these elements have been so successful at propagating copies of themselves that they now comprise ~45% of the human genome. In Chapter 11, we consider the mechanism by which transposable elements move around the genome and how their movement is controlled to prevent integration into genes.

Although we have discussed the nature of the intergenic sequence in the context of the human genome, many of the same features are found in other organisms. For example, comparison of the sequences of several plants with very large genomes (such as maize) indicates that transposable elements are likely to comprise an even larger percentage of these genomes. Similarly, even in the compact genomes of *E. coli* and *S. cerevisiae,* there are examples of transposable elements and microsatellite repeats (see Fig. 7-2). The difference is that these elements have been less successful at occupying the genomes of these simpler organisms. This lack of success is likely a combination of inefficient duplication and more efficient elimination (either by repair events or through selection against organisms in which duplication has occurred).

Although it is tempting to refer to repeated DNA as junk DNA, the stable maintenance of these sequences over hundreds to thousands of generations suggests that intergenic DNA confers a positive value (or selective advantage) to the host organism.

CHROMOSOME DUPLICATION AND SEGREGATION

Eukaryotic Chromosomes Require Centromeres, Telomeres, and Origins of Replication to Be Maintained during Cell Division

There are several important DNA elements in eukaryotic chromosomes that are not genes and are not involved in regulating the expression of genes (Fig. 7-6). These elements include origins of replication that direct the duplication of the chromosomal DNA, centromeres that act as "handles" for the movement of chromosomes into daughter

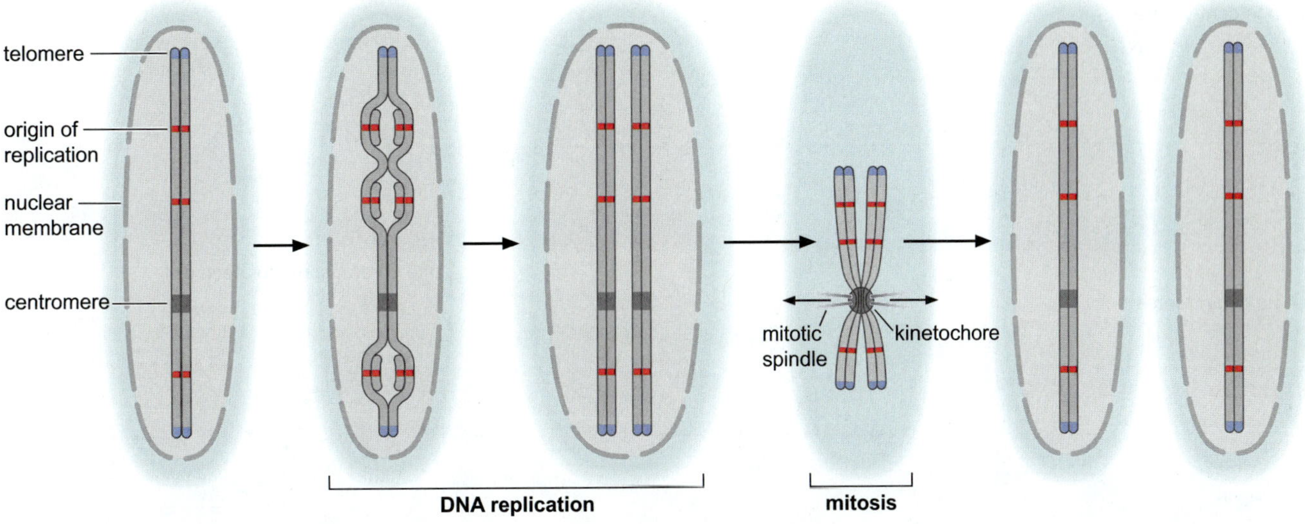

FIGURE 7-6 **Centromeres, origins of the replication, and telomeres are required for eukaryotic chromosome maintenance.** Each eukaryotic chromosome includes two telomeres, one centromere, and many origins of replication. Telomeres are located at both ends of each chromosome. Unlike telomeres, the single centromere found on each chromosome is not in a defined position. Some centromeres are near the middle of the chromosome and others are closer to a telomere. Origins of replication are located throughout the length of each chromosome (e.g., approximately every 30 kb in the budding yeast *S. cerevisiae*).

cells, and telomeres that protect and replicate the ends of linear chromosomes. All of these features are critical for the proper duplication and segregation of the chromosomes during cell division. We now look at each of these elements in more detail.

Origins of replication are the sites at which the DNA replication machinery assembles to initiate replication. They are typically found some 30–40 kb apart throughout the length of each eukaryotic chromosome. Prokaryotic chromosomes also require origins of replication. Unlike their eukaryotic counterparts, prokaryotic chromosomes typically have only a single site of replication initiation. In general, origins of replication are found in noncoding regions. The DNA sequences that are recognized as origins of replication are discussed in detail in Chapter 8.

Centromeres are required for the correct segregation of the chromosomes after DNA replication. The two copies of each replicated chromosome are called sister chromosomes, and during cell division they must be separated with one copy going to each of the two daughter cells. Like origins of replication, centromeres direct the formation of an elaborate protein complex called in this case a **kinetochore**. The kinetochore interacts with the centromere DNA and with protein filaments called microtubules that pull the sister chromosomes away from each other and into the two daughter cells. In contrast to the many origins of replication found on each eukaryotic chromosome, it is critical that each chromosome has *only one* centromere (Fig. 7-7a). In the absence of a centromere, the replicated chromosomes segregate randomly, resulting in daughter cells that either have lost a chromosome or have two copies of a chromosome (Fig. 7-7b). The presence of more than one centromere on each chromosome is equally disastrous. If the associated kinetichores are attached to filaments pulling in opposite directions, this

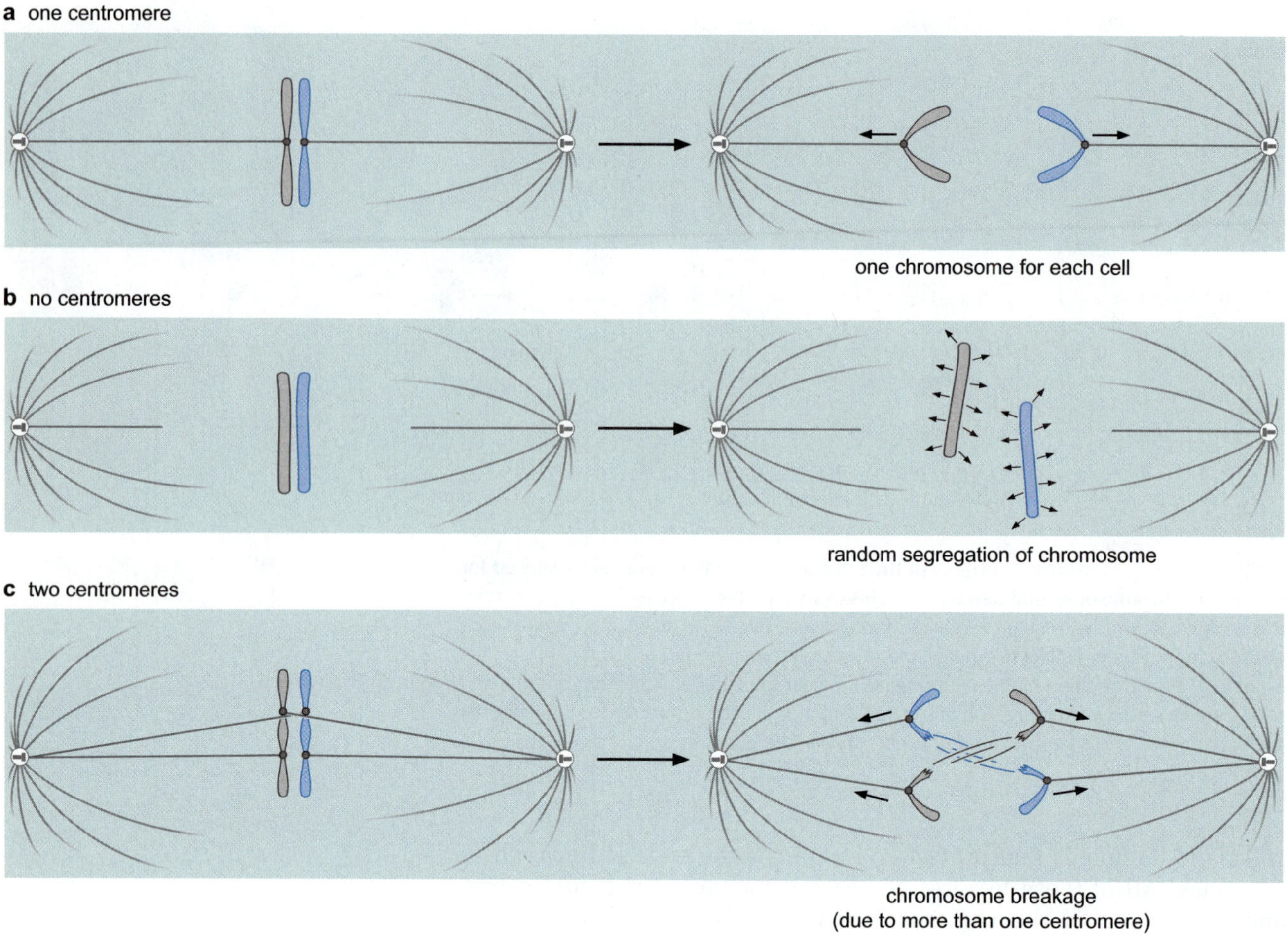

FIGURE 7-7 **More or less than one centromere leads to chromosome loss or breakage.** (a) Normal chromosomes have one centromere. After replication of a chromosome, each copy of the centromere directs the formation of a kinetochore. These two kinetochores then bind to opposite poles of the mitotic spindle and are pulled into the opposite sides of the cell prior to cell division. (b) Chromosomes lacking centromeres are rapidly lost from cells. In the absence of the centromere, the chromosomes do not attach to the spindle and are randomly distributed to the two daughter cells. This leads to frequent events in which one daughter gets two copies of a chromosome and the other daughter cell is missing the same chromosome. (c) Chromosomes with two or more centromeres are frequently broken during segregation. If a chromosome has more than one centromere, it can be bound simultaneously to both poles of the mitotic spindle. When segregation is initiated, the opposing forces of the mitotic spindle frequently break chromosomes attached to both poles.

can lead to chromosome breakage (Fig. 7-7c). Centromeres vary greatly in size. In the yeast *S. cerevisiae,* centromeres are composed of unique sequences that are less than 200 bp in length. In contrast, in the majority of eukaryotes, centromeres are greater than 40 kb and are composed of largely repetitive DNA sequences (Fig. 7-8).

Telomeres are located at the two ends of a linear chromosome. Like origins of replication and centromeres, telomeres are bound by a number of proteins. In this case, the proteins perform two important functions. First, telomeric proteins distinguish the natural ends of the chromosome from sites of chromosome breakage and other DNA breaks in the cell. Ordinarily, DNA ends are the sites of frequent recombination and DNA degradation. The proteins that assemble at

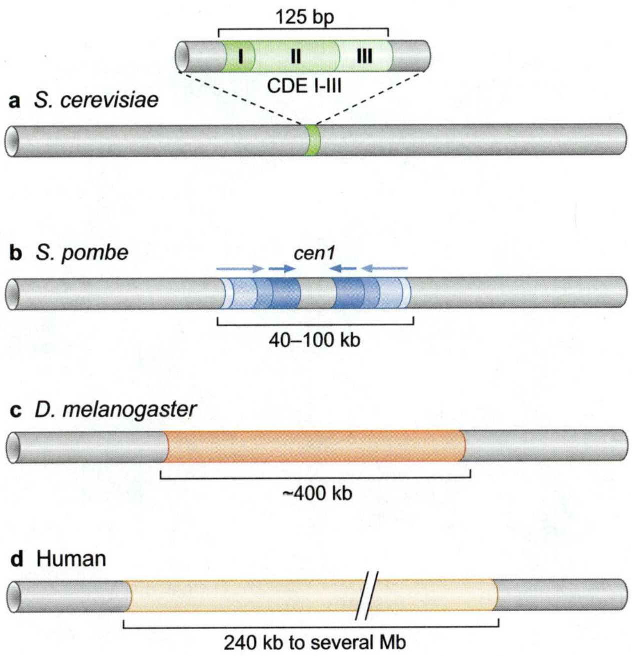

FIGURE 7-8 **Centromere size and composition vary dramatically among different organisms.** *Saccharomyces cerevisiae* centromeres are small and composed of nonrepetitive sequences. In contrast, the centromeres of other organisms such as the fruit fly, *Drosophila melanogaster,* and the fission yeast, *Schizosaccharomyces pombe,* are much larger and are mostly composed of repetitive sequences. Only the central 4–7 kb of the *S. pombe* centromere is nonrepetitive, and the large majority of the *Drosophila* and human centromeres are repetitive DNA.

telomeres form a structure that is resistant to both of these events. Second, telomeres act as a specialized origin of replication that allows the cell to replicate the ends of the chromosomes. For reasons described in detail in Chapter 8, the standard DNA replication machinery cannot completely replicate the ends of a linear chromosome. Telomeres facilitate end replication through the recruitment of an unusual DNA polymerase called **telomerase**.

In contrast to most of the chromosome, a portion of the telomere is maintained in a single-stranded form (Fig. 7-9). Most telomeres have a simple repeating sequence that varies from organism to organism. This repeat is typically composed of a short TG-rich repeat. For example, human telomeres have the repeating sequence of 5′-TTAGGG-3′. As we shall see in Chapter 8, the repetitive nature of telomeres is a consequence of their unique method of replication.

Eukaryotic Chromosome Duplication and Segregation Occur in Separate Phases of the Cell Cycle

During cell division, the chromosomes must be duplicated and segregated into the daughter cells. In bacterial cells, these events occur simultaneously; that is, as the DNA is replicated, the resulting two copies are separated into opposite sides of the cell. Although it is clear that these events are tightly regulated in bacteria, how this regulation is achieved is poorly understood. In contrast, eukaryotic cells duplicate and segre-

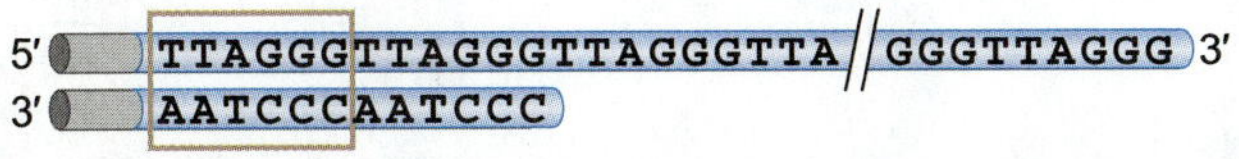

FIGURE 7-9 **Structure of a typical telomere.** The repeated sequence (from human cells) is shown in a representative box. Note that the region of single-stranded DNA at the 3′ end of the chromosome can be hundreds of bases long.

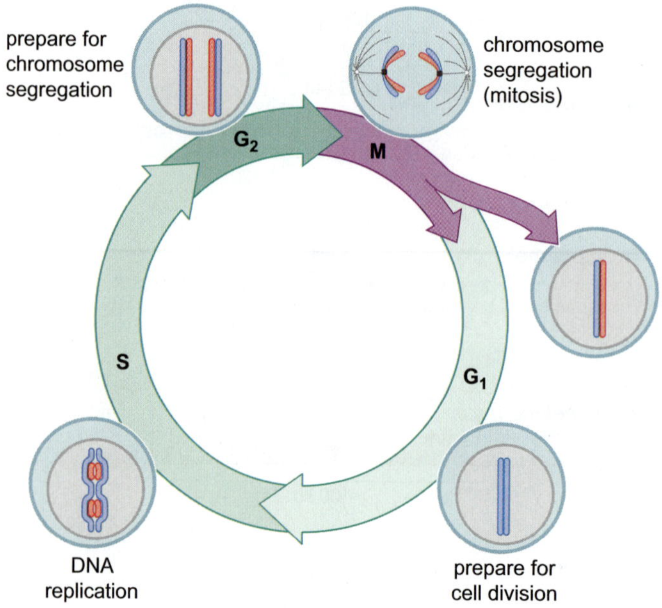

FIGURE 7-10 Eukaryotic mitotic cell cycle. There are four stages of the eukaryotic cell cycle. Chromosomal replication occurs during S phase and chromosome segregation occurs during M phase. The G_1 and G_2 gap phases allow the cell to prepare for the next event in the cell cycle. For example, many eukaryotic cells use the G_1 phase of the cell cycle to establish that the level of nutrients is sufficiently high to allow the completion of cell division.

gate their chromosomes at distinct times during cell division. We focus on these events for the remainder of our discussion of chromosomes.

WEB ANIMATION

The events required for a single round of cell division are collectively known as the **cell cycle** (see Interactive Animation 7-1). Most eukaryotic cell divisions maintain the number of chromosomes in the daughter cells that were present in the parental cell. This type of division is called **mitotic cell division.**

The mitotic cell cycle can be divided into four phases: G_1, S, G_2, and M (Fig. 7-10). Chromosome replication occurs during the **synthesis,** or **S phase,** of the cell cycle, resulting in the duplication of each chromosome (Fig. 7-11). Each chromosome of the duplicated pair is called a **chromatid**, and the two chromatids of a given pair are called **sister chromatids**. Sister chromatids are held together after duplication through the action of a molecule called **cohesin**, which we describe below. The process that holds them together is called **sister-**

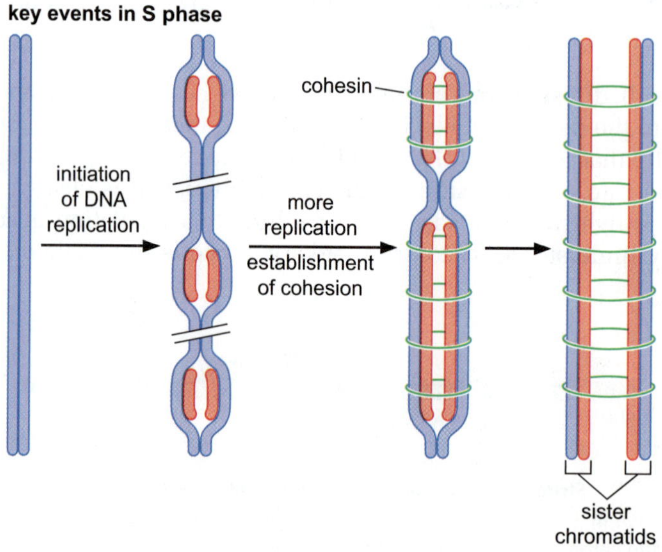

FIGURE 7-11 Events of S phase. Two major chromosomal events occur during S phase. DNA replication copies each chromosome completely, and shortly after replication has occurred, sister-chromatid cohesion is established by ring-shaped cohesin molecules, which are hypothesized to encircle the two copies of the recently replicated DNA. Each blue or red "tube" represents a single-stranded DNA molecule, with red DNA being newly synthesized.

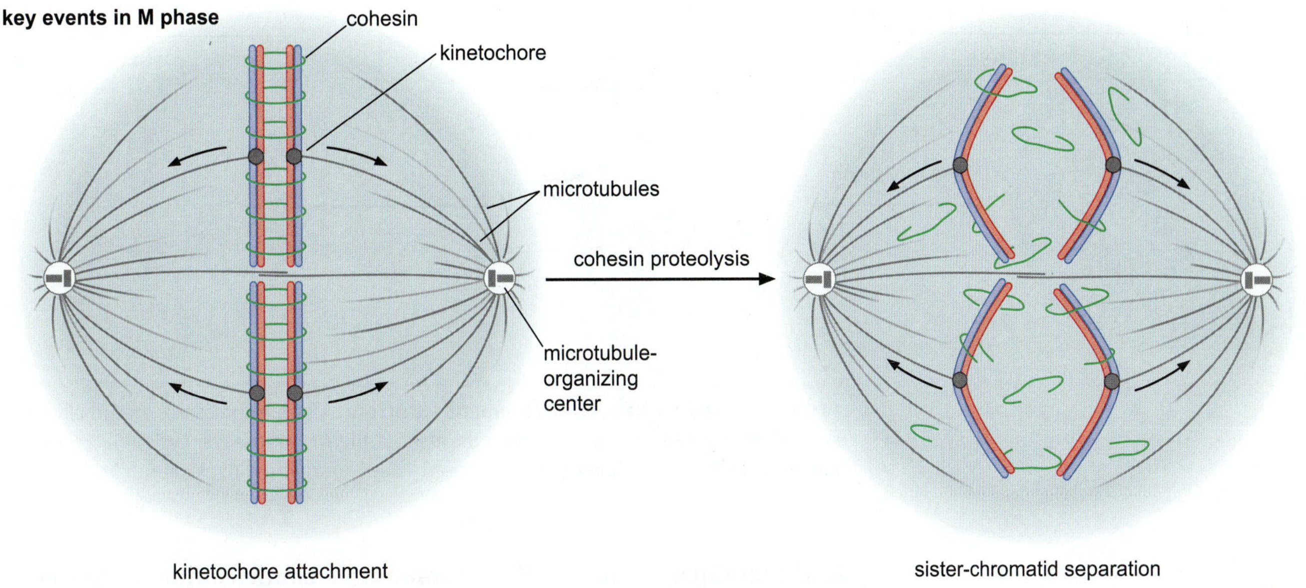

FIGURE 7-12 **Events of mitosis (M phase).** Three major events occur during mitosis. First, the two kinetochores of each linked sister-chromatid pair attach to opposite poles of the mitotic spindle. Once all kinetochores are bound to opposite poles, sister-chromatid cohesion is eliminated by destroying the cohesin ring. Finally, after cohesion is eliminated, the sister chromatids are segregated to opposite poles of the mitotic spindle.

chromatid cohesion, and this tethered state is maintained until the chromosomes segregate from one another.

Chromosome segregation occurs during **mitosis** or the **M phase** of the cell cycle. We consider the overall process of mitosis below, but first we shall focus on three key steps in the process (Fig. 7-12). First, each pair of sister chromatids is bound to a structure called the **mitotic spindle**. This structure is composed of long protein fibers called **microtubules** that are attached to one of the two **microtubule-organizing centers** (also called **centrosomes** in animal cells or **spindle pole bodies** in yeasts and other fungi). The microtubule-organizing centers are located on opposite sides of the cell forming "poles" toward which the microtubules pull the chromatids. Attachment of the chromatids to the microtubules is mediated by the **kinetochore** assembled at each centromere (see Fig. 7-6). Second, the cohesion between the chromatids is dissolved by proteolysis of cohesin. Before cohesion is dissolved, it resists the pulling forces of the mitotic spindle. After cohesion is dissolved, the third major event in mitosis can occur: **sister-chromatid separation.** In the absence of the counterbalancing force of chromatid cohesion, the chromatids are rapidly pulled toward opposite poles of the mitotic spindle. Thus, cohesion between the sister chromatids and attachment of sister-chromatid kinetochores to opposite poles of the mitotic spindle play opposing roles that must be carefully coordinated for chromosome segregation to occur properly.

Chromosome Structure Changes as Eukaryotic Cells Divide

As chromosomes proceed through a round of cell division, their structure is altered numerous times; however, there are two main states for the chromosomes (Fig. 7-13). The chromosomes are in their most compact form as cells segregate their chromsomes. The process that results

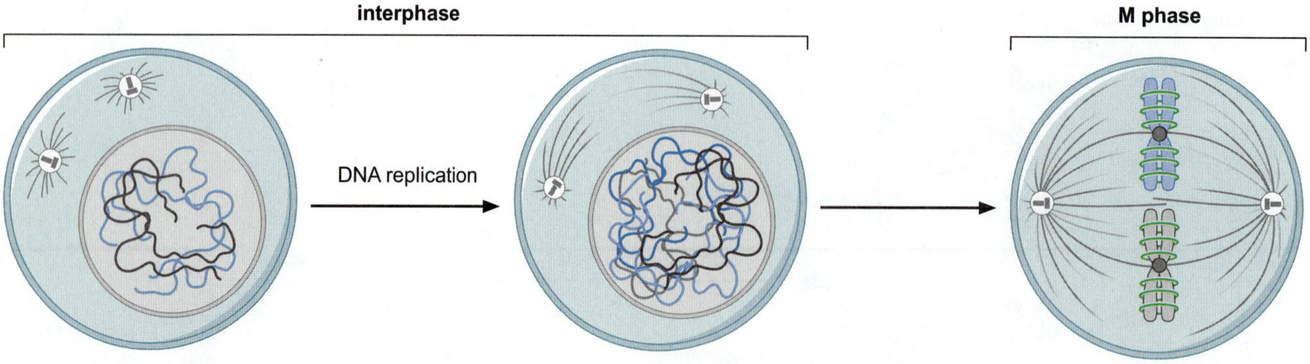

FIGURE 7-13 Changes in chromatin structure. Chromosomes are maximally condensed in M phase and decondensed throughout the rest of the cell cycle (G_1, S, and G_2 in mitotic cells). Together, these decondensed stages are referred to as interphase.

in this compact form is called **chromosome condensation**. In this condensed state, the chromosomes are completely disentangled from one another, greatly facilitating the segregation process.

During phases of the cell cycle when chromosome segregation is not occurring (collectively referred to as interphase), the chromosomes are significantly less compact. Indeed, at these stages of the cell cycle, the chromosomes are likely to be highly intertwined, resembling more of a plate of spaghetti than the organized view of chromosomes during mitosis. Nevertheless, even during these stages, the structures of the chromosomes change. DNA replication requires the nearly complete disassembly and reassembly of the proteins associated with each chromosome. Immediately after DNA replication, sister-chromatid cohesion is established, linking the newly replicated chromatids to one another. As transcription of individual genes is turned on and off or up and down, there are associated changes in the structure of the chromosomes in those regions occurring throughout the cell cycle. Thus, the chromosome is a constantly changing structure that is more like an organelle than a simple string of DNA.

Sister-Chromatid Cohesion and Chromosome Condensation Are Mediated by SMC Proteins

The key proteins that mediate sister-chromatid cohesion and chromosome condensation are related to one another. The structural maintenance of chromosome (SMC) proteins are extended proteins that form defined pairs by interacting through lengthy coiled-coil domains (see Fig. 5-15). Together with non-SMC proteins, they form multiprotein complexes that act to link two DNA helices together. Cohesin is an SMC-protein-containing complex that, as we discussed above, is required to link the two daughter DNA duplexes (sister chromatids) together after DNA replication. It is this linkage that is the basis for sister-chromatid cohesion. The structure of cohesin is thought to be a large ring composed of two SMC proteins and two non-SMC proteins. Although the exact mechanism of sister-chromatid cohesion is still under investigation, a prominent model proposes that chromatid cohesion occurs as the result of both sister chromatids passing through the center of the cohesin protein ring (Fig. 7-14). In this model, proteolytic cleaveage of the non-SMC subunit of cohesin results in the opening of the ring and the loss of cohesin.

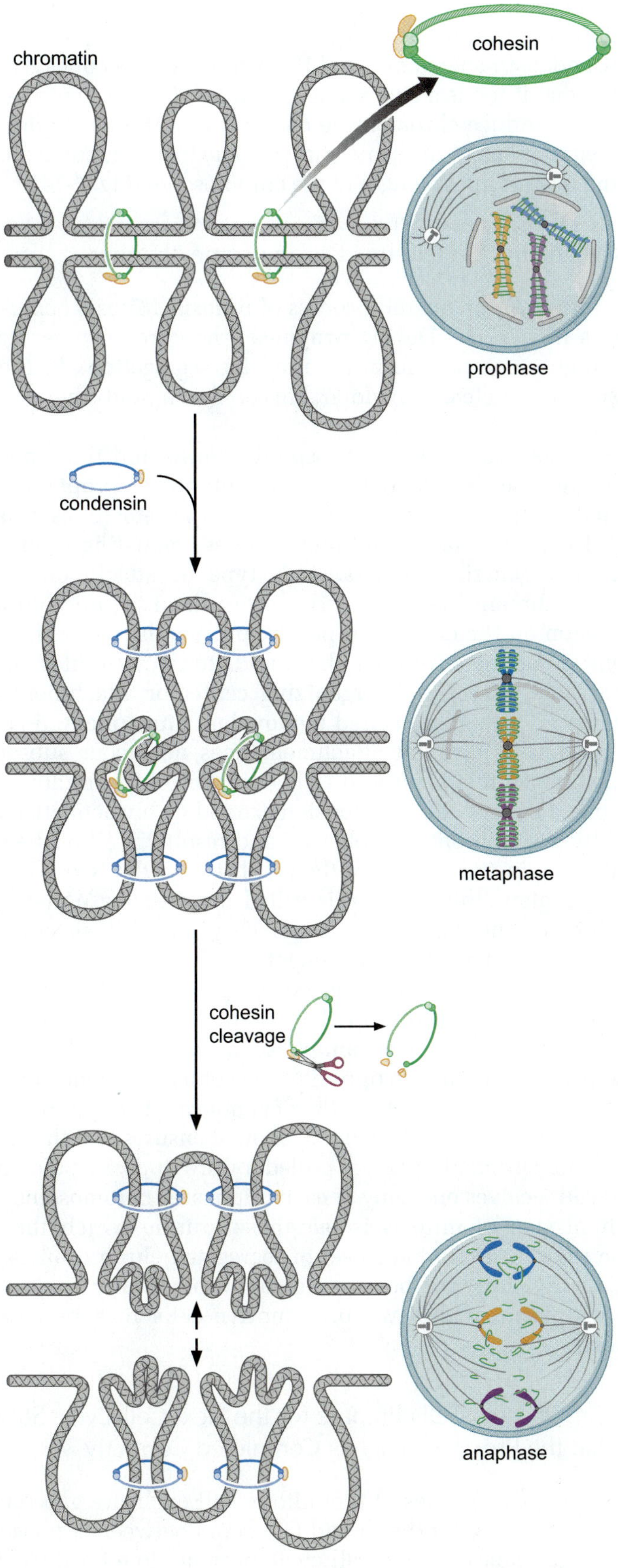

FIGURE 7-14 Model for the structure and function of cohesins and condensins. Cohesins and condensins are ring-shaped protein complexes that include two SMC proteins that play important roles in bringing distant or different regions of DNA together. The proposed ring-shaped structure of these proteins would allow a flexible but strong link between two regions of DNA. In this illustration, the SMC proteins are shown as green (cohesin) or blue (condensin). (Adapted, with permission, from Haering C.H. et al. 2002. *Mol. Cell* 9: 773–788, Fig. 8, p. 785. © Elsevier.)

The chromosome condensation that accompanies chromosome segregation also requires a related SMC-containing complex called **condensin**. Condensin shares many of the features of the cohesin complex, suggesting that it too is a ring-shaped complex. If so, it may use its ring-like nature to induce chromosome condensation. For example, by linking different regions of the same chromosome together, condensin could reduce the overall linear length of the chromosome (Fig. 7-14).

Mitosis Maintains the Parental Chromosome Number

We now return to the overall process of mitosis. Mitosis occurs in several stages (Fig. 7-15). During **prophase**, the chromosomes condense into the highly compact form required for segregation. At the end of prophase, the nuclear envelope breaks down and the cell enters metaphase.

During metaphase, the mitotic spindle forms and the kinetochores of sister chromatids attach to the microtubules. Proper chromatid attachment is only achieved when the two kinetochores of a sister-chromatid pair are attached to microtubules emanating from opposite microtubule-organizing centers. This type of attachment is called **bivalent attachment** (see Fig. 7-15) and results in the microtubules exerting tension on the chromatid pair by pulling the sisters in opposite directions. Attachment of both chromatids to microtubules emanating from the same microtubule-organizing center or attachment of only one chromatid of the pair, called **monovalent attachment**, does not result in tension. If bivalent attachment does not occur subsequently, monovalent attachment can lead to both copies of a chromosome moving into one daughter cell. The tension exerted by bivalent attachment is opposed by sister-chromatid cohesion and results in all of the chromosomes aligning in the middle of the cell between the two microtubule-organizing centers (this position is called the metaphase plate). Importantly, chromosome segregation starts only after all sister-chromatid pairs have achieved bivalent attachment.

Chromosome segregation is triggered by proteolytic destruction of the cohesin molecules, resulting in the loss of sister-chromatid cohesion. This loss occurs as cells enter **anaphase**, during which the sister chromatids separate and move to opposite sides of the cell. Once the two sisters are no longer held together, they cannot resist the outward pull of the microtubule spindle. Bivalent attachment ensures that the two members of a sister-chromatid pair are pulled toward opposite poles and each daughter cell receives one copy of each duplicated chromosome.

The final step of mitosis is **telophase**, during which the nuclear envelope reforms around each set of segregated chromosomes. At this point, cell division can be completed by physically separating the shared cytoplasm of the two presumptive cells in a process called **cytokinesis**.

During Gap Phases, Cells Prepare for the Next Cell Cycle Stage and Check That the Previous Stage Is Completed Correctly

The remaining two phases of the mitotic cell cycle are gap phases. G_1 occurs prior to DNA synthesis and G_2 occurs between S phase and M phase. The gap phases of the cell cycle provide time for the cell to accomplish two goals: (1) to prepare for the next phase of the cell cycle and (2) to check that the previous phase of the cell cycle has been completed appropriately. For example, prior to entry into S phase,

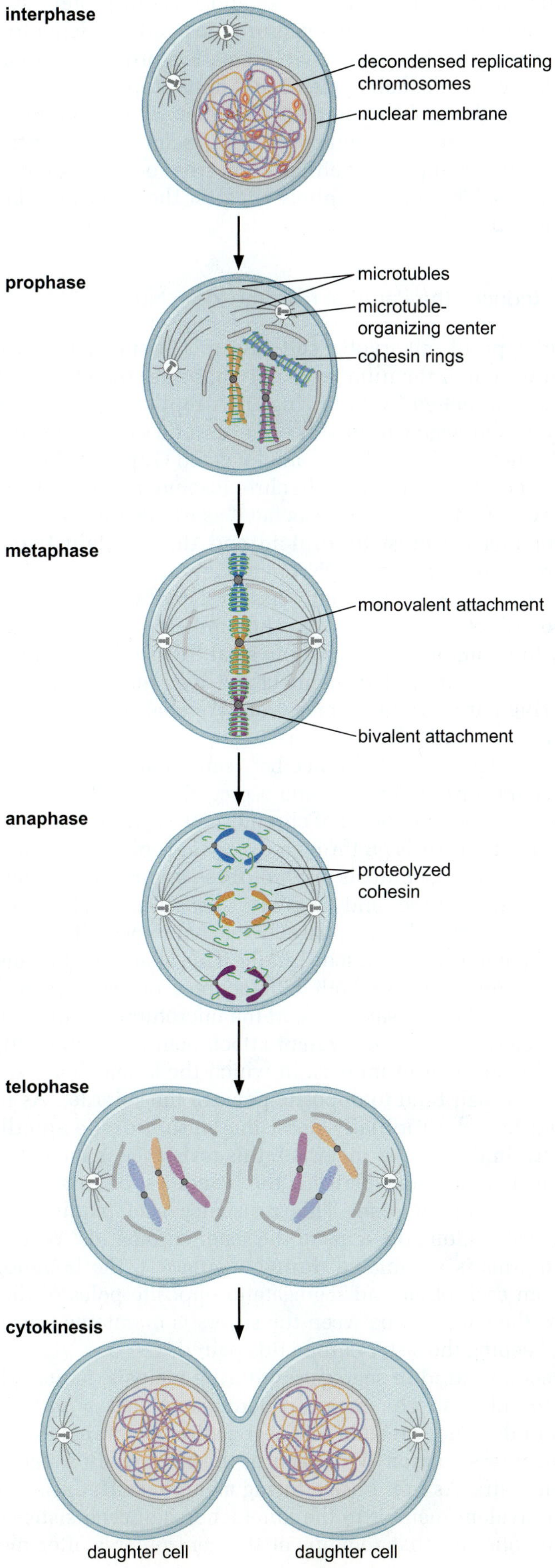

FIGURE 7-15 Mitosis in detail. Prior to mitosis, the chromosomes are in a decondensed state called interphase. During prophase, chromosomes are condensed and detangled in preparation for segregation, and the nuclear membrane surrounding the chromosomes breaks down in most eukaryotes. During metaphase, each sister-chromatid pair attaches to opposite poles of the mitotic spindle. Anaphase is initiated by the loss of sister-chromatid cohesion, resulting in the separation of sister chromatids. Telophase is distinguished by the loss of chromosome condensation and the reformation of the nuclear membrane around the two populations of segregated chromosomes. Cytokinesis is the final event of the cell cycle during which the cellular membrane surrounding the two nuclei constricts and eventually completely separates into two daughter cells. All DNA molecules are double-stranded.

most cells must reach a certain size and level of protein synthesis to ensure that there will be adequate proteins and nutrients to complete the next round of DNA synthesis. If there is a problem with a previous step in the cell cycle, **cell cycle checkpoints** stop the cell cycle to provide time for the cell to complete that step. For example, cells with damaged DNA arrest the cell cycle in G_1 before DNA synthesis or in G_2 before mitosis to prevent either event from occurring with damaged chromosomes. These delays allow time for the damage to be repaired before the cell cycle continues.

Meiosis Reduces the Parental Chromosome Number

A second type of eukaryotic cell division is specialized to produce cells that have half the number of chromosomes than the parental cell. This is accomplished by following DNA replication with two rounds of chromosome segregation. Like the mitotic cell cycle, the **meiotic cell cycle** includes a G_1, S, and an elongated G_2 phase (Fig. 7-16). During the meiotic S phase, each chromosome is replicated, and the daughter chromatids remain associated as in the mitotic S phase. Cells that enter meiosis must be diploid and thus contain two copies of each chromosome prior to DNA replication, one derived from each parent. After DNA replication, these related sister-chromatid pairs, called **homologs**, pair with each other and recombine. Recombination between the homologs creates a physical linkage between the two homologs that is required to connect the two related sister-chromatid pairs during chromosome segregation. We shall discuss the details of meiotic recombination in Chapter 10.

The most significant difference between the mitotic and meiotic cell cycles occurs during chromosome segregation. Unlike mitosis, during which there is a single round of chromosome segregation, chromosomes participating in meiosis go through two rounds of segregation known as meiosis I and II. Like mitosis, each of these segregation events includes a prophase, metaphase, and anaphase stage. During the metaphase of **meiosis I**, also called metaphase I, the homologs attach to opposite poles of the microtubule-based spindle. This attachment is mediated by the kinetochore. Because both kinetochores of each sister-chromatid pair are attached to the same pole of the microtubule spindle, this interaction is referred to as **monovalent attachment** (in contrast to the bivalent attachment seen in mitosis, in which the kinetochores of each sister-chromatid pair bind to opposite poles of the spindle). As in mitosis, the paired homologs initially resist the tension of the spindle pulling them apart. In the case of meiosis I, this resistance is mediated through the physical connections between the homologs, or crossovers, that are induced by recombination. This resistance also requires sister-chromatid cohesion along the arms of the sister chromatids. When cohesion along the arms is eliminated during anaphase I, the homologs are released from each other and segregate to opposite poles of the cell. Importantly, the cohesion between the sisters is maintained near the centromere, keeping the sister chromatids paired.

The second round of segregation during meiosis, **meiosis II**, is very similar to mitosis. The major difference is that a round of DNA replication does not precede this segregation event. Instead, a spindle is formed in association with each of the two newly separated sister-chromatid pairs. As in mitosis, during **metaphase II**, these spindles attach in a bivalent manner to the kinetochores of each sister-chromatid pair. The cohesion that remains at the centromeres after meiosis I is

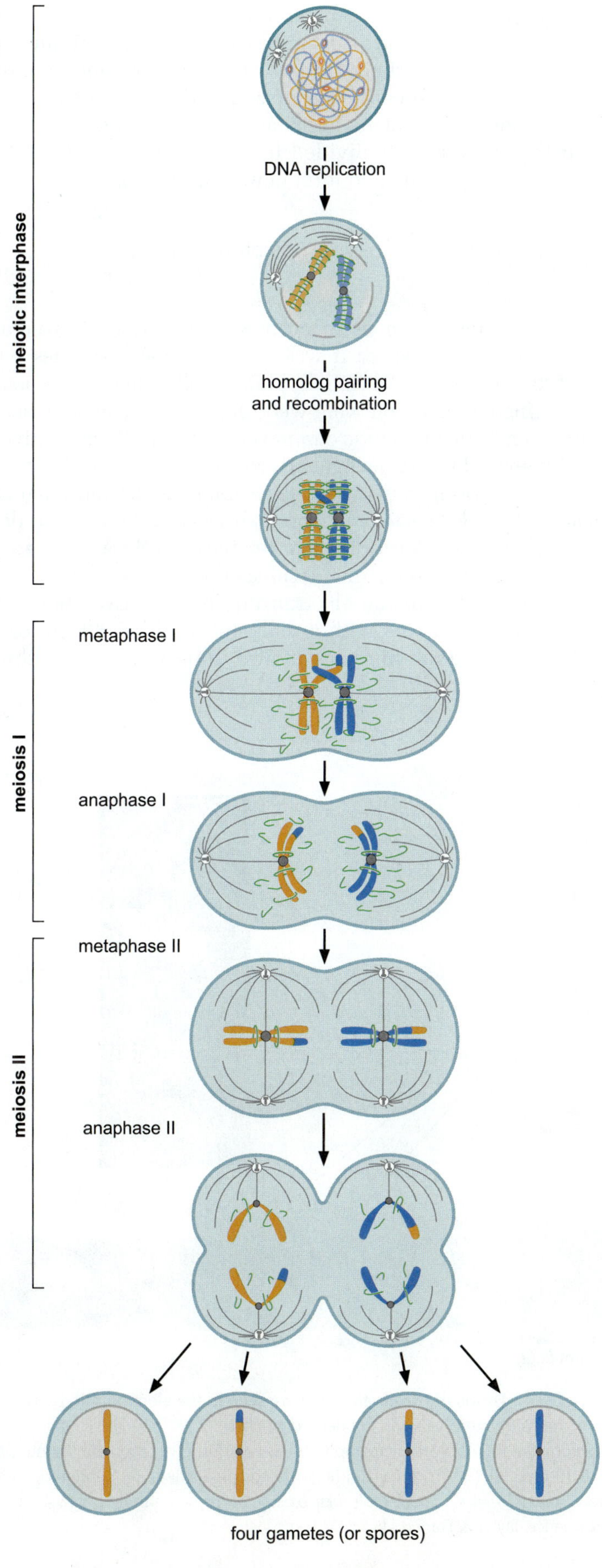

FIGURE 7-16 Meiosis in detail. Like mitosis, meiosis can be divided into discrete stages. After DNA replication, homologous sister chromatids pair with each other to form structures with four related chromosomes. For simplicity, only a single chromosome is shown segregating with the blue copy being from one parent and the yellow copy from the other. During pairing, chromatids from the different homologs recombine to form a link between the homologous chromosomes called a chiasma. During metaphase I, the two kinetochores of each sister-chromatid pair attach to one pole of the meiotic spindle. Homologous sister-chromatid kinetochores attach to opposite poles creating tension that is resisted by the connection between the homologs. Entry into anaphase I is correlated with two events that together result in the separation of the homologous chromosomes from one another. The sister-chromatid cohesion is lost along the arms of the chromosomes, and the chiasma between the homologs is resolved. Together, these events result in the separation of the homologs from one another. The sister chromatids remain attached through cohesion at the centromere. Meiosis II is very similar to mitosis. During meiotic metaphase II, two meiotic spindles are formed. As in mitotic metaphase, the kinetochores associated with each sister-chromatid pair attach to opposite poles of the meiotic spindles. During anaphase II, the remaining cohesion between the sisters is lost and the sister chromatids separate from each other. The four separate sets of chromosomes are then packaged into nuclei and separated into four cells to create four spores or gametes. All DNA molecules are double-stranded. (Adapted, with permission, from Murray A. and Hunt T. 1993. *The cell cycle: An introduction,* Fig. 10.2. © Oxford University Press, Inc.)

critical to oppose the pull of the spindle. The second round of chromosome segregation occurs in **anaphase II** and is initiated by the elimination of centromeric cohesion. At this point, there are four sets of chromosomes in the cell, each of which contains a single copy of each chromosome. A nucleus forms around each set of chromosomes, and then the cytoplasm is divided to form four haploid cells. These cells are now ready to mate to form new diploid cells.

Different Levels of Chromosome Structure Can Be Observed by Microscopy

Microscopy has long been used to observe chromosome structure and function. Indeed, long before it was clear that chromosomes were the source of the genetic information in the cell, their movements and changes during cell division were well-understood. The compact nature of condensed mitotic or meiotic chromosomes makes them relatively easy to visualize even by simple light microscopy. Microscopic analysis of condensed chromosomes is used to determine the chromosomal makeup of human cells and detect such abnormalities as chromosomal deletions or individuals with too few or too many copies of a chromosome.

Outside of mitosis (i.e., in interphase), chromosomal DNA is less compact (Fig. 7-17a). In the electron microscope, two states of chromatin are observed: fibers with a diameter of either 30 nm or 10 nm (Fig. 7-17b). The 30-nm fiber is a more compact version of chromatin

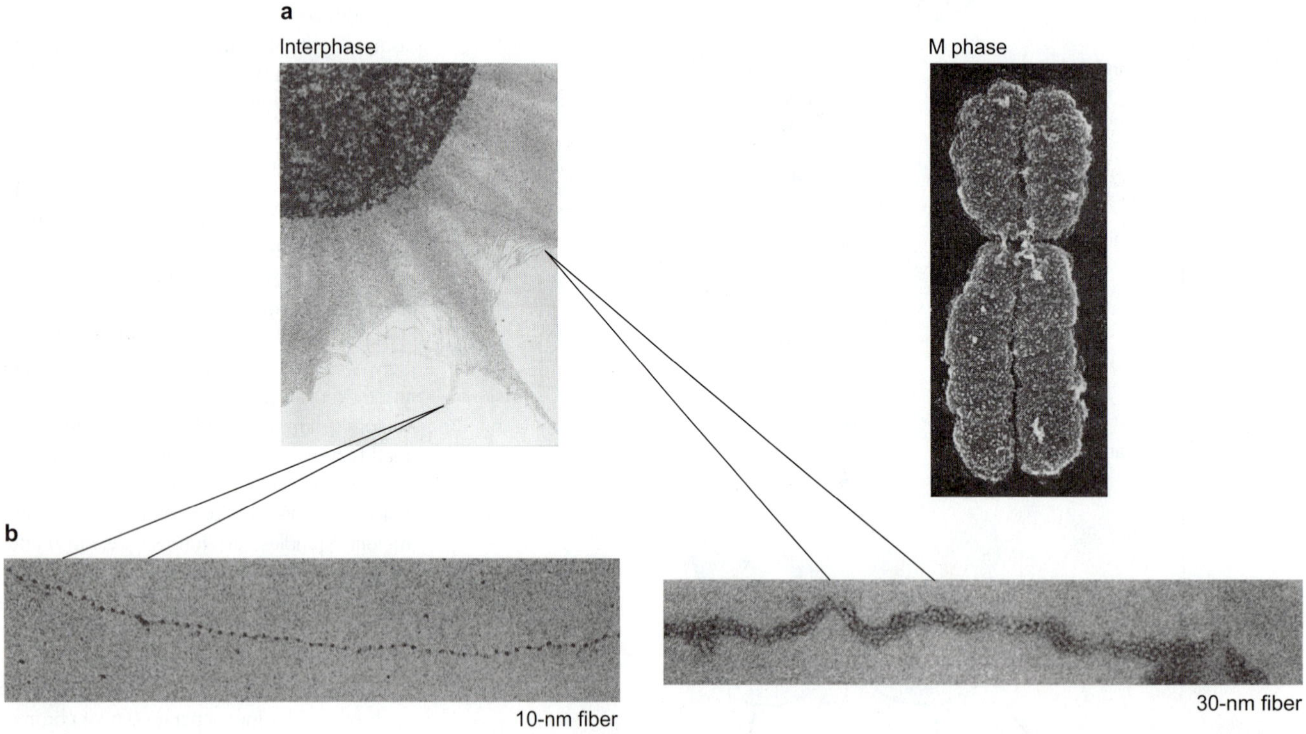

FIGURE 7-17 **Forms of chromatin structure seen in the electron microscope.** (a) Electron micrographs of interphase and condensed M-phase DNA show the changes in the structure of chromatin. (b) Electron micrographs of different forms of chromatin in interphase cells show the 30-nm and 10-nm chromatin fibers (beads on a string). (a, Reprinted, with permission, from Alberts B. et al. 2002. *Molecular biology of the cell,* 4th ed., Figs. 4-21 and 4-23. Garland Science/Taylor & Francis LLC. © V. Foe.)

that is frequently folded into large loops reaching out from a protein core or scaffold. In contrast, the 10-nm fiber is a less compact form of chromatin that resembles a regular series of "beads on a string." These beads are nucleosomes and these protein–DNA structures play a critical role in regulating the structure and function of chromosomes. In the rest of the chapter, we shall first focus on the nature of the nucleosome, including how they are formed, and then describe how nucleosome-dependent structures control the accessibility of nuclear DNA.

THE NUCLEOSOME

Nucleosomes Are the Building Blocks of Chromosomes

The majority of the DNA in eukaryotic cells is packaged into nucleosomes. Each nucleosome is composed of a core of eight histone proteins and the DNA wrapped around them. The DNA between each nucleosome (the "string" in the "beads on a string") is called **linker DNA**. By assembling into nucleosomes, the DNA is compacted approximately sixfold. This is far short of the 1,000–10,000-fold DNA compaction observed in eukaryotic cells. Nevertheless, this first stage of DNA packaging is essential for all the remaining levels of DNA compaction.

The DNA most tightly associated with the nucleosome, called the **core DNA**, is wound ~1.65 times around the outside of the histone octamer like thread around a spool (Fig. 7-18). The length of DNA asso-

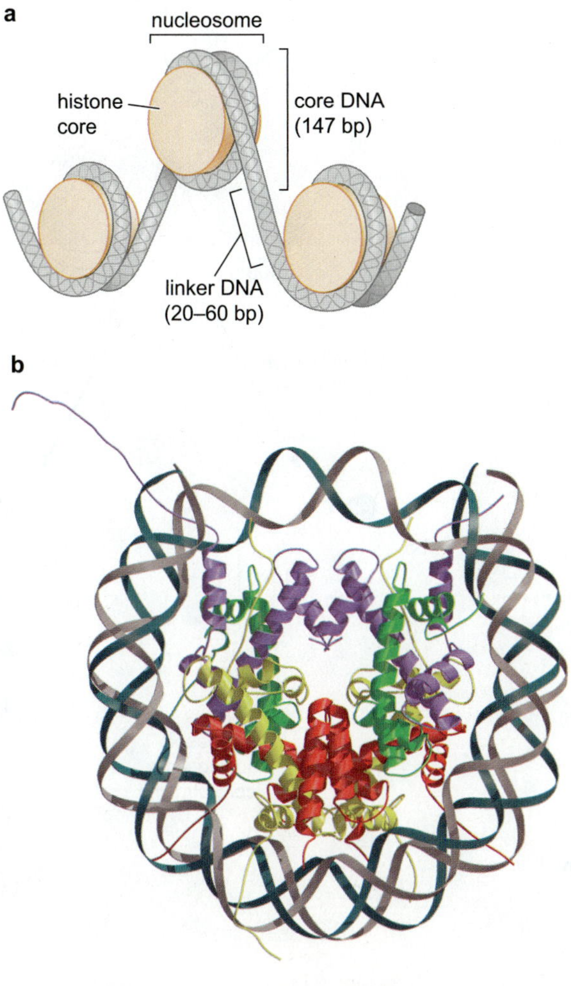

FIGURE 7-18 DNA packaged into nucleosomes. (a) Schematic of the packaging and organization of nucleosomes. (b) Crystal structure of a nucleosome showing DNA wrapped around the histone protein core. (Red) H2A; (yellow) H2B; (purple) H3; (green) H4. Note that the colors of the different histone proteins here and in following structures are the same. (Luger K. et al. 1997. *Nature* 389: 251–260.) Image prepared with MolScript, BobScript, and Raster 3D.

ciated with each nucleosome can be determined using nuclease treatment (Box 7-1, Micrococcal Nuclease and the DNA Associated with the Nucleosome). The ~147-bp length of this DNA is an invariant feature of nucleosomes in all eukaryotic cells. In contrast, the length of the linker DNA between nucleosomes is variable. Typically, this distance is 20–60 bp and each eukaryote has a characteristic average linker DNA length (Table 7-4). The difference in average linker DNA length is likely to reflect the differences in the larger structures formed by nucleosomal DNA in each organism, rather than differences in the nucleosomes themselves (see next section on Higher-Order Chromatin Structure).

KEY EXPERIMENTS

Box 7-1 Micrococcal Nuclease and the DNA Associated with the Nucleosome

Nucleosomes were first purified by treating chromosomes with a sequence-nonspecific nuclease called **micrococcal nuclease (MNase)**. The ability of this enzyme to cleave DNA is primarily governed by the accessibility of the DNA. Thus, MNase cleaves protein-free DNA sequences rapidly and protein-associated DNA sequences poorly. Limited treatment of chromosomes with this enzyme results in a nuclease-resistant population of DNA molecules that are associated with histones. These DNA molecules are between 160 and 220 bp in length and are associated with two copies each of histones H2A, H2B, H3, and H4. On average, these particles include the DNA tightly associated with the nucleosome as well as one unit of linker DNA. More extensive MNase treatment degrades all of the linker DNA. The remaining minimal nucleosome includes only 147 bp of DNA and is called the **nucleosome core particle**.

The average length of DNA associated with each nucleosome can be measured in a simple experiment (Box 7-1 Fig. 1). Chromatin is treated with the enzyme micrococcal nuclease but this time only gently. This results in single cuts in some but not all of the linker DNA. After nuclease treatment, the DNA is extracted from all proteins (including the histones) and subjected to gel electrophoresis to separate the DNA by size. Electrophoresis reveals a "ladder" of fragments that are multiples of the average nucleosome-to-nucleosome distance. A ladder of fragments is observed because the MNase-treated chromatin is only partially digested. Thus, sometimes, multiple nucleosomes will remain unseparated by digestion, leading to DNA fragments equivalent to all of the DNA bound by these nucleosomes. Further digestion would result in all linker DNA being cleaved and the formation of nucleosome core particles and a single ~147-bp fragment.

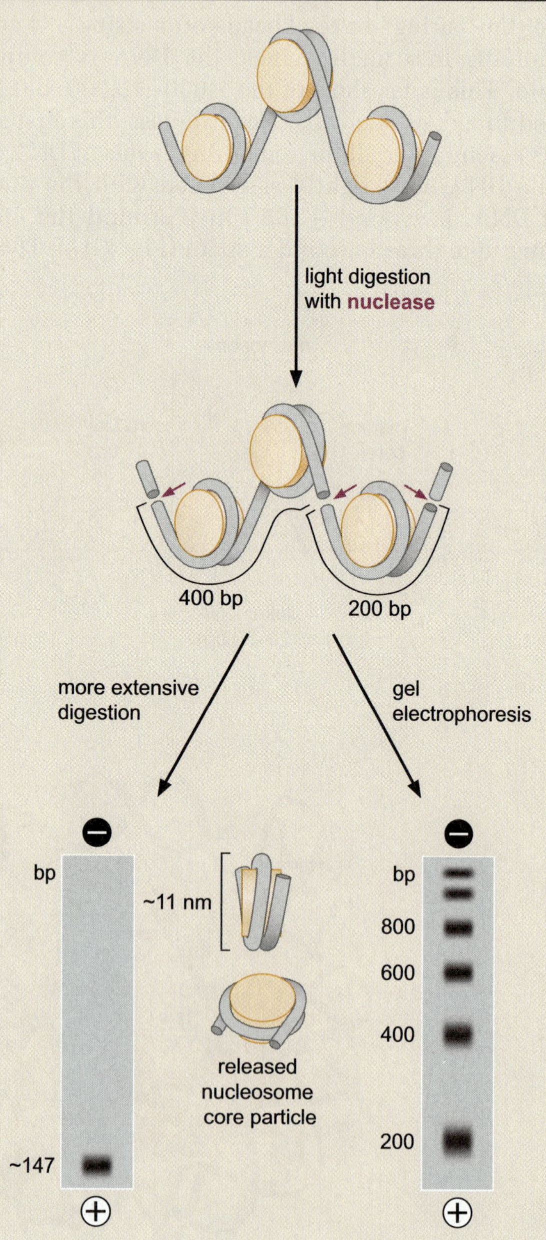

BOX 7-1 FIGURE 1 Progressive digestion of nucleosomal DNA with MNase. (Courtesy of R.D. Kornberg.)

TABLE 7-4 Average Lengths of Linker DNA in Various Organisms

Species	Nucleosome Repeat Length (bp)	Average Linker DNA Length (bp)
Saccharomyces cerevisiae	160–165	13–18
Sea urchin (sperm)	~260	~110
Drosophila melanogaster	~180	~33
Human	185–200	38–53

In any cell, there are stretches of DNA that are not packaged into nucleosomes. Typically, these are regions of DNA engaged in gene expression, replication, or recombination. Although not bound by nucleosomes, these sites are typically associated with nonhistone proteins that are either regulating or participating in these events. We discuss the mechanisms that remove nucleosomes from DNA and maintain such regions of DNA in a nucleosome-free state below and in Chapter 17.

Histones Are Small, Positively Charged Proteins

Histones are by far the most abundant proteins associated with eukaryotic DNA. Eukaryotic cells commonly contain five abundant histones: H1, H2A, H2B, H3, and H4. Histones H2A, H2B, H3, and H4 are the **core histones**, and two copies of each of these histones form the protein core around which nucleosomal DNA is wrapped. Histone H1 is not part of the nucleosome core particle. Instead, it binds to the linker DNA and is referred to as a **linker histone**. The four core histones are present in equal amounts in the cell. H1 is half as abundant as the other histones, which is consistent with the finding that only one molecule of H1 can associate with a nucleosome.

Consistent with their close association with the negatively charged DNA molecule, the histones have a high content of positively charged amino acids (Table 7-5). At least 20% of the residues in each histone are either lysine or arginine. The core histones are also relatively small proteins ranging in size from 11 to 15 kilodaltons (kD). Histone H1 is slightly larger at ~21 kD.

The protein core of the nucleosome is a disk-shaped structure that assembles in an ordered fashion only in the presence of DNA. Without DNA, the core histones form intermediate assemblies in solution. A conserved region found in every core histone, called the **histone-fold domain**, mediates the assembly of these histone-only intermediates (Fig. 7-

TABLE 7-5 General Properties of the Histones

Histone Type	Histone	Molecular Weight (M_r)	% of Lysine and Arginine
Core histones	H2A	14,000	20
	H2B	13,900	22
	H3	15,400	23
	H4	11,400	24
Linker histone	H1	20,800	32

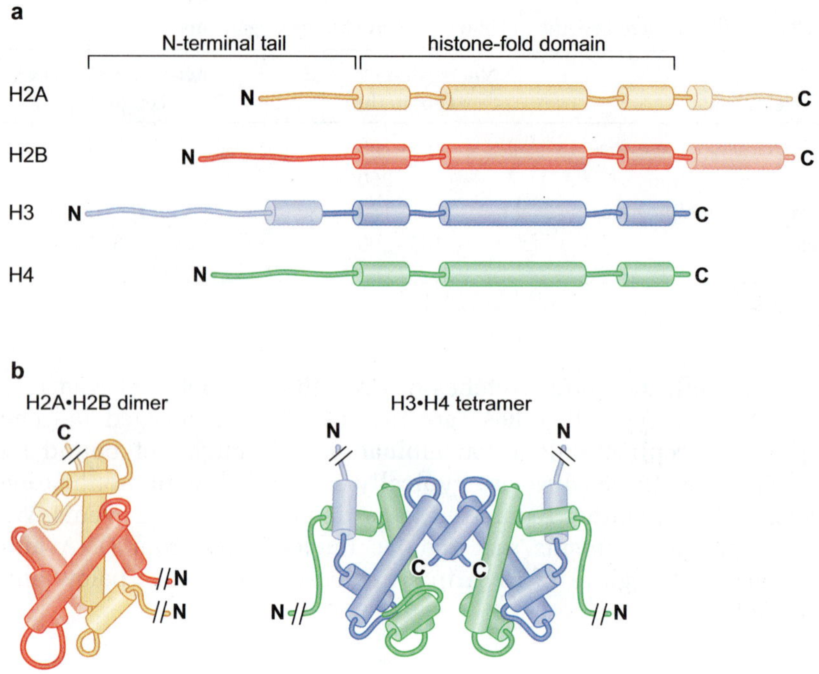

FIGURE 7-19 Core histones share a common structural fold. (a) The four histones are diagramed as linear molecules. The regions of the histone-fold motif that form α helices are indicated as cylinders. Note that there are adjacent regions of each histone that are structurally distinct including additional α-helical regions. (b) The helical regions of two histones (here H2A and H2B) come together to form a dimer. H3 and H4 also use a similar interaction to form $H3_2{\cdot}H4_2$ tetramers. (Adapted, with permission, from Alberts B. et al. 2002. *Molecular biology of the cell,* 4th ed., p. 209, Fig. 4–26. © Garland Science/Taylor & Francis LLC.)

19). The histone-fold domain is composed of three α-helical regions separated by two short unstructured loops. In each case, this domain mediates the formation of head-to-tail heterodimers of specific pairs of histones. H3 and H4 histones first form heterodimers that then come together to form a tetramer with two molecules each of H3 and H4. In contrast, H2A and H2B form heterodimers in solution but not tetramers.

The assembly of a nucleosome involves the ordered association of these building blocks with DNA (Fig. 7-20). First, the H3•H4 tetramer binds to DNA; then two H2A•H2B dimers join the H3•H4-DNA complex to form the final nucleosome. We discuss how and when this assembly process is accomplished in the cell later in the chapter.

The core histones each have an amino-terminal extension, called a "tail," because it lacks a defined structure and is accessible within the intact nucleosome. This accessibility can be detected by treatment of nucleosomes with the protease trypsin (which specifically cleaves proteins after positively charged amino acids). Treatment of nucleosomes with trypsin rapidly removes the accessible amino-terminal tails of the histones but cannot cleave the tightly packed histone-fold regions (Fig. 7-21). The exposed amino-terminal tails are not required for the association of DNA with the histone octamer, as DNA is still tightly associated with the nucleosome after protease treatment. Instead, the tails are the sites of extensive modifications that alter the function of individual nucleosomes. These modifications include phosphorylation, acetylation, and methylation on serine, lysine, and arginine residues. We return to the role of histone tail modification in nucleosome function later. We now turn to the detailed structure of the nucleosome.

The Atomic Structure of the Nucleosome

The high-resolution three-dimensional structure of the nucleosome core particle (see Fig. 7-18b; 147 bp of DNA plus an intact histone octamer) has provided many insights into nucleosome function. The

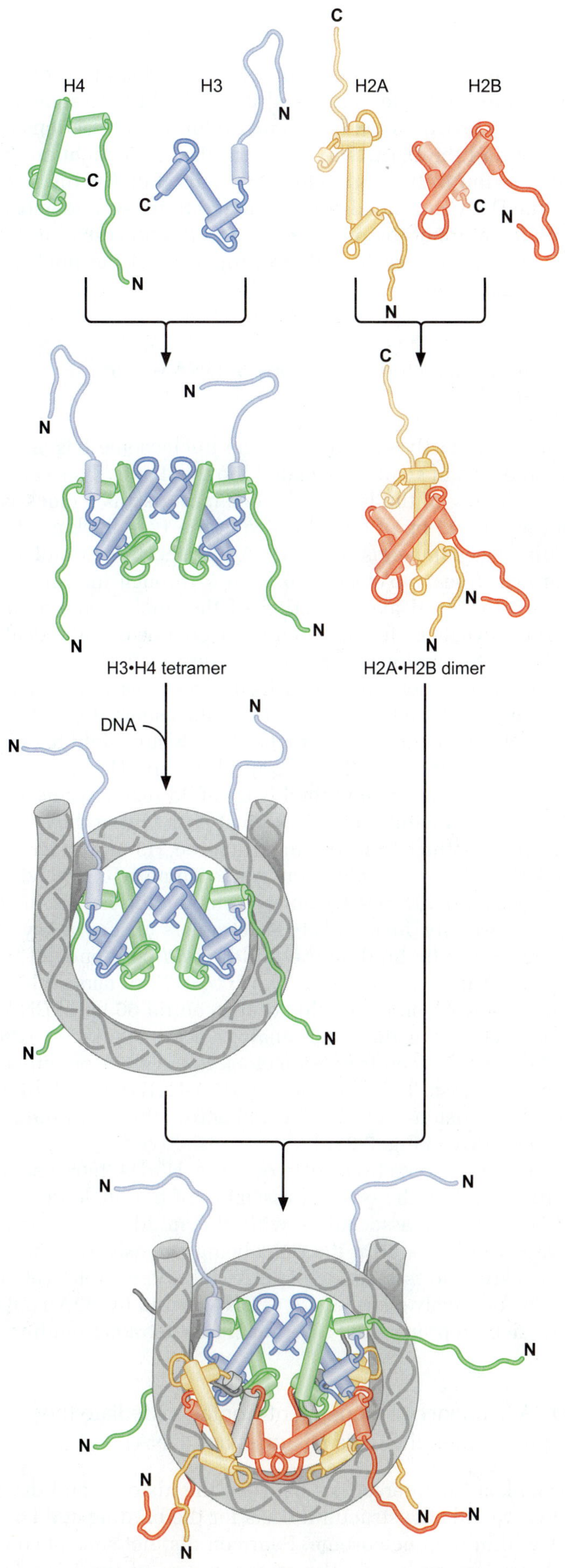

FIGURE 7-20 Assembly of a nucleosome. The assembly of a nucleosome is initiated by the formation of a $H3_2{\bullet}H4_2$ tetramer. The tetramer then binds to double-stranded DNA. The $H3_2{\bullet}H4_2$ tetramer bound to DNA recruits two copies of the H2A•H2B dimer to complete the assembly of the nucleosome. (Adapted, with permission, from Alberts B. et al. 2002. *Molecular biology of the cell,* 4th ed., p. 210, Fig. 4-27. Garland Science/Taylor & Francis LLC, © J. Waterborg.)

high affinity of the nucleosome for DNA, the distortion of the DNA when bound to the nucleosome, and the lack of DNA sequence specificity can each be explained by the nature of the interactions between the histones and the DNA. The structure also sheds light on the function and location of the amino-terminal tails. Finally, the interaction between the DNA and the histone octamer allows an understanding of the dynamic nature of the nucleosome and the process of nucleosome assembly. We discuss each of these properties of the nucleosome in the following sections.

Histones Bind Characteristic Regions of DNA within the Nucleosome

Although not perfectly symmetrical, the nucleosome has an approximate twofold axis of symmetry, called the **dyad axis**. This can be visualized by thinking of the face of the octamer disk as a clock with the midpoint of the 147 bp of DNA located at the 12 o'clock position (Fig. 7-22). This places the ends of the DNA just short of 11 and 1 o'clock. A line drawn from 12 o'clock to 6 o'clock through the middle of the disk defines the dyad axis. Rotation of the nucleosome around this axis by 180° reveals a view of the nucleosome nearly identical to that observed prior to rotation (see Structural Tutorial 7-1).

WEB STRUCTURAL TUTORIAL

The H3•H4 tetramers and H2A•H2B dimers each interact with a particular region of the DNA within the nucleosome (Fig. 7-23). Of the 147 bp of DNA included in the structure, the histone-fold regions of the H3•H4 tetramer interact with the central 60 bp. The amino-terminal region of H3 most proximal to the histone-fold region forms a fourth α helix that interacts with the final 13 bp at each end of the bound DNA (this region is distinct from the unstructured H3 amino-terminal tail described above). If we picture the nucleosome with a clock face as described above, the H3•H4 tetramer forms the top half of the histone octamer. Importantly, histone H3•H4 tetramers occupy a key position in the nucleosome by binding the middle *and* both ends of the DNA (turquoise DNA in Fig. 7-23a). The two H2A•H2B dimers each associate with ~30 bp of DNA on either side of the central 60 bp of DNA bound by H3 and H4. Using the clock analogy again, the DNA associated with H2A•H2B is located from ~5 o'clock to 9 o'clock on either face of the nucleosome disc. Together, the two H2A•H2B dimers form the bottom part of the histone octamer located across the disc from the DNA ends (orange DNA in Fig. 7-23b).

The extensive interactions between the H3•H4 tetramer and the DNA help to explain the ordered assembly of the nucleosome (Fig. 7-24). H3•H4 tetramer association with the middle and ends of the bound DNA would result in the DNA being extensively bent and constrained, making the association of H2A•H2B dimers relatively easy. In contrast, the relatively short length of DNA bound by H2A•H2B dimers is not sufficient to prepare the DNA for H3•H4 tetramer binding.

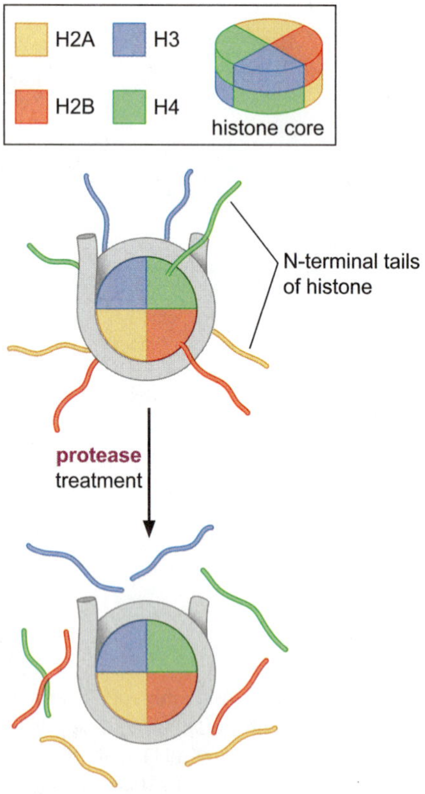

FIGURE 7-21 Amino-terminal tails of the core histones are accessible to proteases. Treatment of nucleosomes with limiting amounts of proteases that cleave after basic amino acids (e.g., trypsin) specifically removes the amino-terminal "tails" leaving the histone core intact.

Many DNA Sequence–Independent Contacts Mediate the Interaction between the Core Histones and DNA

A closer look at the interactions between the histones and the nucleosomal DNA reveals the structural basis for the binding and bending of the DNA within the nucleosome. Fourteen distinct sites of contact are observed, one for each time the minor groove of the DNA faces the

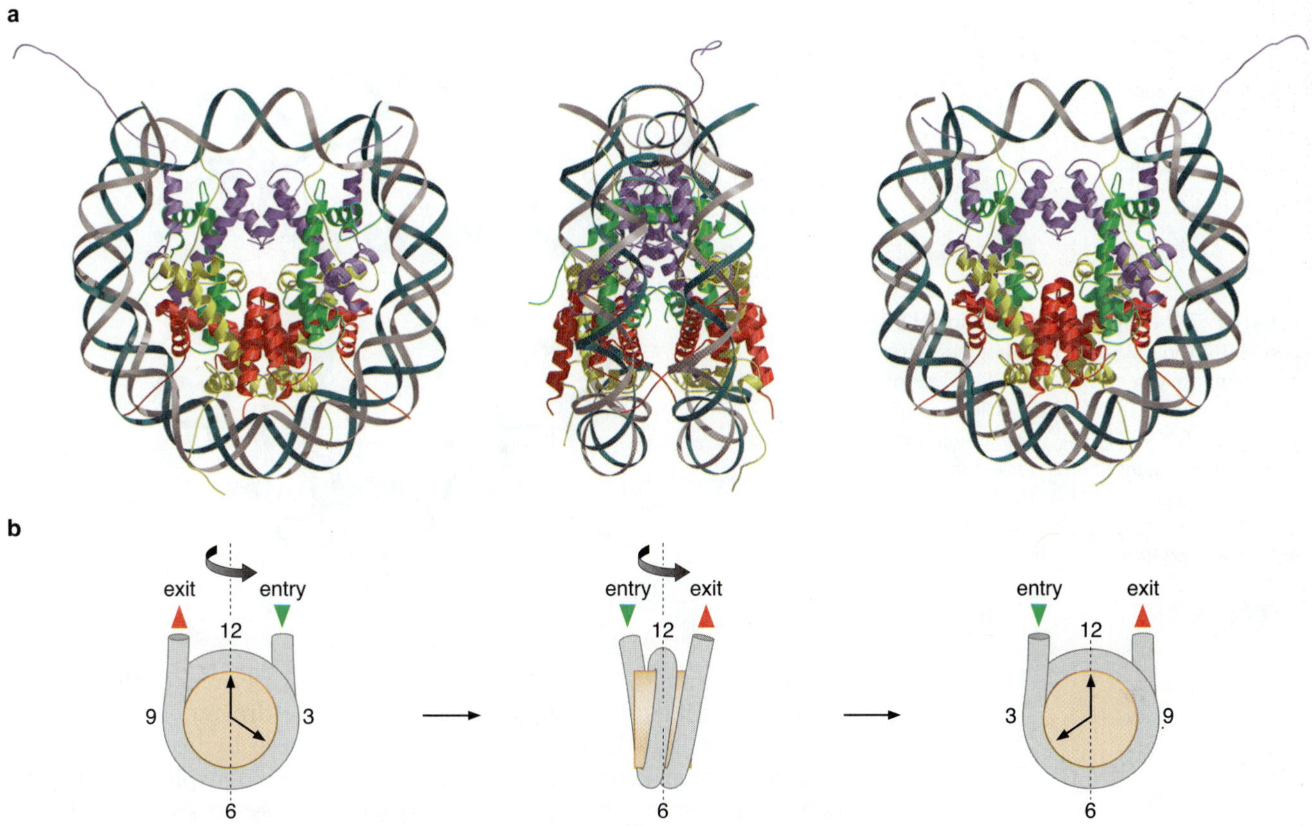

FIGURE 7-22 The nucleosome has an approximate twofold axis of symmetry. (a) Three-dimensional structure. (b) Cartoon illustrating "clock face" analogy to nucleosome. Three views of the nucleosome are shown in each representation. Each view shows a 90° rotation around the axis between 12 and 6 o'clock positions illustrated in the first panel of b. (a, Luger K. et al. 1997. *Nature* 389: 251–260.) Images prepared with MolScript, BobScript, and Raster 3D.

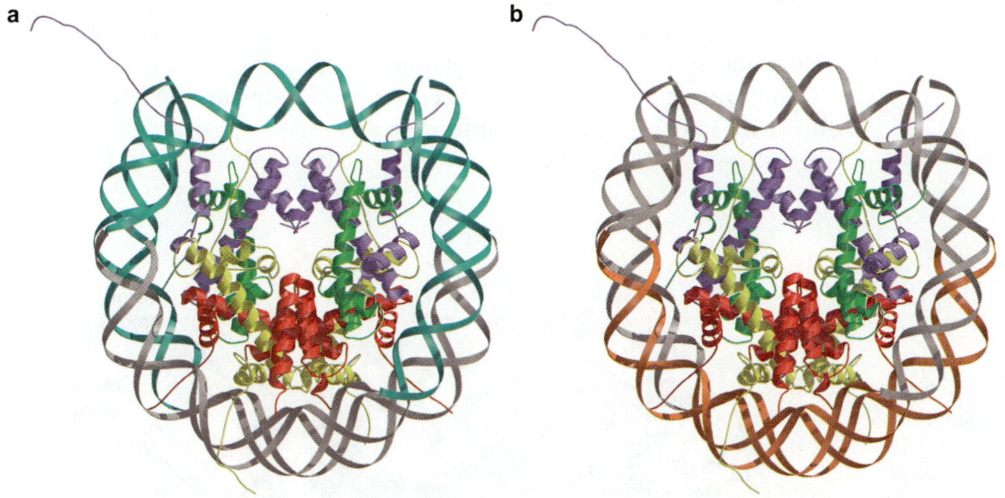

FIGURE 7-23 Interactions of the histones with nucleosomal DNA. (a) H3•H4 bind the middle and the ends of the DNA (shown in turquoise). (b) H2A•H2B bind 30 bp of DNA on one side of the nucleosome (shown in orange). (Luger K. et al. 1997. *Nature* 389: 251–260.) Images prepared with MolScript, BobScript, and Raster 3D.

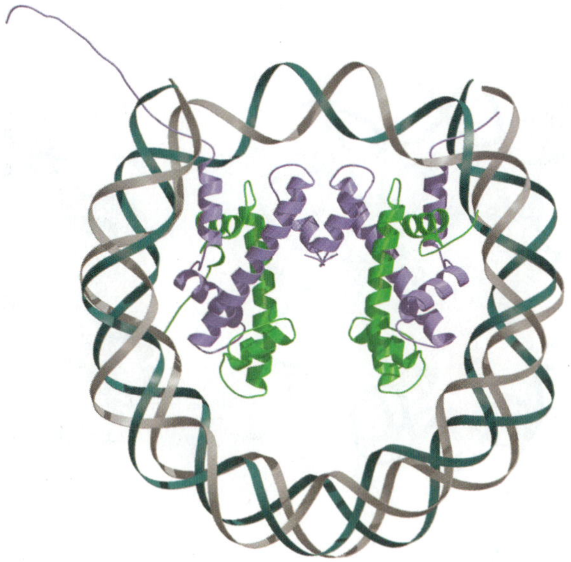

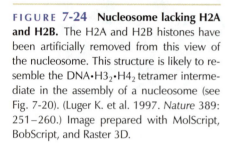

FIGURE 7-24 Nucleosome lacking H2A and H2B. The H2A and H2B histones have been artificially removed from this view of the nucleosome. This structure is likely to resemble the DNA•$H3_2$•$H4_2$ tetramer intermediate in the assembly of a nucleosome (see Fig. 7-20). (Luger K. et al. 1997. *Nature* 389: 251–260.) Image prepared with MolScript, BobScript, and Raster 3D.

histone octamer (Fig. 7-25). The association of DNA with the nucleosome is mediated by a large number (~40) of hydrogen bonds between the histones and the DNA. The majority of these hydrogen bonds are between the proteins and the oxygen atoms in the phosphodiester backbone near the minor groove of the DNA. Only seven hydrogen bonds are made between the protein side chains and the bases, and all of these are made in the minor groove of the DNA.

The large number of these hydrogen bonds (a typical sequence-specific DNA-binding protein only has ~20 hydrogen bonds with DNA) provides the driving force to bend the DNA. The highly basic nature of the histones further facilitates DNA bending by masking the negative charge of the phosphates that ordinarily resists DNA bending. This is because when DNA is bent, the phosphates on the inside of the bend are brought into unfavorably close proximity. The positively charged nature of the histones also facilitates the close juxtaposition of the two adjacent DNA helices necessary to wrap the DNA more than once around the histone octamer.

The finding that all of the sites of contact between the histones and the DNA involve either the minor groove or the phosphate backbone

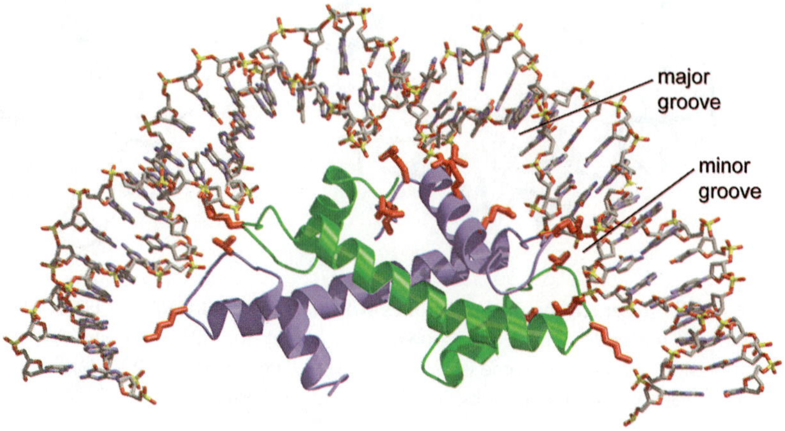

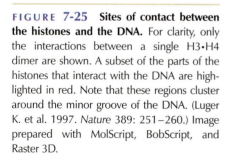

FIGURE 7-25 Sites of contact between the histones and the DNA. For clarity, only the interactions between a single H3•H4 dimer are shown. A subset of the parts of the histones that interact with the DNA are highlighted in red. Note that these regions cluster around the minor groove of the DNA. (Luger K. et al. 1997. *Nature* 389: 251–260.) Image prepared with MolScript, BobScript, and Raster 3D.

is consistent with the non-sequence-specific nature of the association of the histone octamer with DNA. Neither the phosphate backbone nor the minor groove is rich in base-specific information. Moreover, of the seven hydrogen bonds formed with the bases in the minor groove, *none* is with parts of the bases that distinguish between G:C and A:T base pairs (see Chapter 6, Fig. 6-10).

The Histone Amino-Terminal Tails Stabilize DNA Wrapping around the Octamer

The structure of the nucleosome also tells us something about the histone amino-terminal tails. The four H2B and H3 tails emerge from between the two DNA helices. In each case, their path of exit is formed by two adjacent minor grooves, making a "gap" between the two DNA helices just big enough for a polypeptide chain (Fig. 7-26a). Strikingly, the H2B and H3 tails emerge at approximately equal distances from each other around the octamer disk (at ~1 and 11 o'clock for the H3 tails and 4 and 8 o'clock for H2B). Instead of emerging between the two DNA helices, the H2A and H4 amino-terminal tails emerge from either "above" or "below" both DNA helices (Fig. 7-26a). These tails are also distributed around the face of the nucleosome with the H2A tails emerging at 5 and 7 o'clock and the H4 tails at 3 and 9 o'clock (Fig. 7-26b). By emerging both between and on either side of the DNA helices, the histone tails can be thought of as the grooves of a screw, directing the DNA to wrap around the histone octamer disc in a left-handed manner. As we discussed in Chapter 6, the left-handed nature of the DNA wrapping introduces negative supercoils in the DNA. The parts of the tails most proximal to the histone disc (and therefore not subject to the

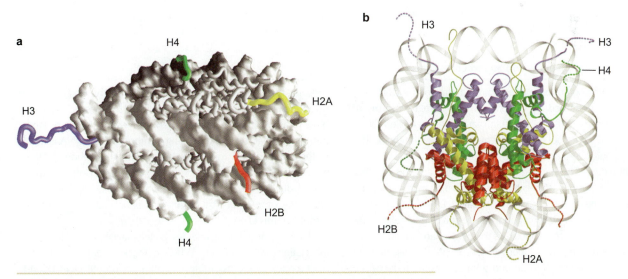

FIGURE 7-26 Histone tails emerge from the core of the nucleosome at specific positions. (a) The side view illustrates that the H3 and H2B tails emerge from between the two DNA helices. In contrast, the H4 and H2A tails emerge either above or below both DNA helices. (Luger K. et al. 1997. *Nature* 389: 251–260.) Image prepared with GRASP. (b) The position of the tails relative to the entry and exit of the DNA is shown here. This view reveals that the histone tails emerge at numerous positions relative to the DNA. (Davey C.A. et al. 2002. *J. Mol. Biol.* 319: 1097–1113.) Image prepared with MolScript, BobScript, and Raster 3D.

protease cleavage discussed above) also make some of the many hydrogen bonds between the histones and the DNA as they pass by the DNA.

Wrapping of the DNA around the Histone Protein Core Stores Negative Superhelicity

Each nucleosome added to a covalently closed circular template changes the linking number of the associated DNA by approximately –1.2. Because the remainder of the DNA is kept relaxed by topoisomerases, the DNA that is packaged into nucleosomes would become negatively supercoiled if nucleosomes were removed from the DNA. Thus, nucleosomes can be viewed as storing or stabilizing negative superhelicity. Why would the cell want to maintain a stockpile of negative superhelicity? There are many instances when it is useful to drive unwinding of DNA in the cell, including initiation of DNA replication, transcription, and recombination. Importantly, negatively supercoiled DNA favors DNA unwinding (see Fig. 6-17). Thus, removal of a nucleosome not only allows increased access to the DNA, but also facilitates DNA unwinding of nearby DNA sequences (Box 7-2, Nucleosomes and Superhelical Density).

If nucleosomes store negative superhelicity in eukaryotic cells, what serves the equivalent function in prokaryotic cells? The answer for many prokaryotic organisms is that the entire genome is maintained in a negatively supercoiled state. This is accomplished by a specialized topoisomerase called **gyrase** that has the ability to introduce negative superhelicity into relaxed DNA by reducing the linking number. For example, in *E. coli* cells, gyrase action results in the genome having an average superhelical density of approximately –0.07. The addition of negative supercoils into otherwise relaxed DNA is an energy-requiring reaction. Consistent with this, gyrase requires the presence of ATP to introduce negative supercoils. In the absence of

KEY EXPERIMENTS

Box 7-2 Nucleosomes and Superhelical Density

Why do nucleosomes alter the topological state of the DNA they include? As described in Chapter 6, there are two forms of writhe that can contribute to the formation of supercoiled DNA: toroidal and interwound (also referred to as plectonemic). The wrapping of DNA around the histone octamer is a form of toroidal writhe. The handedness of the writhe controls whether it introduces positive or negative supercoils (i.e., increases or decreases linking number of the associated DNA). For toroidal writhe, left-handed wrapping induces negative superhelicity (for interwound writhe, the opposite is true; right-handed pitch is associated with negative superhelicity). Thus, the left-handed toroidal wrapping of DNA around the nucleosome reduces the linking number of the associated DNA. For this reason, nucleosomes preferentially incoporate DNA with negative superhelical density. In contrast, assembling nucleosomes on DNA that has a positive superhelical density is very difficult.

The assembly of many nucleosomes on covalently closed circular DNA (cccDNA) requires the presence of a topoisomerase to accommodate changes in the linking number of the DNA bound to histones (see Box 7-2 Fig.1). Without a topoisomerase present, for every nucleosome formed with the cccDNA, the unbound DNA (not associated with nucleosomes) would have to accommodate an equivalent *increase* in linking number (remember that the overall linking number of a cccDNA is fixed in the absence of a topoisomerase). Thus, the unbound DNA would accumulate increased linking number and positive superhelical density. The more positively supercoiled the unbound DNA, the more difficult it is for additional nucleosomes to assemble on this DNA.

Addition of a topoisomerase greatly facilitates nucleosome association with cccDNA. When a topoisomerase is present during nucleosome assembly, it cannot act on the DNA bound to the nucleosome. Instead, the topoisomerase relaxes the DNA not included in nucleosomes, reducing the positive superhelical density in these regions by decreasing the linking number. By maintaining the unbound DNA in a relaxed state, topoisomerases facilitate the binding of histones to the DNA and the formation of additional nucleosomes. Importantly, the overall effect on the plasmid is that the linking number is decreased as more nucleosomes are assembled.

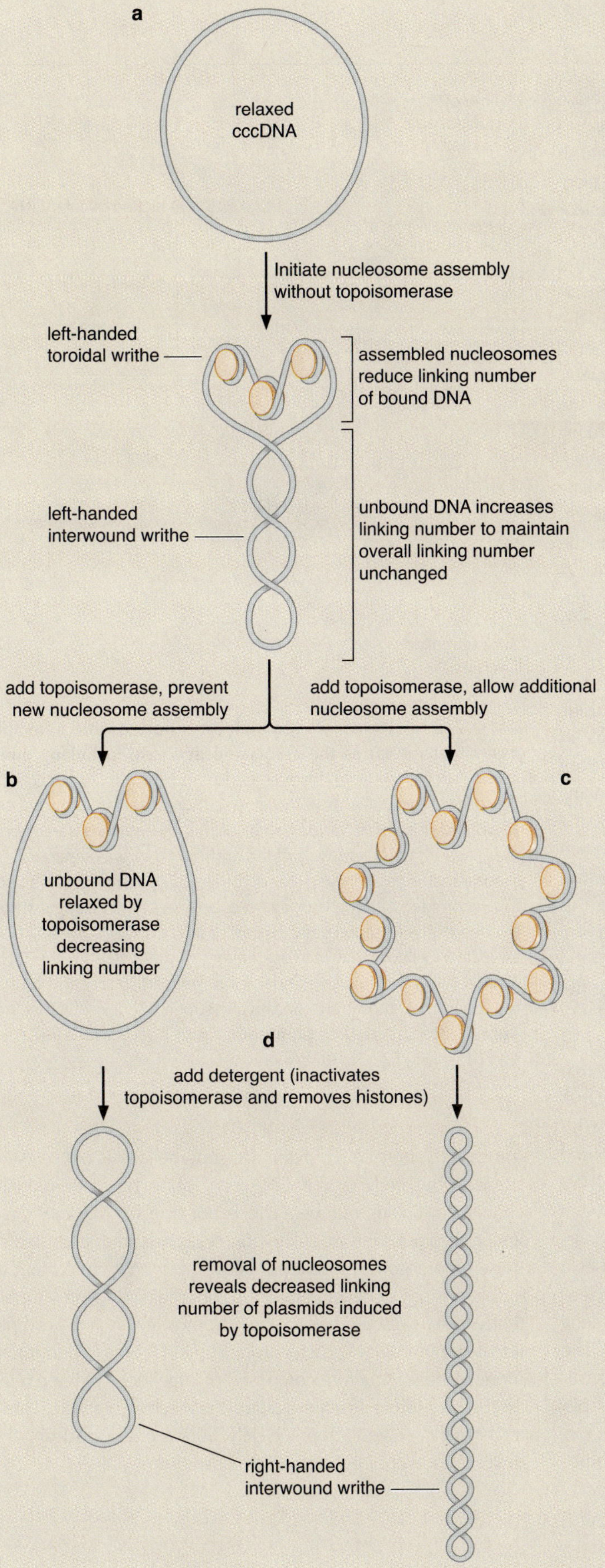
a
relaxed cccDNA
Initiate nucleosome assembly without topoisomerase
left-handed toroidal writhe
assembled nucleosomes reduce linking number of bound DNA
left-handed interwound writhe
unbound DNA increases linking number to maintain overall linking number unchanged
add topoisomerase, prevent new nucleosome assembly
add topoisomerase, allow additional nucleosome assembly
b
c
unbound DNA relaxed by topoisomerase decreasing linking number
d
add detergent (inactivates topoisomerase and removes histones)
removal of nucleosomes reveals decreased linking number of plasmids induced by topoisomerase
right-handed interwound writhe

BOX 7-2 FIGURE 1 Topoisomerase is required for nucleosome assembly using covalently closed circular DNA (cccDNA). (a) Assembly of nucleosomes using cccDNA in the absence of topoisomerase is limited by the accumulation of positive superhelicity in the DNA not associated with nucleosomes. (b) Addition of topoisomerase without additional nucleosome assembly illustrates how topoisomerase reduces linking number to relax the DNA not incorporated into nucleosomes. (c) Additional nucleosome assembly in the presence of topoisomerase. (d) Simultaneous removal of histones and inactivation of topoisomerase (e.g., by addition of a strong detergent) reveals the reduced linking number associated with nucleosomal DNA.

Box 7-2 *(Continued)*

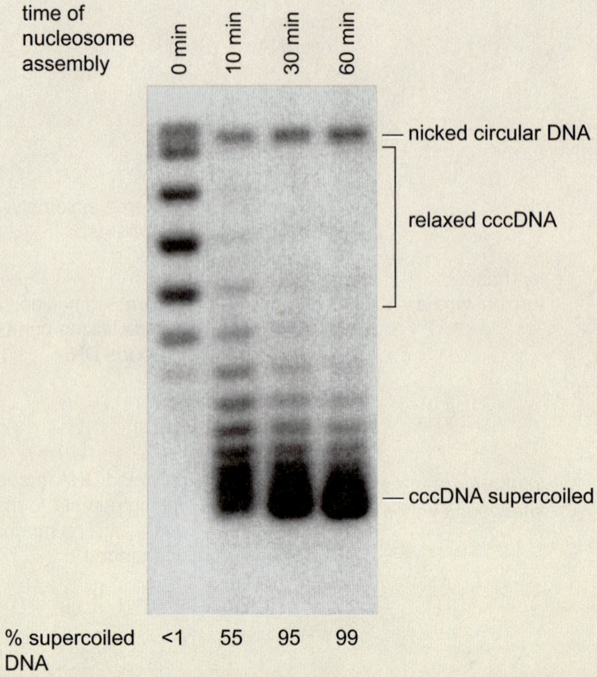

Box 7-2 Figure 2 Example of a nucleosome assembly assay that measures the associated decrease in linking number. Nucleosome assembly was performed on a relaxed cccDNA in the presence of a topoisomerase. Prior to the initiation of assembly (0 min) or at various times during the nucleosome assembly reaction, detergent was added and the DNA was separated on a nondenaturing agarose gel and visualized by staining with ethidium bromide. Although an agarose gel does not distinguish between positively and negatively supercoiled cccDNA, the ability of DNA intercalators that increase linking number to shift the DNA toward the top of the gel (and a more relaxed state) can be used to show that these are negatively supercoiled cccDNAs (not shown). (Reprinted, with permission, from Ito T. et al. 1997. *Cell* 90: 145–155, Fig. 2c. ©Elsevier.)

The decrease in linking number caused by topoisomerase during nucleosome assembly can be used as an assay for this event. The assay takes advantage of the ability to distinguish between relaxed and supercoiled cccDNA by gel electrophoresis (see Fig. 6-27). The first step is to assemble nucleosomes onto a cccDNA in the presence of a topoisomerase. At appropriate times, a strong detergent (e.g., SDS [sodium dodecyl sulfate]) is added to the assembly reaction, inactivating the topoisomerase and removing any histones from the DNA. The resulting DNA is then separated by gel electrophoresis to determine the supercoiled nature of the DNA. Because the detergent inactivates the topoisomerase at the same time as removing the histones from the DNA, the linking number of the DNA assembled into nucleosomes is preserved. On average, the topoisomerase will have decreased the linking number by −1.2 for each nucleosome assembled on the cccDNA. Thus, the more nucleosomes assembled on the cccDNA, the more negatively supercoiled the cccDNA (Box 7-2 Fig. 1c,d). This can easily be observed by the faster migration of supercoiled DNA during gel electrophoresis (Box 7-2 Fig. 2).

Because nucleosomal DNA wraps around the histone protein 1.65 times, the formation of a single nucleosome using covalently closed circular plasmid would create a writhe of −1.65 and thus change the linking number by an equivalent amount. As described above, when the change in linking number associated with each nucleosome was measured, the number was lower than this, approximately −1.2 for each nucleosome added. This discrepancy is referred to as the "nucleosome linking number paradox," and the solution to this conundrum was revealed when the high-resolution crystal structure of the nucleosome was solved. Careful analysis of the DNA associated with the histone protein core showed that the number of bases per turn was reduced relative to naked DNA (from 10.5 to 10.2 bp/turn). A reduction in the number of bp/turn results in an increase in the linking number for that DNA. Consider the example of a 10,500-bp cccDNA described in Chapter 6. Normal B-form DNA will have 10.5 bp/turn resulting in a linking number of +1000 for the plasmid (10,500/10.5). In contrast, the same DNA with a pitch of 10.2 bp/turn will have a linking number of about +1029 (10,500/10.2). Thus, by decreasing the number of base pairs per turn of the helix, binding to the histone octamer causes a slight increase in linking number over the length of the nucleosome-bound DNA. This change reduces the change in linking number per nucleosome assembled from −1.65 to −1.2. The difference of approximately +0.4 per nucleosome can be calculated using the difference in the number of base pairs per turn and the length of DNA associated with a nucleosome.

Are these issues relevant to the linear eukaryotic chromosomes? For short linear fragments, superhelicity is not relevant because the ends of the DNA can rotate to accommodate changes in linking number. This is not true for the very large linear chromosomes of eukaryotic cells. First, the large size of these chromosomes would not allow rapid enough rotation to easily dissipate changes in DNA superhelicity. More importantly, as we discuss below, the chromosome is not a simple linear strand of DNA. Each chromosomal DNA is folded into a more compact structure composed of large loops that are tethered to a protein structure called the nuclear scaffold. These attachments serve to topologically isolate one loop from the next and prevent free rotation of chromosomal DNA.

ATP, gyrase can only relax DNA (e.g., reduce the linking number of positively supercoiled DNA).

Not all bacteria need to maintain their DNA in a negatively supercoiled state. Bacteria that prefer to grow at very high temperatures (>80°C) must expend energy *to prevent* their DNA from unwinding due to thermal denaturation. These organisms have a different topoisomerase called **reverse gyrase.** Consistent with its name, reverse gyrase increases the linking number of relaxed DNA in the presence of ATP. By keeping the genome positively supercoiled, reverse gyrase counteracts the effect of thermal denaturation that would ordinarily result in many regions of the genome being unwound.

HIGHER-ORDER CHROMATIN STRUCTURE

Heterochromatin and Euchromatin

From the earliest observations of chromosomes in the light microscope, it was clear that they were not uniform structures. Early studies of chromosomes divided chromosomal regions into two categories: **euchromatin** and **heterochromatin**. Heterochromatin was characterized by dense staining with a variety of dyes and a more condensed appearance, whereas euchromatin had the opposite characteristics, staining poorly with dyes and having a relatively open structure. As our molecular understanding of genes and their expression advanced, it became clear that heterochromatic regions of chromosomes had very limited gene expression. In contrast, euchromatic regions showed higher levels of gene expression, suggesting that these different structures were connected to global levels of gene expression.

Heterochromatic regions show little gene expression, but this does not mean that these regions are not important. As we shall learn when gene expression is discussed, keeping a gene off can be just as important as turning a gene on. In addition, heterochromatin is associated with particular chromosomal regions, including the telomere and the centromere, and is important for the function of both of these key chromosomal elements.

Over the years, researchers have gained a more complete molecular understanding of heterochromatin and euchromatin structure. It is clear that DNA in both types of chromatin is packaged into nucleosomes. The difference in heterochromatin and euchromatin structure is how the nucleosomes in these different chromosomal regions are (or are not) assembled into larger assemblies. It has become clear that heterochromatic regions are composed of nucleosomal DNA assembled into higher-order structures that result in a barrier to gene expression. In contrast, euchromatic nucleosomes are found to be in much less organized assemblies. In the following sections, we discuss what is known about how nucleosomes are assembled into higher-order structures.

Histone H1 Binds to the Linker DNA between Nucleosomes

Once nucleosomes are formed, the next step in the packaging of DNA is the binding of histone H1. Like the core histones, H1 is a small, positively charged protein (see Table 7-5). H1 interacts with the linker DNA between nucleosomes, further tightening the association of the

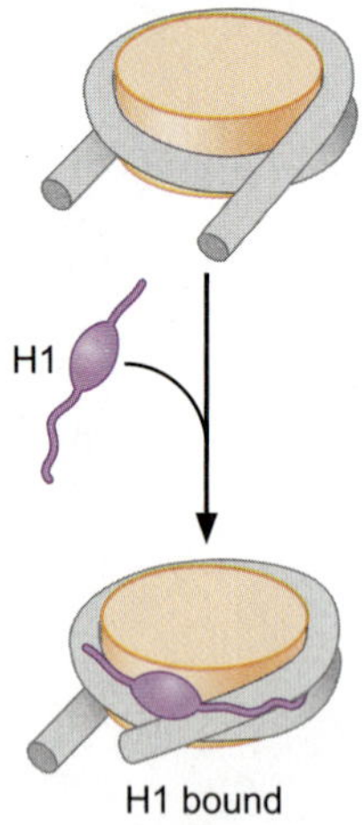

FIGURE 7-27 Histone H1 binds two DNA helices. Upon interacting with a nucleosome, histone H1 binds to the linker DNA at one end of the nucleosome and the central DNA helix of the nucleosome bound DNA (the middle of the 147 bp bound by the core histone octamer).

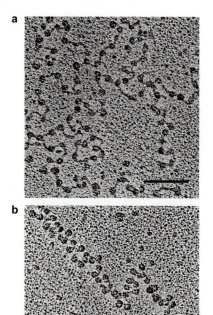

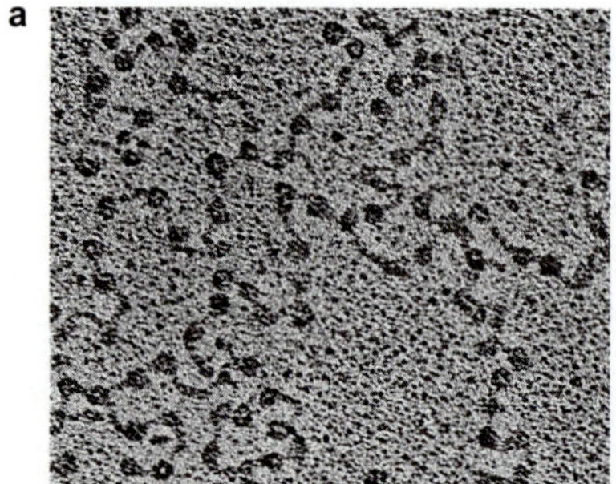

FIGURE 7-28 Addition of H1 leads to more compact nucleosomal DNA. The two images show an electron micrograph of nucleosomal DNA in the absence (a) and presence (b) of histone H1. Note the more compact and defined structure of the DNA in the presence of histone H1. (Reprinted, with permission, from Thoma F. et al. 1979. *J. Cell Biol.* 83: 403–427, Figs. 4. and 6. © Rockefeller University Press.)

DNA with the nucleosome. This can be detected by the increased protection of nucleosomal DNA from micrococcal nuclease digestion. Thus, beyond the 147 bp protected by the core histones, addition of histone H1 to a nucleosome protects an additional 20 bp of DNA from micrococcal nuclease digestion.

Histone H1 has the unusual property of binding two distinct regions of the DNA duplex. Typically, these two regions are part of a single DNA molecule associated with a nucleosome (Fig. 7-27). The sites of H1 binding are located asymmetrically relative to the nucleosome. One of the two regions bound by H1 is the linker DNA at *one* end of the nucleosome. The second site of DNA binding is in the middle of the associated 147 bp (the only DNA duplex present at the dyad axis). Thus, the additional DNA, protected from the nuclease digestion described above, is restricted to linker DNA on only *one* side of the nucleosome. By bringing these two regions of DNA into close proximity, H1 binding increases the length of the DNA wrapped tightly around the histone octamer.

H1 binding produces a more defined angle of DNA entry and exit from the nucleosome. This effect, which can be visualized in the electron microscope (Fig. 7-28), results in the nucleosomal DNA taking on a distinctly zigzag appearance. The angles of entry and exit observed vary substantially depending on conditions (including salt concentration, pH, and the presence of other proteins). If we assume that these angles are ~20° relative to the dyad axis, this would result in a pattern in which nucleosomes would alternate on either side of a central region of linker DNA bound by histone H1 (Fig. 7-29).

Nucleosome Arrays Can Form More Complex Structures: The 30-nm Fiber

Binding of H1 stabilizes higher-order chromatin structures. In the test tube, as salt concentrations are increased, the addition of histone H1 results in the nucleosomal DNA forming a **30-nm fiber**. This structure, which can also be observed in vivo, represents the next level of DNA compaction. Importantly, the incorporation of DNA into this fiber makes the DNA less accessible to many DNA-dependent enzymes (such as RNA polymerases).

There are two models for the structure of the 30-nm fiber. In the **solenoid model**, the nucleosomal DNA forms a superhelix containing approximately six nucleosomes per turn (see Fig. 7-18a). This structure is supported by both electron microscopy and X-ray diffraction studies, which indicate that the 30-nm fiber has a helical pitch of ~11 nm. This distance is also the approximate diameter of the nucleosome disc, suggesting that the 30-nm fiber is composed of nucleosome discs stacked on edge in the form of a helix (Fig. 7-30a). In this model, the flat surfaces on either face of the histone octamer disc are adjacent to each other, and the DNA surface of the nucleosomes forms the accessible surface of the superhelix. The linker DNA is buried in the center of the superhelix, but it never passes through the axis of the fiber. Rather, the linker DNA circles around the central axis as the DNA moves from one nucleosome to the next.

An alternative model for the 30-nm fiber is the "zigzag" model (Fig. 7-30b). This model is based on the zigzag pattern of nucleosomes formed upon H1 addition. In this case, the 30-nm fiber is a compacted form of these zigzag nucleosome arrays. A recent X-ray

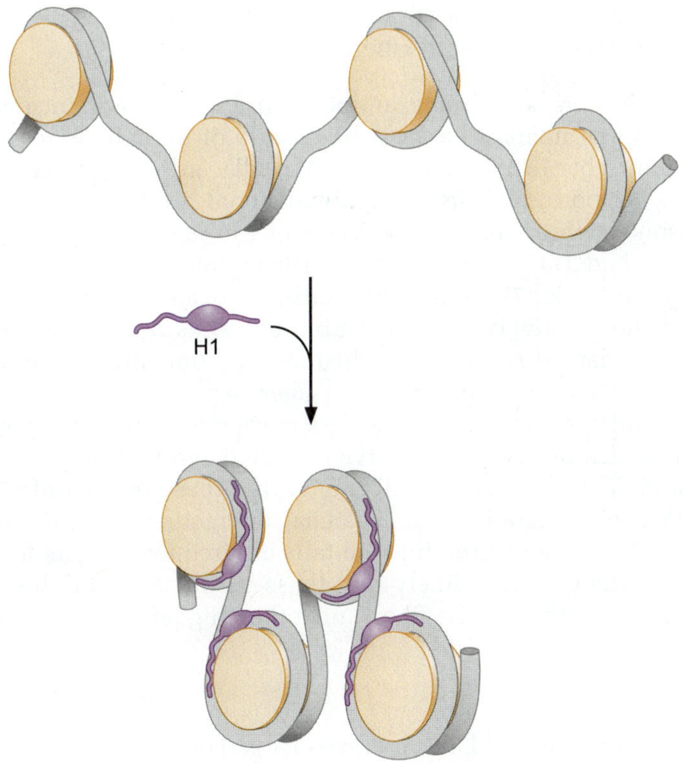

FIGURE 7-29 Histone H1 induces tighter DNA wrapping around the nucleosome. The two illustrations show a comparison of the wrapping of DNA around the nucleosome in the presence and absence of histone H1. One histone H1 can associate with each nucleosome.

structure of a single DNA molecule participating in four nucleosomes and biophysical studies of the spring-like nature of isolated 30-nm fibers support the zigzag model. Unlike the solenoid model, the zigzag conformation requires the linker DNA to pass through the central axis of the fiber in a relatively straight form (see Fig. 7-30b). Thus, longer linker DNA favors this conformation. Because the average linker DNA varies between different species (see Table 7-4), the form of the 30-nm fiber may not always be the same and both models may be correct.

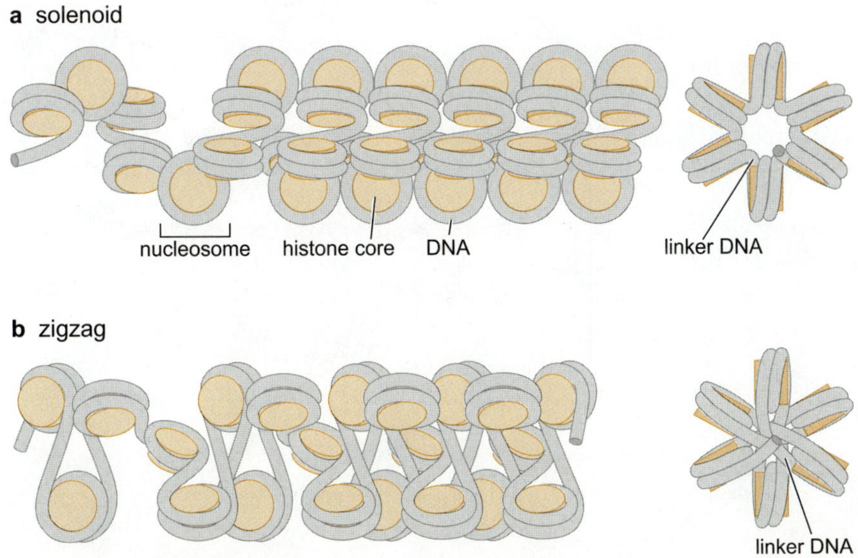

FIGURE 7-30 Two models for the 30-nm chromatin fiber. In each panel, the left-hand view shows the side of the fiber and the right-hand view shows a view down the central axis of the fiber. (a) The solenoid model. Note that the linker DNA does not pass through the central axis of the superhelix and that the sides and entry and exit points of the nucleosomes are relatively inaccessible. (b) The "zigzag" model. In this model, the linker DNA frequently passes through the central axis of the fiber, and the sides and even the entry and exit points are more accessible. (Reproduced, with permission, from Pollard T. and Earnshaw W. 2002. *Cell biology,* 1st ed., Fig. 13–6. © Elsevier.)

The Histone Amino-Terminal Tails Are Required for the Formation of the 30-nm Fiber

Core histones lacking their amino-terminal tails are incapable of forming 30-nm fibers. The most likely role of the tails is to stabilize the 30-nm fiber by interacting with adjacent nucleosomes. This model is supported by the three-dimensional crystal structure of the nucleosome, which shows that each of the amino-terminal tails of H2A, H3, and H4 interacts with adjacent nucleosome cores in the crystal lattice (Fig. 7-31). Recent studies indicate that the interaction between the positively charged amino terminus of histone H4 and a negatively charged region of the histone-fold domain of histone H2A is particularly important for 30-nm fiber formation. The residues of H2A that interact with the H4 tail are conserved across many eukaryotic organisms but are not involved in DNA binding or formation of the histone octamer. One possibility is that these regions of H2A are conserved to mediate internucleosomal interactions with the H4 tail. As we shall see below, the histone tails are frequent targets for modification in the cell. It is likely that these modifications influence the ability to form the 30-nm fiber and other higher-order nucleosome structures.

Further Compaction of DNA Involves Large Loops of Nucleosomal DNA

Together, the packaging of DNA into nucleosomes and the 30-nm fiber results in the compaction of the linear length of DNA by ~40-fold. This is still insufficient to fit 1–2 meters of DNA into a nucleus ~10^{-5} m across. Additional folding of the 30-nm fiber is required to compact the DNA further. Although the exact nature of this folded structure remains unclear, one popular model proposes that the 30-nm fiber forms loops of 40–90 kb that are held together at their bases by a proteinaceous structure referred to as the **nuclear scaffold** (Fig. 7-32). A variety of methods have been developed to identify proteins that are part of this structure, although the true nature of the nuclear scaffold remains mysterious.

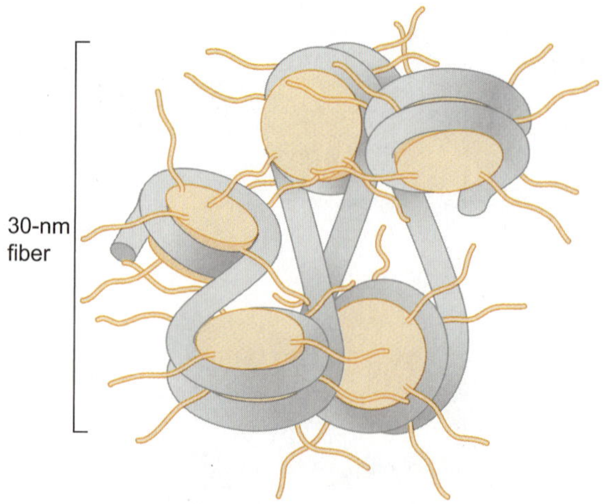

FIGURE 7-31 Speculative model for the stabilization of the 30-nm fiber by histone amino-terminal tails. In this model, the 30-nm fiber is illustrated using the "zigzag" model. Several different tail–histone core interactions are possible. Here, the interactions are shown as between every alternate histone, but they could also be with adjacent or more distant histones.

a

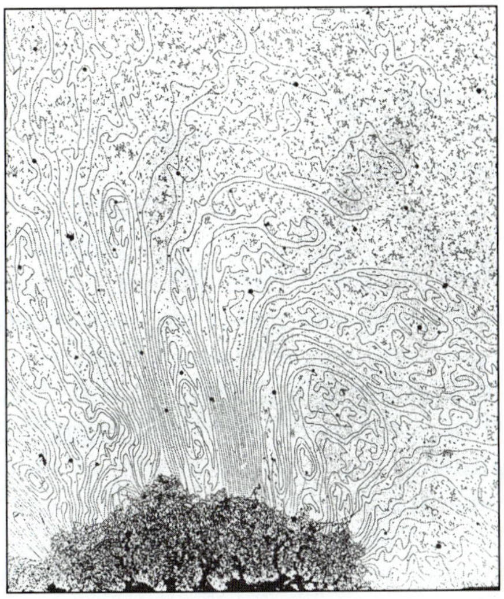

FIGURE 7-32 Higher-order structure of chromatin. (a) A transmission electron micrograph shows chromatin emerging from a central structure of a chromosome. The electron-dense regions are the nuclear scaffold that acts to organize the large amounts of DNA found in eukaryotic chromosomes. (b) A model for the structure of a eukaryotic chromosome shows that the majority of the DNA is packaged into large loops of 30-nm fiber that are tethered to the nuclear scaffold at their base. Sites of active DNA manipulation (e.g., sites of transcription or DNA replication) are in the form of 10-nm fiber or even naked DNA. (a, Courtesy of J.R. Paulson and U.K. Laemmli.)

b

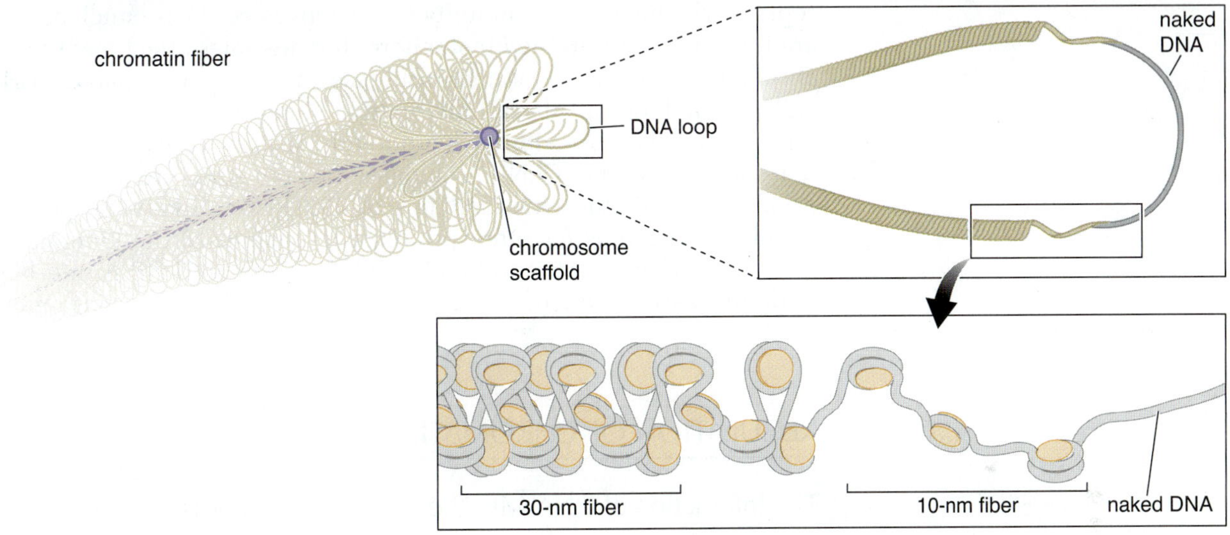

Two classes of proteins that contribute to the nuclear scaffold have been identified. One of these is topoisomerase II (Topo II), which is abundant in both scaffold preparations and purified mitotic chromosomes. Treating cells with drugs that result in DNA breaks at the sites of Topo II DNA binding generates DNA fragments that are ~50 kb in size. This is similar to the size range observed for limited nuclease digestion of chromosomes and suggests that Topo II may be part of the mechanism that holds the DNA at the base of these loops. In addition, the presence of Topo II at the bottom of each loop would ensure that the loops are topologically isolated from one another.

The SMC proteins are also abundant components of the nuclear scaffold. As we discussed earlier (see above section on Chromosome Duplication and Segregation), these proteins are key components of the machinery that condenses and holds sister chromatids together after chromosome duplication. The associations of these proteins with the nuclear scaffold may serve to enhance their functions by providing an underlying foundation for their interactions with chromosomal DNA.

Histone Variants Alter Nucleosome Function

The core histones are among the most conserved eukaryotic proteins; therefore, the nucleosomes formed by these proteins are very similar in all eukaryotes. But there are numerous histone variants found in eukaryotic cells. Such unorthodox histones can replace one of the four standard histones to form alternate nucleosomes. Such nucleosomes may serve to demarcate particular regions of chromosomes or confer specialized functions to the nucleosomes into which they are incorporated. For example, H2A.X is a variant of H2A that is widely distributed in eukaryotic nucleosomes. When chromosomal DNA is broken (referred to as a double-strand break), H2A.X located adjacent to the break is phosphorylated. Phosphorylated H2A.X is specifically recognized by DNA repair enzymes leading to their localization at the site of DNA damage.

A second histone variant, CENP-A, is associated with nucleosomes that include centromeric DNA. In this chromosomal region, CENP-A replaces the histone H3 subunits in nucleosomes. These nucleosomes are incorporated into the kinetochore that mediates attachment of the chromosome to the mitotic spindle (see Fig. 7-12). Compared to H3, CENP-A includes a substantially extended amino-terminal tail region but has an otherwise similar histone-fold region. Thus, it is unlikely that incorporation of CENP-A changes the core structure of the nucleosome. Instead, the extended tail of CENP-A is likely to generate novel binding sites for other protein components of the kinetochore (Fig. 7-33). Consistent with this hypothesis, loss of CENP-A interferes with the association of kinetochore components with centromeric DNA.

REGULATION OF CHROMATIN STRUCTURE

The Interaction of DNA with the Histone Octamer Is Dynamic

As discussed in detail in Chapter 17, the incorporation of DNA into nucleosomes can have a profound impact on the expression of the genome. In many instances, it is critical that nucleosomes can be moved or that their grip on the DNA can be loosened to allow access to particular regions of DNA. Consistent with this requirement, the association of the histone octamer with the DNA is inherently dynamic. In addition, there are factors that act on the nucleosome to increase or decrease the dynamic nature of this association. Together, these properties allow changes in nucleosome position and DNA association in response to the frequently changing needs for DNA accessibility.

Like all interactions mediated by noncovalent bonds, the association of any particular region of DNA with the histone octamer is not

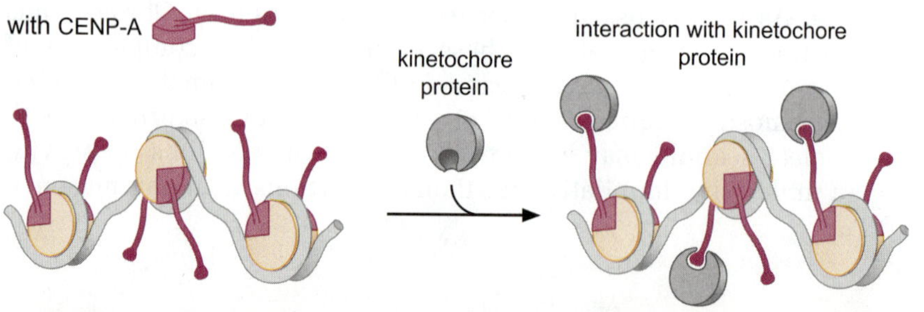

FIGURE 7-33 Alteration of chromatin by incorporation of histone variants. Incorporation of CENP-A in place of histone H3 is proposed to act as a binding site for one or more protein components of the kinetochore.

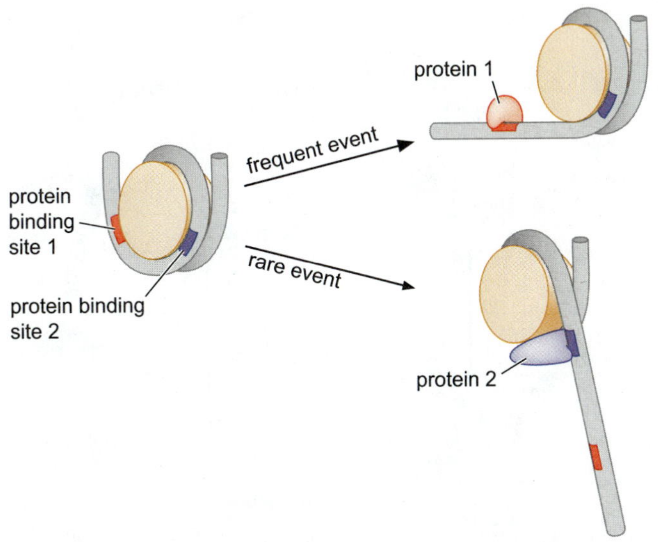

FIGURE 7-34 Model for gaining access to nucleosome-associated DNA. Studies of the ability of sequence-specific DNA-binding proteins to bind nucleosomes suggest that unwrapping of the DNA from the nucleosome is responsible for accessibility of the DNA. DNA sites closest to the entry and exit points are the most accessible and sites closest to the midpoint of the bound DNA are least accessible.

permanent: any individual region of the DNA will transiently be released from tight interaction with the octamer now and then. This release is analogous to the occasional opening of the DNA double helix (as we discussed in Chapter 6). The dynamic nature of DNA binding to the histone core structure is important, because many DNA-binding proteins strongly prefer to bind to histone-free DNA. Such proteins can recognize their binding site only when it is released from the histone octamer or is contained in linker or nucleosome-free DNA.

As a result of intermittent, spontaneous unwrapping of DNA from the nucleosome, a protein can gain access to its DNA-binding sites with a probability of 1 in 50 to 1 in 100,000, depending on where the binding site is within the nucleosome. The more central the binding site, the less frequently it is accessible. Thus, a binding site near position 73 of the 147 bp tightly associated with a nucleosome is least frequently accessible, whereas binding sites near the ends (position 1 or 147) of the nucleosomal DNA are most frequently accessible. These findings indicate that the mechanism of exposure is due to unwrapping of the DNA from the nucleosome, rather than to the DNA briefly coming off the surface of the histone octamer (Fig. 7-34). It is important to note that these studies were performed on a population of individual nucleosomes in a test tube: the ability of DNA to unwrap from the nucleosome may be different for the large stretches of DNA participating in many adjacent nucleosomes (called nucleosome arrays) present in cells. Association of H1 and incorporation of nucleosomes into the 30-nm fiber will also alter these probabilities. Nevertheless, the dynamic nature of nucleosome structure indicates that nucleosomes only look like the structure revealed in the X-ray crystallography studies for short periods of time and instead spend much of their time in other conformations.

Nucleosome-Remodeling Complexes Facilitate Nucleosome Movement

In addition to the intrinsic dynamics exhibited by the nucleosome, the stability of the histone octamer–DNA interaction is influenced by large protein complexes called **nucleosome-remodeling complexes.**

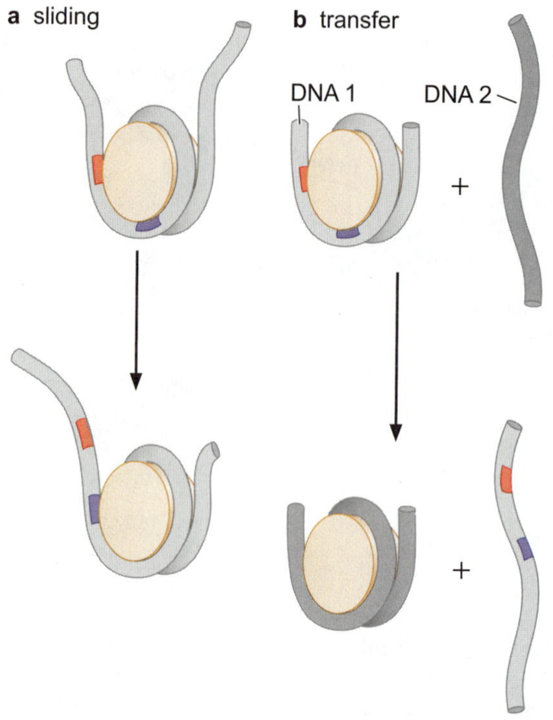

FIGURE 7-35 Nucleosome movement catalyzed by nucleosome-remodeling activities. (a) Nucleosome movement by sliding along a DNA molecule exposes sites for DNA-binding proteins. (b) Nucleosome movement can alternatively occur by transfer of the nucleosome from one strand of DNA to another.

These multiprotein complexes facilitate changes in nucleosome location or interaction with the DNA using the energy of ATP hydrolysis. There are two basic types of nucleosome changes mediated by these enzymes (Fig. 7-35). All nucleosome-remodeling complexes can catalyze the **"sliding"** of DNA along the surface of the histone octamer. A subset of nucleosome-remodeling complexes can catalyze a second, more extreme change in which a histone octamer is **"transferred"** from one DNA helix to another.

Recent studies have begun to reveal how nucleosome-remodeling complexes move DNA on the surface of the histone octamer. Each of these multisubunit enzymes contains an ATP-hydrolyzing subunit that is capable of moving in a directional manner (also called translocating) on double-stranded DNA. In the context of a nucleosome-remodeling complex, these enzymes are thought to bind tightly to the histone octamer and translocate the DNA relative to the histones. Because it would be energetically difficult to move all parts of the DNA relative to the histones simultaneously, it is proposed that the ATP-hydrolyzing subunit moves only a portion of the DNA at a time, resulting in an "inchworm-like" movement of the DNA on the surface of the histone octamer (Fig. 7-36). It is important to keep in mind that different DNA sequences interact with the histone octamer with roughly equal affinities. Thus, a DNA molecule that is sliding across a histone octamer can be viewed as binding to the octamer in many different energetic equivalent states and the nucleosome-remodeling complex is allowing DNA to access these different states more easily.

There are multiple types of nucleosome-remodeling complexes in any given cell (Table 7-6). They can have as few as two subunits or more than ten subunits. Each of these complexes contains a similar ATP hydrolyzing subunit that catalyzes the DNA movement described above and in Figure 7-36. Although the ATP-hydrolyzing sub-

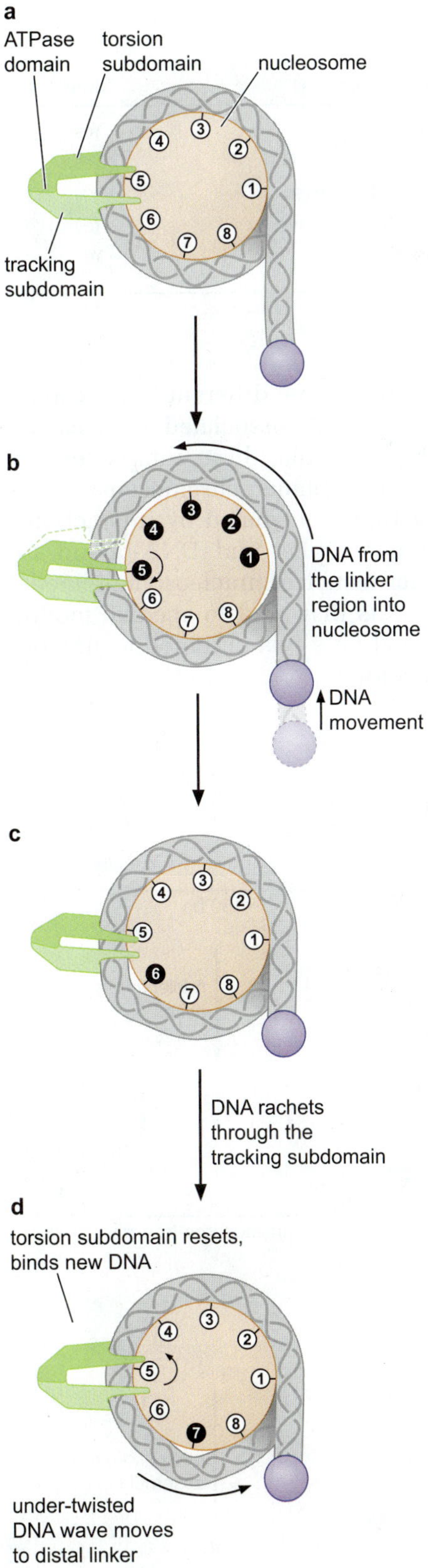

FIGURE 7-36 A model for nucleosomal DNA sliding catalyzed by nucleosome-remodeling complexes. (a) The model proposes that the ATP-hydrolyzing subunit of the nucleosome-remodeling complex binds the nucleosomal DNA two helical turns from the central dyad (e.g., at position 52 out of the total of 147 bp associated with the nucleosome). Other subunits of the nucleosome-remodeling complex bind tightly to the histones. The illustration shows each of the contacts between the DNA and the histones from the dyad to the closest unbound DNA (one contact per helical turn, 7 of the 14). Using the ATP-dependent DNA translocating activity, the nucleosome-remodeling complex first pulls the DNA from the nearest linker domain into the nucleosome (b). This breaks the five histone–DNA contacts between the ATP-hydrolyzing subunit and the linker DNA (broken contacts are shown in black, intact contacts in white). (c) The broken contacts then re-form with the translocated DNA, leaving too much DNA associated with the nucleosome next to the ATP-hydrolyzing subunit (between contacts 5 and 6). (d) To relieve this strain, the model proposes that the additional sequences move like a "wave" across the surface of the histones, breaking one contact at a time (first contact 6 and then 7) until all the contacts have re-formed with the appropriate amount of DNA between them, at which point the excess DNA is no longer present within the histone-associated DNA and the nucleosome has shifted its position on the DNA. (Adapted, with permission, from Saha A. et al. 2006. *Nat. Rev. Mol. Cell Biol.* 7: 437–447, Fig. 4a. © Macmillan.)

TABLE 7-6 ATP-Dependent Nucleosome-Remodeling Complexes

Type	Number of Subunits	Histone-Binding Domains	Slide	Transfer
SWI/SNF	8–11	Bromodomain	Yes	Yes
ISWI	2–4	Bromodomain, SANT domain, PHD finger	Yes	No
CHD	8–10	Chromodomain, PHD finger	Yes	No
SWR1	12–14	Bromodomain, SANT domain	Yes	nd
INO80	10–12	None	Yes	nd

nd, not determined.

unit is similar among the different nucleosome-remodeling complexes, the other subunits associated with each complex modulate their function. For example, these complexes can include subunits that target them to particular chromosomal locations. In some instances, this targeting is mediated by interactions between subunits of the remodeling complex and DNA-bound transcription factors (Fig. 7-37). In other instances, nucleosome-remodeling complexes are localized by subunits that bind to specific modifications of the histone amino-terminal tails (via chromodomains or bromodomains, as we shall discuss below).

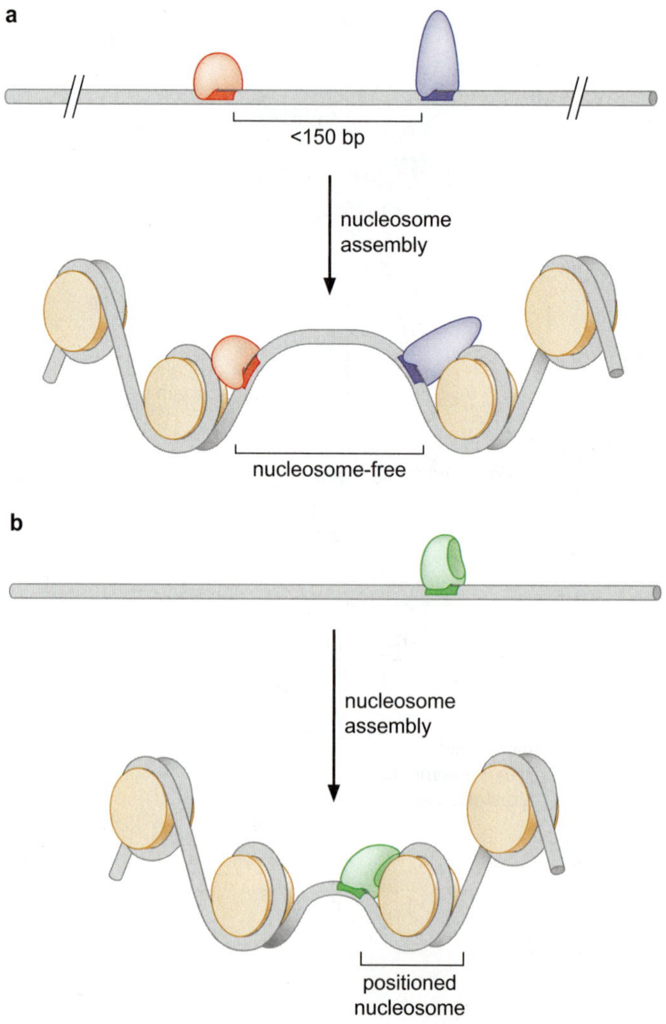

FIGURE 7-37 Two modes of DNA-binding protein-dependent nucleosome positioning. (a) Association of many DNA-binding proteins with DNA is incompatible with the association of the same DNA with the histone octamer. Because a nucleosome requires more than 147 bp of DNA to form, if two such factors bind to the DNA less than this distance apart, the intervening DNA cannot assemble into a nucleosome. (b) A subset of DNA-binding proteins have the ability to bind to nucleosomes. Once bound to DNA, such proteins will facilitate the assembly of nucleosomes immediately adjacent to the protein's DNA-binding site.

Some Nucleosomes Are Found in Specific Positions: Nucleosome Positioning

Because of their sequence-nonspecific and dynamic interactions with DNA, most nucleosomes are not fixed in their locations. But there are occasions when restricting nucleosome location, or **positioning** nucleosomes as it is called, is beneficial. Typically, positioning a nucleosome allows the DNA-binding site for a regulatory protein to remain in the accessible linker DNA region. In many instances, such nucleosome-free regions are larger to allow the binding sites for multiple regulatory proteins to remain accessible.

Nucleosome positioning can be directed by DNA-binding proteins or particular DNA sequences. In the cell, one frequent method involves competition between nucleosomes and DNA-binding proteins. Just as many proteins cannot bind to DNA within a nucleosome, binding of a protein to the DNA can prevent the subsequent association of the core histones with that stretch of DNA. If two such DNA-binding proteins are bound to sites closer than the minimal region of DNA required to assemble a nucleosome (~150 bp), the DNA between the proteins will remain nucleosome-free (Fig. 7-37a). Binding of additional proteins to adjacent DNA can further increase the size of a nucleosome-free region. In addition to this inhibitory mechanism of protein-dependent nucleosome positioning, some DNA-binding proteins interact tightly with adjacent nucleosomes, leading to nucleosomes *preferentially* assembling immediately adjacent to these proteins (Fig. 7-37b).

A second method of nucleosome positioning involves particular DNA sequences that have a high affinity for the nucleosome. Because DNA bound in a nucleosome is bent, nucleosomes preferentially form on DNA that bends easily. A:T-rich DNA has an intrinsic tendency to bend toward the minor groove. Thus, A:T-rich DNA is favored in positions in which the minor groove faces the histone octamer. G:C-rich DNA has the opposite tendency and is therefore favored when the minor groove is facing away from the histone octamer (Fig. 7-38). Each

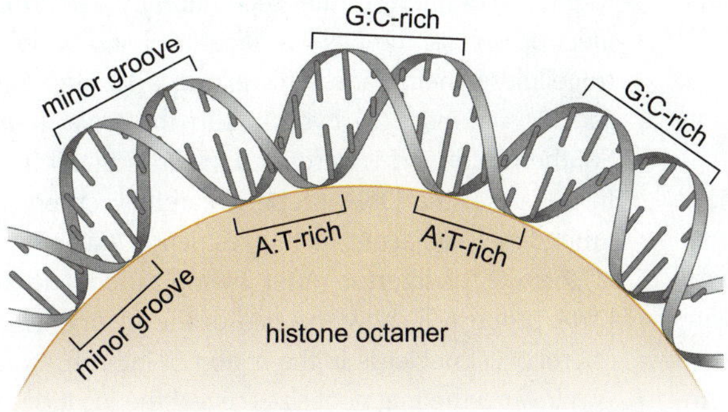

FIGURE 7-38 Nucleosomes prefer to bind bent DNA. Specific DNA sequences can position nucleosomes. Because the DNA is bent severely during association with the nucleosome, DNA sequences that position nucleosomes are intrinsically bent. A:T base pairs have an intrinsic tendency to bend toward the minor groove and G:C base pairs have the opposite tendency. Sequences that alternate between A:T- and G:C-rich sequences with a periodicity of ~5 bp will act as preferred nucleosome-binding sites. (Adapted, with permission, from Alberts B. et al. 2002. *Molecular biology of the cell,* 4th ed., Fig. 4-28. © Garland Science/Taylor & Francis LLC.)

nucleosome will try to maximize this arrangement of A:T-rich and G:C-rich sequences. Recent studies of nucleosome positioning in the yeast *S. cerevisiae* suggest that as many as 50% of tightly positioned nucleosomes can be attributed to preferential binding of the histone core to the sequences they include. It is important to note that, despite being favored, such sequences are not required for nucleosome assembly, and the action of other proteins including chromatin-remodeling and transcription factors can move nucleosomes from such preferred positions.

These mechanisms of nucleosome positioning influence the organization of nucleosomes in the genome. Despite this, many nucleosomes are not tightly positioned. As discussed in the chapters on eukaryotic transcription (Chapters 12 and 17), tightly positioned nucleosomes are most often found at sites directing the initiation of transcription. Although we have discussed positioning primarily as a method to ensure that a regulatory DNA sequence is accessible, a positioned nucleosome can just as easily prevent access to specific DNA sites by being positioned in a manner that overlaps the same sequence. Thus, positioned nucleosomes can have either a positive or negative effect on the accessibility of nearby DNA sequences. An approach to mapping nucleosome locations is described in Box 7-3, Determining Nucleosome Position in the Cell.

KEY EXPERIMENTS

Box 7-3 Determining Nucleosome Position in the Cell

The significance of the location of nucleosomes adjacent to important regulatory sequences has led to the development of methods to monitor the location of nucleosomes in cells. Many of these methods exploit the ability of nucleosomes to protect DNA from digestion by micrococcal nuclease. As described in Box 7-1, micrococcal nuclease has a strong preference to cleave DNA between nucleosomes, rather than DNA tightly associated with nucleosomes. This property can be used to map nucleosomes that are associated with the same position throughout a cell population (Box 7-3 Fig. 1).

To map nucleosome location accurately, it is important to isolate the cellular chromatin and treat it with the appropriate amount of micrococcal nuclease with minimal disruption of the overall chromatin structure. This is typically achieved by gently lysing cells while leaving the nuclei intact. The nuclei are then briefly treated (typically for 1 min) with several different concentrations of micrococcal nuclease, a protein small enough to rapidly diffuse into the nucleus. The goal of the titration is for micrococcal nuclease to cleave the region of interest only once in each cell. Once the DNA has been digested, the nuclei can be lysed, and all of the protein can be removed from the DNA. The sites of cleavage (and, more importantly, the sites not cleaved) leave a record of the protein bound to DNA.

To identify the sites of cleavage in a particular region, it is necessary to create a defined end point for all of the cleaved fragments and exploit the specificity of DNA hybridization. To create a defined end point, the purified DNA from each sample is cut with a restriction enzyme known to cleave adjacent to the site of interest. After separation by size using agarose gel electrophoresis, the DNA is denatured and transferred to a nitrocellulose membrane. This allows a labeled DNA probe of specific sequence to hybridize to the DNA (this is called a Southern blot and is described in more detail in Chapter 21). In this case, the DNA probe is carefully chosen to hybridize immediately adjacent to the restriction enzyme cleavage site at the site of interest. After hybridization and washing, the DNA probe will show the size of the fragments generated by micrococcal nuclease in the region of interest.

How do the fragment sizes reveal the location of positioned nucleosomes? DNA associated with positioned nucleosomes will be resistant to micrococcal nuclease digestion, leaving an ~160–200-bp region of DNA that is not cleaved. This will appear as a large gap in the ladder of DNA bands detected on the Southern blot. Frequently, there are arrays of positioned nucleosomes leading to a 160–200-bp periodicity to sites of cleavage and protection.

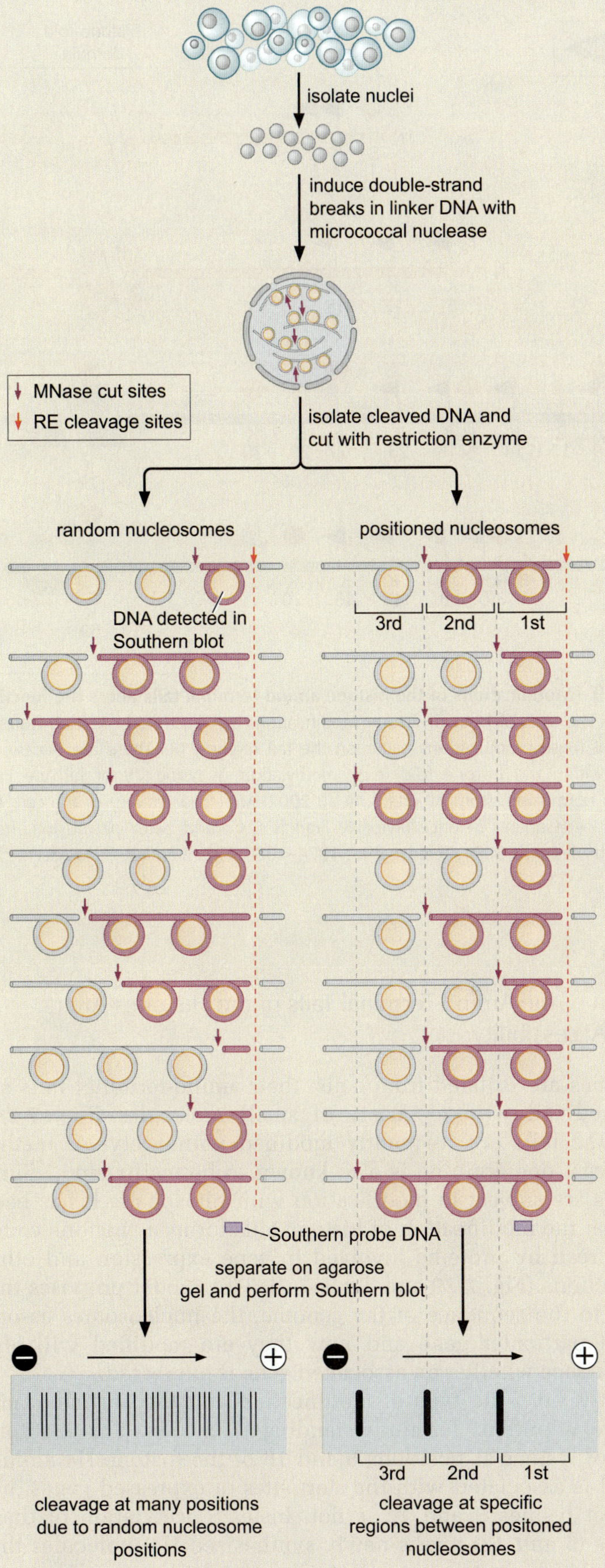
isolate nuclei
induce double-strand breaks in linker DNA with micrococcal nuclease
MNase cut sites
RE cleavage sites
isolate cleaved DNA and cut with restriction enzyme
random nucleosomes
positioned nucleosomes
DNA detected in Southern blot
3rd
2nd
1st
Southern probe DNA
separate on agarose gel and perform Southern blot
3rd
2nd
1st
cleavage at many positions due to random nucleosome positions
cleavage at specific regions between positoned nucleosomes

BOX 7-3 FIGURE 1 Analysis of nucleosome positioning in the cell. The experimental steps in determining nucleosome positioning in the cell are illustrated. See box text for details.

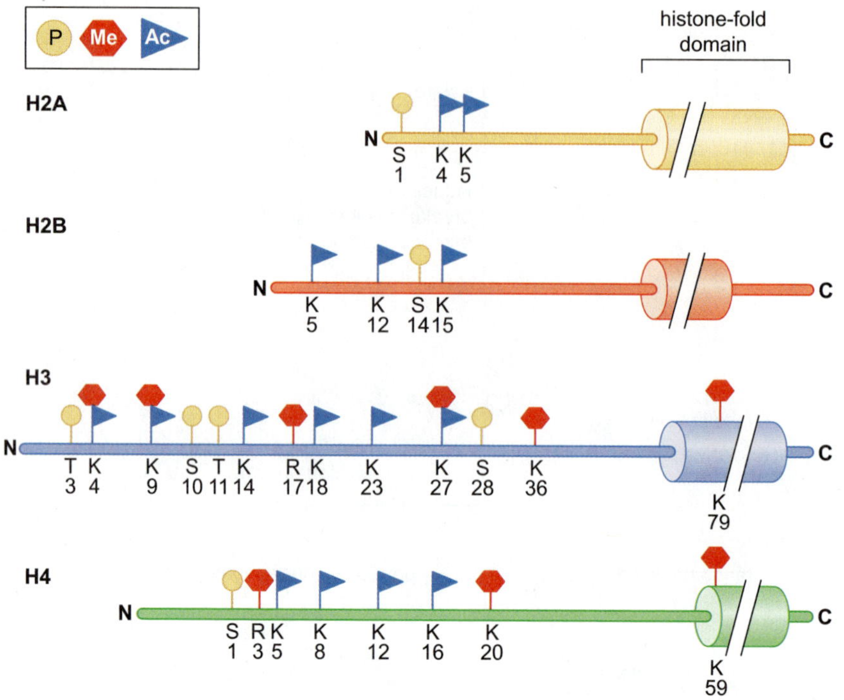

FIGURE 7-39 Modifications of the histone amino-terminal tails alters the function of chromatin. The sites of known histone modifications are illustrated on each histone. The majority of these modifications occur on the tail regions, but there are occasional modifications within the histone fold (e.g., methylation of lysine 79 of histone H3). (Adapted, with permission, from Alberts B. et al. 2002. *Molecular biology of the cell,* 4th ed., Fig. 4–35. © Garland Science/Taylor & Francis LLC; and, with permission, from Jenuwein T. and Allis C.D. 2001. *Science* 293: 1074–1080, Figs. 2 and 3. © AAAS.)

Modification of the Amino-Terminal Tails of the Histones Alters Chromatin Accessibility

When histones are isolated from cells, their amino-terminal tails are typically modified with a variety of small molecules (Fig. 7-39). Lysines in the tails are frequently modified with acetyl or methyl groups (at least one arginine is also known to be methylated). Similarly, serines are subject to modification with phosphate. It has been proposed that the modification of histone tails forms a "histone code" that can be read by proteins involved in gene expression and other DNA transactions (Fig. 7-39 and Table 7-7). This model proposes that in addition to the sequence of the genome, the nucleosomes associated with a particular gene and how they are modified will also strongly influence whether an associated gene is expressed.

Importantly, both the type of modification and the particular site modified are important for understanding this code. For example, acetylation of lysines at positions 8 and 16 of the histone H4 amino-terminal tail is associated with the start sites of expressed genes, but acetylation at lysines 5 and 12 is not. Instead, acetylation of these other lysines (5 and 12) marks newly synthesized H4 molecules that are ready to be deposited onto DNA as part of a new nucleosome. Similarly, methylation of histone tails can also have different mean-

TABLE 7-7 Histone Modifications

Histone Subunit Residue	Modification	Consequence(s)
H2A		
Serine 1	Phosphorylation	Mitosis, transcriptional repression
Lysine 4	Acetylation	Transcriptional activation
Lysine 5	Acetylation	Transcriptional activation
Lysine 7	Acetylation	Transcriptional activation
H2B		
Lysine 5	Acetylation	Transcriptional activation
Lysine 12	Acetylation	Transcriptional activation
Serine 14	Phosphorylation	Apoptosis
Lysine 15	Acetylation	Transcriptional activation
H3		
Threonine 3	Phosphorylation	Mitosis
Lysine 4	Acetylation, methylation	Acetylation: transcriptional activation Methylation: active euchromatin
Lysine 9	Acetylation,methylation	Acetylation: transcriptional activation Methylation: transcriptional repression
Serine 10	Phosphorylation	Transcriptional activation
Threonine 11	Phosphorylation	Mitosis
Lysine 14	Acetylation	Transcriptional activation/elongation
Arginine 17	Methylation	Transcriptional activation
Lysine 18	Acetylation	Transcriptional activation, DNA repair
Lysine 23	Acetylation	Transcriptional activation, DNA repair
Lysine 27	Methylation, acetylation	Transcriptional silencing
Serine 28	Phosphorylation	Mitosis
Lysine 36	Methylation	Transcriptional elongation
Lysine 56	Acetylation	DNA repair
Lysine 79	Methylation	Transcriptional elongation
H4		
Serine 1	Phosphorylation	Mitosis
Arginine 3	Methylation	Transcriptional activation
Lysine 5	Acetylation	Histone deposition
Lysine 8	Acetylation	Transcriptional activation
Lysine 12	Acetylation	Histone deposition, telomeric silencing
Lysine 16	Acetylation	Transcriptional activation, DNA repair
Lysine 20	Methylation	Transcriptional silencing

Adapted from Peterson C.L. and Laniel M.A. 2004. *Curr. Biol.* 14: R546–R551.

A listing of known modifications of vertebrate histones is shown. Most of these modifications are also observed in other organisms, although occasionally the histone structure between organisms varies so that the numbering of various modifications changes.

ings. Methylation of lysines 4, 36, or 79 of histone H3 is associated with expressed genes, whereas methylation of lysines 9 or 27 of the same histone is associated with transcriptional repression. Finally, phosphorylation of the amino-terminal tail of histone H3 is commonly observed in the highly condensed chromatin of mitotic chromosomes.

How does histone modification alter nucleosome function? One obvious change is that acetylation and phosphorylation each acts to reduce the overall positive charge of the histone tails; acetylation of lysine neutralizes its positive charge (Fig. 7-40). This loss of positive charge reduces the affinity of the tails for the negatively charged back-

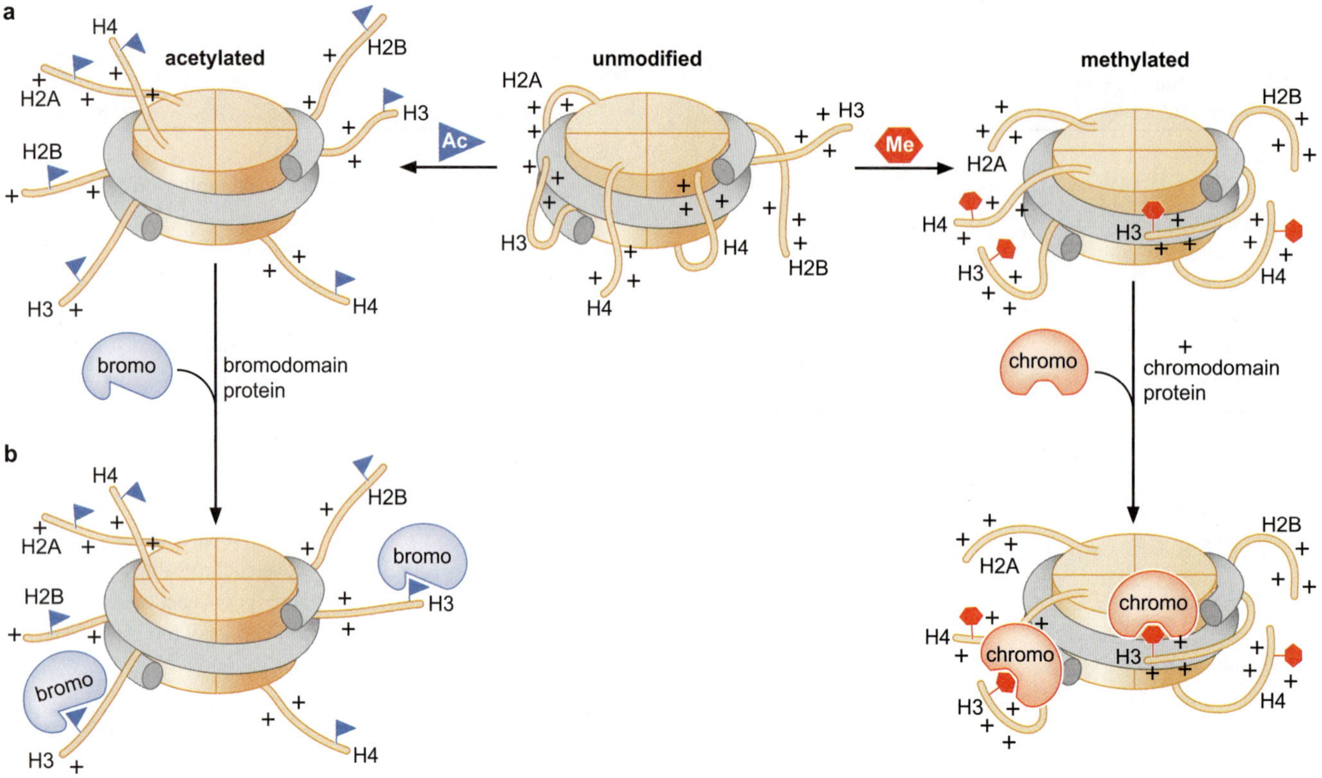

FIGURE 7-40 **Effects of histone tail modifications.** (a) The effect on the association with nucleosome-bound DNA. Unmodified and methylated histone tails are thought to associate more tightly with nucleosomal DNA than acetylated histone tails. (b) Modification of histone tails creates binding sites for chromatin-modifying enzymes.

bone of the DNA. More importantly, modification of the histone tails affects the ability of nucleosome arrays to form more repressive higher-order chromatin structure. As we described above, histone amino-terminal tails are required to form the 30-nm fiber, and modification of the tails modulates this function. For example, consistent with the association of some types of acetylated histones with expressed regions of the genome, acetylation of the H4 amino-terminal tail interferes with the ability of nucleosomes to be incorporated into the repressive 30-nm fiber. As we described earlier, formation of the 30-nm fiber is facilitated by an interaction between the positively charged H4 amino-terminal tail and the negatively charged surface of the H2A histone-fold domain. Acetylation interferes with this association by altering the charge of the H4 tail.

Protein Domains in Nucleosome-Remodeling and -Modifying Complexes Recognize Modified Histones

Modified histone tails can also act to recruit specific proteins to the chromatin (Fig. 7-40b). Protein domains called **bromodomains, chromodomains**, **TUDOR domains**, and **PHD** (for plant homeodomain) **fingers** specifically recognize modified forms of histone tails. Bromodomain-containing proteins interact with acetylated histone tails, and chromodomain-TUDOR domains and PHD-finger-containing proteins interact with methylated histone tails. Yet another protein domain, called a **SANT domain**, has the opposite property. SANT-domain-containing proteins interact preferentially with unmodified histone tails. Consistent with these protein domains being important for interpreting histone modifications, in many instances proteins containing these domains specifically recognize the modified form of only one of the many possible sites of histone modification. For example, the protein HP1 contains a chromodomain that will bind to methylated lysine 9 of histone H3 but not to any other site of histone methylation. Intriguingly, there are proteins that include more than one of these domains, suggesting that they are specialized for recognizing histone tails that are multiply modified. For example, there are proteins that contain a PHD finger specific for methylated lysine 4 of histone H3 immediately next to a bromodomain capable of recognizing an acetylated lysine.

How do the domains that recognize modified histones alter the function of the associated nucleosomes? One important way is that modified histones recruit enzymes that will further modify adjacent nucleosomes. For example, many of the enzymes that acetylate histone tails (called histone acetyltransferases) include bromodomains that recognize the same histone modifications that they create (Table 7-8). In this case, the bromodomain will facilitate the maintenance and propagation of acetylated histones by further modifying regions that are already acetylated (as we shall discuss below).

Modified histones can also recruit other proteins that act on chromatin. Many nucleosome-remodeling complexes include one or more subunits with domains that recognize modified histones (see Table 7-6) allowing modified histones to recruit these enzymes. A number of proteins involved in regulating transcription also include these domains. For example, a key component of the transcription machinery called TFIID includes a bromodomain. This domain directs the transcription machinery to sites of nucleosome acetylation, which is an additional way that histone acetylation contributes to the increased transcriptional activity of the associated DNA. Chromodomains that recognize sites of histone methylation associated with transcriptionally repressed genes are found in several proteins that are important for the establishment of heterochromatin, including the HP1 protein and polycomb proteins.

Specific Enzymes Are Responsible for Histone Modification

The histone modifications we have just described are dynamic and are catalyzed by specific enzymes. Histone acetyltransferases catalyze the addition of acetyl groups to histones, whereas histone deacetylases remove these modifications. Similarly, histone methyltransferases add methyl groups to histones and histone demethylases remove these modifications. A number of different histone acetyltransferases and deacetylases have been identified and are distinguished by their abilities to target a diffferent subset of histones or in some cases specific lysines in one histone. Histone methyltransferases and demethylases appear to be much more specific, always targeting only one of the many lysines or arginines on a specific histone (Table 7-8). Because these different modifications have different effects on nucleosome function, the modification of a nucleosome with different histone acetyltransferases or methyltransferases (or the removal of modifications by histone deacetylase or demethylases) can modulate chro-

TABLE 7-8 Histone-Modifying Enzymes

Histone Acetyltransferase Complexes

Type	Number of Subunits	Catalytic Subunit	Histone-Binding Domains	Target Histones
SAGA	15	Gcn5	Bromodomain, chromodomain	H3 and H2B
PCAF	11	PCAF	Bromodomain	H3 and H4
NuA3	5	Sas3	PHD finger	H3
NuA4	6	Esa1	Chromodomain, SANT Domain, PHD finger	H4 and H2A
P300/CBP	1	P300/CBP	Bromodomain, PHD finger	H2A, H2B, H3, and H4

Histone Deacetylase Complexes

Type	Number of Subunits	Catalytic Subunit(s)	Histone-Binding Domains
NuRD	9	HDAC1/HDAC2	Chromodomain, PHD finger
SIR2 complex	3	Sir2	Neither
Rpd3 large	12	Rpd3	PHD finger
Rpd3 small	5	Rpd3	Chromodomain, PHD finger

Histone Methyltransferases

Name	Histone-Binding Domains	Target Histone
SET1	None	H3 (lysine 4)
SUV39/CLR4	Chromodomain	H3 (lysine 9)
SET2	None	H3 (lysine 36)
DOT1	None	H3 (lysine 79)
PRMT	None	H3 (arginine 3)
SET9/SUV4-20	None	H4 (lysine 20)

Histone Demethylases

Name	Histone-Binding Domains	Target Methylated Histone
LSD1	PHD finger, SANT domain	H3 (lysine 4)
JHDM1	PHD finger	H3 (lysine 36)
JHDM3	PHD finger, TUDOR domain	H3 (lysines 9 and 36)

matin structure and influence a wide array of DNA transactions (see Fig. 7-39).

Like their nucleosome-remodeling complex counterparts, these modifying enzymes are part of large multiprotein complexes. Additional subunits play important roles in recruiting these enzymes to specific regions of the DNA. Similar to the nucleosome-remodeling complexes, these interactions can be with transcription factors bound to DNA or directly with specifically modified nucleosomes. The recruitment of these enzymes to particular DNA regions is responsible for the distinct patterns of histone modification observed along the chromatin and is a major mechanism for modulating the levels of gene expression along the eukaryotic chromosome (see Chapter 17).

Nucleosome Modification and Remodeling Work Together to Increase DNA Accessibility

The combination of amino-terminal tail modifications and nucleosome remodeling can dramatically change the accessibility of the

DNA. As discussed in Chapters 12 and 17, the protein complexes involved in these modifications are frequently recruited to sites of active transcription. Although the order of their function is not always the same, the combined action can result in a profound, but localized, change in DNA accessibility. Modification of amino-terminal tails can reduce the ability of nucleosome arrays to form repressive structures, creating sites that can recruit other proteins, including nucleosome remodelers. Remodeling of the nucleosomes can then further increase the accessibility of the nucleosomal DNA to allow DNA-binding proteins access to their binding sites. In combination with the appropriate DNA-binding proteins or DNA sequences, these changes can result in the positioning or release of nucleosomes at specific sites on the DNA (Fig. 7-41).

NUCLEOSOME ASSEMBLY

Nucleosomes Are Assembled Immediately after DNA Replication

The duplication of a chromosome requires replication of the DNA *and* the reassembly of the associated proteins on each daughter DNA molecule. The latter process is tightly linked to DNA replication to ensure that the newly replicated DNA is rapidly packaged into nucleosomes. In Chapter 8, we discuss the mechanisms of DNA replication in detail. Here, we discuss the mechanisms that direct the assembly of nucleosomes after the DNA is replicated (see Interactive Animation 7-2).

WEB ANIMATION

Although the replication of DNA requires the partial disassembly of the nucleosome, the DNA is rapidly repackaged in an ordered series of events. As discussed earlier, the first step in the assembly of nucleosomes on the DNA is the binding of an H3•H4 tetramer. Once the tetramer is bound, two H2A•H2B dimers associate to form the final nucleosome. H1 joins this complex last, presumably during the formation of higher-order chromatin assemblies.

To duplicate a chromosome, at least half of the nucleosomes on the daughter chromosomes must be newly synthesized. Are all the old histones lost and only new histones assembled into nucleosomes? If not, how are the old histones distributed between the two daughter chromosomes? The fate of the old histones is a particularly important issue given the effects that histone modification can have on the accessibility of the resulting chromatin. If the old histones were lost completely, then chromosome duplication would erase any "memory" of the previously modified nucleosomes. In contrast, if the old histones were retained on a single chromosome, that chromosome would have a distinct set of modifications relative to the other copy of the chromosome.

In experiments that differentially labeled old and new histones, it was found that the old histones are present on both of the daughter chromosomes (Fig. 7-42). Mixing is not entirely random, however. H3•H4 tetramers and H2A•H2B dimers are composed of either all new or all old histones. Thus, as the replication fork passes, nucleosomes are broken down into their component subassemblies. H3•H4 tetramers appear to remain bound to one of the two daughter duplexes at random and are never released from DNA into the free pool of histones. In contrast, the H2A•H2B dimers are released and enter the local pool available for new nucleosome assembly.

The distributive inheritance of old histones during chromosome dupli-

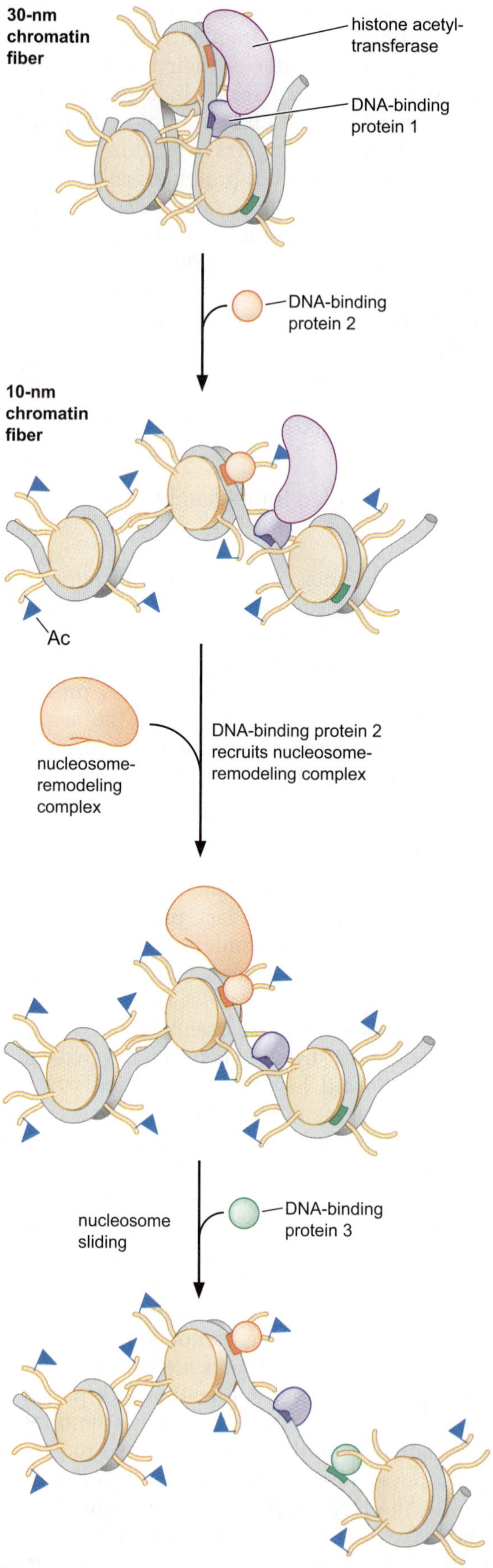

FIGURE 7-41 Chromatin-remodeling and histone-modifying complexes work together to alter chromatin structure. Sequence-specific DNA-binding proteins typically recruit these enzymes to specific regions of a chromosome. In the illustration, the blue DNA-binding protein first recruits a histone acetyltransferase that modifies the adjacent nucleosomes, increasing the accessibility of the associated DNA by locally converting the chromatin fiber from 30-nm fiber to the more accessible 10-nm form. This increased accessibility allows the binding of a second DNA-binding protein (orange) that recruits a nucleosome-remodeling complex. Localization of the nucleosome-remodeling complex facilitates the sliding of the adjacent nucleosomes, which allows the binding site for a third DNA-binding protein (green) to be exposed. For example, this could be the binding site for the TATA-binding protein at a start site of transcription. Although we show the order of association as histone acetylation complex and then nucleosome-remodeling complex, both orders are observed and can be equally effective. It is also true that recruitment of a different histone-modifying complex could result in the formation of more compact and inaccessible chromatin.

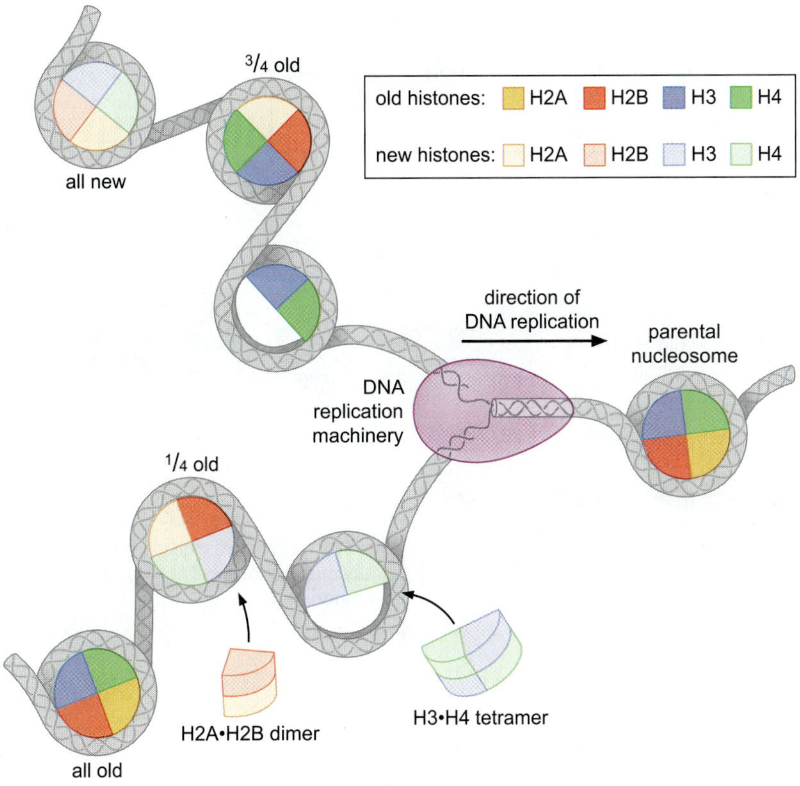

FIGURE 7-42 Inheritance of histones after DNA replication. As the chromosome is replicated, histones that were associated with the parental chromosome are differently distributed. The histone H3•H4 tetramers are randomly transferred to one of the two daughter strands but do not enter into the soluble pool of H3•H4 tetramers. Newly synthesized H3•H4 tetramers form the basis of the nucleosomes on the strand that does not inherit the parental tetramer. In contrast, H2A and H2B dimers are released into the soluble pool and compete for H3•H4 association with newly synthesized H2A and H2B. As a consequence of this type of distribution, on average, every second H3•H4 tetramer on newly synthesized DNA will be derived from the parental chromosome. These tetramers will include all of the modifications added to the parental nucleosomes. The H2A•H2B dimers are more likely to be derived from newly synthesized protein.

cation provides a mechanism for the propagation of the parental pattern of histone modification. By this mechanism, old modified histones will tend to rebind one of the daughter chromosomes at a position near their previous position on the parental chromosome (Fig. 7-43). The old histones have an equal probablility of binding either daughter chromosome; therefore, localized inheritance of modified histones provides a limited number of modifications in similar positions on each daughter chromosome. The ability of these modified histones to recruit enzymes that add similar modifications to adjacent nucleosomes (see the discussion of modified histone-binding domains above) provides a simple mechanism to maintain states of modification after DNA replication has occurred (Fig. 7-43). Such mechanisms are likely to play a critical role in the inheritance of chromatin states from one generation to another. Given the importance of histone modification in controlling gene expression as well as other DNA transactions, the maintenance of such modification states is critical to maintaining cell identity as cells replicate their DNA and divide.

Assembly of Nucleosomes Requires Histone "Chaperones"

The assembly of nucleosomes is not a spontaneous process. Early studies found that the simple addition of purified histones to DNA resulted in little or no nucleosome formation. Instead, the majority of the histones aggregate in a nonproductive form. For correct nucleosome assembly, it was necessary to raise salt concentrations to very high levels (>1 M NaCl) and then slowly reduce the concentration over many hours. Although useful for assembling nucleosomes for in vitro studies (such as for the structural studies of the nucleosome described

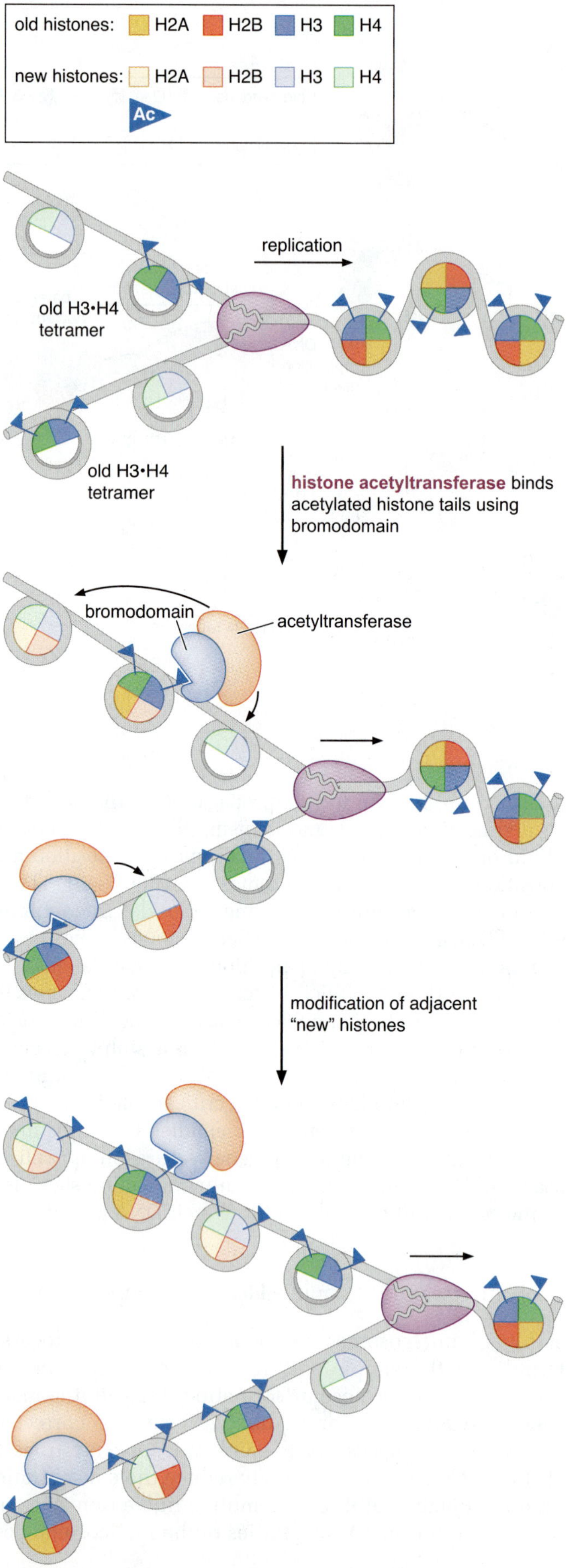

FIGURE 7-43 Inheritance of parental H3·H4 tetramers facilitates the inheritance of chromatin states. As a chromosome is replicated, the distribution of the parental H3·H4 tetramers results in the daughter chromosomes receiving the same modifications as the parent. The ability of these modifications to recruit enzymes that perform the same modifications facilitates the correct propagation of the same state of modification to the two daughter chromosomes. For simplicity, acetylation is shown on the core regions of the histones. In reality, this modification is generally on the amino-terminal tails.

TABLE 7-9 Properties of Histone Chaperones

Name	Number of Subunits	Histones Bound	Interaction with Sliding Clamp
CAF-1	4	H3•H4	Yes
HIRA	4	H3•H4	No
RCAF	1	H3•H4	No
NAP-1	1	H2A•H2B	No

earlier), elevated salt concentrations are not involved in nucleosome assembly in vivo.

Studies of nucleosome assembly under physiological salt concentrations identified factors required to direct the assembly of histones onto the DNA. These factors are negatively charged proteins that form complexes with either H3•H4 tetramers or H2A•H2B dimers (see Table 7-9) and escort them to sites of nucleosome assembly. Because they act to keep histones from interacting with the DNA nonproductively, these factors have been referred to as **histone chaperones** (see Fig. 7-44).

How do the histone chaperones direct nucleosome assembly to sites of new DNA synthesis? Studies of the histone H3•H4 tetramer chaperone CAF-I reveal a likely answer. Nucleosome assembly directed by CAF-I requires that the target DNA be replicating. Thus, replicating DNA is marked in some way for nucleosome assembly. Interestingly, this mark is gradually lost after replication is completed. Studies of CAF-I-dependent assembly have determined that the mark is a ring-shaped sliding clamp protein called PCNA. As we will discuss in detail in Chapter 8, this factor forms a ring around the DNA duplex and is responsible for holding DNA polymerase on the DNA during DNA synthesis. After the polymerase is finished, PCNA is released from the DNA polymerase but still encircles the DNA. In this condition, PCNA is available to interact with other proteins. CAF-I associates with the released PCNA and assembles H3•H4 tetramers preferentially on the PCNA-bound DNA. Thus, by associating with a component of the DNA replication machinery, CAF-I is directed to assemble nucleosomes at sites of recent DNA replication.

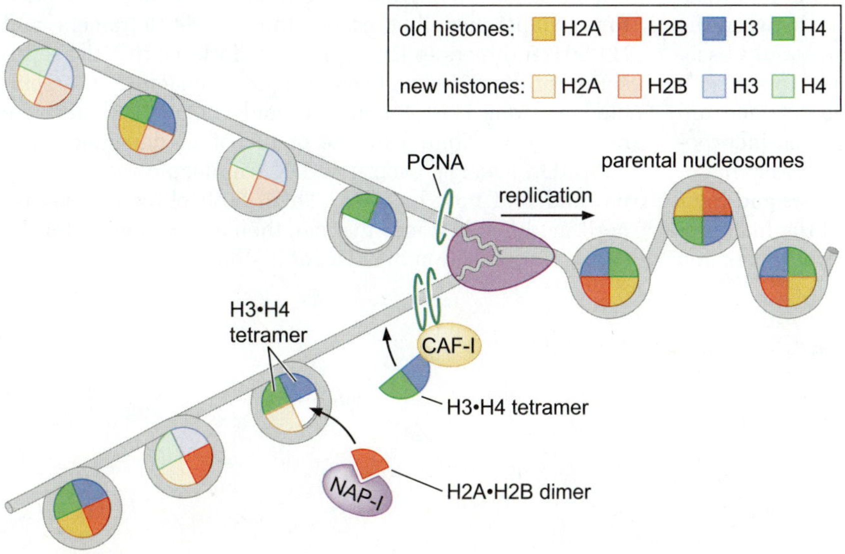

FIGURE 7-44 Chromatin assembly factors facilitate the assembly of nucleosomes. After the replication fork has passed, chromatin assembly factors chaperone free H3•H4 tetramers (e.g., CAF-I) and H2A•H2B dimers (NAP-I) to the site of newly replicated DNA. Once at the newly replicated DNA, these factors transfer their histone contents to the DNA. CAF-I is recruited to the newly replicated DNA by interactions with DNA sliding clamps. These ring-shaped, auxiliary replication factors encircle the DNA and are released from the replication machinery as the replication fork moves. For a more detailed description of DNA sliding clamps and their function in DNA replication, see Chapter 8.

SUMMARY

Within the cell, DNA is organized into large structures called chromosomes. Although the DNA forms the foundations for each chromosome, as much as half of each chromosome is composed of protein. Chromosomes can be either circular or linear; however, each cell has a characteristic number and composition of chromosomes. We now know the sequence of the entire genome of thousands of organisms. These sequences have revealed that the underlying DNA of each organism's chromosomes is used more or less efficiently to encode proteins. Simple organisms tend to use the majority of DNA to encode protein; however, more complex organisms use only a small portion of their DNA to actually encode proteins or RNAs.

Cells must carefully maintain their complement of chromosomes as they divide. Each chromosome must have DNA elements that direct chromosome maintenance during cell division. All chromosomes must have one or more origins of replication. In eukaryotic cells, centromeres play a critical role in the segregation of chromosomes and telomeres help to protect and replicate the ends of linear chromosomes. Eukaryotic cells carefully separate the events that duplicate and segregate chromosomes as cell division proceeds. Chromosome segregation can occur in one of two ways. During mitosis, a highly specialized apparatus ensures that one copy of each duplicated chromosome is delivered to each daughter cell. During meiosis, an additional round of chromosome segregation (without DNA replication) further reduces the number of chromosomes in the resulting daughter cells by half.

The combination of eukaryotic DNA and its associated proteins is referred to as chromatin. The fundamental unit of chromatin is the nucleosome, which is made up of two copies each of the core histones (H2A, H2B, H3, and H4) and ~147 bp of DNA. This protein–DNA complex serves two important functions in the cell: it compacts the DNA to allow it to fit into the nucleus and it restricts the accessibility of the DNA. This latter function is extensively exploited by the cell to regulate many different DNA transactions including gene expression.

The atomic structure of the nucleosome shows that the DNA is wrapped ~1.7 times around the outside of a disc-shaped, histone protein core. The interactions between the DNA and the histones are extensive but uniformly base-nonspecific. The nature of these interactions explain both the bending of the DNA around the histone octamer and the ability of virtually all DNA sequences to be incorporated into a nucleosome. This structure also reveals the location of the amino-terminal tails of the histones and their role in directing the path of the DNA around the histones.

Once DNA is packaged into nucleosomes, it has the ability to form more complex structures that allow additional compaction of the DNA. This process is facilitated by a fifth histone called H1. By binding the DNA associated with the nucleosome, H1 causes the DNA to wrap more tightly around the octamer. A more compact form of chromatin, the 30-nm fiber, is readily formed by arrays of H1-bound nucleosomes. This structure is more repressive than DNA packaged into nucleosomes alone. Current evidence suggests that the incorporation of DNA into this structure results in a dramatic reduction in its accessibility to the enzymes and proteins involved in transcription of the DNA.

The interaction of the DNA with the histones in the nucleosome is dynamic, allowing DNA-binding proteins intermittent access to the DNA. Nucleosome-remodeling complexes increase the accessibility of DNA incorporated into nucleosomes by increasing the mobility of nucleosomes. Two forms of mobilities can be observed: sliding of the histone octamer along the DNA or complete transfer of the histone octamer from one DNA molecule to another. Nucleosome-remodeling complexes are recruited to particular regions of the genome to facilitate alterations in chromatin accessibility. A subset of nucleosomes is restricted to fixed positions in the genome and are said to be "positioned." Nucleosome positioning can be directed by DNA-binding proteins or particular DNA sequences.

Modification of the histone amino-terminal tails also alters the accessibility of chromatin. The types of modifications include acetylation and methylation of lysines and phosphorylation of serines. Acetylation of amino-terminal tails is frequently associated with regions of active gene expression. These modifications alter the properties of the nucleosome itself, as well as acting as binding sites for proteins that influence the accessibility of the chromatin. In addition, these modifications recruit enzymes that perform the same modification, leading to similar modification of adjacent nucleosomes. It is likely that this leads to the stable propagation of regions of modified nucleosomes/chromatin as the chromosomes are duplicated.

Nucleosomes are assembled immediately after the DNA is replicated, leaving little time during which the DNA is unpackaged. This involves the function of specialized histone chaperones that escort the H3•H4 tetramers and H2A•H2B dimers to the replication fork. During the replication of the DNA, nucleosomes are transiently disassembled. Histone H3•H4 tetramers and H2A•H2B dimers are randomly distributed to one or the other daughter molecules. On average, each new DNA molecule receives half old and half new histones. Thus, both chromosomes inherit modified histones that can then act as "seeds" for the similar modification of adjacent histones.

BIBLIOGRAPHY

Books

Allis C.D., Jenuwein T., Reinberg D., and Caparros M.-L., eds. 2007. *Epigenetics.* Cold Spring Harbor Laboratory Press, Cold Spring Harbor, New York.

Brown T.A. 2007. *Genomes 3,* 2nd ed. Garland Science, New York.

Morgan D.O. 2007. *The cell cycle: Principles of control.* New Science Press Ltd., London.

Chromosomes

Bendich A.J. and Drlica K. 2000. Prokaryotic and eukaryotic chromosomes: What's the difference? *Bioessays* **22:** 481–486.

Thanbichler M., Wang S.C., and Shapiro L. 2005. The bacterial nucleoid: A highly organized and dynamic structure. *J. Cell Biochem.* **96:** 506–521.

Nucleosomes

Annunziato A.T. and Hansen J.C. 2000. Role of histone acetylation in the assembly and modulation of chromatin structures. *Gene Expr.* **9:** 37–61.

Belmont A.S., Dietzel S., Nye A.C., Strukov Y.G., and Tumbar T. 1999. Large-scale chromatin structure and function. *Curr. Opin. Cell Biol.* **11:** 307–311.

de la Cruz X., Lois S., Sánchez-Molina S., and Martinez-Balbás M.A. 2005. Do protein motifs read the histone code? *Bioessays* **27:** 164–175.

Eberharter A. and Becker P.B. 2002. Histone acetylation: A switch between repressive and permissive chromatin. *EMBO Rep.* **31:** 224–229.

Gregory P.D., Wagner K., and Horz W. 2001. Histone acetylation and chromatin remodeling. *Exp. Cell. Res.* **265:** 195–202.

Hayes J.J. and Hansen J.C. 2001. Nucleosomes and the chromatin fiber. *Curr. Opin. Genet. Dev.* **11:** 124–129.

Jenuwein T. and Allis C.D. 2001. Translating the histone code. *Science* **293:** 1074–1080.

Klose R.J., Kallin E.M., and Zhang Y. 2006. JmjC-domain-containing proteins and histone demethylation. *Nat. Rev. Genet.* **7:** 715–727.

Luger K. and Richmond T.J. 1998. DNA binding within the nucleosome core. *Curr. Opin. Struct. Biol.* **8:** 33–40.

———. 1998. The histone tails of the nucleosome. *Curr. Opin. Genet. Dev.* **8:** 140–146.

Luger K., Madev A.W., and Richmond R.K. 1997. Crystal structure of the nucleosome core particle at 2.8 Å resolution. *Nature* **389:** 251–260.

Margueron R., Trojer P., and Reinberg D. 2005. The key to development: Interpreting the histone code? *Curr. Opin. Genet. Dev.* **15:** 163–176.

Martin C. and Zhang Y. 2005. The diverse functions of histone lysine methylation. *Nat. Rev.* **6:** 838–849.

Narliker G.J., Fan H.-Y., and Kingston R.E. 2002. Cooperation between complexes that regulate chromatin structure and transcription. *Cell* **108:** 475–487.

Robinson P.J.J. and Rhodes D. 2006. Structure of the '30 nm' chromatin fibre: A key role for the linker histone. *Curr. Opin. Struct. Biol.* **16:** 336–343.

Roth S.Y., Denu J.M., and Allis D. 2001. Histone acetyl transferases. *Annu. Rev. Biochem.* **70:** 81–120.

Saha A., Wittmeyer J., and Cairns B.R. 2006. Chromatin remodelling: The industrial revolution of DNA around histones. *Nat. Rev. Mol. Cell. Biol.* **7:** 437–447.

Thiriet C. and Hayes J.J. 2005. Chromatin in need of a fix: Phosphorylation of H2AX connects chromatin to DNA repair. *Mol. Cell* **18:** 617–622.

Thomas J.O. 1999. Histone H1: Location and role. *Curr. Opin. Cell. Biol.* **11:** 312–317.

Woodcock C.L. and Dimitrov S. 2001. Higher-order structure of chromatin and chromosomes. *Curr. Opin. Genet. Dev.* **11:** 130–135.

CHAPTER

8

The Replication of DNA

WHEN THE DNA DOUBLE HELIX WAS DISCOVERED, the feature that most excited biologists was the complementary relationship between the bases on its intertwined polynucleotide chains. It seemed unimaginable that such a complementary structure would not be utilized as the basis for DNA replication. In fact, it was the self-complementary nature revealed by the DNA structure that finally led most biologists to accept Oswald T. Avery's conclusion that DNA, not some form of protein, was the carrier of genetic information (Chapter 2).

In our discussion of how templates act (Chapter 6), we emphasized that two identical surfaces will not attract each other. Instead, it is much easier to visualize the attraction of groups with opposite shape or charge. Thus, without any detailed structural knowledge, we might guess that a molecule as complicated as the gene could not be copied directly. Instead, replication would involve the formation of a molecule complementary in shape, and this in turn would serve as a template to make a replica of the original molecule. So, in the days before detailed knowledge of protein or nucleic acid structure, some geneticists wondered whether DNA served as a template for a specific protein that in turn served as a template for a corresponding DNA molecule.

But as soon as the self-complementary nature of DNA became known, the idea that protein templates might play a role in DNA replication was discarded. It was immensely simpler to postulate that each of the two strands of every parental DNA molecule served as a template for the formation of a complementary daughter strand. Although from the start this hypothesis seemed too good not to be true, experimental support nevertheless had to be generated. Happily, within 5 years of the discovery of the double helix, decisive evidence emerged for the separation of the complementary strands during DNA replication (see discussion of the Meselson and Stahl experiment in Chapter 2) and firm enzymological proof showed that DNA alone can function as the template for the synthesis of new DNA strands.

With these results, the problem of how genes replicate was in one sense solved. But in another sense, the study of DNA replication had only begun. As we will see in this chapter, the replication of even the simplest DNA molecule is a complex, multistep process, involving many more enzymes than was initially anticipated following the discovery of the first DNA polymerizing enzyme. The replication of the large, linear chromosomes of eukaryotes is still more complex. These

OUTLINE

chromosomes require many start sites of replication to synthesize the entire chromosome in a timely fashion, and the initiation of replication must be carefully coordinated to ensure that all sequences are replicated exactly once. Moreover, because conventional DNA replication cannot completely replicate the chromosome ends (called telomeres), cells have developed a novel method to maintain the integrity of this part of the chromosome.

In this chapter, we first describe the basic chemistry of DNA synthesis and the function of the enzymes that catalyze this reaction. We then discuss how the synthesis of DNA occurs in the context of an intact chromosome at structures called replication forks. An array of additional proteins are required to prepare the DNA for replication at these sites. The last part of the chapter focuses on the initiation and termination of DNA replication. DNA replication is tightly controlled in all cells and initiation is the step that is regulated. We describe how replication initiation proteins unwind the DNA duplex at specific sites in the genome called origins of replication. Together, the proteins involved in DNA replication form a complex machine that performs this critical process with astounding speed, accuracy, and completeness.

THE CHEMISTRY OF DNA SYNTHESIS

DNA Synthesis Requires Deoxynucleoside Triphosphates and a Primer:Template Junction

WEB ANIMATION

For the synthesis of DNA to proceed, two key substrates must be present (see Interactive Animation 8-1). First, new synthesis requires the four deoxynucleoside triphosphates—dGTP, dCTP, dATP, and dTTP (Fig. 8-1a). Nucleoside triphosphates have three phosphoryl groups that are attached via the 5′-hydroxyl of the 2′-deoxyribose. The innermost phosphoryl group (i.e., the group proximal to the deoxyribose) is called the α-phosphate, whereas the middle and outermost groups are called the β- and γ-phosphates, respectively.

The second important substrate for DNA synthesis is a particular arrangement of single-stranded DNA (ssDNA) and double-stranded DNA (dsDNA) called a **primer:template junction** (Fig. 8-1b). As suggested by its name, the primer:template junction has two key compo-

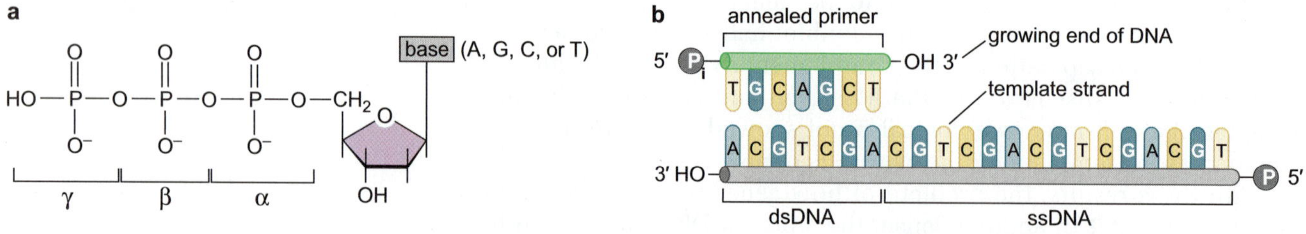

FIGURE 8-1 Substrates required for DNA synthesis. (a) The general structure of the 2′-deoxynucleoside triphosphates. The positions of the α-, β-, and γ-phosphates are labeled. (b) The structure of a generalized primer:template junction. The shorter primer strand is completely annealed to the longer DNA strand and must have a free 3′-OH adjacent to an ssDNA region of the template. The longer DNA strand includes a region annealed to the primer and an adjacent ssDNA region that acts as the template for new DNA synthesis. New DNA synthesis extends the 3′ end of the primer.

nents. The template provides the ssDNA that directs the addition of each complementary deoxynucleotide. The primer is complementary to, but shorter than, the template. The primer must have an exposed 3′-OH adjacent to the single-strand region of the template. It is this 3′-OH that will be extended as new nucleotides are added.

Formally, only the primer portion of the primer:template junction is a substrate for DNA synthesis because only the primer is chemically modified during DNA synthesis. The template provides only the information necessary to pick which nucleotides are added. Nevertheless, both a primer and a template are essential for all DNA synthesis.

DNA Is Synthesized by Extending the 3′ End of the Primer

The chemistry of DNA synthesis requires that the new chain grows by extending the 3′ end of the primer (Fig. 8-2). Indeed, this is a feature of the synthesis of both RNA and DNA. The phosphodiester bond is formed in an S_N2 reaction in which the hydroxyl group at the 3′ end of the primer strand attacks the α-phosphoryl group of the incoming nucleoside triphosphate. The leaving group for the reaction is pyrophosphate, which arises from the release of the β- and γ-phosphates of the nucleotide substrate.

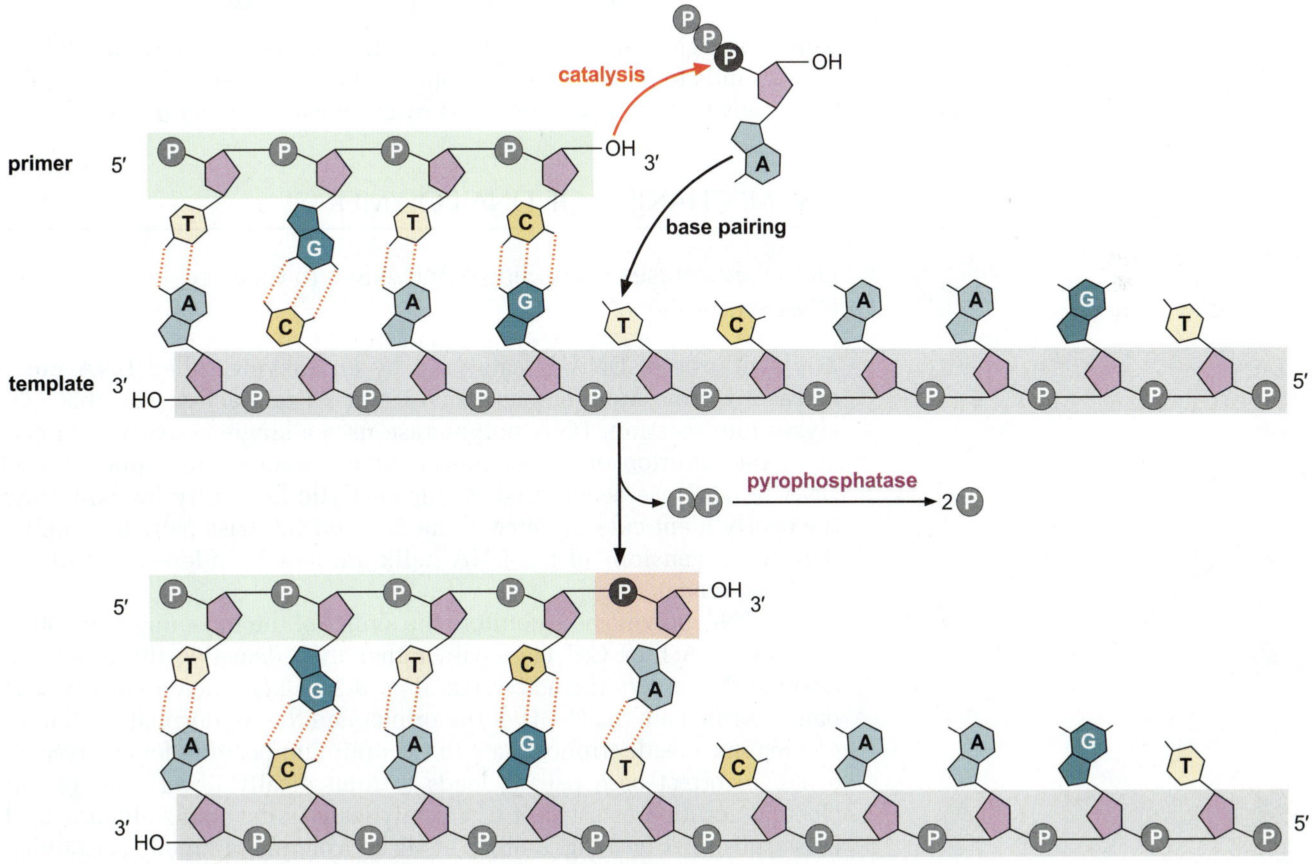

FIGURE 8-2 Diagram of the mechanism of DNA synthesis. DNA synthesis is initiated when the 3′-OH of the primer mediates the nucleophilic attack of the α-phosphate of the incoming dNTP. This results in the extension of the incoming 3′ end of the primer by one nucleotide and the release of one molecule of pyrophosphate. Pyrophosphatase rapidly hydrolyzes released pyrophosphate into two phosphate molecules.

The template strand directs which of the four nucleoside triphosphates is added. The nucleoside triphosphate that base-pairs with the template strand is highly favored for addition to the primer strand. Recall that the two strands of the double helix have an antiparallel orientation. This arrangement means that the template strand for DNA synthesis has the opposite orientation of the growing DNA strand.

Hydrolysis of Pyrophosphate Is the Driving Force for DNA Synthesis

The addition of a nucleotide to a growing polynucleotide chain of length *n* is indicated by the following reaction:

$$XTP + (XMP)_n \rightarrow (XMP)_{n+1} + Ⓟ \sim Ⓟ.$$

But the free energy for this reaction is rather small (ΔG = −3.5 kcal/mole). What then is the driving force for the polymerization of nucleotides into DNA? Additional free energy is provided by the rapid hydrolysis of the pyrophosphate into two phosphate groups by an enzyme known as pyrophosphatase:

$$Ⓟ \sim Ⓟ \rightarrow 2\ Ⓟ_i.$$

The net result of nucleotide addition *and* pyrophosphate hydrolysis is the breaking of two high-energy phosphate bonds. Therefore, DNA synthesis is a coupled process, with an overall reaction of:

$$XTP + (XMP)_n \rightarrow (XMP)_{n+1} + 2\ Ⓟ_i.$$

This is a highly favorable reaction with a ΔG of −7 kcal/mole, which corresponds to an equilibrium constant (K_{eq}) of about 10^5. Such a high K_{eq} means that the DNA synthesis reaction is effectively irreversible.

THE MECHANISM OF DNA POLYMERASE

DNA Polymerases Use a Single Active Site to Catalyze DNA Synthesis

The synthesis of DNA is catalyzed by an enzyme called **DNA polymerase**. Unlike most enzymes, which have one active site that catalyzes one reaction, DNA polymerase uses a single active site to catalyze the addition of any of the four deoxynucleoside triphosphates. DNA polymerase accomplishes this catalytic flexibility by exploiting the nearly identical geometry of the A:T and G:C base pairs (remember that the dimensions of the DNA helix are largely independent of the DNA sequence).

The DNA polymerase monitors the ability of the incoming nucleotide to form an A:T or G:C base pair, rather than detecting the exact nucleotide that enters the active site (Fig. 8-3). *Only* when a correct base pair is formed are the 3′-OH of the primer and the α-phosphate of the incoming nucleoside triphosphate in the optimum position for catalysis to occur. Incorrect base pairing leads to dramatically lower rates of nucleotide addition as a result of a catalytically unfavorable alignment of these substrates (see Fig. 8-3b). This is an example of kinetic selectivity, in which an enzyme favors catalysis using one of several possible substrates by dramatically increasing the rate of bond formation only when the correct substrate is present. Indeed, the rate of incorporation of an incorrect nucleotide is as much as 10,000-fold slower than incorporation when base pairing is correct. A common method to monitor synthesis of new DNA is described in Box 8-1, Incorporation Assays Can Be Used to

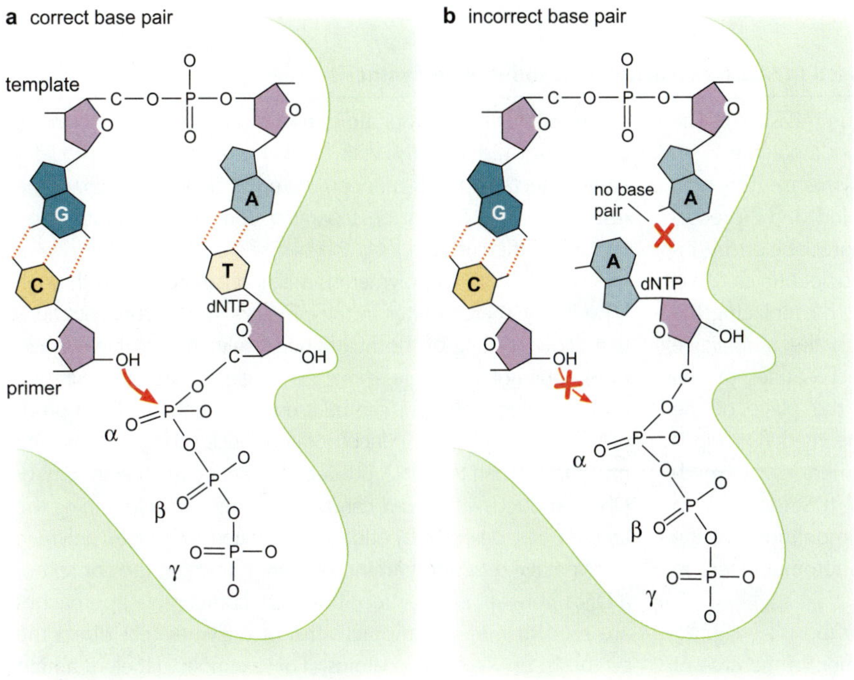

FIGURE 8-3 Correctly paired bases are required for DNA-polymerase-catalyzed nucleotide addition. (a) Schematic diagram of the attack of a primer 3′-OH end on a correctly base-paired dNTP. (b) Schematic diagram of the consequence of incorrect base pairing on catalysis by DNA polymerase. In the example shown, the incorrect A:A base pair displaces the α-phosphate of the incoming nucleotide. This incorrect alignment reduces the rate of catalysis dramatically, resulting in the DNA polymerase preferentially adding correctly base-paired dNTPs. (Adapted, with permission, from Brautigan C.A. and Steitz T.A. 1998. *Curr. Opin. Struct. Biol.* 8: 60, Fig. 4d. © Elsevier.)

Measure Nucleic Acid and Protein Synthesis.

DNA polymerases show an impressive ability to distinguish between ribonucleoside and deoxyribonucleoside triphosphates (rNTPs and dNTPs). Although rNTPs are present at approximately ten-fold higher concentration in the cell, they are incorporated at a rate that is more than 1000-fold lower than dNTPs. This discrimination is mediated by the steric exclusion of rNTPs from the DNA polymerase active site (Fig. 8-4). In DNA polymerase, the nucleotide-binding pocket is too small to allow the presence of a 2′-OH on the incoming nucleotide. This space is occupied by two amino acids that make van der Waals contacts with the sugar ring. Changing these amino acids to other amino acids with

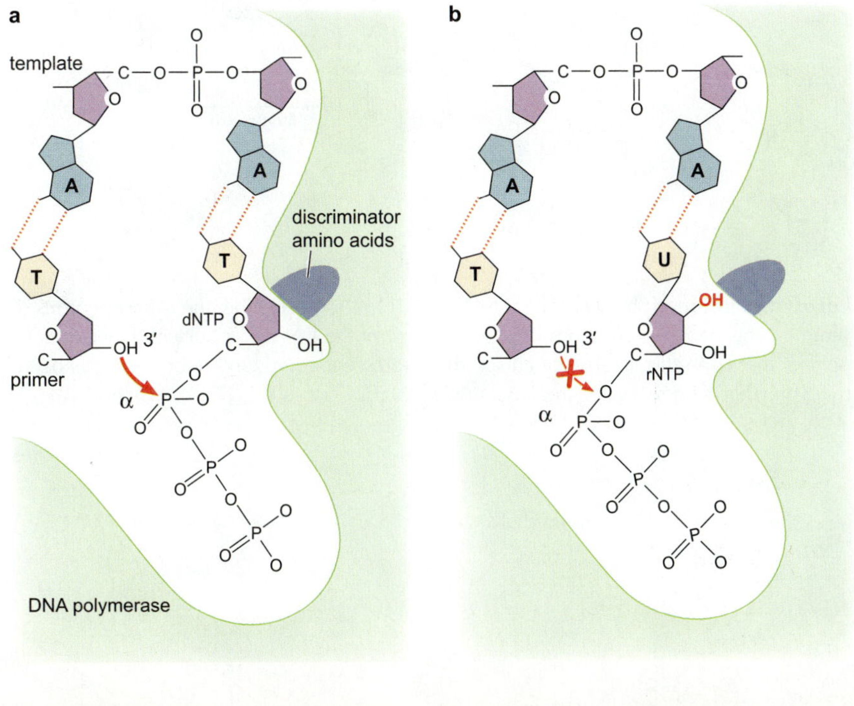

FIGURE 8-4 Schematic illustration of the steric constraints preventing DNA polymerase from using rNTP precursors. (a) Binding of a correctly base-paired dNTP to the DNA polymerase. Under these conditions, the 3′-OH of the primer and the α-phosphate of the dNTP are in close proximity. (b) Addition of a 2′-OH results in a steric clash with amino acids (the discriminator amino acids) in the nucleotide-binding pocket. This results in the α-phosphate of the dNTP being displaced. In this state, the α-phosphate is incorrectly aligned with the 3′-OH of the primer, dramatically reducing the rate of catalysis.

TECHNIQUES

Box 8-1 Incorporation Assays Can Be Used to Measure Nucleic Acid and Protein Synthesis

How can the activity of a DNA polymerase be measured? The simplest assay used to measure the synthesis of a polymer is an incorporation assay. In the case of DNA polymerase, this type of assay measures the incorporation of labeled dNTP precursors into DNA molecules. Typically, dNTPs are labeled by including radioactive atoms in a part of the nucleotide that will be retained in the final DNA product (e.g., by replacing the phosphorous atom in the α phosphate with the radioactive isotope ^{32}P, Box 8-1 Fig. 1a). Alternatively, nucleotides can be synthesized with fluorescent molecules in the place of the methyl group on dTTP (Box 8-1 Fig. 1b). This methyl group is not involved in base pairing, and DNA polymerases can readily accommodate much larger moieties in this location. In either case, these modifications allow easy monitoring of the labeled nucleotide using film or sensitive photomultipliers to detect emitted electrons or photons.

An incorporation assay requires two steps (Box 8-1 Fig. 2). First, the precursor is incorporated into polymer. In the case of DNA polymerase, this is accomplished by incubating the polymerase with a primer:template junction and the labeled dNTP precursors for an appropriate period of time. In most instances, only one of the four dNTPs is labeled as this generally provides easily detectable levels of incorporated nucleotides. Second, the resulting polymers must be separated from unincorporated precursors. In the case of DNA, this can be accomplished in one of two ways. The DNA polymerase reaction can be passed through a positively charged filter in the presence of salt concentrations that allow binding of the highly negatively charged DNA backbone, but not of the less charged single nucleotides. Alternatively, gel electrophoresis can be used to separate the DNA products by size, as the unincorporated nucleotides will migrate much faster than the DNA product. In either case, the amount of DNA product synthesized can be measured by determining the amount of labeled nucleotide incorporated into DNA polymer.

We have described an incorporation assay in the context of a DNA polymerase reaction; however, comparable approaches are used to measure the activities of enzymes that direct the synthesis of RNA or proteins. For example, labeled amino acids can be similarly used to analyze their incorporation into proteins.

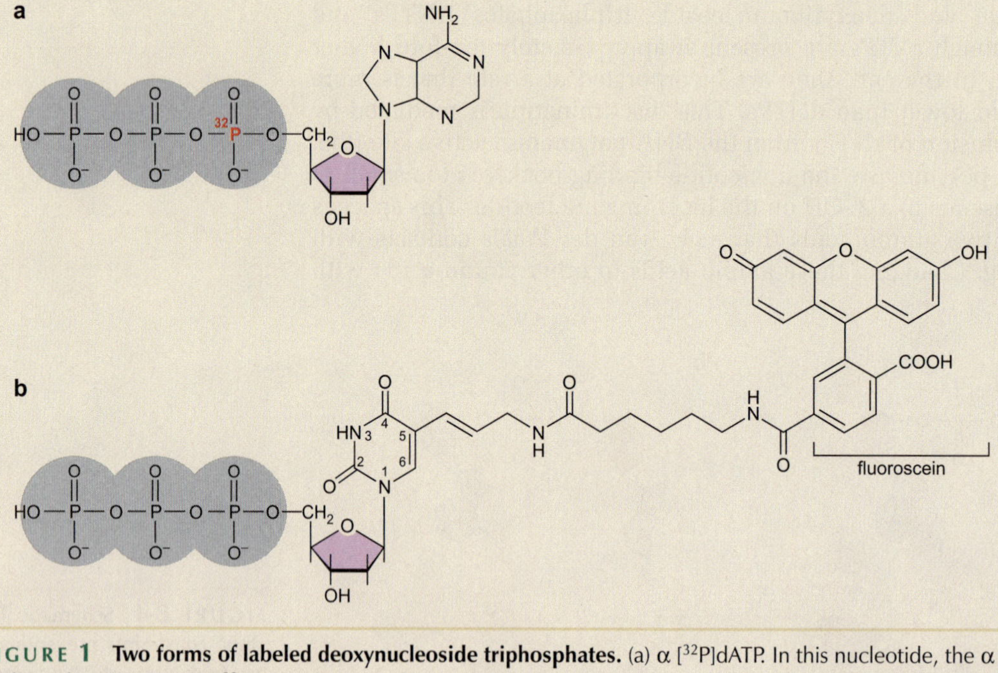

Box 8-1 Figure 1 Two forms of labeled deoxynucleoside triphosphates. (a) α [^{32}P]dATP. In this nucleotide, the α phosphorous is replaced with the radioactive isotope ^{32}P. Note that only this phosphorous atom will become part of the DNA after nucleotide incorporation. (b) Fluorescently labeled thymidine triphosphate analog. In this labeled precursor, the fluorescent compound fluorescein has been attached via a linker to the 5 position of the thymine ring that is normally attached to a methyl group.

Box 8-1 *(Continued)*

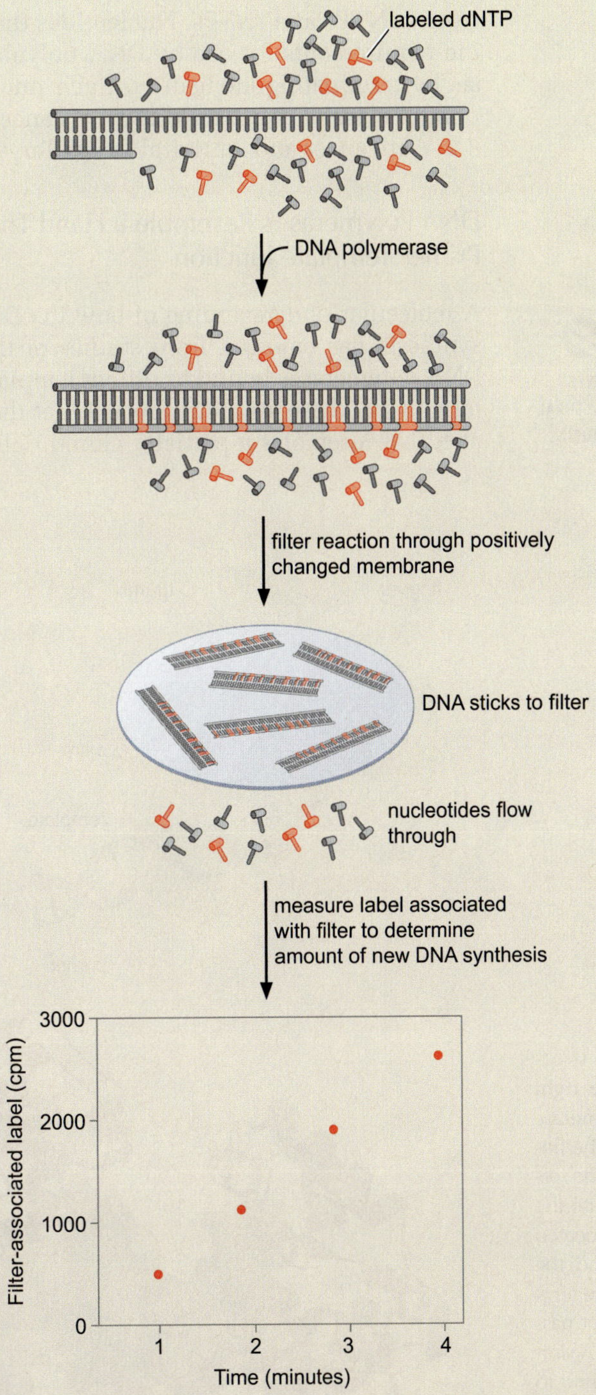
labeled dNTP
DNA polymerase
filter reaction through positively changed membrane
DNA sticks to filter
nucleotides flow through
measure label associated with filter to determine amount of new DNA synthesis
Filter-associated label (cpm)
3000
2000
1000
0
1
2
3
4
Time (minutes)

Box 8-1 Figure 2 Incorporation assay to measure DNA synthesis. In the example shown, filter binding is used to separate unincorporated from DNA-incorporated labeled nucleotides.

smaller side chains (e.g., by changing a glutamate to an alanine) results in a DNA polymerase with significantly reduced discrimination between dNTPs and rNTPs. Nucleotides that can meet some but not all of the requirements for use by DNA polymerase can inhibit DNA synthesis by terminating elongation. Such nucleotides also represent an important class of drugs used to treat cancer and viral infections (see Box 8-2, Anticancer and Antiviral Drugs Target DNA Replication).

DNA Polymerases Resemble a Hand That Grips the Primer:Template Junction

WEB STRUCTURAL TUTORIAL

A molecular understanding of how the DNA polymerase catalyzes DNA synthesis has emerged from studies of the atomic structure of various DNA polymerases bound to primer:template junctions (see Structural Tutorial 8-1). These structures reveal that the DNA substrate sits in a large cleft that resembles a partially closed right hand (Fig. 8-5). On the basis

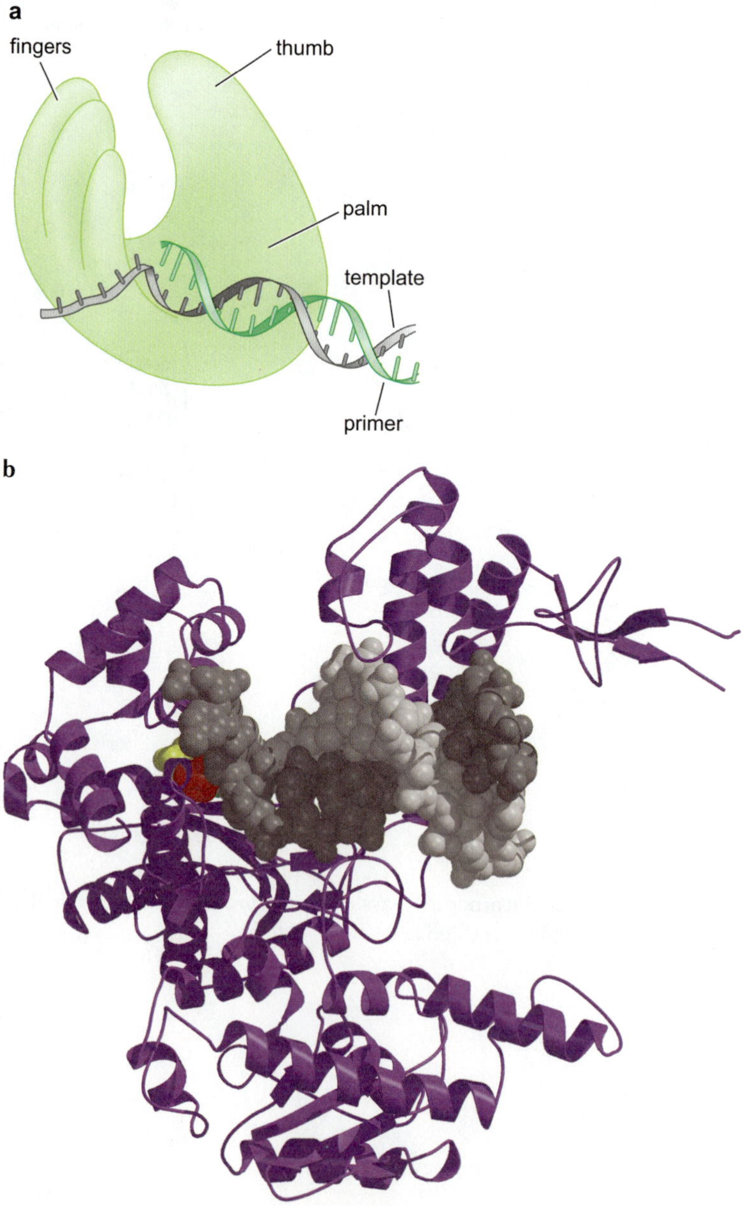

FIGURE 8-5 Three-dimensional structure of DNA polymerase resembles a right hand. (a) Schematic of DNA polymerase bound to a primer:template junction. The fingers, thumb, and palm are noted. The recently synthesized DNA is associated with the palm, and the site of DNA catalysis is located in the crevice between the fingers and the thumb. The single-stranded region of the template strand is bent sharply and does not pass between the thumb and the fingers. (b) A similar view of the T7 DNA polymerase bound to DNA. The DNA is shown in a space-filling manner, and the protein is shown as a ribbon diagram. The fingers and the thumb are composed of α helices. The palm domain is obscured by the DNA. The incoming dNTP is shown in red (for the base and the deoxyribose) and yellow (for the triphosphate moiety). The template strand of the DNA is shown in dark gray, and the primer strand is shown in light gray. (Doublie S. et al. 1998. *Nature* 391: 251.) Image prepared with MolScript, BobScript, and Raster 3D.

MEDICAL CONNECTIONS

Box 8-2 Anticancer and Antiviral Agents Target DNA Replication

The central role of DNA replication during cell division makes it a common target for chemotherapeutic drugs that aim to prevent the growth of tumors. These drugs target DNA replication at various stages.

A number of common chemotherapeutic reagents target the biosynthesis of the nucleotide precursors for DNA, thus starving DNA polymerase for new building blocks. For example, the drugs 5-fluorouracil (5-FU) and 6-mercaptopurine (6-MP) are analogs of nucleotide precursors that inhibit the synthesis of pyrimidine and purine nucleotides respectively (Box 8-2 Fig. 1a,b). 5-FU is the major agent used in the treatment of colorectal cancer and is also used in the treatment of stomach, pancreatic, and advanced breast cancer. 6-MP is primarily used to treat patients with acute leukemia (blood cell cancers).

Other anticancer drugs target DNA synthesis more directly. Cytosine arabinoside (AraC) is a deoxycytidine analog that after conversion to a nucleoside triphosphate is incorporated into DNA in place of dCTP (Box 8-2 Fig. 1c). Once incorporated, the difference between the deoxyribose sugar of dCTP and the arabinose sugar of AraCTP leads to incorrect positioning of the 3′ end of the DNA and termination of elongaton. Like 6-MP, AraC is primarily used in the treatment of acute leukemia.

A third class of chemotherapies damages DNA to block DNA replication. Cisplatin and bischloroethylnitrosourea (BCNU) cause intrastrand and interstrand DNA cross-links when G residues are adjacent to one another (Box 8-2 Fig. 1d,e). These cross-links (particularly the interstrand variety) interfere with DNA elongation. Cisplatin is a major drug used to treat metastatic testicular cancer and BCNU is used to treat brain tumors and leukemias. Similarly, camptothecin and etoposide are inhibitors of topoisomerases that block the ability of these proteins to re-form a phosphodiester bond after cleaving the DNA backbone (see Chapter 6, Fig. 6-24). Treatment with either of these inhibitors leaves a break in the DNA template that terminates DNA replication when DNA polymerase attempts to use it as a template.

As a class these drugs target cells that are replicating their DNA and therefore are frequently dividing. Although the rapidly dividing nature of cancer cells makes them particularly susceptible to such drugs, other cells in the body are affected also. Not surprisingly, these DNA replication inhibitors are also toxic toward rapidly growing host cells such as red and white blood cells, hair cells, and gastrointestinal mucosal cells. Inhibiting the growth of these cells leads to the now familiar side effects of many chemotherapies, including immunosuppression (due to loss of white blood cells), anemia (due to loss of red blood cells), diarrhea (due to gastrointestinal defects), and hair loss.

Replication inhibitors have also been used as antiviral agents. The first drug found to be effective against HIV infection was azidothymidine (AZT), a thymidine analog that inhibits the specialized DNA polymerase (called a reverse transcriptase; see Chapter 11) that copies the RNA genome of HIV into DNA after infection. More recently a guanine nucleoside analog called Acyclovir has replaced AZT as the preferred HIV DNA polymerase inhibitor. This analog is the ribose of a normal nucleoside with an open chain structure (Box 8-2 Fig. If,g). Nevertheless, this analog can be modified to a triphosphate form that can be incorporated by the viral DNA polymerase into DNA. Once incorporated, these analogs act as chain terminators due to their lack of a ribose group and therefore the 3′-OH required for further nucleotide addition. Importantly, these drugs are poorly recognized by cellular DNA polymerases and, thus, have fewer side effects than chemotherapeutic nucleotide analogs.

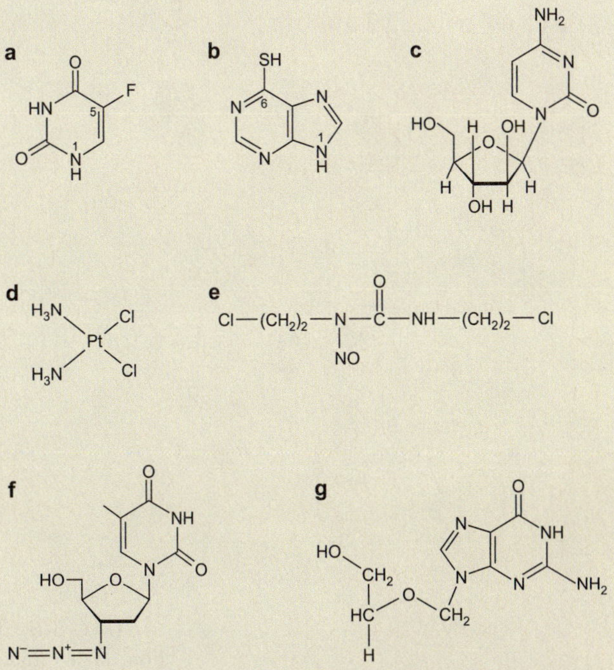

BOX 8-2 FIGURE 1 Structures of common chemotherapeutic reagents that target DNA replication. (a) 5-fluorouracil, (b) 6-mercaptopurine, (c) cytosine arabinoside, (d) cisplatin, (e) bischloroethylnitrosourea, (f) azidothymidine, and (g) acyclovir.

of the analogy to a hand, the three domains of the polymerase are called the thumb, fingers, and palm.

The palm domain is composed of a β sheet and contains the primary elements of the catalytic site. In particular, this region of DNA polymerase binds two divalent metal ions (typically Mg^{+2} or Zn^{+2}) that alter the chemical environment around the correctly base-paired dNTP and the 3′-OH of the primer (Fig. 8-6). One metal ion reduces the affinity of the 3′-OH for its hydrogen. This generates a 3′O^- that is primed for the nucleophilic attack of the α-phosphate of the incoming dNTP. The second metal ion coordinates the negative charges of the β- and γ-phosphates of the dNTP and stabilizes the pyrophosphate produced by joining the primer and the incoming nucleotide.

In addition to its role in catalysis, the palm domain also monitors the base pairing of the most recently added nucleotides. This region of the polymerase makes extensive hydrogen bond contacts with base pairs in the minor groove of the newly synthesized DNA. These contacts are not base-specific but only form if the recently added nucleotides (whichever they may be) are correctly base-paired. Mis-

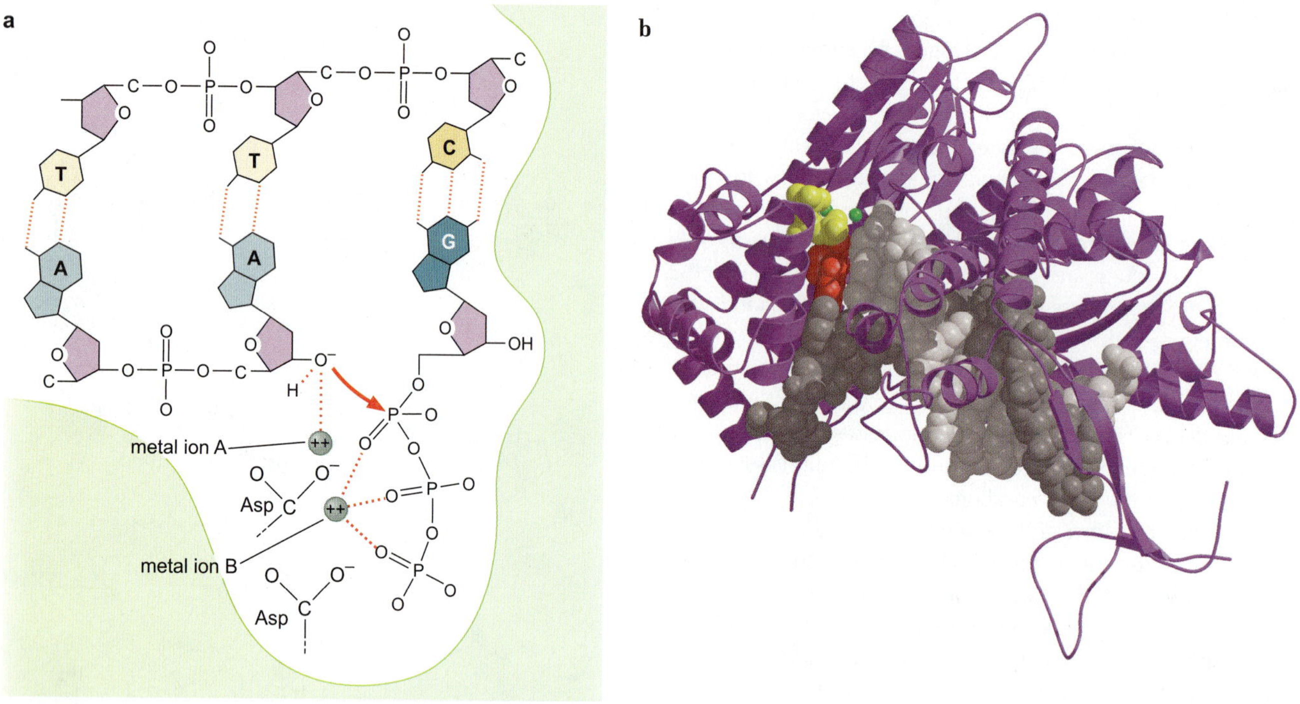

FIGURE 8-6 Two metal ions bound to DNA polymerase catalyze nucleotide addition. (a) Illustration of the active site of a DNA polymerase. The two metal ions (shown in green) are held in place by interactions with two highly conserved aspartate residues. Metal ion A primarily interacts with the 3′-OH, resulting in reduced association between the O and the H. This leaves a nucleophilic 3′O^-. Metal ion B interacts with the triphosphates of the incoming dNTP to neutralize their negative charge. After catalysis, the pyrophosphate product is stabilized through similar interactions with metal ion B (not shown). (b) Three-dimensional structure of the active site metal ions associated with the T7 DNA polymerase, the 3′-OH end of the primer, and the incoming nucleotide. The metal ions are shown in green and the remaining elements are shown in the same colors as those in Figure 8-5b. The view of the polymerase shown here is roughly equivalent to rotating the image shown in Figure 8-5b ~180° around the axis of the DNA helix. (Doublie S. et al. 1998. *Nature* 391: 251.) Image prepared with MolScript, BobScript, and Raster 3D.

matched DNA in this region interferes with these minor groove contacts and dramatically slows catalysis. The combination of the slowed catalysis and reduced affinity for newly synthesized mismatched DNA allows the release of the primer strand from the polymerase active site, and, in many cases, this strand binds and is acted on by a proofreading nuclease that removes the mismatched DNA (see below).

What are the roles of the fingers and the thumb? The fingers are also important for catalysis. Several residues located within the fingers bind to the incoming dNTP. More importantly, once a correct base pair is formed between the incoming dNTP and the template, the finger domain moves to enclose the dNTP (Fig. 8-7). This closed form of the polymerase hand stimulates catalysis by moving the incoming nucleotide in close contact with the catalytic metal ions.

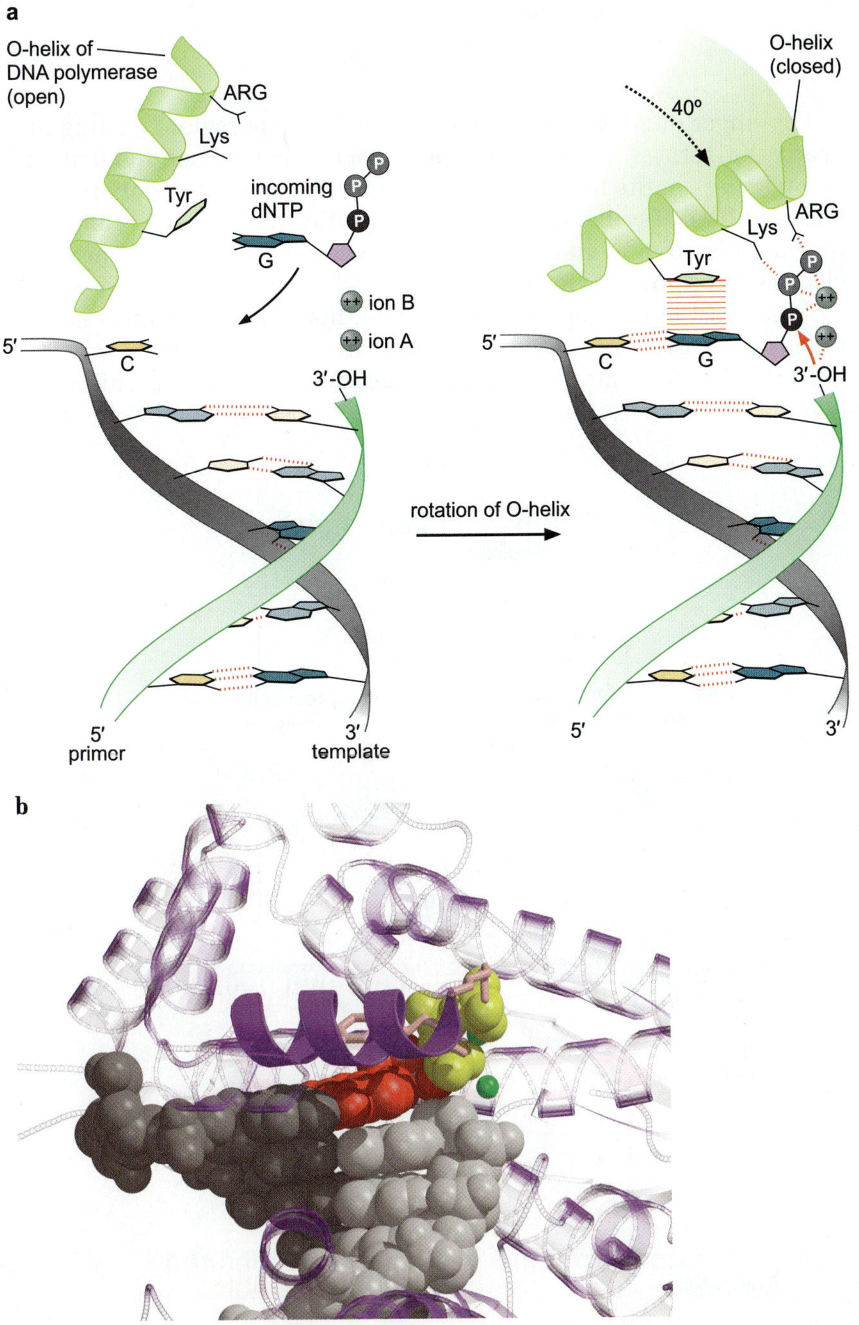

FIGURE 8-7 DNA polymerase "grips" the template and the incoming nucleotide when a correct base pair is made. (a) An illustration of the changes in DNA polymerase structure after the incoming nucleotide base-pairs correctly to the template DNA. The primary change is a 40° rotation of one of the helices in the finger domain called the O-helix. In the open conformation, this helix is distant from the incoming nucleotide. When the polymerase is in the closed conformation, this helix moves and makes several important interactions with the incoming dNTP. A tyrosine makes stacking interactions with the base of the dNTP and two charged residues associate with the triphosphate. The combination of these interactions positions the dNTP for catalysis mediated by the two metal ions bound to the DNA polymerase. (Based on Doublie S., Tabor S., Long A.M., Richardson C.C., and Ellenberger T. 1998. *Nature* 391: 251, Fig. 5. © 1998.) (b) The structure of T7 DNA polymerase bound to its substrates in the closed conformation. The O-helix is shown in purple and the rest of the protein structure is shown as transparent for clarity. The critical tyrosine, lysine, and arginine can be seen behind the O-helix in pink. (Red) The base and the deoxyribose of the incoming dNTP; (light gray) the primer; (dark gray) template strand; (green) the two catalytic metal ions; (yellow) phosphates. (Doublie S. et al. 1998. *Nature* 391: 251.) Image prepared with MolScript, BobScript, and Raster 3D.

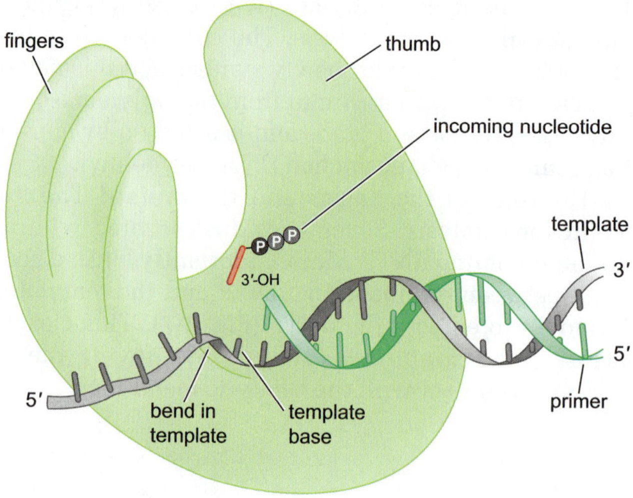

FIGURE 8-8 Illustration of the path of the template DNA through the DNA polymerase. The recently replicated DNA is associated with the palm region of the DNA polymerase. At the active site, the first base of the single-stranded region of the template is in a position expected for dsDNA. As one follows the template strand toward its 5′ end, the phosphodiester backbone abruptly bends 90°. This results in the second and all subsequent single-stranded bases being placed in a position that prevents any possibility of base pairing with a dNTP bound at the active site.

The finger domain associates with the template region, leading to a nearly 90° turn of the phosphodiester backbone between the first and second bases of the template. This bend serves to expose only the first template base after the primer at the catalytic site and avoids any confusion concerning which template base should pair with the next nucleotide to be added (Fig. 8-8).

In contrast to the fingers and the palm, the thumb domain is not intimately involved in catalysis. Instead, the thumb interacts with the DNA that has been most recently synthesized (see Fig. 8-9). This serves two purposes. First, it maintains the correct position of the primer and

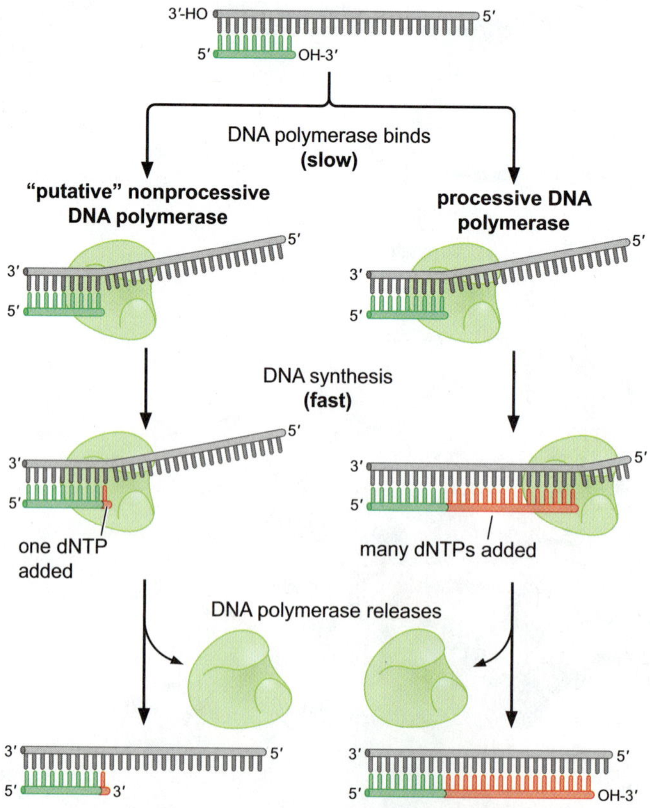

FIGURE 8-9 DNA polymerases synthesize DNA in a processive manner. This illustration shows the difference between a processive and a nonprocessive DNA polymerase. Both DNA polymerases bind the primer:template junction. Upon binding, the nonprocessive enzyme adds a single dNTP to the 3′ end of the primer and then is released from the new primer:template junction. In contrast, a processive DNA polymerase adds many dNTPs each time it binds to the template.

the active site. Second, the thumb helps to maintain a strong association between the DNA polymerase and its substrate. This association contributes to the ability of the DNA polymerase to add many dNTPs each time it binds a primer:template junction (see below).

To summarize, an ordered series of events occurs each time the DNA polymerase adds a nucleotide to the growing DNA chain. The incoming nucleotide base-pairs with the next available template base. This interaction causes the "fingers" of the polymerase to close around the base-paired dNTP. This conformation of the enzyme places the critical catalytic metal ions in a position to catalyze formation of the next phosphodiester bond. Attachment of the base-paired nucleotide to the primer leads to the reopening of the fingers and the movement of the primer:template junction by one base pair. The polymerase is then ready for the next cycle of addition. Importantly, each of these events is strongly stimulated by correct base pairing between the incoming dNTP and the template.

DNA Polymerases Are Processive Enzymes

Catalysis by DNA polymerase is rapid. DNA polymerases are capable of adding as many as 1000 nucleotides per second to a primer strand. The speed of DNA synthesis is largely due to the processive nature of DNA polymerase. **Processivity** is a characteristic of enzymes that operate on polymeric substrates. In the case of DNA polymerases, the degree of processivity is defined as the *average number of nucleotides added each time the enzyme binds a primer:template junction*. Each DNA polymerase has a characteristic processivity that can range from only a few nucleotides to more than 50,000 bases added per binding event (Fig. 8-9).

The rate of DNA synthesis is dramatically increased by adding multiple nucleotides per binding event. It is the initial binding of polymerase to the primer:template junction that is rate-limiting for DNA synthesis. In a typical DNA polymerase reaction, it takes ~1 second for the DNA polymerase to locate and bind a primer:template junction. Once bound, addition of a nucleotide is very fast (in the millisecond range). Thus, a completely nonprocessive DNA polymerase would add ~1 bp per second. In contrast, the fastest DNA polymerases add as many as 1000 nucleotides per second by remaining associated with the template for thousands of rounds of dNTP addition. Consequently, a highly processive polymerase increases the overall rate of DNA synthesis by as much as 1000-fold compared to a completely nonprocessive enzyme.

Increased processivity is facilitated by the ability of DNA polymerases to slide along the DNA template. Once bound to a primer:template junction, DNA polymerase interacts tightly with much of the double-stranded portion of the DNA in a sequence-nonspecific manner. These interactions include electrostatic interactions between the phosphate backbone and the "thumb" domain and interactions between the minor groove of the DNA and the palm domain (described above). The sequence-independent nature of these interactions permits the easy movement of the DNA even after it binds to polymerase. Each time a nucleotide is added to the primer strand, the DNA partially releases from the polymerase (the hydrogen bonds with the minor groove are broken, but the electrostatic interactions with the thumb are maintained). The DNA then rapidly rebinds to the polymerase in a position that is shifted by 1 bp using the same sequence-nonspecific mechanism. Further increases in processivity are

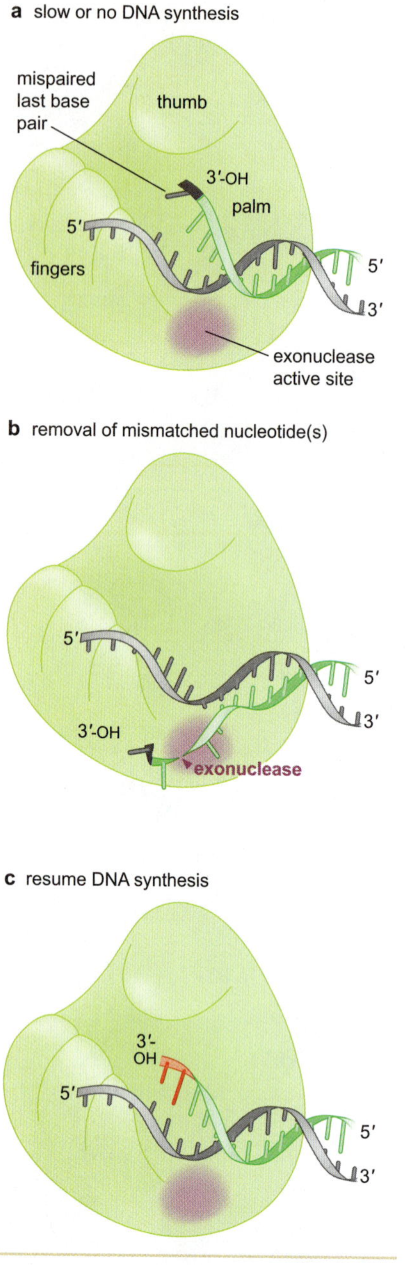

FIGURE 8-10 Proofreading exonucleases removes bases from the 3′ end of mismatched DNA. (a) When an incorrect nucleotide is incorporated into the DNA, the rate of DNA synthesis is reduced because of the incorrect positioning of the 3′-OH. (b) In the presence of a mismatched 3′ end, the last 3–4 nucleotides of the primer become single-stranded, resulting in an increased affinity for the exonuclease active site. Once bound at this active site, the mismatched nucleotide (and frequently an additional nucleotide) is removed from the primer. (c) Once the mismatched nucleotide is removed, a properly base paired primer:template junction is reformed and polymerization resumes (newly synthesized DNA is shown in red). (Adapted, with permission, from Baker T.A. and Bell S.P. 1998. *Cell* 92: 296, Fig. 1b. © Elsevier.)

achieved through interactions between the DNA polymerase and a "sliding clamp" protein that completely encircles the DNA, as we shall discuss further below.

Exonucleases Proofread Newly Synthesized DNA

A system based only on base-pair geometry and the complementarity between the bases cannot reach the extraordinarily high levels of accuracy that are observed for DNA synthesis in the cell (~1 mistake in every 10^{10} bp added). A major limit to DNA polymerase accuracy is the occasional (~1 in 10^5 times) flickering of the bases into the "wrong" tautomeric form (imino or enol; see Fig. 6-5). These alternate forms of the bases allow incorrect base pairs to be correctly positioned for catalysis. Proofreading allows these mistakes to be corrected.

Proofreading of DNA synthesis is mediated by nucleases that remove incorrectly base-paired nucleotides. This type of nuclease was originally identified in the same polypeptide as the DNA polymerase and is referred to as **proofreading exonuclease**. These exonucleases are capable of degrading DNA starting from a 3′ DNA end (i.e., from the growing end of the new DNA strand). (Nucleases that can only degrade from a DNA end are called *exo*nucleases; nucleases that can cut in the middle of a DNA strand are called *endo*nucleases.)

Initially, the presence of a 3′ exonuclease as part of the same polypeptide as a DNA polymerase made little sense. Why would the DNA polymerase need to degrade the DNA it had just synthesized? The role for these exonucleases became clear when it was determined that they have a strong preference to degrade DNA containing incorrect base pairs. Thus, in the rare event that an incorrect nucleotide is added to the primer strand, the proofreading exonuclease removes this nucleotide from the 3′ end of the primer strand. This "proofreading" of the newly added DNA gives the DNA polymerase a second chance to add the correct nucleotide.

The removal of mismatched nucleotides is facilitated by the reduced ability of DNA polymerase to add a nucleotide adjacent to an incorrectly base-paired primer. Mispaired DNA alters the geometry of the 3′-OH and the incoming nucleotide because of poor interactions with the palm region. This altered geometry reduces the rate of nucleotide addition in much the same way that addition of an incorrectly paired dNTP reduces catalysis. Thus, when a mismatched nucleotide is added, it both decreases the rate of new nucleotide addition and increases the rate of proofreading exonuclease activity.

As for processive DNA synthesis, proofreading occurs without releasing the DNA from the polymerase (Fig. 8-10). When a mismatched base pair is present in the polymerase active site, the primer:template junction is destabilized, creating several base pairs of unpaired DNA. The DNA polymerase active site binds such a mismatched template poorly, but the exonuclease active site has a ten-fold higher affinity for single-stranded 3′ ends. Thus, the newly unpaired 3′ end moves from the polymerase active site to the exonuclease active site. The incorrect nucleotide is removed by the exonuclease (an additional nucleotide may also be removed). The removal of the mismatched base allows the primer:template junction to reform and rebind the polymerase active site, enabling DNA synthesis to continue.

In essence, proofreading exonucleases work like a "delete key," removing only the most recent errors. The addition of a proofreading exonuclease greatly increases the accuracy of DNA synthesis. On average,

DNA polymerase inserts one incorrect nucleotide for every 10^5 nucleotides added. Proofreading exonucleases decrease the appearance of incorrect base pairs to 1 in every 10^7 nucleotides added. This error rate is still significantly short of the actual rate of mutation observed in a typical cell (~1 mistake in every 10^{10} nucleotides added). This additional level of accuracy is provided by the postreplication mismatch repair process described in Chapter 9.

THE REPLICATION FORK

Both Strands of DNA Are Synthesized Together at the Replication Fork

Thus far, we have discussed DNA synthesis in a relatively artificial context: that is, at a primer:template junction that is producing only one new strand of DNA. In the cell, both strands of the DNA duplex are replicated at the same time (see Interactive Animation 8-2). This requires separation of the two strands of the double helix to create two template DNAs. The junction between the newly separated template strands and the unreplicated duplex DNA is known as the **replication fork** (Fig. 8-11). The replication fork moves continuously toward the duplex region of unreplicated DNA, leaving in its wake two ssDNA templates that direct the formation of two daughter DNA duplexes.

WEB ANIMATION

The antiparallel nature of DNA creates a complication for the simultaneous replication of the two exposed templates at the replication fork. Because DNA is synthesized only by elongating a 3′ end, only one of the two exposed templates can be replicated continuously as the replication fork moves. On this template strand, the polymerase simply "chases" the moving replication fork. The newly synthesized DNA strand directed by this template is known as the **leading strand**.

Synthesis of the new DNA strand directed by the other ssDNA tem-

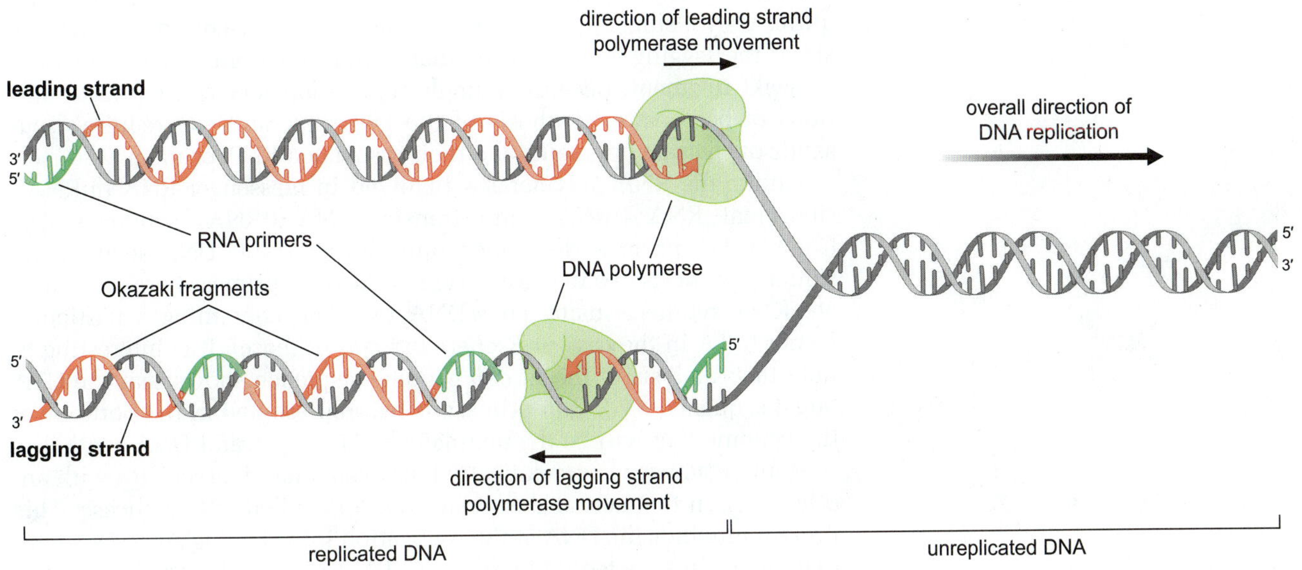

FIGURE 8-11 Replication fork. (Red) Newly synthesized DNA; (green) RNA primers. The Okazaki fragments shown are artficially short for illustrative purposes. In the cell, Okazaki fragments can vary between 100 and greater than 1000 bases.

plate is more complicated. This template directs the DNA polymerase to move in the opposite direction of the replication fork. The new DNA strand directed by this template is known as the **lagging strand**. As shown in Figure 8-11, this strand of DNA must be synthesized in a discontinuous fashion.

Although the leading-strand DNA polymerase can replicate its template as soon as it is exposed, synthesis of the lagging strand must wait for movement of the replication fork to expose a substantial length of template before it can be replicated. Each time a substantial length of new lagging-strand template is exposed, DNA synthesis is initiated and continues until it reaches the 5′ end of the previous newly synthesized stretch of lagging-strand DNA.

The resulting short fragments of new DNA formed on the lagging strand are called **Okazaki fragments** and vary in length from 1000 to 2000 nucleotides in bacteria and from 100 to 400 nucleotides in eukaryotes. Shortly after being synthesized, Okazaki fragments are covalently joined together to generate a continuous, intact strand of new DNA (see below). Okazaki fragments are therefore transient intermediates in DNA replication.

The Initiation of a New Strand of DNA Requires an RNA Primer

As described above, all DNA polymerases require a primer with a free 3′-OH. They cannot initiate a new DNA strand de novo. How, then, are new strands of DNA synthesis started? To accomplish this, the cell takes advantage of the ability of *RNA* polymerases to do what DNA polymerases cannot: start new RNA chains de novo. **Primase** is a specialized RNA polymerase dedicated to making short RNA primers (5–10 nucleotides long) on an ssDNA template. These primers are subsequently extended by DNA polymerase. Although DNA polymerases incorporate only deoxyribonucleotides into DNA, they can initiate synthesis using either an RNA primer or a DNA primer annealed to the DNA template.

Although both the leading and lagging strands require primase to initiate DNA synthesis, the frequency of primase function on the two strands is dramatically different (see Fig. 8-11). Each leading strand requires only a single RNA primer. In contrast, the discontinuous synthesis of the lagging strand means that new primers are needed for each Okazaki fragment. Because a single replication fork can replicate millions of base pairs, synthesis of the lagging strand can require thousands of Okazaki fragments and their associated RNA primers.

Unlike the RNA polymerases involved in messenger RNA (mRNA), ribosomal RNA (rRNA), and transfer RNA (tRNA) synthesis (see Chapter 12), primase does not require an extensive DNA sequence to initiate synthesis of a new RNA primer. Instead, primases prefer to initiate RNA synthesis using an ssDNA template containing a particular trimer (GTA in the case of *Escherichia coli* primase). It is interesting to note that analysis of the *E. coli* genome sequence shows that the GTA target sequence for *E. coli* primase is overrepresented in the portions of the genome that will be the template for lagging-strand DNA synthesis.

Primase activity is dramatically increased when it associates with another protein that acts at the replication fork called DNA helicase. This protein unwinds the DNA at the replication fork creating an ssDNA template that can be acted on by primase. DNA helicase function is considered in more detail below. The requirement for an ssDNA template and association with another replication fork protein ensures that primase is only active at the recently unwound DNA at the replication fork.

RNA Primers Must Be Removed to Complete DNA Replication

To complete DNA replication, the RNA primers used for initiation must be removed and replaced with DNA (Fig. 8-12). Removal of the RNA primers can be thought of as a DNA-repair event, and this process shares many of the properties of excision DNA repair, a process covered in detail in Chapter 9.

To replace the RNA primers with DNA, an enzyme called **RNase H** recognizes and removes most of each RNA primer. This enzyme specifically degrades RNA that is base-paired with DNA (hence, the "H" in its name, which stands for hybrid in RNA:DNA hybrid). RNase H removes all of the RNA primer except the ribonucleotide directly linked to the DNA end. This is because RNase H can cleave only bonds between two ribonucleotides. The final ribonucleotide is removed by an exonuclease that degrades RNA or DNA from their 5′ end.

Removal of the RNA primer leaves a gap in the dsDNA that is an ideal substrate for DNA polymerase—a primer:template junction (see Fig. 8-12). DNA polymerase fills this gap until every nucleotide is base-paired, leaving a DNA molecule that is complete except for a break in the backbone between the 3′-OH and 5′ phosphate of the repaired strand. This "nick" in the DNA can be repaired by an enzyme called **DNA ligase**. DNA ligases use high-energy cofactors (such as ATP) to create a phosphodiester bond between an adjacent 5′ phosphate and 3′-OH. Only after all RNA primers are replaced by DNA and the associated nicks are sealed is DNA synthesis complete.

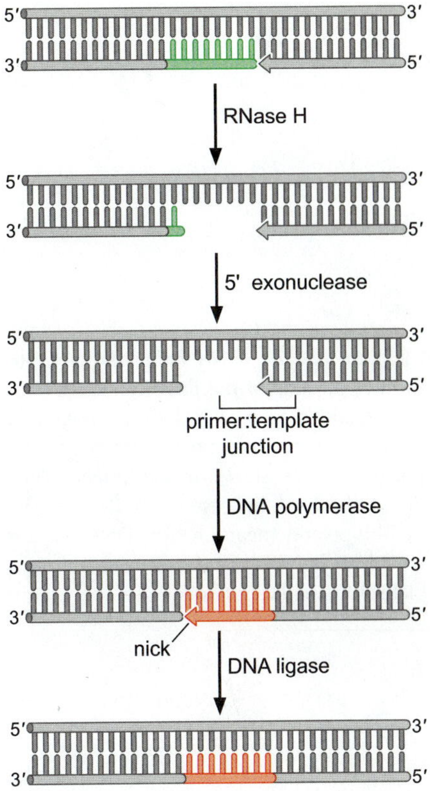

FIGURE 8-12 Removal of RNA primers from newly synthesized DNA. The sequential function of RNase H, 5′ exonuclease, DNA polymerase, and DNA ligase during the removal of RNA primers is illustrated. (Gray) DNA present prior to RNA primer removal; (green) RNA primer; (red) the newly synthesized DNA that replaces the RNA primer.

DNA Helicases Unwind the Double Helix in Advance of the Replication Fork

DNA polymerases are generally poor at separating the two base-paired strands of duplex DNA. Therefore, at the replication fork, a second class of enzymes, called **DNA helicases**, catalyze the separation of the two strands of duplex DNA. These enzymes bind to and move directionally along ssDNA using the energy of nucleoside triphosphate (usually ATP) hydrolysis to displace any DNA strand that is annealed to the bound ssDNA. Typically, DNA helicases that act at replication forks are hexameric proteins that assume the shape of a ring (Fig. 8-13). These ring-shaped protein complexes encircle one of the two single strands at the replication fork near the single-stranded:double-stranded junction.

Like DNA polymerases, DNA helicases act processively. Each time they associate with substrate, they unwind multiple base pairs of DNA. The ring-shaped hexameric DNA helicases found at replication forks exhibit high processivity because they encircle the DNA. Release of the helicase from its DNA substrate therefore requires the opening of the hexameric protein ring, which is a rare event. Alternatively, the helicase can dissociate when it reaches the end of the DNA strand that it has encircled.

Of course, this arrangement of enzyme and DNA poses problems for the binding of the DNA helicase to the DNA substrate in the first place. This problem is most obvious for circular chromosomes where there is no DNA end for the DNA helicase to thread onto. However, because helicases are almost always loaded onto the DNA at internal sites of linear chromsomes, the same problem exists during the replication of these DNAs. Thus, there are specialized mechanisms that open the DNA helicase ring and place it around the DNA before re-

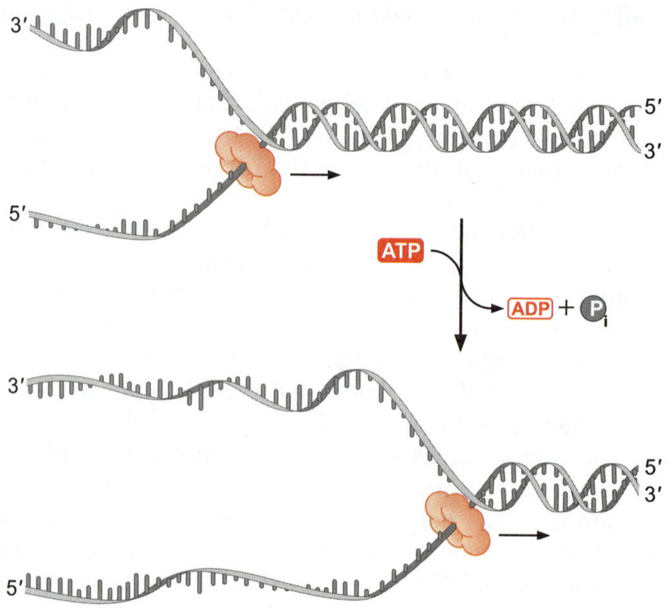

FIGURE 8-13 DNA helicases separate the two strands of the double helix. When ATP is added to a DNA helicase bound to ssDNA, the helicase moves with a defined polarity on the ssDNA. In the instance illustrated, the DNA helicase has a 5′→3′ polarity. This polarity means that the DNA helicase would be bound to the lagging-strand template at the replication fork.

forming the ring (see section on Initiation of Replication below). This topological linkage between proteins involved in DNA replication and their DNA substrates is a common mechanism to increase processivity.

Each DNA helicase moves along ssDNA in a defined direction. This property is a characteristic of each DNA helicase called its **polarity** (see Box 8-3, Determining the Polarity of a DNA Helicase). DNA helicases can have a polarity of either 5′→3′ or 3′→5′. This direction is always defined according to the strand of DNA bound (or encircled for a ring-shaped helicase), rather than the strand that is displaced. In the case of a DNA helicase that functions on the lagging-strand template of the replication fork, the polarity is 5′→3′ to allow the DNA helicase to proceed toward the duplex region of the replication fork (see Fig. 8-13). As is true for all enzymes that move along DNA in a directional manner, movement of the helicase along ssDNA requires the input of chemical energy. For helicases, this energy is provided by ATP hydrolysis.

TECHNIQUES

Box 8-3 Determining the Polarity of a DNA Helicase

The activity of a DNA helicase can be detected by its ability to displace one strand of a DNA duplex from another. In a typical DNA helicase assay, the substrate is composed of one short labeled ssDNA annealed to a second long unlabeled ssDNA (often the label is radioactive ^{32}P incorporated into the short ssDNA). Consider a large circular ssDNA (e.g., 5000 bases) hybridized to a short (200 bases) labeled linear ssDNA molecule (Box 8-3 Fig. 1). A DNA helicase will displace the short linear ssDNA from the large ssDNA circle. Separation of the strands can be detected by a change in electrophoretic mobility of the short labeled ssDNA, in a nondenaturing agarose gel (see Chapter 21). After the gel is exposed to X-ray film to detect only the radiolabeled DNA, the position in the gel that the short DNA occupies can be determined. When it is hybridized to the ssDNA circle, the short ssDNA will comigrate with the large ssDNA circle. In contrast, if the short ssDNA has been displaced from the ssDNA circle by DNA helicase, it will migrate according to its actual size, 200 bases.

Box 8-3 *(Continued)*

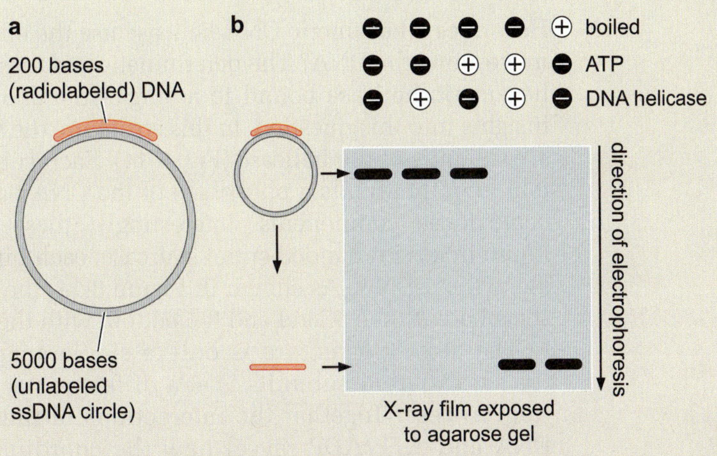

Box 8-3 Figure 1 A biochemical assay for DNA helicase activity. (a) DNA substrate to detect helicase activity. A 5000-bp unlabeled ssDNA circular DNA is annealed to a 200-base radiolabeled DNA. For convenience, the two molecules are not drawn to scale. (b) To detect DNA helicase activity, the DNA substrate is exposed to the DNA helicase (in this case, with and without ATP). After the reaction, the resulting DNA molecules are separated by agarose gel electrophoresis (nondenaturing). When the short radiolabeled DNA is base-paired with the large ssDNA circle, both molecules will comigrate as a large molecule. In contrast, after the DNA helicase has acted, the short radiolabeled ssDNA will migrate at a position consistent with its length. After exposure of the agarose gel to X-ray film, only the position of the radiolabeled DNA will be visible. As a control, the two DNA molecules can be separated by boiling, which also denatures the base-paired region.

A modification of this simple experiment can be used to determine the polarity of a DNA helicase. Suppose there is a restriction enzyme cleavage site located asymmetrically within the base-paired region (Box 8-3 Fig. 2). When this site is cleaved, it will generate a largely single-stranded, linear DNA with two regions of dsDNA of different lengths at each end. Remember that DNA helicases bind to ssDNA, not dsDNA. Thus, the only place that a DNA helicase can bind this new linear substrate is between the two dsDNA regions. Because of the polarity of DNA helicases, any given DNA helicase can displace only one of the two short ssDNAs. Because the two short ssDNA regions are of different lengths, the size of the released fragment will reveal which direction the DNA helicase moved along the ssDNA region of the linear substrate.

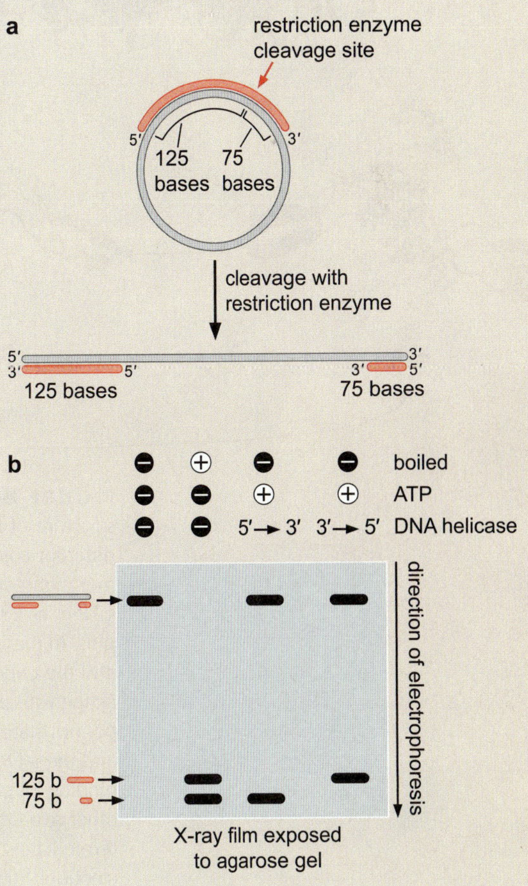

Box 8-3 Figure 2 A biochemical assay for DNA helicase polarity. (a) The DNA substrate. The same DNA substrate illustrated in Box 8-3 Figure 1 is cleaved with a restriction enzyme that leaves fully double-stranded DNA ends. The restriction enzyme is chosen to cleave asymmetrically, leaving 125-base and 75-base radiolabeled ssDNA fragments annealed to the ends of a 5000-base unlabeled ssDNA. The 5′ and 3′ ends of the resulting DNA molecules are indicated. (b) An illustration of an X-ray film exposed to an agarose gel used to separate the DNA products after DNA helicase treatment is shown. The substrate generated in part (a) can be incubated with a DNA helicase to determine its polarity. Results for a 5′→3′ and a 3′→5′ DNA helicase are shown. Boiling of the substrate indicates the consequences of complete denaturation of all base pairing.

DNA Helicase Pulls Single-Stranded DNA through a Central Protein Pore

How does a hexameric DNA helicase use the energy of ATP hydrolysis to move along the DNA? The determination of the atomic structure of a viral hexameric helicase bound to a single-stranded DNA substrate provides insights into this question. In this structure, the ssDNA is encircled by the six subunits of the helicase (Fig. 8-14). Each subunit has a "hairpin" protein loop that binds a phosphate of the DNA backbone and its two adjacent ribose components. Interestingly, these DNA-binding loops are found in a right-handed spiral staircase, each binding the next phosphate along the ssDNA. As shown in Figure 8-14, the top of the staircase is associated with the 5′ end and the bottom with the 3′ end of the ssDNA.

The atomic structure is only a single snapshot; however, each of the six different subunits is at a different stage in the DNA translocation process. Together the interactions of the different subunits with DNA and ATP/ADP reveal how the coordinated movement of these

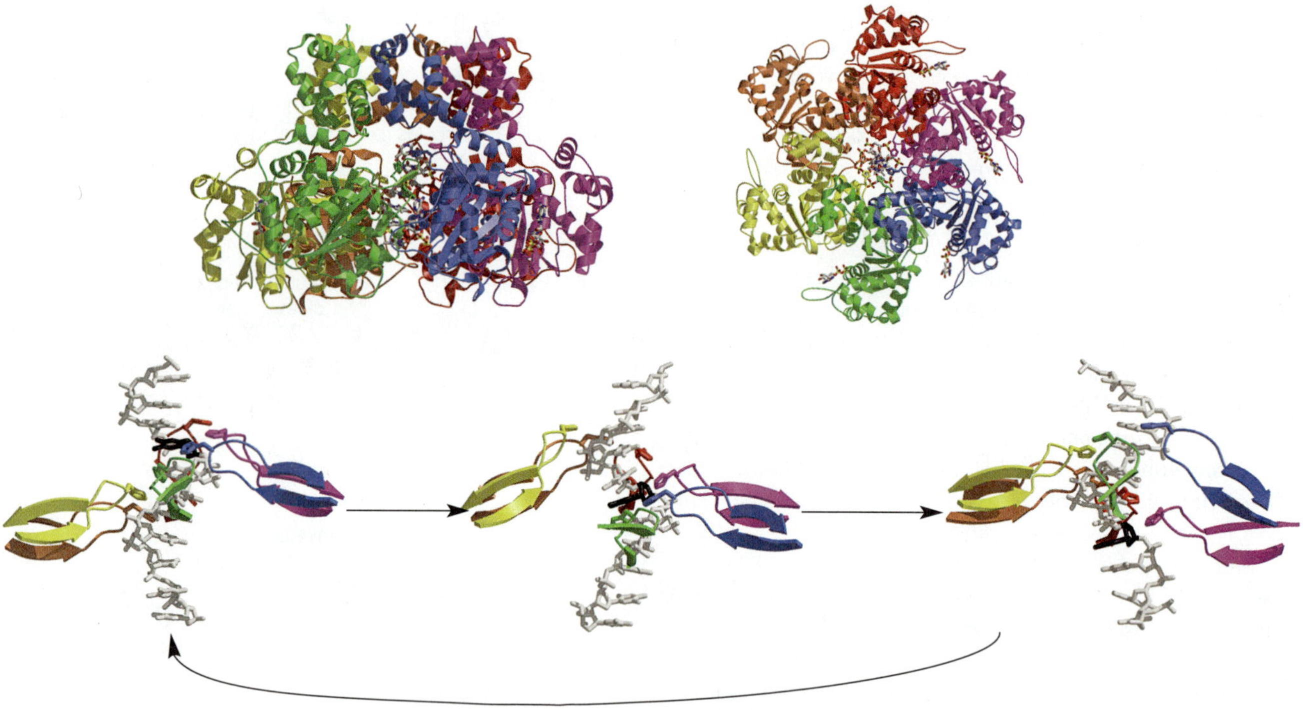

FIGURE 8-14 Structure and proposed mechanisms of a DNA helicase. (a) Overall structure of the bovine papillomavirus E1 hexameric helicase. Each subunit is shown in a different color and the complex is shown looking down the central channel of the hexamer (right) and from the side (left). The protein subunits are shown in ribbon form and the ssDNA as a stick diagram. (b) Illustration of the proposed movement of DNA-binding hairpins. In these views, we are only showing the ssDNA in the central channel of the helicase and the critical hairpins that bind this DNA. The three views show the blue hairpin interacting with the phosphate associated with the black nucleotide moving from the top (ATP-bound) state to the middle (ADP-bound) state to the bottom (nucleotide-unbound) state and the ssDNA moving with the hairpin. Rebinding of ATP is expected to drive the DNA-binding loop back to the top position to allow binding to a new phosphate. Note that the other five DNA-binding loops shown in other colors are also moving through the same intermediates but with different starting points and binding different phosphates/nucleotides. (Enemark E.J. and Joshua-Tor L. 2006. Nature 442: 270–275. PDB Code: 2GXA.) Image prepared with MolScript, BobScript, and Raster 3D.

protein hairpins can pull the ssDNA through the central pore of the helicase. A subunit first binds the ssDNA at the top of the structure (Fig. 8-14b) and the DNA-binding loop moves through successive conformations toward the bottom, bringing the bound DNA along with it. Importantly, each of these conformations is associated with a different nucleotide-bound state; the top conformation is in an ATP-bound state, the middle is in an ADP-bound state, and the bottom lacks a bound nucleotide. Thus, as a single subunit binds, hydrolyzes, and releases ATP, it will cycle through the top, middle, and bottom conformations. Overall, one can think of the helicase as having six hands pulling on a rope in a hand over hand manner.

In addition to translocating along ssDNA, a helicase must also displace the complementary strand to cause DNA unwinding. In the case of this hexameric helicase, the structure of the central channel shows that strand displacement must occur for DNA to pass through the channel. At its narrowest point, the central channel has a diameter of 13 Å, large enough to fit ssDNA, but much too small to fit the 20 Å diameter of dsDNA.

Single-Stranded DNA-Binding Proteins Stabilize ssDNA prior to Replication

After the DNA helicase has passed, the newly generated ssDNA must remain free of base pairing until it can be used as a template for DNA synthesis. To stabilize the separated strands, ssDNA-binding proteins (designated **SSBs**) rapidly bind to the separated strands. Binding of one SSB promotes the binding of another SSB to the immediately adjacent ssDNA (Fig. 8-15). This is called **cooperative binding** and occurs because SSB molecules bound to immediately adjacent regions of ssDNA also bind to each other. This SSB–SSB interaction strongly stabilizes SSB binding to ssDNA and makes sites already occupied by one or more SSB molecules preferred SSB binding sites.

Cooperative binding ensures that ssDNA is rapidly coated by SSB as it emerges from the DNA helicase. (Cooperative binding is a property of many DNA-binding proteins, see Box 16-4, Concentration,

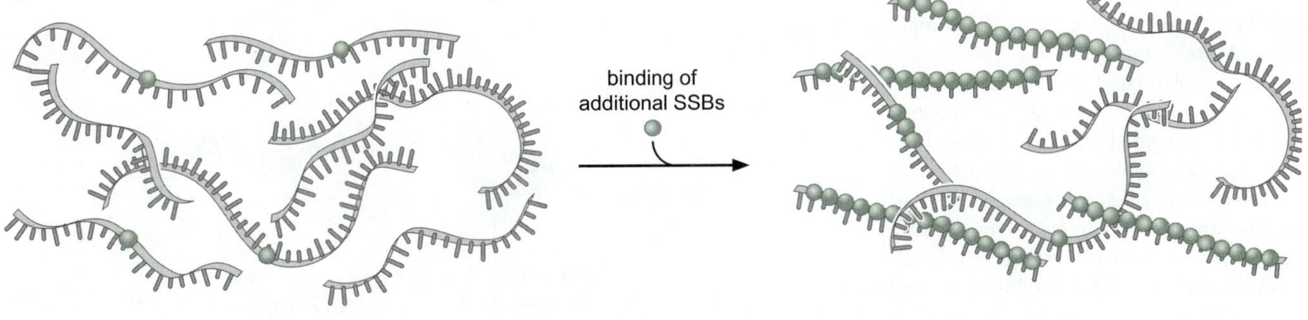

FIGURE 8-15 **Binding of single-stranded binding protein (SSB) to DNA.** (a) A limiting amount of SSB is bound to four of the nine ssDNA molecules shown. (b) As more SSB binds to DNA, it preferentially binds adjacent to previously bound SSB molecules. Only after SSB has completely coated the initially bound ssDNA molecules does binding occur on other molecules. Note that when ssDNA is coated with SSB, it assumes a more extended conformation that inhibits the formation of intramolecular base pairs.

Affinity, and Cooperative Binding.) Once covered with SSB, ssDNA is held in an elongated state that facilitates its use as a template for DNA or RNA primer synthesis.

SSB interacts with ssDNA in a sequence-independent manner. SSBs primarily contact ssDNA through electrostatic interactions with the phosphate backbone and stacking interactions with the DNA bases. In contrast to sequence-specific DNA-binding proteins, SSBs make few, if any, hydrogen bonds to the ssDNA bases.

Topoisomerases Remove Supercoils Produced by DNA Unwinding at the Replication Fork

As the strands of DNA are separated at the replication fork, the dsDNA in front of the fork becomes increasingly positively supercoiled (Fig. 8-16). This accumulation of supercoils is the result of DNA helicase eliminating the base pairs between the two strands. If the DNA strands remain unbroken, there can be no reduction in link-

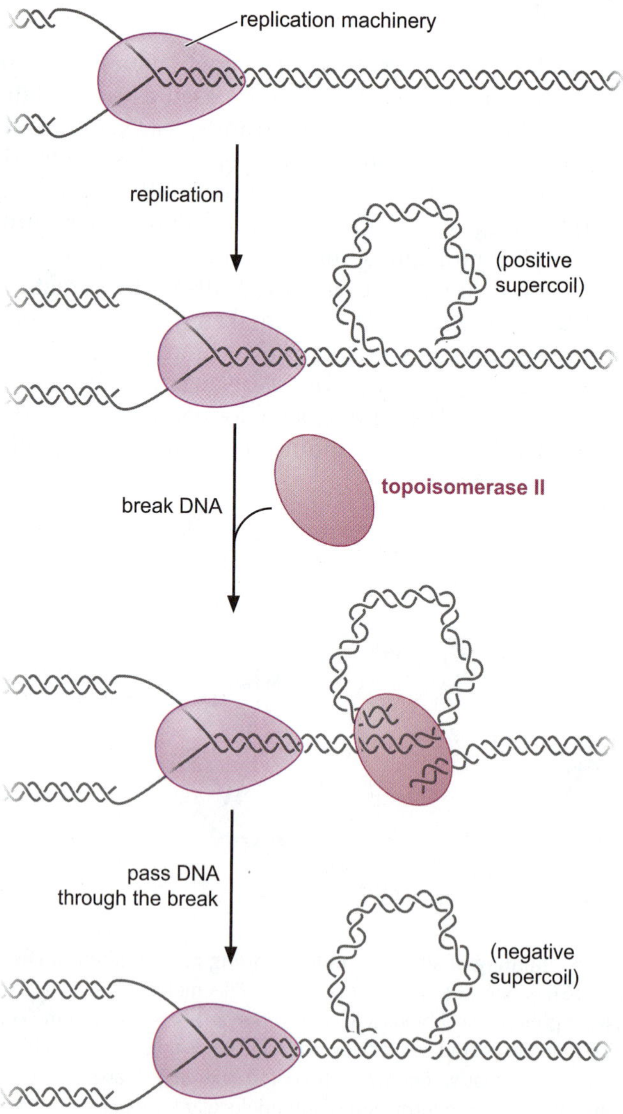

FIGURE 8-16 Action of topoisomerase at the replication fork. As positive supercoils accumulate in front of the replication fork, topoisomerases rapidly remove them. In this diagram, the action of Topo II removes the positive supercoil induced by a replication fork. By passing one part of the unreplicated dsDNA through a double-stranded break in a nearby unreplicated region, the positive supercoils can be removed. It is worth noting that this change would reduce the linking number by two and thus would only have to occur once every 20 bp replicated. Although the action of a type II topoisomerase is illustrated here, type I topoisomerases can also remove the positive supercoils generated by a replication fork. Note that the positive superhelicity in front of the replication fork is shown as right-handed toroidal writhe (one complete turn equals a positive writhe of +1). Passage of one dsDNA molecule through the other at the site of the writhe changes this to one complete left-handed toroidal writhe (equal to a writhe of –1). This illustrates how the linking number is changed by 2 units by a type II topoisomerase (for more information about DNA topology and writhe, see Chapter 6, DNA Topology, and Chapter 7, Box 7-2).

ing number (the number of times the two DNA strands are intertwined) to accommodate this unwinding of the DNA duplex (see Chapter 6). Thus, as the DNA helicase proceeds, the DNA must accommodate the same linking number within a smaller and smaller number of base pairs. Indeed, for the DNA in front of the replication fork to remain relaxed, one DNA link must be removed approximately every 10 bp of DNA unwound. If there were no mechanism to relieve the accumulation of these supercoils, the replication machinery would grind to a halt in the face of mounting strain placed on the DNA in front of the replication fork.

The problem is most clear for the circular chromosomes of bacteria, but it also applies to eukaryotic chromosomes. Because eukaryotic chromosomes are not closed circles, they could in principle rotate along their length to dissipate the introduced supercoils. This is not the case, however: it is simply not possible to rotate a DNA molecule that is millions of base pairs long each time one turn of the helix is unwound.

The supercoils introduced by the action of the DNA helicase are removed by topoisomerases that act on the unreplicated dsDNA in front of the replication fork (Fig. 8-16). These enzymes do this by breaking either one or both strands of the DNA without letting go of the DNA and passing the same number of DNA strands through the break (as we discussed in Chapter 6). This action relieves the accumulation of supercoils. In this way, topoisomerases act as a "swivelase" that rapidly dissipates the accumulation of supercoils ahead of the replication fork.

Replication Fork Enzymes Extend the Range of DNA Polymerase Substrates

On its own, DNA polymerase can only efficiently extend 3′-OH primers annealed to ssDNA templates. The addition of primase, DNA helicase, and topoisomerase dramatically extends the possible substrates for DNA polymerase. Primase provides the ability to initiate new DNA strands on any piece of ssDNA. Of course, the use of primase also imposes a requirement for the removal of the RNA primers to complete replication. Similarly, strand separation by DNA helicase and dissipation of positive supercoils by topoisomerase allow DNA polymerase to replicate dsDNA. Although the names of the proteins change from organism to organism (Table 8-1), the same set of enzymatic activities is used by organisms as diverse as bacteria, yeast, and humans to accomplish chromosomal DNA replication.

It is noteworthy that both DNA helicase and topoisomerase perform their functions without permanently altering the chemical structure of DNA or synthesizing any new molecule. DNA helicase breaks only the hydrogen bonds that hold the two strands of DNA together without

TABLE 8-1 Enzymes That Function at the Replication Fork

	Escherichia coli	*Saccharomyces cerevisiae*	Human
Primase	DnaG	Primase (PRI l/PRI 2)	Primase
DNA helicase	DnaB	Mcm complex	Mcm complex
SSB	SSB	RPA	RPA
Topoisomerases	Gyrase, Topo I	Topo I, II	Topo I, II

breaking any covalent bonds. Although topoisomerases break one or more of DNA's covalent bonds, each bond broken is precisely reformed before the topoisomerase releases the DNA (see Fig. 6-25). Instead of altering the chemical structure of DNA, the action of these enzymes results in a DNA molecule with an altered conformation. Importantly, these conformational alterations are essential for the duplication of the large dsDNA molecules that are the foundation of both bacterial and eukaryotic chromosomes.

The proteins that act at the replication fork interact tightly but in a sequence-independent manner with the DNA. These interactions exploit the features of DNA that are the same regardless of the particular base pair: the negative charge and structure of the phosphate backbone (e.g., the thumb domain of DNA polymerase), the hydrogen-bonding residues in the minor groove (e.g., the palm domain of the DNA polymerase), and the hydrophobic stacking interactions between the bases (e.g., SSB). In addition, the structures of some of these proteins are specialized to encourage processive action by either fully (e.g., DNA helicase) or partially (e.g., DNA polymerase) encircling the DNA.

THE SPECIALIZATION OF DNA POLYMERASES

DNA Polymerases Are Specialized for Different Roles in the Cell

The central role of DNA polymerases in the efficient and accurate replication of the genome requires that cells have evolved multiple specialized DNA polymerases. For example, *E. coli* has at least five DNA polymerases that are distinguished by their enzymatic properties, subunit composition, and abundance (Table 8-2). DNA polymerase III (DNA Pol III) is the primary enzyme involved in the replication of the chromosome. Because the entire 4.6-Mb *E. coli* genome is replicated by two replication forks, DNA Pol III must be highly processive. Consistent with these requirements, DNA Pol III is generally found to be part of a larger complex that confers very high processivity—a complex known as the **DNA Pol III holoenzyme**.

In contrast, **DNA polymerase I (DNA Pol I)** is specialized for the removal of the RNA primers that are used to initiate DNA synthesis. For this reason, this DNA polymerase has a 5′ exonuclease that allows DNA Pol I to remove RNA or DNA immediately *upstream* of the site of DNA synthesis. Unlike DNA Pol III holoenzyme, DNA Pol I is not highly processive, adding only 20–100 nucleotides per binding event. These properties are ideal for RNA primer removal and DNA synthesis across the resulting ssDNA gap. The 5′ exonuclease of DNA Pol I can remove the RNA–DNA linkage that is resistant to RNase H (see Fig. 8-12). The short extent of synthesis by DNA Pol I is ideal for replacing the short region previously occupied by the RNA primers (<10 nucleotides).

Because both DNA Pol I and DNA Pol III are involved in DNA replication, both of these enzymes must be highly accurate. Thus, both proteins carry an associated proofreading exonuclease. The remaining three DNA polymerases in *E. coli* are specialized for DNA repair and lack proofreading activities. These enzymes are discussed in Chapter 9.

Eukaryotic cells also have multiple DNA polymerases, with a typical cell having more than 15. Of these, three are essential to duplicate the genome: DNA Pol δ, DNA Pol ε, and DNA Pol α/primase. Each of these eukaryotic DNA polymerases is composed of multiple subunits (see Table 8-2). DNA Pol α/primase is specifically involved in initiating new DNA strands. This four-subunit protein complex consists of

TABLE 8-2 Activities and Functions of DNA Polymerases

Prokaryotic (*E. coli*)	Number of Subunits	Function
Pol I	1	RNA primer removal, DNA repair
Pol II (Din A)	1	DNA repair
Pol III core	3	Chromosome replication
Pol III holoenzyme	9	Chromosome replication
Pol IV (Din B)	1	DNA repair, translesion synthesis (TLS)
Pol V (UmuC, UmuD$'_2$C)	3	TLS
Eukaryotic	**Number of Subunits**	**Function**
Pol α	4	Primer synthesis during DNA replication
Pol β	1	Base excision repair
Pol γ	3	Mitochondrial DNA replication and repair
Pol δ	2–3	Lagging-strand DNA synthesis; nucleotide and base excision repair
Pol ε	4	Leading-strand DNA synthesis; nucleotide and base excision repair
Pol θ	1	DNA repair of cross-links
Pol ζ	1	TLS
Pol λ	1	Meiosis-associated DNA repair
Pol μ	1	Somatic hypermutation
Pol κ	1	TLS
Pol η	1	Relatively accurate TLS past *cis-syn* cyclobutane dimers
Pol ι	1	TLS, somatic hypermutation
Rev1	1	TLS

Data from Sutton M.D. and Walker G.C. 2001. *Proc. Natl. Acad Sci.* 98: 8342–8349, and references therein.

a two-subunit DNA Pol α *and* a two-subunit primase. After the primase synthesizes an RNA primer, the resulting RNA primer:template junction is immediately handed off to the associated DNA Pol α to initiate DNA synthesis.

Because of its relatively low processivity, DNA Pol α/primase is rapidly replaced by the highly processive DNA Pol δ and Pol ε. The process of replacing DNA Pol α/primase with DNA Pol δ or Pol ε is called **polymerase switching** (Fig. 8-17) and results in three different DNA polymerases functioning at the eukaryotic replication fork. Recent studies indicate that DNA Pol δ and ε are specialized to synthesize different strands at the replication fork, with DNA Pol ε making the leading strand and DNA Pol δ the lagging strand. As in bacterial cells, the majority of the remaining eukaryotic DNA polymerases are involved in DNA repair.

Sliding Clamps Dramatically Increase DNA Polymerase Processivity

High processivity at the replication fork ensures rapid chromosome duplication. As we have discussed, DNA polymerases at the replication fork synthesize thousands to millions of base pairs without re-

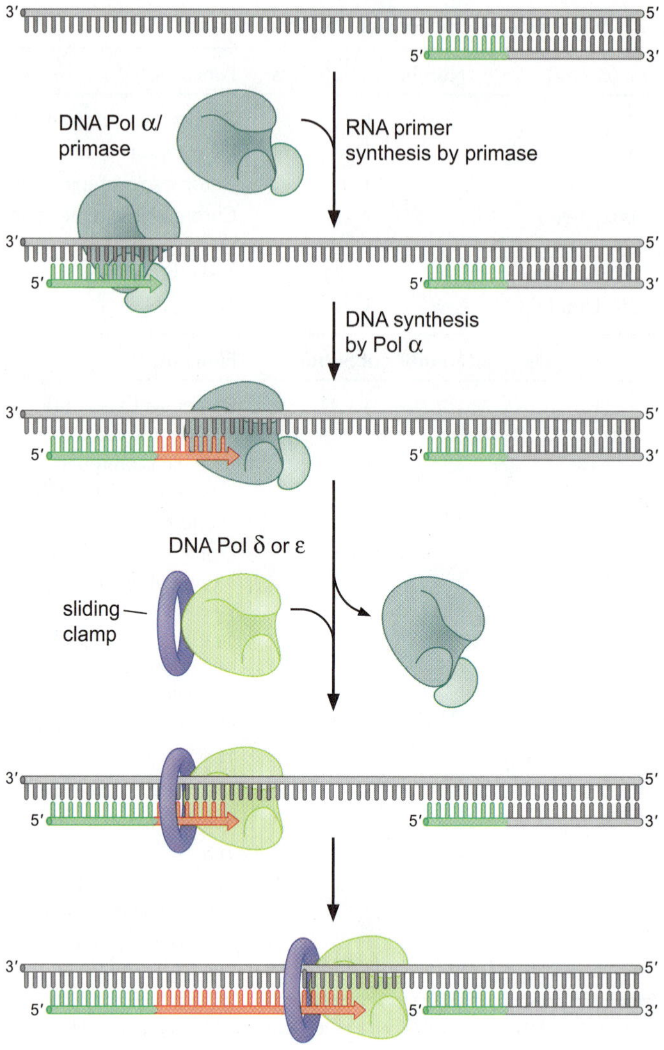

FIGURE 8-17 DNA polymerase switching during eukaryotic DNA replication. The order of eukaryotic DNA polymerase function is illustrated. The length of the DNA synthesized is shorter than in reality for illustrative purposes. Typically, the combined DNA Pol α/primase product is between 50 and 100 bp and the further extension by Pol ε or Pol δ is between 100 and 10,000 nucleotides. Although both DNA Pol δ and ε can substitute for DNA Pol α/primase, recent studies indicate that DNA Pol ε substitutes on the leading-strand template and DNA Pol δ substitutes on the lagging-strand template.

leasing from the template. Despite this, when looked at in the absence of other proteins, the DNA polymerases that act at the replication fork are only able to synthesize 20–100 bp before releasing from the template. How is the processivity of these enzymes increased so dramatically at the replication fork?

One key to the high processivity of the DNA polymerases that act at replication forks is their association with proteins called **sliding DNA clamps**. These proteins are composed of multiple identical subunits that assemble in the shape of a "doughnut." The hole in the center of the clamp is large enough to encircle the DNA double helix and leave room for a layer of one or two water molecules between the DNA and the protein (Fig. 8-18a; see Structural Tutorial 8-2). These properties allow the clamp proteins to slide along the DNA without dissociating from it. Sliding DNA clamps also bind tightly to DNA polymerases at replication forks (Fig. 8-18b). The resulting complex between the polymerase and the sliding clamp moves efficiently along the DNA template during DNA synthesis.

WEB STRUCTURAL TUTORIAL

How does the association with the sliding clamp change the processivity of the DNA polymerase? In the absence of the sliding clamp, a DNA polymerase dissociates and diffuses away from the template DNA on average once every 20–100 bp synthesized. In the presence of

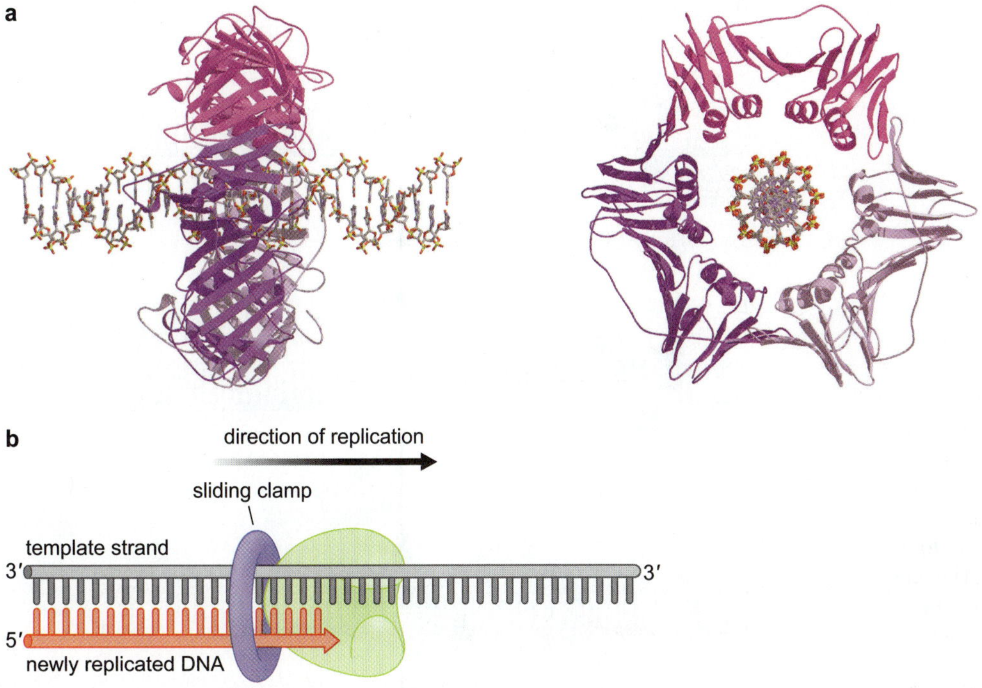

FIGURE 8-18 Structure of a sliding DNA clamp. (a) Three-dimensional structure of a sliding DNA clamp associated with DNA. The opening through the center of the sliding clamp is about 35 Å and the width of the DNA helix is ~20 Å. This provides enough space to allow a thin layer of one or two water molecules between the sliding clamp and the DNA. This is thought to allow the clamp to slide along the DNA easily. (Krishna T.S. et al. 1994. *Cell* 79: 1233.) Image prepared with MolScript, BobScript, and Raster 3D. DNA modeled by Leemor Joshua-Tor. (b) Sliding DNA clamps encircle the newly replicated DNA produced by an associated DNA polymerase. The sliding clamp interacts with the part of the DNA polymerase that is closest to the newly synthesized DNA as it emerges from the DNA polymerase.

the sliding clamp, the DNA polymerase still disengages its active site from the 3′-OH end of the DNA frequently, but the association with the sliding clamp prevents the polymerase from diffusing away from the DNA (Fig. 8-19). By keeping the DNA polymerase in close proximity to the DNA, the sliding clamp ensures that the DNA polymerase rapidly rebinds *the same* primer:template junction, vastly increasing the processivity of the DNA polymerase.

Once an ssDNA template has directed synthesis of its complementary DNA strand, the DNA polymerase must release from the completed dsDNA and the sliding clamp to act at a new primer:template junction. This release is accomplished by a change in the affinity between the DNA polymerase and the sliding clamp that depends on the bound DNA. DNA polymerase bound to a primer:template junction has a high affinity for the clamp. In contrast, when a DNA polymerase reaches the end of an ssDNA template (e.g., at the end of an Okazaki fragment), the presence of dsDNA in its active site results in a change in conformation that reduces the polymerase's affinity for the sliding clamp and the DNA (see Fig. 8-19). Thus, when a polymerase completes the replication of a stretch of DNA, it is released from the sliding clamp so it can act at a new primer:template junction.

Once released from a DNA polymerase, sliding clamps are not immediately removed from the replicated DNA. Instead, other proteins that must function at the site of recent DNA synthesis to perform their function interact with the clamp proteins. As described in Chapter 7,

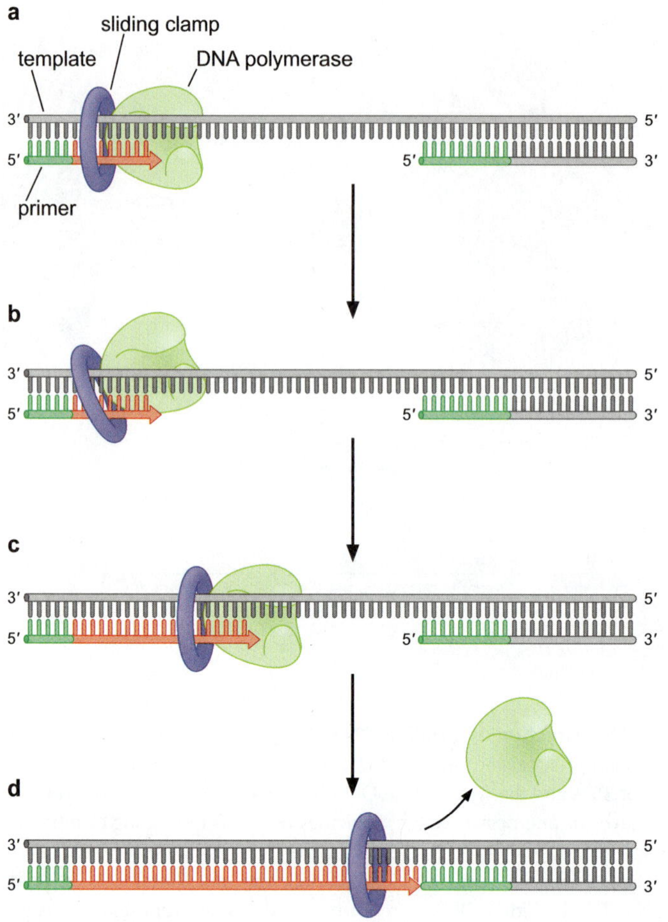

FIGURE 8-19 Sliding DNA clamps increase the processivity of associated DNA polymerases. (a) The sliding DNA clamp encircles the DNA and simultaneously binds the DNA polymerase. (b) The relatively low processivity of DNA polymerases leads to frequent release from the primer:template junction, but the association of the polymerase with the sliding clamp prevents diffusion away from the DNA. (c) The association of DNA polymerase with the sliding clamp ensures that the DNA polymerase rebinds the same primer:template junction and resumes DNA synthesis. (d) After DNA polymerase has completed synthesis of the template, the absence of a primer:template junction causes a change in the DNA polymerase that releases it from the sliding clamp.

enzymes that assemble chromatin in eukaryotic cells are recruited to the sites of DNA replication by an interaction with the eukaryotic sliding DNA clamp (called PCNA). Similarly, eukaryotic proteins involved in Okazaki fragment repair also interact with sliding clamp proteins. In each case, by interacting with sliding clamps, these proteins accumulate at sites of new DNA synthesis where they are needed most.

Sliding clamp proteins are a conserved part of the DNA replication apparatus derived from organisms as diverse as viruses, bacteria, yeast, and humans. Consistent with their conserved function, the structure of sliding clamps derived from these different organisms is also conserved (Fig. 8-20). In each case, the clamp has the same sixfold symmetry and the same diameter. Despite the similarity in overall structure, however, the number of subunits that come together to form the clamp differs.

Sliding Clamps Are Opened and Placed on DNA by Clamp Loaders

The sliding clamp is a closed ring in solution but must open to encircle the DNA double helix. A special class of protein complexes, called **sliding clamp loaders**, catalyze the opening and placement of sliding clamps on the DNA. These enzymes couple ATP binding and hydrolysis to the placement of the sliding clamp around primer:template junctions on the DNA (see Box 8-4, ATP Control of Protein Function: Loading a Sliding Clamp). The clamp loader also removes sliding

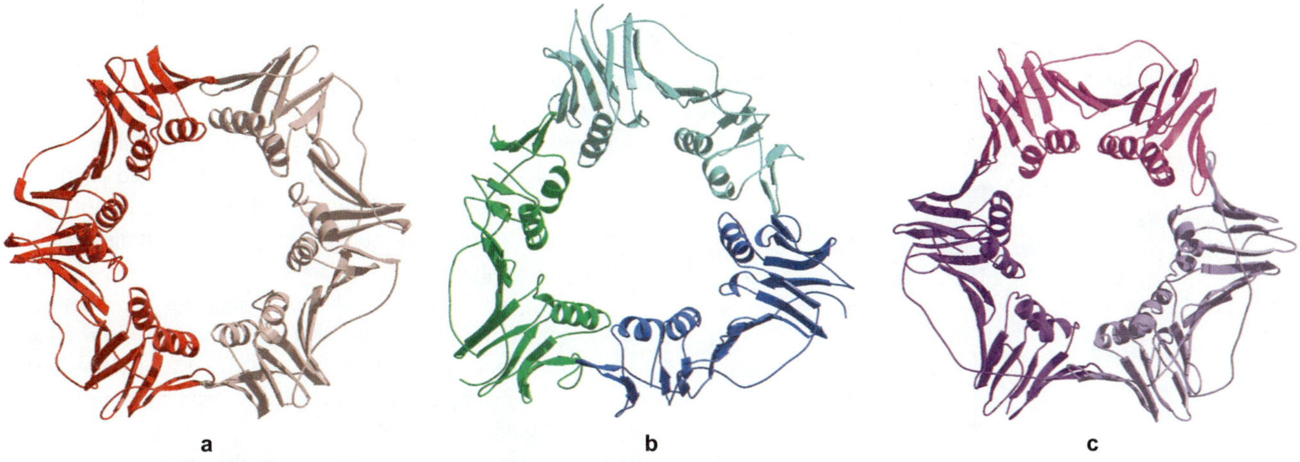

FIGURE 8-20 Three-dimensional structure of sliding DNA clamps isolated from different organisms. Sliding DNA clamps are found across all organisms and share a similar structure. (a) The sliding DNA clamp from *E. coli* is composed of two copies of the β protein. (Kong X.P. et al. 1992. *Cell* 69: 425.) (b) The T4 phage sliding DNA clamp is a trimer of the gp45 protein. (Moarefi I. et al. 2002. *J. Mol. Biol.* 296: 1215.) (c) The eukaryotic sliding DNA clamp is a trimer of the PCNA protein. (Krishna T.S. et al. 1994. *Cell* 79: 1233.) Images prepared with MolScript, BobScript, and Raster 3D.

ADVANCED CONCEPTS

Box 8-4 ATP Control of Protein Function: Loading a Sliding Clamp

The five subunits that come together to form a sliding DNA clamp loader are all members of a large class of proteins that have a related ATP-binding and hydrolysis site called AAA+ proteins. These proteins assemble into multi-AAA+ protein assemblies that use the energy of ATP binding and hydrolysis to alter the structure of target proteins or DNA. Sliding DNA clamp loaders are formed from several different AAA+ proteins, but other AAA+ protein complexes are composed of multiple copies of the same AAA+ protein. Although AAA+ proteins have related ATP-binding and hydrolysis sites, they perform diverse functions. In addition to loading sliding clamps, AAA+ protein complexes unwind DNA, load DNA helicase around substrate DNA (e.g., ORC and DnaC, as we shall see when we discuss the initiation of DNA replication), unfold proteins, and disassemble protein complexes. Indeed, it is the diversity of their functions that led to their name ATPases—associated with various activities.

How is ATP binding and hydrolysis coupled to sliding clamp loading? The simple answer is that ATP binding regulates the ability of the sliding clamp loader to bind its targets: the sliding clamp and DNA. When bound to ATP, the clamp loader can bind and open the sliding clamp ring by causing one of the subunit:subunit interfaces to come apart (Box 8-4 Fig. 1). The now open sliding clamp is brought to the DNA through a high-affinity DNA-binding site on the clamp loader. Like sliding clamp binding, DNA binding requires that the clamp loader be bound to ATP. Consistent with the need for sliding clamps at the sites of DNA synthesis, the clamp loader's DNA-binding site specifically recognizes primer:template junctions. The DNA is bound in such a way that the open sliding clamp is placed around the double-stranded region of the primer:template junction.

The final steps in sliding clamp loading are stimulated by ATP hydrolysis. Binding of the clamp loader to the primer:template junction activates ATP hydrolysis (by the clamp loader). Because the clamp loader can only bind the sliding clamp and DNA when it is bound to ATP (but not ADP), hydrolysis causes the clamp loader to release the sliding clamp and disassociate from the DNA. Once released from the clamp loader, the sliding clamp spontaneously closes around the DNA. The net result of this process is the loading of the sliding clamp at the site of DNA polymerase action—the primer:template junction. Release of ADP and P_i and binding to a new ATP molecule allow the clamp loader to initiate a new cycle of loading.

The function of the clamp loader illustrates several general features of the coupling of ATP binding and hydrolysis to a molecular event. ATP binding to a protein typically is involved in the *assembly stage* of the event: the association of factor with the target molecule. For example, the clamp loader has two target molecules: the sliding clamp and the primer:template junction. ATP is required for the clamp loader to bind to either target. Similarly, ATP binding stimulates the ability of DNA helicases to bind to ssDNA. In each case, the events coupled to ATP binding could be considered the action part of the cycle. For the clamp loader, ATP binding but not ATP hydrolysis is required to open the sliding clamp ring. For the DNA helicase, binding ssDNA is likely to be the key event unwinding DNA. In these cases, binding to ATP stabilizes a conformation of the enzyme that favors interaction with the substrate in a particular conformation.

What is the role of ATP hydrolysis? ATP hydrolysis typically is involved in the *disassembly stage* of the event: releasing the bound targets from the enzyme. Once the ATP-stabilized complex is formed, it must be disassembled. This could occur by simple disassociation; however, more often than not, this process would return the components to their starting situation (e.g., the sliding clamp free in solution), and this process would be slow if the ATP-stabilized complex is tightly associated. To ensure that disassembly occurs at the appropriate time, place, and rate, ATP hydrolysis is used to initiate disassembly. For example, ATP hydrolysis causes the clamp loader to revert back to a state in which it cannot bind either the sliding clamp or DNA. Reversion to this ground state may occur while the enzyme is still bound to the products of ATP hydrolysis (ADP and P_i) or may require their release.

The final key mechanism to couple ATP hydrolysis to a reaction pertains to the *trigger for ATP hydrolysis*. It is critical that the factor not hydrolyze ATP until a desired complex is assembled. Typically, formation of a particular complex triggers ATP hydrolysis. In the case of the clamp loader, this complex is the tertiary complex of the sliding clamp, the clamp loader, and the primer:template junction.

ATP control of these molecular events is thus most directly related to controlling the timing of conformational changes by the enzyme. By requiring the enzyme to alternate between two conformational states in order and requiring the formation of a key intermediate to trigger ATP hydrolysis, the enzyme can accomplish work. In contrast, if the enzyme merely bound and released ATP (without hydrolysis), the reaction would return to the initial state as often as it would proceed forward and little, if any, work would be accomplished.

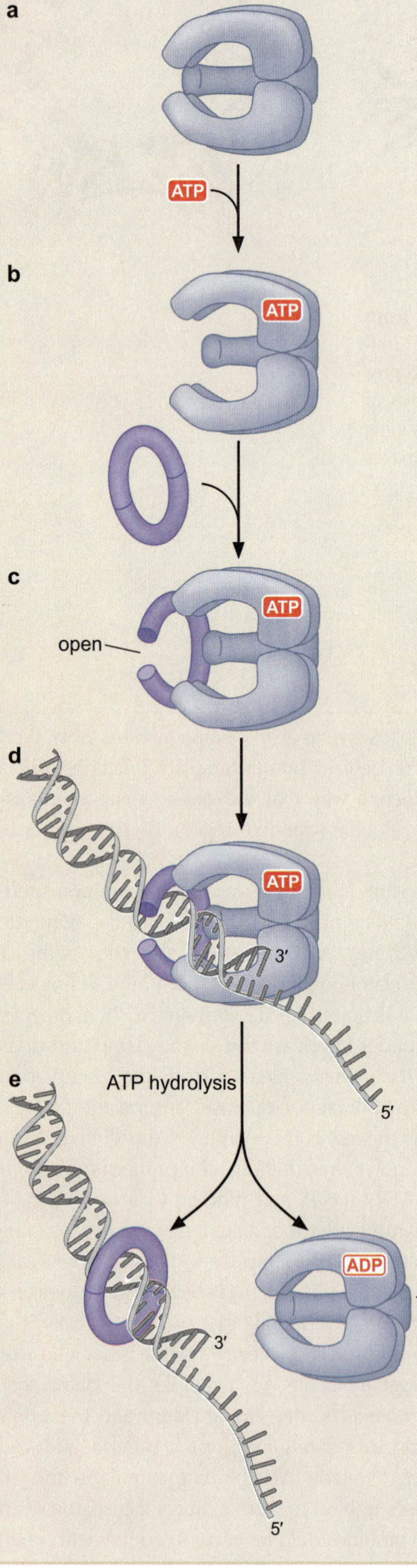

BOX 8-4 FIGURE 1 ATP control of sliding DNA clamp loading. (a) Sliding clamp loaders are five-subunit protein complexes whose activity is controlled by ATP binding and hydrolysis. In *E. coli*, the clamp loader is called the γ complex, and in eukaryotic cells, it is called replication factor C (RF-C). (b) To catalyze the sliding clamp opening, the clamp loader must be bound to ATP. (c) Once bound to ATP, the clamp loader binds the clamp and opens the ring at one of the subunit:subunit interfaces. (d) The resulting complex can now bind to DNA. DNA binding is mediated by the clamp loader, which preferentially binds to primer:template junctions. Correct binding to the DNA has two consequences. First, the opened sliding clamp is positioned so that dsDNA is in what will be the "hole" of the clamp. Second, DNA binding stimulates ATP hydrolysis by the clamp loader. (e) Because only an ATP-bound clamp loader can bind to the clamp and to DNA, the ADP form of the clamp loader rapidly disassociates from the clamp and the DNA, leaving behind a closed clamp positioned around the dsDNA portion of the primer:template junction. (Adapted, with permission, from O'Donnell M. et al. 2001. *Curr. Biol.* 11: R942, Fig. 5. © Elsevier.)

clamps from the DNA when they are no longer in use. Like DNA helicases and topoisomerases, these enzymes alter the conformation of their target (the sliding clamp) but not its chemical composition.

What controls when sliding clamps are loaded and removed from the DNA? Loading of a sliding clamp occurs anytime a primer:template junction is present in the cell. These DNA structures are formed not only during DNA replication, but also during several DNA-repair events (see Chapter 9). A sliding clamp can only be removed from the DNA if it is not bound by another protein. Sliding clamp loaders and DNA polymerases cannot interact with a sliding clamp at the same time because they have overlapping binding sites on the same face of the sliding clamp. Thus, a sliding clamp that is bound to a DNA polymerase is not subject to removal from the DNA. Similarly, nucleosome assembly factors, Okazaki fragment repair proteins, and other DNA-repair proteins all interact with the same region of the sliding clamp as the clamp loader. Thus, sliding clamps are only removed from the DNA once all of the enzymes that interact with them have completed their function.

DNA SYNTHESIS AT THE REPLICATION FORK

At the replication fork, the leading and lagging strands are synthesized simultaneously. This has the important benefit of limiting the amount of ssDNA present in the cell during DNA replication. When an ssDNA region of DNA is broken, there is a complete break in the chromosome that is much more difficult to repair than an ssDNA break in a dsDNA region. Moreover, repair of this type of lesion frequently leads to mutation of the DNA (see Chapter 9). Thus, limiting the time DNA is single-stranded is crucial. To coordinate the replication of both DNA strands, multiple DNA polymerases function at the replication fork.

In *E. coli,* the coordinate action of these polymerases is facilitated by physically linking them together in a large multiprotein complex called the DNA Pol III holoenzyme (Fig. 8-21). Holoenzyme is a general name for a multiprotein complex in which a core enzyme activity is associated with additional components that enhance function. The DNA Pol III holoenzyme includes two copies of the "core" DNA Pol III enzyme and one copy of the five-protein γ complex (the *E. coli* sliding clamp loader). Although present in only one copy, the γ complex binds to both copies of the core DNA Pol III and is essential to the formation of the holoenzyme (see Fig. 8-21). The same subunits of the γ complex that bind to the two copies of the DNA Pol III also interact more weakly with the DNA helicase (see next section).

How do two DNA polymerases remain linked at the replication fork while synthesizing DNA on both the leading and lagging template strands? A model that explains this coupling proposes that the replication machinery exploits the flexibility of DNA (Fig. 8-22). As the helicase unwinds the DNA at the replication fork, the leading-strand template is exposed and acted on immediately by the leading-strand DNA polymerase, which synthesizes a continuous strand of complementary DNA. In contrast, the lagging-strand template is not immediately acted on by DNA polymerase. Instead, it is spooled out as ssDNA that is rapidly bound by SSB. Intermittently, primase interacts with the DNA helicase and is activated to synthesize a new RNA primer on the lagging-strand template. The resulting RNA:DNA hy-

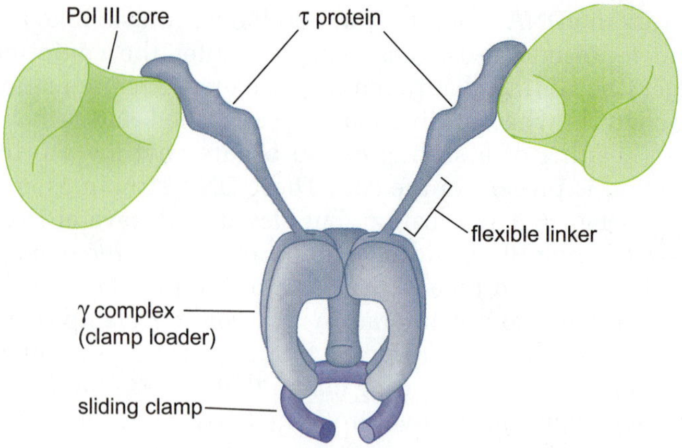

FIGURE 8-21 Composition of the DNA Pol III holoenzyme. There are three enzymes in each copy of the DNA Pol III holoenzyme: two copies of the DNA Pol III core enzyme and one copy of the γ complex. The γ complex includes two copies of the τ protein, each of which includes a domain that interacts with one DNA Pol III core. Analysis of the amino acid sequence of the τ protein indicates that the DNA Pol III-binding region of the protein is separated from the part of the protein involved in clamp loading by an extended flexible linker. This linker is proposed to allow the two polymerases to move in a relatively independent manner that would be necessary for one polymerase to replicate the leading strand and the other to replicate the lagging strand. (Adapted, with permission, from O'Donnell M. et al. 2001. *Curr. Biol.* 11: R943, Fig. 6. © Elsevier.)

brid is recognized as a primer:template junction by the sliding DNA clamp loader and a sliding clamp is assembled at this site.

When the lagging-strand DNA polymerase completes the previous Okazaki fragment, this polymerase is released from the template (recall that once DNA polymerase completes synthesis of an Okazaki fragment, it is released from its associated sliding clamp). Because this polymerase remains tethered to the leading-strand DNA polymerase and the sliding clamp loader, it is in an ideal position to bind the RNA primer:template junction immediately after the addition of a sliding clamp. Indeed, since the time required for the lagging-strand polymerase to synthesize an Okazaki fragment is much shorter than the time required to synthesize an RNA primer and assemble a sliding

FIGURE 8-22 "Trombone" model for coordinating replication by two DNA polymerases at the *E. coli* replication fork. (a) The DNA helicase at the *E. coli* DNA replication fork travels on the lagging-strand template in a 5′→3′ direction. The DNA Pol III holoenzyme interacts with the DNA helicase through the τ subunit, which also binds to both DNA polymerases. One DNA Pol III core is replicating the leading strand and the other DNA Pol III core replicates the lagging strand. SSB coats the ssDNA regions of the DNA (for simplicity, SSB on the lagging strand is only shown in part a). (b) Periodically, DNA primase will associate with the DNA helicase and synthesize a new primer on the lagging-strand template. (c) When the lagging-strand DNA polymerase completes an Okazaki fragment, it is released from the sliding clamp and the DNA. (d) The recently primed lagging-strand DNA is then a target of the clamp loader, which assembles a new sliding clamp at the primer:template junction created by synthesizing a new RNA primer. (e) The primer:template junction with its associated sliding clamp binds to the lagging-strand DNA polymerase, which initiates DNA synthesis on the next Okazaki fragment. Although this description has concentrated on the more complex action occurring during the synthesis of the lagging strand, during this entire process, a new ssDNA template for the leading strand has been generated and rapidly replicated by the leading strand DNA Pol III.

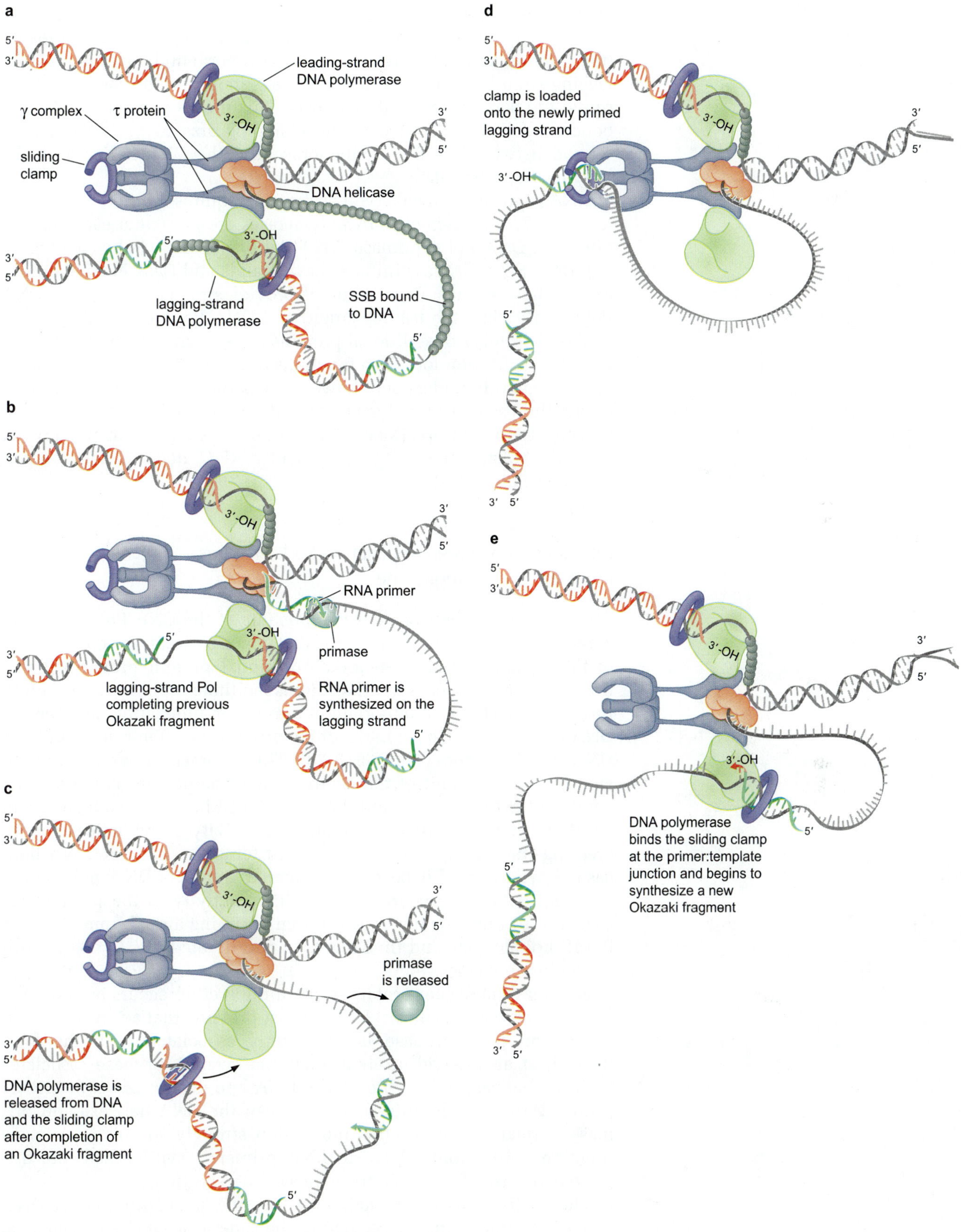
a
leading-strand DNA polymerase
γ complex
τ protein
sliding clamp
DNA helicase
lagging-strand DNA polymerase
SSB bound to DNA
3′-OH
b
RNA primer
primase
lagging-strand Pol completing previous Okazaki fragment
RNA primer is synthesized on the lagging strand
c
primase is released
DNA polymerase is released from DNA and the sliding clamp after completion of an Okazaki fragment
d
clamp is loaded onto the newly primed lagging strand
e
DNA polymerase binds the sliding clamp at the primer:template junction and begins to synthesize a new Okazaki fragment

FIGURE 8-22 (*See facing page for legend.*)

clamp, the lagging-strand DNA polymerase is ready and waiting as the primer:template junction is released by the sliding DNA clamp loader (the γ complex). By binding to the primer:template junction, the lagging-strand polymerase forms a new loop and initiates the next round of Okazaki fragment synthesis. This model is called the "trombone model" in reference to the changing size of the ssDNA loop formed between the DNA polymerase and the DNA helicase on the lagging-strand template.

DNA replication in eukaryotic cells also requires multiple DNA polymerases. *Three* different DNA polymerases are present at each replication fork: DNA Pol α/primase, DNA Pol δ, and DNA Pol ε (see Fig. 8-17). DNA Pol α/primase initiates new strands and DNA Pol δ and Pol ε extend these strands. Recent studies provide strong evidence that DNA Pol ε synthesizes the leading strand and DNA Pol δ makes the lagging strand. A number of additional proteins are known to be part of the eukaryotic replication fork. The functions of these additional proteins are currently poorly understood; however, it is likely that they act to coordinate the three DNA polymerases and couple their action to the eukaryotic DNA helicase (Mcm2-7). Unlike the situation in prokaryotic cells, the eukaryotic sliding clamp loader, RF-C, does not seem to perform these functions.

Interactions between Replication Fork Proteins Form the *E. coli* Replisome

The connections between the components of the DNA Pol III holoenzyme are not the only interactions that occur between the components of the bacterial replication fork. Additional protein–protein interactions between replication fork proteins facilitate rapid replication fork progression. The most important of these is an interaction between the DNA helicase (the hexameric dnaB protein; see Table 8-1) and the DNA Pol III holoenzyme (Fig. 8-23). This interaction, which is mediated by the τ subunit of the clamp loader component of the holoenzyme, holds the helicase and the DNA Pol III holoenzyme together. In addition, this association stimulates the activity of the helicase by increasing the rate of helicase movement tenfold. Thus, the DNA helicase slows down if it becomes separated from the DNA polymerase (see Fig. 8-23). The coupling of helicase activity to the presence of DNA Pol III prevents the helicase from "running away" from the DNA Pol III holoenzyme and thus serves to coordinate these two key replication fork enzymes.

A second important protein–protein interaction occurs between the DNA helicase and primase. Unlike most proteins that act at the *E. coli* replication fork, primase is not tightly associated with the fork. Instead, at an interval of about once per second, primase associates with the helicase and SSB-coated ssDNA and synthesizes a new RNA primer. Although the interaction between the DNA helicase and primase is relatively weak, this interaction strongly stimulates primase function (~1000-fold). After an RNA primer is synthesized, the primase is released from the DNA helicase into solution.

The relatively weak interaction between the *E. coli* primase and DNA helicase is important for regulating the length of Okazaki fragments. A tighter association would result in more frequent primer synthesis on the lagging strand and therefore shorter Okazaki fragments. Similarly, a weaker interaction would result in longer Okazaki fragments.

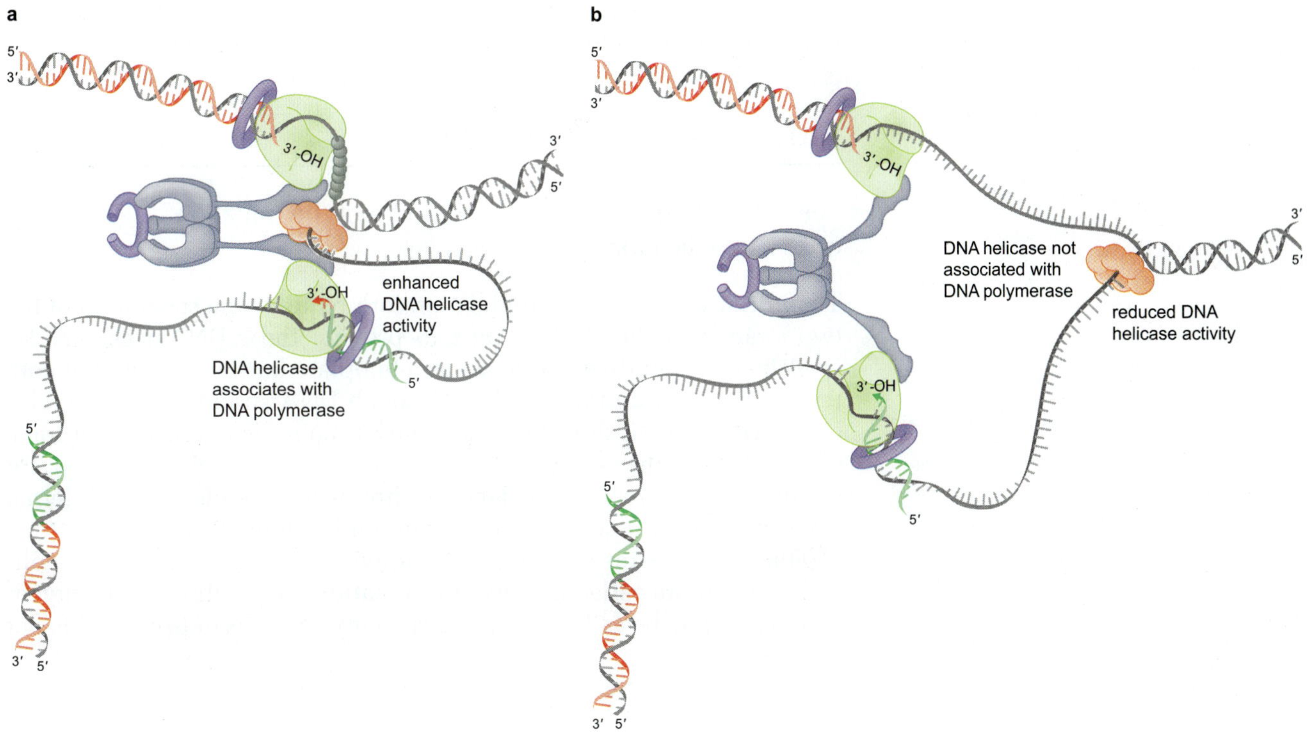

FIGURE 8-23 **Binding of the DNA helicase to DNA Pol III holoenzyme stimulates the rate of DNA strand separation.** The τ subunit of the clamp loader interacts with both the DNA helicase and the DNA polymerase at the replication fork. (a) When this interaction is made, the DNA helicase unwinds the DNA at approximately the same rate as the DNA polymerases replicate the DNA. (b) If the DNA helicase is not associated with DNA Pol III holoenzyme, DNA unwinding slows by tenfold. Under these conditions, the DNA polymerases can replicate faster than the DNA helicase can separate the strands of unreplicated DNA. This allows the DNA Pol III holoenzyme to "catch up" to the DNA helicase and re-form the replisome.

The combination of all of the proteins that function at the replication fork is referred to as the **replisome**. Together, these proteins form a finely tuned factory for DNA synthesis that contains multiple interacting machines. Individually, these machines perform important specific functions. When brought together, their activities are coordinated by the interactions between them. Although these interactions are particularly well-understood in *E. coli* cells, studies of bacteriophage and eukaryotic DNA replication machinery show that a similar coordination between multiple machines is involved in DNA replication in these organisms. Indeed, there are clear parallels between the proteins known to be involved in replication in *E. coli* and those functioning in these other organisms. Table 8-1 presents a list of factors performing analogous functions in phage, prokaryotic, and eukaryotic DNA replication.

To fully appreciate the amazing capabilities of the enzymes that replicate DNA, imagine a situation in which a DNA base is the size of a textbook. Under these conditions, dsDNA would be ~1 m in diameter and the *E. coli* genome would be a large circle, ~500 miles (800 km) in circumference. More importantly, the replisome would be the size of a delivery truck and would be moving at more than 375 mph (600 km/hr)! Replicating the *E. coli* genome would be a 40-minute, 250-mile (400 km) trip for two such machines, each leaving two 1-m

DNA cables in their wake. Impressively, during this trip, the replication machinery would, on average, make only a single error.

INITIATION OF DNA REPLICATION

Specific Genomic DNA Sequences Direct the Initiation of DNA Replication

The initial formation of a replication fork requires the separation of the two strands of the DNA duplex to provide the ssDNA necessary for DNA helicase binding and to act as a template for the synthesis of both the RNA primer and new DNA. Although DNA strand separation (also called DNA unwinding) is most easily accomplished at chromosome ends, DNA synthesis generally initiates at internal regions. Indeed, for circular chromosomes, the lack of chromosome ends makes internal DNA unwinding essential to replication initiation.

The specific sites at which DNA unwinding and initiation of replication occur are called **origins of replication**. Depending on the organism, there may be as few as one or as many as thousands of origins per chromosome.

The Replicon Model of Replication Initiation

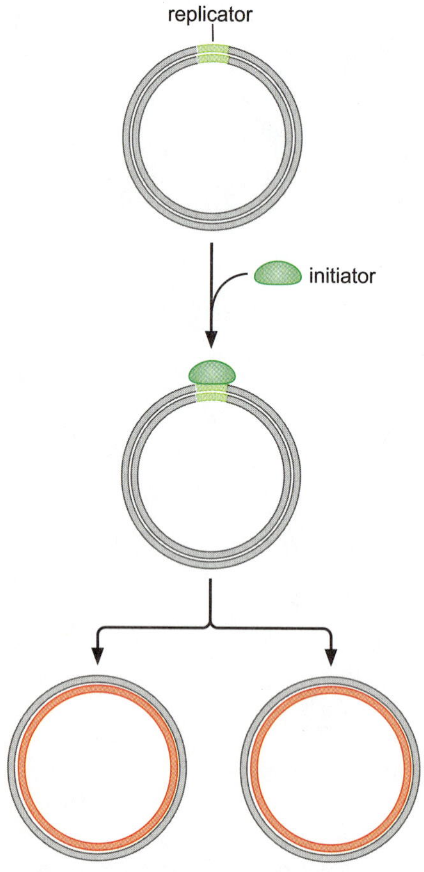

FIGURE 8-24 The replicon model. Binding of the initiator to the replicator stimulates initiation of replication and the duplication of the associated DNA.

In 1963, François Jacob, Sydney Brenner, and Jacques Cuzin proposed a model to explain the events controlling the initiation of replication in bacteria. They defined all the DNA replicated from a particular origin of replication as a **replicon**. For example, because the single chromosome found in *E. coli* cells has only one origin of replication, the entire chromosome is a single replicon. In contrast, the presence of multiple origins of replication divides each eukaryotic chromosome into multiple replicons—one for each origin of replication.

The replicon model proposed two components that controlled the initiation of replication: the replicator and the initiator (Fig. 8-24). The **replicator** is defined as the entire set of *cis*-acting DNA sequences that is *sufficient* to direct the initiation of DNA replication. This is in contrast to the origin of replication, which is the site on the DNA where the DNA is unwound and DNA synthesis initiates. Although the origin of replication is always part of the replicator, sometimes (particularly in eukaryotic cells) the origin of replication is only a fraction of the DNA sequences required to direct the initiation of replication (the replicator). The same distinction can be made between a transcriptional promoter and the start site of transcription, as we shall see in Chapter 12.

The second component of the replicon model is the **initiator** protein. This protein specifically recognizes a DNA element in the replicator and activates the initiation of replication (see Fig. 8-24). Initiator proteins have been identified in many different organisms, including bacteria, viruses, and eukaryotic cells. All initiator proteins select the sites that will become origins of replication, although they recognize DNA using different DNA-binding motifs. Interestingly, all of the known initiator proteins are regulated by ATP binding and hydrolysis and share a common core ATP-binding motif related to, but distinct from, that used by sliding DNA clamp loaders.

As we will see below, the initiator protein is the only sequence-specific DNA-binding protein involved in the initiation of replication.

The remaining proteins required for replication initiation do not bind to a DNA sequence specifically. Instead, these proteins are recruited to the replicator through a combination of protein–protein interactions and affinity for specific DNA structures (e.g., ssDNA or a primer:template junction).

Replicator Sequences Include Initiator Binding Sites and Easily Unwound DNA

The DNA sequences of known replicators share two common features (Fig. 8-25). First, they include a binding site for the initiator protein that nucleates the assembly of the replication initiation machinery. Second, they include a stretch of AT-rich DNA that unwinds readily but not spontaneously. Unwinding of DNA at replicators is controlled by the replication initiation proteins, and the action of these proteins is tightly regulated in most organisms.

The single replicator required for *E. coli* chromosomal replication is called *oriC*. Two repeated motifs are critical for *oriC* function (Fig. 8-25a). The 9-mer motif is the binding site for the *E. coli* initiator, DnaA, and is repeated five times at *oriC*. The 13-mer motif, repeated three times, is the initial site of ssDNA formation during initiation.

Although the specific sequences are different, the overall structures of replicators derived from many eukaryotic viruses and the single-cell eukaryote *Saccharormyces cerevisiae* are similar (Fig. 8-25b,c). The methods used to define origins of replication are described in Box 8-5, The Identification of Origins of Replication and Replicators.

Replicators functioning in multicellular eukaryotes are not well-understood. Their identification and characterization have been

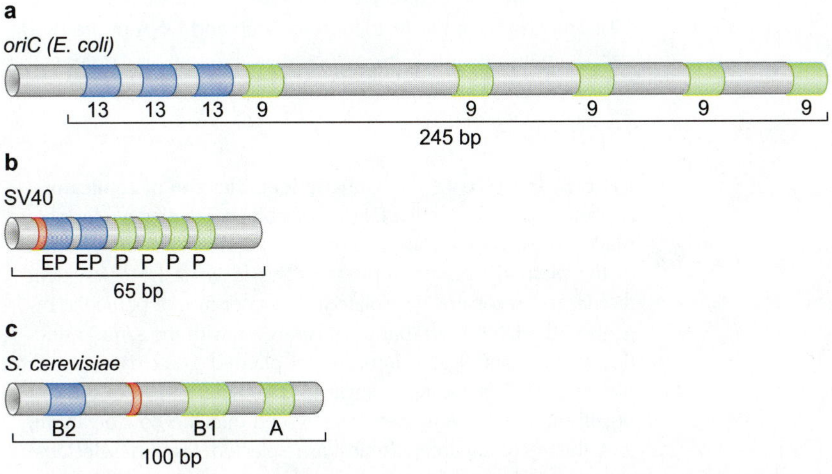

FIGURE 8-25 Structure of replicators. The DNA elements that make up three well-characterized replicators are shown. For each diagram, green represents the initiator DNA-binding site, blue represents DNA elements that facilitate DNA unwinding, and red represents the site of initial DNA synthesis (the site for *oriC* is outside the sequence shown). (a) *oriC* is composed of five "9-mer" DnaA-binding sites and three "13-mer" repeated elements that are the site of initial DNA unwinding. (b) The origin of the eukaryotic virus SV40 is composed of four pentamer binding sites (P) for the initiator protein called large T antigen and a 20-bp early palindrome (EP) that is the site of DNA unwinding. (c) Three elements are commonly found at *S. cerevisiae* replicators. The A and B1 elements bind to the initiator ORC. The B2 element facilitates DNA unwinding and binding of other replication factors.

KEY EXPERIMENTS

Box 8-5 The Identification of Origins of Replication and Replicators

Replicator sequences are typically identified using genetic assays. For example, the first yeast replicators were identified using a DNA transformation assay (Box 8-5 Fig. 1). In these studies, investigators randomly cloned genomic DNA fragments into plasmids lacking a replicator but containing a selectable marker lacking in the host cells. For the plasmid to be maintained in the host cell after transformation, the cloned DNA fragment must contain a yeast replicator. The identified DNA fragments were called **autonomously replicating sequences** (**ARSs**). Although these sequences acted as replicators in the artificial context of a circular plasmid, further evidence was required to demonstrate that these sequences were also replicators in their native chromosomal location.

To demonstrate that ARSs acted as replicators in the chromosome, it was necessary to develop methods to identify the location of origins of replication in the cell. One approach to identify origins takes advantage of the unusual structure of the DNA replication intermediates formed during replication initiation. Unlike either fully replicated or fully unreplicated DNA, DNA that is in the process of being replicated is not linear. For example, a DNA fragment (generated by cleavage of the DNA with a restriction enzyme) that does not contain an origin of replication will take on a variety of "Y-shaped" conformations as it is replicated (Box 8-5 Fig. 2, blue DNA fragments). Similarly, immediately after the initiation of replication, a DNA fragment containing an origin of replication will take on a "bubble" shape. Finally, if the origin of replication is located asymmetrically within the DNA fragment, the DNA will start out as a bubble shape and then convert to a Y shape (Box 8-5 Fig. 2, red DNA fragments). These unusually shaped DNAs can be distinguished from simple linear DNA using two-dimensional agarose gel electrophoresis (Box 8-5 Fig. 3).

To identify DNA that is in the process of replicating, DNA derived from dividing cells is first cut with a restriction enzyme and separated on a two-dimensional agarose gel. In the first dimension, the DNA is separated by size *and shape* and in the second dimension, the DNA is separated primarily by size. This is accomplished by using different density of agarose and electrophoresis rates for each dimension. To separate by size and shape, the agarose gel pores are small (high agarose density) and the rate of electrophoresis is fast. In contrast, to separate primarily by size, the agarose gel pores are larger (low agarose density) and the rate of electrophoresis is slower. Once electrophoresis is complete, the DNA molecules are transferred to nitrocellulose and detected by Southern blotting (see Chapter 21). The choice of the restriction enzyme and DNA probe used can dramatically affect the outcome of the analysis. In general,

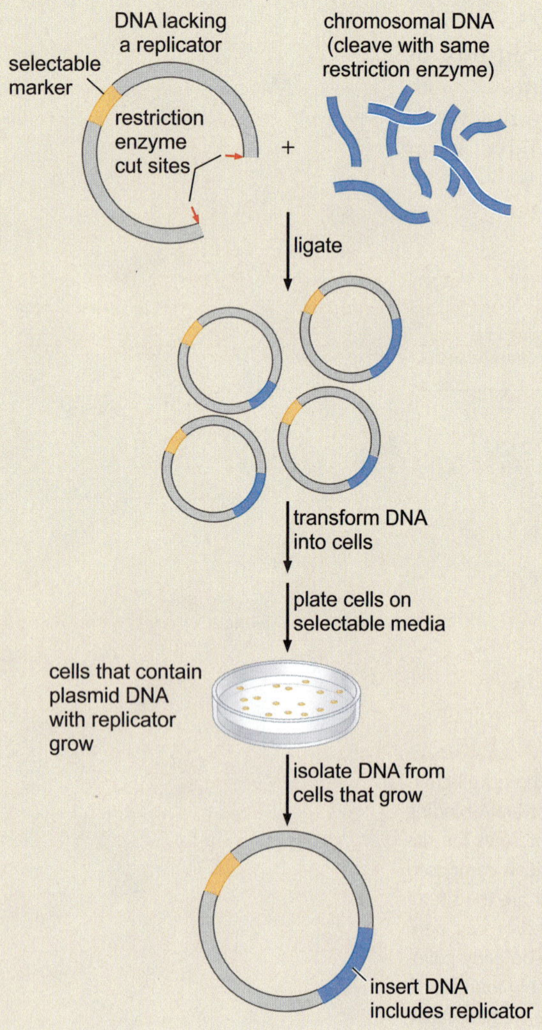

BOX 8-5 FIGURE 1 Genetic identification of replicators. A plasmid (a small circular DNA molecule) containing a selectable marker is cut with a restriction enzyme that results in the excision of the plasmid's normal replicator. This leaves a DNA fragment that lacks a replicator. To isolate a replicator from a particular organism, the DNA from that organism is cut with the same restriction enzyme and ligated into the cut plasmid to recreate circular plasmids each including a single fragment derived from the test organism. This DNA is then transformed into the host organism, and the recombinant plasmids are selected using a selectable marker on the plasmid (e.g., if the marker conferred antibiotic resistance, the cells would be grown in the presence of the antibiotic). Cells that grow are able to maintain the plasmid and its selectable marker, indicating that the plasmid can replicate in the cell and must contain a replicator. Isolation of the plasmid from the host cell and sequencing of the inserted DNA allow the identification of the sequence of the fragment that contains the replicator. Further mutagenesis of the inserted DNA (such as deletion of specific regions of the inserted DNA), followed by a repetition of the assay, allows a more precise definition of the replicator.

Box 8-5 ***(Continued)***

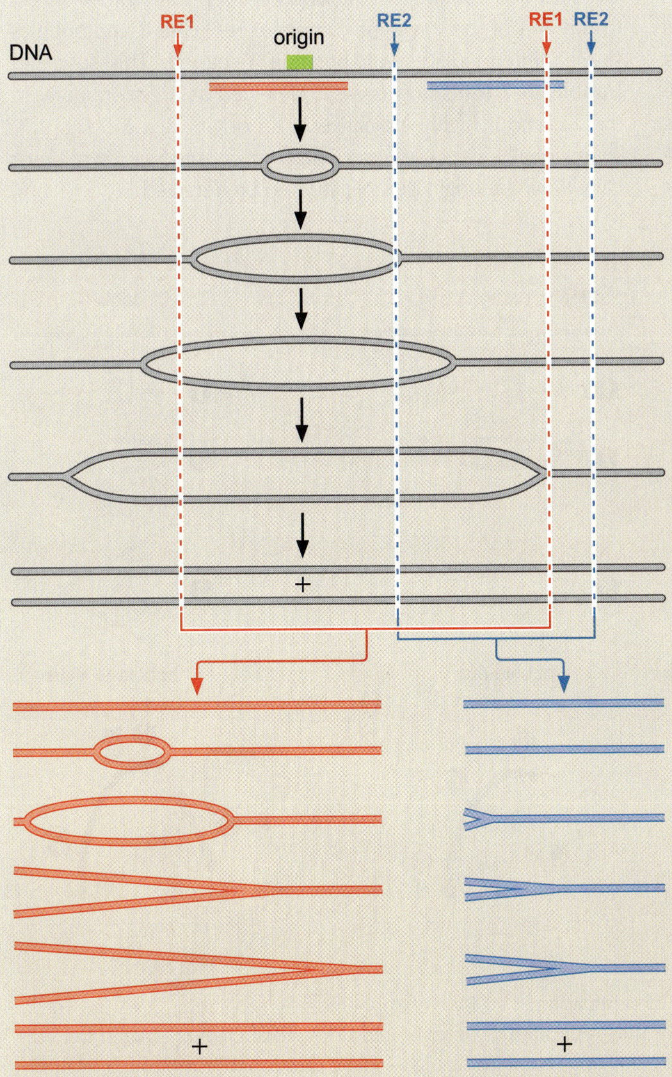

Box 8-5 Figure 2 DNA that is in the process of replication has an unusual structure. Results of restriction enzyme cleavage of DNA in the process of replication are shown. The illustration shows the growth of a "replication bubble" (created by two replication forks progressing away from an origin of replication). The consequences of cutting these replication intermediates is followed by detection by hybridization with the indicated labeled DNA probe. If the red restriction enzyme is used and only the fragments that hybridize to the red DNA probe are examined, the pattern on the left side will be generated. If the blue restriction enzyme and the blue DNA probe is used to detect the resulting DNA fragments, the pattern on the right will be observed. Note that the left-hand pattern starts with a DNA fragment containing a "bubble" and eventually ends with "Y-shaped" molecules. The right-hand pattern never has a "bubble" but does assume a full variety of "Y-shaped" intermediates. Only a DNA fragment containing an origin of replication can produce the pattern on the left.

this method requires that the investigator already know the approximate location of a potential origin of replication. Recently, newer methods have been developed that use DNA microarrays to identify the location of origins and that do not require any prior knowledge concerning origin locations.

How can the two-dimensional gels identify the DNA intermediates associated with a replication origin? The particular pattern of DNA migration can lead to unequivocal evidence of an origin of replication. The most unusual structures migrate most slowly in the first dimension. For example, a Y-shaped molecule that has three equal length arms will migrate the most slowly of all such molecules derived from a particular DNA fragment (Box 8-5 Fig. 3b), and therefore will be at the top of an arc of DNA molecules that are nonlinear. In contrast, a Y-shaped molecule with two very short replicated arms and a large replicated region will migrate very similarly to the unreplicated version of the same DNA fragment. Finally, the Y-shaped molecule that results from the almost completely replicated fragment is similar in shape to a linear molecule two times the size of the unreplicated fragment. Thus, as a DNA molecule is replicated by a single replication fork, it will migrate in positions that vary from a spot that is close to the unreplicated fragment in an arc that eventually reaches a location that a linear molecule twice the size of the unreplicated DNA would be expected to migrate to. This shape is called a **Y-arc** and is indicative that a molecule is in the process of being replicated. Because all DNA molecules are replicated during each round of replication, the majority of DNA fragments will show this type of pattern.

Molecules that contain an origin of replication form bubble-shaped replication intermediates that migrate even more slowly in the first dimension than Y-shaped molecules. The

Box 8-5 *(Continued)*

larger the bubble, the more these molecules migrate differently from linear DNA (Box 8-5 Fig. 3c). Unfortunately, it is difficult to distinguish the arc of intermediates created by a bubble-containing fragment (called a bubble arc) from one created by Y-shaped intermediates (Box 8-5 Fig. 3b,c). This difficulty can be overcome if the origin is located asymmetrically in the DNA fragment. In this instance, the intermediates will start out as bubbles, but when the replication fork closest to the end of the fragment completes replication, the bubble-shaped intermediates will become Y-shaped. This so-called **bubble-to-Y transition** is easily detected as a discontinuity in the arc and is highly indicative of an origin (Box 8-5 Fig. 3d). Thus, ideally, the restriction enzymes chosen will asymmetrically flank the origin of replication to be detected.

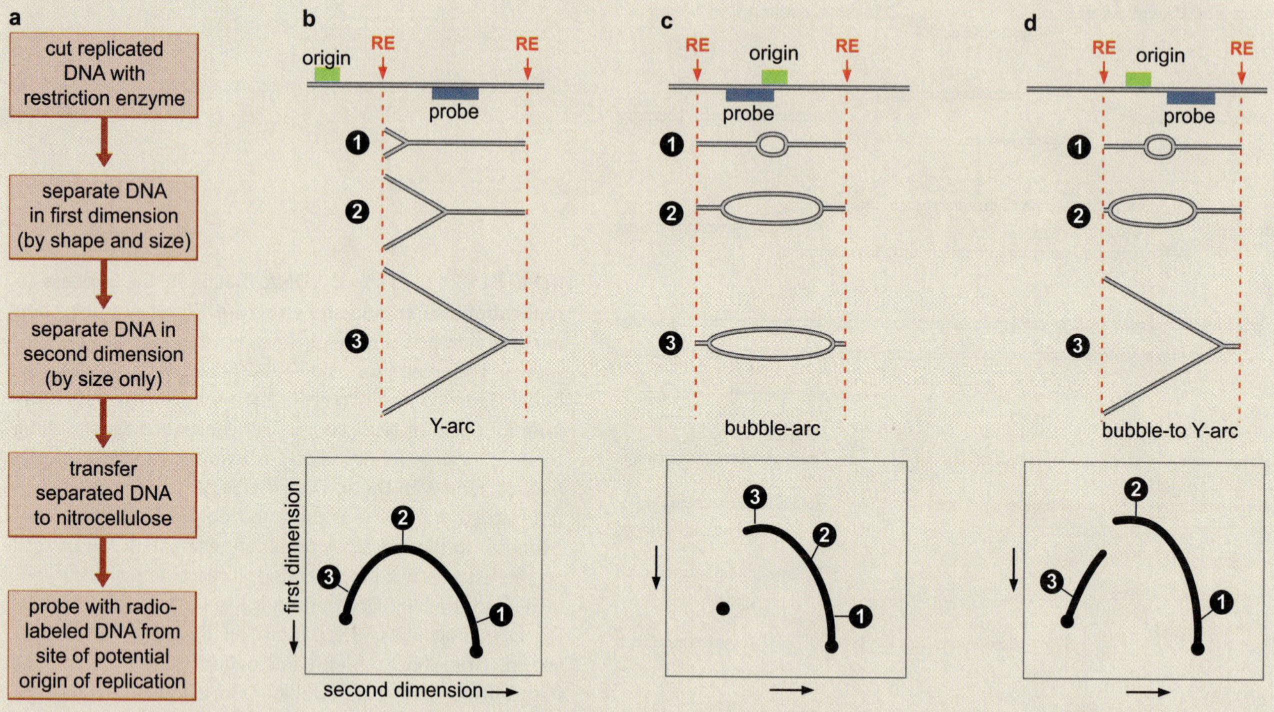

Box 8-5 Figure 3 Molecular identification of an origin of replication. (a) By electrophoretically separating DNA in two dimensions, DNA in the process of replication can be separated from fully replicated or unreplicated DNA. Total DNA is isolated from dividing (and therefore replicating) cells. The DNA is first separated by size and shape (using high-voltage electrophoresis through relatively small pores), the electric field is rotated by 90°, and the DNA is then separated predominantly by size (electrophoresed with low voltage in large pore agarose). Southern blot analysis is used to detect the DNA of interest. The location of the origin (green), restriction enzyme cleavage sites (red), and Southern blot DNA probe (blue) are illustrated for the three different patterns that can be observed. The largest replication bubbles migrate the slowest in the first dimension (c) and Y-shaped molecules with nearly equal length arms migrate the next slowest (b). Because the "Y-arc" and "bubble-arc" patterns are difficult to distinguish, the "bubble-to-Y-arc" pattern (d) is considered the most indicative of an origin.

hampered by the lack of genetic assays for stable propagation of small circular DNA comparable to those used to identify origins in single-cell eukaryotes and bacteria (see Box 8-5). In the few instances in which replicators have been identified, they are found to be much larger than the replicators identified in *S. cerevisiae* and bacterial chromosomes, generally encompassing more than 1000 bp of DNA. Unlike their smaller counterparts, mutations that eliminate the function of these replicators are not readily isolated, perhaps because important elements within these sequences are redundant.

BINDING AND UNWINDING: ORIGIN SELECTION AND ACTIVATION BY THE INITIATOR PROTEIN

Initiator proteins always perform at least two different functions during the initiation of replication (Fig. 8-26). First, these proteins bind a specific DNA sequence within the replicator. Second, initiator proteins interact with additional factors required for replication initiation, thus recruiting them to the replicator. Some but not all initiator proteins perform a third function: they distort or unwind a region of DNA adjacent to their binding site to facilitate the initial opening of the DNA duplex.

Consider, for example, the *E. coli* initiator protein, DnaA. DnaA binds the repeated 9-mer elements in *oriC* (see Fig. 8-25) and is regulated by ATP. When bound to ATP (but not ADP), DnaA also interacts with DNA in the region of the repeated 13-mer repeats of *oriC*. These additional interactions result in the separation of the DNA strands over more than 20 bp within the 13-mer repeat region. This unwound DNA provides an ssDNA template for additional replication proteins to begin the RNA and DNA synthesis steps of replication (see below).

The formation of ssDNA at a site in the chromosome is not sufficient for the DNA helicase and other replication proteins to assemble. Rather, DnaA recruits additional replication proteins to the ssDNA formed at the replicator including the DNA helicase (see the next section). The regulation of *E. coli* replication is linked to the control of DnaA activity and will be discussed later in Box 8-7 (*E. coli* DNA Replication Is Regulated by DnaA·ATP Levels and SeqA).

In eukaryotic cells, the initiator is a six-protein complex called the **origin recognition complex (ORC)**. The function of ORC is best understood in yeast cells. ORC recognizes a conserved sequence found in yeast replicators, called the A element, as well as a second, less-conserved B1 element (see Fig. 8-25). Like DnaA, ORC binds and hydrolyzes ATP. ATP binding is required for sequence-specific DNA binding at the origin, and ATP hydrolysis is required for ORC to participate in the loading of the eukaryotic DNA helicase onto the the replicator DNA. Unlike DnaA, binding of ORC to yeast replicators does not lead to strand separation of the adjacent DNA. Nevertheless, ORC is required to recruit, either directly or indirectly, all of the remaining replication proteins to the replicator (see the section Prereplicative Complex Formation Is the First Step in the Initiation of Replication in Eukaryotes).

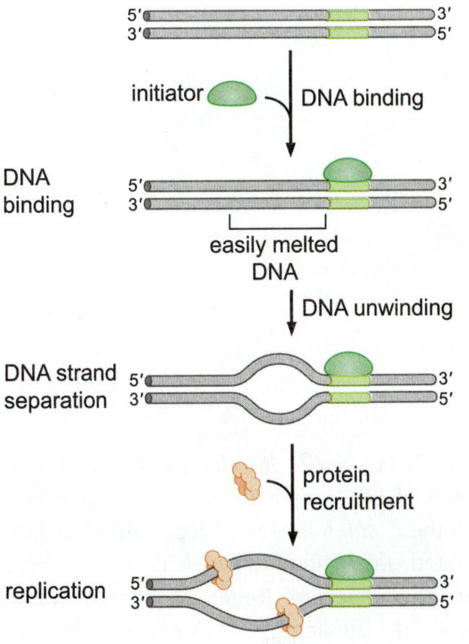

FIGURE 8-26 Functions of the initiator proteins during the initiation of DNA replication. The three common functions of initiator proteins are illustrated: DNA binding, DNA strand separation, and replication protein recruitment. (Here the recruited protein is illustrated as a DNA helicase; however, the recruited proteins differ for each initiator protein.)

Protein–Protein and Protein–DNA Interactions Direct the Initiation Process

Once the initiator binds to the replicator, the remaining steps in the initiation of replication are largely driven by protein–protein interactions and protein–DNA interactions that are sequence-independent. The end result is the assembly of two replisomes that we described earlier. To explore the events that produce these protein machines, we first turn to *E. coli*, in which they are understood in the most detail.

After the initiator (DnaA) has bound to *oriC* and unwound the 13-mer DNA, the combination of ssDNA and DnaA recruits a complex of two proteins: the DNA helicase, DnaB, and helicase loader, DnaC (Fig. 8-27). Both proteins are present in six copies within the complex. The DNA helicase is inactive in the helicase/helicase loader complex to prevent it from functioning at inappropriate sites. Once bound to the ssDNA at the origin, the helicase loader directs the assembly of its associated DNA he-

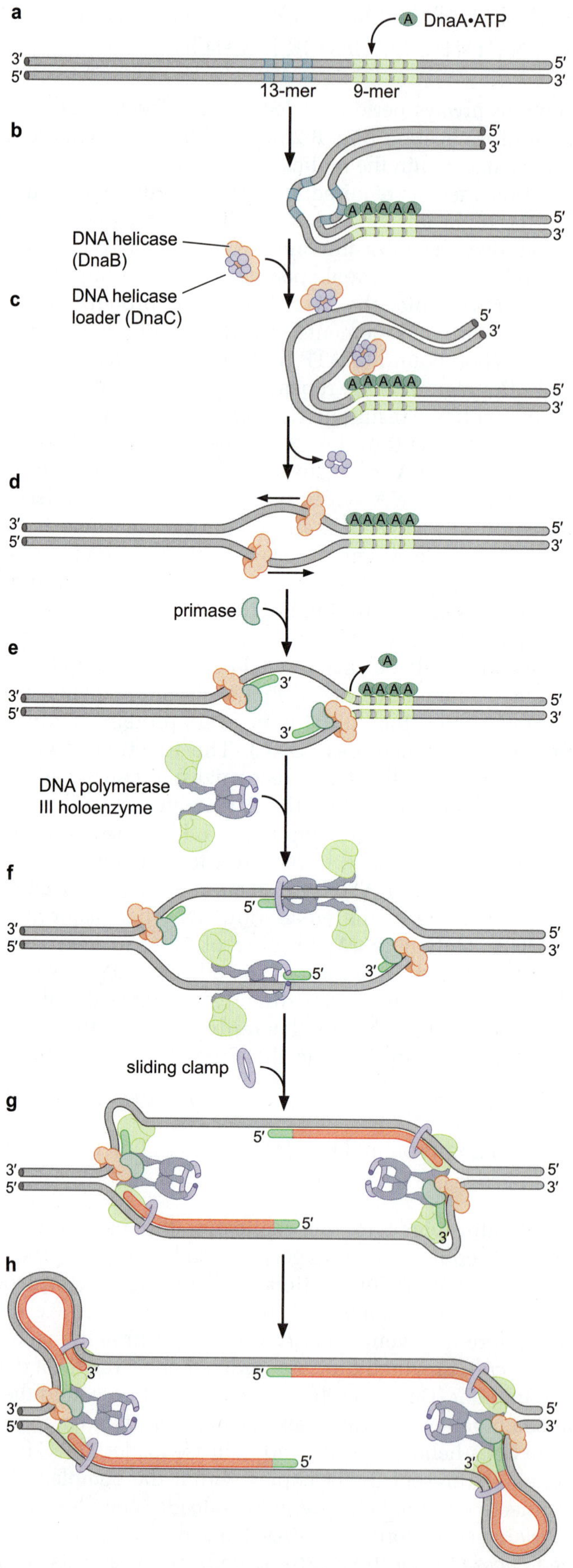

FIGURE 8-27 Model for *E. coli* initiation of DNA replication. The major events in the *E. coli* initiation of replication are illustrated. (a) Multiple DnaA·ATP proteins bind to the repeated 9-mer sequences within *oriC*. (b) Binding of DnaA·ATP to these sequences leads to strand separation within the 13-mer repeats. This is mediated by an ssDNA-binding domain in DnaA·ATP. (c) DNA helicase (DnaB) and the DNA helicase loader (DnaC) associate with the DnaA-bound origin. An ssDNA-binding domain in the helicase loader and protein–protein interactions with DnaA are required to form this complex. (d) DNA helicase loaders catalyze the opening of the DNA helicase protein ring and placement of the ring around the ssDNA at the origin. Loading of the DNA helicase leads to the disassociation of the helicase loader from the replicator and activates the DNA helicases. (e) The DNA helicases each recruit a DNA primase which synthesizes an RNA primer on each template. The movement of the DNA helicases also removes any remaining DnaA bound to the replicator. (f) The newly synthesized primers are recognized by the clamp loader components of two DNA Pol III holoenzymes. Sliding clamps are assembled on each RNA primer, and leading-strand synthesis is initiated by one of the two core DNA Pol III enzymes of each holoenzyme. (g) After each DNA helicase has moved ~1000 bases, a second RNA primer is synthesized on each lagging-strand template and a sliding clamp is loaded. The resulting primer:template junction is recognized by the second DNA Pol III core enzyme in each holoenzyme, resulting in the initiation of lagging-strand synthesis. (h) Leading- and lagging-strand synthesis are now initiated at each replication fork and continue to the end of the template or until another replication fork from an adjacent origin of replication is reached.

licase around the ssDNA (recall that ssDNA passes through the middle of the DnaB helicase's hexameric protein ring). This process is analogous to the assembly of sliding DNA clamps around a primer:template junction, and like the sliding clamp loader, DnaC is a ATP-utilizing AAA+ protein (see Box 8-4). Upon completion of this task, the helicase loader is released, allowing the helicase to become active. One helicase is loaded onto each of the two separated ssDNA strands at the origin, and the orientation of these two helicases is such that they will proceed toward each other as they move with a 5′→3′ polarity along their associated ssDNAs.

The protein–protein interactions between the helicase and other components of the replication fork described above direct the assembly of the rest of the replication machinery (see Fig. 8-27). Helicase recruits DNA primase to the origin DNA, resulting in the synthesis of an RNA primer on each strand of the origin. The DNA Pol III holoenzyme is brought to the origins through interactions with the primer:template junction and the helicase. Once the holoenzyme is present, sliding clamps are assembled on the RNA primers, and the leading-strand polymerases are engaged. As new ssDNA is exposed by the action of the helicase, it is bound by SSB and DNA primase synthesizes the first lagging-strand primers. These new primer:template junctions are targeted by the clamp loaders, which place two additional sliding clamps on the lagging strands. These clamps are recognized by the remaining unengaged core DNA Pol III enzymes, resulting in the initiation of lagging-strand DNA synthesis. At this point, two replication forks have been assembled, and initiation of replication is complete (exactly how the two replication forks are assembled is a matter of debate, see Box 8-6, The Replication Factory Hypothesis).

ADVANCED CONCEPTS

Box 8-6 The Replication Factory Hypothesis

There are two ways to think of the relative motion of the DNA and the replication machinery (Box 8-6 Fig. 1). One simple view is that the replication machinery moves along the DNA in a manner analogous to a train moving along its tracks, replicating both strands of the approaching DNA. In this traditional view, the DNA helicases pass by one another immediately after being loaded at the origin and subsequently act independently from one another at the two new replication forks. An alternative view suggests that the DNA moves while the replication machinery is stationary, similar to film moving into a movie projector. Mechanistically, it has been proposed that this is accomplished when the two DNA helicases do not pass by each other but instead "run into each other" and remain associated for the remainder of the replication process.

The view of replication occurring at stationary sites has become increasingly favored. Studies of bacterial DNA replication clearly indicate that the replication machinery remains in a single location within the cell during DNA synthesis. Instead of the replication machinery moving, the DNA moves in and out of this "replication factory" and in the process is duplicated. Similarly, replication in eukaryotic cells is observed to occur at discrete sites within the cell nucleus. Studies of the helicases that function at replication forks also support a static replication machinery. Several hexameric DNA helicases form double hexamers. This suggests that rather than the two hexameric helicases rapidly separating from each other after initiation (as suggested by the "railroad" model), they remain together throughout the replication process.

These two views of the assembly of the replication fork also have interesting consequences concerning the DNA that is replicated by each DNA Pol III holoenzyme. If the DNA helicases pass by one another immediately after they are loaded, then the closest strands that can be replicated simultaneously by the two polymerases of the DNA Pol III holoenzyme will be the Watson and Crick strands of the most recently unwound DNA (Box 8-6 Fig. 1, left panel). In contrast, if the two helicases remain associated after initiation, then it is possible that the lagging-strand DNA polymerases of the DNA Pol III holoenzyme could associate with either of two primed templates, since they are now both nearby. By most estimations, in this scenario, the choice will be for each DNA Pol III holoenzyme to have the same template strand for the leading- and lagging-strand synthesis; that is, one core enzyme will replicate the "Crick" strand of the DNA and the other will replicate the "Watson" strand of the DNA (Box 8-6 Fig. 1, right panel).

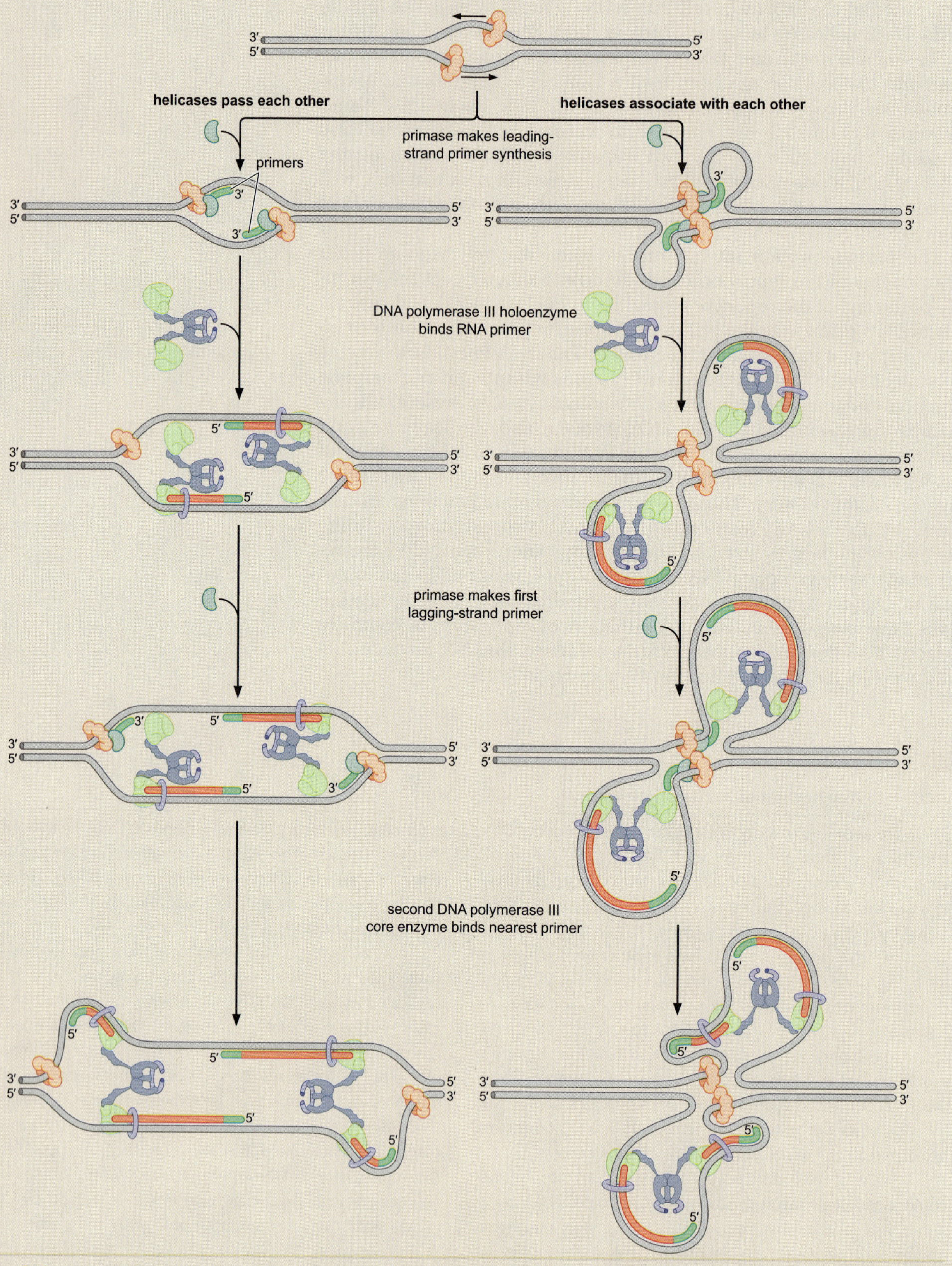

Box 8-6 Figure 1 Two views of how the replication machinery functions. In the left panel, the two DNA helicases function independently. In the right panel, the two DNA helicases remain associated with each other. Note that in the right panel, one DNA Pol III holoenzyme uses only the Watson strand as a template and the other uses only the Crick strand as a template. For simplicity, the DNA Pol III is not shown associated with the DNA helicases.

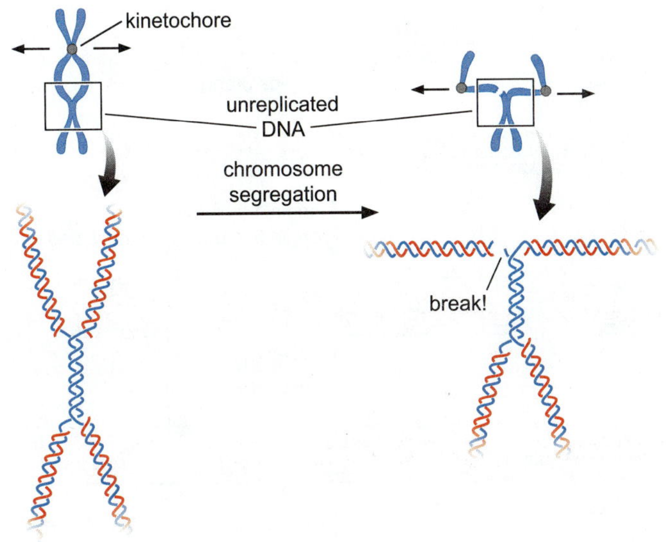

FIGURE 8-28 Chromosome breakage as a result of incomplete DNA replication. This illustration shows the consequences of incomplete replication followed by chromosome segregation. The top of each illustration shows the entire chromosome. The bottom shows the details of the chromosome breakage at the DNA level. (For details of chromosome segregation, see Chapter 7.) As the chromosomes are pulled apart, stress is placed on the unreplicated DNA, resulting in the breakage of the chromosome.

Eukaryotic Chromosomes Are Replicated Exactly Once per Cell Cycle

As discussed in Chapter 7, the events required for eukaryotic cell division occur at distinct times during cell cycle. Chromosomal DNA replication occurs only during the S phase of the cell cycle. During this time, all of the DNA in the cell must be duplicated exactly once. Incomplete replication of any part of a chromosome causes inappropriate links between daughter chromosomes. Segregation of linked chromosomes causes chromosome breakage or loss (Fig. 8-28). Rereplication of even limited amounts of eukaryotic DNA leads to DNA lesions that are difficult for the cell to repair. Attempts to repair such lesions frequently result in amplification of the associated DNA, which can inappropriately increase the expression of the associated genes. Addition of even one or two more copies of critical regulatory genes can lead to catastrophic defects in gene expression, cell division, or the response to environmental signals. Thus, it is critical that every base pair in each chromosome be replicated *once and only once* each time a eukaryotic cell divides.

The need to replicate the DNA once and only once is a particular challenge for eukaryotic chromosomes because they each have many origins of replication. Origins are typically separated by ~30 kb so even a small eukaryotic chromosome may have more than ten origins and a large human chromosome may have thousands. Enough of these origins must be activated to ensure that each chromosome is fully replicated during each S phase. Typically, not all potential origins need to be activated to complete replication, but if too few are activated, regions of the genome will escape replication. On the other hand, although some potential origins may not be used in any given round of cell division, *no* origin of replication can initiate after it has been replicated. Thus, whether an origin is activated to cause its own replication or replicated by a replication fork derived from an adjacent origin, it *must be inactivated* until the next round of cell division (Fig. 8-29). If these conditions were not true, the DNA associated with an origin could be replicated twice in the same cell cycle.

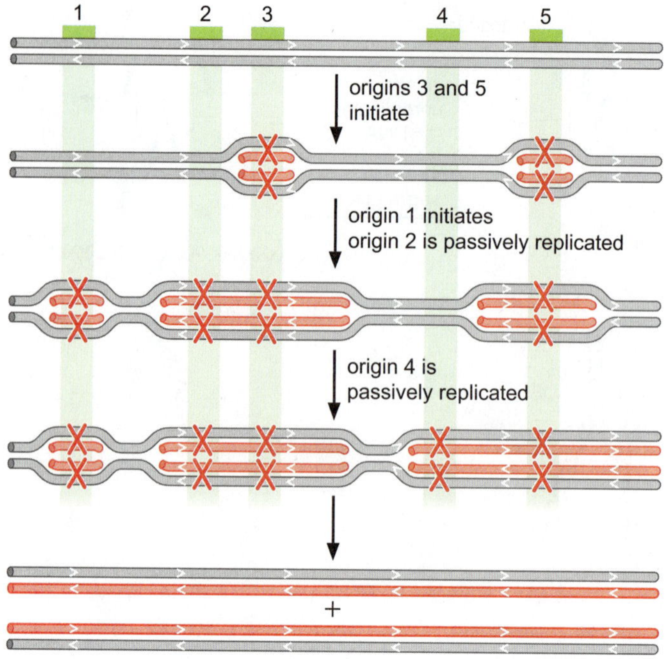

FIGURE 8-29 Replicators are inactivated by DNA replication. A chromosome with five replicators is shown. The replicators labeled 3 and 5 are the first to be activated, leading to the formation of two pairs of bidirectional replication forks. Activation of the parental replicator results in the inactivation of the copies of each replicator on both daughter DNA molecules until the next cell cycle (indicated by a red X). Further extension of the resulting replication forks replicates the DNA overlapping with the number 2 and 4 replicators. When a replicator is copied by a fork derived from an adjacent origin prior to initiation, it is said to have been passively replicated. Although these replicators have not initiated, they are nevertheless inactivated by the act of replicating their DNA. In contrast, replicator 1 is not reached by an adjacent fork prior to initiation and is able to initiate normally. The presence of more replicators than needed to complete DNA replication is a form of redundancy to ensure the complete replication of each chromosome.

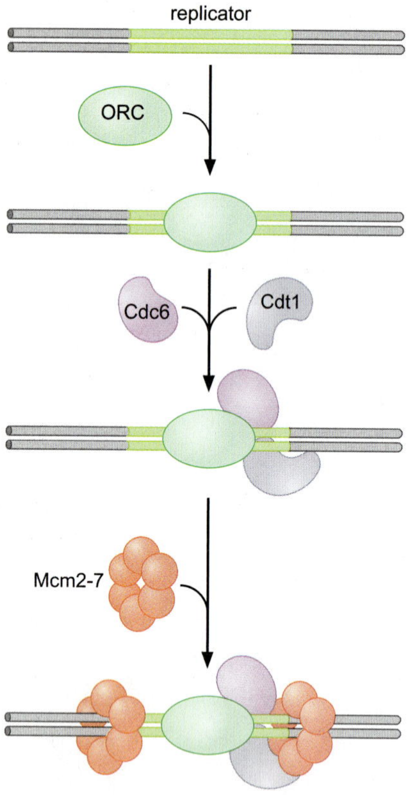

FIGURE 8-30 Steps in the formation of the prereplicative complex (pre-RC). The assembly of the pre-RC is an ordered process that is initiated by the association of the origin recognition complex with the replicator. Once bound to the replicator, ORC recruits at least two additional proteins, Cdc6 and Cdt1. These three proteins function together to assemble the eukaryotic DNA helicase—the Mcm2–7 complex—at the replication and complete the formation of the pre-RC.

Prereplicative Complex Formation Is the First Step in the Initiation of Replication in Eukaryotes

The initiation of replication in eukaryotic cells involves two events that occur at distinct times in the cell cycle (see Chapter 7): replicator selection and origin activation. Replicator selection is the process of identifying sequences that will direct the initiation of replication and occurs in G_1 (prior to S phase). This process leads to the assembly of a multiprotein complex at each replicator in the genome. Origin activation only occurs after cells enter S phase and triggers the replicator-associated protein complex to initiate DNA unwinding and DNA polymerase recruitment.

The separation of replicator selection and origin activation is different from the situation in prokaryotic cells, where the recognition of replicator DNA is intrinsically coupled to DNA unwinding and polymerase recruitment. As we will see below, the temporal separation of these two events during the eukaryotic cell cycle ensures that each chromosome is replicated only once during each cell cycle (bacterial cells solve this problem differently, see Box 8-7).

Replicator selection is mediated by the formation of prereplicative complexes (pre-RCs) (Fig. 8-30). The pre-RC is composed of four separate proteins that assemble in an ordered fashion at each replicator. The first step in the formation of the pre-RC is the recognition of the replicator by the eukaryotic initiator, ORC. Once ORC is bound, it recruits two helicase loading proteins (Cdc6 and Cdt1). Together, ORC and the loading proteins recruit the eukaryotic replication fork helicase, the Mcm 2–7 complex. Interestingly, both ORC and the Cdc6 protein are members of the AAA+ family of proteins like DnaC and the subunits of the sliding clamp loaders. Studies of pre-RC assembly suggest that these proteins use ATP binding and hydrolysis to load the ring-shaped Mcm2–7 complex around dsDNA. Consistent with the Mcm2–7 complex encircling dsDNA, formation of the pre-RC does not lead to the immediate unwinding of origin DNA or the recruitment of DNA polymerases. Instead, the pre-RCs that are formed during G_1 are only activated to unwind DNA and initiate replication after cells pass from the G_1 to the S phase of the cell cycle.

Pre-RCs are activated to initiate replication by two protein kinases: Cdk (cyclin-dependent kinase) and Ddk (Dbf4-dependent kinase) (Fig. 8-31). Protein kinases are proteins that covalently attach phosphate groups to target proteins (see Chapter 5). Each of these kinases is inactive in G_1 and is activated only when cells enter S phase. Once activated, these kinases target the pre-RC and other replication proteins. Phosphorylation of these proteins results in the recruitment of many additional replication proteins to the origin and the initiation of replication (see Fig. 8-31). These new proteins include the three eukaryotic DNA polymerases and a number of other proteins required for their recruitment. Interestingly, the polymerases assemble at the origin in a particular order. DNA Pol δ and Pol ε associate first, followed by DNA Pol α/primase. This order ensures that all three DNA polymerases are present at the origin prior to the synthesis of the first RNA primer (by DNA Pol α/primase).

Only a subset of the proteins that assemble at the origin go on to function as part of the eukaryotic replisome. In addition to the three DNA polymerases, the Mcm complex and many of the factors required for DNA polymerase recruitment become part of the replication fork machinery. Similar to the *E. coli* DNA helicase loader (DnaC), the other factors (such as Cdc6 and Cdt1) are released or destroyed after their role is complete (see Fig. 8-31).

Pre-RC Formation and Activation Are Regulated to Allow Only a Single Round of Replication during Each Cell Cycle

How do eukaryotic cells control the activity of hundreds or even thousands of origins of replication such that *not even one* is activated more than once during a cell cycle? The answer lies in the tight regulation of the formation and activation of pre-RCs by cyclin-dependent kinases (Cdks).

Cdks have two seemingly contradictory roles in regulating pre-RC function (Fig. 8-32). First, as we described above, they are required to activate pre-RCs to initiate DNA replication. Second, Cdk activity *inhibits* the formation of new pre-RCs.

The tight connection between pre-RC function, Cdk levels, and the cell cycle ensures that the eukaryotic genome is replicated only once per cell cycle (Fig. 8-33). Cdk levels are low during G_1, whereas elevated levels of Cdk are present during the remainder of the cell cycle (S, G_2, and M phases). Thus, during each cell cycle, there is *only one* opportunity for pre-RCs to form (during G_1) and *only one* opportunity for those

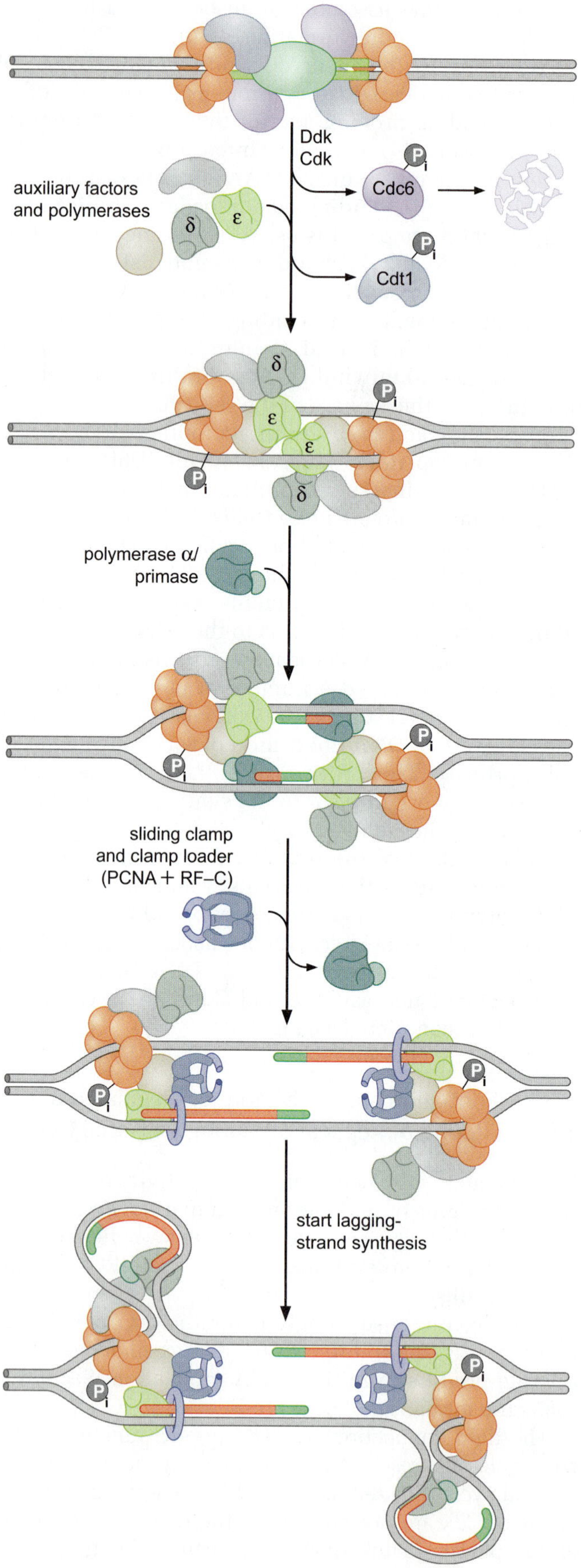

FIGURE 8-31 Activation of the pre-RC leads to the assembly of the eukaryotic replication fork. As cells enter into the S phase of the cell cycle, Cdk and Ddk phosphorylate replication proteins to trigger the initiation of replication. Activation of these kinases leads to the release of Cdc6 and Cdt1. In the case of Cdc6, the phosphorylated form of the protein is subject to rapid proteolysis. The events that lead to DNA unwinding at the origin are poorly understood but are likely to require the activity of the Mcm complex and result in the recruitment of a number of auxiliary replication factors and DNA Pol δ and Pol ε. DNA Pol α/primase is only recruited after DNA Pol δ and Pol ε. Once present at the origin, DNA Pol α/primase synthesizes an RNA primer and briefly extends it. The resulting primer:template junction is recognized by the eukaryotic sliding clamp loader (RF-C), which assembles a sliding clamp (PCNA) at these sites. DNA Pol ε recognizes this primer and begins leading-strand synthesis. After a period of DNA unwinding, DNA Pol α/primase synthesizes additional primers, which allow the initiation of lagging-strand DNA synthesis by DNA Pol δ.

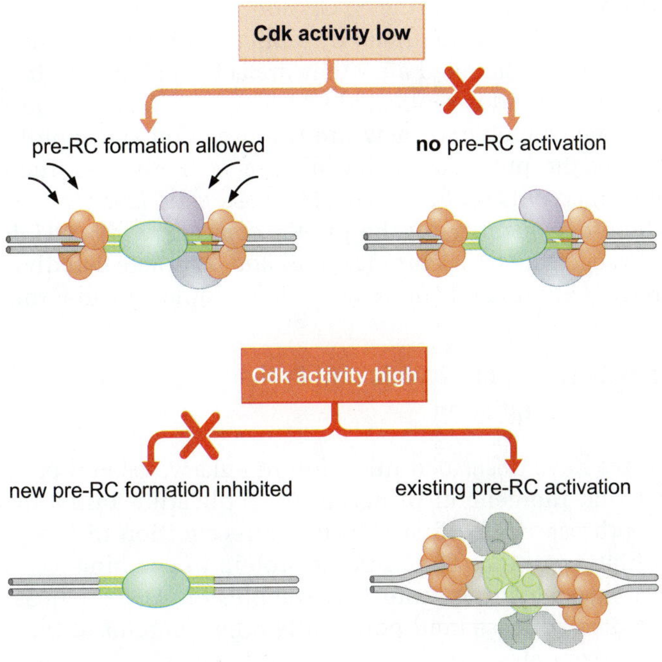

FIGURE 8-32 Effect of Cdk activity on pre-RC formation and activation. High Cdk activity is required for existing pre-RC complexes to initiate DNA replication. These same elevated levels of Cdk activity completely inhibit the formation of new pre-RC complexes. In contrast, low Cdk activity is conducive to new pre-RC formation but is inadequate to trigger DNA replication initiation by the newly formed pre-RC complexes.

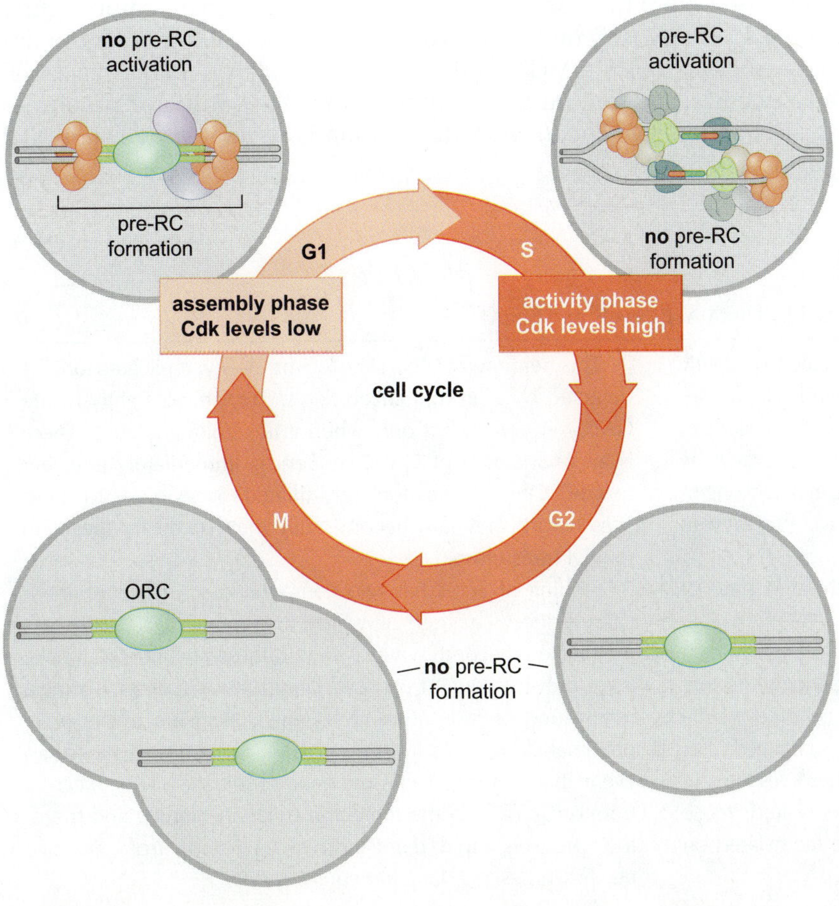

FIGURE 8-33 Cell cycle regulation of Cdk activity and pre-RC formation. In G_1, Cdk levels are low and new pre-RC complexes can form but cannot be activated. During S phase, the elevated levels of Cdk activity trigger the initiation of DNA replication and prevent any new pre-RC complex formation on newly replicated DNA. Once a pre-RC is used for the initiation of replication, it is necessarily dismantled (recall that at least one key component of the pre-RC, the Mcm complex, becomes part of the replication fork). Similarly, replication of pre-RC-associated DNA also destroys the complex (not shown). Because Cdk levels remain high until the end of mitosis, no new pre-RC complexes can be formed until chromosome segregation is complete. Without new pre-RC complexes, reinitiation is impossible.

pre-RCs to be activated (during S, G_2, and M—although in practice, all pre-RCs are activated or disrupted by replication forks in S phase).

Pre-RCs are disassembled after they are activated or after the DNA to which they are bound is replicated (see Fig. 8-29). These exposed replicators are then available for new pre-RC formation and rapidly bind to ORC. Despite the presence of the initiator at these sites, the elevated levels of Cdk activity in S, G_2, and M phase cells prevent the association of the other members of the pre-RC complex with ORC. It is only when cells segregate their chromosomes and complete cell division that Cdk activity is eliminated and new pre-RC complexes can form.

Similarities between Eukaryotic and Prokaryotic DNA Replication Initiation

Now that we have described initiation in eukaryotes and prokaryotes, it is clear that the general principles of replication initiation are the same in both cases. The first step is the recognition of the replicator by the initiator protein. The initiator protein in combination with one or more helicase loading proteins assembles the DNA helicase on the replicator. The helicase (and potentially other proteins at the origin in eukaryotes) generates a region of ssDNA that can act as a template for RNA primer synthesis. Once primers are synthesized, the remaining components of the replisome assemble through interactions with the resulting primer:template junction.

Although the events of initiation are similar, the regulation of replication in bacteria and eukaryotic cells is distinctly different. For example, unlike eukaryotic cells, rapidly dividing bacterial cells initiate replication more than once per cell cycle. The step that is most tightly regulated is also different. Eukaryotic cells focus regulation on the initial association of the MCM helicase with the DNA, whereas bacterial cells focus regulation on the binding of the DnaA initiator protein to the DNA (Box 8-7, *E. coli* DNA Replication Is Regulated by DnaA•ATP Levels and SeqA).

ADVANCED CONCEPTS

Box 8-7 *E. coli* DNA Replication Is Regulated by DnaA·ATP Levels and SeqA

In all organisms, it is critical that replication initiation be tightly controlled to ensure that chromosome number and cell number remain appropriately balanced. Although this balance is most tightly regulated in eukaryotic cells, *E. coli* also prevent runaway chromosome duplication by inhibiting recently initiated origins from reinitiating. Several different mechanisms act to prevent rapid replication reinitiation from *oriC*.

One method exploits changes in the methylated state of the DNA before and after DNA replication (Box 8-7 Fig. 1). In *E. coli* cells, an enzyme called Dam methyltransferase adds a methyl group to the A within every GATC sequence (note that the sequence is a palindrome). Typically, the genome is fully methylated at GATC sequences. This situation is changed after each GATC sequence is replicated. Because the A residues in the newly synthesized DNA strands are unmethylated, those sites that have been recently replicated will be methylated on only one strand (referred to as hemimethylated).

The hemimethylated state of the newly replicated *oriC* is detected by a protein called SeqA. SeqA binds tightly to the GATC sequence, but *only* when it is hemimethylated. There is an abundance of GATC sequences immediately adjacent to *oriC*. Once replication has initiated, SeqA binds to these sites before they can become fully methylated by the Dam methyl transferase.

Binding of SeqA has two consequences. First, it dramatically reduces the rate at which the bound GATC sites are methylated. Second, when bound to these *oriC* proximal sites, SeqA prevents DnaA from associating with *oriC* and initiating a new round of replication. Thus, the conversion of the *oriC*-proximal GATC sites from methylated to hemimethylated (an event that is a direct consequence of initiation of replication from *oriC*) leads to the inhibition of DnaA binding and therefore prevents rapid reinitiation of replication from the two newly synthesized daughter copies of *oriC*.

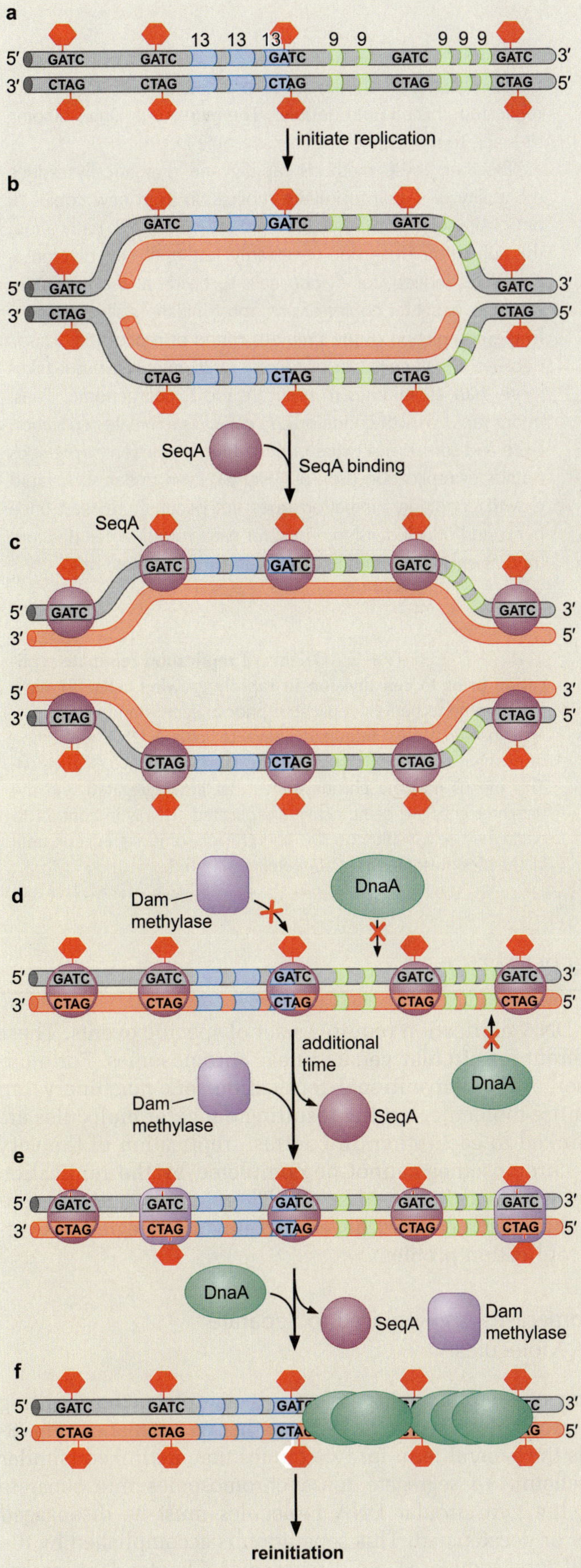

BOX 8-7 FIGURE 1 SeqA bound to hemimethylated DNA inhibits reinitiation from recently replicated daughter origins. (a) Prior to DNA replication, GATC sequences throughout the *E. coli* genome are methylated on both strands ("fully" methylated). Note that throughout the figure, the methyl groups are represented by red hexagons. (b) DNA replication converts these sites to the hemimethylated state (only one strand of the DNA is methylated). (c) Hemimethylated GATC sequences are rapidly bound by SeqA. (d) Bound SeqA protein inhibits the full methylation of these sequences and the binding of *oriC* by DnaA protein (for simplicity, only one of the two daughter molecules is illustrated in parts d, e, and f). (e) When SeqA infrequently disassociates from the GATC sites, the sequences can become fully methylated by Dam DNA methyltransferase, preventing rebinding by SeqA. (f) When the GATC sites become fully methylated, DnaA can bind the 9-mer sequences and direct a new round of replication from the daughter *oriC* replicators.

BOX 8-7 *(Continued)*

DnaA is targeted by other mechanisms that inhibit rapid reinitiation at newly synthesized copies of *oriC*. As described above, only DnaA bound to ATP can direct initiation of replication; however, this bound ATP is converted to ADP during the initiation process. Thus, the process of directing a round of replication initiation inactivates DnaA, preventing its reuse. The process of exchanging the bound ADP for an ATP is a slow one, further delaying the accumulation of replication-competent ATP bound DnaA. The process of replicating nearby sequences also acts to reduce the amount of DnaA available to bind at *oriC*. There are more than 300 DnaA 9-mer binding sites outside of *oriC* (DnaA also acts as a transcriptional regulator at a number of promoters), and as they are replicated, this number doubles. The increase in DnaA-binding sites acts to reduce the levels of available DnaA.

Together, these methods rapidly and dramatically reduce the ability of *E. coli* to initiate replication from new copies of *oriC*. Although these mechanisms prevent rapid reinitiation, this inhibition does not necessarily last until cell division is complete. Indeed, for *E. coli* cells to divide at the maximum rate, the daughter copies of *oriC* must initiate replication prior to the completion of the previous round of replication. This is because *E. coli* cells can divide every 20 minutes, but it takes more than 40 minutes to replicate the *E. coli* genome. Thus, under rapid growth conditions, *E. coli* cells reinitiate replication once and sometimes twice prior to the completion of previous rounds of replication (Box 8-7 Fig. 2). Even under such rapid growth conditions, initiation does not occur more than once per round of cell division. Thus, for each round of cell division, there is only one round of replication initiation from *oriC*.

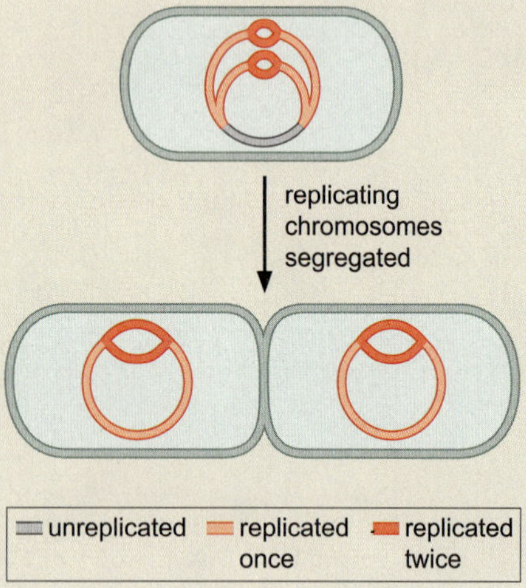

BOX 8-7 FIGURE 2 Origins of replication reinitiate replication prior to cell division in rapidly growing cells. To allow the genome to be fully replicated prior to each round of cell division, bacterial cells frequently have to initiate DNA replication from their single origin prior to the completion of cell division. This means that the chromosomes that are segregated into the daughter cells are being actively replicated. This is in contrast to eukaryotic cells, which do not start chromosome segregation until all the chromosomes are completely replicated.

FINISHING REPLICATION

Completion of DNA replication requires a set of specific events. These events are different for circular versus linear chromosomes. For a circular chromosome, the conventional replication fork machinery can replicate the entire molecule, but the resulting daughter molecules are topologically linked to each other. In contrast, replication of the very ends of linear chromosomes cannot be completed by the replication fork machinery we have discussed so far. Therefore, organisms containing linear chromosomes have developed novel strategies to overcome this end replication problem.

Type II Topoisomerases Are Required to Separate Daughter DNA Molecules

After replication of a circular chromosome is complete, the resulting daughter DNA molecules remain linked together as catenanes (Fig. 8-34). Catenane is the general term for two circles that are linked (similar to links in a chain). To segregate these chromosomes into separate daughter cells, the two circular DNA molecules must be disengaged from each other or decatenated. This separation is accomplished by the action of type II topoisomerases. As discussed in Chapter 6, these enzymes have the ability to break a dsDNA molecule and pass a second dsDNA molecule through this break. This reaction can easily decatenate the two circular daughter chromosomes by breaking one and passing the second through the break, allowing their segregation into separate cells.

Although the importance of this activity for the separation of circular chromosomes is most clear, the activity of type II topoisomerases is also critical to the segregation of large linear molecules. Although there is no inherent topological linkage after the replication of a linear molecule, the large size of eukaryotic chromosomes necessitates the intricate folding of the DNA into loops attached to a protein scaffold (see Chapter 7). These attachments lead to many of the same problems that circular chromosomes have when the two daughter chromosomes must be separated.

Lagging-Strand Synthesis Is Unable to Copy the Extreme Ends of Linear Chromosomes

The requirement for an RNA primer to initiate all new DNA synthesis creates a dilemma for the replication of the ends of linear chromosomes, called the **end replication problem** (Fig. 8-35). This difficulty is not observed during the duplication of the leading-strand template. In that case, a single internal RNA primer can direct the initiation of a DNA strand that can be extended to the extreme 5′ terminus of its template. In contrast, the requirement for multiple primers to complete lagging-strand synthesis means that a complete copy of its template cannot be made. Even if the end of the last RNA primer for Okazaki fragment synthesis anneals to the final base pairs of the

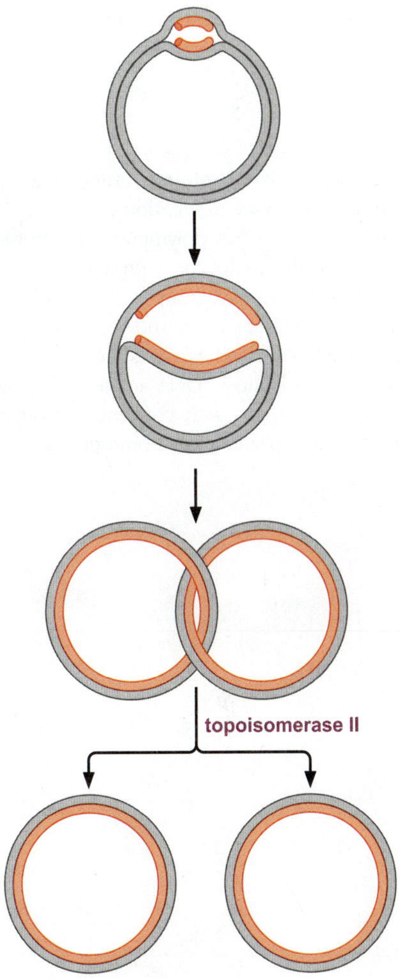

FIGURE 8-34 Toposiomerase II catalyzes the decatenation of replication products. After a circular DNA molecule is replicated, the resulting complete daughter DNA molecules remain linked to each other. Type II DNA topoisomerases can efficiently separate (or decatenate) these DNA circles.

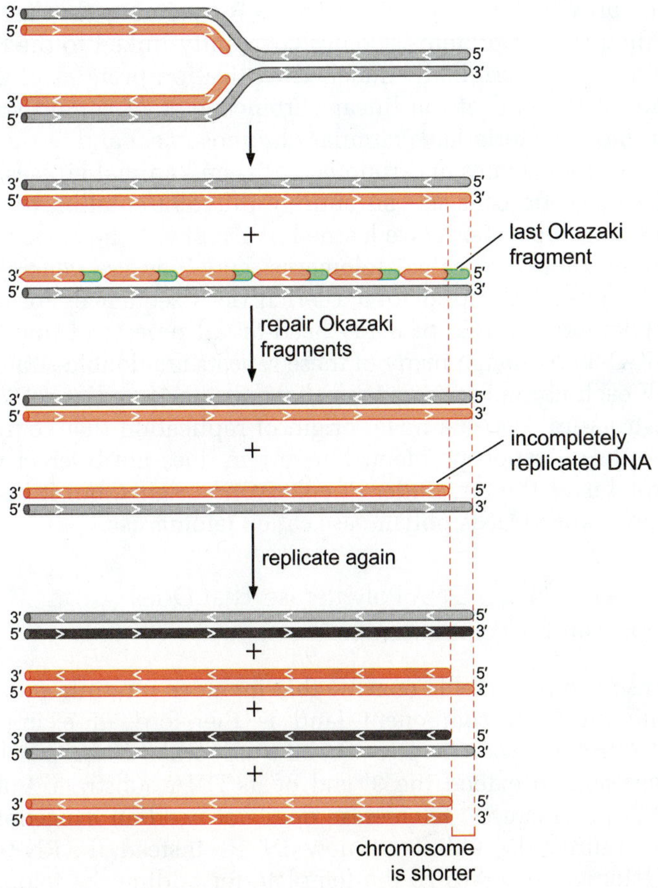

FIGURE 8-35 The end replication problem. As the lagging-strand replication machinery reaches the end of the chromosome, at some point, primase no longer has sufficient space to synthesize a new RNA primer. This results in incomplete replication and a short ssDNA region at the 3′ end of the lagging-strand DNA product. When this DNA product is replicated in the next round, one of the two products will be shortened and will lack the region that was not fully copied in the previous round of replication.

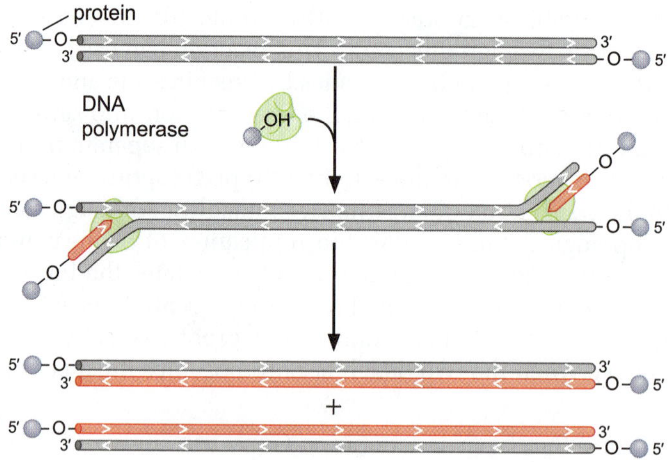

FIGURE 8-36 Protein priming as a solution to the end replication problem. By binding to the DNA polymerase and to the 3′ end of the template, a protein provides the priming hydroxyl group to initiated DNA synthesis. In the example shown, the protein primes all DNA synthesis as is seen for many viruses. For longer DNA molecules, this method combines with conventional origin function to replicate the chromosomes.

lagging-strand template, once this RNA molecule is removed, there will remain a short region of unreplicated ssDNA at the end of the chromosome. This means that each round of DNA replication would result in the shortening of one of the two daughter DNA molecules. Obviously, this scenario would disrupt the complete propagation of the genetic material from generation to generation. Slowly, but surely, genes at the end of the chromosomes would be lost.

Organisms solve the end replication problem in a variety of ways. One solution is to use a protein instead of an RNA as the primer for the last Okazaki fragment at each end of the chromosome (Fig. 8-36). In this situation, the "priming protein" binds to the lagging-strand template and uses an amino acid to provide an OH that replaces the 3′-OH normally provided by an RNA primer. By priming the last lagging strand, the priming protein becomes covalently linked to the 5′ end of the chromosome. Terminally attached replication proteins of this kind are found at the end of the linear chromosomes of certain species of bacteria (most bacteria have circular chromosomes) and at the ends of the linear chromosomes of certain bacterial and animal viruses.

Most eukaryotic cells use an entirely different solution to replicate their chromosome ends. As we learned in Chapter 7, the ends of eukaryotic chromosomes are called **telomeres**, and they are generally composed of head-to-tail repeats of a TG-rich DNA sequence. For example, human telomeres consist of many head-to-tail repeats of the sequence 5′-TTAGGG-3′. Although many of these repeats are double-stranded, the 3′ end of each chromosome extends beyond the 5′ end as ssDNA. This unique structure acts as a novel origin of replication that compensates for the end replication problem. This origin does not interact with the same proteins as the remainder of eukaryotic origins, but it instead recruits a specialized DNA polymerase called **telomerase**.

Telomerase Is a Novel DNA Polymerase That Does Not Require an Exogenous Template

Telomerase is a remarkable enzyme that includes multiple protein subunits *and* an RNA component (and is therefore an example of a ribonucleoprotein, see Chapter 5). Like all other DNA polymerases, telomerase acts to extend the 3′ end of its DNA substrate. But unlike most DNA polymerases, telomerase does not need an exogenous DNA template to direct the addition of new dNTPs. Instead, the RNA component of telomerase serves as the template for adding the telomeric sequence to the 3′ terminus at the end of the chromosome (see Interactive Animation 8-3). Telomerase specifically elongates the 3′-OH of particular ssDNA sequences using its own RNA as a template. As a result of this unusual mechanism, the newly synthesized DNA is single-stranded.

WEB ANIMATION

The key to telomerase's unusual functions is revealed by the RNA component of the enzyme, called telomerase RNA or TER. Depending on the organism, TER varies in size from 150 to 1300 bases. In all organisms, the sequence of the RNA includes a short region that encodes ~1.5 copies of the complement of the telomere sequence (for humans, this sequence is 5′-AAUCCCAAUC-3′). This region of the RNA can anneal to the ssDNA at the 3′ end of the telomere (Fig. 8-37). An-

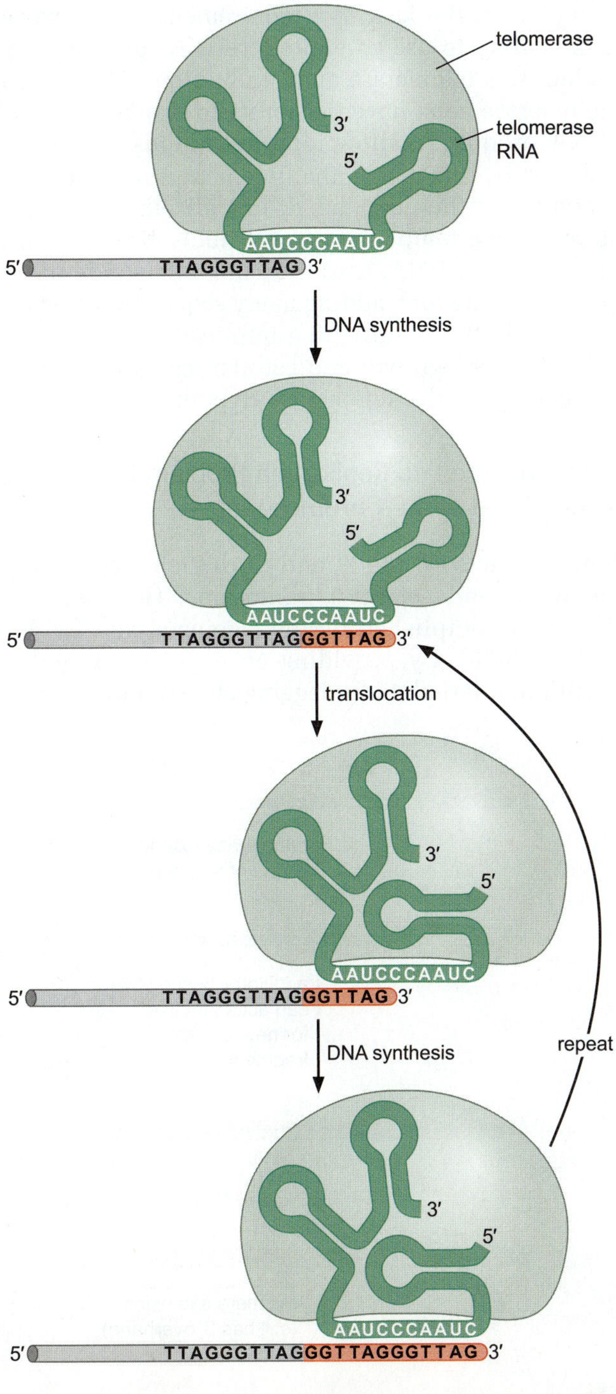

FIGURE 8-37 Replication of telomeres by telomerase. Telomerase uses its RNA component to anneal to the 3′ end of the ssDNA region of the telomere. Telomerase then uses its reverse transcription activity to synthesize DNA to the end of the RNA template. Telomerase then displaces the RNA from the DNA product and rebinds at the end of the telomere and repeats the process.

nealing occurs in such a way that a part of the RNA template remains single-stranded, creating a primer:template junction that can be acted on by telomerase. Interestingly, one of the protein subunits of telomerase is a member of a class of DNA polymerases that use RNA templates called reverse transcriptases (this subunit is called telomerase reverse transcriptase or TERT). As we shall see in Chapter 11, these enzymes "reverse-transcribe" RNA into DNA instead of the more conventional transcription of DNA into RNA. Using the associated RNA template, TERT synthesizes DNA to the end of the TER template region but cannot continue to copy the RNA beyond that point. At this point, the RNA template disengages from the DNA product, reanneals to the last four nucleotides of the telomere, and then repeats this process.

The characteristics of telomerase are in some ways distinct and in other ways similar to those of other DNA polymerases. The inclusion of an RNA component, the lack of a requirement for an exogenous template, and the ability to use an entirely ssDNA substrate to produce an ssDNA product sets telomerase apart from other DNA polymerases. In addition, telomerase must have the ability to displace its RNA template from the DNA product to allow repeated rounds of template-directed synthesis. Formally, this means that telomerase includes an RNA·DNA helicase activity. On the other hand, like all other DNA polymerases, telomerase requires a template to direct nucleotide addition, can only extend a 3′-OH end of DNA, uses the same nucleotide precursors, and acts in a processive manner, adding many sequence repeats each time it binds to a DNA substrate. Intriguing implications of the role of telomerase in regulating cell growth and cellular aging are discussed in Box 8-8, Aging, Cancer, and the Telomerase Hypothesis.

Telomerase Solves the End Replication Problem by Extending the 3′ End of the Chromosome

When telomerase acts on the 3′ end of the telomere, it extends only one of the two strands of the chromosome. How is the 5′ end extended? This is accomplished by the lagging-strand DNA replication machinery (Fig. 8-38). By providing an extended 3′ end, telomerase provides additional template for lagging-strand replication machinery.

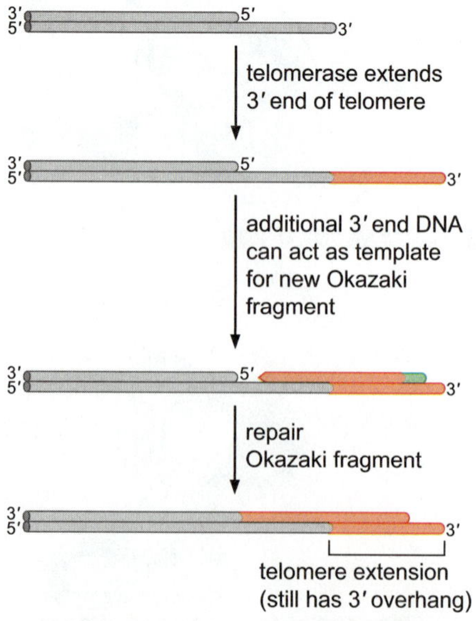

FIGURE 8-38 Extension of the 3′ end of the telomere by telomerase solves the end replication problem. Although telomerase only directly extends the 3′ end of the telomere, by providing an additional template for lagging-strand DNA synthesis, both ends of the chromosome are extended.

MEDICAL CONNECTIONS

Box 8-8 Aging, Cancer, and the Telomere Hypothesis

All organisms are mortal. Whether it is the days or weeks lived by many smaller organisms or the many years that the average human lives, organisms cannot escape their intrinsic mortality. Not surprisingly, researchers (and others) have long studied these limitations hoping to understand them and, perhaps, overcome them and find the mythical "fountain of youth."

When researchers developed ways to grow individual cells outside of the body, they thought that the cells were immortal. This suggested that mortality was a problem of whole organisms, not of cells. This hypothesis was eliminated when Leonard Hayflick studied cell division in culture more carefully. He found that, even in isolation, cells could divide only a limited number of times. Interestingly, Hayflick's studies found that the number of divisions a cell can pass through is characteristic of the source of cells, now known as the Hayflick limit.

Hayflick's studies led to the idea that cells contain an intrinsic countdown clock that limits the number of divisions that a cell can participate in. When the clock reaches zero, a cell would be prevented from dividing again. For years the molecular identity of such a clock was unknown; however, as the nature of telomeres and their role in DNA replication were better understood, it became clear that the telomere could be the long sought after divisional clock. Consistent with this idea, telomere DNA isolated from young people is longer than that isolated from older people. This observation led to the hypothesis that the length of telomeric DNA limited the number of times a cell could divide.

Although the concept is still very much a hypothesis, experimental support for the idea that telomeres are connected to cellular aging has accumulated. For example, for the hypothesis to be viable, normal cells should have little or no telomerase activity. Otherwise these cells would simply continue to extend their telomeres as they shortened. Indeed, many normal cells have limited telomerase activity. In contrast, cells that have increased proliferative capacity, such as stem cells and cells derived from tumors, have higher levels of telomerase activity. Indeed, studies of cancer cells in culture indicate that they can divide indefinitely. A second important experiment in support of the model showed that expression of telomerase in normal cells effectively immortalized the cells.

The finding of elevated telomerase activity in cancer cells has led to the hypothesis that telomeres may represent a method to limit the growth capacity of cells that have lost normal growth control. If true, this may explain why multicellular organisms have not allowed telomerase activity to be present in all cells. Indeed, there are numerous efforts seeking telomerase inhibitors as chemotherapeutic agents. The elevation of telomerase activity in cancer cells also suggests that globally activating telomerase would not be a wise method to seek immortality!

By synthesizing and extending RNA primers using the telomerase extended 3′ end as a template, the cell can effectively increase the length of the 5′ end of the chromosome as well.

Even after the action of the lagging-strand machinery has acted, there remains a short ssDNA region at the end of the chromosome. Indeed, the presence of a 3′ overhang may be important for the end protection function of the telomere (as we discuss below). Nevertheless, the action of telomerase and the lagging-strand replication machinery ensures that the telomere is maintained at sufficient length to protect the end of the chromosome from shortening. Because of the repetitive and non–protein coding nature of the telomeric DNA, variations in the length of the telomere are easily tolerated by the cell.

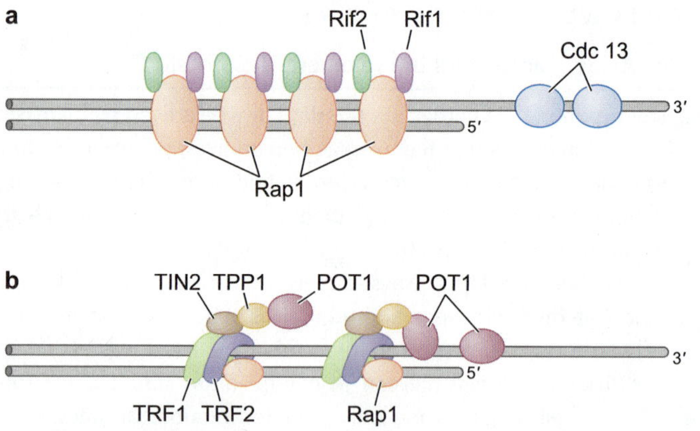

FIGURE 8-39 Telomere-binding proteins. Telomere-binding proteins that regulate telomerase activity are illustrated for *S. cerevisiae* and human cells. (a) *S. cerevisiae* cells. Rap1 directly binds to the double-stranded telomere repeat DNA, whereas Rif1 and Rif2 associate with the telomere indirectly by binding to Rap1. All three proteins have been implicated in the inhibition of telomerase activity. Cdc13 binds to the single-stranded telomere repeat DNA and is involved in telomerase recruitment. (b) Human cells. TRF1 and TRF2 bind directly to the double-stranded telomere repeat DNA. The human homolog of Rap1, TIN2, TPP1, and POT1 all associate with either TRF1 or TRF2. Together these proteins form a complex that is called Shelterin for its ability to "shelter" the telomeres from the action of DNA repair enzymes. POT1 also has the ability to bind directly to the single-stranded telomere repeat DNA.

Telomere-Binding Proteins Regulate Telomerase Activity and Telomere Length

Although extension of telomeres by telomerase could theoretically go on indefinitely, proteins bound to the double-strand regions of the telomere regulate telomere length (Fig. 8-39). These proteins or other proteins that bind to them at the telomere act as weak inhibitors of telomerase activity (Fig. 8-40). When there are relatively few copies of the telomere sequence repeat, few of these proteins are bound to the telomere and telomerase can extend the 3′-OH end of the telomere. As the telomere becomes longer, more of the telomere-binding proteins accumulate and telomerase can extend the 3′-OH end of the telomere. This simple negative feedback loop mechanism (longer telomeres inhibit telomerase) is a robust method to maintain a similar telomere length at the ends of all chromosomes.

Proteins that recognize the single-strand form of the telomere can also modulate telomerase activity. In *S. cerevisiae* cells the Cdc13 protein binds to single-strand regions of the telomere. Studies of this protein indicate that it recruits telomerase to the telomeres. Thus, Cdc13 is a positive activator of telomerase. In contrast, the human protein that binds to single-stranded telomeric DNA, POT1, acts in the opposite manner—that is, as an inhibitor of telomerase activity. In vitro studies show that POT1 binding to single-stranded telomere DNA inhibits telomere activity. Cells that lack this protein show dramatically increased telomere DNA length. Interestingly, this protein interacts indirectly with the double-strand telo-mere-binding proteins in human cells. It has been proposed that as telomeres increase in length, more POT1 is recruited, thereby increasing the likelihood that it binds to the ssDNA ends of the telomere and inhibits telomerase.

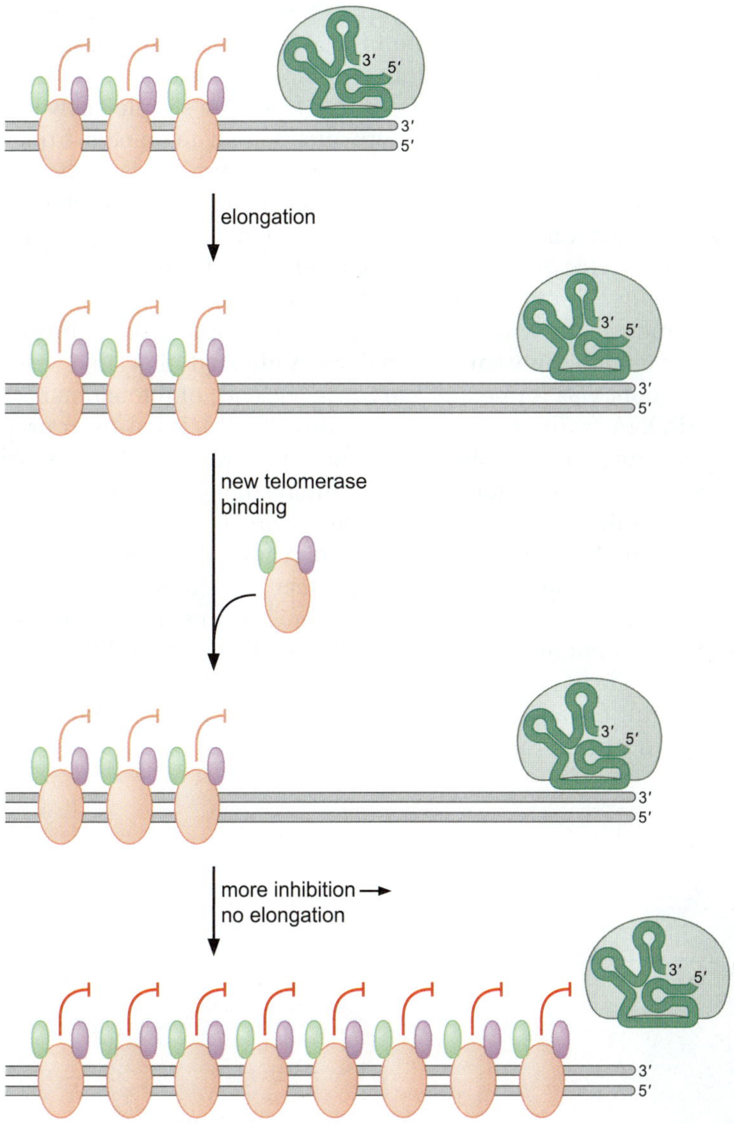

FIGURE 8-40 Telomere length regulation by telomere-binding proteins. When telomeres are relatively short, few telomere-binding proteins will be present and inhibition of telomerase is weak. Under these conditions, telomerase can extend the 3′ end of the telomere. When these regions are made double-stranded by the action of the lagging-strand DNA synthesis machinery, additional telomere-binding proteins can associate with the telomere. This process increases the level of inhibition preventing further elongation by telomerase. (Adapted, with permission, from Smogorzeska A. and de Lange T. 2004. *Annu. Rev. Biochem.* 73: 177–208, Fig. 3a. © Annual Reviews.)

Telomere-Binding Proteins Protect Chromosome Ends

In addition to their role in regulating telomerase function, telomere-binding proteins also play a crucial role in protecting the ends of chromosomes. Ordinarily in a cell, the presence of a DNA end is considered the sign of a double-stranded break in the DNA, which is targeted by the DNA repair machinery (see Chapter 9). The most common outcome of this repair is to initiate recombination with other DNA in the genome (in a diploid cell this recombination is targeted to the intact copy of the broken chromosome). Whereas this response is appropriate for random DNA breaks, it would be disastrous for the telomeres to participate in the same events. Attempts to repair telo-meres in the same manner as double-stranded DNA breaks would lead to chromosome fusion events, which eventually result in random chromosome breaks.

What protects the telomeres from this fate? The simple answer is that the proteins bound at the telomere distinguish telomeres from other DNA ends in the cell. Elimination of these proteins leads to the recognition of the telomeres as normal DNA breaks. It is possible that protection is conferred simply by coating the telomere with binding proteins. Studies of the structure of the human telomere have led to an alternative possibility. Telomeres isolated from human cells were observed by electron microscopy and found to form a loop rather than a linear structure (Fig. 8-41a). Subsequent analysis indicated that this structure, called a **t-loop**, was formed by the 3′ ssDNA end of the telomere invading the dsDNA region of the telomere (Fig. 8-41b). It has been proposed that by forming a t-loop, the end of the telomere is masked and cannot be recognized as a normal DNA end. Interestingly, purified TRF2 is capable of directing t-loop formation with purified telomere DNA.

The t-loop structure may also be relevant to telomere length control. Just as the loop structure may protect the telomere from DNA repair enzymes, it is also likely that telomerase cannot recognize this form of the telomere, as it lacks an obvious single-strand 3′ end. It has been proposed that as telomeres shorten, they would have an increasingly difficult time forming the t-loop, thereby allowing increased access to the 3′ end of the telomere.

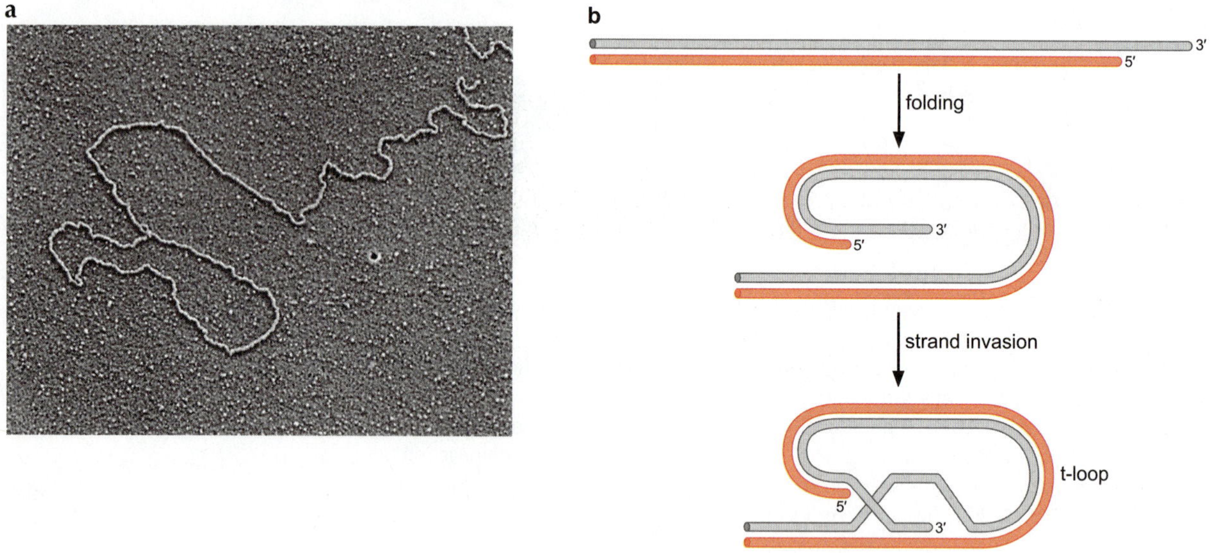

FIGURE 8-41 Telomeres form a looped structure in the cell. (a) An electron micrograph of a telomere isolated from a human cell. The loop found at the end of the DNA included the ssDNA at the end of the telomere and is referred to as a t-loop. The end of the DNA in the upper right-hand corner would be attached to the rest of the chromosome. (b) An illustration of the proposed mechanism of t-loop formation. The first step folds the telomere such that the ssDNA at the end of the telomere can access the dsDNA telomeric repeats. Once the ssDNA end is positioned properly, it can invade the dsDNA repeats and form a helix with the complementary strand, displacing the other strand of the dsDNA. This is called a displacement loop and is a common intermediate in homologous recombination (see Chapter 10). It is likely that telomere-binding proteins and other cellular proteins (e.g., recombination proteins) facilitate this process. Note how the folding process would be increasingly difficult as the telomere become shorter. (a, Reprinted, with permission, from Griffith J.D. et al. 1999. *Cell* 97: 503–514, Fig. 3f. © Elsevier.)

SUMMARY

DNA synthesis is dependent on the presence of two types of substrates: the four deoxynucleoside triphosphates (dATP, dGTP, dCTP, and dGTP) and the template DNA structure, a primer:template junction. The template DNA determines the sequence of incorporated nucleotides. The primer serves as the substrate for deoxynucleotide addition, each being added successively to the OH at its 3′ end.

DNA synthesis is catalyzed by an enzyme called DNA polymerase that uses a single active site to add any of the four dNTP precursors. Structural studies of DNA polymerases reveal that these enzymes resemble a hand that grips the catalytic site. This structure contributes to the extremely accurate nature of the DNA synthesis reaction. DNA polymerases are processive: each time they bind a substrate, they add many nucleotides. Proofreading exonucleases further enhance the accuracy of DNA synthesis by acting like a "delete key" that removes incorrectly added nucleotides.

In the cell, both strands of a DNA template are duplicated simultaneously at a structure called the replication fork. Because the two strands of the DNA are antiparallel, only one of the template DNA strands can be replicated in a continuous fashion (called the leading strand). The other DNA strand (called the lagging strand) must be synthesized first as a series of short DNA fragments, called Okazaki fragments. Each DNA strand is initiated with an RNA primer that is synthesized by an enzyme called primase. These primers must be removed to complete the replication process. After the replacement of the RNA primers with DNA, all of the separately primed lagging-strand DNA fragments are joined together to form one continuous DNA strand by DNA ligase.

An array of proteins in addition to the DNA polymerases coordinate and facilitate the DNA replication reaction. These additional factors facilitate the unwinding of the dsDNA template (DNA helicase), stabilize the ssDNA template (SSB), and remove supercoils generated in front of the replication fork (topoisomerase). DNA polymerases are specialized to perform different events during DNA replication. Some are designed to be highly processive and others, only weakly processive. DNA sliding clamps enhance the processivity of the DNA polymerases that replicate large regions of DNA (such as whole chromosomes). These clamp proteins are topologically linked to DNA, but they are able to slide along the recently synthesized DNA while bound to the DNA polymerase. This effectively prevents the attached DNA polymerase from dissociating from the primer:template junction. Special protein complexes called sliding DNA clamp loaders use the energy of ATP hydrolysis to place sliding clamps on the DNA near primer:template junctions.

Interactions between the proteins at the replication fork have an important role in DNA synthesis. In *E. coli*, the two DNA polymerases are part of a large complex called the DNA Pol III holoenzyme. Binding of the DNA Pol III holoenzyme to the DNA helicase stimulates the rate of DNA unwinding. Similarly, binding of primase to the DNA helicase increases its ability to synthesize RNA primers. Thus, the replication reaction works best when the entire array of replication proteins are present at the replication fork. Together, this set of proteins forms a complex called the replisome.

The initiation of DNA replication is directed by specific DNA sequences called replicators. The physical site of replication initiation is called an origin of replication. The replicator is specifically bound by a protein called the initiator, which stimulates the unwinding of the origin DNA and the recruitment of other proteins required for the initiation of replication (such as DNA helicase). The subsequent events in the initiation of DNA replication are largely driven by either protein–protein or nonspecific protein–DNA interactions.

In eukaryotic cells, the initiation of DNA replication is tightly regulated to ensure that every nucleotide of every chromosome is replicated once and only once per round of cell division. This tight regulation is accomplished by controlling the formation and activation of a multiprotein assembly called the prereplicative complex (pre-RC). Formation of these complexes at replicators is required to recruit the proteins necessary to initiate DNA replication. The ability to form and activate pre-RCs is controlled by a cell-cycle-regulated kinase called cyclin-dependent kinase. During the G_1 phase of the cell cycle, pre-RCs can be formed but cannot direct the initiation of replication. During the remainder of the cell cycle (the S, G_2, and M phases), any existing pre-RCs can initiate DNA replication, but no new pre-RCs can be formed. Thus, any particular pre-RC can only direct one round of initiation per cell cycle, ensuring that the DNA is replicated exactly once.

Finishing DNA replication requires the action of specific enzymes. For circular chromosomes, type II DNA topoisomerases separate the topologically linked circular products from one another. Linear chromosomes also require special proteins to ensure their complete replication. In eukaryotic cells, a specialized DNA polymerase called telomerase allows the ends of the chromosome (called telomeres) to act as a unique origin of replication. By extending the 3′ ends of the telomere, telomerase eliminates the progressive loss of chromosome ends that conventional DNA synthesis by the replication fork machinery would cause. Proteins bound to telomeric DNA act to regulate the activity of telomerase and protect the ends of chromosomes from degradation and recombination.

BIBLIOGRAPHY

Books

DePamphilis M.L. 2006. *DNA replication and human disease.* Cold Spring Harbor Laboratory Press, Cold Spring Harbor, New York.

The Chemistry of DNA Synthesis

Brautigam C.A. and Steitz T.A. 1998. Structural and functional insights provided by crystal structures of DNA polymerases. *Curr. Opin. Struct. Biol.* **8:** 54–63.

Jäger J. and Pata J.D. 1999. Getting a grip: Polymerases and their substrate complexes. *Curr. Opin. Struct. Biol.* **9:** 21–28.

The Mechanism of DNA Polymerase

Doublié S. and Ellenberger T. 1998. The mechanism of action of T7 DNA polymerase. *Curr. Opin. Struct. Biol.* **8:** 704–712.

Steitz T.A. 1998. A mechanism for all polymerases. *Nature* **391:** 231–232.

———. 2006. Visualizing polynucleotide polymerase machines at work. *EMBO J.* **25:** 3458–3468.

The Replication Fork

Corn J.E. and Berger J.M. 2006. Regulation of bacterial priming and daughter strand synthesis through helicase-primase interactions. *Nucleic Acids Res.* **34:** 4082–4088.

Johnson A. and O'Donnell M. 2005. Cellular DNA replicases: Components and dynamics at the replication fork. *Annu. Rev. Biochem.* **74:** 283–315.

O'Donnell M. and Kuriyan J. 2006. Clamp loaders and replication initiation. *Curr. Opin. Struct. Biol.* **16:** 405–415.

Wang J.C. 2002. Cellular roles of DNA topoisomerases. *Nat. Rev. Mol. Cell Biol.* **3:** 430–440.

The Specialization of DNA Polymerases

Lemon K.P. and Grossman A.D. 2000. Movement of replicating DNA through a stationary replisome. *Mol. Cell* **6:** 1321–1330.

Lovett S.T. 2007. Polymerase switching in DNA replication. *Mol. Cell* **27:** 523–526.

Sutton M.D. and Walker G.C. 2001. Managing DNA polymerases: Coordinating DNA replication, DNA repair, and DNA recombination. *Proc. Natl. Acad. Sci.* **98:** 8342–8349.

Initiation of DNA Replication

Bell S.P. and Dutta A. 2002. DNA replication in eukaryotic cells. *Annu. Rev. Biochem.* **71:** 333–374.

Blow J.J. and Dutta A. 2005. Preventing re-replication of chromosomal DNA. *Nat. Rev. Mol. Cell Biol.* **6:** 476–486.

Robinson N.P. and Bell S.D. 2005. Origins of DNA replication in the three domains of life. *FEBS J.* **272:** 3757–3766.

Finishing Replication

Blackburn E.H., Greider C.W., and Szostak J.W. 2006. Telo-meres and telomerase: The path from maize, *Tetrahymena* and yeast to human cancer and aging. *Nat. Med.* **12:** 1133–1138.

Collins K. 2006. The biogenesis and regulation of telomerase holoenzymes. *Nat. Rev. Mol. Cell Biol.* **7:** 484–494.

CHAPTER

9 The Mutability and Repair of DNA

OUTLINE

Visit Web Content for Structural Tutorials and Interactive Animations

THE PERPETUATION OF THE GENETIC MATERIAL from generation to generation depends on maintaining rates of mutation at low levels. High rates of mutation in the germ line would destroy the species, and high rates of mutation in the soma would destroy the individual. Living cells require the correct functioning of thousands of genes, each of which could be damaged by a mutation at many sites in its protein-coding sequence or in flanking sequences that govern its expression or the processing of its messenger RNA (mRNA).

If progeny are to have a good chance at survival, DNA sequences must be passed on largely unchanged in the germ line. Likewise, the specialized cells of the adult organism could not carry out their mission if mutation rates in the soma were high. Cancer, for example, arises from cells that have lost the capacity to grow and divide in a controlled manner as a consequence of damage to genes that govern the cell cycle. If the rates of mutation in the soma were high, the incidence of cancer would be catastrophic and unsustainable.

At the same time, if the genetic material were perpetuated with perfect fidelity, the genetic variation needed to drive evolution would be lacking, and new species, including humans, would not have arisen. Thus, life and biodiversity depend on a happy balance between mutation and its repair. In this chapter, we consider the causes of mutation and the systems that are responsible for reversing or correcting, and thereby minimizing, damage to the genetic material.

Two important sources of mutations are inaccuracy in DNA replication and chemical damage to the genetic material. Replication errors arise from tautomerization, which, as we have seen in Chapter 8, imposes an upper limit on the accuracy of base pairing during DNA replication. The enzymatic machinery for replicating DNA attempts to cope with the misincorporation of incorrect nucleotides through a proofreading mechanism, but some errors escape detection. In addition, DNA is a complex and fragile organic molecule of finite chemical stability. Not only does it suffer spontaneous damage such as the loss of bases, but it is also assaulted by natural and unnatural chemicals and radiation that break its backbone and chemically alter its bases. Simply put, errors in replication and damage to the genetic material from the environment are unavoidable. A third important source of mutation is the class of insertions generated by DNA elements known as transposons. Transposition is a major topic in its own right, which we shall consider in detail in Chapter 11.

Errors in replication and damage to DNA have two consequences. One is, of course, permanent changes to the DNA (mutations), which can alter the coding sequence of a gene or its regulatory sequences. The second consequence is that some chemical alterations to the DNA prevent its use as a template for replication and transcription. The effect of mutations generally become manifest only in the progeny of the cell in which the sequence alteration has occurred, but lesions that impede replication or transcription can have immediate effects on cell function and survival.

The challenge for the cell is twofold. First, it must scan the genome to detect errors in synthesis and damage to the DNA. Second, it must mend the lesions and do so in a way that, if possible, restores the original DNA sequence. Here, we discuss errors that are generated during replication, lesions that arise from spontaneous damage to DNA, and damage that is wrought by chemical agents and radiation. In each case, we consider how the alteration to the genetic material is detected and how it is properly repaired. Among the questions we address are the following: how is the DNA mended rapidly enough to prevent errors from becoming set in the genetic material as mutations? How does the cell distinguish the parental strand from the daughter strand in repairing replication errors? How does the cell restore the proper DNA sequence when, because of a break or severe lesion, the original sequence can no longer be read? How does the cell cope with lesions that block replication? The answers to these questions depend on the kind of error or lesion that needs to be repaired.

We begin by considering errors that occur during replication and how they are repaired. We then consider various kinds of lesions that arise spontaneously or from environmental assaults before turning to the multiple repair mechanisms that allow the cell to mend this damage. We will see that multiple overlapping systems enable the cell to cope with a wide range of insults to DNA, underscoring the investment that living organisms make in the preservation of the genetic material.

REPLICATION ERRORS AND THEIR REPAIR

The Nature of Mutations

Mutations include almost every conceivable change in DNA sequence. The simplest mutations are switches of one base for another. There are two kinds: **transitions**, which are pyrimidine-to-pyrimidine and purine-to-purine substitutions, such as T to C and A to G; and **transversions**, which are pyrimidine-to-purine and purine-to-pyrimidine substitutions, such as T to G or A and A to C or T (Fig. 9-1). Other simple mutations are insertions or deletions of a nucleotide or a small number of nucleotides. Mutations that alter a single nucleotide are called **point mutations**.

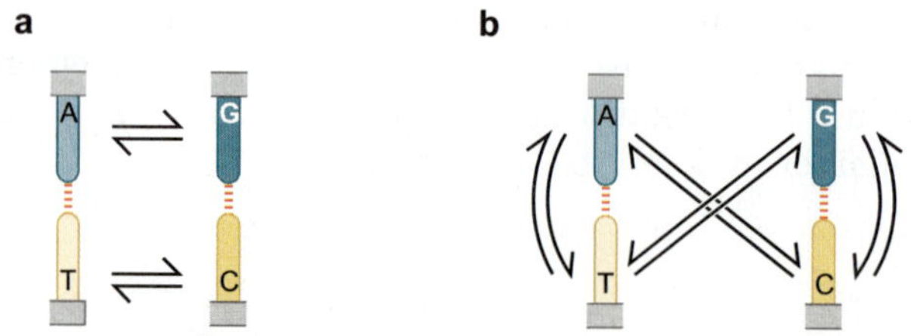

FIGURE 9-1 Base-change substitutions. (a) Transitions. (b) Transversions.

Other kinds of mutations cause more drastic changes in DNA, such as extensive insertions and deletions and gross rearrangements of chromosome structure. Such changes might be caused, for example, by the insertion of a transposon, which typically places many thousands of nucleotides of foreign DNA in the coding or regulatory sequences of a gene (see Chapter 11) or by the aberrant actions of cellular recombination processes. The overall rate at which new mutations arise spontaneously at any given site on the chromosome ranges from about 10^{-6} to 10^{-11} per round of DNA replication, with some sites on the chromosome being "hot spots" where mutations arise at high frequency and other sites undergoing alterations at a comparatively low frequency.

One kind of sequence that is particularly prone to mutation merits special comment because of its importance in human genetics and disease. These mutation-prone sequences are repeats of simple di-, tri-, or tetranucleotide sequences, which are known as **DNA microsatellites.** One well-known example involves repeats of the dinucleotide sequence CA. Stretches of CA repeats are found at many widely scattered sites in the chromosomes of humans and some other eukaryotes. The replication machinery has difficulty copying such repeats accurately, frequently undergoing "slippage." This slippage increases or reduces the number of copies of the repeated sequence. As a result, the CA repeat length at a particular site on the chromosome is often highly polymorphic in the population. This polymorphism provides a convenient physical marker for mapping inherited mutations, such as mutations that increase the propensity to certain diseases in humans (see Box 9-1, Expansion of Triple Repeats Causes Disease).

Some Replication Errors Escape Proofreading

As we have seen, the replication machinery achieves a remarkably high degree of accuracy using a proofreading mechanism, the $3' \rightarrow 5'$ exonuclease component of the replisome, which removes wrongly incorporated nucleotides (as we discussed in Chapter 8). Proofreading

MEDICAL CONNECTIONS

Box 9-1 Expansion of Triple Repeats Causes Disease

Another well-known example of error-prone sequences is repeats of the triplet nucleotide sequences CGG and CAG in certain genes. In humans, such triplet repeats are often found to undergo expansion from one generation to the next, resulting in diseases that are progressively more severe in the children and grandchildren of afflicted individuals. Examples of diseases that are caused by triplet expansion are adult muscular (myotonic) dystrophy; fragile X syndrome, which causes mental retardation; and Huntington's disease, which causes neurodegeneration. CAG is the codon for glutamine, and its expansion in the coding sequence for the huntingtin protein results in an extended stretch of glutamine residues in the mutant protein in patients with Huntington's disease. Recent research indicates that this polyglutamine stretch interferes with the normal interaction between a glutamine-rich patch in a transcription factor called Sp1 and a corresponding glutamine-rich patch in "TAFII130," a subunit of a component of the transcription machinery called TFIID (see Chapter 12). This interference impairs transcription in neurons of the brain, including the transcription of the gene for the receptor of a neurotransmitter. Similar polyglutamine stretches from CAG expansions in other genes may also exert their effects by disrupting interactions between transcription factors and TAFII130.

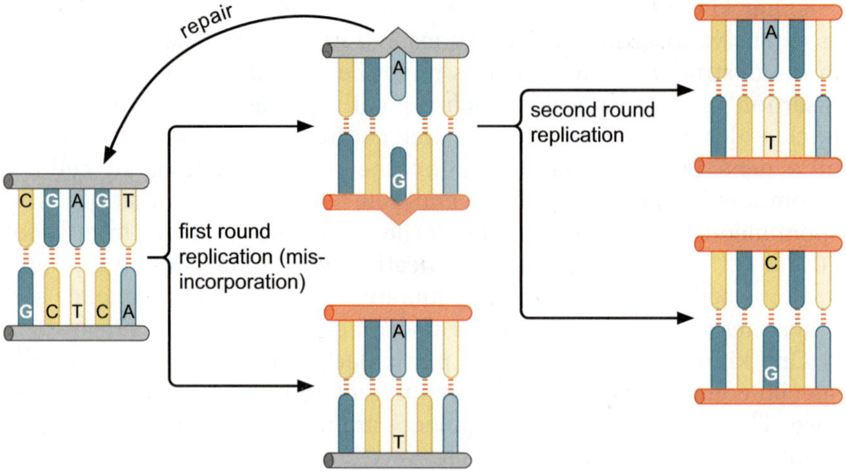

FIGURE 9-2 A mutation can be permanently incorporated by replication. A mutation may be introduced by misincorporation of a base in the first round of replication. In the second round of replication, the mutation becomes permanently incorporated in the DNA sequence.

improves the fidelity of DNA replication by a factor of about 100. The proofreading exonuclease is not, however, foolproof. Some misincorporated nucleotides escape detection and become a mismatch between the newly synthesized strand and the template strand. Three different nucleotides can be misincorporated opposite each of the four kinds of nucleotides in the template strand (e.g., T, G, or C opposite a T in the template) for a total of 12 possible mismatches (T:T, T:G, T:C, and so forth). If the misincorporated nucleotide is not subsequently detected and replaced, the sequence change will become permanent in the genome: during a second round of replication, the misincorporated nucleotide, now part of the template strand, will direct the incorporation of its complementary nucleotide into the newly synthesized strand (Fig. 9-2). At this point, the mismatch will no longer exist; instead, it will have resulted in a permanent change (a mutation) in the DNA sequence.

Mismatch Repair Removes Errors That Escape Proofreading

Fortunately, a mechanism exists for detecting mismatches and repairing them. Final responsibility for the fidelity of DNA replication rests with this **mismatch repair system**, which increases the accuracy of DNA synthesis by an additional two to three orders of magnitude. The mismatch repair system faces two challenges. First, it must scan the genome for mismatches. Because mismatches are transient (they are eliminated following a second round of replication when they result in mutations), the mismatch repair system must rapidly find and repair mismatches. Second, the system must correct the mismatch accurately; that is, it must replace the misincorporated nucleotide in the newly synthesized strand and not the correct nucleotide in the parental strand.

WEB STRUCTURAL TUTORIAL

In *Escherichia coli*, mismatches are detected by a dimer of the mismatch repair protein **MutS** (Fig. 9-3; see Structural Tutorial 9-1). MutS scans the DNA, recognizing mismatches from the distortion they cause in the DNA backbone. MutS embraces the mismatch-containing DNA, inducing a pronounced kink in the DNA and a conformational change in MutS itself (Fig. 9-4). A key to the specificity of MutS is that DNA containing a mismatch is much more readily distorted than properly base-paired DNA. MutS has an ATPase activity that is required for mismatch repair, but its precise role in repair is not understood. The complex of MutS and the mismatch-containing DNA recruits **MutL**, a second pro-

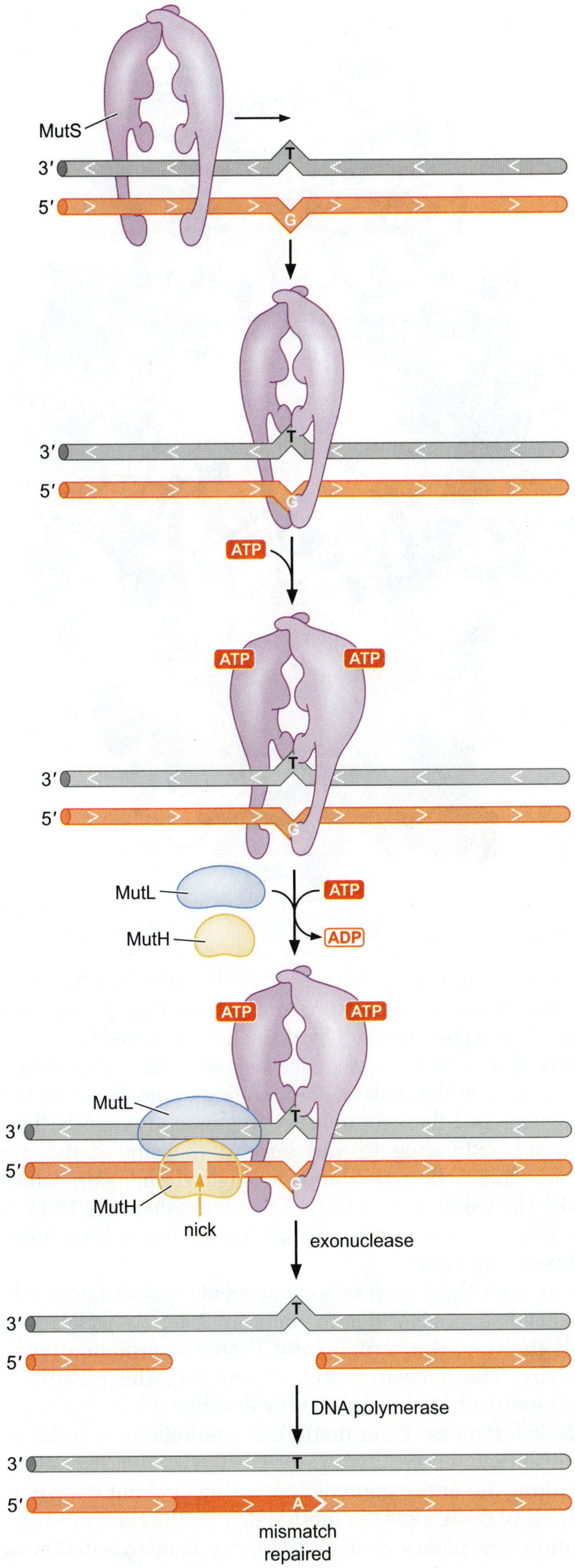

FIGURE 9-3 Mismatch repair pathway for the repair of replication errors. MutS embraces mismatch-containing DNA, inducing a kink (not shown, but see Fig. 9-4). In subsequent steps, MutS recruits MutL and MutH and the ATPase activity of MutS catalyzes the hydrolysis of ATP. MutH is an endonuclease that creates a nick in the DNA near the site of the mismatch. Next, an exonuclease digests the nicked strand moving toward and beyond the mismatch. Finally, the resulting single-strand gap is filled in by DNA polymerase, eliminating the mismatch. (Adapted, with permission, from Junop M.S. et al. 2001. *Mol. Cell* 7: 1–12, Fig. 6b. © Elsevier.)

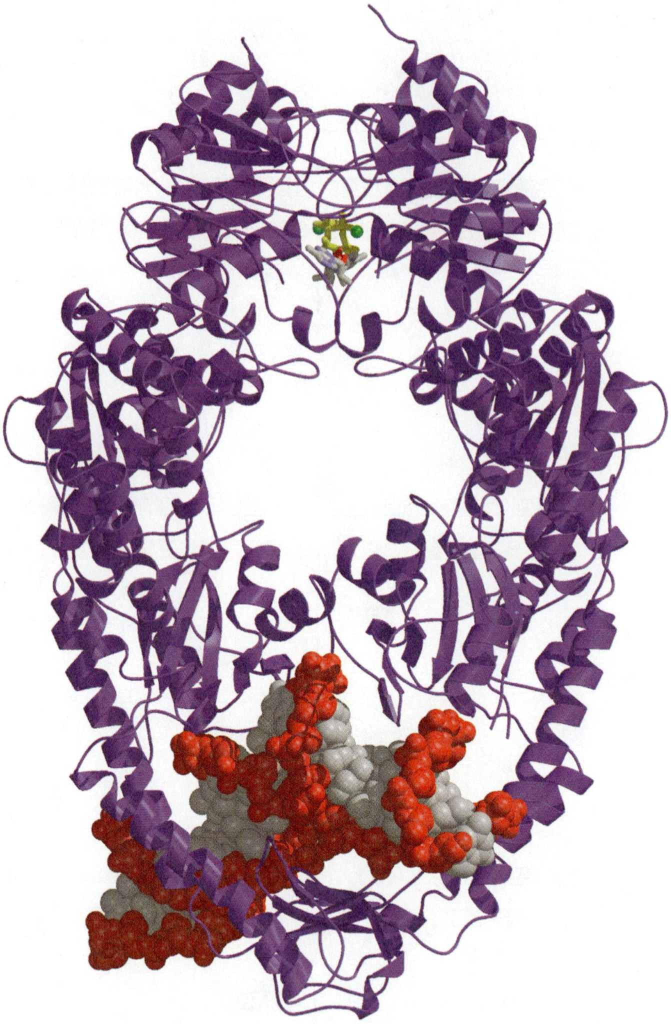

FIGURE 9-4 Crystal structure of the MutS–DNA complex. Notice the kink in the DNA, present near the bottom of the structure. In addition, near the top of the structure of the enzyme is ATP, shown in yellow, green, and red. The DNA is depicted as a space-filling representation with the backbone in red and bases in gray. (Junop M.S. et al. 2002. *Mol. Cell* 7: 1–12.) Image prepared with MolScript, BobScript, and Raster 3D.

tein component of the repair system. MutL, in turn, activates **MutH**, an enzyme that causes an incision or nick on one strand near the site of the mismatch. Nicking is followed by the action of a specific helicase (UvrD) and one of three exonucleases (see below). The helicase unwinds the DNA, starting from the incision and moving in the direction of the site of the mismatch, and the exonuclease progressively digests the displaced single strand, extending to and beyond the site of the mismatched nucleotide. This action produces a single-strand gap, which is then filled in by DNA polymerase III (Pol III) and sealed with DNA ligase. The overall effect is to remove the mismatch and replace it with the correctly base-paired nucleotide.

But how does the *E. coli* mismatch repair system know which of the two mismatched nucleotides to replace? If repair occurred randomly, then half the time the error would become permanently established in the DNA. The answer is that *E. coli* tags the parental strand by transient hemimethylation as we now describe.

The *E. coli* enzyme **Dam methylase** methylates A residues on both strands of the sequence 5′-GATC-3′. The GATC sequence is widely distributed along the entire genome (occurring at about once every 256 bp [4^4]), and all of these sites are methylated by the Dam methylase. When a replication fork passes through DNA that is methylated at GATC sites on both strands (fully methylated DNA), the resulting daughter DNA duplexes will be hemimethylated (i.e., methylated on only the parental

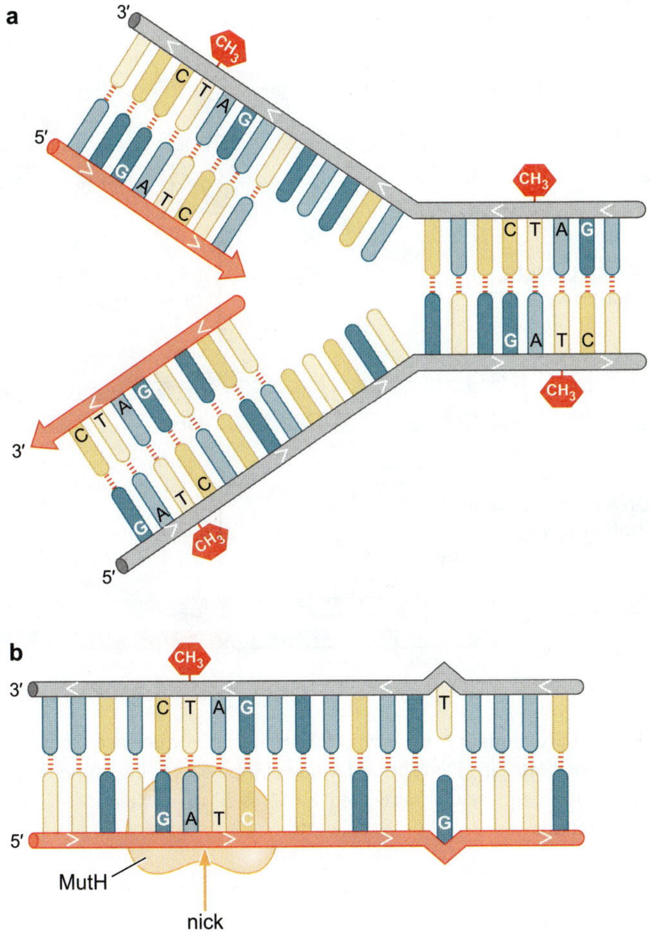

FIGURE 9-5 **Dam methylation at replication fork.** (a) Replication generates hemimethylated DNA in *E. coli*. (b) MutH makes incision in unmethylated daughter strand.

strand). Thus, for a few minutes, until the Dam methylase catches up and methylates the newly synthesized strand, daughter DNA duplexes will be methylated only on the strand that served as a template (Fig. 9-5a). Thus, the newly synthesized strand is marked (it lacks a methyl group) and hence can be recognized as the strand for repair.

The MutH protein binds at such hemimethylated sites, but its endonuclease activity is normally latent. Only when it is contacted by MutL and MutS located at a nearby mismatch (which is likely to be within a distance of a few hundred base pairs) does MutH become activated as we described above. Just how this interaction takes place over distances of up to several hundred base pairs is uncertain, but recent evidence indicates that the MutS–MutL complex leaves the mismatch and moves along the DNA contour to reach MutH at the site of hemimethylation. Once activated, MutH selectively nicks the unmethylated strand, so only newly synthesized DNA in the vicinity of the mismatch is removed and replaced (Fig. 9-5b). Methylation is therefore a "memory" device that enables the *E. coli* repair system to retrieve the correct sequence from the parental strand if an error has been made during replication.

Different exonucleases are used to remove single-stranded DNA between the nick created by MutH and the mismatch, depending on whether MutH cuts the DNA on the 5′ or the 3′ side of the misincorporated nucleotide. If the DNA is cleaved on the 5′ side of the mismatch, then exonuclease VII or RecJ, which degrades DNA in a 5′→3′ direction, removes the stretch of DNA from the MutH-induced cut through the misincorporated nucleotide. Conversely, if the nick is on the 3′ side

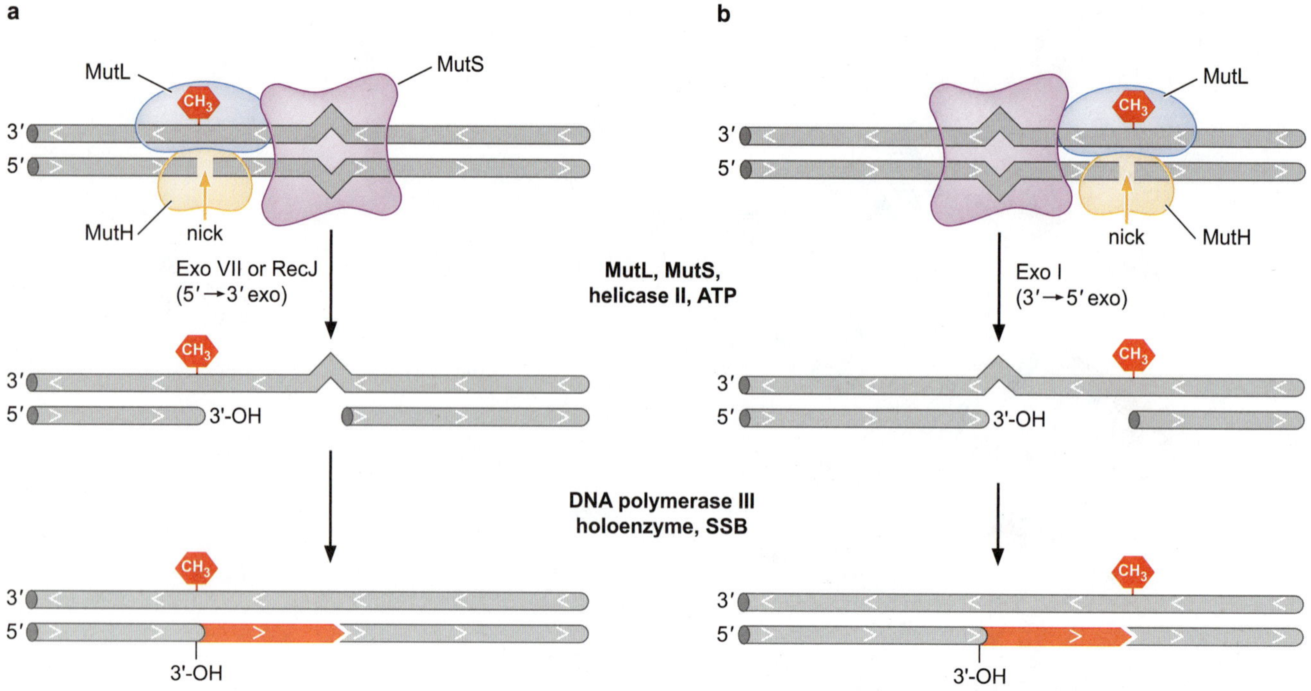

FIGURE 9-6 **Directionality in mismatch repair: Exonuclease removal of mismatched DNA.** For simplicity, DNA-bound MutH is shown as being immediately adjacent to MutS at the mismatch. (a) Unmethylated GATC is 5′ of mutation. (b) Unmethylated GATC is 3′ of mutation.

of the mismatch, then the DNA is removed by exonuclease I, which degrades DNA in a 3′→5′ direction. As we have seen, after removal of the mismatched base, DNA Pol III fills in the missing sequence (Fig. 9-6).

Eukaryotic cells also repair mismatches and do so using homologs to MutS (called MSH proteins for MutS homologs) and MutL (called MLH and PMS). Indeed, eukaryotes have multiple MutS-like proteins with different specificities. For example, one is specific for simple mismatches, whereas another recognizes small insertions or deletions resulting from "slippage" during DNA replication. Dramatic evidence that mismatch repair has a critical role in higher organisms came from the discovery that a genetic predisposition to colon cancer (hereditary nonpolyposis colorectal cancer) is due to a mutation in the genes for human homologs of MutS (specifically the MSH2 homolog) and MutL.

Even though eukaryotic cells have mismatch repair systems, they lack MutH and *E. coli*'s clever trick of using hemimethylation to tag the parental strand. (Indeed, most bacteria lack Dam methylase and are also unable to use hemimethylation to mark the newly synthesized strand.) How then does the mismatch repair system know which of the two strands to correct? Lagging-strand synthesis, as we saw in Chapter 8, takes place discontinuously with the formation of Okazaki fragments that are joined to previously synthesized DNA by DNA ligase. Prior to the ligation step, the Okazaki fragment is separated from previously synthesized DNA by a nick, which can be thought of as being equivalent to the nick created in *E. coli* by MutH on the newly synthesized strand. Indeed, extracts of eukaryotic cells will repair mismatches in artificial templates that contain a nick and do so selectively on the strand that carries the nick. Recent results indicate that human homologs of MutS (MSH) interact with the sliding-clamp component of the replisome (PCNA, which we discussed in Chapter 8), and would thereby be recruited to the site of discontinuous DNA synthesis on the lagging strand. Interaction with the sliding clamp could also recruit mismatch repair proteins to the 3′ (growing) end of the leading strand.

DNA DAMAGE

DNA Undergoes Damage Spontaneously from Hydrolysis and Deamination

Mutations arise not only from errors in replication, but also from damage to the DNA. Some damage is caused, as we shall see, by environmental factors, such as radiation and so-called **mutagens**, which are chemical agents that increase the rate of mutation (see Box 9-2, The Ames Test). But DNA also undergoes spontaneous damage from the action of water. (This is ironic since the proper structure of the double helix depends on an aqueous environment.)

The most frequent and important kind of hydrolytic damage is deamination of the base cytosine (Fig. 9-7a). Under normal physiological conditions, cytosine undergoes spontaneous deamination, thereby generating the unnatural (in DNA) base uracil. Uracil preferentially pairs with adenine and so introduces that base in the opposite strand upon replication, rather than the G that would have been directed by C. Adenine and guanine are also subject to spontaneous deamination. Deamination converts adenine to hypoxanthine, which hydrogen-bonds to cytosine rather than to thymine; guanine is converted to xanthine, which continues to pair with cytosine, although with only two hydrogen bonds. DNA also undergoes **depurination** by spontaneous hydrolysis of the *N*-glycosyl linkage, and this produces an abasic site (i.e., deoxyribose lacking a base) in the DNA (Fig. 9-7b).

Notice that, in contrast to the replication errors discussed above, all of these hydrolytic reactions result in alterations to the DNA that are unnatural. Apurinic sites are, of course, unnatural and each of the deamination reactions generates an unnatural base. This situation allows changes to be recognized by the repair systems described below. This situation also suggests an explanation for why DNA has thymine instead of uracil. If DNA naturally contained uracil instead of thymine, then deamination of cytosine would generate a natural base, which the repair systems could not easily recognize.

The hazard of having deamination generate a naturally occurring base is illustrated by the problem caused by the presence of 5-methylcytosine. Vertebrate DNA frequently contains 5-methylcytosine in place of cytosine as a result of the action of methyltransferases. This modified base has a role in the transcriptional silencing (see Chapter 17). Deamination of 5-methylcytosine generates thymine (Fig. 9-7c), which obviously will not be recognized as an abnormal base and, following a round of DNA replication, can become fixed as a C-to-T transition. Indeed, methylated Cs are hot spots for spontaneous mutations in vertebrate DNA.

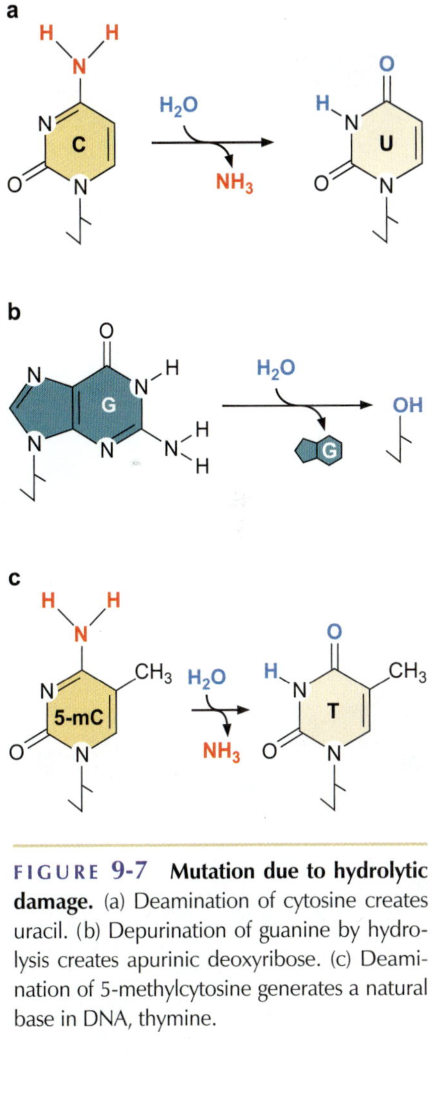

FIGURE 9-7 Mutation due to hydrolytic damage. (a) Deamination of cytosine creates uracil. (b) Depurination of guanine by hydrolysis creates apurinic deoxyribose. (c) Deamination of 5-methylcytosine generates a natural base in DNA, thymine.

DNA Is Damaged by Alkylation, Oxidation, and Radiation

DNA is vulnerable to damage from alkylation, oxidation, and radiation. In alkylation, methyl or ethyl groups are transferred to reactive sites on

MEDICAL CONNECTIONS

Box 9-2 The Ames Test

Determining the potential carcinogenic effects of chemicals in animals is time-consuming and expensive. However, because most tumor-causing agents are mutagens, the potential carcinogenic effects of chemicals can be conveniently assessed from their capacity to cause mutations. Bruce Ames of the University of California at Berkeley devised a simple test for the potential carcinogenic effects of chemicals based on their capacity to cause mutations in the bacterium *Salmonella typhimurium*. The Ames test uses a strain of *S. typhimurium* that is mutant for the operon responsible for the biosynthesis of the amino acid histidine. For example, the mutant operon might contain a missense or a frameshift mutation in one of the genes for histidine biosynthesis. As a consequence, cells of the mutant fail to grow and form colonies on solid medium lacking histidine (Box 9-2 Fig. 1). However, if the mutant cells are treated with a chemical that is mutagenic (and hence potentially carcinogenic), the chemical will cause the missense or frameshift mutation (depending on the nature of the mutagen) to revert in a small number of the mutant cells. This reversal restores the capacity of the cells to grow and form colonies on solid medium lacking histidine. The more potent the mutagen, the greater the number of colonies. Some chemicals that cause cancers are not mutagenic to begin with, but rather are converted into mutagens by the liver, which metabolizes foreign substances. To identify chemicals that are converted into mutagens in the liver, the Ames test treats potential mutagens with a mixture of liver enzymes. Chemicals that are found to be mutagenic in the Ames test can then be tested for their potential carcinogenic effects in animals.

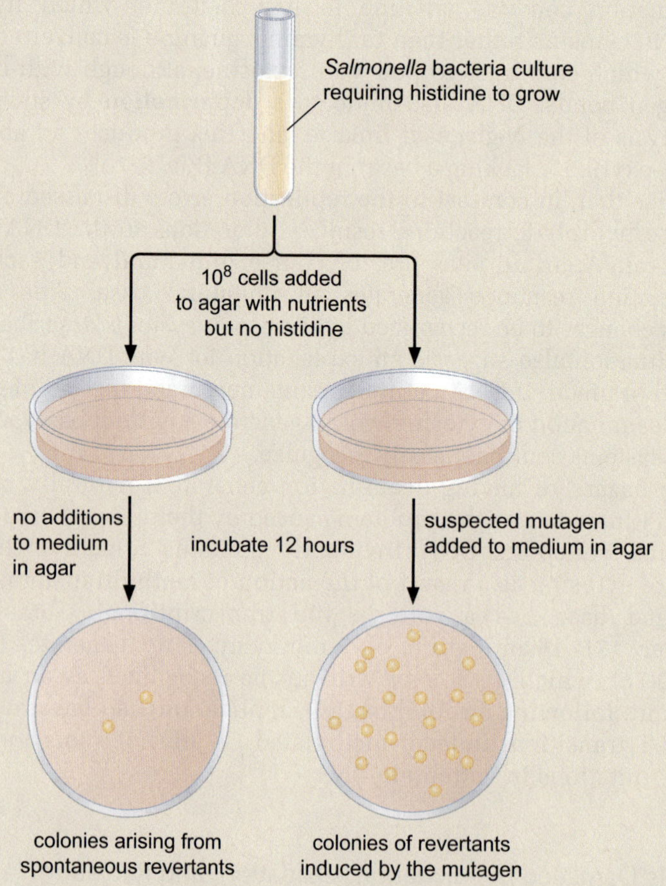

BOX 9-2 FIGURE 1 The Ames test.

the bases and to phosphates in the DNA backbone. Alkylating chemicals include nitrosamines and the very potent laboratory mutagen *N*-methyl-N^1-nitro-*N*-nitrosoguanidine. One of the most vulnerable sites of alkylation is the oxygen of carbon atom 6 of guanine (Fig. 9-8). The product of this methylation, O^6-methylguanine, often mispairs with thymine, resulting in the change of a G:C base pair into an A:T base pair when the damaged DNA is replicated.

DNA is also subject to attack from reactive oxygen species (e.g., O_2^-, H_2O_2, and OH•). These potent oxidizing agents are generated by ionizing radiation and by chemical agents that generate free radicals. Oxidation of guanine, for example, generates 7,8-dihydro-8-oxoguanine or oxoG. The oxoG adduct is highly mutagenic because it can base-pair with adenine as well as with cytosine. If it base-pairs with adenine during replication, it gives rise to a G:C to T:A transversion, which is one of the most common mutations found in human cancers. Thus, perhaps the carcinogenic effects of ionizing radiation and oxidizing agents are partly caused by free radicals that convert guanine to oxoG.

Yet another type of damage to bases is caused by ultraviolet light. Radiation with a wavelength of ~260 nm is strongly absorbed by the bases, one consequence of which is the photochemical fusion of two pyrimidines that occupy adjacent positions on the same polynucleotide chain. In the case of two thymines, the fusion is called a **thymine dimer** (Fig. 9-9), which comprises a **cyclobutane** ring generated by links between carbon atoms 5 and 6 of adjacent thymines. In the case of a thymine adjacent to a cytosine, the resulting fusion is a thymine-cytosine adduct in which the thymine is linked via its carbon atom 6 to the carbon atom 4 of cytosine. These linked bases are incapable of base pairing and cause the DNA polymerase to stop during replication.

Finally, γ radiation and X-rays (ionizing radiation) are particularly hazardous because they cause double-strand breaks in the DNA, which are difficult to repair. Ionizing radiation can directly attack (ionize) the deoxyribose in the DNA backbone. Alternatively, this radiation can attack indirectly by generating reactive oxygen species (described above), which in turn react with the deoxyribose subunits. Because cells require intact chromosomes to replicate their DNA, ionizing radiation is used therapeutically to kill rapidly proliferating cells in cancer treatment. Certain anticancer drugs, such as bleomycin, also cause breaks in DNA. Ionizing radiation and agents like bleomycin that cause DNA to break are said to be **clastogenic** (from the Greek *klastos*, which means "broken").

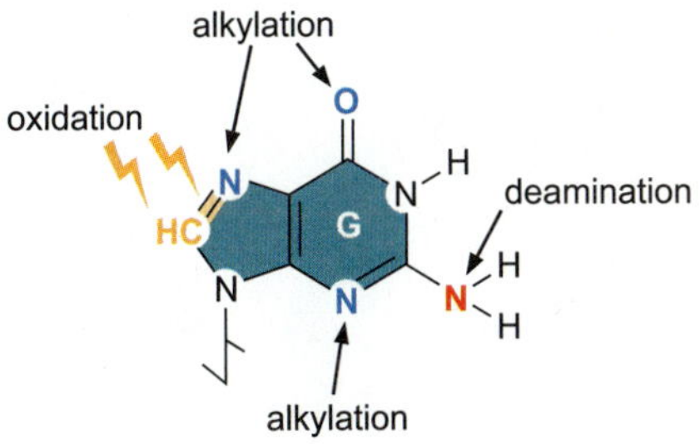

FIGURE 9-8 G modification. The figure shows specific sites on guanine that are vulnerable to damage by chemical treatment, such as alkylation or oxidation, and by radiation. The products of these modifications are often highly mutagenic.

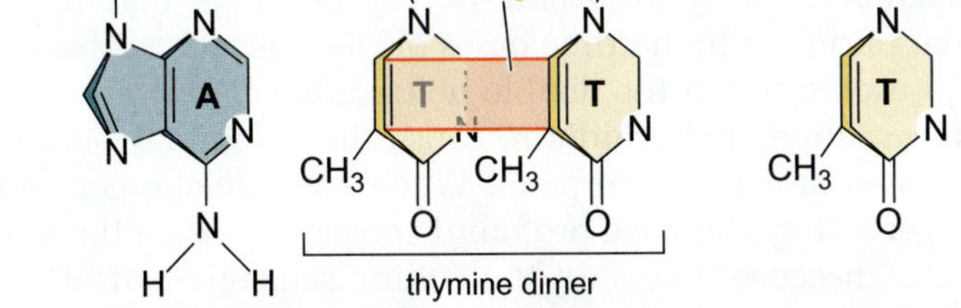

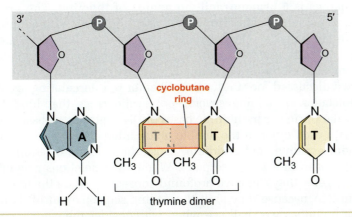

FIGURE 9-9 Thymine dimer. Ultraviolet light induces the formation of a cyclobutane ring between adjacent thymines.

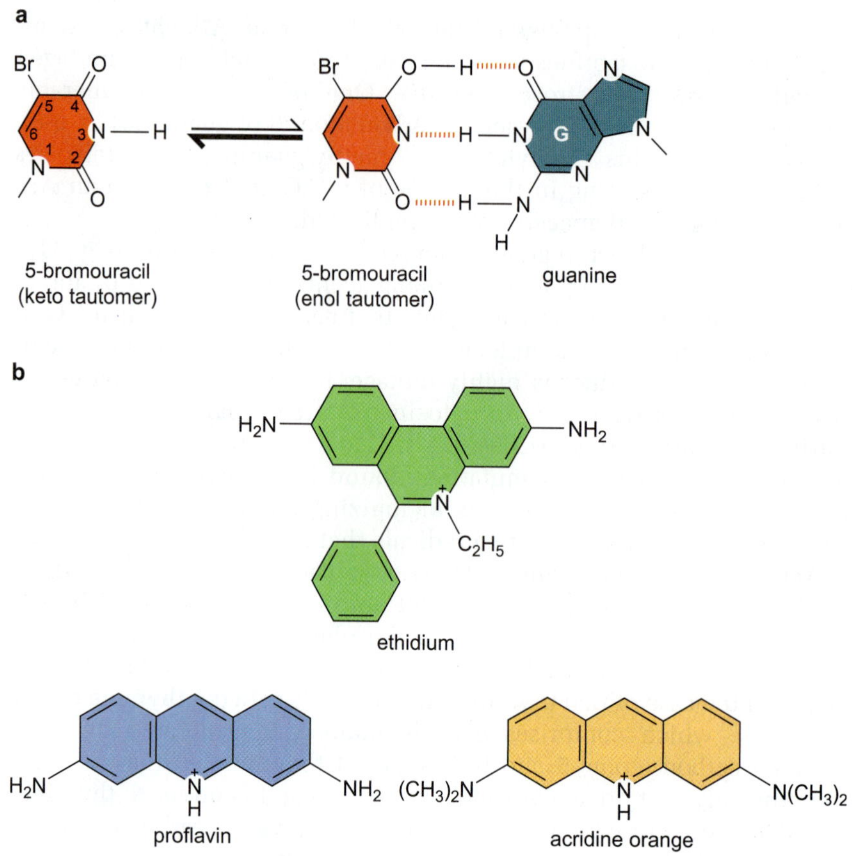

FIGURE 9-10 Base analogs and intercalating agents that cause mutations in DNA. (a) Base analog of thymine, 5-bromouracil, can mispair with guanine. (b) Intercalating agents.

Mutations Are Also Caused by Base Analogs and Intercalating Agents

Mutations are also caused by compounds that substitute for normal bases (**base analogs**) or slip between the bases (**intercalating agents**) to cause errors in replication (Fig. 9-10). Base analogs are structurally similar to proper bases but differ in ways that make them treacherous to the cell. Thus, base analogs are similar enough to the proper bases to get taken up by cells, converted into nucleoside triphosphates, and incorporated into DNA during replication. But, because of the structural differences between these analogs and the proper bases, the analogs base-pair inaccurately, leading to frequent mistakes during the replication process. One of the most mutagenic base analogs is **5-bromouracil**, an analog of thymine. The presence of the bromo substituent allows the base to mispair with guanine via the enol tautomer (see Fig. 9-10a). As we saw in Chapter 6, the keto tautomer is strongly favored over the enol tautomer, but more so for thymine than for 5-bromouracil.

As we discussed for ethidium in Chapter 6, **intercalating agents** are flat molecules containing several polycyclic rings that bind to the equally flat purine or pyrimidine bases of DNA, just as the bases bind or stack with each other in the double helix. Intercalating agents, such as **proflavin**, **acridine**, and **ethidium**, cause the deletion or addition of a base pair or even a few base pairs. When such deletions or additions arise in a gene, they can have profound consequences on the translation of its mRNA because they shift the coding sequence out of its proper reading frame, as we shall see when we consider the genetic code in Chapter 15.

How do intercalating agents cause short insertions and deletions? One possibility in the case of insertions is that, by slipping between the bases in the template strand, these mutagens cause the DNA polymerase to insert an extra nucleotide opposite the intercalated molecule. (The intercalation of one of these structures approximately doubles the typical distance between two base pairs.) Conversely, in the case of deletions, the distortion to the template caused by the presence of an intercalated molecule might cause the polymerase to skip a nucleotide.

REPAIR OF DNA DAMAGE

As we have seen, damage to DNA can have two consequences. Some kinds of damage, such as thymine dimers or nicks and breaks in the DNA backbone, create impediments to replication or transcription. Other kinds of damage create altered bases that have no immediate structural consequence on replication but cause mispairing; these can result in a permanent alteration to the DNA sequence after replication. For example, the conversion of cytosine to uracil by deamination creates a U:G mismatch, which, after a round of replication, becomes a C:G to T:A transition mutation on one daughter chromosome. These considerations explain why cells have evolved elaborate mechanisms to identify and repair damage before it blocks replication or causes a mutation. Cells would not endure long without such mechanisms.

In this section, we consider the systems that repair damage to DNA (Table 9-1). In the most direct of these systems (representing true repair), a repair enzyme simply reverses (undoes) the damage. One more elaborate step involves **excision repair systems**, in which the damaged nucleotide is not repaired but removed from the DNA. In excision repair systems, the other, undamaged, strand serves as a template for reincorporation of the correct nucleotide by DNA polymerase. As we shall see, two kinds of excision repair systems exist, one involving the removal of only the damaged nucleotide and the other involving the removal of a short stretch of single-stranded DNA that contains the lesion.

Yet more elaborate is **recombinational repair**, which is employed when both strands are damaged as when the DNA is broken. In such situations, one strand cannot serve as a template for the repair of the other. Hence, in recombinational repair (known as **double-strand break repair**), sequence information is retrieved from a second undamaged copy of the chromosome. Finally, when progression of a replicating DNA polymerase is blocked by damaged bases, a special **translesion**

TABLE 9-1 DNA Repair Systems

Type	Damage	Enzyme
Mismatch repair	Replication errors	MutS, MutL, and MutH in *E. coli*; MSH, MLH, and PMS in humans
Photoreactivation	Pyrimidine dimers	DNA photolyase
Base excision repair	Damaged base	DNA glycosylase
Nucleotide excision repair	Pyrimidine dimer Bulky adduct on base	UvrA, UvrB, UvrC, and UvrD in *E. coli*; XPC, XPA, XPD, ERCCI-XPF, and XPG in humans
Double-strand break repair	Double-strand breaks	RecA and RecBCD in *E. coli*
Translesion DNA synthesis	Pyrimidine dimer or apurinic site	Y-family DNA polymerases, such as UmuC in *E. coli*

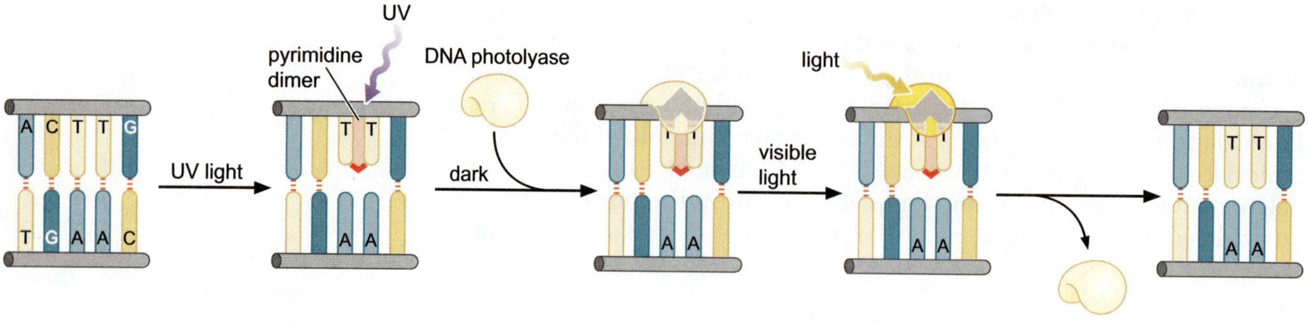

FIGURE 9-11 Photoreactivation. Ultraviolet irradiation causes formation of thymine dimers. Upon exposure to light, DNA photolyase breaks the ring formed between the dimers to restore the two thymine residues.

polymerase copies across the site of the damage in a manner that does not depend on base pairing between the template and newly synthesized DNA strands. This mechanism is a system of last resort because translesion synthesis is inevitably highly error-prone (mutagenic).

Direct Reversal of DNA Damage

An example of repair by simple reversal of damage is **photoreactivation.** Photoreactivation directly reverses the formation of pyrimidine dimers that result from ultraviolet irradiation. In photoreactivation, the enzyme DNA photolyase captures energy from light and uses it to break the covalent bonds linking adjacent pyrimidines (Fig. 9-11). In other words, the damaged bases are mended directly.

Another example of direct reversal is the removal of the methyl group from the methylated base O^6-methylguanine (see above). In this case, a methyltransferase removes the methyl group from the guanine residue by transferring it to one of its own cysteine residues (Fig. 9-12). This is very costly to the cell because the methyltransferase is not catalytic; having once accepted a methyl group, it cannot be used again.

Base Excision Repair Enzymes Remove Damaged Bases by a Base-Flipping Mechanism

The most prevalent way in which DNA is cleansed of damaged bases is by repair systems that remove and replace the altered bases. The two principal repair systems are **base excision repair** and **nucleotide excision repair.** In base excision repair, an enzyme called a **glycosylase** recognizes

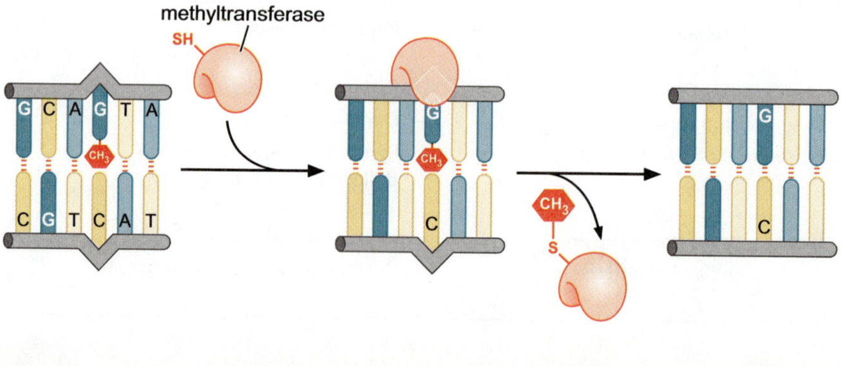

FIGURE 9-12 Methyl group removal. Methyltransferase catalyzes the transfer of the methyl group on O^6-methylguanine to a cysteine residue on the enzyme, thereby restoring the normal G in DNA.

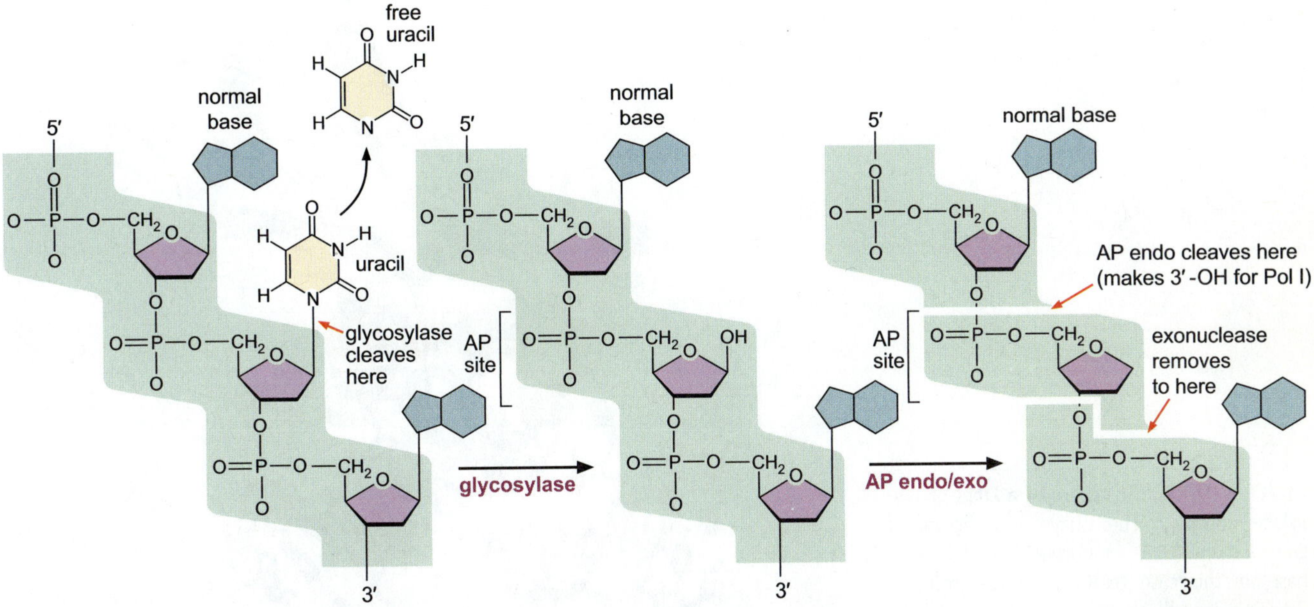

FIGURE 9-13 Base excision pathway: The uracil glycosylase reaction. Uracil glycosylase hydrolyzes the glycosidic bond to release uracil from the DNA backbone to leave an AP site (apurinic or, in this case, apyrimidinic site). AP endonuclease cuts the DNA backbone at the 5′ position of the AP site, leaving a 3′-OH; exonuclease cuts at the 3′ position of the AP site, leaving a 5′-phosphate. The resulting gap is filled in by DNA Pol I.

and removes the damaged base by hydrolyzing the glycosidic bond (Fig. 9-13). The resulting abasic sugar is removed from the DNA backbone in a further endonucleolytic step. Endonucleolytic cleavage also removes apurinic and apyrimidinic sugars that arise by spontaneous hydrolysis. After the damaged nucleotide has been entirely removed from the backbone, a repair DNA polymerase and DNA ligase restore an intact strand using the undamaged strand as a template.

DNA glycosylases are lesion-specific and cells have multiple DNA glycosylases with different specificities. Thus, a specific glycosylase recognizes uracil (generated as a consequence of deamination of cytosine), and another is responsible for removing oxoG (generated as a consequence of oxidation of guanine). A total of eight different DNA glycosylases have been identified in the nuclei of human cells.

Cleansing the genome of damaged bases is a formidable problem because each base is buried in the DNA helix. How do DNA glycosylases detect damaged bases while scanning the genome? Evidence indicates that these enzymes diffuse laterally along the minor groove of the DNA until a specific kind of lesion is detected. But how is the enzyme able to act on the base if it is buried in the helix? The answer to this riddle highlights the remarkable flexibility of DNA. X-ray crystallographic studies reveal that the damaged base is flipped out so that it projects away from the double helix, where it sits in the specificity pocket of the glycosylase (Fig. 9-14). Interestingly, the double helix is able to allow base flipping with only modest distortion to its structure and hence the energetic cost of base flipping may not be great (see Chapter 6 and Fig. 6-8). Nevertheless, it is unlikely that glycosylases flip out every base to check for abnormalities as they diffuse along DNA. Thus, the mechanism by which these enzymes scan for damaged bases remains mysterious.

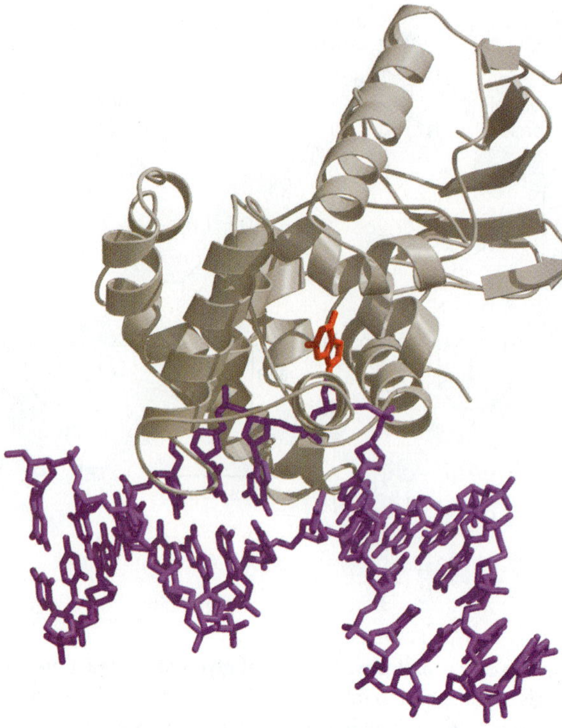

FIGURE 9-14 **Structure of a DNA-glycosylase complex.** The enzyme is shown in gray and the DNA in purple. The damaged base, in this case oxoG (shown in red), is flipped out of the helix and into the catalytic center of the enzyme. (Bruner S.D. et al. 2000. *Nature* 403: 859–866.) Image prepared with MolScript, BobScript, and Raster 3D.

What if a damaged base is not removed by base excision before DNA replication? Does this inevitably mean that the lesion will cause a mutation? In the case of oxoG, which has the tendency to mispair with A, a fail-safe system exists (Fig. 9-15). A dedicated glycosylase recognizes oxoG:A base pairs generated by misincorporation of an A opposite an oxoG on the template strand. In this case, however, the glycosylase removes the A. Thus, the repair enzyme recognizes an A opposite an oxoG as a mutation and removes the undamaged but incorrect base.

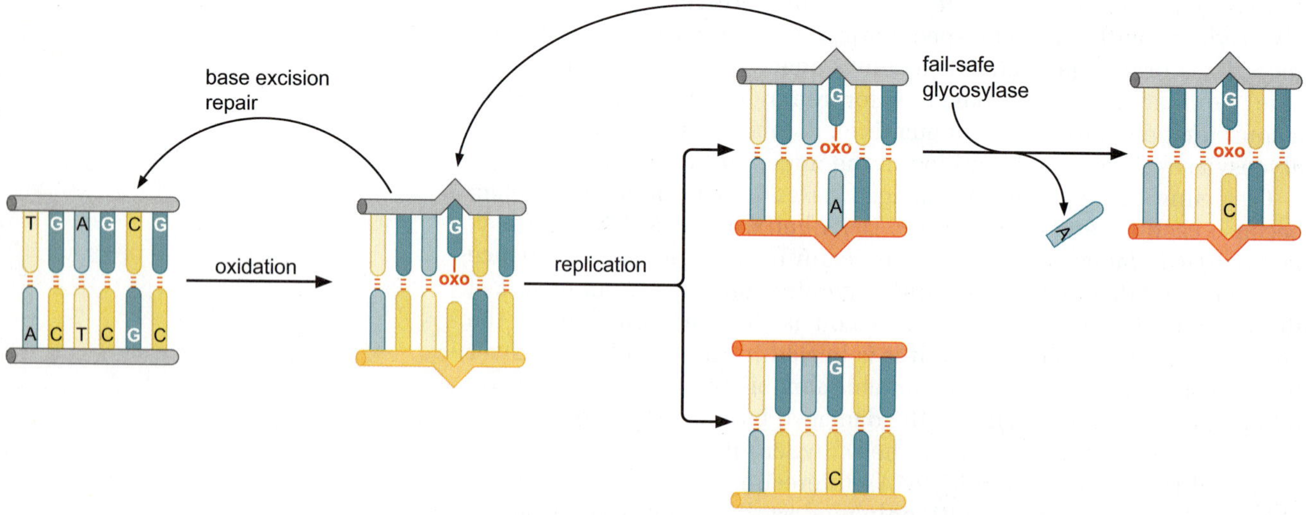

FIGURE 9-15 **oxoG:A repair.** Oxidation of guanine produces oxoG. The modified base can be repaired prior to replication by DNA glycosylase via the base excision pathway. If replication occurs before the oxoG is removed, resulting in the misincorporation of an A, then a fail-safe glycosylase can remove the A, allowing it to be replaced by a C. This provides a second opportunity for the DNA glycosylase to remove the modified base.

Another example of a fail-safe system is a glycosylase that removes T opposite a G. Such a T:G mismatch can arise, as we have seen, by spontaneous deamination of 5-methylcytosine, which occurs frequently in the DNA of vertebrates. Because both T and G are normal bases, how can the cell recognize which is the incorrect base? The glycosylase system assumes, so to speak, that the T in a T:G mismatch arose from deamination of 5-methylcytosine and selectively removes the T so that it can be replaced with a C.

Nucleotide Excision Repair Enzymes Cleave Damaged DNA on Either Side of the Lesion

Unlike base excision repair, the nucleotide excision repair enzymes do not recognize any particular lesion. Rather, this system works by recognizing distortions to the shape of the double helix, such as those caused by a thymine dimer or by the presence of a bulky chemical adduct on a base. Such distortions trigger a chain of events that lead to the removal of a short single-strand segment (or patch) that includes the lesion. This removal creates a single-strand gap in the DNA, which is filled in by DNA polymerase using the undamaged strand as a template and thereby restoring the original nucleotide sequence.

Nucleotide excision repair in *E. coli* is largely accomplished by four proteins: UvrA, UvrB, UvrC, and UvrD (Fig. 9-16). A complex of UvrA and UvrB scans the DNA, with UvrA being responsible for detecting distortions to the helix. Upon encountering a distortion, UvrA exits the complex and UvrB melts the DNA to create a single-stranded bubble around the lesion. Next, UvrB recruits UvrC, and UvrC creates two incisions: one located eight nucleotides away on the 5′ side of the lesion and the other four or five nucleotides away on the 3′ side of the lesion. These cleavages create a 12–13-residue-long single-stranded DNA segment, which is made accessible by the action of the DNA helicase UvrD. Finally, DNA Pol I and DNA ligase fill in the resulting gap.

The principle of nucleotide excision repair in higher cells is much the same as that in *E. coli*, but the machinery for detecting, excising, and repairing the damage is more complicated, involving 25 or more polypeptides. Among these is XPC, which is responsible for detecting distortions to the helix, a function attributed to UvrA in *E. coli*. As in *E. coli*, the DNA is opened to create a bubble around the lesion. Formation of the bubble involves the helicase activities of the proteins XPA and XPD (the equivalent to UvrB in *E. coli*) and the single-strand binding protein RPA. The bubble creates cleavage sites on the 5′ side of the lesion for a nuclease known as ERCC1-XPF and on the 3′ side for the nuclease XPG (representing the function of UvrC). In higher cells, the resulting single-stranded DNA segment is 24–32 nucleotides long. As in bacteria, the DNA segment is released to create a gap that is filled in by the action of DNA polymerase and ligase.

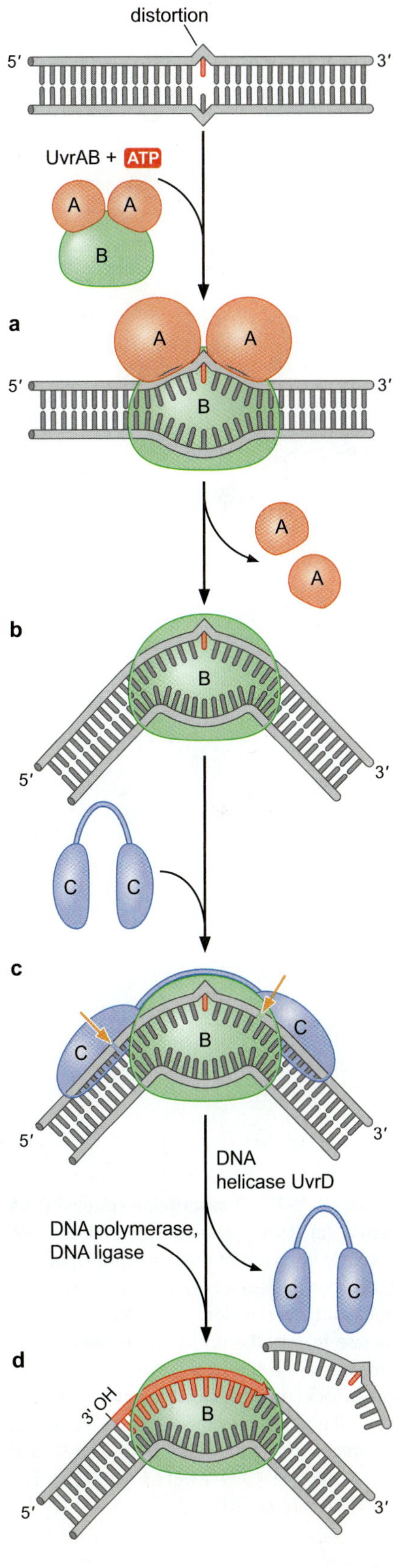

FIGURE 9-16 Nucleotide excision repair pathway. (a) UvrA and UvrB scan DNA to identify a distortion. (b) UvrA leaves the complex, and UvrB melts DNA locally around the distortion. (c) UvrC forms a complex with UvrB and creates nicks to the 5′ side of the lesion and to the 3′ side of the lesion. (d) DNA helicase UvrD releases the single-strand fragment from the duplex, and DNA Pol I and ligase repair and seal the gap. (a–d, Adapted, with permission, from Zou Y. and Van Houten B. 1999. *EMBO J.* 18: 4898, Fig. 7. © Macmillan.)

As their names imply, the UVR proteins are needed to mend damage from ultraviolet light; mutants of the *uvr* genes are sensitive to ultraviolet light and lack the capacity to remove thymine-thymine and thymine-cytosine adducts. In fact, these proteins broadly recognize and repair bulky adducts of many kinds. Nucleotide excision repair is important in humans, too. Humans can exhibit a genetic disease called xeroderma pigmentosum, which renders afflicted individuals highly sensitive to sunlight and results in skin lesions, including skin cancer. Seven genes (referred to as XP genes) have been identified in which mutations give rise to xeroderma pigmentosum. These genes correspond to proteins (such as XPA, XPC, XPD, XPF, and XPG, referred to above) in the human pathway for nucleotide excision repair, underscoring the importance of nucleotide excision repair in mending damage from ultraviolet light.

Not only is nucleotide excision repair capable of mending damage throughout the genome, but it is also capable of rescuing RNA polymerase, the progression of which has been arrested by the presence of a lesion in the transcribed (template) strand of a gene. This phenomenon, known as **transcription-coupled repair**, involves recruitment to the stalled RNA polymerase of nucleotide excision repair proteins (Fig. 9-17). The significance of transcription-coupled repair is that it

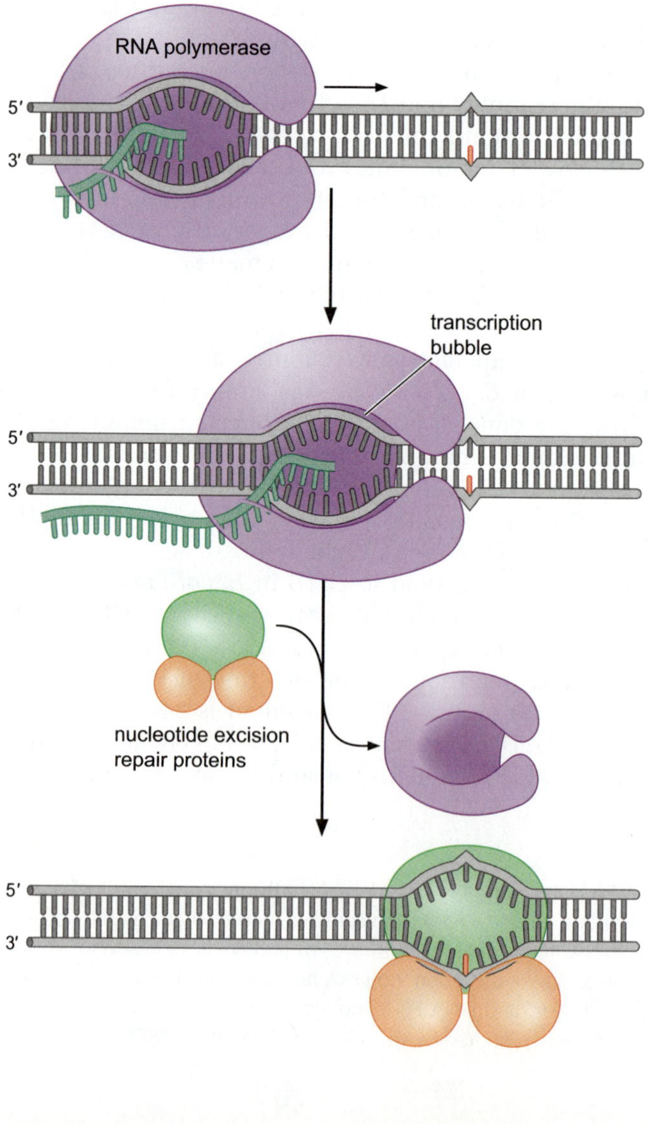

FIGURE 9-17 **Transcription-coupled DNA repair.** (a) RNA polymerase transcribes DNA normally upstream of the lesion. (b) Upon encountering the lesion in DNA, RNA polymerase stalls and transcription stops. (c) RNA polymerase recruits the nucleotide excision repair proteins to the site of the lesion, and then it either backs up or dissociates from the DNA to allow the repair proteins access to the lesion. (Adapted, with permission, from Zou Y. and Van Houten B. 1999. *EMBO J.* 18: 4898, Fig. 7. © Macmillan.)

focuses repair enzymes on DNA (genes) being actively transcribed. In effect, RNA polymerase serves as another damage-sensing protein in the cell. Central to transcription-coupled repair in eukaryotes is the general transcription factor TFIIH. As we shall see in Chapter 12, TFIIH unwinds the DNA template during the initiation of transcription. Subunits of TFIIH include the DNA helix-opening proteins XPA and XPD discussed above. Thus, TFIIH is responsible for two separate functions: its strand-separating helicases melt the DNA around a lesion during nucleotide excision repair (including transcription-coupled repair) and also help to open the DNA template during the process of gene transcription. Systems for coupling repair to transcription also exist in prokaryotes.

Recombination Repairs DNA Breaks by Retrieving Sequence Information from Undamaged DNA

Excision repair uses the undamaged DNA strand as a template to replace a damaged segment of DNA on the other strand. How do cells repair double-strand breaks in DNA in which both strands of the duplex are broken? This is accomplished by the **double-strand break (DSB) repair pathway**, which retrieves sequence information from the sister chromosome. Because of its central role in general homologous recombination as well as in repair, the DSB repair pathway is an important topic in its own right, which we shall consider in detail in Chapter 10.

DNA recombination also helps to repair errors in DNA replication. Consider a replication fork that encounters a lesion in DNA (such as a thymine dimer) that has not been corrected by nucleotide excision repair. The DNA polymerase will sometimes stall attempting to replicate over the lesion. Although the template strand cannot be used, the sequence information can be retrieved from the other daughter molecule of the replication fork by recombination (see Chapter 10). Once this recombinational repair is complete, the nucleotide excision system has another opportunity to repair the thymine dimer. Indeed, mutants defective in recombination are known to be sensitive to ultraviolet light. Consider also the situation in which the replication fork encounters a nick in the DNA template. Passage of the fork over the nick will create a DNA break, repair of which can only be accomplished by the DSB repair pathway. Although we generally consider recombination as an evolutionary device to explore new combinations of sequences, it may be that its original function was to repair damage in DNA.

DSBs in DNA Are Also Repaired by Direct Joining of Broken Ends

A DSB is the most cytotoxic of all kinds of DNA damage. If left unmended, a DNA break can have multiple deleterious consequences, such as blocking replication and causing chromosome loss, which result in cell death or neoplastic transformation. Cells typically have multiple overlapping pathways for coping with DNA damage. It should therefore come as no surprise that cells do not rely on the DSB repair pathway alone for mending DSBs. As we have seen and will consider in further detail in Chapter 10, the DSB repair pathway relies on DNA sequence information in a sister chromosome to repair broken DNA molecules. This is an effective strategy because the sister chromosome provides a template for the precise restoration of the original sequence across the site of the break. In yeast cells, DSB repair is the principal pathway by which breaks are mended. But what happens early in the cell cycle before two sister chromosomes have been generated by DNA replication? If a still-unreplicated chromosome suffers a break, then no sister chromosome is present to serve as a template in the DSB repair pathway. Under such conditions, an alternative system comes into play known as **nonhomologous end joining** or **NHEJ**. NHEJ is a backup system in yeast, but in higher cells it is the principal pathway by which breaks are repaired (see Box 9-3, Nonhomologous End Joining).

The machinery for carrying out NHEJ protects and processes the broken ends and then joins them together, as we shall explain. Because sequence information is lost from the broken ends, the original sequence across the break is not faithfully restored during NHEJ. Thus, NHEJ is mutagenic. Of course, the mutagenic consequences of NHEJ-mediated DNA end joining are far less hazardous to the cell than are the consequences leaving broken DNA unrepaired!

What is the mechanism that joins DNA ends together in NHEJ? As its name implies, NHEJ does not involve extensive strectches of homologous sequences. Instead, the two ends of the broken DNA are joined to each other by misalignment between single strands protruding from the broken ends. This misalignment is believed to occur by pairing between tiny stretches (as short as 1 bp) of complementary bases (serendipitous microhomologies). Single-strand tails are removed by nucleases, and gaps are filled in by DNA polymerase.

MEDICAL CONNECTIONS

Box 9-3 Nonhomologous End Joining

NHEJ can repair DSBs arising from exposure to exogenous agents, such as ionizing radiation, and from cell-intrinsic insults, such as failures in DNA replication. Remarkably, NHEJ is also employed in the entirely normal cell-intrinsic process of adaptive immunity. The immune system produces an enormously diverse group of antibody molecules, which are composed of so-called light and heavy polypeptide chains. The light and heavy chains are generated by a recombinational process that involves the joining, in a bewildering number of combinations, of a large repertoire of protein-coding DNA elements known as V and J segments (and, in the case of the antibody heavy chain, a D segment) for different parts of the polypeptides. As we discuss in Chapter 11, this process is known as **V(D)J recombination**. V(D)J recombination is initiated by the introduction of breaks in the DNA by a process that is specific to lymphocytes and involves an enzyme composed of the proteins RAG1 and RAG2. Once the breaks are created, the NHEJ pathway, which is not lymphocyte-specific, joins the ends together. In this case, however, the ends of the protein-coding segments are not rejoined to their original partners. Rather, the ends are joined to new partners to create the composite coding sequences for the heavy and light chains. NHEJ also participates in a second example of V(D)J recombination that governs the production of an additional category of immunological polypeptides called T-cell receptors as discussed in Chapter 11.

Underscoring the importance of NHEJ in human biology are rare inherited syndromes that are characterized by hypersensitivity to ionizing radiation and DNA-damaging agents and by immunodeficiency, which is attributed to defective V(D)J recombination. Revealingly, patients exhibiting this syndrome harbor mutations in the genes for the Artemis, Ligase IV, or Cernunnos-XLF members of the NHEJ pathway.

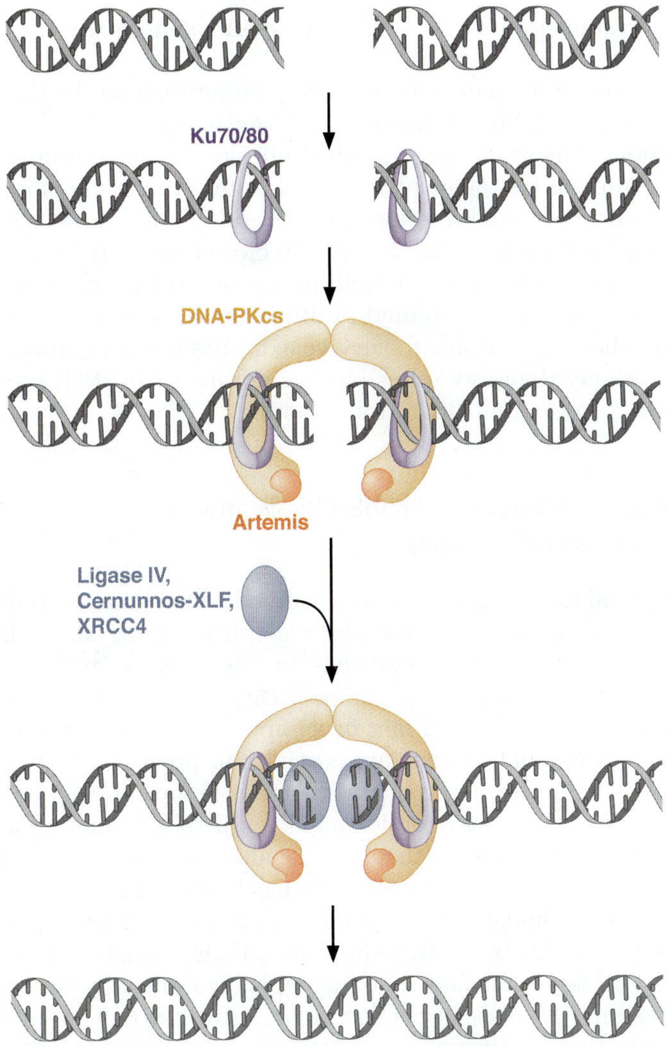

FIGURE 9-18 Mammalian pathway for NHEJ. A heterodimer of Ku70 and Ku80 binds to broken DNA ends and recruits the protein kinase DNA-PKcs. DNA-PKcs in turn recruits Artemis, an enzyme having exonuclease and endonuclease activities, which processes the broken ends. Finally, a complex of Ligase IV with XRCC4 and Cernunnos-XLF joins the broken ends to each other. (Adapted, with permission, from Sekiguchi J.M. and Ferguson D.O. 2006. *Cell* 124: 260–262. © Elsevier.)

A growing number of proteins that mediate NHEJ have been identified. To date, seven components of the NHEJ pathway have been discovered in mammalian cells. These proteins, which have formidable-sounding names, are Ku70, Ku80, DNA-PKcs, Artemis, XRCC4, Cernunnos-XLF, and DNA ligase IV (Fig. 9-18). Ku70 and Ku80 are the most fundamental components of NHEJ. They constitute a heterodimer that binds to the DNA ends and recruits DNA-PKcs, which is a protein kinase. DNA-PKcs, in turn, forms a complex with Artemis. Artemis is both a 5′ to 3′ exonuclease and a latent endonuclease that is activated by phosphorylation by DNA-PKcs. These nucleolytic activities process the broken ends and prepare them for ligation. Ligation is carried out by Ligase IV in a complex with XRCC4 and Cernunnos-XLF.

NHEJ is ubiquitous in eukaryotic organisms, but it occurs less frequently in bacteria. Nevertheless, a fascinating specialized example has been discovered in spores of the bacterium *Bacillus subtilis*. *B. subtilis* produces a Ku-like protein and a DNA ligase when it sporulates and packages the proteins into the mature spore. Ku and the DNA ligase, representing a simple, two-protein NHEJ system, repair DNA breaks when the spore germinates. Mutant spores lacking these proteins are highly susceptible to dry heat, a condition that is known to cause breaks in DNA. Upon germination, heated mutant spores are unable to resume growth because they are unable to rejoin the heat-induced breaks.

That germinating spores rely on NHEJ, rather than on the DSB repair pathway, to mend breaks makes good sense. Spores have only one chromosome. Therefore, they cannot rely on a sister chromosome to use as a template for repair of the break. Interestingly, the spore chromosome is tightly coiled into a remarkable doughnut-like structure that could hold the ends of breaks in DNA in close proximity to each other. This close juxtaposition could facilitate correct rejoining of ends even if the chromosome has sustained multiple breaks. Spores of *B. subtilis* and related bacteria are able to survive extremes of the environment far more effectively than any other kind of dormant cell. NHEJ is part of the basis for this extraordinary robustness.

Translesion DNA Synthesis Enables Replication to Proceed across DNA Damage

In many of the examples we have considered so far, damage to the DNA is mended by excision, followed by resynthesis using an undamaged template. But such repair systems do not operate with complete efficiency and sometimes a replicating DNA polymerase encounters a lesion, such as a pyrimidine dimer or an apurinic site, that has not been repaired. Because such lesions are obstacles to progression of the DNA polymerase, the replication machinery must attempt to copy across the lesion or be forced to cease replicating. Even if cells cannot repair these lesions, there is a fail-safe mechanism that allows the replication machinery to bypass these sites of damage. This mechanism is known as **translesion synthesis**. Although this mechanism is, as we shall see, highly error-prone and thus likely to introduce mutations, translesion synthesis spares the cell the worse fate of an incompletely replicated chromosome.

Translesion synthesis is catalyzed by a specialized class of DNA polymerases that synthesize DNA directly across the site of the damage (Fig. 9-19). Translesion synthesis in *E. coli* is carried out by a complex of the proteins UmuC and UmuD′. UmuC is a member of a distinct family of DNA polymerases found in many organisms known as the Y family of DNA polymerases (Fig. 9-20 and Box 9-4, The Y Family of DNA Polymerases).

An important feature of these polymerases is that although they are template-dependent, they incorporate nucleotides in a manner that is independent of base pairing. This explains how the enzymes can synthesize DNA over a lesion on the template strand. But, because the enzyme is not "reading" sequence information from the template, translesion synthesis is often highly error-prone. Consider the case of an apurinic or apyrmidinic site in which the lesion contains no base-specific information. The translesion polymerase synthesizes across the lesion by inserting nucleotides in a manner that is not guided by base pairing. Nonetheless, the nucleotide incorporated may not be random—some translesion polymerases incorporate specific nucleotides. For example, a human member of the Y family of translesion polymerases correctly inserts two A residues opposite a thymine dimer.

Because of its high error rate, translesion synthesis (like NHEJ) can be considered a system of last resort. It enables the cell to survive what might otherwise be a catastrophic block to replication, but the price that is paid is a higher level of mutagenesis. For this reason, in *E. coli*, the translesion polymerase is not present under normal circumstances.

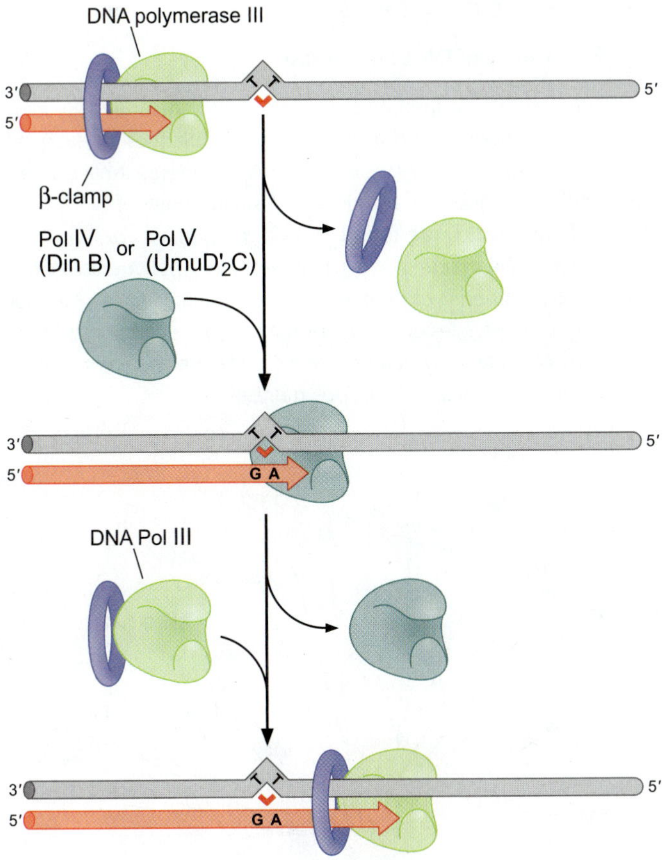

FIGURE 9-19 Translesion DNA synthesis. Upon encountering a lesion in the template during replication, DNA Pol III with its sliding clamp dissociates from the DNA and is replaced by the translesion DNA polymerase, which extends DNA synthesis across the thymine dimer on the template (upper) strand. The translesion polymerase is then replaced by the DNA polymerase III. (Adapted, with permission, from Woodgate R. 1999. *Genes Dev.* 13: 2191–2195, Fig. 1. © Cold Spring Harbor Laboratory Press.)

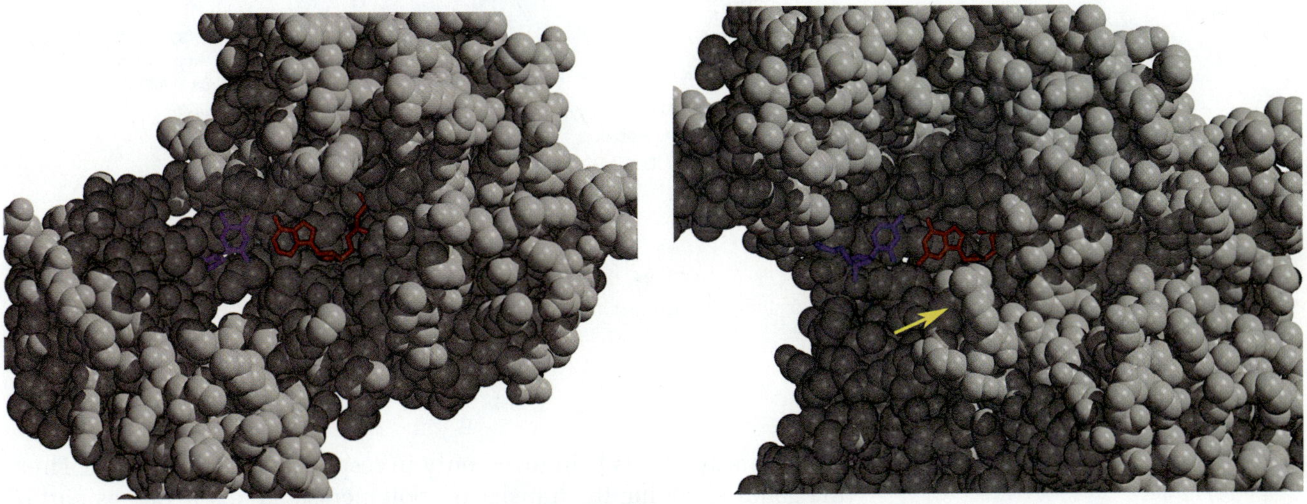

FIGURE 9-20 Crystal structure of a translesion polymerase. The structures shown here represent two different types of DNA polymerases. The structure on the left is a Y-family (lesion bypass) polymerase; that on the right is a typical DNA polymerase from bacteriophage T7. Notice the more open structure around the active site in the Y-polymerase structure, and the absence of the protein region that closes the channel (indicated by the yellow arrow). The incoming nucleotides are in red, and template nucleotides are in blue. (Y polymerase, Ling H. et al. 2001. *Cell* 107: 91. PDB Code: 1JX4. T7 polymerase, Doublié S. et al. 1998. Nature 391: 251. PDB Code: 1T7P.) Images prepared with MolScript, BobScript, and Raster 3D.

ADVANCED CONCEPTS

Box 9-4 The Y Family of DNA Polymerases

DNA polymerases can be grouped into families, shown in various colors in Box 9-4 Figure 1, based on their amino acid sequence similarities to each other. Recently, UmuC and certain other translesion DNA polymerases have been discovered to be founding members of a large and distinct family of DNA polymerases known as the Y family, which are found in all three domains of life: Bacteria, Archaea, and Eukaryota. Members of the Y family of DNA polymerases characteristically carry out DNA synthesis with low fidelity on undamaged DNA templates but have the capacity to bypass lesions in DNA that block replication by members of the other families of DNA polymerases. Box 9-4 Figure 1 shows a phylogenetic tree for the Y family of translesion DNA polymerases.

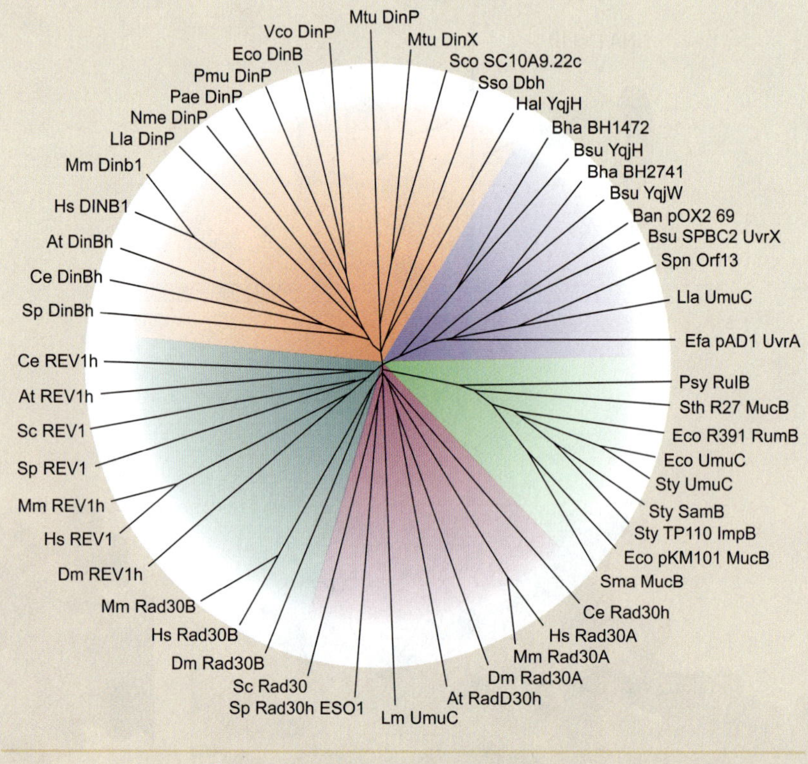

BOX 9-4 FIGURE 1 The phylogenetic tree of the Y family of DNA polymerases. (Adapted, with permission, from Ohmori H. et al. 2001. *Mol. Cell* 8: 7, Fig 1. © Elsevier.)

Rather, its synthesis is induced only in response to DNA damage. Thus, the genes encoding the translesion polymerase are expressed as part of a pathway known as the **SOS response**. Damage leads to the proteolytic destruction of a transcriptional repressor (the LexA repressor) that controls expression of genes involved in the SOS response, including those for UmuC and UmuD, the inactive precursor for UmuD′. Interestingly, the same pathway is also responsible for the proteolytic conversion of UmuD to UmuD′. Cleavage of LexA and UmuD are both stimulated by a protein called RecA, which is activated by single-stranded DNA resulting from DNA damage. RecA is a dual-function protein that is also involved in DNA recombination as we shall see in Chapter 10.

We next address the question of how a translesion polymerase gains access to the stalled replication machinery at the site of DNA damage. In mammalian cells, entry into the translesion repair pathway is triggered by chemical modification of the sliding clamp. As we saw in **Chapter 8**, the sliding clamp, which is known as PCNA in eukaryotes, anchors the replicative polymerase to the DNA template. The chemical modification is the covalent attachment to the sliding clamp of a peptide known as ubiquitin in a process known as ubiquitination. Ubiquitination is widely used in eukaryotic cells to mark proteins for various processes, such as degradation. Its use in triggering translesion synthesis was only recently discovered and adds to the growing list of cellular processes that are governed by tagging proteins with the ubiquitin peptide. Once ubiquitinated, the sliding clamp recruits a translesion polymerase, which contains domains that recognize and bind to ubiquitin. The translesion polymerase, in turn, somehow displaces the replicative polymerase from the 3' end of the growing strand and extends it across the site of the damage. Ubiquitination of the sliding clamp is therefore a distress signal that recruits a translesion polymerase to rescue a replication machine that is stalled at a site of DNA damage.

Finally, several fascinating, but as yet unanswered, questions remain. How exactly does the translesion enzyme replace the normal replicative polymerase in the DNA replication complex? Once DNA synthesis is extended across the lesion, how does the normal replicative polymerase switch back to and replace the translesion enzyme at the replication fork? Translesion polymerases have low processivity, so perhaps they simply dissociate from the template shortly after copying across a lesion. Nonetheless, this explanation still leaves us with the challenge of understanding how the normal processive enzyme is able to reenter the replication machinery.

SUMMARY

Organisms can survive only if their DNA is replicated faithfully and is protected from chemical and physical damage that would change its coding properties. The limits of accurate replication and repair of damage are revealed by the natural mutation rate. Thus, an average nucleotide is likely to be changed by mistake only about once every 10^9 times it is replicated, although error rates for individual bases can vary over a 10,000-fold range. Much of the accuracy of replication is inherent in the way DNA polymerase copies a template. The initial selection of the correct base is guided by complementary pairing. Accuracy is increased by the proofreading activity of DNA polymerase. Finally, in mismatch repair, the newly synthesized DNA strand is scanned by an enzyme that initiates replacement of DNA containing incorrectly paired bases. Despite these safeguards, mistakes of all types occur: base substitutions, small and large additions and deletions, and gross rearrangements of DNA sequences.

Cells have a large repertoire of enzymes devoted to repairing DNA damage that would otherwise be lethal or would alter DNA so as to engender damaging mutations. Some enzymes directly reverse DNA damage, such as photolyases, which reverse pyrimidine dimer formation. A more versatile strategy is excision repair, in which a damaged segment is removed and replaced through new DNA synthesis for which the undamaged strand serves as a template. In base excision repair, DNA glycosylases and endonucleases remove only the damaged nucleotide, whereas in nucleotide excision repair, a short patch of single-stranded DNA containing the lesion is removed. In *E. coli*, excision repair is initiated by the UvrABC endonuclease, which creates a bubble over the site of the damage and cuts out a 12-nucleotide segment of the DNA strand that includes the lesion. Higher cells carry out nucleotide excision repair in a similar manner, but a much larger number of proteins are involved, and the excised, single-stranded DNA is 24–32 residues long.

The most hazardous kind of damage is a DNA break. Breaks are mended by recombinational or DSB repair, in which the sequence across the break is copied from a different but homologous duplex. If no template for repair synthesis is available, breaks in DNA are mended by NHEJ, which rejoins the ends but in an error-prone manner. Another error-prone repair pathway is translesion synthesis. Translesion synthesis enables replication to continue across damage that blocks the progression of a replicating DNA polymerase. Translation synthesis is mediated by a distinct and widespread family of DNA polymerases that are able to carry out DNA synthesis in a manner that, although inaccurate, does not depend on base pairing.

Mutagenesis and its repair are of concern to us because they permanently affect the genes that organisms inherit and because cancer is often caused by mutations in somatic cells.

BIBLIOGRAPHY

Books

Friedberg E.C., Walker G.C., Siede W., Wood R.D., Schultz R.A., and Ellenberger T. 2005. *DNA repair and mutagenesis*. ASM Press, Washington, D.C.

Kornberg A. and Baker T.A. 1992. *DNA replication*, 2nd ed. W.H. Freeman, New York.

Replication Errors and Their Repair

Kunkel T.A. and Erie D.A. 2005. DNA mismatch repair. *Annu. Rev. Biochem.* **76:** 681–710.

Repair of DNA Damage

Bridges B.A. 1999. DNA repair: Polymerases for passing lesions. *Curr. Biol.* **9:** R475–R477.

Citterio E., Vermeulen W., and Hoeijmakers J.H. 2000. Transcriptional healing. *Cell* **101:** 447–450.

Daley J.M., Palmbos P.L., Wu D., and Wilson T.E. 2005. Nonhomologous end joining in yeast. *Annu. Rev. Genet.* **39:** 431–451.

de Laat W.L., Jaspers N.G., and Hoeijmakers J.H. 1999. Molecular mechanism of excision nucleotide repair. *Genes Dev.* **13:** 768–785.

Drapkin R., Reardon J.T., Ansari A., Huang J.C., Zawel L., Ahn K., Sancar A., and Reinberg D. 1994. Dual role of TFIIH in DNA excision repair and in transcription by RNA polymerase II. *Nature* **368:** 769–772.

Kleczkowska H.E., Marra G., Lettieri T., and Jiricny J. 2001. hMSH3 and hMSH6 interact with PCNA and colocalize with it to replication foci. *Genes Dev.* **15:** 724–736.

Sekiguchi J.M. and Ferguson D.O. 2006. DNA double-strand break repair: A relentless hunt uncovers new prey. *Cell* **124:** 260–262.

CHAPTER

10 Homologous Recombination at the Molecular Level

ALL DNA IS RECOMBINANT DNA. GENETIC exchange works constantly to blend and rearrange chromosomes, most obviously during meiosis, when homologous chromosomes pair prior to the first nuclear division. During this pairing, genetic exchange between the chromosomes occurs. This exchange, classically termed **crossing over**, is one of the results of **homologous recombination**. This recombination involves the physical exchange of DNA sequences between the chromosomes. The frequency of crossing over between two genes on the same chromosome depends on the physical distance between these genes, with long distances giving the highest frequencies of exchange. In fact, genetic maps derived from early measurements of crossing-over frequencies gave the first real information about chromosome structure by revealing that genes are arranged in a fixed, linear order.

Sometimes, however, gene order does change: for example, movable DNA segments called **transposons** occasionally "jump" around chromosomes and promote DNA rearrangements, thus altering chromosomal organization. The recombination mechanisms responsible for transposition and other genome rearrangements are distinct from those of homologous recombination. These mechanisms are discussed in detail in Chapter 11.

Homologous recombination is an essential cellular process catalyzed by enzymes specifically made and regulated for this purpose. Besides providing genetic variation, recombination allows cells to retrieve sequences lost through DNA damage by replacing the damaged section with an undamaged DNA strand from a homologous chromosome. Recombination also provides a mechanism to restart stalled or damaged replication forks. Furthermore, special types of recombinations regulate the expression of some genes. For example, by switching specific segments within chromosomes, cells can put otherwise dormant genes into sites where they are expressed.

In addition to providing an explanation for genetic processes, elucidating the molecular mechanisms of recombination has led to the development of methods to manipulate genes. It is, for example, now routine to generate gene "knock-out" and "transgenic" variants in many different experimental organisms (see Chapter 22). These methods for deleting and introducing genes within the context of a whole organism rely on recombination and are exceedingly powerful for determining gene function.

OUTLINE

DNA BREAKS ARE COMMON AND INITIATE RECOMBINATION

Double-stranded breaks (DSBs) in DNA arise frequently. If these breaks are not repaired, the consequence to the cell is disastrous. For example, a single DSB in the *Escherichia coli* chromosome is lethal to a cell that lacks the ability to repair it. The major mechanism used to repair DSBs in most cells is homologous recombination. Some cell types also use a simpler mechanism, such as nonhomologous end joining (NHEJ) to heal their chromosomes. This process is described in Chapter 9.

In bacteria, the major biological role of homologous recombination is to repair DSBs. These broken DNA ends arise by several means (see also Chapter 9). Ionizing radiation and other damaging agents can directly break both strands of the DNA backbone. Many types of DNA damage also indirectly give rise to DSBs by interfering with the progress of a replication fork. For example, an unrepaired nick in one DNA strand will lead to collapse of a passing replication fork (Fig. 10-1). Similarly, a lesion in DNA that makes a strand unable to serve as a template will stop a replication fork. This type of stalled fork can be processed by several different pathways (e.g., fork regression or nuclease digestion; see Fig. 10-1) that give rise to a DNA end with a DSB. These broken DNA ends then initiate recombination with a homologous DNA molecule, a process that will, in turn, heal the break.

In addition to repairing DSBs in chromosomal DNA, homologous recombination promotes genetic exchange in bacteria. This exchange occurs between the chromosome of one cell and the DNA that enters that cell via phage-mediated transduction or cell-to-cell conjugation (as we shall see in Chapter 21). In these cases, the new DNA enters the cell as a linear molecule and thus provides the critical "broken" DNA end needed to initiate recombination.

In eukaryotic cells, homologous recombination is critical for repairing DNA breaks and collapsed replication forks. This role of chromosome repair and replication restart is the principal function of homologous recombination in many somatic cells in complex organisms as well as in vegetatively growing single cellular eukaryotes. However, there are other times when recombination for genetic exchange and chromosome maintenance is specifically needed. As described below, recombination is *essential* to the process of chromosome pairing during meiosis. In this case, as cells enter meiosis, they produce a specific protein to introduce DSBs into the DNA and thereby initiate the recombination pathway. Thus, although they arise from many different sources, the appearance of DSBs in DNA is a key early event in homologous recombination.

MODELS FOR HOMOLOGOUS RECOMBINATION

Elegant early experiments using heavy isotopes of atoms incorporated into DNA provided the first molecular view of the process of homologous recombination. This is the same approach used by Matthew Meselson and Franklin W. Stahl to show that DNA replicates in a semiconservative manner (see Chapter 2). In their experiments, Meselson and Stahl demonstrated that the products of replication contain one old and one newly synthesized DNA strand. In contrast, this same

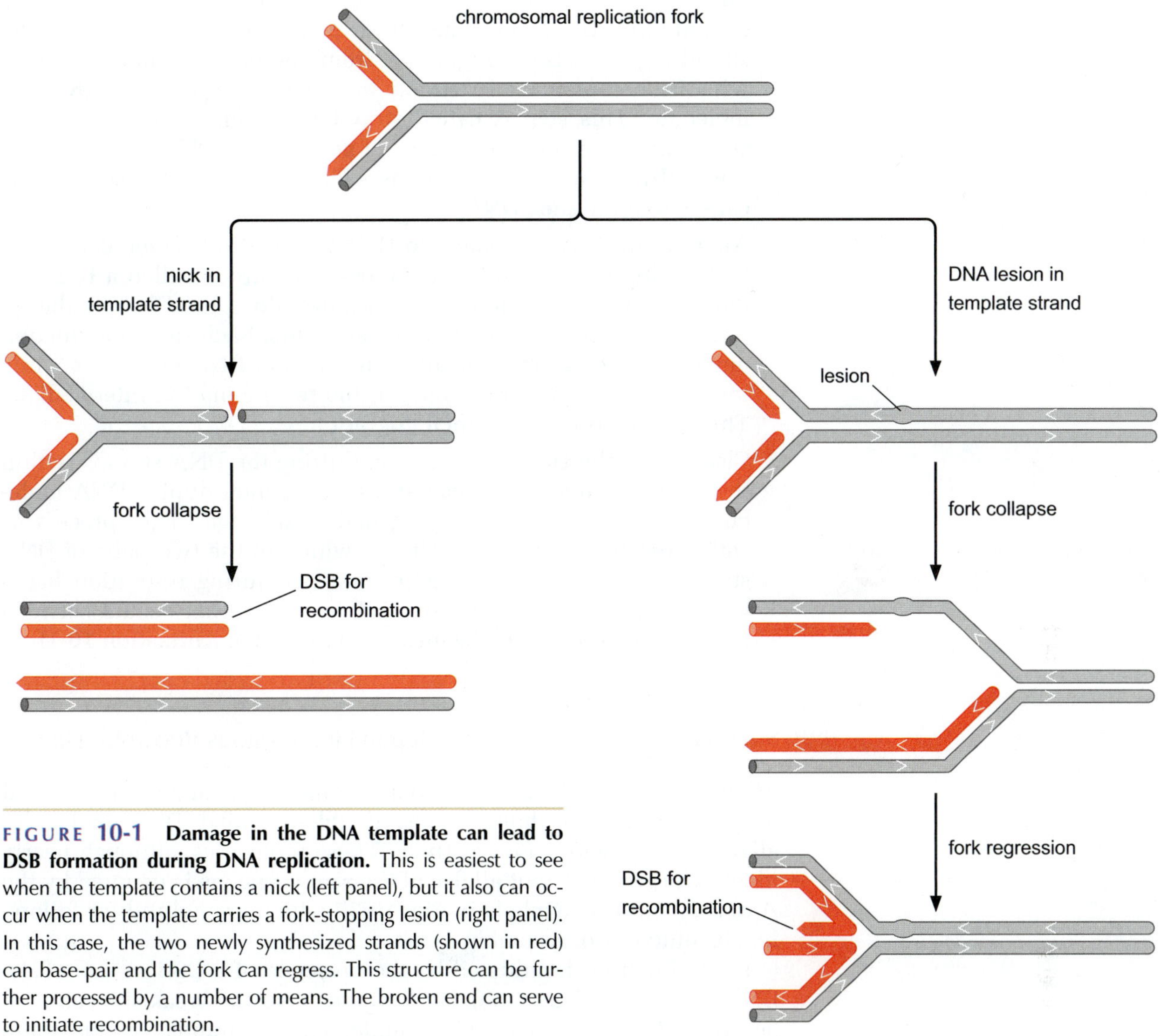

FIGURE 10-1 Damage in the DNA template can lead to DSB formation during DNA replication. This is easiest to see when the template contains a nick (left panel), but it also can occur when the template carries a fork-stopping lesion (right panel). In this case, the two newly synthesized strands (shown in red) can base-pair and the fork can regress. This structure can be further processed by a number of means. The broken end can serve to initiate recombination.

experimental approach revealed that the recombination process being studied involved the direct breakage and rejoining of DNA molecules. As we shall see in the following sections, we now understand that breakage and joining of DNA is a central aspect of homologous recombination. But recombination usually involves also at least the limited destruction and resynthesis of DNA strands. In the years since these initial experiments, numerous models have been proposed to explain the molecular mechanism of genetic exchange. Key steps of homologous recombination present in these models include:

1. Alignment of two homologous DNA molecules. By homologous we mean that the DNA sequences are identical or nearly identical for a region of at least 100 bp or so. Despite this high degree of similarity, DNA molecules can have small regions of sequence difference and may, for example, carry different sequence variants, known as **alleles**, of the same gene.
2. Introduction of breaks in the DNA. These breaks are then further processed to generate regions of single-stranded DNA.

3. Formation of initial short regions of base pairing between the two recombining DNA molecules. This pairing occurs when a single-strand region of DNA originating from one parental molecule pairs with its complementary strand in the homologous duplex DNA molecule. This step is called **strand invasion**. As a result of the strand invasion process, regions of new duplex DNA are generated; this DNA, which often contains some mismatched base pairs, is called **heteroduplex DNA**.
4. After strand invasion, the two DNA molecules become connected by crossing DNA strands. This cross structure is called a **Holliday junction**. A Holliday junction can move along the DNA by the repeated melting and formation of base pairs. Each time the junction moves, base pairs are broken in the parental DNA molecules while identical base pairs are formed in the recombination intermediate. This process is called **branch migration**.
5. Cleavage of the Holliday junction. Cutting the DNA strands within the Holliday junction regenerates two separate duplex DNA molecules and therefore finishes genetic exchange. This process is called **resolution**. As we shall see, which of the two pairs of DNA strands in the Holliday junction are cut during resolution has a large impact on the extent of DNA exchange that occurs between the two recombining molecules (see Interactive Animation 10-1).

WEB ANIMATION

Strand Invasion Is a Key Early Step in Homologous Recombination

When illustrating the steps of homologous recombination, it is useful to picture the two homologous, double-stranded DNA molecules aligned, as shown in Figure 10-2a. These molecules, although nearly identical, carry different alleles of the same gene (as is denoted by the *A*/*a*, *B*/*b*, and *C*/*c* symbols in Fig. 10-2), which are helpful for following the outcome of recombination.

Recombination is initiated by the presence of a DSB in one of the DNA molecules (Fig. 10-2b). DNA strands near the break site can then be "peeled" away from their complementary strands, freeing these strands to invade and ultimately base-pair with the homologous duplex (Fig. 10-2c). Processing of the strands near the break site is described in more detail below. Strand invasion is the central step in homologous recombination, as it is this invasion and then pairing of complementary strands between the two homologous duplexes that establishes the stable pairing between these molecules. This process also initiates the exchange of DNA strands between the two "parental" DNAs. As we shall see below, the enzymes that catalyze strand invasion are called **strand-exchange proteins** because they promote this critical reaction.

Strand invasion generates a Holliday junction that can then move along the DNA by branch migration. This migration increases the length of the DNA exchanged. If the two DNA molecules are not identical—but, for example, carry a few small sequence differences, as is often true between two alleles of the same gene—branch migration through these regions of sequence difference generates DNA duplexes carrying one or a few sequence mismatches (see *B* and *b* alleles in Fig. 10-2d and the inset). Repair of these mismatches in the heteroduplex DNA can have important genetic consequences, a point we return to at the end of the chapter.

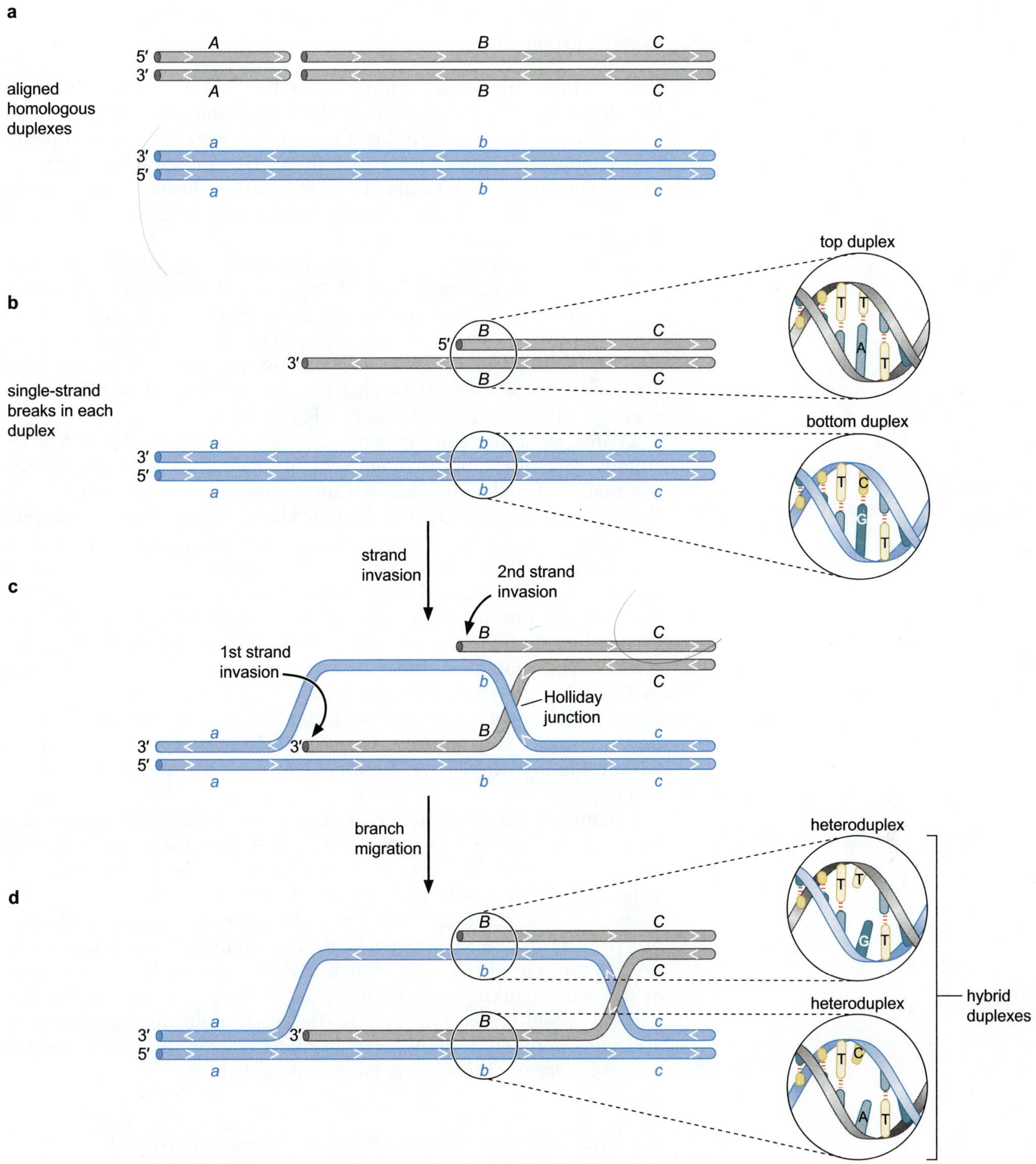

FIGURE 10-2 **Holliday model through the steps of branch migration.** The small arrowheads on the DNA single strands point in the 5´ to 3´ direction. Note that *A* and *a*, *B* and *b*, and *C* and *c* specify different alleles and have slightly different DNA sequences. Therefore, heteroduplex DNA containing those genes (shown in the expanded section in panel d) will have some mismatches.

Resolving Holliday Junctions Is a Key Step to Finishing Genetic Exchange

Finishing recombination requires resolution of the Holliday junction by cutting the DNA strands near the site of the cross; this reaction separates the two recombining DNA molecules and thus completes the genetic exchange. Figure 10-3 shows two homologous DNA duplexes connected by a single Holliday junction. Resolution occurs in one of two ways, and, therefore, gives rise to two distinct classes of DNA products, as we now describe.

Figure 10-3 illustrates where the alternative pairs of DNA cut sites occur on this simple branched DNA generated by exchange between two similar duplex DNA molecules. To make these cut sites easier to visualize, the Holliday junction is "rotated" to give a square, planar structure with no crossing strands. The two strands with the same sequence and polarity must be cleaved; the two alternative choices for cleavage sites are marked 1 and 2 in Figure 10-3.

In this example, the cut sites marked 1 occur in the two DNA strands that are composed entirely of DNA from one of the two parental DNA molecules (e.g., the solid blue and solid gray strands). If these strands are now cut and then covalently joined (the second reaction catalyzed by DNA ligase as discussed below), the resulting DNA molecules will have the structure and sequence shown on the left in the bottom of the figure. These products are referred to as "splice" recombination products, because the two original duplexes are now "spliced together" such that regions from the parental DNA molecules are covalently joined together by a region of hybrid duplex. As seen by following the allele markers, generation of splice products results in reassortment of genes that flank the site of recombination. Therefore, this type of recombinant is also called the **crossover product**, as, within this DNA molecule, crossing over has occurred between the *A* and *C* genes.

In contrast, the alternative pair of cut sites in the Holliday junction (marked 2 in Fig. 10-3) is in the two DNA strands that *contain regions of sequence from both parental molecules* (e.g., both blue and gray segments). After resolution and covalent joining of the strands at these sites, the resulting DNA molecules contain a region or "patch" of hybrid DNA. These molecules are thus known as the **patch products**. In these products, recombination does not result in reassortment of the genes flanking the site of initial cleavage (see fate of the *A/a* and *C/c* allele markers in Fig. 10-3). These molecules are therefore also known as the **noncrossover products**. Factors that influence the site and polarity of resolution are discussed below.

The Double-Strand Break–Repair Model Describes Many Recombination Events

Homologous recombination is often initiated by DSBs breaks in DNA. A common model describing this type of genetic exchange is the **double-strand break–repair pathway** (Fig. 10-4). This pathway starts with the introduction of a DSB in one of two homologous duplex DNA molecules (Fig. 10-4a). The other DNA duplex remains intact. The asymmetric initial breakage of the two DNA molecules in the DSB-repair model necessitates that later stages in the recombination process are also asymmetric (i.e., the two duplexes are treated differently, as we shall see).

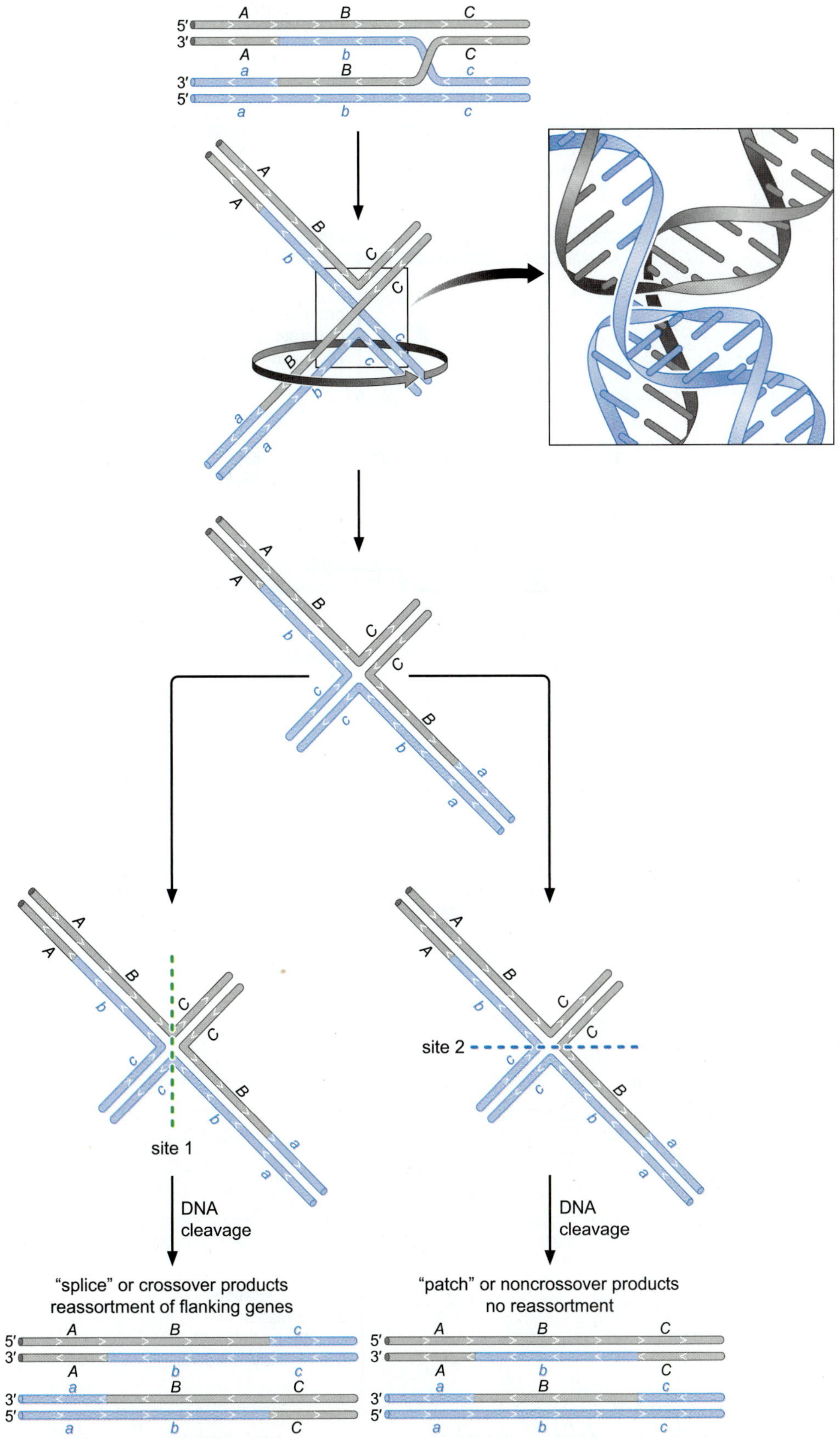

FIGURE 10-3 Holliday junction cleavage. Two alternative pairs of DNA sites can be cut during resolution. Cleavage at one pair of sites generates the "splice" or crossover products. Cleavage at the second pair of sites yields the "patch" or noncrossover products. The inset shows a Holliday junction DNA structure. Notice that the DNA is completely base-paired in this structure.

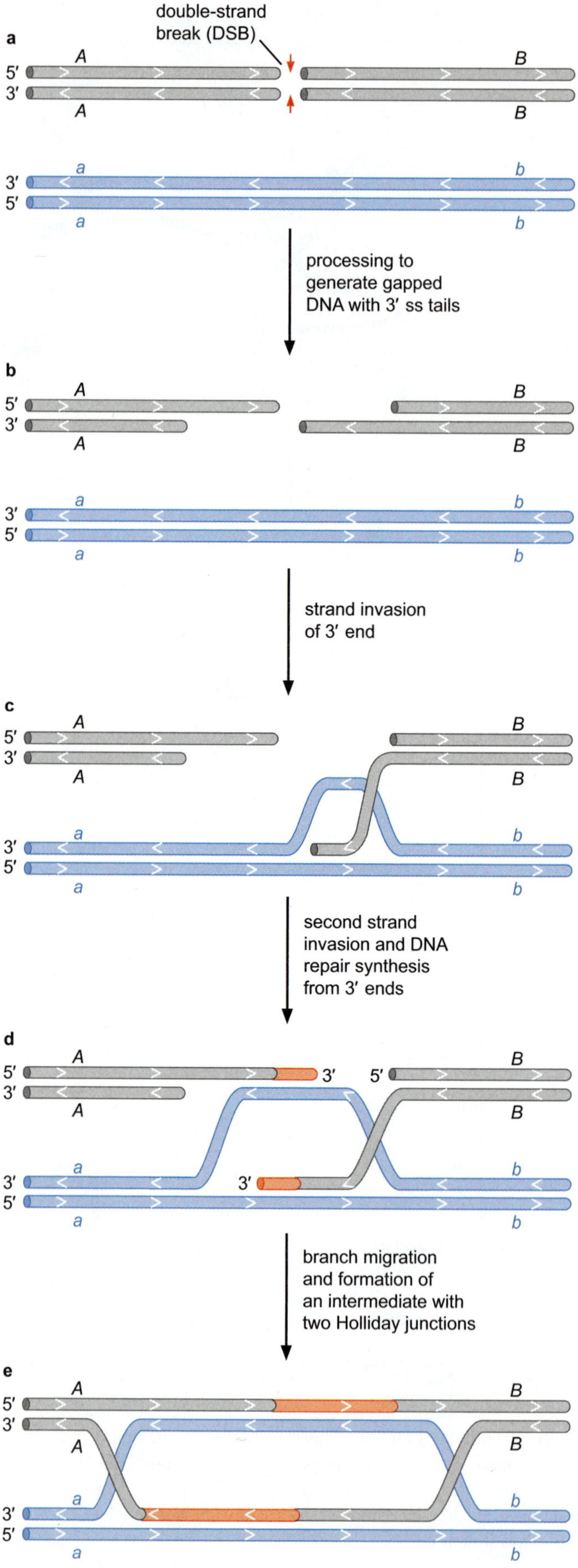

FIGURE 10-4 DSB-repair model for homologous recombination. The figure shows the steps leading to generation of a recombination intermediate with two Holliday junctions.

After introduction of the DSB, a DNA-cleaving enzyme sequentially degrades the broken DNA molecule to generate regions of single-stranded DNA (ssDNA) (Fig. 10-4b). This processing creates single-strand extensions, known as ssDNA tails, on the broken DNA molecules; these ssDNA tails terminate with 3′ ends. In some cases, both strands at a DSB are processed, whereas in other cases, only the 5′-terminating strand is degraded.

The ssDNA tails generated by this process then invade the unbroken homologous DNA duplex (Fig. 10-4c). This panel of the figure shows one strand invasion, as likely occurs initially, whereas the next panel shows the two invading strands. In each case, the invading strand base-pairs with its complementary strand in the other DNA molecule. Because the invading strands end with 3′ termini, they can serve as primers for new DNA synthesis. Elongation from these DNA ends—using the complementary strand in the homologous duplex as a template—serves to regenerate the regions of DNA that were destroyed during the processing of the strands at the break site (Fig. 10-4d,e).

If the two original DNA duplexes were not identical in sequence near the site of the break (e.g., having single-base-pair changes as described above), sequence information could be lost during recombination by the DSB-repair pathway. In the recombination event shown in Figure 10-4, sequence information lost from the gray DNA molecule as a result of DNA processing is replaced by the sequence present on the blue duplex as a result of DNA synthesis. This nonreciprocal step in DSB repair sometimes leaves a genetic trace—giving rise to a **gene conversion** event—a point that we shall return to at the end of the chapter.

The two Holliday junctions found in the recombination intermediates generated by this model move by branch migration and ultimately are resolved to finish recombination. Once again, the strands that are cleaved during resolution of these Holliday junctions determine whether or not the product DNA molecules will contain reassorted genes in the regions flanking the site of recombination (i.e., result in crossing over). The different ways to resolve a recombination intermediate containing two Holliday junctions are explained in Box 10-1, How to Resolve a Recombination Intermediate with Two Holliday Junctions.

HOMOLOGOUS RECOMBINATION PROTEIN MACHINES

Organisms from all branches of life encode enzymes that catalyze the biochemical steps of recombination. In some cases, members of homologous protein families provide the same function in all organisms. In contrast, other recombination steps are catalyzed by different classes of proteins in different organisms but with the same general outcome (see Interactive Animation 10-2). Our most detailed understanding of the mechanism of recombination comes from studies of *E. coli* and its phage. Thus, in the following sections, we first focus on the proteins that promote recombination in *E. coli* via a major DSB-repair pathway, known as the **RecBCD pathway**. Homologous recombination in eukaryotic cells and the proteins involved in these events are considered in later sections.

WEB ANIMATION

Table 10-1 lists the proteins that catalyze critical recombination steps in bacteria as well as those that serve these same functions in eukaryotes (the budding yeast *Saccharomyces cerevisiae* is the best-understood example). These proteins provide activities needed to complete important steps in the DSB-repair pathway. In addition to

ADVANCED CONCEPTS

Box 10-1 How to Resolve a Recombination Intermediate with Two Holliday Junctions

How the Holliday junctions present in a recombination intermediate are cleaved has a huge impact on the structure of the product DNA molecules. Products will either have the DNA flanking the site of recombination reassorted (in the splice/crossover products) or not (in the patch/noncrossover products) depending on how resolution is achieved. Because the intermediates generated by the DSB-repair pathway contain two Holliday junctions, it can be difficult to see which products are generated by the different possible combinations of Holliday junction–cleavage events. In fact, there is a simple pattern that determines whether crossover or noncrossover products are generated.

To explain the different possible ways these intermediates can be resolved, consider the two junctions (labeled x and y) in Box 10-1 Figure 1. For each junction, there are two possible cleavage sites (labeled site 1 and site 2). The simple rule that determines whether or not resolution will result in crossover versus noncrossover products is as follows. If both junctions are cleaved *in the same way*, that is, either both at site 1 or both at site 2, then noncrossover products will be generated. An example of this type of product is shown in panel b of the figure; these are the molecules generated when both Holliday junctions are cleaved at site 2. Note that the allele markers *A/B* and *a/b* are still on the same DNA molecules as they were in the parental chromosomes. Cleavage of both junctions at site 1 also generates noncrossover products.

In contrast, when the two Holliday junctions are cleaved *using different sites*, then the crossover products are generated. An example of this type of resolution is shown in panel c of Box 10-1 Figure 1. Here junction x was cleaved at site 1, whereas junction y was cleaved at site 2. Note that now gene *A* is linked to gene *b*, whereas gene *a* is linked to gene *B*; thus, reassortment of the flanking genes has occurred. Cleavage of junction x at site 2 and junction y at site 1 also generates crossover products.

Why is the simple rule true? To understand this, compare the junctions shown here to the single Holliday junction shown in Figure 10-3. It can be seen that, at a single junction, cleavage at site 1 would give the splice products, whereas cleavage at site 2 would generate patch products. So when the results of cleavage at the two junctions are combined, this is what happens:

- Cleavage of both junctions at site 2 will give a patch product (patch + patch = patch, noncrossover products).
- Cleavage at both junctions at site 1 also gives a patch product (splice + splice = patch) because the second splice-type resolution essentially "undoes" the rearrangement caused by the first cleavage.
- Cleavage of one junction at site 1, but the other at site 2 therefore generates crossover products (splice + patch = splice), because the rearrangement caused by the site 1 cleavage is retained in the final product.

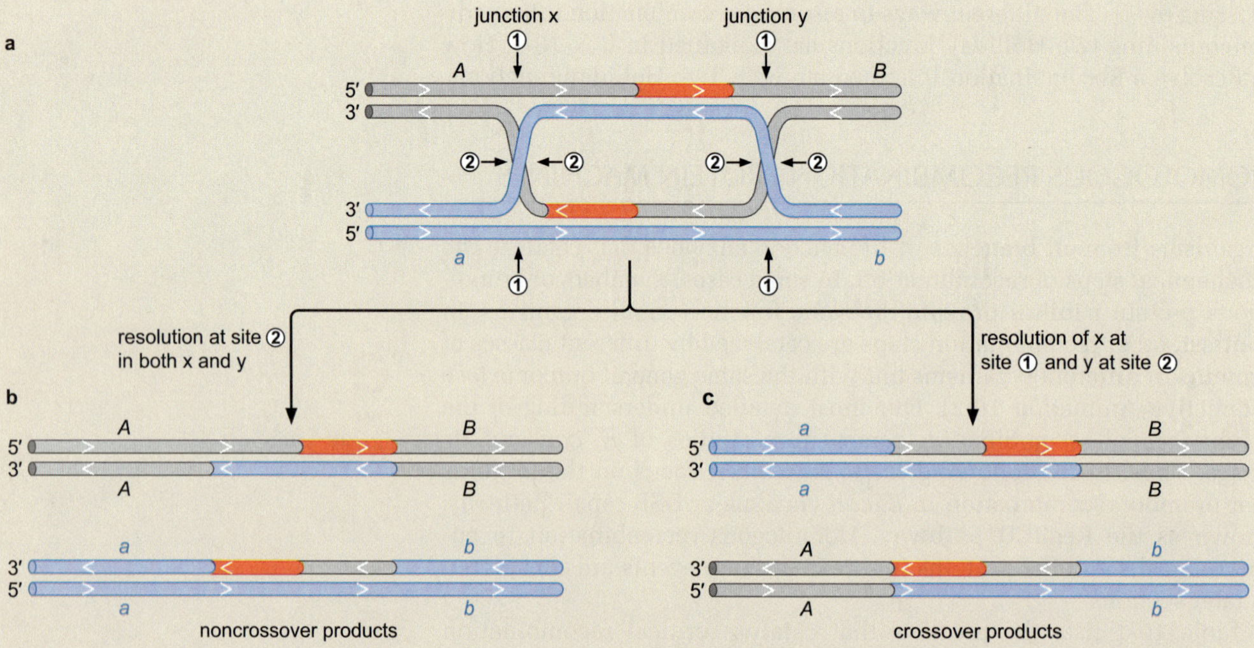

BOX 10-1 FIGURE 1 Two possible ways of resolving an intermediate from the DSB-repair pathway. The parental DNA molecules were like those in Fig. 10-4. The regions of red DNA are those that were resynthesized during recombination.

TABLE 10-1 Prokaryotic and Eukaryotic Factors That Catalyze Recombination Steps

Recombination Step	*E. coli* Protein Catalyst	Eukaryotic Protein Catalyst
Pairing homologous DNAs and strand invasion	RecA protein	Rad51 Dcm1 (in meiosis)
Introduction of DSB	None	Spo11 (in meiosis) HO (for mating-type switching)
Processing DNA breaks to generate single strands for invasion	RecBCD helicase/nuclease	MRX protein (also called Rad50/58/60 nuclease)
Assembly of strand-exchange proteins	RecBCD and RecFOR	Rad52 and Rad59
Holliday junction recognition and branch migration	RuvAB complex	Unknown
Resolution of Holliday junctions	RuvC	Perhaps Rad51c-XRCC3 complex and others

these dedicated recombination proteins, DNA polymerases, ssDNA-binding proteins, topoisomerases, and ligases also have critical roles in the process of genetic exchange.

Note that absent from the list in Table 10-1 is an *E. coli* protein that introduces DSBs in DNA, despite the fact that recombination via the RecBCD pathway requires a DSB on one of the recombining two DNA molecules. As discussed above, in bacteria, no specific protein has been found that carries out this task. Rather, breaks generated as a result of DNA damage or failure of a replication fork are the major source of these initiating events in chromosomal DNA. Alternatively, during genetic exchange reactions, such as phage-mediated transduction (which we shall consider in Chapter 22), the infecting DNA segment carries broken DNA ends.

The following sections describe the *E. coli* recombination proteins and how they perform their functions during recombination by the DSB-repair pathway. These proteins are discussed in the order in which they appear during the reaction pathway. First, we see how the RecBCD enzyme processes DNA at the site of the DSB to generate single-strand regions. Next, the structure and mechanism of RecA, the strand-exchange protein, are described. RecA, after assembling on the ssDNA, finds regions of sequence homology in the DNA molecules and generates new base-pairing partners between these regions. The RuvA and RuvB proteins that drive DNA branch migration are then described. Finally, the Holliday junction–resolving enzyme, RuvC, is considered.

The RecBCD Helicase/Nuclease Processes Broken DNA Molecules for Recombination

DNA molecules with ssDNA extensions or tails are the preferred substrate for initiating strand exchange between regions of homologous sequence. The RecBCD enzyme processes broken DNA molecules to generate these regions of ssDNA. RecBCD also helps load the RecA strand-exchange protein onto these ssDNA ends. In addition, as we shall see, the multiple enzymatic activities of RecBCD provide a means for cells to "choose" whether to recombine with or destroy DNA molecules that enter a cell.

RecBCD is composed of three subunits (the products of the *recB*, *recC*, and *recD* genes) and has both DNA helicase and nuclease activities. It binds to DNA molecules at the site of a DSB and tracks along DNA using the energy of ATP hydrolysis. As a result of its action, the DNA is unwound, with or without the accompanying nucleolytic destruction of one or both of the DNA strands. The activities of RecBCD are controlled by specific DNA sequence elements known as **chi sites** (for crossover hot spot instigator). Chi sites were discovered because they stimulate the frequency of homologous recombination.

Figure 10-5 shows a schematic of RecBCD processing a DNA molecule containing a single chi site to activate this DNA for recombination. RecBCD enters the DNA at the site of the DSB and moves along the DNA, unwinding the strands. The RecB and RecD subunits are both DNA helicases, that is, enzymes that use ATP hydrolysis to melt DNA base pairs (see Chapter 8). RecB is a 3′ to 5′ helicase and also has a multifunctional DNA nuclease domain (see below), whereas RecD is a 5′ to 3′ helicase. RecC functions to recognize the chi sites. The nuclease activity of RecBCD frequently cleaves each strand during unwinding and thereby destroys the DNA.

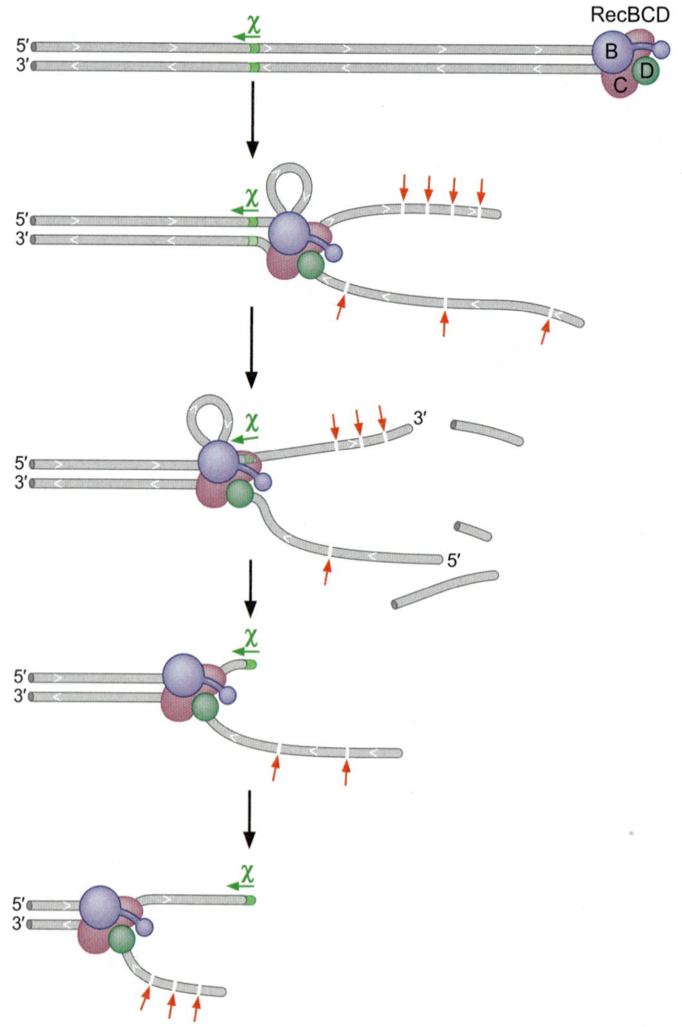

FIGURE 10-5 Steps of DNA processing by RecBCD. Note that RecBCD protein could have entered this DNA molecule from either or both broken ends. However, chi sites function only in one orientation. On the DNA molecule shown, the chi site is oriented such that it will only modify a RecBCD enzyme that is moving from right to left. The RecBCD enzyme has two DNA helicases: RecD, which moves rapidly on the 5′-ending strand (bottom strand), and RecB, which moves slowly on the 3′-ending strand (top strand). Because these two subunits travel at different speeds, the DNA molecules accumulate a single-stranded DNA loop on the top strand during unwinding. After the enzyme has encountered the chi site, RecD is lost or inactivated.

Upon encountering the chi sequence, the nuclease activity of the RecBCD enzyme is altered. As RecBCD moves into the sequence distal to the chi site (with respect to the broken DNA site at which the enzyme entered), it no longer cleaves the DNA strand with 3′→5′ polarity. Furthermore, after the encounter with the chi site, the other DNA strand (the one with the 5′→3′ polarity) is cleaved even more frequently than it was prior to the chi site. As a result of this change in activity, a duplex DNA molecule is converted into one with a 3′ single-strand extension terminating with the chi sequence at the 3′ end. This structure is ideal for assembly of RecA and initiation of strand exchange (see below). The molecular basis of the change in RecBCD's enzyme activity after the encounter with a chi site is due to inactivation of the RecD subunit and a change in the way the DNA travels through the multisubunit RecBCD complex as we shall now describe.

The structure of the RecBCD complex bound to DNA provides further insight into how this three-subunit machine functions, and how its activity changes upon encountering a chi site (see Structural Tutorial 10-1). As shown in Figure 10-6, the protein complex has an overall triangular shape, with duplex DNA entering the protein from the top point of the triangle. Here, the DNA encounters a "pin" structure protruding from the RecC subunit that splits the duplex and guides the two individual strands of DNA to the two motors within the enzyme. The RecC subunit channels the 3′ strand to the RecB motor and the 5′ strand to the RecD motor. In this manner, RecC, which is not itself a helicase, contributes to the overall efficiency of the helicase activity of the complex. The organization of the channels within the enzyme causes the 3′ DNA tail to be fed along a groove that emerges at the nuclease active site on the RecB subunit. As a result, this strand is efficiently and processively degraded, prior to the enzyme encountering a chi site. The 5′ DNA tail also moves by the nuclease active site upon leaving the RecD motor, but it is digested less frequently than the 3′

WEB STRUCTURAL TUTORIAL

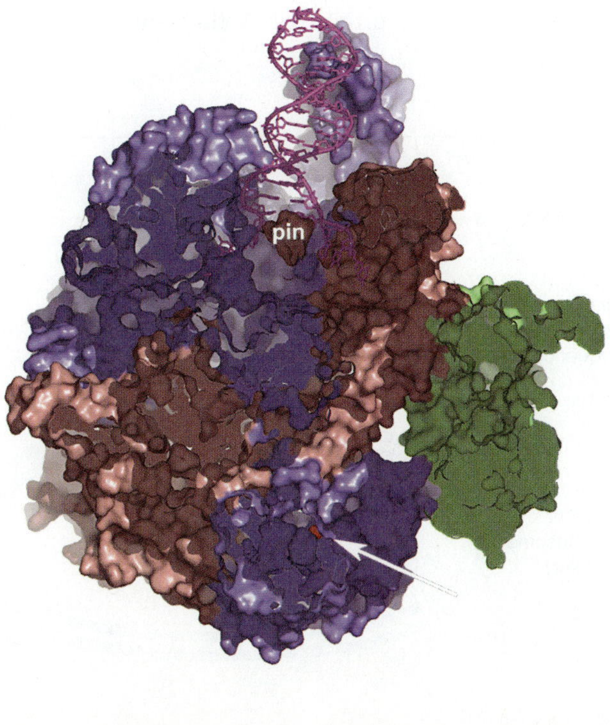

FIGURE 10-6 Structure of the RecBCD–DNA complex. Here RecB is shown in blue, RecC in magenta, and RecD in green. The bound DNA entering through the "top" of the complex is colored purple and the white arrow points to a bound calcium ion (red) in the RecB nuclease active site. The structure shows a cutaway view to reveal the DNA. RecC contacts both DNA strands and splits them, directing the 3′ strand to RecB on the left and the 5′ tail to RecD on the right. (Singleton M.R. et al. 2004. *Nature* 432: 7015; PDB Code: 1W36.) Image prepared with PyMOL (DeLano W.L. 2002. *The PyMOL molecular graphics system*. DeLano Scientific, Palo Alto, California. http:www.pymol.org).

tail because it must compete with the more favorably positioned 3′ strand. However, upon encountering a chi site, this situation changes. RecC recognizes and binds tightly to this DNA site, and once this 3′ end is bound, it is prevented from entering the nuclease. This binding therefore both prevents further digestion of the 3′ tail and promotes digestion of the 5′ tail, by removing its competitor.

The ssDNA tail generated by RecBCD must be coated by the RecA protein for recombination to occur. However, cells also contain ssDNA-binding protein (SSB) that can bind to this DNA. To ensure that RecA, rather than SSB, binds these ssDNA tails, RecBCD interacts directly with RecA and promotes its assembly. This loading activity involves a direct protein–protein interaction between the nuclease domain of the RecB subunit and the RecA protein and serves to load RecA on the DNA with the 3′ tail.

Chi Sites Control RecBCD

Chi sites increase the frequency of recombination about tenfold. This stimulation is most pronounced directly adjacent to the chi site. Although elevated recombination frequencies are observed for ~10 kb distal to the chi site, they drop off gradually over this distance (Fig. 10-7). The observation that recombination is stimulated specifically only on one "side" of the chi site was initially puzzling. It is now clear, however, why this pattern is observed: the DNA between the DSB (where RecBCD enters) and the chi site is cut into small pieces by the enzyme and is therefore not available for recombination. In contrast, DNA sequences met by RecBCD after its encounter with chi are preserved in a recombinogenic, single-strand form and are specifically loaded with RecA.

The ability of chi sites to control the nuclease activity of RecBCD also helps bacterial cells protect themselves from foreign DNA that may enter via phage infection or conjugation. The eight-nucleotide chi site (GCTGGTGG) is highly overrepresented in the *E. coli* genome: whereas it is predicted to occur only once every 65 kb, or about 80 times, the chromosomal sequence reveals the presence of 1009 chi sites! Because of this overrepresentation, *E. coli* DNA that enters an *E. coli* cell is

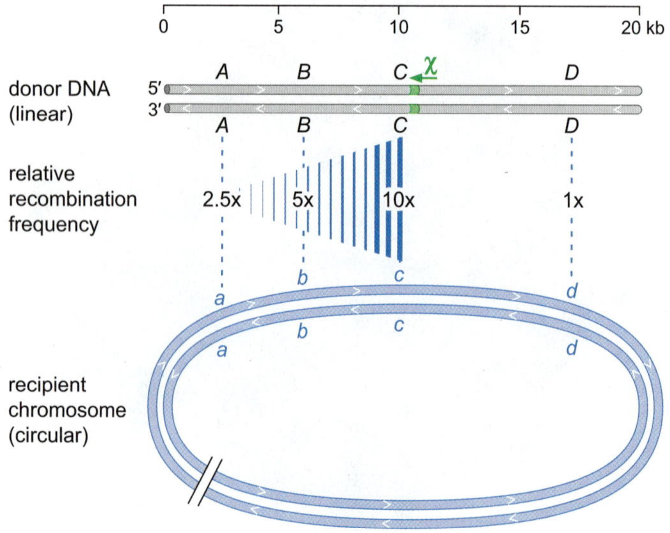

FIGURE 10-7 Polar action of chi. This schematic shows that a chi site specifically elevates recombination frequencies directly at the site, as well as in the distal sequences. The recombination event shown represents exchange between a transferred linear DNA segment introduced into a cell by transduction or conjugation and the bacterial chromosome. The actual DNA segments participating may be much longer. For example, phage transduction often delivers an ~80-kb segment of DNA. The *E. coli* chromosome is ~5 Mb.

likely to be processed by RecBCD in a manner that generates the 3′ ss-DNA tails, and thus to be activated for recombination. In contrast, DNA from another species (in which *E. coli* chi sites are not overrepresented) will lack frequent chi sites. RecBCD action on this DNA will lead to its extensive degradation, rather than activation for recombination.

In summary, the DNA-degradation activity of RecBCD has multiple consequences: this degradation is needed to process DNA at a break site for the subsequent steps of RecA assembly and strand invasion. In this manner, RecBCD promotes recombination. However, because RecBCD degrades DNA to activate it, the overall process of homologous recombination must also involve DNA synthesis to regenerate the degraded strands. In addition, RecBCD sometimes functions simply to destroy DNA, as it does when foreign DNA lacking frequent chi sites enters cells. In this way, RecBCD can protect cells from the potentially deleterious consequences of taking up foreign sequences, which, for example, may carry a bacteriophage or other harmful agent.

RecA Protein Assembles on Single-Stranded DNA and Promotes Strand Invasion

RecA is the central protein in homologous recombination. It is the founding member of a family of enzymes called strand-exchange proteins. These proteins catalyze the pairing of homologous DNA molecules. Pairing involves both the search for sequence matches between two molecules and the generation of regions of base pairing between these molecules.

The DNA pairing and strand-exchange activities of RecA can be observed using simple DNA substrates in vitro; examples of DNA pairing and strand-exchange reactions useful for demonstrating the biochemical activities of RecA are shown in Figure 10-8. The important features of these DNA molecules are (1) DNA sequence complementarity between the two partner molecules, (2) a region of ssDNA on at least one molecule to allow RecA assembly, and (3) the presence of a

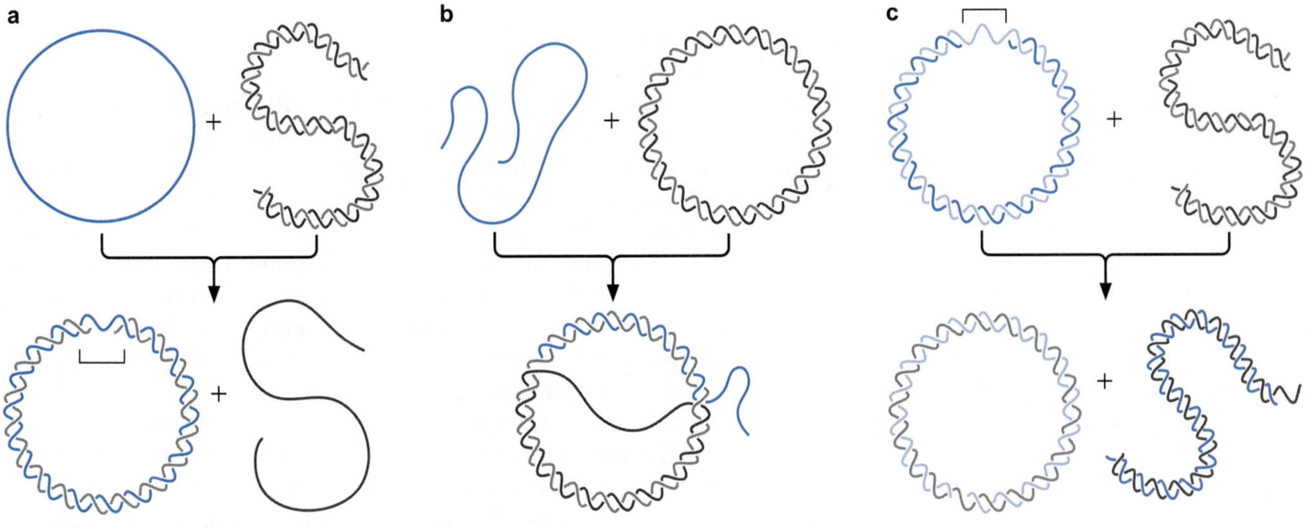

FIGURE 10-8 Substrates for RecA strand exchange. Note that the brackets in parts a and c show the location of a gap in one of the strands.

a

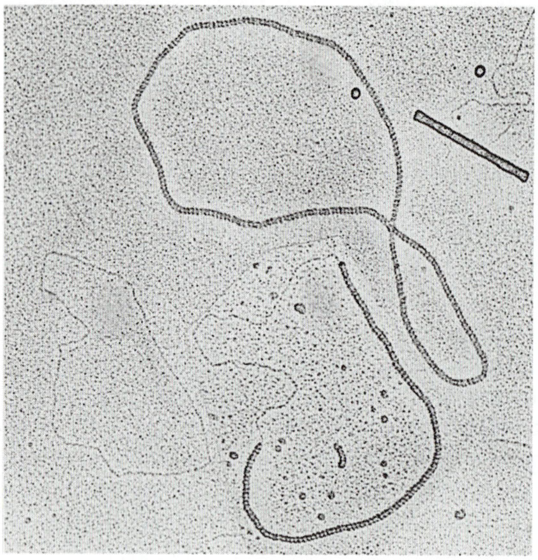

b

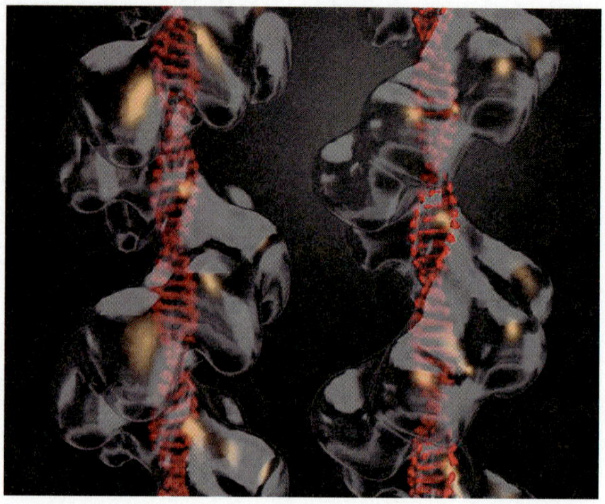

c

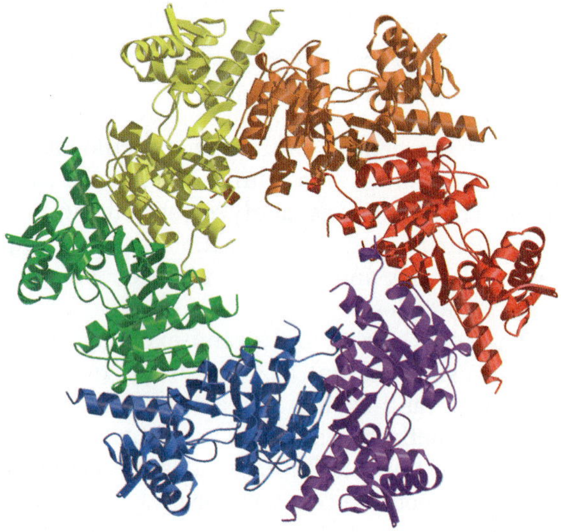

FIGURE 10-9 Three views of the RecA filament. (a) Electron micrograph (EM) of circular DNA molecules that are fully or partially coated with RecA. An uncoated DNA molecule is also shown to illustrate how the DNA is elongated upon RecA binding. (Reprinted, with permission, from Stasiak A. and Egelman E.H. 1988. *Genetic recombination* [ed. R. Kucherlapati and G. Smith], pp. 265–307, Fig. 3. © ASM Press.) (b) A higher-resolution view of the filament generated by averaging many EM images. The picture on the left is *E. coli* RecA, whereas the one on the right is the related strand-exchange protein Rad51 from yeast. (Image provided by Edward Egelman, University of Virginia.) (c) A higher-resolution view generated by X-ray crystallography. Here, one turn of the helical filament is shown from a top-down view. Individual subunits are colored; the red subunit is closest to the viewer. (Story R.M. et al. 1992. *Nature* 355: 318–325.) Image prepared with MolScript, BobScript, and Raster 3D.

DNA end within the region of complementarity, enabling the DNA strands in the newly formed duplex to intertwine.

The active form of RecA is a protein–DNA filament (Fig. 10-9). Unlike most proteins involved in molecular biology that function in smaller discrete protein units, such as monomers, dimers, or hexamers, the RecA filament is huge and variable in size; filaments that contain approximately 100 subunits of RecA and 300 nucleotides of DNA are common. The filament can accommodate one, two, three, or even four strands of DNA. As described below, filaments with either one or three bound strands are most common in recombination intermediates.

The structure of DNA within the filament is highly extended compared to either uncoated ssDNA or a standard B-form helix. On average, the distance between adjacent bases is 5 Å, rather than the 3.4-Å spacing normally observed (Chapter 6). Thus, upon RecA binding, the length of a DNA molecule is extended ~1.5-fold (Fig. 10.9a). It is within this RecA filament that the search for homologous DNA sequences is conducted and the exchange of DNA strands executed.

To form a filament, subunits of RecA bind cooperatively to DNA. RecA binding and assembly are much more rapid on ssDNA than on

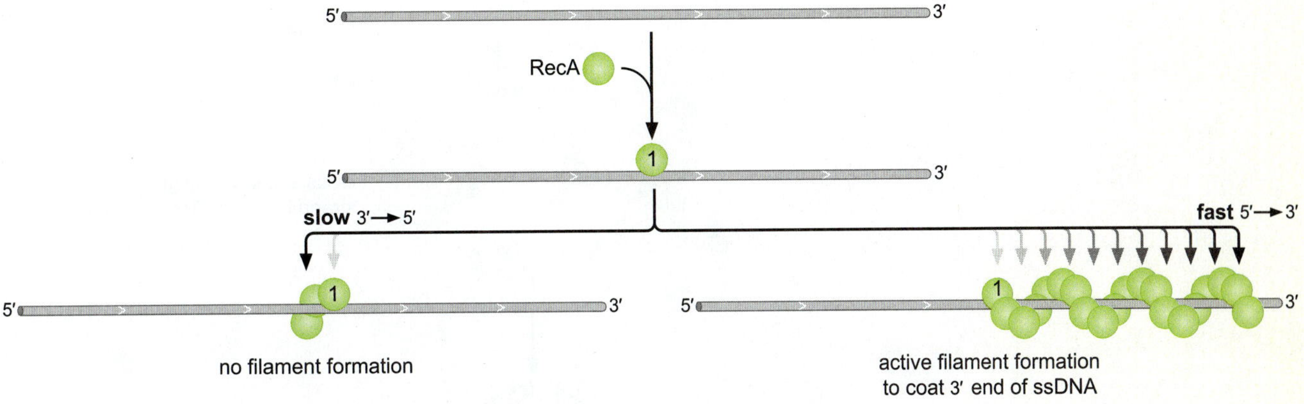

FIGURE 10-10 Polarity of RecA assembly. Note that new subunits of RecA join the filament on the DNA 3′ side to an existing subunit much faster than these subunits join on the 5′ side. Because of this polarity of assembly, DNA molecules with 3′ ssDNA extensions will be efficiently coated with RecA. In contrast, molecules with 5′ ssDNA extensions would not serve as substrates for filament assembly.

double-stranded DNA, thus explaining the need for regions of ssDNA in strand-exchange substrates. The filament grows by the addition of RecA subunits in the 5′ to 3′ direction, such that a DNA strand that terminates in 3′ ends is most likely to be coated by RecA (Fig. 10-10). Note that in the DSB-repair model for recombination, it is DNA molecules with just this configuration that participate in strand invasion.

Newly Base-Paired Partners Are Established within the RecA Filament

RecA-catalyzed strand exchange can be divided into distinct reaction stages. First, the RecA filament must assemble on one of the participating DNA molecules. Assembly occurs on a molecule containing a region of ssDNA, such as an ssDNA tail. This RecA–ssDNA complex is the active form that participates in the search for a homology. During this search, RecA must "look" for base-pair complementarity between the DNA within the filament and a new DNA molecule.

This homology search is promoted by RecA because the filament structure has two distinct DNA-binding sites: a primary site (bound by the first DNA molecule) and a secondary site (Fig. 10-11). This secondary DNA-binding site can be occupied by double-stranded DNA. Binding to this site is rapid, weak, transient, and—importantly—independent of DNA sequence. In this way, the RecA filament can bind and rapidly "sample" huge stretches of DNA for sequence homology.

How does the RecA filament sense sequence homology? Details of this mechanism are not yet clear. The DNA in the secondary binding site is transiently opened and tested for complementarity with the ssDNA in the primary site. This "testing" is presumably via base-pairing interactions, although it occurs initially without disrupting the global base pairing between the two strands of the DNA in the secondary site. In support of this idea, experiments suggest that the initial alignment may involve base flipping of some of the bases in the DNA duplex (see Chapter 9 for a discussion of base flipping during DNA re-

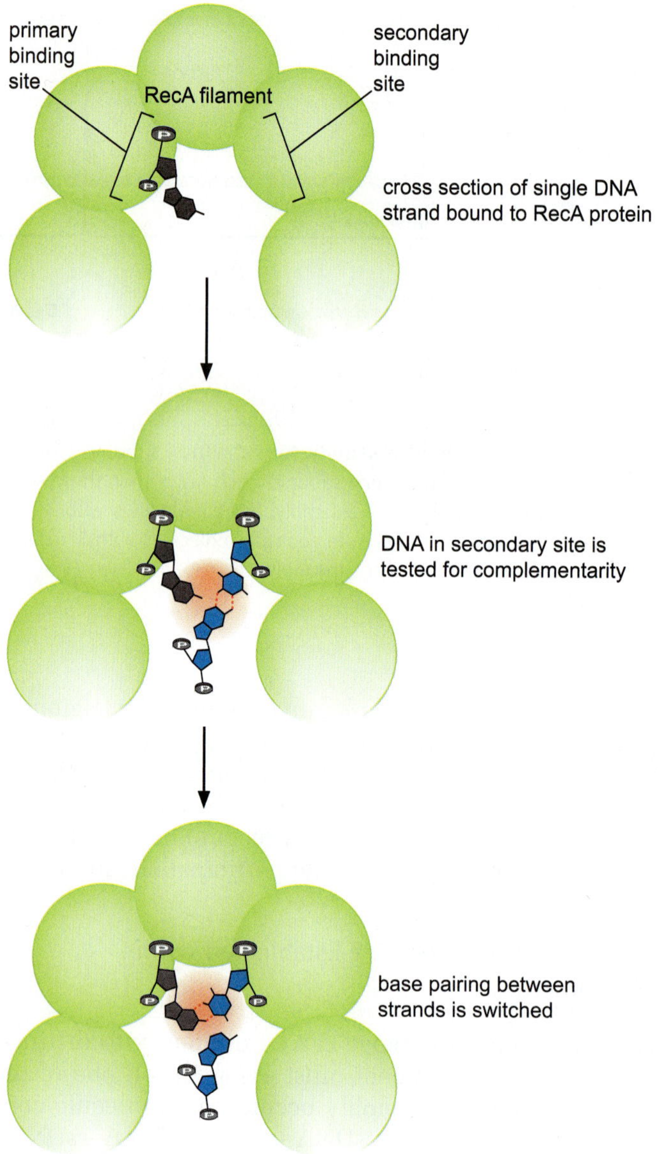

FIGURE 10-11 Model of two steps in the search for homology and DNA strand exchange within the RecA filament. Here, the RecA filament is represented from a top-down view as in Fig. 10-8c. The incoming DNA duplex is shown in blue. (Adapted from Howard-Flanders et al. 1984. *Nature* 309: 215–220.)

pair). In vitro experiments indicate that a sequence match of just 15 bp provides a sufficient signal to the RecA filament that a match has been found and thereby triggers strand exchange.

Once a region of base-pair complementarity is located, RecA promotes the formation of a stable complex between these two DNA molecules. This RecA-bound three-stranded structure is called a **joint molecule** and usually contains several hundred base pairs of hybrid DNA. It is within this joint molecule that the actual exchange of DNA strands occurs. The DNA strand in the primary binding site becomes base-paired with its complement in the DNA duplex bound in the secondary site. Strand exchange thus requires the breaking of one set of base pairs and the formation of a new set of identical base pairs. Completion of strand exchange also requires that the two newly paired strands be intertwined to form a proper double helix. RecA binds preferentially to the DNA products after strand exchange has occurred, and it is this binding energy that actually drives the exchange reaction toward the new DNA configuration.

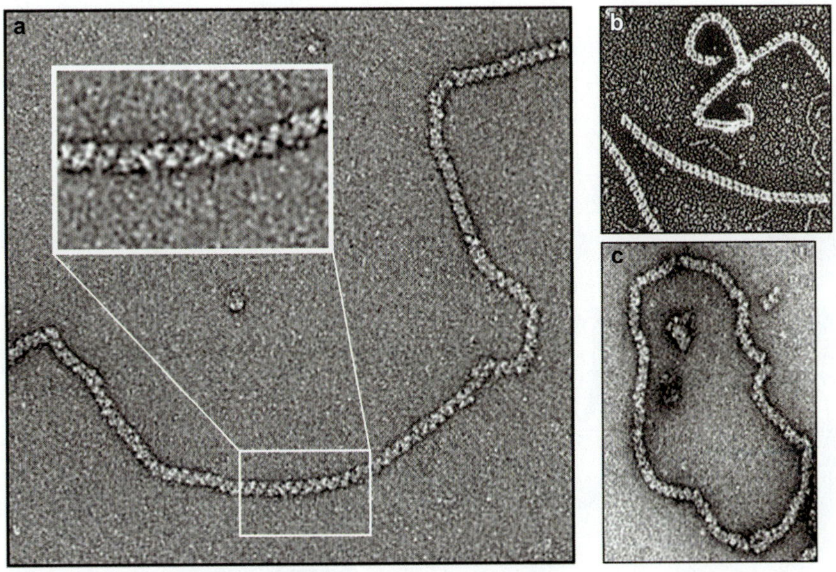

FIGURE 10-12 **RecA-like proteins in three branches of life.** Nucleoprotein filaments are shown for (a) human Rad51, (b) *E. coli* RecA, and (c) *Archaeoglobus fulgidus* RadA proteins. The Rad51 and RecA proteins are also shown in Fig. 10-8. Notice the similar helical structure of the filaments revealed by the stripes in these EM images. (Reprinted, with permission, from West S.C. et al. 2003. *Nat. Rev. Mol. Cell Biol.* 4: 435–445. © Macmillan. Images provided by A. Stasiak, University of Lausanne, Switzerland.)

RecA Homologs Are Present in All Organisms

Strand-exchange proteins of the RecA family are present in all forms of life. The best-characterized members are RecA from Eubacteria, RadA from Archaea, Rad51 and Dmc1 from Eukaryota, and the bacteriophage T4 UvsX protein. These proteins form filaments similar to those made by RecA (Fig. 10-12) and likely function in an analogous manner (although some features of the proteins are specifically tailored for their specific cellular roles and interaction partners). We discuss the roles of Rad51 and Dmc1 recombination in eukaryotic cells below.

The RuvAB Complex Specifically Recognizes Holliday Junctions and Promotes Branch Migration

After the strand invasion step of recombination is complete, the two recombining DNA molecules are connected by a DNA branch known as a Holliday junction (see above). Movement of the site of this branch requires exchange of DNA base pairs between the two homologous DNA duplexes. Cells encode proteins that greatly stimulate the rate of branch migration.

RuvA protein is a Holliday junction–specific DNA-binding protein that recognizes the structure of the DNA junction, regardless of its specific DNA sequence (see Structural Tutorial 10-2). RuvA recognizes and binds to Holliday junctions and recruits the RuvB protein to this site. RuvB is a hexameric ATPase, similar to the hexameric helicases involved in DNA replication (see Chapter 8). The RuvB ATPase provides the energy to drive the exchange of base pairs that move the DNA branch. This energy is needed to move the branch rapidly and in one direction. Structural models for RuvAB com-

WEB STRUCTURAL TUTORIAL

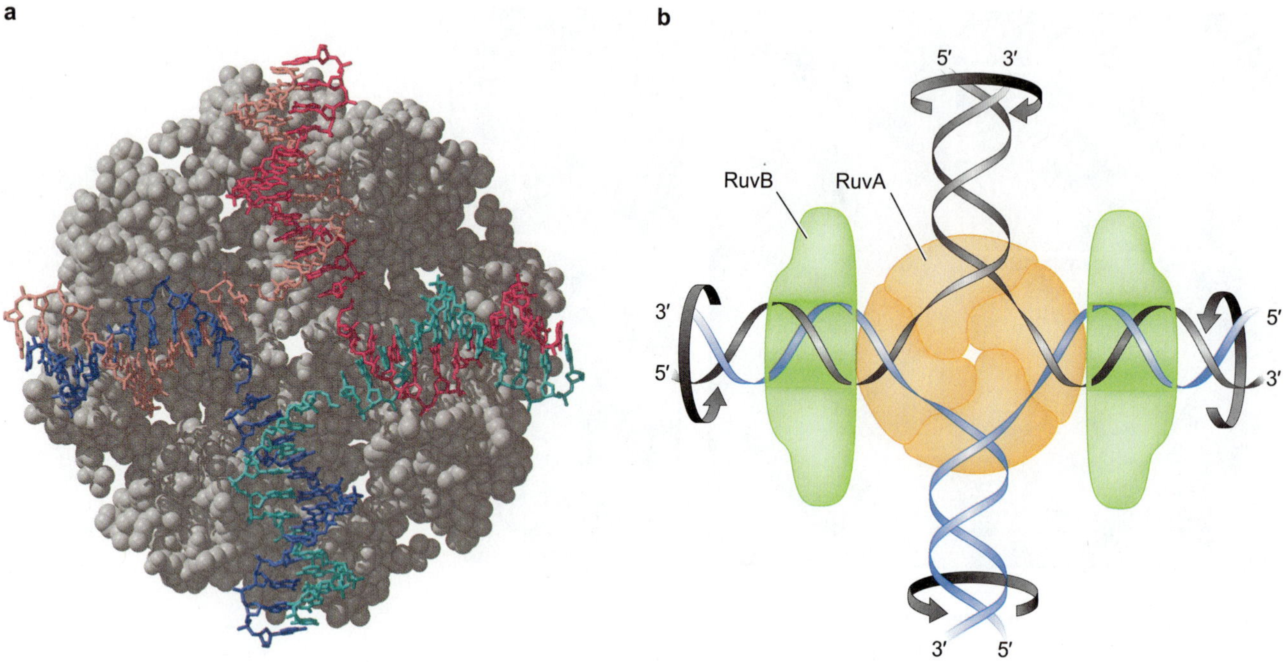

FIGURE 10-13 High-resolution structure of RuvA and schematic model of the RuvAB complex bound to Holliday junction DNA. (a) The crystal structure of the RuvA tetramer shows the fourfold symmetry of the protein. (Ariyoshi M et al. 2000. *Proc. Natl. Acad. Sci.* 97: 8257–8262.) Image prepared with MolScript, BobScript, and Raster 3D. (b) A schematic model of the crystal structure is shown with two RuvB hexamers. Notice how a tetramer of RuvA binds with fourfold symmetry to the junction. Two hexamers of RuvB bind on opposite sides of RuvA and function as a motor to pump DNA through the junction. The RuvB hexamers are shown in cross sections, so that the DNA threading through these complexes can be seen. (Redrawn from Yamada K. et al. 2002. *Mol. Cell* 10: 671–681, Fig. 4.)

plexes at a Holliday junction show how a tetramer of RuvA together with two hexamers of RuvB work together to power this DNA-exchange process (Fig. 10-13).

RuvC Cleaves Specific DNA Strands at the Holliday Junction to Finish Recombination

Completion of recombination requires that the Holliday junction (or junctions) between the two recombining DNA molecules be resolved. In bacteria, the major Holliday junction resolving endonuclease is RuvC. RuvC was discovered and purified based on its ability to cut DNA junctions made by RecA in vitro. Evidence indicates that it functions in concert with RuvA and RuvB.

Resolution by RuvC occurs when RuvC recognizes the Holliday junction—likely in a complex with RuvA and RuvB—and specifically nicks two of the homologous DNA strands that have the same polarity. This cleavage results in DNA ends that terminate with 5′-phosphates and 3′-OH groups that can be directly joined by DNA ligase. Depending on which pair of strands is cleaved by RuvC, the resulting ligated recombination products will be of either the "splice" (crossover) or "patch" (noncrossover) type. The structure of RuvC and a model schematic proposing how it may interact with junction DNA are shown in Figure 10-14.

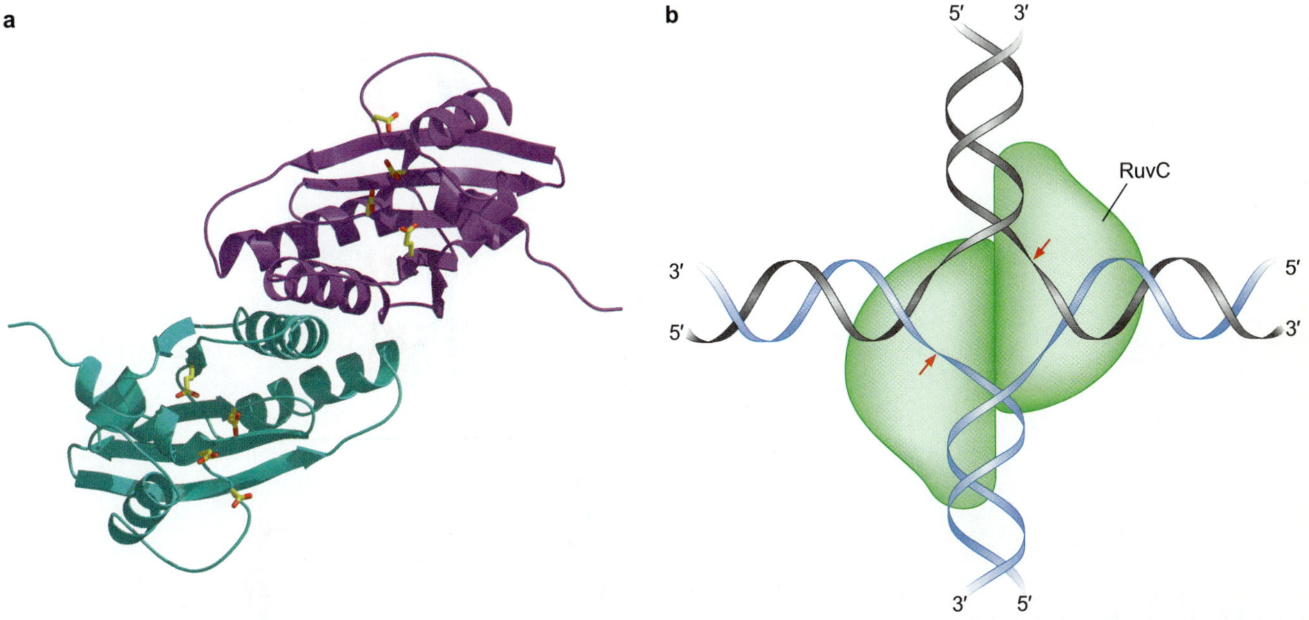

FIGURE 10-14 High-resolution structure of the RuvC resolvase and schematic model of the RuvC dimer bound to Holliday junction DNA. (a) The crystal structure of the RuvC protein. (Ariyoshi M. et al. 1994. *Cell* 78: 1063–1072.) Image prepared with MolScript, BobScript, and Raster 3D. (b) Model for binding of a RuvC dimer to a Holliday junction. Notice how, in this model, a dimer of RuvC can bind the Holliday junction and introduce symmetrical cleavages into the two identical DNA strands. (Adapted, with permission, from Rafferty J.B. et al. 1996. *Science* 274: 416–421, Fig. 1b. © AAAS.)

Despite recognizing a structure rather than a specific sequence, RuvC cleaves DNA with modest sequence specificity. Cleavage takes place only at sites conforming to the consensus 5′A/T-T-T-G/C. Cleavage occurs after the second T in this sequence. Sequences with this consensus are found frequently in DNA, averaging once every 64 nucleotides. This modest sequence selectivity ensures that at least some branch migration occurs before resolution. Without this sequence selectivity, RuvC might simply cleave Holliday junctions as soon as they are formed, thereby restricting the region of DNA that participates in strand exchange.

HOMOLOGOUS RECOMBINATION IN EUKARYOTES

Homologous Recombination Has Additional Functions in Eukaryotes

As we have just described, homologous recombination in bacteria is required to repair DBSs in DNA, to restart collapsed replication forks, and to allow a cell's chromosomal DNA to recombine with DNA that enters via phage infection or conjugation. Homologous recombination is also required for DNA repair and the restarting of collapsed replication forks in eukaryotic cells. This requirement is illustrated by the fact that cells with defects in the proteins that promote recombination are hypersensitive to DNA-damaging agents, especially those that break DNA strands. Furthermore, animals carrying mutations that interfere with homologous recombination are predisposed to certain types of

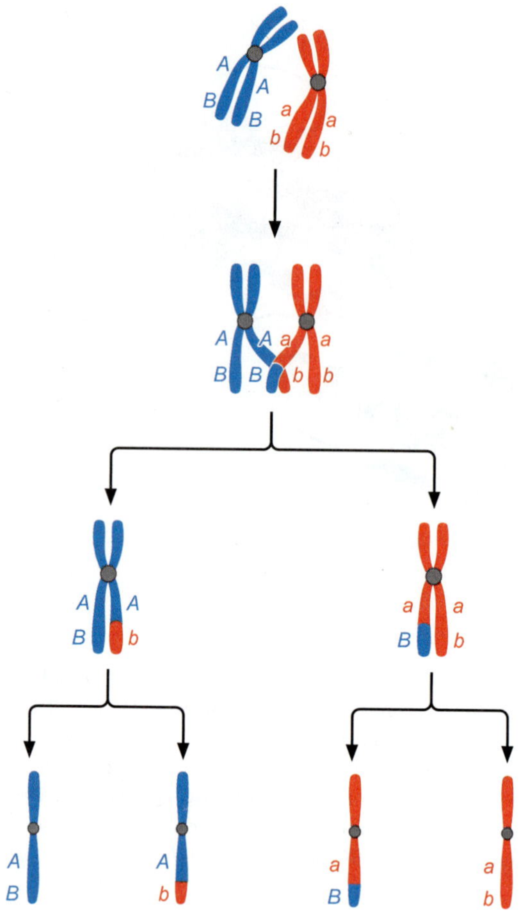

FIGURE 10-15 DNA dynamics during meiosis. Here, only one type of chromosome is shown for clarity. The two homologs are shown, in red and blue, after they have been duplicated by a round of DNA replication. Homologous recombination is required to pair these homologous chromosomes in preparation for the first nuclear division. This recombination can also lead to crossing over, as is shown here between the *A* and *B* genes.

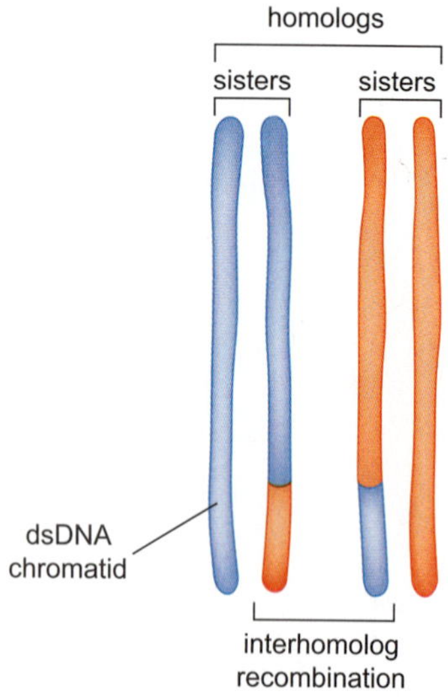

FIGURE 10-16 Meiotic recombination between homologous chromatids. Each structure shown is a replicated, double-stranded DNA molecule called a chromatid. The pairs are called sister chromatids, and recombination that occurs between nonsister pairs is mediated by Dmc1 (see Fig. 10-19).

cancer. However, as we discuss below, homologous recombination plays important additional roles in eukaryotic organisms. Most importantly, homologous recombination is critical for meiosis. During meiosis, homologous recombination is *required* for proper chromosome pairing and thus for maintaining the integrity of the genome. This recombination also reshuffles genes between the parental chromosomes, ensuring variation in the sets of genes passed to the next generation.

Homologous Recombination Is Required for Chromosome Segregation during Meiosis

As we saw in Chapter 7, meiosis involves two rounds of nuclear division, resulting in a reduction of the DNA content from the normal content of diploid cells (2N) to the content present in gametes (1N). Figure 10-15 schematically shows how the chromosomes are configured during these two division cycles. Before division, the cell has two copies of each chromosome (the homologs), one each that was inherited from its two parents. During S phase, these chromosomes are replicated to give a total DNA content of 4N (Fig. 10-16). The products of replication—that is, the sister chromatids—stay together. Then, in preparation for the first nuclear division, these *duplicated homologous chromosomes must pair* and align at the center of the cell. It is this pairing of homologs that requires homologous recombination (Fig. 10-15). These events are carefully timed. Recombination must be complete before the first nuclear di-

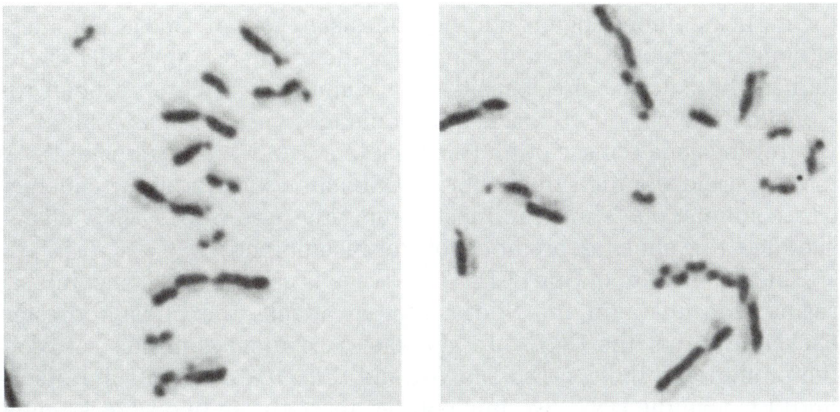

FIGURE 10-17 Cytological view of crossing over. Reciprocal crossing over directly visualized in hamster cells in tissue culture. Chromosomes whose DNA contains bromodeoxyuridine in place of thymidine in both strands appear light after treatment with Giemsa stain, whereas those containing DNA substituted in only one strand appear dark. After two generations of growth in bromodeoxyuridine, one newly replicated chomatid has only one of its strands substituted, whereas its sister has both substituted. Thus, sister chromatids can be distinguished by staining. Then, crossovers are easily detected as alternating lengths of light and dark (left). Similar recombinant chromosomes are also seen when mitotically growing cells are treated with a DNA-damaging agent (right). (Courtesy of Sheldon Wolff and Jody Bodycote.)

vision to allow the homologs to properly align and then separate. During this process, sister chromatids remain paired (Fig. 10-16; see also Fig. 7-16). Then, in the second nuclear division, it is the sister chromatids that separate. The products of this division are the four gametes, each with one copy of each chromosome (i.e., the 1N DNA content).

In the absence of recombination, chromosomes often fail to align properly for the first meiotic division, and, as a result, there is a high incidence of chromosome loss. This improper segregation of chromosomes, called **nondisjunction**, leads to a large number of gametes without the correct chromosome complement. Gametes with either too few or too many chromosomes cannot develop properly once fertilized; thus, a failure in homologous recombination is often reflected in poor fertility. The homologous recombination events that occur during meiosis are called **meiotic recombination**.

Meiotic recombination also frequently gives rise to crossing over between genes on the two homologous parental chromosomes. This genetic exchange, shown schematically in Figure 10-16, can be observed cytologically (Fig. 10-17, left panel). An important consequence is that the alleles present on the parental DNA molecules are reassorted for the next generation.

Programmed Generation of Double-Stranded DNA Breaks Occurs during Meiosis

The developmental program needed for cells to successfully complete meiosis involves turning on the expression of many genes that are not needed during normal growth. One of these is *SPO11*. This gene encodes a protein that introduces DBSs in chromosomal DNA to initiate meiotic recombination.

The Spo11 protein cuts the DNA at many chromosomal locations, with little sequence selectivity, but at a very specific time during meiosis. Spo11-mediated DNA cleavage occurs right around the time when the replicated homologous chromosomes start to pair. Spo11 cut sites, although frequent, are not randomly distributed along the DNA. Rather, the cut sites are located most commonly in chromosomal regions that are not tightly packed with nucleosomes, such as promoters controlling gene transcription (see Chapters 7 and 17). Regions of DNA that experience a high frequency of DSBs also show a high frequency of recombination. Thus, the most commonly used Spo11 DNA cleavage sites, like chi sites, are hot spots for recombination.

The mechanism of Spo11 DNA cleavage is as follows. A specific tyrosine side chain in the Spo11 protein attacks the phosphodiester backbone to cut the DNA and generate a covalent complex between the protein and the severed DNA strand (Fig. 10-18). Two subunits of Spo11 cleave the DNA two nucleotides apart on the two DNA strands to make a staggered DSB. Spo11 shares this DNA cleavage mechanism with the DNA topoisomerases and the site-specific recombinases (see Chapters 6 and 11). Protein sequence comparisons reveal that Spo11 appears to be a distant cousin of these enzymes.

The fact that Spo11 cleavage involves a covalent protein–DNA complex has two consequences. First, the 5´ ends of the DNA at the site of Spo11 cleavage are covalently bound to the enzyme. It is these Spo11-linked 5´ DNA ends that are the initial sites of DNA processing

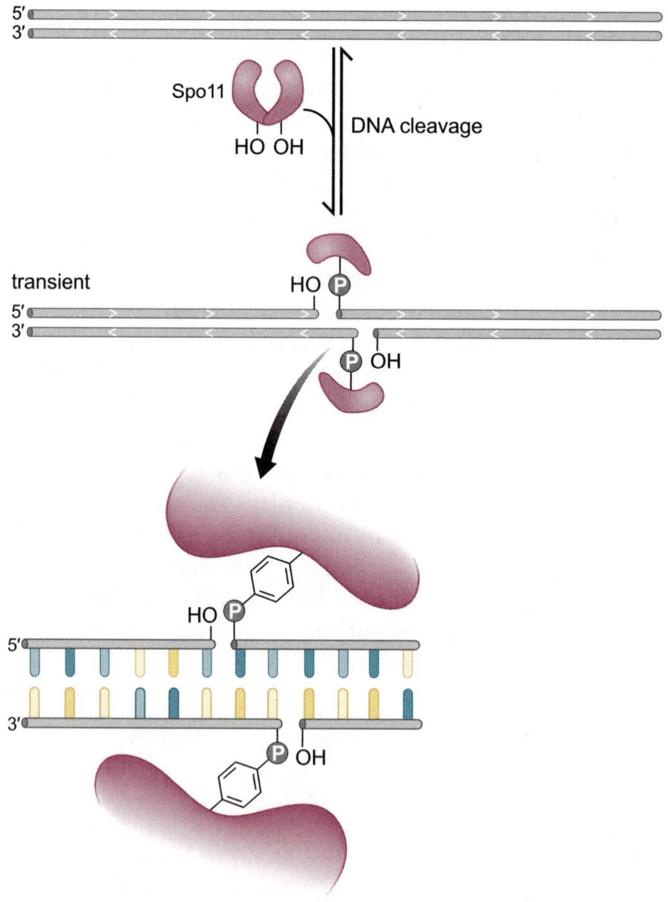

FIGURE 10-18 Mechanism of cleavage by Spo11. The OH group of a tyrosine in the Spo11 protein attacks the DNA to form a covalent protein–DNA linkage. Two subunits of Spo11 are required to generate a double-stranded DNA break, one to attack each of the two DNA strands. Note that because of this cleavage mechanism, the DSB can be resealed by the simple reversal of the cleavage reaction.

to create the ssDNA tails required for assembly of RecA-like proteins and initiation of DNA strand invasion (see below). Second, the energy of the cleaved DNA phosphodiester bond is stored in the bound protein–DNA linkage, and so the DNA strands can be resealed by a simple reversal of the cleavage reaction (see Fig. 11-5 for chemical mechanism). This resealing can occur when cells receive a signal to stop proceeding with meiosis.

MRX Protein Processes the Cleaved DNA Ends for Assembly of the RecA-like Strand-Exchange Proteins

The DNA at the site of the Spo11-catalyzed DSB is processed to generate single-strand regions needed for assembly of the RecA-like strand-exchange proteins. As was observed in the RecBCD pathway from bacteria, this processing generates long segments of ssDNA that terminate in 3´ ends (Fig. 10-19). During meiotic recombination, the MRX-en-

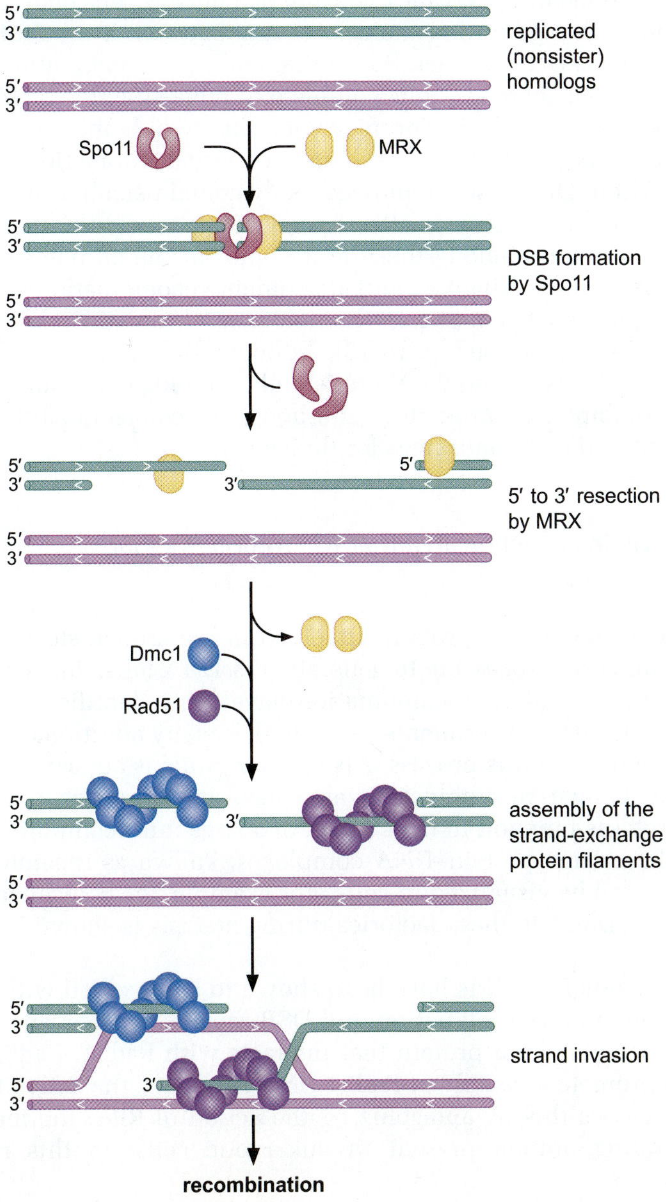

FIGURE 10-19 Overview of meiotic recombination pathway. Formation of the DSBs during meiosis requires the presence of both Spo11 and the MRX complex. This observation suggests that DSB formation and subsequent strand processing are normally coupled by the coordinated action of several proteins. MRX protein is responsible for resection of the 5´-ending strands at the break site. The strand-exchange proteins Dmc1 and Rad51 then assemble on the ssDNA tails. Both proteins participate in recombination, but how they work together is not known. They are shown forming separate filaments for clarity. (Redrawn, with permission, from Lichten M. 2001. *Curr. Biol.* 11: R253–R256, Fig 2. © Elsevier.)

zyme complex is responsible for this DNA-processing event. This complex, although not homologous to RecBCD, is also a multisubunit DNA nuclease. MRX is composed of protein subunits called Mre11, Rad50, and Xrs2; the first letters of these subunits give the complex its name.

Processing of the DNA at the break site occurs exclusively on the DNA strand that terminates with a 5′ end—that is, the strands covalently attached to the Spo11 protein (as described above). The strands terminating with 3′ ends are not degraded. This DNA-processing reaction is therefore called 5′ to 3′ resection. The MRX-dependent 5′ to 3′ resection generates the long ssDNA tails with 3′ ends that are often 1 kb or longer. The MRX complex is also thought to remove the DNA-linked Spo11.

Dmc1 Is a RecA-like Protein That Specifically Functions in Meiotic Recombination

Eukaryotes encode two well-characterized homologs of the bacterial RecA protein: Rad51 and Dmc1. Both proteins function in meiotic recombination. Whereas Rad51 is widely expressed in cells dividing mitotically and meiotically, Dmc1 is expressed only as cells enter meiosis.

Strand exchange during meiosis occurs between a particular type of homologous DNA partner. Recall that meiotic recombination occurs at a time when there are four complete, double-stranded DNA molecules representing each chromosome: the two homologs each of which have been copied to generate two sister chromatids (see Fig. 10-16). Although the two homologs likely contain small sequence differences and carry distinct alleles for various genes, the majority of the DNA sequence among these four copies of the chromosome will be identical. Interestingly, Dmc1-dependent recombination is preferentially between the *nonsister* homologous chromatids, rather than between the sisters (see Fig. 10-16). Although the mechanistic basis of this selectivity is unknown, there is a clear biological rationale: meiotic recombination promotes interhomolog connections to assist alignment of the chromosomes for division.

Many Proteins Function Together to Promote Meiotic Recombination

As we have described, proteins involved in the critical stages of DSB formation, DNA processing to generate 3′ ssDNA tails, and strand exchange during meiotic recombination have been identified and characterized. Genetic experiments indicate that many additional proteins also participate in this process. Also, many proteins appear to interact with the known recombination enzymes, and it seems likely that these proteins function in the context of a large multicomponent complex. These large protein–DNA complexes, known as **recombination factories**, can be visualized in cells. For example, the colocalization of Rad51 and Dmc1 to these factories during meiosis is shown in Figure 10-20.

Various other proteins have been shown to be involved with Rad51 to help promote recombination and DSB repair. Rad52 is another essential recombination protein that interacts with Rad51. Rad52 functions to promote assembly of Rad51 DNA filaments, the active form of Rad51. It does this by antagonizing the action of RPA, the major ssDNA-binding protein present in eukaryotic cells. In this respect,

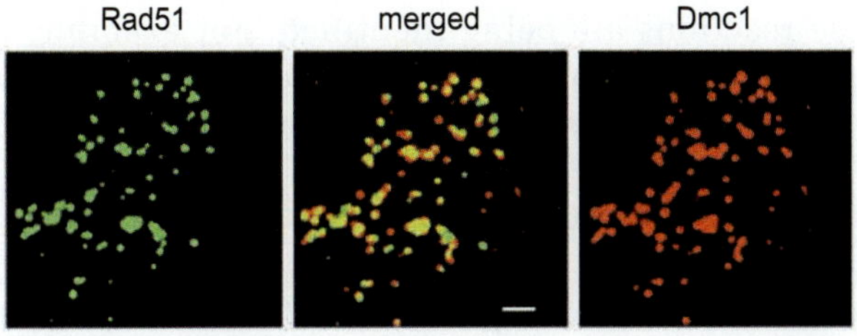

FIGURE 10-20 Colocalizations of the Rad51 and Dmc1 proteins to "recombination factories" in cells undergoing meiosis. Proteins were detected by immunostaining with fluorescently labeled antibodies to Rad51 (green) and Dmc1 (red). When the two proteins colocalize, the merged image appears yellow. (Reprinted, with permission, from Shinohara M. et al. 2000. *Proc. Natl. Acad. Sci.* 97: 10814–10819, Fig. 1A. © National Academy of Sciences.)

Rad52 shares an activity with the *E. coli* RecBCD protein, which, as we learned, helps RecA load onto ssDNA that would otherwise have been bound by SSB. Rad52 protein also promotes the annealing and base pairing of complementary ssDNA molecules, and this activity may also play a role in the strand-pairing reactions that occur during initiation of recombination. The product of the *BRCA2* gene also participates in Rad51-mediated DSB repair (see Box 10-2, The Product of the Tumor Suppressor Gene *BRCA2* Interacts with Rad51 Protein and Controls Genome Stability).

By analogy with bacteria, we expect that eukaryotic cells encode proteins that promote the branch migration and Holliday junction resolution steps of recombination. In fact, enzymes capable of pro-

MEDICAL CONNECTIONS

Box 10-2 The Product of the Tumor Suppressor Gene *BRCA2* Interacts with Rad51 Protein and Controls Genome Stability

The *BRCA2* gene is important for maintaining genome stability. In humans, mutations in *BRCA2* are thought to be responsible for half the familial breast cancers. This cancer predisposition appears attributable, at least in part, to a direct role of the BRCA2 protein in Rad51-mediated DSB repair. When cells are subjected to agents that damage DNA, Rad51 foci assemble as an apparent prerequisite to activation of the repair functions. Cells with defects in BRCA2 fail to assemble these foci in the nucleus of the damaged cells and have a corresponding defect in repair of DNA breaks. BRCA2 makes direct protein–protein contacts with Rad51 (see Box 10-2 Fig. 1), and these interactions are likely important for recruiting Rad51 to the proper cellular location for repair, as well as modulating the activity of the protein. The strong phenotype associated with the *BRCA2* mutations therefore illustrates the central importance of DSB repair and Rad51-dependent homologous recombination in eukaryotes, including humans.

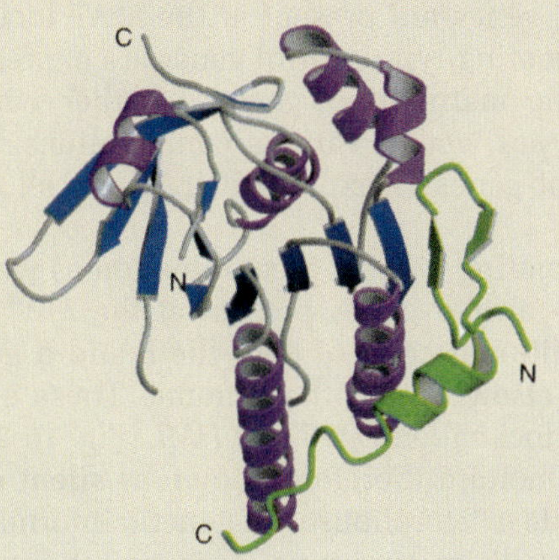

BOX 10-2 FIGURE 1 Structure of the complex between Rad51 and BRCA2 repeat motif. Various biochemical and structural studies have shown that specific regions within BRCA2, conserved repeat sequences known as the BRC motifs, are the major sites of interaction with Rad51. One of these motifs, BRC repeat 4 (BRC4), has been shown to bind Rad51 with high affinity. Structural analysis has revealed more precisely how BRC4 forms a complex with Rad51. In this view, the α helices of Rad51 are shown in purple and the β strands in blue, whereas the peptide BRC repeat sequence is shown in green. The amino and carboxyl termini are marked for each sequence. (Reprinted, with permission, from Pellegrini L. et al. 2002. *Nature* 420: 287–293, Fig. 1a. © Macmillan.)

moting these reactions are being identified. For example, a complex containing a Rad51-like protein, called Rad51C, and a second protein, called XRCC3, has been found to contain Holliday junction resolvase activity.

As we have seen, meiotic recombination aligns homologous chromosomes and promotes genetic exchange between them. These recombination reactions often lead to crossing over between the parental chromosomes. Recall, however, that depending on how the Holliday junctions in the recombination intermediates are resolved, recombination via the DSB-repair pathway can also give rise to noncrossover products (see above). These events may provide the essential chromosome-pairing function needed for a successful meiotic division, yet leave no detectable change in the genetic makeup of the chromosomes.

But even noncrossover recombination can have genetic consequences, such as giving rise to a **gene conversion** event. Gene conversion happens when an allele of a gene is lost and replaced by an alternative allele. Examples of how gene conversion can occur both in mitotically growing cells and during meiosis are described in the following sections.

MATING-TYPE SWITCHING

In addition to promoting DNA pairing, DNA repair, and genetic exchange, homologous recombination can also serve to change the DNA sequence at a specific chromosomal location. This type of recombination is sometimes used to regulate gene expression. For example, recombination controls the mating type of the budding yeast *S. cerevisiae* by switching which mating-type genes are present at a specific location that is being expressed in that organism's genome.

S. cerevisiae is a single-cell eukaryote that can exist as any of three different cell types (see Chapter 22). Haploid *S. cerevisiae* cells can be either of two mating types, **a** or α. In addition, when an **a** and α cell come in close proximity, they can fuse (i.e., "mate") to form an **a**/α diploid cell. The **a**/α cell may then go through meiosis to form two haploid **a** cells and two haploid α cells.

The mating-type genes encode transcriptional regulators. These regulators control expression of target genes whose products define each cell type. The mating-type genes expressed in a given cell are those found at the mating-type locus (***MAT* locus**) in that cell (Fig. 10-21). Thus, in **a** cells the **a**1 gene is present at the *MAT* locus, whereas in α cells, the α1 and α2 genes are present at the *MAT* locus. In the diploid cell, both sets of mating-type control genes are expressed. The regulators encoded by the mating-type genes, together with others found in all three cell types, act in various combinations to ensure that the correct pattern of genes is expressed in each cell type (see Chapter 17).

Cells can switch their mating type by recombination as we now describe. In addition to the **a** or α genes present at the *MAT* locus in each cell, there is an additional copy of both the *a* and α genes present (but not expressed) elsewhere in the genome. These additional silent copies are found at loci called *HMR* and *HML* (Fig. 10-21).

These *HMR* and *HML* loci are therefore known as **silent cassettes**. Their function is to provide a "storehouse" of genetic information that can be used to switch a cell's mating type. This switch requires the

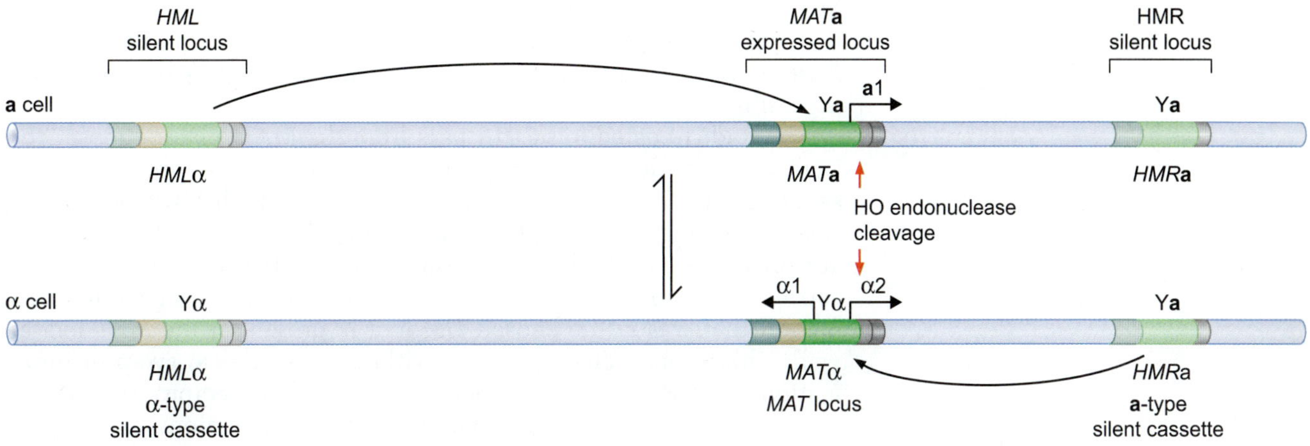

FIGURE 10-21 Genetic loci encoding mating-type information. Although chromosome III carries three mating-type loci, only the genes at the *MAT* locus are expressed. *HML* encodes a silent copy of the α genes, whereas *HMR* encodes a silent copy of the **a** genes. When recombination occurs between *MAT* and *HML*, **a** cells switch to α cells. When recombination occurs between *MAT* and *HMR*, α cells switch to **a** cells. (Adapted, with permission, from Haber J.E. 1998. *Annu. Rev. Genet.* 32: 561–599, Fig. 3. © Annual Reviews.)

transfer of genetic information from the *HM* sites to the *MAT* locus via homologous recombination.

Mating-Type Switching Is Initiated by a Site-Specific Double-Strand Break

Mating-type switching is initiated by the introduction of a DSB at the *MAT* locus. This reaction is performed by a specialized DNA-cleaving enzyme, called the **HO endonuclease**. Expression of the HO gene is tightly regulated to ensure that switching occurs only when it should. The mechanisms responsible for this regulation are discussed in Chapter 17. HO is a sequence-specific endonuclease; the only sites in the yeast chromosome that carry HO recognition sequences are the mating-type loci. HO cutting introduces a staggered break in the chromosome. In contrast to Spo11 cleavage, HO simply hydrolyzes the DNA and does not remain covalently linked to the cut strands.

5′ to 3′ resection of the DNA at the site of the HO-induced break occurs by the same mechanism used during meiotic recombination. Thus, resection depends on the MRX-protein complex and is specific for the strands that terminate with 5′ ends. In contrast, the strands terminating with 3′ ends are very stable and not subjected to nuclease digestion. Once the long 3′ ssDNA tails have been generated, this DNA becomes coated by the Rad51 and Rad52 proteins (as well as other proteins that help the assembly of the recombinogenic protein–DNA complex). These Rad51 protein–coated strands then search for homologous chromosomal regions to initiate strand invasion and genetic exchange.

Mating-type switching is unidirectional. That is, sequence information (although not the actual DNA segment) is "moved" to the *MAT* locus, from *HMR* and *HML*, but information never "goes" in the other direction. Thus, the cut *MAT* locus is always the "recipient" partner during recombination, and the *HMR* and *HML* sites remain unchanged by the recombination process. This directionality stems from the fact that HO endonuclease cannot cleave its recognition sequence at either *HML* or *HMR* because the chromatin structure renders these sites inaccessible to this enzyme.

The Rad51-coated 3´ ssDNA tails from the MAT locus "choose" the DNA at either the *HMR* or *HML* locus for strand invasion. If the DNA sequence at *MAT* is **a**, then invasion will occur with *HML*, which carries the "storage" copy of the α sequences. In contrast, if the α genes are present at *MAT*, then invasion occurs with *HMR*, the locus that carries the stored **a** sequences. After recombination, the genetic information that was at the chosen *HM* loci is present at the *MAT* loci as well. This genetic change occurs without a reciprocal swap of information from *MAT* to the *HR* loci. This type of nonreciprocal recombination event is a specialized example of gene conversion.

Mating-Type Switching Is a Gene Conversion Event and Not Associated with Crossing Over

Although the DSB-repair pathway could explain the mechanism of mating-type switch recombination, substantial experimental evidence indicates that after the strand invasion step, this recombination pathway diverges from the DSB-repair mechanism. One hint that the mechanism is distinct is that the crossover class of recombination products is never observed during mating-type switching. Recall that in the DSB-repair pathway, resolution of the Holliday junction intermediates gives two classes of products: the splice, or crossover class, and the patch, or noncrossover, class (see Fig. 10-3). According to the DSB-repair model, these two types of products are predicted to occur at a similar frequency, yet, in mating-type switching, crossover products are never observed. Therefore, models for recombination that do not involve resolution of Holliday junction intermediates better explain mating-type switching.

To explain gene conversion without crossing over, a new recombination model termed **synthesis-dependent strand annealing (SDSA)** has been proposed. Figure 10-22 shows how mating-type switching can occur using this mechanism. The initiating event is, as described above, the introduction of a DSB at the recombination site (Fig. 10-22a). After strand invasion (Fig. 10-22b), the invading 3´ end serves as the primer to initiate new DNA synthesis (Fig, 10-22c,d). Remarkably, in contrast to what occurs during the DSB-repair pathway, a complete replication fork is assembled at this site. Both leading-strand and lagging-strand DNA synthesis occurs. In contrast to normal DNA replication, however, the newly synthesized strands are displaced from the template. As a result, a new double-stranded DNA segment is synthesized, joined to the DNA site that was originally cut by HO, and resected by MRX. This new segment has the sequence of the DNA segment used as the template (*HMR***a** in Fig. 10-22).

Completing recombination requires that the other "old" DNA strand present at *MAT* (the 3´-ending strand not cleaved by MRX) be removed (the bottom strand in Fig. 10-22d). Then, the newly synthesized DNA—an exact copy of the information in the partner DNA molecule—replaces the information that was originally present. This mechanism nicely explains how gene conversion occurs without the need to cleave a Holliday junction. Thus, by this model, the absence of crossover products during mating-type recombination is no longer mysterious.

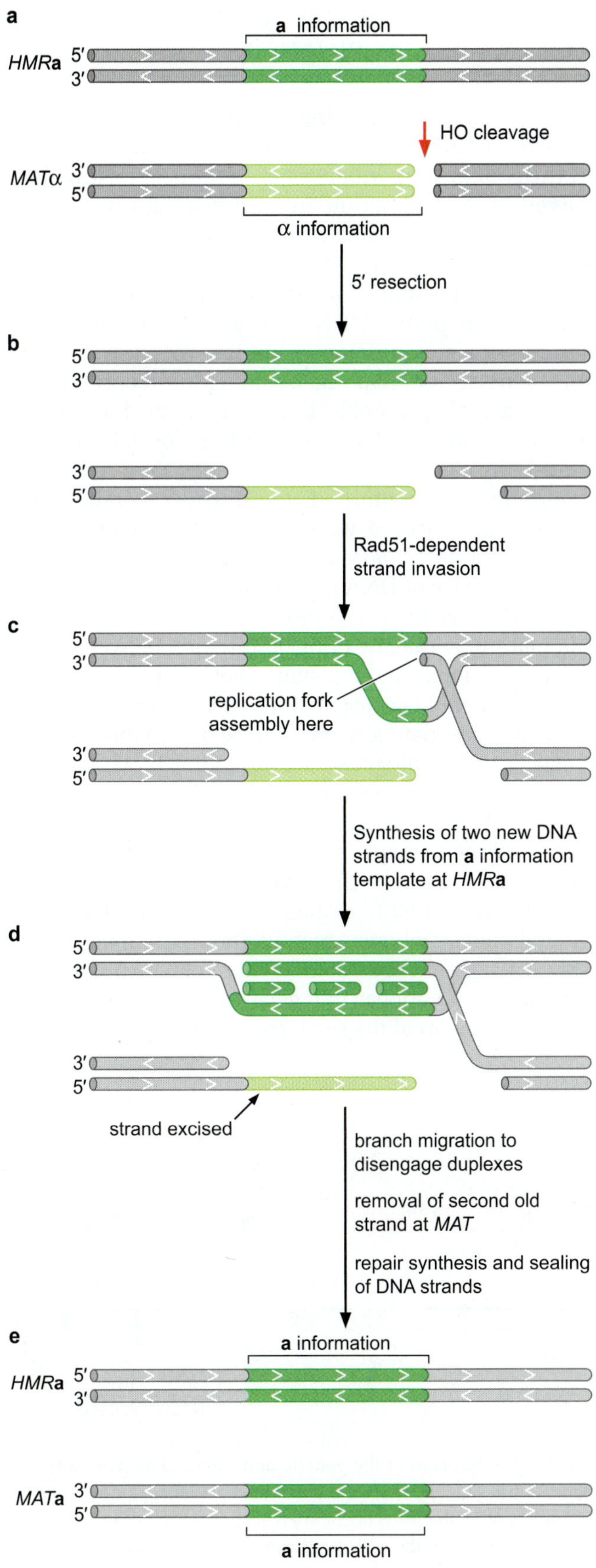

FIGURE 10-22 Recombination model for mating-type switching: Synthesis-dependent strand annealing (SDSA). Shown are the steps leading to gene conversion at the *MAT* locus. The *HMR* and *MAT* regions are shown in green; the region of *HMR* encoding the **a** information is represented in dark green, whereas the region of *MAT* encoding the α information is shown in lime green. Upon completion of process of SDSA, the α region originally present at *MAT* has been replaced by (i.e., converted to) the **a** information present in the *HMR* region.

GENETIC CONSEQUENCES OF THE MECHANISM OF HOMOLOGOUS RECOMBINATION

As we discussed at the beginning of this chapter, initial models for the mechanism of homologous recombination were formulated largely to explain the genetic consequences of the process. Now that the basic steps involved in recombination are understood, it is useful to review how the process of homologous recombination alters DNA molecules and thereby generates specific genetic changes.

A central feature of homologous recombination is that it can occur between any two regions of DNA, regardless of the sequence, provided these regions are sufficiently similar. We now understand why this is true; none of the steps in homologous recombination require recognition of a highly specific DNA sequence. For steps that have some sequence preference (such as the transformation of RecBCD by chi sites and DNA cleavage by RuvC protein), the preferred sequences are very common. The committed step during recombination between two DNA molecules occurs when a strand-exchange protein of the RecA family successfully pairs the molecules, a process dictated only by the normal capacity of DNA strands to form proper base pairs.

A corollary of the fact that recombination is generally independent of sequence is that the frequency of recombination between any two genes is generally proportional to the distance between those genes. This proportionality is observed because regions of DNA are, in general, equally likely to be used to initiate a successful recombination event. This fundamental aspect of homologous recombination is what makes it possible to use recombination frequencies to generate useful genetic maps that display the order and spacing of genes along a chromosome.

Distortions in genetic maps compared to physical maps occur when a region of DNA does not have the "average" probability of participating in recombination (Fig. 10-23). Regions with a higher than average probability are "hot spots," whereas regions that participate less commonly than an average segment are "cold." Therefore, two genes that have a hot spot between them appear in a genetic map to be farther apart than is true in a physical map of the same region. In contrast, genes separated

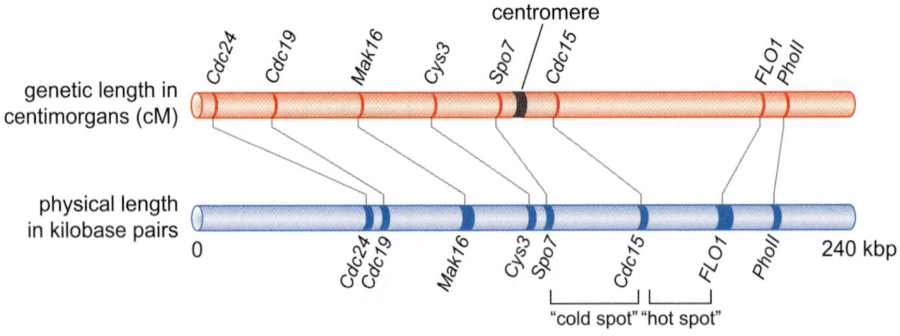

FIGURE 10-23 Comparison of the genetic and physical maps of a typical region of a yeast chromosome. Markers show the location of various genes. Notice in the region between *Spo7* and *Cdc15* that the genetic map is contracted because of a low frequency of crossing over. In contrast, in the region between *Cdc15* and *FL01* the genetic map is expanded because of a high frequency of crossing over. (Adapted, with permission, from Alberts B. et al. 2002. *Molecular biology of the cell*, 4th ed., p. 1138, Fig. 20-14. © Garland Science/Taylor & Francis LLC.)

by a "cold" interval appear by genetic mapping to be closer together than is true from their physical distance. We have encountered two examples for the molecular explanation of hot and cold spots in chromosomes. Regions near chi sites and Spo11 cleavage sites have a higher than average probability of initiating recombination and are "hot," whereas regions having few such sites are correspondingly "cold."

One Cause of Gene Conversion Is DNA Repair during Recombination

Another genetic consequence of homologous recombination is gene conversion. We have introduced the concept of gene conversion during the specialized recombination events responsible for mating-type switching in yeast. However, gene conversion is also commonly observed during normal homologous recombination events, such as those responsible for genetic exchange in bacteria and for pairing chromosomes during meiosis.

To illustrate gene conversion during meiotic recombination, consider a cell undergoing meiosis that has the *A* allele on one homolog and the *a* allele on the other. After DNA replication, four copies of this gene are present and the genotype would be *A A a a*. In the absence of gene conversion, two gametes carrying the *A* allele and two gametes carrying the *a* allele would be generated. If instead the gametes with genotypes *A*, *a*, *a*, *a* (or *A*, *A*, *A*, *a*) are formed, then a gene conversion event has occurred, in which one copy of the A gene has been converted into a (or vice versa). How might this conversion arise?

There are two ways that gene conversion can occur during the DSB-repair pathway. First, consider what would happen if the *A* gene was very close to the site of the DSB. In this case, when the 3′ ssDNA tails invade the homologous duplexes and are elongated, they may copy the *a* information, which could replace the *A* information in the product chromosome upon completion of recombination (see Fig. 10-4d).

The second mechanism of gene conversion involves the repair of base-pair mismatches that occur in the recombination intermediates. For example, if either strand invasion or branch migration includes the *A*/*a* gene, a segment of heteroduplex DNA carrying the *A* sequence on one strand and the *a* sequence on the other strand would be formed (Fig. 10-24; see also Fig. 10-2d inset). This region of DNA carrying base-pair mismatches could be recognized and acted upon by the cellular mismatch repair enzymes (which we discussed in Chapter 9). These enzymes are specialized for fixing base-pair mismatches in DNA. When they detect a mismatched base pair, these enzymes excise a short stretch of DNA from one of the two strands. A repair DNA polymerase then fills in the gap, now with the properly base-paired sequence. When working on recombination intermediates, the mismatch repair enzymes will likely choose randomly which strand to repair. Therefore, after their action, both strands will carry the sequence encoding either the *A* information or the *a* information (depending on which strand was "fixed" by the repair enzymes), and gene conversion will be observed.

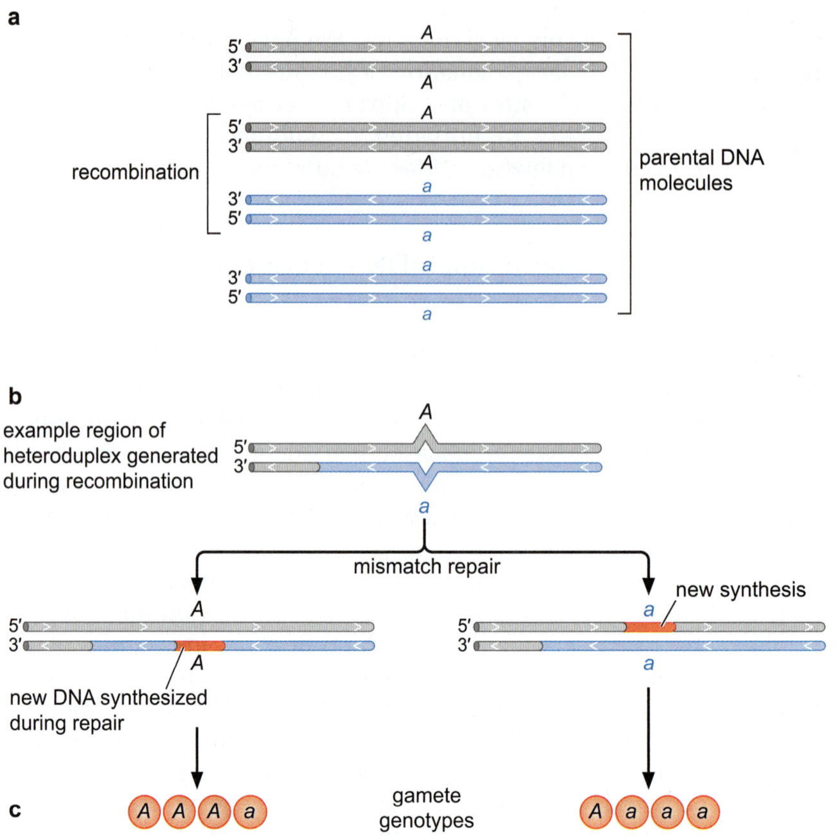

FIGURE 10-24 Mismatch repair of heteroduplex DNA within recombination intermediates can give rise to gene conversion.

SUMMARY

Homologous recombination occurs in all organisms, allowing for genetic exchange, the reassortment of genes along chromosomes, and the repair of broken DNA strands and collapsed replication forks. The recombination process involves the breaking and rejoining of DNA molecules. The double-strand repair pathway of homologous recombination well describes many recombination events. By this model, initiation of exchange requires that one of the two homologous DNA molecules have a double-strand break. The broken DNA ends are processed by DNA-degrading enzymes to generate single-stranded DNA segments. These single-strand regions participate in DNA pairing with the homologous partner DNA. Once pairing occurs, the two DNA molecules are joined by a branched structure in the DNA called a Holliday junction. Cutting the DNA at the Holliday junction resolves the junction and terminates recombination. Holliday junctions can be cut in two alternative ways. One way generates crossover products, in which regions from two parental DNA molecules are now covalently joined. The alternative way of cleaving the junction generates a "patch" of recombined DNA but does not result in crossing over.

Cells encode enzymes that catalyze all of the steps in homologous recombination. Key enzymes are the strand-exchange proteins. Of these, *E. coli* RecA is the premier example; RecA-like proteins are found in all organisms. RecA-like strand-exchange proteins promote the search for homologous sequences between two DNA molecules and the exchange of DNA strands within the recombination intermediate. RecA functions as a large protein–DNA complex, known as the RecA filament. Eukaryotic cells encode two strand-exchange proteins, called Rad51 and Dmc1. Other important recombination enzymes are the DNA-cleaving enzymes that generate double-strand breaks in DNA to initiate recombination; these proteins appear to be found only in eukaryotes and include Spo11 and HO. Nucleases that process the DNA at the break site to generate the required single-strand regions include the RecBCD enzyme in prokaryotes and the MRX-enzyme complex in eukaryotes. Additional enzymes promote the movement (branch migration) and cleavage (resolution) of Holliday junctions.

During meiosis, recombination is essential for the proper homologous pairing of chromosomes prior to the first nuclear division. Therefore, recombination is highly regulated to ensure that it occurs on all chromosomes. The Spo11 DNA-cutting enzyme and the Dmc1 strand-exchange protein are both specifically involved in these recombination reactions. Homologous recombination is also sometimes used to control gene expression. The mating-type switching of yeast is an excellent example of this type of regulation; it is also an example of gene conversion. Analysis of the mechanism of mating-type switching has generated a new class of models to describe some homologous recombination events called synthesis-dependent strand annealing. This mechanism gives rise to the gene-conversion-type genetic exchange products but does not result in crossing over.

BIBLIOGRAPHY

Books

Brown T.A. 2007. *Genomes*, 3rd ed. Garland Science, New York.

Griffiths A.J.F., Miller J.H., Suzuki D.T., Lewontin R.C., and Gelbart W.M. 2000. *An introduction to genetic analysis*, 7th ed. W.H. Freeman, New York.

Recombination in Bacteria

Court D.L., Sawitzke J.A., and Thomason L.C. 2002. Genetic engineering using homologous recombination. *Annu. Rev. Genet.* **36:** 361–388.

Cox M.M. 2001. Recombinational DNA repair of damaged replication forks in *Escherichia coli:* Questions. *Annu. Rev. Genet.* **35:** 53–82.

Kowalczykowski S.C., Dixon D.A., Eggleston A.K., Lauder S.D., and Rehrauer W.M. 1994. Biochemistry of homologous recombination in *Escherichia coli. Microbiol. Rev.* **58:** 401–465.

Lusetti S.L. and Cox M.M. 2002. The bacterial RecA protein and the recombinatorial DNA repair of stalled replication forks. *Annu. Rev. Biochem.* **71:** 71–100.

Smith G.R. 2001. Homologous recombination near and far from DNA breaks: Alternative roles and contrasting views. *Annu. Rev. Genet.* **35:** 243–274.

Recombination in Eukaryotes

Eichler E.E. and Sankoff D. 2003. Structural dynamics of eukaryotic chromosome evolution. *Science* **301:** 793–797.

Keeney S. 2001. Mechanism and control of meiotic recombination initiation. *Curr. Top. Dev. Biol.* **52:** 1–53.

Page S.L. and Hawley R.S. 2003. Chromosome choreography: The meiotic ballet. *Science* **301:** 785–789.

Pâques F. and Haber J.E. 1999. Multiple pathways of recombination induced by double-strand breaks in *Saccharomyces cerevisiae. Microbiol. Mol. Biol. Rev.* **63:** 349–404.

Pastink A., Eeken J.C., and Lohman P.H. 2001. Genomic integrity and the repair of double-strand DNA breaks. *Mutat. Res.* **480-481:** 37–50.

Prado F., Cortes-Ledesma F., Huertas P., and Aguilera A. 2003. Mitotic recombination in *Saccharomyces cerevisiae. Curr. Genet.* **42:** 185–198.

Symington L.S. 2002. Role of *RAD52* epistasis group genes in homologous recombination and double-strand break repair. *Microbiol. Mol. Biol. Rev.* **66:** 630–670.

van den Bosch M., Lohman P.H., and Pastink A. 2002. DNA double-strand break repair by homologous recombination. *Biol. Chem.* **383:** 873–892.

West S.C. 2003. Molecular views of recombination proteins and their control. *Nat. Rev. Mol. Cell Biol.* **4:** 435–445.

Mating-Type Switching in Yeast

Haber J.E. 2002. Switching of *Saccharomyces cerevisiae* mating-type genes. In *Mobile DNA II* (ed. N.L. Craig et al.), pp. 927–952. ASM Press, Washington, D.C.

CHAPTER

11 Site-Specific Recombination and Transposition of DNA

DNA IS A VERY STABLE MOLECULE. DNA replication, repair, and homologous recombination, as we have learned in the previous chapters, all occur with high fidelity. These processes serve to ensure that the genomes of an organism are nearly identical from one generation to the next. Importantly, however, there are also genetic processes that rearrange DNA sequences and thus lead to a more dynamic genome structure. These processes are the subject of this chapter.

Two classes of genetic recombination, **conservative site-specific recombination (CSSR)** and **transpositional recombination** (generally called **transposition**), are responsible for many important DNA rearrangements. CSSR is recombination between two defined sequence elements (Fig. 11-1). Transposition, in contrast, is recombination between specific sequences and nonspecific DNA sites. The biological processes promoted by these recombination reactions include the insertion of viral genomes into the DNA of the host cell during infection, the inversion of DNA segments to alter gene structure, and the movement of **transposable elements**—often called "jumping" genes—from one chromosomal site to another.

The impact of these DNA rearrangements on chromosome structure and function is profound. In many organisms, transposition is the major source of spontaneous mutation, and nearly half the human genome consists of sequences derived from transposable elements (although most elements are currently inactive). Furthermore, as we shall see, both viral infection and development of the vertebrate immune system depend critically on these specialized DNA rearrangements.

Conservative site-specific recombination and transposition share key mechanistic features. Proteins known as **recombinases** recognize specific sequences where recombination will occur within a DNA molecule. The recombinases bring these specific sites together to form a protein–DNA complex bridging the DNA sites, known as the **synaptic complex**. Within the synaptic complex, the recombinase catalyzes the cleavage and rejoining of the DNA molecules either to invert a DNA segment or to move a segment to a new site. One recombinase protein is usually responsible for all of these steps. Both types of recombination are also carefully controlled such that the danger to the cell of introducing breaks in the DNA, and rearranging DNA segments in an unintended manner, is minimized. As we shall see, however, the two types of recombinations also have key mechanistic differences.

In the following sections, the simpler site-specific recombination reactions are introduced first, followed by a discussion of transposi-

OUTLINE

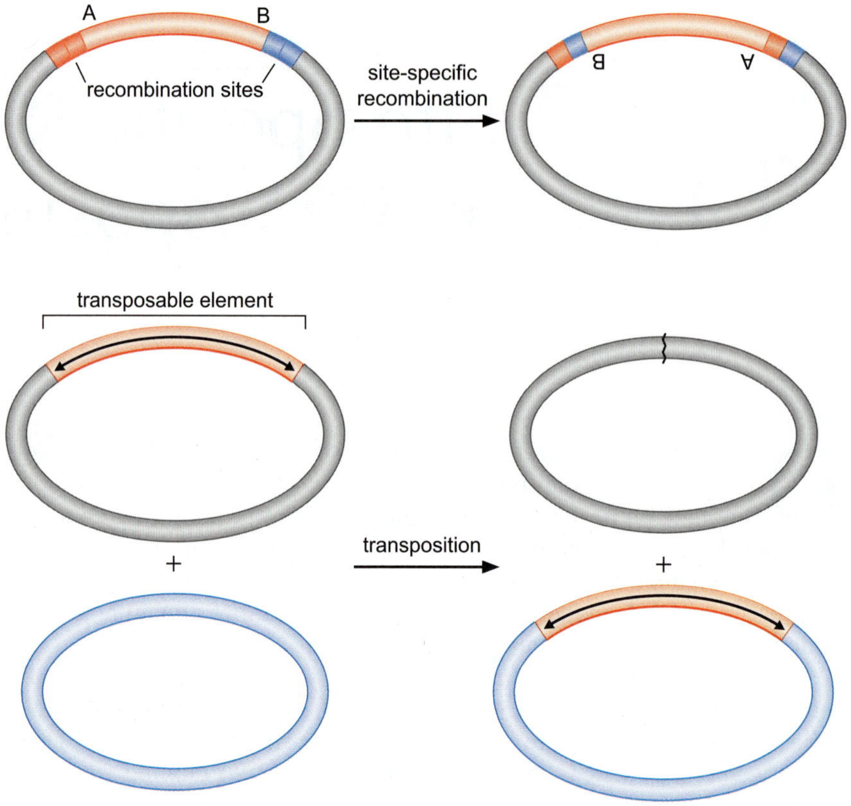

FIGURE 11-1 Two classes of genetic recombination. (Top panel) Example of site-specific recombination. Here, recombination between the red and blue recombination sites inverts the DNA segment carrying the A and B genes. (Bottom panel) Example of transposition in which the red transposable element excises from the gray DNA and inserts into an unrelated site in the blue DNA.

tion. Each of these sections is organized to describe general features of the mechanism first and then to provide some specific examples.

CONSERVATIVE SITE-SPECIFIC RECOMBINATION

Site-Specific Recombination Occurs at Specific DNA Sequences in the Target DNA

CSSR is responsible for many reactions in which a defined segment of DNA is rearranged. A key feature of these reactions is that the segment of DNA that will be moved carries specific short sequence elements, called **recombination sites**, where DNA exchange occurs. An example of this type of recombination is the integration of the bacteriophage λ genome into the bacterial chromosome (Fig. 11-2 and Chapter 22).

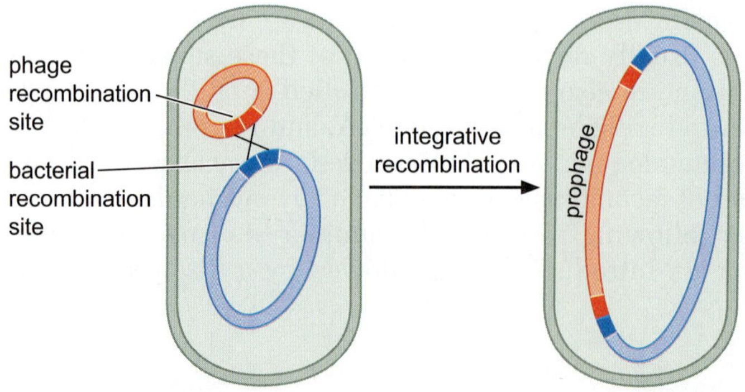

FIGURE 11-2 Integration of the λ genome into the chromosome of the host cell. DNA exchange occurs specifically between the recombination sites on the two DNA molecules. The relative lengths of the λ and cellular chromosomes are not shown to scale.

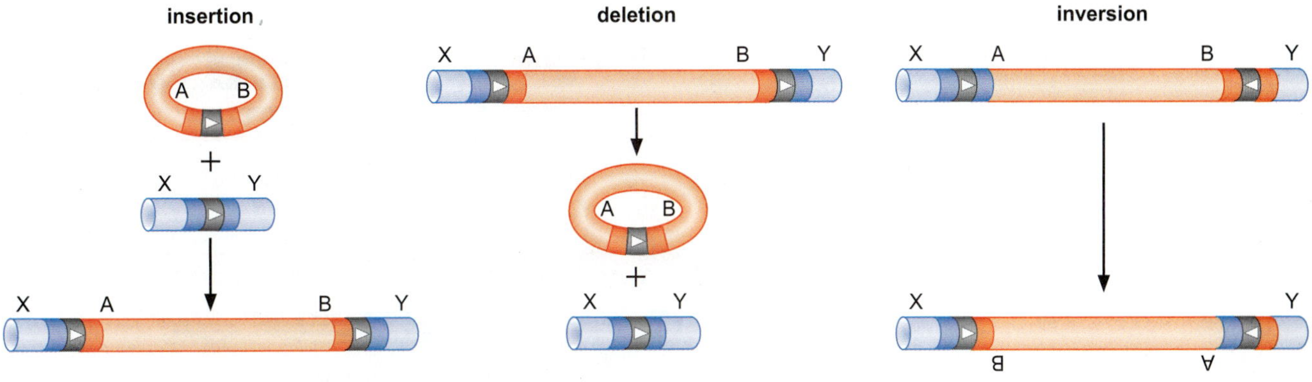

FIGURE 11-3 Three types of CSSR recombination. In each case, it is the red segment of DNA that is moved or rearranged during recombination. A, B, X, and Y denote genes that lie within the different segments of DNA. The darker red and blue boxes are the recombinase recognition sequences and the black arrows are the crossover regions. These sequence elements together form the recombination sites.

During λ integration, recombination always occurs at exactly the same nucleotide sequence within two recombination sites, one on the phage DNA and the other on the bacterial DNA. Recombination sites carry two classes of sequence elements: sequences specifically bound by the recombinases and sequences where DNA cleavage and rejoining occur. Recombination sites are often quite short, 20 bp or so, although they may be much longer and carry additional sequence motifs and protein-binding sites. Examples of the more complex recombination sites are discussed when we consider specific recombination examples.

CSSR can generate three different types of DNA rearrangements (Fig. 11-3): (1) insertion of a segment of DNA into a specific site (as occurs during bacteriophage λ DNA integration), (2) deletion of a DNA segment, or (3) inversion of a DNA segment. Whether recombination results in DNA insertion, deletion, or inversion depends on the organization of the recombinase recognition sites on the DNA molecule or molecules that participate in recombination.

To understand how the organization of recombination sites determines the type of DNA rearrangement, we must look at the sequence elements within the recombination sites in more detail (Fig. 11-4). Each recombination site is organized as a pair of **recombinase recognition sequences**, positioned symmetrically. These recognition sequences flank a central short asymmetric sequence, known as the **crossover region**, where DNA cleavage and rejoining occurs.

Because the crossover region is asymmetric, a given recombination site always has a defined polarity. The orientation of two sites present on a single DNA molecule will be related to each other in either an **inverted repeat** or a **direct repeat** manner. Recombination between a pair of inverted sites will invert the DNA segment between the two sites (Fig. 11-3, right panel). In contrast, recombination using the identical mechanism but occurring between sites organized as direct repeats deletes the DNA segment between the two sites (Fig. 11-3, middle panel). Finally, insertion specifically occurs when recombination sites on two different molecules are brought together for DNA exchange (Fig. 11-3, left panel). Examples of each of these three types of rearrangements are considered below, after a general discussion of the recombinases.

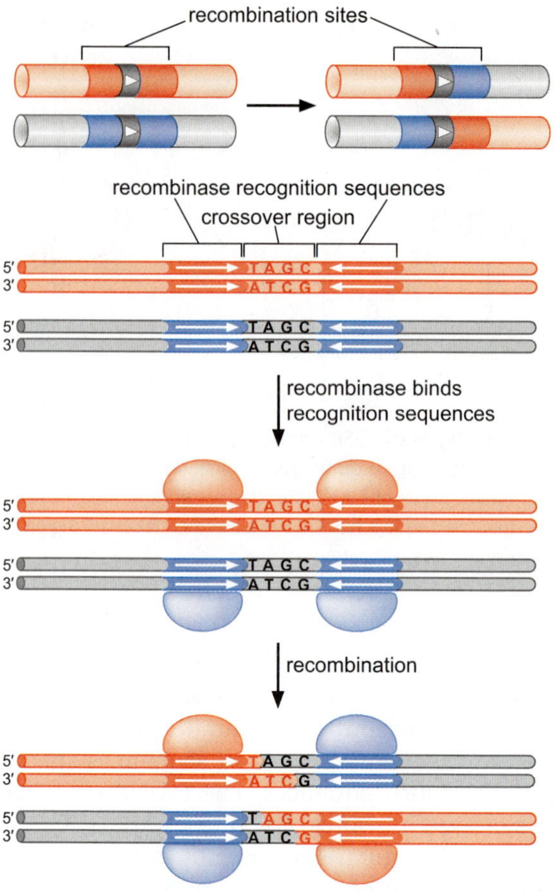

FIGURE 11-4 Structures involved in CSSR. The pair of symmetric recombinase recognition sequences flank the crossover region where recombination occurs. The subunits of the recombinase bind these recognition sites. Note that the sequence of the crossover region is not palindromic, resulting in an intrinstic asymmetry to the recombination sites. (Adapted, with permission, from Craig N. et al. 2002. *Mobile DNA II*, p. 4, Fig. 1. © ASM Press.)

Site-Specific Recombinases Cleave and Rejoin DNA Using a Covalent Protein–DNA Intermediate

There are two families of conservative site-specific recombinases: the **serine recombinases** and the **tyrosine recombinases**. Fundamental to the mechanism used by both families is that when they cleave the DNA, a covalent protein–DNA intermediate is generated. For the serine recombinases, the side chain of a serine residue within the protein's active site attacks a specific phosphodiester bond in the recombination site (Fig. 11-5). This reaction introduces a single-strand break in the DNA and simultaneously generates a covalent linkage between the serine and a phosphate at this DNA cleavage site. Likewise, for the tyrosine recombinases, it is the side chain of the active-site tyrosine that attacks and then becomes joined to the DNA. Table 11-1 classifies a number of important recombinases by family and biological function.

The covalent protein–DNA intermediate conserves the energy of the cleaved phosphodiester bond within the protein–DNA linkage. As a result, the DNA strands can be rejoined by reversal of the cleavage process. For reversal, an OH group from the cleaved DNA attacks the covalent bond that links the protein to the DNA. This process covalently seals the DNA break and regenerates the free (non–DNA bound) recombinase (see Fig. 11-5).

It is this mechanistic feature that contributes "conservative" to the CSSR name: it is called conservative because every DNA bond that is broken during the reaction is resealed by the recombinase. No external

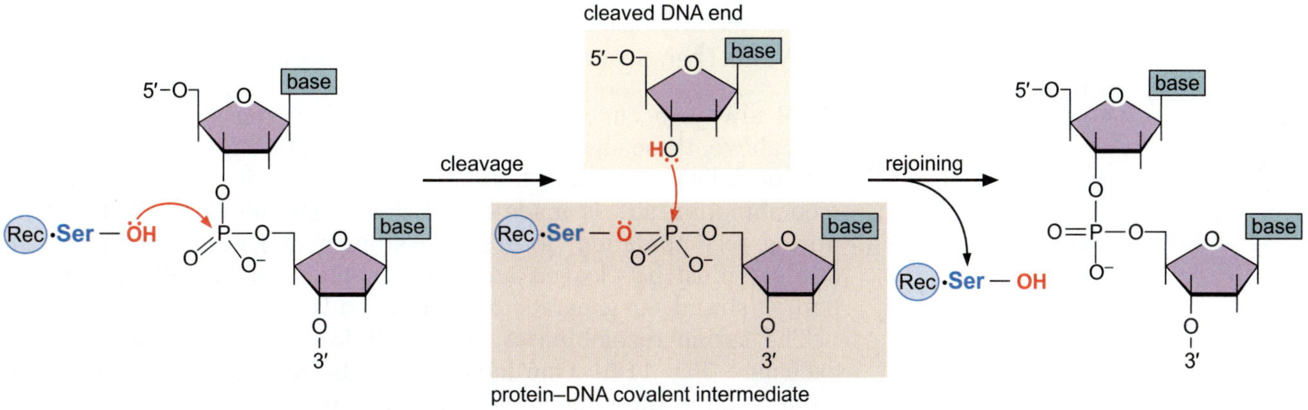

FIGURE 11-5 Covalent-intermediate mechanism used by the serine and tyrosine recombinases. Here, an OH group from an active-site serine is shown to attack the phosphate and thereby introduce a single-strand break at the site of recombination. The liberated OH group on the broken DNA can then reattack the protein–DNA covalent bond to reverse this cleavage reaction, reseal the DNA, and release the protein. The recombinase, labeled Rec, is shown in blue.

energy, such as that released by ATP hydrolysis, is needed for DNA cleavage and joining by these proteins. This cleavage mechanism, with its covalent intermediate, is not unique to the recombinases. Both DNA topoisomerases (see Chapter 6) and Spo11, the protein that introduces double-strand breaks into DNA to initiate homologous recombination during meiosis (see Chapter 10), use this mechanism.

TABLE 11-1 Recombinases by Family and by Function

Recombinase	Function
Serine family	
Salmonella Hin invertase	Inverts a chromosomal region to flip a gene promoter by recognizing *hix* sites. Allows expression of two distinct surface antigens.
Transposon Tn*3* and γδ resolvases	Promotes a DNA deletion reaction to resolve the DNA fusion event that results from replicative transposition. Recombination sites are called *res* sites.
Tyrosine family	
Bacteriophage λ integrase	Promotes DNA integration and excision of the λ genome into, and out of, a specific sequence on the *E. coli* chromosome. Recombination sites are called *att* sites.
Phage P1 Cre	Promotes circularization of the phage DNA during infection by recognizing sites (called *lox* sites) on the phage DNA.
Escherichia coli XerC and XerD	Promotes several DNA deletion reactions that convert dimeric circular DNA molecules into monomers. Recognizes both plasmid-borne sites (*cer*) and chromosomal sites (*dif*).
Yeast FLP	Inverts a region of the yeast 2μ plasmid to allow for a DNA amplification reaction called rolling circle replication. Recombination sites are called *frt* sites.

Serine Recombinases Introduce Double-Strand Breaks in DNA and Then Swap Strands to Promote Recombination

CSSR always occurs between two recombination sites. As we have seen above, these sites may be on the same DNA molecule (for inversion or deletion) or on two different molecules (for integration). Each recombination site is made up of double-stranded DNA. Therefore, during recombination, four single strands of DNA (two from each duplex) must be cleaved and then rejoined—now with a different partner strand—to generate the rearranged DNA.

The serine recombinases cleave all four strands prior to strand exchange (Fig. 11-6). One molecule of the recombinase protein promotes each of these cleavage reactions; therefore, a minimum of four subunits (i.e., a tetramer) of the recombinase is required.

These double-stranded DNA breaks in the parental DNA molecules generate four double-stranded DNA segments (marked by the proteins bound to them as R1, R2, R3, and R4 in Fig. 11-6). For recombination to occur, the R2 segment of the top DNA molecule must recombine with the R3 segment of the bottom DNA molecule. Likewise, the R1 segment of the top molecule must recombine with the R4 segment of the bottom DNA molecule. Once this DNA

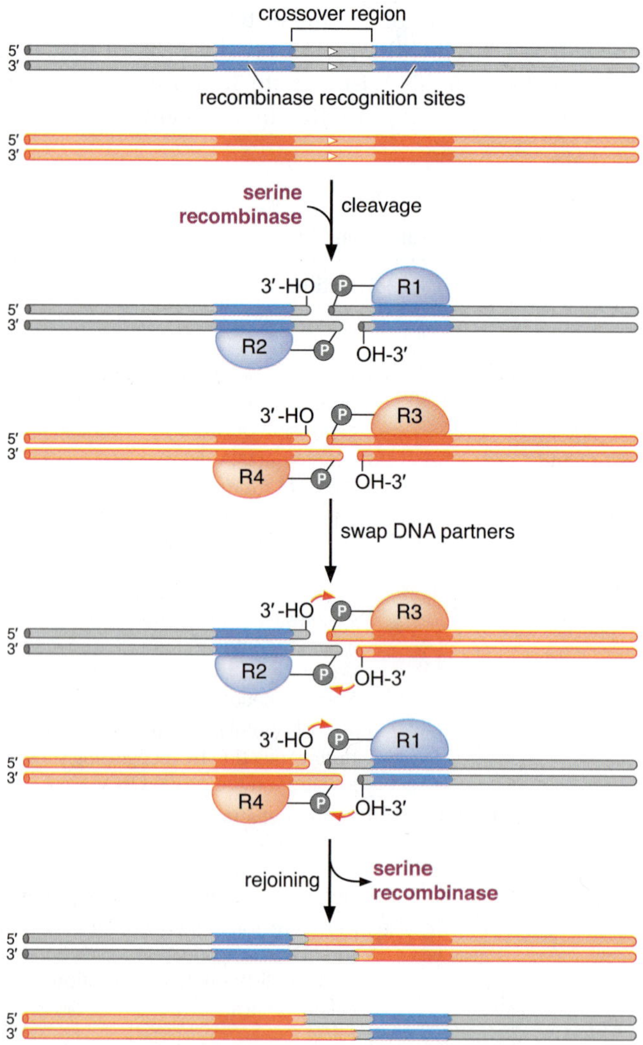

FIGURE 11-6 Recombination by a serine recombinase. Each of the four DNA strands is cleaved within the crossover region by one subunit of the protein. These subunits are labeled R1, R2, R3, and R4. Cleavage of the two individual strands of one duplex is staggered by two bases. This two-base region forms a hybrid duplex in the recombinant products. The recombination sites are similar to those shown in Fig. 11-4.

"swap" has occurred, the 3′-OH ends of each of the cleaved DNA strands can attack the recombinase–DNA bond in their new partner segment. As discussed above, this reaction liberates the recombinase and covalently seals the DNA strands to generate the rearranged DNA product.

Structure of the Serine Recombinase–DNA Complex Indicates That Subunits Rotate to Achieve Strand Exchange

The structure of a serine recombinase–DNA complex in the process of recombination provides a snapshot of how the exchange of DNA strands is physically coordinated. The complex contains four subunits of the recombinase, and two cleaved, double-stranded DNA molecules. The covalent linkage between the active-site serine in each recombinase subunit and 5′-phosphates in the DNA of each recombination half-site is clearly visible. Each of these linkages, in turn, leaves a free 3′-OH DNA end that can participate in strand exchange.

The most dramatic feature of the structure is the large, flat interface between the "top" and "bottom" recombinase dimers (Fig. 11-7). This structure is largely hydrophobic, and slippery, providing little barrier to impede rotation of the top and bottom halves of the complex around each other. However, some regions of complementary positive and negative charge can serve to stabilize the structure specifically in the initial and the 180° rotated orientation. Thus, analysis of this complex strongly supports the model that the mechanism of recombination is (1) DNA cleavage to form the covalent enzyme-DNA intermediate, (2) an 180° rotation of the dimers in the protein–DNA complex, and (3) attack of the 3′-OH DNA ends on the resolvase-DNA linkages to join the strands in the new, recombined configuration.

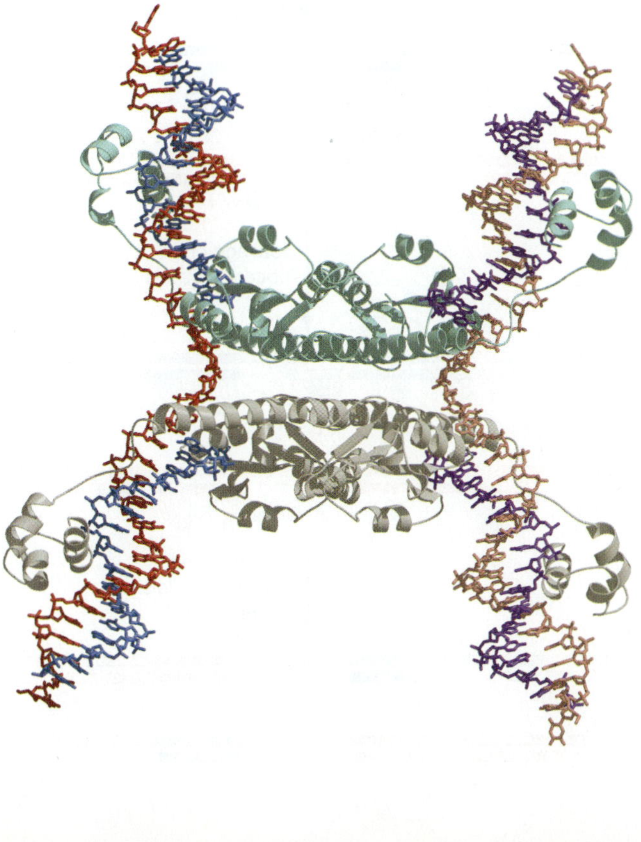

FIGURE 11-7 The structure of serine recombinase. The structure shows the large, flat tetramer interface that is the site of rotation. For the full recombination reaction, all four DNA strands must be cleaved (as in Fig. 11-6). In this structure, however, only two are cleaved. (Li W. et al. 2005. *Science* 309: 1210. PDB Code: 1ZR4.) Image prepared with MolScript, BobScript, and Raster 3D.

Tyrosine Recombinases Break and Rejoin One Pair of DNA Strands at a Time

In contrast to the serine recombinases, the tyrosine recombinases cleave and rejoin two DNA strands first and only then cleave and rejoin the other two strands (Fig. 11-8). Consider two DNA molecules with their recombination sites aligned. Here also, four molecules of the recombinase are needed, one to cleave each of the four individual DNA strands. To start recombination, the subunits of recombinase bound to the left recombinase-binding sites (marked as R1 and R3 in Fig. 11-8a) each cleave the top strand of the DNA molecule to which they are bound. This cleavage occurs at the first nucleotide of the crossover region. Next, the right top strand from the top (gray) DNA molecule and the right top strand from the bottom (red) DNA molecule "swap" partners. These two DNA strands are then joined, now in the recombined configurations. This "first-strand" exchange reaction generates a branched DNA intermediate known as a Holliday junction (see Chapter 10) (Fig. 11-8b).

Once the first-strand exchange is complete, two more recombinase subunits (those marked R2 and R4) cleave the bottom strands of each

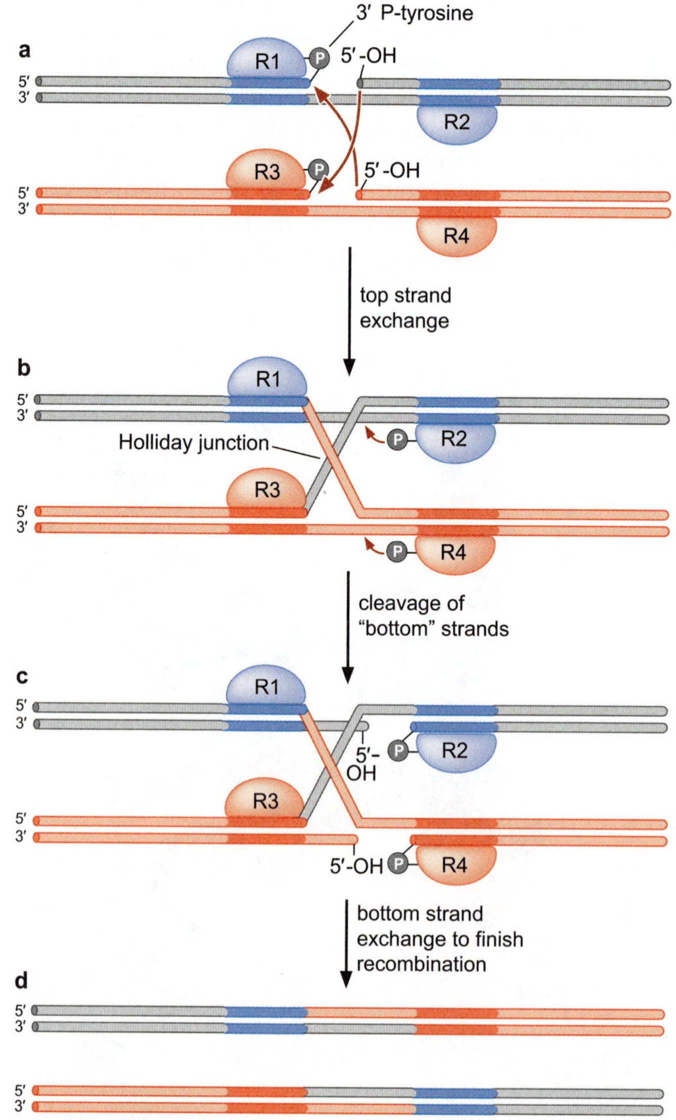

FIGURE 11-8 Recombination by a tyrosine recombinase. Here, the R1 and R3 subunits cleave the DNA in the first step (a); in the example shown, the protein becomes linked to the cut DNA by a 3′ P-tyrosine bond. Exchange of the first pair of strands occurs when the two 5′-OH groups at the break sites each attack the protein–DNA bond on the other DNA molecule (b). The second-strand exchange occurs by the same mechanism, using the R2 and R4 subunits (c and d). (Adapted from Craig N. et al. 2002. *Mobile DNA II*, Color Plate 1, Chapter 2. © ASM Press.)

DNA molecule (Fig. 11-8c). These strands again switch partners and then are joined by the reversal of the cleavage reaction. This "second–strand" exchange reaction "undoes" (i.e., resolves; see Chapter 10) the Holliday junction, to yield the rearranged DNA products. In the next section, we discuss how these chemical steps occur in the context of the recombinase protein–DNA complex.

Structures of Tyrosine Recombinases Bound to DNA Reveal the Mechanism of DNA Exchange

The mechanism of site-specific recombination is best understood for the tyrosine recombinases. Several structures of members of this protein class have been solved, and these structures reveal the recombinases "caught in the act" of recombination. One beautiful example is the structure of the Cre recombinase bound to two different configurations of the recombining DNA. Insights into the mechanisms derived from these structures are explained below. Cre is an enzyme encoded by phage P1, which functions to circularize the linear phage genome during infection. The recombination sites on the DNA, where Cre acts, are called *lox* sites. Cre-*lox* is a simple example of recombination by the tyrosine recombinase family; only Cre protein and the *lox* sites are needed for complete recombination. Cre is also widely used as a tool in genetic engineering (see Box 11-1, Application of Site-specific Recombination to Genetic Engineering).

The Cre-*lox* structures reveal that recombination requires four subunits of Cre, with each molecule bound to one binding site on the sub-

MEDICAL CONNECTIONS

Box 11-1 Application of Site-Specific Recombination to Genetic Engineering

Because some site-specific recombination systems are so simple, they have become widely used as tools in experimental genetics. Cre recombinase and its close relative FLP recombinase are both used experimentally to delete genes in eukaryotic organisms (also see example in Chapter 22).

An example of the usefulness of this strategy becomes clear when we consider the following hypothetical example. A researcher is interested in the role of a specific gene in the development of lung cancer and wishes to study this process using the mouse as a model organism (see Chapter 22). When the gene of interest is disrupted or "knocked out" (see Fig. 22-27), however, the mice all die during early embryogenesis. Apparently, the gene is required very early in development. How can its role in lung cancer be studied in the adult animal?

Site-specific recombination can often provide the answer. Using routine methods, researchers can introduce recombination sites recognized by Cre (or FLP) flanking the gene of interest. These sites will have no effect on the gene's function, unless the recombinase is also present. Therefore, the Cre protein (or FLP protein) can be introduced into the same organism, under the control of a promoter that can be carefully regulated (see Chapter 17). The mice can therefore be allowed to develop in the absence of the recombinase, but then after birth, Cre expression can be "turned on." The presence of the recombinase causes deletion of the gene of interest. In this case, the propensity of the Cre-treated mice (in which the gene is deleted) for lung cancer can now be compared with their "normal" littermates, in which the gene of interest is still intact. Thus, recombination using Cre allows the potential functions of the genes to be uncovered in different stages of development.

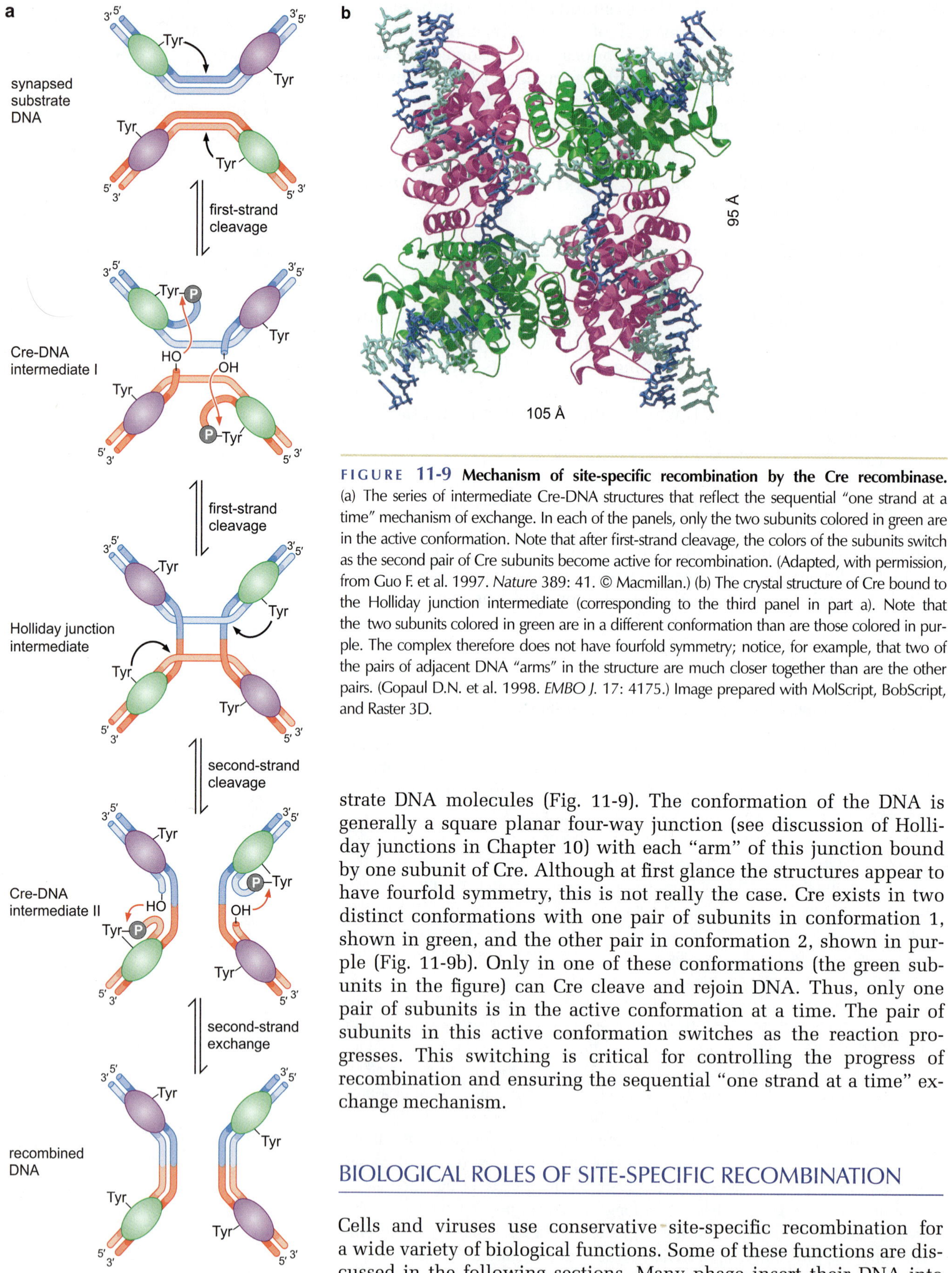

FIGURE 11-9 Mechanism of site-specific recombination by the Cre recombinase. (a) The series of intermediate Cre-DNA structures that reflect the sequential "one strand at a time" mechanism of exchange. In each of the panels, only the two subunits colored in green are in the active conformation. Note that after first-strand cleavage, the colors of the subunits switch as the second pair of Cre subunits become active for recombination. (Adapted, with permission, from Guo F. et al. 1997. *Nature* 389: 41. © Macmillan.) (b) The crystal structure of Cre bound to the Holliday junction intermediate (corresponding to the third panel in part a). Note that the two subunits colored in green are in a different conformation than are those colored in purple. The complex therefore does not have fourfold symmetry; notice, for example, that two of the pairs of adjacent DNA "arms" in the structure are much closer together than are the other pairs. (Gopaul D.N. et al. 1998. *EMBO J.* 17: 4175.) Image prepared with MolScript, BobScript, and Raster 3D.

strate DNA molecules (Fig. 11-9). The conformation of the DNA is generally a square planar four-way junction (see discussion of Holliday junctions in Chapter 10) with each "arm" of this junction bound by one subunit of Cre. Although at first glance the structures appear to have fourfold symmetry, this is not really the case. Cre exists in two distinct conformations with one pair of subunits in conformation 1, shown in green, and the other pair in conformation 2, shown in purple (Fig. 11-9b). Only in one of these conformations (the green subunits in the figure) can Cre cleave and rejoin DNA. Thus, only one pair of subunits is in the active conformation at a time. The pair of subunits in this active conformation switches as the reaction progresses. This switching is critical for controlling the progress of recombination and ensuring the sequential "one strand at a time" exchange mechanism.

BIOLOGICAL ROLES OF SITE-SPECIFIC RECOMBINATION

Cells and viruses use conservative site-specific recombination for a wide variety of biological functions. Some of these functions are discussed in the following sections. Many phage insert their DNA into the host chromosome during infection using this recombination mechanism. In other cases, site-specific recombination is used to alter gene expression. For example, inversion of a DNA segment can allow two alternative genes to be expressed. Site-specific recombination is also widely used to help maintain the structural integrity of circular DNA molecules during cycles of DNA replication, homologous recombination, and cell division.

A comparison of site-specific recombination systems reveals some general themes. All reactions depend critically on the assembly of the recombinase protein on the DNA and the bringing together of the two recombination sites. For some recombination reactions, this assembly is very simple, requiring only the recombinase and its DNA recognition sequences as just described for Cre. In contrast, other reactions require accessory proteins. These accessory proteins include so-called **architectural proteins** that bind specific DNA sequences and bend the DNA. They organize DNA into a specific shape and thereby stimulate recombination. Architectural proteins can also control the direction of a recombination reaction, for example, to ensure that integration of a DNA segment occurs while preventing the reverse reaction—DNA excision. Clearly, this type of regulation is essential for a logical biological outcome. Finally, we will also see that recombinases can be regulated by other proteins to control when a particular DNA rearrangement takes place and coordinate it with other cellular events.

λ Integrase Promotes the Integration and Excision of a Viral Genome into the Host-Cell Chromosome

When bacteriophage λ infects a host bacterium, a series of regulatory events result either in establishment of the quiescent **lysogenic state** or in phage multiplication, a process called **lytic growth** (see Chapters 16 and 22). Establishment of a lysogen requires the integration of the phage DNA into the host chromosome. Likewise, when the phage leaves the lysogenic state to replicate and make new phage particles, it must excise its DNA from the host chromosome. The analysis of this integration/excision reaction provided the first molecular insights into site-specific recombination.

To integrate, the λ integrase protein (λInt) catalyzes recombination between two specific sites, known as the *att,* or attachment, sites. The *attP* site is on the phage DNA (*P* for phage), and the *attB* site is in the bacterial chromosome (*B* for bacteria; see Fig. 11-2). λInt is a tyrosine recombinase, and the mechanism of strand exchange follows the pathway described above for the Cre protein. Unlike Cre recombination, however, λ integration requires accessory proteins to help the required protein–DNA complex to assemble. These proteins control the reaction to ensure that DNA integration and DNA excision occur at the right time in the phage life cycle. We first consider the integration pathway and then look at how excision is triggered.

Important to the regulation of λ integration is the highly asymmetric organization of the *attP* and *attB* sites (Fig. 11-10). Both sites carry a central core segment (~30 bp). These core recombination sites each consist of two λInt-binding sites and a crossover region where strand exchange occurs (as described above). Whereas *attB* consists only of this central core region, *attP* is much longer (240 bp) and carries several additional protein-binding sites.

Flanking each side of the core region of *attP* are DNA regions known as the "arms." These arms carry a variety of protein-binding sites, including additional sites bound by λInt (labeled as P_1, P_2, and P′, in Fig. 11-10). λInt is an unusual protein because it has two domains involved in sequence-specific DNA binding: one domain binds

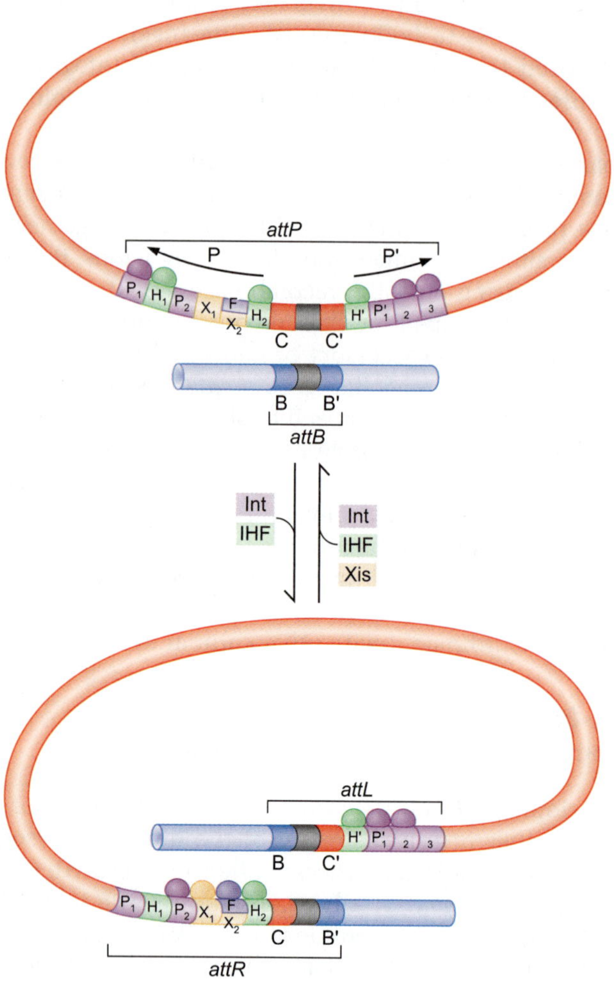

FIGURE 11-10 Recombination sites involved in λ integration and excision showing the important sequence elements. C, C′, B, and B′ are the core λInt-binding sites. The additional protein-binding sites are on *attP* and flank the C and C′ sites. These regions are called the "arms"; the sequences on the left are called the P arm and those on the right are called the P′ arm. The small purple boxes labeled P_1, P_2, and P_1' are the arm λInt-binding sites. Sites marked H are the integration host factor (IHF) binding sites, and sites marked X are the sites which bind Xis. F is the site bound by Fis, another architectural protein not discussed further here. The gray regions are the crossover regions. For clarity, λInt is not shown bound to the core sites. Note that not all protein-binding sites are filled during either integrative or excisive recombination. After recombination, the P arm is part of *attL*, whereas the P′ arm becomes part of *attR*.

to the arm recombinase recognition sites and the other binds to the core recognition sites. In addition, the arms of *attP* carry sites bound by several architectural proteins. Binding of these proteins governs the directionality and efficiency of recombination.

Integration requires *attB, attP,* λInt, and an architectural protein called **integration host factor (IHF).** IHF is a sequence-dependent DNA-binding protein that introduces large bends (>160°) in DNA (Fig. 11-11). The arms of *attP* carry three IHF binding sites (labeled H_1, H_2, and H′ in Fig. 11-10). The function of IHF is to bring together the λInt sites on the DNA arms (where λInt binds strongly) with the sites present at the central core (where λInt binds only weakly). Thus, bending of the DNA, mediated by IHF, allows λInt to find the weak core sites and to catalyze recombination.

When recombination is complete, the circular phage genome is stably integrated into the host chromosome. As a result, two new hybrid sites are generated at the junctions between the phage and the host DNA. These sites are called *attL* (left) and *attR* (right) (see Fig. 11-10). Both of these sites contain the core region, but the two arm regions are now separated from each other (see the location of the P and P′ regions in Fig. 11-10). Thus, neither of the two core regions in this new arrangement is competent to assemble an active λInt recombinase complex via the mechanism that was used to generate this complex during integration; the DNA sites important for assembly are simply not in the right place.

Bacteriophage λ Excision Requires a New DNA-Bending Protein

How does λ excise? An additional architectural protein, this one phage-encoded, is essential for excisive recombination. This protein, called Xis (for excise), binds to specific DNA sequences and introduces bends in the DNA. In this manner, Xis is similar in function to IHF. Xis recognizes two sequence motifs present in one arm of *attR* (and also present in *attP*—marked X_1 and X_2 in Fig. 11-10). Binding these sites introduces a large bend (>140°), and together Xis, λInt, and IHF stimulate excision by assembling an active protein–DNA complex at *attR*. This complex then interacts productively with proteins assembled at *attL* and recombination occurs.

In addition to stimulating excision (recombination between *attL* and *attR*), DNA binding by Xis also inhibits integration (recombination between *attP* and *attB*). The DNA structure created upon Xis binding to *attP* is incompatible with proper assembly of λInt and IHF at this site. Xis is a phage-encoded protein and is only made when the phage is triggered to enter lytic growth. Xis expression is described in detail in Chapter 16. Its dual action as a stimulatory cofactor for excision and an inhibitor of integration ensures that the phage genome will be free, and remain free, from the host chromosome when Xis is present.

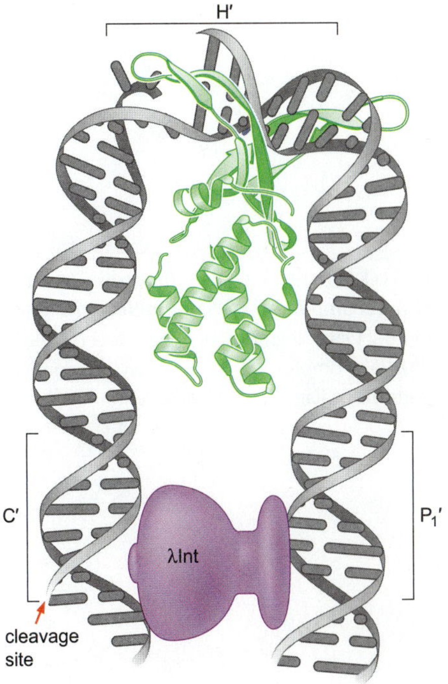

FIGURE 11-11 Model for IHF bending DNA to bring DNA-binding sites together. The λInt and IHF-binding sites from the P′ arm of *attP* are shown. IHF binding to the H′ site bends the DNA to allow one molecule of λInt to bind both the P_1′ and C′ sites. The break in the DNA within the H′ site reflects a nick that was present in the DNA used for structural analysis of the IHF-DNA complex. (Adapted, with permission, from Rice P. et al. 1996. *Cell* 87: 1295–1306, Fig. 8. © Elsevier.)

The Hin Recombinase Inverts a Segment of DNA Allowing Expression of Alternative Genes

The *Salmonella* Hin recombinase inverts a segment of the bacterial chromosome to allow expression of two alternative sets of genes. Hin recombination is an example of a class of recombination reactions, relatively common in bacteria, known as programmed rearrangements. These reactions often function to "pre-adapt" a portion of a population to a sudden change in the environment. In the case of Hin inversion, recombination is used to help the bacteria evade the host immune system as we shall now explain.

The genes that are controlled by the inversion process encode two alternative forms of flagellin (called the H1 and H2 forms), the protein component of the flagellar filament. Flagella are on the surface of the bacteria and are thus a common target for the immune system (Fig. 11-12). By using Hin to switch between these alternative forms,

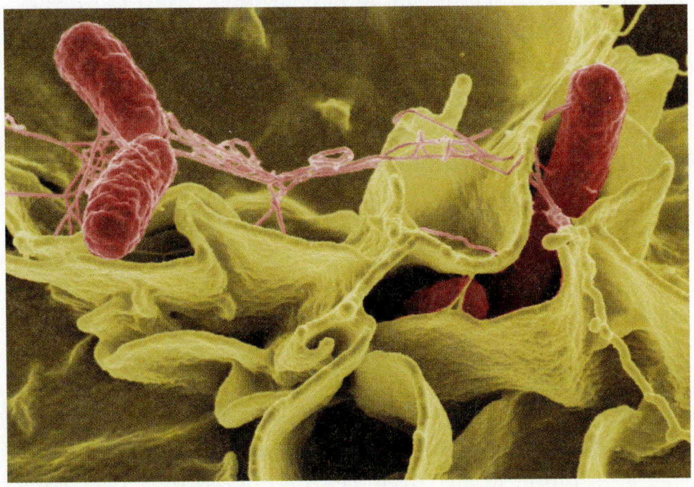

FIGURE 11-12 Micrograph of bacteria (*Salmonella*) showing flagella. The color-enhanced scanning electron micrograph shows *Salmonella typhimurium* (red) invading cultured human cells. The hair-like protrusions on the bacteria are the flagella. (Courtesy of the Rocky Mountain Laboratories, NIAID, NIH.)

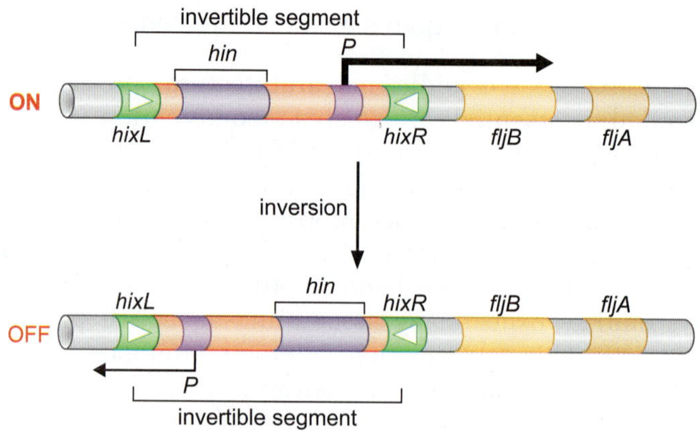

FIGURE 11-13 DNA inversion by the Hin recombinase of *Salmonella*. Inversion of the DNA segment between the *hix* sites flips a promoter (*P*) to give two alternative patterns of flagellin gene expression.

at least some individuals in the bacterial population can avoid recognition of this surface structure by the immune system.

The chromosomal region inverted by Hin is ~1000 bp and is flanked by specific recombination sites called *hixL* (on the left) and *hixR* (on the right) (Fig. 11-13). These sequences are in inverted orientation with respect to one another. Hin, a serine recombinase, promotes inversion using the basic mechanism described above for this enzyme family. The invertible segment carries the gene encoding Hin, as well as a promoter, which in one orientation is positioned to express the genes located outside the invertible segment directly adjacent to the *hixR* site. When the invertible segment is in the "on" orientation, these adjacent genes are expressed, whereas when the segment is flipped into the "off" orientation, the genes cannot be transcribed, because they lack a functional promoter.

The two genes under control of this "flipping" promoter are *fljB,* which encodes the H2 flagellin, and *fljA,* which encodes a transcriptional repressor of the gene for the H1 flagellin. The H1 flagellin gene is located at a distant site. Thus, in the "on" orientation, H2 flagellin and the H1 repressor are expressed. These cells have exclusively H2-type flagella on their surface. In the "off" orientation, however, neither H2 nor the H1 repressor is synthesized, and the H1-type flagella are present.

Hin Recombination Requires a DNA Enhancer

Hin recombination requires a sequence in addition to the *hix* sites. This short (~60 bp) sequence is an enhancer that stimulates the rate of recombination ~1000-fold. Like enhancer sequences that stimulate transcription (see Chapter 17), this sequence can function even when located quite a distance from the recombination sites. Enhancer function requires the bacterial Fis protein (named because it was discovered as a **factor for inversion stimulation**). Like IHF, Fis is a site-specific DNA-bending protein. In addition, it makes protein–protein contacts with Hin that are important for recombination.

The Fis–enhancer complex activates the catalytic steps of recombination. Hin can actually assemble and pair the *hix* recombination sites to form a synaptic complex in the absence of the Fis–enhancer complex (Fig. 11-14). This contrasts with the role of IHF in λ integration, where the accessory protein is essential for assembly of the recombinase–DNA complex. For Fis activation of Hin, the three DNA sites (*hixL, hixR,* and enhancer) need to come together. Formation of this three-way complex is greatly facilitated by negative DNA supercoiling (see Chapter 6),

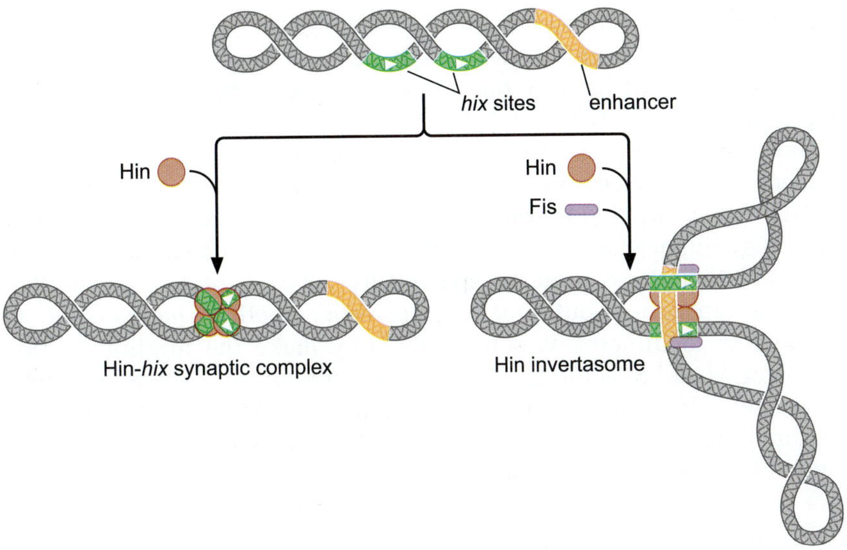

FIGURE 11-14 Complexes formed during Hin-catalyzed recombination. Hin protein alone recognizes and pairs the two *hix* sites. When Fis protein is also present, the three-segment complex can form. This complex is called the invertasome and is the most active complex for promoting recombination. (Adapted, with permission, from Craig N. et al. 2002. *Mobile DNA II*, p. 246, Fig. 9. © ASM Press.)

which stabilizes the association of the distant DNA sites. Another bacterial architectural protein, HU, also facilitates assembly of this invertasome complex. HU is a close structural homolog of IHF, yet in contrast to IHF, it binds DNA in a sequence-independent manner.

What is the biological rationale for control of Hin inversion by the Fis–enhancer complex? The principal function is to ensure that recombination occurs only between *hix* sites that are present on the same DNA molecule. This selectivity ensures that the invertible segment is flipped frequently but also that intermolecular DNA rearrangements, which could disrupt the integrity of the bacterial chromosome, are avoided.

In contrast to integration and excision of bacteriophage λ, Hin-catalyzed inversion is not highly regulated. Rather, inversion occurs stochastically, such that within a population of cells, there will always be some cells that carry the invertible segment in each orientation.

Recombinases Convert Multimeric Circular DNA Molecules into Monomers

Site-specific recombination is critical to the maintenance of circular DNA molecules within cells. The chromosomes of most bacteria are circular, as are most plasmids in both prokaryotic and eukaryotic cells. Some viral genomes are also circular. An intrinsic problem with circular DNA molecules is that they sometimes form dimers and even higher multimeric forms during the process of homologous recombination. Site-specific recombination can be used to convert these DNA multimers back into monomers.

Consider what happens when a DNA crossover occurs between two identical circular molecules. This process is shown occurring between two copies of a bacterial chromosome during replication (Fig. 11-15) (for a discussion of homologous recombination, see Chapter 10). A single homologous recombination event can generate one large circu-

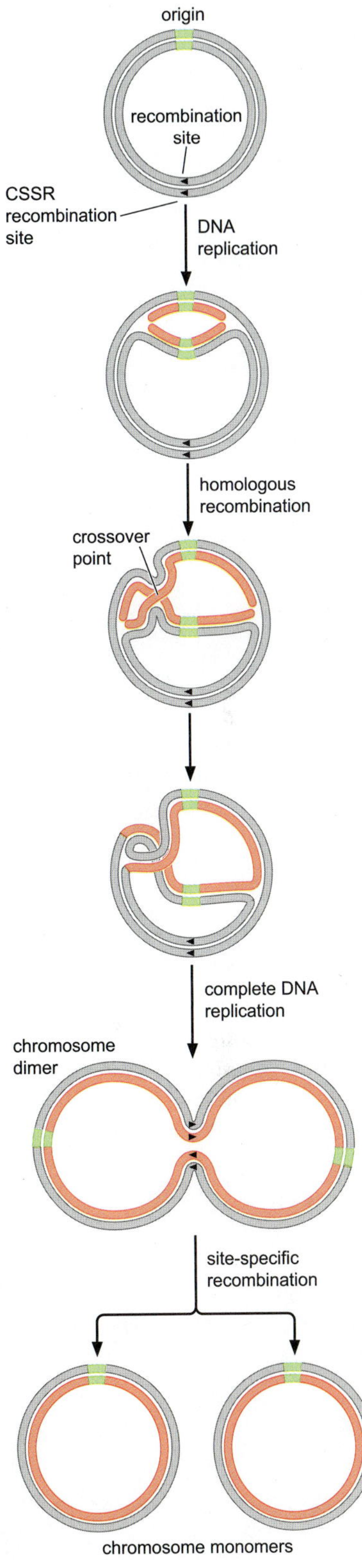

FIGURE 11-15 Circular DNA molecules can form multimers. Homologous recombination between the two daughter DNA molecules during DNA replication generates a dimeric chromosome (or plasmid). Site-specific recombination by the XerCD recombinase is then needed to generate the monomeric DNA molecules needed for cell division.

lar chromosome with two copies of all the genes (i.e., a dimeric chromosome). At the time of cell division, this dimer poses a major problem, as there will be only one rather than two DNA molecules to be segregated into the two daughter cells.

Because of this multimerization problem, many circular DNA molecules carry sequences recognized by site-specific recombinases. Proteins that function at these sequences are called **resolvases,** as they "resolve" dimers (and larger multimers) into monomers. Clearly, it is essential that these proteins specifically catalyze resolution (a DNA deletion reaction) but not the reverse reaction (conversion of monomers to dimers), which would only make the multimerization problem worse! Specific mechanisms are in place to enforce this directional selectivity on the recombination process (see Box 11-2, The Xer Recombinase Catalyzes the Monomerization of Bacterial Chromosomes and of Many Bacterial Plasmids).

There Are Other Mechanisms to Direct Recombination to Specific Segments of DNA

Although we have limited our discussion to conservative site-specific recombination, other recombination events occur at specific sequences and serve similar biological functions. Some of these reactions, for example, mating-type switching in yeast, occur by a targeted gene-conversion event, as we described in Chapter 10. The gene rearrangements responsible for assembly of gene segments encoding critical proteins for the vertebrate immune system—known as V(D)J recombination—also occur at specific sites. This reaction is mechanistically similar to transposition and therefore is considered later in this chapter.

TRANSPOSITION

Some Genetic Elements Move to New Chromosomal Locations by Transposition

Transposition is a specific form of genetic recombination that moves certain genetic elements from one DNA site to another. These mobile genetic elements are called **transposable elements** or **transposons.** Movement occurs through recombination between the DNA sequences at the very ends of the transposable element and a sequence in the DNA of the host cell (Fig. 11-16); movement can occur with or without duplication of the element, as we shall see. In some cases, the recombination reaction involves a transient RNA intermediate.

When transposable elements move, they often show little sequence selectivity in their choice of insertion sites. As a result, transposons can insert within genes, often completely disrupting gene function. They can also insert within the regulatory sequences of a gene where their presence may lead to changes in how that gene is expressed. It was these disruptions in gene function and expression that led to the discovery of transposable elements (see Box 11-4, Maize Elements and the Discovery of Transposons, later in this chapter). Perhaps not surprisingly, therefore, transposable elements are the most common source of new mutations in many organisms. In fact, these elements are an important cause of mutations leading to genetic disease in hu-

Box 11-2 The Xer Recombinase Catalyzes the Monomerization of Bacterial Chromosomes and of Many Bacterial Plasmids

Xer is a member of the tyrosine recombinase family, and its mechanism for promoting recombination is very similar to that described above for Cre. Xer is a heterotetramer, containing two subunits of a protein called XerC and two subunits of a protein called XerD. Both XerC and XerD are tyrosine recombinases, but they recognize different DNA sequences. Therefore, the recombination sites used by the Xer recombinase must carry recognition sequences for each of these proteins. The recombination sites in bacterial chromosomes, called *dif* sites, have a XerC recognition sequence on one side and an XerD recognition sequence on the other side of the crossover region (Box 11-2 Fig. 1). There is one *dif* site on the chromosome. It is located within the region where DNA replication terminates (see Chapter 8). When the chromosome is dimeric, this dimer will of course have two *dif* sites (see Fig. 11-15).

How do cells make sure that Xer-mediated recombination at *dif* sites will convert a chromosome dimer into monomers without ever promoting the reverse reaction? This directional regulation is achieved through the interaction between the Xer recombinase and a cell division protein called FtsK. This regulation is shown in Box 11-2 Figures 1 and 2 and occurs as follows. When FtsK is unavailable for interaction with the XerCD complex at the *dif* site, the recombinase complex adopts a conformation in which only the two XerC subunits are active. As a result, XerC will promote exchange of one pair of DNA strands to form the Holliday junction intermediate (see the discussion of the tyrosine recombinase mechanism above). Because XerD is never activated, recombination is never completed. Instead, reversal of the XerC cleavage reaction often occurs. This reversal simply regenerates the original DNA arrangement (see Box 11-2 Fig. 1).

In contrast, when the FtsK protein is available and interacts with the XerCD complex, it alters the conformation of the complex and activates XerD protein. In this case, XerD promotes recombination of the first pair of strands to generate the Holliday junction intermediate. Once this reaction is completed, XerC promotes the second pair of strand-exchange reactions, yielding the recombined DNA products (see Box 11-2 Fig. 1).

FtsK is an ATPase that tracks along DNA. It functions as a "DNA-pumping protein machine" similar to the RuvB protein that promotes DNA branch migration during homologous recombination (discussed in Chapter 10). FtsK is also a membrane-bound protein that is localized in the cell at the site where cell division occurs. It moves DNA away from the center of the cell prior to division so that the cell can divide at this site (Box 11-2 Fig. 2).

This localization of FtsK to the division site is key to how the cells ensure that XerD is activated specifically when a dimeric chromosome is present. In this case, the chromosome will be "stuck" in the middle of the dividing cell as one-half of the chromosome dimer is moved into each daughter cell. FtsK also interacts with specific polar DNA sequences (called KOPS) that are arranged asymmetrically around the *dif* site. As a result, FtsK translocates *dif* sites toward the septum and toward each other. This movement therefore facilitates their pairing, as well as activating XerD recombination. In this manner, site-specific recombination is regulated to occur at the right time and place within the cell division cycle.

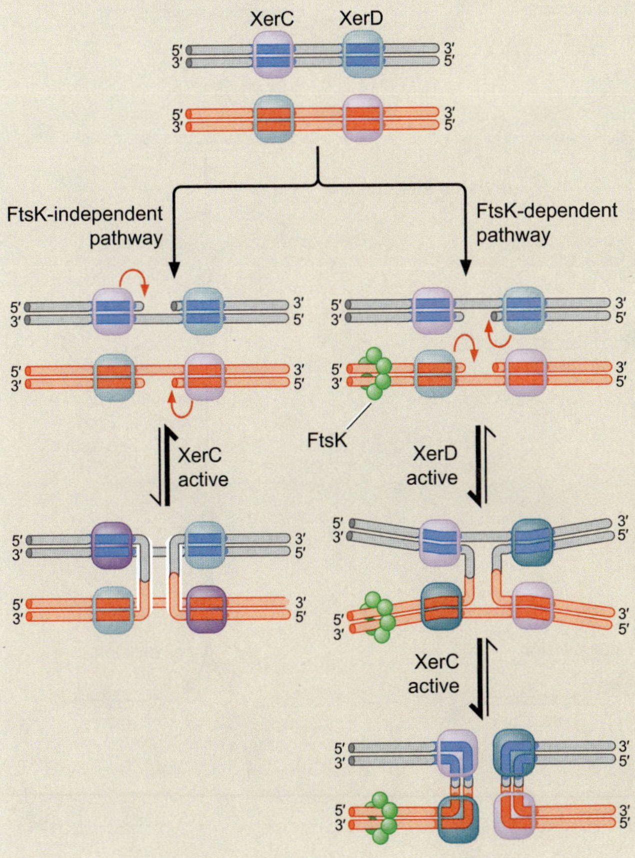

BOX 11-2 FIGURE 1 Pathways for Xer-mediated recombination at *dif*. In the absence of FtsK (FtsK-independent pathway shown in the left panel), only XerC is active to promote strand exchange to form a Holliday junction intermediate. In this case (because XerD is not active), recombination is not completed and the XerC reaction is frequently reversed. In the presence of FtsK (FtsK-dependent pathway shown in the right panel), XerD, now active, catalyzes formation of the Holliday junction intermediate, and XerC promotes second-strand exchange to complete the recombination event and generate chromosome monomers. (Adapted, with permission, from Aussel L. et al. 2002. *Cell* 108: 195–205, Fig. 6. © Elsevier.)

BOX 11-2 (*Continued*)

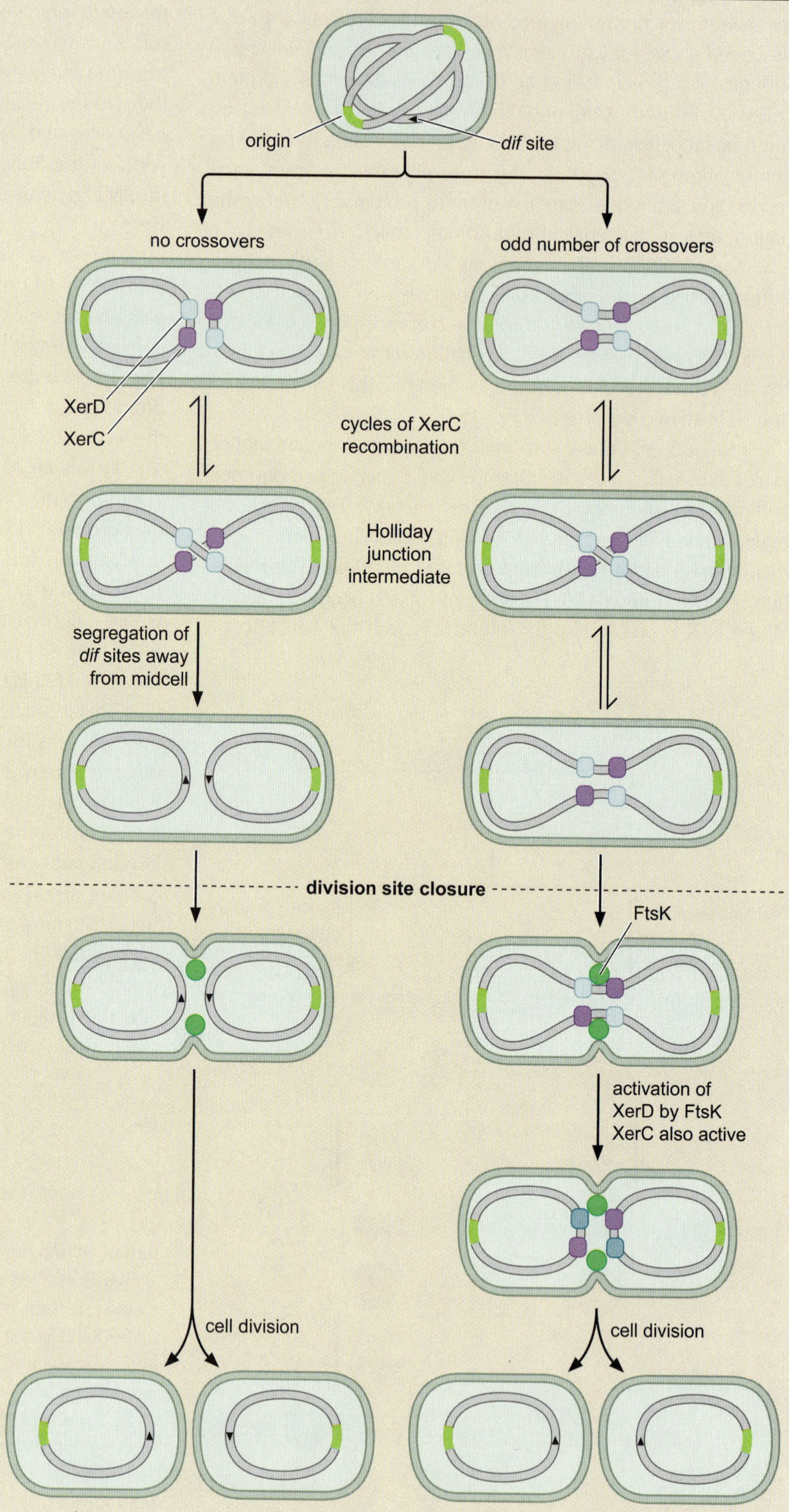

BOX 11-2 FIGURE 2 Regulation of chromosome segregation by FtsK. Just before cell division, the newly replicated origins, shown in green, move to the poles of the cell, whereas the replication terminus that includes *dif,* shown as a triangle, typically remains localized at the midcell. When the *dif* site is replicated, the two daughter *dif* sites can recombine to form a Holliday junction, which is resolved by XerC. If the replicated chromosome forms monomers, segregation will break the synaptic complex and the *dif* sites will move away from the midcell location before division. In contrast, if the chromosome forms a dimer (right panel), the synaptic complex remains trapped at midcell and allows access to FtsK, which is localized to the cell division site. FtsK then activates XerD. XerD-mediated recombination, followed by XerC-mediated recombination, then allows resolution of the dimers into monomers for cell division. (Adapted, with permission, from Barre F.-X. et al. 2001. *Proc. Natl. Acad. Sci.* 98: 8189–8195, Fig. 5. © National Academy of Sciences.)

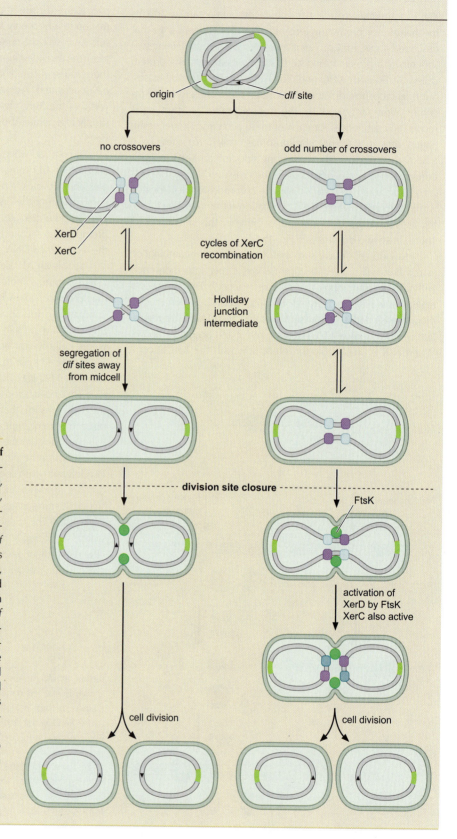

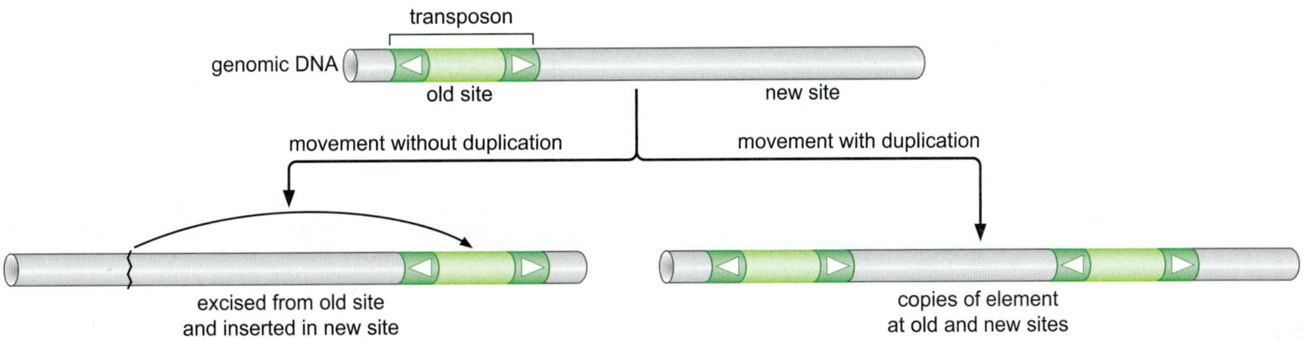

FIGURE 11-16 Transposition of a mobile genetic element to a new site in the host DNA. Recombination, in some cases, involves excision of the transposon from the old DNA location (left). In other cases, one copy of the transposon stays at the old location and another copy is inserted into the new DNA site (right).

mans. The ability of transposable elements to insert so promiscuously in DNA has also led to their modification and use as mutagens and DNA-delivery vectors in experimental biology.

Transposable elements are present in the genomes of all life-forms. The comparative analysis of genome sequences reveals two fascinating observations. First, transposon-related sequences can make up huge fractions of the genome of an organism. For example, more than 50% of both the human and maize genomes are composed of transposon-related sequences (including fragments of transposons or "dead" elements that have been inactivated by mutations). This contribution is in sharp contrast to the small percentage of the sequence that actually encodes cellular proteins (<2% in human). Second, the transposon content in different genomes is highly variable (Fig. 11-17). For example, compared to humans or maize, the fly and yeast genomes are very "gene-rich" and "transposon-poor."

There are many different types of transposable elements. These elements can be divided into families that share common aspects of structure and recombination mechanism. In the following sections, we introduce three major families of transposable elements and the recombination mechanism associated with each family. Some of the best-studied individual elements are then described. In the description of individual elements, we focus on how transposition is regulated to balance the maintenance and propagation of these elements with their potential to disrupt or misregulate genes within the host organism.

The genetic recombination mechanisms responsible for transposition are also used for functions other than the movement of transposons. For example, many viruses use a recombination pathway nearly identical to transposition to integrate into the genome of the host cell during infection. These viral integration reactions will therefore be considered together with transposition. Likewise, some DNA rearrangements used by cells to alter gene expression occur using a mechanism very similar to DNA transposition. V(D)J recombination, a reaction required for development of a functional immune system in vertebrates, is a well-understood example. V(D)J recombination is discussed at the end of this chapter.

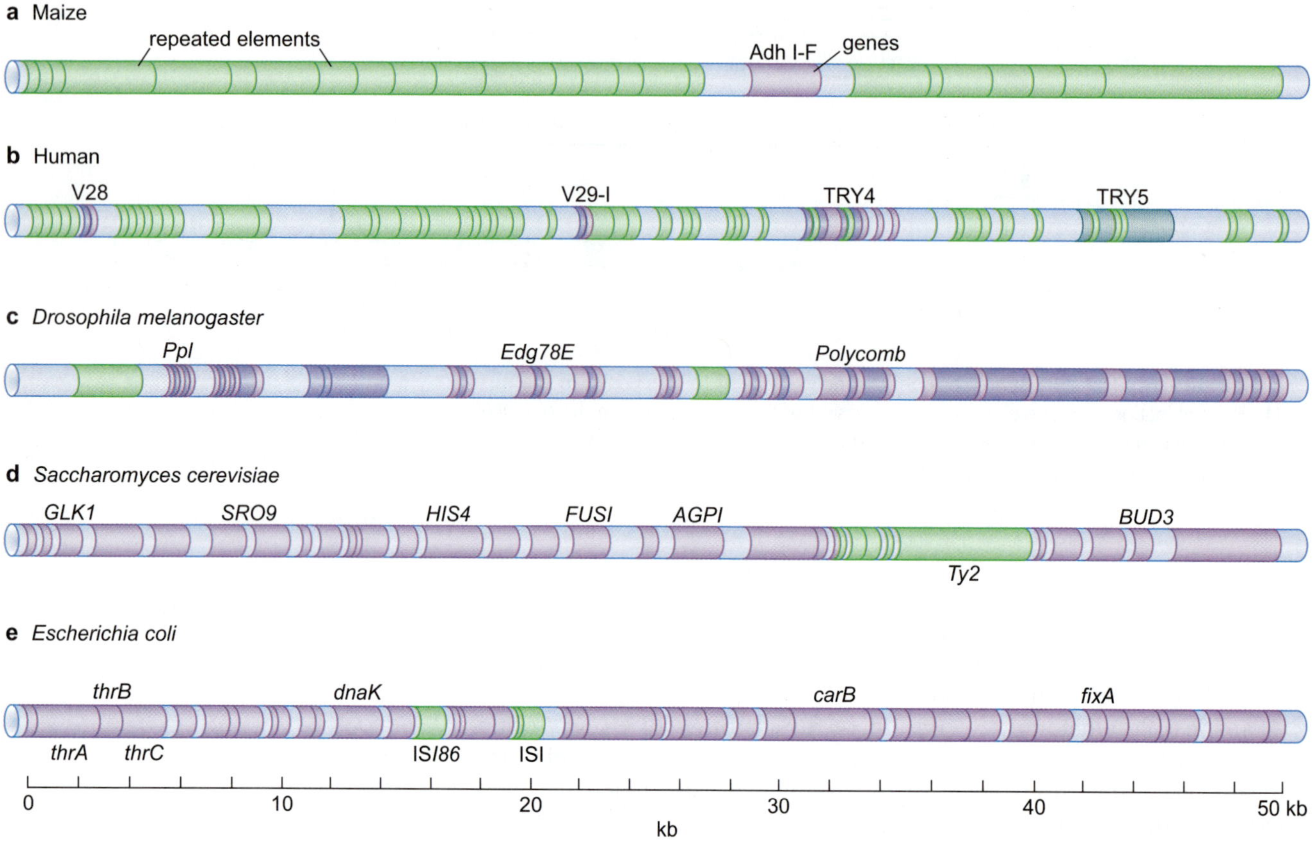

FIGURE 11-17 Transposons in genomes: Occurrence and distribution. (Green) Repeated elements, mostly composed of transposons or transposon-related sequences (such as truncated elements); (purple) cellular genes. (a) Maize; (b) human; (c) *Drosophila*; (d) budding yeast; (e) *E. coli*. (From Brown T.A. 2002. *Genomes*, 2nd ed., p. 34, Fig. 2.2, and references therein.)

There Are Three Principal Classes of Transposable Elements

Transposons can be divided into the following three families on the basis of their overall organization and mechanism of transposition:

1. **DNA transposons.**
2. **Virus-like retrotransposons.** This class includes the retroviruses. These elements are also called **long terminal repeat (LTR)** retrotransposons.
3. **Poly-A retrotransposons.** These elements are also called nonviral retrotransposons.

Figure 11-18 shows a schematic of the general genetic organization of each of these element families. DNA transposons remain as DNA throughout a cycle of recombination. They move using mechanisms that involve the cleavage and rejoining of DNA strands, and in this way, they are similar to elements that move by conservative site-specific recombination. Both types of retrotransposons move to a new DNA location using a transient RNA intermediate.

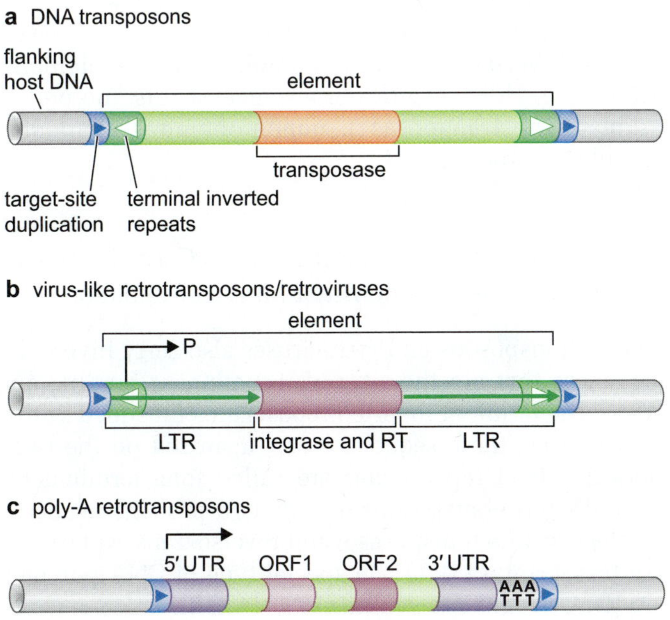

FIGURE 11-18 Genetic organization of the three classes of transposable elements. (a) DNA transposons. The element includes the terminal inverted-repeat sequences (shown in green with white arrows), which are the recombination sites, and a gene encoding transposase. (b) Virus-like retrotransposons and retroviruses. The element includes two LTR sequences that flank a region encoding two enzymes: integrase and reverse transcriptase (RT). (c) Poly-A retrotransposons. The element terminates in the 5′ and 3′ untranslated region (UTR) sequences and encodes two enzymes: an RNA-binding enzyme (ORF1) and an enzyme having both reverse transcriptase and endonuclease activities (ORF2).

DNA Transposons Carry a Transposase Gene, Flanked by Recombination Sites

DNA transposons carry both DNA sequences that function as recombination sites and genes encoding proteins that participate in recombination (Fig. 11-18a). The recombination sites are at the two ends of the element and are organized as inverted-repeat sequences. These terminal inverted repeats vary in length from ~25 to a few hundred base pairs, are not exact sequence repeats, and carry the recombinase recognition sequences. The recombinases responsible for transposition are usually called **transposases** (or, sometimes, **integrases**).

DNA transposons carry a gene encoding their own transposase. They may carry a few additional genes, sometimes encoding proteins that regulate transposition or provide a function useful to the element or its host cell. For example, many bacterial DNA transposons carry genes encoding proteins that promote resistance to one or more antibiotic(s). The presence of the transposon therefore causes the host cell to be resistant to that antibiotic.

The DNA sequences immediately flanking the transposon have a short (2–20 bp) segment of duplicated sequence. These segments are organized as direct repeats, are called **target-site duplications**, and are generated during the process of recombination as we shall discuss below.

Transposons Exist as Both Autonomous and Nonautonomous Elements

DNA transposons that carry a pair of terminal inverted repeats and a transposase gene have everything they need to promote their own transposition. These elements are called **autonomous transposons**. However, genomes also contain many even simpler mobile DNA segments known as **nonautonomous transposons**. These elements carry only the terminal inverted repeats, that is, the *cis*-acting sequences needed for transposition. In a cell that also carries an autonomous transposon, encoding a transposase that will recognize these terminal inverted repeats, the nonautonomous element will be able to transpose. However, in the absence of this "helper" transposon (to donate the transposase), nonautonomous elements remain frozen, unable to move.

Virus-like Retrotransposons and Retroviruses Carry Terminal Repeat Sequences and Two Genes Important for Recombination

Virus-like retrotransposons and retroviruses also carry inverted terminal repeat sequences that are the sites of recombinase binding and action (Fig. 11-18b). The terminal inverted repeats are embedded within longer repeated sequences; these sequences are organized on the two ends of the element as direct repeats and are called long terminal repeats or LTRs. Virus-like retrotransposons encode two proteins needed for their mobility: integrase (the transposase) and reverse transcriptase.

Reverse transcriptase (RT) is a special type of DNA polymerase that can use an RNA template to synthesize DNA. This enzyme is needed for transposition because an RNA intermediate is required for the transposition reaction. Because these elements convert RNA into DNA, the reverse of the normal pathway of biological information flow (DNA to RNA), they are known as "retro" elements. The distinction between virus-like retrotransposons and retroviruses is that the genome of a retrovirus is packaged into a viral particle, escapes its host cell, and infects a new cell. In contrast, the retrotransposons can move only to new DNA sites within a cell but can never leave that cell. Like the DNA transposons, these elements are flanked by short target-site duplications that are generated during recombination.

Poly-A Retrotransposons Look Like Genes

The poly-A retrotransposons do not have the terminal inverted repeats present in the other transposon classes. Instead, the two ends of the element have distinct sequences (Fig. 11-18c). One end is called the 5′ UTR, whereas the other end has a region called the 3′ UTR followed by a stretch of A:T base pairs called the **poly-A sequence**. These elements are also flanked by short target-site duplications.

Retrotransposons carry two genes, known as *ORF1* and *ORF2*. *ORF1* encodes an RNA-binding protein. *ORF2* encodes a protein with both reverse transcriptase activity and an endonuclease activity. This protein, although distinct from the transposases and integrases encoded by the other classes of elements, has essential roles during recombination. Like their DNA and virus-like transposon counterparts, poly-A retrotransposons exist commonly in both autonomous and nonautonomous forms. Furthermore, genome sequence analysis reveals that there are many truncated elements that do not have a complete 5′ UTR sequence and have lost their ability to transpose.

DNA Transposition by a Cut-and-Paste Mechanism

DNA transposons, virus-like retrotransposons, and retroviruses all use a similar mechanism of recombination to insert their DNA into a new site. First, let us consider the simplest transposition reaction: the movement of a DNA transposon by a nonreplicative mechanism. This recombination pathway involves the excision of the transposon from its initial location in the host DNA, followed by integration of this excised transposon into a new DNA site. This mechanism is therefore called **cut-and-paste transposition** (Fig. 11-19).

To initiate recombination, the transposase binds to the terminal inverted repeats at the end of the transposon. Once the transposase recognizes these sequences, it brings the two ends of the transposon

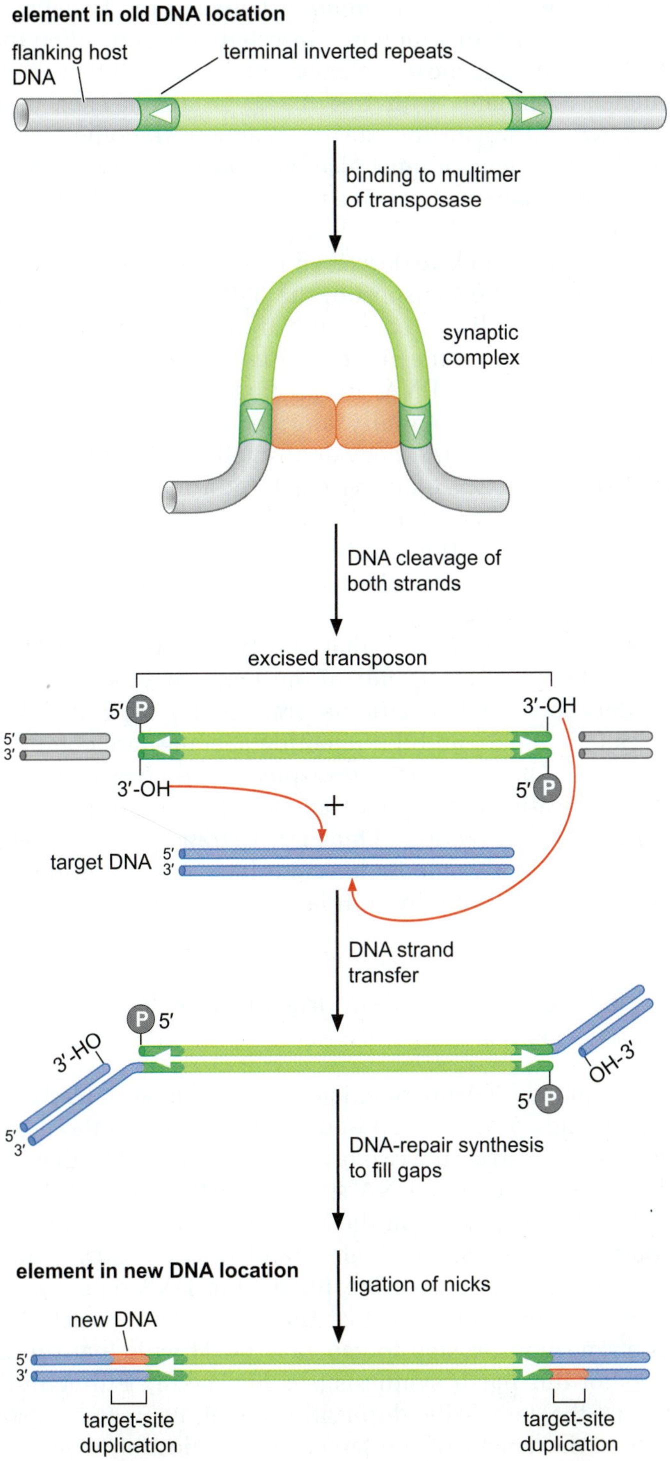

FIGURE 11-19 The cut-and-paste mechanism of transposition. The figure shows movement of a transposon from a target site in the gray host DNA to a new site in the blue DNA. Note the staggered cleavage sites on the target DNA during the DNA strand transfer reaction that give rise to short repeated sequences at the new target site (the target-site duplications). The DNA at the original insertion site (here in gray) will be left with a double-stranded DNA break as a result of transposon excision. This break can be repaired by nonhomologous end joining or homologous recombination (see Chapters 9 and 10). Note that the chemical steps are shown without the bound protein for clarity.

DNA together to generate a stable protein–DNA complex. This complex is called the **synaptic complex** or **transpososome**. It contains a multimer of transposase—usually two or four subunits—and the two DNA ends (see below). This complex functions to ensure that the DNA cleavage and joining reactions needed to move the transposon occur simultaneously on the two ends of the element's DNA. It also protects the DNA ends from cellular enzymes during recombination. The next step is the excision of the transposon DNA from its original location in the genome. To achieve this, the transposase subunits within the transpososome first cleave one DNA strand at each end of the transposon, exactly at the junction between the transposon DNA and the host sequence in which it is inserted (a region called the **flanking host DNA**). The transposase cleaves the DNA such that the transposon sequence terminates with free 3′-OH groups at each end of the element's DNA. To finish the excision reaction, the other DNA strand at each end of the element must also be cleaved. Different transposons use different mechanisms to cleave these "second" DNA strands (those strands that terminate with 5′ ends at the transposon host DNA junction). These mechanisms are described in a following section.

After excision of the transposon, the 3′-OH ends of the transposon DNA—the ends first liberated by the transposase—attack the DNA phosphodiester bonds at the site of the new insertion. This DNA segment is called the **target DNA**. Recall that for most transposons, the target DNA can have essentially any sequence. As a result of this attack, the transposon DNA is covalently joined to the DNA at the target site. During each DNA-joining reaction, a nick is also introduced into the target DNA (Fig. 11-19). This DNA-joining reaction occurs by a one-step transesterification reaction that is called **DNA strand transfer** (Fig. 11-20). A similar mechanism for joining nucleic acid strands is used for RNA splicing (see Chapter 13).

The transpososome ensures that the two ends of the transposon DNA attack the two DNA strands of the same target site together. The sites of attack on the two strands are usually separated by a few nucleotides (e.g., 2-, 5-, and 9-nucleotide spacings are common). This distance is fixed for each type of transposon and gives rise to the short target-site duplications that flank transposed copies of the element (as explained in the next section). Once DNA strand transfer is complete, the job of the transpososome is also complete. The remaining recombination steps are carried out by cellular DNA-repair proteins.

The Intermediate in Cut-and-Paste Transposition Is Finished by Gap Repair

The structure of the DNA intermediate generated after DNA strand transfer has the 3′ ends of the transposon DNA attached to the target DNA. This structure also carries the two nicks in the target DNA that were generated during the process of DNA strand transfer. The fact that the two sites of DNA strand transfer on the two strands are separated by a few nucleotides results in short single-stranded DNA gaps flanking the joined transposon. These gaps are filled by a DNA-repair polymerase encoded by the host cell. Note that the target DNA is cleaved during the DNA strand transfer step to generate 3′-OH ends that can serve as the primers for this repair synthesis (see Fig. 11-18). Filling in the gaps gives rise to the target-site duplications that flank transposons (see above). Thus, the length of the target-site duplication reveals the distance between the sites attacked on the two strands of the target DNA

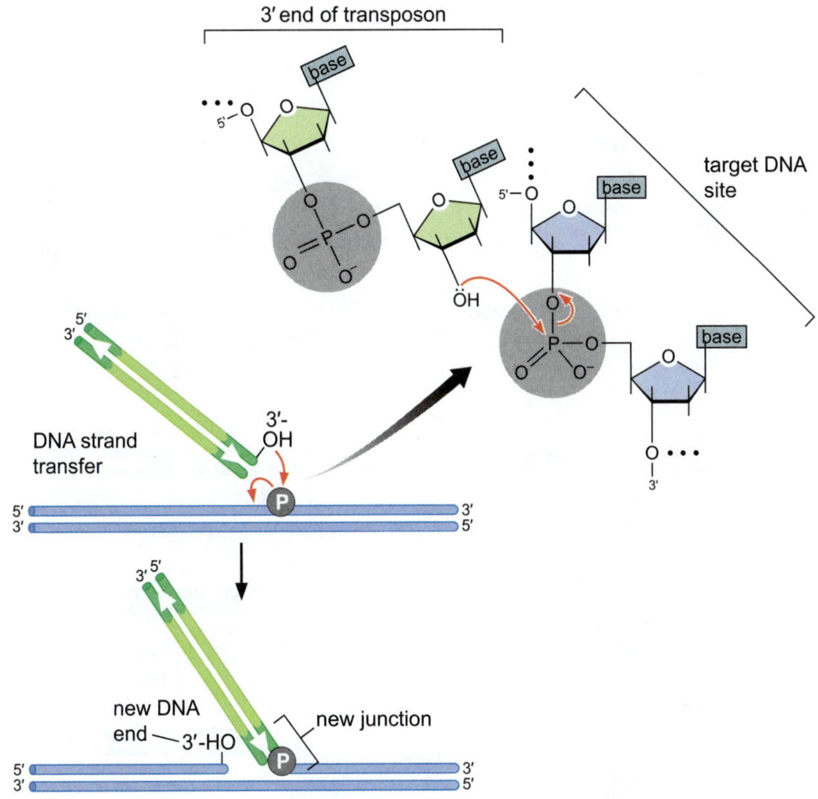

FIGURE 11-20 Close-up view of the chemical step of DNA strand transfer. In the inset, only one strand is shown for the transposon and for the target DNA for clarity.

during DNA strand transfer. After gap-repair synthesis, DNA ligase is needed to seal the DNA strands.

Cut-and-paste transposition also leaves a double-strand break in the DNA at the site of the "old" insertion, which must be repaired to maintain the integrity of the host cell's genome. Repair of double-stranded DNA breaks by homologous recombination is described in Chapter 10. These breaks are also sometimes more directly rejoined, as we will see below in the discussion of the Tc*1*/*mariner* family of transposons.

There Are Multiple Mechanisms for Cleaving the Nontransferred Strand during DNA Transposition

As described above, the transposase cleaves the 3′ ends of the element DNA and promotes DNA strand transfer to catalyze cut-and-paste transposition. However, transposons that move by this mechanism also need to cleave the 5′-terminating strands at the junctions between the transposon and the flanking host DNA. These DNA strands are called the **nontransferred strands**, as their 5′ ends are not directly linked to the target DNA during the DNA strand transfer reaction. Different transposons use different mechanisms to catalyze this second-strand cleavage reaction (Fig. 11-21). Two methods are described here.

An enzyme other than the transposase can be used to cleave the nontransferred strand (Fig. 11-21). For example, the bacterial transposon Tn*7* encodes a specific protein (called TnsA) that does this job (Fig. 11-21a). TnsA has a structure very similar to that of a restriction endonuclease. TnsA assembles with the Tn*7*-encoded transposase (the

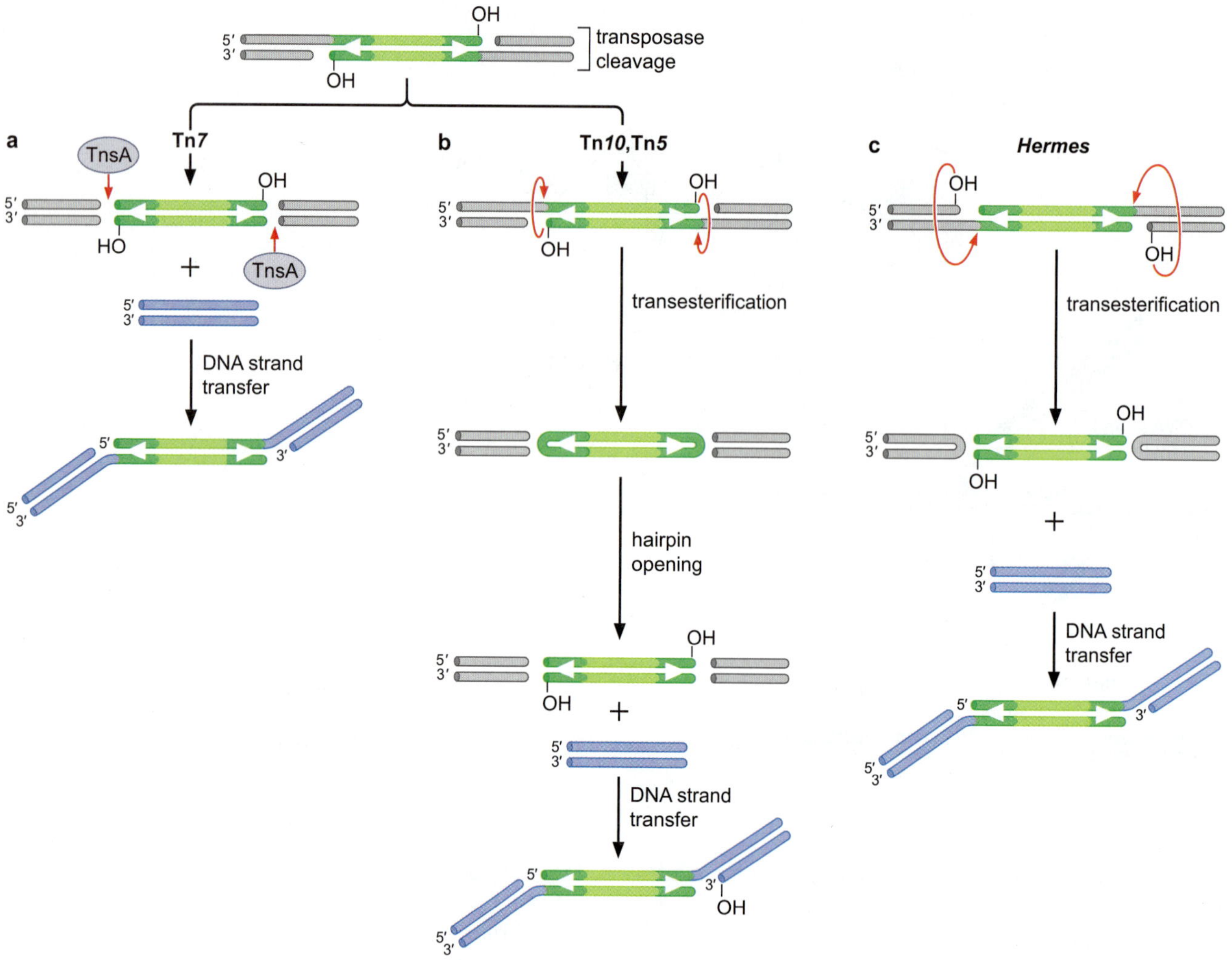

FIGURE 11-21 Three mechanisms for cleaving the nontransferred strand. (a) An enzyme other than transposase is used. (b) The transposase catalyzes the attack of one DNA strand on the opposite strand to form the DNA–hairpin intermediate. In this case, attack is of the transferred strand on the nontransferred strand. The two hairpin ends are subsequently hydrolyzed by the transposase. (c) The *Hermes* transposon uses a second mechanism of second-strand cleavage by hairpin formation. In this case, cleavage of the top strand (nontransferred strand) occurs first, and the hairpins are generated on the original insertion site DNA, rather than at the transposon ends.

TnsB protein). By working together, the transposase and TnsA excise the transposon from its original target site.

The other way of cleaving the nontransferred strand is promoted by the transposase itself, using a DNA transesterification mechanism that is similar to DNA strand transfer. For example, the transposons Tn*5* and Tn*10* cleave the nontransferred strand by generating a structure known as a "DNA hairpin." To form this hairpin, the transposase uses the initially cleaved 3′-OH end of the transposon DNA to attack a phosphodiester bond directly across the DNA duplex on the opposite strand (Fig. 11-21b). This reaction both cleaves the attacked DNA strand and covalently joins the 3′ end of the transposon DNA to one side of the break. As a result, the two DNA strands are covalently joined by a looped end, reminiscent in shape to a hairpin.

This hairpin DNA end is then cleaved (i.e., "opened") by the transposase to generate a standard double-strand break in the DNA. The opening reaction occurs on both ends of the transposon DNA. Once these steps are complete, the 3′-OH ends of the element DNA are ready to be joined to a new target DNA by the DNA strand transfer reaction.

The *Hermes* transposon, a member of the *hAT* family of elements, also uses DNA-hairpin intermediates to excise the transposon from the old DNA insertion site. However, in this case, the order of the strand cleavage and transesterification reactions is different, such that the DNA hairpins are formed in the host cell's DNA, rather than on the end of the transposable element (Fig. 11-21c). As we will see later in the chapter, this pathway of DNA cleavage and joining reactions is highly reminiscent of that observed during the early steps of V(D)J recombination. This mechanistic similarity strongly supports the hypothesis that V(D)J recombination arose from the capture and "taming" of a transposon by a host organism during vertebrate evolution.

Although not shown in Figure 11-21, DNA cleavage via a transesterification reaction can also occur *between* the two ends of the transposon. In this case, one cleaved 3′-OH end attacks the DNA strand at the opposite end of the element's DNA and the resulting DNA intermediate is further processed to generate the excised transposon. IS*3* family transposons use this mechanism.

Why might transposases use transesterification as a cleavage mechanism? It is probably an economic solution. Transposases have the intrinsic ability to promote (1) site-specific hydrolysis of the 3′ ends of the transposon DNA and (2) transesterification of this end into a nonspecific DNA site. These same activities, with the transesterification reaction simply applied to a new DNA site (e.g., the strand opposite the initial cleavage site), can allow the transposase to promote transposon excision. This mechanism therefore avoids the need for the transposon to encode a second enzyme to cleave the nontransferred strand.

DNA Transposition by a Replicative Mechanism

Some DNA transposons move using a mechanism called **replicative transposition**, in which the element DNA is duplicated during each round of transposition. Although the products of the transposition reaction are clearly different, as we shall now see, the mechanism of recombination is very similar to that used for cut-and-paste transposition (Fig. 11-22).

The first step of replicative transposition is the assembly of the transposase protein on the two ends of the transposon DNA to generate a transpososome. As we saw for cut-and-paste transposition, transpososome formation is essential to coordinate the DNA cleavage and joining reactions on the two ends of the transposon's DNA.

The next step is DNA cleavage at the ends of the transposon DNA. This reaction is catalyzed by the transposase within the transpososome. The transposase introduces a nick into the DNA at each of the two junctions between the transposon sequence and the flanking host DNA (see Fig. 11-22). This cleavage liberates two 3′-OH DNA ends on the transposon sequence. In contrast to cut-and-paste transposition, the transposon DNA is not excised from the host sequences at this stage. This is the major difference between replicative and cut-and-paste transposition.

The 3′-OH ends of the transposon DNA are then joined to the target DNA site by the DNA strand transfer reaction. The mechanism is the

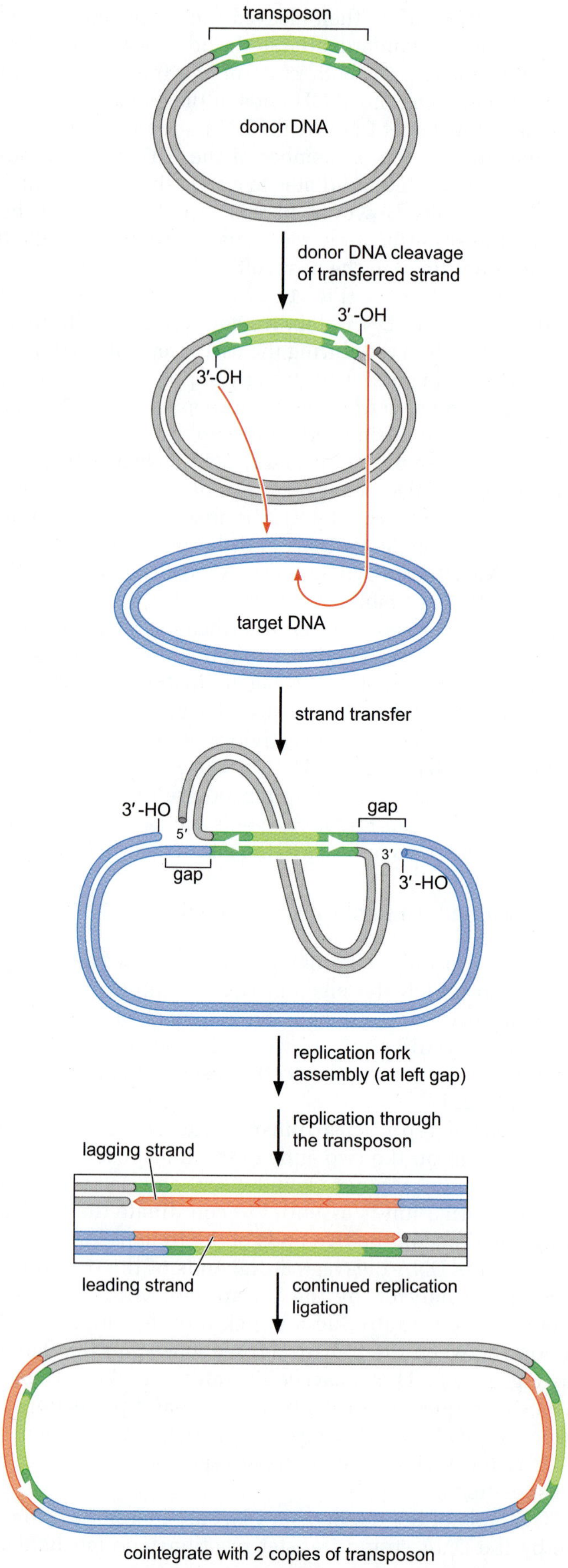

FIGURE 11-22 Mechanism for replicative transposition. The transpososome introduces a single-strand nick at each of the ends of the transposon DNA. This cleavage generates a 3′-OH group at each end. These OH groups then attack the target DNA and become joined to the target by DNA strand transfer. Note that at each end of the transposon, only one strand is transferred into the target at this point, resulting in the formation of a doubly branched DNA structure. The replication apparatus assembles at one of these "forks" (the left one in the figure). Replication continues through the transposon sequence. The resulting product, called a cointegrate, has the two starting circular DNA molecules joined by two copies of the transposon. The single-stranded DNA gaps in the branched intermediate give rise to the target-site duplications. These duplications are not shown in the cointegrate for clarity.

same as discussed above for cut-and-paste transposition. However, the intermediate generated by DNA strand transfer is in this case a doubly branched DNA molecule (see Fig. 11-22). In this intermediate, the 3′ ends of the transposon are covalently joined to the new target site, whereas the 5′ ends of the transposon sequence remain joined to the old flanking DNA.

The two DNA branches within this intermediate have the structure of a replication fork (see Chapter 8). After DNA strand transfer, the DNA replication proteins from the host cell can assemble at these forks. In the best-understood example of replicative transposition (phage Mu, which we discuss below), this assembly specifically occurs at only one of the two forked structures (see Fig. 11-22 bottom panels). The 3′-OH end in the cleaved target DNA serves as a primer for DNA synthesis. Replication proceeds through the transposon sequence and stops at the second fork. This replication reaction generates two copies of the transposon DNA. These copies are flanked by the short direct target-site duplications.

Replicative transposition frequently causes chromosomal inversions and deletions that can be highly detrimental to the host cell. This propensity to cause rearrangements may put replicative transposons at a selective disadvantage. Perhaps this is why so many elements have developed ways to excise completely from their original DNA location prior to joining to a new DNA site. By excision, transposons avoid generating these major disruptions to the host genome. As we describe below, transposition via an RNA intermediate also avoids generating these disruptive rearrangements.

Virus-like Retrotransposons and Retroviruses Move Using an RNA Intermediate

Virus-like retrotransposons and retroviruses insert into new sites in the genome of the host cell, using the same steps of DNA cleavage and DNA strand transfer we have described for the DNA transposons. In contrast to the DNA transposons, however, recombination for these retroelements involves an RNA intermediate.

A cycle of transposition starts with transcription of the retrotransposon (or retroviral) DNA sequence into RNA by a cellular RNA polymerase. Transcription initiates at a promoter sequence within one of the LTRs (Fig. 11-23) and continues across the element to generate a nearly full-length RNA copy of the element's DNA. The RNA is then reverse-transcribed to generate a double-stranded DNA molecule. This DNA molecule is called the **cDNA** (for copied DNA) and is free from any flanking host DNA sequences.

It is the cDNA that is recognized by the integrase protein (a protein highly related to the transposases of DNA elements, as we shall see below) for recombination with a new target DNA site. Integrase assembles on the ends of this cDNA and then cleaves a few nucleotides off the 3′ end of each strand. This cleavage reaction is identical to the DNA cleavage step of DNA transposition. As the direct precursor DNA for integration is generated from the RNA template by reverse transcription, it is already in the form of an excised transposon. Therefore, a mechanism to cleave the second strand is unnecessary for these elements. Integrase then catalyzes the insertion of these cleaved 3′ ends into a DNA target site in the host-cell genome using the DNA strand transfer reaction. As we discussed above, this target site can have essentially any DNA sequence. Host-cell DNA-

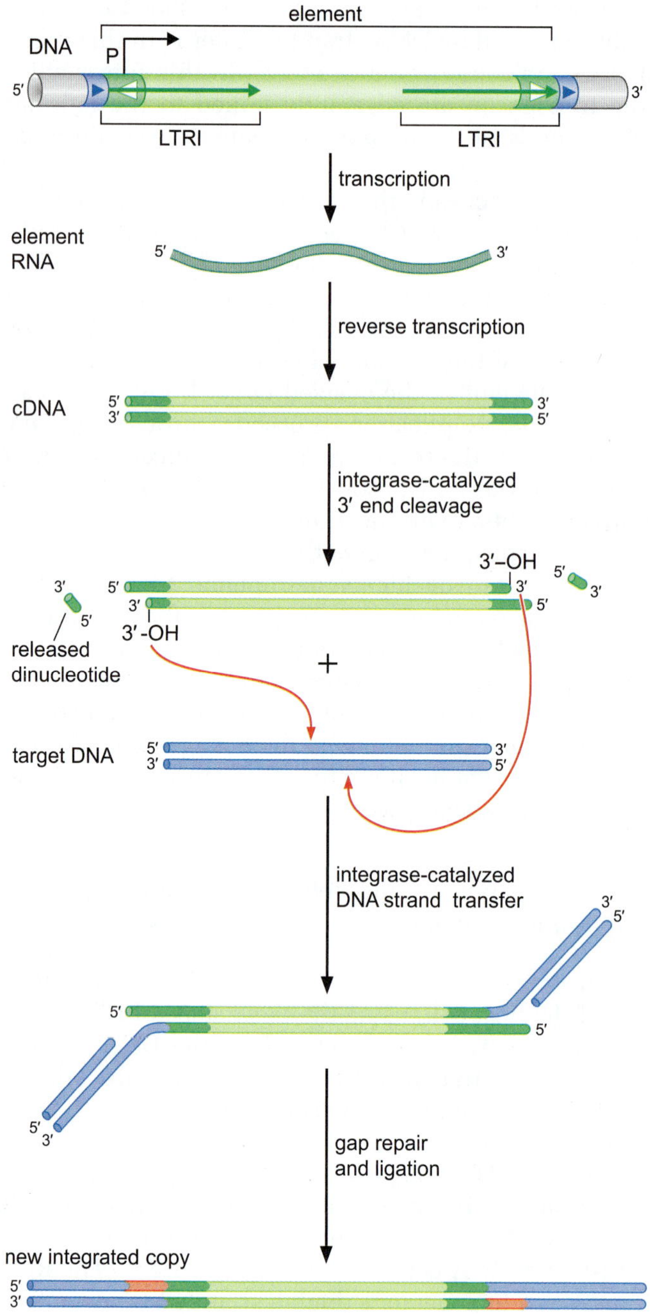

FIGURE 11-23 Mechanism of retroviral integration and transposition of virus-like retrotransposons. The top panel shows integrated provirus. For a more detailed view of the LTR sequences, see the figures in Box 11-3. The promoter for transcription of the viral RNA is embedded in the left LTR as shown. cDNA synthesis from this viral RNA is explained in Box 11-3. The integrase-catalyzed DNA cleavage and DNA strand transfer steps are shown.

repair proteins fill the gaps at the target site generated during DNA strand transfer to complete recombination. This gap-repair reaction generates the target-site duplications.

Because transcription to generate the RNA intermediate initiates within one of the LTRs, this RNA does not carry the entire LTR sequence; the sequence between the transcription start site and the end of the element is missing. Therefore, a special mechanism is needed to regenerate the full-length element sequence during reverse transcription. The pathway of reverse transcription involves two internal priming events and two strand switches (for details of the process, see Box 11-3, The Pathway of Retroviral cDNA Formation). These

ADVANCED CONCEPTS

Box 11-3 The Pathway of Retroviral cDNA Formation

To understand the process of retroviral reverse transcription (or that of the virus-like retrotransposons), we first need to look in more detail at the structure of the LTR sequences. Each LTR is constructed of three sequence elements. These are called U3 (for unique 3′ end), R (for repeat), and U5 (for unique 5′ end). Transcription from the integrated copy of the retroviral genome generates the viral RNA with the R sequence at each end (Box 11-3 Fig. 1). Therefore, during the process of reverse transcription, one additional U3 and U5 region must be synthesized. As explained below, this duplication happens because priming of DNA synthesis occurs at internal sites within the RNA genome, and the R sequence allows two "strand switches" to occur during the replication process.

It is the viral RNA that is packaged into viral particles, and this RNA enters the new cell during infection. The viral RNA is packaged with a cellular tRNA molecule (see Chapter 14) that serves as the primer for synthesis of the first cDNA strand. This tRNA forms base pairs with a specific sequence near the U5 region, known as the **primer-binding site (PBS)** (Box 11-3 Fig. 2a). DNA synthesis by the reverse transcriptase enzyme then copies the U5 region and the first R segment (Box 11-3 Fig. 2b).

Reverse transcriptase has two enzymatic activities that are important for cDNA formation: a DNA polymerase activity and an RNase H activity. RNase H enzymes degrade RNA that is base-paired with DNA (as we discussed in Chapter 8). During reverse transcription, RNase H removes the template RNA strands. When this step occurs on the first RNA–DNA hybrid intermediate (see Box 11-3 Fig. 2b,c), the U5-R DNA strand is released in a single-stranded form.

This U5-R DNA strand can then base-pair with the R region on the other end of the viral RNA molecule (Box 11-3 Fig. 2d). This step is the first of the two strand switches. Once this switching occurs, reverse transcriptase continues DNA synthesis to copy the remainder of the RNA template (Box 11-3 Fig. 2e). The resulting DNA strand ends with the PBS sequence at its 3′ terminus. The RNA template strand is removed, as before, by RNase H (Box 11-3 Fig. 2d,e).

RNase H–mediated degradation of the viral RNA also generates an RNA fragment that serves as the primer for synthesis of the second cDNA strand. This region of RNA remains base-paired with a sequence called the **polypurine tract (PPT)** at the edge of the U3 sequence (Box 11-3 Fig. 2e,f). Elongation of this primer copies the U3, R, U5, and PBS sequences into DNA.

Once the tRNA primer is removed from the first cDNA strand, the second-strand switch occurs. The complementary sequence of the PBS on the 3′ ends allows base-pairing interactions between the two DNA strands and formation of a circular intermediate. Elongation of each of the 3′ DNA ends present in this intermediate to the end of the other strand generates the double-stranded cDNA with two complete LTR sequences. This DNA molecule is then ready to be integrated into the cell's genome by the integrase protein.

Reverse transcriptase is a virus-encoded (or retrotransposon-encoded) enzyme and serves no essential cellular function. It is, however, absolutely essential for retroviral replication. Thus, it is a common target of antiviral drugs, including many of the drugs that have been used to fight the AIDS epidemic.

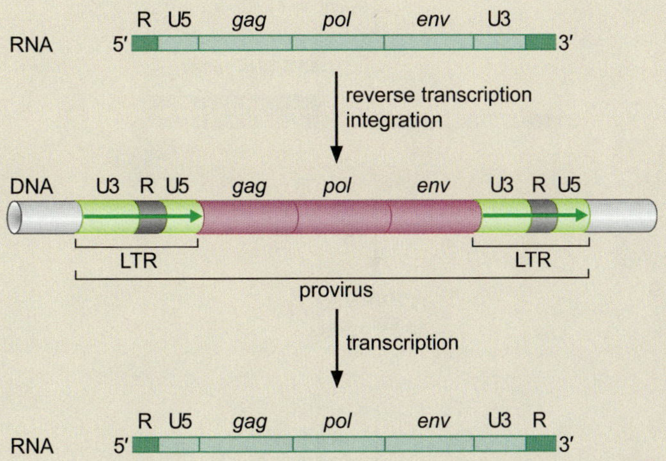

BOX 11-3 FIGURE 1 Detailed view of the sequence elements near the ends of the retroviral RNA and cDNA. Virus-like retrotransposons have a very similar sequence organization. The *pol* gene encodes both reverse transcriptase (including the RNase H activity) and integrase.

Box 11-3 (*Continued*)

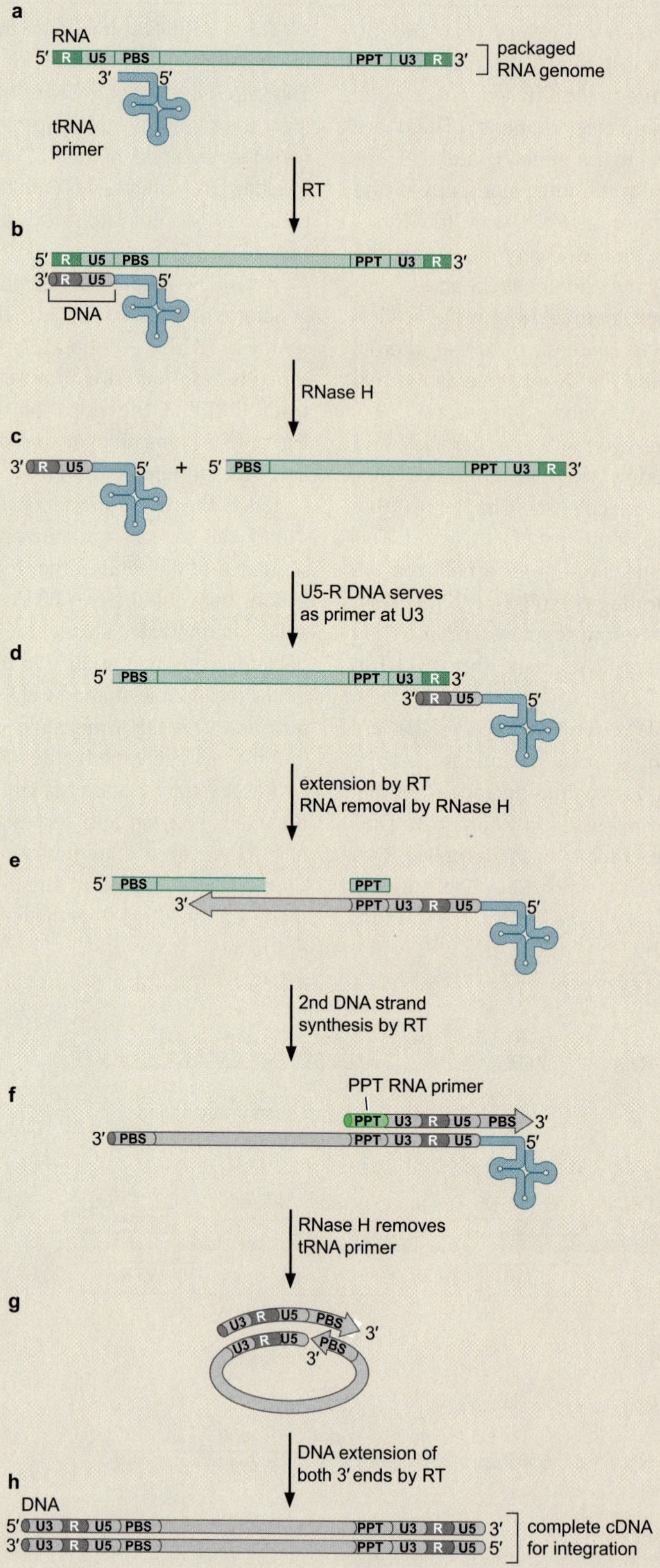

Box 11-3 Figure 2 **Pathway of reverse transcription to generate the cDNA copy of the retroviral or retrotransposon RNA.**

switching events result in the duplication of sequences at the ends of the cDNA. Thus, the cDNA has complete reconstructed LTR sequences to compensate for regions of sequence lost during transcription. This reconstruction of the LTRs is essential for recognition of the cDNA by integrase and for subsequent recombination.

DNA Transposases and Retroviral Integrases Are Members of a Protein Superfamily

As we have seen, DNA cleavage of the 3′ ends of the transposon DNA (or cDNA) and DNA strand transfer are common steps used for DNA transposition and the movement of virus-like retrotransposons and retroviruses. This conserved recombination mechanism is reflected in the structure of the transposase/integrase proteins (Fig. 11-24). High-resolution structures reveal that many different transposases and integrases carry a catalytic domain that has a common three-dimensional shape. This catalytic domain contains three evolutionarily invariant acidic amino acids: two aspartates (D) and a glutamate (E). Therefore, recombinases of this class are referred to as DDE-motif transposase/integrase proteins. These acidic amino acids form part of the active site and coordinate divalent metal ions (such as Mg^{2+} or Mn^{2+}) that are required for activity (as described for the DNA polymerases, see Chapter 8). An unusual feature of the transposase/integrase proteins is that they use this same active site to catalyze both the DNA cleavage and DNA strand transfer, rather than having two active sites, each specialized for one chemical reaction.

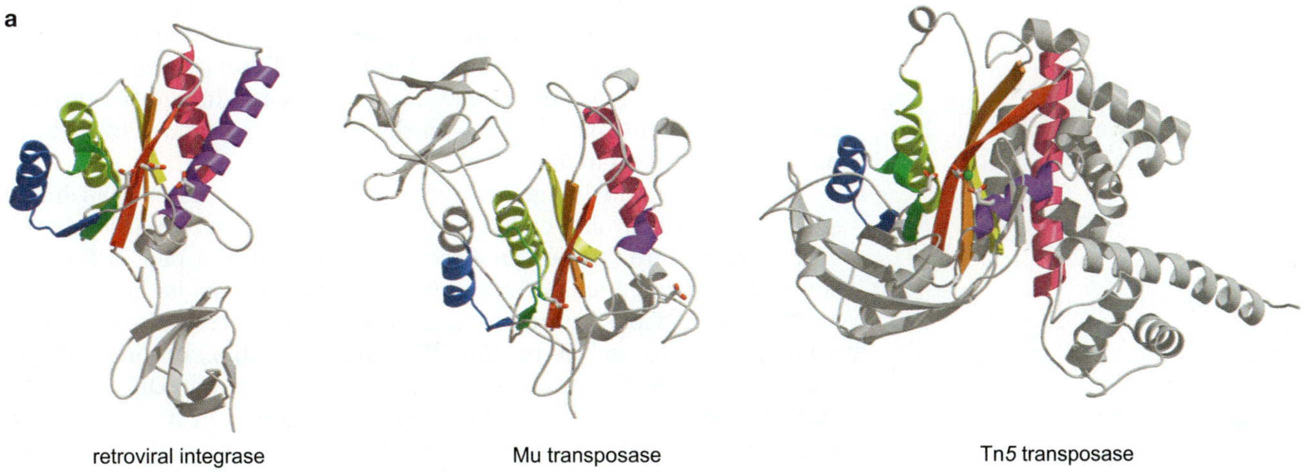

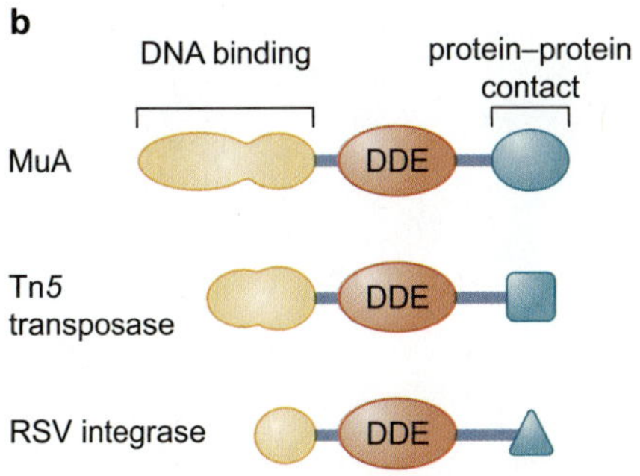

FIGURE 11-24 Similarities of catalytic domains of transposases and integrases. (a) Structures of the conserved core domains (shown right to left) of Tn5 transposase (Davies D.R. et al. 2000. *Science* 289: 77–85), of phage Mu transposase (Rice P. and Mizuuchi K. 1995. *Cell* 82: 209–220), and of RSV integrase (Chook Y.M. et al. 1994. *J. Mol. Biol.* 240: 476–500). Common secondary structure elements are shown in the same colors. The DDE-motif active-site residues are shown in stick form. Images prepared with MolScript, BobScript, and Raster 3D. (b) Schematic of the domain organization of the three proteins shown in part a. The amino-terminal domains bind to the element DNA. The middle domains contain the catalytic regions shown in a. The carboxy-terminal domains are involved in protein–protein contacts needed to assemble the transpososome and/or to interact with other proteins that regulate transposition. (Derived from Rice P.A. and Baker T.A. 2001. *Nat. Struct. Biol.* 8: 302–307.)

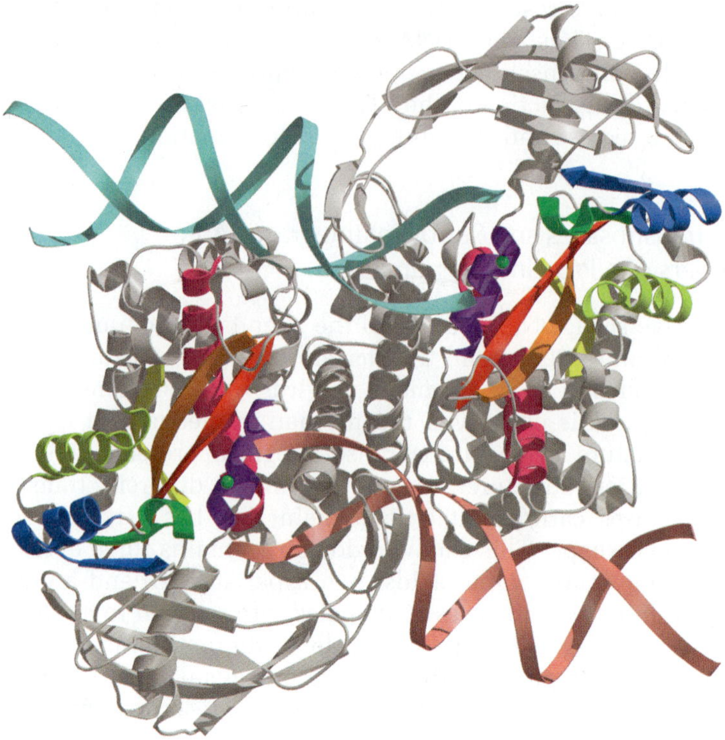

FIGURE 11-25 Cocrystal of Tn*5* bound to substrate DNA. The complex contains a dimer of transposase. The catalytic domains are colored as in Fig. 11-24. The green balls are divalent metal ions bound in the protein's active site. Note that the subunit bound via its DNA-binding domain to one transposon end donates the catalytic domain for recombination on the other DNA end. The DNA is shown in light blue and pink. (Davies D.R. et al. 2000. *Science* 289: 77–85.) Image prepared with MolScript, BobScript, and Raster 3D with additional modeling of the DNA by Leemor Joshua-Tor.

In contrast to the highly conserved structure of the catalytic domains, the remaining regions of proteins in this family are not conserved. These regions encode site-specific DNA-binding domains and regions involved in protein–protein interactions needed to assemble the protein–DNA complex specific for each individual element. Thus, these unique domains ensure that transposases and integrases catalyze recombination specifically only on the element that encoded them or on a very highly related element.

Transposases and integrases are only active when assembled into a synaptic complex, also called a transpososome, on DNA (see above). The cocrystal structure of Tn*5* transposase bound to a pair of transposon end DNA fragments provides insight into why this is the case (Fig. 11-25). The transposase subunit that is bound to the recombinase recognition sequences on one of these DNA fragments (i.e., on one transposon end) donates the catalytic domain that promotes the DNA cleavage and DNA strand transfer reactions on the other end of the transposon. Because of this subunit organization, the transposase will be properly positioned for recombination only when two subunits and a pair of DNA ends are present together in the complex.

Poly-A Retrotransposons Move by a "Reverse Splicing" Mechanism

The poly-A retrotransposons (e.g., human LINE elements) move using an RNA intermediate but use a mechanism different from that used by the virus-like elements. This mechanism is called **target-site-primed reverse transcription** (Fig. 11-26). The first step is transcription of the DNA of an integrated element by a cellular RNA polymerase (Fig. 11-26a). Although the promoter is embedded in the 5′ UTR, it can in this

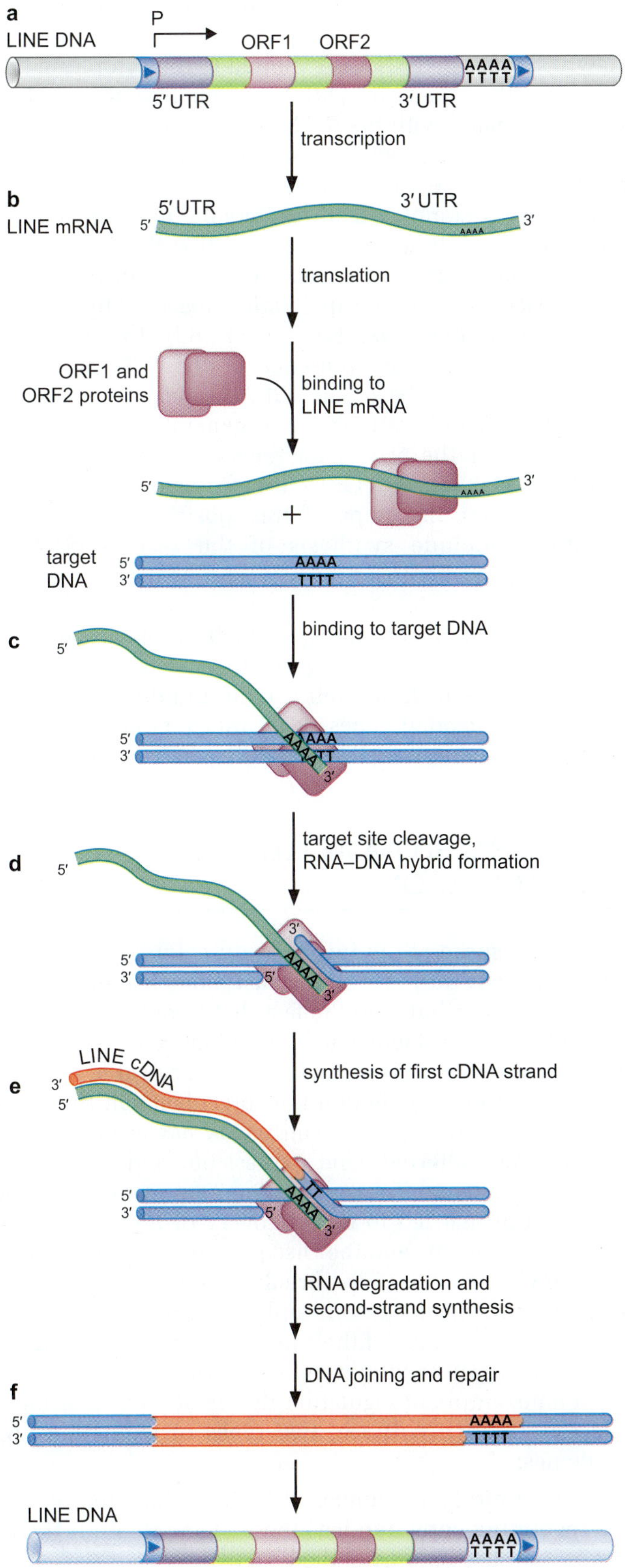

FIGURE 11-26 Transposition of a poly-A retrotransposon by target-site-primed reverse transcription. The figure outlines a model for the movement of a LINE element. (a) A cellular RNA polymerase initiates transcription of an integrated LINE sequence. (b) The resulting mRNA is translated to produce the products of the two encoded ORFs that then bind to the 3′ end of their mRNA. (c) The protein–mRNA complex then binds to a T-rich site in the target DNA. (d) The proteins initiate cleavage in the target DNA, leaving a 3′-OH at the DNA end and forming an RNA:DNA hybrid. (e) The 3′-OH end of the target DNA serves as a primer for reverse transcription of the element RNA to produce cDNA (first-strand synthesis). (f) The final steps of the transposition reaction include second-strand synthesis and DNA joining and repair to create a newly inserted LINE element.

case direct RNA synthesis to begin at the first nucleotide of the element's sequence.

This newly synthesized RNA is exported to the cytoplasm and translated to generate the ORF1 and ORF2 proteins (see above). These proteins remain associated with the RNA that encoded them (Fig. 11-26b). In this way, an element promotes its own transposition and does not donate proteins to competing elements.

The protein–RNA complex then reenters the nucleus and associates with the cellular DNA (Fig. 11-26c). Recall that the ORF2 protein has both a DNA endonuclease activity and a reverse transcriptase activity. The endonuclease initiates the integration reaction by introducing a nick in the chromosomal DNA (see Fig. 11-26d). T-rich sequences are preferred cleavage sites. The presence of these Ts at the cleavage site permits the DNA to base-pair with the poly-A tail sequence of the element RNA. The 3′-OH DNA end generated by the nicking reaction then serves as the primer for reverse transcription of the element RNA (Fig. 11-26e). The ORF2 protein also catalyzes this DNA synthesis. The remaining steps of transposition, although not yet well-understood, include synthesis of the second cDNA strand, repair of DNA gaps at the insertion site, and ligation to seal the DNA strands.

Many of the poly-A retrotransposons that have been detected by large-scale genomic sequencing are truncated elements. Most of these are missing regions from their 5′ ends and do not have complete copies of element-encoded genes or an intact promoter. These truncated elements have therefore lost the ability to transpose.

EXAMPLES OF TRANSPOSABLE ELEMENTS AND THEIR REGULATION

Transposons have successfully invaded and colonized the genomes of all life-forms. Clearly, they are very robust biological entities. Some of this success can be attributed to the fact that transposition is regulated in ways that help to establish a harmonious coexistence with the host cell. This coexistence is essential for the survival of the element as transposons cannot exist without a host organism. On the other hand, as introduced above, transposons can wreak havoc in a cell, causing insertion mutations, altering gene expression, and promoting large-scale DNA rearrangements. These disruptions are particularly noticeable in plants, a feature that led to the discovery of transposons in maize (Box 11-4, Maize Elements and the Discovery of Transposons).

In the following sections, we briefly describe some of the best-understood individual transposons and transposon families. (A larger list of transposons and some of their important features is summarized in Table 11-2.) Each subsection provides a brief overview of a specific element and an example of regulation that is of particular importance to that element. As we will see, two types of regulation appear as recurring themes:

- Transposons control the number of their copies present in a given cell. By **regulating copy number**, these elements limit their deleterious impact on the genome of the host cell.
- Transposons control target-site choice. Two general types of target-site regulations are observed. In the first, some elements preferentially insert into regions of the chromosome that tend not to be harmful to the host cell. These regions are called **safe havens** for

TABLE 11-2 Major Types of Transposable Elements

Type	Structural Features	Mechanism of Movement	Examples
DNA-mediated transposition			
Bacterial replicative transposons	Terminal inverted repeats that flank antibiotic-resistance and transposase genes	Copying of element DNA accompanying each round of insertion into a new target site	Tn*3*, γδ, phage Mu
Bacterial cut-and-paste transposons	Terminal inverted repeats that flank antibiotic-resistance and transposase genes	Excision of DNA from old target site and insertion into new site	Tn*5*, Tn*10*, Tn*7*, IS*911*, Tn*917*
Eukaryotic transposons	Inverted repeats that flank coding region with introns	Excision of DNA from old target site and insertion into new site	P elements (*Drosophila*), *hAT* family elements, Tc*1*/*Mariner* elements
RNA-mediated transposition			
Virus-like retrotransposons	~250- to 600-bp direct terminal repeats (LTRs) flanking genes for reverse transcriptase, integrase, and retrovirus-like Gag protein	Transcription into RNA from promoter in left LTR by RNA polymerase II followed by reverse transcription and insertion at target site	Ty elements (yeast), *Copia* elements (*Drosophila*)
Poly-A retrotransposons	3′ A-T-rich sequence and 5′ UTR flank genes encoding an RNA-binding protein and reverse transcriptase	Transcription into RNA from internal promoter; target-primed reverse transcription initiated by endonuclease cleavage	F and G elements (*Drosophila*), LINE and SINE elements (mammals), *Alu* sequences (humans)

transposons. In the second type of regulation, some transposons specifically avoid transposing into their own DNA. This phenomenon is called **transposition target immunity**.

IS*4* Family Transposons Are Compact Elements with Multiple Mechanisms for Copy Number Control

The bacterial transposon Tn*10* is a well-characterized representative of the IS*4* family, which also includes Tn*5*. Tn*10* is a compact element of 9 kb and encodes a gene for its own transposase and genes imparting resistance to the antibiotic tetracycline (Fig. 11-27).

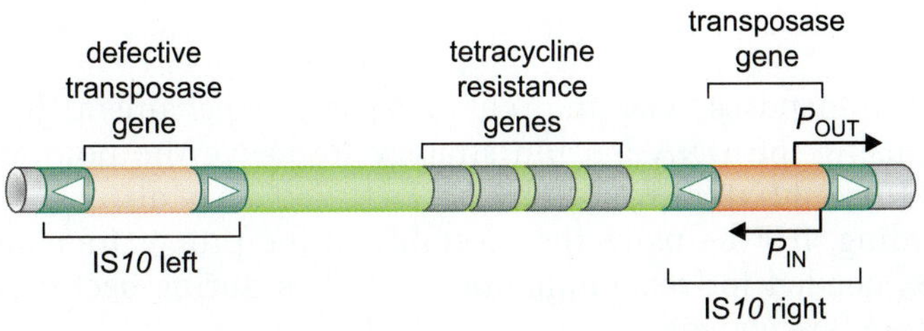

FIGURE 11-27 Genetic organization of bacterial transposon Tn*10*. The map shows the functional elements in the bacterial transposon Tn*10*. Tn*10*, like many bacterial transposons, actually carries two "minitransposons" at its termini. For Tn*10*, these elements are called IS*10L* (left) and IS*10R* (right). Both types of IS*10* elements can transpose and are found in DNA separately from Tn*10*. The white triangles show the inverted repeat sequences at the ends of the IS elements and Tn*10*. Although these four copies are not exactly the same in sequence, all are recognized by the Tn*10* transposase and are used as recombination sites.

KEY EXPERIMENTS

Box 11-4 Maize Elements and the Discovery of Transposons

Plant genomes are very rich in transposons. Furthermore, the ability of transposable elements to alter gene expression can often be readily observed as dramatic variation in the coloration of the plant (Box 11-4 Fig. 1). Thus, it is not surprising that transposable elements, and many of their salient features, were first discovered in plants.

Barbara McClintock discovered "controlling elements" in maize in the late 1940s. It was actually the ability of transposable elements to break chromosomes that first came to McClintock's attention. She found that some strains experienced broken chromosomes very frequently, and she named the genetic element responsible for these chromosome breaks *Ds* (dissociator). Surprisingly, she observed that the sites of these "hot spots" for chromosome breaks were different in different strains and could even be in different chromosomal locations in the descendents of an individual plant. This observation provided the first insight that genetic elements could move (i.e., "transpose") within chromosomes.

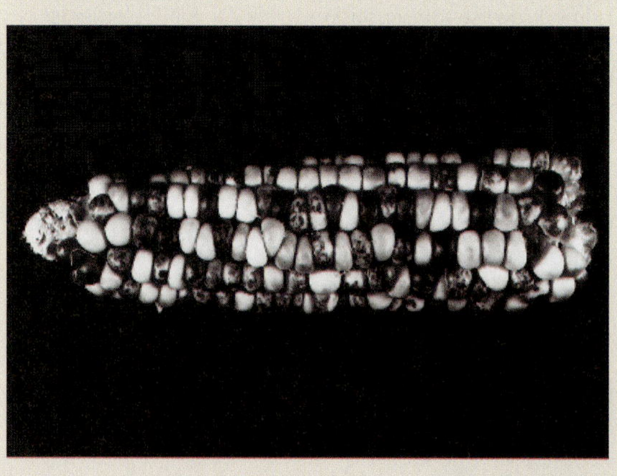

BOX 11-4 FIGURE 1A Example of corn (maize) cob showing color variegation due to transposition. (Photograph taken by Barbara McClintock; image courtesy Cold Spring Harbor Laboratory Archives.)

BOX 11-4 FIGURE 1B Example of color variegation in snapdragon flowers due to Tam3 transposition. The size of white patches is related to the frequency of transposition. (Reprinted, with permission, from Chatterjee M. and Martin C. 1997. Plant J. 11: 759–771, Fig. 2a. © Blackwell Publishing.)

Ds, in fact, is a nonautonomous DNA transposon that moves by cut-and-paste transposition. *Ds* movement requires the *Ac* (activator) element (also discovered by McClintock) to be present in the same cell and provide the transposase protein. *Ac* is now recognized to be part of a large family of DNA transposons called the *hAT* family, named for the *hobo* elements from flies, the *Ac* elements from maize, and the *Tam* elements from snapdragon. The *Hermes* element from housefly is also a member of this family and has proved amenable to mechanistic analysis.

Tn*10* transposes via the cut-and-paste mechanism (described above), using the DNA hairpin strategy to cleave the nontransferred strands (Figs. 11-18 and 11-21). The Tn*10* sequence also has a site for IHF binding. IHF helps in the assembly of the proper transpososome complex needed for recombination as it does during bacteriophage λ integration (see above).

Tn*10* is organized into three functional modules. This organization is relatively common, and elements that have it are called **composite transposons**. The two outermost modules, called IS*10L* (left) and IS*10R* (right), are actually minitransposons. "IS" stands for **insertion sequence**. IS*10R* encodes the gene for the transposase that recognizes the terminal inverted repeat sequences of IS*10R*, IS*10L*, and Tn*10*. IS*10L*, although very similar in sequence to IS*10R*, does not encode a functional transposase. Thus, both IS*10R* and Tn*10* are autonomous, whereas IS*10L* is a nonautonomous transposon. Both types of IS*10* elements are found, as expected considering their own mobility, unassociated with Tn*10* in genomes.

Tn*10* limits its copy number in any given cell by strategies that restrict its transposition frequency. One mechanism is the use of an **antisense RNA** to control the expression of the transposase gene (see Fig. 11-29) (for a discussion of antisense RNA regulation, see Chapters 17 and 18). Near the end of IS*10R* are two promoters that direct the synthesis of RNA by the host cell's RNA polymerase. The promoter that directs RNA synthesis inward (called P_{IN}) is responsible for the expression of the transposase gene. The promoter that directs transcription outward (P_{OUT}), in contrast, serves to regulate transposase expression by making an antisense RNA, as follows. The RNAs synthesized from P_{IN} and P_{OUT} overlap (by 36 bp) and therefore can pair by hydrogen bonding between these overlapping (complementary) regions. This pairing prevents binding of ribosomes to the P_{IN} transcript, and thus synthesis of the transposase protein.

By this mechanism, cells that carry more copies of Tn*10* will transcribe more of the antisense RNA, which in turn will limit expression of the transposase gene (Fig. 11-28, see legend for more details). The transposition frequency will therefore be very low in such a strain. In contrast, if there is only one copy of Tn*10* in the cell, the level of antisense RNA will be low, synthesis of the transposable protein will be efficient, and transposition will occur at a higher frequency.

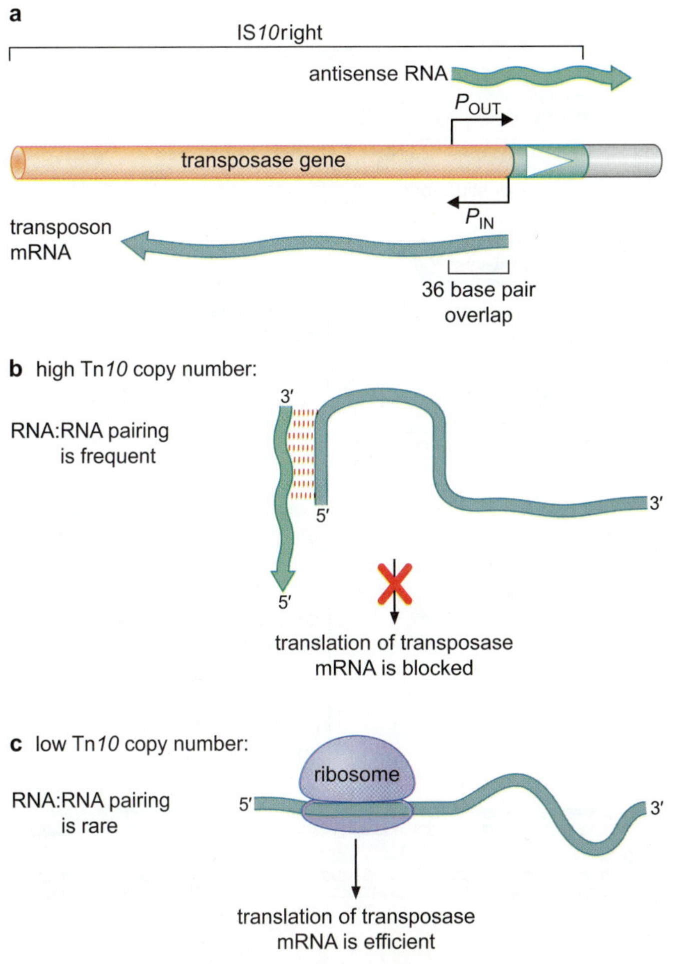

FIGURE 11-28 Antisense regulation of Tn*10* expression. (a) A map of the overlapping promoter regions is shown. The leftward promoter (P_{IN}) promotes expression of the transposase gene; the rightward promoter (P_{OUT}), which lies 36 bases to the left of P_{IN}, promotes expression of an antisense RNA. The first 36 bases of each transcript are complementary to one another. Note that in cells, the antisense transcript initiated at P_{OUT} is longer-lived than is the mRNA initiated at P_{IN}. (b) In cells having a high copy number of Tn*10*, the RNA:RNA pairing occurs frequently and blocks translation of the transposase mRNA (thereby eventually reducing the copy number of the element). (c) In cells having a low copy number of the transposon, RNA:RNA pairing is rare; the translation of transposase mRNA is efficient, and the copy number in the cell is increased.

Tn*10* Transposition Is Coupled to Cellular DNA Replication

Tn*10* also couples transposition to cellular DNA replication. Recall that bacteria such as *E. coli* (a common host for Tn*10*) methylate their DNA at GATC sites (see Chapter 8, Box 8-7). This methylation occurs after DNA replication, such that GATC sites are hemimethylated for the few minutes between passage of the replication fork and recognition of these sequences by the methylase enzyme.

It is during this brief period—when the Tn*10* DNA is hemimethylated—that transposition is most likely to occur. This coupling of transcription to the methylation state is due to the presence of two critical GATC sites in the transposon sequence. One of these sites is in the promoter for the transposase gene; the second is in the binding site for the transposase within one of the inverted terminal repeats. Both RNA polymerase and transposase bind more tightly to the hemimethylated sequences than to their fully methylated versions. As a result, when the DNA is hemimethylated, the transposase gene is most efficiently expressed, and the transposase protein binds most efficiently to the DNA. Therefore, transposition of Tn*10* occurs at its highest frequency during this brief phase of the cell cycle just after its DNA has been replicated (Fig. 11-29).

Regulation of Tn*10* transposition by DNA methylation serves to limit the overall frequency of transposition. It also restricts transposition specifically to actively dividing cells. This timing ensures that there are two copies of the chromosome present to "heal" the double-stranded DNA break left in the old target site as a result of transposon

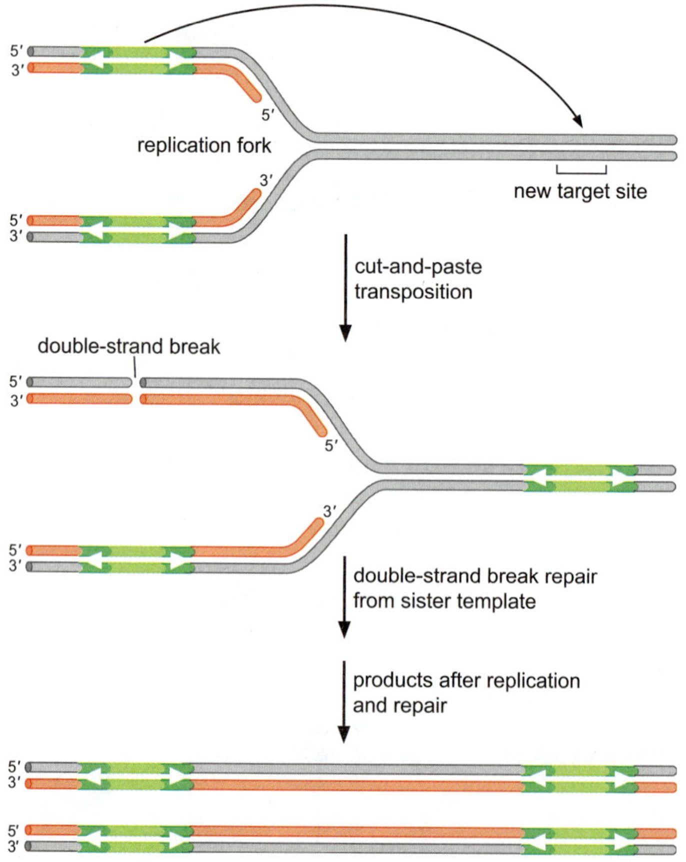

FIGURE 11-29 Transposition of Tn*10* after passage of a replication fork. Transposition is activated by the hemimethylated DNA that exists just after DNA replication (methylation sites are not shown). During transposition, a double-strand break is made in the chromosomal DNA where the element excised. This break can be repaired by the double-strand break repair pathway (see Chapter 10), a process that regenerates a copy of Tn*10* at the site of excision. By this mechanism, transposition may appear to be "replicative" in nature, although the actual recombination process goes through the cut-and-paste (nonreplicative) pathway.

excision. Because "empty target sites" are repaired via homologous recombination by the double-strand break repair pathway, it is important for cell survival that two copies of the chromosomal region be present during transcription (see Chapter 10).

Phage Mu Is an Extremely Robust Transposon

Phage Mu, like bacteriophage λ, is a lysogenic bacteriophage (see Chapter 22). Mu is also a large DNA transposon. This phage uses transposition to insert its DNA into the genome of the host cell during infection and in this way is similar to the retroviruses (discussed above). Mu also uses multiple rounds of replicative transposition to amplify its DNA during lytic growth. During the lytic cycle, Mu completes approximately 100 rounds of transposition per hour, making it the most efficient transposon known. Furthermore, even when present as a quiescent lysogen, the Mu genome transposes quite frequently, compared to traditional transposons such as Tn*10*. The name Mu is short for mutator and stems from this ability to transpose promiscuously: cells carrying an inserted copy of the Mu DNA frequently accumulate new mutations due to insertion of the phage DNA into cellular genes.

The Mu genome is ~40 kb and carries more than 35 genes, but only two encode proteins with dedicated roles in transposition. These are the *A* and *B* genes, which encode the proteins MuA and MuB. MuA is the transposase and is a member of the DDE protein superfamily discussed above. MuB is an ATPase that stimulates MuA activity and controls the choice of the DNA target site (Fig. 11-30). This process is explained in the next section.

Mu Uses Target Immunity to Avoid Transposing into Its Own DNA

Mu, like many transposons, shows very little sequence preference at its target sites. As a result, "good" target sites occur very frequently in DNA including the DNA of the Mu genome itself. Given this nearly random sequence preference, how does Mu avoid transposing into its own DNA, a situation that would likely result in serious disruptions of the phage's genes?

This problem is solved because Mu transposition is regulated by a process called **transposition target immunity** (see Box 11-5, Mechanism of Transposition Target Immunity). DNA sites surrounding a copy of the Mu element, including the element's own DNA, are rendered very poor targets for a new transposition event.

Transposition target immunity is observed for a number of different transposable elements and can work over very long distances. For Mu, sequences within ~15 kb of an existing Mu insertion are immune to new insertions. For some elements—for example, Tn*3* and Tn*7*—target immunity occurs over distances greater than 100 kb. Target immunity protects an element from transposing into itself or from having another new copy of the same type of element insert into its genome. Furthermore, this type of regulation of target DNA selection also provides a driving force for elements to move to new locations "far" from where they are initially inserted, a feature that may also be advantageous for their overall propagation and survival.

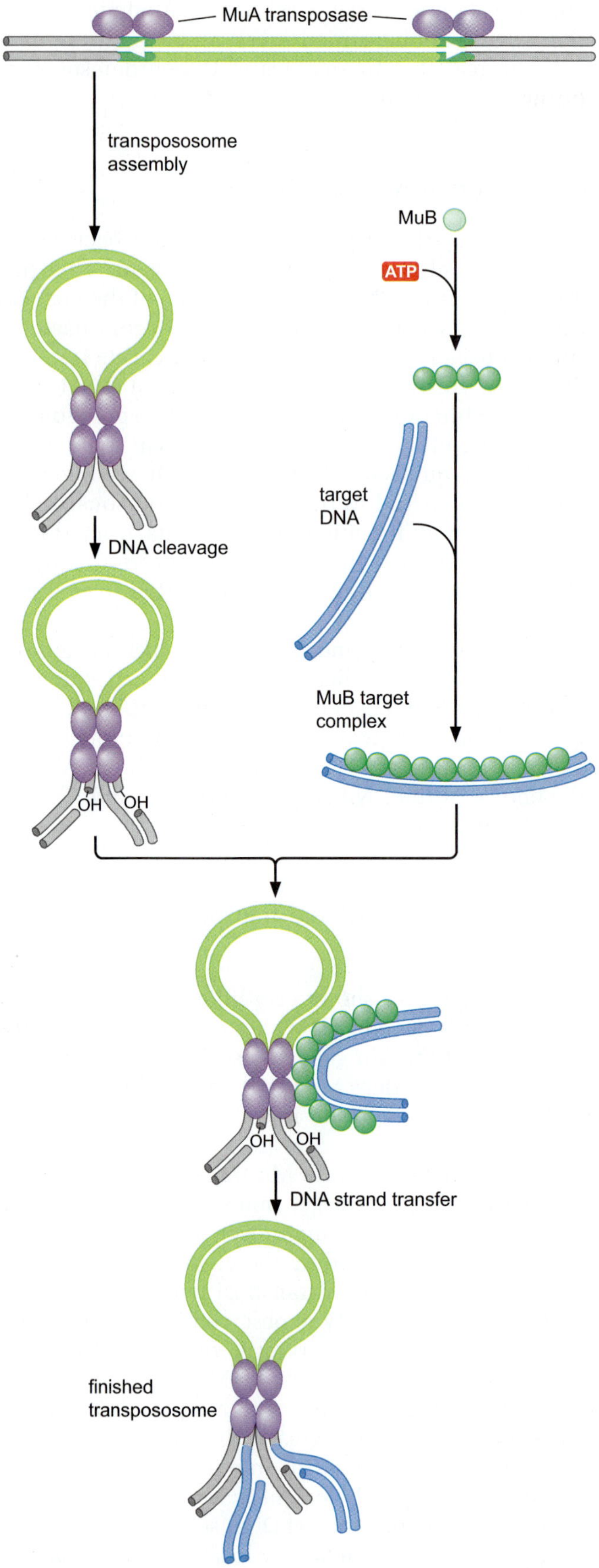

FIGURE 11-30 Overview of the early steps of Mu transposition. Four subunits of the MuA transposase assemble on the ends of Mu DNA. MuB binds ATP and then binds to DNA of any sequence. A protein–protein interaction between MuA and MuB brings the MuA DNA–transpososome complex to a new DNA target site. MuB is not shown in the final panel because, after DNA strand transfer, it is no longer needed and probably leaves the complex.

ADVANCED CONCEPTS

Box 11-5 Mechanism of Transposition Target Immunity

Interplay between the MuA transposase and the MuB ATPase is at the center of the mechanism of transposition target immunity. MuA–MuB interactions prevent MuB from binding to the DNA near where MuA is bound. The interactions responsible for this interplay are listed below.

- MuA inhibits MuB from binding to nearby DNA sites. This inhibition requires ATP hydrolysis.
- MuB helps MuA find a target site for transposition.

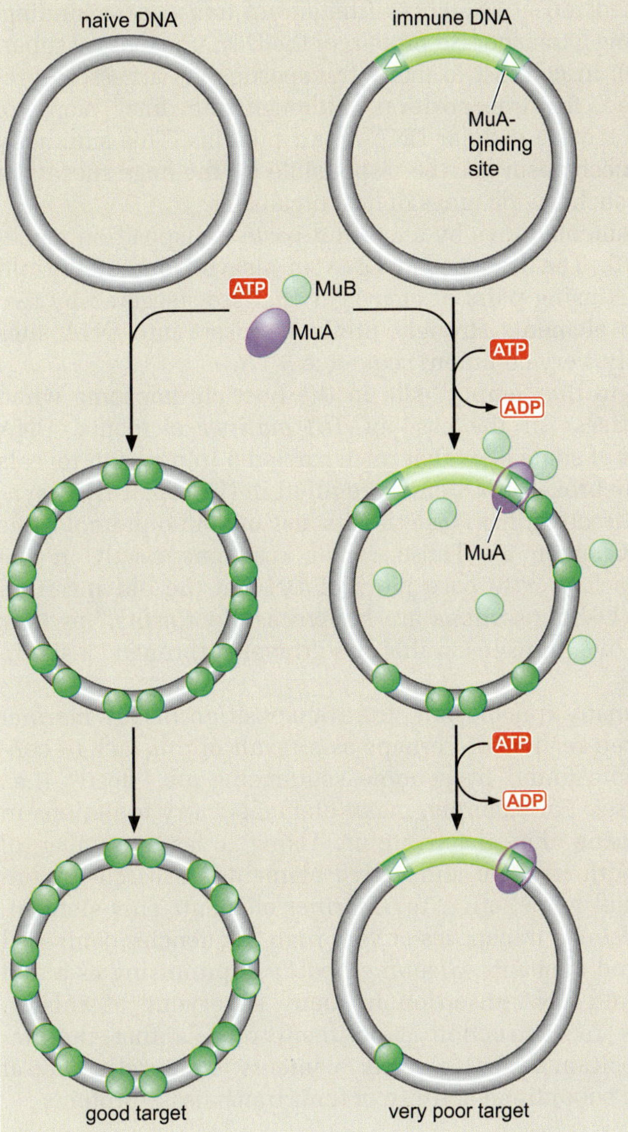

To see how individual protein-protein and protein–DNA interactions function together to generate target immunity, consider transposition into two candidate DNA segments: one is any representative segment of DNA, whereas the second has a copy of Mu already inserted (see Box 11-5 Fig. 1). We call the first DNA segment the naïve region and the second DNA segment the immune region.

What happens at each of these DNA regions as Mu prepares to transpose? First, we consider events at the naïve region. MuB, in complex with ATP (MuB•ATP), will bind the DNA, using its nonspecific DNA-binding activity. At the same time, MuA transposase will assemble a transpososome on the Mu DNA. This MuA in the transpososome can then make protein–protein contacts with the MuB•DNA complex at the naïve region. As a result of this interaction, MuB delivers this DNA to MuA for use as a target site.

In contrast, both MuA and MuB bind to DNA in the immune region. MuA interacts with its specific binding sites on the Mu genome that is already present; MuB•ATP again binds using its affinity for any DNA sequence. However, when both MuA and MuB are bound to this region, they will interact. As a result, MuA stimulates ATP hydrolysis by MuB and the dissociation of MuB from this DNA. MuB therefore does not accumulate on this immune DNA segment. By this means, the Mu transposition proteins use the energy stored in ATP to protect the Mu genome from becoming the target of transposition. As expected from this mechanism, even a single MuA-binding site within a DNA molecule is sufficient to impart target immunity.

BOX 11-5 FIGURE 1 The interplay between MuA and MuB on DNA leads to the development of an immune target DNA. The MuA-binding sites are in the terminal inverted repeats on the ends of the transposon (shown in dark green). MuA is shown bound to only one of the two repeat regions for clarity. Every time MuB hydrolyzes ATP, it dissociates from the DNA (MuB bound to ATP is shown in the darker green); MuA–MuB contact stimulates this hydrolysis reaction. Although shown contacting only two molecules of MuB, MuA will preferentially contact all the MuB bound within close proximity to its DNA-binding site. DNA lengths of 5–15 kb can be rendered "immune" by a single MuA-bound terminal inverted repeat sequence.

Tc*1/mariner* Elements Are Extremely Successful DNA Elements in Eukaryotes

Recognizable members of the Tc*1/mariner* family of elements are widespread in both invertebrate and vertebrate organisms. Elements in this family are the most common DNA transposons present in eukaryotes. Although these elements are clearly related, members isolated from different organisms have distinguishing features and are named differently. For example, elements from the worm *Caenorhabditis elegans* are called Tc elements, whereas the original element named *Mariner* was isolated from a *Drosophila* species.

Tc*1/mariner* elements are among the simplest autonomous transposons known. Typically, they are 1.5–2.5 kb long and carry only a pair of terminal inverted repeat sequences (the site of transposase binding) and a gene encoding a transposase protein of the DDE transposase superfamily (see above). In contrast to many transposons, no accessory proteins are required for transposition, although the final steps of recombination do require cellular DNA-repair proteins. This simplicity in structure and mechanism may be responsible for the huge success of these elements in such a wide range of host organisms.

Tc*1/mariner* elements move by a cut-and-paste transposition mechanism (Fig. 11-19). The transposon DNA is cleaved out of the old flanking host DNA using pairs of cleavages that are staggered by two base pairs. These elements strongly prefer to insert into DNA sites with the (obviously, very common) sequence 5′TA.

What happens to the "empty" site in the host chromosome when a transposon excises? In the case of Tc*1/mariner* elements, DNA sequence analysis of some sites that once carried a transposon reveals that sometimes the broken DNA ends are filled in (by repair DNA synthesis) and then directly joined (see the discussion on nonhomologous end joining in Chapter 9). These repair reactions result in the incorporation of a few extra base pairs of DNA at the old insertion site. These small DNA insertions are known as "footprints," as they are the traces left by a transposon that has "traveled through" a site in the genome.

In contrast to many transposons, the transposition of Tc*1/mariner* elements is not well-regulated. Perhaps as a result of this lack of control, many elements found by genome sequencing are "dead" (i.e., unable to transpose). For example, many elements carry mutations in the transposase gene that inactivate it. Using a large number of sequences from both inactive and active elements, researchers constructed an artificial hyperactive Tc*1/mariner* element. This element, named *Sleeping Beauty,* transposes at very high frequencies compared to naturally isolated elements. *Sleeping Beauty* is promising as a tool for mutagenesis and DNA insertion in many eukaryotic organisms. Furthermore, this reconstruction experiment reveals that the frequency of transposition by Tc*1/mariner* elements is naturally kept at bay because of the suboptimal activity of their transposase proteins.

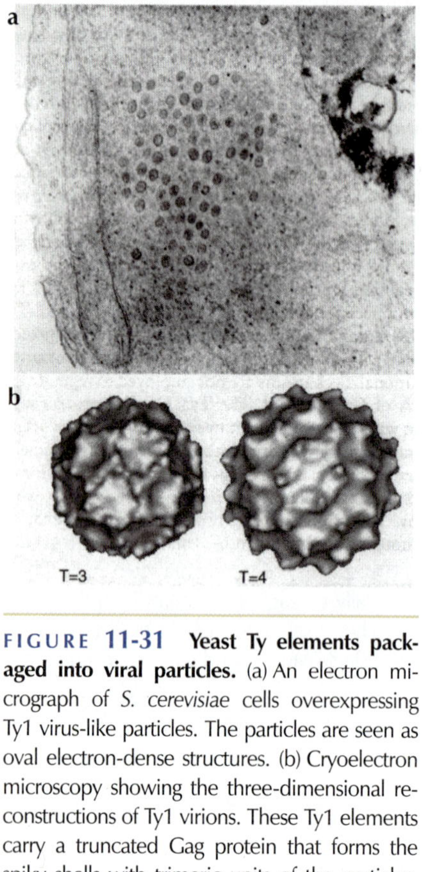

FIGURE 11-31 Yeast Ty elements packaged into viral particles. (a) An electron micrograph of *S. cerevisiae* cells overexpressing Ty1 virus-like particles. The particles are seen as oval electron-dense structures. (b) Cryoelectron microscopy showing the three-dimensional reconstructions of Ty1 virions. These Ty1 elements carry a truncated Gag protein that forms the spiky shells with trimeric units of the particles. (Reprinted, with permission, from Craig N. et al. 2002. *Mobile DNA II,* © ASM Press. b, Also courtesy of H. Saibil.)

Yeast Ty Elements Transpose into Safe Havens in the Genome

The Ty elements (transposons in yeast), prominent transposons in yeast, are virus-like retrotransposons. In fact, their similarity to retroviruses extends beyond their mechanism of transposition: Ty RNA is found in cells packaged into virus-like particles (Fig. 11-31). Thus, these elements seem to be viruses that cannot escape one cell and

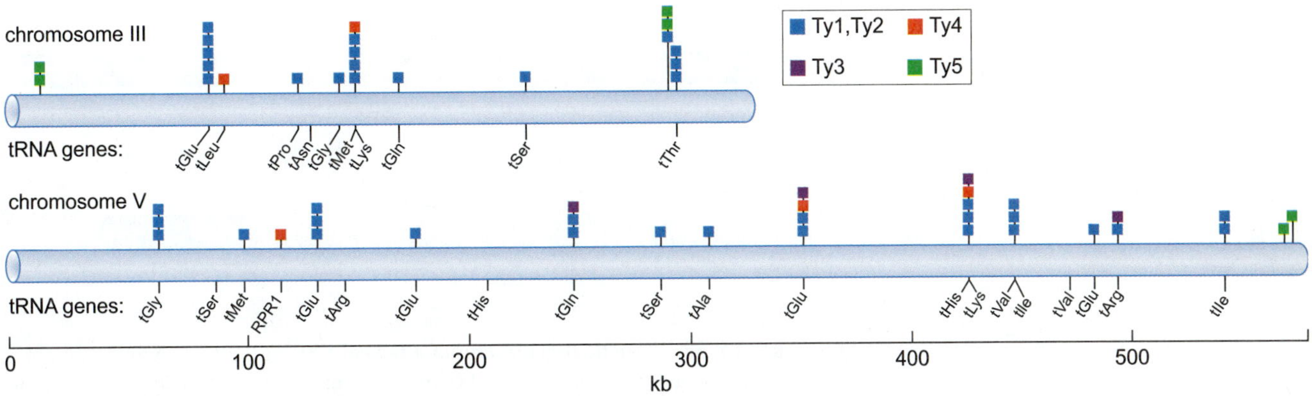

FIGURE 11-32 Clustered integration sites observed for Ty elements. Each colored box represents a known site for transposon insertion. Note that Ty1, Ty2, Ty3, and Ty4 insertions are near tRNA genes, which are transcribed by the cellular RNA polymerase III. Insertion occurs upstream of the actual gene and therefore does not disrupt expression. Ty1 and Ty2 are closely related elements and therefore are grouped together. Ty5 is found near the ends of chromosomes and near the mating-type loci (see Chapter 10) that are "silenced" (i.e., not highly transcribed). (Courtesy of Dan Voytas.)

infect new cells. There are many types of well-studied Ty elements; for example, *S. cerevisiae* carries members of the Ty1, Ty3, Ty4, and Ty5 classes (although the Ty5 elements in this yeast species all appear to be inactive). Each of these classes of Ty elements promotes its own mobility but does not mobilize elements of another class.

Ty elements preferentially integrate into specific chromosomal regions (Fig. 11-32). For example, Ty1 elements nearly always transpose into DNA within ~200 bp upstream of a start site for transcription by the host RNA polymerase III (Pol III) enzyme (see Chapter 12). RNA Pol III specifically transcribes tRNA genes, and most Ty1 insertions are near these genes. Ty3 integration is also tightly linked to Pol III promoters. In this case, integration is precisely targeted to the start site of transcription (±2 bp). In contrast, Ty5 preferentially integrates into regions of the genome that are in a silenced, transcriptionally quiescent state. Silenced regions targeted by Ty5 include the telomeres and the silent copies of the mating-type loci (see Chapter 10). In all of these cases, the mechanism of regional target-site selection involves the formation of specific protein–protein complexes between the element's integrase—bound in a complex to the cDNA—and host-specific proteins bound to these chromosomal sites. For example, Ty5 integrase forms a specific complex with the DNA-silencing protein Sir4 (see Chapter 17).

Why do Ty elements exhibit this regional target-site preference? It is proposed that this target specificity enables the transposons to persist in a host organism by focusing most of their insertions away from important regions of the genome that are involved directly in coding for proteins. The use of this type of targeted transposition may be especially important in organisms with small gene-rich genomes, such as yeast.

LINEs Promote Their Own Transposition and Even Transpose Cellular RNAs

The autonomous poly-A retrotransposons known as LINEs are abundant in the genomes of vertebrate organisms. In fact, about 20% of the human

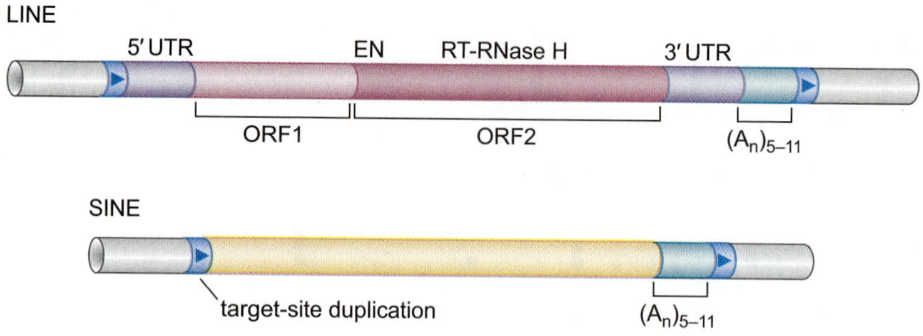

FIGURE 11-33 Genetic organization of a typical LINE and SINE. Note the variable-length poly-A sequence at the right end of the elements. This is a defining feature of the poly-A retrotransposons. These elements are also flanked by target-site duplications that are variable in length (blue arrows). Sequence elements are not shown to scale. Both types of elements also carry promoter sequences; see Figs. 11-20 and 11-28. (Adapted, with permission, from Bushman F. 2002. *Lateral DNA transfer*, p. 251, Fig. 8.4. © Cold Spring Harbor Laboratory Press.)

genome is composed of LINE sequences. These elements were first recognized as a family of repeat sequences. Their name is derived from this initial identification: LINE is the acronym for long interspersed nuclear element. L1 is one of the best understood LINEs in humans. In addition to promoting their own mobility, LINEs also donate the proteins needed to reverse-transcribe and integrate another related class of repeat sequences, the nonautonomous poly-A retrotransposons, known as short interspersed nuclear elements (SINEs). Genome sequences reveal, once again, the presence of huge numbers of these elements, which are typically only between 100 and 400 bp in length. The *Alu* sequence is an example of a widespread SINE in the human genome. A comparison of the structures of typical LINE and SINE elements is shown in Figure 11-33. The sequences of LINEs and SINEs look like simple genes. In fact, the *cis*-acting sequences important for transposition simply include a promoter, to direct transcription of the element into RNA, and a poly-A sequence. Recall that these A residues pair with the DNA at the target site to help generate the primer terminus for reverse transcription (see Fig. 11-23).

These simple sequence requirements for transposition pose a problem for LINEs: how do they avoid transposing cellular mRNA molecules? All genes have a promoter, and most are transcribed into an mRNA that will carry a poly-A sequence at the 3′ end of the molecule (Chapter 12). Thus, any mRNA should be an attractive "substrate" for transposition. In fact, genome sequences provide clear evidence for transposition of cellular RNA via the target-primed reverse transcription mechanism.

For many cellular genes, there are additional copies of a highly related sequence in the genome. These copies appear to have lost their promoter and their introns (regions of sequence present within a gene but removed from the mRNA by RNA splicing; see Chapter 13) and often carry truncations near their 5′ ends. These sequences are known as **processed pseudogenes** and usually are not expressed by the cell. These pseudogenes are often flanked by short repeats in the target DNA. This structure is exactly that expected of LINE-promoted transposition of a cellular mRNA.

Although transposition of cellular RNAs can occur, it is a rare event. The principal mechanism used to avoid this process is that the

LINE-encoded proteins bind immediately to their own RNA during translation (see Fig. 11-23). Thus, they show a strong bias to catalyzing reverse transcription and integration of the RNA that encoded them.

V(D)J RECOMBINATION

We have seen that transposition is involved in the movement of many different genetic elements. Cells, however, have also harnessed this recombination mechanism for functions that directly help the organism. The best example is V(D)J recombination, which occurs in the cells of the vertebrate immune system.

The immune system of vertebrates has the job of recognizing and fending off invading organisms, including viruses, bacteria, and pathogenic eukaryotes. Vertebrates have two specialized cell types dedicated to recognizing these invaders: B cells and T cells. B cells produce **antibodies** that circulate in the bloodstream, whereas T cells produce cell surface–bound receptor proteins (called **T-cell receptors**). Recognition of a "foreign" molecule by either of these classes of proteins starts a cascade of events focused on destruction of the invader. To fulfill their functions successfully, antibodies and T-cell receptors must be able to recognize an enormously diverse group of molecules. The principal mechanism cells use to generate antibodies and T-cell receptors with such diversity relies on a specialized set of DNA rearrangement reactions known as **V(D)J recombination**.

Antibody and T-cell receptor genes are composed of gene segments that are assembled by a series of sequence-specific DNA rearrangements. To understand how this recombination generates the needed diversity, we need to look at the structure of an antibody molecule (Fig. 11-34); T-cell receptors have a similar modular structure. A genomic re-

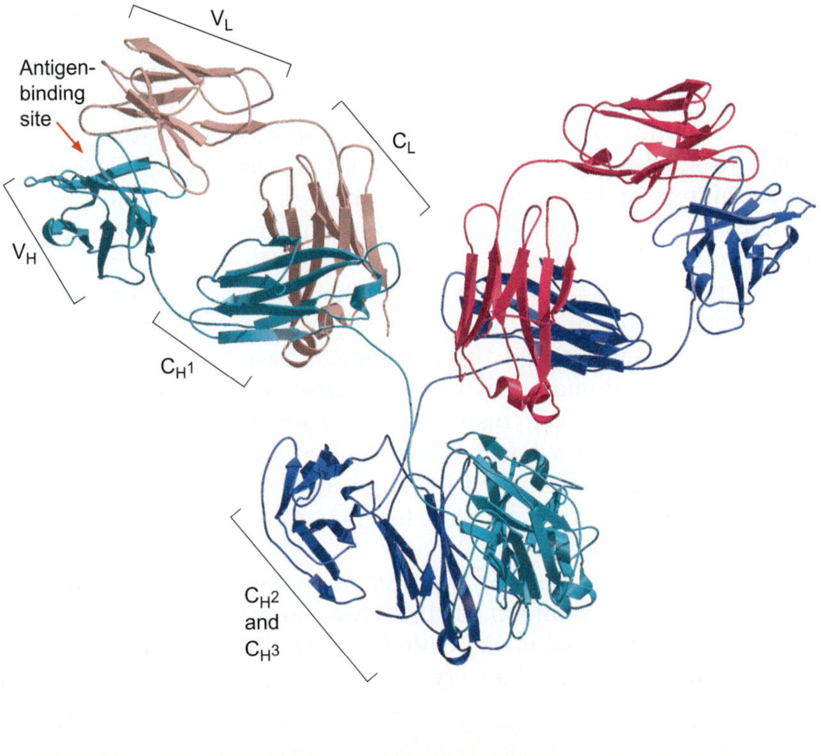

FIGURE 11-34 Structure of an antibody molecule. (Pink) The two light chains; (blue) the heavy chains. The variable and constant regions are labeled on the left side of the molecule only. Note that the antigen-binding region is formed at the interface between the V_L and V_H domains. (Harris L.J. et al. 1998. *J. Mol. Biol.* 275: 861–872.) Image prepared with MolScript, BobScript, and Raster 3D.

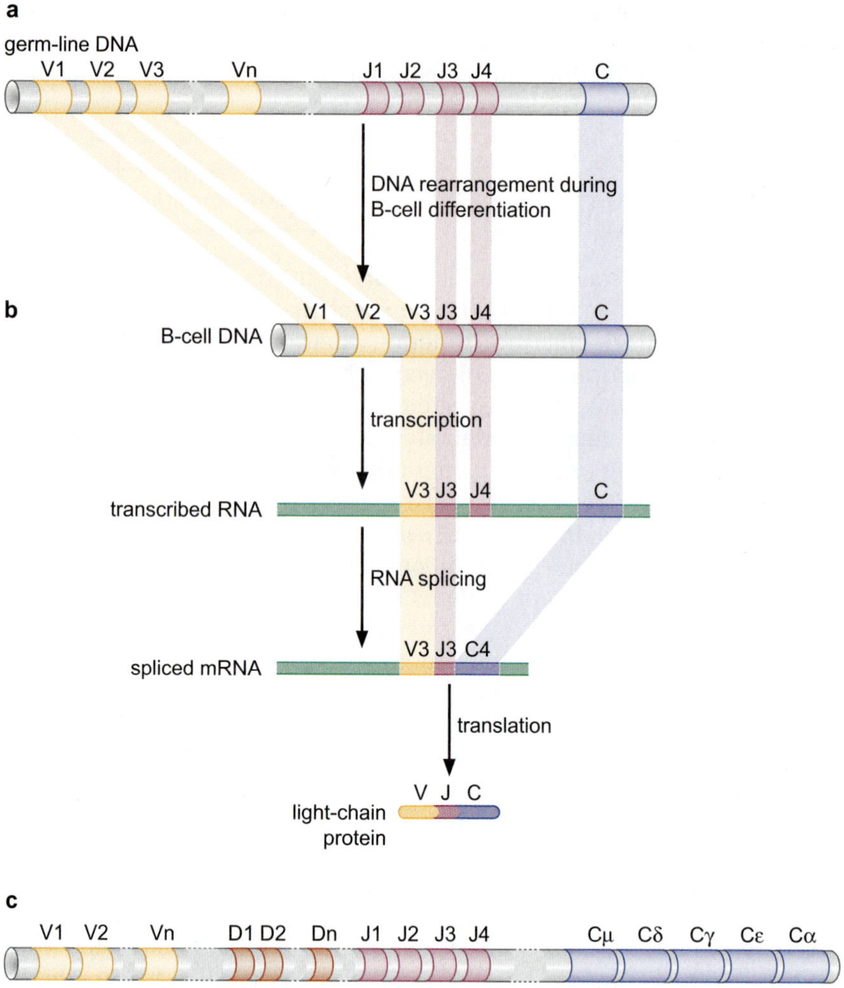

FIGURE 11-35 Overview of the process of V(D)J recombination. The top panels show the steps involved in producing the light chain of an antibody protein. (a) The genetic organization of part of the light-chain DNA in cells that have not experienced V(D)J recombination (germ-line DNA). (b) Recombination between two specific gene segments (V3 and J3) as occurs during B-cell development. This is only one of the many types of recombination events that can occur in different pre-B cells. The recombined locus is then transcribed and the RNA spliced (Chapter 13) to juxtapose a constant region gene segment. This mRNA is then translated to generate the light-chain protein. (c) Schematic of the even more complex heavy-chain genetic region, with its additional "D" gene segments and multiple types of constant region segments ($C\mu$, $C\gamma$, etc.). (Adapted, with permission, from Bushman F. 2002. *Lateral DNA transfer*, p. 345, Fig. 11.3. © Cold Spring Harbor Laboratory Press.)

gion encoding an antibody molecule is shown in Figure 11-35. Antibodies are constructed of two copies each of a light chain and a heavy chain. The part of the protein that interacts with foreign molecules is called the **antigen-binding site**. This binding region is constructed from V_L and V_H domains of the antibody molecule, shown in Figure 11-34. The "V" signifies that the protein sequence in this region is highly variable. The remaining domains of the antibody are called "C," or constant, regions and do not differ among different antibody molecules.

Figure 11-35a shows the genomic region encoding an antibody light chain (from a mouse), called the kappa locus. This region carries ~300 gene segments coding for different versions of the light-chain V_L protein region. There are also four gene segments encoding a short region of protein sequence called the J region, followed by a single coding region for the C_L domain. By the mechanism we shall describe below, V(D)J recombination can fuse the DNA between any pair of V and J segments. Thus, as a result of recombination, 1200 variants of the antibody light chain can be produced from this single genomic region. These segments are then brought together with the C_L-coding region by RNA splicing (Chapter 13).

The situation for assembly of the gene segments encoding the antibody heavy chain is similar. In this case, however, there is an additional type of gene segment, called D (for diversity) (Fig. 11-35c). Heavy-chain genes can be very complex. For example, a specific heavy-chain locus in a mouse has more than 100 V regions, 12 D regions, and 4 J regions. V(D)J recombination can assemble this gene to generate more than 4800 different protein sequences. Because functional antibodies can be constructed from any pair of light and heavy chains, the diversity generated by recombination at the light and heavy loci have a multiplicative impact on protein structure.

The Early Events in V(D)J Recombination Occur by a Mechanism Similar to Transposon Excision

Recombination sequences, called **recombination signal sequences**, flank the gene segments that are assembled by V(D)J recombination. These signals all have two highly conserved sequence motifs, one 7 bp (the 7-mer) and the second 9 bp (the 9-mer) in length (Fig. 11-36). These motifs are bound by the recombinase (see below). The recombination signal sequences come in two classes. One class has the 7-mer and 9-mer motifs spaced by 12 bp of sequence, whereas the second class has these motifs spaced by 23 bp (Fig. 11-36a). Recombination always occurs between a pair of recombination signal sequences in which one partner has the 12-bp "spacer" and the other partner has the 23-bp "spacer." These pairs of recombination signal sequences are organized as inverted repeats flanking the DNA segments that are destined to be joined (Fig. 11-36b).

The recombinase responsible for recognizing and cleaving the recombination signal sequences is composed of two protein subunits called **RAG1** and **RAG2** (RAG for recombination-activating gene). These proteins function in a manner very similar to that of a transposase (Fig. 11-37). They recognize the recombination signal sequences and pair the two sites to form a protein–DNA synaptic complex.

The RAG1 proteins within this complex then introduce single-strand breaks in the DNA at each of the junctions between the recombination signal sequence and the gene segment that will be rearranged (Fig. 11-37a). The site of cleavage is such that the protein-coding segment now has a free 3′-OH DNA end (Fig. 11-37b). Then, as we have seen previ-

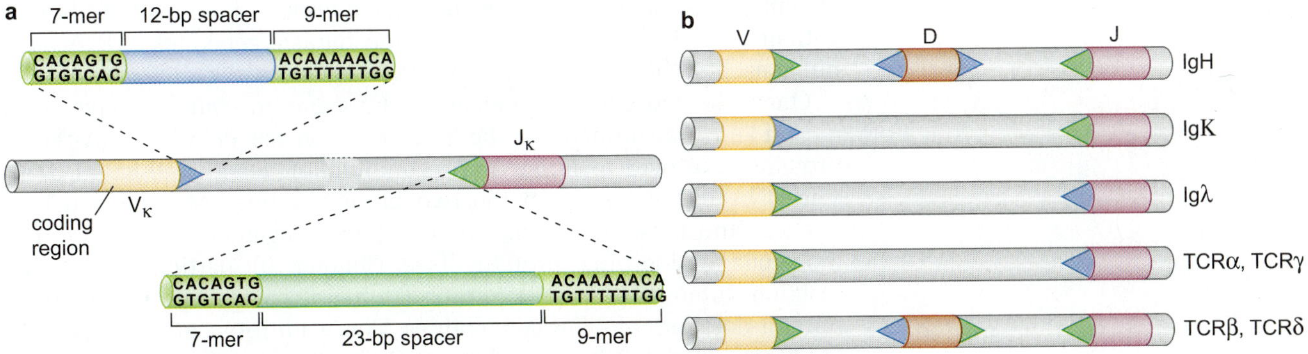

FIGURE 11-36 Recombination signal sequences recognized in V(D)J recombination. (a) Close-up of the two types of recombination signal sequences (RSSs). (Blue) The 12-bp spacer; (green) the 23-bp spacer; (light green) conserved 7-mer and 9-mer sequence elements, shared by both types of sequences. The nucleotide sequence in the spacer region is not important. The length, however, is critical. (b) Examples of RSS arrangements in the genetic regions encoding antibodies (Ig genes) and T-cell receptor proteins (TCR genes). (a, Adapted, with permission, from Bushman F. 2002. *Lateral DNA transfer*, p. 346, Fig. 11.5. © Cold Spring Harbor Laboratory Press.)

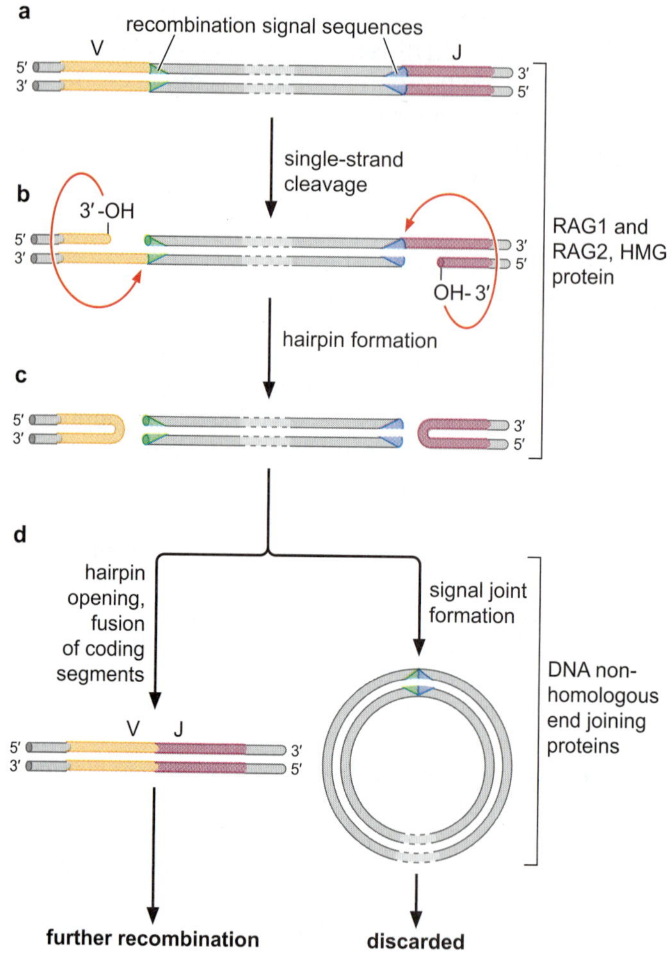

FIGURE 11-37 The V(D)J recombination pathway: Cleavages occur by a mechanism similar to transposon excision. The recombinases catalyze single-strand cleavage at the ends of the signal sequences, leaving a free 3′-OH. Each 3′-OH then initiates attack on the opposite strands to form a hairpin intermediate (see Fig. 11-23b). The hairpin structures are subsequently hydrolyzed and then joined together to form a coding joint between the V and J regions. The two ends carrying the recombination signal sequences are also joined to form a signal joint. The former structure undergoes further recombination, whereas the latter is discarded. (Adapted, with permission, from Bushman F. 2002. *Lateral DNA transfer*, p. 348, Fig. 11.6. © Cold Spring Harbor Laboratory Press.)

ously for some transposon excision reactions (especially in the *Hermes* pathway, Fig. 11-21), this 3′-OH DNA end attacks the opposite strand of the DNA double helix. This attack results in the coupled DNA cleavage and joining reaction that generates a hairpin DNA end. It is the protein-coding sequence segments that have the DNA hairpin ends, whereas the recombination signal sequences now have normal double-strand breaks at their ends (Fig. 11-37c). This same mechanism generates a DNA hairpin at each of the two recombining DNA segments.

Once the two DNA sequences in the synaptic complex have been nicked and "hairpinned" by the RAG recombinase, cellular DNA-repair proteins take over to finish the recombination reaction (Fig. 11-37d). The DNA hairpin ends on the two protein-coding segments must be opened, and these ends must then be joined together. Cellular nonhomologous end-joining proteins (see Chapter 9) participate in this reaction. Interestingly, DNA joining is often accompanied by the addition (or deletion) of a few nucleotides. These additions are analogous to the "footprints" left in the old target DNA when transposons excise, as we described for the Tc*1*/*mariner* transposons. The added nucleotides contribute an extra component to the sequence diversity of the resulting protein molecule. The pair of cleaved recombination signal sequences are also joined together during recombination. This event generates a circular DNA molecule that is usually discarded by the cell.

The similarities between the mechanism of DNA cleavage to initiate V(D)J recombination and transposon excision are remarkable. In fact, the recombination signal sequences also look similar to the terminal inverted repeats found at the ends of a transposon, and the RAG1 protein has some sequence similarity to the DDE transposase protein family. In fact, genomic analysis has recently uncovered a transposon family called *Transib* that are the likely source of both RAG1 and the recombination signal sequences. These observations, together with many others, provide overwhelming evidence for the proposal that V(D)J recombination, now a critical feature of the immune system of higher animals, evolved from a DNA transposon. This conclusion speaks to the critical importance of transposable elements in the evolution of cellular genomes.

SUMMARY

Although DNA is normally thought of as a very static molecule that archives the genetic material, it is also subject to numerous types of rearrangements. Two classes of genetic recombinations—conservative site-specific recombination and transposition—are responsible for many of these events.

Conservative site-specific recombination occurs at defined sequence elements in the DNA. Recombinase proteins recognize these sequence elements and act to cleave and join DNA strands to rearrange DNA segments containing the recombination sites. Three types of rearrangements are common: DNA insertion, DNA deletion, and DNA inversion. These rearrangements have many functions, including insertion of a viral genome into that of the host cell during infection, resolving DNA multimers, and altering gene expression.

The organization of the recombination sites on the DNA and the participation of DNA architectural proteins dictate the outcome of a specific recombination reaction. The architectural proteins function to bend DNA segments and can have a large influence on the reactions occurring on a specific region of DNA.

There are two families of conservative site-specific recombinases. Both families cleave DNA using a protein–DNA covalent intermediate. For the serine recombinases, this linkage is via an active-site serine residue; for the tyrosine recombinases, it is via a tyrosine. Structures of the tyrosine recombinases yield many insights into the details of the recombination mechanism.

Transposition is a class of recombination that moves mobile genetic elements, called transposons, to new genomic sites. There are three major classes of transposons: DNA transposons, virus-like retrotransposons, and poly-A retrotransposons. The DNA transposons exist as DNA throughout a cycle of transposition. They move either by a cut-and-paste recombination mechanism, which involves an excised transposon intermediate, or by a replicative mechanism. The two classes of retrotransposons move using an RNA intermediate. These "retro" elements require the RNA-dependent DNA polymerase, called reverse transcriptase, as well as a recombinase protein for mobility.

Transposons are present in the genomes of all organisms, where they can constitute a huge fraction of the total DNA sequence. They are a major cause of mutations and genome rearrangements. Transposition is often regulated to help ensure that transposons do not cause too much of a disruption to the genome of the host cell. Control of transposon copy number and regulation of the choice of new insertion sites are commonly observed.

Finally, a transposition-like mechanism can be used for other types of DNA rearrangement reactions. The prime example of this is the V(D)J recombination reaction, responsible for assembly of gene fragments during development of the vertebrate immune system.

BIBLIOGRAPHY

Books

Bushman F. 2002. *Lateral DNA transfer: Mechanisms and consequences.* Cold Spring Harbor Laboratory Press, Cold Spring Harbor, New York.

Craig N.L., Craigie R., Gellert M., and Lambowitz A.M., eds. 2002. *Mobile DNA II.* American Society for Microbiology, Washington, D.C.

Site-Specific Recombination

Chen Y. and Rice P.A. 2003. New insight into site-specific recombination from FLP recombinase-DNA structures. *Annu. Rev. Biophys. Biomol. Struct.* **32:** 135–159.

Grindley N.D.F., Whiteson K.L., and Rice P.A. 2006. Mechanisms of site-specific recombination. *Annu. Rev. Biochem.* **75:** 567–605.

Hallet B. and Sherratt D.J. 1997. Transposition and site-specific recombination: Adapting DNA cut-and-paste mechanisms to a variety of genetic rearrangements. *FEMS Microbiol. Rev.* **21:** 157–178.

Smith M.C. and Thorpe H.M. 2002. Diversity in the serine recombinases. *Mol. Microbiol.* **44:** 299–307.

Stark W.M., Boocock M.R., and Sherratt D.J. 1992. Catalysis by site-specific recombinases. *Trends Genet.* **8:** 432–439.

Transposition

Haren L., Ton-Hoang B., and Chandler M. 1999. Integrating DNA: Transposases and viral integrases. *Annu. Rev. Microbiol.* **53:** 245–281.

Plasterck R. 1995. The Tc*1*/*mariner* transposon family. *Curr. Top. Microbiol. Immunol.* **204:** 125–143.

Prak E.T.L. and Kazazian H.H., Jr. 2000. Mobile elements in the human genome. *Nat. Rev. Genet.* **1:** 134–144.

Rice P.A. and Baker T.A. 2001. Comparative architecture of transposase and integrative complexes. *Nat. Struct. Biol.* **8:** 302–307.

Smit A.F.A. 1999. Interspersed repeats and other mementos of transposable elements in mammalian genomes. *Curr. Opin. Genet. Dev.* **9:** 657–663.

Williams T.L. and Baker T.A. 2000. Transposase team puts a headlock on DNA. *Science* **289:** 73–74.

Transposition and V(D)J Recombination

Gueguen E., Rousseau P., Duval-Valentin G., and Chandler M. 2005. The transpososome: Control of transposition at the level of catalysis. *Trends Microbiol.* **13:** 543–549.

Oettinger M.A. 2004. Hairpins at split ends in DNA. *Nature* **432:** 960–961.

V(D)J Recombination

Fugmann S.D., Lee A.I., Schockett P.E., Villey I.J., and Schatz D.G. 2000. The RAG proteins and V(D)J recombination: Complexes, ends, and transposition. *Annu. Rev. Immunol.* **18:** 495–527.

Gellert M. 2002. V(D)J recombination: RAG proteins, repair factors, and regulation. *Annu. Rev. Biochem.* **71:** 101–132.

chain of a protein. Chapter 14 describes the four principal participants in translation: the coding sequence in messenger RNA; adaptor molecules known as tRNAs; enzymes known as aminoacyl-tRNA synthetases that load amino acids onto the tRNA adaptors; and the protein-synthesizing factory itself, the ribosome, which is composed of RNA and protein. The remainder of the chapter describes how these four components, with help from a number of key auxiliary factors, manage the remarkable process of recognizing the protein-coding sequences and converting the nucleotide code of a given mRNA into the correct order of amino acids in its protein product.

Finally, Chapter 15 describes the classic experiments that led to the elucidation of the genetic code and lays out the rules by which the code is translated. The nucleotide sequence information is based on a three-letter code, whereas the protein sequence information is based on 20 different amino acids. The code is degenerate with two or more codons (in most cases) specifying the same amino acid. There are also specific codons that indicate where translation should start and where it should stop.

PHOTOS FROM THE COLD SPRING HARBOR LABORATORY ARCHIVES

Richard Roberts, 1977 Symposium on Chromatin. Much of Roberts' research has focused on the function and diversity of restriction enzymes (Chapter 21), but he was also a codiscoverer of "split genes," for which he shared the Nobel Prize in Physiology or Medicine with Phillip Sharp in 1993. Shown here with him are, left to right, Yasha Gluzman, the tumor virologist; Ahmad Bukhari, who worked on phage Mu transposition (Chapter 11); and James Darnell, whose work focuses on signal transduction in gene regulation (Chapter 17).

Sydney Brenner and James Watson, 1975 Symposium on The Synapse. Brenner, shown here with Watson, contributed to the discoveries of mRNA and the nature of the genetic code (Chapters 2 and 15); his share of a Nobel Prize, in 2002, however, was for establishing the worm, *Caenorhabditis elegans,* as a model system for the study of developmental biology (Chapter 22).

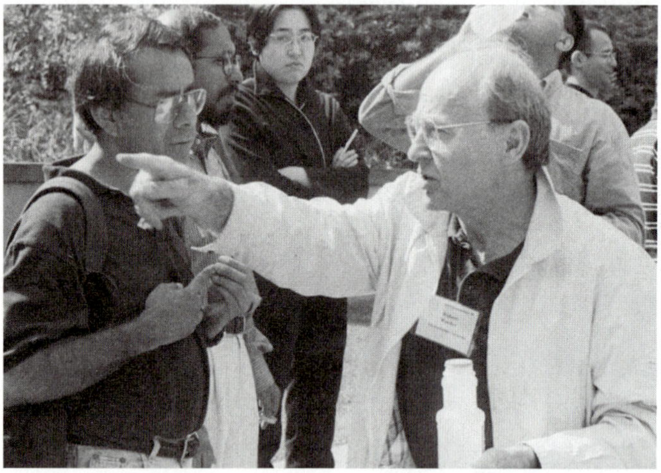

Robert Roeder, 1998 Symposium on Mechanisms of Transcription. Roeder discovered the three eukaryotic RNA polymerases—Pol I, II, and II—purifying all three enzymes and other factors they each need to initiate transcription from their respective promoters (Chapter 12). On the left, looking on skeptically, is Camilo Parada, at the time a postdoc in Roeder's lab.

Roger D. Kornberg, 1977 Symposium on Chromatin. Having earlier worked on the structure of the nucleosome (Chapter 7), Kornberg won the Nobel Prize in Chemistry in 2006 for his structural studies of RNA polymerase II (Chapter 12). His father is Arthur Kornberg, whose picture is on page 98.

Francis Crick, 1963 Symposium on Synthesis and Structure of Macromolecules. In addition to his role in solving the structure of DNA, Crick was an intellectual driving force in the development of molecular biology during the field's critical early years. His "adaptor hypothesis" (published in the RNA Tie Club newsletter) predicted the existence of molecules required to translate the genetic code of RNA into the amino acid sequence of proteins. Only later were tRNAs found to do just that (Chapter 14).

Phillip Sharp, 1974 Symposium on Tumor Viruses. Sharp and Richard Roberts shared the 1993 Nobel Prize in Physiology or Medicine for discovering that many eukaryotic genes are "split"—that is, their coding regions are interrupted by stretches of noncoding DNA. The noncoding regions are removed from the RNA copy by "splicing" (Chapter 13). Sharp is shown here with his wife Ann.

Paul Zamecnik, 1969 Symposium on The Mechanism of Protein Synthesis. Zamecnik developed in vitro systems of protein synthesis that proved critical to understanding how the genetic code works and how cells manufacture proteins (Chapters 2 and 15). Together with Mahlon Hoagland, he also discovered tRNAs, a key component in that process (Chapter 14.)

David Baltimore, François Jacob, and Walter Gilbert, 1985 Symposium on the Molecular Biology of Development. Baltimore codiscovered, with Howard Temin, the enzyme reverse transcriptase, which makes DNA using RNA as a template (Chapter 11). Jacob, with Jacques Monod, proposed the basic model for how gene expression is regulated (Chapter 16) and also proposed a model for how DNA replication is regulated (Chapter 8). Gilbert provided biochemical validation for aspects of the Jacob and Monod model of gene regulation; he also invented a chemical method for sequencing DNA (Chapter 20). They all separately shared in Nobel Prizes, in 1975 (in Physiology or Medicine), 1965 (in Physiology or Medicine), and 1980 (in Chemistry), respectively.

David Allis and Emily Bernstein, 2004 Symposium on Epigenetics. Allis was the first to identify an enzyme that modifies histones—a histone acetyltransferase from *Tetrahymena* (Chapter 7). Since that discovery, a whole field has grown up examining the range of histone modifications that exist and their effects on gene expression. Allis is here shown with Bernstein, a postdoc in his lab at the time the photo was taken and formerly a graduate student in Greg Hannon's lab, where she identified the Dicer enzyme involved in RNAi (Chapter 18).

CHAPTER

12 Mechanisms of Transcription

OUTLINE

Up to this point, we have been considering maintenance of the genome—that is, how the genetic material is organized, protected, and replicated. We now turn to the question of how that genetic material is *expressed*—that is, how the series of bases in the DNA directs the production of the RNAs and proteins that perform cellular functions and define cellular identity. In the next few chapters, we describe the basic processes responsible for gene expression: transcription, RNA processing, and translation.

Transcription is, chemically and enzymatically, very similar to DNA replication (Chapter 8). Both involve enzymes that synthesize a new strand of nucleic acid complementary to a DNA template strand. There are some important differences, of course; most notably, in the case of transcription, the new strand is made from ribonucleotides rather than deoxyribonucleotides (see Chapter 6). Other mechanistic features of transcription that differ from that of replication include the following:

- **RNA polymerase** (the enzyme that catalyzes RNA synthesis) does not need a primer; rather, it can initiate transcription de novo (although in vivo, initiation is permitted only at certain sequences, as we shall see).
- The RNA product does not remain base-paired to the template DNA strand; rather, the enzyme displaces the growing chain only a few nucleotides behind where each ribonucleotide is added (Fig. 12-1). This displacement is critical for the RNA to carry out its functions (for example, as is most often the case, to be translated to produce its protein product). Furthermore, because this release follows so closely behind the site of polymerization, multiple RNA polymerase molecules can transcribe the same gene at the same time, each following closely behind another. Thus, a cell can synthesize large numbers of transcripts from a single gene (or other DNA sequence) in a short time. It is important to note that as the RNA product dissociates from the DNA template just behind each advancing RNA polymerase, the two DNA strands reanneal (Fig. 12-1).
- Transcription, although very accurate, is less accurate than replication (one mistake occurs in 10,000 nucleotides added, compared to one in 10,000,000 for replication). This difference reflects the lack of extensive proofreading mechanisms for transcription, although two forms of proofreading for RNA synthesis do exist.

  It makes sense for the cell to worry more about the accuracy of replication than of transcription. DNA is the molecule in which the genetic material is stored, and DNA replication is the process by

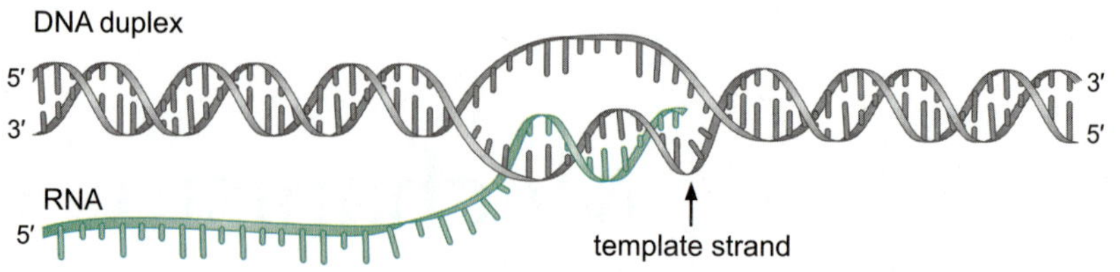

FIGURE 12-1 Transcription of DNA into RNA. The figure shows, in the absence of the enzymes involved, how the DNA double helix is unwound and an RNA strand is built on the template strand. It also shows how the RNA transcript dissociates from the DNA template a few nucleotides behind the point of synthesis, and how the DNA strands reanneal. In the figure, transcription proceeds from left to right.

which that genetic material is passed on. Any mistake that arises during replication can therefore easily be catastrophic: it becomes permanent in the genome of that individual and gets passed on to subsequent generations. Transcription, in contrast, produces only transient copies and normally several from each transcribed region. Thus, a mistake during transcription will rarely do more harm than render one out of many transient transcripts defective.

Beyond these mechanistic differences between DNA replication and transcription, one profound difference reflects the different purposes served by these processes. Transcription selectively copies only certain parts of the genome and makes anywhere from one to several hundred, or even thousand, copies of any given section. In contrast, replication must copy the entire genome and do so once (and only once) every cell division (as we saw in Chapter 8). The choice of which regions to transcribe is not random: each typically includes one or more genes, and there are specific DNA sequences that direct the initiation of transcription at the start of each region and others at the end that terminate transcription.

Not only are different parts of the genome transcribed to different extents, but the choice of which part to transcribe, and how extensively, can also be regulated. Thus, in different cells, or in the same cell at different times, different sets of genes might be transcribed. So, for example, two genetically identical cells in a human will, in many cases, transcribe different sets of genes, leading to differences in the character and function of those two cells (e.g., one might be a muscle cell and the other, a neuron). Or a given bacterial cell will transcribe a different set of genes, depending on the medium in which it is growing. These questions of transcriptional regulation are dealt with in Part 4.

RNA POLYMERASES AND THE TRANSCRIPTION CYCLE

RNA Polymerases Come in Different Forms but Share Many Features

RNA polymerase performs essentially the same reaction in all cells, from bacteria to humans. It is thus not surprising that the enzymes from these organisms share many features, especially in those parts of the enzyme directly involved with catalyzing the synthesis of RNA. From bacteria to mammals, the cellular RNA polymerases are made up of multiple subunits (although some phage and organelles do encode single-subunit enzymes that perform the same task, as we shall see in Box 12-2). Table 12-1 shows the numbers and sizes of subunits found in each case and also shows which subunits are conserved at the sequence level between different enzymes.

TABLE 12-1 The Subunits of RNA Polymerases

Prokaryotic		Eukaryotic		
Bacterial	**Archaeal**	**RNAP I**	**RNAP II**	**RNAP III**
Core	**Core**	**(Pol I)**	**(Pol II)**	**(Pol III)**
β′	A′/A″	RPA1	RPB1	RPC1
β	B	RPA2	RPB2	RPC2
α^{I}	D	RPC5	RPB3	RPC5
α^{II}	L	RPC9	RPB11	RPC9
ω	K	RPB6	RPB6	RPB6
	[+6 others]	[+9 others]	[+7 others]	[+11 others]

Adapted, with permission, from Ebright R.H. 2000. *J. Mol. Biol.* 304: 687–698, Fig. 1, p. 688. © Elsevier.
The subunits in each column are listed in order of decreasing molecular weight.

As can be seen from the table, bacteria have only a single RNA polymerase, whereas eukaryotic cells have three: RNA polymerases I, II, and III (RNA Pol I, II, and III). **Pol II** is the enzyme we focus on when dealing with eukaryotic transcription in the second half of this chapter because it is the most studied of these enzymes. It is also the polymerase responsible for transcribing most genes—indeed, essentially all protein-encoding genes. **Pol I** and **Pol III** are each involved in transcribing specialized, RNA-encoding genes. Specifically, Pol I transcribes the large RNA precursor gene, whereas Pol III transcribes tRNA genes, some small nuclear RNA genes, and the 5*S* rRNA gene. We return to these enzymes at the end of the chapter.

The bacterial RNA polymerase **core enzyme** alone is capable of synthesizing RNA and comprises two copies of the α subunit and one each of the β, β′, and ω subunits. This enzyme is closely related to the eukaryotic polymerases (see Table 12-1). Specifically, the two large subunits, β and β′, are homologous to the two large subunits found in RNA Pol II (RPB1 and RPB2). The α subunits are homologous to RPB3 and RPB11, and ω is homologous to RPB6. The structure of a bacterial RNA polymerase core enzyme is similar to that of the yeast Pol II enzyme. These are shown side by side in Figure 12-2. Later, we describe some of the structural details that shed light on how these enzymes work. For now, we just highlight some of the general features.

The bacterial and yeast enzymes share an overall shape and organization; indeed, they are more alike than the comparison of the subunit sequences would predict. This is particularly true of the internal parts, near the active site, and less so on the peripheries. This distribution of similarities and differences makes sense: the internal parts of the enzyme are involved in synthesis of RNA on a DNA substrate—the same in all organisms; many of the peripheral regions of the enzyme, however, are involved in interactions with other proteins, and these differ in eukaryotic cells compared to prokaryotic cells, as we shall see.

Overall, the shape of each enzyme resembles a crab claw. This is reminiscent of the "hand" structure of DNA polymerases described in Chapter 8 (Fig. 8-5). The two pincers of the crab claw are made up predominantly of the two largest subunits of each enzyme (β′ and β for the bacterial case and RPB1 and RPB2 for the eukaryotic enzyme). The active site, which is made up of regions from both these subunits, is found at the base of the pincers within a region called the "active center cleft" (see Fig. 12-2). The active site works according to the two-metal ion catalytic mechanism for nucleotide addition proposed

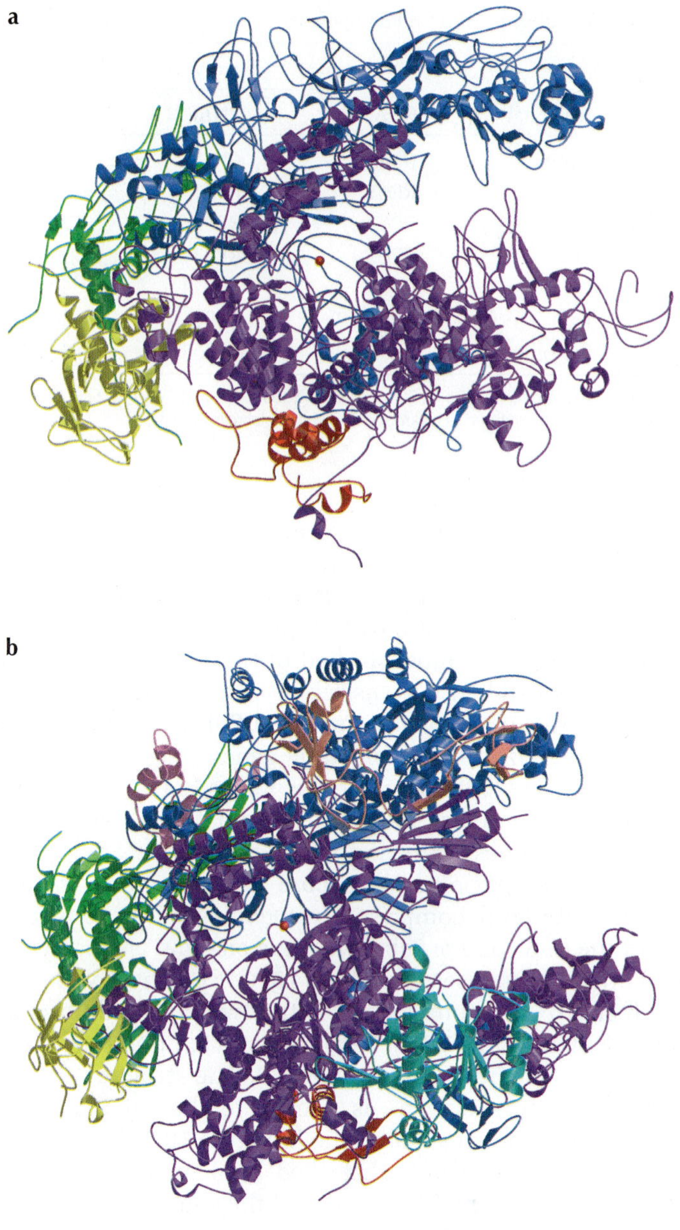

FIGURE **12-2** **Comparison of the crystal structures of prokaryotic and eukaryotic RNA polymerases.** (a) Structure of RNA polymerase core enzyme from *Thermus aquaticus*. The subunits are colored as follows: (blue) β; (purple) β'; (yellow and green) the two α subunits; (red) ω. The Mg^{2+} ion, represented as a red ball, marks the active site here and in part b (Seth Darst, The Rockefeller University, personal communication). (b) Structure of RNA Pol II from yeast *Saccharomyces cerevisiae*. The subunits are colored to show their relatedness to those in the bacterial enzyme (see Table 12-1). Thus, RPB1 and RPB2 are shown in purple and blue, respectively; RPB3 and RPB11 are shown in green and yellow, respectively; and RPB6 is shown in red. (Cramer P. et al. 2001. *Science* 292: 1863.) Images prepared with MolScript, BobScript, and Raster 3D.

for all types of polymerase (see Chapter 8). In this case, however, the active site contains only one tightly bound Mg^{2+} ion, and the second Mg^{2+} is brought in with each new nucleotide in the addition cycle and released with the pyrophosphate.

There are various channels that allow DNA, RNA, and ribonucleotides into and out of the enzyme's active center cleft. These we discuss later when considering the mechanisms of transcription.

Transcription by RNA Polymerase Proceeds in a Series of Steps

To transcribe a gene, RNA polymerase proceeds through a series of well-defined steps grouped into three phases: **initiation**, **elongation**, and **termination**. Here, and in Figure 12-3, we summarize the basic features of each phase.

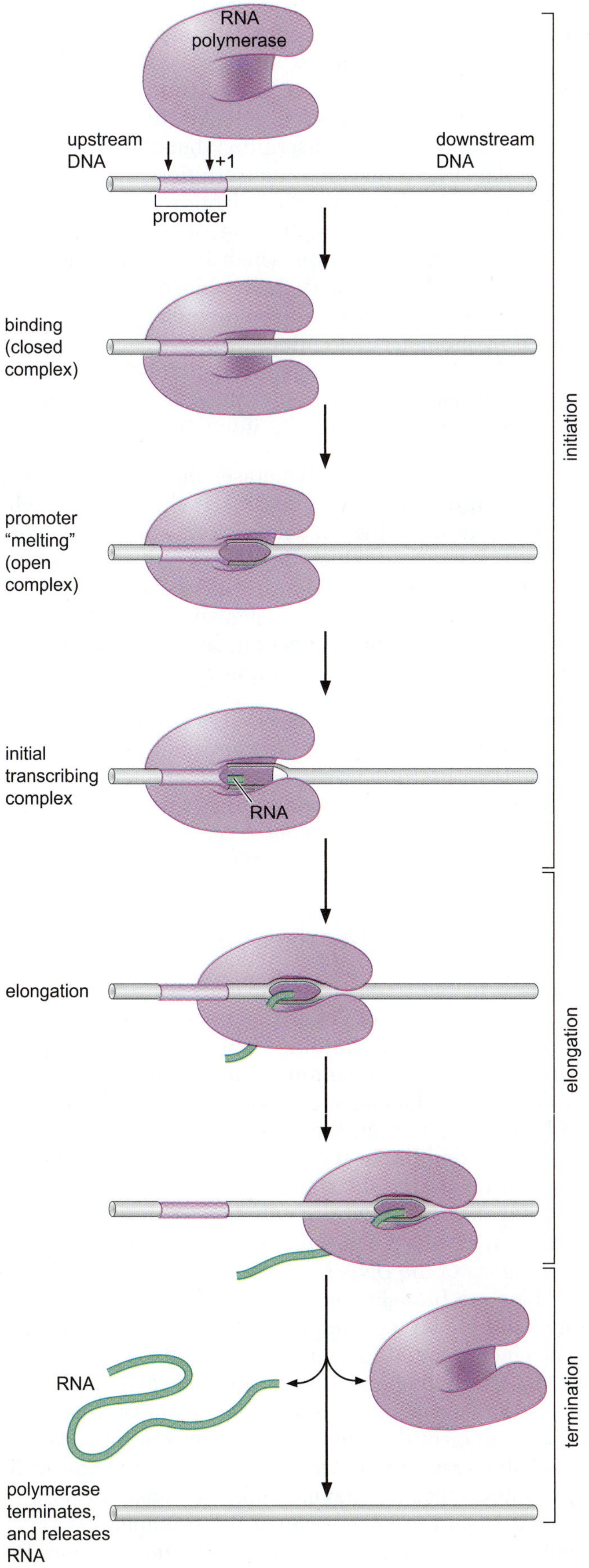

FIGURE 12-3 The phases of the transcription cycle: Initiation, elongation, and termination. The figure shows the general scheme for the transcription cycle. The features shown hold for both bacterial and eukaryotic cases. Other factors required for initiation, elongation, and termination are not shown here but are described later in the text. The DNA nucleotide encoding the beginning of the RNA chain is called the transciption start site and is designated the "+1" position. Sequences in the direction in which transcription proceeds are referred to as downstream from the start site. Likewise, sequences preceding the start site are referred to as upstream sequences. When referring to a specific position in the upstream sequence, this is given a negative value. Downstream sequences are allotted positive values.

Initiation A **promoter** is the DNA sequence that initially binds the RNA polymerase (together with any initiation factors required). Once formed, the promoter–polymerase complex undergoes structural changes required for initiation to proceed. As in replication initiation, the DNA around the point where transcription will start unwinds: the base pairs are disrupted, producing a "transcription bubble" of single-stranded DNA. Again, like DNA replication, transcription always occurs in a 5′ to 3′ direction; that is, the new ribonucleotide is added to the 3′ end of the growing chain. Unlike replication, however, only one of the DNA strands acts as a template on which the RNA strand is built. As RNA polymerase binds promoters in a defined orientation, the same strand is always transcribed from a given promoter.

The choice of promoter determines which stretch of DNA is transcribed and is the main step at which regulation is imposed; that is, deciding whether or not to initiate transcription of a given gene is chiefly the way a cell regulates which proteins it will make at any given time.

Elongation Once the RNA polymerase has synthesized a short stretch of RNA (approximately ten bases), it shifts into the elongation phase. During elongation, the enzyme performs an impressive range of tasks in addition to the catalysis of RNA synthesis. It unwinds the DNA in front and reanneals it behind, it dissociates the growing RNA chain from the template as it moves along, and it performs proofreading functions. Recall that during replication, in contrast, several different enzymes are required to catalyze a similar range of functions.

Termination Once the polymerase has transcribed the length of the gene (or genes), it must stop and release the RNA product (as well as dissociating from the DNA itself). This step is called termination. In some cells, specific, well-characterized, sequences trigger termination; in others, it is less clear what instructs the enzyme to cease transcribing and dissociate from the template.

Transcription Initiation Involves Three Defined Steps

The first phase in the transcription cycle—initiation—can itself be broken down into a series of defined steps (as indicated in Fig. 12-3). The first step is the initial binding of polymerase to a promoter to form what is called a **closed complex**. In this form, the DNA remains double-stranded, and the enzyme is bound to one face of the helix. In the second step of initiation, the closed complex undergoes a transition to the **open complex** in which the DNA strands separate over a distance of ~14 bp around the start site to form the transcription bubble. In the next stage of initiation, polymerase enters the phase of initial transcription followed by promoter escape, as we now describe.

The opening up of the DNA frees the template strand. The first two ribonucleotides are brought into the active site, aligned on the template strand, and joined together. In the same way, subsequent ribonucleotides are incorporated into the growing RNA chain. Incorporation of the first ten or so ribonucleotides is a rather inefficient process, and at that stage, the enzyme often releases short transcripts (each of less than ten or so nucleotides) and then begins synthesis again. In this phase, the polymerase–promoter complex is called the **initial transcribing complex**. Once an enzyme makes a transcript longer than 10 nucleotides, it is said to have **escaped** the promoter. At this point, it has formed a stable ternary complex, containing enzyme, DNA, and RNA. This is the transition to the elongation phase.

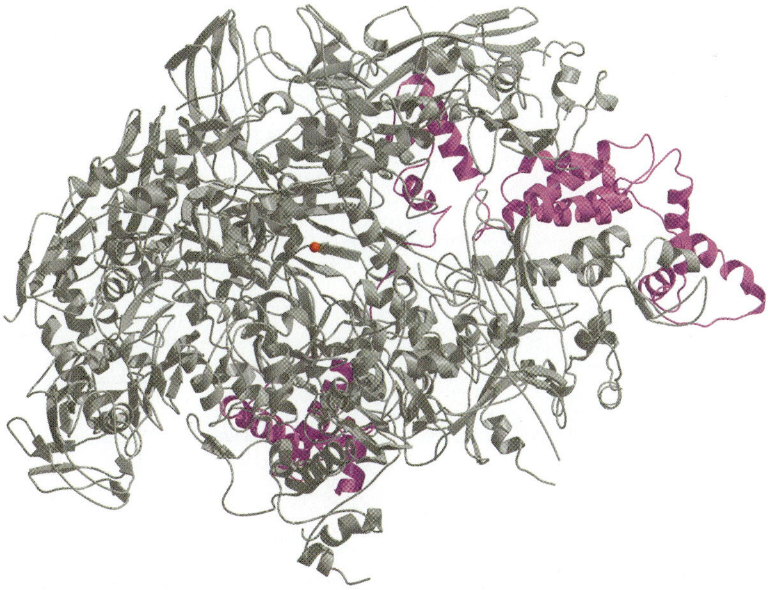

FIGURE 12-4 RNA polymerase holoenzyme from *Thermus aquaticus*. Shown in gray is the core enzyme (the same enzyme shown in Fig. 12-2a). In purple is the σ^{70} subunit (regions 2, 3, and 4; see Fig. 12-6). On the right is region 2, at the top region 3, and at the bottom region 4. As described later in the text, it is σ regions 2 and 4 that recognize the –10 and –35 regions of the promoter, respectively. (Murakami K.S. et al. 2002. *Science* 296: 1280.) Image prepared with MolScript, BobScript, and Raster 3D.

In the remainder of this chapter, we describe the transcription cycle in more detail—first for the bacterial case and then for eukaryotic systems.

THE TRANSCRIPTION CYCLE IN BACTERIA

Bacterial Promoters Vary in Strength and Sequence but Have Certain Defining Features

The bacterial core RNA polymerase can, in principle, initiate transcription at any point on a DNA molecule, and this can be shown in vitro using purified core enzyme. In cells, however, polymerase initiates transcription only at promoters. It is the addition of an initiation factor called σ that converts core enzyme ($\alpha_2\beta\beta'\omega$) into the form that initiates only at promoters. This form of the enzyme is called the RNA polymerase **holoenzyme** (Fig. 12-4).

In the case of *Escherichia coli*, the predominant σ factor is called σ^{70} (we consider other alternative σ factors in Chapter 16). Promoters recognized by polymerase containing σ^{70} share the following characteristic structure: two conserved sequences, each of six nucleotides, are separated by a nonspecific stretch of 17–19 nucleotides (Fig. 12-5a). The two defined sequences are centered, respectively, at ~10 bp and at ~35 bp upstream of the site where RNA synthesis starts. The sequences are thus called the **–35** (minus 35) and **–10** (minus 10) **regions**, or **elements**, according to the numbering scheme described in Figure 12-3, in which the DNA nucleotide encoding the beginning of the RNA chain is designated +1.

Although the vast majority of σ^{70} promoters contain recognizable –35 and –10 regions, the sequences are not identical. By comparing many different promoters, a **consensus sequence** can be derived (for a discussion of how these are derived, see Box 12-1, Consensus Sequences). The consensus sequence reflects preferred –10 and –35 regions, separated by the optimum spacing (17 bp). Very few promoters have this exact sequence, but most differ from it only by a few nucleotides.

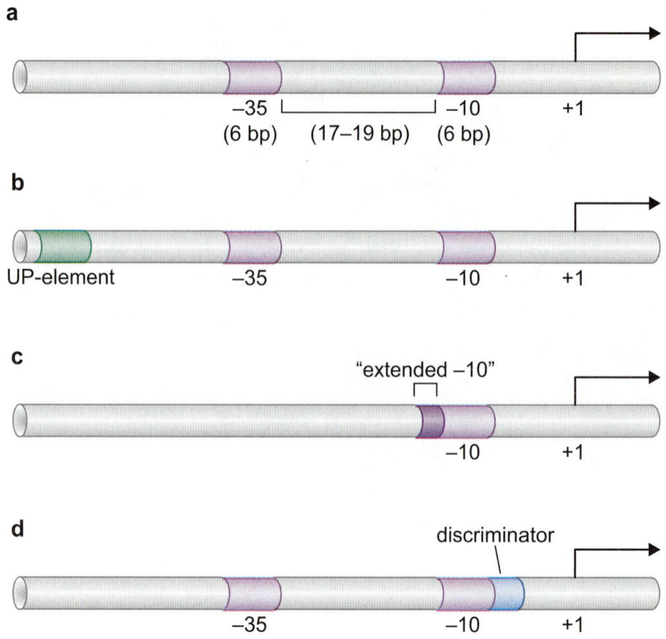

FIGURE 12-5 Features of bacterial promoters. Various combinations of bacterial promoter elements are shown. Details of how each element contributes to polymerase binding and function are described in the text.

Promoters with sequences closer to the consensus are generally "stronger" than those that match less well. By the strength of a promoter, we mean how many transcripts it initiates in a given time. That measure is influenced by how well the promoter binds polymerase initially, how efficiently it supports isomerization, and how readily the polymerase can then escape. The correlation between promoter strength and sequence explains why promoters are so heterogeneous: some genes need to be expressed more highly than others and the former are likely to have sequences closer to the consensus.

An additional DNA element that binds RNA polymerase is found in some strong promoters, for example, those directing expression of the ribosomal RNA (rRNA) genes. This is called an **UP-element** (see Fig. 12-5b) and increases polymerase binding by providing an additional specific interaction between the enzyme and the DNA.

Another class of σ^{70}-promoters lacks a –35 region and instead has a so-called "extended –10" element (see Fig. 12-5c). This comprises a standard –10 region with an additional short sequence element at its upstream end. Extra contacts made between polymerase and this additional sequence element compensate for the absence of a –35 region. The *E. coli gal* genes (whose products direct metabolism of the sugar galactose; see Chapter 16) use such a promoter.

More recently, an additional DNA element that binds RNA polymerase has been found just downstream from the –10 element. This new element is called the **discriminator** and is shown in Figure 12-5d. The strength of the interaction between the discriminator and polymerase influences the stability of the complex between the enzyme and the promoter.

The σ Factor Mediates Binding of Polymerase to the Promoter

The σ^{70} factor can be divided into four regions called σ region 1 through σ region 4 (see Fig. 12-6). The regions that recognize the –10 and –35 elements of the promoter are region 2 and 4, respectively.

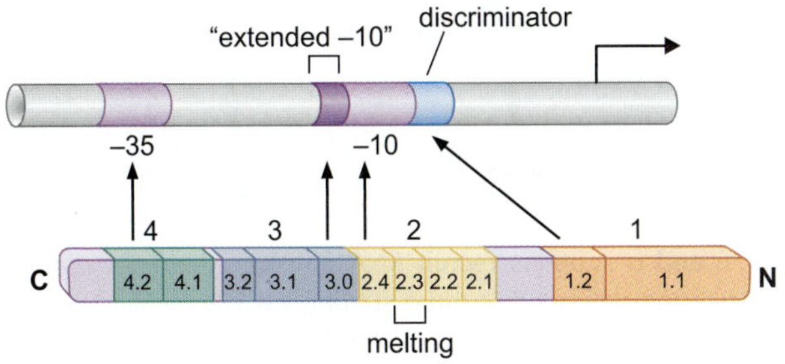

FIGURE 12-6 **Regions of σ.** Those regions of σ factor that recognize specific regions of the promoter are indicated by arrows. Region 2.3 is responsible for melting the DNA. For a schematic view of σ recruiting RNA polymerase core enzyme to a standard promoter, see Fig. 12-7. (Adapted, with permission, from Young B.A. et al. 2002. *Cell* 109: 417–420, Fig. 1. © Elsevier.)

Two helices within region 4 form a common DNA-binding motif called a **helix-turn-helix**. One of these helices inserts into the major groove and interacts with bases in the −35 region; the other lies across the top of the groove, making contacts with the DNA backbone. This structural motif is found in many DNA-binding proteins—for example, almost all transcriptional activators and repressors found in bacterial cells (described in Chapter 16)—and was discussed earlier in Chapter 5 (Fig. 5-20).

The −10 region is also recognized by an α helix. But in this case, the interaction is less well-characterized and is more complicated for the following reason: whereas the −35 region simply provides binding energy to secure polymerase to the promoter, the −10 region has a more elaborate role in transcription initiation, because it is within that element that DNA melting is initiated in the transition from the closed to open complex. Thus, the region of σ that interacts with the −10 region is doing more than simply binding DNA. In keeping with this expectation, the α helix involved in recognition of the −10 region contains several essential aromatic amino acids that can interact with bases on the nontemplate strand in a manner that stabilizes the melted DNA. In Chapter 8, we described a similar role for the single-strand binding protein (SSB) during DNA replication.

The extended −10 element, where present, is recognized by an α helix in σ region 3. This helix makes contact with the two specific base pairs that constitute that element. The discriminator is recognized by σ region 1.2.

Unlike the other elements within the promoter, the UP-element is not recognized by σ but is recognized by a carboxy-terminal domain of the α subunit, called the **αCTD** (Fig. 12-7). The αCTD is connected to the αNTD by a flexible linker. Thus, although the αNTD is embedded in the body of the enzyme, the αCTD can reach the upstream element and can do so even when that element is not located immediately adjacent to the −35 region, but further upstream.

The σ subunit is positioned within the holoenzyme structure in such a way as to make feasible the recognition of various promoter elements. Thus, the DNA-binding regions point away from the body of the enzyme, rather than being embedded. Moreover, the spacing between those regions is consistent with the distance between the DNA elements they recognize. Thus, σ regions 2 and 4 are separated by about 75 Å when σ is bound in the holoenzyme, and this is about the same distance as that between the centers of the −10 and −35 elements of a typical σ^{70} promoter (see Fig. 12-7). This rather large spac-

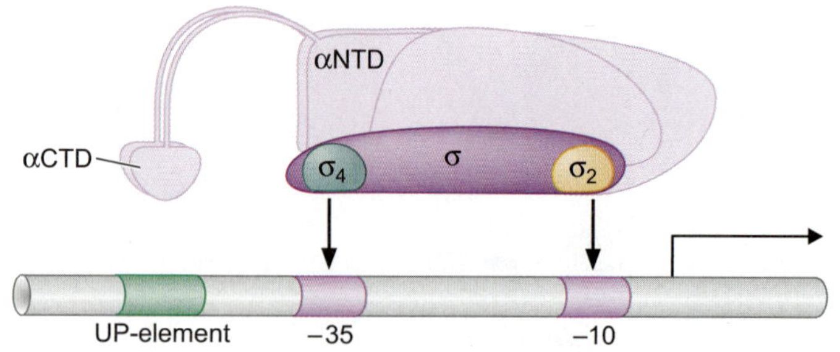

FIGURE 12-7 σ and α subunits recruit RNA polymerase core enzyme to the promoter. The carboxy-terminal domain of the α subunit (αCTD) recognizes the UP-element (where present), whereas σ regions 2 and 4 recognize the −10 and −35 regions, respectively (see Fig. 12-6). In this figure, RNA polymerase is shown in a schematic representation rather different from that presented in earlier figures. This representation is particularly useful for indicating surfaces that touch DNA and regulating proteins, and we use it again in some figures in Chapter 16 when we consider regulation of transcription in bacteria.

ing of the protein domains is accommodated by the region between σ regions 2 and 4, that is, by region 3—especially region 3.2, also called the $\sigma_{3/4}$ linker (see Figs. 12-4 and 12-6).

Transition to the Open Complex Involves Structural Changes in RNA Polymerase and in the Promoter DNA

The initial binding of RNA polymerase to the promoter DNA in the closed complex leaves the DNA in double-stranded form. The next stage in initiation requires the enzyme to become more intimately engaged with the promoter, in the open complex. The transition from closed to open complex involves structural changes in the enzyme and the opening of the DNA double helix to reveal the template and nontemplate strands. This "melting" occurs between positions −11 and +3, with respect to the transcription start site.

In the case of the bacterial enzyme bearing σ^{70}, this transition, often called **isomerization**, does not require energy derived from ATP hydrolysis and is instead the result of a spontaneous conformational change in the DNA–enzyme complex to a more energetically favorable form. Isomerization is essentially irreversible and, once complete, typically guarantees that transcription will subsequently initiate (although regulation can still be imposed after this point in some cases). Formation of the closed complex, in contrast, is readily reversible: polymerase can as easily dissociate from the promoter as make the transition to the open complex.

To picture the structural changes that accompany isomerization, we need to examine the structure of the holoenzyme in more detail. A channel runs between the pincers of the claw-shaped enzyme, as we described earlier (see Fig. 12-2). The active site of the enzyme, which is made up of regions from both the β and β′ subunits, is found at the base of the pincers within the "active center cleft."

There are five channels into the enzyme, as shown in the illustration of the open complex in Figure 12-8. The NTP-uptake channel (not shown in the figure; see figure caption) allows ribonucleotides to enter the active center. The RNA-exit channel allows the growing

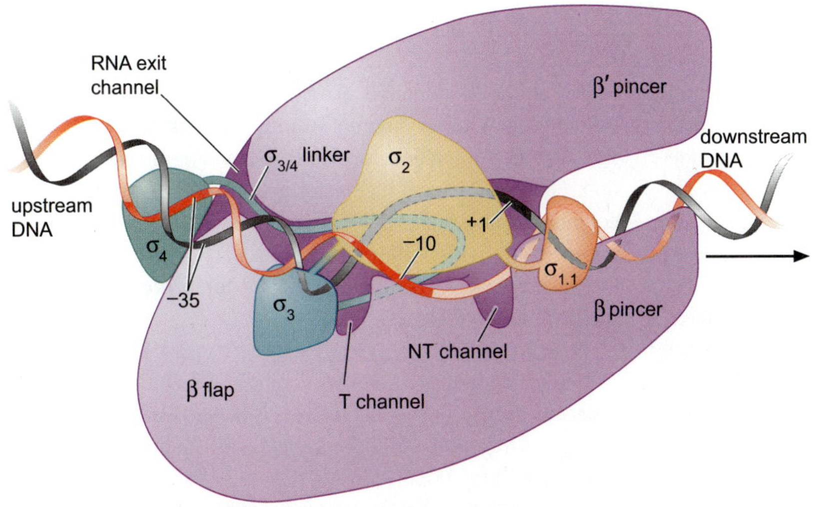

FIGURE 12-8 Channels into and out of the open complex. This figure shows the relative positions of the DNA strands (template strand in gray, nontemplate strand in orange), the four regions of σ, the –10 and –35 regions of the promoter, and the start site of transcription (+1). The channels through which DNA and RNA enter or leave the RNA polymerase enzyme are also shown. The only channel not shown here is the nucleotide entry (NTP-uptake) channel, through which nucleotides enter the active-site cleft for incorporation into the RNA chain as it is made. As drawn, that channel would enter the active site from the back of the page at about the position shown as "+1" on the DNA. Where a DNA strand passes underneath a protein, it is drawn as a dotted ribbon. σ region 3/4 linker—also called $\sigma_{3.2}$—is the linker region between $\sigma_{3.1}$ and σ_4. (Original figure design by Richard Ebright.)

RNA chain to leave the enzyme as it is synthesized during elongation. The remaining three channels allow DNA entry and exit from the enzyme, as follows.

The downstream DNA (i.e., DNA ahead of the enzyme, yet to be transcribed) enters the active center cleft in double-stranded form through the downstream DNA channel (between the pincers). Within the active center cleft, the DNA strands separate from position +3. The nontemplate strand exits the active center cleft through the nontemplate-strand (NT) channel and travels across the surface of the enzyme. The template strand, in contrast, follows a path through the active center cleft and exits through the template-strand (T) channel. The double helix re-forms at –11 in the upstream DNA behind the enzyme.

Two striking structural changes are seen in the enzyme upon isomerization from the closed to open complex. First, the pincers at the front of the enzyme clamp down tightly on the downstream DNA. Second, there is a major shift in the position of the amino-terminal region of σ (region 1.1) as we now describe. When not bound to DNA, σ region 1.1 lies within the active center cleft of the holoenzyme, blocking the path that, in the open complex, is followed by the template DNA strand. In the open complex, region 1.1 shifts some 50 Å and is now found on the outside of the enzyme, allowing the DNA access to the cleft (see Fig. 12-8). Region 1.1 of σ is highly negatively charged (just like DNA). Thus, in the holoenzyme, region 1.1 acts as a **molecular mimic** of DNA. The space in the active center cleft, which may be occupied either by region 1.1 or by DNA, is highly positively charged.

TECHNIQUES

Box 12-1 Consensus Sequences

The DNA sequences of binding sites recognized by a given protein may not always be exactly the same. Likewise, a stretch of amino acids that bestows upon a protein a particular function may be slightly different in different proteins. A consensus sequence is, in each case, a version of the sequence having at each position the nucleotide (or amino acid) most commonly found there in different examples. Thus, the consensus sequence for promoters in *E. coli* recognized by RNA polymerase containing σ^{70} is shown in the figure (Box 12-1 Fig. 1). This consensus sequence was derived by aligning 300 sequences known to function as σ^{70} promoters and ascertaining the most common base found at each position in the –35 and –10 hexamers. That nucleotide is then chosen as the nucleotide of choice at that position in the consensus; its relative frequency and the frequencies with which the other three nucleotides occur at each position are portrayed in the graph. Note that there is no significant consensus among the 17–19 nucleotides that lie in the region between –35 and –10.

In that example, each individual promoter sequence had previously been identified, so aligning the sequences is trivial. But consider a rather different example. In this case, no binding site has been identified for the DNA-binding protein in question. However, several regions of a chromosome are known to contain binding sites somewhere within their lengths. A computer algorithm is employed that scans each of the sequences of these chromosomal regions, searching for a potential binding site common to them all.

A second approach to deriving the consensus sequence for a DNA-binding protein when the binding site is not already known takes advantage of chemical methods for synthesizing vast sets of short DNA fragments of random sequence. The protein of interest is mixed with the population of DNA molecules and those DNAs to which it binds are retrieved and sequenced. A comparison of the sequences bound reveals the consensus readily, because each of the fragments is very short. This last method (often called SELEX) is widely used to define binding sites for previously uncharacterized DNA-binding proteins. SELEX is described in more detail in Chapter 21.

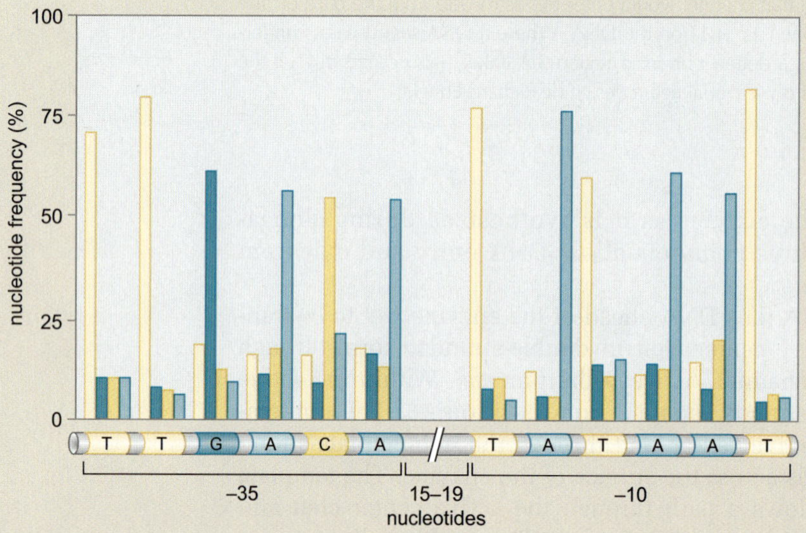

BOX 12-1 FIGURE 1 Promoter consensus sequence and spacing consensus. (Redrawn, with permission, from Alberts B. et al. 2002. *Molecular biology of the cell*, 4th ed., p. 308, Fig. 6.12. © Garland Science/Taylor & Francis Books LLC.)

Transcription Is Initiated by RNA Polymerase without the Need for a Primer

Recall from Chapter 8 that DNA polymerase does not synthesize new DNA strands de novo; that is, it can only extend an existing polynucleotide chain. For this reason, replication always requires a primer strand. The primer is typically a short piece of RNA that binds to the DNA template strand to form a short hybrid double-strand region; DNA polymerase then adds nucleotides to the 3′ end of the primer.

RNA polymerase can initiate a new RNA chain on a DNA template and thus does not need a primer. This impressive feat requires that the initiating ribonucleotide be brought into the active site and held stably on the template while the next NTP is presented with correct geometry for the chemistry of polymerization to occur. This is particularly difficult because RNA polymerase starts most transcripts with an A, and that ribonucleotide binds the template nucleotide (T) with only two hydrogen bonds (as opposed to the three between C and G).

Thus the enzyme has to make specific interactions with one or both of the initiating ribonucleotide and the second ribonucleotide, holding one (or both) rigidly in the correct orientation to allow chemical attack on the incoming NTP. The requirement for such specific interactions between the enzyme and the initiating nucleotide probably explains why most transcripts start with the same nucleotide. It is believed that the interactions are provided by various parts of polymerase holoenzyme, including σ region 3/4 linker. Consistent with this, in experiments using an RNA polymerase containing a σ^{70} derivative lacking this part of σ, initiation requires much higher than normal concentrations of one or both of the first two ribonucleotides.

During Initial Transcription, RNA Polymerase Remains Stationary and Pulls Downstream DNA into Itself

As we have outlined, during initial transcription, RNA polymerase produces and releases short RNA transcripts of less than ten nucleotides (abortive synthesis) before escaping the promoter, entering the elongation phase, and synthesizing the proper transcript. It has long been unclear how the enzyme's active site translocates along the DNA template during initial abortive cycles of transcription. Three general models were proposed (as shown in Fig. 12-9 and described below):

1. The so-called "transient excursion" model proposes transient cycles of forward and reverse translocation of RNA polymerase. Thus, polymerase is thought to leave the promoter and translocate a short way along the DNA template, synthesizing a short transcript before aborting transcription, releasing the transcript, and returning to its original location on the promoter.
2. "Inchworming" invokes a flexible element within the polymerase that allows a module at the front of the enzyme, containing the active site, to move downstream, synthesizing a short transcript before aborting and retracting to the body of the enzyme still at the promoter.
3. "Scrunching" proposes that DNA downstream from the stationary, promoter-bound, polymerase is pulled into the enzyme. The DNA thus accumulated within the enzyme is accommodated as single-stranded bulges.

It is now believed that the third model—scrunching—reflects what happens. This conclusion is based on a number of findings, including experiments using single-molecule analyses that allow the positions of different parts of polymerase to be measured relative to each other and to the template DNA during initial transcription. These experiments show that during initial transcription, the polymerase remains stationary on the promoter, unwinds downstream DNA, and pulls that DNA into itself. Only the scrunching model is consistent with these results.

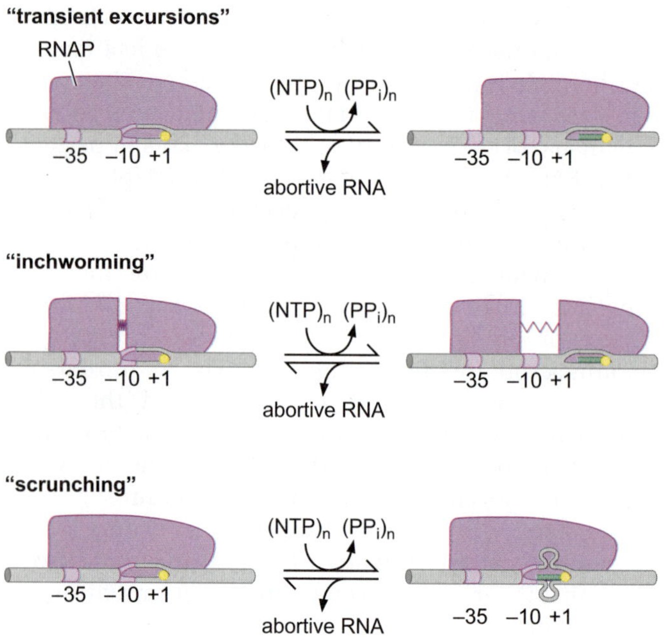

FIGURE 12-9 Mechanism of initial transcription. During initial transcription, the active center of RNA polymerase is translocated forward relative to the DNA template and synthesizes short transcripts before aborting, then repeats this cycle until it escapes the promoter. Three models have been proposed to account for this and are shown in the figure. According to the first of these—transient excursions (shown at the top)—polymerase moves along the DNA. In the second—inchworming (shown in the middle)—the front part of the enzyme moves along the DNA, but because of a flexible region within the enzyme, the back part of the enzyme can remain stationary at the promoter. In the third model—scrunching (shown at the bottom)—the enzyme remains stationary and pulls the DNA into it. The differences between these models are explained in the text, as is the evidence supporting scrunching as the true picture of what goes on. (Modified, with permission, from Kapanidis A.N. et al. 2006. *Science* 314: 1144–1147, Fig. 1a. © AAAS.)

Promoter Escape Involves Breaking Polymerase–Promoter Interactions and Polymerase Core–σ Interactions

As we have seen, during initial transcription, the process of abortive initiation takes place, and short—nine nucleotides or shorter—transcripts are generated and released. Polymerase manages to escape from the promoter and enter the elongation phase only once it has managed to synthesize a transcript of a threshold length of ten or more nucleotides. Once this length, the transcript cannot be accommodated within the region where it hybridizes to the DNA and must start threading into the RNA exit channel (Fig. 12-8). Promoter escape is associated with the breaking of all interactions between polymerase and promoter elements and between polymerase and any regulatory proteins operating at the given promoter (Chapter 16).

It is not clear why RNA polymerase must undergo this period of abortive initiation before achieving escape, but once again a region of the σ factor appears to be involved, acting as a molecular mimic. In this case, it is the region 3/4 linker, and it mimics RNA. This region of σ lies in the middle of the RNA exit channel in the open complex (see Fig. 12-8), and for an RNA chain to be made longer than about ten nucleotides, this region of σ must be ejected from that location, a process that can take the enzyme several attempts.

The ejection of σ region 3/4 linker probably accounts for σ being more weakly associated with the elongating enzyme than it is with the open complex; indeed, it is often lost altogether from the elongating complex.

Scrunching is reversed upon escape. That is, the DNA unwound during scrunching is rewound, with concomitant collapse of the transcription bubble from a size of 22–24 nucleotides back down to 12–14 nucleotides (Fig. 12-3). It is believed that this process provides the energy required by polymerase to break the polymerase–promoter and core–σ interactions associated with escape. Thus, scrunching is a way to store and mobilize energy during transcription initiation, and its release upon escape is what enables polymerase to break free of the promoter and dislodge σ factor from core.

In Box 12-2, The Single-Subunit RNA Polymerases, we see how these simple RNA polymerases, despite lacking a σ subunit, undergo a structurally comparable shift in transition from the initiating to the elongating complex.

The Elongating Polymerase Is a Processive Machine That Synthesizes and Proofreads RNA

DNA passes through the elongating enzyme in a manner very similar to its passage through the open complex (Fig. 12-8). Thus, double-stranded DNA enters the front of the enzyme between the pincers. At the opening of the catalytic cleft, the strands separate to follow different paths through the enzyme before exiting via their respective channels and re-forming a double helix behind the elongating polymerase. Ribonucleotides enter the active site through their defined channel and are added to the growing RNA chain under the guidance of the template DNA strand. Only eight or nine nucleotides of the growing RNA chain remain base-paired to the DNA template at any given time; the remainder of the RNA chain is peeled off and directed out of the enzyme through the RNA exit channel. See Figure 12-10 for a schematic diagram of the elongating complex.

During elongation, the enzyme adds one nucleotide at a time to the growing RNA transcript. In addition, the size of the bubble, that is, the length of DNA that is not double-helical, remains constant throughout elongation: as 1 bp is separated ahead of the processing enzyme, 1 bp is formed behind it.

In addition to all this, RNA polymerase carries out two proofreading functions as well. The first of these is called **pyrophosphorolytic editing**. In this, the enzyme uses its active site, in a simple back-reaction, to catalyze the removal of an incorrectly inserted ribonucleotide, by reincorporation of PPi. The enzyme can then incorporate another ribonucleotide in its place in the growing RNA chain. Note that the enzyme can remove either correct or incorrect bases in this manner, but spends longer hovering over mismatches than matches, and so removes the former more frequently. In the second proofreading mechanism, called **hydrolytic editing**, the polymerase backtracks by one or more nucleotides (see Fig. 12-10d) and cleaves the RNA product, removing the error-containing sequence.

Hydrolytic editing is stimulated by Gre factors, which both enhance hydrolytic editing function and serve as elongation stimulating factors; that is, they ensure that polymerase elongates efficiently and helps overcome "arrest" at sequences that are difficult to transcribe. This combination of functions is comparable to those imposed on the

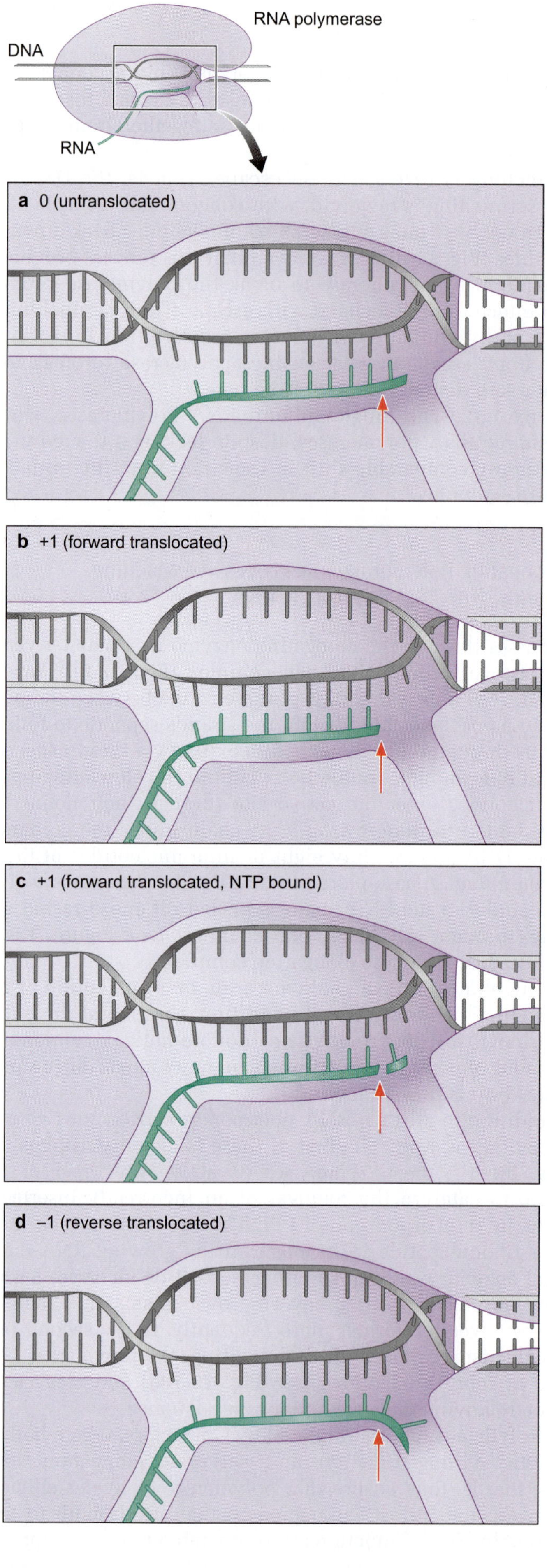

FIGURE 12-10 Template and transcript within the RNA polymerase elongating complex. The figure shows schematic diagrams of the relative positions of RNA and the DNA template within RNA polymerase at various states of the transcription process. (a) Untranslocated polymerase (0) shows the RNA chain paired with the template DNA strand for a nine-base stretch. (b) Forward translocated polymerase (+1) shows the situation when the enzyme has translocated one base forward. (c) Forward translocated polymerase with NTP bound shows the DNA and RNA in the same position as in b with the incoming NTP bound. (d) Reverse translocated polymerase (–1) shows the situation when the enzyme is translocated backward one base as it does during hydrolytic editing. The red arrow indicates a set position within the polymerase, the same in all parts of the figure. See text for more details. For clarity, the polymerase shown here is in a different orientation from that in Fig. 12-8, with the RNA exit channel downward. (Figures based on images courtesy of Richard Ebright.)

ADVANCED CONCEPTS

Box 12-2 The Single-Subunit RNA Polymerases

In the text, we discuss the multisubunit RNA polymerases found in bacteria and eukaryotic cells. But there are several examples of single-subunit RNA polymerases that are capable of performing the same basic reaction as their more complex multicellular counterparts. Thus, many bacteriophage, for example, the *E. coli* phage T7, encode polymerases of this type with which, upon infection, they transcribe most of their genes. Similarly, the majority of mitochondrial and chloroplast genes are transcribed by polymerases closely related to the single-subunit phage enzymes. It is remarkable that evolution has produced these relatively simple enzymes capable of carrying out transcription, a task that we, in the text, emphasize as an impressive achievement even for the much larger and more complicated multisubunit enzymes.

The T7 polymerase is the most widely studied of the single-subunit enzymes. It has a molecular mass of 100 kD—compared to 400 kD for the bacterial core enzyme (without σ factor)—and a structure shown in Box 12-2 Figure 1. Overall, it looks like the Pol I family of DNA polymerases that we considered in Chapter 8. Thus, the T7 RNA polymerase resembles a right hand, with the fingers, thumb, and palm representing domains arranged around a central cleft, within which lies the active site.

Although it is not structurally related to the cellular RNA polymerases (and instead is structurally related to the DNA polymerases), the T7 enzyme does have features functionally analogous to the cellular RNA polymerases as well, features that have become more apparent since the structure of the T7 and bacterial enzymes have been compared in complex with their templates. As we saw in the text, the bacterial enzyme has various channels into and out of the active center cleft (see Fig. 12-8). One of these, for example, allows the NTPs access to the active site and template, where they are polymerized, under the influence of the template, into the growing RNA chain. Another channel provides the growing RNA chain an exit from the enzyme. Analogous channels are seen in the structure of the phage polymerase as well.

The initiation and elongation complexes of the bacterial and T7 polymerases have been compared. These comparisons highlight one striking example of how an analogous functional transition can be achieved through different kinds of structural change in the two cases. We noted in the text that in the bacterial case the transition from initiation to elongation involves a significant shift in the location of a domain of the σ factor. This movement opens up the RNA-exit channel, thereby allowing production of transcripts larger than ten nucleotides in length. The T7 enzyme has no σ factor; but a comparable structural change in the body of that single-subunit enzyme mediates the transition from the initiating to elongating complex, and this structural change is required to form the RNA exit channel.

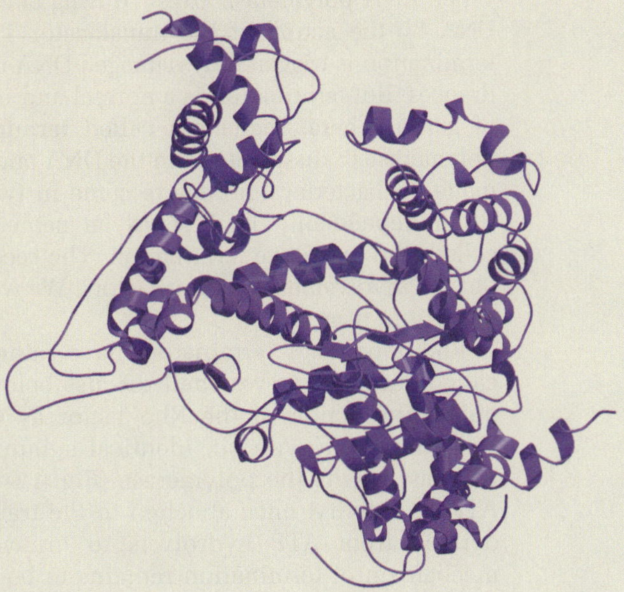

BOX 12-2 FIGURE 1 **Bacteriophage T7 RNA polymerase.** (Jeruzalmi D. and Steitz T.A. 1998. *EMBO J.* 17: 4101.) Image prepared with MolScript, BobScript, and Raster 3D.

eukaryotic RNA polymerase II by the transcription factor TFIIS (see below). Another group of proteins—the Nus proteins—joins polymerase in the elongation phase and promotes, in still rather undefined ways, the processes of elongation and termination (for examples of regulation during elongation, see Chapter 16).

RNA Polymerase Can Become Arrested and Need Removing

Under certain circumstances, an elongating RNA polymerase can become arrested and cease transcribing. One common cause of arrest is a damaged DNA strand. The consequences of arrest can be catastrophic if the gene being transcribed is essential: no product will be made by the arrested polymerase and that same enzyme will cause a roadblock to other polymerases attempting to transcribe the same gene.

To deal with this situation, the cell has machinery that removes the arrested polymerase and at the same time recruits repair enzymes (in particular, the endonuclease Uvr(A)BC); the repair that follows is called transcription-coupled repair (which we discussed in Chapter 9). Both polymerase removal and repair enzyme recruitment are carried out by a single protein called TRCF.

TRCF has an ATPase activity. It binds double-stranded DNA upstream of the polymerase and uses the ATPase motor to translocate along the DNA until it encounters the stalled RNA polymerase. The collision pushes polymerase forward, either allowing it to restart elongation or, more often, causing dissociation of the ternary complex of RNA polymerase, template DNA, and RNA transcript. This terminates transcription by that enzyme, but it makes way for another polymerase.

Transcription Is Terminated by Signals within the RNA Sequence

We have already seen one way in which transcription can be terminated. When RNA polymerase arrests during elongation, it can be knocked off DNA by the action of the translocator TRCF (discussed above). This termination is triggered by damaged DNA or by other unanticipated hindrances. But termination is a normal and important function at the ends of genes. There, sequences called **terminators** trigger the elongating polymerase to dissociate from the DNA and release the RNA chain it has made. In bacteria, terminators come in two types: **Rho-dependent** and **Rho-independent**. The first, as its name suggests, requires a protein called Rho to induce termination. The second causes termination without the involvement of other factors. We will deal with each kind of terminator in turn.

Rho-dependent terminators have rather ill-defined RNA elements called **rut** sites, as we shall discuss below, and for them to work requires the action of the Rho factor as well. Rho, which is a ring-shaped protein with six identical subunits, binds to single-stranded RNA as it exits the polymerase (Fig. 12-11). The protein also has an ATPase activity: once attached to the transcript, Rho uses the energy derived from ATP hydrolysis to induce termination. The precise mechanism of termination remains to be determined, and models include the following: Rho pushes polymerase forward relative to the DNA and RNA, resulting in termination in a manner analogous to termination by TRCF (described above); Rho pulls RNA out of the polymerase, resulting in termination; or Rho induces a conformational change in polymerase, causing the enzyme to terminate.

How is Rho directed to a particular RNA molecule? First, there is some specificity in the sites it binds (the rut sites, for Rho utilization

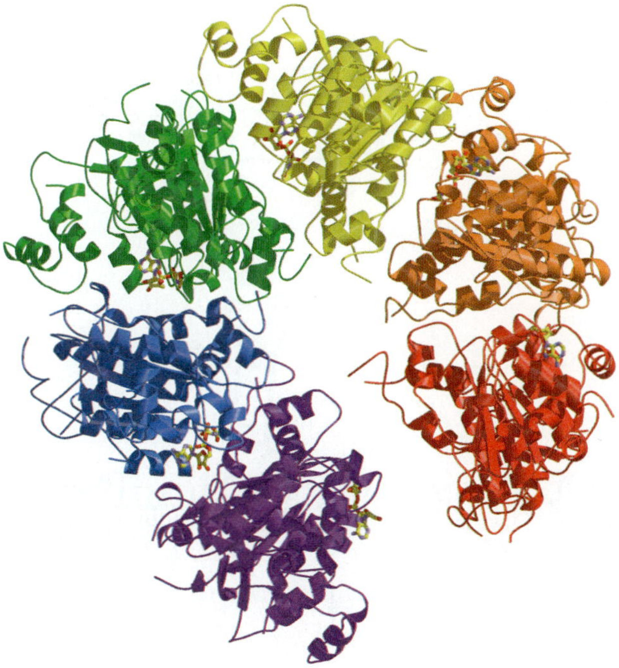

FIGURE 12-11 The Rho transcription termination factor. The crystal structure of the Rho termination factor is shown in a top-down view. It consists of a hexamer of Rho protein, each monomer shown in a different color. The six monomers form an open ring. The ring is not flat; the sixth subunit is further down in the plane of the page than the first. The gap between the two subunits is 12 Å, and the helical pitch between them is 45 Å. The RNA transcript on which Rho acts (not shown) is believed to bind along the bottom of each subunit and then thread through the middle of the ring. (Skordalakes E. and Berger J.M. 2003. *Cell* 114: 135.) Image prepared with MolScript, BobScript, and Raster 3D.

mentioned above). Optimally, these sites consist of stretches of ~40 nucleotides that do not fold into a secondary structure (i.e., they remain largely single-stranded); they are also rich in C residues.

The second level of specificity is that Rho fails to bind any transcript that is being translated (i.e., a transcript bound by ribosomes). In bacteria, transcription and translation are tightly coupled—translation initiates on growing RNA transcripts as soon as they start exiting polymerase, while they are still being synthesized. Thus, Rho typically terminates only those transcripts still being transcribed beyond the end of a gene or operon.

Rho-independent terminators, also called **intrinsic terminators**, because they need no other factors to work, consist of two sequence elements: a short inverted repeat (of approximately 20 nucleotides) followed by a stretch of about eight A:T base pairs (Fig. 12-12). These

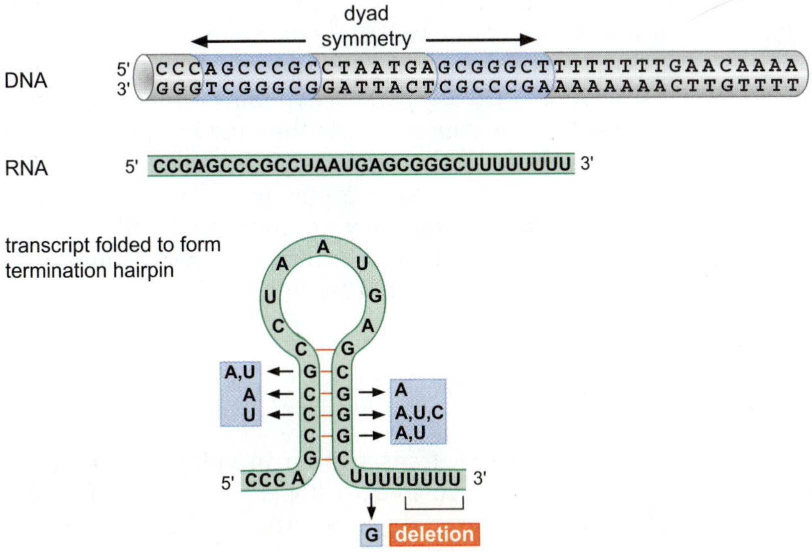

FIGURE 12-12 Sequence of a Rho-independent terminator. At the top is the sequence, in the DNA, of the terminator. Below is shown the sequence of the RNA, and the bottom shows the structure of the terminator hairpin. The terminator in question is from the trp attenuator, discussed in Chapter 16. The boxes show mutations isolated in the sequence that disrupt the terminator. (Adapted from Yanofsky C. 1981. *Nature* 289: 751–758.)

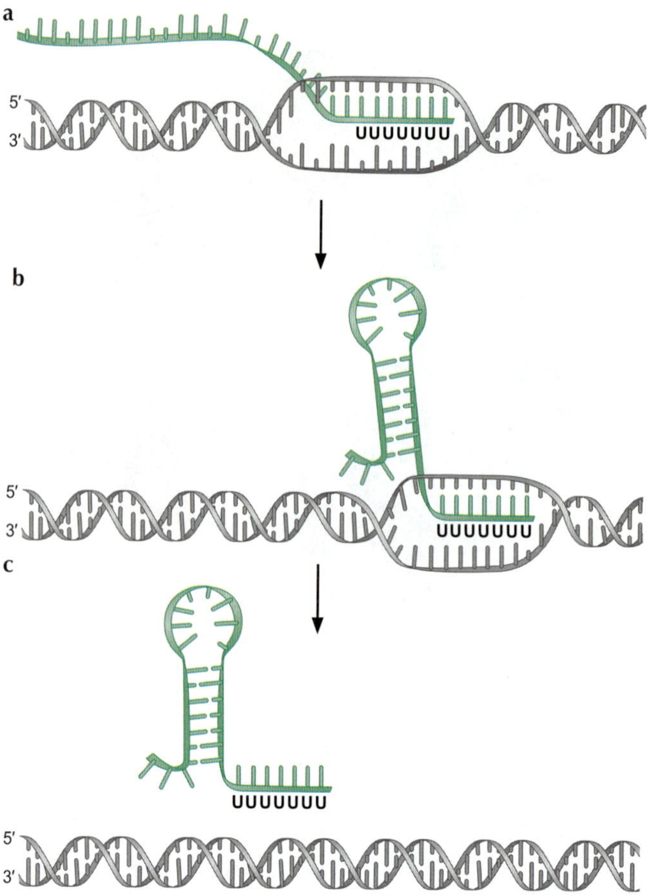

FIGURE 12-13 Transcription termination. Shown is a model for how the Rho-independent terminator might work. (a) The hairpin forms in the RNA (Fig. 12-10) as soon as that region has been transcribed by polymerase (the enzyme is not shown here). (b) That RNA structure disrupts polymerase just as the enzyme is transcribing the AT-rich stretch of DNA downstream. (c) Exactly how the hairpin disrupts the transcribing polymerase is not clear (see text for alternative models), but the weak interactions between the transcript and the template DNA (Us in the transcript and As in the template) appear to make release of that transcript easier. (Adapted from Platt T. 1981. *Cell* 24: 10–23.)

elements do not affect the polymerase until they have been transcribed—that is, they function in the RNA rather than in the DNA. When polymerase transcribes an inverted repeat sequence, the resulting RNA can form a stem-loop structure (often called a "hairpin") by base-pairing with itself (see Chapter 6). Formation of the hairpin causes termination by disrupting the elongation complex. As with Rho-dependent termination, the mechanism remains to be determined and current models are much the same as those proposed for Rho. That is, the hairpin induces termination by either pushing polymerase forward relative to the DNA and RNA, wresting the transcript from polymerase, or inducing a conformational change in polymerase.

The hairpin works as an efficient terminator only when it is followed by a stretch of A:U base pairs, as we have described. This is because, under those circumstances, at the time the hairpin forms, the growing RNA chain will be held on the template at the active site by only A:U base pairs. As A:U base pairs are the weakest of all base pairs (weaker even than A:T base pairs), they are more easily disrupted by the effects of the stem loop on the transcribing polymerase, and so the RNA will more readily dissociate (Fig. 12-13).

TRANSCRIPTION IN EUKARYOTES

As we have already discussed, transcription in eukaryotes is undertaken by polymerases closely related to RNA polymerases found in prokaryotes. This is hardly surprising: the process of transcription itself is identical in the two cases. There are, however, differences in the machinery used in each case. One we have already seen: eukaryotes have three different polymerases (Pol I, II, and III), whereas bacteria have only one. Also, whereas bacteria require only one additional initiation factor (σ), several initiation factors are required for efficient and promoter-specific initiation in eukaryotes. These are called the **general transcription factors (GTFs)**.

In vitro, the general transcription factors are all that are required, together with Pol II, to initiate transcription on a DNA template (without histones). In vivo, however, the DNA template in eukaryotic cells is incorporated into nucleosomes, as we have seen in Chapter 7. Under these circumstances, the general transcription factors are not alone sufficient to bind promoter sequences and elicit significant expression. Rather, additional factors are required, including DNA-binding regulatory proteins, the so-called Mediator complex, and chromatin-modifying enzymes.

We first consider the basic mechanism by which Pol II and the general transcription factors assemble at a promoter to initiate transcription in vitro. We then consider the roles of the additional components required to promote transcription in vivo.

RNA Polymerase II Core Promoters Are Made Up of Combinations of Four Different Sequence Elements

The eukaryotic **core promoter** refers to the minimal set of sequence elements required for accurate transcription initiation by the Pol II machinery, as measured in vitro. A core promoter is typically approximately 40–60 nucleotides long, extending either upstream or downstream from the transcription start site. Figure 12-14 shows the location, relative to the transcription start site, of elements found in Pol II core promoters. These are the TFIIB recognition element (BRE), the TATA element (or box), the initiator (Inr), and the downstream promoter elements (known as DPE, DCE, and MTE). Typically, a promoter includes some subset of these elements. Thus, for example, promoters typically have either a TATA element or a DPE element, not both. Often, a TATA-containing promoter also contains a DCE. The Inr is the most common element, found in combination with both TATA and DPEs. The consensus sequence for each element and the general transcription factor that binds it are also shown, and these features are described in more detail in coming sections.

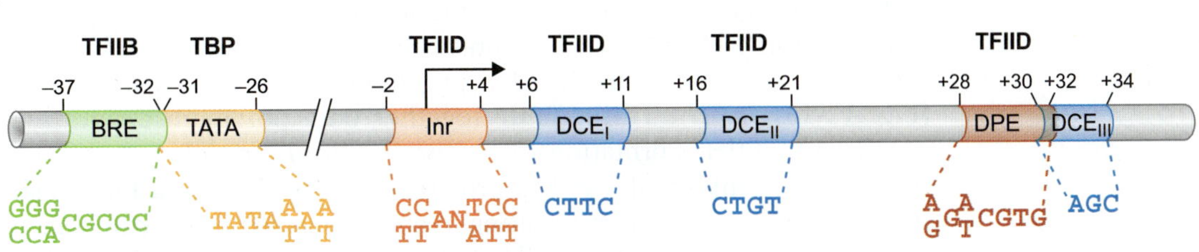

FIGURE 12-14 Pol II core promoter. The figure shows the positions of various DNA elements relative to the transcription start site (indicated by the arrow above the DNA). These elements, described in the text, are as follows: BRE (TFIIB recognition element); TATA (TATA Box); Inr (initiator element); DPE (downstream promoter element), and DCE (downstream core element). Another element, MTE (motif ten element), described in the text, is not shown in this figure but is located just upstream of the DPE. Also shown are the consensus sequences for each element (determined in the same way as described for the bacterial promoter elements, see Box 12-1) and (above) the name of the general transcription factor that recognizes each element.

Beyond—and typically upstream of—the core promoter, there are other sequence elements required for efficient transcription in vivo. Together, these elements constitute the **regulatory sequences** and can be grouped into various categories, reflecting their location, and the organism in question, as much as their function. These elements include promoter proximal elements, upstream activator sequences (UASs), enhancers, and a series of other elements called silencers, boundary elements, and insulators. All of these DNA elements bind regulatory proteins (activators and repressors), which help or hinder transcription from the core promoter, and are the subject of Chapter 17. Some of these regulatory sequences can be located many tens or even hundreds of kilobases from the core promoters on which they act.

RNA Polymerase II Forms a Preinitiation Complex with General Transcription Factors at the Promoter

The general transcription factors collectively perform the functions performed by σ in bacterial transcription, and some of these factors share regions of weak homology with different regions of σ. Thus, the general transcription factors help polymerase bind to the promoter and melt the DNA (comparable to the transition from closed to open complex in the bacterial case). They also help polymerase escape from the promoter and embark on the elongation phase. The complete set of general transcription factors and polymerase, bound together at the promoter and poised for initiation, is called the **preinitiation complex**.

As we described above (and in Fig. 12-14), many Pol II promoters contain a so-called TATA element (some 30 bp upstream of the transcription start site). This is where preinitiation complex formation begins. The TATA element is recognized by the general transcription factor called **TFIID**. (The nomenclature "TFII" denotes a transcription factor for Pol II, with individual factors distinguished as A, B, and so on.) Like many of the general transcription factors, TFIID is in fact a multisubunit complex. The component of TFIID that binds to the TATA DNA sequence is called **TBP** (TATA-binding protein). The other subunits in this complex are called **TAFs**, for TBP-associated factors. Some TAFs recognize other core promoter elements such as the Inr, DPE, and DCE, although the strongest binding is between TBP and TATA. Thus, TFIID is a critical factor in promoter recognition and preinitiation complex establishment.

Upon binding DNA, TBP extensively distorts the TATA sequence (we shall discuss this event in more detail presently). The resulting TBP–DNA complex provides a platform to recruit other general transcription factors and polymerase itself to the promoter. In vitro, these proteins assemble at the promoter in the following order (Fig. 12-15): TFIIA, TFIIB, TFIIF together with polymerase, and then TFIIE and TFIIH. Formation of the preinitiation complex containing these components is followed by promoter melting. In contrast to the situation in bacteria, promoter melting in eukaryotes requires hydrolysis of ATP and is mediated by TFIIH. It is a helicase-like activity of that factor that stimulates unwinding of promoter DNA.

Promoter Escape Requires Phosphorylation of the Polymerase "Tail"

Just as we have seen in the bacterial case, there now follows a period of abortive initiation before the polymerase escapes the promoter and enters the elongation phase. Recall that during abortive initiation, the

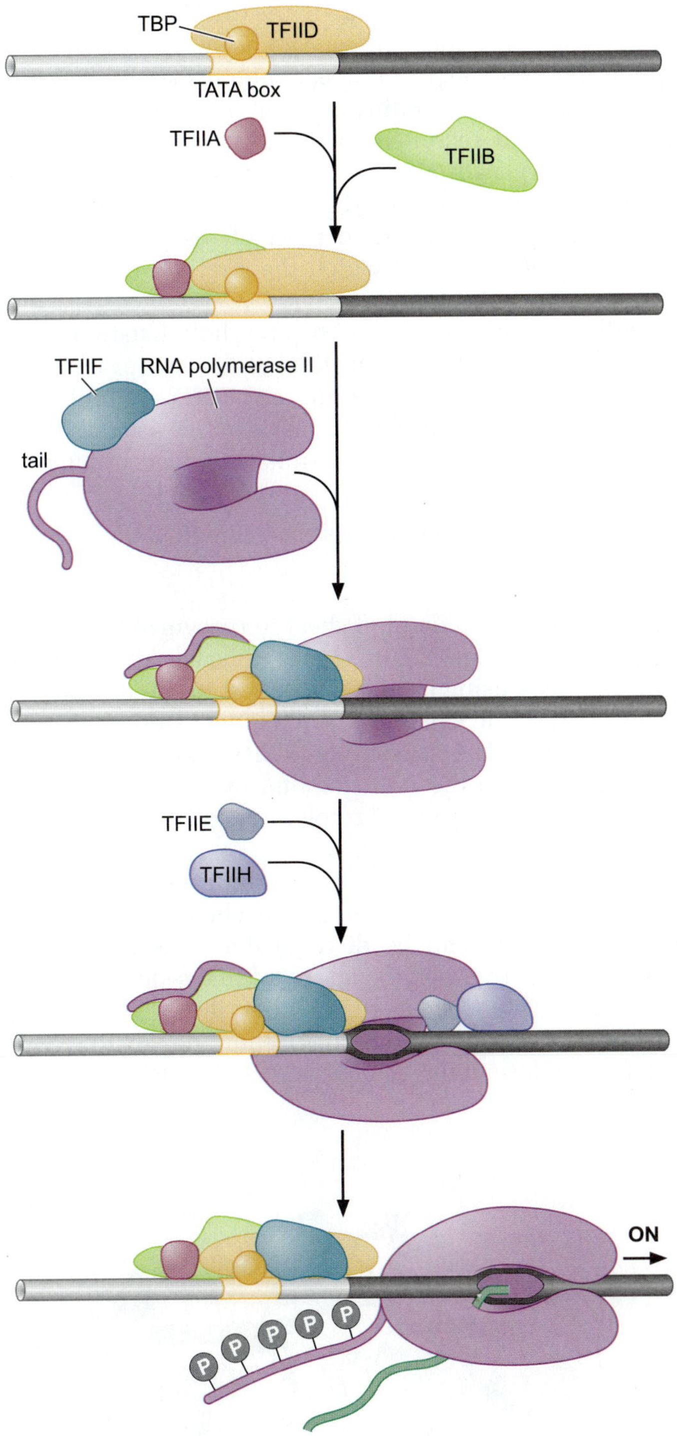

FIGURE 12-15 Transcription initiation by RNA Pol II. The stepwise assembly of the Pol II preinitiation complex is shown here and described in detail in the text. Once assembled at the promoter, Pol II leaves the preinitiation complex upon addition of the nucleotide precursors required for RNA synthesis and after phosphorylation of serine resides within the enzyme's "tail." The tail contains multiple repeats of the heptapeptide sequence: Tyr-Ser-Pro-Thr-Ser-Pro-Ser (see Fig. 12-19).

polymerase synthesizes a series of short transcripts. In eukaryotes, promoter escape involves two steps not seen in the bacterial case: one is ATP hydrolysis (in addition to the earlier ATP hydrolysis needed for DNA melting), and the other is phosphorylation of the polymerase, as we now describe.

The large subunit of Pol II has a carboxy-terminal domain (CTD), which is referred to as the "tail" (see Fig. 12-15). The CTD contains a series of repeats of the heptapeptide sequence: Tyr-Ser-Pro-Thr-Ser-Pro-Ser. There are 27 of these repeats in the yeast Pol II CTD, 32 in the worm *Caenorhabditis elegans*, 45 in the fly *Drosophila*, and 52 in humans. Indeed, the number of repeats seems to correlate with the complexity of the genome. Each repeat contains sites for phosphorylation by specific kinases, including one that is a subunit of TFIIH.

The form of Pol II recruited to the promoter initially contains a largely unphosphorylated tail, but the species found in the elongation complex bears multiple phosphoryl groups on its tail. Addition of these phosphates helps polymerase shed most of the general transcription factors used for initiation, and which the enzyme leaves behind as it escapes the promoter.

As we will see, regulating the phosphorylation state of the CTD of Pol II controls later steps—elongation and even processing of the RNA—as well. Indeed, in addition to TFIIH, a number of other kinases have been identified that act on the CTD, as well as a number of phosphatases that remove the phosphates added by those kinases.

TBP Binds to and Distorts DNA Using a β Sheet Inserted into the Minor Groove

TBP uses an extensive region of β sheet to recognize the minor groove of the TATA element (Fig. 12-16). This is unusual. More typically, proteins recognize DNA using α helices inserted into the major groove of DNA, as we have seen in Chapters 5 and 6 and also for the σ factor earlier in this chapter. The reason for TBP's unorthodox recognition mechanism is linked to the need for that protein to distort the local DNA structure. But this mode of recognition raises a problem: how is specificity achieved?

We have seen in Chapter 6 that compared to the major groove, the minor groove of DNA is less rich in the chemical information that would enable base pairs to be distinguished. Instead, to select the TATA sequence, TBP relies on the ability of that sequence to undergo a specific structural distortion, as we now describe.

When it binds DNA, TBP causes the minor groove to be widened to an almost flat conformation; it also bends the DNA by an angle of

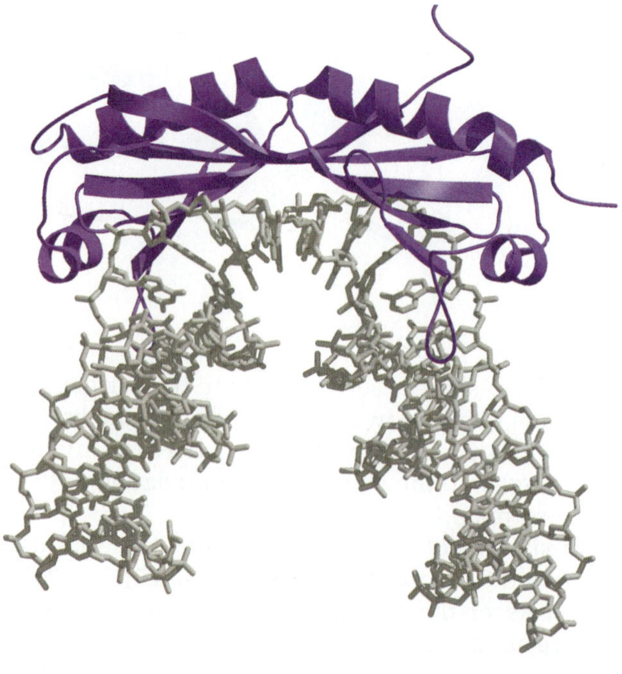

FIGURE 12-16 TBP–DNA complex. TBP is shown here in purple complexed with the DNA TATA sequence (shown in gray) found at the start of many Pol II genes. The details of this interaction are described in the text. (Nikolov D.B. et al. 1995. *Nature* 377: 119.) Image prepared with MolScript, BobScript, and Raster 3D. Extended DNA on either side of image modeled by Leemor Joshua-Tor.

~80°. The interaction between TBP and DNA involves only a limited number of hydrogen bonds between the protein and the edges of the base pairs in the minor groove. Instead, much of the specificity is imposed by two pairs of phenylalanine side chains that intercalate between the base pairs at either end of the recognition sequence and drive the strong bend in the DNA.

A:T base pairs are thus favored because they are more readily distorted to allow the initial opening of the minor groove. There are also extensive interactions between the phosphate backbone and basic residues in the β sheet, adding to the overall binding energy of the interaction.

The Other General Transcription Factors Also Have Specific Roles in Initiation

We do not know in detail the functions of all the other general transcription factors. As we have noted, some of these factors are in fact complexes made up of two or more subunits (shown in Table 12-2). Below we comment on a few structural and functional characteristics.

TABLE 12-2 The General Transcription Factors of RNA Polymerase II

GTFs	Number of Subunits
TBP	1
TFIIA	2
TFIIB	1
TFIIE	2
TFIIF	3
TFIIH	10
TAFs	11

The numbers shown are for yeast but are similar for other eukaryotes, including humans. There are some differences, however—for example, human TFIIF has only two subunits and its TFIIA has three.

TAFs TBP is associated with about ten TAFs. Two of the TAFs bind DNA elements at the promoter; for example, the initiator element (Inr) and the downstream promoter elements (see Fig. 12-14). Several of the TAFs have structural homology with histone proteins, and it has been proposed that they might bind DNA in a similar manner, although evidence for such a form of DNA binding has not been obtained. For example, TAF42 and TAF62 from *Drosophila* have been shown to form a structure similar to that of the H3•H4 tetramer (see Chapter 7). These histone-like TAFs are found not only in the TFIID complex, but also associated with some histone modification enzymes, such as the yeast SAGA complex (see Table 7-7).

Another TAF appears to regulate the binding of TBP to DNA. It does this using an inhibitory flap that binds to the DNA-binding surface of TBP, another example of molecular mimicry. This flap must be displaced for TBP to bind TATA.

TFIIB This protein, a single polypeptide chain, enters the preinitiation complex after TBP (Fig. 12-14). The crystal structure of the ternary complex of TFIIB–TBP–DNA shows specific TFIIB–TBP and TFIIB–DNA contacts (Fig. 12-17). These include base-specific interactions with the major groove upstream (to the BRE; see Fig. 12-14) and the minor groove downstream from the TATA element. The asymmetric binding of TFIIB to the TBP-TATA complex accounts for the asymmetry in the rest of the assembly of the preinitiation complex and the unidirectional transcription that results. TFIIB also contacts Pol II in the preinitiation complex. Thus, this protein appears to bridge the TATA-bound TBP and polymerase. Structural studies suggest that segments of TFIIB insert into the RNA-exit channel and active-center cleft of Pol II in a manner analogous to σ region 3/4 linker in the bacterial case.

TFIIF This two-subunit (in humans) factor associates with Pol II and is recruited to the promoter together with that enzyme (and other factors). Binding of Pol II-TFIIF stabilizes the DNA–TBP–TFIIB complex and is required before TFIIE and TFIIH are recruited to the preinitiation complex (Fig. 12-15). In yeast, this factor includes a third subunit (as shown in Table 12-2), but the function of the third subunit is not known.

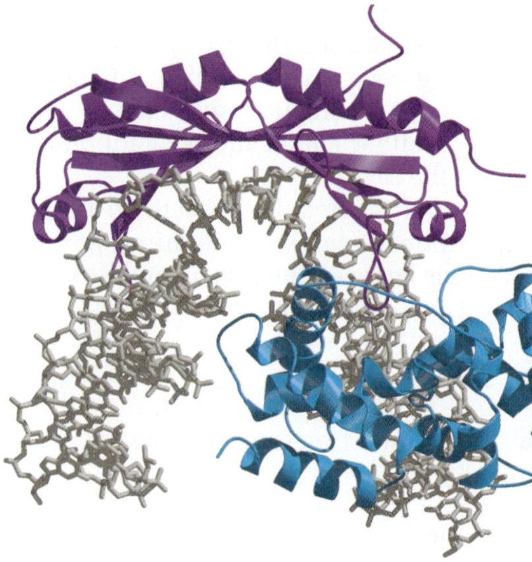

FIGURE 12-17 TFIIB–TBP–promoter complex. This structure shows the TBP protein bound to the TATA sequence, just as we have seen in the previous figure. Here, the general transcription factor TFIIB (shown in turquoise) has been added. This tripartite complex forms the platform to which other general transcription factors, and Pol II itself, are recruited during preinitiation complex assembly. (Nikolov D.B. et al. 1995. *Nature* 377: 119.) Image prepared with MolScript, BobScript, and Raster 3D. Extended DNA on either side of image modeled by Leemor Joshua-Tor.

TFIIE and TFIIH TFIIE, which, like TFIIF, consists of two subunits, binds next and has roles in the recruitment and regulation of TFIIH. TFIIH controls the ATP-dependent transition of the preinitiation complex to the open complex. It is also the largest and most complex of the general transcription factors; it has ten subunits and a molecular mass comparable to that of the polymerase itself! Within TFIIH are two subunits that function as ATPases and another that is a protein kinase, with roles in promoter melting and escape, as described above. Together with other factors, the ATPase subunits are also involved in nucleotide excision repair (see Chapter 9).

In Vivo, Transcription Initiation Requires Additional Proteins, Including the Mediator Complex

Thus far, we have described what is needed for Pol II to initiate transcription from a naked DNA template in vitro. But we have already noted that high, regulated levels of transcription in vivo require, additionally, transcriptional regulatory proteins, the Mediator complex, and nucleosome-modifying enzymes (which are themselves often parts of large protein complexes) (Fig. 12-18). For characteristics of various modifying complexes, see Table 7.8.

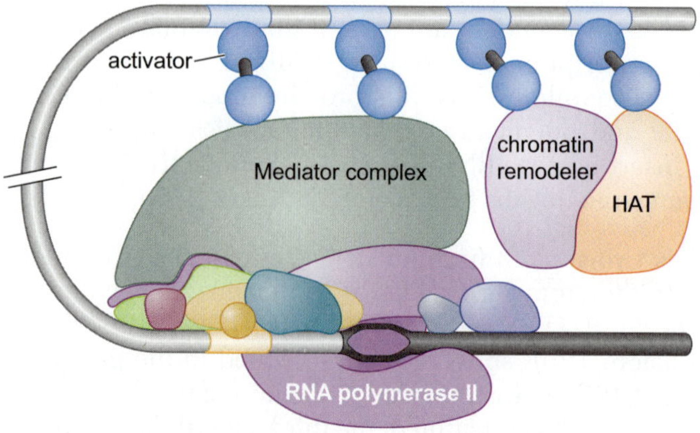

FIGURE 12-18 Assembly of the preinitiation complex in the presence of Mediator, nucleosome modifiers and remodelers, and transcriptional activators. In addition to the general transcription factors shown in Fig. 12-15, transcriptional activators bound to sites near the gene recruit nucleosome-modifying and -remodeling complexes and the Mediator complex, which together help form the preinitiation complex.

One reason for these additional requirements is that the DNA template in vivo is packaged into chromatin, as we discussed in Chapter 7. This condition complicates binding to the promoter of polymerase and its associated factors. Transcriptional regulatory proteins called activators help recruit polymerase to the promoter, stabilizing its binding there. This recruitment is mediated through interactions between DNA-bound activators, chromatin modifying and remodeling factors, and parts of the transcription machinery. One such interaction is with the Mediator complex (hence, its name). Mediator is associated with the CTD "tail" of the large polymerase subunit through one surface, while presenting other surfaces for interaction with DNA-bound activators. This explains the need for Mediator to achieve significant transcription in vivo.

Despite this central role in transcriptional activation, deletion of individual subunits of Mediator often leads to loss of expression of only a small subset of genes, different for each subunit (it is made up of many subunits). This result likely reflects the fact that different activators are believed to interact with different Mediator subunits to bring polymerase to different genes. In addition, Mediator aids initiation by regulating the CTD kinase in TFIIH.

The need for nucleosome modifiers and remodelers also differs at different promoters or even at the same promoter under different circumstances. When and where required, these complexes are also typically recruited by the DNA-bound activators.

We discuss the role of Mediator and modifiers in stimulating transcription in Chapter 17. We now consider some of the structural and functional properties of Mediator.

Mediator Consists of Many Subunits, Some Conserved from Yeast to Human

As shown in Figure 12-19, the yeast and human Mediator each include more than 20 subunits, of which 7 show significant sequence homology between the two organisms. (The names of the subunits were initially different in each case, reflecting the experimental approaches that led to their identification, but recently a convention has been established so that equivalent subunits in different organisms take the same name. It is these that are given in Fig. 12-19.) Very few of these subunits have any identified function. Only one (Srb4/Med17) is essential for transcription of essentially all Pol II genes in vivo. Low-resolution structural comparisons suggest that both Mediators have a similar shape, and both are very large, even bigger than RNA polymerase itself.

The Mediator from both yeast and humans is organized in modules, each a subset of the subunits shown (see Fig, 12-19). These modules can be dissociated from one another under certain conditions in vitro. This observation, together with the fact that human Mediator varies in its composition (and size) depending on how it is isolated, has led to the idea that there are various forms of Mediator (particularly in metazoans), each containing subsets of Mediator subunits. Furthermore, it has been argued that the different forms are involved in regulating different subsets of genes or responding to different groups of regulators (activators and repressors). It is equally possible, however, that the variations seen in subunit composition are artifacts, simply reflecting different methods of isolation.

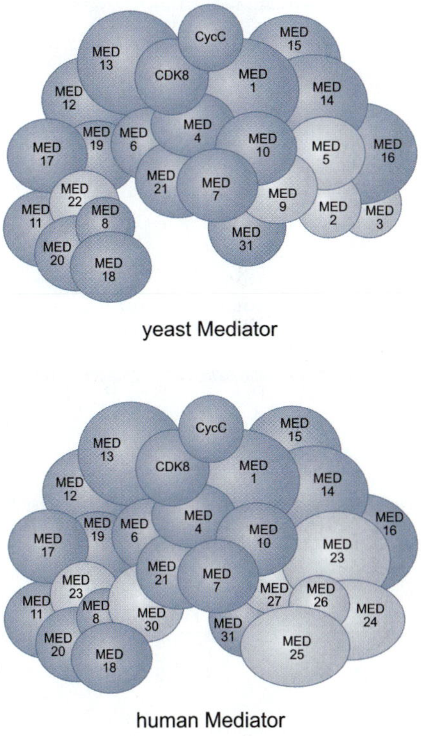

FIGURE 12-19 Comparison of the yeast and human Mediators. The majority of the subunits are found in both cases, but differences are indicated by paler shading. (Yeast Mediator: modified from Guglielmi B. et al. 2004. *Nucleic Acids Res.* 32: 5379–5391, Fig. 8B; human Mediator: modified, with permission, from Malik S. and Roeder R.G. 2005. *Trends Biochem. Sci.* 30: 256–263, Fig. 1a. © Elsevier.)

A New Set of Factors Stimulate Pol II Elongation and RNA Proofreading

Once polymerase has escaped the promoter and initiated transcription, it shifts into the elongation phase, as we have discussed. This transition involves the Pol II enzyme shedding most of its initiation factors—for example, the general transcription factors and Mediator. In their place, another set of factors is recruited. Some of these (such as TFIIS and hSPT5) are **elongation factors** (i.e., factors that stimulate elongation). Others are required for RNA processing. The enzymes involved in RNA processing (described in detail later) are, like several of the initiation factors we have discussed, recruited to the carboxy-terminal (CTD) tail of the large subunit of Pol II (Fig. 12-20). In this case, however, the factors favor the phosphorylated form of the CTD. Thus, phosphorylation of the CTD leads to an exchange of initiation factors for those factors required for elongation and RNA processing.

As is evident from the crystal structure of yeast Pol II, the polymerase CTD lies directly adjacent to the channel through which the

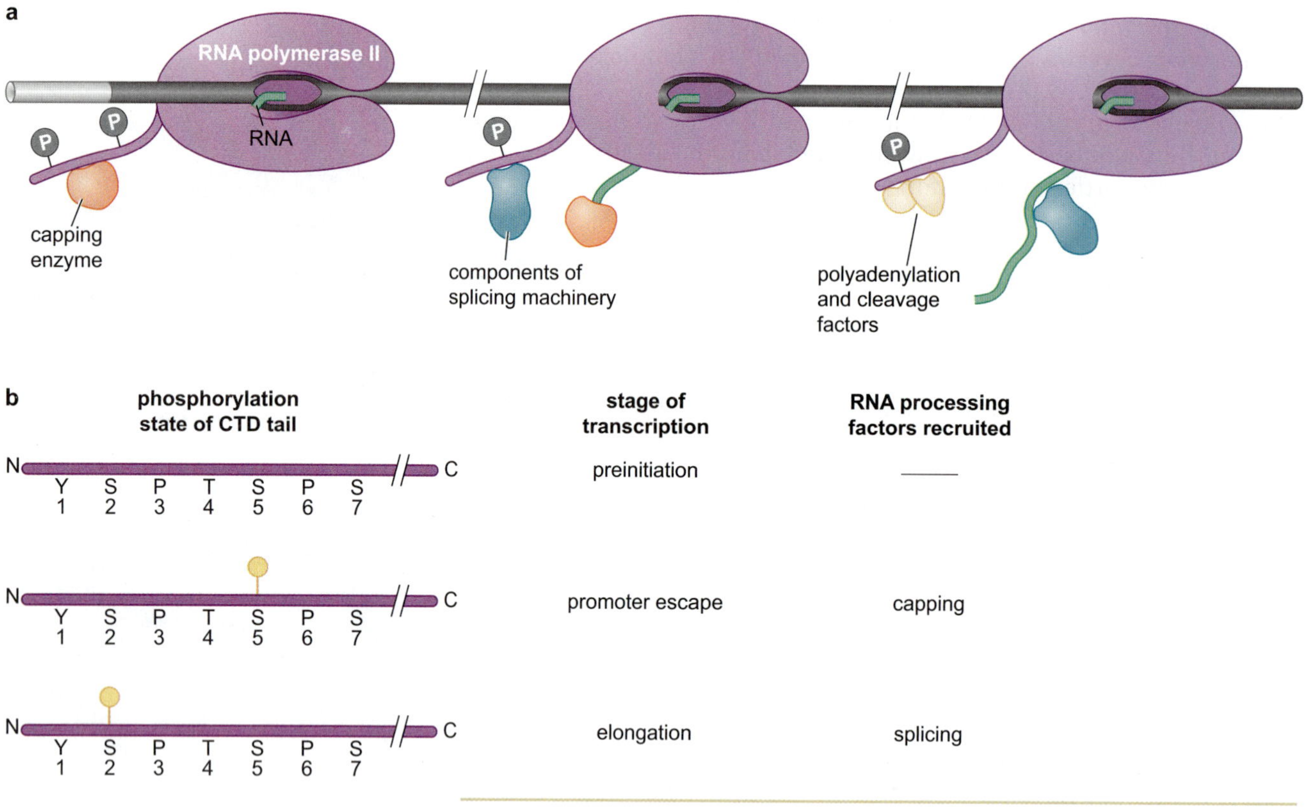

FIGURE 12-20 RNA processing enzymes are recruited by the CTD tail of polymerase. (a) Various factors involved in RNA processing recruited by the CTD tail of polymerase. Different factors are recruited depending on the phosphorylation state of the tail. Those factors are then transferred to the RNA as they are needed (see next section in text). (b) A schematic of the tail, with the sequence of one copy of the heptapeptide repeat shown in the top line. The positions of serine residues that get phosphorylated are indicated in lines 2 and 3. Phosphorylation of serine at position 5 is seen upon promoter escaper and is associated with recruitment of capping factors, whereas phosphorylation of serine at position 2 is seen during elongation and is associated with recruitment of splicing factors. Recruitment of factors involved in elongation of transcription and in RNA processing overlap. Thus, elongation factor hSPT5 is recruited to the tail phosphorylated on Ser5.

newly synthesized RNA exits the enzyme. It is also very long (it could potentially extend ~800 Å from the body of the enzyme—that is, about seven times the length of the rest of the enzyme). Together, these features allow the tail to bind several components of the elongation and processing machinery and deliver them to the emerging RNA.

Various proteins are thought to stimulate elongation by Pol II. One of these, the kinase P-TEFb, is recruited to polymerase by transcriptional activators. Once bound to Pol II, this protein phosphorylates the serine residue at position 2 of the CTD repeats as described earlier. That phosphorylation event correlates with elongation. In addition, P-TEFb phosphorylates and thereby activates another protein, called hSPT5, itself an elongation factor. Finally, TAT-SF1, yet another elongation factor, is recruited by P-TEFb. Thus, P-TEFb stimulates elongation in three separate ways.

Yet another class of elongation factor is the so-called ELL family. These also bind to elongating polymerase and suppress transient pausing by the enzyme; such pausing otherwise occurs at many sites along the DNA. The first human ELL protein was originally identified as the product of a gene that undergoes translocations in acute myeloid leukemia. Even the initiation factor TFIIF has an additional role in stimulating elongation.

Another factor that does not affect initiation, but stimulates elongation, is TFIIS. This factor, like ELL, stimulates the overall rate of elongation by limiting the length of time that polymerase pauses when it encounters sequences that would otherwise tend to slow the enzyme's progress. It is a feature of polymerase that it does not transcribe through all sequences at a constant rate. Rather, it pauses periodically, sometimes for rather long periods, before resuming transcription. In the presence of TFIIS, the length of time that polymerase pauses at any given site is reduced.

TFIIS has another function: it contributes to proofreading by polymerase. We saw at the start of the chapter how polymerases are able, inefficiently, to remove misincorporated bases using the active site of the enzyme to perform the reverse reaction to nucleotide incorporation. In addition, TFIIS stimulates an inherent RNase activity in polymerase (not part of the active site), allowing an alternative approach to removing misincorporated bases through local limited RNA degradation. This feature is comparable to the hydrolytic editing in the bacterial case stimulated by the Gre factors we discussed there. Figure 12-21 shows how TFIIS and GreB, although structurally unrelated (and unrelated in sequence, too), nevertheless interact with the yeast and bacterial polymerases, respectively, in comparable ways, to stimulate the same reactions.

Elongating RNA Polymerase Must Deal with Histones in Its Path

As with initiation of transcription, elongation also takes place in the presence of histones, because the DNA template is incorporated into nucleosomes. How does RNA polymerase transcribe through these potential barriers?

Experiments in vitro comparing transcription on naked DNA and on DNA incorporated in chromatin revealed that chromatin greatly impedes transcription. This experimental setup provided the assay for identifying factors that facilitate transcription in the presence of chromatin. In this way, a factor called FACT (facilitates chromatin transcription) was identified in human cell extracts. As its name suggests, this factor makes transcription on chromatin templates much more ef-

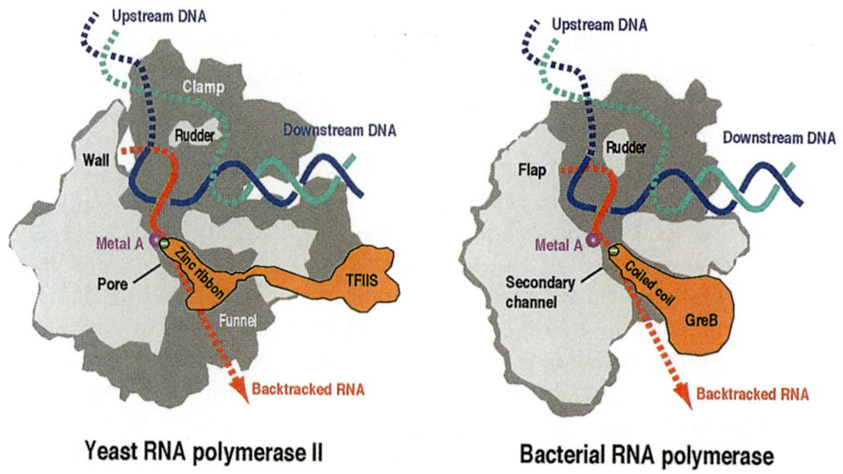

FIGURE 12-21 TFIIS and GreB act in analogous ways. The figure shows cutaway views of the major features of the complexes of arrested RNA polymerase II and TFIIS (left) and bacterial RNA polymerase and GreB (right). TFIIS (in orange) is inserted into the RNA polymerase II core, and GreB (also in orange) is inserted into the bacterial RNA polymerase channel. In each case, the primary catalytic magnesium ion is designated as Metal A (in pink) and the positions of the two conserved acidic residues are indicated by green circles. Thus we see that although the two proteins are so different, they act in essentially the same way. The dashed arrows indicate the presumed locations of the backtracked RNAs (see also Fig. 12-10). (Reprinted, with permission, from Conaway R.C. et al. 2003. *Cell* 114: 272–274, Fig. 1. © Elsevier.)

ficient. FACT is a heterodimer of two well-conserved proteins: Spt16 and SSRP1. The yeast homolog of the former had already been linked to chromatin modulation from genetic studies, and a role for FACT in elongation was established through genetic interactions between this complex and known elongation factors, including TFIIS.

How does FACT work? Recall from Chapter 7 that nucleosomes are octomers, made up of H2A, H2B, H3, and H4 histone subunits and DNA (see Fig. 7-20). These histones are arranged in two modules: the H2A/H2B dimers and the H3/H4 tetramer. Spt16 binds to the former; SSRP1 binds to the latter. Strikingly, FACT can both dismantle histones, by removing one H2A/H2B dimer, and reassemble them by restoring that dimer.

Thus evolved a picture of how FACT works during elongation (Fig. 12-22). Ahead of a transcribing RNA polymerase, FACT removes one H2A/H2B dimer. This allows polymerase to pass that nucleosome (in vitro, it has been shown that removing H2A/H2B from a template allows transcription). FACT also has histone chaperone activity; this allows it to restore the H2A/H2B dimer to the histone hexamer immediately behind the processing polymerase. In this way, FACT allows polymerase to elongate and at the same time maintains the integrity of the chromatin through which the enzyme is transcribing.

Elongating Polymerase Is Associated with a New Set of Protein Factors Required for Various Types of RNA Processing

Once transcribed, eukaryotic RNA has to be processed in various ways before being exported from the nucleus where it can be translated. These processing events include the following: capping of the 5′ end of the RNA, splicing, and polyadenylation of the 3′ end of the RNA. The most

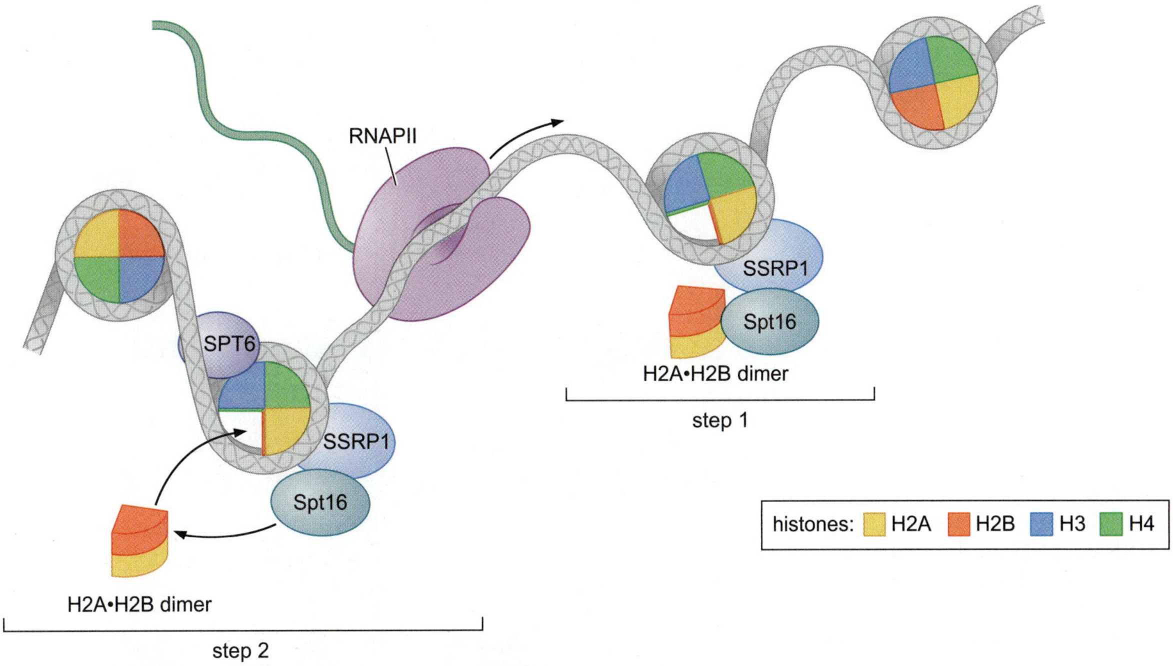

FIGURE 12-22 A model for FACT-aided elongation through nucleosomes. As described in the text, FACT, shown as the heterodimer of Spt 16 and SSRP1, is able to dismantle nucleosomes ahead of the transcribing RNA polymerase (Step 1) and reassemble them behind (Step 2). Specifically, it removes the H2A/H2B dimer. SPT6 binds histone H3 and is believed to aid in nucleosome reassembly. (Adapted, with permission, from Reinberg D. and Sims R. 2006. *J. Biol. Chem.* 281: 23297–23301, Fig. 2b. © American Society for Biochemistry and Molecular Biology.)

complicated of these is splicing—the process whereby noncoding introns are removed from RNA to generate the mature mRNA. The mechanisms and regulation of that process and others, such as RNA editing, are the subject of Chapter 13. We consider the other two processes—capping and polyadenylating the transcript—here.

Strikingly, there is an overlap in proteins involved in elongation and those required for RNA processing. In one case, for example, an elongation factor mentioned above (hSPT5) also helps to recruit the 5′-capping enzyme to the CTD tail of polymerase (phosphorylated at serine position 5 [Fig. 12-20b]). The hSPT5 stimulates the 5′-capping enzyme activity. In another case, elongation factor TAT-SF1 recruits components of the splicing machinery to polymerase with Ser2 phosphorylated tail (Fig. 12-20b). Thus, elongation, termination of transcription, and RNA processing are interconnected, presumably to ensure their proper coordination.

The first RNA processing event is capping. This involves the addition of a modified guanine base to the 5′ end of the RNA. Specifically, it is a methylated guanine, and it is joined to the RNA transcript by an unusual 5′–5′ linkage involving three phosphates (this structure is shown in the last step at the bottom of Fig. 12-23).

The 5′ cap is created in three enzymatic steps, as detailed in Figure 12-23 and described in detail in the legend. In the first step, a phosphate group is removed from the 5′ end of the transcript. Then, in the second step, the GMP moiety is added. In the final step, that nucleotide is modified by the addition of a methyl group. The RNA is capped as soon as it emerges from the RNA-exit channel of polymerase. This happens when the transcription cycle has progressed only to the transition between the

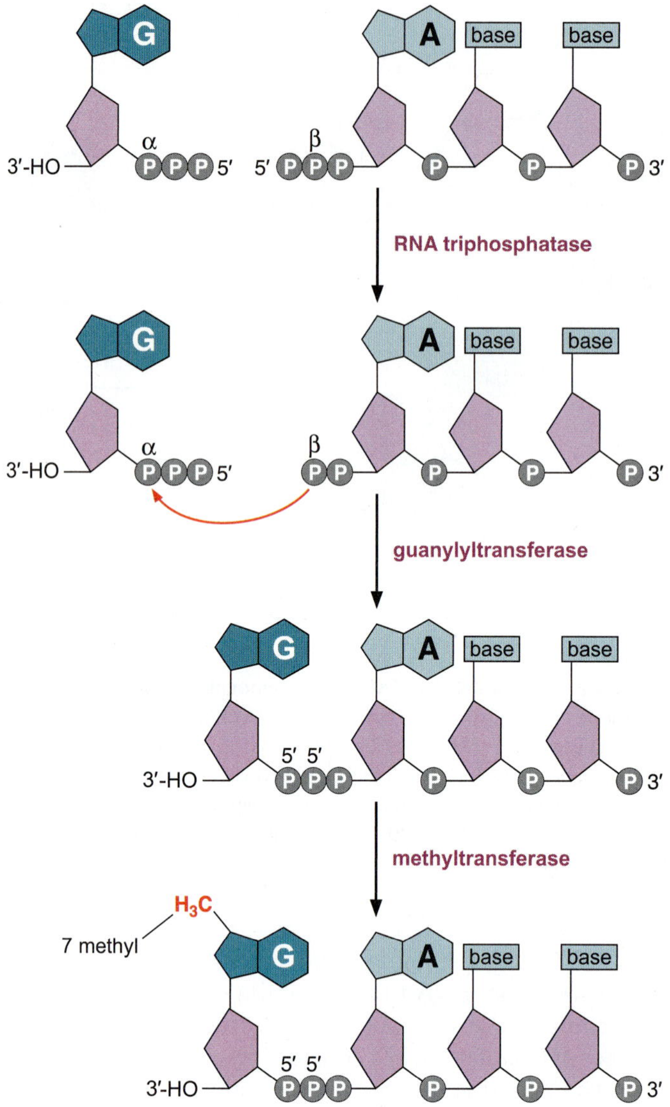

FIGURE 12-23 The structure and formation of the 5′ RNA cap. In the first step, the γ-phosphate at the 5′ end of the RNA is removed by an enzyme called RNA triphosphatase (the initiating nucleotide of a transcript initially retains its α-, β-, and γ-phosphates). In the next step, the enzyme guanylyltransferase adds a GMP moiety to the resulting terminal β-phosphatase. This is a two-step process: first, an enzyme–GMP complex is generated from GTP with release of the β- and γ-phosphates of that GTP, and then the GMP from the enzyme is transferred to the β-phosphate of the 5′ end of the RNA. Once this linkage is made, the newly added guanine and the purine at the original 5′ end of the mRNA are further modified by the addition of methyl groups by methyltransferase. The resulting 5′ cap structure later recruits the ribosome to the mRNA for translation to begin (see Chapter 14).

initiation and elongation phases. After capping, dephosphorylation of Ser-5 within the tail repeats may be responsible for dissociation of the capping machinery, and further phosphorylation (this time of Ser-2 within the tail repeats) causes recruitment of the machinery needed for RNA splicing (see Fig. 12-20b).

The final RNA processing event, polyadenylation of the 3′ end of the mRNA, is intimately linked with the termination of transcription (Fig. 12-24). Just as with capping and splicing, the polymerase CTD tail is involved in recruiting some of the enzymes necessary for polyadenylation (Fig. 12-20). Once polymerase has reached the end of a gene, it encounters specific sequences that, after being transcribed into RNA, trigger the transfer of the polyadenylation enzymes to that RNA, leading to four events: cleavage of the message; addition of many adenine residues to its 3′ end; degradation of the RNA remaining associated with RNA polymerase by a 5′ to 3′ ribonuclease; and, subsequently, termination of transcription. This series of events unfold as follows.

Two protein complexes are carried by the CTD of polymerase as it approaches the end of the gene: CPSF (cleavage and polyadenylation

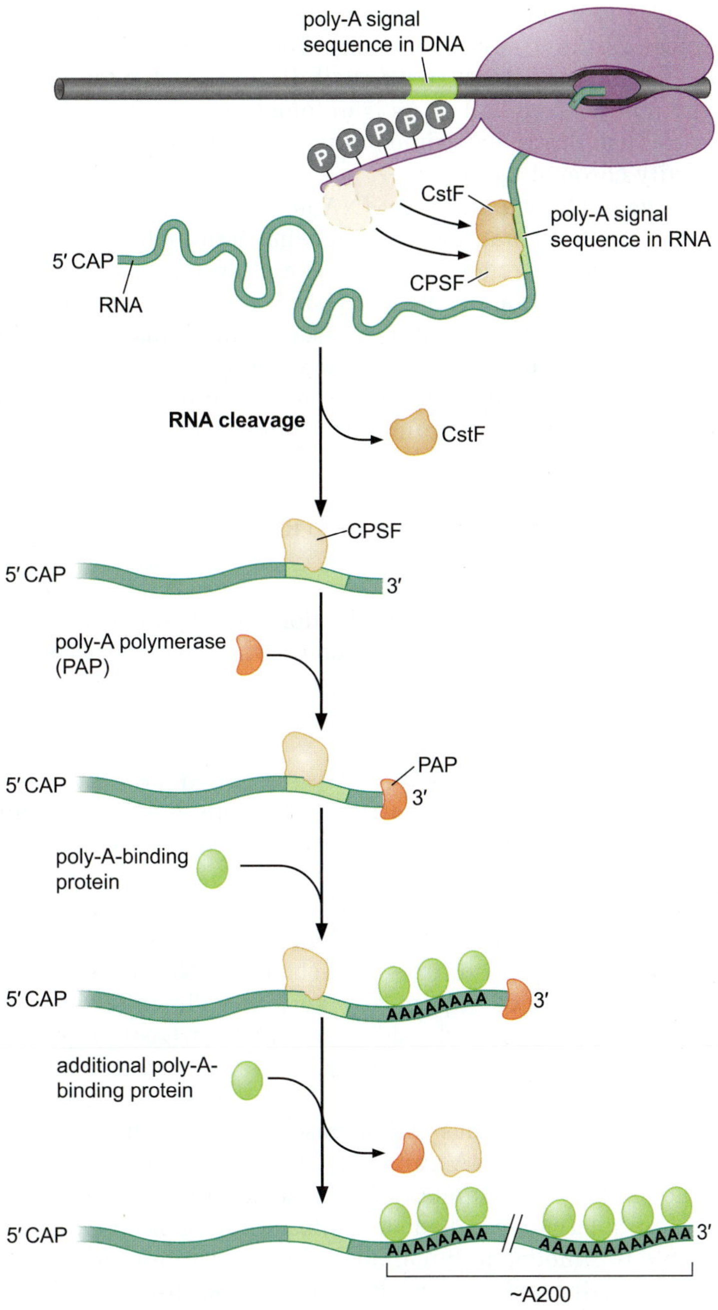

FIGURE 12-24 Polyadenylation and termination. The various steps in this process are described in the text.

specificity factor) and CstF (cleavage stimulation factor). The sequences that, once transcribed into RNA, trigger transfer of these factors to the RNA are called poly-A signals and their operation is shown in Figure 12-24. Once CPSF and CstF are bound to the RNA, other proteins are recruited as well, leading initially to RNA cleavage and then polyadenylation.

Polyadenylation is mediated by an enzyme called poly-A polymerase, which adds approximately 200 adenines to the RNA's 3′ end produced by the cleavage. This enzyme uses ATP as a precursor and adds the nucleotides using the same chemistry as RNA polymerase. But it does so without a template. Thus, the long tail of As is found in the RNA but not the DNA. It is not clear what determines the length of the poly-A tail, but this process involves other proteins that bind specifically to the poly-A sequence. The mature mRNA is then transported from the nucleus, as we shall discuss in Chapter 13. It is noteworthy that the long tail of As is unique to transcripts made by Pol II, a feature that allows experimental isolation of protein-coding mRNAs by affinity chromatography.

We thus see how a mature mRNA is released from polymerase once the gene has been transcribed. But what terminates transcription by polymerase? In fact, the enzyme does not terminate immediately after the RNA is cleaved and polyadenylated. Rather, it continues to move along the template, generating a second RNA molecule. The polymerase can continue transcribing for several thousand nucleotides before terminating and dissociating from the template. We now describe current models for how termination might happen.

Transcription Termination Is Linked to RNA Destruction by a Highly Processive RNase

Polyadenylation is linked to termination, although exactly how is still not quite clear. Recently, however, an enzyme that degrades the second RNA as it emerges from the polymerase has been identified, and this enzyme may itself trigger termination. This is called the "torpedo model" of termination (Fig. 12-25a).

The free end of the second RNA is uncapped, and thus can be distinguished from genuine transcripts. This new RNA is recognized by an RNase called, in yeast, Rat1 (in humans Xrn2) that is loaded onto the end of the RNA by another protein (Rtt103) that binds the CTD of RNA polymerase. The Rat1 enzyme is very processive and quickly degrades the RNA in a 5′ to 3′ direction, until it catches up to the still-transcribing polymerase from which the RNA is being spewed. Termination may not require any very specific interaction between Rat1 and polymerase and might in fact be triggered in a manner rather similar to that described earlier in the chapter for Rho-dependent termination in bacteria—that is, the highly processing RNase polymerase either pushes polymerase forward and/or pulls the remains of the nascent RNA transcript from the enzyme.

Although the torpedo model for termination is now the favored one, there is an alternative called the "allosteric model" (Fig. 12-25b). According to this model, termination depends on a conformational change in the elongating polymerase that reduces the processivity of the enzyme leading to spontaneous termination soon afterward. This conformational change would be linked to polyadenylation and could, for example, be triggered by the transfer of the 3′ processing enzymes from the CTD tail of polymerase to the RNA or by the subsequent binding to the CTD tail of other factors that induce a conformational change.

TRANSCRIPTION BY RNA POLYMERASES I AND III

RNA Pol I and Pol III Recognize Distinct Promoters, Using Distinct Sets of Transcription Factors, but Still Require TBP

We have already mentioned that eukaryotes have two other polymerases—Pol I and Pol III—in addition to Pol II. These enzymes are related to Pol II and even share several subunits (Table 12-2), but

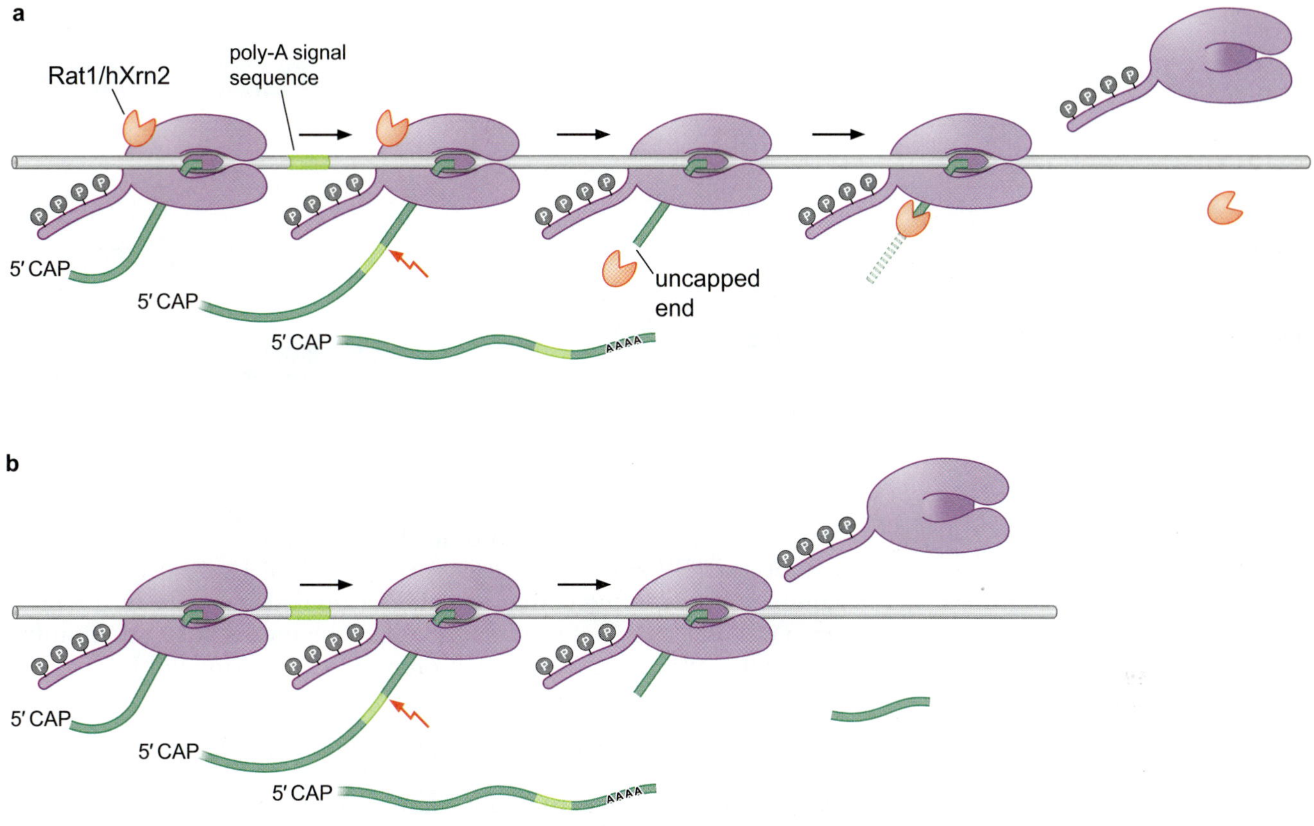

FIGURE 12-25 Models of termination: Torpedo and allosteric. As described in the text, there are two proposed models for how transcription by eukaryotic RNA Pol II terminates after transcribing a gene. In the figure, the poly-A site is marked by the light green stretch in the DNA and is located just downstream from the gene. It is also light green in the transcript. The dotted green line indicates degraded transcript. (a) In the torpedo model, RNA transcribed downstream from the poly-A site is attacked by the 5′ to 3′ RNase (the torpedo), which is loaded onto this transcript from polymerase itself. When this exonuclease catches up with polymerase, it triggers dissociation from the DNA template and termination of transcription. (b) In the allosteric model, the polymerase is highly processive within the gene, and then, once the poly-A signal is passed, becomes less processive. This alteration could be due to a modification or a conformational change. Even in the allosteric model, the second RNA would be degraded by the RNase, but that would not be the cause of termination. In this case, RNA degradation is not shown in the figure to emphasize the different mechanisms of termination in these two models. (Adapted, with permission, from Luo W. and Bentley D. 2004. *Cell* 119: 911–914, Fig. 1. © Elsevier.)

they initiate transcription from distinct promoters and transcribe distinct genes. These genes encode specialized RNAs rather than proteins, as we discussed earlier in the chapter. Each of these enzymes also works with its own unique set of general transcription factors. TBP, however, is universal—it is involved in initiating transcription by Pol I and Pol III, as well as Pol II.

Pol I is required for the expression of only one gene, that encoding the rRNA precursor. There are many copies of that gene in each cell, and, indeed, it is expressed at far higher levels than any other gene, perhaps explaining why it has its own dedicated polymerase.

The promoter for the rRNA gene comprises two parts: the core element and the UCE (upstream control element) as shown in Figure 12-26. The former is located around the start site of transcription, and the

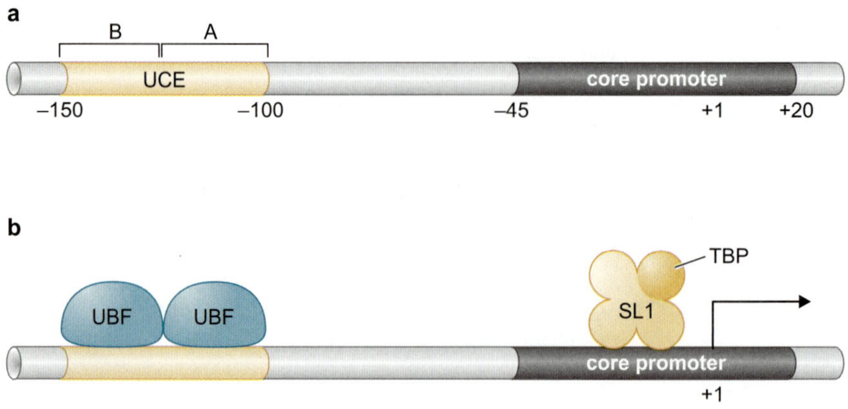

FIGURE 12-26 Pol I promoter region. (a) Structure of the Pol I promoter. (b) Pol I transcription factors. The case shown here is for humans. The set of proteins involved in helping Pol I transcription in yeast is rather different.

latter between 100 and 150 bp upstream (in humans). In addition to Pol I, initiation requires two other factors, called SL1 and UBF. SL1 comprises TBP and three TAFs specific for Pol I transcription. This complex binds to the core element. SL1 binds DNA only in the presence of UBF. This factor binds to UCE, bringing in SL1 and stimulating transcription from the core promoter by recruiting Pol I.

Pol III Promoters Are Found Downstream of Transcription Start Site

Pol III promoters come in various forms, and the vast majority have the unusual feature of being located *downstream* from the transcription start site. Some Pol III promoters (e.g., those for the tRNA genes) consist of two regions, called Box A and Box B, separated by a short element (Fig. 12-27); others contain Box A and Box C (e.g., the 5*S* rRNA gene); and still others contain a TATA element like those of Pol II.

Just as with Pol II and Pol I, transcription by Pol III requires transcription factors in addition to polymerase. In this case, the factors are called TFIIIB and TFIIIC for the tRNA genes and those plus TFIIIA for the 5*S* rRNA gene.

Figure 12-27 shows the tRNA promoter. Here, the TFIIIC complex binds to the promoter region. This complex recruits TFIIIB to the DNA just upstream of the start site, where it in turn recruits Pol III to the start site of transcription. The enzyme then initiates, presumably displacing TFIIIC from the DNA template as it goes. As with the other two classes of polymerase, Pol III uses TBP. In this case, that ubiquitous factor is found within the TFIIIB complex.

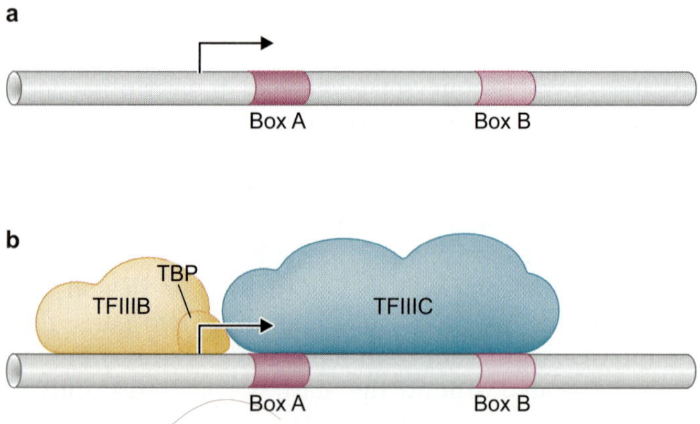

FIGURE 12-27 Pol III core promoter. Shown here is the promoter for a yeast tRNA gene. The order of events leading to transcription initiation is described in the text.

SUMMARY

Gene expression is the process by which the information in the DNA double helix is converted into the RNAs and proteins whose activities bestow upon a cell its morphology and functions. Transcription is the first step in gene expression and involves copying DNA into RNA. This process, catalyzed by the enzyme RNA polymerase, is in many ways similar to the process of DNA replication discussed in Chapter 8. In both cases, a new chain of nucleotides is synthesized upon a DNA template, and both DNA and RNA syntheses proceed in a 5′ to 3′ direction (i.e., the enzyme adds each successive nucleotide to the 3′ end of the growing chain). But there are several critical differences between these two processes, some mechanistic and others reflecting the different roles they serve.

For example, in DNA replication, the entire genome is duplicated once and only once each cell division. In transcription, only some regions of the genome are transcribed, and the regions chosen vary in different cells or in the same cell at different times. Different regions can be transcribed to different extents; that is, anywhere from one to several thousand transcripts can be made of a given region in a single cell.

Mechanistic differences between transcription and replication include the following: the nucleotides used to build a new DNA chain are deoxyribonucleotides, whereas in transcription they are ribonucleotides. In addition, whereas DNA polymerase can only elongate existing polynucleotide chains, and thus requires a primer, RNA polymerase can initiate RNA synthesis de novo.

RNA polymerases from bacteria to humans are highly conserved. Eukaryotes have three different polymerases each; bacteria have just one. The three eukaryotic enzymes are called RNA Pol I, II, and III. Of these, in this chapter, we focused primarily on Pol II, as this is the enzyme that transcribes the vast majority of genes in the cell and all of the protein-coding genes.

The basic enzyme from *E. coli*, called the core enzyme, has one copy of each of three subunits—β, β', and ω—and two copies of α. All of these subunits have homologs in the eukaryotic enzymes. The structures of the bacterial and yeast Pol II enzyme are also similar. Both resemble a crab claw in shape, the pincers being made up of the largest subunits, β and β' in the case of the bacterial enzyme. The active site is at the base of the pincers, and access to and from the active site is afforded through five channels: one channel allows double-stranded DNA to enter between the pincers at the front of the enzyme; two other channels allow the two single strands—the template and nontemplate strands—to leave the enzyme behind the active site; another channel provides the route by which NTPs enter the active site; and the RNA product, which peels off the DNA template a short distance behind the site of polymerization, exits the enzyme through the fifth channel.

Pol II differs from the bacterial enzyme in one important way. The former has a so-called "tail" at the carboxy-terminal end of the large subunit, and this is absent from the bacterial enzyme. This tail is made up of multiple repeats of a heptapeptide sequence.

A round of transcription proceeds through three phases called initiation, elongation, and termination. Although RNA polymerases can synthesize RNA unaided, other proteins—called initiation factors—are required for accurate and efficient initiation. These factors ensure that the enzyme initiates transcription only from appropriate sites on the DNA, called promoters. In bacteria, there is only one initiation factor, σ, whereas in eukaryotes there are several, collectively called the general transcription factors. In eukaryotes, the DNA is wrapped within nucleosomes, and, in vivo, efficient initiation requires additional proteins, including the Mediator complex and nucleosome modifying enzymes. Transcriptional activator proteins are also needed (see Chapter 17).

During initiation, RNA polymerase (together with the initiation factors) binds to the promoter in a closed complex. In that state, the DNA remains in a double-stranded form. This closed complex then undergoes isomerization to the open complex. In that form, the DNA around the transcription start site is unwound, disrupting the base pairs, and forming a bubble of single-stranded DNA. This transition allows access to the template strand, which determines the order of bases in the new RNA strand. This phase of initiation is followed by promoter escape: once the enzyme has synthesized a series of short RNAs, called abortive initiation, it manages to make a transcript that grows beyond 10 bp. At this point, the enzyme leaves the promoter and enters the elongation phase. During this phase, polymerase moves along the gene while the enzyme performs several functions: it opens the DNA downstream and reseals it upstream (behind) the active site; it adds ribonucleotides to the 3′ end of the growing transcript; it peels the newly formed RNA off the template some 8 or 9 bp behind the point of polymerization; and it also proofreads the transcript checking for (and replacing) incorrectly inserted nucleotides.

Transcription in both bacteria and eukaryotes follows these same steps. There are differences in the two cases, however. For example, in bacteria, isomerization to the open complex occurs spontaneously and does not require ATP hydrolysis. In eukaryotes, this step does require ATP hydrolysis. More strikingly, in eukaryotes, promoter escape is regulated by the phosphorylation state of the CTD tail. Thus, the form of Pol II that binds the promoter in the preinitiation complex has an unphosphorylated CTD. This domain becomes phosphorylated by one or more kinases, including the kinase that is part of one of the general transcription factors, TFIIH.

Once phosphorylated, the CTD tail of the Pol II frees itself from the other proteins at the promoter, releasing polymerase into the elongation phase. The CTD then binds factors involved in transcriptional elongation and RNA processing. Thus, there is an exchange of initiation for elongation and processing factors as the polymerase moves away from the promoter and starts transcribing the gene. There are also interactions between the elongation factors and those involved in processing, ensuring proper coordination of these events. Another difference between bacteria and eukaryotes is that the latter must deal with nucleosomes during elongation. This requires yet another complex that can dismantle nucleosomes ahead of, and reassemble them behind, the advancing polymerase.

Termination also works differently in bacteria and eukaryotes. Thus, in bacteria, there are two kinds of terminators: intrinsic (Rho-independent) and Rho-dependent. Intrinsic terminators consist of two sequence elements that operate once transcribed into RNA. One element is an inverted repeat that forms a stem loop in the RNA, disrupting the elongating polymerase. In combination with a string of U nucleotides (which bond only weakly with the template strand), this leads to release of the transcript. Rho-dependent terminators require the ATPase Rho, a protein that hops on elongating transcripts and translocates along them until they reach polymerase, triggering termination. In eukaryotes, termination is closely linked to an RNA processing event called 5′ polyadenylation. But in these organisms too, termination is believed to involve another protein—in this case, an RNase enzyme—traveling along a nascent transcript until it collides with polymerase, triggering termination.

In this chapter, we considered capping of the 5′ end of the RNA transcripts, polyadenylation of the 3′ end, and the link between the last of these and transcriptional termination. Splicing is described in the next chapter.

BIBLIOGRAPHY

Books

Cold Spring Harbor Symposia on Quantitative Biology. 1998. Volume 63: Mechanisms of transcription. Cold Spring Harbor Laboratory Press, Cold Spring Harbor, New York.

Ptashne M. and Gann A. 2002. *Genes and signals.* Cold Spring Harbor Laboratory Press, Cold Spring Harbor, New York.

White R.J. 2001. *Gene transcription: Mechanisms and control.* Blackwell Science, Malden, Connecticut.

RNA Polymerase

Cramer P. 2004. RNA polymerase II structure: From core to functional complexes. *Curr. Opin. Genet. Dev.* **14:** 218–226.

Darst S. 2004. New inhibitors targeting bacterial RNA polymerase. *Trends Biochem. Sci.* **29:** 159–162.

Ebright R.H. 2000. RNA polymerase: Structural similarities between bacterial RNA polymerase and eukaryotic RNA polymerase II. *J. Mol. Biol.* **304:** 687–698.

Murakami K.S. and Darst S.A. 2003. Bacterial RNA polymerases: The whole story. *Curr. Opin. Struct. Biol.* **13:** 31–39.

Paget M.S. and Helmann J.D. 2003. The σ70 family of sigma factors. *Genome Biol.* **4:** 203.

Promoters

Butler J.E. and Kadonaga J.T. 2002. The RNA polymerase II core promoter: A key component in the regulation of gene expression. *Genes Dev.* **16:** 2583–2592.

Juven-Gershon T., Hsu J.-Y., and Kadonaga J.T. 2006. Perspectives on the RNA polymerase II core promoter. *Biochem. Soc. Trans.* **34:** 1047–1050.

Transcription Initiation

Boeger H., Bushnell D.A., Davis R., Griesenbeck J., Lorch Y., Strattan J.S., Westover K.D., and Kornberg R.D. 2005. Structural basis of eukaryotic gene transcription. *FEBS Lett.* **579:** 899–903.

Conaway R.C., Sato S., Tomomori-Sato C., Yao T., and Conaway J.W. 2005. The mammalian Mediator complex and its role in transcriptional regulation. *Trends Biochem. Sci.* **30:** 250–255.

Malik S. and Roeder R.G. 2005. Dynamic regulation of pol II transcription by the mammalian Mediator complex. *Trends Biochem. Sci.* **30:** 256–263.

Myers L.C. and Kornberg R.D. 2000. Mediator of transcriptional regulation. *Annu. Rev. Biochem.* **69:** 729–749.

Roberts J.W. 2006. RNA polymerase, a scrunching machine. *Science* **314:** 1097–1098.

Woychik N.A. and Hampsey M. 2002. The RNA polymerase II machinery: Structure illuminates function. *Cell* **108:** 453–463.

Young B.A., Gruber T.M., and Gross C.A. 2002. Views of transcription initiation. *Cell* **109:** 417–420.

Elongation and RNA Processing

Armache K.-J., Kettenberger H., and Cramer P. 2005. The dynamic machinery of mRNA elongation. *Curr. Opin. Struct. Biol.* **15:** 197–203.

Bentley D. 2005. Rules of engagement: Co-transcriptional recruitment of pre-mRNA processing factors. *Curr. Opin. Cell. Biol.* **17:** 251–256

Maniatis T. and Reed R. 2002. An extensive network of coupling among gene expression machines. *Nature* **416:** 499–506.

Reinberg D. and Smith R.J., III 2006. de FACTo nucleosome dynamics. *J. Biol. Chem.* **281:** 23297–23301.

Saunders A., Core J.C., and Lis J.T. 2006. Breaking barriers to transcription elongation. *Nat. Rev. Mol. Cell. Biol.* **7:** 557.

Termination

Luo W. and Bartley D. 2004. A ribonucleolytic rat torpedoes RNA polymerase II. *Cell* **119:** 911–914.

Richardson J.P. 2003. Loading Rho to terminate transcription. *Cell* **114:** 157–159.

———. 2006. How Rho exerts its muscle as RNA. *Mol. Cell* **23:** 711–712.

Rosonina E., Kaneko S., and Manley J.L. 2006. Terminating the transcript: Breaking up is hard to do. *Genes Dev.* **20:** 1050–1056.

RNA Polymerases I and III

Grummt I. 2003. Life on a planet of its own: Regulation of RNA polymerase I. *Genes Dev.* **17:** 1691–1702.

Schramm L. and Hernandez N. 2002. Recruitment of RNA polymerase III to its target promoters. *Genes Dev.* **16:** 2593–2620.

White R.J. 2005. RNA polymerase I and III, growth control and cancer. *Nat. Rev. Mol. Cell. Biol.* **6:** 69–78.

CHAPTER

13 RNA Splicing

THE CODING SEQUENCE OF A GENE IS A SERIES of three-nucleotide codons that specify the linear sequence of amino acids in its polypeptide product. Thus far, we have tacitly assumed that the coding sequence is contiguous: the codon for one amino acid is immediately adjacent to the codon for the next amino acid in the polypeptide chain. This is true in the vast majority of cases in bacteria and their phage. But it is rarely so for eukaryotic genes. In those cases, the coding sequence is periodically interrupted by stretches of noncoding sequence.

Many eukaryotic genes are thus mosaics, consisting of blocks of coding sequences separated from each other by blocks of noncoding sequences. The coding sequences are called **exons** and the intervening sequences are called **introns.** As a consequence of this alternating pattern of exons and introns, genes bearing noncoding interruptions are often said to be "in pieces" or "split." Technically, the term exon applies to any region retained in a mature RNA, whether or not it is coding. Noncoding exons include the 5′ and 3′ untranslated regions of an mRNA (Chapter 12); all portions of spliced, stable noncoding RNAs such as the X chromosome inactivation regulator *Xist* (Chapter 18); and regions that give rise to functional RNAs such as the micro-RNAs we shall also encounter in Chapter 18.

Figure 13-1 shows a typical eukaryotic gene in which the coding region is interrupted by three introns, splitting it into four exons. The number of introns found within a gene varies enormously—from one in the case of most intron-containing yeast genes (and a few human

OUTLINE

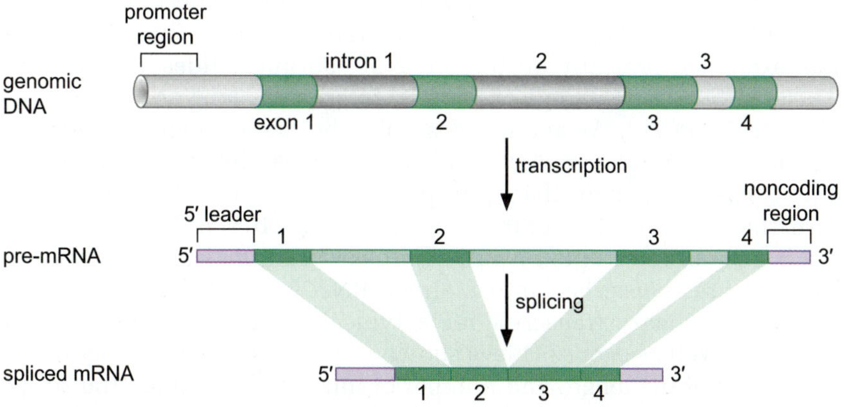

FIGURE 13-1 **A typical eukaryotic gene.** The depicted gene contains four coding exons separated by three introns. Transcription from the promoter generates a pre-mRNA, shown in the middle line, that contains all the exons and introns. Splicing removes the introns and fuses the exons to generate the mature mRNA that, once processed further (see polyadenylation, Chapter 12) and exported from the nucleus, can be translated to give a protein product. Technically, the 5′ leader and 3′ noncoding regions are also exons because they are retained in the mature mRNA. They are shown here in purple to indicate their status as noncoding exons.

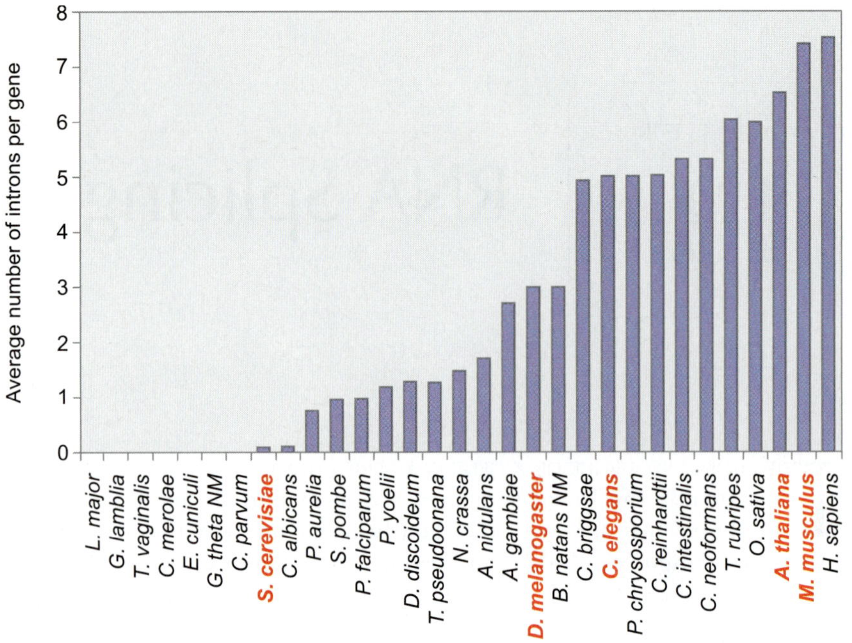

FIGURE 13-2 Number of introns per gene in various eukaryotic species. The average number of introns per gene is shown for a selection of eukaryotic species. The names in red are those of the common model organisms (Chapter 22): the yeast (*Saccharomyces cerevisiae*), the fruit fly (*Drosophila melanogaster*), the roundworm (*Caenorhabditis elegans*), the plant (*Arabidopsis thaliana*), and the mouse (*Mus musculus*). The other species shown are *Anopheles gambiae*; *Aspergillus nidulans*; *Bigelowiella natans* Nucleomorph; *Caenorhabditis briggsae*; *Candida albicans*; *Chlamydomonas reinhardtii*; *Ciona intestinalis*; *Cryptococcus neoformans*; *Cryptosporidium parvum*; *Cyanidioschyzon merolae*; *Dictyostelium discoideum*; *Encephalitozoon cuniculi*; *Giardia lamblia*, *Guillardia theta* Nucleomorph; *Homo sapiens*; *Leishmania major*; *Neurospora crassa*; *Oryza sativa*; *Paramecium aurelia*; *Phanerochaete chrysosporium*; *Plasmodium falciparum*; *Plasmodium yoelii*; *Schizosaccharomyces pombe*; *Takifugu rubripes*; *Thalassiosira pseudonana*; and *Trichomonas vaginalis*. (Redrawn, with permission, from Roy S.W. and Gilbert W. 2006. *Nat. Rev. Genet.* 7: 212, Fig. 1. © Macmillan.)

genes), to 50 in the case of the chicken *proα2* collagen gene, to as many as 363 in the case of the *Titin* gene of humans. Figure 13.2 shows the average number of introns per gene for a range of organisms sequenced to date. Clearly, the average number increases as one looks from simple single-celled eukaryotes, such as yeast, through higher organisms such as worms and flies, all the way up to humans.

The sizes of the exons and introns vary as well. Indeed, introns are very often much longer than the exons they separate. Thus, for example, exons are typically on the order of 150 nucleotides, whereas introns—although they too can be short—can be as long as 800,000 nucleotides (800 kb). As another example, the mammalian gene for the enzyme dihydrofolate reductase is more than 31 kb long, and within it are dispersed six exons that correspond to 2 kb of mRNA. Thus, in this case, the coding portion of the gene is less than 10% of its total length.

Like the uninterrupted genes of prokaryotes, the split genes of eukaryotes are transcribed into a single RNA copy of the entire gene. Thus, the primary transcript for a typical eukaryotic gene contains introns as well as exons. This is shown in the middle part of Figure 13-1. Because of the length and number of introns, the primary transcript (or **pre-mRNA**) can be very long indeed. In the extreme case of the human *dystrophin* gene, RNA polymerase must traverse 2400 kb of DNA to copy the entire gene into RNA. (Given that transcription proceeds at a rate of 40 nucleotides per second, it can readily be seen that it takes a staggering 17 hours to make a single transcript of this gene!)

Despite this seemingly odd gene organization, the protein-synthesizing machinery of the cell (Chapter 14) is equipped only to translate mRNAs containing a contiguous stretch of codons; it has no way of identifying and skipping over a block of noncoding sequence. Thus, the primary transcripts of split genes must have their introns removed before they can be translated into proteins.

Introns are removed from the pre-mRNA by a process called **RNA splicing.** This process converts the pre-mRNA into mature mRNA and must occur with great precision to avoid the loss, or addition, of even a single nucleotide at the sites at which the exons are joined. As we shall see in Chapters 14 and 15, the triplet-nucleotide codons of mRNA are translated in a fixed reading frame that is set by the first codon in the protein-coding sequence. Lack of precision in splicing—if, for example, a base were lost or gained at the boundary between two exons—would throw the reading frames of exons out of register; downstream codons would be incorrectly selected and the wrong amino acids incorporated into proteins.

Some pre-mRNAs can be spliced in more than one way, generating alternative mRNAs. So, for example, mRNAs containing different selections of exons can be generated from a given pre-mRNA. This is called **alternative splicing,** and, by this strategy, a gene can give rise to more than one polypeptide product. These alternative products are called **isoforms**. It is estimated that up to 75% of the genes in the human genome are spliced in alternative ways to generate more than one isoform.

The number of different variants a given gene can encode in this way varies from two to hundreds or even thousands. For example, the *Slo* gene from rat, which encodes a potassium channel expressed in neurons, has the potential to encode 500 alternative versions of that product. And, as we shall see, one particular *Drosophila* gene can encode as many as 38,000 possible products as a result of alternative splicing.

In this chapter, we discuss not only the mechanisms and regulation of RNA splicing, but also ideas about why eukaryotic genes have interrupted coding regions. We also describe RNA editing, another way initial transcripts can be altered to change what they encode.

Splicing was discovered in studies of gene expression in the mammalian adenovirus, as described in Box 13-1, Adenovirus and the Discovery of Splicing.

THE CHEMISTRY OF RNA SPLICING

Sequences within the RNA Determine Where Splicing Occurs

WEB ANIMATION

We now consider the molecular mechanisms of the splicing reaction (see Interactive Animation 13-1). How are the introns and exons distinguished from each other? How are introns removed? How are exons joined with high precision? The borders between introns and exons are marked by specific nucleotide sequences within the pre-mRNAs. These sequences delineate where splicing will occur. Thus, as shown in Figure 13-3, the exon–intron boundary—that is, the boundary at the 5′ end of the intron—is marked by a sequence called the **5′ splice site.** The intron–exon boundary at the 3′ end of the intron is marked by the **3′ splice site.** (The 5′ and 3′

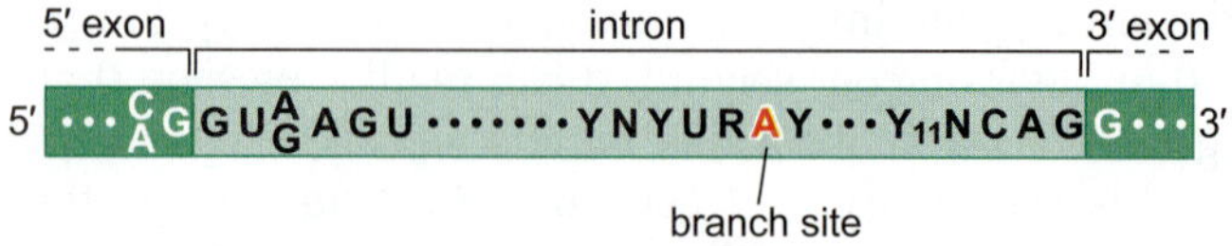

FIGURE 13-3 Sequences at intron–exon boundaries. Shown in the figure are the consensus sequences for both the 5′ and 3′ splice sites, and also the conserved A at the branch site. As in other cases of consensus sequences, where two alternative bases are similarly favored, those bases are both indicated at that position. In this figure, the consensus sequences shown are for humans. This is true for all other figures in this chapter, unless otherwise stated.

splice sites were sometimes referred to as the **donor** and **acceptor** sites, respectively, but this nomenclature is rarely used today.)

The figure shows a third sequence necessary for splicing. This is called the **branch point site** (or branch point sequence). It is found entirely within the intron, usually close to its 3′ end, and is followed by a polypyrimidine tract (Py tract).

The consensus sequence for each of these elements is shown in Figure 13-3. The most highly conserved sequences are the GU in the 5′ splice site, the AG in the 3′ splice site, and the A at the branch site. These highly conserved nucleotides are all found within the intron itself—perhaps not surprisingly, as the sequence of most exons, in contrast to the introns, is constrained by the need to encode the specific amino acids of the protein product.

The Intron Is Removed in a Form Called a Lariat as the Flanking Exons Are Joined

Let us begin by considering the chemistry of splicing. An intron is removed through two successive **transesterification** reactions in which phosphodiester linkages within the pre-mRNA are broken and new ones are formed (Fig. 13-4). The first reaction is triggered by the 2′-OH of the conserved A at the branch site. This group acts as a nucleophile to attack the phosphoryl group of the conserved G in the

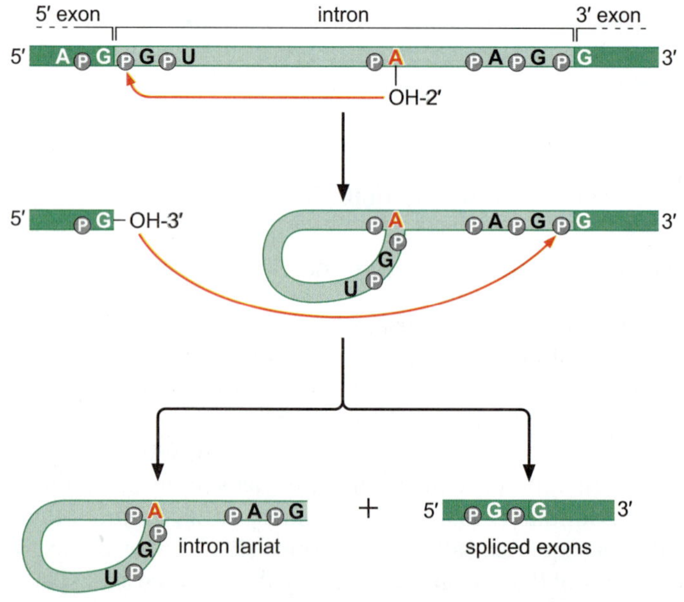

FIGURE 13-4 The splicing reaction. Shown are the two steps of the splicing reaction described in the text. In the first step, the RNA forms a loop structure, which is shown in detail in the next figure.

KEY EXPERIMENTS

Box 13-1 Adenovirus and the Discovery of Splicing

Studies with bacteria and their phage led to the view that the mRNA is an exact replica in terms of nucleotide sequence of the gene from which it is transcribed (see Chapter 15). It therefore came as a shock when, in 1977, it was discovered that certain (and, as we now know, most) eukaryotic mRNAs are spliced together in patchwork fashion from much longer primary transcripts. How was this startling discovery made?

In an effort to understand gene transcription in eukaryotes, scientists focused on the human DNA virus called adenovirus. This virus was intended to serve as a model for understanding the molecular biology of the eukaryotic gene just as phage T4 and λ had done for the prokaryotic gene (see Chapter 22). The virion of adenovirus is composed of several different virus-encoded proteins, and the mRNAs for these proteins were purified with the hope that their 5′ termini would pinpoint the transcription initiation sites for each gene on the viral genome. Instead, all of the mRNAs, even though they encoded different proteins, were found to have identical 5′ sequences. We now know that all of the mRNAs for the virion proteins of adenovirus arise from a single promoter known as the major late promoter. Initiation from this promoter generates long transcripts that span the coding sequences for multiple proteins (Box 13-1 Fig. 1). This transcript then undergoes alternative splicing to generate separate mRNAs for individual virion components such as the hexon and fiber proteins. All of the mRNAs share the same 5′ sequence, which is stitched together from three short non-protein-coding sequences known as the tripartite leader. The leader is then alternatively spliced to the coding sequences for the hexon, fiber, and other virion proteins to generate each of the late viral mRNAs.

That these messengers are spliced together from RNAs arising from several regions of the genome emerged from a variety of experiments—one of which is known as R-loop mapping (Box 13-1 Fig. 2). When RNA is incubated, under the appropriate conditions, with a double-stranded DNA containing a stretch of sequence identical to that of the RNA, the RNA anneals to its complement, displacing a stretch of the noncomplementary strand in the form of a loop (Box 13-1 Fig. 2a). Following the staining procedure used to visualize nucleic acids, this R loop can be observed in the electron microscope, as RNA–DNA and DNA–DNA duplexes appear thicker than single-stranded nucleic acids. When such an experiment was perfomed with adenovirus messengers, the resulting R loops were found not to be fully contiguous with a single region of DNA. Instead, and depending on which fragment of viral DNA was used, one or both ends of the RNA were found to protrude from the RNA loops as single-strand tails (Box 13-1 Fig. 2b). In other cases, one of the tails is seen to anneal with a DNA fragment from a different region of the viral genome (Box 13-1 Fig. 2c). Clearly, these mRNAs were composite molecules that had been joined together from sequences complementary to noncontiguous regions of the genome. These and other kinds of DNA–RNA annealing experiments were used to deduce the pattern of alternative splicing shown in Box 13-1 Figure 1.

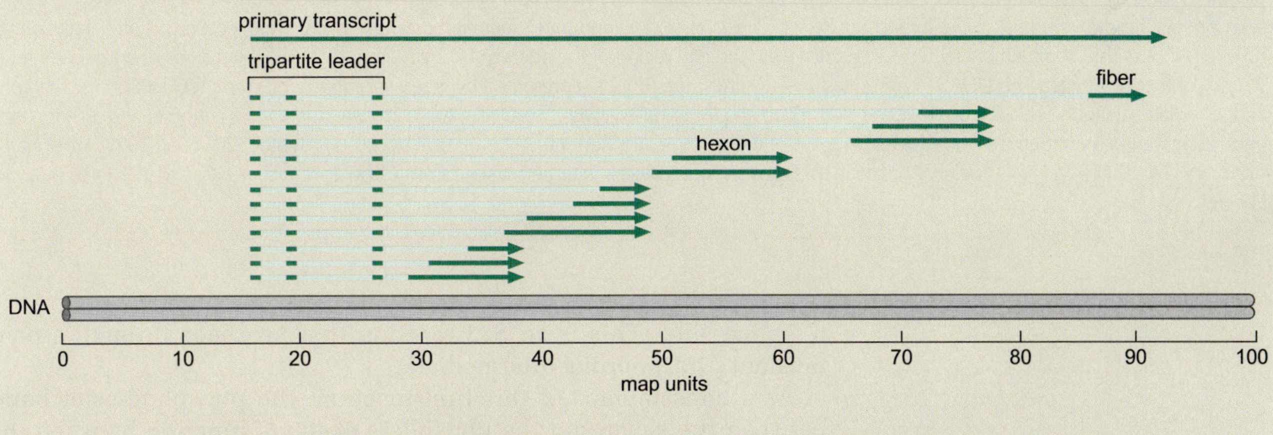

BOX 13-1 FIGURE 1 Map of the human adenovirus-2 genome. The map shows the transcription patterns of the late mRNAs, including the primary transcript (shown as a long dark green arrow at the top); the tripartite leader sequences found at positions 16.6, 19.6, and 26.6 (shown as green bars); and the map positions of the DNA sequences that encode the various late mRNAs (the late mRNAs are shown as short dark green arrows).

Box 13-1 *(Continued)*

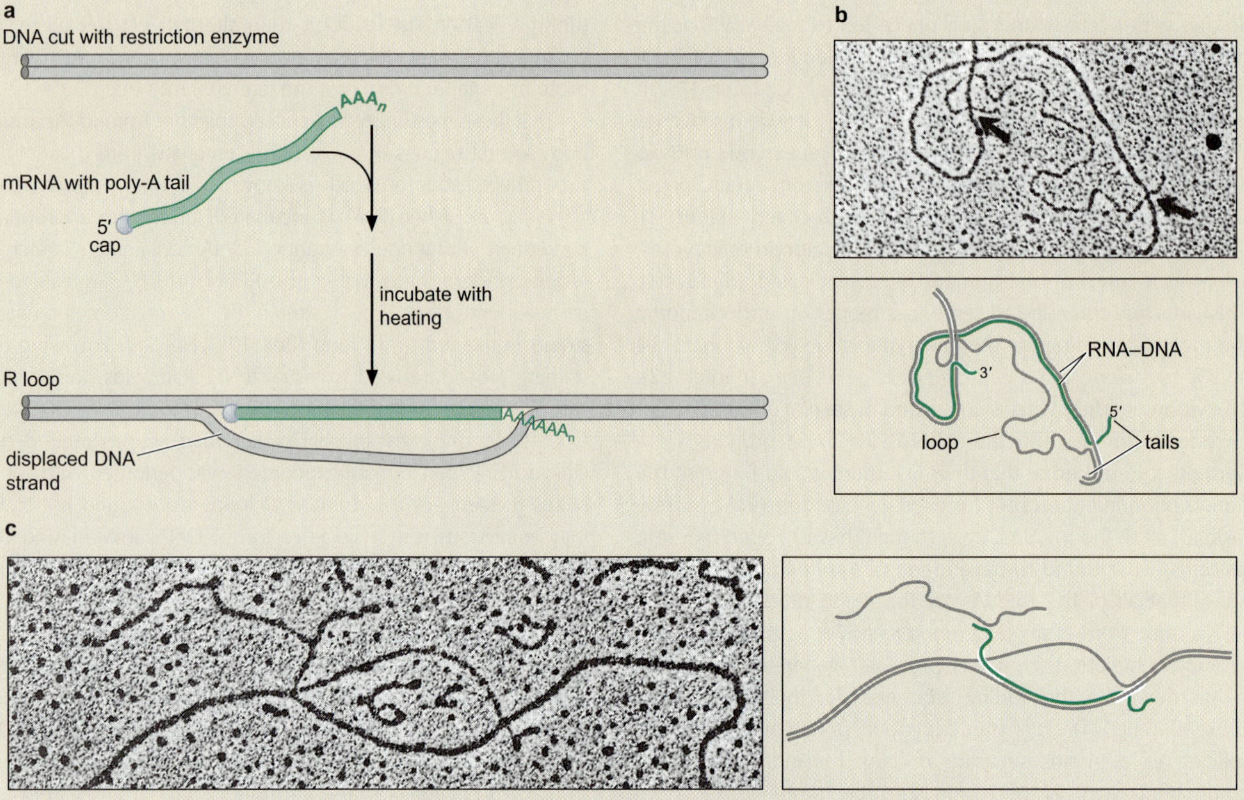

Box 13-1 Figure 2 R-loop mapping of the adenovirus-2 late messenger RNAs. (a) The schematic shows the formation of an R-loop structure. A double-stranded DNA fragment generated by digestion with a restriction endonuclease is incubated with mRNA and heated to just above the melting temperature of the DNA in 80% formamide. The hybrid formed between the messenger and its complementary DNA sequence results in displacement of the second DNA strand. The poly-A tail of the mRNA (not encoded by DNA; see Chapter 12) is seen projecting from the end of the hybrid duplex. (b) Electron micrograph and schematic diagram of an R loop observed after incubating hexon mRNA with a complementary DNA sequence from the late region of the adenovirus-2 genome. Note the extensions of both the 5′ and 3′ ends of the messenger. The DNA is represented by gray lines; the RNA is represented by green lines in the diagram. (c) Electron micrograph and schematic diagram of an R loop observed after incubating fiber mRNA with two DNAs, the complete adenovirus genome, and a restriction endonuclease fragment derived from the early region of the genome. (b, Reprinted, with permission, from Berget S.M. et al. 1977. *Proc. Natl. Acad. Sci.* 74: 3171–3175. © National Academy of Sciences; c, reprinted with permission, from Chow L.T. et al. 1977. *Cell* 12: 1–8, p. 2. © Elsevier.)

5′ splice site. (This is an S_N2 reaction that proceeds through a pentavalent phosphorous intermediate.)

As a consequence of this first reaction, the phosphodiester bond between the sugar and the phosphate at the 5′ junction between the intron and the exon is cleaved. The freed 5′ end of the intron is joined to the A within the branch site. Thus, in addition to the 5′ and 3′ backbone linkages, a third phosphodiester extends from the 2′-OH of that A to create a three-way junction (hence its description as a branch point). The structure of the three-way junction is shown in Figure 13-5.

Note that the 5′ exon is a leaving group in the first transesterification reaction. In the second reaction, the 5′ exon (more precisely, the newly liberated 3′-OH of the 5′ exon) reverses its role and becomes a nucleophile that attacks the phosphoryl group at the

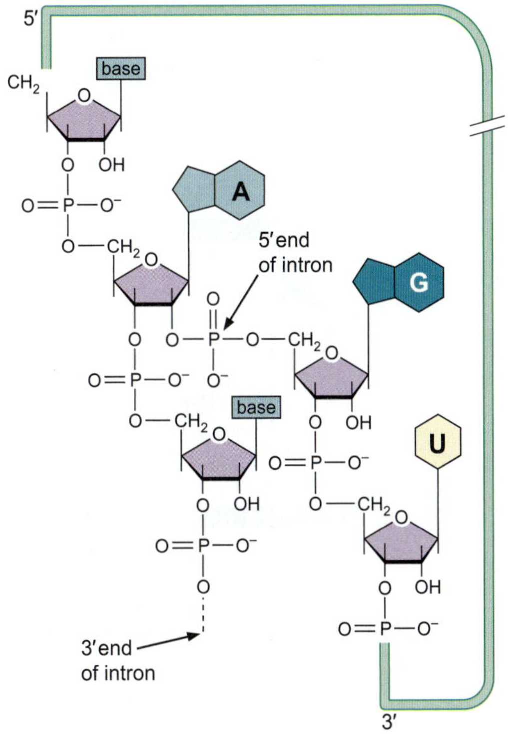

FIGURE 13-5 The structure of the three-way junction formed during the splicing reaction.

3′ splice site (Fig. 13-4). This second reaction has two consequences. First, and most importantly, it joins the 5′ and 3′ exons; thus, this is the step in which the two coding sequences are actually "spliced" together. Second, this same reaction liberates the intron, which serves as a leaving group. Because the 5′ end of the intron had been joined to branch point A in the first transesterification reaction, the newly liberated intron has the shape of a **lariat.**

In the two reaction steps, there is no net gain in the number of chemical bonds—two phosphodiester bonds are broken, and two new ones made. As it is just a question of shuffling bonds, no energy input is demanded by the chemistry of this process. But, as we see below, a large amount of ATP is consumed during the splicing reaction. This energy is required, not for the chemistry, but to properly assemble and operate the splicing machinery.

Another point about the splicing reaction is direction: what ensures that splicing only goes forward—that is, toward the products shown in Figure 13-4? Two features that could contribute to this are as follows. First, the forward reaction involves an increase in entropy—a single pre-mRNA molecule is split into two molecules, the mRNA and the liberated lariat. Second, the excised intron is rapidly degraded after its removal and so is not available to partake in the reverse reaction.

Exons from Different RNA Molecules Can Be Fused by *trans*-Splicing

In our description of splicing above, we assumed that the 5′ splice site of one exon is joined to the 3′ splice site of the exon that immediately follows it. This is not always the case. In alternative splicing, exons can be skipped, and a given exon is joined to one further downstream (as we see later in the text). In some cases, two exons

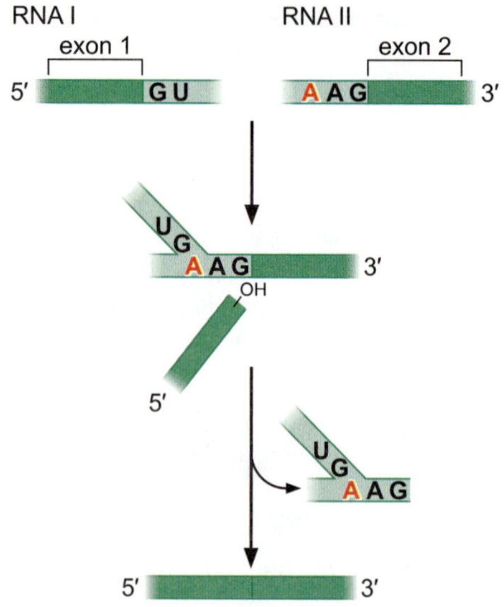

FIGURE 13-6 ***trans*-splicing.** In *trans*-splicing, two exons, initially found in two separate RNA molecules, are spliced together into a single mRNA. The chemistry of this reaction is the same as that of the standard splicing reaction described previously, and the spliced product is indistinguishable. The only difference is that the other product—the lariat in the standard reaction—is, in *trans*-splicing, a Y-shaped branch structure instead. This is because the initial reaction brings together two RNA molecules rather than forming a loop within a single molecule.

carried on different RNA molecules can be spliced together in a process called ***trans*-splicing.** Although generally rare, *trans*-splicing occurs in almost all the mRNAs of trypanosomes. In the nematode worm (*Caenorhabditis elegans*), all mRNAs undergo *trans*-splicing (to attach a 5′ leader sequence), and many of them undergo *cis*-splicing as well. Figure 13-6 shows how the basic splicing reaction just described is adapted to carry out *trans*-splicing.

THE SPLICEOSOME MACHINERY

RNA Splicing Is Carried Out by a Large Complex Called the Spliceosome

WEB ANIMATION

The transesterification reactions just described are mediated by a huge molecular "machine" called the **spliceosome** (see Interactive Animation 13-1). This complex comprises about 150 proteins and 5 RNAs and is similar in size to a ribosome—the machine that translates mRNA into protein (Chapter 14). In carrying out even a single splicing reaction, the spliceosome hydrolyzes several molecules of ATP. Strikingly, it is believed that many of the functions of the spliceosome are carried out by its RNA components rather than the proteins, again reminiscent of the ribosome. Thus, RNAs locate the sequence elements at the intron–exon borders and likely participate in catalysis of the splicing reaction itself.

The five RNAs (U1, U2, U4, U5, and U6) are collectively called **small nuclear RNAs (snRNAs)**. Each of these RNAs is between 100 and 300 nucleotides long in most eukaryotes and is complexed with several proteins. These RNA–protein complexes are called **small nuclear ribonuclear proteins (snRNPs**—pronounced "snurps"). In Chapter 5, we saw the crystal structure of a section of the U1 snRNA bound to one of the proteins of the U1 snRNP (Fig. 5-23).

The spliceosome is the large complex made up of these snRNPs, but the exact makeup differs at different stages of the splicing reaction: different snRNPs come and go at different times, each carrying out particular functions in the reaction. There are also many proteins within the spliceosome that are not part of the snRNPs, and others besides that are only loosely bound to the spliceosome.

The snRNPs have three roles in splicing. They recognize the 5′ splice site and the branch site; they bring those sites together as required; and they catalyze (or help to catalyze) the RNA cleavage and joining reactions. To perform these functions, RNA–RNA, RNA–protein, and protein–protein interactions are all important. We start by considering some of the RNA–RNA interactions. These operate within individual snRNPs, between different snRNPs, and between snRNPs and the pre-mRNA.

Thus, for example, Figure 13-7a shows the interaction, through complementary base pairing, of the U1 snRNA and the 5′ splice site in the pre-mRNA. Later in the reaction, that splice site is recognized by the U6 snRNA. In another example, shown in Figure 13-7b, the branch site is recognized by the U2 snRNA. A third example, in Figure 13-7c, shows an interaction between U2 and U6 snRNAs. This brings the 5′ splice site and the branch site together. It is these and other similar interactions, and the rearrangements they lead to, that drive the splicing reaction and contribute to its precision, as we shall see a little later.

Some non-snRNPs are involved in splicing as mentioned above. One example, U2AF (U2 auxilliary factor), recognizes the polypyrimidine (Py) tract/3′ splice site and, in the initial step of the splicing reaction, helps another protein, branch-point-binding protein (BBP), bind to the branch site. BBP is then displaced by the U2 snRNP, as shown in Figure 13-7d. Other proteins involved in the splicing reaction include RNA-annealing factors, which help load snRNPs onto the mRNA, and DEAD-box helicase proteins. The latter use their ATPase activity to dissociate given RNA–RNA interactions, allowing alternative pairs to form and thereby driving the rearrangements that occur through the splicing reaction.

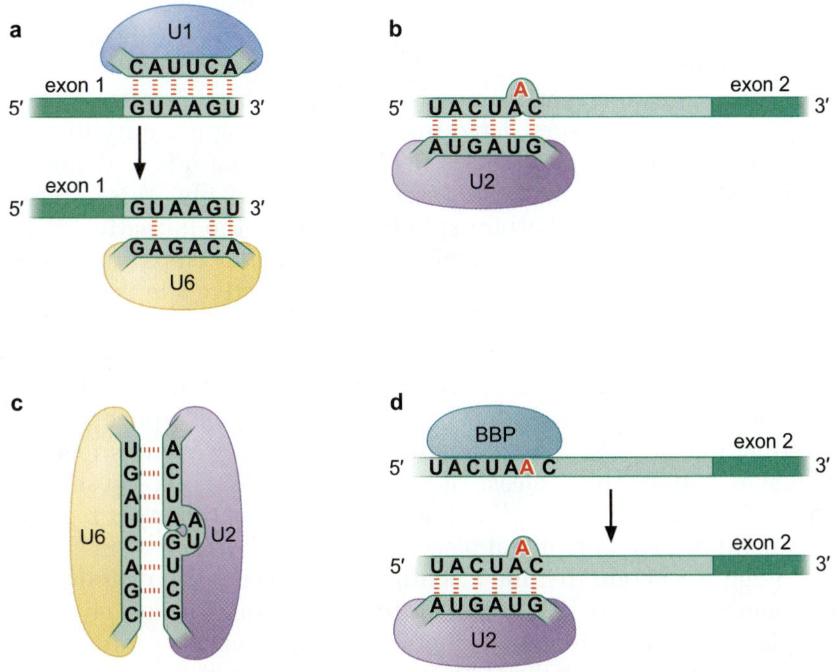

FIGURE 13-7 Some RNA–RNA hybrids formed during the splicing reaction. In some cases, (a) different snRNPs recognize the same (or overlapping) sequences in the pre-mRNA at different stages of the splicing reaction, as shown here for U1 and U6 recognizing the 5′ splice site. (b) snRNP U2 is shown recognizing the branch site. (c) The RNA:RNA pairing between the snRNPs U2 and U6 is shown. Finally, (d), the same sequence within the pre-mRNA is recognized by a protein (not part of an snRNP) at one stage and displaced by an snRNP at another. Each of these changes accompanies the arrival or departure of components of the spliceosome and a structural rearrangement that is required for the splicing reaction to proceed. The sequences in this figure are from yeast.

SPLICING PATHWAYS

Assembly, Rearrangements, and Catalysis within the Spliceosome: The Splicing Pathway

The steps of the splicing pathway are shown in Figure 13-8. Initially, the 5′ splice site is recognized by the U1 snRNP (using base pairing between its snRNA and the pre-mRNA, shown in Fig. 13-7). One subunit of U2AF binds to the Py tract and the other to the 3′ splice site. The former subunit interacts with BBP and helps that protein bind to the branch site. This arrangement of proteins and RNA is called the early (E) complex.

U2 snRNP then binds to the branch site, aided by U2AF and displacing BBP. This arrangement is called the A complex. The base pairing between the U2 snRNA and the branch site is such that the branch site A residue is extruded from the resulting stretch of double-helical RNA as a single-nucleotide bulge as shown in Figure 13-7b. This A residue is thus unpaired and available to react with the 5′ splice site.

The next step is a rearrangement of the A complex to bring together all three splice sites. This is achieved as follows: the U4 and U6 snRNPs, along with the U5 snRNP, join the complex. Together, these three snRNPs are called the tri-snRNP particle, within which the U4 and U6 snRNPs are held together by complementary base pairing between their RNA components, and the U5 snRNP is more loosely associated through protein–protein interactions. With the entry of the tri-snRNP, the A complex is converted into the B complex.

In the next step, U1 leaves the complex, and U6 replaces it at the 5′ splice site. This requires that the base pairing between the U1 snRNA and the pre-mRNA be broken, allowing the U6 RNA to anneal with the same region (in fact, to an overlapping sequence, as shown in Fig. 13-7a).

Those steps complete the assembly pathway. The next rearrangement triggers catalysis, and occurs as follows: U4 is released from the complex, allowing U6 to interact with U2 (through the RNA:RNA base pairing shown in Fig. 13-7c). This arrangement, called the C complex, produces the active site. That is, the rearrangement brings together within the spliceosome those components—believed to be solely regions of the U2 and U6 RNAs—that together form the active site. The same rearrangement also ensures the substrate RNA is properly positioned to be acted upon. It is striking not only that the active site is primarily formed of RNA, but also that it is only formed at this stage of spliceosome assembly. Presumably, this strategy lessens the chance of aberrant splicing. Linking the formation of the active site to the successful completion of earlier steps in spliceosome assembly makes it highly likely that the active site is available only at legitimate splice sites.

Formation of the active site juxtaposes the 5′ splice site of the pre-mRNA and the branch site, facilitating the first transesterification reaction. The second reaction, between the 5′ and 3′ splice sites, is aided by the U5 snRNP, which helps to bring the two exons together. The final step involves release of the mRNA product and the snRNPs. The snRNPs are initially still bound to the lariat, but they get recycled after rapid degradation of that piece of RNA.

It might seem odd that the machinery and mechanism of splicing is so complicated. How did it evolve that way? Would it not have been simpler to fuse the exons in a single reaction, rather than undergo the two reactions just described? To consider this question, we turn to a

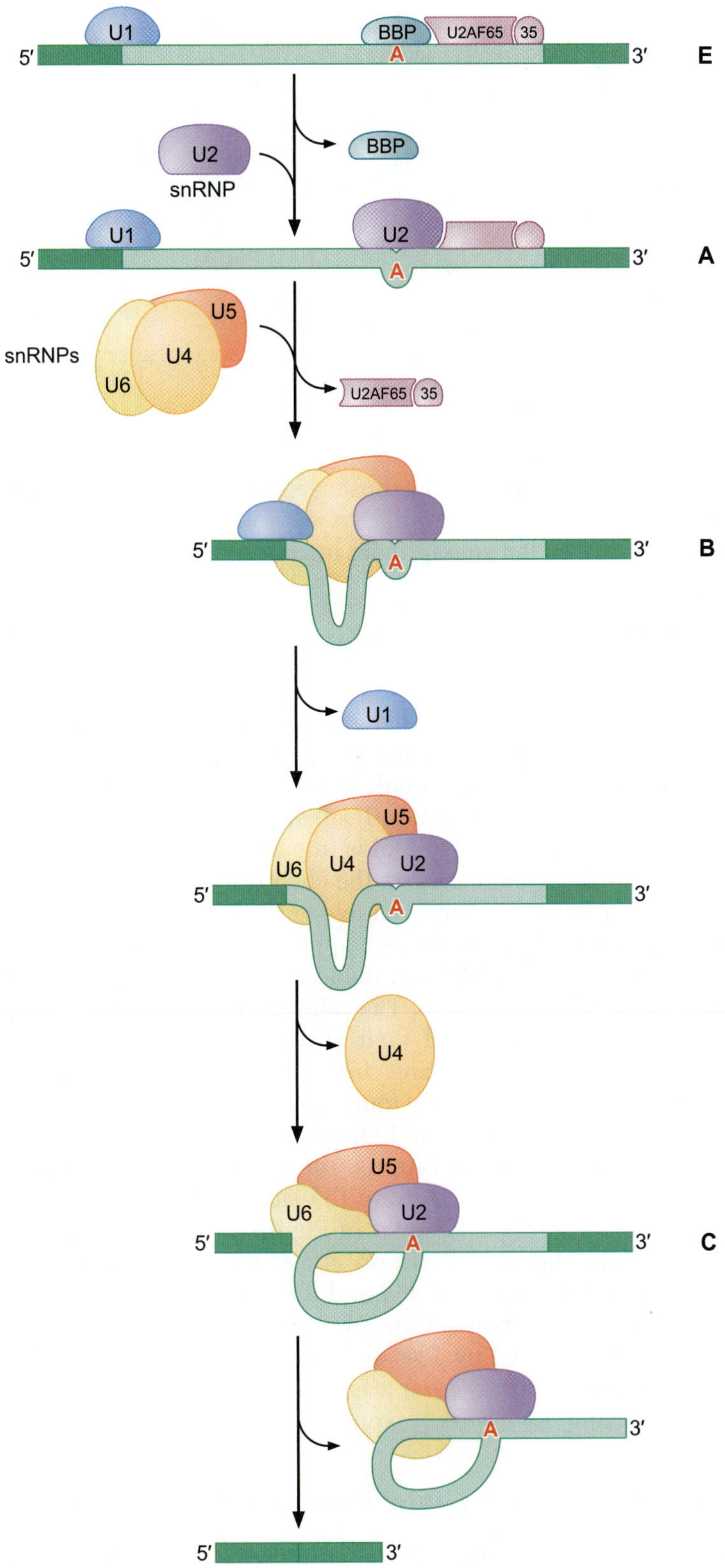

FIGURE 13-8 Steps of the spliceosome-mediated splicing reaction. The assembly and action of the spliceosome are shown, and the details of each step are described in the text. Components of the splicing machinery arrive or leave the complex at each step, changes that are associated with structural rearrangements necessary for the splicing reaction to proceed. Note that the name of each complex is shown to the right. There is evidence to suggest that some of the components shown do not arrive or leave precisely when indicated in this figure; they may, for example, remain present but weaken their association with the complex rather than dissociating completely. It is also not possible to be sure of the order of some changes shown, particularly the two steps involving changes in U6 pairing: when it takes over from U1 at the 5′ splice site, compared to when it takes over from U4 in binding U2. Despite these uncertainties, the critical involvement of different components of the machinery at different stages of the splicing reaction and the general dynamic nature of the spliceosome are as shown.

TABLE 13-1 Three Classes of RNA Splicing

Class	Abundance	Mechanism	Catalytic Machinery
Nuclear pre-mRNA	Very common; used for most eukaryotic genes	Two transesterification reactions; branch site A	Major and minor spliceosomes
Group II introns	Rare; some eukaryotic genes from organelles and prokaryotes	Same as pre-mRNA	RNA enzyme encoded by intron (ribozyme)
Group I introns	Rare; nuclear rRNA in some eukaryotes, organelle genes, and a few prokaryotic genes	Two transesterification reactions; branch site G	Same as group II

group of introns that—unlike those we have considered thus far—can splice themselves out of pre-mRNA without the need for the spliceosome. They are called **self-splicing introns.**

Self-Splicing Introns Reveal That RNA Can Catalyze RNA Splicing

The three classes of splicing found in cells (not including tRNA processing, which we discuss in Chapter 14) are shown in Table 13-1. Thus far, we have dealt only with nuclear pre-mRNA splicing, that mediated by the spliceosome found in all eukaryotes. Also shown in Table 13-1 are the so-called **group I** and **group II** self-splicing introns. By self-splicing, we mean that the intron itself folds into a specific conformation within the precursor RNA and catalyzes the chemistry of its own release (recall that we discussed the general features of RNA enzymes in Chapter 6). In terms of a practical definition, self-splicing refers to introns that can remove themselves from RNAs in the test tube in the absence of any proteins or other RNA molecules. The self-splicing introns are grouped into two classes on the basis of their structure and splicing mechanism. Strictly speaking, self-splicing introns are not enzymes (catalysts) because they mediate only one round of RNA processing (as we consider in Box 13-2, Converting Group I Introns into Ribozymes).

In the case of group II introns, the chemistry of splicing and the RNA intermediates produced are the same as those for nuclear pre-mRNAs. For example, as shown in Figure 13-9, the intron uses an A residue within the branch site to attack the phosphodiester bond at the boundary between its 5′ end and the end of the 5′ exon—that is, at the 5′ splice site. This reaction produces the branched lariat, as seen above, and is followed by a second reaction in which the newly freed 3′-OH of the exon attacks the 3′ splice site, releasing the intron as a lariat and fusing the 3′ and 5′ exons.

Group I Introns Release a Linear Intron Rather Than a Lariat

WEB ANIMATION

Group I introns splice by a different pathway (Fig. 13-9c and Interactive Animation 13-2). Instead of a branch point A residue, they use a free G nucleotide or nucleoside. This G species is bound by the RNA and its 3′-OH group is presented to the 5′ splice site. The same type of transesterification reaction that leads to the lariat formation in the earlier examples here fuses the "G" to the 5′ end of the intron. The second reaction now proceeds just as it does in the earlier examples: the freed

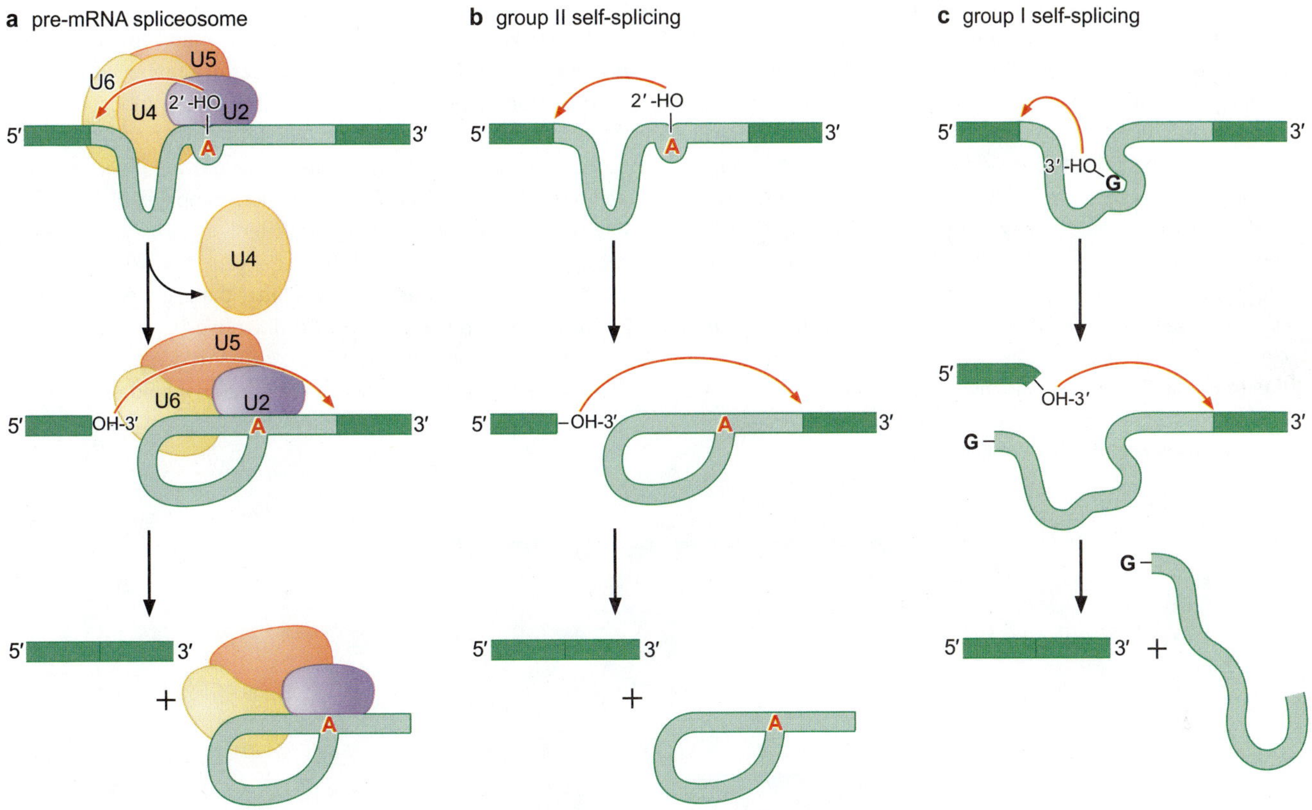

FIGURE 13-9 Group I and group II introns. This figure compares the reaction of the self-splicing group I and II introns and the spliceosome-mediated reaction already described. The chemistry in the case of group II introns is essentially the same as in the spliceosome case, with a highly reactive adenine within the intron initiating splicing and leading to the formation of a lariat product. In the case of the group I intron, the RNA folds in a way that forms a guanine-binding pocket, which allows the molecule to bind a free guanine nucleotide and use that to initiate splicing. Although these introns can splice themselves out of RNA molecules unaided by proteins in vitro, in vivo they typically do require protein components to stimulate the reaction. (Adapted, with permission, from Cech T.R. 1986. *Cell* 44: 207–210, Fig. 1. © Elsevier.)

3′ end of the exon attacks the 3′ splice site. This fuses the two exons and releases the intron, although, in this case, the intron is linear rather than a lariat structure.

Group I introns, which are smaller than group II introns, share a conserved secondary structure (RNA folding is discussed in Chapter 6). The structure of group I introns includes a binding pocket that will accommodate any guanine nucleotide or nucleoside as long as it is a ribose form. In addition to the nucleotide-binding pocket, group I introns contain an "internal guide sequence" that base-pairs with the 5′ splice site sequence and thereby determines the precise site at which nucleophilic attack by the G nucleotide takes place (see Box 13-2).

A typical self-splicing intron is between 400 and 1000 nucleotides long, and, in contrast to introns removed by spliceosomes, much of the sequence of a self-splicing intron is critical for the splicing reaction. This sequence requirement holds because the intron must fold into a precise structure to perform the reaction chemistry. In addition, in vivo, the intron is complexed with a number of proteins that help stabilize the correct structure—partly by shielding regions of the backbone from each other. Thus, the folding requires certain sections of the RNA back-

Box 13-2 Converting Group I Introns into Ribozymes

Once a group I self-splicing intron has been spliced out, the active site it contains remains intact. So what prevents this splicing reaction from reversing itself? One factor is the high cellular concentration of G nucleotides—this strongly favors the forward reaction. But in addition, the intron undergoes a further reaction that effectively prevents it from participating in the back reaction. Conveniently, at the extreme 3′ end of the intron is a G, which can bind in the G-binding pocket. Meanwhile, the 5′ end of the intron can bind along the internal guide sequence. Thus, a third transesterification reaction can occur to cyclize the intron. The new bond formed with the terminal G is labile and hydrolyzes spontaneously. As a consequence, the intron is relinearized, but it is truncated and so precluded from the back-splicing reaction.

As explained earlier in the text, group I (and II) introns are not enzymes because they have a turnover number of only 1. But they can be readily converted into enzymes (ribozymes) in the following way (Box 13-2 Fig. 1): the relinearized intron described above retains its active site. If we provide it with free G and a substrate that includes a sequence complementary to the internal guide sequence, it will repeatedly catalyze cleavage of substrate molecules. We will have converted a group I intron into a ribozyme, similar to the way that the self-cleaving hammerhead could be converted to a ribozyme by separating the active site from the substrate (Chapter 6). We can go a step further by changing the sequence of the internal guide sequence and thereby generate tailor-made ribonucleases that cleave RNA molecules of our choice.

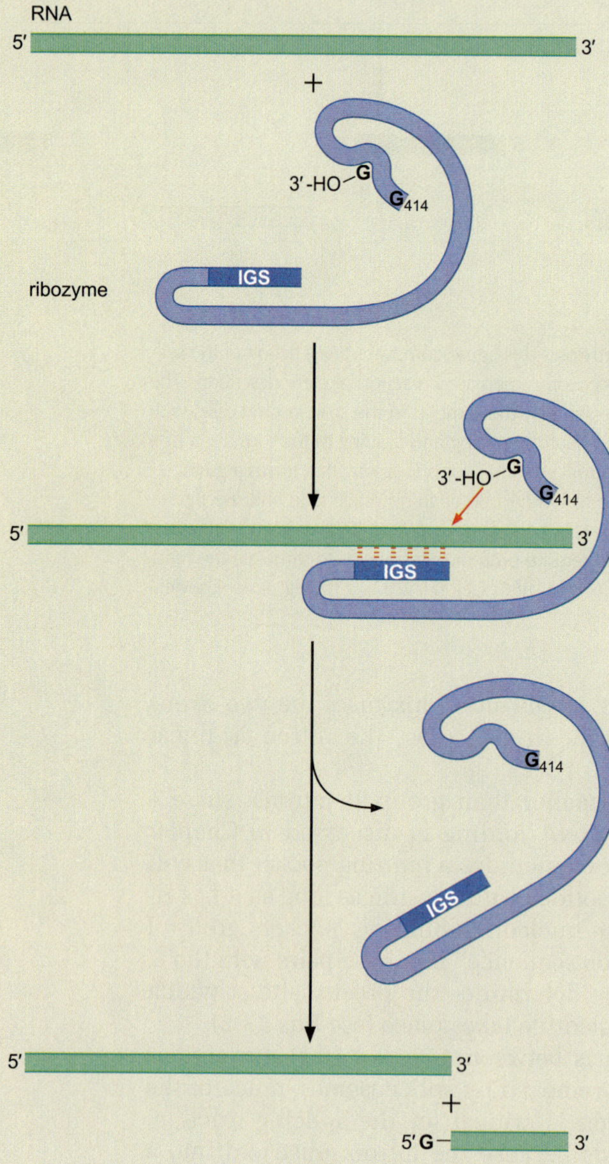

BOX 13-2 FIGURE 1 Group I introns can be converted into true ribozymes.

bone to be in close proximity to other sections, and the negative charges provided by the phosphates in those backbone regions would repel each other if not shielded. In vitro, high salt concentrations (and thus positive ions) compensate for the absence of these proteins. This is how we know that the proteins are not needed for the splicing reaction itself.

The similar chemistry seen in self- and spliceosome-mediated splicing is believed to reflect an evolutionary relationship. Perhaps ancestral group II–like self-splicing introns were the starting point for the evolution of modern pre-mRNA splicing. The catalytic functions provided by the RNA were retained, but the requirement for extensive sequence specificity within the intron itself was relieved by having the snRNAs and their associated proteins provide most of those functions in *trans*. In this way, introns had only to retain the minimum of sequence elements required to target splicing to the correct places. Thus, many more and varied sizes and sequences of introns were permitted.

The structure of the catalytic region that performs the first transesterification reaction is very similar in the group II intron and the pre-mRNA/snRNP complex (Fig. 13-10). This observation fuels the broader speculation (discussed in Chapter 6) that early in the evolution of modern organisms, many catalytic functions in the cell were carried out by RNAs and that these functions have, on the whole, since been replaced by proteins. In the case of the spliceosome and the ribosome,

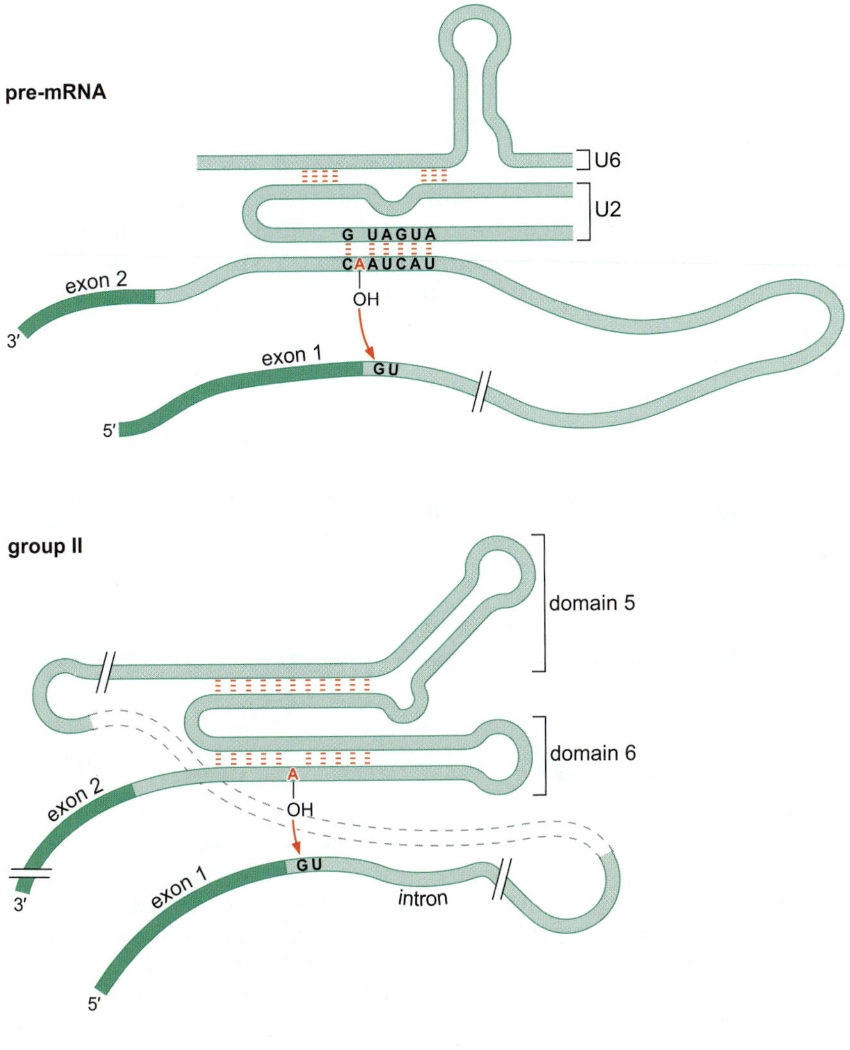

FIGURE 13-10 Proposed folding of the RNA catalytic regions for splicing of group II introns and pre-mRNAs. The dotted regions of the RNA in the group II case replace an additional four folded domains not shown in this depiction.

however, these activities have not been entirely replaced by proteins. Rather, the vestigial RNA-catalyzed mechanisms remain at the heart of the present complex machinery.

How Does the Spliceosome Find the Splice Sites Reliably?

We have already seen one mechanism that guards against inappropriate splicing: the active site of the spliceosome is only formed on RNA sequences that pass the test of being recognized by multiple elements during spliceosome assembly. Thus, for example, the 5′ splice site must be recognized initially by the U1 snRNP and then by the U6 snRNP. It is unlikely both would recognize an incorrect sequence, and so selection is stringent. Yet, the problem of appropriate splice-site recognition in the pre-mRNA remains formidable.

Consider the following. The average human gene has seven or eight exons and can be spliced in three alternative forms. But there is one human gene with 363 exons and one *Drosophila* gene that can be spliced in 38,000 alternative ways, a case we describe in detail in the next section. If the snRNPs had to find the correct 5′ and 3′ splice sites on a complete RNA molecule and bring them together in the correct pairs, unaided, it seems inevitable that many errors would occur. Remember also that the average exon is only some 150 nucleotides long, whereas the average intron is approximately 3000 nucleotides long (and as we have seen, some introns can be as long as 800,000 nucleotides). Thus, the exons must be identified within a vast ocean of intronic sequences.

Splice-site recognition is prone to two kinds of errors (Fig. 13-11). First, splice sites can be skipped, with components bound at, for example, a given 5′ splice site pairing with those at a 3′ site beyond the correct one.

Second, other sites, close in sequence but not legitimate splice sites, could be mistakenly recognized. This is easy to appreciate when one recalls that the splice site consensus sequences are rather loose. And so, for example, components at a given 5′ splice site might pair with components bound incorrectly at such a "pseudo" 3′ splice site (see Fig. 13-11b).

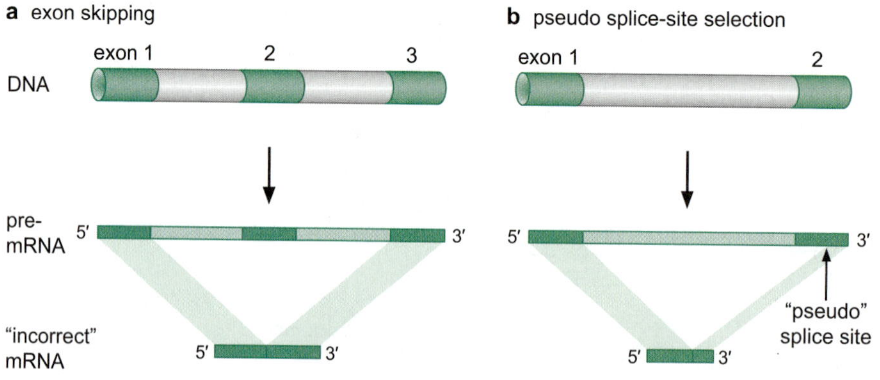

FIGURE 13-11 Errors produced by mistakes in splice-site selection. (a) The consequence of skipping an exon. This happens if the spliceosome components bound at the 5′ splice site of one exon interact with spliceosome components bound at the 3′ splice site of not the next exon, but one beyond. (b) The effect of spliceosome components recognizing "pseudo" splice sites—sequences that resemble (but are not) legitimate splice sites. In the case shown, the pseudo site is within an exon and leads to regions near the 5′ end of that exon being mistakenly spliced out along with the intron.

Two ways in which the accuracy of splice-site selection can be enhanced are as follows. First, as we saw in Chapter 12, while transcribing a gene to produce the RNA, RNA polymerase II carries with it various proteins with roles in RNA processing (see Fig. 12-19). These include proteins involved in splicing. When a 5′ splice site is encountered in the newly synthesized RNA, the factors that recognize that site are transferred from the polymerase carboxy-terminal "tail" (that part of the enzyme where they hitch a ride) onto the RNA. Once in place, the 5′ splice site components are poised to interact with those other factors that bind to the next 3′ splice site to be synthesized. Thus, the correct 3′ splice site can be recognized before any competing sites further downstream have been transcribed. This cotranscriptional loading process greatly diminishes the likelihood of exon skipping.

It is worth noting that even though much of the splicing machinery assembles while the gene is being transcribed—and on individual introns in the order they are transcribed—this does not mean the introns are themselves spliced out in that order. Thus, in contrast to many other activities we have heard about—transcription, replication, and so on—there appears to be no "tracking" mechanism involved, whereby the machinery assembles at one end of the gene or message and acts as it tracks to the other end.

A second mechanism guards against the use of incorrect sites by ensuring that splice sites close to exons (and thus likely to be authentic) are recognized preferentially. So-called SR (serine arginine–rich) proteins bind to sequences called **exonic splicing enhancers (ESEs)** within the exons. SR proteins bound to these sites recruit the splicing machinery to the nearby splice sites. In this way, the machinery binds more efficiently to those nearby splice sites than to incorrect sites not close to exons. Specifically, the SR proteins recruit the U2AF proteins to the 3′ splice site and U1 snRNP to the 5′ site (Fig. 13-12). As we saw earlier, these factors demarcate the splice sites for the rest of the machinery to assemble correctly. This recruitment is either through direct interaction between the SR proteins and proteins within the spliceosome, or, as has recently been suggested, through interaction with, and stablization of, RNA:RNA hybrids formed during spliceosome assembly and action.

SR proteins are essential for splicing. They not only ensure the accuracy and efficiency of constitutive splicing (as we have just seen), but also regulate alternative splicing (as we shall see presently). They come in many varieties, some controlled by physiological signals, others constitutively active. Some are expressed preferentially in certain cell

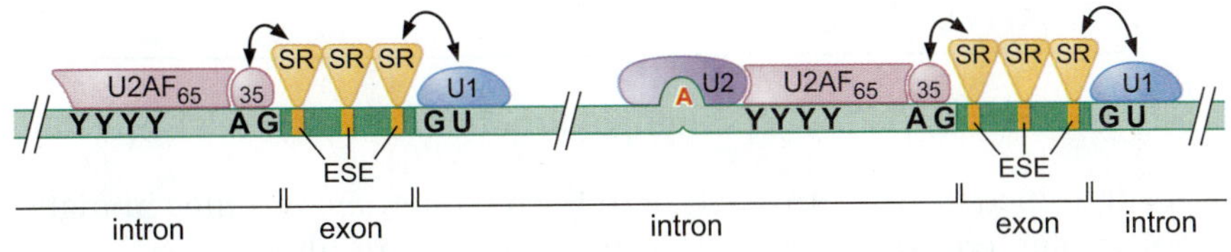

FIGURE 13-12 SR proteins recruit spliceosome components to the 5′ and 3′ splice sites. Legitimate splice sites are recognized by the splicing machinery by virtue of being close to exons. Thus, SR proteins bind to sequences within the exons (exonic splicing enhancers [ESEs]), and from there recruit U2AF and U1snRNP to the downstream 5′ and upstream 3′ splice sites, respectively. This initiates the assembly of the splicing machinery on the correct sites and splicing can proceed as outlined earlier. (Adapted, with permission, from Maniatis T. and Tasic B. 2002. *Nature* 418: 236–243. © Macmillan.)

types and control splicing in cell-type-specific patterns. We discuss some specific examples of the roles of SR proteins in the next section.

A Small Group of Introns Are Spliced by an Alternative Spliceosome Composed of a Different Set of snRNPs

Higher eukaryotes (including mammals, plants, and so on) use the major splicing machinery we have discussed thus far to direct splicing of the majority of their pre-mRNAs. But in these organisms (unlike in yeast), some pre-mRNAs are spliced by an alternative, low-abundance form of spliceosome. This rare form contains some components common to the major spliceosome, but it contains other unique components as well. Thus, U11 and U12 components of the alternative splicesome have the same roles in the splicing reaction as U1 and U2 of the major form, but they recognize distinct sequences. U4 and U6 have equivalent counterparts in both spliceosome forms—although these snRNPs are distinct, they share the same names. Finally, the identical U5 component is found in both the major and the alternative—so-called "minor"—spliceosome.

The minor spliceosome recognizes rarely occurring introns having consensus sequences distinct from the sequences of most pre-mRNA introns. The minor form is also known as the AT-AC spliceosome, because the termini of the originally identified rare introns contain AU at the 5′ splice site and AC at the 3′ site (in RNA or AT and AC in DNA). Later it transpired that many introns spliced by this pathway have GT-AG termini (like mainstream introns), but otherwise their consensus sequences are distinct from those of the major pathway.

Despite the different splice site and branch site sequences recognized by the two systems, these major and minor forms of spliceosomes both remove introns using the same chemical pathway (Fig. 13-13). Consistent with this conserved mechanism, the differences in splice-site sequences recognized by these snRNPs are mirrored by complementary differences in the sequences of their snRNAs. Thus, it is the ability of the snRNAs and splice site sequences to base-pair that is conserved, not any particular sequence within either.

It is also worth noting that AT-AC introns might fit into the evolutionary scheme discussed earlier. Thus, as we mentioned, it has been proposed that the group II introns represent the oldest form of introns. Furthermore, it is suggested that the AT-AC introns evolved from the group II introns and eventually gave rise to the major pre-mRNA introns.

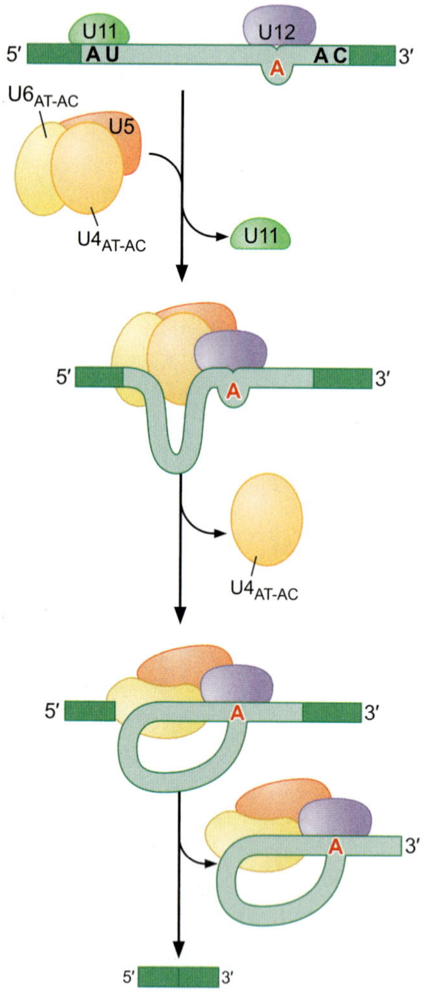

FIGURE 13-13 The AT-AC (minor) spliceosome catalyzed splicing. This minor spliceosome works on a minority of exons (e.g., perhaps 1 in 1000 in humans), and those have distinct splice-site sequences. Regardless, the chemistry is the same, and so are some of the spliceosome components, and others are closely related.

ALTERNATIVE SPLICING

Single Genes Can Produce Multiple Products by Alternative Splicing

As described in the introduction to this chapter, many genes in higher eukaryotes encode RNAs that can be spliced in alternative ways to generate two or more different mRNAs and thus different protein products (or isoforms). Microarray analyses now suggest that 40% of *Drosophila* genes and as many as 75% of human genes undergo alternative splicing. Many of these generate only two alternative products, but in some cases, the number of potential alternatives that can be generated from a single gene is breathtaking—hundreds (e.g., in the

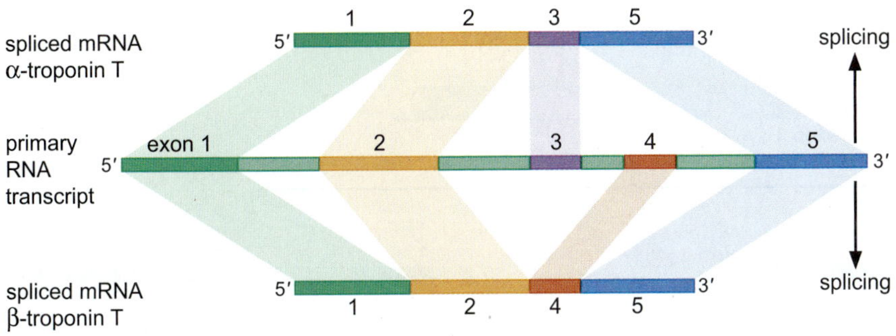

FIGURE 13-14 Alternative splicing in the troponin T gene. A region of the troponin T gene encoding five exons that generates two alternatively spliced forms as indicated. One contains exons 1, 2, 4, and 5; the other contains exons 1, 2, 3, and 5.

human *Slo* gene) or even many thousands (for the *Drosophila Dscam* gene).

For a simple case, consider the gene for the mammalian muscle protein troponin T. Shown in Figure 13-14 is a region of the pre-mRNA made from this gene that contains five exons. This pre-RNA is spliced to form two alternative mature mRNAs, each containing four exons. A different exon is eliminated from each of the two mRNAs, so the two messages have three exons in common, as well as each carrying one unique exon.

But, as shown in Figure 13-15, alternative splicing can occur in a number of ways. Thus, in addition to alternative exons, exons can be extended (by selecting an alternative downstream 5′, or upstream 3′, splice site). In other cases, exons can be skipped (deliberately), or *introns* can

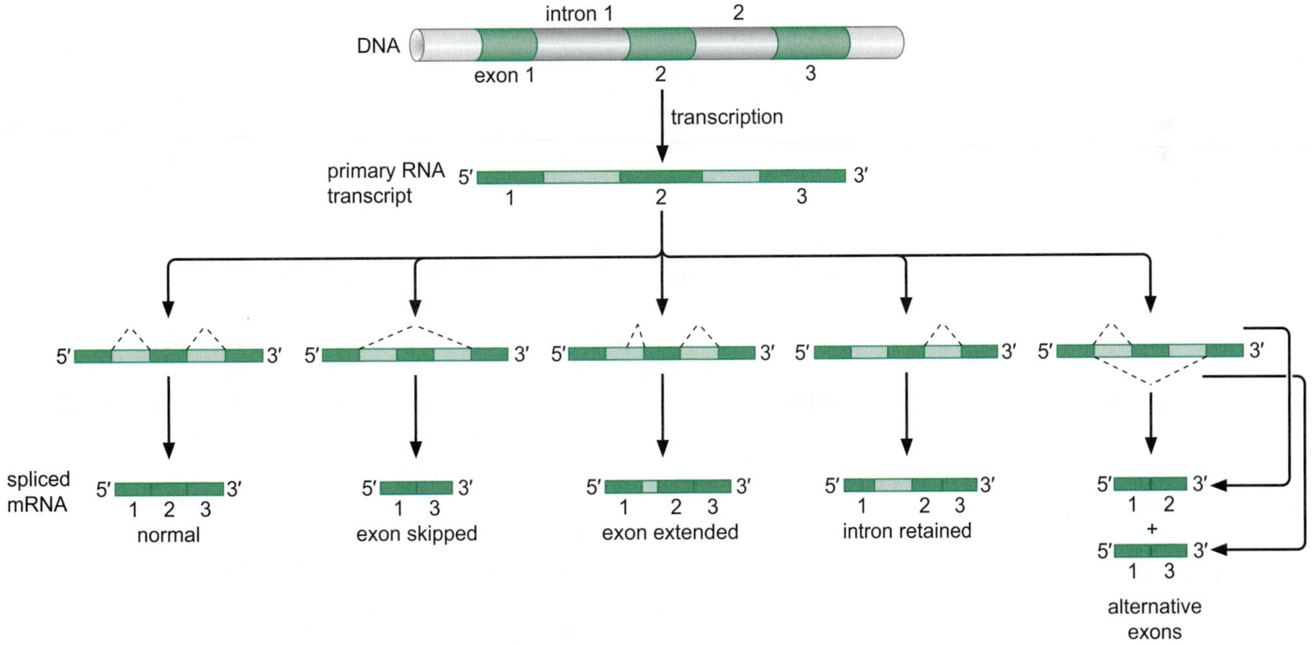

FIGURE 13-15 Five ways to splice an RNA. At the top is shown a gene encoding three exons. This is transcribed into a pre-mRNA, shown in the middle, and then spliced by five different alternative pathways. Thus, by including all exons, an mRNA containing all three exons is generated. Exon skipping gives an mRNA containing just exons 1 and 3. By exon extension, part of intron 1 is included together with the three exons. In another case, a complete intron is retained in the mature mRNA. Finally, exons 2 and 3 might be used as alternatives, generating a mixture of mRNAs, each including exon 1 and either exon 2 or 3.

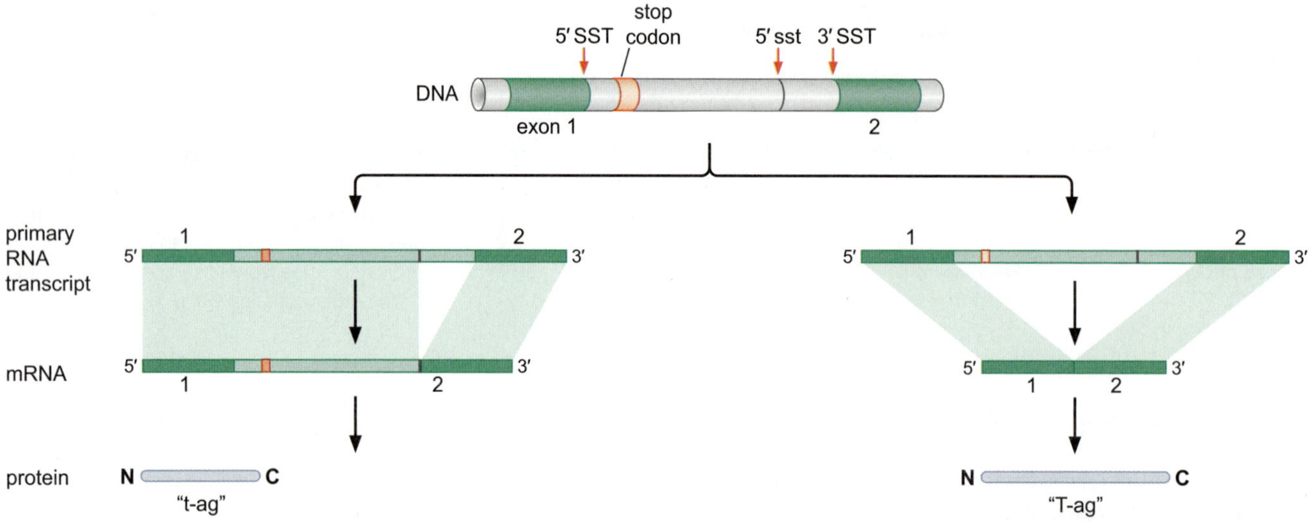

FIGURE 13-16 Alternative splicing of SV40 T-antigen. Splicing of the SV40 T-antigen RNA is shown. Both forms are typically produced, and both proteins made, upon infection. The small t antigen is encoded by the longer of the two mRNAs; that message contains an in-frame stop codon upstream of exon 2. 5′ SST refers to the 5′ splice site used to generate the large T mRNA; 5′ sst refers to the 5′ splice site used for small t. 3′ SST is the 3′ splice site used in generating both mRNAs.

be retained in the mature message. Some alternative spicing results from transcription of a gene from alternative promoters, allowing one transcript to include a 5′ exon not present in the other. Similarly, alternative poly-A sites allow 3′ terminal exons to be extended or alternative 3′ terminal exons to be used in some transcripts of a given gene. There are even cases of alternative *trans*-splicing (see Fig. 13.6).

In an example of an extended exon, Figure 13-16 shows the case of the T antigen of the monkey virus SV40. The T-antigen gene encodes two protein products—the large T antigen (T-ag) and the small t antigen (t-ag). The two proteins result from alternative splicing of the pre-mRNAs from the same gene. Thus, as shown in Figure 13-16, the gene has two exons and different mature mRNAs result from the use of two different 5′ splice sites. In the mRNA encoding T-ag, exon 1 is spliced directly to exon 2, deleting the intron that lies between. The mRNA for t-ag, on the other hand, is formed using the alternative 5′ splice site within the intron. Thus, in this case, the mRNA includes some of the intron as well. (It is therefore an example of the "extended exon" shown in Fig. 13-15.) The reason this larger message encodes the smaller protein is because there is an in-frame stop codon within the region of the intron retained in this mRNA.

Both forms of T antigens are made in a cell infected by SV40 but have different functions. Large T induces transformation and cell cycle reentry, whereas small t blocks the apoptotic response of cells forced down that path. The ratio of the two forms produced differs depending on the level of the splicing regulator SF2/ASF. When present at high levels, this protein directs the machinery to favor use of the 5′ splice site that generates more of the t-ag mRNA. SF2/ASF is an SR protein and, when abundant, presumably binds sites within exon 2 and helps the spliceosome assemble there.

In genome-wide studies, the most commonly seen forms of alternative splicing are cases in which complete exons are included or excluded from the mature message. Such exons are often called cassette exons. In about 10% of cases, cassette exons come in pairs, only one of which is included in the spliced message, just as we saw in the case of α-troponin T (Fig. 13-14). In these cases, there must be mechanisms that ensure that the exons are spliced in a mutually exclusive fashion.

Several Mechanisms Exist to Ensure Mutually Exclusive Splicing

There are several mechanisms to ensure that selection of alternative exons is mutually exclusive—that is, that when one is chosen, the other is not (or, to refer again to the α-troponin T example, when exon 3 is chosen, exon 4 is always excluded, and vice versa). We deal with each of these mechanisms here and then, in the next section, discuss an extreme case where a special mechanism is required.

Steric Hindrance Consider two alternative exons separated by an intron. If the splice sites within the intron are too close together, splicing factors cannot bind to both sites at the same time. Thus, Figure 13-17 shows a case where the binding of U1 snRNP to the 5′ splice site of the intron between two alternative exons (exons 2 and 3) prevents the binding of U2 snRNP to the branch point within that same intron (Fig. 13-17b). Alternatively, binding of U2 snRNP to the branch point excludes use of the 5′ splice site (Fig. 13-17c). The splicing of exons 3 and 4 of α-troponin is made mutually exclusive by this mechanism.

This arrangement can arise through the relative positions of the splice sites within an intron or because the intron is simply too small to work; in *Drosophila*, any exon under 59 nucleotides falls into that category.

Combinations of Major and Minor Splice Sites As we saw earlier, there is a form of the spliceosome—called the minor spliceosome—that recognizes splice sites distinct from those recognized by the major spliceosome. Neither spliceosome can remove an intron that contains

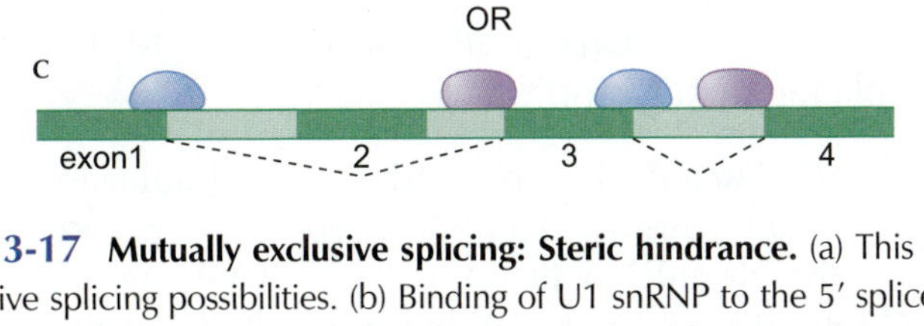

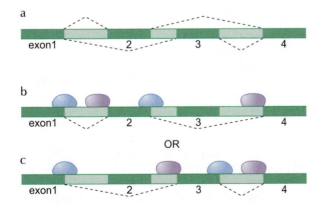

FIGURE 13-17 Mutually exclusive splicing: Steric hindrance. (a) This view shows the alternative splicing possibilities. (b) Binding of U1 snRNP to the 5′ splice site of the second intron excludes binding of U2 snRNP to the branch point of the same intron; binding of U2 to the following intron results in exclusion of exon 3. (c) Here, binding of the U2 snRNP to the branch point of the second intron excludes binding of U1 to the 5′ splice site of the same intron. In this case, binding of U1 to the 5′ splice site of the first intron results in exclusion of exon 2. (Courtesy of Brenton Graveley.)

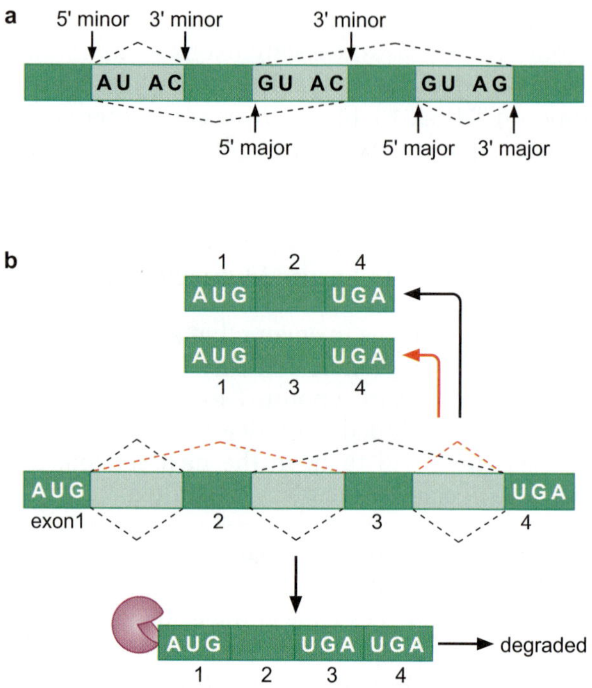

FIGURE 13-18 Mutually exclusive splicing. (a) Splice sites recognized by the major and minor spliceosomes. (b) Nonsense-mediated decay. (Design courtesy of Brenton Graveley.)

a combination of sites (i.e., a 5′ splice site of one type and 3′ of the other). Thus, by judicious arrangement of 5′ and 3′ splice sites recognized by these alternative spliceosomes, mutual exclusion can be achieved, as shown in Figure 13-18a. The human JNK1 gene is an example of this.

Nonsense-Mediated Decay Rather than forcing the splicing machinery to splice in a mutually exclusive fashion, this mechanism instead ensures that only messages that have one or another exon (never both and never neither) survive. In other words, although not ensuring mutually exclusive splicing, the consequences of this mechanism amount to the same thing. Nonsense-mediated decay (NMD) results from the fact that including both exons produces an mRNA that contains a premature termination codon (Fig. 13-18b). These messages are destroyed by NMD, the details of which are described in Chapter 14 (see Fig. 14-50).

The Curious Case of the *Drosophila Dscam* Gene: Mutually Exclusive Splicing on a Grand Scale

The *Drosophila Dscam* (Down syndrome cell-adhesion molecule) gene potentially encodes 38,016 protein isoforms. As shown in Figure 13-19, each possible mRNA made from this gene contains 24 exons, 20 of which are always the same, but 4 of which (exons 4, 6, 9, and 17) come in multiple alternative forms in the pre-mRNA. Thus, there are 12 possible versions of exon 4, 48 of exon 6, 33 of exon 9, and two of exon 17. The permutations these allow (12 x 48 x 33 x 2) give rise to the huge number of possible forms.

The proteins encoded by this gene are cell-surface proteins of the

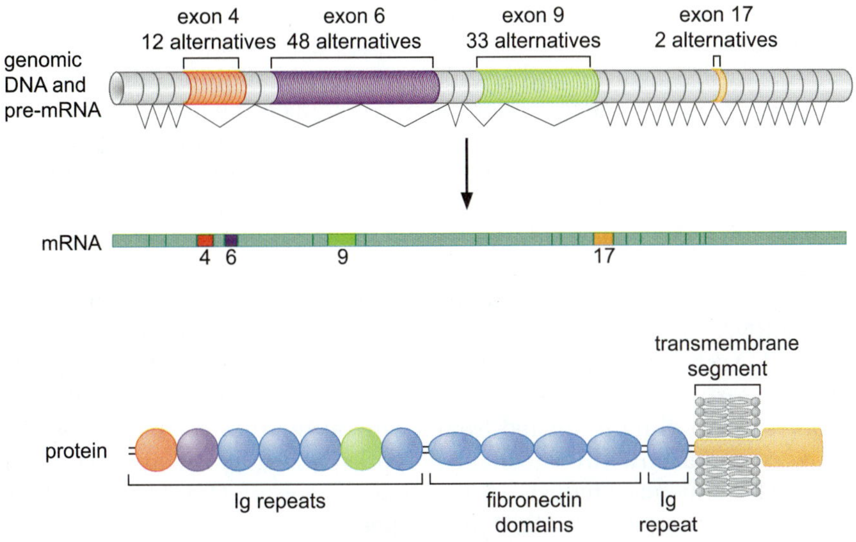

FIGURE 13-19 The multiple exons of the *Drosophila Dscam* gene. The *Dscam* gene (shown at the top) is 61.2 kb long; once transcribed and spliced, it produces one or more versions of a 7.8-kb, 24-exon mRNA (the figure shows the generic structure of those mRNAs). As shown, there are several mutually exclusive alternatives for exons 4, 6, 9, and 17. Thus, each mRNA will contain one of 12 possible alternatives for exon 4 (in red), 1 of 48 for exon 6 (purple), 1 of 33 for exon 9 (blue), and 1 of 2 for exon 17 (yellow). Exons 4, 6, and 9 encode parts of three Ig domains, depicted in the corresponding colors, and exon 17 encodes the transmembrane domain. If all possible combinations of these exons are used, the *Dscam* gene produces 38,016 different mRNAs and proteins. (Adapted, with permission, from Schmucker D. 2000. *Cell* 101: 671, Fig. 8. © Elsevier.)

immunoglobulin (Ig) superfamily. A generic form of the protein is shown at the bottom of Figure 13-19. The molecule has a transmembrane segment (encoded by exon 17, and thus coming in two alternative forms); fibronectin domains that are identical in all isoforms; and Ig domains, parts of three of which are encoded by the highly variable exons 4, 6, and 9. Thus, it is in these Ig domains that the vast majority of the variation from isoform to isoform resides.

The Dscam protein has two disparate functions in the fly—it acts in neural patterning in the brain and also recognizes antigens as part of the innate immune system. In its neuronal function, the Dscam protein mediates specific cell–cell interactions. Any given isoform of the protein interacts with itself but not with other isoforms. This selectivity is believed to enable a given axon to distinguish between other axons on the basis of whether or not they express the same or different isoforms. As such, it is thought to play a key part in establishing neural networks in the brain.

In the immune system, the different isoforms recognize different antigens, much as vertebrate antibodies do. The evolutionary pressure driving diversity is thought to come from selection on this function.

Mutually Exclusive Splicing of Dscam Exon 6 Cannot Be Accounted for by Any Standard Mechanism and Instead Uses a Novel Strategy

As we have seen, exon 6 is one of the four alternatively spliced exons of the *Dscam* gene—and in this case there are 48 alternatives to chose from. The scale of this selection is beyond the scope of the mechanisms discussed above. For example, although steric hindrance could be responsible for adjacent exons not being spliced, it cannot explain how others, further away, could also be excluded. In addition, all the splice sites in the *Dscam* gene are for the major spliceosome, and so the dual spliceosome mechanism is not an option. NMD also cannot explain the mutually exclusive splicing of exon 6: even if frameshifts resulted from including none, two, or three exons, an mRNA with, say, four exons would have the same reading frame as the message with only one. The same would be true of an mRNA that included seven exons, and so on.

So how does the cell ensure only one exon 6 variant is included in the mRNA? The novel mechanism hinges on the formation of alternative RNA:RNA base-paired structures within the pre-mRNA. Each alternative structure ensures that one, and only one, of the exon 6 variants is at any time protected from a general repression of splicing. We now consider how this mechanism works, and how it was discovered through sequence analysis of the *Dscam* gene of *Drosophila* and its various counterparts in other insect species.

The basic model is shown in Figure 13-20. Two classes of conserved sequence element are shown. One, the **docking site**, is located between exon 5 and the first alternative exon 6 variant (exon 6.1). A copy of the second type of element—the **selector sequence**—is found in front of each exon 6 variant (in the figure, exon 6.21 is shown as an example). Each selector sequence is different, but, as shown in Figure 13-21, each can base-pair with the docking site. The regions they each bind in the docking site overlap, and so binding of the different selector sequences to the one docking site is mutually exclusive: only one selector can bind at a time. And the selector sequence that does bind brings its associated exon 6 variant close to exon 5, ensuring it is the exon 6 variant chosen.

In addition to bringing the chosen exon 6 variant close to exon 5, the hybridization of the selector sequence and the docking site also ensures that the chosen exon 6 variant is free from a general repression mechanism that inhibits splicing of other possible exon 6 variants. A protein

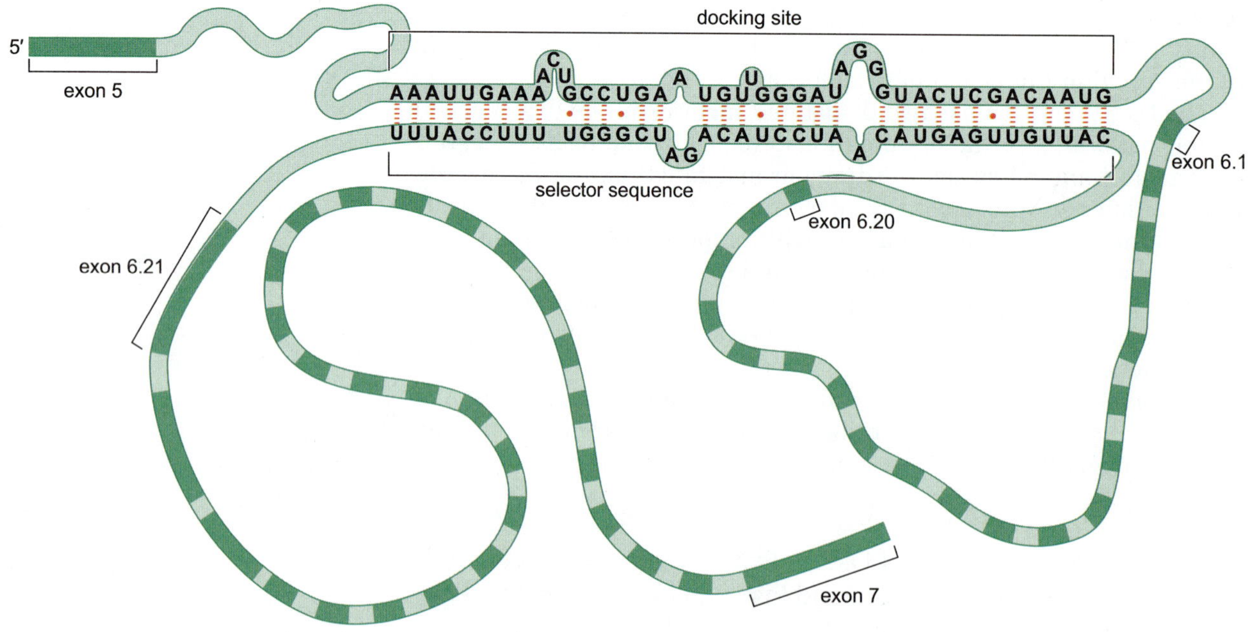

FIGURE 13-20 The docking site:selector sequences. Base pairing of the selector sequence for exon 6.21 with the docking site. (Design courtesy of Brenton Graveley.)

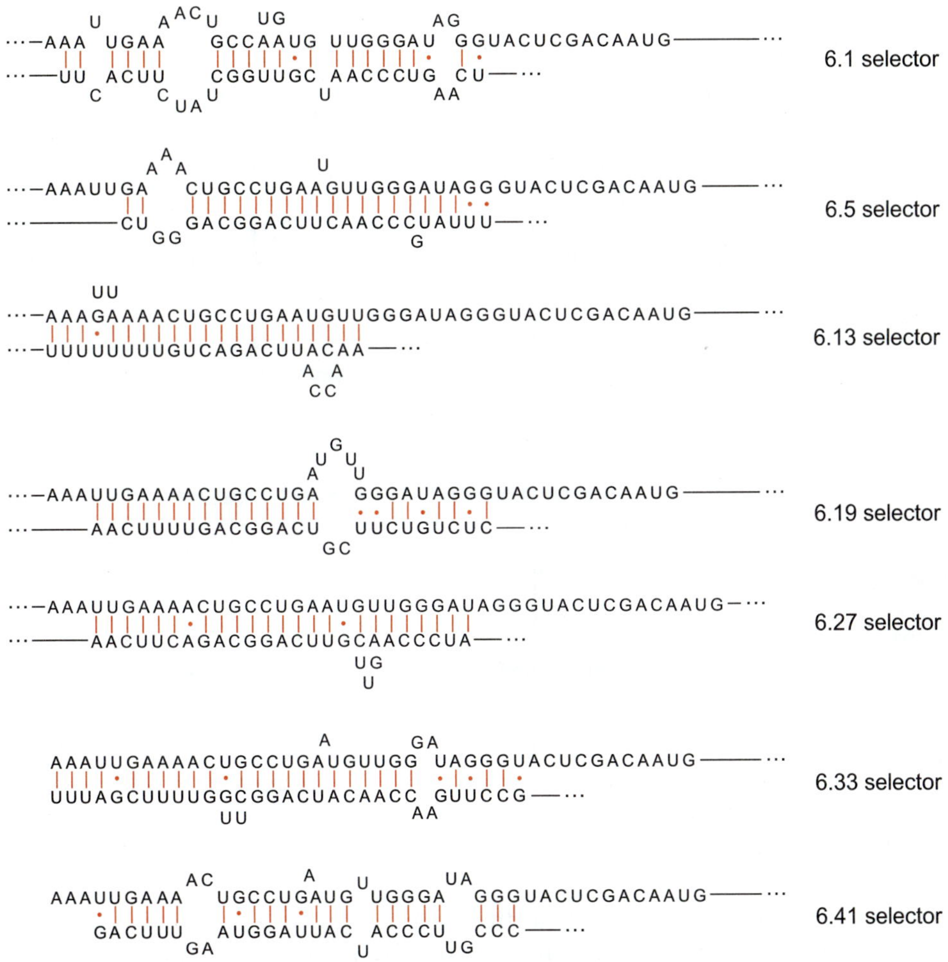

FIGURE 13-21 The selector sequences for six exon 6 variants, each bound to the exon 6 docking sequence. As is evident, each selector sequence base-pairs to a slightly different region of the docking sequence, but their binding to that docking sequence is nevertheless mutually exclusive. (Courtesy of Brenton Graveley.)

(Hrp36) acts as a general repressor of splicing by coating the other exons and inhibiting their inclusion in that mRNA. This local relief from inhibition afforded by the RNA hybridization may occur either as a direct result of the RNA secondary structure; or by the creation of a structure in the RNA recognized by a protein that removes the repressor; or by bringing the chosen exon close to activators within exon 5 that can then overcome the repression.

The docking site and selector sequences were discovered through sequence comparisons in an example of bioinformatic analysis, as described in Box 13-3, Identification of Docking Site and Selector Sequences.

Alternative Splicing Is Regulated by Activators and Repressors

Proteins that regulate splicing bind to specific sites called **exonic** (or **intronic**) **splicing enhancers** (**ESE** or **ISE**) or **silencers** (**ESS** and **ISS**). The former enhance, and the latter repress, splicing at nearby splice sites. We have already encountered enhancers and the SR proteins that bind to them (Fig. 13-12). Indeed, these elements and

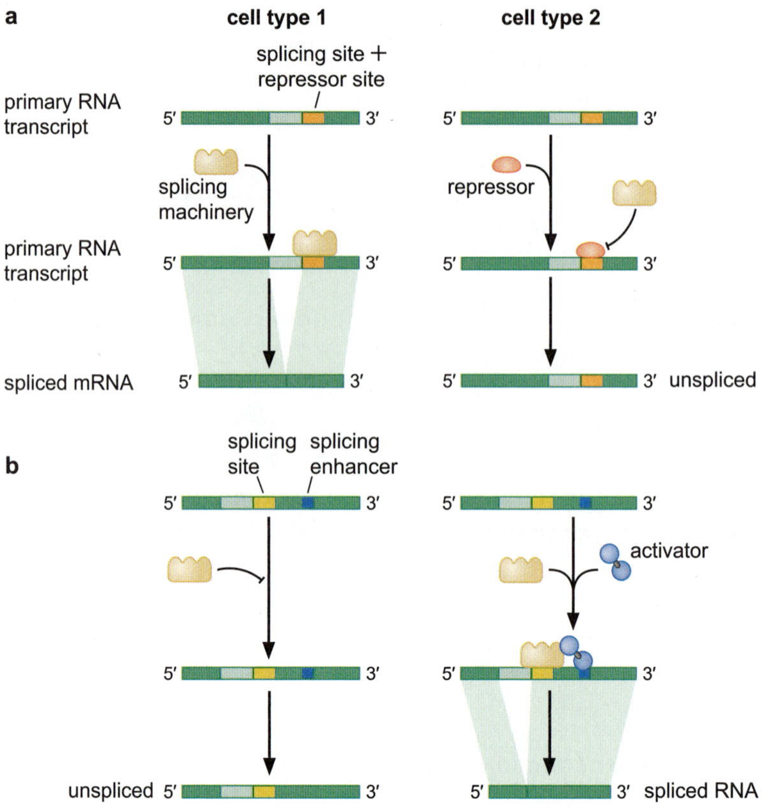

FIGURE 13-22 Regulated alternative splicing. Some alternatively spliced exons appear in mRNAs unless prevented from doing so by a repressor protein (a). Others appear only if a specific activator promotes their inclusion (b). Either mechanism can be used to regulate splicing such that in one cell type a particular exon is included in an mRNA, whereas in another it is not.

proteins are important in directing the splicing machinery to many exons, even when alternative splicing is not involved. In addition, in the example of T-antigen splicing described earlier (Fig. 13-16), it was an SR protein that ensured that alternative splicing occurred. But this protein family—which is large and diverse—has specific roles in *regulated* alternative splicing as well, directing the splicing machinery to different splice sites under different conditions. Thus, the presence or activity of a given SR protein can determine whether a particular splice site is used in a particular cell type or at a particular stage of development. Figure 13-22 shows hypothetical cases of regulated splicing by an activator bound to a splicing enhancer and a repressor bound to a splicing silencer.

The SR proteins bind RNA using one domain—for example, the well-characterized RNA-recognition motif (RRM) described in Chapter 5 (Figs. 5-23 and 5-24). Each SR protein has another domain, rich in arginine and serine, called an **RS domain.** The RS domain, found at the carboxy-terminal end of the protein, mediates interactions between the SR protein and proteins within the splicing machinery, recruiting that machinery to a nearby splice site.

An example of an activator that promotes a particular alternative splicing event in a specific tissue type is the *Drosophila* Half-pint protein. This activator regulates the alternative splicing of a set of pre-mRNAs in the fly ovary. It works by binding to sites near the 3′ splice site of specific exons in those pre-mRNAs and recruiting the U2AF splicing factor.

Most silencers are recognized by members of the heterogeneous nuclear ribonucleoprotein (hnRNP) family. These bind RNA but lack the RS domains and so cannot recruit the splicing machinery. Instead, by

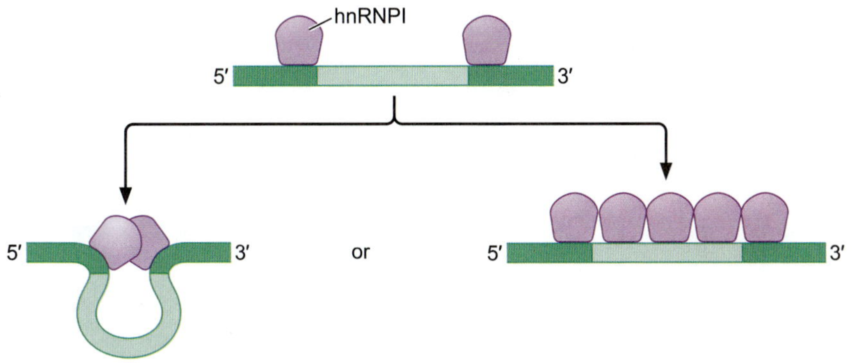

FIGURE 13-23 Inhibition of splicing by hnRNPI. Two models are presented. In one, the protein coats the entire exon. In the other, it binds at each end of the exon and conceals it within a loop.

blocking specific splice sites, they repress the use of those sites. We saw a function like this in the *Dscam* example earlier where Hrp36 inhibits inclusion of exon 6 variants in the mRNA. Another example is hnRNPA1, which binds to an exonic silencer element within an exon of the human immunodeficiency virus (HIV) *tat* pre-RNA and represses the inclusion of that exon in the final mRNA. By binding to its site, the repressor blocks binding of the activator SC35 (an SR protein) to a nearby enhancer element. This blocking is not direct—the two binding sites do not overlap—but hnRNPA1 promotes cooperative binding of additional molecules of hnRNPA1 to adjacent sequences, spreading over the enhancer site. When present, another SR protein (SF2/ASF) can overcome this repression, because it has a higher affinity for the enhancer sequence than does SC35 and therefore displaces the repressors bound there. We will see similar themes of cooperative and competitive binding in examples of transcriptional regulation in Chapters 16 and 17.

Another mammalian splicing repressor is the hnRNPI protein. In some cases, this protein blocks the binding of the basic splicing machinery by binding directly to the Py tract (explaining why hnRNPI is also called the polypyrimidine tract–binding protein). In other cases, it excludes a given exon from the mature mRNA by binding to sequences that flank that exon. This exclusion occurs either because molecules of hnRNPI at each end of the exon interact and loop out the exon, which is then passed over by the spliceosome, or because the molecules of hnRNPI at each end bind cooperatively with other molecules of hnRNPI, coating the RNA across the whole exon. This too would render the exon invisible to the splicing machinery (Fig. 13-23).

Regulation of Alternative Splicing Determines the Sex of Flies

We now consider a particularly elaborate example of regulated alternative splicing—that involving the *double-sex* gene of *Drosophila*. The sex of a given fly depends on which of two alternative splicing variants of this mRNA it produces. The regulation of a splicing cascade by repressors and activators lies at the heart of sex determination in *Drosophila*, as we now briefly describe.

The sex of a fly is determined by the ratio of *X* chromosomes to autosomes. A female results from a ratio of 1 (two *X*s and two sets of autosomes) and a male from a ratio of 0.5. This ratio is initially measured at the level of transcriptional regulation using two activators, called SisA and SisB (we consider mechanisms of transcriptional regulation in detail in Chapters 16 and 17). The genes encoding these regulators are both on the *X* chromosome, and so, in the early embryo, the

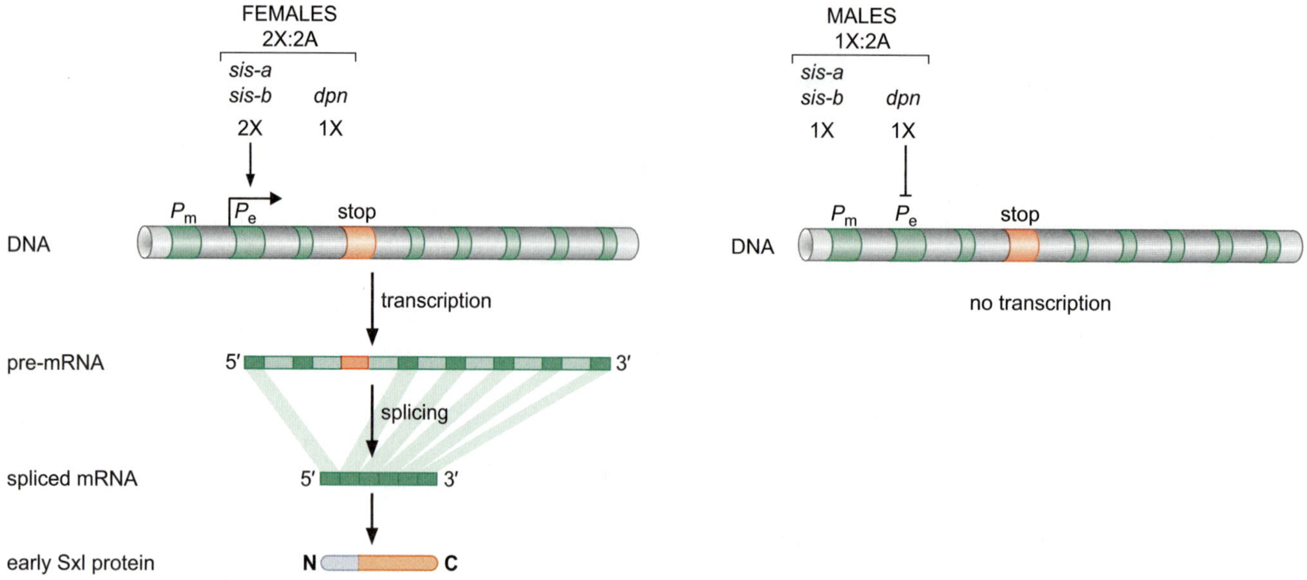

FIGURE 13-24 Early transcriptional regulation of *Sxl* in male and female flies. The *sisA* and *sisB* genes are found on the *X* chromosome and encode transcriptional activators that control expression of the *Sxl* gene. Dpn, a repressor of *Sxl*, is encoded by a gene on chromosome 2. Although both males and females express the same amount of the autosomally encoded Dpn, females make twice as much of the activators as males (because females have two *X* chromosomes and males have only one). The difference in ratio of activators to repressor ensures that *Sxl* is expressed in females but not males. The Sxl protein then autoregulates its own expression as described in the text and the next figure. (Adapted from Estes P. A. et al. 1995. *Mol. Cell. Biol.* 15: 904–917.)

prospective female makes twice as much of their products as does the male (Fig. 13-24).

These activators bind to sites in the regulatory sequence upstream of the gene *Sex-lethal* (*Sxl*). Another regulator that binds to and controls the *Sxl* gene is a repressor called Dpn (Deadpan); this is encoded by a gene found on one of the autosomes (chromosome 2). Thus, the ratio of activators to repressor differs in the two sexes, and this makes the difference between the *Sxl* gene being activated (in females) and repressed (in males).

The *Sxl* gene is expressed from two promoters, P_e and P_m. The former (promoter for establishment) is the one controlled by SisA and SisB

KEY EXPERIMENTS

Box 13-3 Identification of Docking Site and Selector Sequences

The docking site is 66 nucleotides long in *Drosophila melanogaster*. It is 90–100% conserved in ten other *Drosophila* species examined. Even when the comparison includes non-*Drosophila* insect species—mosquito, silkworm, and honeybee, for example—the central 24 nucleotides of the docking site are still very highly conserved. In fact, it is the most conserved sequence in the whole *Dscam* gene (which is more than 60 kb long)! Initial identification of the docking sequence was based entirely on this conservation (Box 13-3 Fig. 1).

The selector sequences were also discovered through sequence comparisons, even though they are less highly conserved than the docking site. Thus, selector sequences turned up as relatively conserved sequences in the introns upstream of exon 6 variants. An alignment of the 48 selector sequences from the exon 6 variants of *D. melanogaster* revealed a 28-nucleotide consensus sequence that was complementary to the docking sequence (Box 13-2 Fig. 2). When each individual selector sequence was compared with the docking site, each was seen to base-pair with it, each in a unique, but overlapping, manner. Some examples are shown in Figure 13-21.

Box 13-3 ***(Continued)***

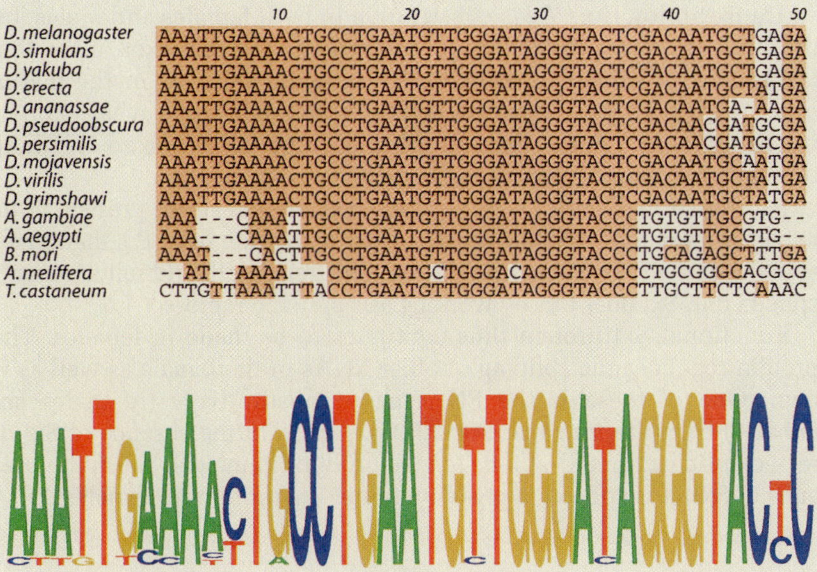

10
20
30
40
50
D. melanogaster
D. simulans
D. yakuba
D. erecta
D. ananassae
D. pseudoobscura
D. persimilis
D. mojavensis
D. virilis
D. grimshawi
A. gambiae
A. aegypti
B. mori
A. meliffera
T. castaneum
AAATTGAAAACTGCCTGAATGTTGGGATAGGGTACTCGACAATGCTGAGA
AAATTGAAAACTGCCTGAATGTTGGGATAGGGTACTCGACAATGCTGAGA
AAATTGAAAACTGCCTGAATGTTGGGATAGGGTACTCGACAATGCTGAGA
AAATTGAAAACTGCCTGAATGTTGGGATAGGGTACTCGACAATGCTATGA
AAATTGAAAACTGCCTGAATGTTGGGATAGGGTACTCGACAATGA-AAGA
AAATTGAAAACTGCCTGAATGTTGGGATAGGGTACTCGACAACGATGCGA
AAATTGAAAACTGCCTGAATGTTGGGATAGGGTACTCGACAACGATGCGA
AAATTGAAAACTGCCTGAATGTTGGGATAGGGTACTCGACAATGCAATGA
AAATTGAAAACTGCCTGAATGTTGGGATAGGGTACTCGACAATGCTATGA
AAATTGAAAACTGCCTGAATGTTGGGATAGGGTACTCGACAATGCTATGA
AAA---CAAATTGCCTGAATGTTGGGATAGGGTACCCTGTGTTGCGTG--
AAA---CAAATTGCCTGAATGTTGGGATAGGGTACCCTGTGTTGCGTG--
AAAT---CACTTGCCTGAATGTTGGGATAGGGTACCCTGCAGAGCTTTGA
--AT--AAAA---CCTGAATGCTGGGACAGGGTACCCCTGCGGGCACGCG
CTTGTTAAATTTACCTGAATGTTGGGATAGGGTACCCTTGCTTCTCAAAC

Box 13-3 Figure 1 The nucleotide sequence alignment of the docking sites of 15 insects. The insects analyzed include ten species of *Drosophila*; two of mosquito, *Anopheles gambiae* (malaria mosquito) and *Aedes aegypti* (yellow fever mosquito); the Lepidopteran *Bombyx mori* (silkworm); the Hymenopteran *Apis mellifera* (honeybee); and the Coleopteran *Tribolium castaneum* (red flour beetle). The most common nucleotide at each position is shaded, and the docking site consensus sequence is represented below as a pictogram. The height of each letter represents the frequency of each nucleotide at that position. (Modified, with permission, from Graveley B.R. 2005. *Cell* 123: 65–73, Fig. 2. © Elsevier.)

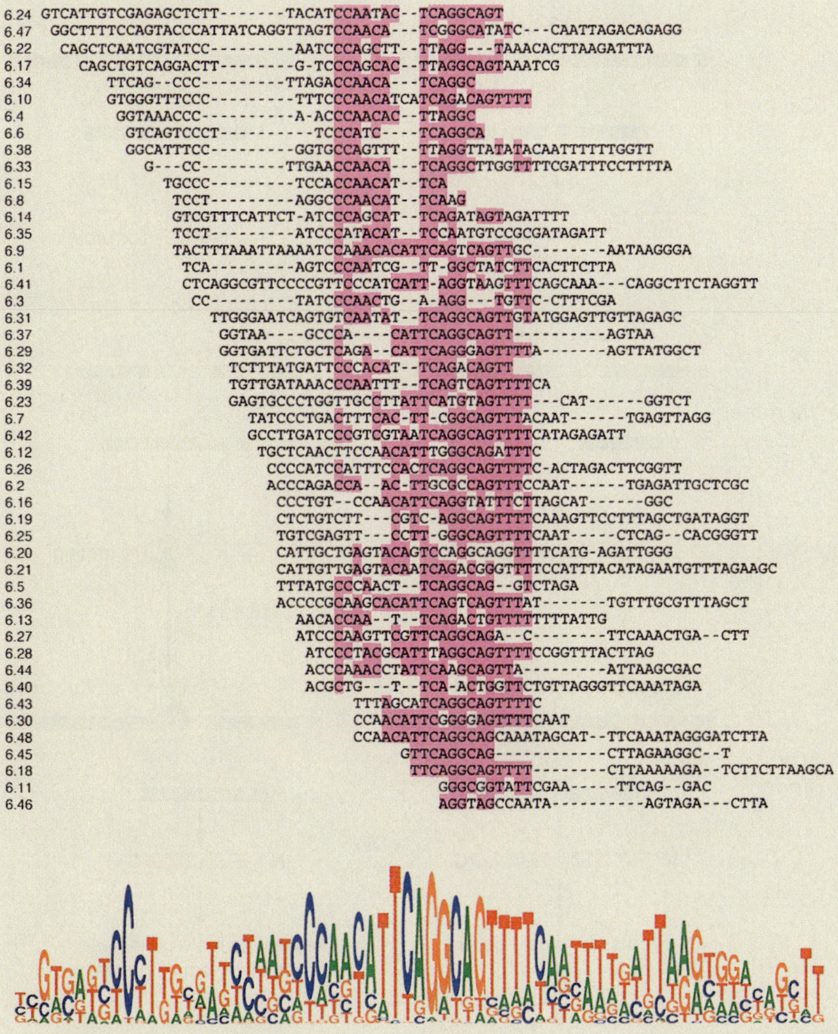

Box 13-3 Figure 2 *Drosophila melanogaster* selector sequence consensus. (Top panel) The 48 selector sequences and flanking sequence were aligned. The most frequent nucleotides in the central portion of the alignment are highlighted. (Bottom panel) The alignment was used to generate a selector sequence consensus. (Reprinted, with permission, from Graveley B.R. 2005. *Cell* 123: 65–73, Fig. 4. © Elsevier.)

(and hence expressed in females only). Later in development, this promoter is switched off permanently. In female embryos, expression of *Sxl* is maintained by expression from P_m (promoter for maintenance).

Transcription from P_m is constitutive in both females and males, but the RNA produced from this promoter contains one exon more than the transcript produced from P_e. If that exon remains in the mature message, it fails to produce an active protein, which is what happens in the male. But in the female, splicing removes that exon and functional Sxl protein continues to be produced.

As shown in Figure 13-25, it is the Sxl protein itself, present in the female but not the male (thanks to earlier expression from P_e), that directs splicing of the RNA made from P_m and ensures that the inhibitory exon is spliced out. Sxl does this by working as a splicing repressor.

Functional Sxl protein thus continues to be made in females. That protein regulates the splicing of other RNAs in the female as well as its own. One of these is the RNA made constitutively (in males and females) from the *tra* gene (Fig. 13-25). Again, in the absence of Sxl-directed splicing, this RNA fails to give protein (in males), but in the presence of Sxl, it is spliced to give functional Tra protein (in females).

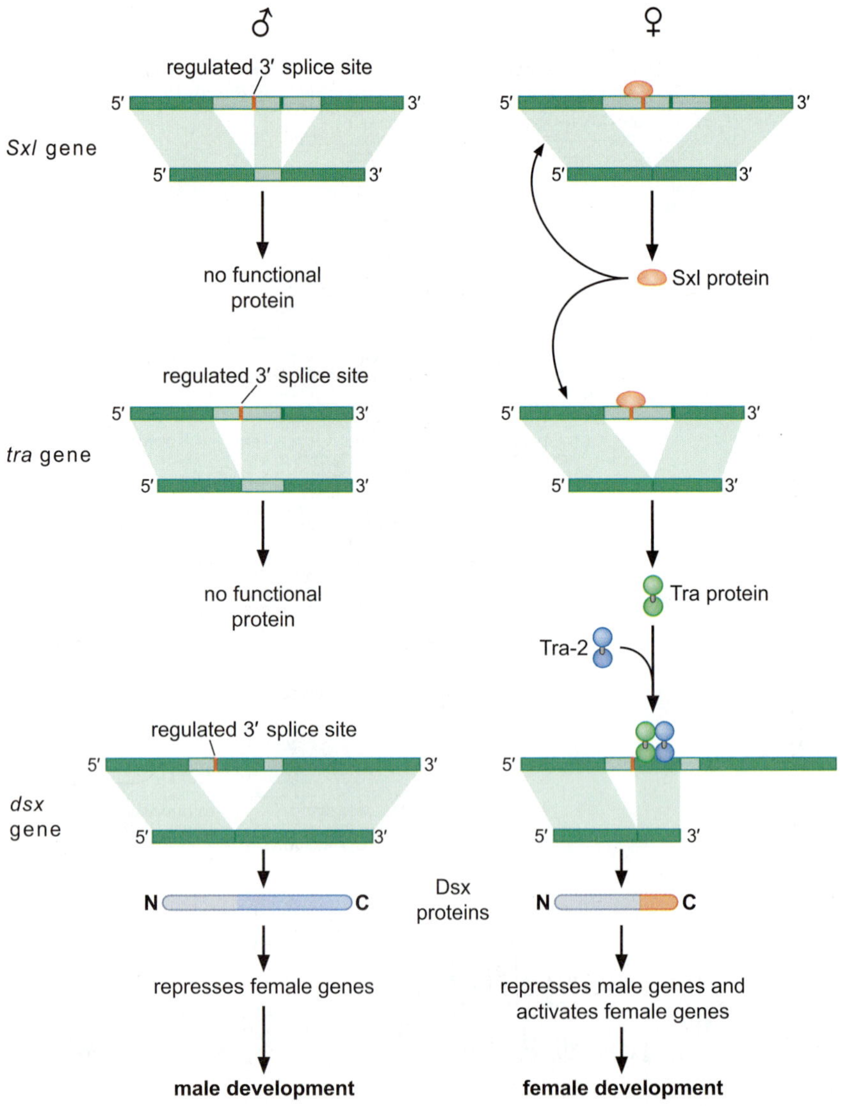

FIGURE 13-25 A cascade of alternative splicing events determines the sex of a fly. As described in detail in the text, the Sex-lethal protein is produced in flies that will develop into females (shown on the right of the figure) but not those that will develop into males (shown on the left). The presence of that protein is maintained by autoregulation of the splicing of its own message. In the absence of this regulation, no functional protein is produced (in males). Sex-lethal also controls splicing of the *tra* gene, producing functional Tra protein in females (but not males). Tra is itself a splicing regulator. It acts on pre-mRNA from the *doublesex* gene. When the *dsx* mRNA is spliced in response to Tra protein, a version of Doublesex protein is produced (in females) with a stretch of 30 amino acids at its carboxy-terminal end that distinguishes it from the form of the protein produced in the absence of the Tra regulator (in males). The female form of Dsx activates genes required for female development and represses those for male development. The male form, which has a stretch of 150 amino acids at the carboxy-terminal end, represses genes that direct female development. Sxl protein acts as a splicing repressor by binding to the pyrimidine tract at the 3′ splice site (see Fig. 13-3). The Tra protein, in contrast, acts as a splicing activator. It binds to an enhancer sequence in one of the exons of *dsx* RNA (see Fig. 13-12).

Tra protein is also a splicing regulator. Whereas Sxl is a splicing repressor, Tra is an activator (Fig. 13-25). One of its targets is RNA made from the gene encoding Doublesex (Dsx). This RNA is spliced in two alternative forms, both encoding regulatory proteins but with different activities. Thus, in the presence of *tra*, *dsx* RNA is spliced in a manner that gives rise to a protein that represses expression of male-specific genes. In the absence of Tra protein, the form of Dsx produced represses female-specific genes.

We have considered the various ways in which splicing is carried out in eukaryotic systems as well as the diversity of components involved. The loss of function in any of these components may lead to serious consequences, as we describe in Box 13-4, Defects in Pre-mRNA Splicing Cause Human Disease.

MEDICAL CONNECTIONS

Box 13-4 Defects in Pre-mRNA Splicing Cause Human Disease

As discussed in the text, the vast majority of human genes contain introns. Indeed, the large majority of human genes contain multiple introns. It is therefore not surprising that many point mutations that cause disease in humans turn out to be nucleotide substitutions that impair pre-mRNA splicing. In fact, estimates indicate that at least 15% of all point mutations that cause human disease alter recognition sequences for splicing. A classic example is β-thalassemia. This human genetic disorder is characterized by a defect in the production of β-globin, a subunit of hemoglobin. One kind of β-thalassemia is due to a mutation in the first intron of the β-globin gene that changes the sequence TT**G**GT to TT**A**GT. This mutation creates a sequence that resembles a normal 3′ splice site (Py tract AG/G; see Fig. 13-3). As a result, the splicing of β-globin pre-mRNA in afflicted individuals predominantly occurs at the mutationally created 3′ splice site rather than at the normal site.

An inherited disease known as "familial isolated growth hormone deficiency type II" is caused by a defect in the splicing of the pre-mRNA for growth hormone, resulting in individuals who are short in stature. Frasier syndrome is a urogenital disorder that is attributed to a defect in pre-mRNA splicing for a gene known to be important for kidney and gonad development. Two additional examples are a kind of dementia that is due to a splicing defect in the mRNA for a cytoskeleton protein and a form of cystic fibrosis.

Yet other disorders are caused by mutations that impair the splicing machinery itself. One example is retinitis pigmentosa, which is characterized by progressive degeneration of the retina and eventually blindness. Mutations at many genes cause retinitis pigmentosa, and most of these genes have retina-specific functions. Some of these mutations, however, are in genes for components of the spliceosome. Because afflicted individuals have one normal copy of the gene as well as the mutant copy, the splicing protein is produced but is present in lower than normal amounts.

Why is the effect of a lower than normal level of a splicing component manifest in a specific tissue, the retina? One possible explanation stems from the fact that the photopigment rhodopsin of the retina undergoes a high level of turnover. Thus, the splicing machinery must meet the very high demand for opsin (the protein component of rhodopsin) production to replace that lost from degradation. Hence, the retina might be more sensitive to a partial impairment of splicing than other tissues that do not have the burden of producing a specific protein at high levels.

Spinal muscular atrophy is one of the most common genetic causes of mortality in children. The disease, which is characterized by the progressive loss of spinal neurons, results from a mutation in a gene for a ubiquitous component known as SMN (survival motor neuron) of the splicing machinery whose precise function is not well understood. As in the case of retinitis pigmentosa, we are left with the mystery of why the effect of the splicing defect is principally manifest in motor neurons.

The examples considered thus far are inherited disorders. But disease-causing mutations that impair splicing also arise somatically. An example comes from mutants of the gene for the cell cycle regulator p73. The p73 protein exists in multiple forms as a result of alternative splicing of its mRNA. Mutations that cause faulty alternative splicing of the p73 pre-mRNA have been implicated in a kind of cancer known as squamous cell carcinoma. It is likely that somatic mutations that impair splicing or cause faulty alternative splicing for many other pre-mRNAs also contribute to the etiology of cancer.

It is sometimes said that medicine is the greatest teacher of biology. Certainly, and as we have seen, this adage aptly applies to the field of pre-mRNA splicing where the study of human genetic disorders has provided a wealth of insights into the sequences that govern splicing and the machinery that carries it out. Indeed, the very discovery of snRNPs, the most fundamental components of the pre-mRNA splicing machinery, arose from studies of a form of the autoimmune disease lupus in which afflicted individuals produce antibodies against these ribonuclear protein particles. It seems likely that the continued study of human genetic disorders will lead to additional insights into the mechanisms of pre-mRNA maturation.

EXON SHUFFLING

Exons Are Shuffled by Recombination to Produce Genes Encoding New Proteins

As we have noted, all eukaryotes have introns, and yet these elements are rare—almost nonexistent—in bacteria. There are two likely explanations for this situation.

First, in the so-called **introns early model**, introns existed in all organisms but had been lost from bacteria. If introns originally did exist in bacteria, why might they subsequently have been lost? The argument is that these "gene-rich" organisms (see Chapters 7 and 11) have streamlined their genomes in response to selective pressure to increase the rate of chromosome replication and cell division. (Recall also that among eukaryotes, the yeast, which are unicellular and rapidly growing, have far fewer introns than do complex multicellular organisms.)

In the alternative view, introns never existed in bacteria but rather arose later in evolution. According to this so-called **introns late model,** introns were inserted into genes that previously had no introns, perhaps by a transposon-like mechanism (see Chapter 11).

Irrespective of which explanation is true—and at this stage, it is impossible to decide the matter unambiguously—there is the second, perhaps more interesting, question: why have the introns been retained in eukaryotes and, in particular, in the extensive form seen in multicellular eukaryotes? One clear advantage is that the presence of introns, and the need to remove them, allows for alternative splicing that can generate multiple protein products from a single gene. But, on an even grander scale, another advantage afforded these organisms is believed to be the following: having the coding sequence of genes divided into several exons allows new genes to be created by reshuffling exons. Three observations strongly suggest that this process actually occurs:

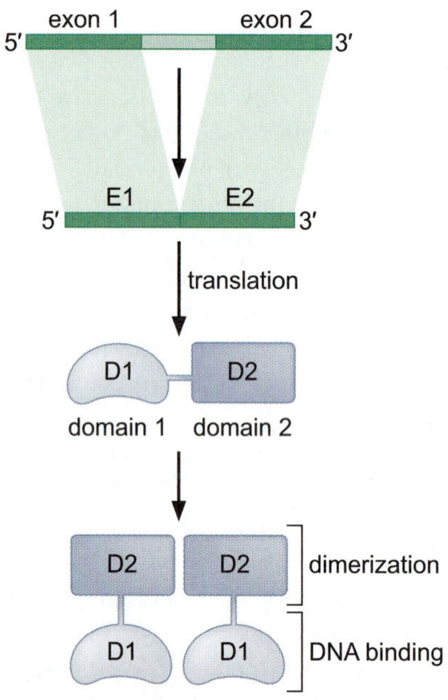

FIGURE 13-26 Exons encode protein domains. In this example, the DNA-binding domain of a protein is encoded by one exon, whereas the dimerization domain of that same protein is encoded by a separate exon. Protein domains fold independently of the rest of the protein in which they are found and often carry out a single function (as we discussed in Chapter 5). Thus, exons can often be exchanged between proteins productively.

- First, the borders between exons and introns within a given gene often coincide with the boundaries between domains (see Chapter 5) within the protein encoded by that gene. That is, it seems that each exon very often encodes an independently folding unit of protein (often corresponding to an independent function). For example, consider the DNA-binding protein depicted in Figure 13-26. Like most DNA-binding proteins, this one has two domains—the DNA recognition domain and the dimerization domain. As shown in the figure, these domains (D1 and D2) are encoded by separate exons (E1 and E2) within the gene.
- Second, many genes, and the proteins they encode, have apparently arisen during evolution in part via exon duplication and divergence. Proteins made up of repeating units (such as immunoglobulins) have probably arisen this way (see Fig. 11-35). The presence of introns between each exon makes the duplication more likely.
- Third, related exons are sometimes found in otherwise unrelated genes. That is, there is evidence that exons really have been reused in genes encoding different proteins. As an example, consider the low-density lipoprotein (LDL) receptor gene (Fig. 13-27). This gene contains some exons that are clearly evolutionarily related to exons found in the gene encoding the epidermal growth factor (EGF) precursor. At the same time, it has other exons that are clearly related to exons from the C9 complement gene (Fig. 13-27). More extensive examples of exon accretion are apparent from the complete sequences of genomes. As shown in Figure 13-28, there are

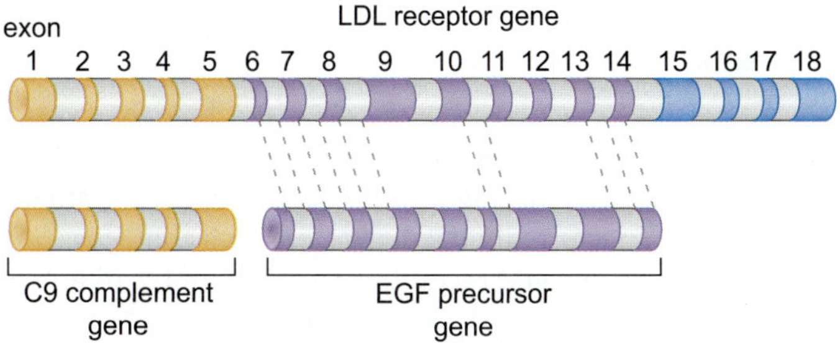

FIGURE 13-27 Genes made up of parts of other genes. The LDL receptor (the plasma low-density lipoprotein receptor) gene contains a stretch of six exons closely related to six exons from the C9 complement gene and eight closely related to eight from the EGF (epidermal growth factor) precursor gene. Thus, the LDL receptor gene is made up of exons shuffled between other genes; although not shown here, these same parts appear in yet other genes as well. The introns are, in many cases, not positioned in exactly the same positions within the EGF precursor gene and the comparable region of the LDL receptor gene. When they *are* in the same place, this is indicated by dotted lines.

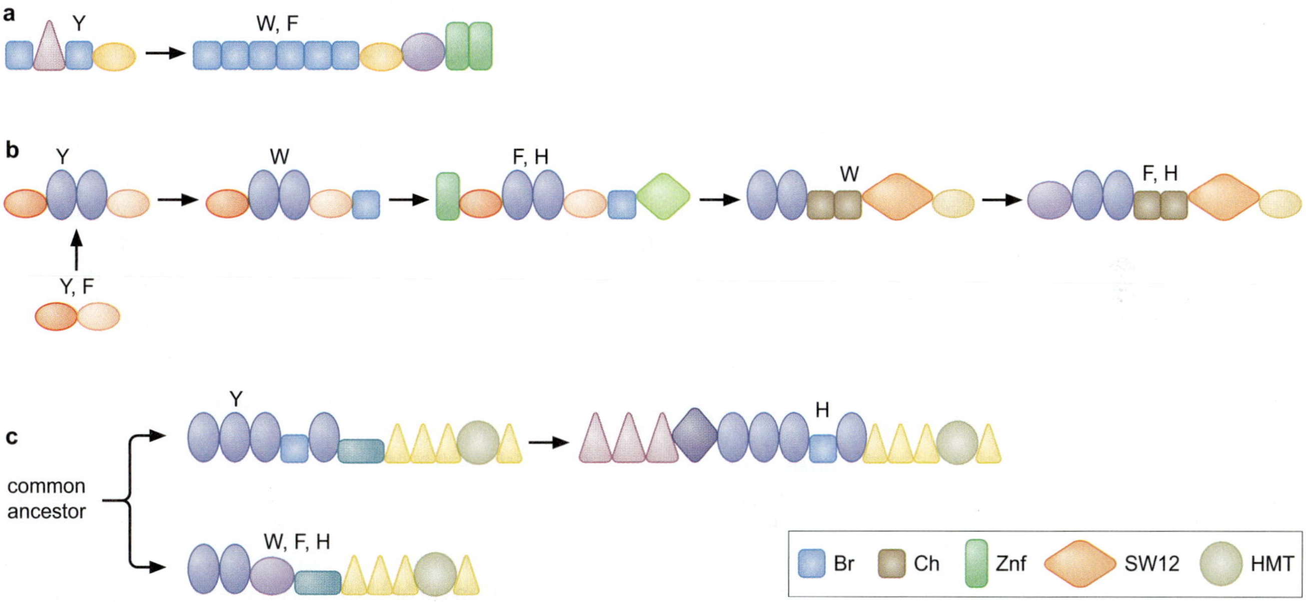

FIGURE 13-28 Accumulation, loss, and reshuffling of domains during the evolution of a family of proteins. Proposed routes whereby different related proteins might have evolved by gain and loss of specific domains. Three examples are given; in each case, the proteins in question are chromatin-modifying enzymes (Chapter 7) from yeast (Y), worms (W), flies (F), and humans (H). Each protein is depicted by a series of differently colored and shaped domains, and above each protein is shown the organism(s) in which proteins are found containing the domain arrangement shown. Some arrangements are found in more than one organism, and in some cases a given organism has more than one related arrangement of similar domains. A few of the domains—those whose functions we discussed in Chapters 7 or 17—are identified and are as follows: bromodomain (Br); chromodomain (Ch); a histone methyltransferase domain (HMT); an ATPase activity associated with chromatin-remodeling enzymes (SW12); and a zinc finger domain (Znf). (Adapted, with permission, from Lander et al. 2001. *Nature* 409: 906, Fig. 42. © Macmillan.)

numerous examples of proteins made up of highly related domains used in various combinations, encoded by genes made up of shuffled exons.

As we have seen, exons tend to be rather short (~150 nucleotides or so), whereas introns vary in length and can be very long indeed (up to several hundred kilobases). The size ratio ensures that, for the average gene in a higher eukaryote, recombination is more likely to occur within the introns than within the exons. Thus, exons are more likely to be reshuffled than disrupted. The mechanism of splicing—the use of the 5′ and 3′ splice sites—guarantees that almost all recombinant genes will be expressed, because the splice sites in different genes are largely interchangeable. In addition, alternative splicing can allow new exons to be tried without discarding the original gene product—that is, both the new and old products can be made initially.

RNA EDITING

RNA Editing Is Another Way of Altering the Sequence of an mRNA

RNA editing, like RNA splicing, can change the sequence of an RNA after it has been transcribed. Thus, the protein produced upon translation is different from that predicted from the gene sequence, but this example is perhaps even more dramatic than the case of splicing—instead of stretches of the mRNA being reassorted, during editing, individual bases are either inserted, deleted, or changed. That is, the coding information in the RNA is altered. There are two mechanisms that mediate editing: site-specific deamination of adenines or cytosines and guide RNA–directed uridine insertion or deletion. We consider each in turn.

In one form of site-specific **deamination,** a specifically targeted cytosine residue within mRNA is converted into uridine by deamination. For a given mRNA species that undergoes editing, that process typically occurs only in certain tissues or cell types and in a regulated manner. Figure 13-29 shows the mammalian apolipoprotein-B gene. This gene has several exons, within one of which is a particular CAA codon that is targeted for editing; it is the C within this codon that gets deaminated. That deamination, carried out by the enzyme **cytidine deaminase**, converts the C to a U (Fig. 13-30). In this example, the deamination occurs in a tissue-specific manner: messages are edited in intestinal cells but not in liver cells.

The CAA codon, which is translated as glutamine in the unedited message in the liver, is thus converted to UAA—a stop codon—in the intestine. The result is that the full-length protein (of some 4500 amino acids) is produced in the liver, but a truncated polypeptide of only about 2100 amino acids is made in the intestine (see Fig. 13-29).

The two forms of apolipoprotein B are both involved in lipid metabolism. The longer form, found in the liver, is involved in the transport of endogenously synthesized cholesterol and triglycerides. The smaller version, found in the intestines, is involved in the transport of dietary lipids to various tissues.

Other examples of mRNA editing by enzymatic deamination include adenosine deamination. This reaction, carried out by the enzyme **ADAR (adenosine deaminase acting on RNA)**—of which there are three in humans—produces inosine. Inosine can base-pair with cytosine, and

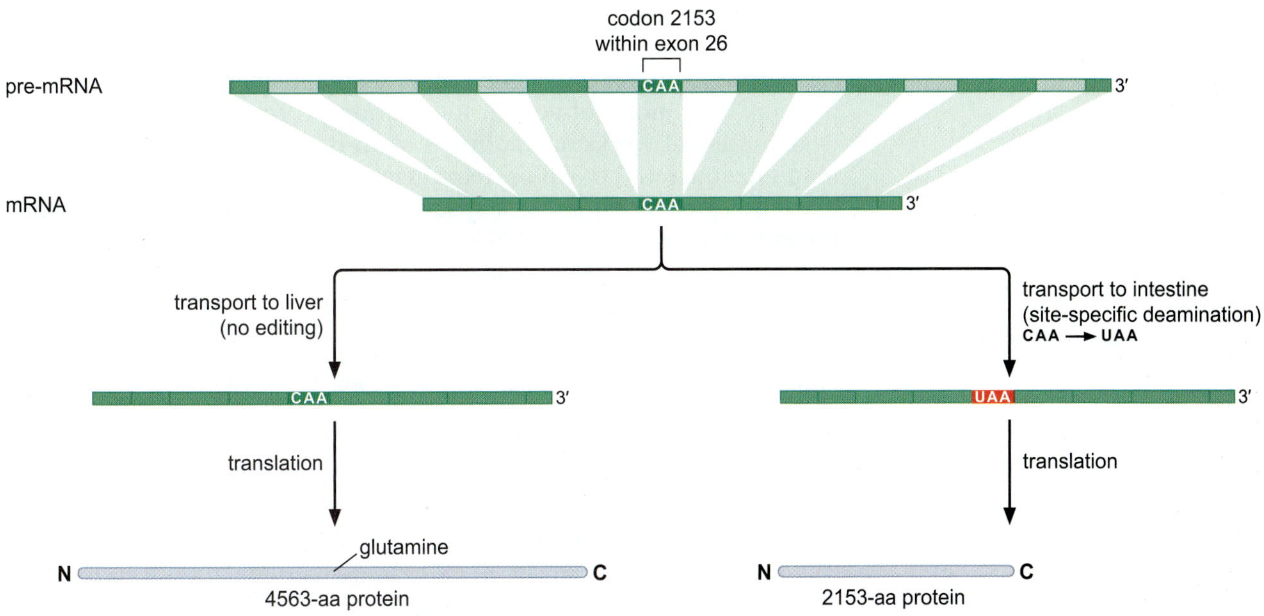

FIGURE 13-29 RNA editing by deamination. The RNA made from the human apolipoprotein gene is edited in a tissue-specific manner by deamination of a specific cytidine to generate a uridine. This event occurs in RNAs in the intestine, but not in those found in the liver. The result, as described in the text, is that a stop codon introduced into the intestinal mRNA generates a shorter protein than that produced in the liver. The figure is not drawn to scale: thus the edited exon is exon 26, and the codon marked as filling it is in reality only a very short part of that exon.

so this change can readily alter the sequence of the protein encoded by the mRNA. An ion channel expressed in mammalian brains is the target of this type of editing. A single edit in its mRNA elicits a single-amino-acid change in the protein, which in turn alters the Ca^{2+} permeability of the channel. In the absence of this editing, brain development is seriously impaired.

This type of editing—enzymatic deamination—seems to be quite rare, but important. In *Drosophila*, it has been estimated that there may be as few as 20 cytosines targeted for deamination, but all are in genes involved in neurotransmitter production or activity.

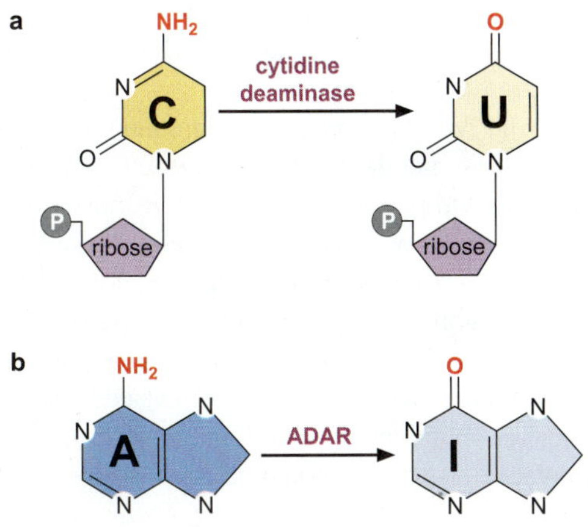

FIGURE 13-30 The deamination of cytosine and adenine to produce uracil and inosine. (a) The amino group on the nucleotide ring is removed by the cytidine deaminase enzyme. (b) In the case of adenine deamination, the same chemical group is removed from the adenine by ADAR to generate inosine.

It is not entirely clear how the deaminase enzymes work so specifically: their active sites could act on any cytosine. Often the proteins are part of complexes in which other components might influence the proteins' specificity of action. In addition, in the case of the cytosine deaminase that works on apolipo-protein-B, the enzyme bears an RNA-binding domain that helps recognize the specific site for deamination—by recognizing either a specific sequence or perhaps a particular secondary structure in the RNA.

Another role for deaminase enzymes in the cellular defenses against HIV infection is described in Box 13-5, Deaminases and HIV.

Guide RNAs Direct the Insertion and Deletion of Uridines

A very different form of RNA editing is found in the RNA transcripts that encode proteins in the mitochondria of trypanosomes. In this case, multiple Us are inserted into specific regions of mRNAs after transcription (or, in other cases, Us may be deleted). These insertions can be so extensive that, in an extreme case, they amount to as many as half the nucleotides of the mature mRNA. The addition of Us to the message changes codons and reading frames, completely altering the "meaning" of the message. As an example, consider the trypanosome *coxII* gene. In a specific region of the mRNA of this gene, four Us are inserted between adjacent bases at three sites (two Us at one site and one U at each of two additional sites). These additions alter some codons and cause a "–1" change in the reading frame, a

MEDICAL CONNECTIONS

Box 13-5 Deaminases and HIV

Deamination of the human apolipoprotein-B mRNA described in the text is undertaken by an enzyme called APOBEC1 (apolipoprotein-B editing enzyme, catalytic polypeptide-like 1). This is a member of a family of enzymes that direct deamination of cytidines in both RNA and DNA. Another member of the family—APOBEC3G (A3G)—is a potent inhibitor of infection by a range of retroviruses, including HIV.

As viral particles, retroviruses such as HIV carry RNA genomes. Upon infection, the RNA is converted to a cDNA copy by reverse transcription (see Box 11-2). It is the minus strand of the cDNA produced during reverse transcription that is attacked by the A3G enzyme. The enzyme deaminates Cs to produce Us in that DNA strand, leading to hypermutation at levels the virus cannot accommodate, or even to destruction of the damaged strand by DNA glycosylase and apurinic-apyrimidinic endonuclease (Chapter 9).

To counter this, wild-type HIV produces a protein called Vif (viral infectivity factor) that directs proteosomal degradation of the A3G enzyme, thereby excluding it from viral particles and protecting the virus in its next round of infection. Vif is required by the virus to grow in all its biologically relevant target cells in vivo. Some cell lines used in laboratories to grow virus were found to support growth of HIV lacking the Vif function. Such cells were called **permissive**. Heterokaryons (cells made by fusing two other cell types) made from permissive and nonpermissive cells had the nonpermissive character. This revealed that the nonpermissive cells make a factor that countered viral replication. That factor was shown to be deaminase A3G, and permissive cells could be made nonpermissive simply by expressing A3G.

shift that is required to generate the correct open reading frame, as shown in Figure 13-31a.

How are these additional bases inserted? Us are inserted into the message by so-called **guide RNAs (gRNAs)**, as shown in Figure 13-31. These gRNAs range from 40 to 80 nucleotides in length and are encoded by genes distinct from those that encode the mRNAs they act on. Each gRNA is divided into three regions. The first, at the 5′ end, is called the "anchor" and directs the gRNA to the region of the mRNA it will edit; the second determines exactly where the Us will be inserted within the edited sequence; and the third, at the 3′ end, is a poly-U stretch. We now look more closely at how the gRNAs direct editing.

The anchor region of the gRNA contains a sequence that can base-pair with a region of the message immediately beside (3′ to) the region that will be edited (Fig. 13-31b). This is followed by the editing "instruc-

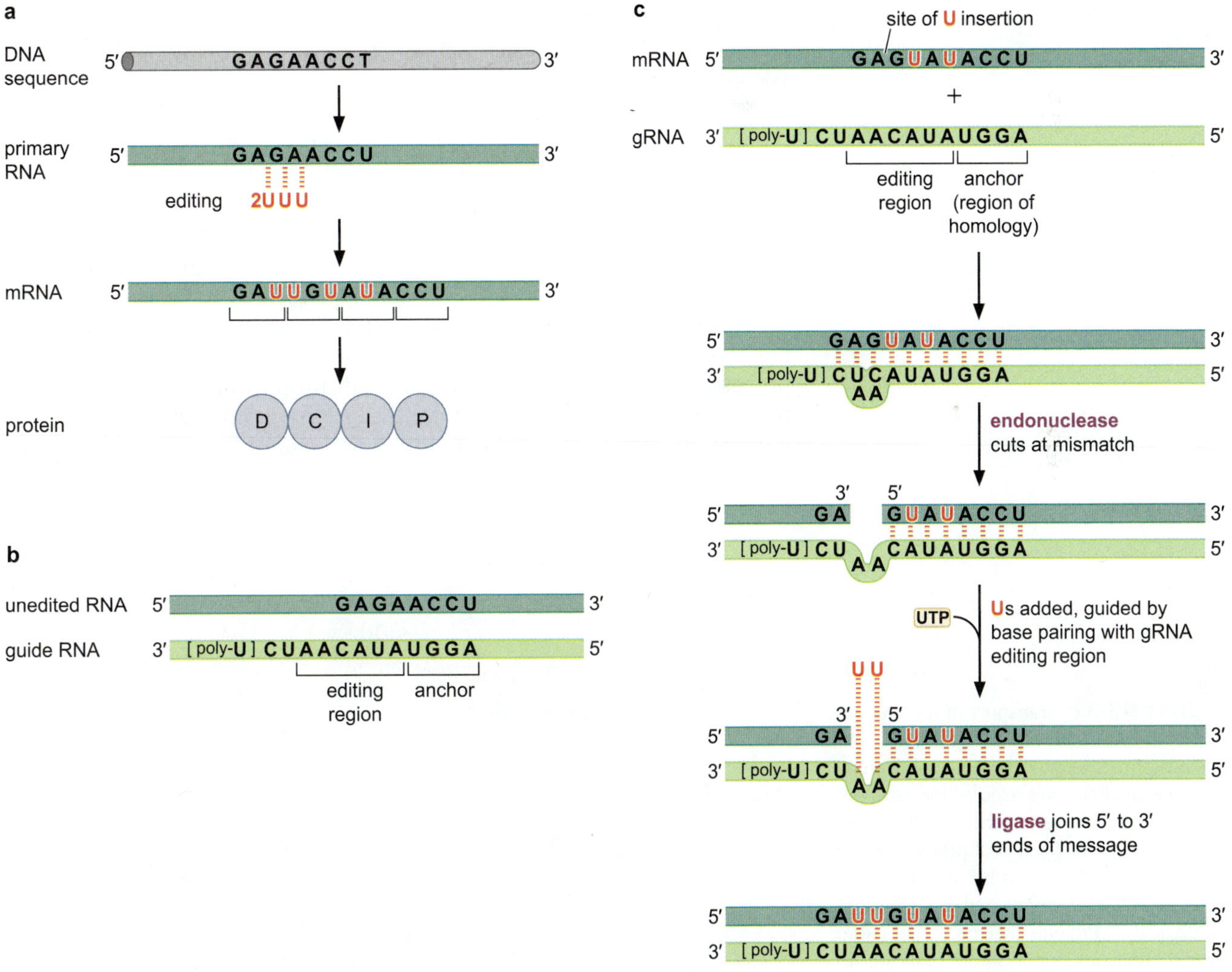

FIGURE 13-31 **RNA editing by guide RNA–mediated U insertion.** Editing of the trypanosome *coxII* gene RNA. (a) The positions of the four U nucleotides inserted into the pre-mRNA of the *coxII* gene. These generate the correct reading frame and coding information in the mRNA. (b) The sequence of the guide RNA that determines the U insertion pattern, and the sequence of the unedited stretch of mRNA. (c) The editing reaction itself.

tions": a stretch of gRNA complementary to the region in the message to be edited but containing additional As. The As are at positions in the gRNA opposite where Us will be inserted into the mRNA. At the 3′ end of the gRNA is the poly-U region. The role of the nucleotides in this region is unclear, although it is proposed that they tether the gRNA to purine-rich sequences in the mRNA upstream of (5′ to) the edited region.

As shown in Figure 13-31c, the gRNA and mRNA form an RNA–RNA duplex with looped-out single-stranded regions opposite where Us will be inserted. An endonuclease recognizes and cuts the mRNA opposite these loops. Editing involves the transfer of Us into the gap in the message. This process is catalyzed by the enzyme 3′ terminal uridylyl transferase (TUTase).

After the addition of Us, the two halves of the mRNA are joined by an RNA ligase, and the "editing" region of the gRNA continues its action along the mRNA in a 3′ to 5′ direction. A single gRNA can be responsible for inserting several Us at different sites (as is the case for the one shown in Fig. 13-31). Furthermore, in some cases, several different gRNAs work on different regions of the same message.

mRNA TRANSPORT

Once Processed, mRNA Is Packaged and Exported from the Nucleus into the Cytoplasm for Translation

Once it has been fully processed—capped, spliced, and polyadenylated—an mRNA is transported out of the nucleus and into the cytoplasm (Fig. 13-32), where it is translated to give its protein product (Chapter 14). Movement from the nucleus to the cytoplasm is not a passive process. Indeed, it must be carefully regulated: the fully processed mRNAs represent only a small proportion of the RNA found in the nucleus, and many of the other RNAs would be detrimental to the cell if exported. These include, for example, damaged or mis-

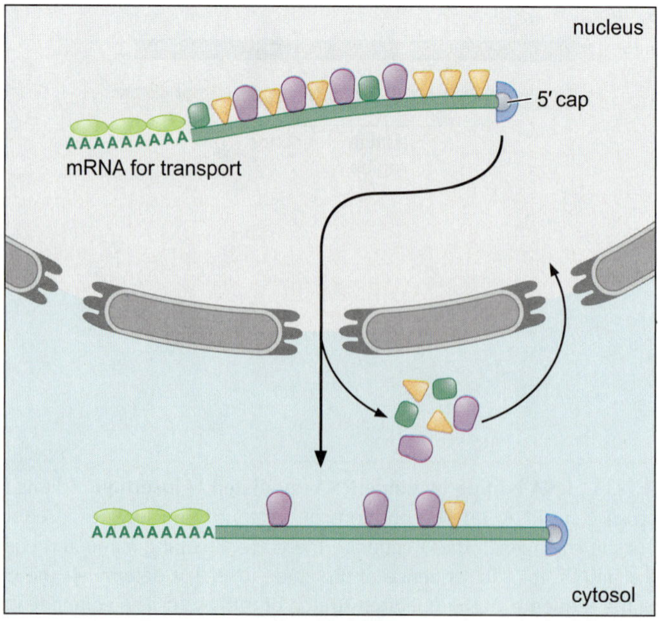

FIGURE 13-32 Transport of mRNAs out of the nucleus. RNA export from the nucleus is an active process, and only certain (appropriate) RNAs are selected for transport. To be selected for transport, the RNA must have the correct collection of proteins bound to it. These will distinguish it from other RNAs, which must be retained in the nucleus or destroyed. Proteins that recognize exon: exon boundaries, for example, indicate an mRNA that has been appropriately spliced, whereas proteins that bind introns indicate an RNA that should be retained in the nucleus. Once in the cytoplasm, some proteins are shed and others are taken on in readiness for translation (Chapter 14).

processed RNAs and liberated introns (which, being, as they tend to be, so much larger than the exons, represent a larger population of RNA than do the mature mRNAs).

How are RNAs selected and transported? As we have emphasized in this and the previous chapter, from the moment an RNA molecule starts to be transcribed, it becomes associated with proteins of various sorts: initially proteins involved in capping, then splicing factors, and finally the proteins that mediate polyadenylation. Some of these proteins are replaced at various steps along the processing path, but others (including, for example, some SR proteins—the serine-argine-rich splicing regulators [Fig. 13-12]) are not; moreover, additional proteins join. As a result, a typical mature mRNA carries a collection of proteins that identifies it as being mRNA destined for transport. Other RNAs not only lack the particular signature collection required for transport, but have their own alternative sets of proteins that actively block export. Thus, for example, excised introns will often carry hnRNPs (repressors of splicing seen in Fig. 13-23), and these probably mark such an RNA for nuclear retention and destruction.

In addition to residual SR proteins, mature mRNAs carry another group of proteins that bind specifically to exon–exon junctions (which are only found in spliced species of course). The mRNAs do also carry some hnRNPs, but fewer than are typically bound to introns, and also in a different context of course. This emphasizes the fact that it is the set of proteins, not any individual kind of protein, that marks RNAs for either export or retention in the nucleus.

Export takes place through a special structure in the nuclear membrane called the **nuclear pore complex.** Small molecules—those under about 50 kD—can pass through these pores unaided, but larger molecules and complexes, including mRNAs and their associated proteins, require active transport. (Other molecules—proteins made in the cytoplasm but with functions in the nucleus, for example—are transported in the other direction, from the cytoplasm into the nucleus, through these same pores.)

The mechanisms of nuclear transport are beyond the scope of this volume. Suffice it to say that some of the proteins associated with the RNA carry nuclear export signals that are recognized by export receptors that guide the RNA out through the pore. Once in the cytoplasm, the proteins are discarded and are then recognized for import back into the nucleus where they associate with another mRNA and repeat the cycle (Fig. 13-32).

Export requires energy, and this is supplied by hydrolysis of GTP by a GTPase protein called Ran. Like other GTPases, Ran exists in two conformations depending on whether complexed with GTP or GDP, and the transition from one state to the other drives movement into or out of the nucleus.

SUMMARY

Most genes encode proteins, and the sequence of amino acids within any given protein is determined by the sequence of "codons" in its gene. Each codon is made up of a group of three adjacent nucleotides. In almost all bacterial and phage genes, the open reading frame is a single stretch of codons with no break. But the coding sequence of many eukaryotic genes is split into stretches of codons interrupted by stretches of noncoding sequence.

The coding stretches in these split genes are called exons (for "expressed sequences"), and the noncoding stretches are called introns (for "intervening sequences"). Some noncoding regions are also included in mature mRNAs—5′ and 3′ untranslated regions of mRNA and entire noncoding RNAs such as microRNAs (Chapter 18). Such regions are therefore also classified as exons. The numbers and sizes of the introns and exons vary enormously from gene to gene. Thus, in yeast, only a relatively small proportion of genes have introns, and where they occur, they tend to be short and few in number (one or occasionally two per gene). In multicellular organisms such as humans, the number of genes containing introns is much larger, as is the number of introns per gene (up to 362 in an extreme case). The sizes of exons do vary but are often about 150 nucleotides; introns, on the other hand, vary from 61 bp to as much as a staggering 800 kb.

When a gene containing introns is transcribed, the RNA initially contains those introns. These are then removed to produce the mature mRNA. The process of intron removal is called splicing.

Many intron-containing genes give rise to a unique mRNA species. That is, in each case, all of the introns are removed from the original RNA, leaving an mRNA composed of all the exons. But in other cases, splicing can produce a number of different mRNAs from the same gene by splicing the original RNA in different patterns. Thus, for example, some genes contain alternative versions of some of their exons, and only one of these variants ends up in a given mRNA. In other cases, a given exon might be removed (along with the introns) from some copies of the RNA—again producing alternative versions of mRNA from the same gene. We considered in detail one extreme example of alternative splicing—the *Dscam* gene of *Drosophila*. In this case, one of its exons comes in 48 variants, all of which are found in the pre-mRNA, but one (and only one) region is found in each mRNA. How such mutually exclusive selection is achieved was worked out by comparing the sequences of selected genes from several species in an elegant example of bioinformatics analysis.

Sequences found at the boundary between introns and exons allow the cell to identify introns for removal. These splicing sequences are almost exclusively within the introns (where there are no restrictions imposed by the need to encode amino acids, as there are in exons). These sequences are called the 3′ and 5′ splice sites, denoting their relative locations at one end of the intron or the other end. To splice out an intron also requires a sequence element, called the branch site, near the 3′ end of the intron.

Intron removal proceeds via two transesterification reactions. In the first, an A in the branch site attacks a G in the 5′ splice site. In the second, the liberated 5′ exon attacks the 3′ splice site. These reactions have two consequences. First and foremost, they fuse the two exons. Second, they release the intron in the form of a branched structure called a lariat.

Splicing of nuclear pre-mRNAs requires a large complex of proteins and RNAs called the spliceosome. This is made up of so-called snRNPs, of which there are five—U1, U2, U4, U5, and U6 snRNPs. Each of these comprises an RNA molecule, called the U1 to U6 snRNA, respectively, and a number of proteins, the majority of which are different in each case. The RNA components have a central role in recognizing introns and catalyzing their removal. The spliceosome is a very dynamic structure. That is, at different steps during the process of splicing, the spliceosome constitution alters—different subunits of the machine join and leave the complex, each performing a particular function.

Thus, early on, U1 snRNP recognizes the 5′ splice site, whereas the U2 snRNP recognizes the branch site. U4 and U6 then join, together with U5, bringing the branch site and 5′ splice site together and stimulating the first reaction concomitant with U1 and U4 leaving. Finally, the 3′ and 5′ splice sites are brought together and exons are fused.

There are a few rare introns that can remove themselves from within RNA molecules by a process known as self-splicing. Although not strictly an enzymatic reaction, the RNA of the intron nevertheless mediates the chemistry of removal. These self-splicing introns come in two classes, one of which (group II) splices by the same chemical pathway as that mediated by the spliceosome. These introns probably represent the evolutionary origin of modern spliceosomal introns, and the two-step chemical pathway used by both reflects that evolutionary relationship (and perhaps explains why the spliceosome does not remove introns by a more direct single-step mechanism).

The splice sites are defined by rather short sequences with low levels of conservation. It thus represents a significant challenge for the splicing machinery to recognize and splice only at correct sites. There are various mechanisms by which the spliceosome enhances accuracy. First, it assembles on the sites soon after they have been synthesized. This ensures they are selected before other downstream sites are available to compete. Second, there are other proteins—SR proteins—that bind near legitimate splice sites and help recruit the splicing machinery to those sites. In this way, authentic sites effectively have a higher affinity for the machinery than do so-called pseudo sites of similar sequence.

There are a large variety of SR proteins. Each binds RNA with one surface and stimulates binding of the splicing machinery with another. Some SR proteins regulate splicing. That is, a given SR protein may be found only in one cell type and mediate a particular splicing event only in that cell type. Other SR proteins are only active in the presence of specific physiological signals, and so a given splicing event only occurs in response to that signal. In this way, SR proteins resemble transcriptional activators, as we shall see in later chapters. Also, analogous to transcriptional regulation, there are repressors of splicing that exclude splicing of specific introns under certain circumstances. We considered in detail the case of sex determination in *Drosophila* where a cascade of regulated, alternative splicing events determines whether the fly develops as a male or female.

Together with the other modifications dealt with in Chapter 12, splicing is required before mRNAs can be transported out of the nucleus through nuclear pores. This too can be regulated.

It is believed that a given exon typically encodes an independently folding (and functional) protein domain. Thus, such an exon can readily function in combination with other different exons. This suggests it has been relatively easy, through evolution, to generate new proteins by shuffling existing exons between genes.

RNA editing is another mechanism that allows an RNA to be changed after transcription so as to encode a different protein from that encoded by the gene. Two mechanisms for editing are enzymatic modification of bases and the insertion or deletion of multiple U nucleotides within the message.

BIBLIOGRAPHY

Books

Gesteland R.F., Cech T.R., and Atkins J.F., eds. 2006. *The RNA world*, 3rd ed. Cold Spring Harbor Laboratory Press, Cold Spring Harbor, New York.

Mechanisms of Splicing and the Spliceosome

Crick F. 1979. Split genes and RNA splicing. *Science* **204:** 264–271.

Hastings M.L. and Krainer A.R. 2001. Pre-mRNA splicing in the new millennium. *Curr. Opin. Cell Biol.* **13:** 302–309.

Hertel K.J. and Graveley B.R. 2005. RS domains contact the pre-mRNA throughout spliceosome assembly. *Trends Biochem. Sci.* **30:** 115–118.

Maniatis T. and Reed R. 2002. An extensive network of coupling among gene expression machines. *Nature* **416:** 499–506.

Staley J.P. and Guthrie C. 1998. Mechanical devices of the spliceosome: Motors, clocks, springs, and things. *Cell* **92:** 315–326.

Tange T.Ø., Nott A., and Moore M.J. 2004. The ever-increasing complexities of the exon junction complex. *Curr. Opin. Cell Biol.* **16:** 279–284.

Tarn W.Y. and Steitz J.A. 1997. Pre-mRNA splicing: The discovery of a new spliceosome doubles the challenge. *Trends Biochem. Sci.* **22:** 132–137.

Self-Splicing

Cech T.R. 1990. Nobel lecture. Self-splicing and enzymatic activity of an intervening sequence RNA from *Tetrahymena*. *Biosci. Rep.* **10:** 239–261.

Alternative Splicing and Regulation

Blencowe B.J. 2006. Alternative splicing: New insights from global analyses. *Cell* **126:** 37–47.

Graveley B.R. 2001. Alternative splicing: Increasing diversity in the proteomic world. *Trends Genet.* **17:** 100–107.

Hughes T.A. 2006. Regulation of gene expression by alternative untranslated regions. *Trends Genet.* **22:** 119–122.

Kornblihtt A.R. 2005. Promoter usage and alternative splicing. *Curr. Opin. Cell Biol.* **17:** 262–268.

Ladd A.N. and Cooper T.A. 2002. Finding signals that regulate alternative splicing in the post-genomic era. *Genome Biol.* **3:** reviews0008.1-0008.16.

Maniatis T. and Tasic B. 2002. Alternative pre-mRNA splicing and proteome expansion in metazoans. *Nature* **418:** 236–243.

Smith C.W.J. 2005. Alternative splicing—When two's a crowd. *Cell* **123:** 1–3.

Xing Y. and Lee C. 2006. Alternative splicing and RNA selection pressure—Evolutionary consequences for eukaryotic genomes. *Nat. Rev. Genet.* **7:** 499–509.

mRNA Transport

Dreyfuss G., Kim V.N., and Kataoka N. 2002. Messenger-RNA-binding proteins and the messages they carry. *Nat. Rev. Mol. Cell Biol.* **3:** 195–205.

Evolution

Roy S.W. and Gilbert W. 2006. The evolution of spliceosomal introns: Patterns, puzzles and progress. *Nat. Rev. Genet.* **7:** 211–221.

RNA Editing

Blanc V. and Davidson N.O. 2003. C-to-U RNA editing: Mechanisms leading to genetic diversity. *J. Biol. Chem.* **278:** 1395–1398.

Chiu Y.-L. and Greene W.C. 2006. Multifaceted antiviral actions of APOBEC3 cytidine deaminases. *Trends Immunol.* **27:** 291–297.

———. 2006. APOBEC3 cytidine deaminases: Distinct antiviral actions along the retroviral life cycle. *J. Biol. Chem.* **281:** 8309–8312.

Decatur W.A. and Fournier M.J. 2003. RNA-guided nucleotide modification of ribosomal and other RNAs. *J. Biol. Chem.* **278:** 695–698.

Keegan L.P., Leroy A., Sproul D., and O'Connell M.A. 2004. Adenosine deaminases acting on RNA (ADARs): RNA-editing enzymes. *Genome Biol.* **5:** 209.

Madison-Antenucci S., Grams J., and Hajduk S.L. 2002. Editing machines: The complexities of trypanosome RNA editing. *Cell* **108:** 435–438.

Sattelle D.B., Jones A.K., Sattelle B.M., Matsuda K., Reenan R., and Biggin P.C. 2005. Edit, cut and paste in the nicotinic acetylcholine receptor gene family of *Drosophila melanogaster*. *BioEssays* **27:** 366–376.

Simpson L., Sbicego S., and Aphasizhev R. 2003. Uridine insertion/deletion RNA editing in trypanosome mitochondria: A complex business. *RNA* **9:** 265–276.

Stuart K.D., Schnaufer A., Ernst N.L., and Panigrahi A.K. 2004. Complex management: RNA editing in trypanosomes. *Trends Biochem. Sci.* **30:** 97–105.

Turelli P. and Trono D. 2005. Editing at the crossroad of innate and adaptive immunity. *Science* **307:** 1061–1065.

Splicing and Disease

Faustino N.A. and Cooper T.A. 2003. Pre-mRNA splicing and human disease. *Genes Dev.* **17:** 419–437.

Licatalosi D.D. and Darnell R.B. Splicing regulation in neurologic disease. *Neuron* **52:** 93–101.

Srebrow A. and Kornblihtt A.R. 2006. The connection between splicing and cancer. *J. Cell Sci.* **119:** 2635–2641.

Yeo G.W.-M. 2005. Splicing regulators: Targets and drugs. *Genome Biol.* **6:** 240.

CHAPTER

14 Translation

THE CENTRAL QUESTION ADDRESSED IN THIS CHAPTER, as well as the next chapter, is how genetic information contained within the order of nucleotides in messenger RNA (mRNA) is interpreted to generate the linear sequences of amino acids in proteins. This process is known as **translation**. Of the events we have discussed, translation is among the most highly conserved across all organisms and among the most energetically costly for the cell. In rapidly growing bacterial cells, up to 80% of the cell's energy and 50% of the cell's dry weight are dedicated to protein synthesis. Indeed, the synthesis of a single protein requires the coordinated action of well over 100 proteins and RNAs. Consistent with the more complex nature of the translation process, we have divided our discussion into two chapters. In this first chapter, we describe the events that allow decoding of the mRNA, and in Chapter 15, we describe the nature of the genetic code and its recognition by transfer RNAs.

Translation is a much more formidable challenge in information transfer than the transcription of DNA into RNA. Unlike the complementarity between the DNA template and the ribonucleotides of the mRNA, the side chains of amino acids have little or no specific affinity for the purine and pyrimidine bases found in RNA. For example, the hydrophobic side chains of the amino acids alanine, valine, leucine, and isoleucine cannot form hydrogen bonds with the amino and keto groups of the nucleotide bases. Likewise, it is hard to imagine that several different combinations of three bases of RNA could form surfaces with unique affinities for the aromatic amino acids phenylalanine, tyrosine, and tryptophan. Thus, it seemed unlikely that direct interactions between the mRNA template and the amino acids could be responsible for the specific and accurate ordering of amino acids in a polypeptide.

With these considerations in mind, in 1955 Francis H. Crick proposed that prior to their incorporation into polypeptides, amino acids must attach to a special adaptor molecule that is capable of directly interacting with and recognizing the three-nucleotide-long coding units of the mRNA. Crick imagined that the adaptor would be an RNA molecule because it would need to recognize the code by Watson–Crick base-pairing rules. Just two years later, Paul C. Zamecnik and Mahlon B. Hoagland demonstrated that prior to their incorporation into proteins, amino acids are attached to a class of RNA molecules (representing 15% of all cellular RNA). These RNAs are called transfer RNAs (or tRNAs) because their attached amino acid is subsequently transferred to the growing polypeptide chain.

OUTLINE

The machinery responsible for translating the language of mRNAs into the language of proteins is composed of four primary components: **mRNAs**, **tRNAs**, **aminoacyl-tRNA synthetases**, and the **ribosome**. Together, these components accomplish the extraordinary task of translating a code written in a four-base alphabet into a second code written in the language of the 20 amino acids. The mRNA provides the information that must be interpreted by the translation machinery and is the template for translation. The protein-coding region of the mRNA consists of an ordered series of three-nucleotide-long units called **codons** that specify the order of amino acids. The tRNAs provide the physical interface between the amino acids being added to the growing polypeptide chain and the codons in the mRNA. Enzymes called aminoacyl-tRNA synthetases couple amino acids to specific tRNAs that recognize the appropriate codon. The final major player in translation is the ribosome, a remarkable, multi-megadalton machine composed of both RNA and protein. The ribosome coordinates the correct recognition of the mRNA by each tRNA and catalyzes peptide bond formation between the growing polypeptide chain and the amino acid attached to the selected tRNA.

We shall first consider the key attributes of each of these four components. We then describe how these components work together to accomplish translation. Recent progress in elucidating the structure of the components of the translational machinery make this an exciting area—one that is rich in mechanistic insights. Among the questions we will ask are: what is the organization of nucleotide sequence information in mRNA? What is the structure of tRNAs, and how do aminoacyl-tRNA synthetases recognize and attach the correct amino acids to each tRNA? Finally, how does the ribosome orchestrate the decoding of nucleotide sequence information and the addition of amino acids to the growing polypeptide chain?

MESSENGER RNA

Polypeptide Chains Are Specified by Open Reading Frames

The translation machinery decodes only a portion of each mRNA. As we saw in Chapter 2 and will consider in detail in Chapter 15, the information for protein synthesis is in the form of three-nucleotide codons, which each specify one amino acid. The protein-coding region(s) of each mRNA is composed of a contiguous, nonoverlapping string of codons called an **open reading frame** (commonly known as an **ORF**). Each ORF specifies a single protein and starts and ends at internal sites within the mRNA. That is, the ends of an ORF are distinct from the ends of the mRNA.

Translation starts at the 5′ end of the ORF and proceeds one codon at a time to the 3′ end. The first and last codons of an ORF are known as the **start** and **stop codons**. In bacteria, the start codon is usually 5′-AUG-3′ but 5′-GUG-3′ and sometimes even 5′-UUG-3′ are also used. Eukaryotic cells always use 5′-AUG-3′ as the start codon. The start codon has two important functions. First, it specifies the first amino acid to be incorporated into the growing polypeptide chain. Second, it defines the reading frame for all subsequent codons. Because each codon is immediately adjacent to (but not overlapping with) the next codon, and because codons are three nucleotides long, any stretch of mRNA could be translated in three different reading frames (Fig. 14-

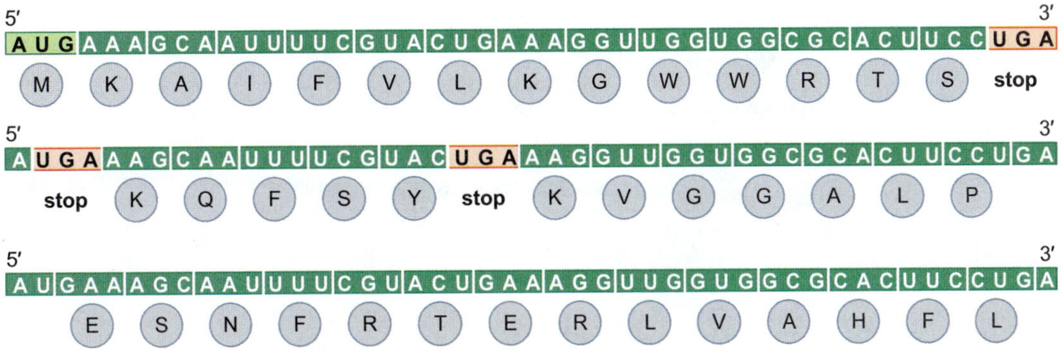

FIGURE 14-1 Three possible reading frames of the *Escherichia coli trp* leader sequence. Start codons are shaded in green and stop codons are shaded in red. The amino acid sequence encoded by each reading frame is indicated in the single-letter code below each codon.

1). Once translation starts, however, the reading frame is determined. Thus, by setting the location of the first codon, the start codon determines the location of all following codons.

Stop codons, of which there are three (5′-UAG-3′, 5′-UGA-3′, and 5′-UAA-3′), define the end of the ORF and signal termination of polypeptide synthesis. We can now fully appreciate the origin of the term *open* reading frame. It is a contiguous stretch of codons "read" in a particular frame (as set by the first codon) that is "open" to translation because it lacks a stop codon (i.e., until the last codon in the ORF).

mRNAs contain at least one ORF. The number of ORFs per mRNA is different between eukaryotes and prokaryotes. Eukaryotic mRNAs almost always contain a single ORF. In contrast, prokaryotic mRNAs frequently contain two or more ORFs and hence can encode multiple polypeptide chains. mRNAs containing multiple ORFs are known as **polycistronic mRNAs**, and those encoding a single ORF are known as **monocistronic mRNAs**. As we learned in Chapter 12, the polycistronic mRNAs found in bacteria often encode proteins that perform related functions, such as different steps in the biosynthesis of an amino acid or nucleotide. The structures of typical prokaryotic and eukaryotic mRNAs are shown in Figure 14-2.

Prokaryotic mRNAs Have a Ribosome-Binding Site That Recruits the Translational Machinery

For translation to occur, the ribosome must be recruited to the mRNA. To facilitate binding by a ribosome, many prokaryotic ORFs contain a short sequence upstream (on the 5′ side) of the start codon called the **ribosome-binding site (RBS)**. This element is also referred to as a **Shine–Dalgarno sequence** after the scientists who discovered it by comparing the sequences of multiple mRNAs. The RBS, typically located three to nine base pairs on the 5′ side of the start codon, is complementary to a sequence located near the 3′ end of one of the ribosomal RNA components, the 16*S* ribosomal RNA (rRNA, see Fig. 14-2a). The RBS base-pairs with this RNA, thereby aligning the ribosome with the beginning of the ORF. The core of this region of the 16S rRNA has the sequence 5′-CCUCCU-3′. Not surprisingly, prokaryotic RBS are most often a subset of the sequence 5′-AGGAGG-3′. The extent of complementarity and the spacing between the RBS and the

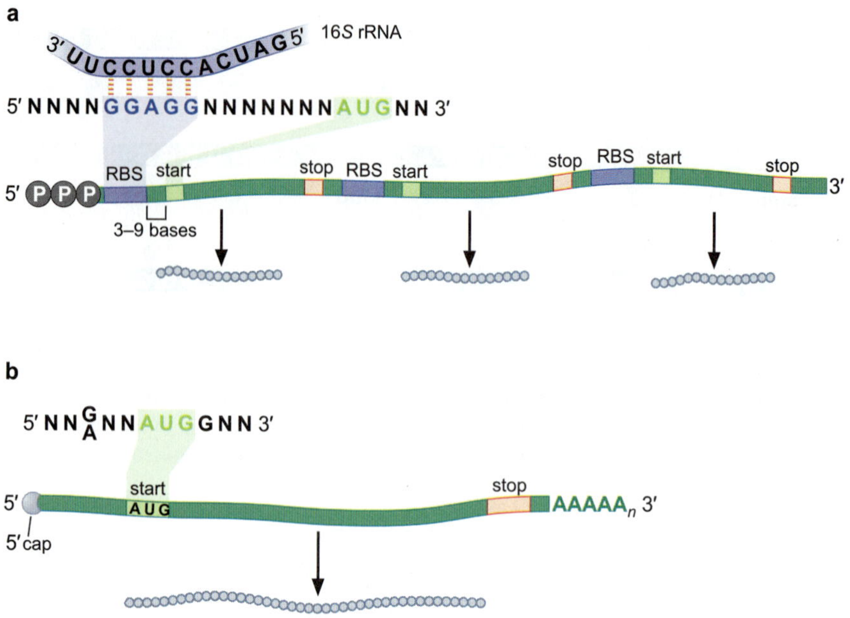

FIGURE 14-2 Structure of messenger RNA. (a) A polycistronic prokaryotic message. Each ribosome-binding site is indicated by RBS. (b) A monocistronic eukaryotic message. The 5′ cap is indicated by a "ball" at the end of the mRNA.

start codon has a strong influence on how actively a particular ORF is translated: high complementarity and proper spacing promote active translation, whereas limited complementarity and/or poor spacing generally support lower levels of translation.

Some prokaryotic ORFs lack a strong RBS but are nonetheless actively translated. These ORFs are not found as the first ORF in an mRNA but instead are located just after another ORF in a polycistronic message (not all prokaryotic mRNAs are polycistronic). In these cases, the start codon of the downstream ORF often overlaps the 3′ end of the upstream ORF (most often as the sequence 5′-AUGA-3′, which contains a start and a stop codon). Thus, a ribosome that has just completed translating the upstream ORF is well positioned to begin translating from the start codon for the downstream ORF. This arrangement circumvents the need for an RBS to recruit the ribosome. This phenomenon of linked translation between overlapping ORFs is known as **translational coupling**. It is important to note that in this situation translation of the downstream ORF requires translation of the upstream ORF. Indeed, with two translationally coupled genes, a stop mutation in the upstream ORF also prevents translation of the downstream ORF.

Eukaryotic mRNAs Are Modified at Their 5′ and 3′ Ends to Facilitate Translation

Unlike their prokaryotic counterparts, eukaryotic mRNAs recruit ribosomes using a specific chemical modification called the **5′ cap**, which is located at the extreme 5′ end of the message (see Chapter 12 and Fig. 14-2b). The 5′ cap is a methylated guanine nucleotide that is joined to the 5′ end of the mRNA via an unusual 5′ to 5′ linkage. Created in three steps (see Fig. 12-23 in Chapter 12), the guanine nucleotide of the 5′ cap is connected to the 5′ end of the mRNA through three phosphate groups. The resulting 5′ cap is required to recruit the ribosome to the mRNA. Once bound to the mRNA, the ribosome moves in a 5′ → 3′ direction until it encounters a 5′-AUG-3′ start codon, a process called **scanning**.

Two other features of eukaryotic mRNAs stimulate translation. One feature is the presence, in some mRNAs, of a purine three bases upstream of the start codon and a guanine immediately downstream (5′-G/ANNAUGG-3′). This sequence was originally identified by Marilyn Kozak and is referred to as the Kozak sequence. Many eukaryotic mRNAs lack these bases, but their presence increases the efficiency of translation. In contrast to the situation in prokaryotes, these bases are thought to interact with the initiator tRNA, not with an RNA component of the ribosome. A second feature that contributes to efficient translation is the presence of a poly-A tail at the extreme 3′ end of the mRNA. As we saw in Chapter 12, this tail is added enzymatically by the enzyme poly-A polymerase. Despite its location at the 3′ end of the mRNA, the poly-A tail enhances the level of translation of the mRNA by promoting efficient recycling of ribosomes (as we shall discuss below).

TRANSFER RNA

tRNAs Are Adaptors between Codons and Amino Acids

The heart of protein synthesis is the "translation" of nucleotide sequence information (in the form of codons) into amino acids. This is accomplished by tRNA molecules, which act as adaptors between codons and the amino acids they specify. There are many types of tRNA molecules, but each is attached to a specific amino acid and each recognizes a particular codon, or codons, in the mRNA (most tRNAs recognize more than one codon, as we shall discuss in Chapter 15). tRNA molecules are between 75 and 95 ribonucleotides in length. Although the exact sequence varies, all tRNAs have certain features in common. First, all tRNAs end at the 3′ terminus with the sequence 5′-CCA-3′ (see Box 14-1, CCA-adding Enzymes: Synthesizing RNA without a Template). Consistent with this absolute conservation, the 3′ end of this sequence (and of the tRNA) is the site that is attached to the cognate amino acid.

A second striking aspect of tRNAs is the presence of several unusual bases in their primary structure. These unusual features are created posttranscriptionally by enzymatic modification of normal bases in the polynucleotide chain. For example, **pseudouridine** (ΨU) is derived from uridine by an isomerization in which the site of attachment of the uracil base to the ribose is switched from the nitrogen at ring position 1 to the carbon at ring position 5 (Fig. 14-3). Likewise, **dihydrouridine** (**D**) is derived from uridine by enzymatic reduction of the double bond between the carbons at positions 5 and 6. Other unusual bases found in tRNA include hypoxanthine, thymine, and methylguanine. These modified bases are not essential for tRNA function, but cells lacking

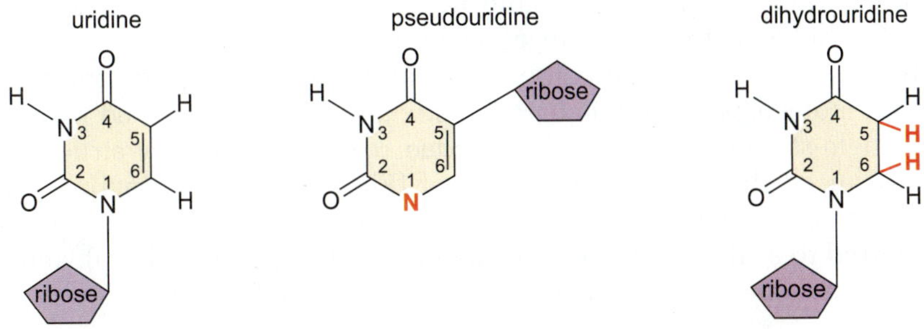

FIGURE 14-3 A subset of modified nucleosides found in tRNA. Uridine and two uridine-related nucleotides are shown.

ADVANCED CONCEPTS

Box 14-1 CCA-Adding Enzymes: Synthesizing RNA without a Template

As we have described, the 5´-CCA-3´ end is universally conserved for all tRNAs and is absolutely required for protein synthesis. Oddly, when the genes that encode tRNAs were cloned, it was found that many do not encode the CCA end. Instead, these genes end three nucleotides short of the 3´ ends found in the mature tRNA. In fact, the genes encoding many bacterial tRNAs and almost all eukaryotic tRNAs lack this final three-base sequence. How then do these tRNAs acquire their CCA ends? The answer is provided by a specialized RNA polymerase called a **CCA-adding enzyme**. As its name indicates, this enzyme adds the terminal CCA to tRNAs that initially lack this sequence. Surprisingly, there is no nucleic acid component to this enzyme; that is, CCA-adding enzymes add a specific sequence to the end of the tRNA without an RNA or DNA template.

How do CCA-adding enzymes add a specific sequence without a template? A series of three-dimensional structures of these enzymes has begun to reveal the solution. First, like other RNA and DNA polymerases, CCA-adding enzymes have only one active site that uses a similar two-metal ion mechanism of catalysis (see Fig. 8-6). Within this active site, an amino acid and a phosphate from the terminal tRNA nucleotide form hydrogen bonds with A and C bases but not with G or U. This specificity can be understood by observing the pattern of hydrogen-bond donors and acceptors on each of the bases (see Figs. 6-5 and 6-6). The patterns of A and C are overlapping but G and U (or T) have opposite and complementary patterns. Indeed, it is this complementarity that is responsible for the specificity of base pairing within the double-stranded DNA helix. This hydrogen bonding pattern explains the specificity of the enzyme for C and A.

Specificity for the addition of C versus A is controlled by changes in the active site as each base is added. Unlike other polymerases, the tRNA template does not change its position as each additional nucleotide is added. Instead, the template tRNA is held firmly in place and each added nucleotide alters the structure of the active site. The result of these changes is that the active site is specific for C when a CCA-less tRNA binds but is altered to be specific for A after two C residues have been added. Once the A residue is added, the active site is no longer accessible to additional bases and the tRNA with its newly added CCA end is released.

these modified bases show reduced rates of growth. This observation suggests that the modified bases lead to improved tRNA function. For example, as we will see in Chapter 15, hypoxanthine plays an important role in the process of codon recognition by certain tRNAs.

tRNAs Share a Common Secondary Structure That Resembles a Cloverleaf

As we saw in Chapter 6, RNA molecules typically contain regions of self-complementarity that enable them to form limited stretches of double helix that are held together by base pairing. Other regions of RNA molecules have no complement and hence are single-stranded. tRNA molecules exhibit a characteristic and highly conserved pattern of single-stranded and double-stranded regions (secondary structure) that can be illustrated as a cloverleaf (Fig. 14-4). The principal features of the tRNA cloverleaf are an acceptor stem, three stem loops (referred to as the ΨU loop, the D loop, and the anticodon loop), and a fourth variable loop. Descriptions of each of these features follows:

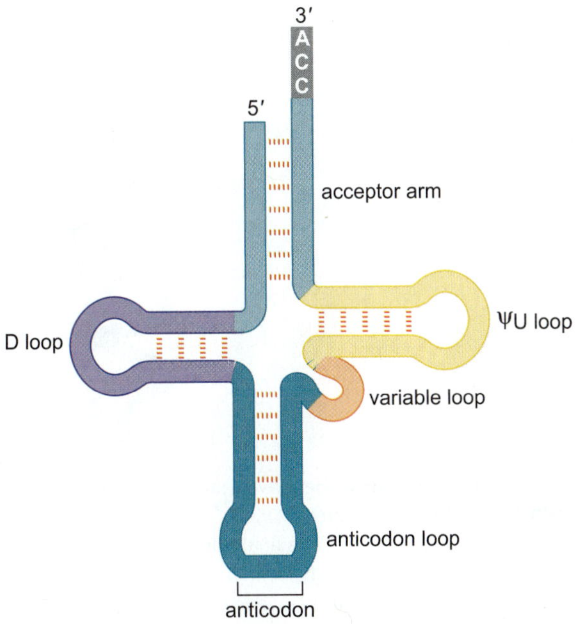

FIGURE 14-4 Cloverleaf representation of the secondary structure of tRNA. In this representation of a tRNA, the base pairings between different parts of the tRNA are indicated by the dotted red lines.

- The acceptor stem, so-named because it is the site of attachment of the amino acid, is formed by pairing between the 5´ and 3´ ends of the tRNA molecule. The 5´-CCA-3´ sequence at the extreme 3´ end of the molecule in a single-strand region that protrudes from this double-strand stem.
- The ΨU loop is so-named because of the characteristic presence of the unusual base ΨU in the loop. The modified base is often found within the sequence 5´-TΨUCG-3´.
- The D loop takes its name from the characteristic presence of dihydrouridines in the loop.
- The anticodon loop, as its name implies, contains the anticodon, a three-nucleotide-long sequence that is responsible for recognizing the codon by base pairing with the mRNA. The anticodon is always bracketed on the 3´ end by a purine and on its 5´ end by uracil.
- The variable loop sits between the anticodon loop and the ΨU loop and, as its name implies, varies in size from 3 to 21 bases.

tRNAs Have an L-shaped Three-Dimensional Structure

The cloverleaf reveals regions of self-complementarity within tRNAs. What is the actual three-dimensional configuration of this adaptor molecule? X-ray crystallography reveals an L-shaped tertiary structure in which the terminus of the acceptor stem is at one end of the molecule and the anticodon loop is ~70 Å away at the other end (Fig. 14-5c). To understand the relationship of this L-shaped structure to the cloverleaf, consider the following: the acceptor stem and the stem of the ΨU loop form an extended helix in the final tRNA structure (Fig. 14-5b). Similarly, the anticodon stem and the stem of the D loop form a second extended helix. These two extended helices align at a right angle to each other, with the D loop and the ΨU loop coming together.

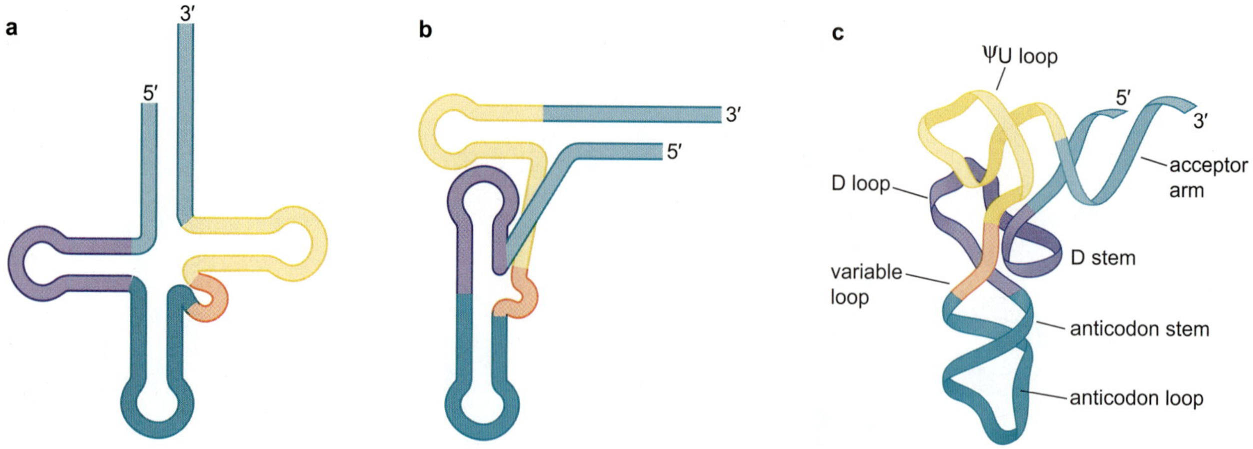

FIGURE 14-5 Conversion between the cloverleaf and the actual three-dimensional structure of a tRNA. (a) Cloverleaf representation. (b) L-shaped representation showing the location of the base-paired regions of the final folded tRNA. (c) Ribbon representation of the actual folded structure of a tRNA. Note that although this diagram illustrates how the actual tRNA structure is related to the cloverleaf representation, a tRNA does not attain its final structure by first base pairing and then folding into an L shape.

Three kinds of interactions stabilize this L-shaped structure. First, the formation of the two extended regions of base pairing result in base-stacking interactions similar to those seen in double-stranded DNA. Second, hydrogen bonds are formed between bases in different helical regions that are brought near each other in three-dimensional space by the tertiary structure. These base–base interactions are generally unconventional (non-Watson–Crick) bonding. Finally, there are interactions between the bases and the sugar–phosphate backbone.

ATTACHMENT OF AMINO ACIDS TO tRNA

tRNAs Are Charged by the Attachment of an Amino Acid to the 3′-Terminal Adenosine Nucleotide via a High-Energy Acyl Linkage

tRNA molecules to which an amino acid is attached are said to be charged, and tRNAs that lack an amino acid are said to be **uncharged**. Charging requires an acyl linkage between the carboxyl group of the amino acid and the 2′- or 3′-hydroxyl group (see below) of the adenosine nucleotide that protrudes from the acceptor stem at the 3′ end of the tRNA. This acyl linkage is a high-energy bond because its hydrolysis results in a large change in free energy. This is significant for protein synthesis: the energy released when the bond is broken helps drive the formation of the peptide bonds that link amino acids to each other in polypeptide chains.

Aminoacyl-tRNA Synthetases Charge tRNAs in Two Steps

All aminoacyl-tRNA synthetases attach an amino acid to a tRNA in two enzymatic steps (Fig. 14-6). Step one is **adenylylation** in which the amino acid reacts with ATP to become adenylylated with the con-

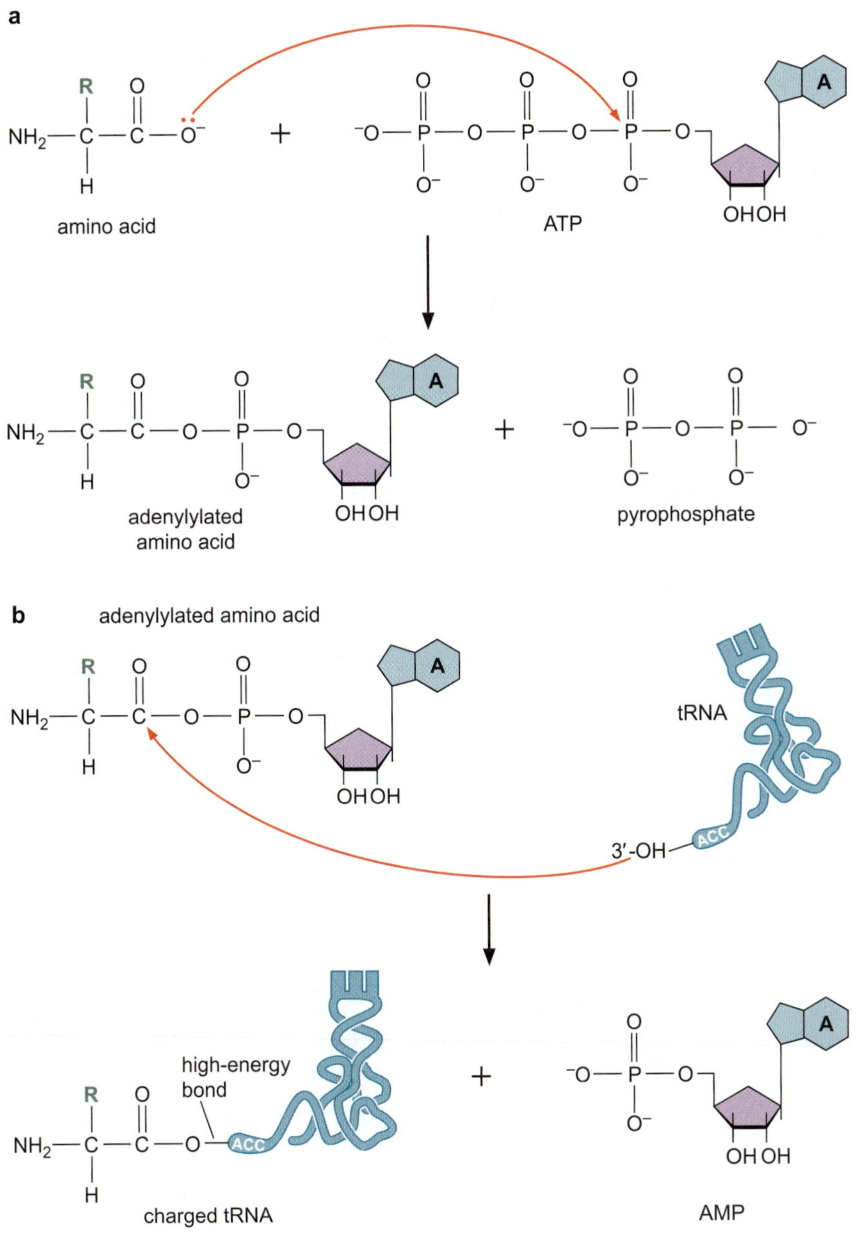

FIGURE 14-6 The two steps of aminoacyl-tRNA charging. (a) Adenylylation of amino acid. (b) Transfer of the adenylylated amino acid to tRNA. The process shown is for a class II tRNA synthetase (which attaches the amino acid to the 3′-OH).

comitant release of pyrophosphate. Adenylylation refers to transfer of AMP, as opposed to adenylation, which would indicate the transfer of adenine. As we have seen in the case of polynucleotide synthesis (see Chapter 8), the principal driving force for the adenylylation reaction is the subsequent hydrolysis of pyrophosphate by pyrophosphatase. As a result of adenylylation, the amino acid is attached to adenylic acid via a high-energy ester bond in which the carbonyl group of the amino acid is joined to the phosphoryl group of AMP. Step two is **tRNA charging** in which the adenylylated amino acid, which remains tightly bound to the synthetase, reacts with tRNA. This reaction results in the transfer of the amino acid to the 3′ end of the tRNA via the 2′- or 3′-hydroxyl and the release of AMP.

There are two classes of tRNA synthetases (Table 14-1). Class I enzymes attach the amino acid to the 2′-OH of the tRNA and are generally monomeric. Class II enzymes attach the amino acid to the 3′-OH

TABLE 14-1 Classes of Aminoacyl-tRNA Synthetases

Class II	Quarternary Structure	Class I	Quarternary Structure
Gly	$(\alpha_2\beta_2)$	Glu	(α)
Ala	(α_4)	Gln	(α)
Pro	(α_2)	Arg	(α)
Ser	(α_2)	Cys	(α_2)
Thr	(α_2)	Met	(α_2)
His	(α_2)	Val	(α)
Asp	(α_2)	Ile	(α)
Asn	(α_2)	Leu	(α)
Lys	(α_2)	Tyr	(α)
Phe	$(\alpha_2\beta_2)$	Trp	(α)

Adapted, with permission, from Delarue M. 1995. *Curr. Opin. Struct. Biol.* 5: 48–55, Table 1. © Elsevier.
Class I enzymes are generally monomeric, whereas class II enzymes are dimeric or tetrameric, with residues from two subunits contributing to the binding site for a single tRNA. α and β refer to subunits of the tRNA synthetases and the subscripts indicate their stoichiometry.

of the tRNA and are typically dimeric or tetrameric. Although the initial coupling between the tRNA and the amino acid is different, once released from the synthetase, the amino acid rapidly equilibrates between attachment at the 3′-OH and the 2′-OH.

Each Aminoacyl-tRNA Synthetase Attaches a Single Amino Acid to One or More tRNAs

Each of the 20 amino acids is attached to the appropriate tRNA by a single, dedicated tRNA synthetase. Because most amino acids are specified by more than one codon (see Chapter 15), it is not uncommon for one synthetase to recognize and charge more than one tRNA (known as isoaccepting tRNAs). Nevertheless, the same tRNA synthetase is responsible for charging all tRNAs for a particular amino acid. Thus, one and only one tRNA synthetase attaches each amino acid to all of the appropriate tRNAs.

Most organisms have 20 different tRNA synthetases, but this is not always the case. For example, some bacteria lack a synthetase for charging the tRNA for glutamine ($tRNA^{Gln}$) with its cognate amino acid. Instead, a single species of aminoacyl-tRNA synthetase charges $tRNA^{Gln}$ as well as $tRNA^{Glu}$ with glutamate. A second enzyme then converts (by amination) the glutamate moiety of the charged $tRNA^{Gln}$ molecules to glutamine. That is, Glu-$tRNA^{Gln}$ is aminated to Gln-$tRNA^{Gln}$ (the prefix identifies the attached amino acid and the superscript identifies the type of codon the tRNA recognizes). The presence of this second enzyme removes the need for a glutamine tRNA synthetase. Nevertheless, an aminoacyl-tRNA synthetase can never attach more than one kind of amino acid to a given tRNA.

tRNA Synthetases Recognize Unique Structural Features of Cognate tRNAs

As we can see from the above considerations, aminoacyl-tRNA synthetases face two important challenges: they must recognize the cor-

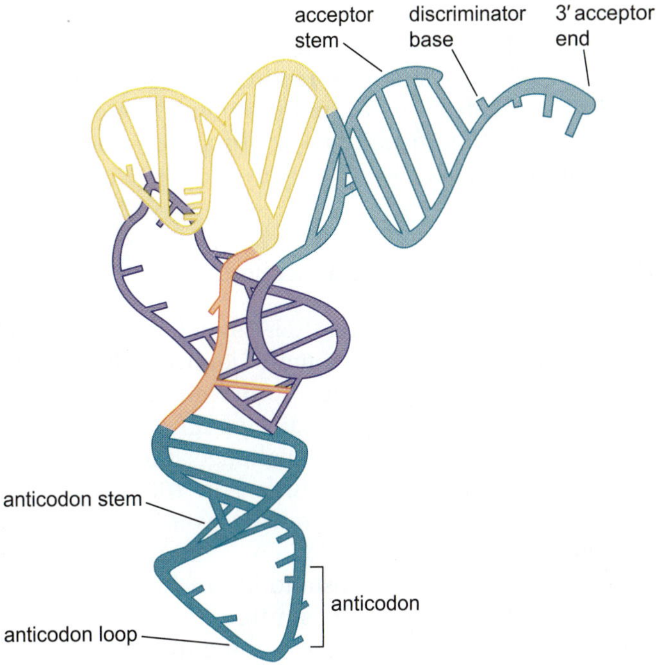

FIGURE 14-7 **Structure of tRNA: Elements required for aminoacyl synthetase recognition.**

rect set of tRNAs for a particular amino acid and they must charge all of these isoaccepting tRNAs with the correct amino acid. Both processes must be carried out with high fidelity.

Let us first consider the specificity of tRNA recognition: what features of the tRNA molecule enable a synthetase to discriminate the correct set of isoaccepting tRNAs from the tRNAs for the other 19 amino acids? Genetic, biochemical, and X-ray crystallographic evidence indicate that the specificity determinants are clustered at two distant sites on the molecule: the acceptor stem and the anticodon loop (Fig. 14-7). The acceptor stem is an especially important determinant for the specificity of tRNA synthetase recognition. In some cases, changing a single base in the acceptor stem (known as the **discriminator base**) is sufficient to convert the recognition specificity of a tRNA from one synthetase to another. Nonetheless, the anticodon loop frequently contributes to discrimination as well. The synthetase for glutamine, for example, makes numerous contacts both in the acceptor stem and across the anticodon loop, including the anticodon itself (Fig. 14-8).

One might expect that the anticodon would always be used for recognition by tRNA synthetases because it is the ultimate defining feature of a tRNA—the anticodon dictates the amino acid that the tRNA is responsible for incorporating into the growing polypeptide chain. However, because each amino acid is usually specified by more than one codon, recognition of the anticodon cannot be used in many cases. For example, the amino acid serine is specified by six codons, including 5′-AGC-3′ and 5′-UCA-3′, which are completely different from one another. Hence, the tRNAs for serine necessarily have a variety of different anticodons, which could not be easily recognized by a single tRNA synthetase. So, to recognize its tRNAs, the synthetase for serine must rely on determinants that lie outside of the anticodon.

The set of tRNA determinants that enable synthetases to discriminate among tRNAs is sometimes referred to as the "second genetic

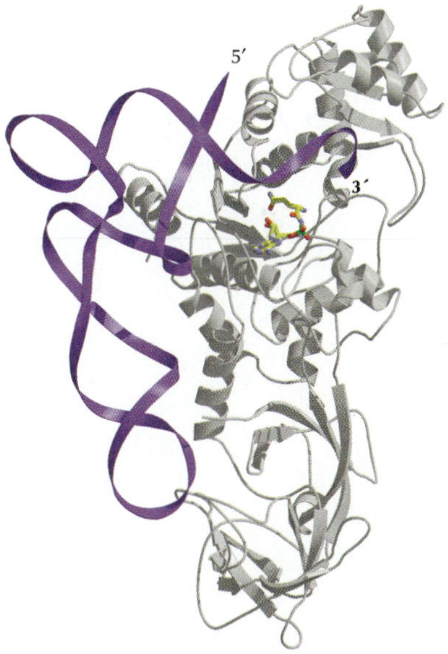

FIGURE 14-8 **Cocrystal structure of glutaminyl aminoacyl-tRNA synthetase with $tRNA^{Gln}$.** The enzyme is shown in gray and $tRNA^{Gln}$ is shown in purple. The yellow, red, and green molecule is glutaminyl-AMP. Note the proximity of this molecule to the 3′ end of the tRNA and the points of contact between the tRNA and the synthetase. (Rath V.L. et al. 1998. *Structure* 6: 439–449.) Image prepared with MolScript, BobScript, and Raster 3D.

code" because of its central importance in information flow. As we discussed above, this code is significantly more complex than the "first genetic code" and cannot be readily tabulated. Without such a code, however, synthetases could not distinguish one tRNA from another, and the translation machinery would not produce polypeptides with a reproducible sequence.

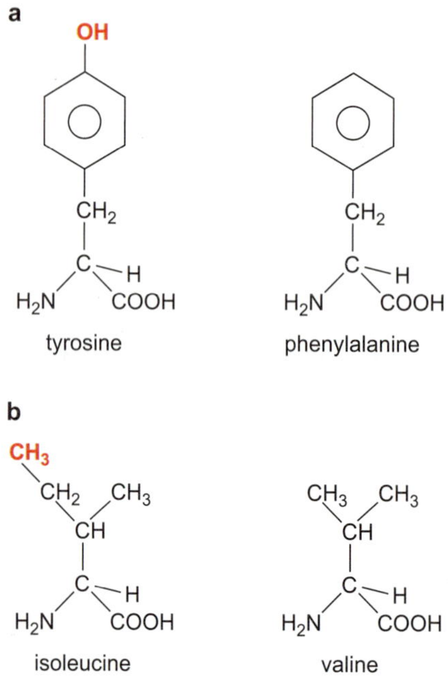

FIGURE 14-9 Distinguishing features of similar amino acids.

Aminoacyl-tRNA Formation Is Very Accurate

The challenge faced by aminoacyl-tRNA synthetases in selecting the correct amino acid is perhaps even more daunting than the challenge the enzyme faces in recognizing the appropriate tRNA (Fig. 14-9). The reason for this is the relatively small size of amino acids and, in some cases, their similarity. Despite this challenge, the frequency of mischarging is very low; typically, less than 1 in 1000 tRNAs is charged with the incorrect amino acid. In certain cases it is easy to understand how this high accuracy is achieved. For example, the amino acids cysteine and tryptophan differ substantially in size, shape, and chemical groups. Even in the case of the similar-looking amino acids tyrosine and phenylalanine (see Fig. 14-9a), the opportunity for forming a strong and energetically favorable hydrogen bond with the hydroxyl moiety of the former but not the latter allows the synthetase for tyrosine (tyrosyl-tRNA synthetase) to discriminate effectively against phenylalanine.

It is more challenging to understand the case of isoleucine and valine, which differ by only a single methylene group (see Fig. 14-9b). Valyl-tRNA synthetase can sterically exclude isoleucine from its catalytic pocket because isoleucine is larger than valine. In contrast, valine should slip easily into the catalytic pocket of the isoleucyl-tRNA synthetase. Although both amino acids will fit into the isoleucyl-tRNA synthetase amino acid–binding site, interactions with the extra methylene group on isoleucine will provide an extra −2 to −3 kcal/moles of free energy (see Table 3-1). As we described in Chapter 3, even this relatively small difference in free energy will make binding to isoleucine ~100-fold more likely than binding to valine if the two amino acids are present at equal concentrations. Thus, valine would be attached to isoleucine tRNAs ~1% of the time; however, this is an unacceptably high rate of error. As we have discussed, the actual frequency of misincorporation is <0.1%. How is this additional level of fidelity achieved?

Some Aminoacyl-tRNA Synthetases Use an Editing Pocket to Charge tRNAs with High Accuracy

One common mechanism to increase the fidelity of an aminoacyl-tRNA synthetase is to proofread the products of the charging reaction as we have seen for DNA polymerases in Chapter 8. For example, in addition to its catalytic pocket (for adenylylation), isoleucyl-tRNA synthetase has a nearby editing pocket (a deep cleft in the enzyme) that allows it to proofread the product of the adenylylation reaction. AMP-valine (as well as adenylylates of other small amino acids, such as alanine) can fit into this editing pocket, where it is hydrolyzed and released as free valine and AMP. In contrast, AMP-isoleucine is too large to enter the editing pocket and is therefore not subject to hydrolysis. As a consequence, isoleucyl-tRNA synthetase discriminates against valine twice: in the initial binding and adenylylation of the amino acid (discriminating by a factor of ~100), and then in the editing of the adenylylated amino acid (again discriminating by a factor of ~100), for an overall selectivity of ~10,000-fold (i.e., an error rate of ~0.01%).

The Ribosome Is Unable to Discriminate between Correctly and Incorrectly Charged tRNAs

The reason that so much responsibility falls on aminoacyl-tRNA synthetases to couple the proper amino acid with its cognate tRNA is that the ribosome cannot distinguish between correctly and incorrectly charged tRNAs. In other words, the ribosome "blindly" accepts any charged tRNA that exhibits a proper codon–anticodon interaction, whether or not the tRNA is charged with the correct amino acid.

This conclusion is supported by two kinds of experiments: one genetic and the other biochemical. The genetic experiment involves the isolation of a mutant tRNA that carries a nucleotide substitution in the anticodon. Recall that tRNA synthetases frequently do not rely on interaction with the anticodon to recognize cognate tRNAs. Hence, a subset of tRNAs can be mutated in their anticodons but still be charged with their usual cognate amino acids. As a consequence of the anticodon mutation, however, the mutant tRNA delivers its amino acid to the wrong codon. In other words, the ribosome and the auxiliary proteins that work in conjunction with the ribosome (which we shall discuss shortly) primarily check that the charged tRNA makes a proper codon–anticodon interaction with the mRNA. The ribosome and these proteins do little to prevent an incorrectly charged tRNA from adding an inappropriate amino acid to the growing polypeptide.

A classic biochemical experiment nicely illustrates the point that the ribosome recognizes tRNA and not the amino acid that it is carrying. Consider the charged tRNA cysteinyl-$tRNA^{Cys}$ (remember that the prefix identifies the amino acid and the superscript identifies the nature of the tRNA). The cysteine attached to cysteinyl-$tRNA^{Cys}$ can be converted to an alanine by chemical reduction to give alanine-$tRNA^{Cys}$ (Fig. 14-10). When added to a cell-free protein-synthesizing system, alanine-$tRNA^{Cys}$ introduces alanines at codons that specify insertion of cysteine. Thus, the translation machinery relies on the high fidelity of the aminoacyl-tRNA synthetases to ensure the accurate decoding of each mRNA (see Box 14-2, Selenocysteine).

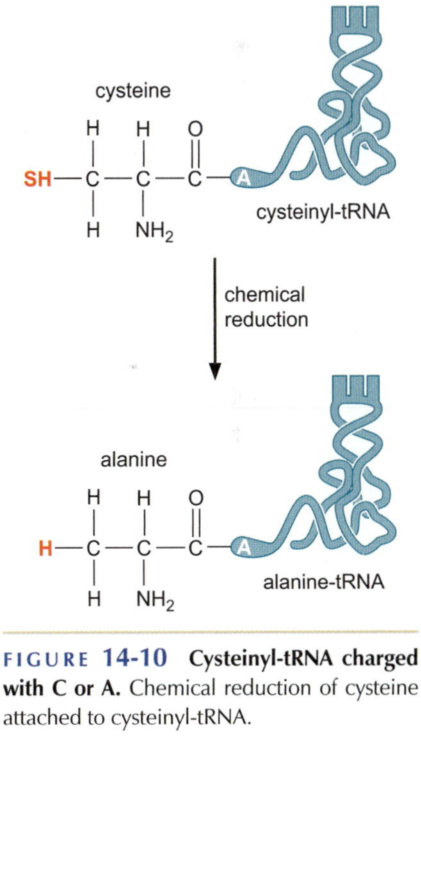

FIGURE 14-10 Cysteinyl-tRNA charged with C or A. Chemical reduction of cysteine attached to cysteinyl-tRNA.

THE RIBOSOME

The ribosome is the macromolecular machine that directs the synthesis of proteins. Consistent with the additional challenges of translating a nucleic acid code into an amino acid code, the ribosome is larger and more complex than the minimal machinery required for DNA or RNA synthesis. Indeed, single polypeptides can perform DNA or RNA synthesis (although DNA replication and transcription are more frequently mediated by larger multisubunit complexes). In contrast, the machinery for polymerizing amino acids is composed of at least three RNA molecules and more than 50 different proteins, with an overall molecular mass of greater than 2.5 megadaltons. Compared to the

ADVANCED CONCEPTS

Box 14-2 Selenocysteine

Certain proteins, such as the enzymes glutathione peroxidase and formate dehydrogenase, contain an unusual amino acid called selenocysteine, which is part of the catalytic center of the enzymes. Selenocysteine contains the trace element selenium in place of the sulfur atom of cysteine (Box 14-2 Fig. 1). Interestingly, selenocysteine is not incorporated into proteins by chemical modification after translation (as is true for certain other unusual amino acids, such as hydroxyproline, which is found in collagen). Instead, selenocysteine is generated enzymatically from serine carried on a special tRNA that is charged by serine-tRNA synthetase. This altered tRNA is used to incorporate selenocysteine directly into enzymes such as glutathione peroxidase as they are synthesized. A dedicated (EF-Tu-like; see below) translation elongation factor delivers selenocysteinyl-tRNA to the ribosome at a codon (UGA) that would normally be recognized as a stop codon. Incorporation of selenocysteine at UGA codons requires the presence of a special sequence element elsewhere in the mRNA. Thus, selenocysteine can be thought of as a 21st amino acid that is incorporated into proteins by a modification of the standard translation machinery of the cell.

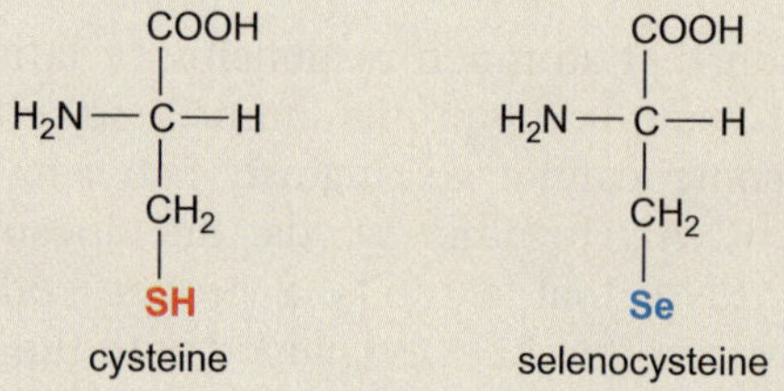

BOX 14-2 FIGURE 1 The structures of cysteine and selenocysteine.

speed of DNA replication—200 to 1000 nucleotides per second—translation takes place at a rate of only 2–20 amino acids per second.

In prokaryotes, the transcription machinery and the translation machinery are located in the same compartment. Thus, the ribosome can commence translation of the mRNA as it emerges from the RNA polymerase. This situation allows the ribosome to proceed in tandem with the RNA polymerase as it elongates the transcript (Fig. 14-11). Recall that the 5′ end of an RNA is synthesized first, and thus the ribosome, which begins translation at the 5′ end of the mRNA, can start translating a nascent transcripts as soon as it emerges from the RNA polymerase. Interestingly, there are several instances in which the coupling of transcription and translation is exploited during the regulation of gene expression, as we shall see in Chapter 18.

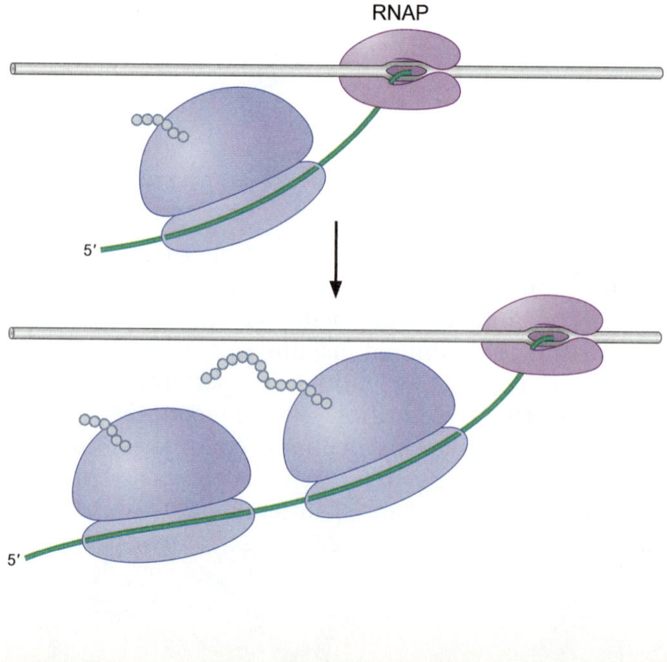

FIGURE 14-11 Prokaryotic RNA polymerase and ribosomes at work on the same mRNA.

Although slow relative to DNA synthesis in prokaryotes, the ribosome is capable of keeping up with the transcription machinery. The typical prokaryotic rate of translation of 20 amino acids per second corresponds to the translation of 60 nucleotides (20 codons) of mRNA per second. This is similar to the rate of 50–100 nucleotides per second synthesized by RNA polymerase.

In contrast to the situation in prokaryotes, translation in eukaryotes is completely separate from transcription. These events occur in separate compartments of the cell: transcription occurs in the nucleus, whereas translation occurs in the cytoplasm. Perhaps because of the lack of coupling to transcription, eukaryotic translation proceeds at the more leisurely speed of 2–4 amino acids per second.

The Ribosome Is Composed of a Large and a Small Subunit

The ribosome is composed of two subassemblies of RNA and protein known as the large and small subunits. The large subunit contains the **peptidyl transferase center**, which is responsible for the formation of peptide bonds. The small subunit contains the **decoding center** in which charged tRNAs read or "decode" the codon units of the mRNA.

By convention, the large and small subunits are named according to the velocity of their sedimentation when subjected to a centrifugal force (Fig. 14-12). The unit used to measure sedimentation velocity is the **Svedberg** (*S*; the larger the *S* value the faster the sedimentation velocity and the larger the molecule), which is named after the inventor of the ultracentrifuge, Theodor Svedberg. In bacteria, the large subunit has a sedimentation velocity of 50 Svedberg units and is accordingly known as the 50*S* subunit, whereas the small subunit is called the 30*S* subunit. The intact prokaryotic ribosome is referred to as the 70*S* ribosome. Note that 70*S* is less than the sum of 50*S* and 30*S*! The explanation for this apparent discrepancy is that sedimentation velocity is determined by both shape and size and hence is not an exact measure of mass. The eukaryotic ribosome is somewhat larger, composed of 60*S* and 40*S* subunits, which together form an 80*S* ribosome.

The large and small subunits are each composed of one or more RNAs (known as ribosomal RNAs or rRNAs), and many ribosomal proteins (Fig. 14-13). Svedberg units are once again used to distinguish among the rRNAs. Thus, in bacteria, the 50*S* subunit contains a 5*S* rRNA and a 23*S* rRNA, whereas the 30*S* subunit contains a single 16*S* rRNA. Although there are far more ribosomal proteins than rRNAs in each subunit, more than two-thirds of the mass of the prokaryotic ribosome is RNA. This is true because the ribosomal proteins are small (the average molecular mass of a ribosomal protein in

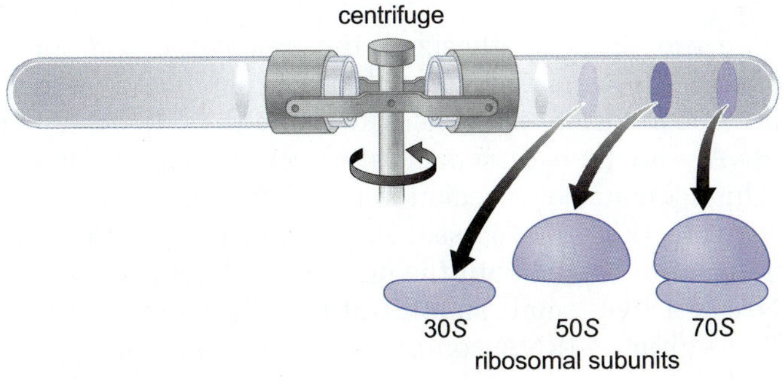

FIGURE 14-12 **Sedimentation by ultracentrifugation separates individual bacterial ribosome subunits from the full ribosome.**

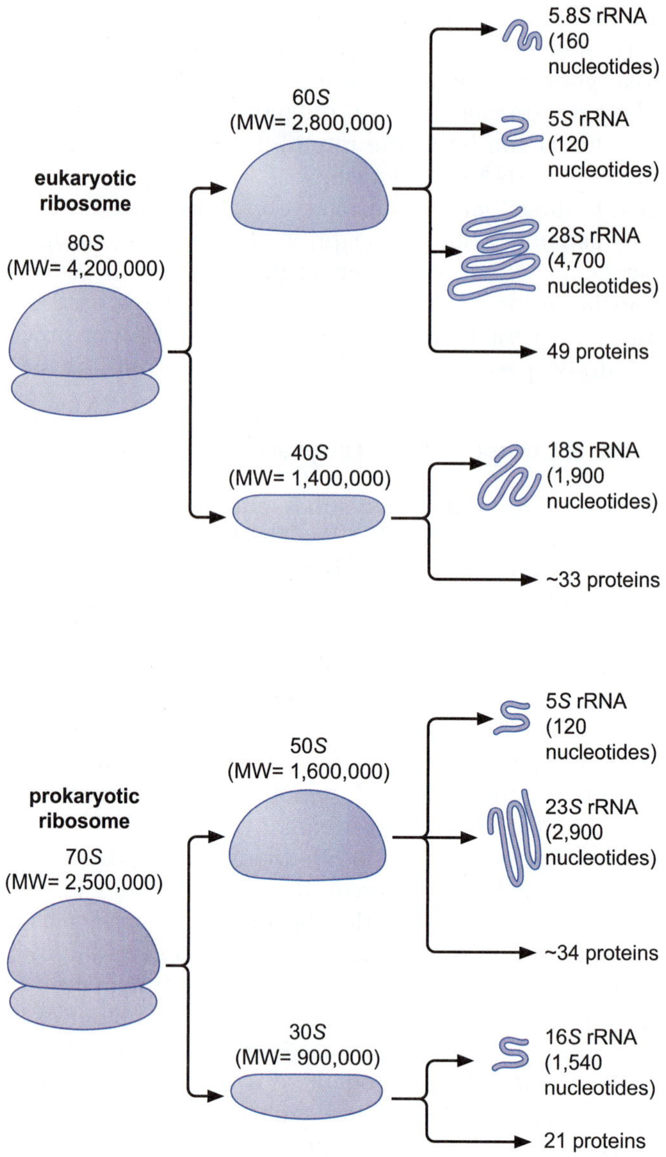

FIGURE 14-13 **Composition of the prokaryotic and eukaryotic ribosomes.** The rRNA and protein composition of the different subunits are indicated. The length of the rRNA and the number of ribosomal proteins are indicated for each subunit.

the bacterial small subunit is ~15 kD). In contrast, the 16*S* and 23*S* rRNAs are large. Recall that, on average, a single nucleotide has a molecular mass of 330 D; therefore, on its own, the 2900-nucleotide-long 23*S* rRNA has a molecular mass of almost 1000 kD.

The Large and Small Subunits Undergo Association and Dissociation during Each Cycle of Translation

WEB ANIMATION

Each time a protein is synthesized, the translation components undergo a specific series of events in which the small and large subunits of the ribosome associate with each other and the mRNA, translate the target mRNA, and then dissociate after completing synthesis of the protein. This sequence of association and dissociation is known as the **ribosome cycle** (Fig. 14-14; see also Interactive Animation 14-1). Briefly, translation begins with the binding of the mRNA and an initiating tRNA to a free, small subunit of the ribosome. The small subunit–mRNA–initiator–tRNA complex then recruits a large subunit to

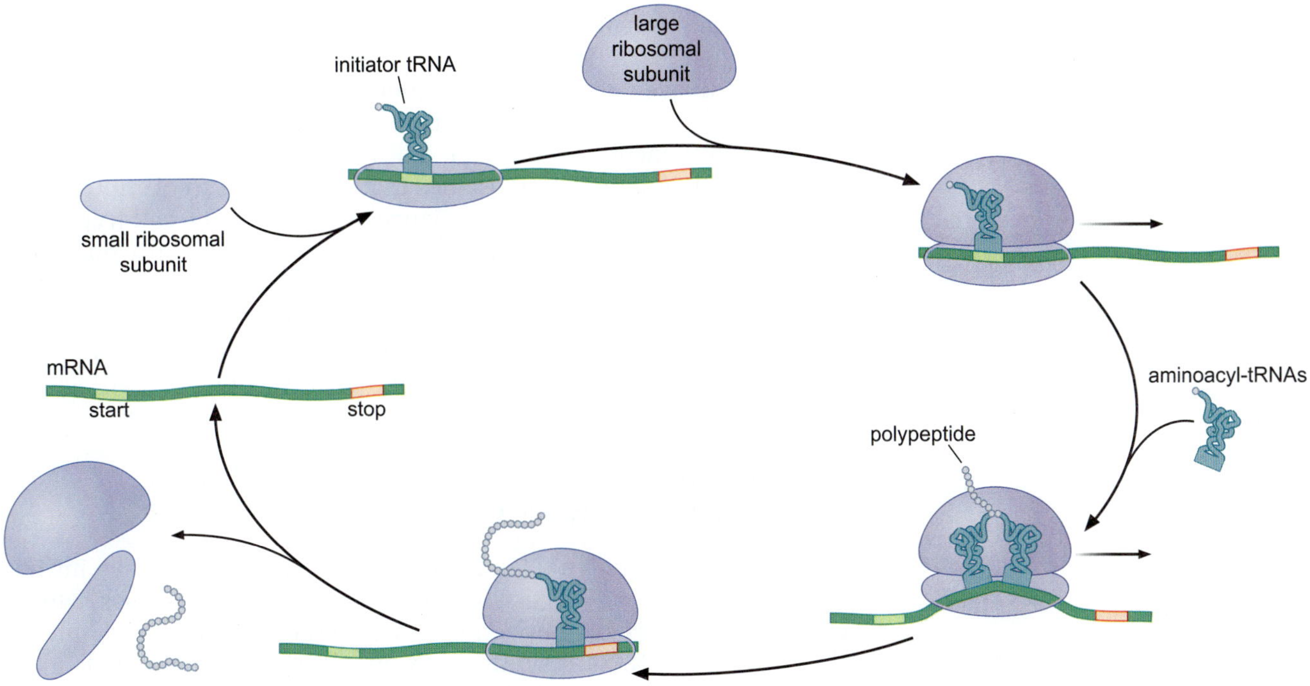

FIGURE 14-14 Overview of the events of translation: The ribosome cycle.

create an intact ribosome with the mRNA sandwiched between the two subunits. Protein synthesis is initiated in the next step, commencing at the start codon at the 5′ end of the message and progressing toward the 3′ end of the mRNA. As the ribosome translocates from codon to codon, one charged tRNA after another is slotted into the decoding and peptidyl transferase centers of the ribosome. When the elongating ribosome encounters a stop codon, the now completed polypeptide chain is released, and the ribosome dissociates from the mRNA as separate large and small subunits. The separated subunits are now available to bind to a new mRNA molecule and repeat the cycle of protein synthesis.

Although a ribosome can synthesize only one polypeptide at a time, each mRNA can be translated simultaneously by multiple ribosomes (for simplicity, let us assume that the message we are considering is monocistronic). An mRNA bearing multiple ribosomes is known as a **polyribosome** or a **polysome** (Fig. 14-15). A single ribo-

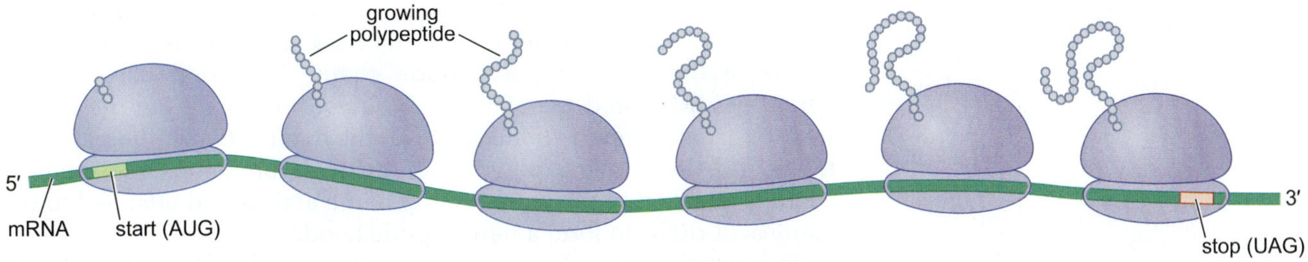

FIGURE 14-15 A polyribosome.

some contacts ~30 nucleotides of mRNA, but the large size of the ribosome only allows a density of 1 ribosome for every 80 nucleotides of mRNA. Still, even a small ORF of 1000 bases (which would encode a protein of ~35 kD) can bind more than ten ribosomes and therefore direct the simultaneous synthesis of multiple polypeptides.

The ability of multiple ribosomes to function on a single mRNA explains the relatively limited abundance of mRNA in the cell (typically 1–5% of total RNA). If an mRNA could be translated by only one ribosome at a time, then as few as 10% of the ribosomes would be engaged in protein synthesis in a typical cell. Instead, the association of multiple ribosomes with each mRNA ensures that the majority of the ribosomes are engaged in translation at any given time.

New Amino Acids Are Attached to the Carboxyl Terminus of the Growing Polypeptide Chain

As we know, both polynucleotide and polypeptide chains have intrinsic polarities. Thus, for each of these molecules, we can ask which end of the chain is synthesized first. We learned in Chapters 8 and 12 that DNA and RNA are synthesized by adding each new nucleotide triphosphate to the 3′ end of the growing polynucleotide chain (often referred to as synthesis in the 5′ → 3′ direction).

What is the order of synthesis of a growing polypeptide chain? This was first determined in a classic experiment performed by Dintzis that is described in Chapter 2. This experiment found that each new amino acid must be added to the carboxyl terminus of the growing polypeptide chain (often referred to as synthesis in the amino- to carboxy-terminal direction). As described in the next section, this directionality is a direct result of the chemistry of protein synthesis.

Peptide Bonds Are Formed by Transfer of the Growing Polypeptide Chain from One tRNA to Another

The ribosome catalyzes a single chemical reaction: the formation of a peptide bond. This reaction occurs between the amino acid residue at the carboxy-terminal end of the growing polypeptide and the incoming amino acid to be added to the chain. Both the growing chain and the incoming amino acid are attached to tRNAs; as a result, during peptide bond formation, the growing polypeptide is continuously attached to a tRNA.

The actual substrates for each round of amino acid addition are two charged species of tRNAs—an aminoacyl-tRNA and a **peptidyl-tRNA**. As we discussed earlier in this chapter (see the section on Attachment of Amino Acids to tRNAs) the aminoacyl-tRNA is attached at its 3′ end to the carboxyl group of the amino acid. The peptidyl-tRNA is attached in exactly the same manner (at its 3′ end) to the carboxyl-terminus of the growing polypeptide chain. The bond between the aminoacyl-tRNA and the amino acid *is not* broken during the formation of the next peptide bond. Instead, the bond between the peptidyl-tRNA and the growing polypeptide chain is broken as the growing chain is attached to the amino group of the amino acid attached to the aminoacyl-tRNA to form a new peptide bond.

To catalyze peptide bond formation, the 3′ ends of these two tRNAs are brought into close proximity on the ribosome. This positioning allows the amino group of the aminoacyl-tRNA to attack the carbonyl group of the most carboxy-terminal amino acid attached to the peptidyl-tRNA to form a new peptide bond (Fig. 14-16). There are two consequences of this method of polypeptide synthesis. First, this mechanism of peptide bond formation requires that the amino terminus of the protein be synthesized before the carboxyl terminus. Second, the growing polypeptide chain is transferred from the peptidyl-tRNA to the aminoacyl-tRNA. For this reason, the reaction to form a new peptide bond is called the **peptidyl transferase reaction**.

Interestingly, peptide bond formation takes place without the simultaneous hydrolysis of a nucleoside triphosphate. This is because peptide bond formation is driven by breaking the high-energy acyl bond that joins the growing polypeptide chain to the tRNA. Recall that this bond was created during the tRNA synthetase–catalyzed reaction that is responsible for charging tRNA. The charging reaction involves the hydrolysis of a molecule of ATP. Thus, the energy for peptide bond formation originates from the molecule of ATP that was hydrolyzed during the tRNA charging reaction (Fig. 14-6).

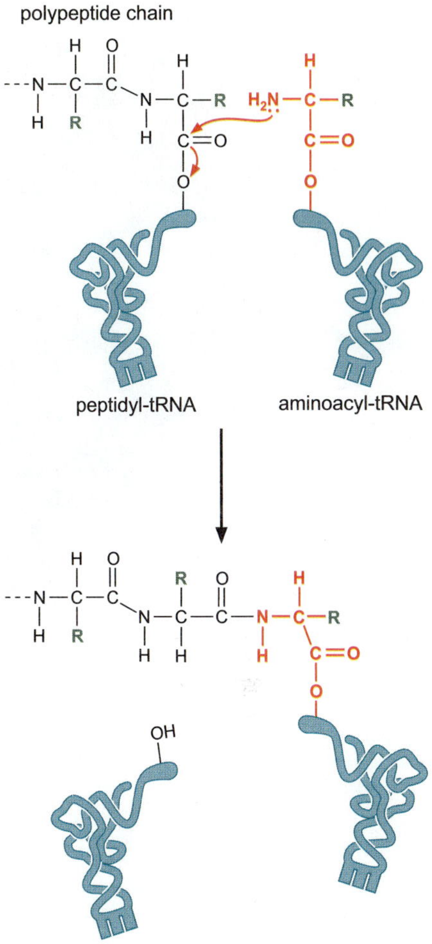

FIGURE 14-16 The peptidyl transferase reaction.

Ribosomal RNAs Are Both Structural and Catalytic Determinants of the Ribosome

Although the ribosome and its basic functions were discovered more than 40 years ago, the determination of numerous high-resolution, three-dimensional structures of the ribosome has vastly increased our understanding of the workings of this molecular machine (Fig. 14-17). Perhaps the most important outcome of these studies is the finding that rRNAs are much more than structural components of the ribosome. Rather, they are directly responsible for the key functions of the ribosome. The most obvious example of this is the demonstration that the peptidyl transferase center is composed almost entirely of RNA, as we will discuss in detail below. RNA also plays a central role in the function of the small subunit of the ribosome. The anticodon loops of the charged tRNAs and the codons of the mRNA contact the 16S rRNA, not the ribosomal proteins of the small subunit.

A further indication of the importance of RNA in the structure and function of the ribosome is that most ribosomal proteins are on the periphery of the ribosome, not in its interior (Fig. 14-17; see also Structural Tutorial 14-1). The core functional domains of the ribosome (the peptidyl transferase center and the decoding center) are composed either entirely or mostly from RNA. Portions of some ribosomal proteins do reach into the core of the subunits, where their function seems to be to stabilize the tightly packed rRNAs by shielding the negative charges of their sugar–phosphate backbones. Indeed, it is likely that the contemporary ribosome evolved from a primitive protein-synthesizing machine that was composed entirely of RNA and that the ribosomal proteins were added to enhance the function of this primordial RNA machine.

WEB STRUCTURAL TUTORIAL

The Ribosome Has Three Binding Sites for tRNA

To carry out the peptidyl transferase reaction, the ribosome must be able to bind at least two tRNAs simultaneously. In fact, the ribosome contains three tRNA-binding sites, called the A, P, and E sites (Figs. 14-18 and 14-19). The **A site** is the binding site for the aminoacylated-

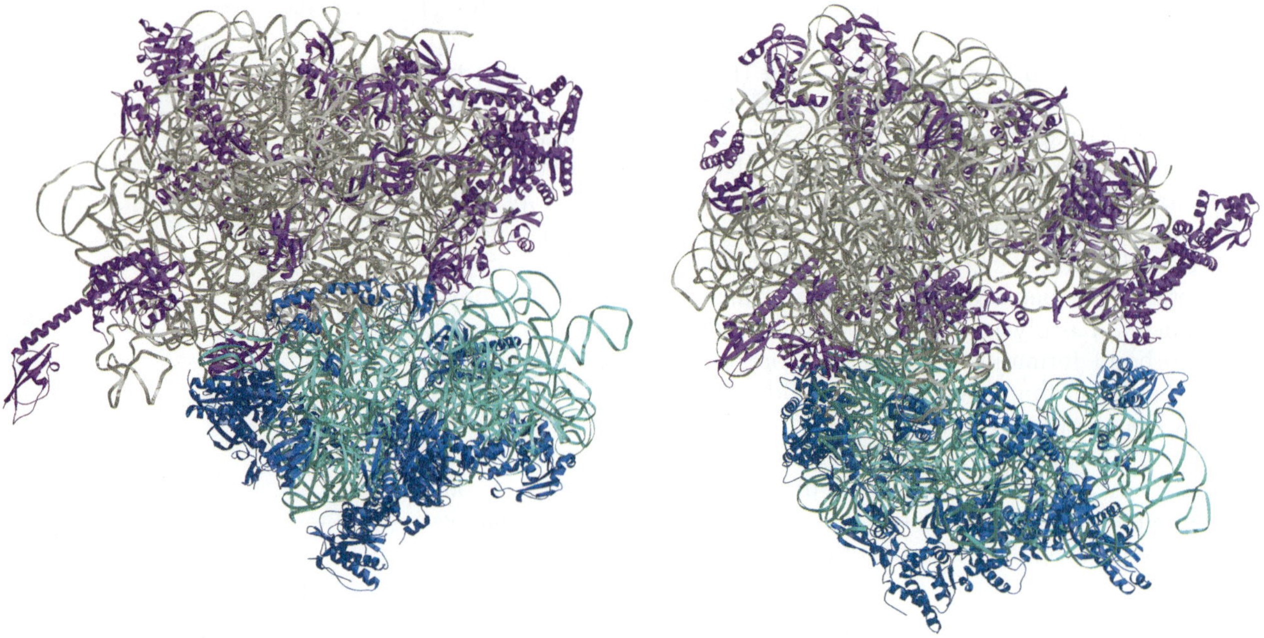

FIGURE 14-17 **Two views of the ribosome.** The 50*S* subunit is above the 30*S* subunit in both views. The cavity between the 50*S* and 30*S* subunits in the right-hand image represents the site of tRNA association (see Fig. 14-19b). The RNA component of the 50*S* subunit is shown in gray and the protein component is shown in purple. The RNA component of the 30*S* subunit is shown in light blue and the protein component in dark blue. (Yusupov M.M. et al. 2001. *Science* 292: 883–896.) Images prepared with MolScript, BobScript, and Raster 3D.

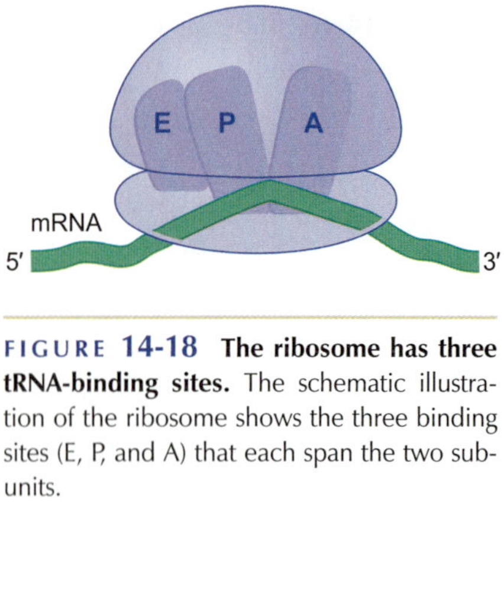

FIGURE 14-18 **The ribosome has three tRNA-binding sites.** The schematic illustration of the ribosome shows the three binding sites (E, P, and A) that each span the two subunits.

tRNA, the **P site** is the binding site for the peptidyl-tRNA, and the **E site** is the binding site for the tRNA that is released after the growing polypeptide chain has been transferred to the aminoacyl-tRNA (E is for exit).

Each tRNA-binding site is formed at the interface between the large and the small subunits of the ribosome (Fig. 14-19a,b). In this way, the bound tRNAs can span the distance between the peptidyl transferase center in the large subunit (Fig. 14-19c) and the decoding center in the small subunit (Fig. 14-19d). The 3′ ends of the tRNAs that are coupled to the amino acid or to the growing peptide chain are adjacent to the large subunit. The anticodon loops of the bound tRNAs are located adjacent to the small subunit.

Channels through the Ribosome Allow the mRNA and Growing Polypeptide to Enter and/or Exit the Ribosome

Both the decoding center and the peptidyl transferase center are buried within the intact ribosome. Yet, mRNA must be threaded through the decoding center during translation, and the nascent polypeptide chain must escape from the peptidyl transferase center. How do these polymers enter (in the case of mRNA) and exit the ribosome? The answer is provided by the structure of the ribosome, which reveals "tunnels" in and out of the ribosome.

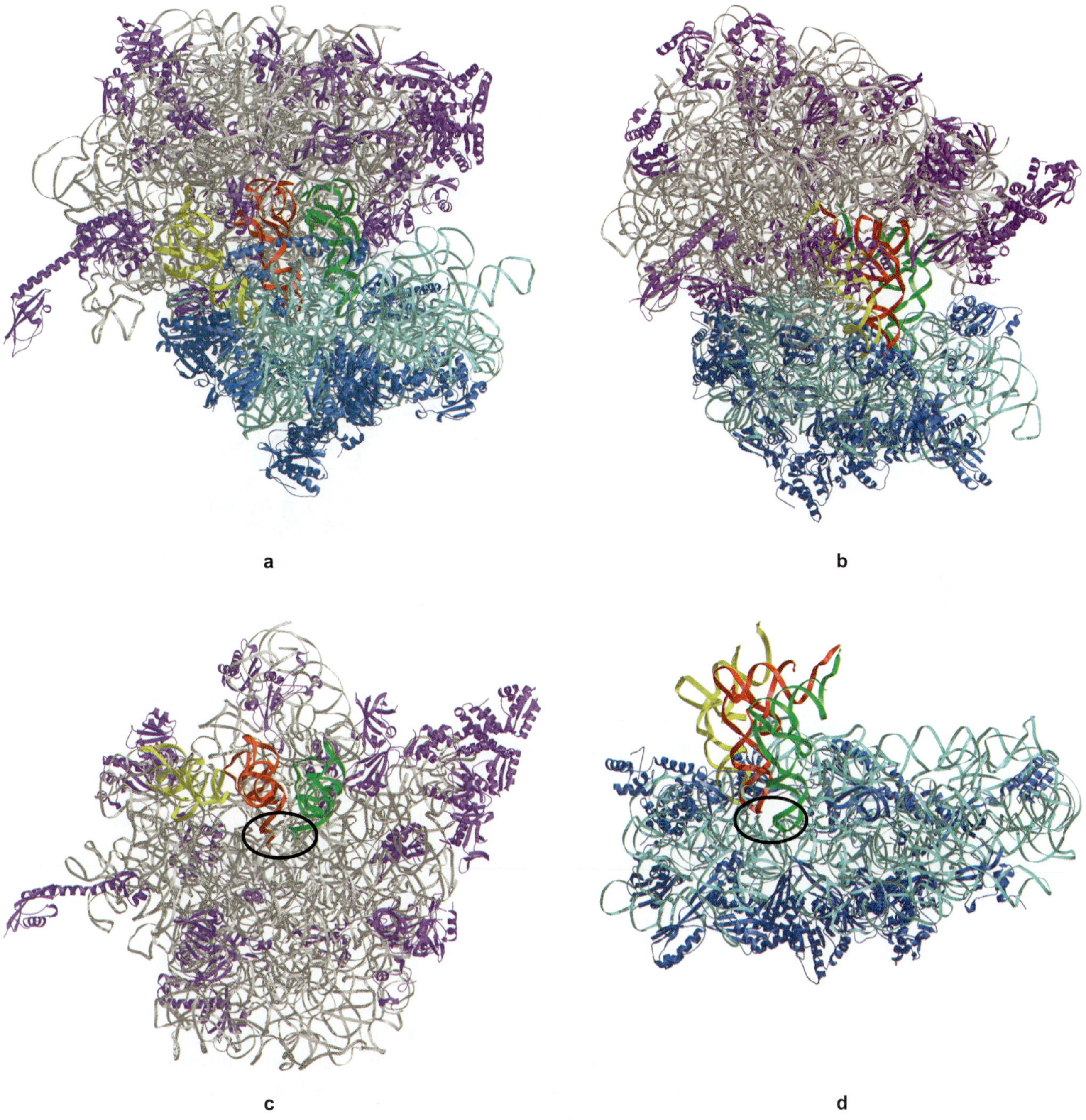

FIGURE 14-19 Views of the three-dimensional structure of the ribosome including three bound tRNAs. The E-, P-, and A-site tRNAs are shown in yellow, red, and green, respectively. The colors representing the RNA and protein components of the small and large subunits are the same as those in Fig. 14-17. (a and b) Two views of the ribosome bound to the three tRNAs in the E, P, and A sites. Note that the left (a) and right (b) views shown here correspond to those views of the ribosome shown in Fig. 14-17. (c) The isolated 50*S* subunit with tRNAs as seen in the full ribosome (this view is as if you were looking up at the large subunit from the small subunit). The peptidyl transferase center is circled. (d) The isolated 30*S* subunit with tRNAs as seen in the full ribosome. The decoding center is circled. (Yusupov M.M. et al. 2001. *Science* 292: 883–896.) Images prepared with MolScript, BobScript, and Raster 3D.

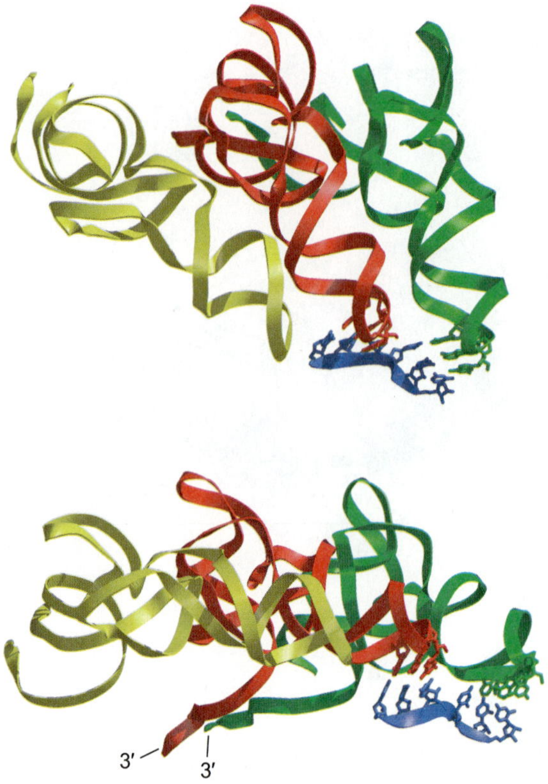

FIGURE 14-20 The interaction between the A-site and P-site tRNAs and the mRNA within the ribosome. Two views of the structure of the mRNA and tRNAs are shown as they are found in the ribosome. For clarity, the ribosome is not shown. The E-, P-, and A-site tRNAs are shown in yellow, red, and green, respectively, and the mRNA is shown in blue. Only the bases involved in the codon–anticodon interaction are shown. The strong kink in the mRNA clearly distinguishes between the A-site and P-site codons. The close proximity of the 3′ ends of the A-site and P-site tRNAs can be seen in the lower image. (Yusupov M.M. et al. 2001. *Science* 292: 883–896.) Image prepared with MolScript, BobScript, and Raster 3D.

The mRNA enters and exits the decoding center through two narrow channels in the small subunit. The entry channel is only wide enough for unpaired RNA to pass through. This feature ensures that the mRNA is in a single-stranded form as it enters the decoding center by removing any intramolecular base-pairing interactions that may have formed in the mRNA. In between the two channels is a region that is accessible to tRNAs and where adjacent codons can bind to the aminoacyl-tRNA and peptidyl-tRNA in the A and P sites, respectively. Interestingly, there is a pronounced kink in the mRNA between the two codons that facilitates maintenance of the correct reading frame (Fig. 14-20). This kink places the vacant A-site codon in a distinctive position that ensures the incoming aminoacyl-tRNA does not have access to bases immediately adjacent to the codon.

A second channel through the large subunit provides an exit path for the newly synthesized polypeptide chain (Fig. 14-21). As with the mRNA channel, the size of the peptide exit channel limits the conformation of the growing polypeptide chain. In this case, a polypeptide can form an α helix within the channel, but other secondary struc-

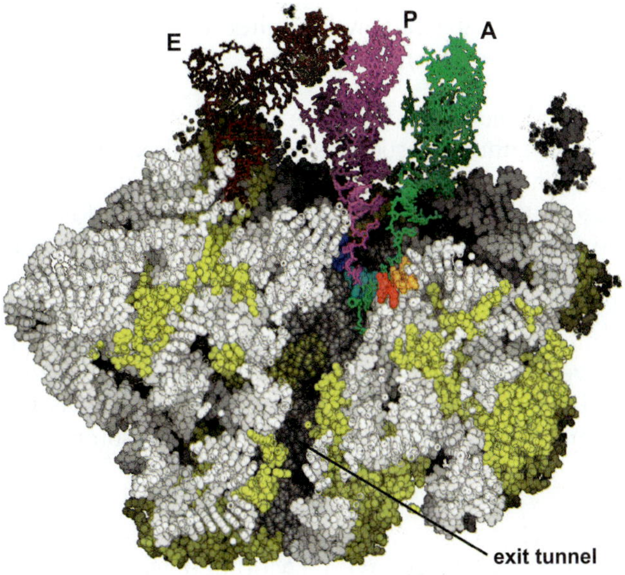

FIGURE 14-21 **The polypeptide exit tunnel.** In this image, the 50*S* subunit is cut in half to reveal the polypeptide exit tunnel. The rRNA is shown in white and the ribosomal proteins are shown in yellow. The three bound tRNAs are colored as follows: E site (brown), P site (purple), and A site (green). The red and gold parts of the rRNA adjacent to the A-site tRNA are components of the peptidyl transferase center. (Courtesy of T. Martin Schmeing and Thomas Steitz; adapted from Schmeing T.M. et al. 2002. *Nat. Struct. Biol.* 9: 225–230.)

tures (such as β sheets) and tertiary interactions can form only after the polypeptide exits the large ribosomal subunit. For this reason, the final three-dimensional structure of a newly synthesized protein is not attained until after it is released from the ribosome.

Now that we have described the four primary components of the translation process, the remainder of the chapter will focus on the individual stages of translation. Our description will proceed in order through the three stages of translation: initiation of the synthesis of a new polypeptide chain, elongation of the growing polypeptide, and termination of polypeptide synthesis. As we shall see, there are important similarities and differences between prokaryotes and eukaryotes in the strategies they employ to carry out these events. We shall consider the nature of the translation machinery from both kinds of cells in each of the following sections. As we have seen for DNA and RNA synthesis, although the ribosome is the center of activity, auxiliary factors play critical functions in each of the steps of translation and are required for protein synthesis to occur in a rapid and accurate fashion.

INITIATION OF TRANSLATION

For translation to be successfully initiated, three events must occur (Fig. 14-22): the ribosome must be recruited to the mRNA; a charged tRNA must be placed into the P site of the ribosome; and the ribosome must be precisely positioned over the start codon. The correct positioning of the ribosome over the start codon is critical, because this establishes the reading frame for the translation of the mRNA. Even a one-base shift in the location of the ribosome would result in the synthesis of a completely unrelated polypeptide (see the discussion of mRNA above and in Chapter 15). The dissimilar structures of prokaryotic and eukaryotic mRNAs result in distinctly different means of accomplishing these events. We start by addressing the initiation events in prokaryotes and then discuss the differences observed in eukaryotic cells.

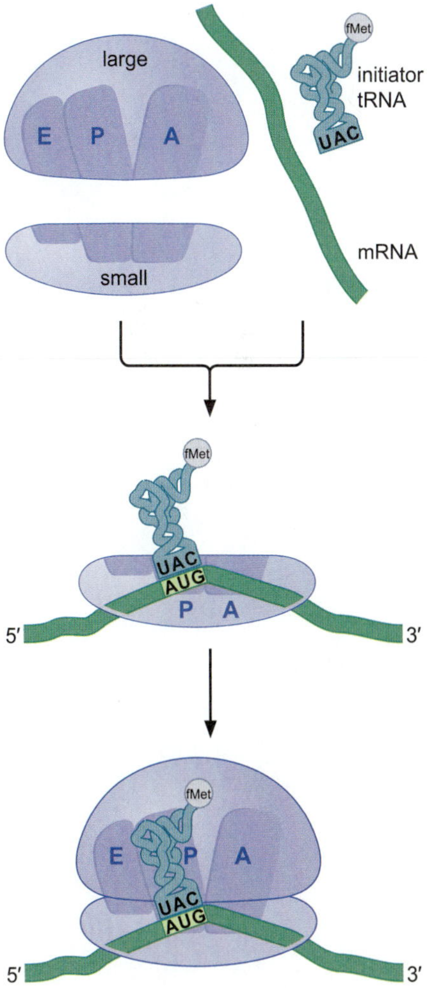

FIGURE 14-22 **An overview of the events of translation initiation.**

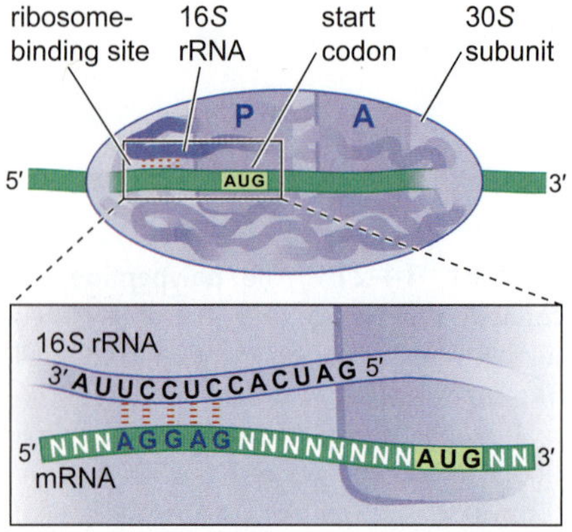

FIGURE 14-23 The 16S rRNA interacts with the RBS to position the AUG in the P site. This illustration shows an mRNA with the ideal separation between the RBS and the initiating AUG. This spacing places the AUG in the region of the P site. Many mRNA have nonideal spacings leading to a reduced rate of translation. Other mRNA lack an RBS completely.

Prokaryotic mRNAs Are Initially Recruited to the Small Subunit by Base Pairing to rRNA

The assembly of the ribosome on an mRNA occurs one subunit at a time. The small subunit associates with the mRNA first. As described during our discussion of mRNA structure (see Fig. 14-2), in prokaryotes, the association of the small subunit with the mRNA is mediated by base-pairing interactions between the RBS and the 16*S* rRNA (Fig. 14-23). For ideally positioned RBSs, the small subunit is positioned on the mRNA such that the start codon will be in the P site when the large subunit joins the complex. The large subunit joins its partner only at the very end of the initiation process, just prior to the formation of the first peptide bond. Thus, many of the key events of translation initiation occur in the absence of the full ribosome.

A Specialized tRNA Charged with a Modified Methionine Binds Directly to the Prokaryotic Small Subunit

Typically charged tRNAs enter the ribosome in the A site and only reach the P site after a round of peptide bond synthesis. During initiation, however, a charged tRNA enters the P site directly. This event requires a special tRNA known as the **initiator tRNA**, which base-pairs with the start codon—usually AUG or GUG. AUG and GUG have a different meaning when they occur within an ORF, where they are read by tRNAs for methionine ($tRNA^{Met}$) and valine ($tRNA^{Val}$), respectively (see Chapter 15). Neither methionine nor valine is attached to the initiator tRNA. Instead, it is charged with a modified form of methionine (***N*-formyl methionine**) that has a formyl group attached to its amino group (Fig. 14-24). The charged initiator tRNA is referred to as **$fMet\text{-}tRNA_i^{fMet}$**.

Because *N*-formyl methionine is the first amino acid to be incorporated into a polypeptide chain, one might think that all prokaryotic proteins have a formyl group at their amino terminus. This is not the case, however, as an enzyme known as a **deformylase** removes the formyl group from the amino terminus during or after the synthesis of the polypeptide chain. In fact, many mature prokaryotic proteins do not even start with a methionine; aminopeptidases often remove the amino-terminal methionine as well as one or two additional amino acids.

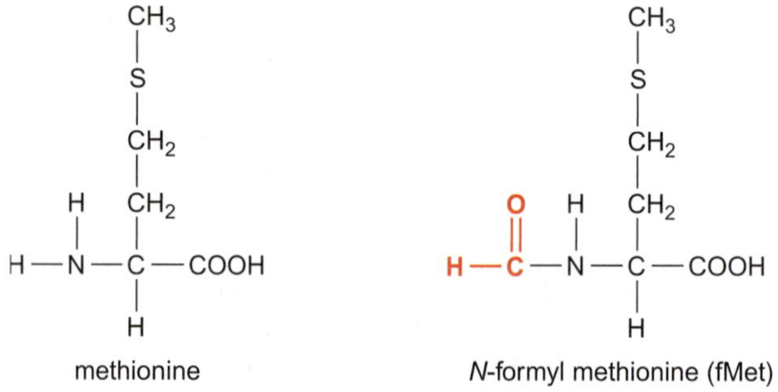

FIGURE 14-24 Methionine and *N*-formyl methionine.

Three Initiation Factors Direct the Assembly of an Initiation Complex That Contains mRNA and the Initiator tRNA

The initiation of prokaryotic translation commences with the small subunit and is catalyzed by three **translation initiation factors** called **IF1**, **IF2**, and **IF3**. Each factor facilitates a key step in the initiation process:

- IF1 prevents tRNAs from binding to the portion of the small subunit that will become part of the A site.
- IF2 is a GTPase (a protein that binds and hydrolyzes GTP) that interacts with three key components of the initiation machinery: the small subunit, IF1, and charged initiator tRNA (fMet-$tRNA_i^{fMet}$). By interacting with these components, IF2 facilitates the association of fMet-$tRNA_i^{fMet}$ with the small subunit and prevents other charged tRNAs from associating with the small subunit.
- IF3 binds to the small subunit and blocks it from reassociating with a large subunit. Because initiation requires a free small subunit, the binding of IF3 is critical for a new cycle of translation. IF3 becomes associated with the small subunit at the end of a previous round of translation when it helps to dissociate the 70*S* ribosome into its large and small subunits.

Each of the initiation factors binds at, or near, one of the three tRNA-binding sites on the small subunit. Consistent with its role in blocking the binding of charged tRNAs to the A site, IF1 binds directly to the portion of the small subunit that will become the A site. IF2 binds to IF1 and reaches over the A site into the P site to contact the fMet-$tRNA_i^{fMet}$. Finally, IF3 occupies the part of the small subunit that will become the E site. Thus, of the three potential tRNA binding sites on the small subunit, only the P site is capable of binding a tRNA in the presence of the initiation factors.

With all three initiation factors bound, the small subunit is prepared to bind to the mRNA and the initiator tRNA (Fig. 14-25). These two RNAs can bind in either order and independently of each other. As discussed above, binding to the mRNA involves base pairing between the RBS and the 16*S* rRNA in the small subunit. Meanwhile, binding fMet-$tRNA_i^{fMet}$ to the small subunit is facilitated by its interactions with IF2 bound to GTP and (once the mRNA is bound) base pairing between the anticodon and the start codon of the mRNA. Similarly, base pairing between the fMet-$tRNA_i^{fMet}$ and the mRNA serves to position the start codon in the P site.

The last step of initiation involves the association of the large subunit to create the **70*S* initiation complex**. When the start codon and fMet-$tRNA_i^{fMet}$ base-pair, the small subunit undergoes a change in conformation. This altered conformation results in the release of IF3. In the absence of IF3, the large subunit is free to bind to the small subunit with its cargo of IF1, IF2, mRNA, and fMet-$tRNA_i^{fMet}$. Binding of the large subunit stimulates the GTPase activity of IF2•GTP, causing it to hydrolyze GTP. IF2 bound to GDP has reduced affinity for the ribosome and the initiator tRNA leading to the release of IF2•GDP as well as IF1 from the ribosome. Thus, the net result of initiation is the formation of an intact (70*S*) ribosome assembled at the start site of the mRNA with fMet-$tRNA_i^{fMet}$ in the P site and an empty A site. The ribosome–mRNA complex is now poised to accept a charged tRNA into the A site and commence polypeptide synthesis.

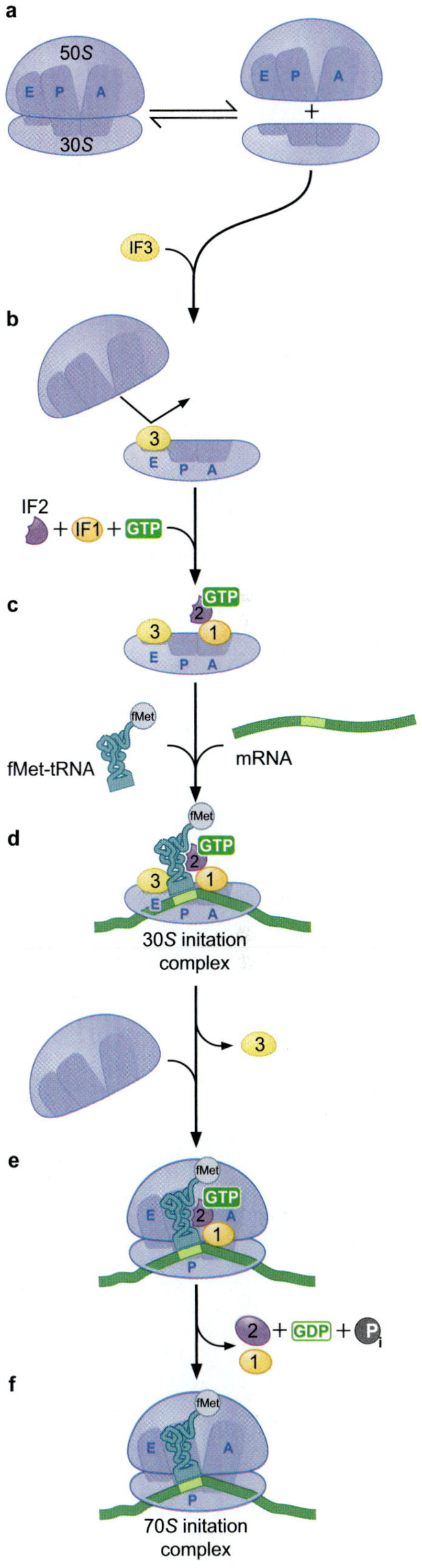

FIGURE 14-25 A summary of translation initiation in prokaryotes.

Eukaryotic Ribosomes Are Recruited to the mRNA by the 5′ Cap

Initiation of translation in eukaryotes is similar to prokaryotic initiation in many ways. Both use a start codon and a dedicated initiator tRNA, and both use initiation factors to form a complex with the small ribosomal subunit that assembles on the mRNA prior to addition of the large subunit. Nevertheless, eukaryotes use a fundamentally distinct method to recognize the mRNA and the start codon, which has important consequences for eukaryotic translation.

In eukaryotes, the small subunit is already associated with an initiator tRNA when it is recruited to the capped 5′ end of the mRNA. It then "scans" along the mRNA in a 5′→ 3′ direction until it reaches the first 5′-AUG-3′ (see the discussion of the Kozak sequence in the preceding section on mRNA), which it recognizes as the start codon. Thus, in most instances (see Box 14-3, uORFs and IRESs: Exceptions That Prove the Rule), only the first AUG can be used as the start site of translation in eukaryotic cells. Note that this method of initiation is consistent with the fact that the vast majority of eukaryotic RNAs encode a single polypeptide (monocistronic); recognition of an internal start codon is generally neither required nor possible.

As we have seen for other molecular processes (such as promoter recognition during transcription), eukaryotic cells require more auxiliary proteins to drive the initiation process than do prokaryotes. The events of initiation can be broken down into four steps. First, in contrast to the situation in prokaryotes, in eukaryotic cells, binding of the initiator tRNA to the small subunit *always* precedes association with the mRNA (Fig. 14-26a). Second, a separate set of auxiliary factors mediates the recognition of the mRNA. Third, the ribosome bound to the initiator tRNA scans the mRNA for the first AUG sequence. Finally, the large subunit of the ribosome is recruited after the initiator tRNA base pairs with the start codon. We now describe these events in detail.

As the eukaryotic ribosome completes a cycle of translation, it dissociates into free large and small subunits when four initiation factors—eIF1, eIF1A, eIF3, and eIF5—bind to the small subunit. In yeast cells, three of these factors (eIF1, eIF3, and eIF5) are assembled into a multifactor complex before they bind the small subunit, and similar interactions are likely to occur in other species. Together, these four factors act in an analogous manner to the prokaryotic initiation factors IF3 and IF1 to prevent both large subunit binding and tRNA binding to the A site. The initiator tRNA is escorted to the small subunit by the three-subunit GTP-binding protein eIF2. Like IF2, eIF2 will bind the initiator tRNA only in the GTP-bound state. This resulting complex is called the **ternary complex** (TC). For eukaryotes the initiator tRNA is charged with methionine, *not N*-formyl methionine, and is referred to as Met-$tRNA_i^{Met}$. eIF2 positions the Met-$tRNA_i^{Met}$ in the future P site of the small subunit, resulting in the formation of the **43*S* preinitiation complex**.

In a separate series of reactions, the mRNA is prepared for recognition by the small subunit. This process begins with recognition of the 5′ cap by the cap-binding protein eIF4E. A series of additional initiation factors is then recruited. eIF4G binds to both eIF4E and the mRNA, whereas eIF4A binds eIF4G and the mRNA (see Fig. 14-26b). The association of eIF4G with eIF4E is particularly important—the overall level of translation in the cell is controlled at this step by a family of proteins that compete with eIF4G-binding called eIF4E binding proteins (see Regulation of Translation later in this chapter). This complex is joined by eIF4B which activates the RNA helicase ac-

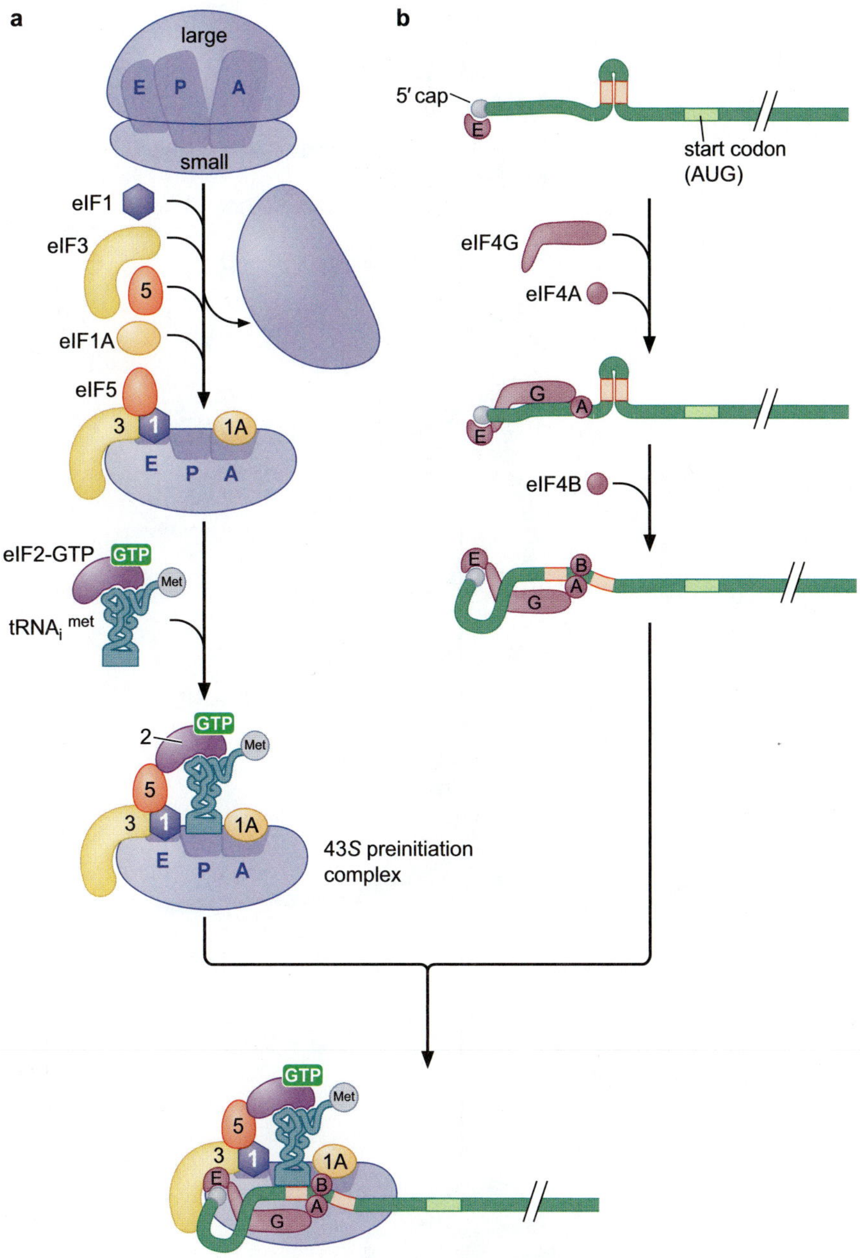

FIGURE 14-26 **Assembly of the eukaryotic small ribosomal subunit and initiator tRNA onto the mRNA.**

tivity of eIF4A. The helicase unwinds any secondary structures (such as hairpins) that may have formed at the end of the mRNA. Removal of secondary structures is critical as the 5′ end of mRNA must be unstructured to bind to the small subunit. Finally, interactions between the eIF4 factors bound to the unstructured mRNA and the initiation factors bound to the small subunit recruit the 43*S* preinitiation complex to the mRNA to form the **48*S* preinitiation complex**.

The Start Codon Is Found by Scanning Downstream from the 5′ End of the mRNA

Once assembled at the 5′ end of the mRNA, the small subunit and its associated factors move along the mRNA in a 5′ → 3′ direction in an ATP-dependent process that is stimulated by the eIF4A/B-associated RNA helicase (Fig. 14-27). During this movement, the small subunit

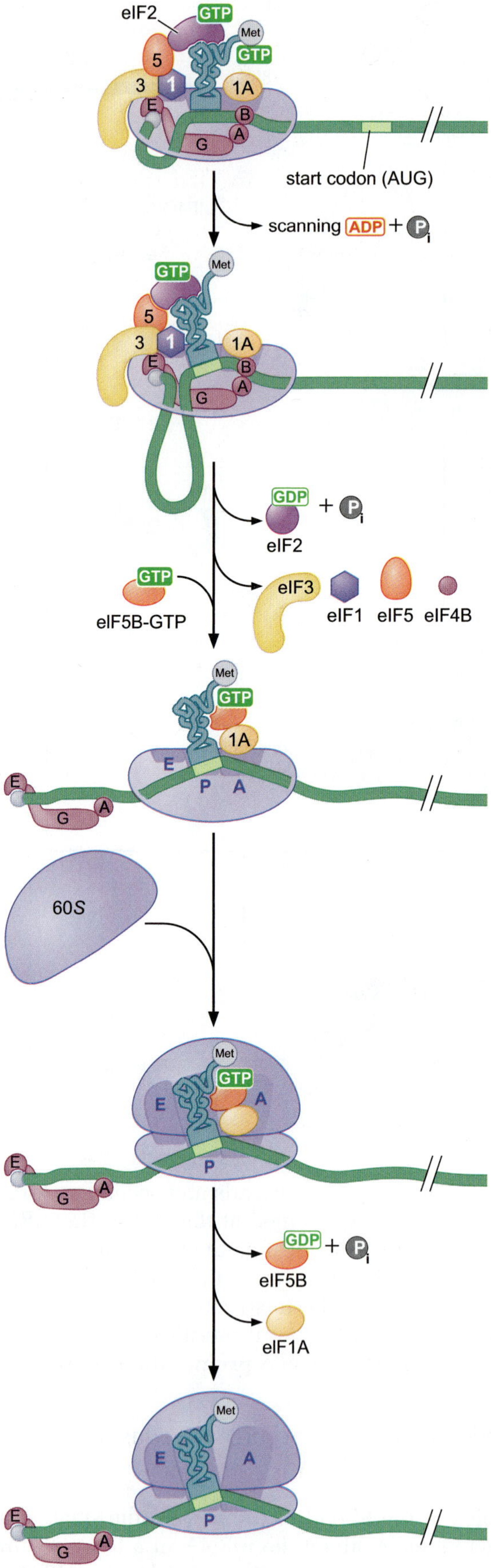

FIGURE 14-27 Identification of the initiating AUG by the eukaryotic small ribosomal subunit.

ADVANCED CONCEPTS

Box 14-3 uORFs and IRESs: Exceptions That Prove the Rule

Not all eukaryotic polypeptides are encoded by an ORF that starts with the AUG that is most proximal to the 5′ terminus. In some cases, the first AUG is not in a proper sequence context, resulting in its bypass. In other cases, short, upstream ORFs (uORFs, encoding peptides less than ten amino acids long) are found upstream of the principal ORF, that encodes a large polypeptide (Box 14-3 Fig. 1). In these cases, the uORFs act to regulate the extent of translation of a larger, downstream ORF. In general, these uORFs reduce but do not eliminate translation of the long downstream ORF. We discuss a specific example of this later in this chapter.

A more extreme example of initiating translation at sites downstream from the most 5′ proximal AUG are internal ribosome entry sites (IRESs). IRESs are RNA sequences that function like prokaryotic RBSs. They recruit the small subunit to bind and initiate at an internal site in the mRNA (Box 14-3 Fig. 2). These are relatively rare in eukaryotic transcripts and are most often encoded in viral mRNAs that often lack a 5′ cap end and have a need to exploit the sequences of their genome maximally. By using an IRES, a viral mRNA can encode more than one protein, reducing the need for extended transcriptional regulatory sequences for each protein-coding sequence. Different IRES sequences work by different mechanisms. At least one viral IRES directly binds to eIF4G, mimicking the normal recruitment of this protein through interactions with the 5′ cap–binding protein eIF4E. Other IRESs interact directly with the small subunit and its associated initiation factors (e.g., eIF3).

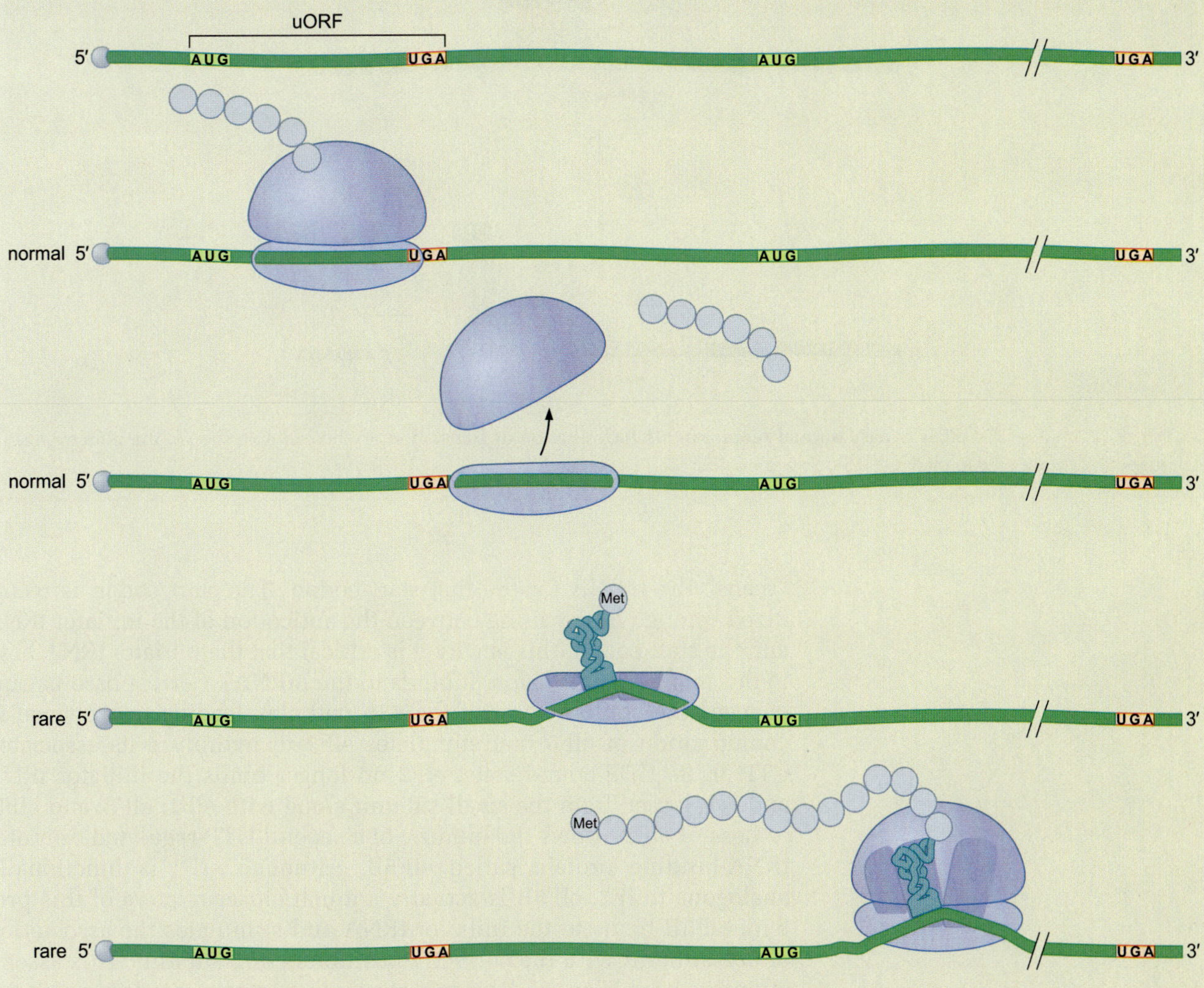

BOX 14-3 FIGURE 1 uORFs regulate translation of downstream ORFs. In some cases, after a ribosome translates a uORF, the small subunit remains on the mRNA and resumes scanning for a second AUG. It can only identify a second AUG when it binds a new initiator rRNA.

BOX 14-3 *(Continued)*

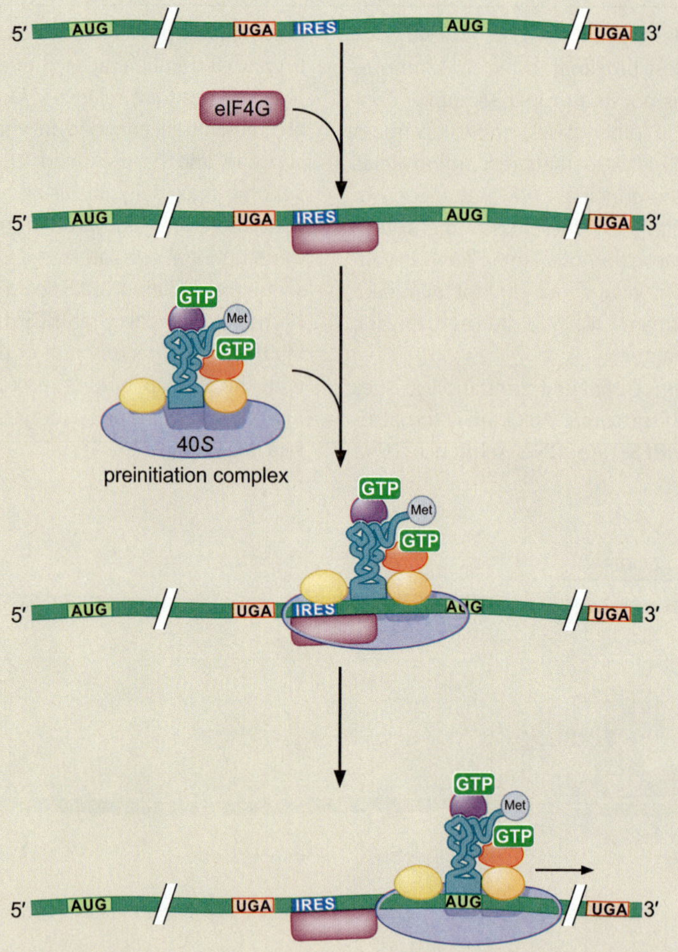

BOX 14-3 FIGURE 2 IRESs bypass normal requirements for initiation of translation. In the example shown, the IRES bypasses the requirement for the 5′ cap by directly binding to eIF4G.

"scans" the mRNA for the first start codon. The start codon is recognized through base pairing between the anticodon of the initiator tRNA and the start codon (this is why it is critical that the initiator tRNA bind to the small subunit *before* it binds to the mRNA). Correct base pairing changes the conformation of the 43*S* complex leading to a change in conformation of eIF5 that stimulates eIF2 to hydrolyze its associated GTP. In its GDP-bound state, eIF2 no longer binds the initiator tRNA and is released from the small subunit along with eIF1, eIF3, and eIF5.

Loss of eIF2 allows the binding of a second GTP-regulated, initiator tRNA-binding protein called eIF5B. Although eIF2 is functionally analogous to IF2, eIF5B is actually a much closer relative of this protein. eIF5B binds to the initiator tRNA and stimulates the association of 60*S* subunit with the correctly positioned 40*S* subunit. This association is possible as the factors that previously prevented this association (eIF1, eIF3, and eIF5) have been released. As in the prokaryotic situation, binding of the large subunit leads to the release of the remaining initiation factors by stimulating GTP hydrolysis by the IF2 analog, eIF5B. As a result of these events, the Met-$tRNA_i^{Met}$ is placed

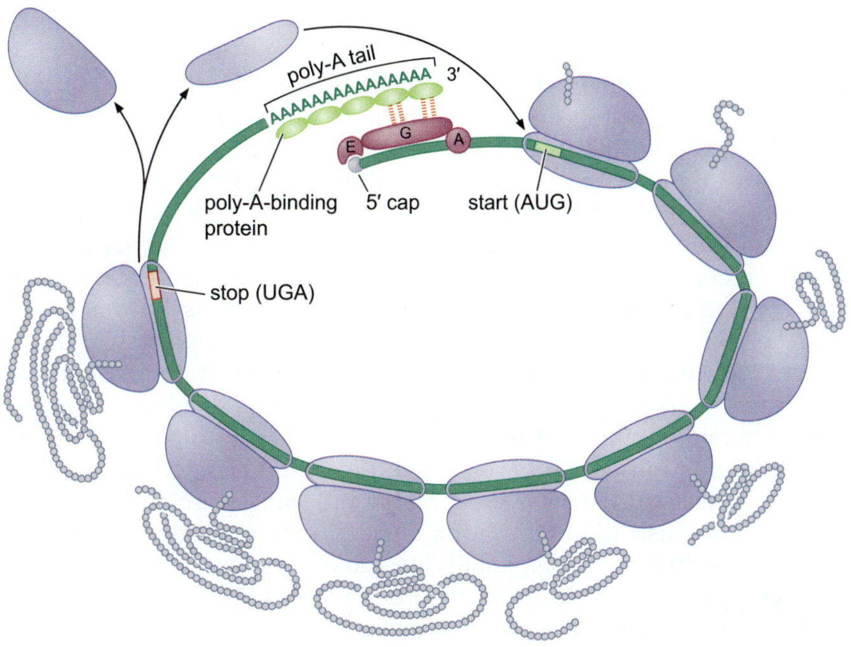

FIGURE 14-28 A model for the circularization of eukaryotic mRNA. Circularization is proposed to be mediated by an interaction between eIF4G and the poly-A-binding protein.

in the P site of the resulting **80*S* initiation complex**. With the start codon and Met-tRNA$_i^{Met}$ placed in the P site, the eukaryotic ribosome is now poised to accept a charged tRNA into its A site and carry out the formation of the first peptide bond.

Translation Initiation Factors Hold Eukaryotic mRNAs in Circles

In addition binding to the 5′ end of eukaryotic mRNAs, the initiation factors are closely associated with the 3′ end of the mRNA through its poly-A tail (Fig. 14-28). This is mediated by an interaction between eIF4G and the **poly-A-binding protein** that coats the poly-A tail. A consistent interaction between the two ends occurs because both eIF4G and the poly-A-binding protein are bound to the mRNA through multiple rounds of translation. The interaction between these proteins results in the mRNA being held in a circular configuration via a protein bridge between the 5′ and 3′ ends of the molecule. It has long been known that the poly-A tail contributes to efficient translation of mRNA. The finding that translation initiation factors "circularize" mRNA in a poly-A-dependent manner provides a simple rationale for this observation: once a ribosome finishes translating an mRNA that is circularized via its poly-A tail, the newly released ribosome is well positioned to reinitiate translation on the same mRNA.

TRANSLATION ELONGATION

Once the ribosome is assembled with the charged initiator tRNA in the P site, polypeptide synthesis can begin. There are three key events that must occur for the correct addition of each amino acid (Fig. 14-29). First, the correct aminoacyl-tRNA is loaded into the A site of the

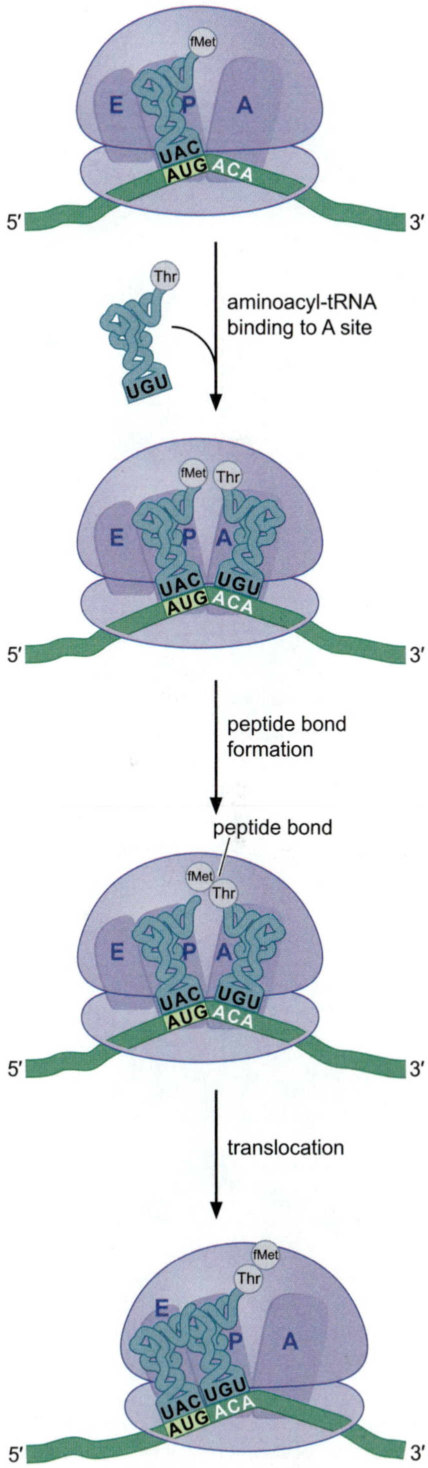

FIGURE 14-29 Summary of the steps of translation.

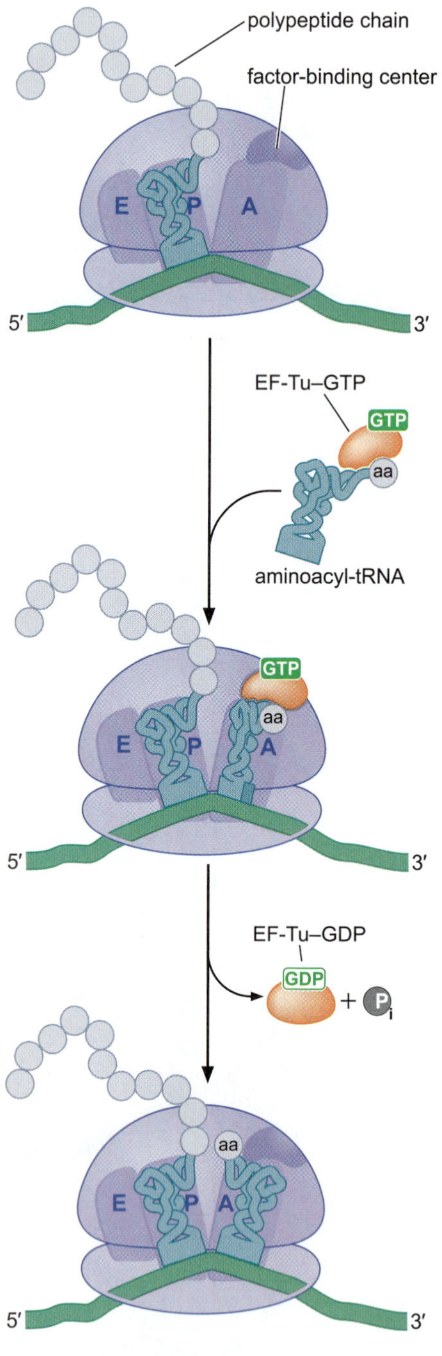

FIGURE 14-30 EF-Tu escorts aminoacyl-tRNA to the A site of the ribosome. Charged tRNAs are bound to EF-Tu–GTP as they first interact with the A site of the ribosome. When the correct codon–anticodon interaction occurs, EF-Tu interacts with the factor-binding center, hydrolyzes its bound GTP, and is released from the tRNA and the ribosome.

ribosome as dictated by the A-site codon. Second, a peptide bond is formed between the aminoacyl-tRNA in the A site and the peptide chain that is attached to the peptidyl-tRNA in the P site. This peptidyl transferase reaction, as we have seen, results in the transfer of the growing polypeptide from the tRNA in the P site to the amino acid moiety of the charged tRNA in the A site. Third, the resulting peptidyl-tRNA in the A site and its associated codon must be **translocated** to the P site so that the ribosome is poised for another cycle of codon recognition and peptide bond formation. As with the original positioning of the mRNA, this shift must occur precisely to maintain the correct reading frame of the message. Two auxiliary proteins known as **elongation factors** control these events. Both of these factors use the energy of GTP binding and hydrolysis to enhance the rate and accuracy of ribosome function.

Unlike the initiation of translation, the mechanism of elongation is highly conserved between prokaryotic and eukaryotic cells. We will limit our discussion to translation elongation in prokaryotes, which is understood in the greatest detail, but the events that occur in eukaryotic cells are similar to those in prokaryotes, both in the factors involved and in their mechanism of action.

Aminoacyl-tRNAs Are Delivered to the A Site by Elongation Factor EF-Tu

Aminoacyl-tRNAs do not bind to the ribosome on their own. Instead, they are "escorted" to the ribosome by the elongation factor **EF-Tu** (Fig. 14-30). Once a tRNA is aminocylated, EF-Tu binds to the tRNA's 3′ end, masking the coupled amino acid. This interaction prevents the bound aminoacyl-tRNA from participating in peptide bond formation until it is released from EF-Tu.

Like the initiation factor IF2, the elongation factor EF-Tu binds and hydrolyzes GTP, and the type of guanine nucleotide bound governs its function. EF-Tu can only bind to an aminoacyl-tRNA when it is associated with GTP. EF-Tu bound to GDP, or lacking any bound nucleotide, shows little affinity for aminoacyl-tRNAs. Thus, when EF-Tu hydrolyzes its bound GTP, any associated aminoacyl-tRNA is released. On its own, EF-Tu bound to an aminoacyl-tRNA does not hydrolyze GTP at a significant rate. Instead, the EF-Tu GTPase is activated when it associates with the same domain on the large subunit of the ribosome that activates the IF2 GTPase when the large subunit joins the initiation complex. This domain is known as the **factor-binding center**. EF-Tu only interacts with the factor binding center after the tRNA enters the A site *and* a correct codon-anticodon match is made. At this point, EF-Tu hydrolyzes its bound GTP and is released from the ribosome (Fig. 14-30). As we discuss below, control of GTP hydrolysis by EF-Tu is critical to the specificity of translation.

The Ribosome Uses Multiple Mechanisms to Select against Incorrect Aminoacyl-tRNAs

The error rate of translation is between 10^{-3} to 10^{-4}. That is, no more than 1 in every 1000 amino acids incorporated into protein is incorrect. The ultimate basis for the selection of the correct aminoacyl-tRNA is the base pairing between the charged tRNA and the codon displayed in the A site of the ribosome. Despite this, the energy difference between a correctly formed codon–anticodon pair and that of a near match cannot account for this level of accuracy. In many instances, only one of the three possible base pairs in the anticodon–codon interaction is mismatched, yet the ribosome rarely allows such mismatched aminoacyl-tRNAs to continue in the translation process. At least three different mechanisms contribute to this specificity (see Fig. 14-31). In each case, these mechanisms select *against* incorrect codon–anticodon pairings.

One mechanism that contributes to the fidelity of codon recognition involves two adjacent adenine residues in the 16*S* rRNA component located within the A site of the small subunit. These bases form hydrogen bonds with the minor groove of each correct base pair formed between the anticodon and the first two bases of the codon in the A site (Fig. 14-31a). Recall (see Fig. 6-10) that the hydrogen bonding properties of a Watson–Crick G:C and A:U base pair are very similar in the minor groove. Thus, the adjacent A residues in the 16S rRNA do not discriminate between G:C or A:U base pairs and recognize either as correct. In contrast, non-Watson–Crick base pairs form a minor groove that cannot be recognized by these bases, resulting in significantly reduced affinity for incorrect tRNAs. The net result of these interactions is that correctly paired tRNAs exhibit a much lower rate of dissociation from the ribosome than do incorrectly paired tRNAs.

A second mechanism that helps to ensure correct codon–anticodon pairing involves the GTPase activity of EF-Tu (see Fig. 14-31b). As described above, release of EF-Tu from the tRNA requires GTP hydrolysis, which is highly sensitive to correct codon–anticodon base pairing. Even a single mismatch in the codon–anticodon base pairing alters the position of EF-Tu, reducing its ability to interact with the factor-binding center. This, in turn, leads to a dramatic reduction in EF-Tu GTPase activity. This mechanism is an example of kinetic selectivity and is related to the mechanisms used to ensure correct base pairing during DNA synthesis (see Chapter 8). In both cases, formation of correct base pairing interactions dramatically enhances the rate of a critical biochemical step. For the DNA polymerase, this step was the formation of the phosphodiester bond. In this case, it is the hydrolysis of GTP by EF-Tu.

A third mechanism that ensures pairing accuracy is a form of proofreading that occurs after EF-Tu is released. When the charged tRNA is first introduced into the A site in a complex with EF-Tu-GTP, its 3′ end is distant from the site of peptide bond formation. To participate successfully in the peptidyl transferase reaction, the tRNA must rotate into the peptidyl transferase center of the large subunit in a process called **accommodation** (Fig. 14-31c). During accommodation, the 3′ end of the aminoacylated tRNA moves almost 70 Å. Incorrectly paired tRNAs frequently dissociate from the ribosome during accommodation. It is hypothesized that the rotation of the tRNA places a strain on the codon–anticodon interaction and that only a correctly paired anticodon can sustain this strain. Thus, mispaired tRNAs are more likely to dissociate from the ribosome prior to participating in the peptidyl transferase reaction.

In summary, in addition to the codon–anticodon interactions, the ribosome exploits minor groove interactions and two phases of proofreading to ensure that a correct aminoacyl-tRNA binds in the A site. Each of these three additional selectivity mechanisms inhibits retention of aminoacyl-tRNAs that do not form correct codon–anticodon interactions.

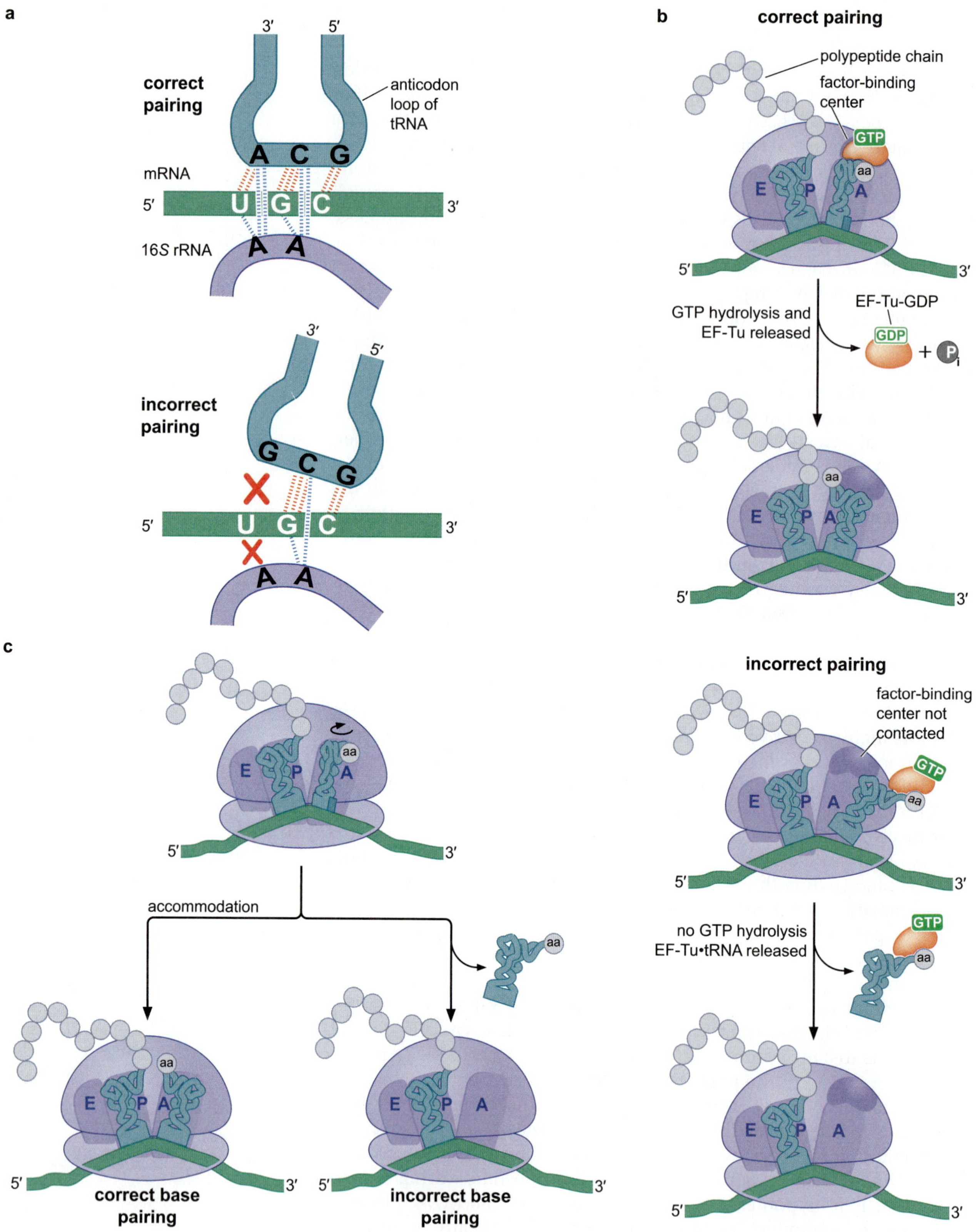

FIGURE 14-31 Three mechanisms to ensure correct pairing between the tRNA and the mRNA. (a) Additional hydrogen bonds are formed between two adenine residues of the 16*S* rRNA and the minor groove of the anticodon–codon pair only when they are correctly base-paired. (b) Correct base pairing allows EF-Tu bound to the aminoacyl-tRNA to interact with the factor-binding center inducing GTP hydrolysis and EF-Tu release. (c) Only correctly base-paired aminoacyl-tRNAs remain associated with the ribosome as they rotate into the correct position for peptide bond formation. This rotation is referred to as tRNA accommodation.

The Ribosome Is a Ribozyme

Once the correctly charged tRNA has been placed in the A site and has rotated into the peptidyl transferase center, peptide bond formation takes place. This reaction is catalyzed by RNA, specifically the 23*S* rRNA component of the large subunit. Early evidence for this came from experiments in which it was shown that a large subunit that had been largely stripped of its proteins was still able to carry out peptide bond formation. In support of this view, structural studies of the large subunit of the ribosome from one prokaryotic species showed that there was no amino acid with 18 Å of the active site (Fig. 14-32).

Recent determination of the three-dimensional structure of the entire *Escherichia coli* ribosome with bound mRNA and tRNAs has revealed that the very amino terminus of one protein (L27) does reach into the active site. This finding suggested a role for this protein in catalysis. To test this possibility, the nine amino acids at the L27 amino terminus that were in close proximity to the active site were eliminated by mutation. The resulting cells produced ribosomes with reduced but detectable peptidyl transferase activity, clearly indicating that this region of the L27 protein contributes to peptidyl transferase activity. The mutant ribosomes, however, still synthesized proteins at 30–50% of wild-type levels and cells containing them continued to grow and divide. The ribosome promotes a 10^7-fold increase in the rate of peptide bond formation relative to the rate observed with substrates (aminoacyl-tRNAs) alone in solution. Clearly, the vast majority of this increase is retained, even without the presence of L27 in the active site. Thus, although this protein facilitates peptide bond formation, it is not essential for peptide transferase activity. Like other ribosomal proteins, the most likely role for L27 is to correctly position one or more of the RNA components of the active site. More importantly, because this protein is the only one close enough to act catalytically, the rRNA component of the ribosome must be primarily responsible for catalyzing peptide bond formation.

How then does the 23*S* rRNA catalyze peptide bond formation? The exact mechanism remains to be determined, but some answers to

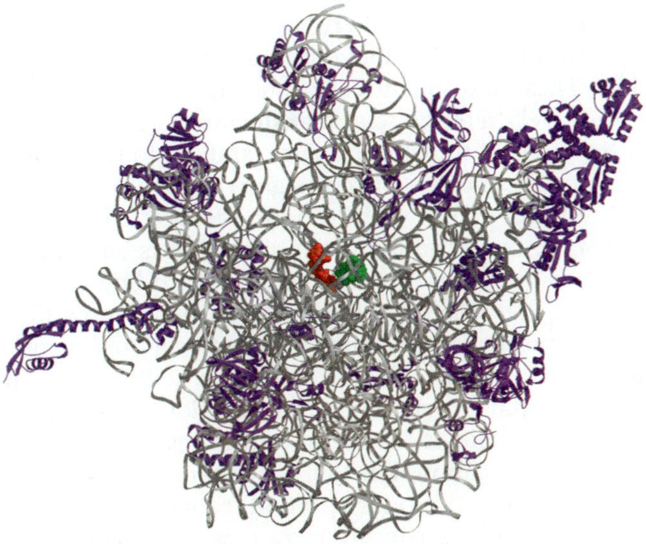

FIGURE 14-32 RNA surrounds the peptidyl transferase center of the large ribosomal subunit. The three-dimensional structure of the bacterial 50*S* subunit is shown. The rRNAs are shown in gray and the ribosomal proteins are shown in purple. The 3′ ends of the A-site and P-site tRNAs that are immediately adjacent to the peptidyl transferase center are shown in green and red, respectively. (Yusupov M.M. et al. 2001. *Science* 292: 883–896.) Image prepared with MolScript, BobScript, and Raster 3D.

this question are beginning to emerge. First, base pairing between the 23*S* rRNA and the CCA ends of the tRNAs in the A and P sites help to position the α-amino group of the aminoacyl-tRNA to attack the carbonyl group of the growing polypeptide attached to the peptidyl-tRNA. These interactions are also likely to stabilize the aminoacyl-tRNA after accommodation. This type of catalytic mechanism is called entropic catalysis. That is, the enzyme works by bringing the substrates together in a manner that stimulates catalysis.

Because close proximity of substrates is rarely sufficient to generate high levels of catalysis, it is likely that other elements of the rRNA contribute to catalysis. Indeed, alterations that eliminate the 2´-OH of a highly conserved residue in the 23*S* rRNA (A2451 in the *E. coli* 23*S* rRNA) reduce the rates of catalysis by at least tenfold. Recent studies have implicated a second unexpected RNA as being critical for catalysis: the P site tRNA. Mutations that remove the 2´-OH of the A residue at the 3´ end of the P-site tRNA result in a *10^6-fold* reduction in catalysis rates. This "substrate-assisted catalysis" is a particularly interesting finding as it indicates that the peptidyl-tRNAs themselves carry critical catalytic elements. This finding suggests that, before the evolution of the ribosome, tRNAs may have provided critical elements to allow them to catalyze protein synthesis on their own.

On the basis of a number of considerations, it has been proposed that the 2´-OH of the P-site tRNA may act as part of a "proton shuttle" (Fig. 14-33). In this model, the 2´-OH donates a hydrogen to the 3´-OH of the peptidyl-tRNA and accepts a proton from the attacking α-amino group of amino acid attached to the A-site tRNA. Importantly, both of these findings strongly support the hypothesis that it is RNA and not protein that catalyzes peptide bond formation. Nevertheless, there is still much to be learned about how the ribosome catalyzes peptide bond formation.

Peptide Bond Formation and the Elongation Factor EF-G Drive Translocation of the tRNAs and the mRNA

Once the peptidyl transferase reaction has occurred, the tRNA in the P site is deacetylated (no longer attached to an amino acid) and the growing polypeptide chain is linked to the tRNA in the A site. For a new round of peptide chain elongation to occur, the P-site tRNA must move to the E site and the A-site tRNA must move to the P site. At the same time, the mRNA must move by three nucleotides to expose the next codon. These movements are coordinated within the ribosome and are collectively referred to as **translocation**.

The initial steps of translocation are coupled to the peptidyl transferase reaction (Fig. 14-34). Once the growing peptide chain has been transferred to the A-site tRNA, the A- and P-site tRNAs are said to be in a hybrid state. The 3´ end of the A-site tRNA is bound to the growing polypeptide chain, but the anticodon is still bound to the codon in the A site of the decoding center. Similarly, the now deacetylated P-site tRNA is no longer attached to the growing polypeptide chain, but its anticodon is still bound to the P-site codon. Thus, translocation in the large subunit precedes translocation in the small subunit and the tRNAs are said to be in "hybrid states." Their 3´ ends have shifted into a new location, but their anticodon ends are still in their pre–peptidyl transfer position (Fig. 14-34, panel 1).

The completion of translocation requires the action of a second elongation factor called **EF-G**. EF-G can only bind to the ribosome

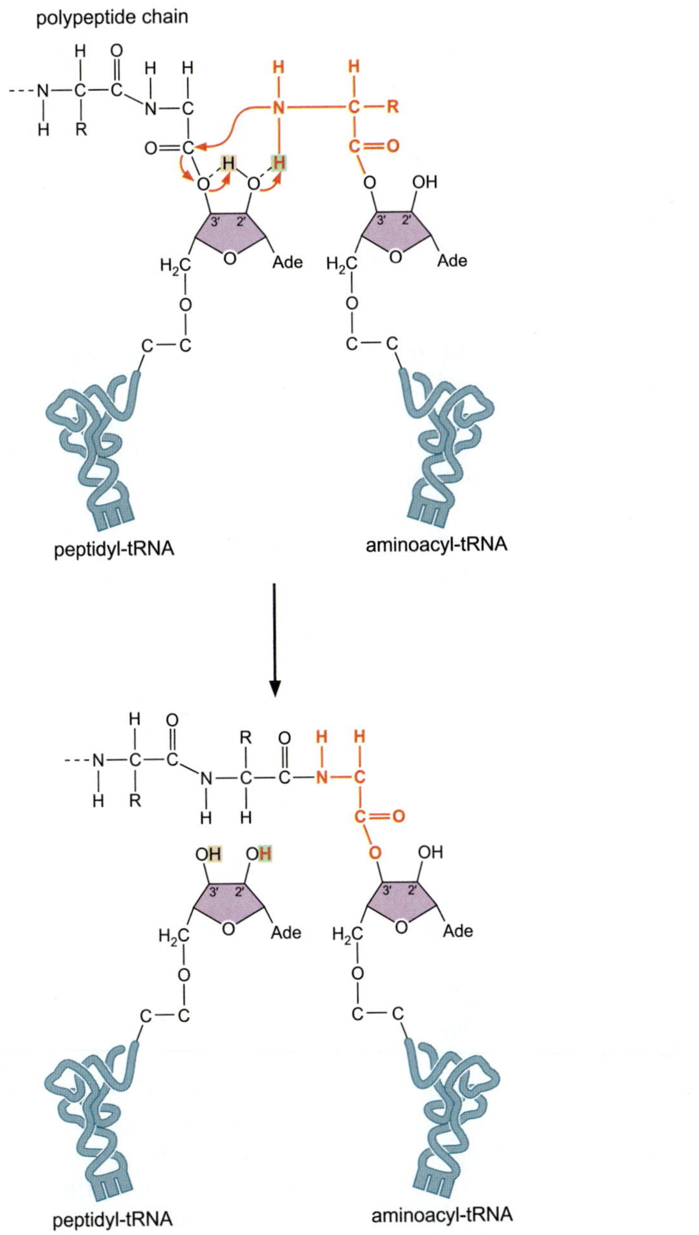

FIGURE 14-33 Proposed role for the 2′-OH of the P-site tRNA in peptide bond formation. The 2′-OH of the final "A" in the peptidyl-tRNA is critical for peptide bond formation. Based on this finding, it has been proposed that the hydrogen component of the 2′-OH participates in a "proton shuttle" illustrated here. In this model, as the bond between the peptidyl-tRNA and the polypeptide chain is broken, the 3′ oxygen extracts a hydrogen (yellow highlight) from the 2′-hydroxyl and the 2′ oxygen, in turn, extracts a hydrogen (green highlight) from the amino group attacking the carbonyl. The red arrows show the proposed direction of electron movement during peptide bond formation.

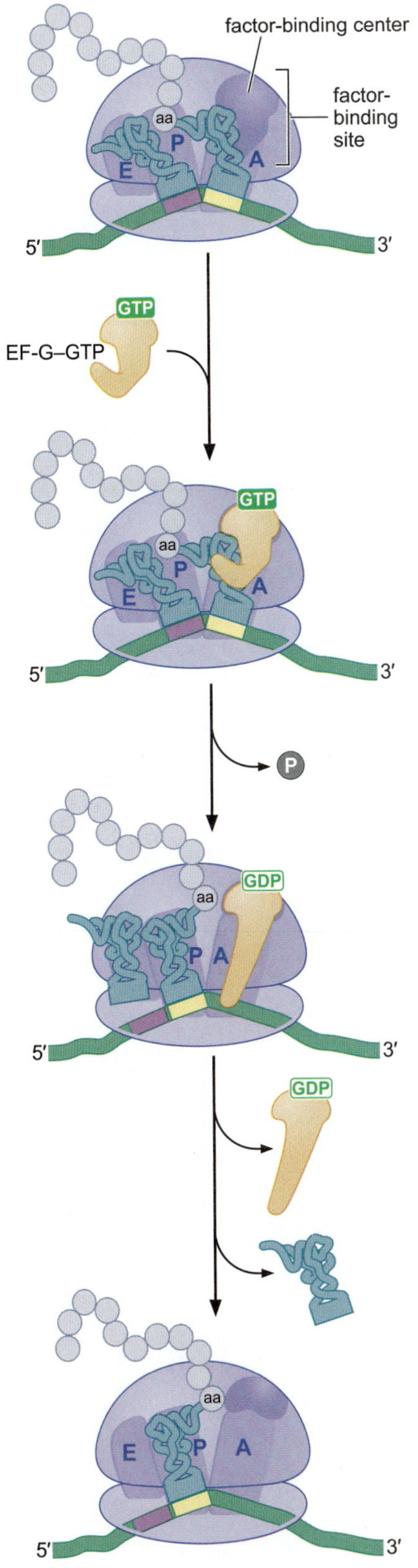

FIGURE 14-34 EF-G stimulation of translocation requires GTP hydrolysis.

when associated with GTP. After the peptidyl transferase reaction, the lack of a peptide bound to the P-site tRNA and the shift in the location of the A-site tRNA uncover a binding site for EF-G in the large subunit portion of the A site. When EF-G–GTP binds, it contacts the factor-binding center of the large subunit, which stimulates GTP hydrolysis. GTP hydrolysis changes the conformation of EF-G, allowing it to reach into the small subunit and trigger translocation of the A-site tRNA (Fig. 14-34, panel 3). When translocation is complete, the resulting ribosome structure has dramatically reduced affinity for EF-G–GDP, allowing the elongation factor to release from the ribosome. Together, these events result in the translocation of the A-site tRNA into the P site, the P-site tRNA into the E site, and the movement of the mRNA by exactly 3 bp (Fig. 14-34, panel 4). The ribosome is now ready for a new cycle of amino acid addition to begin.

EF-G Drives Translocation by Displacing the tRNA Bound to the A Site

The exact means by which EF-G induces translocation is not clear, but part of the mechanism involves the ability of EF-G–GDP to occupy the A-site portion of the decoding center. By interacting with the decoding center, EF-G–GDP displaces the A-site tRNA into the P site. Like dominoes, the displacement of the A-site tRNA into the P site means that the P-site tRNA must move into the E site. During the movement of the tRNAs, the mRNA is shifted by 3 bp. Movement of the mRNA is mediated by base pairing between the moving A-site tRNA and the mRNA, which is maintained during translocation. Essentially, the mRNA is pulled along with the moving A-site tRNA. Indeed, rare "frameshifting" tRNAs that have four-nucleotide-long anticodons (and can therefore compensate for certain frameshift mutations) move the mRNA by four nucleotides instead of three. In contrast to A-site tRNA movement, movement of the P-site tRNA into the E site disrupts base pairing of the tRNA with the mRNA. Hence, the now uncharged tRNA in the E site is free to dissociate from the ribosome and to become recharged with a fresh amino acid by aminoacyl-tRNA synthetase.

Changes in the subunits of the ribosome also contribute to translocation. For example, a counterclockwise rotation of the small subunit relative to the large subunit occurs during translocation. In addition, there must be changes in the structure of the small subunit to allow the release of EF-G–GDP after translocation is complete. Structral studies also reveal that there are "gates" that separate the the A, P, and E sites. Thus, for the tRNAs to translocate to their new positions, these gates must open. The occupancy of the A-site decoding center by EF-G–GDP ensures that, after translocation, the formerly A-site tRNA cannot move back into the A site. When EF-G is released, the "gates" presumably have closed again.

How does EF-G–GDP interact with the A site of the decoding center so effectively? Crystal structures of EF-G and EF-Tu bound to tRNA reveal a clear answer to this question. EF-G–GDP and EF-Tu–GTP–tRNA have a very similar structure (Fig. 14-35). Recall that EF-Tu–GTP–tRNA also binds to the A-site decoding center. What is most remarkable about this similarity is that even though EF-G is composed of a single polypeptide, its structure mimics that of a *tRNA* bound to a protein. This is an example of "molecular mimicry" in which a protein takes on the appearance of a tRNA to facilitate association with the same binding site. Intriguingly, structural studies of the eukaryotic analog of EF-G (called eEF-2) have identified two dramatically different conformations of the protein (one of which is bound to the antibiotic sordarin). One conformation is similar to the structure of the EF-G shown in Figure 14-35, whereas the second conformation results from a dramatic movement of the tRNA mimic region relative to the GTP-binding region. It is likely that such a conformational change is important for the function of EF-G during translocation.

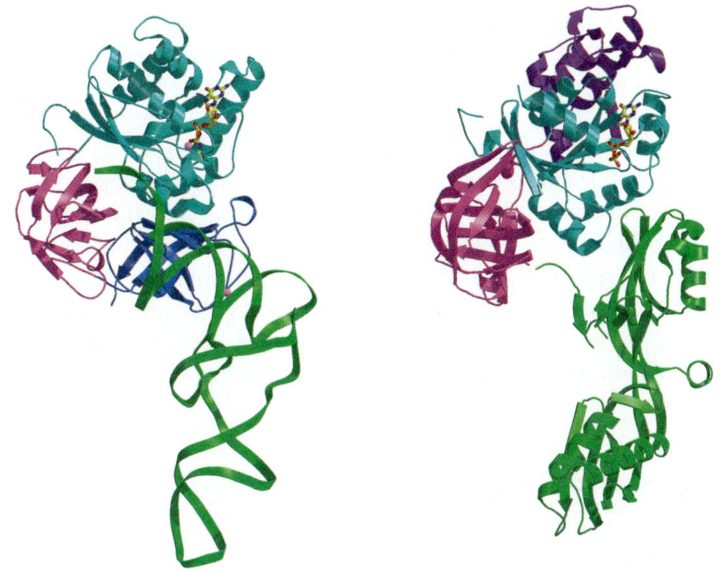

FIGURE 14-35 **Structural comparison of elongation factors.** EF-Tu–GDPNP–Phe– tRNA is shown on the left and EF-G–GDP is shown on the right. GDPNP is an analog of GTP that cannot be hydrolyzed that is used to lock the molecule in the GTP-bound conformation during the determination of the three-dimensional structure. Note the similarity between the structure of the green domain in EF-G and the tRNA bound to EF-Tu (also shown in green). (Left structure: Nissen P. et al. 1995. *Science* 270: 1464–1472. Right structure: al-Karadaghi S. et al. 1996. *Structure* 4: 555–565.) Images prepared with MolScript, BobScript, and Raster 3D.

EF-Tu–GDP and EF-G–GDP Must Exchange GDP for GTP prior to Participating in a New Round of Elongation

EF-Tu and EF-G are catalytic proteins that are used once for each round of tRNA loading onto the ribosome, peptide bond formation, and translocation. After GTP hydrolysis, both proteins must release their bound GDP and bind a new molecule of GTP. For EF-G, this is a simple process, as GDP has a lower affinity for EF-G than does GTP. Thus, after GTP hydrolysis, GDP and phosphate are released and the unbound EF-G rapidly binds a new GTP molecule. In the case of EF-Tu, a second protein is required to exchange GDP for GTP. The elongation factor **EF-Ts** acts as a **GTP exchange factor** for EF-Tu (Fig. 14-36). After EF-Tu–GDP is released from the ribosome, a molecule of EF-Ts binds to EF-Tu, causing the displacement of GDP. Next, GTP binds to the resulting EF-Tu-EF-Ts complex, causing its dissociation into free EF-Ts and EF-Tu–GTP. Finally, EF-Tu–GTP binds a molecule of charged tRNA, regenerating the EF-Tu–GTP aminoacyl-tRNA complex, which is once again ready to deliver a charged tRNA to the ribosome.

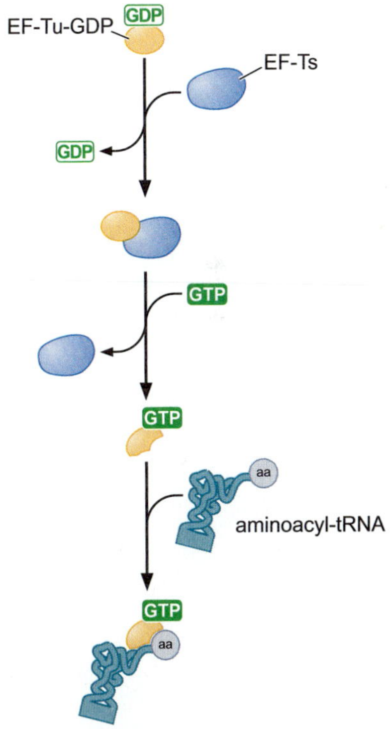

FIGURE 14-36 **EF-Ts stimulates release of GDP from EF-Tu.** GDP bound to EF-Tu is released very slowly in isolation. EF-Ts binds EF-Tu–GDP and causes the rapid release of GDP. GTP binding to EF-Tu in the EF-Tu–EF-Ts complex displaces EF-Ts and leaves EF-Tu–GTP, which can then bind a new aminoacyl-tRNA for delivery to the ribosome.

A Cycle of Peptide Bond Formation Consumes Two Molecules of GTP and One Molecule of ATP

We conclude our discussion of elongation by accounting for the energy spent. How many molecules of nucleoside triphosphate does it cost per round of peptide bond formation (setting aside the energetics of amino acid biosynthesis and the energetics of initiation and termination)? Recall that one molecule of nucleoside triphosphate (ATP) is consumed by the aminoacyl-tRNA synthetase in creating the high-energy acyl bond that links the amino acid to the tRNA. The breakage of this high-energy bond drives the peptidyl transferase reaction that creates the peptide bond. A second molecule of nucleoside triphosphate (GTP) is consumed in the delivery of a charged tRNA to the A site of the ribosome by EF-Tu and in ensuring that correct codon–anticodon recognition had taken place. Finally, a third nucleoside triphosphate is consumed in the EF-G-mediated process of translocation. Thus, making a peptide bond costs the cell two molecules of GTP and one of ATP, with one nucleoside triphosphate being consumed for each step in the translation elongation process. Interestingly, of the three molecules, only one (ATP) is energetically connected to peptide bond formation. The energy of the other two molecules (GTP) is spent to ensure the accuracy and order of events during translation (see Box 14-4, GTP-binding Proteins, Conformational Switching, and the Fidelity and Ordering of the Events of Translation).

Throughout the discussion of translation elongation, we have not distinguished between prokaryotes and eukaryotes. Although the eukaryotic factors analogous to EF-Tu (eEF1) and EF-G (eEF2) are named differently, their functions are remarkably similar to their prokaryotic counterparts.

TERMINATION OF TRANSLATION

Release Factors Terminate Translation in Response to Stop Codons

The ribosome's cycle of aminoacyl-tRNA binding, peptide bond formation, and translocation continues until one of the three stop codons enters the A site. It was initially postulated that there would be one or more chain-terminating tRNAs that would recognize these codons. However, this is not the case. Instead, stop codons are recognized by proteins called **release factors (RFs)** that activate the hydrolysis of the polypeptide from the peptidyl-tRNA.

There are two classes of release factors. Class I release factors recognize the stop codons and trigger hydrolysis of the peptide chain from the tRNA in the P site. Prokaryotes have two class I release factors called RF1 and RF2. RF1 recognizes the stop codon UAG, and RF2 recognizes the stop codon UGA. The third stop codon, UAA, is recognized by both RF1 and RF2. In eukaryotic cells, there is a single class I release factor called eRF1 that recognizes all three stop codons. Class II release factors stimulate the dissociation of the class I factors from the ribosome after release of the polypeptide chain. Prokaryotes and eukaryotes have only one class II factor called RF3 and eRF3, respectively. Like EF-G, IF2, and EF-Tu, class II release factors are regulated by GTP binding and hydrolysis.

Short Regions of Class I Release Factors Recognize Stop Codons and Trigger Release of the Peptidyl Chain

How do release factors recognize stop codons? Because release factors are composed entirely of protein, recognition of stop codons must be mediated by a protein–RNA interaction. Experiments in which short coding regions were genetically swapped between RF1 and RF2 (which have different stop-codon specificity) identified a three-amino-acid sequence that is critical for release factor specificity. Exchange of these three amino acids between RF1 and RF2 swaps the stop-codon specificity of the two complexes. For this reason, this three-amino-acid sequence is called a *peptide* anticodon and must interact with and recognize stop codons. A three-dimensional stucture of RF1 bound to the ribosome confirms that RF1 binds to the A site of the ribosome (Fig. 14-37a). In this structure, the peptide anticodon is located very near the anticodon, but it is likely that there are additional protein regions that contribute to codon recognition (Fig. 14-37b).

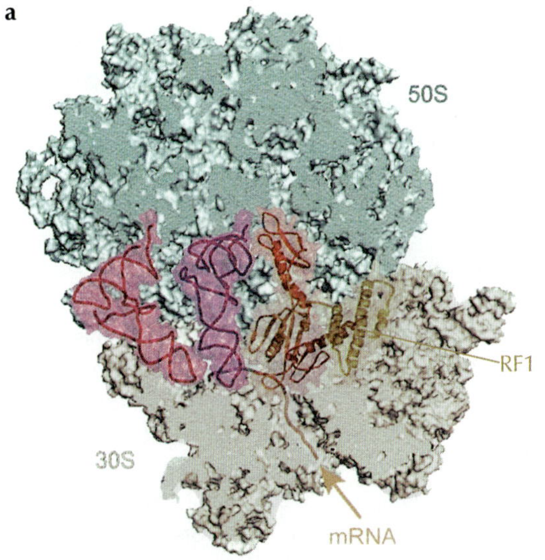

FIGURE 14-37 Three-dimensional stuctures of RF1 bound to the ribosome. (a) This view shows RF1 binding to the A site of the ribosome. (b) This structure shows that the peptide anticodon is located very near the anticodon. (c) In this view, the structure of RF1 bound to the ribosome shows the GGQ motif located close to the 3′ end of the P-site tRNA and the peptidyl transferase center. (Adapted, with permission, from Petry et al. 2005. *Cell* 123: 1255–1266. © Elsevier.)

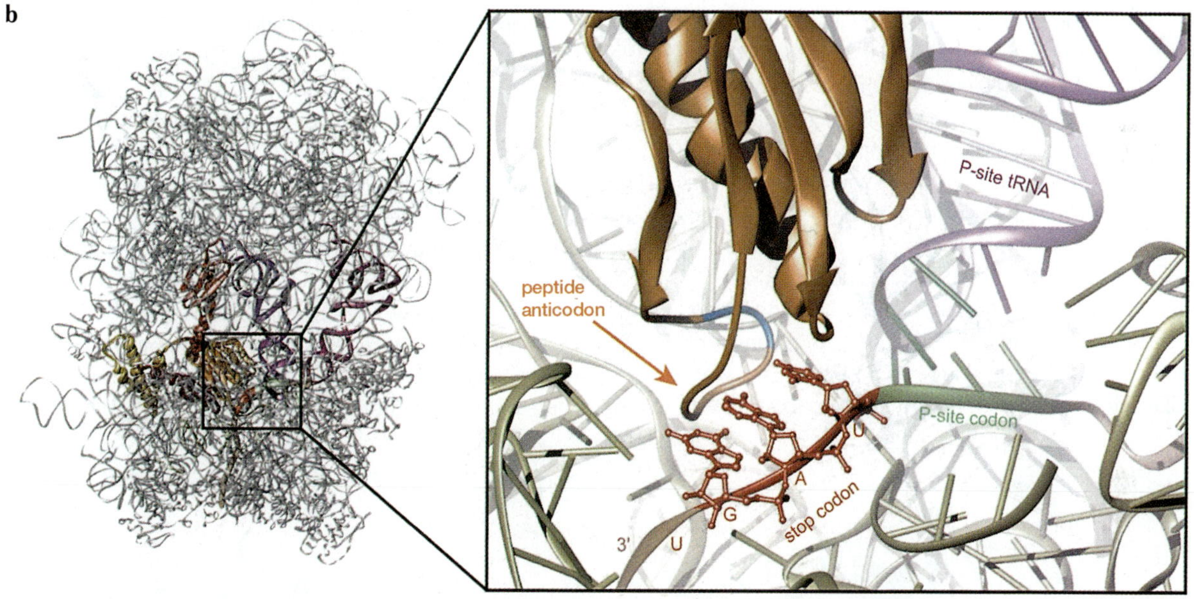

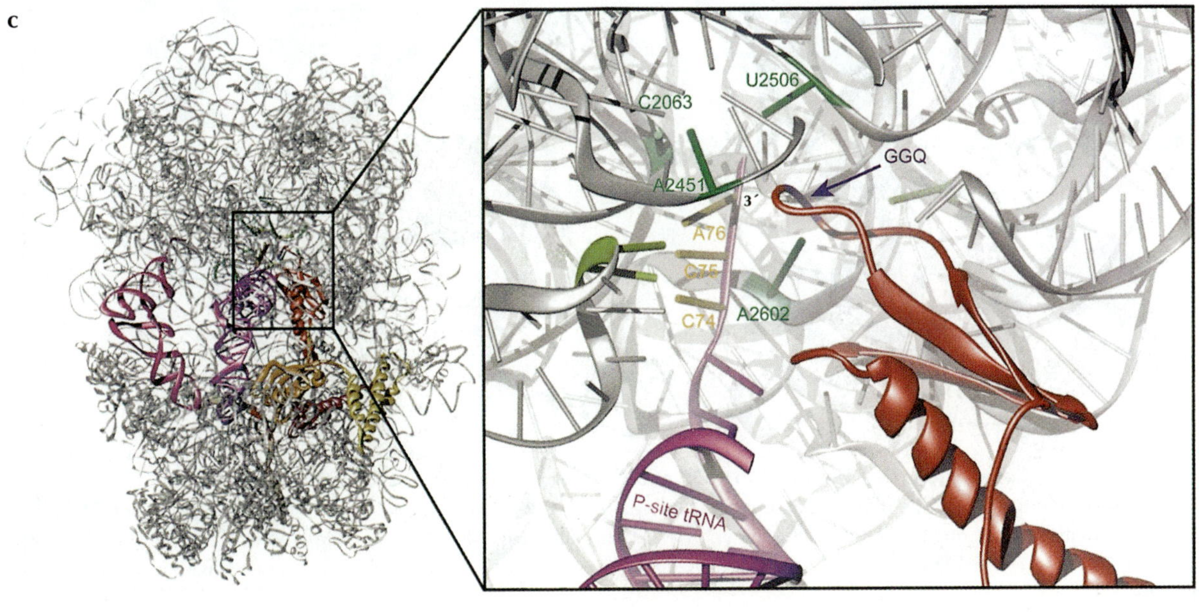

Box 14-4 GTP-Binding Proteins, Conformational Switching, and the Fidelity and Ordering of the Events of Translation

GTP is used throughout translation to control key events. The energy of GTP hydrolysis is not coupled to chemical modification as ATP is in the coupling of amino acids to tRNAs. Instead, the energy of GTP hydrolysis is used to control the order and fidelity of events during translation. How is this accomplished?

A key feature of the GTP-binding proteins involved in translation is that their conformation changes depending on the guanine nucleotide (such as GDP vs. GTP) to which they are bound. This can be seen for EF-Tu in Box 14-4 Figure 1, which shows the three-dimensional structure of EF-Tu bound to GTP or GDP. EF-Tu undergoes a major conformational change when it binds to GTP that results in the formation of its tRNA-binding site. In particular, one domain of EF-Tu (shown in magenta in Box 14-4 Fig. 1) shifts its location relative to the other domains of the protein depending on the nucleotide that is bound. This change in domain location, as well as changes in the conformation of the other two domains (shown in turquoise and dark blue), results in the formation of a new surface on EF-Tu that binds tightly to charged tRNAs (see EF-Tu bound to a tRNA in Fig. 14-35). Thus, depending on the form of guanine nucleotide bound, these factors can have different functions or bind to different proteins/RNAs. For example, EF-Tu–GTP can bind to an aminoacyl-tRNA but EF-Tu–GDP cannot.

By coupling GTP hydrolysis to the completion of key events in translation, the order of these events can be tightly controlled. For EF-Tu, the GTP-dependent association of EF-Tu with aminoacyl-tRNAs ensures that peptide bond formation does not occur prior to correct codon–anticodon pairing. Formation of the correct base pairs triggers GTP hydrolysis. Once bound to GDP, EF-Tu is released from the aminoacyl-tRNA allowing peptide bond formation to ensue.

The mechanism that activates GTP hydrolysis by each of the GTP-regulated auxiliary proteins is the same. In each case, GTPase activity is stimulated through an interaction with a specific region of the large subunit called the factor-binding center. This interaction is not of sufficient affinity to occur in isolation. Instead, each GTP-controlled translation factor must make several other critical interactions with the ribosome to stabilize the precise association with the factor-binding center that leads to GTPase activation. Indeed, as we have seen for EF-Tu, this interaction is highly sensitive to the exact nature of the interactions between EF-Tu, the aminoacyl-tRNA, the mRNA, and the ribosome. Thus, the interaction with the factor-binding center monitors all the other interactions of these proteins and RNAs with the ribosome. Only when correct codon–anticodon pairing is achieved does the GTP-binding site interact productively with the factor-binding center, leading to GTP hydrolysis and the associated changes in protein conformation.

The use of GTP during translation is analogous to the use of ATP by the sliding clamp loaders (see Chapter 8, Box 8-4). Recall that in that case, ATP binding was required to assemble an initial complex with the sliding clamp, but ATP hydrolysis and release of the sliding clamp could only occur when the clamp loader bound the primer:template junction. In translation, GTP is required for the initial association of the GTP-regulated factors with the ribosome (and in some instances, other RNAs and proteins), and GTP hydrolysis occurs only when the factor has correctly interacted with the ribosome. As in the case of the sliding clamp, GTP hydrolysis generally results in the release of the factor from the ribosome.

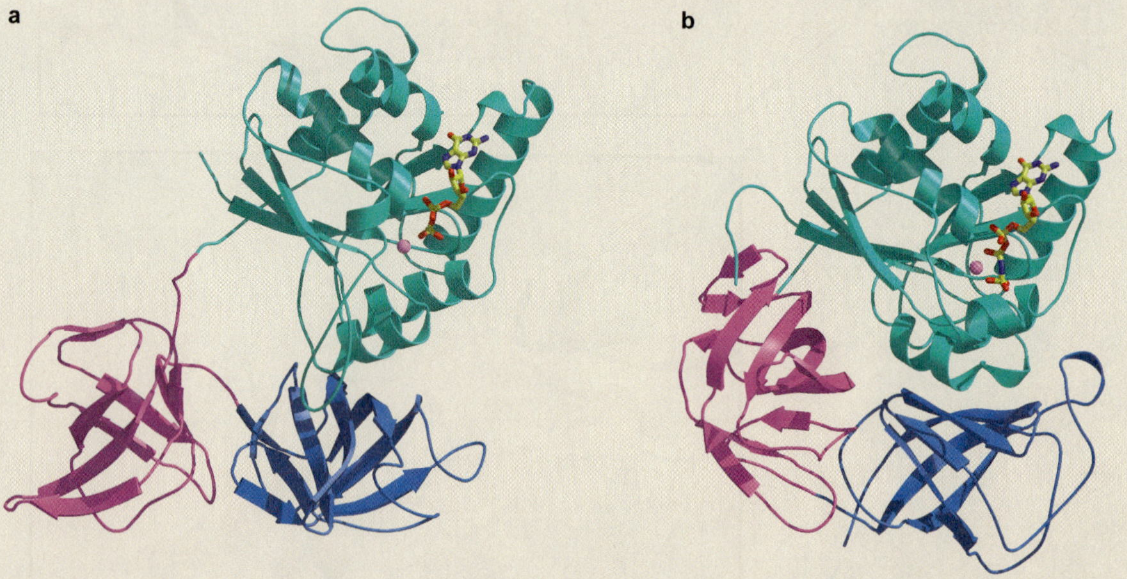

BOX 14-4 FIGURE 1 **Comparison of EF-Tu bound to GDP and GTP.** (a) EF-Tu bound to GDP. (b) EF-Tu bound to GTP. The GTP-binding domain is shown in turquoise. The rotation of the magenta domain and the changes in the structure of the turquoise and blue domains lead to the formation of a strong tRNA-binding site when GTP is bound (see Fig. 14-35). GTP is depicted in stick representation. (a, Polekhina G. et al. 1996. *Structure* 4: 1141–1151; b, Kjeldgard M. et al. 1993. *Structure* 1: 35–50.) Images prepared with MolScript, BobScript, and Raster 3D.

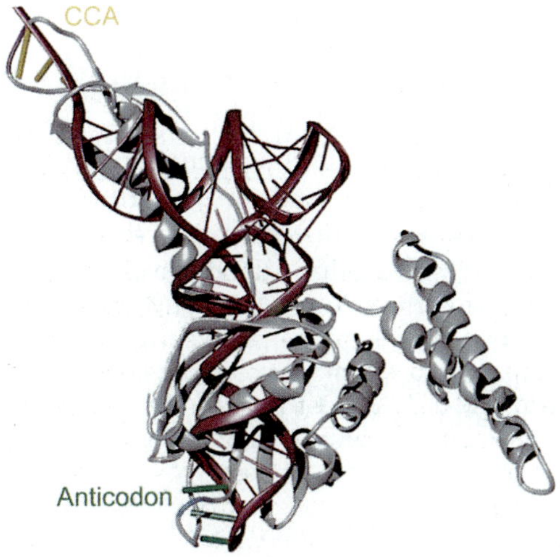

FIGURE 14-38 Comparison of the structures of RF1 to a tRNA. The tRNA is shown in dark red and RF1 is shown occupying the same space in gray. (Redrawn, with permission, from Petry et al. 2005. *Cell* 123: 1255–1266, Fig. 3E. © Elsevier.)

A region of class I release factors that contributes to polypeptide release has also been identified. All class I factors share a conserved three-amino-acid sequence (glycine glycine glutamine, GGQ) that is essential for polypeptide release. Moreover, the structure of RF1 bound to the ribosome confirms that the GGQ motif is located in close proximity to the peptidyl transferase center (Fig. 14-38c). It remains unclear whether the GGQ motif is directly involved in the hydrolysis of the polypeptide from the peptidyl-tRNA or if it induces a change in the peptidyl transferase center that allows the center itself to catalyze hydrolysis. Studies of the conserved bases found adjacent to the CCA ends in the petidyl transferase center (e.g., A2541 or A2602) indicate that several of these residues are required for peptide hydrolysis. Indeed, these bases appear to play a more important role in peptide release than they do in peptide bond formation. A likely explanation for this difference is that only proximal RNA residues can position a small water molecule for hydrolysis, but residues at many sites in the ribosomes can help position the much larger tRNA for catalysis.

Together, these studies have led to the hypothesis that class I release factors functionally mimic a tRNA, having a peptide anticodon that interacts with the stop codon and a GGQ motif that reaches into the peptidyl transferase center. Comparison of the structure of RF1 to a tRNA reveals how the protein functionally mimics a tRNA (Fig. 14-38). Just as the CCA 3′ terminus and the anticodon loop occupy extreme ends of each tRNA, the GGQ and the peptide anticodon loop occupy extreme ends of RF1.

GDP/GTP Exchange and GTP Hydrolysis Control the Function of the Class II Release Factor

Once the class I release factor has triggered the hydrolysis of the peptidyl-tRNA linkage, it must be removed from the ribosome (Fig. 14-39). This step is accomplished by the class II release factor, RF3 (or, in eukaryotic cells, eRF3). RF3 is a GTP-binding protein but, unlike the other GTP-binding proteins involved in translation, this factor has a higher affinity for GDP than GTP. Thus, free RF3 is predominantly in the GDP-bound form. RF3-GDP binds to the ribosome in a manner

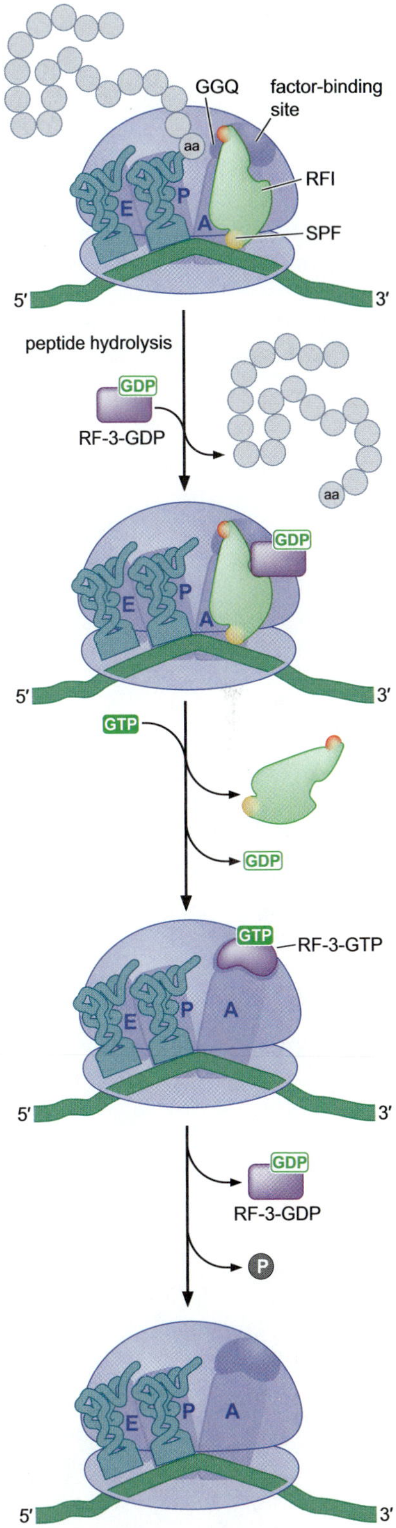

FIGURE 14-39 Polypeptide release is catalyzed by two release factors. The class I release factor (shown here as RF1) recognizes the stop codon and stimulates polypeptide release through a GGQ motif that is localized to the peptidyl transferase center. The class II release factor (RF3) binds only after polypeptide release and drives the dissociation of the class I release factor.

that depends on the presence of a class I release factor. After the class I release factor stimulates polypeptide release, a change in the conformation of the ribosome and the class I release factor stimulates RF3 to exchange its bound GDP for a GTP. That is, these factors act as a GTP exchange factor for RF3 in much the same way that EF-Ts does for EF-Tu. The binding of GTP to RF3 leads to the formation of a high-affinity interaction with the ribosome that displaces the class I factor from the ribosome. This change also allows RF-3 to associate with the factor-binding center of the large subunit. As with other GTP-binding proteins involved in translation, this interaction stimulates the hydrolysis of GTP. In the absence of a bound class I factor, the resulting RF3•GDP has a low affinity for the ribosome and is released.

The Ribosome Recycling Factor Mimics a tRNA

After the release of the polypeptide chain and the release factors, the ribosome is still bound to the mRNA and two deacylated tRNAs (in the P and E sites). To participate in a new round of polypeptide synthesis, the tRNAs and the mRNA must be removed from the ribosome and the ribosome must dissociate into its large and small subunits. Collectively, these events are referred to as **ribosome recycling**.

In prokaryotic cells, a factor known as the **ribosome recycling factor (RRF)** cooperates with EF-G and IF3 to recycle ribosomes after polypeptide release (Fig. 14-40). RRF binds to the empty A site of the ribosome, where it mimics a tRNA. RRF also recruits EF-G–GTP to the ribosome and, in events that mimic EF-G function during elongation, the EF-G stimulates the release of the uncharged tRNAs bound in the P and E sites. Although exactly how this release occurs is unclear, it is thought that RRF is displaced from the A site by EF-G in a manner similar to the displacement of a tRNA from the A site during elongation. Once the tRNAs are removed, EF-G–GDP and RRF are released from the ribosome along with the mRNA. IF3 (the initiation factor) may also participate in the release of the mRNA and is required to separate the two ribosomal subunits from each other. The final outcome of these events is a small subunit bound to IF3 (but not tRNA or mRNA) and a free large subunit. The released ribosome can now participate in a new round of translation.

Reinforcing the view that the RRF is a mimic of tRNA, RRF in fact resembles a tRNA in its three-dimensional structure. Despite this similarity, RRF interacts with the ribosome in a manner very different from that of a tRNA. RRF is closely associated only with the large subunit portion of the A site. We can rationalize this difference between the recycling factor and tRNAs in the following way. If the ribosome recycling factor precisely mimicked an A-site tRNA, then the P-site tRNA would be moved into the E site by EF-G. Instead, EF-G and the recycling factor lead to the release of the P-site tRNA from the ribosome directly from the P site. It is likely that EF-G and the ribosome recycling factor cause a more dramatic change in the structure of the ribosome than normally occurs during translocation, allowing both the mRNA and the tRNAs to be released.

Like initiation and elongation, the termination of translation is mediated by an ordered series of interdependent factor binding and release events. This ordered nature of translation ensures that no one step occurs before the previous step is complete. For example, EF-Tu cannot escort a new tRNA into the A site until EF-G completes translocation.

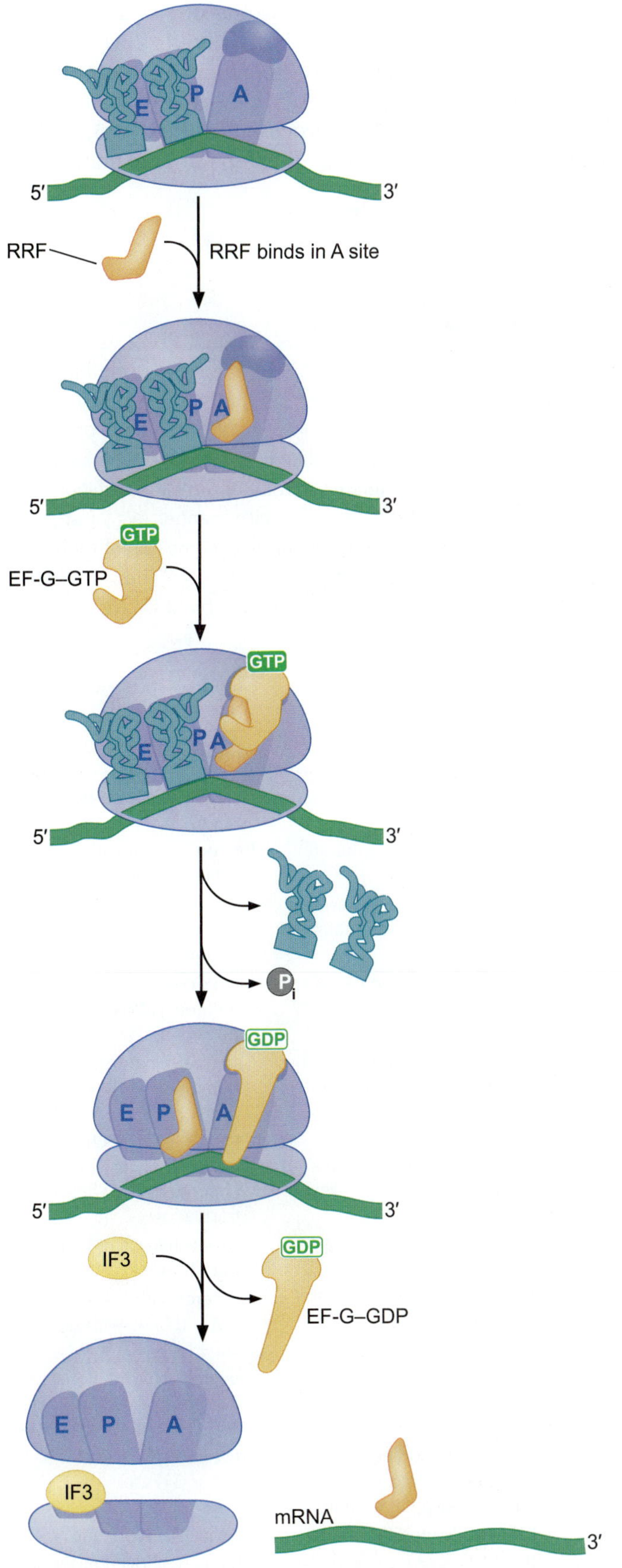

FIGURE 14-40 **RRF and EF-G combine to stimulate the release of tRNA and mRNA from a terminated ribosome.**

Similarly, RF3 cannot bind to the ribosome unless a class I release factor has already recognized a stop codon. There is a weakness to this orderly approach to translation: if any step cannot be completed, then the entire process stops. It is just this Achilles' heel that antibiotics exploit when they target the translation process (see Box 14-5, Antibiotics Arrest Cell Division by Blocking Specific Steps in Translation).

MEDICAL CONNECTIONS

Box 14-5 Antibiotics Arrest Cell Division by Blocking Specific Steps in Translation

Antibiotics represent a powerful tool to fight disease. The most widely used antibiotics in medicine kill bacteria but have little or no effect on eukaryotic cells and hence are not toxic to the patient. Since their discovery in the first half of the last century, antibiotics have helped make previously untreatable infections such as tuberculosis, bacterial pneumonia, syphilis, and gonorrhea largely curable (although the emergence of antibiotic-resistant bacteria is becoming an increasing obstacle to effective treatment). Antibiotics have many different kinds of targets in the bacterial cell, but ~40% of the known antibiotics are inhibitors of the translation machinery (Box 14-5 Table 1). In general, these antibiotics bind a component of the translation apparatus and inhibit its function. Because different antibiotics arrest translation at different steps and do so in a precise manner (e.g., just prior to EF-Tu release), these agents have become useful tools in studies of the mechanism of protein synthesis. Thus, in addition to their obvious medical benefits, antibiotics have come to play an

BOX 14-5 TABLE 1 Antibiotics: Targets and Consequences

Antibiotic/Toxin	Target Cells	Molecular Target	Consequence
Tetracycline	Prokaryotic cells	A site of 30*S* subunit	Inhibits aminoacyl-tRNA binding to A site
Hygromycin B	Prokaryotic and eukaryotic cells	Near A site of 30*S* subunit	Prevents translocation of A-site tRNA to P site
Paromycin	Prokaryotic cells	Adjacent to A-site codon–anticodon interaction site in 30*S* subunit	Increases error rate during translation by decreasing selectivity of codon–anticodon pairing
Chloramphenicol	Prokaryotic cells	Peptidyl transferase center of 50*S* subunit	Blocks correct positioning of A site aminoacyl-tRNA for peptidyl transfer reaction
Puromycin	Prokaryotic and eukaryotic cells	Peptidyl transferase center of large ribosomal subunit	Chain terminator; mimics 3′ end of aminoacyl-tRNA in A site and acts as acceptor for nascent polypeptide chain
Erythromycin	Prokaryotic cells	Peptide exit tunnel of 50*S* subunit	Blocks exit of growing polypeptide chain from the ribosome; arrests translation
Fusidic acid	Prokaryotic cells	EF-G	Prevents release of EF-G–GDP from the ribosome
Thiostrepton	Prokaryotic cells	Factor-binding center of 50*S* subunit	Interferes with the association of IF2 and EF-G with factor-binding center
Kirromycin		EF-Tu	Prevents conformational changes associated with GTP hydrolysis and therefore EF-Tu release
Ricin and α-Sarcin (protein toxins)	Prokaryotic and eukaryotic the cells	Chemically modifies RNA in factor-binding center of large ribosomal subunit	Prevents activation of translation factor GTPases
Diptheria toxin	Eukaryotic cells	Chemically modifies EF-Tu	Inhibits EF-Tu function
Cycloheximide	Eukaryotic cells	Peptidyl transferase center of 60*S* subunit	Inhibits peptidyl transferase activity

Box 14-5 (*Continued*)

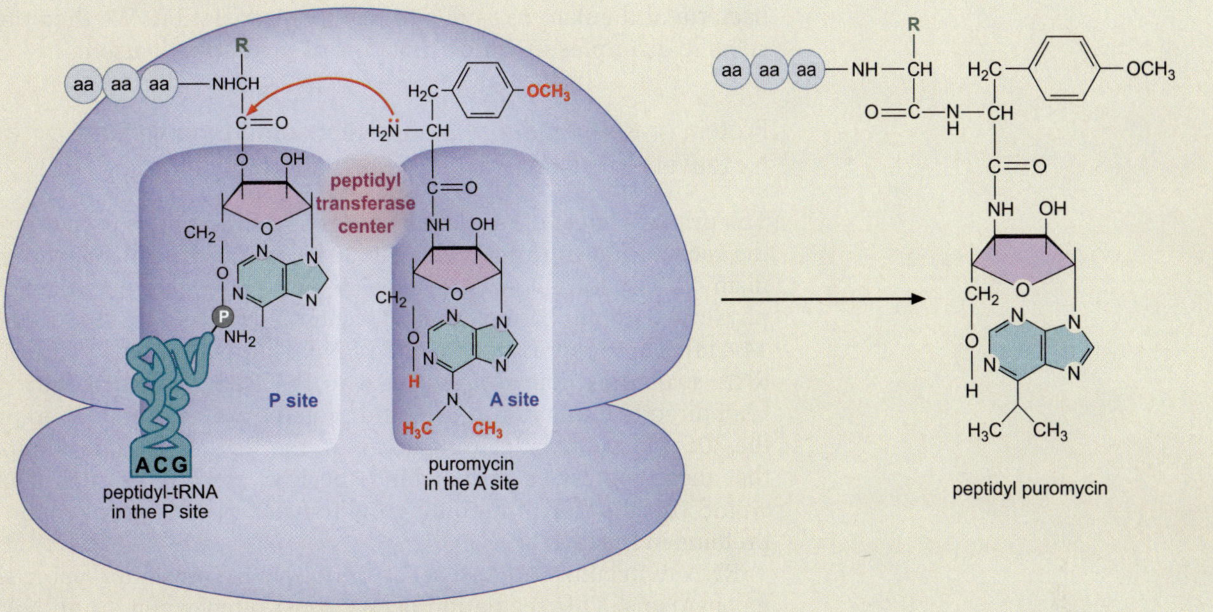

BOX 14-5 FIGURE 1 Puromycin terminates translation by mimicking a tRNA in the A site. Puromycin binds in the A site and participates in peptide bond formation. Once completed, puromycin and any associated polypeptide diffuse out of the ribosome.

important role in helping us understand the workings of the translation machinery.

Puromycin is one antibiotic commonly used in studies of translation. It binds to the large subunit region of the A site. Once bound, puromycin can substitute for an aminoacyl-tRNA in the peptidyl transferase reaction (Box 14-5 Fig. 1). Because puromycin is very small compared to a tRNA, its binding to the A site is not sufficient to retain the polypeptide chain on the ribosome. Thus, peptidyl chains that are transferred to puromycin dissociate from the ribosome as an incomplete, puromycin-bound polypeptide. In other words, puromycin causes polypeptide synthesis to terminate prematurely. Other antibiotics target other features of the ribosome, such as the peptide exit tunnel, the peptidyl transferase center, the factor-binding center, the decoding center, and regions critical for translocation.

Yet other antibiotics are inhibitors of translation factors. For example, kirromycin and fusidic acid are inhibitors of the elongation factors EF-Tu and EF-G, respectively (Box 14-5 Table 1). In both cases, the antibiotic interacts with the GTP-bound form of the translation factor and prevents changes in conformation that would normally occur after GTP hydrolysis. Thus, kirromycin arrests ribosomes with bound EF-Tu•GDP aminoacyl-tRNA. Similarly, fusidic acid arrests ribosomes with bound EF-G•GDP. In both cases, the next step in translation is prevented by the failure to release the elongation factor.

REGULATION OF TRANSLATION

Although the expression of most genes is regulated at the level of mRNA transcription, in some instances, it is more effective for the cell to regulate gene expression at the level of protein synthesis. One advantage of control of translation over transcription is the ability to respond very rapidly to external stimuli. Regulation at the level of protein synthesis eliminates the time required to alter the levels of mRNA transcription (and in eukaryotes also mRNA processing and transport to the cytoplasm), thereby allowing a more rapid change in protein levels. As with other types of regulation, translational control typically functions at the level of initiation. It is generally more efficient to regulate a pathway at an earlier step rather than starting a process and then stopping it. In the case of translation, regulation at the level of initiation also eliminates the production of incomplete proteins that might have altered function.

In this section, we first describe general mechanisms used to by bacteria and eukaryotic cells to regulate translation. We then describe specific examples in which this type of regulation is used.

Protein or RNA Binding Near the Ribosome-Binding Site Negatively Regulates Bacterial Translation Initiation

The primary target of regulators of bacterial initiation is to interfere with the recognition of the RBS by the 30*S* subunit. In general, the mechanism of these inhibitors is to associate with sequences near the RBS and physically inhibit base pairing between the RBS and the 16*S* rRNA (Fig. 14-41a). These repressors are often RNA-binding proteins that recognize RNA structures that form adjacent to the RBS. Although they do not bind directly to the RBS, the bound proteins are large enough to prevent the 30*S* subunit from gaining access to the RBS. Indeed, it is important that these repressors do not bind the RBS directly as such a protein would run the risk of inhibiting the translation of a large proportion of proteins in the cell.

RNA molecules can also act as inhibitors of translation using similar mechanisms. This regulation occurs most often when an mRNA base-pairs with itself to mask one or more RBSs (Fig. 14-41b). This masking

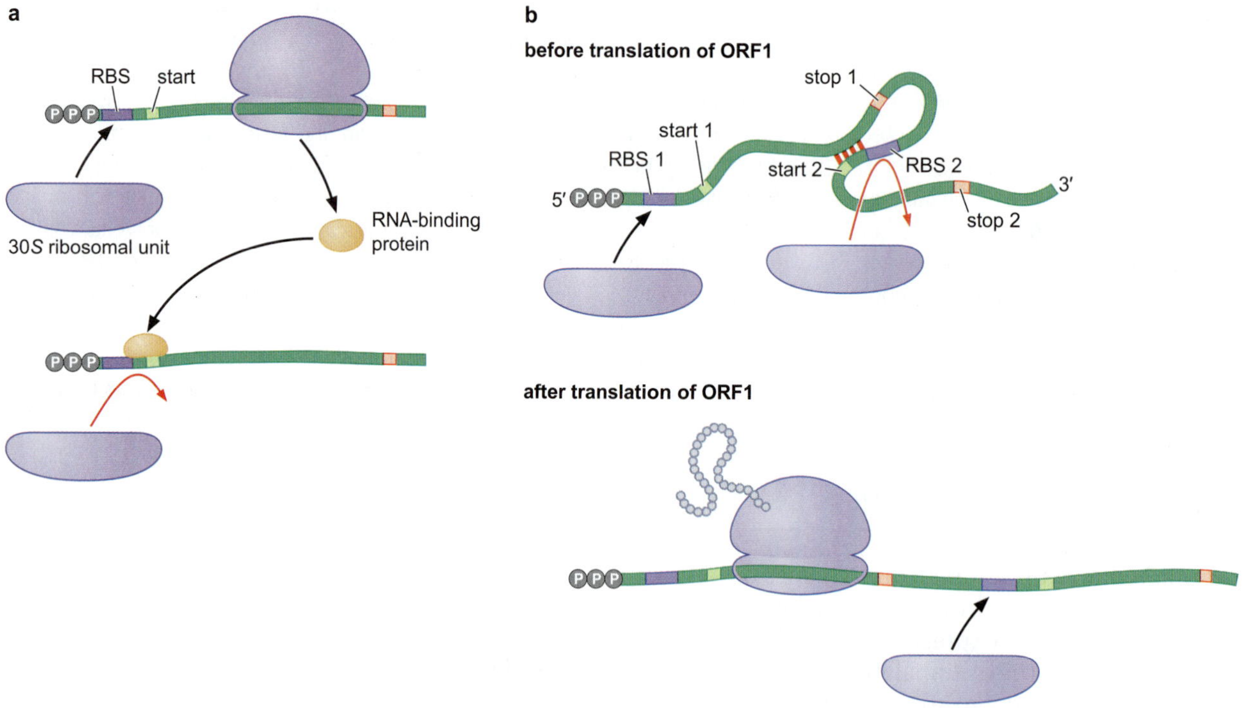

FIGURE 14-41 Regulation of bacterial translation initiation by inhibiting 30*S* subunit binding. (a) Protein binding to sites near the RBS prevents access of the 16*S* rRNA to the RBS. In this case, the protein encoded by the mRNA binds to its own RBS. (b) Intramolecular base pairing of the mRNA can interfere with base pairing by the 16*S* rRNA. In many cases, this inhibition is modulated by the translation of other genes in the same operon. If the region of the mRNA that is base pairing to the RBS proximal region is within an ORF, when that ORF is translated, the interfering base pairing is disrupted allowing a second ribosome to recognize the previously blocked RBS.

can prevent translation of the associated ORF until the interaction is disrupted. In many instances, disruption occurs as a consequence of translating another gene in the operon. In this case, the region of the mRNA that is interacting with the RBS proximal region is within another ORF and the passage of the ribosome disrupts the base pairing, thereby allowing another ribosome to recognize the unmasked RBS.

Regulation of Prokaryotic Translation: Ribosomal Proteins Are Translational Repressors of Their Own Synthesis

We now present an example of regulation of translation in bacteria that illustrates how the cell uses these mechanisms to control correct expression of ribosomal protein genes. Coordinating the expression of ribosomal proteins with rRNA expression poses an interesting regulatory problem for the cell. As we discussed earlier, each ribosome contains more than 50 distinct proteins that should be produced at the same rate as the rRNAs to which they bind. Furthermore, the rate at which a cell makes protein and the number of ribosomes it needs are closely tied to the cell's growth rate. Changes in growth conditions quickly lead to an appropriate increase or decrease in the rate of synthesis of all ribosomal components. How is this coordinated regulation accomplished?

Coordinate regulation of ribosomal protein genes is simplified by their organization into several operons, each containing genes for up to 11 ribosomal proteins (Fig, 14-42). As with other operons, these gene clusters are regulated at the level of RNA synthesis (as we discussed in Chapter 16); however, the most important control of riboso-

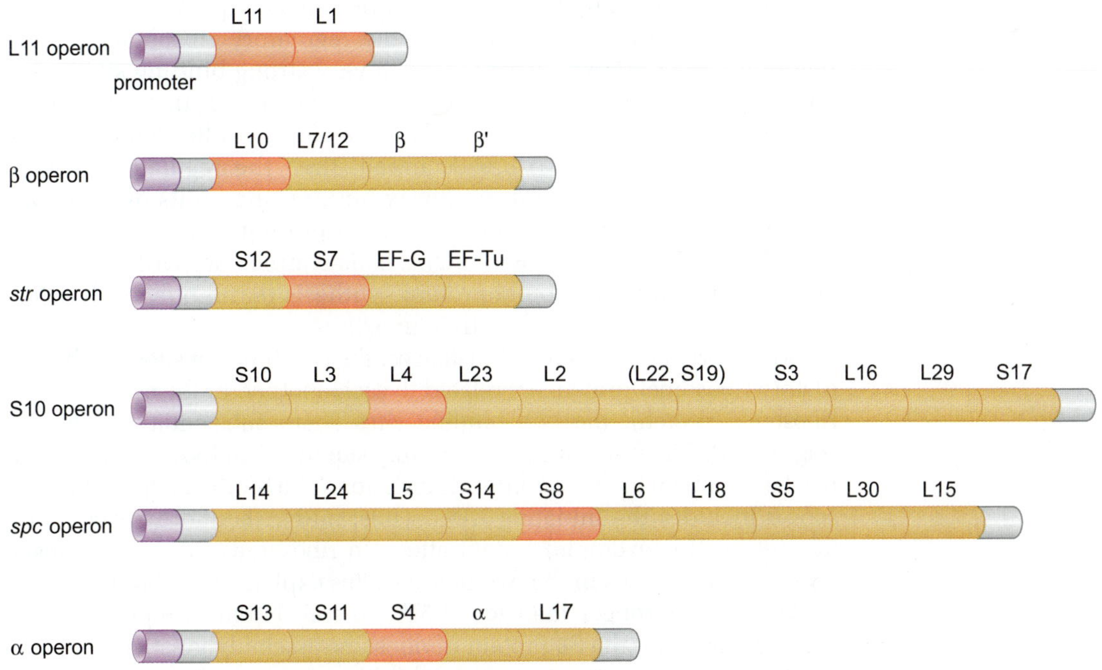

FIGURE 14-42 ***E. coli*** **ribosomal protein operons.** The protein that acts as a translational repressor of the other proteins is shaded red. The promoter is shown in purple, and each ORF is labeled according to the ribosomal protein encoded (e.g., L14 is large ribosomal protein 14). (Adapted, with permission, from Nomura M. et al. 1984. *Annu. Rev. Biochem.* 53: 75–117. © Annual Reviews.)

mal protein synthesis is at the level of *translation* of the mRNA. This can be illustrated by a simple experiment. When extra copies of a ribosomal protein operon are introduced into the cell, the amount of mRNA increases correspondingly, but synthesis of ribosomal proteins stays nearly the same. Thus, the cell compensates for extra mRNA by reducing its use as a template for protein synthesis.

The tight control of the translation of ribosomal protein mRNAs is the result of autorepression. For each ribosomal protein operon, one (or a complex of two) of the encoded ribosomal proteins binds that operon's mRNA near the translation initiation sequence of one of the most 5´-proximal genes. Binding of the ribosomal protein sterically inhibits association of the ribosomal small subunit with the nearby RBS, thereby inhibiting translation initiation.

It is easy to see how ribosomal protein binding prevents translation of the initial gene in the operon. But how does this affect the downstream genes that, in some cases, have their own RBS? Such "polar" effects can occur through multiple mechanisms. As we discussed earlier in the chapter, translational coupling may occur when the stop codon of an upstream gene is located very close to the start codon of a downstream gene. This proximity can create a situation in which translation of the upstream gene is required for translation of the downstream gene. A second mechanism exploits the folding of mRNAs into particular structures. The ribosomal protein operon mRNAs frequently are folded into structures that only allow recognition of internal RBSs if earlier genes in the mRNA are being translated. For example, suppose a region in the coding region of the first gene in the mRNA were to base-pair with a site near the RBS of the second gene. Under these circumstances, the 16*S* rRNA could only recognize this RBS after the inhibitory base pairing is disrupted by a ribosome translating through the first coding region (Fig. 14-43).

How is expression of the ribosomal proteins coupled to the amount of rRNA in the cell? In each case, the regulatory ribosomal protein that binds the mRNA also recognizes a very strong binding site on the appropriate rRNA. If this binding site is unoccupied, then the ribosomal protein will preferentially bind there. On the other hand, if all of these rRNA-binding sites are occupied, then the regulatory protein will bind to the second, lower-affinity binding site on its own mRNA. Thus, only when the ribosomal protein is present in excess to its target rRNA will it bind its own mRNA. This simple competitive binding event ensures that ribosomal protein synthesis is inhibited only when the regulatory ribosomal protein is in excess.

Not surprisingly, in several instances, the two binding sites for the regulatory ribosomal protein are related to each another. In the case of the S8 ribosomal protein, the two binding sites share substantial similarities (Fig. 14-44). The sequence of the binding site in the mRNA reveals a clear mechanism by which S8 inhibits translation. The binding site in the messenger includes the initiating AUG. Thus, mRNA bound by excess protein S8 (in this example) cannot attach to ribosomes to initiate translation. The differences in the two binding sites explains how binding to the rRNA can be stronger than to mRNA, so translation is repressed only when the need for the S8 protein in ribosome assembly is satisfied.

This strategy for translational inhibition is not restricted to ribosomal proteins. Other RNA-binding proteins regulate their expression by binding to their own mRNAs, including some aminoacyl-tRNA synthetases. In addition, there are instances in which mRNAs fold into different structures that favor or inhibit translation depending on

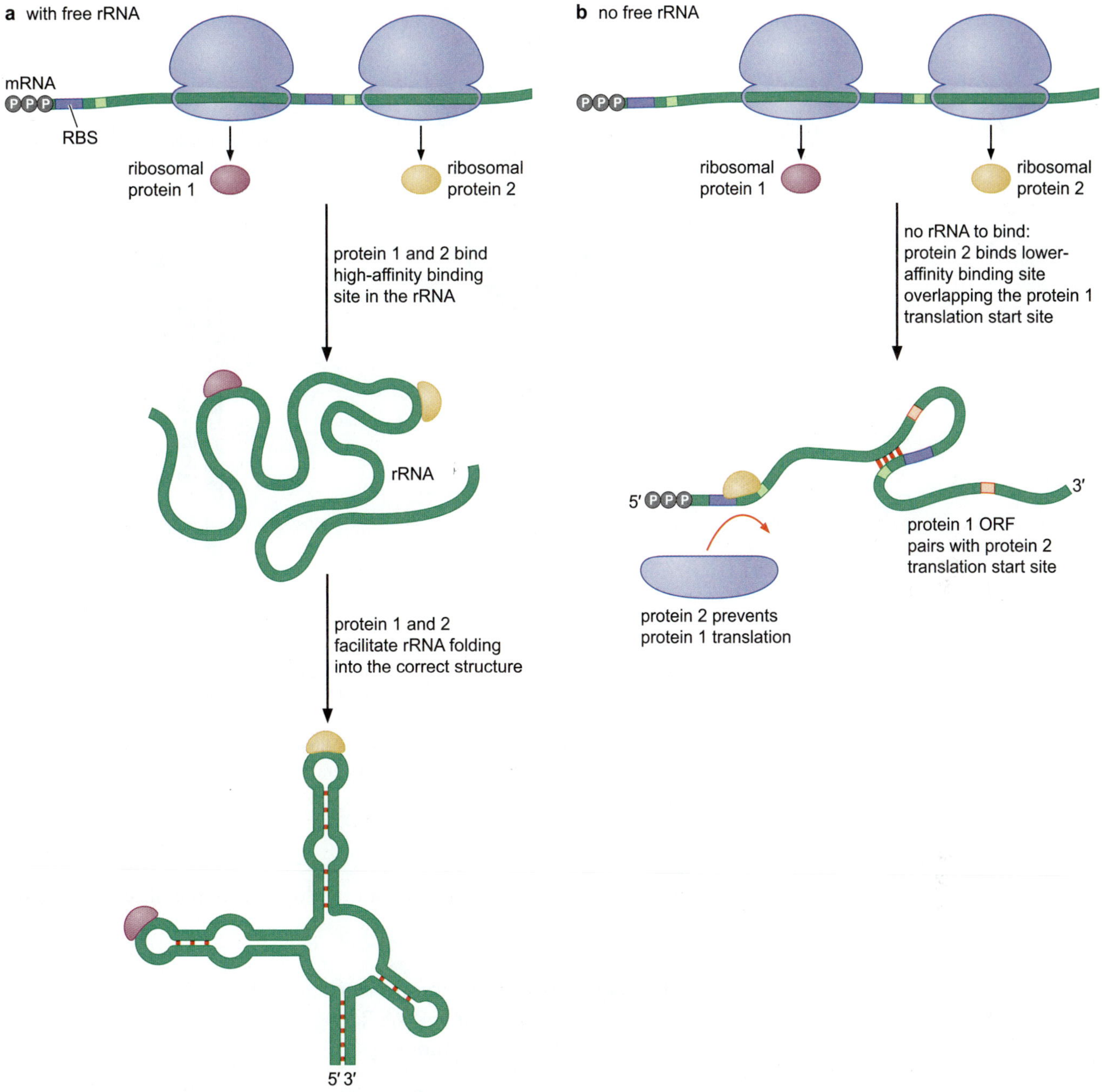

FIGURE 14-43 Regulation of ribosomal protein expression. In this example, we describe a simplified two-protein ribosomal protein gene operon. (a) In the presence of unbound rRNA, both ribosomal proteins expressed from the operon bind to the rRNA and assist in the proper assembly of the ribosome. Under this circumstance, ribosomal protein 2 does not bind to the lower-affinity binding site adjacent to the ribosomal protein 1 RBS. (b) When there is no available rRNA for the ribosomal proteins to associate with, ribosomal protein 2 binds adjacent to the ribosomal protein 1 RBS inhibiting translation of this ORF. In the absence of translation of ribosomal protein 1, sequences that are complementary to sequences adjacent to the ribosomal protein 2 RBS base-pair and prevent binding of 30S subunit to this site. As discussed in the text, translational coupling could also act to inhibit translation of downstream ORFs.

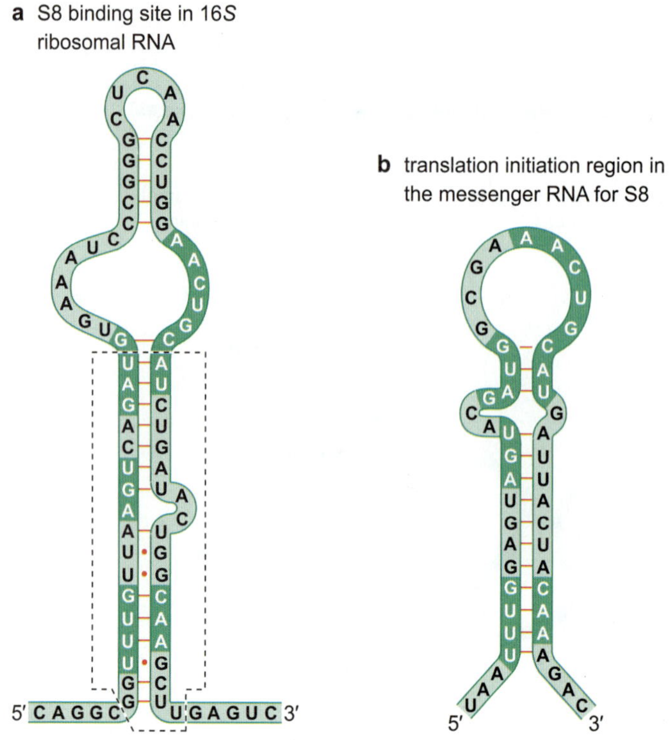

FIGURE 14-44 Ribosomal protein S8 binds 16*S* rRNA and its own mRNA. A comparison of the region where ribosomal protein S8 (encoded by the *spc* operon; Fig. 14-43) binds 16*S* rRNA in the ribosome, with the translation initiation site in its mRNA. Similar sequences are shaded in dark green. The dashed lines box off that region of the 16*S* rRNA protected by the S8 protein. (Adapted, with permission, from Cerretti D.P. et al. 1988. *J. Mol. Biol.* 204: 309–329. © Elsevier.)

the cellular conditions (e.g., temperature or metabolite levels, see Chapter 18).

Global Regulators of Eukaryotic Translation Target Key Factors Required for mRNA Recognition and Initiator tRNA Ribosome Binding

Under conditions of reduced nutrients or other cellular stresses, it is often useful for eukaryotic cells to reduce translation globally. In these instances, two early steps in eukaryotic translation initiation are targeted for inhibition: recognition of the mRNA or initiator tRNA binding to the 40*S* subunit. Recall from our earlier discussion of initiation of eukaryotic translation that these events occur independently of one another but inhibition of either eliminates new protein synthesis. In each case, the mechanism of inhibition is controlled by phosphorylation.

One common mechanism of inhibition is mediated by phosphorylation of eIF2. Recall that eIF2 bound to GTP is required to deliver the initiator tRNA to the P site of the 40*S* subunit of the eukaryotic ribosome. A number of protein kinases have been identified that phosphorylate the α subunit of eIF2. Phosphorylation of this subunit inhibits the action of a GTP-exchange factor for eIF2, called eIF2B, leading to reduced levels of eIF2–GTP. Similar to the action of EF-Ts on EF-Tu–GDP, eIF2B stimulates eIF2–GDP to release its bound GDP and bind GTP. Because eIF2 bound to GTP is required to escort the initiator tRNA to the 40*S* subunit, reduced levels of eIF2–GTP limits initiation of translation. The known eIF2α kinases are activated by a number of different cellular conditions including amino acid starvation (see below), viral infection, and elevated temperature.

A second mechanism to globally inhibit translation initiation targets the 5′ cap–binding protein: eIF4E. Recall that after binding to the 5′ cap, eIF4E binds to eIF4G. The short domain of eIF4G that is recognized by eIF4E is also found in a small family of proteins called eIF4E-binding proteins or 4E-BPs. These proteins compete with eIF4G for binding to eIF4E and therefore act as general inhibitors of translation initiation (Fig. 14-45). Like eIF2, the 4E-BPs are also regulated by phosphorylation. In their unphosphorylated state, 4E-BPs bind to eIF4E tightly and inhibit translation. In contrast, phosphorylation of 4E-BPs inhibits their binding to eIF4E (Fig. 14-45).

Phosphorylation of 4E-BPs is mediated by a key cellular protein kinase called mTor. Growth factors, hormones, and other factors that stimulate cell division activate this kinase and therefore increase the overall translational capacity of the cell. These observations have led to the hypothesis that the control of translation capacity is carefully coordinated with cell proliferation. Indeed, overexpression of eIF4E can result in cancerous transformation of cells, and inhibitors of mTor

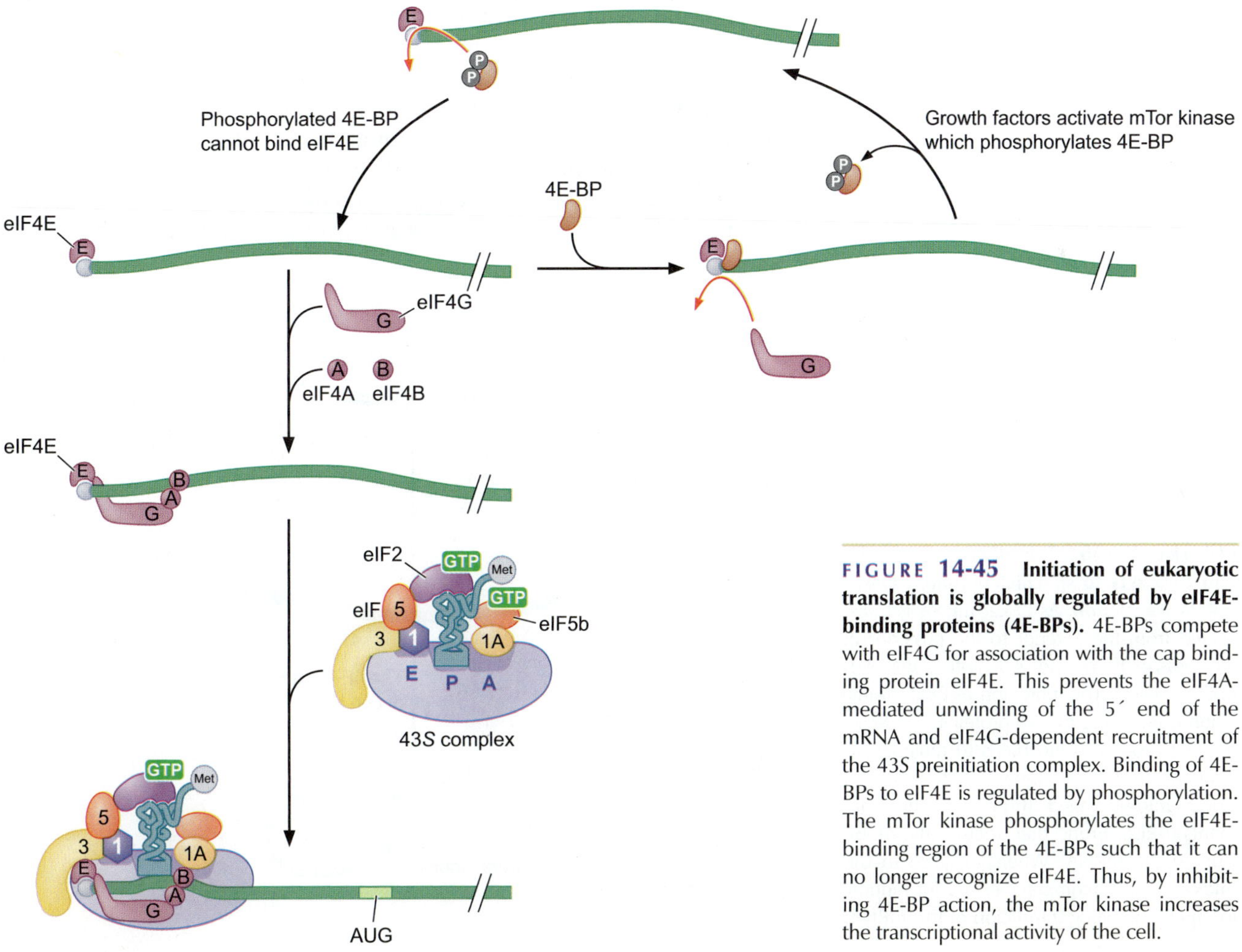

FIGURE 14-45 Initiation of eukaryotic translation is globally regulated by eIF4E-binding proteins (4E-BPs). 4E-BPs compete with eIF4G for association with the cap binding protein eIF4E. This prevents the eIF4A-mediated unwinding of the 5′ end of the mRNA and eIF4G-dependent recruitment of the 43*S* preinitiation complex. Binding of 4E-BPs to eIF4E is regulated by phosphorylation. The mTor kinase phosphorylates the eIF4E-binding region of the 4E-BPs such that it can no longer recognize eIF4E. Thus, by inhibiting 4E-BP action, the mTor kinase increases the transcriptional activity of the cell.

(e.g., rapamycin) are effective chemotherapy agents. Although we have discussed these regulatory mechanisms in the context of the global control of translation, both are also used to regulate the translation of specific mRNAs in the cell as we shall see below.

Spatial Control of Translation by mRNA-Specific 4E-BPs

In addition to globally regulating translation, binding to eIF4E is also used to regulate the translation of specific mRNAs. For example, the correct establishment of the anterior–posterior axis of the *Drosophila melanogaster* oocyte (egg) and developing embryo requires the correct localization of many proteins within a large shared cytoplasm (see Chapter 19 for a complete description of these events). In several instances, spatially restricted translation of these critical regulatory proteins plays a key role in controlling their localization.

The Oskar protein is carefully localized to the posterior regions of the oocyte prior to fertilization. Despite this, Oskar mRNA is synthesized by attached nurse cells of the ovary of the mother fly and deposited into the anterior of the oocyte prior to fertilization. Oskar mRNA is then transported to the posterior region of the oocyte. For the cell to restrict Oskar expression to the posterior region, it is critical that Oskar mRNA not be translated as it moves from the anterior to the posterior region of the oocyte.

The action of a 4E-BP called Cup is critical to specifically repressing translation of Oskar mRNA (Fig. 14-46). The Oskar mRNA contains several sequences in the 3′-untranslated region that specifically bind to a protein called Bruno. Bruno, in turn, binds to Cup, recruiting this 4E-BP to Oskar mRNA. When localized to Oskar mRNA, Cup outcompetes eIF4G for binding to eIF4E, thus inhibiting translation of the mRNA. Cup is not abundant enough to act generally on all translation as do the global 4E-BPs described above. Nevertheless, when localized to a particular mRNA, Cup becomes a very effective inhibitor of translation. This mechanism is not exclusive to Oskar. The Nanos

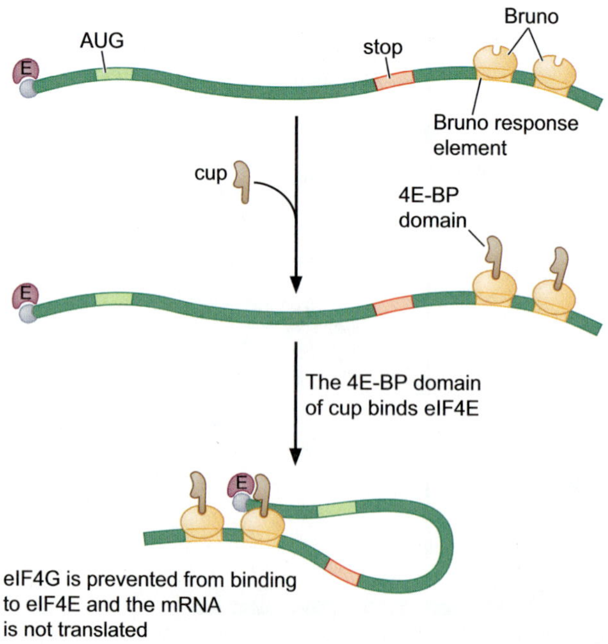

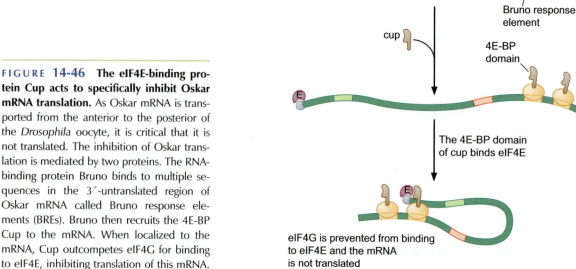

FIGURE 14-46 The eIF4E-binding protein Cup acts to specifically inhibit Oskar mRNA translation. As Oskar mRNA is transported from the anterior to the posterior of the *Drosophila* oocyte, it is critical that it is not translated. The inhibition of Oskar translation is mediated by two proteins. The RNA-binding protein Bruno binds to multiple sequences in the 3′-untranslated region of Oskar mRNA called Bruno response elements (BREs). Bruno then recruits the 4E-BP Cup to the mRNA. When localized to the mRNA, Cup outcompetes eIF4G for binding to eIF4E, inhibiting translation of this mRNA.

protein in *Drosophila* is also regulated by recruitment of Cup to its mRNA. Similarly, an mRNA-binding protein called CPEB recruits a 4E-BP called Maskin to a number of mRNAs whose translation is inhibited during vertebrate oocyte development.

An Iron-Regulated, RNA-Binding Protein Controls Translation of Ferritin

Regulating iron levels in the human body is critical. Many proteins use iron as a cofactor, including the oxygen transport proteins hemoglobin and myoglobin, as well as many of the proteins involved in oxidative phosphorylation. Consistent with the important role of iron in oxygen transport and energy production, a shortage of iron in the human body (called anemia) results in an overall feeling of weakness. On the other hand, excess iron is toxic to cells and can contribute to liver damage, heart failure, and diabetes. Thus, it is critical to properly regulate iron levels in the human body.

The iron-binding protein ferritin is the major regulator of iron levels in the human body. Ferritin stores and releases iron in a controlled manner, thereby maintaining proper iron homeostasis. Thus, it is critical that the levels of ferritin respond rapidly to the levels of free iron in the body. The need to respond rapidly to changes in the levels of free iron has resulted in the regulation of ferritin expression at the level of protein synthesis.

Ferritin translation is regulated by iron-binding protein called iron regulatory proteins (IRPs). These proteins are also RNA-binding proteins that recognize a specific hairpin structure formed at the 5′ end of the ferritin mRNA called the iron regulatory element (IRE; Fig. 14-47). Impor-

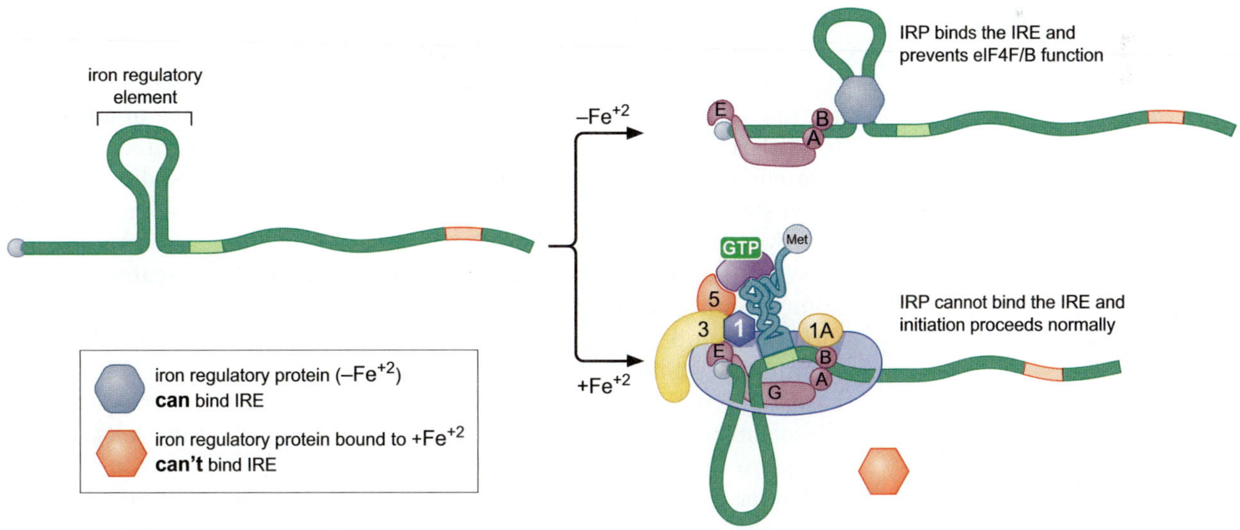

FIGURE 14-47 Regulation of ferritin translation by iron. The 5′-untranslated region of the ferritin genes includes a stem-loop structure called the iron regulatory element (IRE). The iron regulatory protein (IRP) binds tightly to this site when it is not bound to Fe^{+2}. By stabilizing the stem-loop structure of the IRE, IRP prevents eIF4A from removing this structure from the end of the ferritin mRNA. Under these conditions, association of the 43*S* preinitiation complex with the mRNA cannot occur and the ferritin genes are not translated. When iron levels are elevated and ferritin protein is needed, IRP binds to Fe^{+2}, which inhibits its ability to bind to the IRE and, therefore, allows translation of the ferritin protein.

tantly, the ability of these proteins to recognize the IRE is controlled by the levels of iron in the cell. In iron-deficient cells, the concentration of iron is too low to bind the IRPs. In the absence of bound iron, these proteins bind tightly to the IRE and inhibit the ability of eIF4A/B to unwind the IRE hairpin structure. The continued presence of the hairpin acts as a steric block to 43*S* complex binding to the mRNA. In contrast, when the concentration of free iron in the cell is elevated, the IRPs become iron-bound. When bound to iron, the IRPs lose their ability to bind to the IRE and therefore to inhibit translation. In summary, the levels of ferritin mRNA translation are tightly controlled by the affinity of the IRPs for free iron and the resulting effect on IRP binding to the IRE.

Translation of the Yeast Transcriptional Activator Gcn4 Is Controlled by Short Upstream ORFs and Ternary Complex Abundance

Gcn4 is a yeast transcriptional activator that regulates the expression of genes encoding enzymes that direct amino acid biosynthesis. Although it is a transcriptional activator, Gcn4 is itself regulated at the level of translation. In the presence of low levels of amino acids, Gcn4 mRNA is translated (and so the biosynthetic enzymes are expressed). But in the presence of high levels of amino acids, Gcn4 mRNA is not translated. How is this regulation achieved?

Unlike the structure of the typical eukaryotic messenger, the mRNA encoding the Gcn4 protein contains four small open reading frames (called uORFs) upstream of the coding sequence for Gcn4. The most upstream of these short ORFs (uORF1) is efficiently recognized by ribosomes that scan along the message from the 5′ end. Once they have translated uORF1, a unique property of this ORF allows 50% of the small subunits of the ribosome to remain bound to the RNA and resume scanning for downstream initiation (AUG) codons (Fig. 14-48).

Which downstream AUG is recognized by the scanning small subunit is controlled by when the scanning subunit binds to eIF2 complexed with an initiator tRNA. Recall that in the absence of an initiator tRNA in the P site, the 40*S* subunit cannot recognize an AUG sequence in the mRNA. Thus, before initiating translation at any downstream ORF, the scanning 40*S* ribosome subunit must bind eIF2•Met-$tRNA_i^{Met}$ (hereafter referred to as TC for ternary complex).

When amino acids are not limiting, TC rebinds the scanning ribosomes soon after they complete translation of uORF1 (Fig. 14-48a). Once rebound by TC, the small subunits can recognize an AUG and reinitiate translation at one of the other uORFs (2, 3, or 4). Unlike the case with uORF1, after translating these uORFs, the ribosome fully dissociates from the mRNA and fails to translate the Gcn4 ORFs. Thus, no Gcn4 protein is made.

Under conditions of amino acid starvation, a combination of events reduce the rate at which TC binds to the 40*S* subunit. Limited amino acids leads to an abundance of uncharged tRNAs, which in turn activates an eIF2α kinase called Gcn2. As described for the global control of translation by eIF2α kinases, when Gcn2 phosphorylates eIF2, its ability to bind GTP is reduced. Because eIF2 can only bind Met-$tRNA_i^{Met}$ in the presence of GTP, these conditions lead to less TC and therefore a reduced rate of TC binding to the 40*S* subunit (Fig. 14-48b). The reduced rate of binding means that 40*S* subunit scanning

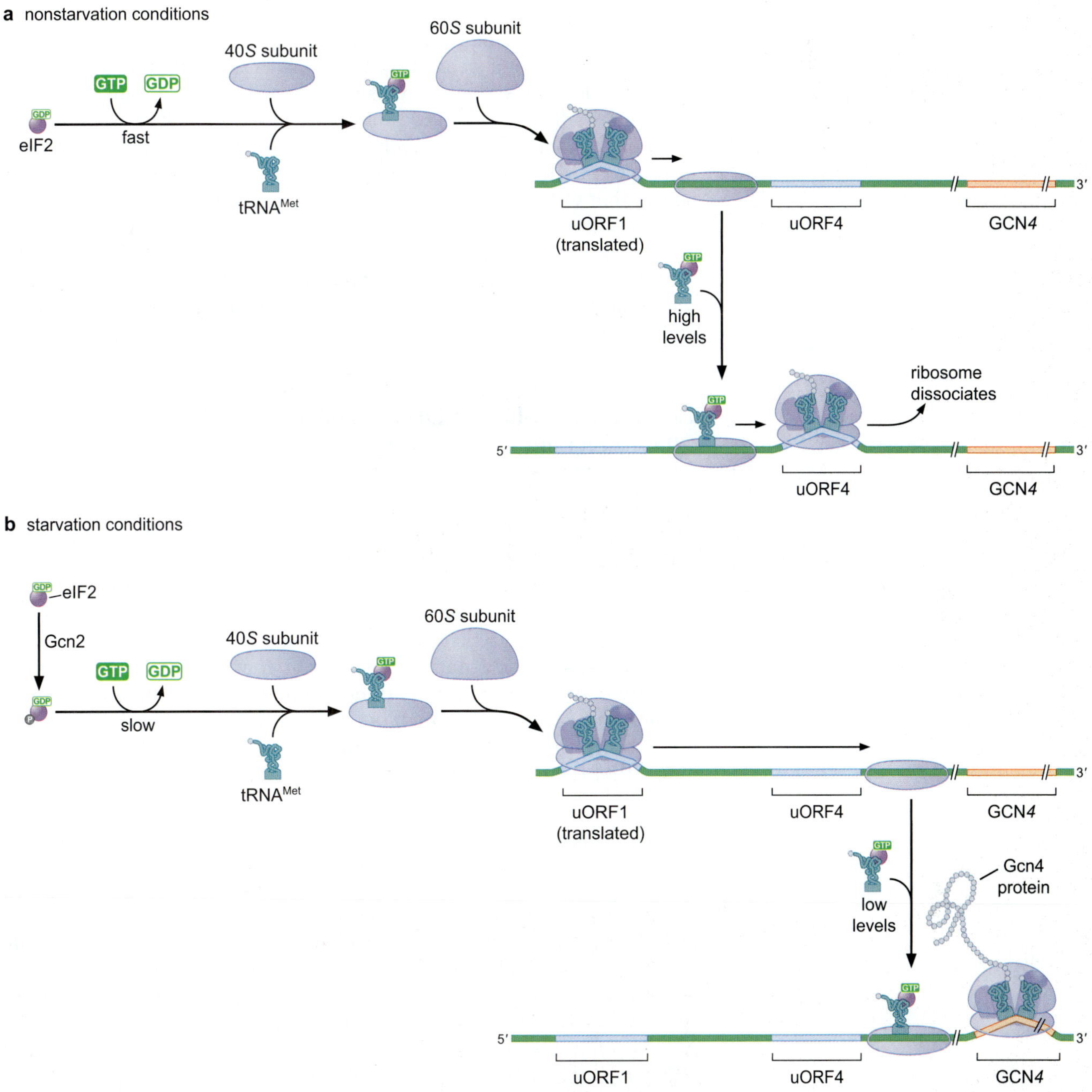

FIGURE 14-48 Translational control of Gcn4 in response to amino acid starvation. As described in detail in the text, the ORF encoding the yeast activator Gcn4 is preceded by four short ORFs called uORFs (here, only uORF1 and uORF4 are shown). The first of these upstream ORFs is translated initially and, because of special properties of this ORF, approximately half of the 40*S* subunits are retained after translation termination to continue scanning the Gcn4 mRNA. (a) When amino acids are abundant, eIF2B stimulates eIF2 to exchange GDP for GTP rapidly. This allows for rapid binding of eIF2-GTP-Met-tRNA$^{Met}_{i}$ to the 40*S* subunit and the ability to recognize one of the three other short ORFs. Translation of any one of these uORFs results in full termination of translation. (b) Under starvation conditions, phosphorylation of eIF2 by the eIF2α kinase Gcn2 reduces the ability of eIF2B to stimulate GTP binding to eIF2. Reduced levels of eIF2-GTP result in slower binding of eIF2-GTP-Met-tRNA$^{Met}_{i}$ to the 40*S* subunit. This reduced rate of initiator tRNA binding increases the chance that the scanning ribosome will pass the remaining uORFs prior to being able to recognize an AUG and therefore favors the translation of Gcn4. (Modified, with permission, from Hinnebusch A.G. 1997. *J. Biol. Cell* 272: 21661–21664, Fig. 1. © American Society for Biochemistry & Molecular Biology.)

continues farther along the mRNA without the ability to detect an AUG. If the ribosome scans through the AUGs for uORF2–4 before rebinding eIF2-tRNA^{Met}, then these ORFs will not be translated. The start codons for uORF2–4 are relatively close to uORF1, whereas the AUG of the Gcn4 ORF is much further downstream. This additional distance provides much more time for TC to bind to the small subunit before that ORF is encountered, thereby increasing the odds that the ribosome translates it. Indeed, removing the spacer RNA between the uORFs and the Gcn4 start codon results in progressively less Gcn4 protein expression. Thus, in the presence of limiting TC, Gcn4 is produced and can switch on the genes needed to synthesize additional amino acids in the cell.

TRANSLATION-DEPENDENT REGULATION OF mRNA AND PROTEIN STABILITY

At some frequency, mRNAs will be made that are mutant or damaged. Such defective mRNAs can arise from mistakes in transcription or from damage that occurs after they are synthesized. For example, because they are single-stranded, mRNAs are more susceptible to breakage. Such damaged mRNAs have the possibility of making incomplete or incorrect proteins that could have negative effects on the cell. In some cases, such as point mutations that change only a single amino acid, there is little that can be done to eliminate the mutant mRNA or its protein product. However, in other cases described below, the process of translation is used to detect defective mRNAs and eliminate either them or their protein products.

The SsrA RNA Rescues Ribosomes That Translate Broken mRNAs

Normally, a stop codon is required to release the ribosome from an mRNA. But what happens to a ribosome that initiates translation of an mRNA fragment that lacks a termination codon in the appropriate reading frame? Such an mRNA can be generated by incomplete transcription or nuclease action. Translation of this type of mRNA can initiate normally and continue until the 3′ end of the mRNA is reached. At this point, the ribosome cannot proceed. There is no codon to bind either an aminoacyl-tRNA or a release factor. Without some mechanism to release them from these defective mRNAs, many ribosomes would be permanently trapped, removing them from polypeptide synthesis. In prokaryotic cells, such stalled ribosomes are rescued by the action of a chimeric RNA molecule that is part tRNA and part mRNA, called a **tmRNA**.

SsrA is a 457-nucleotide tmRNA that includes a region at its 3′ end that strongly resembles tRNA^{Ala} (Fig. 14-49). This similarity allows the SsrA RNA to be charged with alanine and to bind EF-Tu–GTP. When a ribosome is stalled at the 3′ end of an mRNA, the SsrA^{Ala}–EF-Tu–GTP complex binds to the A site of the ribosome and participates in the peptidyl transferase reaction, as would any other tRNA. Translocation of the peptidyl-SsrA RNA results in the release of the broken mRNA. Remarkably, translocation of the SsrA RNA also results in a portion of this RNA entering the mRNA-binding channel of the ribosome. This portion of the SsrA RNA extends the ORF of the incomplete mRNA by ten codons followed by a stop codon.

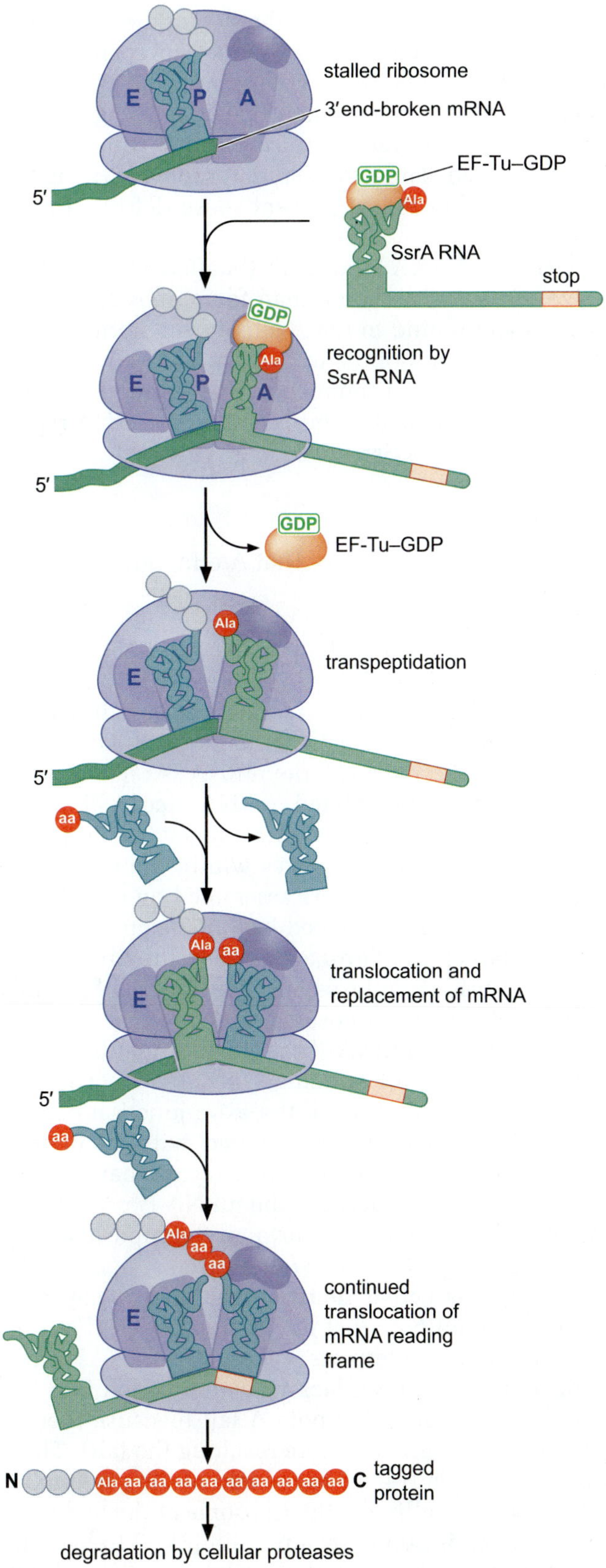

FIGURE 14-49 The tmRNA SsrA rescues ribosomes stalled on prematurely terminated mRNAs. The SsrA RNA mimics a tRNA but can only bind a ribosome that is stalled at the 3′ end of an mRNA. Once bound, the SsrA RNA substitutes part of its sequence to act as a new "mRNA."

The net result of SsrA binding is that when the defective mRNA is released from the ribosome, the incomplete polypeptide is fused to a ten-amino-acid "tag" at its carboxyl terminus and the ribosome is recycled. Interestingly, the ten-amino-acid tag is recognized by cellular proteases that rapidly degrade the tag and the truncated polypeptide to which it is attached. Thus, translation products arising from broken mRNAs are rapidly cleared to prevent these defective proteins from harming the cell.

How does the SsrA RNA bind to only stalled ribosomes? Because of the large size of SsrA (it is more than four times longer than a standard tRNA), it cannot bind to the A site during normal elongation. In contrast, when the 3′ end of the mRNA is missing, additional room is created in the A site to accommodate the larger RNA (Fig. 14-49). Thus, only ribosomes stalled at the 3′ end of an mRNA represent a potential binding site for the SsrA RNA.

Eukaryotic Cells Degrade mRNAs That Are Incomplete or Have Premature Stop Codons

Translation is tightly linked to the process of mRNA decay in eukaryotic cells (Fig. 14-50). This linkage is illustrated by two mechanisms that monitor the integrity of mRNAs that are being translated. For example, when an mRNA contains a premature stop codon (known as a nonsense codon; see Chapter 15), the mRNA is rapidly degraded by a process called **nonsense-mediated mRNA decay** (Fig. 14-50b). In mammals, recognition of mRNAs with premature stop codons relies on the assembly of protein complexes within the ORF of the mRNA. These exon-junction complexes are assembled on the mRNA as a consequence of splicing and are located just upstream of each exon–exon boundary (see Chapter 13). Ordinarily, when the first ribosome translates an mRNA, these complexes are displaced as the mRNA enters the decoding center of the ribosome. However, if a premature stop codon is present in the mRNA (because of mutation of the gene or mistakes in transcription or splicing), then the ribosome is released prior to the displacement of all of the exon-junction complexes. Under these conditions, the complexes interact with the prematurely terminating ribosome, which activates an enzyme that removes the cap at the 5′ end of the mRNA. Because the mRNA is ordinarily protected from degradation by the 5′ cap, removal of the cap causes rapid degradation of the mRNA by a 5′ → 3′ exonuclease.

A different process called **nonstop-mediated decay** rescues ribosomes that translate mRNAs that lack a stop codon (Fig. 14-51c). Unlike their prokaryotic counterparts, eukaryotic mRNAs terminate with a poly-A tail. When an mRNA lacking a stop codon is translated, the ribosome translates through the poly-A tail (because there is no stop codon to cause it to terminate before reaching the tail). This results in the addition of multiple lysines to the end of the protein (AAA is the codon for lysine) and stalling of the ribosome at the end of the mRNA. The stalled ribosome is bound by a protein (Ski7) (related to the class II release factor eRF3) that stimulates ribosome dissociation and recruits a 3′ → 5′ exonuclease that degrades the "nonstop" mRNA. In addition, proteins that contain polylysine at their carboxyl terminus are unstable, leading to the rapid degradation of proteins derived from nonstop mRNAs. Thus, like the situation in prokaryotes, proteins syn-

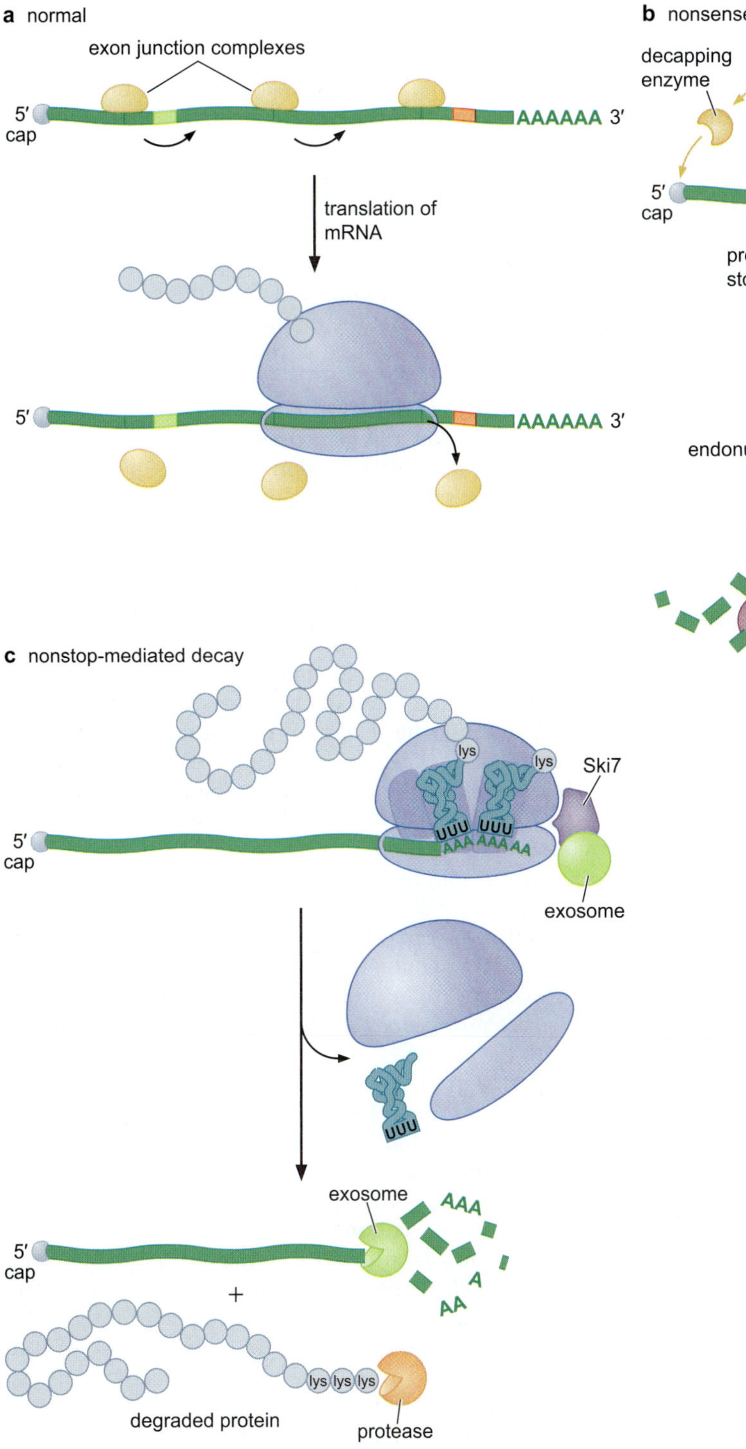

FIGURE 14-50 Eukaryotic mRNAs with premature or no stop codons are targeted for degradation. (a) Translation of a normal mRNA displaces all of the exon junction complexes. (b) Nonsense-mediated decay. Translation of an mRNA with a premature stop codon does not displace one or more of the exon junction complexes. This results in the recruitment of the Upf1, Upf2, and Upf3 proteins to the ribosome. Once bound to the ribosome, these proteins activate a decapping enzyme that removes the 5′ cap of the mRNA. The uncapped mRNA is then rapidly degraded by 5′ to 3′ exonucleases that are normally unable to degrade the mRNA because of the presence of the 5′ cap. (c) Nonstop-mediated decay. In the absence of a stop codon, the poly-A tail of the mRNA is translated. A complex that includes the Ski7 protein and a 3′ to 5′ exonuclease called the exosome binds any ribosome stalled at the 3′ end of the poly-A tail. This results in the release of the ribosome from the mRNA and its degradation. Similar to SsrA-mediated nonstop decay, the polylysine found at the end of proteins derived from such mRNAs targets the protein for degradation.

thesized from mRNAs lacking stop codons are rapidly removed from the cell.

A fascinating feature of nonsense-mediated mRNA decay and nonstop-mediated decay is that both processes of mRNA degradation require translation of the damaged mRNA. In the absence of translation, the damaged mRNAs are not rapidly degraded and have normal stability. Thus, although indirect, eukaryotic cells rely on translation as a mechanism to proofread their mRNAs.

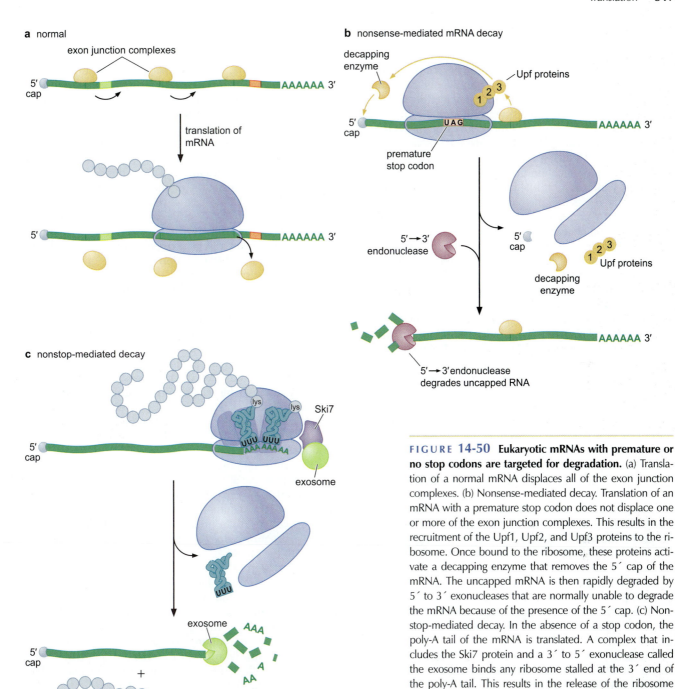

SUMMARY

Proteins are synthesized on RNA templates known as messenger RNAs (mRNAs) in a process known as translation. Translation involves the decoding of nucleotide sequence information into the linear sequence of amino acids of the polypeptide chain. The machinery for protein synthesis consists of four principal components: the mRNA; adaptor RNAs known as transfer RNAs (tRNAs); aminoacyl-tRNA synthetases that attach amino acids to the tRNAs; and the ribosome, which is a multisubunit complex of protein and RNA that catalyzes peptide bond formation.

The mRNA contains the coding sequence for protein and recognition elements for the initiation and termination of translation. The coding sequence is known as an open reading frame (ORF), and consists of a series of three-nucleotide-long units known as codons that are in register with each other. An ORF specifies a single polypeptide chain. Each ORF begins with a start codon and ends with a stop codon. The start codon is usually AUG or GUG in prokaryotes and always AUG in eukaryotes. In prokaryotes, the start codon is preceded by a region of sequence complementarity to the 16*S* rRNA component of the ribosome, which is responsible for aligning the ribosome over the start codon. In eukaryotes, the mRNA contains a special structure at its 5′ terminus known as the 5′ cap, which is responsible for recruiting the ribosome. Eukaryotic mRNAs terminate in a string of A residues known as the poly-A tail, which enhances the efficiency of translation. Prokaryotic mRNAs often contain two or more ORFs; they are referred to as being polycistronic. Eukaryotic mRNAs usually contain only a single ORF and are called monocistronic.

tRNAs are a physical interface between codons in the mRNA and the amino acids that are added to the growing polypeptide chain. tRNAs are L-shaped molecules with a loop at one end that displays the anticodon and a 3′-protruding 5′-CCA-3′ sequence at the other end. The anticodon is complementary to the codon, which it recognizes by base pairing. Amino acids are attached to the terminal residue of the 5′-CCA-3′ via an acyl linkage between the carbonyl group of the amino acid and the 3′-hydroxyl of the terminal ribose.

Aminoacyl-tRNA synthetases attach amino acids to tRNAs in a two-step process known as charging. A single aminoacyl tRNA synthetase is responsible for charging all tRNAs for a specific amino acid. Synthetases recognize the correct tRNAs by interactions with both ends of these L-shaped molecules. Synthetases are responsible for charging their cognate tRNAs with the correct amino acid and do so with high fidelity. Some aminoacyl-tRNA synthetases achieve increased accuracy by means of a proofreading mechanism.

The ribosome consists of a large subunit, which contains the site of peptide bond formation (the peptidyl transferase center), and a small subunit, which contains the site of mRNA decoding (decoding center). Each subunit is composed of one or more RNAs and multiple proteins. The RNAs not only are a principal structural feature of the subunits, but also are responsible for the principal functions of the ribosome. The intact ribosome contains three tRNA-binding sites that reach between the two subunits: an A site where the charged tRNA enters the ribosome, a P site that contains the peptidyl-tRNA, and an E site, where deacylated tRNAs exit the ribosome.

Translation of one protein involves a cycle of association and dissociation of the small and large subunits. In this ribosome cycle, the small and large subunits assemble at the beginning of an ORF and then dissociate into free subunits when translation of the ORF is complete. The mRNA is translated starting at the 5′ end of the ORF and the polypeptide chain is synthesized in an amino-terminal to carboxy-terminal direction.

Translation takes place in three principal steps: initiation, elongation, and termination. Initiation in prokaryotes involves the recruitment of the small ribosomal subunit to the mRNA through the interaction of the ribosome-binding site (RBS) with the 16*S* rRNA. This interaction is facilitated by three auxiliary proteins (called initiation factors IF1, IF2, and IF3) that help to keep the two ribosomal subunits apart and recruit a special initiator tRNA to the start codon. Pairing between the anticodon of the charged initiator tRNA and the start codon triggers the recruitment of the large subunit, the release of the initiation factors, and the placement of the charged initiator tRNA in the P site. This is the prokaryotic initiation complex, and it is poised to accept a charged tRNA into the A site and carry out the formation of the first peptide bond.

Eukaryotic mRNAs recruit the small subunit through recognition of the 5′ cap and the action of numerous auxiliary initiation factors. The small subunit then scans downstream until it encounters an AUG, which it recognizes as the start codon. As in prokaryotes, only when the starting AUG is recognized does the large ribosomal subunit associate with the mRNA.

The first step of the elongation phase of translation is the introduction of a charged tRNA into the A site. This is catalyzed by the GTP-binding protein EF-Tu in prokaryotes and its equivalent in eukaryotes. Multiple mechanisms ensure that proper base pairing has taken place between the codon and the anticodon before the aminoacyl group is allowed to enter the peptidyl transferase center. Next, peptide bond formation takes place through the transfer of the peptidyl chain from the tRNA in the P site to the aminoacyl-tRNA in the A site. Peptide bond formation is catalyzed by RNA in the peptidyl transferase center of the large subunit, as well as the 3′-OH of the P-site tRNA. This ribozyme stimulates the nucleophilic attack of the amino group of the aminoacyl-tRNA in the A site on the carbonyl group that attaches the growing polypeptide chain to the tRNA in the P site. Finally, the ribosome translocates to the next vacant codon in a process that is driven by both the peptidyl transferase reaction and the action of the elongation factor EF-G (or its eukaryotic equivalent). As a result of translocation, the deacylated tRNA in the P site is shifted into the E site where it exits the ribosome, and the peptidyl-tRNA in the A site is shifted into the now vacant P site. The adjacent codon in the mRNA is shifted into the now vacant A site, which is poised to accept the delivery of a charged tRNA by EF-Tu.

Translation terminates when the ribosome encounters a stop codon, which is recognized by one of two class I release factors in prokaryotes and a single class I release factor in eukaryotes. The release factor triggers the hydrolysis of the polypeptide from the peptidyl-tRNA and hence the release of the completed polypeptide. Finally, a class II release factor, a ribosome recycling factor, and an initiation factor (IF3 in prokaryotes) complete termination by causing the release of the mRNA and the deacylated tRNAs and the dissociation of the ribosome into its large and small subunits. The ribosome cycle is now complete and the small subunit is ready to commence a new cycle of polypeptide synthesis.

The expression of many genes is controlled at the level of transcription; however, some genes are also regulated at the level of translation initiation. In bacterial cells, this regulation generally occurs by inhibiting binding of the small subunit to the RBS. This inhibition can be mediated by either protein or RNA binding to mRNA sequences near the RBS. Global levels of eukaryotic translation are regulated by 4E-BP proteins, which bind to eIF4E and compete for its ability to bind eIF4G, and eIF2α kinases, which inhibit the ability of eIF2 to bind GTP. Regulation of the translation of specific eukaryotic mRNAs is sometimes mediated by small uORFs that limit access of the small subunit to a downstream ORF. In the case of Gcn4 translation, the cell exploits the dependence of eIF2•Met-tRNA$_i^{Met}$ on amino acid levels to regulate translation of Gcn4. Only when eIF2•Met-tRNA$_i^{Met}$ is limiting can the scanning small subunit reach the Gcn4 start codon without dissociating first.

Translation is also used by both bacteria and eukaryotic cells to monitor the integrity of mRNAs and eliminate such RNAs and their protein products. mRNAs lacking stop codons result in the synthesis of proteins that are recognized by cellular proteases and degraded. In eukaryotic cells, these mRNAs are also degraded. Similarly, eukaryotic mRNAs with a premature stop codon are detected by the failure of the ribosome to displace proteins bound to the mRNA. If these proteins are not displaced, they recruit enzymes that direct the degradation of the mRNA.

BIBLIOGRAPHY

Books

Mathews M.B., Sonenberg N., and Hershey J.W.B., eds. 2007. *Translational control in biology and medicine.* Cold Spring Harbor Laboratory Press, Cold Spring Harbor, New York.

tRNA and Aminoacyl-tRNA Synthetases

Arnez J.G. and Moras D. 1997. Structural and functional considerations of the aminoacylation reaction. *Trends Biochem. Sci.* **2:** 189–232.

Ibba M. and Soll D. 2000. Aminoacyl-tRNA synthesis. *Annu. Rev. Biochem.* **69:** 617–650.

The Ribosome

Koronstelev A., Trakhanov S., Laurberg M., and Noller H.F. 2006. Crystal structure of a 70*S* ribosome-tRNA complex reveals functional interactions and rearrangements. *Cell* **126:** 1065–1077.

Moore P.B. and Steitz T.A. 2005. The ribosome revealed. *Trends Biochem. Sci.* **30:** 281–283.

Poehlsgaard J. and Douthwaite S. 2005. The bacterial ribosome as a target for antibiotics. *Nat. Rev. Microbiol.* **3:** 870–881.

Ramakrishnan V. 2002. Ribosome structure and the mechanism of translation. *Cell* **108:** 557–572.

Rodnina M.V., Beringer M., and Wintermeyer W. 2006. How ribosomes make peptide bonds. *Trends Biochem. Sci.* **32:** 20–26.

Selmer M., Dunham C.M., Murphy F.V., Weixlbaumer A., Petry S., Kelley A.C., Weir J.R., and Ramakrishnan V. 2006. Structure of the 70*S* ribosome complexed with mRNA and tRNA. *Science* **313:** 1935–1942.

Translation

Broderson D.E. and Ramakrishnan V. 2003. Shapes can be seductive. *Nat. Struct. Biol.* **10:** 78–80.

Laursen B.S., Sorenson H.P., Mortenson K.K., and Sperling-Peterson H.U. 2005. Initiation of protein synthesis in bacteria. *Microbiol. Mol. Biol. Rev.* **60:** 101–123.

Nilsson J. and Nissen P. 2005. Elongation factors on the ribosome. *Curr. Opin. Struct. Biol.* **15:** 349–354.

Nissen P., Kjeldgaard M., and Nyborg J. 2000. Macromolecular mimicry. *EMBO J.* 19: 489–495.

Weinger J.S., Parnell K.M., Forner S., Green R., and Strobel S.A. 2004. Substrate-assisted catalysis of peptide bond formation by the ribosome. *Nat. Struct. Mol. Biol.* **11:** 1101–1106.

Regulation of Translation

Gebauer F and Hentze M.W. 2004. Molecular mechanisms of translational control. *Nat. Rev. Mol. Cell Biol.* **5:** 827–835.

Richter J.D. and Sonenberg N. 2006. Regulation of cap-dependent translation by eIF4E inhibitory proteins. *Nature* **433:** 477–480.

CHAPTER

15 The Genetic Code

OUTLINE

AT THE VERY HEART OF THE CENTRAL DOGMA is the concept of information transfer from the linear sequence of the four-letter alphabet of the polynucleotide chain into the 20-amino-acid language of the polypeptide chain. As we have seen, the translation of genetic information into amino acid sequences takes place on ribosomes and is mediated by special adaptor molecules known as transfer RNAs (tRNAs). These tRNAs recognize groups of three consecutive nucleotides known as codons. With four possible nucleotides at each position, the total number of permutations of these triplets is 64 (4 x 4 x 4), a value well in excess of the number of amino acids. Which of these triplet codons are responsible for specifying which amino acids, and what are the rules that govern their use? In this chapter, we discuss the nature and underlying logic of the genetic code, how the code was "cracked," and the effect of mutations on the coding capacity of messenger RNA.

THE CODE IS DEGENERATE

Table 15-1 lists all 64 permutations, with the left-hand column indicating the base at the 5′ end of the triplet, the row across the top specifying the middle base, and the right-hand column identifying the base in the 3′ position. One of the most striking features of the code is that 61 of the 64 possible triplets specify an amino acid, with the remaining three triplets being chain-terminating signals (see below). This means that many amino acids are specified by more than one codon, a phenomenon called **degeneracy.** Codons specifying the same amino acid are **synonyms.** For example, UUU and UUC are synonyms for phenylalanine, whereas serine is encoded by the synonyms UCU, UCC, UCA, UCG, AGU, and AGC. In fact, when the first two nucleotides are identical, the third nucleotide can be either cytosine or uracil and the codon will still code for the same amino acid. Often, adenine and guanine are similarly interchangeable. However, not all degeneracy is based on equivalence of the first two nucleotides. Leucine, for example, is coded by UUA and UUG, as well as by CUU, CUC, CUA, and CUG (Fig. 15-1). Codon degeneracy, especially the frequent third-place equivalence of cytosine and uracil or guanine and adenine, explains how there can be great variation in the AT/GC ratios in the DNA of various organisms without correspondingly large changes in the relative proportion of amino acids in their proteins. (For example, the genomes of certain bacteria display

TABLE 15-1 The Genetic Code

first position (5′ end)	second position: U	C	A	G	third position (3′ end)
U	UUU Phe	UCU Ser	UAU Tyr	UGU Cys	U
U	UUC	UCC	UAC	UGC	C
U	UUA Leu	UCA	UAA* stop	UGA* stop	A
U	UUG	UCG	UAG* stop	UGG Trp	G
C	CUU Leu	CCU Pro	CAU His	CGU Arg	U
C	CUC	CCC	CAC	CGC	C
C	CUA	CCA	CAA Gln	CGA	A
C	CUG	CCG	CAG	CGG	G
A	AUU Ile	ACU Thr	AAU Asn	AGU Ser	U
A	AUC	ACC	AAC	AGC	C
A	AUA	ACA	AAA Lys	AGA Arg	A
A	AUG† Met	ACG	AAG	AGG	G
G	GUU Val	GCU Ala	GAU Asp	GGU Gly	U
G	GUC	GCC	GAC	GGC	C
G	GUA	GCA	GAA Glu	GGA	A
G	GUG	GCG	GAG	GGG	G

* Chain-terminating or "nonsense" codons.

† Also used in bacteria to specify the initiator formyl-Met-$tRNA^{fMet}$.

vastly different AT/GC ratios and yet are closely related enough to encode proteins of highly similar amino acid sequences.)

Perceiving Order in the Makeup of the Code

Inspection of the distribution of codons in the genetic code suggests that the code evolved in such a way as to minimize the deleterious effects of mutations. For instance, mutations in the first position of a codon will often give a similar (if not the same) amino acid. Furthermore, codons with pyrimidines in the second position specify mostly hydrophobic amino acids, whereas those with purines in the second position correspond mostly to polar amino acids (see Table 15-1 and Chapter 5, Fig. 5-4). Hence, because transitions (A:T to G:C or G:C to A:T substitutions) are the most common type of point mutations, a change in the second position of a codon will usually replace one amino acid with a very similar one. Finally, if a codon suffers a transition mutation in the third position, rarely will a different amino acid be specified. Even a transversion mutation in this position will have no consequence about half the time.

Another consistency noticeable in the code is that whenever the first two positions of a codon are both occupied by G or C, each of the four nucleotides in the third position specifies the same amino acid (such as proline, alanine, arginine, or glycine). On the other hand, whenever the first two positions of the codon are both occupied by A or U, the identity of the third nucleotide does make a

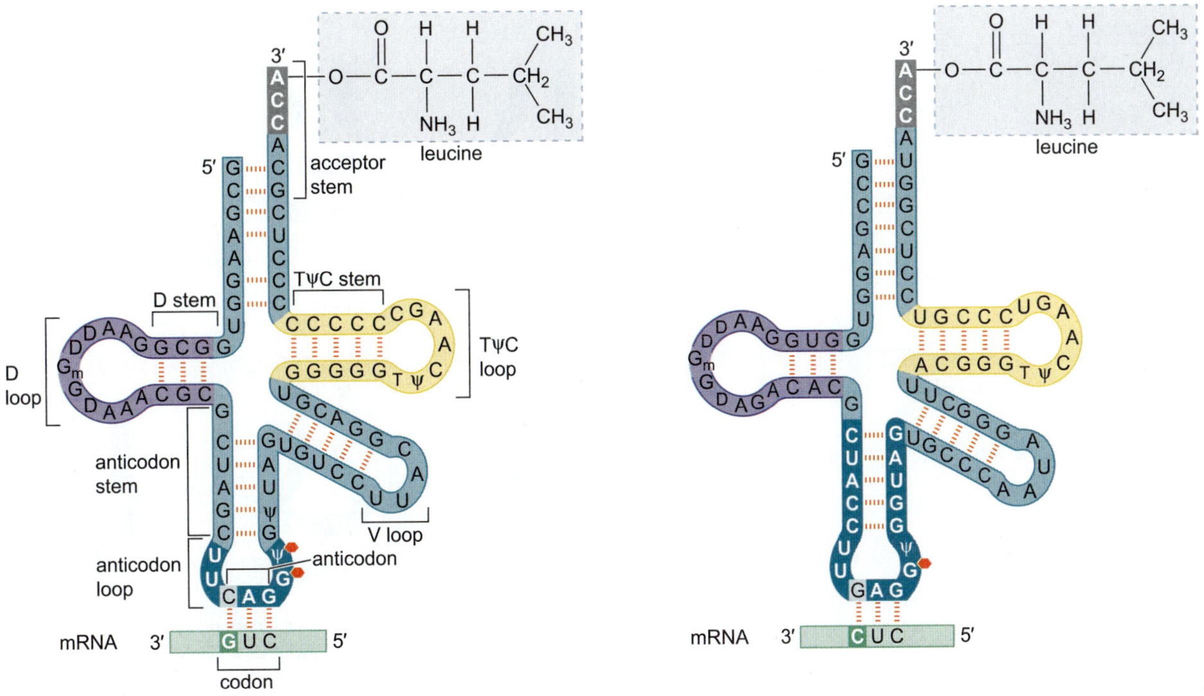

FIGURE 15-1 Codon–anticodon pairing of two tRNA Leu molecules. Critical stem and loop regions of the tRNA structure are labeled (see Chapter 14). The red hexagons linked to the G (3′ to the anticodon) denote methylation at the N1 positions of the base. Note that the codon is shown in a 3′ to 5′ orientation.

difference. Since G:C base pairs are stronger than A:U base pairs, mismatches in pairing the third codon base are often tolerated if the first two positions make strong G:C base pairs. Thus, having all four nucleotides in the third position specify the same amino acid may have evolved as a safety mechanism to minimize errors in the reading of such codons.

Wobble in the Anticodon

It was first proposed that a specific tRNA anticodon would exist for every codon. If that were the case, at least 61 different tRNAs, possibly with an additional 3 for the chain-terminating codons, would be present. Evidence began to appear, however, that highly purified tRNA species of known sequence could recognize several different codons. Cases were also discovered in which an anticodon base was not one of the four regular ones, but a fifth base, inosine. Like all the other minor tRNA bases, inosine arises through enzymatic modification of a base present in an otherwise completed tRNA chain. The base from which it is derived is adenine, whose carbon 6 is deaminated to give the 6-keto group of inosine. (Inosine is actually a nucleoside composed of ribose and the base hypoxanthine, but it has come to be referred to as a base in common usage and we do so here.)

In 1966, Francis Crick devised the **wobble concept** to explain these observations. It states that the base at the 5′ end of the anticodon is not as spatially confined as the other two, allowing it to form hydrogen bonds with any of several bases located at the 3′ end of a codon. Not all combinations are possible, with pairing restricted to those shown in Table 15-2. For example, U at the wobble position can

TABLE 15-2 Pairing Combinations with the Wobble Concept

Base in Anticodon	Base in Codon
G	U or C
C	G
A	U
U	A or G
I	A, U, or C

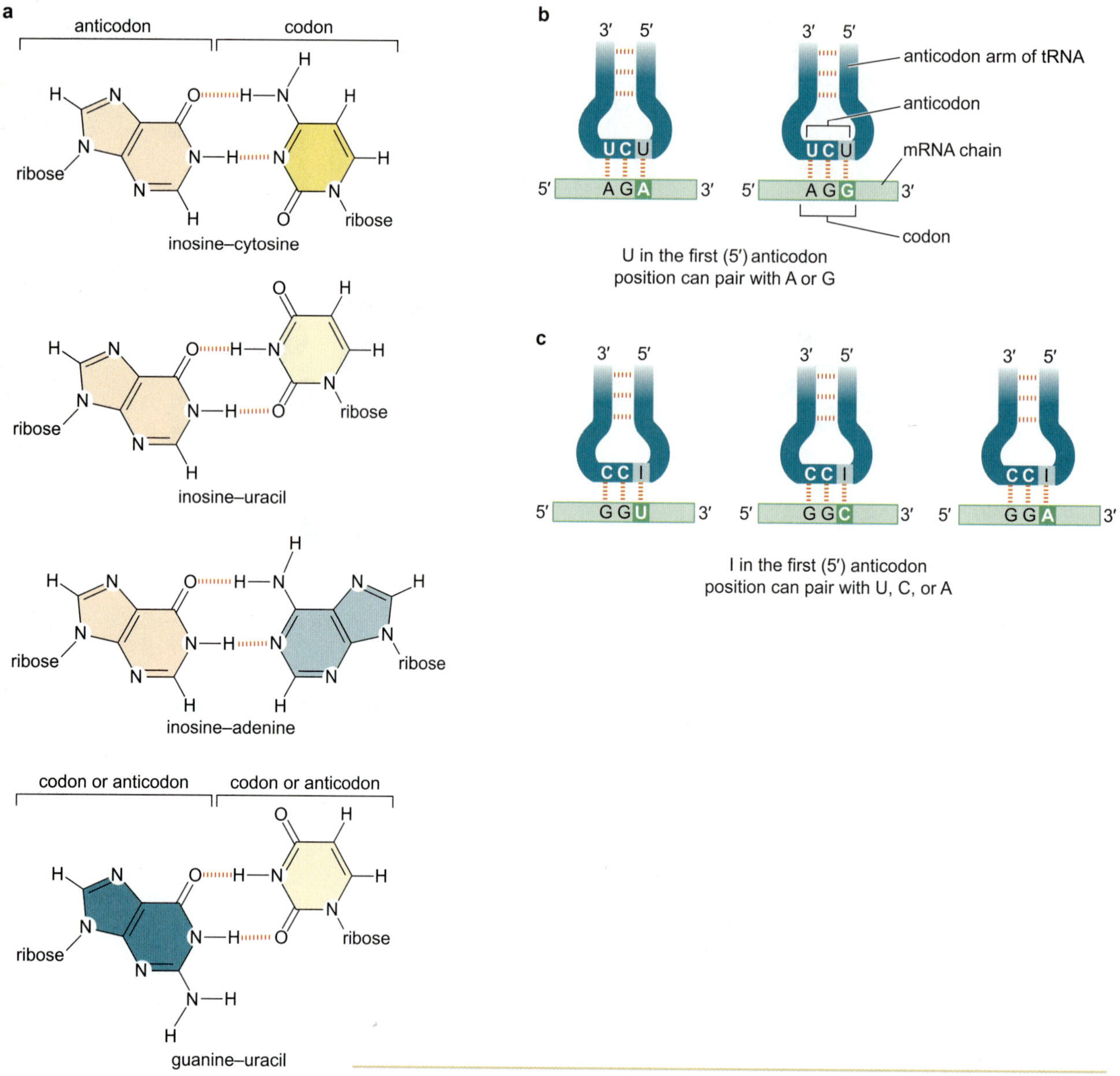

FIGURE 15-2 Wobble base pairing. Note that the ribose–ribose distances for all the wobble pairs are close to those of the standard A:U or G:C base pairs.

pair with either adenine or guanine, while I can pair with U, C, or A (Fig. 15-2). The pairings permitted by the wobble rules are those that give ribose–ribose distances close to that of the standard A:U or G:C base pairs. Purine–purine (with the exception of I:A pairs) or pyrimidine–pyrimidine pairs would give ribose–ribose distances that are too long or too short, respectively.

The wobble rules do not permit any single tRNA molecule to recognize four different codons. Three codons can be recognized only when inosine occupies the first (5′) position of the anticodon.

Almost all the evidence gathered since 1966 supports the wobble concept. For example, the concept correctly predicted that at least three tRNAs exist for the six serine codons (UCU, UCC, UCA, UCG, AGU, and AGC). The other two amino acids (leucine and arginine) that are encoded by six codons also have different tRNAs for the sets of codons that differ in the first or second position.

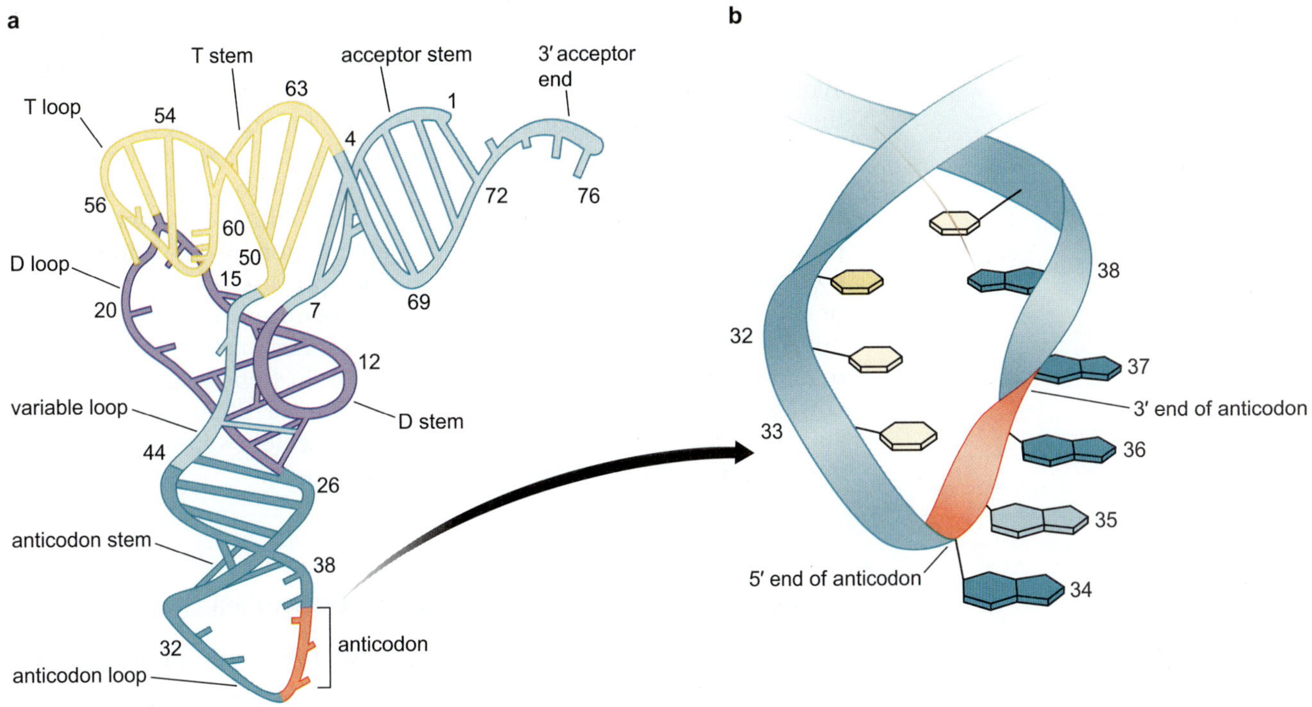

FIGURE 15-3 Structure of yeast tRNA^{Phe}. (a) A view of the L-shaped molecule based on X-ray diffraction data. (b) An enlargement of the anticodon loop. Bases in the anticodon (34–36) are shown in red. The anticodon and the following two bases (37 and 38) on the 3′ side are partially stacked. It can be seen that the base at the 5′ end of the anticodon is freer to wobble than is the fully stacked base at the 3′ end of the anticodon. (Adapted from Kim S-H. et al. 1974. *Proc. Natl. Acad. Sci.* 71: 4970.)

In the three-dimensional structure of tRNA, the three anticodon bases—as well as the two following (3′) bases in the anticodon loop—all point in roughly the same direction, with their exact conformations largely determined by stacking interactions between the flat surfaces of the bases (Fig. 15-3). Thus, the first (5′) anticodon base is at the end of the stack and is perhaps less restricted in its movements than the other two anticodon bases—hence, wobble in the third (3′) position of the codon. By contrast, not only does the third (3′) anticodon base appear in the middle of the stack, but the adjacent base is always a bulky modified purine residue. Thus, restriction of its movements may explain why wobble is not seen in the first (5′) position of the code.

Three Codons Direct Chain Termination

As we have seen, three codons do not correspond to any amino acid. Instead, they signify chain termination. As we discussed in Chapter 14, these chain-terminating codons, UAA, UAG, and UGA, are read not by special tRNAs but by specific proteins known as release factors (RF1 and RF2 in bacteria and eRF1 in eukaryotes). Release factors enter the A site of the ribosome and trigger hydrolysis of the peptidyl-tRNA occupying the P site, resulting in the release of the newly synthesized protein.

How the Code Was Cracked

The assignment of amino acids to specific codons is one of the great achievements in the history of molecular biology (see Chapter 2 for an historic account). How were these assignments made? By 1960, the general outline of how messenger RNA (mRNA) participates in protein synthesis had been established. Nevertheless, there was little optimism that we would soon have a detailed understanding of the genetic code itself. It was believed that identification of the codons for a given amino acid would require exact knowledge of both the nucleotide sequences of a gene and the corresponding amino acid order in its protein product. At that time, the elucidation of the amino acid sequence of a protein, although a laborious process, was already a very practical one. On the other hand, the then-current methods for determining DNA sequences were very primitive. Fortunately, this apparent roadblock did not hold up progress. In 1961, just one year after the discovery of mRNA, the use of artificial messenger RNAs and the availability of cell-free systems for carrying out protein synthesis began to make it possible to crack the code (see Chapter 2).

Stimulation of Amino Acid Incorporation by Synthetic mRNAs

Biochemists found that extracts prepared from cells of *Escherichia coli* that were actively engaged in protein synthesis were capable of incorporating radioactively labeled amino acids into proteins. Protein synthesis in these extracts proceeded rapidly for several minutes and then gradually came to a stop. During this interval, there was a corresponding loss of mRNA owing to the action of degradative enzymes present in the extract. However, the addition of fresh mRNA to extracts that had stopped making protein caused an immediate resumption of synthesis.

The dependence of cell extracts on externally added mRNA provided an opportunity to elucidate the nature of the code using synthetic polyribonucleotides. These synthetic templates were created using the enzyme polynucleotide phosphorylase, which catalyzes the reaction

$$[\text{XMP}]_n + \text{XDP} \rightleftharpoons [\text{XMP}]_{n+1} + \text{Ⓟ}, \qquad \textbf{[Equation 15-1]}$$

where X represents the base and $[\text{XMP}]_n$ represents RNA of length n nucleotides.

Polynucleotide phosphorylase is normally responsible for breaking down RNA and under physiological conditions favors the degradation of RNA into nucleoside diphosphates. By use of high nucleoside diphosphate concentrations, however, this enzyme can be made to catalyze the formation of internucleotide $3' \rightarrow 5'$ phosphodiester bonds and thus make RNA molecules (Fig. 15-4). No template DNA or RNA is required for RNA synthesis with this enzyme; the base composition of the synthetic product depends entirely on the ratio of the various ribonucleoside diphosphates added to the reaction mixture. For example, when only adenosine diphosphate is used, the resulting RNA contains only adenylic acid and is thus called **polyadenylic acid** or **poly-A.** It is likewise possible to make poly-U, poly-C, and poly-G. Addition of two or more different diphosphates produces mixed copolymers such as poly-AU, poly-AC, poly-CU, and poly-AGCU. In all these mixed polymers, the base sequences are approximately random, with the nearest-neighbor frequencies determined solely by the relative concentrations of the reactants. For example, poly-AU molecules with two times as much A as U have sequences like UAAUAUAAAUAAUAAAAUAUU. . . .

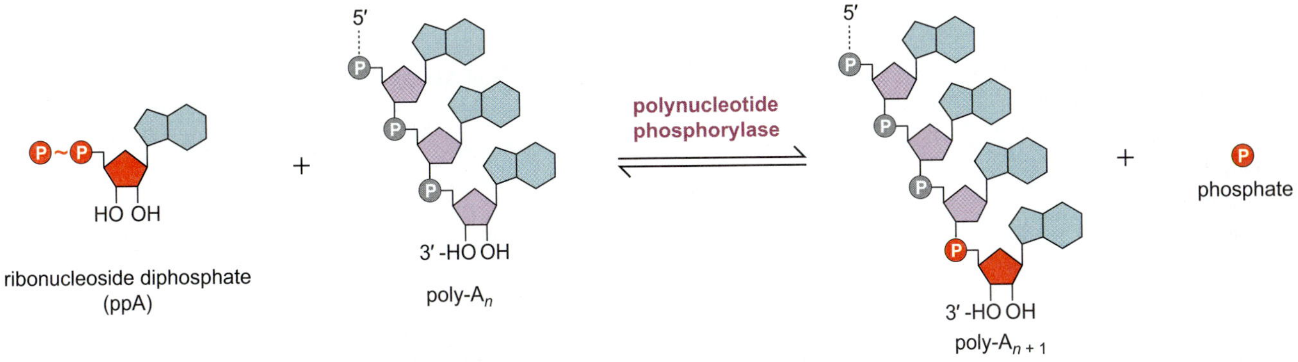

FIGURE 15-4 Polynucleotide phosphorylase reaction. The figure shows the reversible reactions of synthesis or degradation of polyadenylic acid catalyzed by the enzyme polynucleotide phosphorylase.

Poly-U Codes for Polyphenylalanine

Under the right conditions in vitro, almost all synthetic polymers will attach to ribosomes and function as templates. Luckily, high concentrations of magnesium were used in the early experiments. A high magnesium concentration circumvents the need for initiation factors and the special initiator fMet-tRNA, allowing chain initiation to take place without the proper signals in the mRNA. Poly-U was the first synthetic polyribonucleotide discovered to have mRNA activity. It selects phenylalanyl tRNA molecules exclusively, thereby forming a polypeptide chain containing only phenylalanine (polyphenylalanine). Thus, we know that a codon for phenylalanine is composed of a group of three uridylic acid residues, UUU. (That a codon has three nucleotides was known from genetic experiments, as indicated in Chapters 2 and 21, and below.) On the basis of analogous experiments with poly-C and poly-A, CCC was assigned as a proline codon and AAA as a lysine codon. Unfortunately, this type of experiment did not tell us what amino acid GGG specifies. The guanine residues in poly-G firmly hydrogen-bond to each other and form multistranded triple helices that do not bind to ribosomes.

Mixed Copolymers Allowed Additional Codon Assignments

Poly-AC molecules can contain eight different codons, CCC, CCA, CAC, ACC, CAA, ACA, AAC, and AAA, whose proportions vary with the copolymer A/C ratio. When AC copolymers attach to ribosomes, they cause the incorporation of asparagine, glutamine, histidine, and threonine—in addition to the proline previously assigned to CCC codons and the lysine previously assigned to AAA codons. The proportions of these amino acids incorporated into polypeptide products depend on the A/C ratio. Thus, since an AC copolymer containing much more A than C promotes the incorporation of many more asparagine than histidine residues, we conclude that asparagine is coded by two As and one C and that histidine is coded by two Cs and one A (Table 15-3). Similar experiments with other copolymers allowed several additional assignments. Such experiments, however, did not reveal the order of the different nucleotides within

TABLE 15-3 Amino Acid Incorporation into Proteins

Amino Acid	Observed Amino Acid Incorporation	Tentative Codon Assignments	Calculated Triplet Frequency				Sum of Calculated Triplet Frequencies
			3A	2A1C	1A1C	3C	
Poly-AC (5:1)							
Asparagine	24	2A1C		20			20
Glutamine	24	2A1C		20			20
Histidine	6	1A2C			4.0		4
Lysine	100	3A	100				100
Proline	7	1A2C, 3C			4.0	0.8	4.8
Threonine	26	2A1C, 1A2C		20	4.0		24
Poly-AC (1:5)							
Asparagine	5	2A1C		3.3			3.3
Glutamine	5	2A1C		3.3			3.3
Histidine	23	1A2C			16.7		16.7
Lysine	1	3A	0.7				0.7
Proline	100	1A2C, 3C			16.7	83.3	100
Threonine	21	2A1C, 1A2C		3.3	16.7		20

The amino acid incorporation into proteins was observed after adding random copolymers of A and C to a cell-free extract. The incorporation is given as a percentage of the maximal incorporation of a single amino acid. The copolymer ratio was then used to calculate the frequency with which a given codon would appear in the polynucleotide product. The relative frequencies of the codons are a function of the probability that a particular nucleotide will occur in a given position of a codon. For example, when the A/C ratio is 5:1, the ratio of AAA/AAC = 5 x 5 x 5:5 x 5 x 1 = 125:25. If we thus assign to the 3A codon a frequency of 100, then the 2A and 1C codon is assigned a frequency of 20. By correlating the relative frequencies of amino acid incorporation with the calculated frequencies with which given codons appear, tentative codon assignments can be made.

a codon. There is no way of knowing from random copolymers whether the histidine codon containing two Cs and one A is ordered CCA, CAC, or ACC.

Transfer RNA Binding to Defined Trinucleotide Codons

A direct way of ordering the nucleotides within some of the codons was developed in 1964. This method utilized the fact that even in the absence of all the factors required for protein synthesis, specific aminoacyl-tRNA molecules can bind to ribosome–mRNA complexes. For example, when poly-U is mixed with ribosomes, only phenylalanyl tRNA will attach. Correspondingly, poly-C promotes the binding of prolyl-tRNA. Most importantly, this specific binding does not demand the presence of long mRNA molecules. In fact, the binding of a trinucleotide to a ribosome is sufficient. The addition of the trinucleotide UUU results in phenylalanyl-tRNA attachment, whereas if AAA is added, lysyl-tRNA specifically binds to ribosomes. The discovery of this trinucleotide effect provided a relatively easy way of determining the order of nucleotides within many codons. For example, the trinucleotide 5′-GUU-3′ promotes valyl-tRNA binding, 5′-UGU-3′ stimulates cysteinyl-tRNA binding, and 5′-UUG-3′ causes leucyl-tRNA binding (Table 15-4). Although all 64 possible trinucleotides were synthesized with the hope of definitely assigning the order of every codon, not all codons were determined in this way. Some trinucleotides bind to ribosomes much less efficiently than UUU or GUU, making it impossible to know whether they code for specific amino acids.

TABLE 15-4 Binding of Aminoacyl-tRNA Molecules to Trinucleotide-Ribosome Complexes

Trinucleotide						AA-tRNA Bound
5′-UUU-3′	UUC					Phenylalanine
UUA	UUG	CUU	CUC	CUA	CUG	Leucine
AAU	AUC	AUA				Isoleucine
AUG						Methionine
GUU	GUC	GUA	GUG	UCU[a]		Valine
UCU	UCC	UCA	UCG			Serine
CCU	CCC	CCA	CCG			Proline
AAA	AAG					Lysine
UGU	UGC					Cysteine
GAA	GAG					Glutamic acid

AA, aminoacyl.
[a]Note that this codon was misassigned by this method.

Codon Assignments from Repeating Copolymers

At the same time that the trinucleotide binding technique became available, organic chemical and enzymatic techniques were being used to prepare synthetic polyribonucleotides with known repeating sequences (Fig. 15-5). Ribosomes start protein synthesis at random points along these regular copolymers; yet they incorporate specific amino acids into polypeptides. For example, the repeating sequence CUCUCUCU . . . is the messenger for a regular polypeptide in which leucine and serine alternate. Similarly, UGUGUG . . . promotes the synthesis of a polypeptide containing two amino acids, cysteine and valine. And ACACAC . . . directs the synthesis of a polypeptide alternating threonine and histidine. The copolymer built up from repetition of the three-nucleotide sequence AAG (AAGAAGAAG) directs the synthesis of three types of polypeptides: polylysine, polyarginine, and polyglutamic acid. Poly-AUC behaves in the same way, acting as a template for polyisoleucine, polyserine, and polyhistidine (Table 15-5). Further codon assignments were obtained from repeating tetranucleotide sequences.

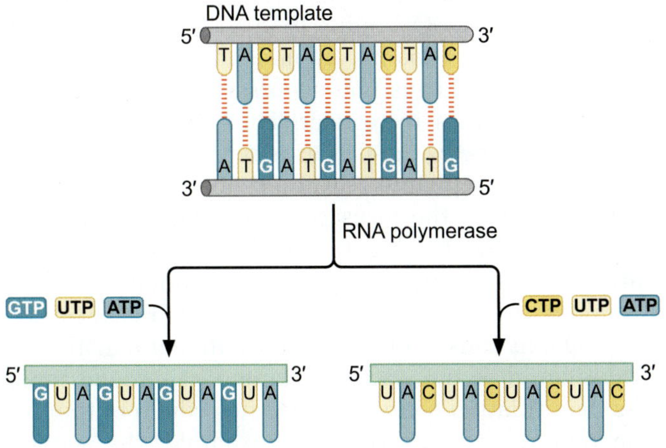

FIGURE 15-5 Preparing oligoribonucleotides. Using a combination of organic synthesis and copying by DNA polymerase I, double-stranded DNA with simple repeating sequences can be generated. RNA polymerase will then synthesize long polyribonucleotides corresponding to one or the other DNA strand, depending on the choice of ribonucleoside triphosphate added to the reaction mixture.

TABLE 15-5 Assignment of Codons Using Repeating Copolymers Built from Two or Three Nucleotides

Copolymer	Codons Recognized	Amino Acids Incorporated or Polypeptide Made	Codon Assignment
$(CU)_n$	CUC|UCU|CUC . . .	Leucine	5′-CUC-3′
		Serine	UCU
$(UG)_n$	UGU|GUG|UGU . . .	Cysteine	UGU
		Valine	GUG
$(AC)_n$	ACA|CAC|ACA . . .	Threonine	ACA
		Histidine	CAC
$(AG)_n$	AGA|GAG|AGA . . .	Arginine	AGA
		Glutamine	GAG
$(AUC)_n$	AUC|AUC|AUC . . .	Polyisoleucine	AUC
	UCA|UCA|UCA . . .	Polyserine	UCA
	CAU|CAU|CAU . . .	Polyhistidine	CAU

The sum of all these observations permitted the assignments of specific amino acids to 61 out of the possible 64 codons (see Table 15-1), with the remaining three chain-terminating codons, UAG, UAA, and UGA, not specifying any amino acid. (Note, as discussed in the previous chapter, that in the special context of translation initiation in *E. coli*, AUG is used as a start codon to specify *N*-formyl methionine rather than its usual codon assignment of methionine.)

THREE RULES GOVERN THE GENETIC CODE

The genetic code is subject to three rules that govern the arrangement and use of codons in messenger RNA. The first rule holds that codons are read in a 5′ to 3′ direction. Thus, in principle and as an example, the coding sequence for the dipeptide NH_2-Thr-Arg-COOH could be written as 5′-ACGCGA-3′ (where 5′-ACG-3′ is a threonine codon and 5′-CGA-3′ an arginine codon) or as 3′-GCAAGC-5′ wherein the codons are written in the same order as before but oppositely to their original orientations. Because messenger RNA is translated in a 5′ to 3′ direction, however, only the former is the correct coding sequence; if the latter were translated in a 5′ to 3′ direction, then the resulting peptide would be NH_2-Arg-Thr-COOH rather than NH_2-Thr-Arg-COOH.

The second rule is that codons are nonoverlapping and the message contains no gaps. This means that successive codons are represented by adjacent trinucleotides in register. Thus, the coding sequence for the tripeptide NH_2-Thr-Arg-Ser-COOH is represented by three contiguous and nonoverlapping triplets in the sequence 5′-ACGCGAUCU-3′.

The final rule is that the message is translated in a fixed reading frame, which is set by the initiation codon. As you will recall from Chapter 14, translation starts at an initiation codon, which is located at the 5′ end of the protein-coding sequence. Because codons are nonoverlapping and consist of three consecutive nucleotides, a stretch of nucleotides could be translated in principle in any of three reading frames. It is the initiation codon that dictates which of the three possible reading frames is used. Thus, for example, the sequence 5′. . .

ACGACGACGACGACGACGACG . . . 3′ could be translated as a series of threonine codons (5′-ACG′-3′), a series of arginine codons (5′-CGA-3′), or a series of asparate codons (5′-GAC-3′) depending on the frame of the upstream start codon.

Three Kinds of Point Mutations Alter the Genetic Code

Now that we have considered the nature of the genetic code, it is instructive to revisit the issue of how the coding sequence of a gene is altered by point mutations (see Chapter 9). An alteration that changes a codon specific for one amino acid to a codon specific for another amino acid is called a **missense mutation.** As a consequence, a gene bearing a missense mutation produces a protein product in which a single amino acid has been substituted for another, as in the classic example of the human genetic disease sickle-cell anemia, in which glutamate 6 in the β-globin subunit of hemoglobin has been replaced with a valine.

A more drastic effect results from an alteration causing a change to a chain-termination codon, which is known as a **nonsense** or **stop mutation.** When a nonsense mutation arises in the middle of a genetic message, an incomplete polypeptide is released from the ribosome owing to premature chain termination. The size of the incomplete polypeptide chain depends on the location of the nonsense mutation. Mutations occurring near the beginning of a gene result in very short polypeptides, whereas mutations near the end produce polypeptide chains of almost normal length. As we saw in Chapter 14, mRNAs that contain a premature stop codon are rapidly degraded in eukaryotic cells by a process known as nonsense-mediated mRNA decay.

The third kind of point mutation is a **frameshift mutation.** Frameshift mutations are insertions or deletions of one or a small number of base pairs that alter the reading frame. Consider a tandem repeat of the sequence GCU in a frame that would be read as a series of alanine codons (the codons are artificially set apart from each other by a gap for clarity but are, of course, contiguous in a real messenger RNA):

Ala	Ala	Ala	Ala	Ala	Ala	Ala	Ala
5′-GCU	GCU	GCU	GCU	GCU	GCU	GCU	GCU-3′

Now imagine the insertion of an A in the message, thereby generating a serine codon (AGC) at the site of the insertion. The resulting frameshift causes triplets downstream of the insertion to be read as cysteines:

Ala	Ala	Ser	Cys	Cys	Cys	Cys	Cys
5′-GCU	GCU	**A**GC	UGC	UGC	UGC	UGC	UGC-3′

Thus, the insertion (or for that matter the deletion) of a single base drastically alters the coding capacity of the message not only at the site of the insertion but for the remainder of the messenger as well. Likewise, the insertion (or deletion) of two bases would have the effect of throwing the entire coding sequence, at and downstream of the insertions, into a different reading frame.

Finally, consider the instructive case of an insertion of three extra bases at nearby positions in a message. It is obvious that the stretch of message, at and between the three insertions, will be drastically altered. But because the code is read in units of three, mRNA downstream of the three inserted bases will be in its proper reading frame and, hence, completely unaltered:

Ala Ala Ser Cys Met Leu His Ala Ala Ala

5′-GCU GCU **A**GC UGC **A**UG CUG C**A**U GCU GCU GCU-3′

Genetic Proof That the Code Is Read in Units of Three

The preceding example is the logic of a classic experiment by Francis Crick, Sydney Brenner, and their coworkers, involving bacteriophage T4 that established that the code is read in units of three and did so purely on the basis of a genetic argument (i.e., without any biochemical or molecular evidence). Genetic crosses were carried out to create a mutant phage harboring three inferred single-base-pair insertion mutations at nearby positions in a single gene. Of course, the three insertions would have scrambled a short stretch of codons but the protein encoded by the gene in question (called *rII*) was able to tolerate the local alteration to its amino acid sequence. This finding indicated that the overall coding capacity of the gene had been chiefly left unaltered despite the presence of three mutations, each of which alone, or any two of which alone, would have drastically altered the reading frame of the gene's message (and rendered its protein product inactive). Because the gene could tolerate three insertions but not one or two (or, for that matter, four), the genetic code must be read in units of three. See Chapters 2 and 22 for a discussion of the historic figures who showed that the code is read in units of three and for a description of the role of bacteriophage T4 as a model system for elucidating the nature of the code.

SUPPRESSOR MUTATIONS CAN RESIDE IN THE SAME OR A DIFFERENT GENE

Often, the effects of harmful mutations can be reversed by a second genetic change. Some of these subsequent mutations are easy to understand, being simple **reverse (back) mutations,** which change an altered nucleotide sequence back to its original arrangement. More difficult to understand are the mutations occurring at different locations on the chromosome that suppress the change due to a mutation at site A by producing an additional genetic change at site B. Such **suppressor mutations** fall into two main categories: those occurring within the same gene as the original mutation, but at a different site in this gene (**intragenic suppression**) and those occurring in another gene (**intergenic suppression**). Genes that cause suppression of mutations in other genes are called **suppressor genes.** Both of the types of suppression that we are considering here work by causing the production of good (or partially good) copies of the protein made inactive by the original harmful mutation. For example, if the first mutation caused the production of inactive copies of one of the enzymes involved in making arginine, then the suppressor mutation allows arginine to be made by restoring the synthesis of some good copies of this same enzyme. However, the mechanisms by which intergenic and intragenic suppressor mutations cause the resumption of the synthesis of good proteins are completely different.

As an example of intragenic supression, consider the case of a missense mutation. Its effect can sometimes be reversed through an

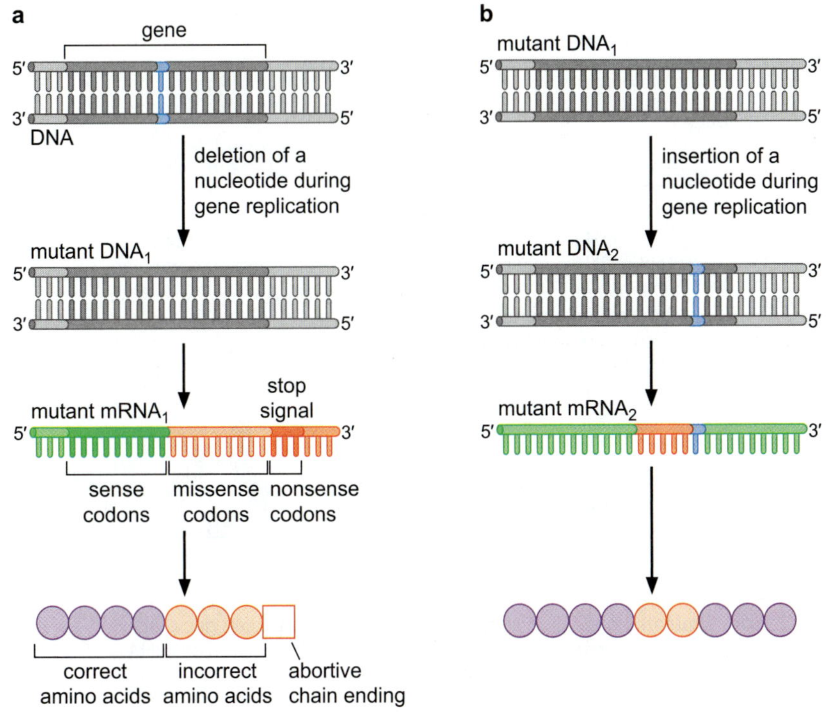

FIGURE 15-6 Suppression of frameshift mutations. (a) A deletion in the nucleotide coding sequence can result in an incomplete, inactive polypeptide chain. (b) The effect of the deletion, shown in panel a, can be overcome by a second mutation, an insertion in the coding sequence. This insertion results in the production of a complete polypeptide chain having two amino acid replacements. Depending on the change in sequence, the protein may have partial or full activity.

additional missense mutation in the same gene. In such cases, the original loss of enzymatic activity is due to an altered three-dimensional configuration resulting from the presence of an incorrect amino acid in the encoded protein sequence. A second missense mutation in the same gene can bring back biological activity if it somehow restores the original configuration around the functional part of the molecule. Figure 15-6 shows another example of intragenic suppression, this time for the case of a frameshift mutation.

Intergenic Suppression Involves Mutant tRNAs

Suppressor genes do not act by changing the nucleotide sequence of a mutant gene. Instead, they change the way the mRNA template is read. One of the best known examples of suppressor mutations are mutant tRNA genes that suppress the effects of nonsense mutations in protein-coding genes (but mutant tRNAs that suppress missense mutations and even frameshift mutations are also known). In *E. coli*, suppressor genes are known for each of the three stop codons. They act by reading a stop codon as if it were a signal for a specific amino acid. There are, for example, three well-characterized genes that suppress the UAG codon. One suppressor gene inserts serine, another glutamine, and a third tyrosine at the nonsense position. In each of the three UAG suppressor mutants, the anticodon of a tRNA species specific for one of these amino acids has been altered. For example, the tyrosine suppressor arises by a mutation within a $tRNA^{Tyr}$ gene that changes the anticodon from GUA (3′-AUG-5′) to CUA (3′-AUC-5′), thereby enabling it to recognize UAG codons (Fig. 15-7). The serine and glutamine suppressor tRNAs also arise by single base changes in their anticodons.

The discovery that cells with nonsense suppressors contain mutationally altered tRNAs raised the question of how their codons correspond-

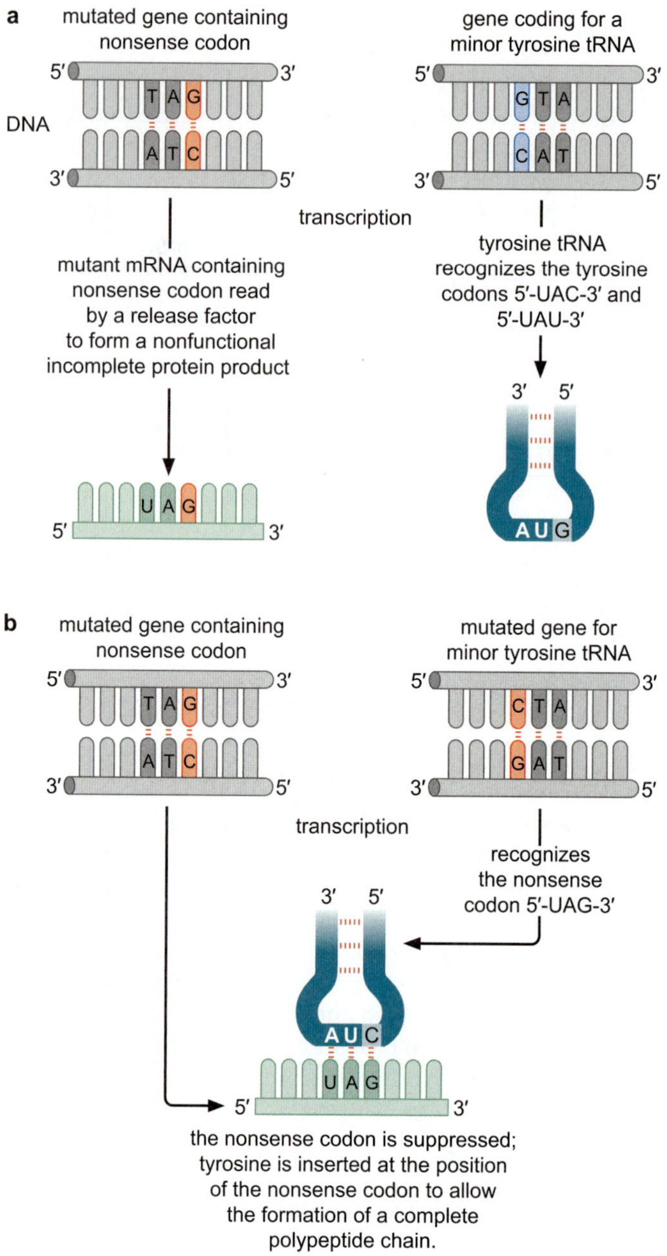

FIGURE 15-7 **Nonsense suppression.** The figure shows how a minor tyrosine tRNA species acts to suppress the nonsense codon in mRNA.

ing to these tRNAs could continue to be read normally. In the case of the tyrosine UAG suppressor, the answer comes from the discovery that three separate genes code for $tRNA^{Tyr}$. One codes for the major $tRNA^{Tyr}$ species, whereas the other two are duplicate genes coding for a species present in smaller amounts. One or the other of the two duplicate genes is always the site of the suppressor mutation. No such dilemma exists for UGA suppression, which is mediated by a mutant form of $tRNA^{Trp}$; the suppressing $tRNA^{Trp}$ retains its capacity to read UGG (tryptophan) codons while also recognizing UGA stop codons. This is possible because the anticodon was changed from CCA (3′-ACC-5′) in the wild type to UCA (3′-ACU-5′) in the mutant $tRNA^{Trp}$, and wobble rules, as we have seen, allow recognition of A or G in the 3′ position of the codon by U in the 5′ position of an anticodon.

Nonsense Suppressors Also Read Normal Termination Signals

The act of nonsense suppression can be viewed as a competition between the suppressor tRNA and the release factor. When a stop codon comes into the ribosomal A site, either readthrough or polypeptide chain termination will occur, depending on which arrives first. Suppression of UAG codons is efficient. In the presence of the suppressor tRNA, more than half of the chain-terminating signals are read as specific amino acid codons. *E. coli* can tolerate this misreading of the UAG stop codon because UAG is used infrequently as a chain-terminating codon at the end of open reading frames. In contrast, suppression of the UAA codon usually averages between 1% and 5% and mutant cells producing UAA-suppressing tRNAs grow poorly. This is expected from the fact that UAA is frequently used as a chain-terminating codon and its recognition by a suppressor tRNA would be expected to result in the production of many more aberrantly long polypeptides.

Proving the Validity of the Genetic Code

The code was cracked, as we have seen, by means of biochemical methods involving the use of cell-free systems for carrying out protein synthesis. But molecular biologists are generally suspicious of a method that relies on in vitro analysis alone. So how do we know definitively that the code as depicted in Table 15-1 is true in living cells? Of course, in the modern era of large-scale DNA sequencing, in which the entire nucleotide sequences of the genomes of diverse organisms ranging from microbes to man have been determined, the genetic code has not only been validated but shown to be universal or nearly so (see below). Nonetheless, a classic and instructive experiment in 1966 helped to validate the genetic code well before DNA sequencing was possible. The experiment was based on the construction by genetic recombination of a mutant gene of phage T4 that harbored a mutually suppressing pair of insertion and deletion mutations (similar to the example given in Fig. 15-6). The gene in question encoded a cell-wall-degrading enzyme called lysozyme, chosen because it is small, easy to purify, and its complete amino acid sequence was known. The experimental strategy was to compare the amino acid sequence of the doubly mutant protein with that of wild-type lysozyme.

When the amino acid sequences of the mutant (. . . NH_2—Thr Lys **Val His His Leu Met** Ala Ala Lys—COOH . . .) and wild type (. . . NH_2—Thr Lys **Ser Pro Ser Leu Asn** Ala Ala Lys—COOH . . .) were compared, they were found to differ by a stretch of five amino acids (highlighted in bold). This observation suggested that the insertion and deletion mutations had scrambled a short stretch of codons in the message of the mutant. Knowing the consequent effect of the scrambled codons on the amino acid sequence of the protein imposed important constraints on the nature of the genetic code. Specifically, if the genetic code as elucidated in biochemical experiments is valid, then it should be possible to identify a set of codons for the wild-type sequence Ser Pro Ser Leu Asn that, when properly aligned and bracketed with an insertion at one end and a deletion at the other, would specify the mutant amino acid sequence. Indeed, such a solution exists, which requires a deletion of a nucleotide at the 5′ end of the coding sequence and the insertion of a nucleotide at the 3′ end:

NH_2—Lys	**Ser**	**Pro**	**Ser**	**Leu**	**Asn**	Ala—COOH
5′—AAA	**AGU**	CCA	UCA	CUU	AAU	GC—3′
5′—AAA	GUC	CAU	CAC	UUA	AUG	**GC—3′**
NH_2—Lys	**Val**	**His**	**His**	**Leu**	**Met**	Ala—COOH

As you can see, the solution verifies several codon assignments and demonstrates that more than one synonymous codon is used to specify the same amino acid in vivo (e.g., 5′-CAU-3′ and 5′-CAC-3′ for histidine). Lastly, and importantly, you should be able to convince yourself from the solution that translation proceeds in a 5′ to 3′ direction. (Hint: see if you can account for the two amino acid sequences in their proper NH_2 to COOH order when you align each of the codons in your solution in a 3′ to 5′ orientation.)

THE CODE IS NEARLY UNIVERSAL

The results of large-scale sequencing of genomes have largely confirmed the expected universality of the genetic code. The universality of the code has had a huge impact on our understanding of evolution as it made it possible to directly compare protein-coding sequences among all organisms for which a genome sequence is available. As we shall see in Chapter 21, powerful computer programs are available that can search for and identify similarities among predicted coding sequences from a wide range of organisms. The universality of the code also helped to create the field of genetic engineering by making it possible to express cloned copies of genes encoding useful protein products in surrogate host organisms, such as the production of human insulin in bacteria (see Chapter 21).

To understand the conservative nature of the code, consider what might happen if a mutation changed the genetic code. Such a mutation might, for example, alter the sequence of the serine tRNA molecule of the class that corresponds to UCU, causing them to recognize UUU sequences instead. This would be a lethal mutation in haploid cells containing only one gene directing the production of $tRNA^{Ser}$, for serine would not be inserted into many of its normal positions in proteins. Even if there were more than one gene for $tRNA^{Ser}$ (as in a diploid cell), this type of mutation would still be lethal since it would cause the simultaneous replacement of many phenylalanine residues by serine in cell proteins.

In view of what we have just said, it was completely unexpected to find that in certain subcellular organelles, the genetic code is in fact slightly different from the standard code. This realization came during the elucidation of the entire DNA sequence of the 16,569-bp human mitochondrial genome but is observed for mitrochondria in yeast, the fruit fly, and higher plants. Sequences of the regions known to specify proteins have revealed the following differences between the standard and mitochondrial genetic codes (Table 15-6).

- UGA is not a stop signal but codes for tryptophan. Hence, the anticodon of mitochondrial $tRNA^{Trp}$ recognizes both UGG and UGA, as if obeying the traditional wobble rules.
- Internal methionine is encoded by both AUG and AUA.
- In mammalian mitochondria, AGA and AGG are not arginine codons (of which there are six in the "universal" code) but specify

TABLE 15-6 Genetic Code of Mammalian Mitochondria

first position (5′ end)	second position: U		C		A		G		third position (3′ end)
U	UUU	Phe (GAA)†	UCU	Ser (UGA)	UAU	Tyr (GUA)	UGU	Cys (GCA)	U
	UUC		UCC		UAC		UGC		C
	UUA	Leu (UAA)	UCA		UAA	stop	UGA	Trp (UCA)	A
	UUG		UCG		UAG	stop	UGG		G
C	CUU	Leu (UAG)	CCU	Pro (UGG)	CAU	His (GUG)	CGU	Arg (UCG)	U
	CUC		CCC		CAC		CGC		C
	CUA		CCA		CAA	Gln (UUG)	CGA		A
	CUG		CCG		CAG		CGG		G
A	AUU	Ile (GAU)	ACU	Thr (UGU)	AAU	Asn (GUU)	AGU	Ser (GCU)	U
	AUC		ACC		AAC		AGC		C
	AUA	Met (CAU)‡	ACA		AAA	Lys (UUU)	AGA	stop	A
	AUG		ACG		AAG		AGG	stop	G
G	GUU	Val (UAC)	GCU	Ala (UGC)	GAU	Asp (GUC)	GGU	Gly (UCC)	U
	GUC		GCC		GAC		GGC		C
	GUA		GCA		GAA	Glu (UUC)	GGA		A
	GUG		GCG		GAG		GGG		G

* Differences between the mitochrondial and "universal" genetic code (Table 15-1) are shown by green shading.

† Each group of codons is shaded in gray and is read by a single tRNA whose anticodon, written 5′⟶ 3′ is in parentheses. Each four-codon group is read by a tRNA having a U in the first (5′) position of the anticodon. Two-codon groups with codons ending in either U/C or A/G are read with GU wobble by tRNAs, with G or U, respectively, in the first position of the anticodon. The anticodons often contain modified bases.

‡ Note that the C in the first anticodon position engages in unusual pairing.

chain termination. Thus, there are four stop codons (UAA, UAG, AGA, and AGG) in the mammalian mitochondrial code.

- In fruit fly mitochondria, AGA and AGG are also not arginine codons but specify serine.

Perhaps not surprisingly, mitochondrial tRNAs are likewise unusual with respect to the rules by which they decode mitochondrial messages. Only 22 tRNAs are present in mammalian mitochondria, whereas a minimum of 32 tRNA molecules are required to decode the "universal" code according to the wobble rules. Consequently, when an amino acid is specified by four codons (with the same first and second positions), only a single mitochondrial tRNA is involved. (Recall that a minimum of two tRNAs would be required by nonmitochondrial systems.) Such mitochondrial tRNAs all have in the 5′ (wobble) position of their anticodons a U residue, which is able to engage in pairing with any of the four nucleotides in the third codon position. In cases where purines in the third position of the codon correspond to different amino acids from pyrimidines in that position, a modified U in the first position of the anticodon of the mitochondrial tRNA restricts wobble to pairing with the two purines only.

Exceptions to the "universal" code are not limited to mitochondria but are also found in several prokaryotic genomes and in the nuclear genomes of certain eukaryotes. The bacterium *Mycoplasma capricolum* uses UGA as a tryptophan codon rather than a chain-termination codon. Likewise, some unicellular protozoa use UAA and UAG, which are stop codons in the "universal" code, as glutamine codons. Finally, a codon (CUG) for one amino acid (leucine) in the "universal" code has become a codon for another amino acid (serine) in the yeast *Candida*.

SUMMARY

In the "universal" genetic code used by every organism from bacteria to humans, 61 codons signify specific amino acids; the remaining three are chain-termination codons. The code is highly degenerate, with several codons (synonyms) usually corresponding to a single amino acid. A given tRNA can sometimes specifically recognize several codons. This ability arises from wobble in the base at the 5′ end of the anticodon. The stop codons UAA, UAG, and UGA are read by specific proteins, not specialized tRNA molecules.

The genetic code is subject to three principal rules. Codons are read in a 5′ to 3′ direction, codons are nonoverlapping and the message contains no gaps, and the message is translated in a fixed reading frame, which is set by the initiation codon.

The genetic code was cracked through the study of protein synthesis in cell-free extracts. Addition of new mRNA to an extract depleted of its original messenger component results in the production of new proteins whose amino acid sequences are determined by the externally added mRNA. The first (and probably most important) step in cracking the genetic code occurred when the synthetic polyribonucleotide poly-U was found to code specifically for polyphenylalanine. Use of other synthetic polyribonucleotides, both homogeneous (poly-C, and so on) and mixed (poly-AU, and so on), then allowed assignment of codons for the various amino acids. Determination of the exact order of nucleotides in codons subsequently came from a study of specific trinucleotide–tRNA–ribosome interactions and the use of regular copolymers as messengers.

Point mutations that alter the code are missense mutations, which change the codon for one amino acid into the codon for another amino acid; nonsense mutations, which cause protein synthesis to terminate prematurely; and frameshift mutations, which alter the reading frame of the message. In some cases the effects of missense, nonsense, and frameshift mutations can be partially suppressed by extragenic suppressors. For example, mutant tRNAs read stop codons generated by nonsense mutations as if they were codons for a specific amino acid.

A slightly different genetic code is utilized in mitochondria and in the principal genomes of certain prokaryotes and protozoa, such as the use of UGA, a stop codon in the "universal code," as a tryptophan codon.

BIBLIOGRAPHY

Books

Celis J.E. and Smith J.D., eds. 1979. *Nonsense mutations and tRNA suppressors.* Academic Press, New York.

Clark B. and Petersen H., eds. 1984. Gene expression: The translational step and its control. *Alfred Benzon Symposium*, vol. 19. Copenhagen, Munksgaard.

Cold Spring Harbor Symposia on Quantitative Biology. 1966. Volume 31: *The genetic code.* Cold Spring Harbor Laboratory, Cold Spring Harbor, New York.

Söll D.G., Abelson J.N., and Schimmel P.R., eds. 1980. *Transfer RNA: Biological aspects.* Cold Spring Harbor Laboratory, Cold Spring Harbor, New York.

Ycas M. 1969. *The biological code.* Wiley (Interscience), New York.

Features of the Genetic Code

Crick F.H.C. 1966. Codon–anticodon pairing: The wobble hypothesis. *J. Mol. Biol.* **19:** 548–555.

Kohli J. and Grosjean H. 1981. Usage of the three termination codons: Compilation and analysis of the known eukaryotic and prokaryotic translation termination sequences. *Mol. Gen. Genet.* **182:** 430–439.

Lagerkvist U. 1981. Unorthodox codon reading and the evolution of the genetic code. *Cell* **23:** 305–306.

How the Code Was Cracked

Crick F.H.C. 1963. The recent excitement in the coding problem. *Prog. Nucleic Acid Res.* **1:** 164.

Khorana H.G. 1968. *Polynucleotide synthesis and the genetic code. Harvey Lecture Series 1966–67.* Vol. 62. Academic Press, New York.

Nirenberg M. and Leder P. 1964. The effect of trinucleotides upon the binding of sRNA to ribosomes. *Science* **145:** 1399–1407.

Speyer J.F., Lengyel P., Basilio C., Wahba A.J., Gardner R.S., and Ochoa S. 1963. Synthetic polynucleotides and the amino acid code. *Cold Spring Harbor Symp. Quant. Biol.* **28:** 559–568.

Three Rules of the Genetic Code

Brenner S., Stretton A.O.W., and Kaplan S. 1965. Genetic code: The nonsense triplets for chain termination and their suppression. *Nature* **206:** 994–998.

Crick F.H.C., Barnett L., Brenner S., and Watts-Tobin R.J. 1961. General nature of the genetic code for proteins. *Nature* **192:** 1227–1232.

Garen A. 1968. Sense and nonsense in the genetic code. *Science* **160:** 149–159.

Terzaghi E., Okada Y., Streisinger G., Emrich J., Inouye M., and Tsugita A. 1966. Change of a sequence of amino acids in phage T4 lysozyme by acridine-induced mutations. *Proc. Natl. Acad. Sci.* **56:** 500–507.

Suppression

Buckingham R.H. and Kurland C.G. 1980. Interactions between UGA-suppressor tRNA^{Trp} and the ribosome: Mechanisms of tRNA selection. In *Transfer RNA: Biological aspects* (ed. D. Söll et al.), pp. 421–426. Cold Spring Harbor Laboratory, Cold Spring Harbor, New York.

Ozeki H., Inokuchi H., Yamao F., Kodaira M., Sakano H., Ikemura T., and Shimura Y. 1980. Genetics of nonsense suppressor of tRNAs in *Escherichia coli*. In *Transfer RNA: Biological aspects* (ed. D. Söll et al.), pp. 341–349. Cold Spring Harbor Laboratory, Cold Spring Harbor, New York.

Steege D.A. and Söll D.G. 1979. Suppression. In *Biological regulation and development I* (ed. R.F. Goldberger), pp. 433–486. Plenum, New York.

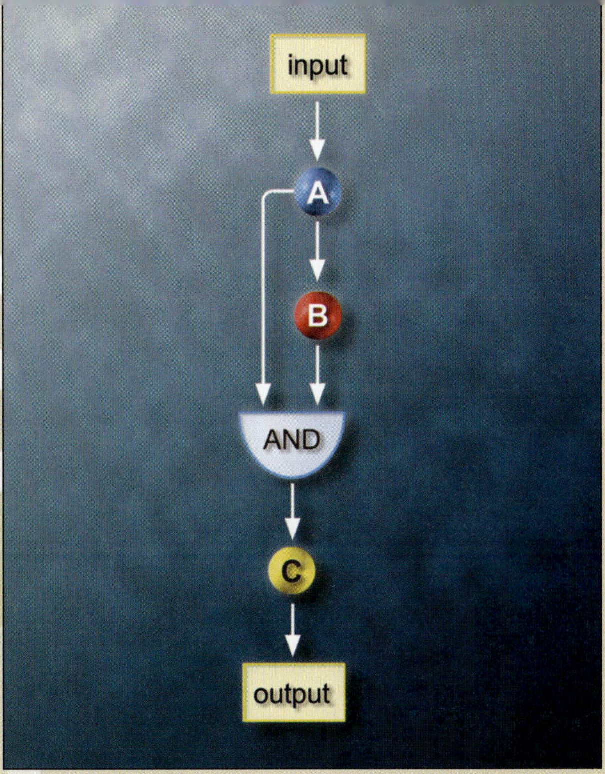

PART

4

REGULATION

IN PART 3, WE CONSIDERED HOW THE GENETIC information encoded in the DNA is expressed. This involves the transcription of DNA sequences into an RNA form, which is then used as a template for translation into protein.

But not all genes are expressed in all cells all the time. Indeed, much of life depends on the ability of cells to express their genes in different combinations at different times and in different places. Even a lowly bacterium expresses only some of its genes at any given time, thus ensuring it can, for example, make the enzymes needed to metabolize the nutrients it encounters while not making enzymes for other nutrients that are not available at that time. Development of multicellular organisms offers an even more striking example of this so-called "differential gene expression." Essentially all the cells in a human contain the same genes, but the set of genes expressed in forming one cell type is different from that expressed in forming another. Thus, a muscle cell expresses a set of genes different (at least in part) from that expressed by a neuron, a skin cell, and so on. By and large, these differences occur at the level of transcription—most commonly, the initiation of transcription.

In the following chapters, we look at how transcription is regulated. We start in Chapter 16 with how this is done in bacteria. It is here that the basic mechanisms can most readily be appreciated. Thus, we deal with simple cases that illustrate different mechanisms of transcriptional regulation. These include the case of the *lac* operon, which is a group of genes that encode proteins needed for metabolism of the sugar lactose—genes that are transcribed only when that sugar is available in the growth medium. In this case we learn how genes can be activated (switched on) and repressed (switched off) in reponse to different signals. We then look at other examples: some where regulation is similar to the *lac* genes and some that illustrate rather different mechanisms of transcriptional regulation. Finally in this chapter, we describe how transcriptional regulation of alternative sets of genes in phage λ underpins the ability of that virus to choose between alternative development pathways upon infection of a bacterial cell.

In Chapter 17, we consider basic mechanisms of transcriptional regulation in eukaryotes, from yeast to higher eukaryotes. Mechanisms of transcriptional activation and repression are compared to those in bacteria, and we see where mechanisms are conserved and where there are additional features—most notably the effects of chromatin modifications of the type discussed in Chapter 7. We also discuss the meaning and mechanisms of epigenetic gene regulation.

Up to this point, the regulation we have discussed is driven by protein regulators—activators and repressors, and proteins they recruit to genes. In Chapter 18, we look at the growing field of regulatory RNAs. Here we describe how RNA molecules can activate, or more commonly repress, expression of genes in bacteria and eukaryotes. This includes long-understood mechanisms, such as attenuation of the Tryptophan operon, and also more recently uncovered mechanisms, such as RNA interference and the role of microRNAs in higher eukaryotes.

In Chapter 19, we consider gene regulation in the context of developmental biology and evolution. We look at how genes are regulated to bestow cell type specificity (differentiation) and pattern formation (morphogenesis) on a group of genetically identical cells—for example, those found in a developing embryo. We also discuss diversity among closely related organisms and see how, in many of these, the differences in morphology or behavior result not from changes in the genes but from differences in where and when those genes are expressed within each organism during development. The most striking finding to come from the various genome sequences elucidated over recent years is that most animals (for example) have essentially the same genes—be they mice, men, or even flies. This observation again underscores the general role of gene regulation—most of it transcriptional regulation—in defining what each genome produces.

This leads us to the last chapter in this section of the book—Genome Analysis and Systems Biology. In the first half, we consider genome sequences and what they have taught us, and in the second half we discuss some aspects of the emerging field of systems biology. Although the field remains rather ill-defined and seems to embrace a range of different areas, in the current context we deal with just one aspect, which is relevant to the theme of this section of the book—namely, gene regulatory networks. We present the nomenclature and basic ideas behind newly defined ways of thinking about how networks of genes are regulated. A new generation of molecular biologists—many with backgrounds in computing or physics—are describing such networks in terms of the logic of information flow rather than molecular mechanisms that underlie their operation. This field has even developed its own representations—for example, the illustration on the cover of this book.

PHOTOS FROM THE COLD SPRING HARBOR LABORATORY ARCHIVES

Edward B. Lewis, Carl C. Lindegren, Alfred D. Hershey, and Joshua Lederberg, 1951 Symposium on Genes and Mutations. Lewis instigated the genetic analysis of development, using the fruit fly as his model (Chapter 19). He shared, with Eric F. Wieschaus and Christiane Nüsslein-Volhard, the 1995 Nobel Prize in Physiology or Medicine for his work. Lindegren was a pioneer of yeast genetics (Chapter 22). Hershey was, together with Max Delbrück and Salvador E. Luria, the leader of the group that used phage as their model system in the early days of molecular biology (Chapter 22); the three of them shared the 1969 Nobel Prize in Physiology or Medicine. Lederberg discovered that DNA could pass between bacteria by a mating process called conjugation (Chapter 22), for which he shared, with George Beadle and Edward Tatum, in the 1958 Nobel Prize in Physiology or Medicine.

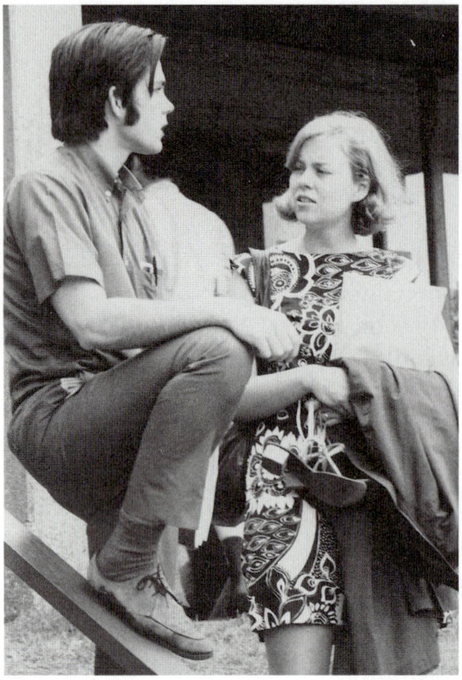

Jeffrey W. Roberts and Ann B. Burgess, 1970 Symposium on Transcription of Genetic Material. Roberts' research has focused on regulators of gene expression in bacteria and phage, particularly antiterminators in phage λ (Chapter 16). Burgess became a biology educator and is involved in national efforts to improve science education. Roberts was an author of the fourth edition of this book, and Burgess has a cousin among the current authors (TB).

Christiane Nüsslein-Volhard, 1996 Meeting on Zebrafish Development and Genetics. Mutant screens carried out in fruit flies by Nüsslein-Volhard and her colleague Eric F. Wieschaus identified many genes critical to the early embryonic development of that organism, and probably all animals (Chapter 19). For this the two of them shared in the 1995 Nobel Prize in Physiology or Medicine with Edward B. Lewis.

Mark Ptashne and Joseph Goldstein, 1988 Symposium on Molecular Biology of Signal Transduction. Ptashne was instrumental in taking the early ideas of Jacob and Monod about how gene expression is regulated, and describing how these work at a molecular level, first in phage λ, and then in yeast (Chapters 16 and 17). Goldstein, with his long-time collaborator Michael S. Brown, worked out the signal transduction pathways (Chapter 17) that control expression of genes involved in cholesterol metabolism, for which they won the 1985 Nobel Prize in Physiology or Medicine.

Mrs. I.H. Herskowitz with sons, Ira and Joel, 1947 Symposium on Nucleic Acids and Nucleoproteins. Ira Herskowitz pioneered the use of the yeast *Saccharomyces cerevisiae* as a model organism for molecular biology (Chapter 22) and made major contributions to ideas about gene regulation in this organism as he had, earlier, in bacteriophage λ (Chapters 16 and 17). His father, Irwin, later the author of a genetics textbook, was attending the symposium that year.

Craig Mello, 2004 Symposium on Epigenetics. Together with Andrew Fire, Mello found that by simply introducing dsRNAs into cells, genes with homology to that RNA can be silenced. From this observation, which they called RNA interference, the whole field of RNAi exploded (Chapter 18). They shared the 2006 Nobel Prize in Physiology or Medicine for their work.

Scott Emmons, Gary Ruvkun, and Barbara Meyer, 2004 Symposium on Epigenetics. While studying the genetics of development in worms, Victor Ambros and Ruvkun identified the first miRNA and target gene (Chapter 18). The NASA T-shirt is a clue to another of Ruvkun's many research interests: the quest for life on Mars. Emmons studies behavior in worms, at all levels from gene expression to the neurobiology, and Meyer, who as a graduate student contributed much to elucidating the phage λ genetic switch (Chapter 16), now works on sex determination and dosage compensation in the worm (Chapter 18).

Jacques Monod and Leo Szilard, 1961 CSH Laboratory. Monod, together with Françoise Jacob, formulated the operon model for the regulation of gene expression (Chapter 16). The two of them, together with their colleague André Lwoff, shared the 1965 Nobel Prize in Physiology or Medicine for this achievement. Leo Szilard was a wartime nuclear physicist who turned to molecular biology after taking the phage course at Cold Spring Harbor in 1947. He ran a lab with Aaron Novick in Chicago. (Courtesy of Esther Bubley.)

Richard Jorgensen and David Baulcombe, 2006 Symposium on Regulatory RNAs. Jorgensen found that overexpression of the petunia pigment gene could generate flowers that had white rather than dark purple flowers (Chapter 18). Although unknown at the time, this effect was caused by RNAi. The small interfering RNAs—the critical intermediates in this process—were later identified by Baulcombe (Chapter 18).

CHAPTER

16 Transcriptional Regulation in Prokaryotes

IN CHAPTER 12, WE SAW HOW DNA IS TRANSCRIBED into RNA by the enzyme RNA polymerase. We also described the sequence elements that constitute a promoter—the region at the start of a gene where the enzyme binds and initiates transcription. In bacteria, the most common form of RNA polymerase (that bearing σ^{70}) recognizes promoters formed from various sequence elements—the three major ones being "–10", "–35", and "UP" elements—and we saw that the strength of any given promoter is determined by which elements it possesses and how well they match optimum "consensus" sequences. In the absence of regulatory proteins, these elements determine the efficiency with which polymerase binds to the promoter and, once bound, how readily it initiates transcription.

Now we turn to the mechanisms that regulate expression—that is, those mechanisms that increase or decrease expression of a given gene as the requirement for its product varies. There are various stages at which expression of a gene can be regulated. The most common is transcription initiation, and the bulk of this chapter focuses on the regulation of that step in bacteria. We start with an overview of general mechanisms and principles and proceed to some well-studied examples that demonstrate how the basic mechanisms are used in various combinations to control genes in specific biological contexts. We also consider mechanisms of transcriptional regulation that operate at steps after initiation, specifically during elongation and termination. Other examples of transcriptional regulation in prokaryotes—those mediated by RNA structures—are considered in Chapter 18, Regulatory RNAs. An example of prokaryotic gene regulation at the level of translation was discussed in Chapter 14.

OUTLINE

PRINCIPLES OF TRANSCRIPTIONAL REGULATION

Gene Expression Is Controlled by Regulatory Proteins

As described in the introduction to this section, genes are very often controlled by extracellular signals; in the case of bacteria, this typically means molecules present in the growth medium. These signals are communicated to genes by regulatory proteins, which come in two types: positive regulators, or **activators**, and negative regulators, or **repressors**. Typically, these regulators are DNA-binding proteins that recognize specific sites at or near the genes they control. An activator increases transcription of the regulated gene; repressors decrease or eliminate that transcription.

How do these regulators work? Recall the steps in transcription initiation described in Chapter 12 (see Fig. 12-3). First, RNA polymerase binds to the promoter in a closed complex (in which the DNA strands remain together). The polymerase–promoter complex then undergoes a transition to an open complex in which the DNA at the start site of transcription is unwound and the polymerase is positioned to initiate transcription. This is followed by promoter escape, the step in which polymerase leaves the promoter and starts transcribing. Polymerase then proceeds through the elongation phase before finally terminating. Which steps are stimulated by activators and inhibited by repressors depends on the promoter and regulators in question.

Most Activators and Repressors Act at the Level of Transcription Initiation

Although we shall see cases where gene expression is regulated at essentially every step from the gene to its product, the most common step at which regulation impinges is the initiation of transcription—the focus of this chapter. There are two reasons why this might make sense. First, transcription initiation is the most energetically efficient step to regulate. By this we mean that deciding whether or not to express a gene at the first step ensures that no energy or resources are wasted making, for example, part or all of an mRNA that will not then be used (e.g., be translated). Second, regulation at this first step is easier to do well. There is only a single copy of each gene (in a haploid genome), and so typically only a single promoter on a single DNA molecule must be regulated to control expression of a given gene. In contrast, to regulate that gene at the point of translation, for example, each of several mRNA molecules must be acted on.

Why then is not all regulation focused on the step of transcription initiation? Regulating later steps can have two advantages. First, it allows for more inputs: if a gene is regulated at more than one step, more signals can modulate its expression, or the same signals can do so even more effectively. Second, regulation at steps later than transcription initiation can reduce the response time. Thus, consider again the example of translational regulation (see Fig. 14-43 for an example). If a signal relieves repression of this step, the protein product encoded by the gene will be produced immediately upon receipt of that signal. This reduced response time might obviously be advantageous in some situations. But, as we have said, it is the initiation of transcription that is most often regulated, and we now consider, in general terms, how activators and repressors regulate transcription initiation (see Interactive Animation 16-1).

WEB ANIMATION

Many Promoters Are Regulated by Activators That Help RNA Polymerase Bind DNA and by Repressors That Block That Binding

At many promoters, in the absence of regulatory proteins, RNA polymerase binds only weakly. This is because one or more of the promoter elements discussed above is absent or imperfect. When polymerase does occasionally bind, however, it spontaneously undergoes a transition to the open complex and initiates transcription. This gives a low level of **constitutive** expression called the **basal** level. Binding of

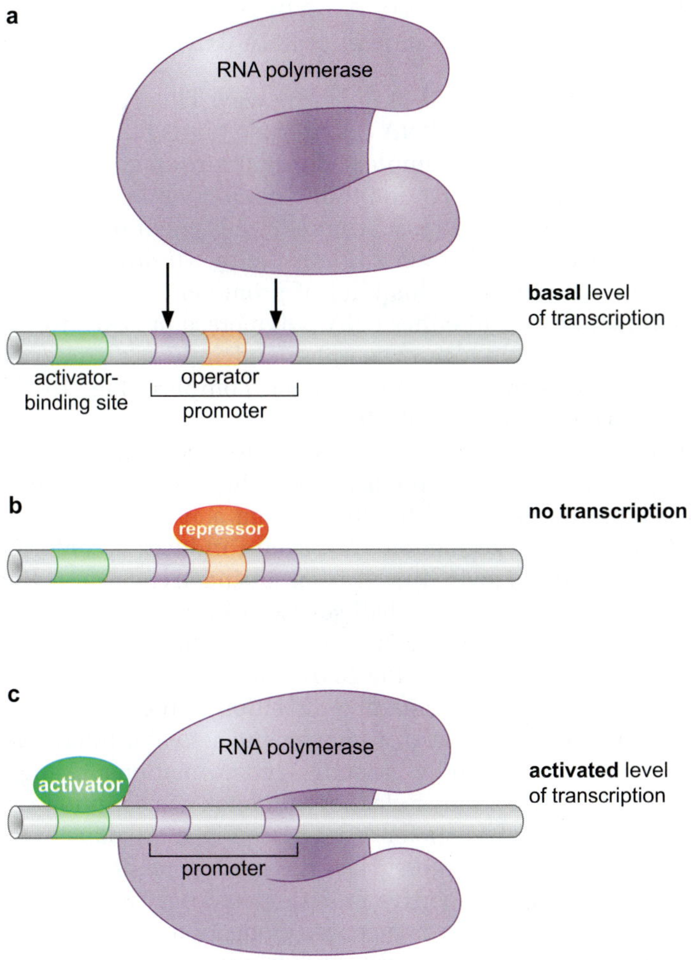

FIGURE 16-1 Activation by recruitment of RNA polymerase. (a) In the absence of both activator and repressor, RNA polymerase occasionally binds the promoter spontaneously and initiates a low level (basal level) of transcription. (b) Binding of the repressor to the operator sequence blocks binding of RNA polymerase and so inhibits transcription. (c) Recruitment of RNA polymerase by the activator gives high levels of transcription. RNA polymerase is shown recruited in the closed complex (see Fig. 12-13). It then spontaneously isomerizes to the open complex and initiates transcription. If both the repressor and activator are present and functional, the action of the repressor typically overcomes that of the activator. (This case is not shown in the figure.)

RNA polymerase is the rate-limiting step in this case (Fig. 16-1a).

To control expression from such a promoter, a repressor need only bind to a site overlapping the region bound by polymerase. In that way, the repressor blocks polymerase binding to the promoter, thereby preventing transcription (Fig. 16-1b), although it is important to note that repression can work in other ways as well. The site on DNA where a repressor binds is called an **operator**.

To activate transcription from this promoter, an activator can just help the polymerase bind the promoter. Typically, this is achieved as follows: the activator uses one surface to bind to a site on the DNA near the promoter; with another surface, the activator simultaneously interacts with RNA polymerase, bringing the enzyme to the promoter (Fig. 16-1c). This mechanism, often called **recruitment**, is an example of **cooperative binding** of proteins to DNA (a process we describe in more detail later, particularly in Box 16-3). The interactions between the activator and polymerase, and between activator and DNA, serve merely "adhesive" roles: the enzyme is active and the activator simply brings it to the nearby promoter. Once there, it spontaneously isomerizes to the open complex and initiates transcription.

The *lac* genes of *Escherichia coli* are transcribed from a promoter that is regulated by an activator and a repressor working in the simple way outlined above. We describe this case in detail later in the chapter.

Some Activators and Repressors Work by Allostery and Regulate Steps in Transcriptional Initiation after RNA Polymerase Binding

Not all promoters are limited in the same way. Thus, consider another class of promoter in which RNA polymerase binds efficiently unaided and forms a stable closed complex. But that closed complex does not spontaneously undergo transition to the open complex (Fig. 16-2a). At this promoter, an activator must stimulate the transition from a closed to open complex, since that transition is the rate-limiting step.

Activators that stimulate this kind of promoter work by triggering a conformational change in either RNA polymerase or DNA; that is, they interact with the stable closed complex and induce a conformational change that causes transition to the open complex (Fig. 16-2b). This mechanism is an example of **allostery**.

In Chapter 5, we encountered allostery as a general mechanism for controlling the activities of proteins. One of the examples we considered there was a protein (a cyclin) binding to, and activating, a kinase (Cdk) involved in cell cycle regulation. The cyclin induces a conformational change in the kinase, switching it from an inactive to an active state (Fig. 5-28). In this chapter, we shall see two examples of transcriptional activators working by allostery. In one case (at the *glnA* promoter), the activator (NtrC) interacts with the RNA polymerase bound in a closed complex at the promoter, stimulating transition to the open complex. In the other example (at the *merT* promoter), the activator (MerR) achieves the same effect but does so by inducing a conformational change in the promoter DNA. In still another class of promoter, transcription initiation is limited at the step of promoter escape (see Fig. 12-3). One example of such a promoter directs expression of the *malT* gene. In the absence of an activator, it undergoes abortive initiation, and only in the presence of an activator will it efficiently escape into elongation.

In a similar vein, repressors can work in ways other than just blocking the binding of RNA polymerase. For example, some repressors interact with polymerase at the promoter and inhibit transition to the open complex, or promoter escape. We consider examples of these later in the chapter (e.g., the Gal repressor).

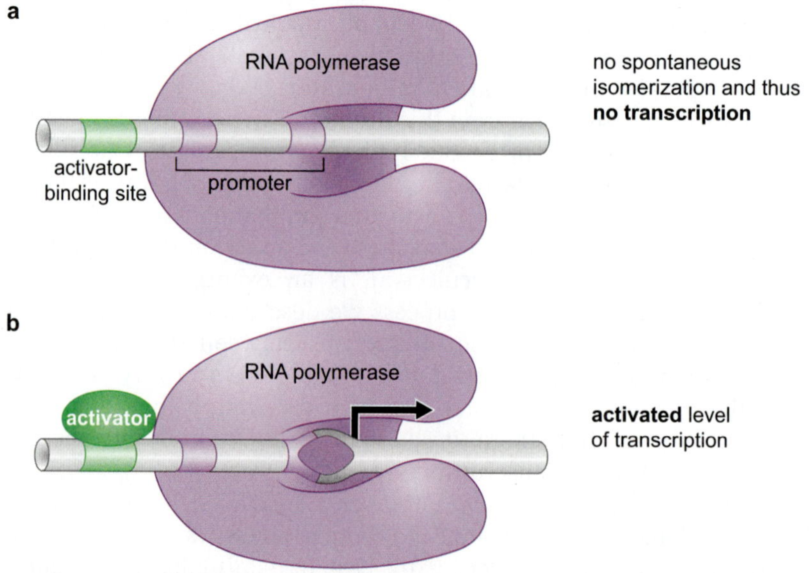

FIGURE 16-2 Allosteric activation of RNA polymerase. (a) Binding of RNA polymerase to the promoter in a stable closed complex. (b) The activator interacts with polymerase to trigger transition to the open complex and high levels of transcription. The representations of the closed and open complexes are shown diagrammatically; for a more complete description of those states, see Chapter 12.

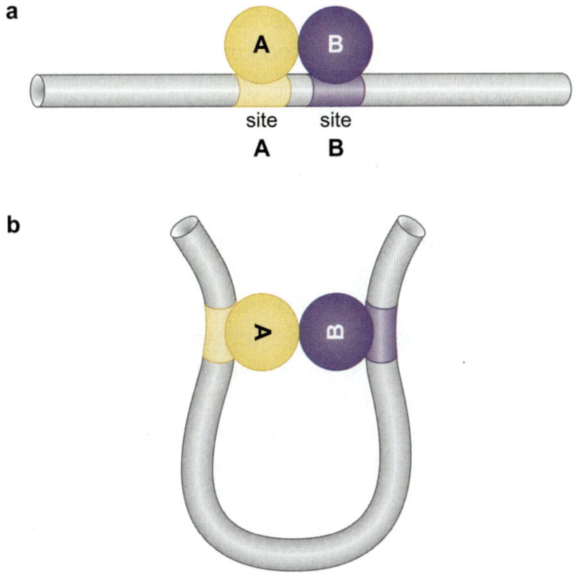

FIGURE 16-3 Interactions between proteins bound to DNA. (a) Cooperative binding of proteins to adjacent sites. (b) Cooperative binding of proteins to separated sites.

Action at a Distance and DNA Looping

Thus far we have tacitly assumed that DNA-binding proteins that interact with each other bind to adjacent sites (e.g., RNA polymerase and activator in Figs. 16-1 and 16-2). This is often the case. But some proteins interact with each other even when bound to sites well separated on the DNA. To accommodate this interaction, the DNA between the sites loops out, bringing the sites into proximity with one another (Fig. 16-3).

We will encounter examples of this kind of interaction in bacteria. Indeed, one of the activators we have already mentioned (NtrC) activates "from a distance": its binding sites are normally located about 150 bp upstream of the promoter, and the activator works even when those sites are placed further away (a kilobase or more). We will also consider repressors that interact to form DNA loops of up to 3 kb. In the next chapter—on eukaryotic transcriptional regulation—we are faced with more numerous and more dramatic examples of this "action at a distance."

Distant DNA sites can be brought closer together to help loop formation. In bacteria, for example, there are cases in which a protein binds between an activator-binding site and the promoter and helps the activator interact with polymerase by bending the DNA (Fig. 16-4). Such

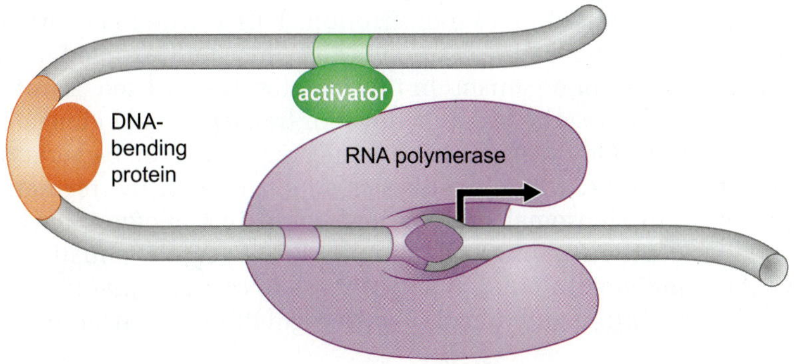

FIGURE 16-4 A DNA-bending protein can facilitate interaction between distantly bound DNA-binding proteins. A protein that bends DNA binds to a site between the activator-binding site and the promoter. If the direction of the bend is favorable, this action brings the two sites closer together in space and thereby helps the interaction between the DNA-bound activator and polymerase.

"architectural" proteins facilitate interactions between proteins in other processes as well (e.g., site-specific recombination; see Chapter 11).

Cooperative Binding and Allostery Have Many Roles in Gene Regulation

We have already pointed out that gene activation can be mediated by simple cooperative binding: the activator interacts simultaneously with DNA and with polymerase and so recruits the enzyme to the promoter. We have also described how activation can, in other cases, be mediated by allosteric events: an activator interacts with polymerase already bound to the promoter and, by inducing a conformational change in the enzyme or the promoter, stimulates transcription initiation. Both cooperative binding and allostery have additional roles in gene regulation.

For example, groups of regulators often bind DNA cooperatively; that is, two or more activators and/or repressors interact with each other and with DNA and thereby help each other bind near a gene they all regulate. As we shall see, this kind of interaction can produce sensitive switches that allow a gene to go from completely off to fully on in response to only small changes in conditions. Cooperative binding of activators can also serve to integrate signals; that is, some genes are activated only when multiple signals (and thus multiple regulators) are simultaneously present. A particularly striking and well-understood example of cooperativity in gene regulation is provided by bacteriophage λ. The basic mechanism and consequences of cooperative binding are considered in more detail when we discuss that example later in the chapter and also in Box 16-3.

Allostery, for its part, is not only a mechanism of gene activation, but also often the way regulators are controlled by their specific signals. Thus, a typical bacterial regulator can adopt two conformations: in one, it can bind DNA; in the other, it cannot. Binding of a signal molecule locks the regulatory protein in one or another conformation, thereby determining whether or not it can act. An example of this was seen in Chapter 5 (Fig. 5-26), where we also considered the basic mechanism of allostery in some detail; in this and the next chapter, we will see several examples of allosteric control of regulators by their signals.

Antitermination and Beyond: Not All of Gene Regulation Targets Transcription Initiation

As stated at the beginning of this chapter, the bulk of gene regulation takes place at the initiation of transcription. This is true in eukaryotes just as it is in bacteria. But regulation is certainly not restricted to that step in either class of organism. In this chapter, we will see examples in bacteria of gene regulation at the level of transcriptional elongation and termination. Other examples of gene regulation in bacteria are found in Chapter 14, where we discuss an example of the regulation of translation of ribosomal protein genes, and in Chapter 18, where we consider cases involving regulation by RNAs (e.g., attenuation, riboswitches, and small RNAs). Some of these RNA cases involve regulation of transcription and others involve regulation of translation.

REGULATION OF TRANSCRIPTION INITIATION: EXAMPLES FROM PROKARYOTES

Having outlined basic principles of transcriptional regulation, we turn to some examples that show these principles in action in real cases. First, we consider the genes involved in lactose metabolism in *E. coli*. Here, we see how an activator and a repressor regulate expression in response to two signals. We also describe some of the experiments that reveal how these regulators work.

An Activator and a Repressor Together Control the *lac* Genes

The three *lac* genes—*lacZ*, *lacY*, and *lacA*—are arranged adjacently on the *E. coli* genome and are together called the ***lac* operon** (Fig. 16-5). The *lac* promoter, located at the 5′ end of *lacZ*, directs transcription of all three genes as a single mRNA (called a polycistronic message because it includes more than one gene); this mRNA is translated to give the three protein products. The *lacZ* gene encodes the enzyme β-galactosidase, which cleaves the sugar lactose into galactose and glucose, both of which are used by the cell as energy sources. The *lacY* gene encodes the lactose permease, a protein that inserts into the cell membrane and transports lactose into the cell. The *lacA* gene encodes thiogalactoside transacetylase, which rids the cell of toxic thiogalactosides that also get transported in by *lacY*.

These genes are expressed at high levels only when lactose is available, and glucose—the preferred energy source—is not. Two regulatory proteins are involved: one is an activator called **CAP**, and the other is a repressor called the **Lac repressor**. The Lac repressor is encoded by the *lacI* gene, which is located near the other *lac* genes, but transcribed from its own (constitutively expressed) promoter. The name CAP stands for catabolite activator protein, but this activator is also known as CRP (for cAMP receptor protein, for reasons that will be explained later). The gene encoding CAP is located elsewhere on the bacterial chromosome, not linked to the *lac* genes. Both CAP and the Lac repressor are DNA-binding proteins and each binds to a specific site on DNA at or near the *lac* promoter (the CAP site and the operator, respectively; see Fig. 16-5).

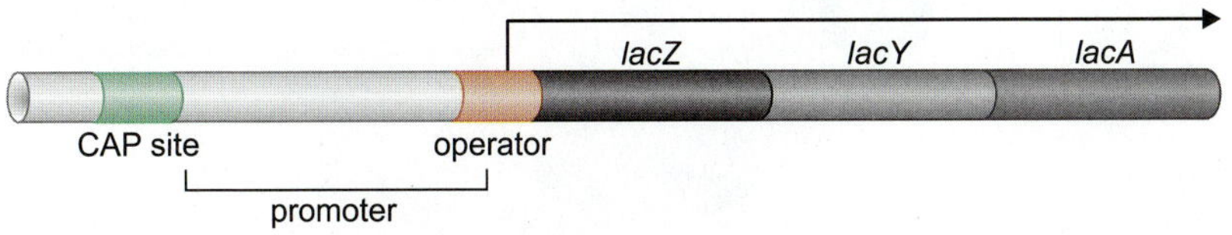

FIGURE 16-5 The *lac* operon. The three genes (*lacZ, lacY,* and *lacA*) are transcribed as a single mRNA from the promoter (as indicated by the arrow). The CAP site and the operator (the site bound by Lac repressor) are each about 20 bp. The operator lies within the region bound by RNA polymerase at the promoter, and the CAP site lies just upstream of the promoter (see Fig. 16-8 for more details of the relative arrangements of these binding sites and the text for a description of the proteins that bind to them). The picture is simplified in that there are two additional, weaker, *lac* operators located nearby (see Fig. 16-13), but we do not need to consider those at present.

Each of these regulatory proteins responds to one environmental signal and communicates it to the *lac* genes. Thus, CAP mediates the effect of glucose, whereas Lac repressor mediates the lactose signal. This regulatory system works in the following way (and as shown in Fig. 16-6). Lac repressor can bind DNA and repress transcription only in the absence of lactose. In the presence of that sugar, the repressor is inactive and the genes derepressed (expressed). CAP can bind DNA and activate the *lac* genes only in the *absence* of glucose. Thus, the combined effect of these two regulators ensures that the genes are expressed at significant levels only when lactose is present and glucose absent.

CAP and Lac Repressor Have Opposing Effects on RNA Polymerase Binding to the *lac* Promoter

As we have seen, the site bound by the Lac repressor is called the ***lac* operator**. This 21-bp sequence is twofold symmetric and is recognized by two subunits of Lac repressor, one binding to each half-site (see

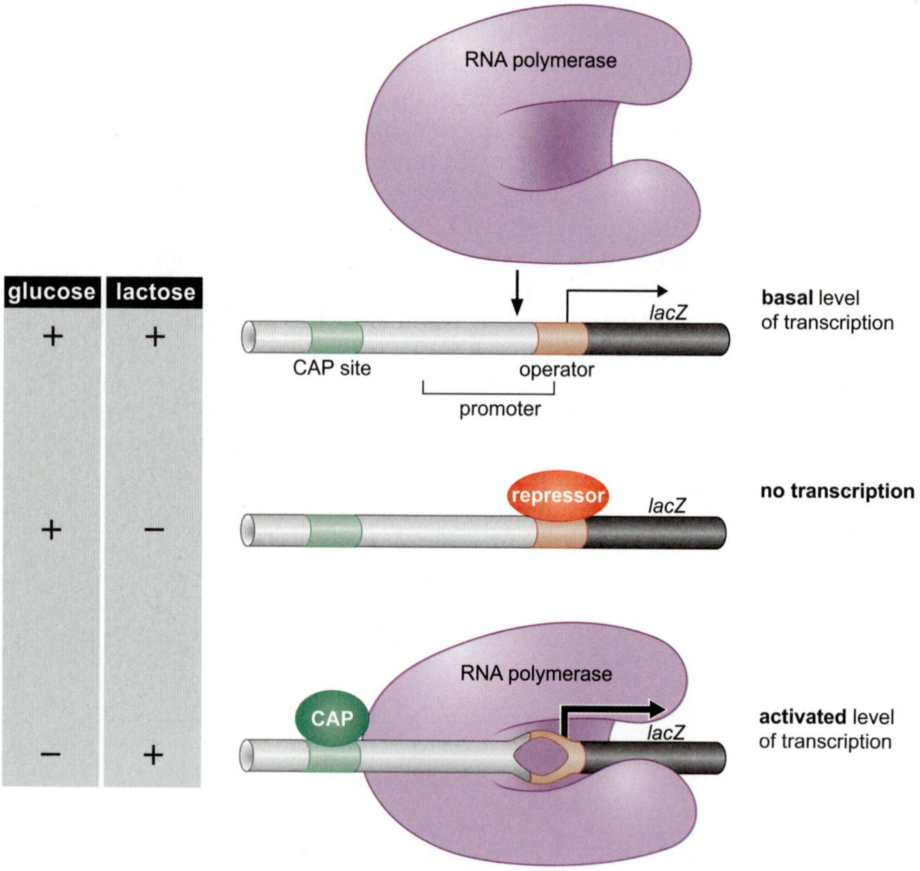

FIGURE 16-6 Expression of the *lac* genes. The presence or absence of the sugars lactose and glucose control the level of expression of the *lac* genes. High levels of expression require the presence of lactose (and hence the absence of functional Lac repressor) and absence of the preferred energy source, glucose (and hence presence of the activator CAP). When bound to the operator, Lac repressor excludes polymerase whether or not active CAP is present. CAP and Lac repressor are shown as single units, but CAP actually binds DNA as a dimer, and Lac repressor binds as a tetramer (see Fig. 16-13). CAP recruits polymerase to the *lac* promoter where it spontaneously undergoes isomerization to the open complex (the state shown in the bottom line).

Fig. 16-7). We discuss that binding in more detail later in this chapter, in the section "CAP and Lac Repressor Bind DNA Using a Common Structural Motif." How does the repressor, when bound to the operator, repress transcription?

The *lac* operator overlaps the promoter, and so the repressor bound to the operator physically prevents RNA polymerase from binding to the promoter and thus initiating RNA synthesis (see Fig. 16-8). Protein-binding sites in DNA can be identified, and their location mapped, using DNA-footprinting and gel-mobility assays as described in Chapter 21.

As we have seen, RNA polymerase binds the *lac* promoter poorly in the absence of CAP, even when there is no active repressor present. This is because the sequence of the −35 region of the *lac* promoter is not optimal for its binding, and the promoter lacks an UP-element (see Fig. 12-5, Box 12-1, and Fig. 16-8). This is typical of promoters that are controlled by activators.

CAP binds as a dimer to a site similar in length to that of the *lac* operator, but different in sequence. This site is located some 60 bp upstream of the start site of transcription (see Fig. 16-8). When CAP binds to that site, the activator helps polymerase bind to the promoter by interacting with the enzyme and recruiting it to the promoter (see Fig. 16-6). This cooperative binding stabilizes the binding of polymerase to the promoter. We now look at CAP-mediated activation in more detail.

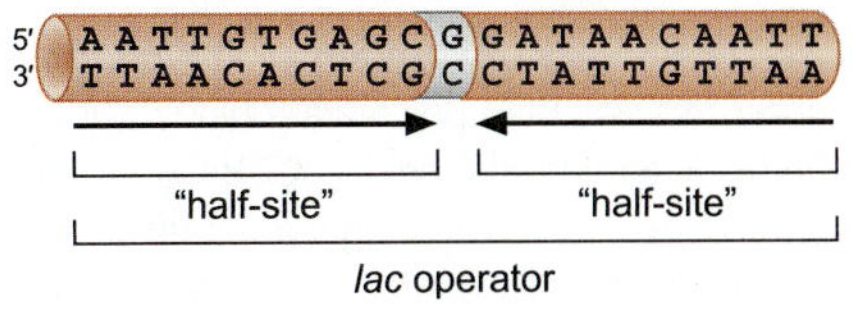

FIGURE 16-7 The symmetric half-sites of the *lac* operator.

CAP Has Separate Activating and DNA-Binding Surfaces

Various experiments support the view that CAP activates the *lac* genes by simple recruitment of RNA polymerase. Mutant versions of CAP have been isolated that bind DNA but do not activate transcription. The existence of these so-called **positive control (*pc*)** mutants demonstrates that to activate transcription, the activator must do more than simply bind DNA near the promoter. Thus, activation is not caused by, for example, the activator changing local DNA structure. The amino acid substitutions in the positive control mutants identify the region of CAP that touches polymerase, called the **activating region.**

Where does the activating region of CAP touch RNA polymerase when activating the *lac* genes? This site is revealed by mutant forms of polymerase that can transcribe most genes normally, but cannot be activated by CAP at the *lac* genes. These mutants have amino acid

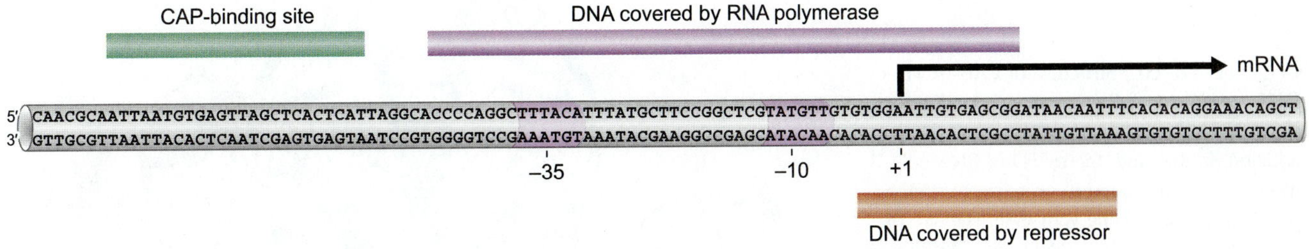

FIGURE 16-8 The control region of the *lac* operon. The nucleotide sequence and organization of the *lac* operon control region are shown. The colored bars above and below the DNA show regions covered by RNA polymerase and the regulatory proteins. Note that the Lac repressor covers more DNA than that sequence defined as the minimal operator-binding site and RNA polymerase more than that defined by the sequences that make up the promoter.

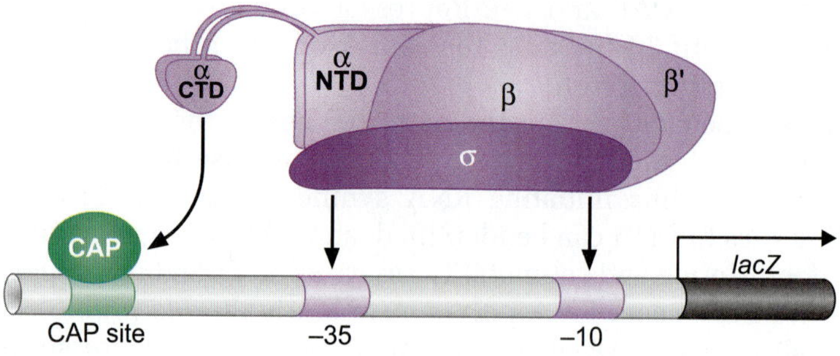

FIGURE 16-9 **Activation of the *lac* promoter by CAP.** RNA polymerase binding at the *lac* promoter with the help of CAP. CAP is recognized by the CTDs of the α subunits. The αCTDs also contact DNA, adjacent to the CAP site, when interacting with CAP. As discussed in Chapter 12, we use this representation of RNA polymerase when indicating specific points of contact between an activator and its target site on polymerase, or between regions of polymerase and the promoter.

substitutions in the **carboxy-terminal domain** (CTD) of the α **subunit** of RNA polymerase. As we saw in Chapter 12, this domain is attached to the amino-terminal domain (NTD) of α by a flexible linker. The αNTD is embedded in the body of the enzyme, but the αCTD extends out from it and binds the UP-element of the promoter (when that element is present) (see Fig. 12-7).

At the *lac* promoter, where there is no UP-element, αCTD binds to CAP and adjacent DNA instead (Fig. 16-9). This picture is supported by a crystal structure of a complex containing CAP, αCTD, and a DNA oligonucleotide duplex containing a CAP site and an adjacent UP-element (Fig. 16-10; see also Structural Tutorial 16-1). In Box 16-1, Activator Bypass Experiments, we describe an experiment showing that activation of the *lac* promoter requires no more than polymerase recruitment.

WEB STRUCTURAL TUTORIAL

Having seen how CAP activates transcription at the *lac* operon, and how the Lac repressor counters that effect, we now look more closely at how these regulators recognize their DNA-binding sites.

CAP and Lac Repressor Bind DNA Using a Common Structural Motif

X-ray crystallography has been used to determine the structural basis of DNA binding for a number of bacterial activators and repressors, including CAP and the Lac repressor. Although the details differ, the basic mechanism of DNA recognition is similar for most bacterial regulators.

In the typical case, the protein binds as a homodimer to a site that is

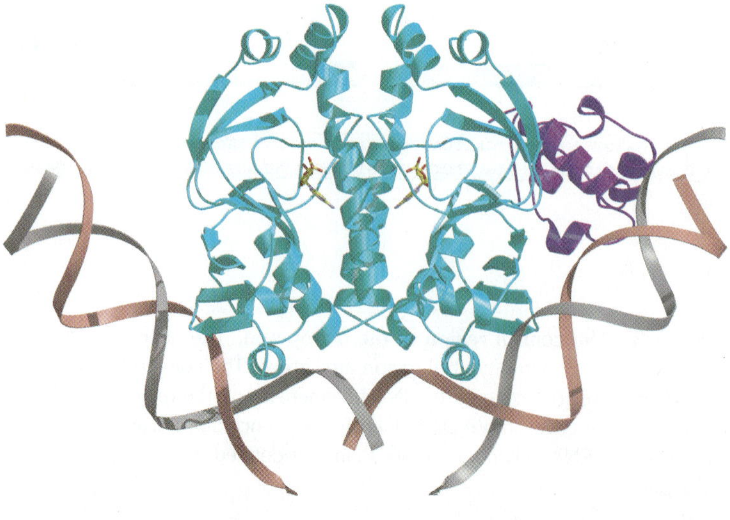

FIGURE 16-10 **Structure of CAP–αCTD–DNA complex.** CAP is shown bound as a dimer to its site just as we saw in Fig. 5-18. In addition, in this case, the αCTD of RNA polymerase is shown bound to an adjacent stretch of DNA and interacting with CAP. The site of interaction on each protein involves the residues identified genetically. In this figure, CAP is shown in turquoise and the αCTD of polymerase in purple. One molecule of cAMP is shown bound to each monomer of CAP. (Benoff B. et al. 2002. *Science* 297: 1562.) Image prepared with MolScript, BobScript, and Raster 3D.

KEY EXPERIMENTS

Box 16-1 Activator Bypass Experiments

If an activator has only to recruit polymerase to the gene, then other methods of bringing the polymerase to the gene should work just as well. This turns out to be true of the *lac* genes, as shown by the following experiments (Box 16-1 Fig. 1).

In one experiment, another protein–protein interaction is used in place of that between CAP and polymerase. This is done by taking two proteins known to interact with each other, attaching one to a DNA-binding domain, and, with the other, replacing the carboxy-terminal domain of the polymerase α subunit (αCTD). The modified polymerase can be activated by the makeshift "activator" as long as the appropriate DNA-binding site is introduced near the promoter. In another experiment, the αCTD of polymerase is replaced with a DNA-binding domain (e.g., that of CAP). This modified polymerase efficiently initiates transcription from the *lac* promoter in the absence of any activator, as long as the appropriate DNA-binding site is placed nearby. A third experiment is even simpler: polymerase can transcribe the *lac* genes at high levels in vitro in the absence of any activator if the enzyme is present at high concentration. So we see that either recruiting polymerase artificially or supplying it at a high concentration is sufficient to produce activated levels of expression of the *lac* genes. These experiments are consistent with the activator having only to help polymerase bind to the promoter. For an explanation of why simply increasing the concentration of a protein (e.g., RNA polymerase) helps it bind to a site on DNA (in this case the promoter), see Box 16-3. The results discussed in this box would not be expected if the activator had to induce a specific allosteric change in polymerase to activate transcription.

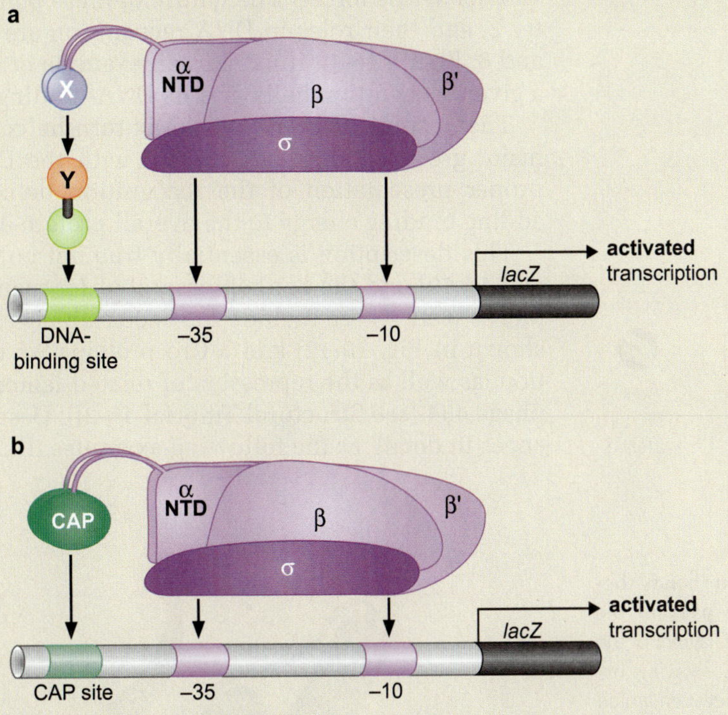

BOX 16-1 FIGURE 1 **Two activator bypass experiments.** (a) The αCTD is replaced by a protein X, which interacts with protein Y. Protein Y is fused to a DNA-binding domain, and the site recognized by that domain is shown placed near the *lac* genes. (b) The αCTD is replaced by the DNA-binding portion of CAP.

an inverted repeat (or near repeat). One monomer binds each half-site, with the axis of symmetry of the dimer lying over that of the binding site (as for the Lac repressor, Fig. 16-7). Recognition of specific DNA sequences is achieved using a conserved region of secondary structure called a **helix-turn-helix** (Fig. 16-11). This domain is composed of two α helices, one of which—the **recognition helix**—fits into the major groove of the DNA. As discussed in Chapter 5, an α helix is just the right size to fit into the major groove, allowing amino acid residues on

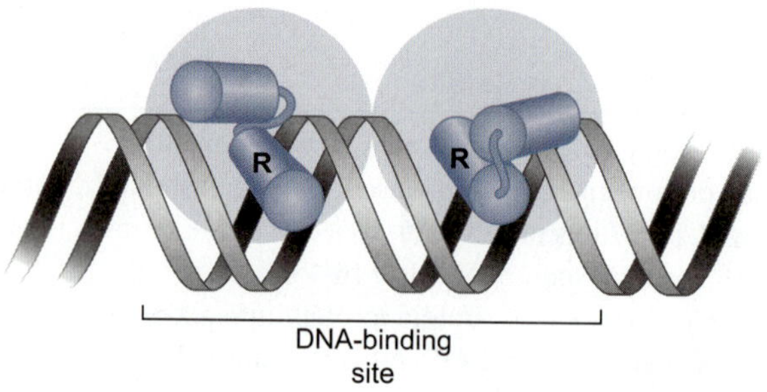

FIGURE 16-11 Binding of a protein with a helix-turn-helix domain to DNA. The protein, as is typically the case, binds as a dimer, and the two subunits are indicated by the shaded circles. The helix-turn-helix motif on each monomer is indicated; the "recognition helix" is labeled R.

its outer face to interact with chemical groups on the edges of base pairs. And recall that in Chapter 6, we saw how each base pair presents a characteristic pattern of hydrogen bonding acceptors and donors (Fig. 6-10). Thus, a protein can distinguish different DNA sequences in this way without unwinding the DNA duplex (see Fig. 16-11).

The contacts made between the amino acid side chains protruding from the recognition helix and the edges of the bases can be mediated by direct H bonds, indirect H bonds (bridged by water molecules), or van der Waals forces. The nature of these bonds is discussed in Chapter 3, and their roles in DNA recognition are discussed in Chapters 5 and 6. Figure 16-12 illustrates an example of the interactions made by a given recognition helix and its DNA-binding site.

The second helix of the helix-turn-helix domain sits across the major groove and makes contact with the DNA backbone, ensuring proper presentation of the recognition helix and at the same time adding binding energy to the overall protein–DNA interaction.

This description is essentially true not only for CAP (see Fig. 5-18 and 16-10) and the Lac repressor, but for many other bacterial regulators as well. These include the bacteriophage λ repressor (the example shown in Fig. 16-12) and λ Cro proteins we encounter in a later section, as well as the repressors of related lambdoid phages (e.g., that of phage 434 [see Structural Tutorial 16-2]). Despite this, there are differences in detail, as the following examples illustrate.

WEB STRUCTURAL TUTORIAL

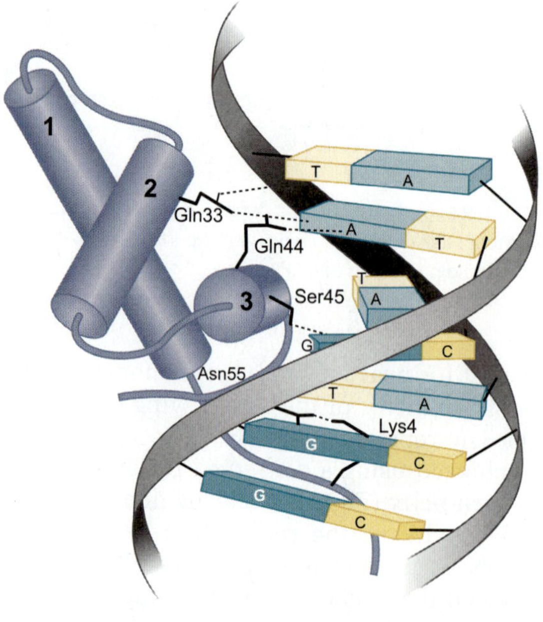

FIGURE 16-12 Hydrogen bonds between λ repressor and base pairs in the major groove of its operator. Diagram of the λ repressor–operator complex, showing hydrogen bonds (in dotted lines) between amino acid side chains and bases in the consensus half-site. We will see more of this regulator and its function later in this chapter. Only the relevant amino acid side chains are shown. In addition to Gln44 and Ser45 in the recognition helix, Asn55 in the loop following the recognition helix also makes contact with a specific base. Furthermore (and unusual to this case, see later in the text), Lys4 in the amino-terminal arm of the protein makes a contact in the major groove on the opposite face of the DNA helix. Gln33 contacts the backbone. (Redrawn, with permission, from Jordan S. and Pabo C. 1988. *Science* 242: 896, Fig. 3B. © AAAS.)

- Lac repressor binds as a tetramer, not a dimer. Nevertheless, each operator is contacted by only two of these subunits. Thus, the different oligomeric form does not alter the mechanism of DNA recognition. The other two monomers within the tetramer can bind one of two other *lac* operators, located 400 bp downstream and 90 bp upstream of the primary operator. In such cases, the intervening DNA loops out to accommodate the reaction (Fig. 16-13).
- In some cases, other regions of the protein, outside the helix-turn-helix domain, also interact with the DNA. The λ repressor, for example, makes additional contacts using amino-terminal arms. These reach around the DNA and interact with the minor groove on the back face of the helix (see Fig. 16-12).
- In many cases, binding of the protein does not alter the structure of the DNA. In some cases, however, various distortions are seen in the protein–DNA complex. For example, CAP induces a dramatic bend in the DNA, partially wrapping it around the protein. This is caused by other regions of the protein, outside the helix-turn-helix domain, interacting with sequences outside the operator. In other cases, binding results in twisting of the operator DNA.

Not all prokaryotic repressors bind using a helix-turn-helix. A few have been described that employ quite different approaches. A striking example is the Arc repressor from phage P22 (a phage related to λ but that infects *Salmonella*). The Arc repressor binds as a dimer to an inverted repeat operator, but instead of an α helix, it recognizes its binding site using two antiparallel β strands inserted into the major groove.

The Activities of Lac Repressor and CAP Are Controlled Allosterically by Their Signals

When lactose enters the cell, it is converted to allolactose. It is allolactose (rather than lactose itself) that controls the Lac repressor. Paradoxically, the conversion of lactose to allolactose is catalyzed by β-galactosidase, itself encoded by one of the *lac* genes. How is this possible?

The answer is that expression of the *lac* genes is leaky: even when they are repressed, an occasional transcript gets made. This happens

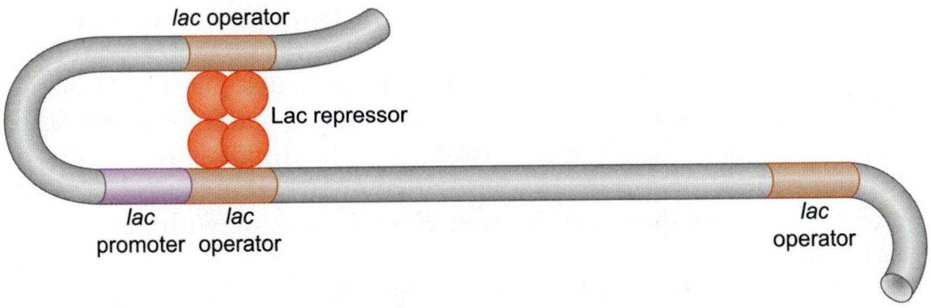

FIGURE 16-13 Lac repressor binds as a tetramer to two operators. The loop shown is between the Lac repressor bound at the primary operator and the upstream auxiliary one. A similar loop can alternatively form with the downstream operator. The primary operator—the one shown against the promoter—is the operator referred to in discussion of regulation of *lac* gene expression. In this figure, each repressor dimer is shown as two circles, rather than as a single oval (as used in earlier figures) to emphasize its oligomeric structure.

because every so often, RNA polymerase will manage to bind the promoter in place of the Lac repressor. This leakiness ensures that there is a low level of β-galactosidase in the cell even in the absence of lactose, and so there is enzyme poised to catalyze the conversion of lactose to allolactose.

Allolactose binds to the Lac repressor and triggers a change in the shape (conformation) of that protein. In the absence of allolactose, the repressor is present in a form that binds its site on DNA (and so keeps the *lac* genes switched off). Once allolactose has altered the shape of the repressor, the protein can no longer bind DNA, and so the *lac* genes are no longer repressed. In Chapter 5, we described the structural basis of this allosteric change in the Lac repressor (Fig. 5-26). An important point to emphasize is that allolactose binds to a part of the Lac repressor distinct from its DNA-binding domain.

CAP activity is regulated in a similar manner. Glucose lowers the intracellular concentration of a small molecule, cAMP. This molecule is the allosteric effector for CAP: only when CAP is complexed with cAMP does the protein adopt a conformation that binds DNA (thus also explaining CAP's alternative name, CRP). And so, only when glucose levels are low (and cAMP levels high) does CAP bind DNA and activate the *lac* genes. The part of CAP that binds the effector, cAMP, is separate from the part of the protein that binds DNA.

The *lac* operon of *E. coli* is one of the two systems used by French biologists François Jacob and Jacques Monod in formulating the early ideas about gene regulation. In Box 16-2, Jacob, Monod, and the Ideas behind Gene Regulation, we provide a brief description of those early studies and why the ideas they generated have proved so influential.

Combinatorial Control: CAP Controls Other Genes As Well

The *lac* genes provide an example of **signal integration**: their expression is controlled by two signals, each of which is communicated to the genes via a single regulator—the Lac repressor and CAP, respectively.

Consider another set of *E. coli* genes, the *gal* genes. These genes encode enzymes involved in galactose metabolism. As with the *lac* genes, the *gal* genes are only expressed when their substrate sugar, in this case galactose, is present, and the preferred energy source, glucose, is absent. Again, analogous to *lac*, the two signals are communicated to the genes via two regulators—an activator and a repressor. The repressor, encoded by the *galR* gene, mediates the effects of the inducer galactose, but the activator of the *gal* genes is again CAP. Thus, a regulator (CAP) works together with different repressors at different genes. This is an example of **combinatorial control**. In fact, CAP acts at more than 100 genes in *E. coli,* working with an array of partners.

Combinatorial control is a characteristic feature of gene regulation. Thus, when the same signal controls multiple genes, it is typically communicated to each of those genes by the same regulatory protein. This regulator will be communicating just one of perhaps several signals involved in regulating each gene; the other signals, different in most cases, will each be mediated by a separate regulator. More complex organisms—higher eukaryotes in particular—tend to have more signal integration, and there we will see greater and more elaborate examples of combinatorial control (Chapter 17).

KEY EXPERIMENTS

Box 16-2 Jacob, Monod, and the Ideas behind Gene Regulation

The idea that the expression of a gene can be controlled by the product of another gene—that there exist regulatory genes the sole function of which is regulating the expression of other genes—was one of the great insights from the early years of molecular biology. It was proposed by a group of scientists working in Paris in the 1950s and early 1960s, in particular François Jacob and Jacques Monod. They sought to explain two apparently unrelated phenomena: the appearance of β-galactosidase in *E. coli* grown in lactose, and the behavior of the bacterial virus (bacteriophage) λ upon infection of *E. coli*. Their work culminated in publication of their operon model in 1961 (and the 1965 Nobel Prize in Physiology or Medicine, which they shared with their colleague, Andre Lwoff).

It is difficult to appreciate the magnitude of their achievement now that we are so familiar with their ideas and have such direct ways of testing their models. To put it in perspective, consider what was known at the time they began their classic experiments: β-galactosidase activity appeared in *E. coli* cells only when lactose was provided in the growth medium. It was not clear that the appearance of this enzyme involved switching on expression of a gene. Indeed, one early explanation was that the cell contained a general (generic) enzyme and that enzyme took on whatever properties were required by the circumstances. Thus, when lactose was present, the generic enzyme took on the appropriate shape to metabolize lactose, using the sugar itself as a template!

Jacob, Monod, and their coworkers dissected the problem genetically. We will not go through their experiments in any detail, but a brief summary gives a taste of their ingenuity.

First, they isolated mutants of *E. coli* that made β-galactosidase irrespective of whether lactose was present (i.e., mutants in which the enzyme was produced **constitutively**). These mutants came in two classes: in one, the gene encoding the Lac repressor was inactivated; in the other, the operator site was defective. These two classes could be distinguished using a *cis-trans* test, as described below.

Jacob and Monod constructed partially diploid cells in which a section of the chromosome from a wild-type cell carrying the *lac* genes (i.e., the Lac repressor gene, *LacI*, the genes of the *lac* operon, and their regulatory elements) was introduced (on a plasmid called an F′) into a cell carrying a mutant version of the *lac* genes on its chromosome. (This genetic trick is described more fully in the Bacteria section in Chapter 22.) This transfer resulted in the presence of two copies of the *lac* genes in the cell, making it possible to test whether the wild-type copy could complement any given mutant copy. When the chromosomal genes were expressed constitutively because of a mutation in the *lacI* gene (encoding repressor), the wild-type copy on the plasmid restored repression (and inducibility); that is, β-galactosidase was once again only made when lactose was present (Box 16-2 Fig. 1). This result is gained because the repressor made from the wild-type *lacI* gene on the plasmid can diffuse to the chromosome (i.e., it can act in *trans*).

When the mutation causing constitutive expression of the chromosomal genes was in the *lac* operator, it could not be complemented in *trans* by the wild-type genes (Box 16-2 Fig. 2). The operator functions only in *cis* (i.e., it only acts on the genes directly linked to it on the same DNA molecule).

These and other results led Jacob and Monod to propose that genes were expressed from specific sites called promoters found at the start of the gene and that this expression was regulated by repressors that act through operator sites located on the DNA beside the promoter.

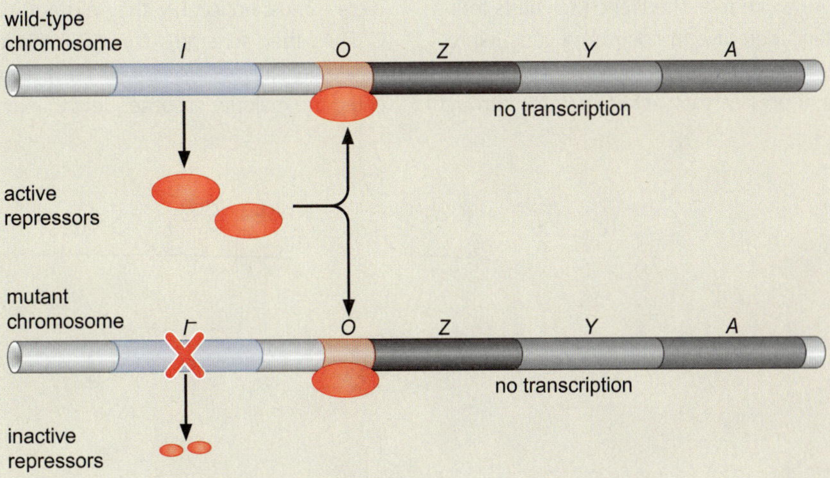

BOX 16-2 FIGURE 1 Partial diploid cells show that functional repressors work in *trans*. In the absence of lactose, the *lac* genes are not expressed, and thus no significant level of β-galactosidase is made in these cells.

BOX 16-2 **(*Continued*)**

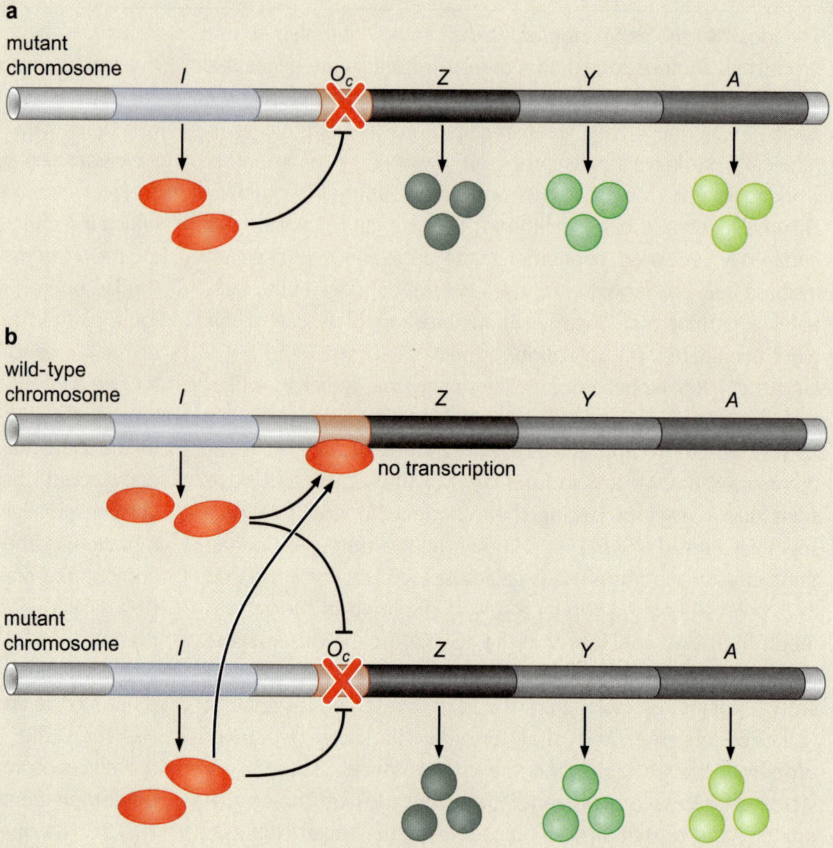

BOX 16-2 FIGURE 2 Partial diploid cells show that operators work only in *cis*. (a) Haploid cell containing mutant operator (O_c). (b) Partially diploid cell containing a normal operator (*O*) and a mutant operator (O_c). The *lac* genes (*Z, Y,* and *A*) attached to the mutant operator continue to be expressed constitutively even in the presence of a wild-type operator on another chromosome in the same cell. Thus, the operator only works in *cis*.

But these experiments with the *lac* system were not carried out in isolation; in parallel, Jacob and Monod did similar experiments on bacteriophage λ (a system we consider in detail later in this chapter). Bacteriophage λ can propogate through either of two life cycles. Which one is chosen depends on which of the relevant phage genes are expressed. The French scientists found they could isolate mutants defective in controlling gene expression in this system just as they had in the *lac* case. These mutations again defined a repressor that acted in *trans* through *cis*-acting operator sites. The similarity of these two regulatory systems (despite the very different biology) convinced Jacob and Monod that they had identified a fundamental mechanism of gene regulation and that their model would apply throughout nature. As we will see, although their description was not complete—most noticeably, they did not include activators (such as CAP) in their scheme—the basic model they proposed of *cis*-regulatory sites recognized by *trans*-regulatory factors has dominated the majority of subsequent thinking about gene regulation.

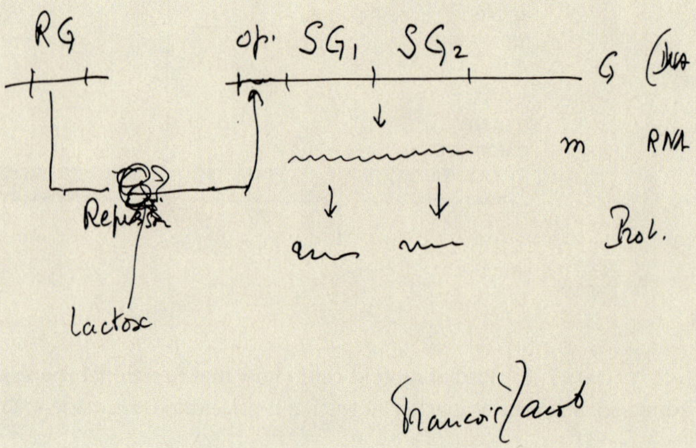

BOX 16-2 FIGURE 3 This drawing, showing the *lac* operon and its regulation, was rendered by François Jacob, 2002. (Courtesy of Jan Witkowski.)

Alternative σ Factors Direct RNA Polymerase to Alternative Sets of Promoters

Recall from Chapter 12 that it is the σ subunit of RNA polymerase that recognizes the promoter sequences (Fig. 12-6). The *lac* promoter that we have been discussing, along with the bulk of other *E. coli* promoters, is recognized by RNA polymerase bearing the σ^{70} subunit. But *E. coli* encodes several other σ subunits that can replace σ^{70} under certain circumstances and direct the polymerase to alternative promoters.

One of these alternatives is the heat shock σ factor, σ^{32}. Thus, when *E. coli* is subject to heat shock, the amount of this new σ factor increases in the cell, it displaces σ^{70} from a proportion of RNA polymerases, and it directs those enzymes to transcribe genes whose products protect the cell from the effects of heat shock. The level of σ^{32} is increased by two mechanisms: first, its translation is stimulated, that is, its mRNA is translated with greater efficiency after heat shock than it was before; and second, the protein is transiently stabilized. Another example of an alternative σ factor, σ^{54}, is considered in the next section. σ^{54} is associated with a small fraction of the polymerase molecules in the cell and directs that enzyme to genes involved in nitrogen metabolism.

Sometimes, a series of alternative sigmas directs a particular program of gene expression. Two examples are found in the bacterium *Bacillus subtilis*. We consider the most elaborate of these, which controls sporulation in that organism, in Chapters 19 and 20. The other we describe briefly here.

Bacteriophage SPO1 infects *B. subtilis,* where it grows lytically to produce progeny phage. This process requires that the phage expresses its genes in a carefully controlled order. That control is imposed on polymerase by a series of alternative σ factors. Thus, upon infection, the bacterial RNA polymerase (bearing the *B. subtilis* version of σ^{70}) recognizes so-called "early" phage promoters, which direct transcription of genes that encode proteins needed early in infection. One of these genes (called gene 28) encodes an alternative σ. This displaces the bacterial σ factor and directs the polymerase to a second set of promoters in the phage genome, those associated with the so-called "middle" genes. One of these genes, in turn, encodes the σ factor for the phage "late" genes (Fig. 16-14).

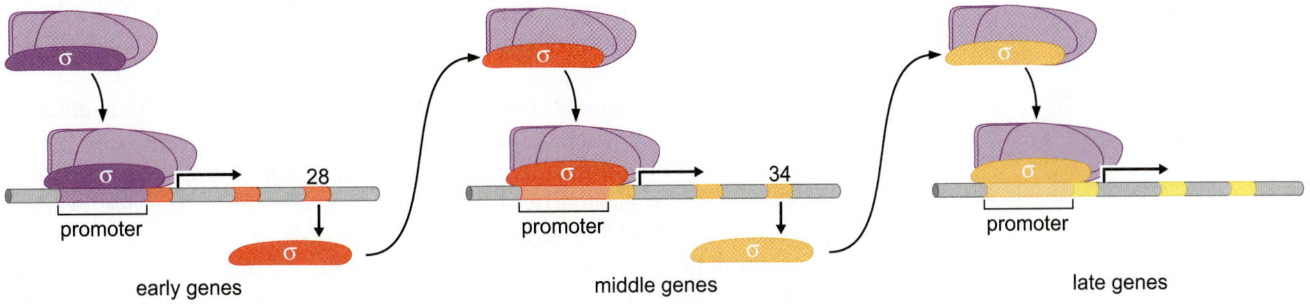

FIGURE 16-14 **Alternative σ factors control the ordered expression of genes in a bacterial virus.** The bacterial phage SPO1 uses three σ factors in succession to regulate expression of its genome. This ensures that viral genes are expressed in the order in which they are needed. (Adapted, with permission, from Alberts B. et al. 2002. *Molecular biology of the cell,* 4th ed., p. 415, Fig. 7-63. © Garland Science/Taylor & Francis LLC.)

NtrC and MerR: Transcriptional Activators That Work by Allostery Rather than by Recruitment

Although the majority of activators work by recruitment, there are exceptions. Two examples of activators that work not by recruitment but by allosteric mechanisms are NtrC and MerR. Recall what we mean by an allosteric mechanism. Activators that work by recruitment simply bring an active form of RNA polymerase to the promoter. In the case of activators that work by allosteric mechanisms, polymerase initially binds the promoter in an inactive complex. To activate transcription, the activator triggers an allosteric change in that complex.

NtrC controls expression of genes involved in nitrogen metabolism, such as the *glnA* gene. At the *glnA* gene, RNA polymerase is prebound to the promoter in a stable closed complex. The activator NtrC induces a conformational change in the enzyme, triggering transition to the open complex. Thus, the activating event is an allosteric change in RNA polymerase (see Fig. 16-2).

MerR controls a gene called *merT,* which encodes an enzyme that makes cells resistant to the toxic effects of mercury. MerR also acts on an inactive RNA polymerase–promoter complex. Like NtrC, MerR induces a conformational change that triggers open complex formation. In this case, however, the allosteric effect of the activator is on the DNA, rather than on the polymerase.

NtrC Has ATPase Activity and Works from DNA Sites Far from the Gene

As with CAP, NtrC has separate activating and DNA-binding domains and binds DNA only in the presence of a specific signal. In the case of NtrC, this signal is low nitrogen levels. Under these conditions, NtrC is phosphorylated by a kinase, NtrB, and, as a result, undergoes a conformational change that reveals the activator's DNA-binding domain. Once active, NtrC binds four sites located approximately 150 bp upstream of the promoter. NtrC binds to each of its sites as a dimer and, through protein–protein interactions between the dimers, binds to the four sites in a highly cooperative manner.

The form of RNA polymerase that transcribes the *glnA* gene contains the σ^{54} subunit. This enzyme binds to the *glnA* promoter in a stable closed complex in the absence of NtrC. Once active, NtrC (bound to its sites upstream) interacts directly with σ^{54}. This requires that the DNA between the activator-binding sites and the promoter form a loop to accommodate the interaction (Fig. 16-15). If the NtrC-binding sites are moved further upstream (as much as 1–2 kb), the activator can still work.

NtrC itself has an enzymatic activity—it is an ATPase. This activity provides the energy needed to induce a conformational change in polymerase. This conformational change triggers polymerase to initiate transcription. Specifically, it stimulates conversion of the stable inactive closed complex to an active open complex.

At some genes controlled by NtrC, there is a binding site for another protein, called IHF, located between the NtrC-binding sites and the promoter. Upon binding, IHF bends DNA; when the IHF-binding site, and hence the DNA bend, are in the correct register, this event increases activation by NtrC. The explanation is that by bending the DNA, IHF brings the DNA-bound activator closer to the promoter, helping the activator interact with the polymerase bound there (see Fig. 16-4; for a closer look at how IHF bends DNA, see Fig. 11-11).

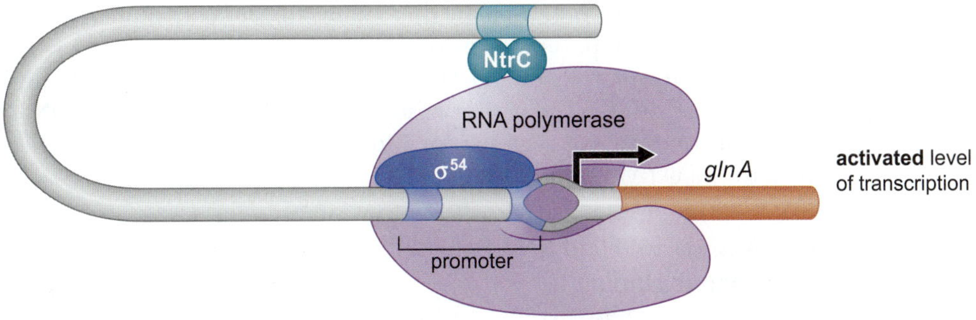

FIGURE 16-15 Activation by NtrC. The promoter sequence recognized by σ^{54}-containing holoenzyme is different from that recognized by σ^{70}-containing holoenzyme. Although not specified in the figure, NtrC contacts the σ^{54} subunit of polymerase. NtrC is shown as a dimer, but in fact forms a higher-order complex on DNA.

MerR Activates Transcription by Twisting Promoter DNA

When bound to a single DNA-binding site, in the presence of mercury, MerR activates the *merT* gene. As shown in Figure 16-16, MerR binds to a sequence located between the –10 and –35 regions of the *merT* promoter (this gene is transcribed by σ^{70}-containing polymerase). MerR binds on the opposite face of the DNA helix from that bound by RNA polymerase, and so polymerase can (and does) bind to the promoter at the same time as MerR.

The *merT* promoter is unusual. The distance between the –10 and –35 elements is 19 bp instead of the 15–17 bp typically found in an efficient σ^{70} promoter (see Chapter 12, Box 12-1). As a result, these two sequence elements recognized by σ are neither optimally separated nor aligned; they are somewhat rotated around the face of the helix with respect to each other. Furthermore, the binding of MerR (in the absence of Hg^{2+}) locks the promoter in this unpropitious conformation: polymerase can bind, but not in a manner that allows it to initiate transcription. Therefore, there is no basal transcription.

When MerR binds Hg^{2+}, however, the protein undergoes a conformational change that causes the DNA in the center of the promoter to twist. This structural distortion restores the disposition of the –10 and –35 regions to something close to that found at a strong σ^{70} promoter. In this new configuration, RNA polymerase can efficiently initiate transcription. The structures of promoter DNA in the "active" and "inactive" states have been determined (for another promoter regulated in this manner) and are shown in Figure 16-17.

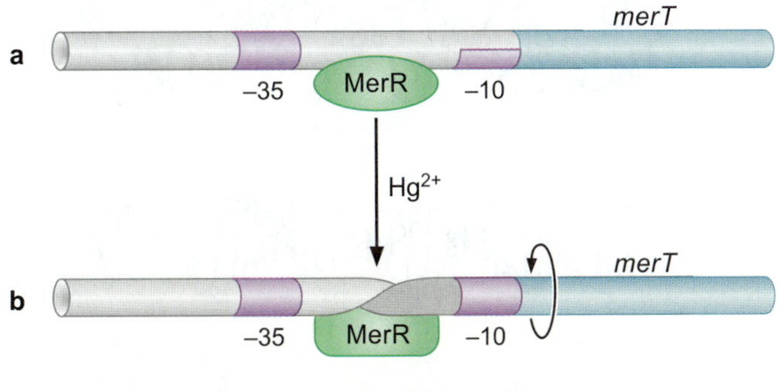

FIGURE 16-16 Activation by MerR. The –10 and –35 elements of the *merT* promoter lie on nearly opposite sides of the helix. (a) In the absence of mercury, MerR binds and stabilizes the inactive form of the promoter. (b) In the presence of mercury, MerR twists the DNA so as to properly align the promoter elements.

It is important to note that in this example, the activator does not interact with RNA polymerase to activate transcription, but instead alters the conformation of the DNA in the vicinity of the prebound enzyme. Thus, unlike the earlier cases, there is no separation of DNA-binding and activating regions: for MerR, DNA binding is intimately linked to the activation process.

Some Repressors Hold RNA Polymerase at the Promoter Rather than Excluding It

The Lac repressor works in the simplest possible way: by binding to a site overlapping the promoter, it blocks RNA polymerase binding. Many repressors work in that same way. In the MerR case, we saw a different form of repression: the protein holds the promoter in a conformation incompatible with transcription initiation. There are other ways repressors can work, one of which we now consider.

Some repressors work from binding sites that do not overlap the promoter. These repressors do not block polymerase binding; rather they bind to sites beside a promoter, interact with polymerase bound at that promoter, and inhibit initiation. One is the *E. coli* Gal repressor, which we mentioned earlier. The Gal repressor controls genes that encode enzymes involved in galactose metabolism; in the absence of galactose, the repressor keeps the genes off. In this case, the repressor interacts with the polymerase in a manner that inhibits transition from the closed to open complex.

Another example is provided by the P_4 protein from a bacteriophage (ϕ29) that grows on the bacterium *B. subtilis*. This regulator binds to a site adjacent to one promoter—a weak promoter called P_{A3}—and, by interacting with polymerase, serves as an activator. The interaction is with the αCTD, just as we saw with CAP. But this activator also binds at another promoter—a strong promoter called P_{A2c}. Here, it makes the same contact with polymerase as at the weak promoter, but the result is repression. It seems that whereas in the former case, the extra binding energy helps recruit polymerase and hence activates the gene, in the latter case, the overall binding energy—provided by the strong interactions between the polymerase and the promoter and the additional interaction provided by the activator—is so strong that the polymerase is unable to escape the promoter.

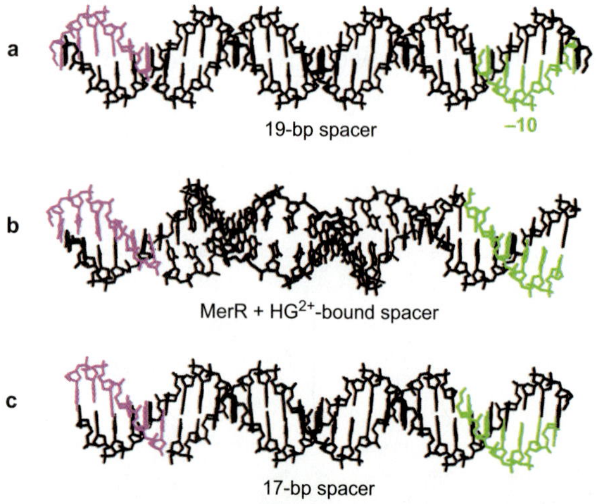

FIGURE 16-17 Structure of a *merT*-like promoter. (a) Promoter with a 19-bp spacer. (b) Promoter with a 19-bp spacer when in complex with active activator. (c) Promoter with a 17-bp spacer. The promoter shown in parts a and b is from the *bmr* gene of *B. subtilis,* which is controlled by the regulator BmrR. BmrR works as an activator when complexed with the drug tetraphenylphosphonium (TPP). The –35 (TTGACT) and –10 (TACAGT) elements of one strand are shown in pink and green, respectively. (Adapted, with permission, from Zheleznova Heldwein E.E. and Brennan R.G. 2001. *Nature* 409: 378; Fig. 3 b–d. © Macmillan.)

AraC and Control of the *araBAD* Operon by Antiactivation

The promoter of the *araBAD* operon from *E. coli* is activated in the presence of arabinose and the absence of glucose and directs expression of genes encoding enzymes required for arabinose metabolism. Unlike the cases of *lac* and *gal* genes, where a repressor and an activator work together, here two activators work together: AraC and CAP. When arabinose is present, AraC binds that sugar and adopts a configuration that allows it to bind DNA as a dimer to the adjacent half-sites, $araI_1$ and $araI_2$ (Fig. 16-18a). Just upstream of these (but not shown in the figure) is a CAP site: in the absence of glucose, CAP binds here and helps activation.

In the absence of arabinose, the *araBAD* genes are not expressed. This is because when not bound to arabinose, AraC adopts a different conformation and binds DNA in a different way: one monomer still binds the $araI_1$ site, but the other monomer binds a distant half-site called $araO_2$, as shown in Figure 16-18b. As these two half-sites are 194 bp apart, when AraC binds in this fashion, the DNA between the two sites forms a loop. In addition, when bound in this way, there is no monomer of AraC at $araI_2$, and as this is the position from which activation of *araBAD* promoter is mediated, there is no activation in this configuration.

The magnitude of induction of the *araBAD* promoter by arabinose is very large, and for this reason, the promoter is often used in **expression vectors**. Expression vectors are DNA constructs in which efficient synthesis of any protein can be ensured by fusing its gene to a strong promoter (see Chapter 21). In this case, fusing a gene to the *araBAD* promoter allows expression of the gene to be controlled by arabinose alone: the gene can be kept off when its expression is undesirable, and then both "derepressed" and "induced" when its product is wanted, simply by addition of arabinose. This allows expression even of genes

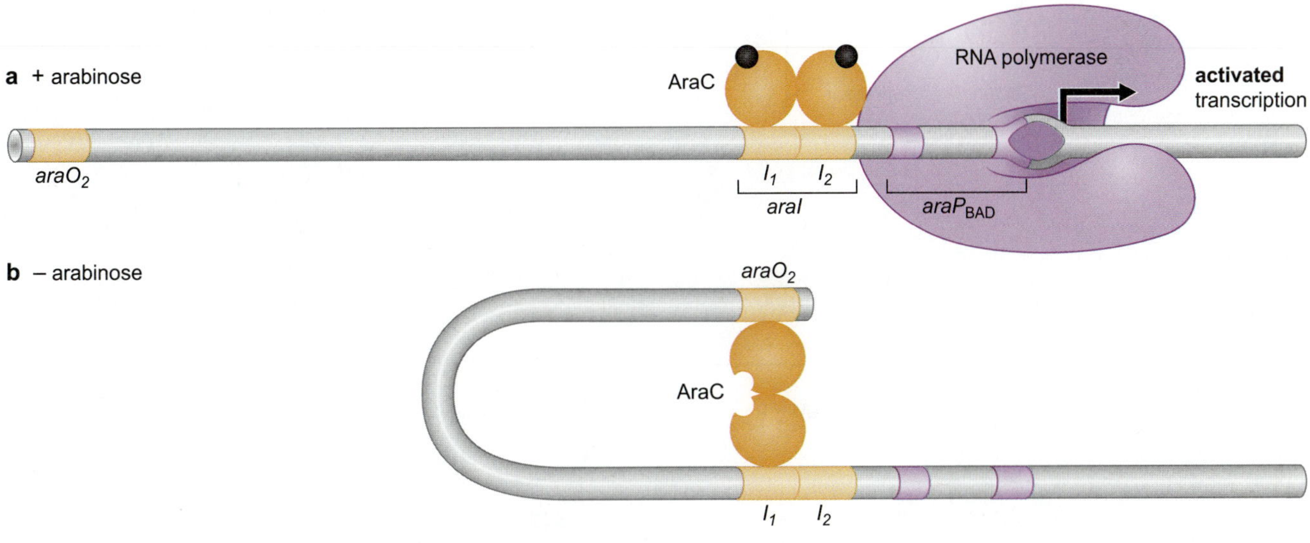

FIGURE 16-18 Control of the *araBAD* operon. (a) Arabinose binds to AraC, changing the shape of that activator so that it binds as a dimer to $araI_1$ and $araI_2$. This places one monomer of AraC close to the promoter from which it can activate transcription. (b) In the absence of arabinose, the AraC dimer adopts a different conformation and binds to $araO_2$ and $araI_1$. In this position, there is no monomer at site $araI_2$, and so the protein cannot activate the *araBAD* promoter ($araP_{BAD}$). This promoter is also controlled by CAP (not shown in this figure).

with products that are toxic to the bacterial cells—that is, genes that must be kept very tightly repressed when not induced.

THE CASE OF BACTERIOPHAGE λ: LAYERS OF REGULATION

Bacteriophage λ is a virus that infects *E. coli.* Upon infection, the phage can propagate in either of two ways: **lytically** or **lysogenically**, as illustrated in Figure 16-19. Lytic growth requires replication of the phage DNA and synthesis of new coat proteins. These components combine to form new phage particles that are released by lysis of the host cell. Lysogeny—the alternative propagation pathway—involves integration of the phage DNA into the bacterial chromosome where it is passively replicated at each cell division, as though it were a legitimate part of the bacterial genome.

A lysogen is extremely stable under normal circumstances, but the phage dormant within it—the **prophage**—can efficiently switch to lytic growth if the cell is exposed to agents that damage DNA (and thus threaten the host cell's continued existence). This switch from lysogenic to lytic growth is called **lysogenic induction**.

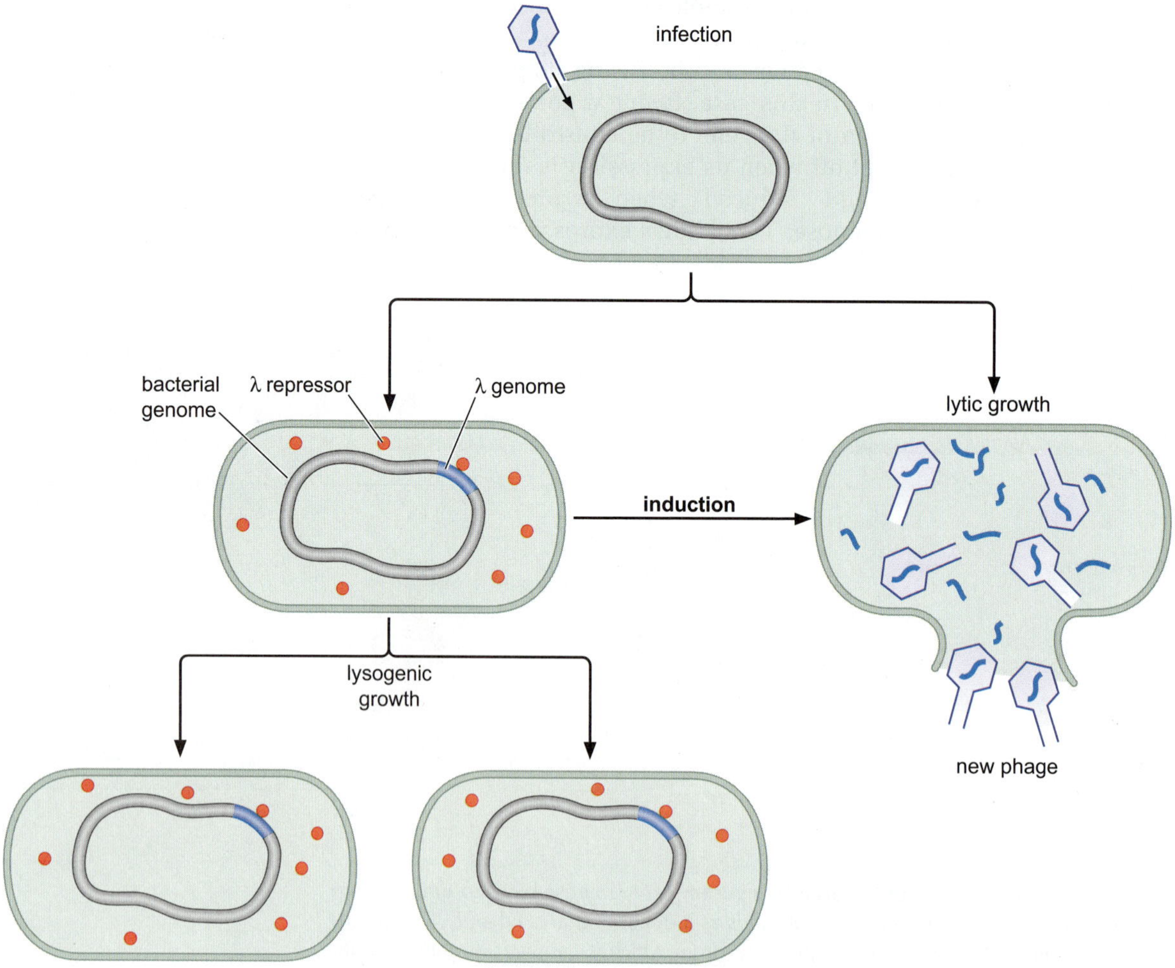

FIGURE 16-19 **Growth and induction of λ lysogen.** Upon infection, λ can grow either lytically or lysogenically. A lysogen can be propogated stably for many generations or it can be induced. Following induction, the lytic genes are expressed in proper order, leading to the production of new phage particles.

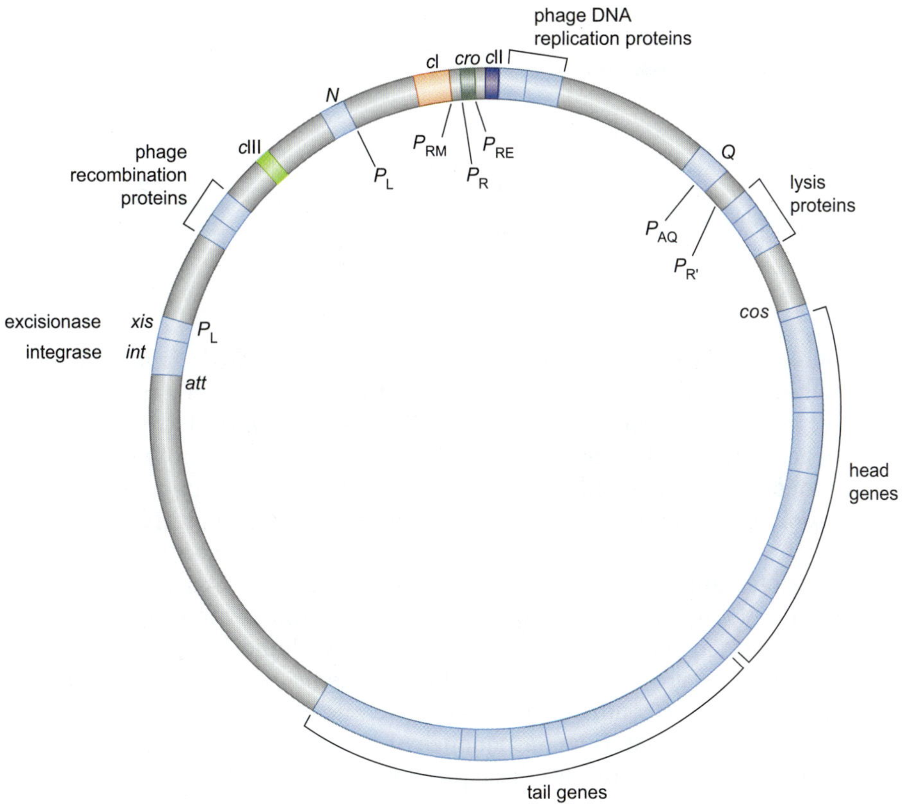

FIGURE 16-20 Map of bacteriophage λ in the circular form. λ genome is linear in the phage head, but, upon infection, circularizes at the *cos* site. When integrated into the bacterial chromosome, the phage genome is again linearized, but this time the ends are at the *att* site (see Chapter 11 for a description of integration).

The choice of developmental pathway depends on which of two alternative programs of gene expression is adopted in that cell. The program responsible for the lysogenic state can be maintained stably for many generations but then, upon induction, switch over to the lytic program with great efficiency.

Alternative Patterns of Gene Expression Control Lytic and Lysogenic Growth

Bacteriophage λ has a 50-kb genome and approximately 50 genes. Most of these genes encode coat proteins, proteins involved in DNA replication, recombination, and lysis (Fig. 16-20). The products of these genes are important in making new phage particles during the lytic cycle, but our concern here is restricted to the regulatory proteins, and where they act. We can therefore concentrate on just a few of them and start by considering a very small area of the genome shown in Figure 16-21.

The depicted region contains two genes (*c*I and *cro*) and three promoters (P_R, P_L, and P_{RM}). All of the other phage genes (except one minor

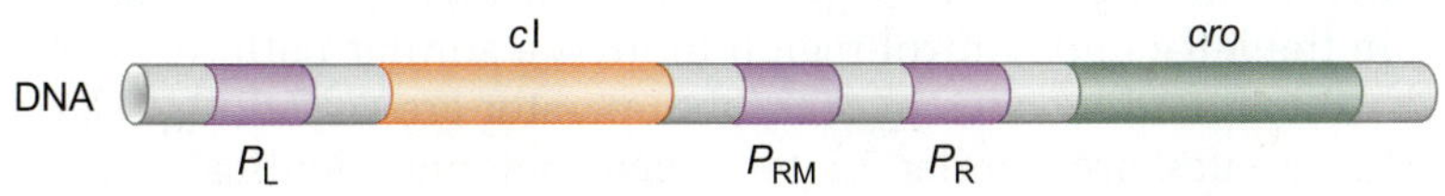

FIGURE 16-21 Promoters in the right and left control regions of bacteriophage λ.

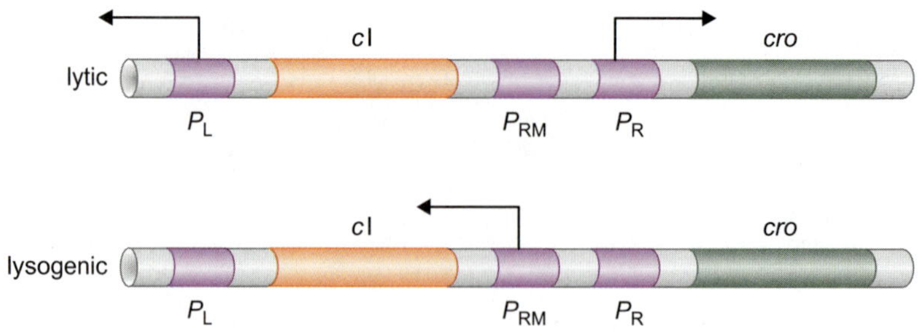

FIGURE 16-22 Transcription in the λ control regions in lytic and lysogenic growth. Arrows indicate which promoters are active at the decisive period during lytic and lysogenic growth, respectively. The arrows also show the direction of transcription from each promoter.

one) are outside this region and are transcribed directly from P_R and P_L (which stand for rightward and leftward promoter, respectively), or from other promoters whose activities are controlled by products of genes transcribed from P_R and P_L. P_{RM} (promoter for repressor maintenance) transcribes only the *c*I gene. P_R and P_L are strong, constitutive promoters; that is, they bind RNA polymerase efficiently and direct transcription without help from an activator. P_{RM}, in contrast, is a weak promoter and only directs efficient transcription when an activator is bound just upstream. P_{RM} resembles the *lac* promoter in this regard.

Two arrangements of gene expression are depicted in Figure 16-22: one renders growth lytic, the other lysogenic. Lytic growth proceeds when P_L and P_R remain switched on while P_{RM} is kept off. Lysogenic growth, in contrast, is a consequence of P_L and P_R being switched off and P_{RM} switched on. How are these promoters controlled?

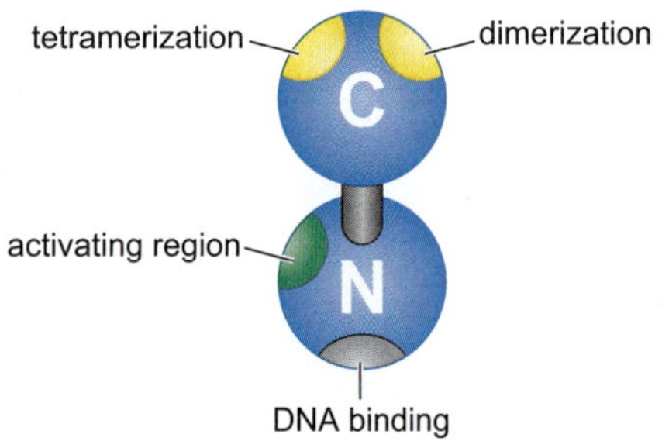

FIGURE 16-23 λ repressor. The figure shows a monomer of λ repressor, indicating various surfaces involved in different activities carried out by the protein. N indicates the amino domain, C the carboxyl domain. "Tetramerization" denotes the region where two dimers interact when binding cooperatively to adjacent sites on DNA. (Adapted, with permission, from Ptashne M. and Gann A. 2002. *Genes & signals,* p. 36, Fig. 1.17. © Cold Spring Harbor Laboratory Press.)

Regulatory Proteins and Their Binding Sites

The *c*I gene encodes λ repressor, a protein of two domains joined by a flexible linker region (Fig. 16-23). The amino-terminal domain contains the DNA-binding region (a helix-turn-helix domain, as we saw earlier). As with the majority of DNA-binding proteins, λ repressor binds DNA as a dimer; the main dimerization contacts are made between the carboxy-terminal domains. A single dimer recognizes a 17-bp DNA sequence, each monomer recognizing one half-site, again just as we saw in the *lac* system. (We have already looked at the details of DNA recognition by λ repressor in Fig. 16-12.)

Despite its name, λ repressor can both activate and repress transcription. When functioning as a repressor, it works in the same way as the Lac repressor: it binds to sites that overlap the promoter and excludes RNA polymerase. As an activator, λ repressor works like CAP—by recruitment. λ repressor's activating region is in the amino-terminal domain of the protein. Its target on polymerase is a region of the σ subunit adjacent to the part of σ that recognizes the –35 region of the promoter (region 4, see Chapter 12, Fig. 12-6).

Cro (which stands for *c*ontrol of *r*epressor and *o*ther things) only represses transcription, like the Lac repressor. It is a single-domain protein and again binds as a dimer to 17-bp DNA sequences, using a helix-turn-helix motif.

λ repressor and Cro can each bind to any one of six operators. These sites are recognized with different affinities by each of the proteins. Three of these sites are found in the left-hand control region and three in the right. We focus on the binding of λ repressor and Cro to the sites in the right-hand region shown in Figure 16-24. Binding to sites in the left-hand control region follows a similar pattern.

The three binding sites in the right operator are called O_{R1}, O_{R2}, and O_{R3}; these sites are similar in sequence, but not identical, and each

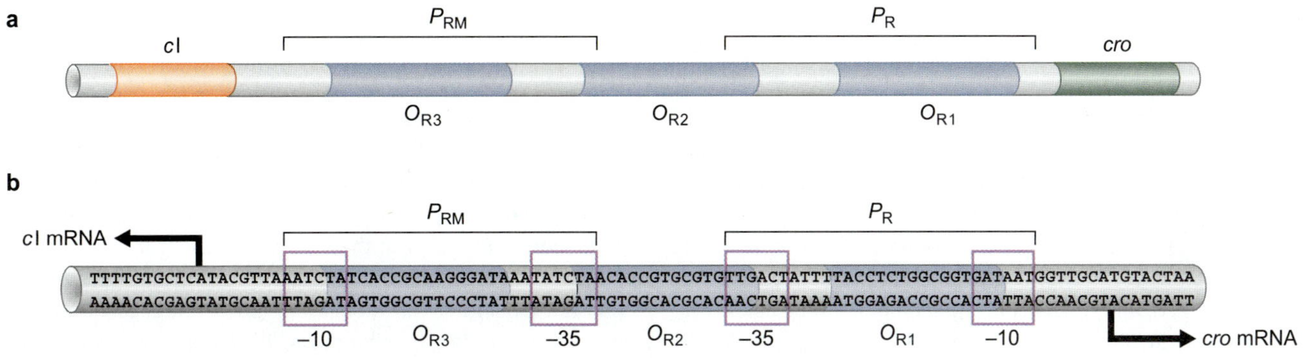

FIGURE 16-24 Relative positions of promoter and operator sites in O_R. Note that O_{R2} overlaps the –35 region of P_R by 3 bp, and that of P_{RM} by 2 bp. This difference is enough for P_R to be repressed and P_{RM} activated by repressor bound at O_{R2}. (b, Adapted, with permission, from Ptashne M. 1992. *A genetic switch: Phage and higher organisms,* 2nd ed. © Blackwell Science.)

one—if isolated from the others and examined separately—can bind either a dimer of repressor or a dimer of Cro. The affinities of these various interactions, however, are not all the same. Thus, repressor binds O_{R1} tenfold better than it binds O_{R2}. In other words, ten times more repressor—a tenfold higher concentration—is needed to bind O_{R2} than to bind O_{R1}. O_{R3} binds repressor with about the same affinity as does O_{R2}. Cro, on the other hand, binds O_{R3} with highest affinity, and only binds O_{R2} and O_{R1} when present at tenfold higher concentration. The significance of these differences will become apparent presently.

λ Repressor Binds to Operator Sites Cooperatively

λ repressor binds DNA cooperatively. This is critical to its function and occurs as follows. Consider repressor binding to sites in O_R. In addition to providing the dimerization contacts, the carboxy-terminal domain of λ repressor mediates interactions *between* dimers (the point of contact is the patch marked "tetramerization" in Fig. 16-23). In this way, two dimers of repressor can bind cooperatively to adjacent sites on DNA.

For example, the repressor at O_{R1} helps the repressor bind to the lower-affinity site O_{R2} by cooperative binding. Repressor thus binds both sites simultaneously and does so at a concentration that would be sufficient to bind only O_{R1} were the two sites tested separately (Fig. 16-25). (Recall that, without cooperativity, a tenfold higher concentration of repressor would be needed to bind O_{R2}.) O_{R3} is not bound: repressor bound cooperatively at O_{R1} and O_{R2} cannot simultaneously make contact with a third dimer at that adjacent site.

We have already discussed the idea of cooperative binding and seen an example: activation of the *lac* genes by CAP. As in that case, cooperative binding of repressors is a simple consequence of their touching each other while simultaneously binding to sites on the same DNA molecule.

For a more detailed discussion of the causes and effects of cooperative binding, see Box 16-3, Concentration, Affinity, and Cooperative Binding. Cooperative binding of regulatory proteins is used to ensure that changes in the level of expression of a given gene can be dramatic even in response to small changes in the level of a signal that controls that gene. The lysogenic induction of λ, discussed below, provides an

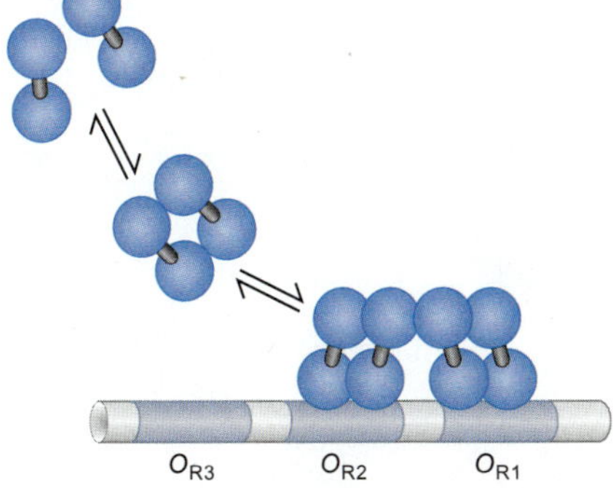

FIGURE 16-25 Cooperative binding of λ repressor to DNA. The λ repressor monomers interact to form dimers, and those dimers interact to form tetramers. These interactions ensure that binding of repressor to DNA is cooperative. That cooperative binding is helped further by interactions between repressor tetramers at O_R interacting with others at O_L (see later in text and Fig. 16-27).

ADVANCED CONCEPTS

Box 16-3 Concentration, Affinity, and Cooperative Binding

What do we mean when we talk about "strong" and "weak" binding sites? When we say two molecules recognize each other, or interact with each other, such as a protein and its site on DNA, we mean that they have some affinity for each other. Whether they are actually found bound together at any given time depends on (1) how high that affinity is (i.e., how tightly they interact), and (2) the concentration of the molecules.

As we emphasized in Chapters 3 and 5, the molecular interactions that underpin regulation in biological systems are reversible: when interacting molecules find each other, they stick together for a period of time and then separate. The higher the affinity, the tighter the two molecules stick together and, in general, the longer they remain together before parting. The higher the concentration, the more often they will find each other in the first place. Thus, higher affinity and higher concentration have similar effects: they both result in the two molecules, in general, spending more time bound to each other.

Cooperativity Visualized

Cooperativity can be expressed in terms of increased affinity. Repressor has a higher affinity for O_{R1} than for O_{R2}. But once repressor is bound to O_{R1}, repressor can bind O_{R2} more tightly because it interacts not only with O_{R2}, but with repressor bound at O_{R1} as well. Neither of these interactions is very strong alone, but when combined, they substantially increase the affinity of binding of that second repressor. As discussed in Chapter 4, the relationship between binding energy and equilibrium is an exponential one (see Table 4-1). Thus, increasing the binding energy as little as twofold increases affinity by one order of magnitude.

Another way to picture how cooperativity works is to think of it as increasing the local concentration of repressor. Picture repressor bound cooperatively at O_{R1} and O_{R2}. Although repressor at O_{R2} periodically lets go of DNA, it is holding on to repressor at O_{R1} and so remains in the proximity of O_{R2}. This effectively increases the local concentration of repressor in the vicinity of that site and ensures that repressor rebinds frequently.

If we dispense with cooperativity and just increase the concentration of repressor in the cell, when repressor falls off O_{R2}, it will not be held nearby by repressor at O_{R1} and will usually drift away before it can rebind O_{R2}. But at the higher concentrations of repressor, another molecule of repressor will likely be close to O_{R2} and bind there. Thus, even if each repressor dimer only sits on O_{R2} for a short time, by either holding it nearby or increasing the number of possible replacements, the likelihood of repressor being bound will increase at any given time.

Yet another way of thinking about cooperative binding is as an entropic effect. When a protein goes from being free in solution to being constrained on a DNA-binding site, the entropy of the system decreases. But repressor held close to O_{R2} by interaction with repressor at O_{R1} is already constrained compared to its free state. Rebinding of that constrained repressor has less entropic cost than does binding of free repressor.

We thus see three ways in which cooperativity can be pictured. We should also consider some of the consequences of cooperative binding that make it so useful in biology. For example, cooperativity not only enables a weak site to be filled at a lower concentration of protein than its inherent affinity would predict, it also changes the steepness of the curve describing the filling of that site with changes in concentration. To understand what is meant by that, consider as an example a protein binding cooperatively to two weak sites, A and B. These sites will go from essentially completely empty to almost completely filled over a much narrower range of protein concentration than would a single site (Box 16-3 Fig. 1). In fact, the cooperativity in the λ system is even greater than one might expect because a large fraction of free repressor (i.e., that not bound to DNA) is found as monomer in the cell; thus, it is in essence a coopera-

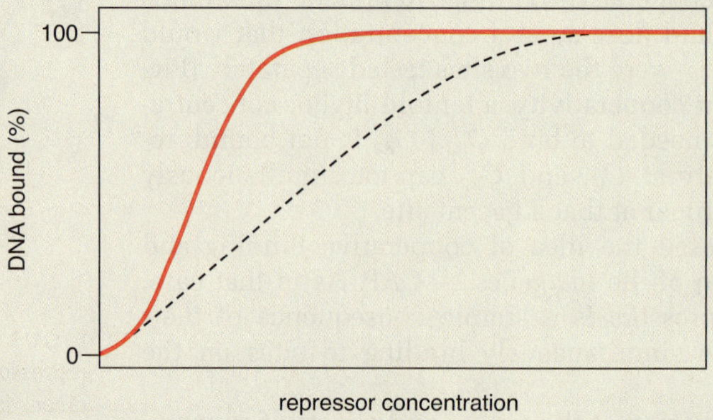

BOX 16-3 FIGURE 1 Cooperative binding reaction. The dashed line shows the curve that describes binding of a protein to a single site. The steeper sigmoid curve shows cooperative binding of, for example, λ repressor to its operator sites. (Adapted, with permission, from Ptashne M. 1992. *A genetic switch: Phage and higher organisms,* 2nd ed. © Blackwell Science.)

Box 16-3 *(Continued)*

tive binding of four monomers, rather than two stable dimers, adding to the concerted nature of complex formation on DNA, and so adding to the steepness of the curve. But why does cooperativity make the binding curve steeper?

We have already seen how the site is filled at a lower concentration of repressor than its affinity would suggest; but how is it that as repressor concentration decreases, binding falls away so quickly? Consider interactions between components of any system: as the concentration of the components is reduced, any given interaction between two of them will occur less frequently. If the system requires multiple interactions between several different components, this will become very rare at lower concentrations. Thus, binding of four monomers of a protein to two sites requires several (in fact, seven) interactions; the chance of the individual components coming together is drastically reduced as their individual concentrations decrease.

Cooperativity and DNA-binding Specificity

A final important aspect of cooperative binding is that it imposes specificity on DNA binding. CAP activation of the *lac* promoter shows this. CAP brings RNA polymerase to promoters that bear CAP sites specifically (as opposed to other promoters of comparable affinity that lack CAP sites). Likewise, λ repressor at O_{R1} directs another molecule of repressor to bind to the weak site adjacent to it, not some other site of equal affinity elsewhere in the cell. In fact, cooperativity is vital to ensuring that proteins can bind with sufficient specificity for life to work as we know it.

To illustrate this, consider a protein binding to a site on DNA. This protein has a high affinity for its correct site. But the DNA within the cell represents a huge number of potential (but incorrect) binding sites for that protein. What is important, therefore, is not simply the absolute affinity of the protein for its correct site, but its affinity for that site compared to its affinity for all the other incorrect sites. And remember, those incorrect sites are at a much higher concentration than the correct site (representing, as they do, all of the DNA in the cell except the correct site). So even if the affinity for the incorrect sites is lower than that for the correct site, the higher concentration of the incorrect sites ensures that the protein will often sample them while attempting to reach its correct site.

What is needed is a strategy that increases affinity for the correct site without aiding interactions with the incorrect sites. Increasing the number of contacts between the protein and its DNA site (e.g., by making the protein larger) does not necessarily help because it also tends to increase binding to the incorrect sites. Once affinity for the incorrect sites gets too high, the protein essentially never finds its correct site; it spends too much time sampling incorrect sites. Thus, a kinetic problem replaces the specificity one and it can be just as disruptive.

Cooperativity solves the problem. By binding to two adjacent sites cooperatively, a protein increases dramatically its affinity for those sites, without increasing affinity for other sites. The reason it does not increase affinity for the incorrect sites is simply because the chance of two molecules of protein binding incorrect sites close together at the same time (allowing cooperativity to stabilize that binding) is extremely remote. Only when they find the correct sites do they remain bound long enough to give a second protein a chance to turn up.

Cooperativity and Allostery

Although in this chapter we use the term *cooperativity* to refer to a particular mechanism of cooperative binding, the term is also used in other contexts where different mechanisms apply. In general, we might say that cooperativity describes any situation in which two ligands bind to a third molecule in such a way that the binding of one of those ligands helps the binding of the other. Thus, for the DNA-binding proteins we considered here, cooperativity is mediated by simple adhesive interactions, but in other situations, cooperativity can be mediated by allosteric events. Perhaps the best example of that is the binding of oxygen molecules to hemoglobin.

Hemoglobin is a homotetramer, and each subunit binds one molecule of oxygen. This binding is cooperative: when the first oxygen binds, it causes a conformational change that fixes the binding site for the next oxygen in a conformation with a higher affinity for that ligand. Thus, in this case, there is no direct interaction between the ligands, but by triggering an allosteric transition, one ligand increases affinity for a second.

excellent example of this sensitive aspect of control. In some systems, cooperative binding between activators is also the basis of signal integration (see the discussion on β-interferon in Chapter 17).

Repressor and Cro Bind in Different Patterns to Control Lytic and Lysogenic Growth

How do repressor and Cro control the different patterns of gene expression associated with the different ways λ can replicate? As shown in Fig-

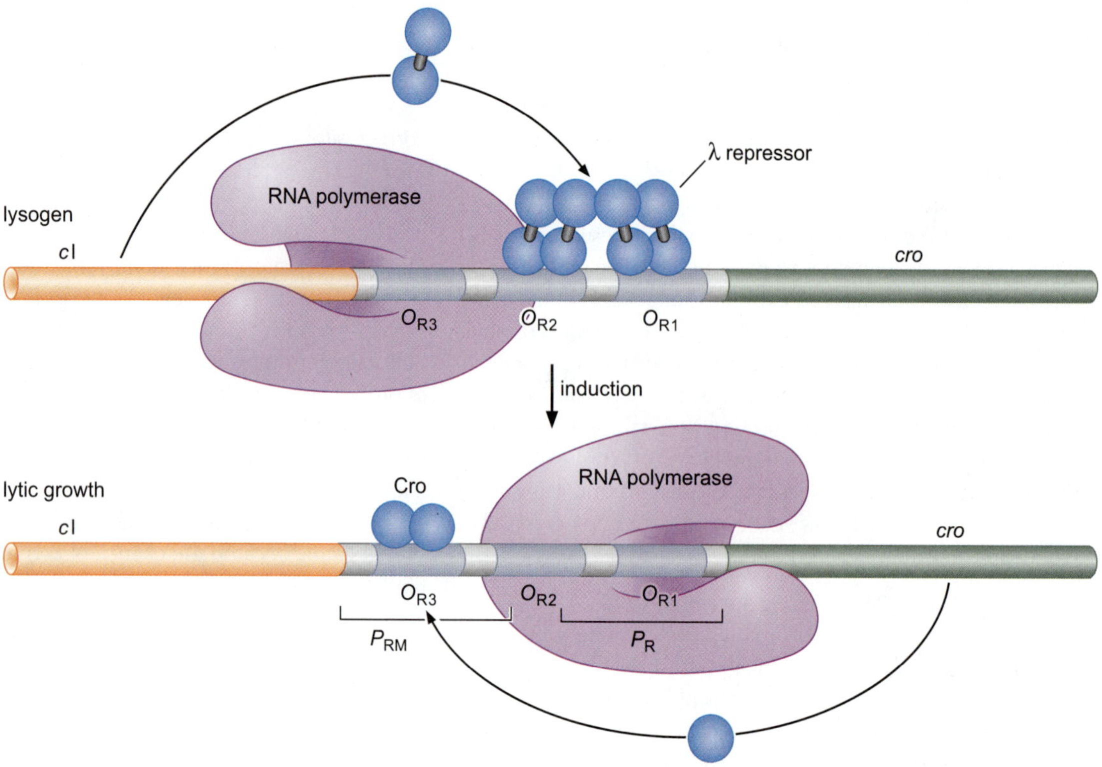

FIGURE 16-26 The action of λ repressor and Cro. Repressor bound to O_{R1} and O_{R2} turns off transcription from P_R. Repressor bound at O_{R2} contacts RNA polymerase at P_{RM}, activating expression of the cI (repressor) gene. O_{R3} lies within P_{RM}; Cro bound there represses transcription of cI. (Adapted, with permission, from Ptashne M. and Gann A. 2002. *Genes & signals,* p. 30, Fig. 1.13. © Cold Spring Harbor Laboratory Press.)

ure 16-26, for lytic growth, a single Cro dimer is bound to O_{R3}; this site overlaps P_{RM} and so Cro represses that promoter (which would only work at a low level anyway in the absence of activator because the promoter is weak) (Fig. 16-26). As neither repressor nor Cro is bound to O_{R1} and O_{R2}, P_R binds RNA polymerase and directs transcription of lytic genes; P_L does likewise. Recall that both P_R and P_L are strong promoters that need no activator.

During lysogeny, P_{RM} is on while P_R (and P_L) are off. Repressor bound cooperatively at O_{R1} and O_{R2} blocks RNA polymerase binding at P_R, repressing transcription from that promoter (Fig. 16-26). But repressor bound at O_{R2} *activates* transcription from P_{RM}.

We will shortly return to the question of how the phage chooses between these alternative pathways. But first we consider induction—how the lysogenic state outlined above switches to the alternative lytic state when the cell is threatened.

Lysogenic Induction Requires Proteolytic Cleavage of λ Repressor

E. coli senses and responds to DNA damage. It does this by activating the function of a protein called RecA. This enzyme is involved in recombination (which accounts for its name; see Chapter 10), but it has another function: it stimulates the proteolytic autocleavage of certain proteins. The primary substrate for this activity is a bacterial repressor protein called LexA that represses genes encoding DNA-repair enzymes. Activated RecA stimulates autocleavage of LexA, releasing repression of those genes. This is called the SOS response (see Chapter 9).

If the cell is a lysogen, it is in the best interests of the prophage to escape under these threatening circumstances. To this end, λ repressor has evolved to resemble LexA, ensuring that λ repressor too undergoes autocleavage in response to activated RecA. The cleavage reaction removes the carboxy-terminal domain of repressor, and so dimerization and cooperativity are immediately lost. As these functions are critical for repressor binding to O_{R1} and O_{R2} (at concentrations of repressor found in a lysogen), loss of cooperativity ensures that the repressor dissociates from those sites (as well as from O_{L1} and O_{L2}). Loss of repression triggers transcription from P_R and P_L, leading to lytic growth. Transcription from P_R quickly produces Cro, which binds O_{R3} and blocks any further synthesis of repressor from P_{RM}. This action ensures that the decision to induce is irreversible.

For induction to work efficiently, the level of repressor in a lysogen must be tightly regulated. If levels were to drop too low, under normal conditions, the lysogen might spontaneously induce; if levels rose too high, appropriate induction would be inefficient. The reason for the latter is that more repressor would have to be inactivated (by RecA) for the concentration to drop enough to vacate O_{R1} and O_{R2}. We have already seen how repressor ensures that its level never drops too low: it activates its own expression, an example of **positive autoregulation**. But how does it ensure levels never get too high? Repressor also regulates itself negatively.

This **negative autoregulation** works as follows. As drawn, Figure 16-26 shows P_{RM} being activated by repressor (at O_{R2}) to make more repressor. But if the concentration gets too high, repressor will bind to O_{R3} as well and repress P_{RM} (in a manner analogous to Cro binding O_{R3} and repressing P_{RM}). This prevents synthesis of new repressor until its concentration falls to a level at which it vacates O_{R3}.

As an aside, it is interesting to note that the term "induction" is used to describe both the switch from lysogenic growth to lytic growth in λ and the switching on of the *lac* genes in response to lactose. This common usage stems from the fact that both phenomena were studied in parallel by Jacob and Monod (see Box 16-2). It is also worth noting that, just as lactose induces a conformational change in Lac repressor to relieve repression of the *lac* genes, so too the inducing signals of λ work by causing a structural change (in this case, proteolytic cleavage) in λ repressor.

Negative Autoregulation of Repressor Requires Long-Distance Interactions and a Large DNA Loop

We have discussed cooperative binding of repressor dimers to adjacent operators such as O_{R1} and O_{R2}. There is yet another level of cooperative binding seen in the prophage of a lysogen, one critical to proper negative autoregulation. Repressor dimers at O_{R1} and O_{R2} interact with repressor dimers bound cooperatively at O_{L1} and O_{L2}. These interactions produce an octomer of repressor; each dimer within the octamer is bound to a separate operator.

To accommodate the long-distance interaction between repressors at O_R and O_L, the DNA between those operator regions (about 3.5 kb, including the *c*I gene itself) must form a loop (Fig. 16-27). When the loop is formed, O_{R3} is held close to O_{L3}. This allows another two

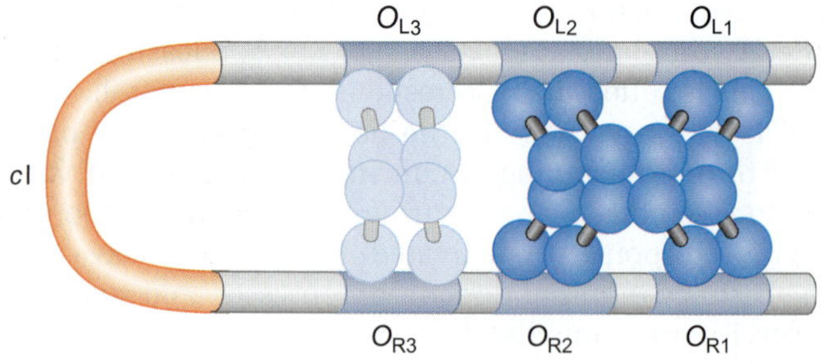

FIGURE 16-27 Interaction of repressors at O_R and O_L. Repressors at O_R and O_L interact as shown. These interactions stabilize binding. In this way, the interactions increase repression of P_R and P_L and allow repressor to bind O_{R3} at a lower concentration than it otherwise could. The repressors bound at O_{L3} and O_{R3} are here shown in a lighter shade to indicate that they will be bound only when the concentration of repressor rises above a certain level, as described in the text. (Adapted, with permission, from Ptashne M. and Gann A. 2002. *Genes & signals,* p. 35, Fig. 1.16. © Cold Spring Harbor Laboratory Press.)

dimers of repressor to bind cooperatively to these two sites. This cooperativity means O_{R3} binds repressor at a lower concentration than it otherwise would—indeed, at a concentration only just a little higher than that required to bind O_{R1} and O_{R2}. Thus, repressor concentration is very tightly controlled: small decreases are compensated for by increased expression of its gene, and increases by switching the gene off. This explains why lysogeny can be so stable while also ensuring that induction is very efficient.

The structure of the carboxy-terminal domain of λ repressor, interpreted in light of earlier genetic studies, reveals the basis of dimer formation, but it also shows how two dimers interact to form the tetrameric form (as occurs when repressor is bound cooperatively to O_{R1} and O_{R2}). Moreover, the structure reveals the basis for the octomer form and shows that this is the highest-order oligomer repressor can form (Fig. 16-28).

In Box 16-4, Evolution of the λ Switch, we discuss how the control circuits that govern lysogenic and lytic growth, and the process of induc-

FIGURE 16-28 Interactions between the carboxy-terminal domain of λ repressors. The figure shows, at the top, a schematic representation of two dimers of the carboxy-terminal domain of λ repressor. Indicated are the two patches here called B and R on the surface of that domain that mediate interactions between two dimers to give a tetramer, in the first instance, and then between two tetramers to give an octamer (the form found when repressor is bound cooperatively to the four sites, O_{R1}, O_{R2}, O_{L1}, and O_{L2}). Once the octamer has formed, there is no space left for a further dimer to enter the complex, and so the octamer is the highest-order structure that forms. (Modified, with permission, from Bell et al. 2000. *Cell* 101: 801–811, Figs. 4a,b and 5a–c. © Elsevier.)

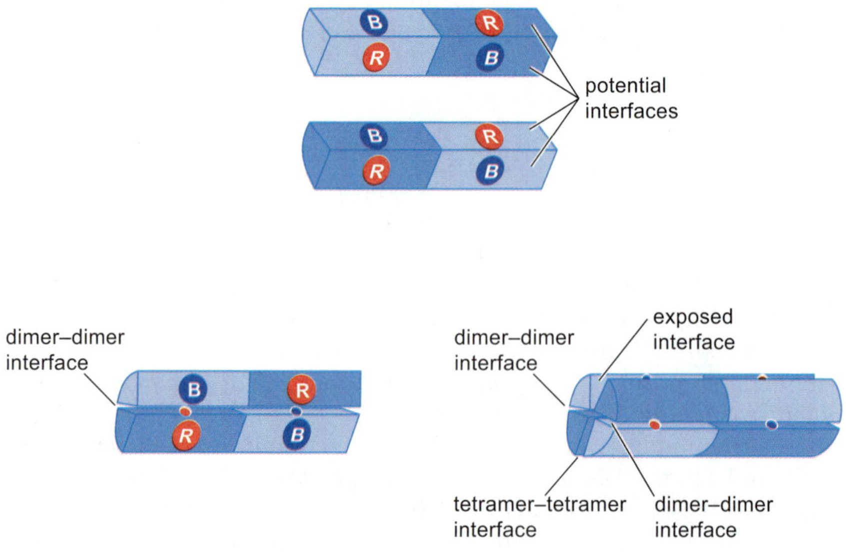

tion, might have evolved. Specifically, we discuss how the interactions between repressor and Cro, their binding sites, and the promoters they regulate could have evolved to their current elaborate form in small steps from an earlier rudimentary system.

Another Activator, λ CII, Controls the Decision between Lytic and Lysogenic Growth upon Infection of a New Host

We have seen how λ repressor and Cro control lysogenic and lytic growth and the switch from one to the other upon induction. Now we turn to those early events of infection that determine which pathway the phage chooses in the first place. Critical to this choice are the products of two other λ genes, *c*II and *c*III. We need only expand slightly our map of the regulatory region of λ to see where *c*II and *c*III lie: *c*II is on the right of *c*I and is transcribed from P_R; *c*III, on the left of *c*I, is transcribed from P_L (Fig. 16-29).

Like the λ repressor, the CII protein is a transcriptional activator. It binds to a site upstream of a promoter called P_{RE} (for repressor establishment) and stimulates transcription of the *c*I (repressor) gene from that promoter. Thus, the repressor gene can be transcribed from two different promoters (P_{RE} and P_{RM}).

P_{RE} is a weak promoter because it has a very poor −35 sequence. The CII protein binds to a site that overlaps the −35 region but is located on the opposite face of the DNA helix; by directly interacting with polymerase, CII helps polymerase bind to the promoter.

Only when sufficient repressor has been made from P_{RE} can that repressor bind to O_{R1} and O_{R2} and direct its own synthesis from P_{RM}. Thus, we see that repressor synthesis is **established** by transcription from one promoter (stimulated by one activator) and then **maintained** by transcription from another promoter (under its own control—positive autoregulation).

We can now see in summary how CII orchestrates the choice between lytic and lysogenic development. Upon infection, transcription is immediately initiated from the two constitutive promoters P_R and P_L. P_R directs synthesis of both Cro and CII. Cro expression favors lytic development: once Cro reaches a certain level, it will bind O_{R3} and block P_{RM}. CII expression, on the other hand, favors lysogenic growth by directing transcription of the repressor gene (Fig. 16-30). For successful lysogeny, repressor must then bind to O_{R1} and O_{R2} and activate P_{RM}.

The efficiency with which CII directs transcription of the *c*I gene, and hence the rate at which repressor is made, is the critical step in deciding how λ will develop. What determines how efficiently CII works in any given infection?

These *c*I, *c*II, and *c*III genes were isolated in elegant genetic screens outlined in Box 16-5, Genetic Approaches That Identified Genes Involved in the Lytic/Lysogenic Choice.

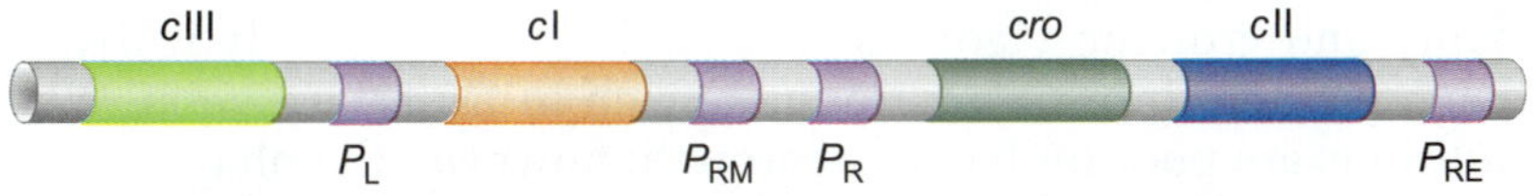

FIGURE 16-29 Genes and promoters involved in the lytic/lysogenic choice. Not shown here is the gene *N*, which lies between P_L and cIII (see Fig. 16-20).

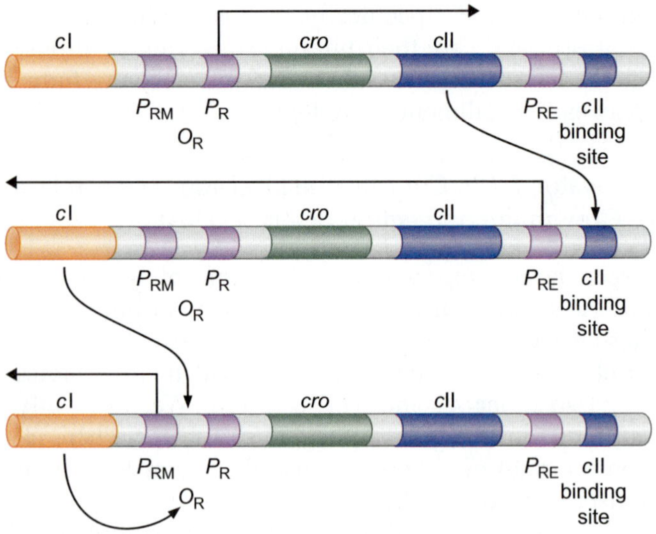

FIGURE 16-30 Establishment of lysogeny. The cI gene is transcribed from P_{RE} when establishing lysogeny and from P_{RM} when maintaining that state. Repressor bound at O_{R1} and O_{R2} not only activates the maintenance mode, but also turns off the establishment mode of expression. Note that P_R controls not only lytic genes, but also expression of cII and is thus important in lysogeny as well as lytic development. Similarly, although not shown in the figure, P_L, which controls many lytic genes, also controls the cIII gene which helps establish lysogeny (see text). (Adapted, with permission, from Ptashne M. and Gann A. 2002. *Genes & signals,* p. 31, Fig. 1.14. © Cold Spring Harbor Laboratory Press.)

The Number of Phage Particles Infecting a Given Cell Affects Whether the Infection Proceeds Lytically or Lysogenically

Multiplicity of infection (or moi) is a measure of how many phage particles infect a given bacterial cell within a population. If the average number is one or fewer phage particles per cell, the infection is more likely to result in lysis. If the number of phage particles is two or more, it is more likely to produce lysogney. And as the numbers of phage per cell become lower and lower, the tendency toward lytic infection increases, and as it becomes higher and higher, the likelihood of lysogeny similarly increases.

Mechanistically, this makes sense. The more phage genomes that enter the cell and start transcribing from P_R and P_L, the more CII and CIII gets made, and the greater the chance that at least one of those phage genomes will establish repressor synthesis and integrate into the bacterial chromosome. As long as one of the infecting phage does this, the others will subsequently be blocked from further lytic development.

One can speculate as to why λ is set up to respond this way—why it would rather develop lysogenically when in a population of many phage and few bacteria, for example. If there are few bacterial cells, then availability of host cells for the next round of infection will be limited, and so the phage might benefit from becoming dormant within a lysogen rather than risk finding no further host cells after a round of lytic infection. The growth conditions of the bacterial cells also influence the outcome of an infection as described below.

Growth Conditions of *E. coli* Control the Stability of CII Protein and thus the Lytic/Lysogenic Choice

When the phage infects a population of bacterial cells that are healthy and growing vigorously, it tends to propagate lytically, releasing progeny into an environment rich in fresh host cells. When conditions are poor for bacterial growth, however, the phage is more likely to form lysogens and sit tight; again, there will likely be few host cells in the vicinity for any progeny phage to infect. These different growth conditions impinge on CII as follows.

KEY EXPERIMENTS

Box 16-4 Evolution of the λ Switch

We have emphasized many of the intricacies that underlie the mechanisms of decision-making by bacteriophage λ: how it chooses between lytic and lysogenic development and how it can efficiently switch from a stable prophage to a lytically replicating virus. Many of the subtleties that give the system these characteristics have been discussed: cooperative binding, autopositive and negative regulation, the use of repressors with opposing effects, and so on. Emphasizing the intricate interplay of these features, and how interdependent they are in the phage we see today, rather begs the question of how such a system could have evolved in simple steps from an earlier primitive version. This is an important question when considering all biological systems, and we address it here for λ.

A proposed step-by-step model of how the λ switch might have evolved from a rudimentary version is shown (Box 16-4 Fig. 1). In each step, one simple addition has been made to a system that already works, to produce one that works a little better.

In the last few years, a series of experiments has explored the issues raised in this scheme. These studies point toward how relatively easily evolution might have molded the λ switch. Thus, each apparently critical feature of the existing switch has been eliminated by mutation, rendering the phage defective in various behaviors; the mutant phage might, for example, lysogenize less efficiently or form lysogens that are unstable, or perhaps too stable, making induction too easy or too difficult.

Other mutations were then found that compensated for the original defect in each case. These experiments revealed that far from the irreducible complexity that might on the surface seem to exist for this system, loss of any individually "essential" feature could be compensated for, at least partially, by alteration of another. For example, positive autoregulation was eliminated by introducing a *pc* mutation into the cI gene of the phage. A *pc* mutation, as we discussed earlier for CAP, eliminates the activation function of an activator. Thus, in this case, the mutant phage would make repressor that can still bind DNA and repress transcrption but cannot activate expression of more repressor from P_{RM}. This mutant phage can form lysogens, but they are very unstable because repressor levels are low. Introducing other changes that strengthen the promoter P_{RM} compensates for this to a great extent, making the lysogens more stable and more like those produced by wild-type λ. The strengthened P_{RM} can direct expression of more repressor without being activated by the existing repressor. It seems that having autopositive regulation of repressor gives the wild-type phage an advantage (explaining why all known lambdoid phage have this feature), but it is not completely essential for the system to work fairly well. Thus, one can see an intermediate step in the evolution of the modern system.

In another example, cooperative binding by repressor was also shown to be a feature that, although advantageous, is not completely necessary for the phage to function in a rudimentary way. Thus, cooperate binding was substantially weakened by introducing mutations that had previously been shown to disrupt cooperative interactions between repressor dimers. Phage carrying this mutant repressor gene were unable to form lysogens. But addition of other modifications—one again strengthening P_{RM} and the other strengthening the binding site O_{R2} for repressor—together generated a phage that now could form lysogens, albeit less efficiently than wild-type λ.

In a further set of remarkable experiments, the λ switch was dismantled and reassembled in ways that test critical ideas about both how it functions and how it arose. In the most recent and ambitious of these, the repressor gene was replaced by a gene for a bacterial repressor protein, the Tet repressor, and in the same phage, the gene for Cro was substituted by *lacI*, the gene encoding Lac repressor. In addition, operator sites within the phage were modified to allow these two bacterial repressors to bind in patterns that mimic some of the critical binding patterns of λ repressor and Cro in wild-type λ.

Phage built from these heterologous pieces could recapitulate some of the behaviors of wild-type λ. Because the binding of both the repressors employed in the modified phage can be titrated precisely by small molecules (*lac* and *tet* inducers), further subtle manipulations can now be used to investigate further the workings, and possible origins, of the λ system.

Taken together, these various experimental approaches make clear two points. First, the λ switch could easily have evolved through a series of steps, each adding a new level of regulation to a system that worked less well, but did work, before. This is what would be expected of any system that evolved through natural selection. Second, there are alternative ways any given behavior can be achieved. Understanding the details of the solution that finally appeared can make the problem of how it evolved seem much more difficult than it necessarily was; that is, the final solution was only one of a variety that would have worked had they arisen.

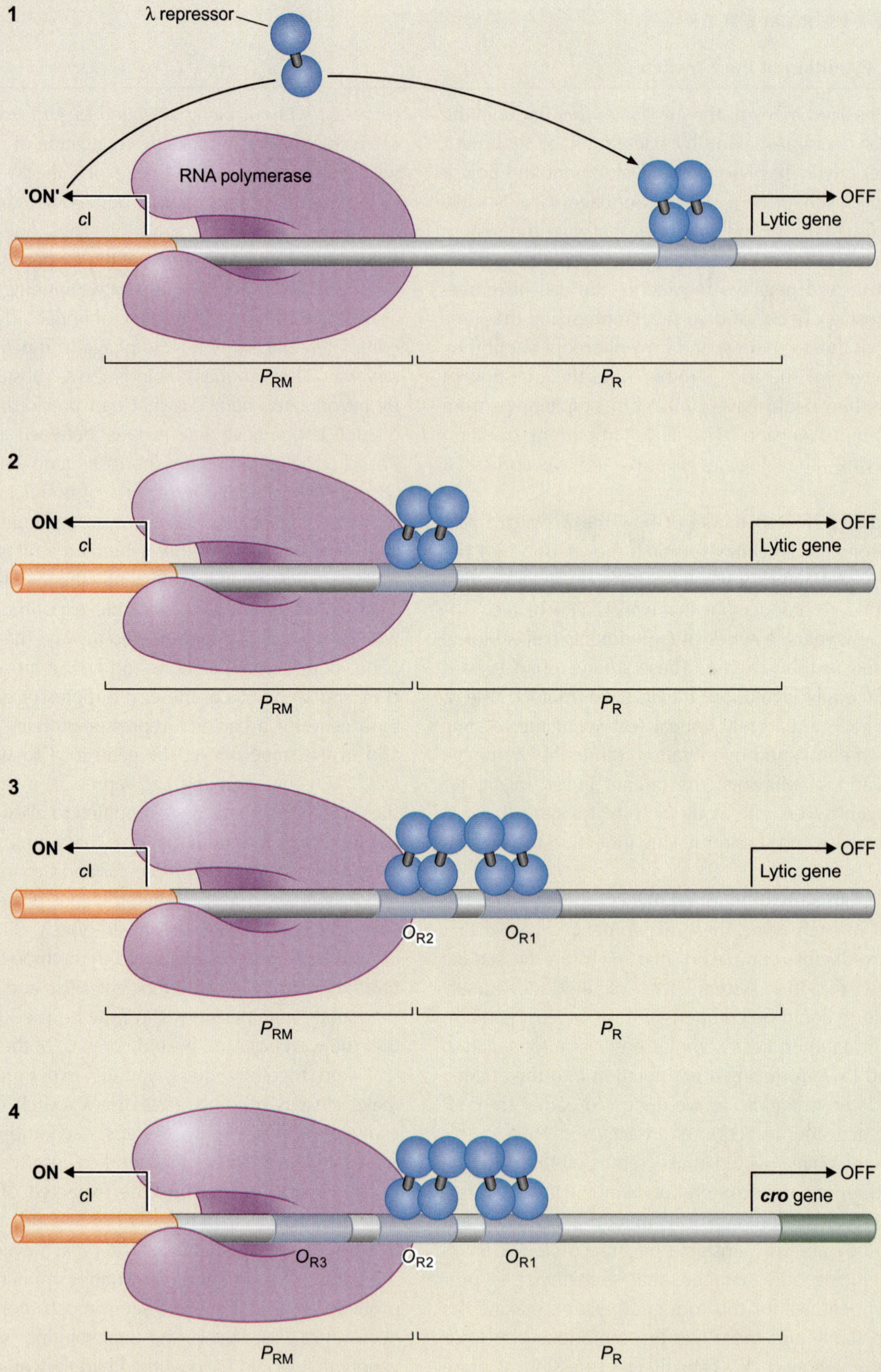

BOX 16-4 FIGURE 1 **Hypothetical stages in the evolution of the λ switch.** (Stage 1) The primitive λ genome bears two promoters, one directing expression of the lytic genes (P_R) and one for the repressor gene (P_{RM}). A single λ repressor-binding site overlaps P_R; when bound to this site, repressor turns off the lytic genes, but its own synthesis is unregulated. (Stage 2) Here, the single repressor-binding site has moved close to P_{RM} (now in the position of O_{R2}), so that bound repressor contacts polymerase at P_{RM} and thereby stimulates that promoter while repressing P_R. (Stage 3) A second repressor-binding site has been introduced (in the position of O_{R1}). In addition, a new protein–protein interaction surface has been introduced, allowing cooperative binding of repressor dimers to these adjacent sites. These features contribute additional aspects of cooperativity to the system that increases the efficiency of the switch mechanism. (Stage 4) The third repressor-binding site (O_{R3}) is introduced. When bound to this site, repressor negatively regulates its own synthesis such that its concentration remains below a critical level and ensures an efficient switch mechanism. (Adapted, with permission, from Ptashne M. and Gann A. 1998. *Curr. Biol.* 8: R812–R822. © Elsevier.)

KEY EXPERIMENTS

Box 16-5 Genetic Approaches That Identified Genes Involved in the Lytic/Lysogenic Choice

Genes involved in lytic/lysogenic choice were identified by screening for λ mutants that efficiently grow only either lytically or lysogenically. To understand how these mutants were found, we need to consider how phage are grown in the laboratory (see Chapter 21). Bacterial cells can be grown as a confluent, opaque lawn across an agar plate. A lytic phage, grown on that lawn, produces clear plaques, or holes (Fig. 21-3). Each plaque is typically initiated by a single phage infecting a bacterial cell. The progeny phage from that infection then infect surrounding cells, and so on, killing off (lysing) the bacterial cells in the vicinity of the original infected cell and causing a clear cell-free zone in the otherwise opaque lawn of bacterial cells.

Bacteriophage λ forms plaques too, but they are turbid (or cloudy)—that is, the region within the plaque is clearer than the uninfected lawn, but only marginally so. The reason for this is that λ, unlike a purely lytic phage, kills only a proportion of the cells it infects; the others survive as lysogens. Lysogens are resistant to subsequent infection and so can grow within the plaque unharmed by the mass of phage particles found there. The reason for this "immunity" is quite simple: in a lysogen, the integrated phage DNA (the prophage) continues making repressor from P_{RM}. Any new λ genome entering that cell will at once be bound by repressor, giving no chance of lytic growth.

In one classic study, mutants of λ that formed clear plaques were isolated. These mutant phage are unable to form lysogens but still grow lytically. The λ clear mutations identified the three phage genes, called cI, cII, and cIII (for clear I, II, and III). In other studies, so-called virulent (*vir*) mutations were isolated. These mutations define the operator sites where λ repressor binds and were isolated by virtue of the fact that such phage can grow on lysogens. By analogy to the *lac* system, the cI mutants are comparable to the Lac repressor (*lacI*) mutants; *vir* mutants are the equivalent of the *lac* operator (*lacO*) mutants (see Box 16-2). Another revealing mutation was identified in a different experiment, this one a mutation in a host gene. The mutant is called *hfl* for high frequency of lysogeny. When infected with wild-type λ, this strain almost always forms lysogens, very rarely allowing the phage to grow lytically. This bacterial strain lacks the protease that degrades the λ CII protein (see text).

CII is a very unstable protein in *E. coli;* it is degraded by a specific protease called FtsH (HflB), encoded by the *hfl* gene (described in Box 16-5). The speed with which CII can direct synthesis of repressor is thus determined by how quickly it is being degraded by FtsH. Cells lacking the *hfl* gene (and thus FtsH) almost always form lysogens upon infection by λ: in the absence of the protease, CII is stable and directs synthesis of ample repressor. FtsH activity is itself regulated by the growth conditions of the bacterial cell, and, although it is not understood exactly how this is achieved, we can state the following. If growth is good, FtsH is very active, CII is destroyed efficiently, repressor is not made, and the phage tend to grow lytically. Under poor growth conditions, the opposite happens: low FtsH activity, slow degradation of CII, repressor accumulation, and a tendency toward lysogenic development. Levels of CII are also modulated by the phage protein CIII. CIII stabilizes CII, probably because it acts as an alternative (and thus competing) substrate for FtsH.

A second CII-dependent promoter, P_I, has a sequence similar to that of P_{RE} and is located in front of the phage gene *int* (see Fig. 16-20); this gene encodes the integrase enzyme that catalyzes site-specific recombination of λ DNA into the bacterial chromosome to form the prophage (see Chapter 11). A third CII-dependent promoter, P_{AQ}, located in the middle of gene *Q*, acts to retard lytic development and thus to promote lysogenic development. This is because the P_{AQ} RNA acts as an antisense message, binding to the Q message and promoting its degradation. Q is another regulator, one that promotes the late stages of lytic growth, as discussed the next section.

Transcriptional Antitermination in λ Development

Two examples of transcriptional regulation *after* initiation are found in λ development, as we now describe. We start with a type of positive transcriptional regulation called **antitermination.**

The transcripts controlled by the λ N and Q proteins are initiated perfectly well in the absence of those regulators. But the transcripts terminate a few hundred to a thousand nucleotides downstream from the promoter unless RNA polymerase has been modified by the regulator; λ N and Q proteins are therefore called antiterminators.

N protein regulates early gene expression by acting at three terminators: one to the left of the *N* gene itself, one to the right of *cro*, and one between genes *P* and *Q* (Figs. 16-20 and 16-31). Q protein has one target, a terminator 200 nucleotides downstream from the late gene promoter, $P_{R'}$, located between the *Q* and *S* genes (see Fig. 16-31). The late gene operon of λ, transcribed from $P_{R'}$, is remarkably large for a prokaryotic transcription unit: about 26 kb, a distance that takes about 10 minutes for RNA polymerase to traverse.

Our understanding of how antiterminators work is incomplete. Like other regulatory proteins, N and Q only work on genes that carry sequences specific for each regulator. Thus, N protein prevents termination in the early operons of λ, but not in other bacterial or phage operons. The specific recognition sequences for antiterminators are not found in the terminators where they act, but instead occur somewhere between the promoter and the terminator. For N, those sites are called *nut* (for *N utilization*) sites which are 60 and 200 nucleotides downstream from P_L and P_R (see Fig. 16-31). But N does not bind to these sequences within DNA. Rather, N binds to RNA transcribed from DNA containing a *nut* sequence. Thus, once RNA polymerase has passed a *nut* site, N binds to the RNA and from there is loaded on to the polymerase itself. In this state, the polymerase is resistant to the terminators found just beyond the *N* and *cro* genes. λ N works together with the products of the bacterial genes *nusA*, *nusB*, *nusE*, and *nusG*. The NusA

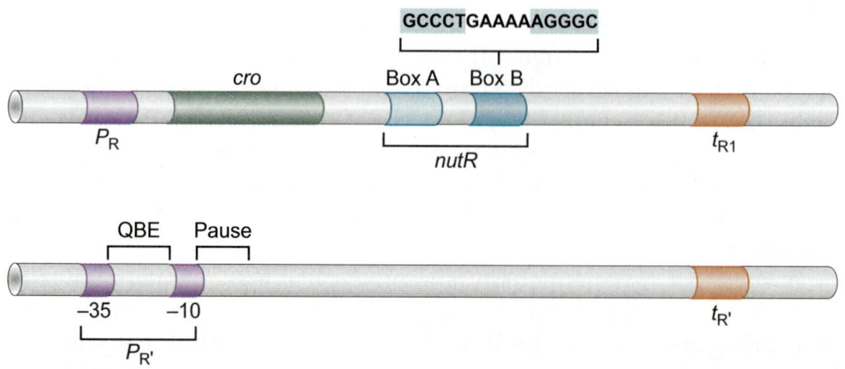

FIGURE 16-31 Recognition sites and sites of action of the λ N and Q transcription antiterminators. The upper line shows the early rightward promoter P_R and its initial terminator, t_{R1}. The *nut* site is divided into two regions, called Box A (7 bp) and Box B, separated by a spacer region of 8 bp. The sequence of Box B has dyad symmetry and forms a stem-loop structure once transcribed into RNA. The sequence of the RNA-like strand of *nutR* is shown above. The lower line shows the promoter $P_{R'}$, the sequences essential for Q protein function, and the terminator at which Q protein acts.

protein is an important cellular transcription factor. NusE is the small ribosomal subunit protein S10, but its role in N protein function is unknown. No cellular function of NusB protein is known. These proteins form a complex with N at the *nut* site, but N can work in their absence if present at high concentration, suggesting that it is N itself that promotes antitermination.

Unlike N protein, the λ Q protein recognizes DNA sequences (QBE) between the −10 and −35 regions of the late gene promoter ($P_{R'}$) (see Fig. 16-31). In the absence of Q, polymerase binds $P_{R'}$ and initiates transcription, only to pause after a mere 16 or 17 nucleotides; it then continues but terminates when it reaches the terminator ($t_{R'}$) about 200 bp downstream. If Q is present, it binds to QBE once the polymerase has left the promoter and transfers from there to the nearby paused polymerase. With Q on board, the polymerase is then able to transcribe through $t_{R'}$.

Recently, it has become clear that the σ factor of polymerase is involved in Q function (see Fig. 16-32). First, the reason polymerase pauses just after initiation at $P_{R'}$ is because it encounters a sequence resembling the "−10" element of a promoter. Region 2 of σ typically recognizes that sequence, binding to base pairs in the nontemplate strand as described Chapter 12. It does the same at this pause site, halting polymerase progress temporarily. At the same time, the nascent transcript exiting the RNA channel of the enzyme facilitates

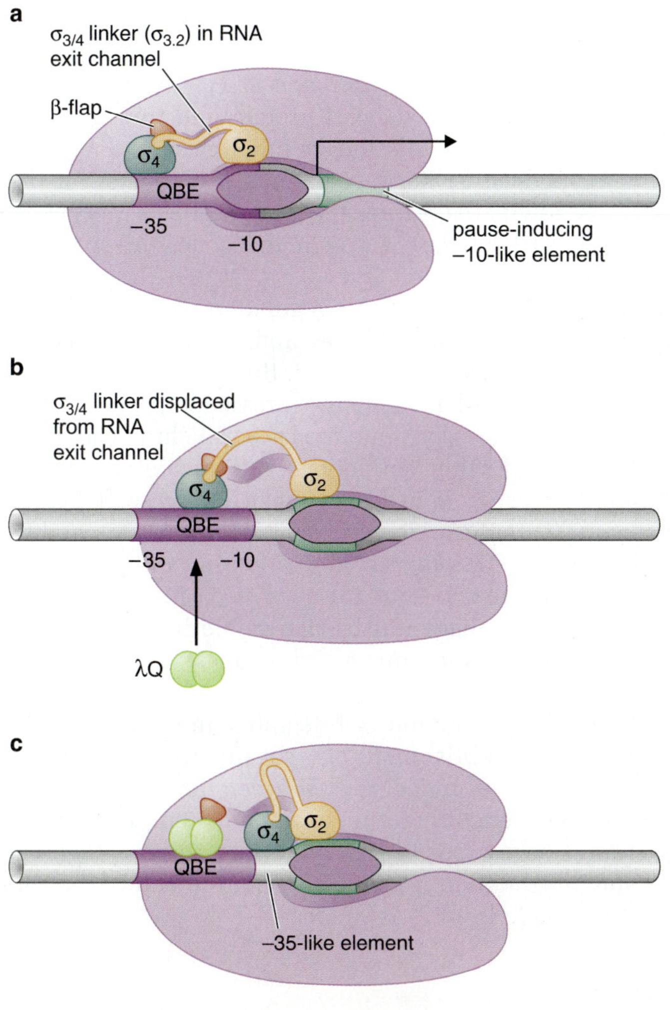

FIGURE 16-32 How λ Q engages RNA polymerase during early elongation. The sequence of events at λ P_R. (a) The organization of polymerase elements in the initiation complex bound to λ P_R. $\sigma_{3/4}$ linker ($\sigma_{3.2}$) is shown within the RNA exit channel. (b) The paused complex. The nascent transcript (not shown) has displaced $\sigma_{3/4}$ linker from the exit channel. This is just before binding of λ Q. (c) λ Q is shown bound to the paused elongation complex. Further details of the process are provided in the text. (Courtesy of Ann Hochschild.)

rearrangements of the interface between σ and the core enzyme, revealing part of σ region 4 that was previously buried (as described in Chapter 12). This surface of σ is then bound by Q.

Why this new complex of polymerase and Q is impervious to the downstream terminator is still unclear. But clearly σ can be involved in regulation downstream from initiation and that σ region 4 can be a target for regulators working at initiation and afterward as well.

Retroregulation: An Interplay of Controls on RNA Synthesis and Stability Determines *int* Gene Expression

The CII protein activates the promoter P_I that directs expression of the *int* gene, as well as the promoter P_{RE} responsible for repressor synthesis (see Fig. 16-20). The Int protein is the enzyme that integrates the phage genome into that of the host cell during formation of a lysogen (see Chapter 11). Therefore, upon infection, conditions favoring CII protein activity give rise to a burst of both repressor and integrase enzyme.

But the *int* gene is transcribed from P_L as well as from P_I, so one would have thought that integrase should be made even in the absence of CII protein. This does not happen. The reason is that *int* mRNA initiated at P_L is degraded by cellular nucleases, whereas mRNA initiated at P_I is stable and can be translated into integrase protein. This occurs because the two messages have different structures at their 3′ ends.

RNA initiated at P_I stops at a terminator about 300 nucleotides after the end of the *int* gene; it has a typical stem-and-loop structure followed by six uridine nucleotides (Fig. 16-33; see Chapter 12, Fig. 12-12). When RNA synthesis is initiated at P_L, on the other hand, RNA polymerase is modified by the N protein and thus goes through and beyond the terminator. This longer mRNA can form a stem that is a substrate for nucleases. Because the site responsible for this negative regulation is downstream from the gene it affects, and because degradation proceeds backward through the gene, this process is called **retroregulation**.

The biological function of retroregulation is clear. When CII activity is low and lytic development is favored, there is no need for integrase enzyme; thus, its mRNA is destroyed. But when CII activity is high and lysogeny is favored, the *int* gene is expressed to promote recombination of the repressed phage DNA into the bacterial chromosome.

There is yet a further subtlety in this regulatory device. When a prophage is induced, it needs to make integrase (together with another enzyme, called excisionase; see Chapter 11) to catalyze reformation of free phage DNA by recombination out of the bacterial DNA; it must do this whether or not CII activity is high. Thus, under these circumstances, the phage must make stable integrase mRNA from P_L despite the antitermination activity of N protein. How is this achieved?

When the phage genome is integrated into the bacterial chromosome during the establishment of lysogeny, the phage attachment site at which recombination occurs is *between* the end of the *int* gene and those sequences encoding the extended stem from which mRNA degradation is begun (see Fig. 16-20). Thus, in the integrated form, the site causing degradation is removed from the end of the *int* gene, and so *int* mRNA made from P_L is stable.

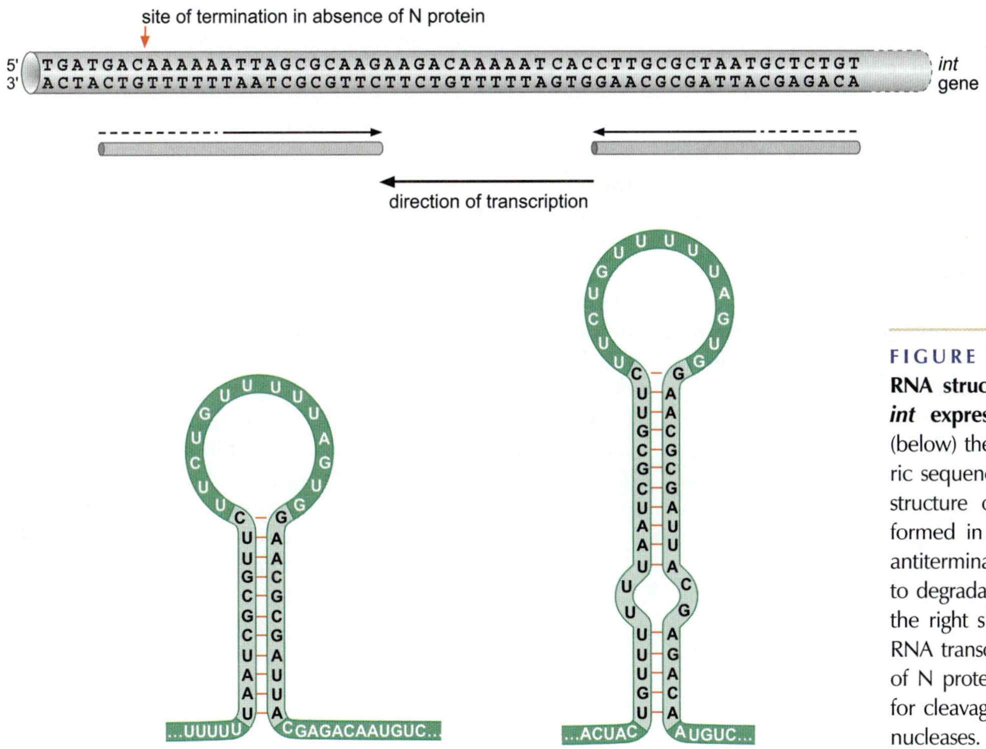

FIGURE 16-33 DNA site and transcribed RNA structures active in retroregulation of *int* expression. (Top) The DNA sequence; (below) the small cylinders show the symmetric sequences that form hairpins in RNA. The structure on the left shows the terminator formed in RNA transcribed from $P_{I'}$ without antitermination by N protein, which is resistant to degradation by nucleases. The structure on the right shows an extended loop formed in RNA transcribed from P_L under the influence of N protein antiterminator, which is a target for cleavage by RNase III and degradation by nucleases.

SUMMARY

A typical gene is switched on and off in response to the need for its product. This regulation is predominantly at the level of transcription initiation. Thus, for example, in *E. coli*, a gene encoding the enzyme that metabolizes lactose is transcribed at high levels only when lactose is available in the growth medium. Furthermore, when glucose (a better energy source) is also available, the gene is not expressed even when lactose is present.

Signals, such as the presence of a specific sugar, are communicated to genes by regulatory proteins. These are of two types: *activators*, positive regulators that switch genes on, and *repressors*, negative regulators that switch genes off. Typically, these regulators are DNA-binding proteins that recognize specific sites at or near the genes they control.

Activators, in the simplest (and most common) cases, work on promoters that are inherently weak; that is, RNA polymerase binds to the promoter (and thus initiates transcription) poorly in the absence of any regulator. An activator binds to DNA with one surface and with another surface binds polymerase and recruits it to the promoter. This process is an example of cooperative binding and is sufficient to stimulate transcription.

Repressors can inhibit transcription by binding to a site that overlaps the promoter, thereby blocking RNA polymerase binding. Repressors can work in other ways as well: for example, by binding to a site beside the promoter and by interacting with polymerase bound at the promoter, inhibiting initiation.

The *lac* genes of *E. coli* are controlled by an activator and a repressor that work in the simplest way just outlined. CAP, in the absence of glucose, binds DNA near the *lac* promoter and, by recruiting polymerase to that promoter, activates expression of those genes. The Lac repressor binds a site that overlaps the promoter and shuts off expression in the absence of lactose.

Another way in which RNA polymerase is recruited to different genes is by the use of alternative σ factors. Thus, different σ factors can replace the most prevalent one (σ^{70} in *E. coli*) and direct the enzyme to promoters of different sequences. Examples include σ^{32}, which directs transcription of genes in response to heat shock, and σ^{54}, which directs transcription of genes involved in nitrogen metabolism. Phage SPO1 uses a series of alternative σs to control the ordered expression of its genes during infection.

There are, in bacteria, examples of other kinds of transcriptional activation as well. Thus, at some promoters, RNA polymerase binds efficiently unaided and forms a stable, but inactive, closed complex. This closed complex does not spontaneously undergo transition to the open complex and initiate transcription. At such a promoter, an activator must stimulate the transition from a closed to open complex.

Activators that stimulate this kind of promoter work by allostery: they interact with the stable, closed complex and induce a conformational change that causes transition to the open complex. In this chapter, we saw two examples of transcriptional activators working by allostery. In one case, the activator (NtrC) interacts with the RNA polymerase (bearing σ^{54}) bound in a stable closed complex at the *glnA* promoter, stimulating transition to the open complex. In the other example, the activator (MerR) induces a conformational change in the *merT* promoter DNA.

In all the cases that we have considered, the regulators themselves are controlled allosterically by signals; that is, the shape of the regulator changes in the presence of its signal. In one state, it can bind DNA, and in the other state, it cannot. Thus, for example, the Lac repressor is controlled by the ligand allolactose (a product made from lactose). When allolactose binds repressor, it induces a change in the shape of that protein; in that state, the protein cannot bind DNA.

Gene expression can be regulated at steps after transcription initiation. For example, regulation can be at the level of transcriptional elongation. Examples considered in this chapter were antitermination by the N and Q proteins of bacteriophage λ. The λ proteins N and Q load on to RNA polymerases initiating transcription at certain promoters in the phage genome. Once modified in this way, the enzyme can pass through certain transcriptional terminator sites that would otherwise block expression of downstream genes.

We concluded this chapter with a detailed discussion of how bacteriophage λ chooses between two alternative modes of propagation. Several of the strategies of gene regulation encountered in this system turn out to operate in other systems as well, including, as discussed in later chapters, those that govern the development of animals—for example, the use of cooperative binding to give stringent on/off switches and the use of separate pathways for establishing and maintaining expression of genes. We also considered how complex and intricate gene networks like that found in λ might have evolved from more rudimentary earlier versions.

BIBLIOGRAPHY

Books

Baumberg S., ed. 1999. *Prokaryotic gene expression.* Oxford University Press, Oxford.

Müller-Hill B. 1996. *The* lac *operon.* de Gruyter, Berlin.

Ptashne M. 2005. *A genetic switch: Phage lambda revisited,* 3rd ed. Cold Spring Harbor Laboratory Press, Cold Spring Harbor, New York.

Ptashne M. and Gann A. 2002. *Genes & signals.* Cold Spring Harbor Laboratory Press, Cold Spring Harbor, New York.

Activation and Repression

Adhya S., Geanacopoulos M., Lewis D.E., Roy S., and Aki T. 1998. Transcription regulation by repressosome and by RNA polymerase contact. *Cold Spring Harbor Symp. Quant. Biol.* **63:** 1–9.

Busby S. and Ebright R.H. 1999. Transcription activation by catabolite activator protein (CAP). *J. Mol. Biol.* **293:** 199–213.

Dodd I.B., Shearwin K.E, and Egan J.B. 2005. Revisited gene regulation in bacteriophage λ. *Curr. Opin. Genet. Dev. Biol.* **15:** 145–152.

Dove S.L., Darst, S.E., and Hochschild A. 2003. Region 4 of σ as a target for transcription regulation. *Mol. Microbiol.* **48:** 863–874.

Gottesmann M. and Wesiberg R. 2004. Little lambda, who made thee? *Microbiol. Mol. Biol. Rev.* **68:** 796–813.

Hochschild A. and Dove S.L. 1998. Protein–protein contacts that activate and repress prokaryotic transcription. *Cell* **92:** 597–600.

Huffman J.L. and Brennan R.G. 2002. Prokaryotic transcription regulators: More than just the helix-turn-helix motif. *Curr. Opin. Struct. Biol.* **12:** 98–106.

Jacob F. and Monod J. 1961. Genetic regulatory mechanisms in the synthesis of proteins. *J. Mol. Biol.* **3:** 318–356.

Lawson C.L., Swigon D., Murakami K.S., Darst S.A., Berman H.M., and Ebright R.H. 2004. Catabolite activator protein: DNA binding and transcription activation. *Curr. Opin. Struct. Biol.* **14:** 10–20.

Magasanik B. 2000. Global regulation of gene expression. *Proc. Natl. Acad. Sci.* **97:** 14044–14045.

Müller-Hill B. 1998. Some repressors of bacterial transcription. *Curr. Opin. Microbiol.* **1:** 145–151.

Murray N.E. and Gann A. 2007. What has phage lambda ever done for us? *Curr. Biol.* **17:** R305–R312.

Oppenheim A.B., Oren Kobiler O., Stavans J., Court D.L., and Adhya S. 2005. Switches in bacteriophage lambda development. *Ann. Rev. Genet.* **39:** 409–429.

Ptashne M. 2006. Lambda's switch: Lesson from a module swap. *Curr. Biol.* **16:** R459–R462.

Ptashne M. and Gann A. 1997. Transcriptional activation by recruitment. *Nature* **386:** 569–577.

Rojo F. 2001. Mechanisms of transcriptional repression. *Curr. Opin. Microbiol.* **4:** 145–151.

Rombel I., North A., Hwang I., Wyman C., and Kustu S. 1998. The bacterial enchancer-binding protein NtrC as molecular machine. *Cold Spring Harbor Symp. Quant. Biol.* **63:** 157–166.

Roy S., Garges S., and Adhya S. 1998. Activation and repression of transcription by differential contact: Two sides of a coin. *J. Biol. Chem.* **273:** 14059–14062.

Schleif R. 2003. AraC protein: A love–hate relationship. *Bioessays* **25:** 274–282.

Wigneshweraraj S.R., Burrows P.C., Bordes P., Schumacher J., Rappas M., Finn R.D., Cannon W.V., Zhang X., and Buck M. 2005. The second paradigm for activation of transcription. *Prog. Nucleic Acid Res. Mol. Biol.* **79:** 339–369.

Xu H. and Hoover T.R. 2001. Transcriptional regulation at a distance in bacteria. *Curr. Opin. Microbiol.* **4:** 138–144.

DNA Binding, Cooperativity, and Allostery

Bell C.E. and Lewis M. 2001. The Lac repressor: A second generation of structural and functional studies. *Curr. Opin. Struct. Biol.* **11:** 19–25.

Hochschild A. 2002. The switch: *c*I closes the gap in autoregulation. *Curr. Biol.* **12:** R87–R89.

Luscombe N.M., Austin S.E., Berman H.M., and Thornton J.M. 2000. An overview of the structures of protein–DNA complexes. *Genome Biol.* **1:** REVIEWS001.

Monod J. 1966. From enzymatic adaptation to allosteric transitions. *Science* **154:** 475–483.

Vilar J.M.G. and Saiz L. 2005. DNA looping in gene regulation: From the assembly of macromolecular complexes to the control of transcriptional noise. *Curr. Opin. Genet. Dev. Biol.* **15:** 136–144.

Antitermination

Artsimovitch I. 2005. Control of transcription termination and antitermination. In *The bacterial chromosome* (ed. N.P. Higgins), Chapter 17. ASM Press, Washington, D.C.

Friedman D.I. and Court D.L. 2001. Bacteriophage λ: Alive and well and still doing its thing. *Curr. Opin. Microbiol.* **4:** 201–207.

Gottesman M. 1999. Bacteriophage λ: The untold story. *J. Mol. Biol.* **293:** 177–180.

Greenblatt J., Mah T.F., Legault P., Mogridge J., Li. J., and Kay L.E. 1998. Structure and mechanism in transcriptional antitermination by the bacteriophage λ N protein. *Cold Spring Harbor Symp. Quant. Biol.* **63:** 327–336.

Roberts J.W., Yarnell W., Bartlett E., Guo J., Marr M., Ko D.C., Sun H., and Roberts C.W. 1998. Antitermination by bacteriophage λ Q protein. *Cold Spring Harbor Symp. Quant. Biol.* **63:** 319–325.

Weisberg R.A. and Gottesman M.E. 1999. Processive antitermination. *J. Bacteriol.* **181:** 359–367.

CHAPTER

17 Transcriptional Regulation in Eukaryotes

OUTLINE

In eukaryotic cells, expression of a gene can be regulated at all those steps seen in bacteria and a few additional steps as well. Most striking among the additional steps is **splicing**, as we saw in Chapter 13. In many cases, a given transcript can be spliced in alternative ways to generate different products, and this too can be regulated.

But just as in bacteria, it is the initiation of transcription that is the most pervasively regulated step. Indeed, many of the principles we encountered when considering how transcription is regulated in bacteria apply to regulation of transcription in eukaryotes as well. These principles are laid out in the first few pages of the chapter on prokaryotic transcriptional regulation (Chapter 16) and in the summary at the end of that chapter. We urge readers who have not previously (or recently) read that chapter to look at those passages before continuing with this chapter.

We have also already seen that the eukaryotic transcriptional machinery is more elaborate than its bacterial counterpart (Chapter 12). This is particularly true of the RNA polymerase II machinery—that which transcribes protein-encoding genes. Despite this added complexity, transcription is once again regulated by activators and repressors, DNA-binding proteins that help or hinder transcription initiation at specific genes in response to appropriate signals. There are, however, additional features of eukaryotic cells and genes that complicate the actions of these regulatory proteins. We begin by summarizing the two most significant of these additional complexities.

Nucleosomes and their modifiers: As discussed in Chapter 7, the genome of a eukaryote is wrapped in proteins called histones to form nucleosomes. Thus, the transcriptional machinery is presented with a partially concealed substrate. This condition reduces the expression of many genes in the absence of regulatory proteins. Eukaryotic cells also contain a number of enzymes that rearrange, or chemically modify, histones; these modifications alter nucleosomes in ways that affect how easily the transcriptional machinery—and DNA-binding proteins in general—can bind and operate. Thus, nucleosomes present a problem not faced in bacteria, but their modification also offers new opportunities for regulation.

More regulators and more extensive regulatory sequences: A further difference between eukaryotes and prokaryotes is the number of regulatory proteins that control a typical gene. This is reflected in the number and arrangement of regulator binding sites associated with a typical gene. As in bacteria, individual regulators bind short sequences, but in eukaryotes, these binding sites are often more numerous and positioned further from the start site of transcription than they are in bacteria. We call the region at the gene where the transcriptional machinery binds the **promoter**, the individual binding sites **regulator binding sites**, and the stretch of DNA encompassing the complete collection of regulator binding sites for a given gene the **regulatory sequences.**

The expansion of regulatory sequences—that is, the increase in the number of binding sites for regulators at a typical gene—is most striking in multicellular organisms such as *Drosophila* and mammals. This situation reflects the more extensive signal integration found in those organisms: that is, the tendency for more signals to be required to switch a given gene on at the right time and place. We saw examples of signal integration in bacteria (Chapter 16), but those examples typically involved just two different regulators integrating two signals to control a gene (glucose and lactose at the *lac* genes, for example). Yeast have less signal integration than multicellular organisms—indeed, they are not so different from bacteria in this regard—and their genes have less extensive regulatory sequences than those of multicellular eukaryotes (Fig. 17-1).

In multicellular organisms, regulatory sequences can spread thousands of nucleotides from the promoter—both upstream and downstream—and can be made up of tens of regulator binding sites. Often, these binding sites are grouped in units called **enhancers,** and a given enhancer binds regulators responsible for activating the gene at a given time and place. Alternative enhancers bind different groups of regulators and control expression of the same gene at different times and places in response to different signals.

Having more extensive regulatory sequences means that some regulators bind sites far from the genes they control, in some cases 50 kb or more. How can regulators act from such a distance? In bacteria, we encountered DNA-binding proteins that communicate over a range of a few kilobases: λ repressors at O_R interacting with those at O_L; and NtrC, which can activate the *glnA* gene from sites placed 1 kb or more upstream. In those examples of "action at a distance," the intervening DNA loops out to accommodate the interaction between the proteins. The same mechanism explains action at a distance in many, if not all, eukaryotic cases as well, although in some cases the distances over which proteins work are very large and it is not clear how the looping occurs.

Activation at a distance raises another problem. When bound at an enhancer, there may be several genes within range of an activator, yet

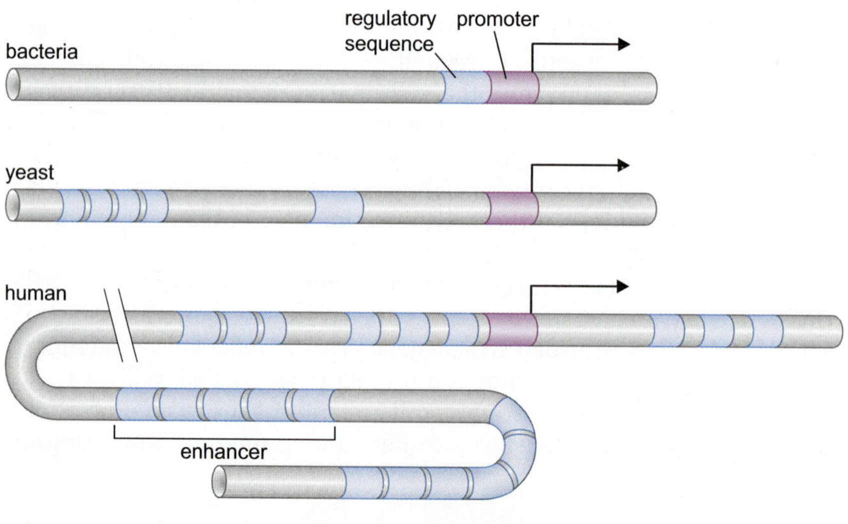

FIGURE 17-1 The regulatory elements of bacterial, yeast, and human genes. Illustrated is the increasing complexity of regulatory sequences from a simple bacterial gene controlled by a repressor to a human gene controlled by multiple activators and repressors. In each case, a promoter is shown at the site where transcription is initiated. Although this is accurate for the bacterial case, in the eukaryotic examples, transcription initiates somewhat downstream from where the transcription machine binds (see Chapter 12). Some groups of regulatory binding sites in the human regulatory sequences represent enhancers, as shown in one case.

a given enhancer typically regulates only one gene. Other regulatory sequences—called **insulators** or **boundary elements**—are found between enhancers and some promoters. Insulators block activation of the promoter by activators bound at the enhancer. These elements, although still poorly understood, ensure that activators do not work indiscriminately.

CONSERVED MECHANISMS OF TRANSCRIPTIONAL REGULATION FROM YEAST TO MAMMALS

In this chapter, we consider transcriptional regulation in organisms ranging from single-celled yeast to mammals. All of these organisms have both the more elaborate transcriptional machinery and the nucleosomes and their modifiers typical of eukaryotes. So it is not surprising that many of the basic features of gene regulation are the same in all eukaryotes. As yeast are the most amenable to a combination of genetic and biochemical dissection, much of the information about how activators and repressors work comes from that organism. In addition, critical to the generality of the conclusions, when expressed in a mammalian cell, a typical yeast activator can stimulate transcription. This is tested using a **reporter gene**. The reporter gene consists of binding sites for the yeast activator inserted upstream of the promoter of a gene whose expression level is readily measured (as we discuss below).

We shall see that the typical eukaryotic activator works in a manner similar to that of the simplest bacterial case: it has separate DNA-binding and activating regions and activates transcription by recruiting protein complexes to specific genes. In contrast, repressors work in a variety of ways, some different from anything we encountered in bacteria. These novel repression mechanisms include examples of what is called **gene silencing**, in which nucleosome and DNA modifiers are recruited to regions of the genome where they act to keep genes switched off, sometimes over large stretches of DNA.

Despite having so much in common, not all details of gene regulation are the same in all eukaryotes. Most importantly, as we have mentioned, a typical yeast gene has less-extensive regulatory sequences than its multicellular counterpart. So we must look to higher organisms to see how the basic mechanisms of gene regulation are extended to accommodate more complicated cases of signal integration and combinatorial control. In this chapter, we restrict discussion to transcriptional regulation mediated by proteins (and their modifications). In the following chapter, we discuss regulation of gene expression mediated by RNA molecules.

Activators Have Separate DNA-Binding and Activating Functions

In bacteria, we saw that a typical activator, such as CAP, has separate DNA-binding and activating functions. We described the genetic demonstration of this: positive control (or *pc*) mutants bind DNA normally, but they are defective in activation. Eukaryotic activators have separate DNA-binding and activating regions as well. Indeed, in this case, the two surfaces are very often on separate domains of the protein.

We take as an example the most studied eukaryotic activator, Gal4 (Fig. 17-2). This protein activates transcription of the galactose genes in the yeast *Saccharomyces cerevisiae*. These genes, like their bacterial counterparts, encode enzymes required for galactose metabolism. One

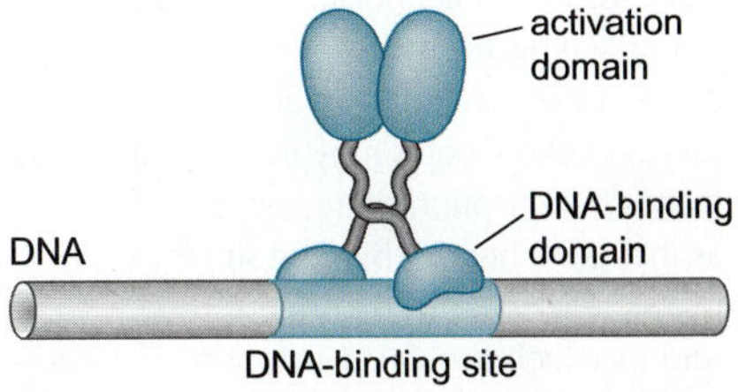

FIGURE 17-2 Gal4 bound to its site on DNA. The yeast activator Gal4 binds as a dimer to a 17-bp site on DNA. The DNA-binding domain of the protein is separate from the region of the protein containing the activating region (the activation domain).

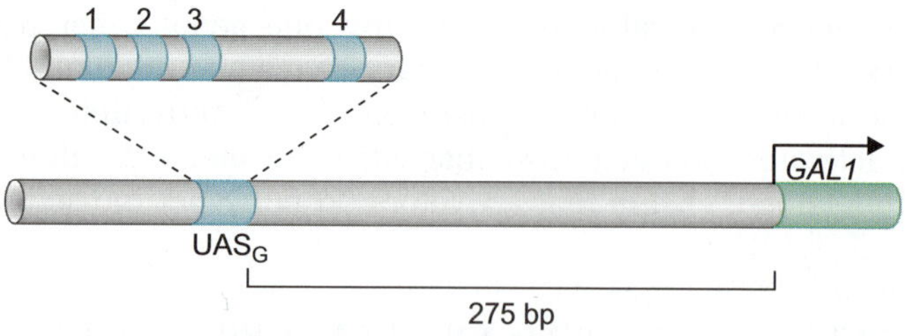

FIGURE 17-3 The regulatory sequences of the yeast *GAL1* gene. The UAS_G (upstream activating sequence for *GAL*) contains four binding sites, each of which binds a dimer of Gal4 as shown in Fig. 17-2. Although not shown here, there is another site between these and the *GAL1* gene that binds a repressor called Mig1, which is discussed later in the chapter (see Fig. 17-21).

such gene is *GAL1*. Gal4 binds to four sites located 275 bp upstream of *GAL1* (Fig. 17-3). When bound there, in the presence of galactose, Gal4 activates transcription of the *GAL1* gene 1000-fold.

The separate DNA-binding and activating regions of Gal4 were revealed in two complementary experiments. In one experiment, expression of a fragment of the *GAL4* gene—encoding the amino-terminal one-third of the activator—produced a protein that bound DNA normally but did not activate transcription. This protein contained the DNA-binding domain but lacked the activating region and was therefore formally comparable to the *pc* mutants of bacterial activators (Fig. 17-4a).

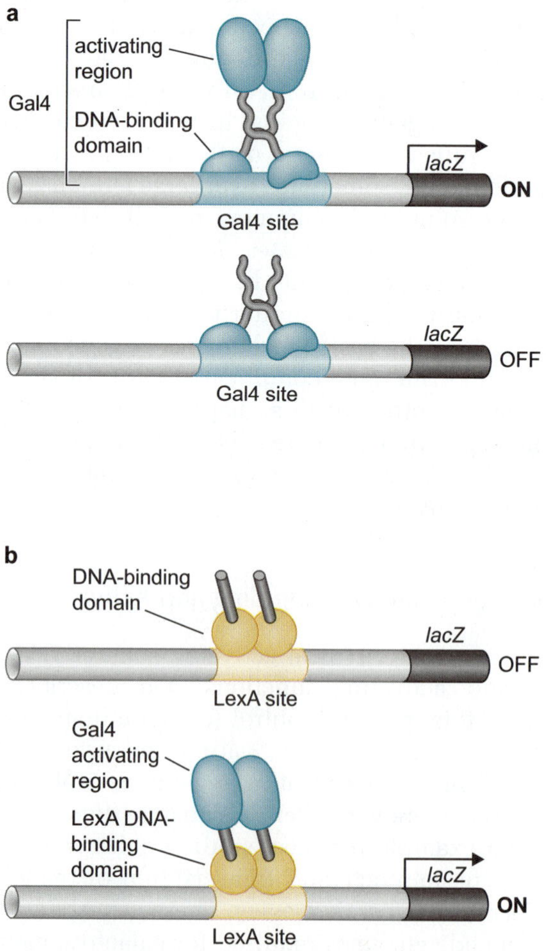

FIGURE 17-4 Domain swap experiment. (a) The DNA-binding domain of Gal4, without that protein's activation domain, can still bind DNA but cannot activate transcription. In another experiment (not shown), the activation domain, without the DNA-binding domain, also does not activate transcription. (b) Attaching the activation domain of Gal4 to the DNA-binding domain of the bacterial protein LexA creates a hybrid protein that activates transcription of a gene in yeast as long as that gene bears a binding site for LexA. Expression is measured using a reporter plasmid in which the *GAL1* promoter is fused to the *Escherichia coli lacZ* gene whose product (β-galactosidase) is readily assayed in yeast cells. Levels of expression from the *GAL1* promoter in response to the various activator constructs can therefore easily be measured. Similar reporter plasmids are used in many experiments in this chapter.

In a second experiment, a hybrid gene was constructed that encoded the carboxy-terminal three-quarters of Gal4 fused to the DNA-binding domain of a bacterial repressor protein, LexA. The fusion protein was expressed in yeast together with a reporter plasmid bearing LexA-binding sites upstream of the *GAL1* promoter. The fusion protein activated transcription of this reporter (Fig. 17-4b). This experiment shows that activation is not mediated by DNA binding alone, as it was in one of the alternative mechanisms we encountered in bacteria—activation by MerR (Fig. 16-16). Instead, the DNA-binding domain serves merely to tether the activating region to the promoter just as in the most common mechanism we saw in bacteria.

Many other eukaryotic activators have been examined in similar experiments, and whether from yeast, flies, or mammals, the same story typically holds: DNA-binding domains and activating regions are separable. In some cases, they are even carried on separate polypeptides: one has a DNA-binding domain and the other has an activating region, and they form a complex on DNA. An example of this is the herpes virus activator VP16, which interacts with the Oct1 DNA-binding protein found in cells infected by that virus. Another example is the *Drosophila* activator Notch, described in Chapter 19. The separable nature of DNA-binding and activating regions of eukaryotic activators is the basis for a widely used assay to detect protein–protein interactions (see Box 17-1, The Two-Hybrid Assay).

Eukaryotic Regulators Use a Range of DNA-Binding Domains, but DNA Recognition Involves the Same Principles as Found in Bacteria

The experiments described above show that a bacterial DNA-binding domain can function in place of the DNA-binding domain of a eukaryotic activator. That result suggests there is no fundamental difference in the ways DNA-binding proteins from these organisms recognize their sites.

Recall from the previous chapter that most bacterial regulators bind as dimers to DNA target sequences that are twofold rotationally symmetric; each monomer inserts an α helix into the major groove of the DNA over one-half of the site and detects the edges of base pairs found there. Binding typically requires no significant alteration in the structure of either the protein or the DNA. The vast majority of bacterial regulatory proteins use the so-called helix-turn-helix motif. This motif, as we saw, consists of two α helices separated by a short turn. One helix (the recognition helix) fits in the major groove of the DNA and recognizes specific base pairs. The other helix makes contacts with the DNA backbone, positioning the recognition helix properly and increasing the strength of binding (see Fig. 5-20).

The same basic principles of DNA recognition are used in most eukaryotic cases, despite variations in detail. Thus, proteins often bind as dimers and recognize specific DNA sequences using an α helix inserted into the major groove. One class of eukaryotic regulatory protein presents the recognition helix as part of a structure very like the helix-turn-helix domain; others present the recognition helix within quite different domain structures. In a variation not seen in prokaryotes, several of the regulatory proteins we encounter in eukaryotes bind DNA as **heterodimers,** and in some cases, even as monomers (although often only when binding cooperatively with other proteins). Heterodimers extend the range of

TECHNIQUES

Box 17-1 The Two-Hybrid Assay

This assay is used to identify proteins that interact with each other. Thus, in the case shown in Box 17-1 Figure 1, activation of a reporter gene depends on the fact that protein A interacts with protein B (even though these proteins need not themselves normally have a role in transcriptional activation). The assay is predicated on the finding, discussed in the text, that the DNA-binding domain and activating region can be on separate proteins, as long as those proteins interact, and the activating region is thereby tethered to the DNA near the gene to be activated. Practically, the assay is carried out as follows. The gene encoding protein A is fused to a DNA fragment encoding the DNA-binding domain of Gal4. The gene for a second protein (B) is fused to a fragment encoding an activating region. Neither protein alone, when expressed in a yeast cell, activates the reporter gene carrying Gal4-binding sites (as shown in the first two lines of the figure). When both hybrid genes are expressed together in a yeast cell, however, the interaction between proteins A and B generates a complete activator, and the reporter is expressed, as shown in the bottom line of the figure. In a widely used elaboration of this simple assay, the two-hybrid assay is employed to screen a library of candidates to find any protein that will interact with a known starting protein. So now, protein A in the figure would be the starting protein (called the "bait"), whereas protein B (the "prey") represents one of many alternatives encoded by the library (see Chapter 21 for a description of how libraries are made). Yeast cells are transfected with the construct encoding protein A fused to the DNA-binding domain, together with the library encoding many unknown proteins fused to the activating region. Thus, each transfected yeast cell contains protein A tethered to DNA and one or another alternative protein B fused to an activating region. Any cell containing a combination of A and B that interact will activate the reporter gene. Such a cell will form a colony that can be identified by plating on suitable indicator medium. Typically, the reporter gene would be *lacZ,* and positive colonies (those comprising cells expressing the reporter gene) would be blue on appropriate indicator plates.

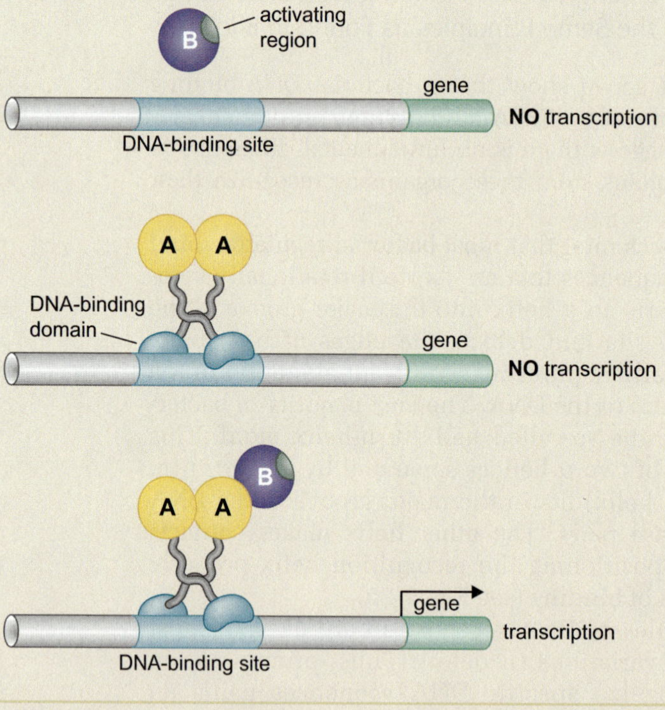

BOX 17-1 FIGURE 1 How the two-hybrid assay works. The reporter gene used in such an assay would typically be *lacZ* or some other gene that makes an easily assayed product.

DNA-binding specificities available: when each monomer has a different DNA-binding specificity, the site recognized by the heterodimer is different from that recognized by either homodimer. Below is a brief survey of some eukaryotic DNA-binding domains.

Homeodomain Proteins The homeodomain is a class of helix-turn-helix DNA-binding domain and recognizes DNA in essentially the same way as those bacterial proteins (Fig. 17-5). Homeodomains from different proteins are structurally very similar: not only is the recognition helix similar, but so is the surrounding protein structure that presents that helix to the DNA. In contrast, as we saw in the previous chapter, the detailed structures of helix-turn-helix domains vary to a greater extent. Homeodomain proteins are found in all eukaryotes. They were discovered in *Drosophila* where they control many basic developmental programs, just as they do in higher eukaryotes; we consider their functions in that regard in Chapter 19. Homeodomain proteins are also found in yeast; some of the mating-type control genes discussed below encode homeodomain proteins. Indeed, it is the structure of one of these that is shown in Figure 17-5. Many homeodomain proteins bind DNA as heterodimers.

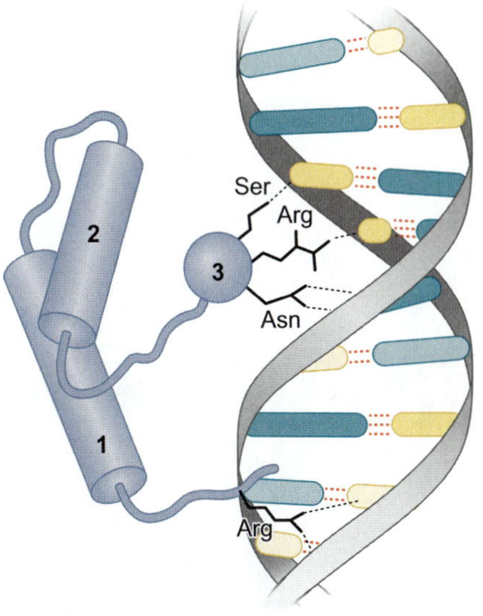

FIGURE 17-5 DNA recognition by a homeodomain. The homeodomain consists of three α helices, of which two (helices 2 and 3 in the figure) form the structure resembling the helix-turn-helix motif (compare this figure with Fig. 16-12, for example). Thus, helix 3 is the recognition helix, and, as shown, it is inserted into the major groove of DNA. Amino acid residues along its outer edge make specific contacts with base pairs. In the case shown, the yeast α2 transcriptional repressor, an arm extending from helix 1, makes additional contacts with base pairs in the minor groove. (Adapted, with permission, from Wolberger C. et al. 1991. *Cell* 67: 517–528. © Elsevier.)

Zinc-Containing DNA-Binding Domains There are various forms of DNA-binding domains that incorporate a zinc atom(s). These include the classically defined **zinc finger** proteins (such as the general transcription factor TFIIIA (Chapter 12) that is involved in the expression of a ribosomal RNA gene) and the related **zinc cluster** domain found in the yeast activator Gal4. In these cases, the zinc atom interacts with cysteine and histidine residues and serves a structural role essential for integrity of the DNA-binding domain (Fig. 17-6). The DNA is again recognized by an α helix inserted into the major groove. Some proteins contain two or more zinc finger domains linked end to end. Each finger inserts an α helix into the major groove, extending—with each additional finger—the length of the DNA sequence recognized and thus the affinity of binding.

There are other DNA-binding domains that use zinc. In those cases, the zinc is coordinated by four cysteine residues and stabilizes a

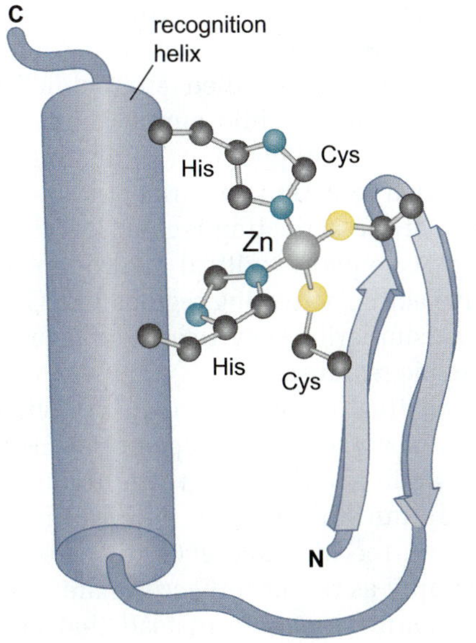

FIGURE 17-6 Zinc finger domain. The α helix on the left of the structure is the recognition helix, and it is presented to the DNA by the β sheet on the right. The zinc is coordinated by the two histidine residues in the α helix and two cysteine residues in the β sheet as shown. This arrangement stabilizes the structure and is essential for DNA binding. (Adapted from Lee M.S. et al. 1989. *Science* 245: 635–637.)

rather different DNA-recognition motif, one resembling a helix-turn-helix. An example of this is found in the glucorticoid receptor, which regulates genes in response to certain hormones in mammals.

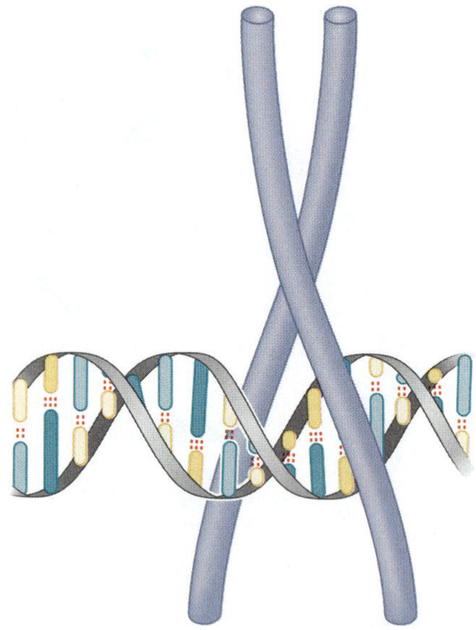

FIGURE 17-7 Leucine zipper bound to DNA. Two large α helices, one from each monomer, form both the dimerization and DNA-binding domain at different sections along their length. Thus, as shown, toward the top, the two helices interact to form a coiled-coil that holds the monomers together; further down, the helices separate enough to embrace the DNA, inserting into the major groove on opposite sides of the DNA helix. Once again, specificity is provided by contacts made between amino acid side chains on the α helices and the edge of base pairs in the major groove. An example of this is found in the yeast transcriptional activator, GCN4 (Fig. 5-15). (Adapted, with permission, from Ellenberger T.G. et al. 1992. *Cell* 71: 1223. © Elsevier.)

Leucine Zipper Motif This motif combines dimerization and DNA-binding surfaces within a single structural unit. As shown in Figure 17-7, two long α helices form a pincer-like structure that grips the DNA, with each α helix inserting into the major groove half a turn apart. Dimerization is mediated by another region within those same α helices: in this region, they form a short stretch of coiled-coil, wherein the two helices are held together by hydrophobic interactions between appropriately spaced leucine (or other hydrophobic) residues. We discussed this protein–protein interaction in more detail in Chapter 5 (Fig. 5-15). Leucine zipper–containing proteins often form heterodimers as well as homodimers. This is also true of our final category, the so-called helix-loop-helix proteins (HLH proteins).

Helix-Loop-Helix Proteins As in the example of the leucine zipper, an extended α-helical region from each of two monomers inserts into the major groove of the DNA. As shown in Figure 17-8, the dimerization surface is formed from two helical regions: the first is part of the same helix involved in DNA recognition; the other is a shorter α helix. These two helices are separated by a flexible loop that allows them to pack together (and gives the motif its name). Leucine zipper and HLH proteins are often called **basic zipper** and **basic HLH proteins:** this is because the region of the α helix that binds DNA contains basic amino acid residues.

Activating Regions Are Not Well-Defined Structures

In contrast to DNA-binding domains, activating regions do not always have well-defined structures. They have been shown to form helical structures when interacting with their targets within the transcriptional machinery, but it is believed that these structures are "induced" by that binding. As we shall see, the lack of defined structure is consistent with the idea that activating regions are adhesive surfaces capable of interacting with several protein surfaces.

Instead of being characterized by structure, therefore, activating regions are grouped on the basis of amino acid content. The activating region of Gal4, for example, is called an "acidic" activating region, reflecting a preponderance of acidic amino acids. The importance of these acidic residues is highlighted by mutations that increase the activator's potency: such mutations invariably increase the overall acidity (negative charge) of the activating region. But despite this, the activating region contains equally critical hydrophobic residues. Many other activators have acidic activating regions like Gal4. Although these show little sequence similarity, they retain the characteristic pattern of acidic and hydrophobic residues.

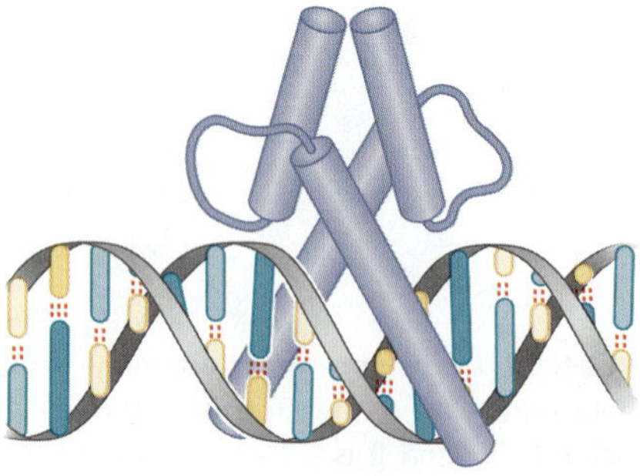

FIGURE 17-8 Helix-loop-helix motif. In this case, we again see a long α helix involved in both DNA recognition and, in combination with a second, shorter, α-helix, dimerization. (Adapted, with permission, from Ma P.C. et al. 1994. *Cell* 77: 451, Fig. 2A. © Elsevier.)

It is believed that activating regions consist of reiterated small units, each of which has a weak activating capacity on its own. Each unit is a short sequence of amino acids. The greater the number of units, and the more acidic each unit, the stronger the resulting activating region. This is consistent with the idea that activating regions lack an overall structure and act simply as rather indiscriminate "sticky" surfaces. (To understand this reasoning, imagine instead that an activating region folded into a precise, stable three-dimensional structure, comparable to, for example, a DNA-binding domain. Under those circumstances, fragments of that domain would not be expected to retain a fraction of the DNA-binding activity of the intact domain; rather the entire domain is needed for any significant activity. But if each activating region is simply a general adhesive surface, it is easy to imagine it being made up of smaller, weaker units.)

There are other kinds of activating regions. These include glutamine-rich activating regions such as that found on the mammalian activator SP1. In addition, proline-rich activating regions have been described—for example, on another mammalian activator CTF1. These too lack defined structure. In general, whereas acidic activating regions are typically strong and work in any eukaryotic organism in which they have been tested, other activating regions are weaker and work less universally than members of the acidic class.

RECRUITMENT OF PROTEIN COMPLEXES TO GENES BY EUKARYOTIC ACTIVATORS

Activators Recruit the Transcriptional Machinery to the Gene

We saw in bacteria that in the most common case, an activator stimulates transcription of a gene by binding to DNA with one surface, and with another, interacting with RNA polymerase and recruiting the enzyme to that gene (see Fig. 16-1). Eukaryotic activators also work this way, but rarely, if ever, through a direct interaction between the activator and RNA polymerase. Instead, the activator recruits polymerase indirectly or recruits other factors needed after polymerase has bound. Thus, the activator can interact with parts of the transcriptional machinery other than polymerase and, by recruiting them, recruit polymerase as well. In addition, activators can recruit nucleosome modifiers that alter chromatin in the vicinity of a gene and thereby help initiation. Finally, activators can recruit factors needed for polymerase to initiate or elongate. In all of these functions, the activator is merely recruiting proteins to the promoter. In bacteria, RNA polymerase is the only protein that needs to be recruited; this is not the case in eukaryotes. Indeed, in eukaryotes a given activator might work in all three ways. We first consider recruitment of the transcriptional machinery.

The eukaryotic transcriptional machinery contains numerous proteins in addition to RNA polymerase, as seen in Chapter 12. Many of these proteins come in preformed complexes such as the **Mediator** and the **TFIID complex** (see Table 12-2 and Fig. 12-16). Activators interact with one or more of these complexes and recruit them to the gene (Fig. 17-9). Other components that are not directly recruited by the activator bind cooperatively with those that are recruited.

Many proteins in the transcriptional machinery have been shown to bind to activating regions in vitro. For example, a typical acidic activating region can interact with components of the Mediator and with subunits of TFIID.

Recruitment can be visualized using the technique called **chromatin immunoprecipitation (ChIP),** described in Chapter 21. This technique reveals when a given protein binds to a defined region of DNA within a cell. At most genes (although not all, as we shall see presently), the transcriptional machinery appears at the promoter only upon

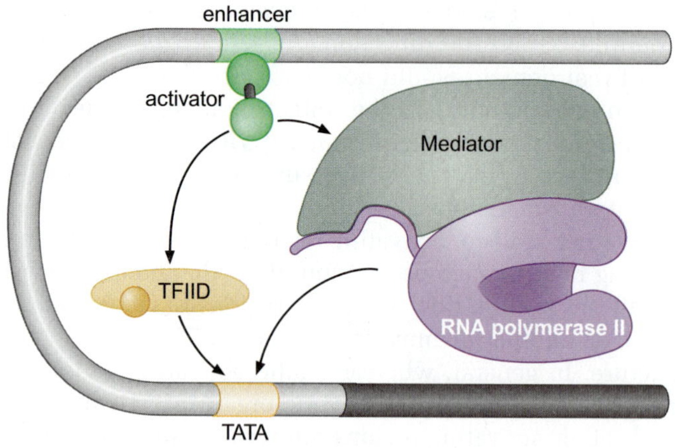

FIGURE 17-9 Activation of transcription initiation in eukaryotes by recruitment of the transcription machinery. A single activator is shown recruiting two possible target complexes: the Mediator and, through that, RNA polymerase II, as well as the general transcription factor TFIID. Other general transcription factors are either recruited as part of the Mediator, Pol II, or TFIID complexes or recruited separately by the activator, or they can bind spontaneously in the presence of the recruited components. These are not shown here. In reality, this recruitment would usually be mediated by more than one activator bound upstream of the gene.

activation of the gene. That is, the machinery is not prebound, confirming that the role of the activator is to recruit it.

In bacteria, we saw that genes activated by recruitment (such as the *lac* genes) can be activated in so-called activator bypass experiments (Box 16-1). In such an experiment, activation is observed when RNA polymerase is recruited to the promoter without using the natural activator-polymerase interaction. Similar experiments work in yeast. Thus, the *GAL1* gene (normally activated by Gal4) can be activated equally well by a fusion protein containing the DNA-binding domain of the bacterial protein LexA fused directly to a component of the Mediator complex (Fig. 17-10).

It is important to note that these experiments do not exclude the possibility that at least some activators not only recruit parts of the transcriptional machinery but also induce allosteric changes in them. Such changes might stimulate the efficiency of transcription initiation. Nevertheless, the recruitment of the machinery to one or another gene is the basis of specificity; that is, which gene is activated depends on which gene has the machinery recruited to it. In addition, the success of the activator bypass experiments suggests that any allosteric events that occur during initiation do not require the *activator* to do anything beyond recruiting proteins to the gene.

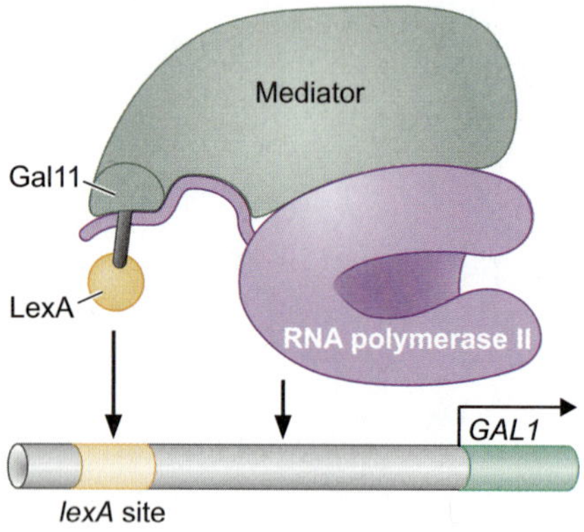

FIGURE 17-10 Activation of transcription through direct tethering of mediator to DNA. This is an example of an activator bypass experiment, as described in Box 16-1. In this case, the *GAL1* gene is activated, in the absence of its usual activator Gal4, by the fusion of the DNA-binding domain of LexA to a component of the Mediator complex (Gal11/Med15; see Fig. 12-19). Activation depends on LexA DNA-binding sites being inserted upstream of the gene. Other components required for transcription initiation—TFIID, etc.—presumably bind together with Mediator and Pol II.

Activators Also Recruit Nucleosome Modifiers That Help the Transcriptional Machinery Bind at the Promoter or Initiate Transcription

In addition to direct recruitment of the transcriptional machinery, recruitment of nucleosome modifiers can help activate a gene packaged within **chromatin.** As discussed in Chapter 7, nucleosome modifiers come in two types: those that add chemical groups to the tails of

HSF binds to specific sites at the promoter and recruits a kinase, P-TEF (positive transcription elongation factor), to the stalled initiated machinery. The kinase phosphorylates the carboxy-terminal domain (CTD) of the largest subunit of RNA polymerase (the so-called CTD "tail") freeing the enzyme from the stall and allowing transcription to proceed through the gene.

We saw in Chapter 12 that phosphorylation of the CTD tail is an important step in the early stages of transcription at all genes, and the kinase TFIIH can perform that phosphorylation. Whether P-TEF is also needed at most genes is not clear. A strong acidic activator like Gal4 is able to recruit P-TEF along with the rest of the machinery. It may be that only at certain genes is the recruitment of the machinery partitioned between regulators in the way we see at this *HSP70* gene, allowing an extra layer of control.

The human immunodeficiency virus (HIV), which causes AIDS, transcribes its genes from a promoter controlled by P-TEF. Again, polymerase initiates transcription at that promoter, under the control of the activator SP1, but stalls soon afterward. In this case, P-TEF is brought to the stalled polymerase by an RNA-binding protein, not a DNA-bound one. The protein responsible is called TAT. TAT recognizes a specific sequence near the start of the HIV RNA and present in the transcript made by the stalled polymerase. Another domain of TAT interacts with P-TEF and recruits it to the stalled polymerase.

Action at a Distance: Loops and Insulators

Many eukaryotic activators—particularly in higher eukaryotes—work from a distance. Thus, in a mammalian cell, for example, enhancers can be found several tens or even hundreds of kilobases upstream (or downstream) of the genes they control. We saw in bacteria that proteins bound to separated sites on DNA can nevertheless interact, a reaction accommodated by DNA looping. But in those cases, we were considering proteins binding only a few hundred base pairs apart. Under that condition, the proteins are bound sufficiently close to each other that their chance of interacting is much higher on DNA than off it. Once the sites to which they bind are separated by more than a few kilobases, this advantage is largely lost.

Mechanisms exist to help communication between distantly bound proteins. Recall, from bacteria, one way this can be done. The "architectural" protein IHF (integration host factor) binds to sites on DNA and bends it. At some genes controlled by NtrC, IHF sites are found between the activator-binding sites and the promoter. By bending the DNA, IHF helps the DNA-bound activator reach RNA polymerase at the promoter (see Fig. 16-4).

Various models have been proposed to explain how proteins binding between enhancers and promoters might help activation in the cells of higher eukaryotes. In *Drosophila*, the *cut* gene is activated from an enhancer some 100 kb away. A protein called Chip (nothing to do with the technique of that name!) aids communication between enhancer and gene. Thus, mutants in the gene encoding Chip affect the strength of activation. How Chip works is still not clear, but one model is that it binds to multiple DNA sites between the enhancer and the promoter and, by interacting with itself, forms multiple mini-loops in the intervening DNA, the cumulative effect of which is to bring the promoter and enhancer into closer proximity.

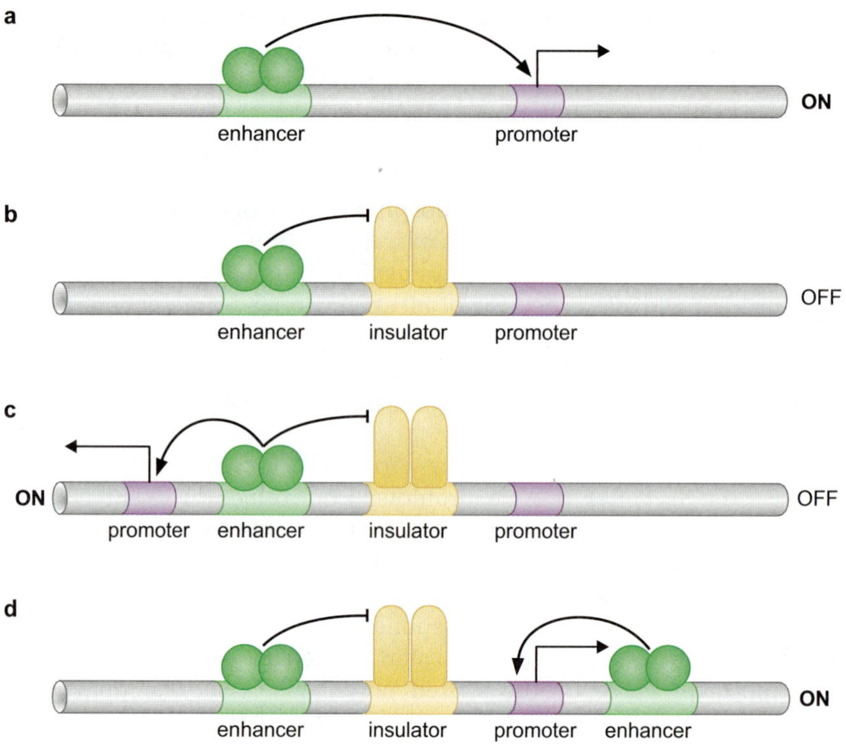

FIGURE 17-12 Insulators block activation by enhancers. (a) A promoter activated by activators bound to an enhancer. (b) An insulator is placed between the enhancer and the promoter. When bound by appropriate insulator-binding proteins, activation of the promoter by the enhancer is blocked, despite activators binding to the enhancer. (c,d) Neither the activators at the enhancer nor the promoter is inactivated by the action of the insulator. Thus, the activator can activate another promoter nearby (c), and the original promoter can be activated by another enhancer placed downstream (d).

There are other models. In eukaryotes, the DNA is wrapped in nucleosomes as we have seen, and the histones within these nucleosomes are subject to various modifications that affect their disposition, compactness, and perhaps the flexibility of the chromatin structure. Thus, sites separated by many base pairs may not, in effect, be as far apart in the cell as might have been thought. In addition, chromatin may in some places form special structures that actively bring enhancers and promoters closer together. In the next section, we describe a few cases of distant regulatory sequences and how they might work. But first, we consider another problem of long-distance activation.

If an enhancer activates a specific gene 50 kb away, what stops it from activating other genes whose promoters are within that range? Specific elements called **insulators** control the actions of activators. When placed between an enhancer and a promoter, an insulator inhibits activation of the gene by that enhancer. As shown in Figure 17-12, the insulator does not inhibit activation of that same gene by a different enhancer, one placed downstream from the promoter; nor does the insulator inhibit the original activator from working on a different gene. Thus, the proteins that bind insulators do not actively repress the promoter, nor do they inhibit the activities of the activators. Rather, they block communication between the two.

In other assays, insulators also inhibit the spread of chromatin modifications. As we have seen, the modification state of local chromatin influences gene expression. We shall see below that propagation of certain repressing histone modifications over stretches of chromatin lies at the heart of a phenomenon called transcriptional **silencing.** Silencing is a specialized form of repression that can spread along chromatin, switching off multiple genes without the need for each to bear binding sites for specific repressors. Insulator elements can block this spreading, and thus insulators protect genes from both indiscriminate activation and repression.

This situation has consequences for some experimental manipulations. A gene inserted at random into the mammalian genome is often "silenced" because it becomes incorporated into a particularly dense form of chromatin called heterochromatin. But if insulators are placed up- and downstream from that gene, they protect it from silencing.

Appropriate Regulation of Some Groups of Genes Requires Locus Control Regions

The human globin genes are expressed in red blood cells of adults and in various cells in the lineage that forms red blood cells during development. There are five different globin genes in humans (Fig. 17-13a). Although clustered, these genes are not all expressed at the same time; rather, the different genes are expressed at different stages of development starting with ε, then the γ genes, followed by δ and β. How is their expression regulated?

Each gene has its own collection of regulatory sites needed to switch that gene on at the right time during development and in the proper tissues. Thus, the β-globin gene (which is expressed in adult bone marrow) has two enhancers: one upstream of the promoter and the other downstream. Only in adult bone marrow are the correct regulators all active and present in appropriate concentrations to bind these enhancers. But more than this is required to switch on the various globin genes in the correct order.

A group of regulatory elements collectively called the **locus control region,** or **LCR,** is found 30–50 kb upstream of the whole cluster of globin genes. A similar situation is seen with the *HoxD* gene cluster in mice. These genes are involved in patterning the developing limbs and are expressed in a precise manner in the embryo (Chapter 19). The *HoxD* genes are controlled by an element called the GCR (global control region) that works like the LCR.

The LCR is made up of multiple sequence elements. Some of these have the properties of enhancers: that is, if these sequences are attached experimentally upstream of a reporter gene, they can activate

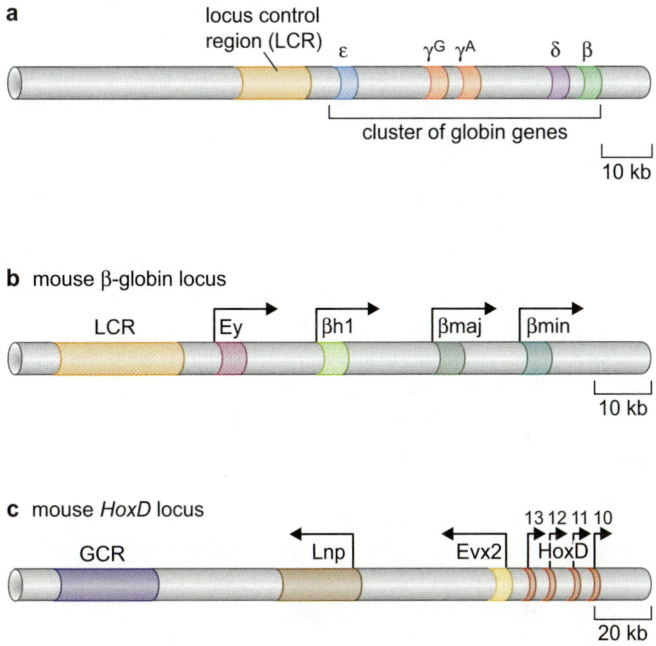

FIGURE 17-13 Regulation by LCRs. (a) The human globin genes and the LCR that ensures their ordered expression are shown. Not shown is the α-globin gene, which is expressed throughout development; its product combines with each of the globins shown here, in turn, to produce different forms of hemoglobins at different stages of development. (b) The globin genes from mice, which are also regulated by an LCR, are shown. (c) The *HoxD* gene cluster from the mouse controlled by an element called the GCR, which like the LCRs, appears to impose ordered expression on the gene cluster.

that gene. Other parts of the LCR act more like insulator elements and still others seem to have properties of promoters. This diversity of elements has led to numerous models for how LCRs might work. One model proposes that the entire transcriptional machinery is recruited to the LCR and from there transcribes all the way through the locus, opening up the chromatin as it goes and freeing up the local control elements in front of each gene. These individual promoters would then produce high level expression of each gene as required.

Recent experiments have used techniques that allow the locations of the LCR and promoter to be visualized in cells during activation. These studies have now been performed with a number of genes controlled by LCRs or LCR-like elements. In all cases, the results show that regulatory proteins bound to the upstream regulatory sequences are found in close proximity to the promoter as that promoter is activated. This is consistent with the idea that proteins bound at the LCR interact with others at the promoter with the intervening DNA looping out to accommodate the interaction. For more detail on how these long-distance mechanisms might work, see Box 17.2, Long-Distance Interactions on the Same and Different Chromosomes.

Activation by LCRs is associated with substantial chromatin modification. How this is linked to activation remains unclear. It might help "open up" the chromatin around the LCR itself or around the promoter. It might also alter the chromatin between the two in a manner that helps loop formation.

KEY EXPERIMENTS

Box 17-2 Long-Distance Interactions on the Same and Different Chromosomes

The colocalization of the LCR and promoters is shown in various ways, including chromosome conformation capture (3C) and RNA-TRAP (tagging and recovering of associated proteins), as well as DNA and RNA fluorescence in situ hybridization (FISH).

The first is a variation on the ChIP technique discussed in Chapter 21, which cross-links proteins to the DNA sites to which they are bound. After cross-linking, the DNA is sheared, and whatever short piece remains attached to the protein can then be identified. In 3C, the same basic procedure is followed, but in this case, the proteins bound to the enhancer (or LCR) are cross-linked not only to the DNA, but also to any other proteins with which they interact. If this includes proteins bound to other DNA sites (e.g., the promoter), then these DNA sites can also be identified as shown in Box 17-2 Figure 1.

There are other examples of activation at a distance where the regulatory sequences are not just distant from the promoters they control but are on a separate chromosome! These cases are also detected by the techniques mentioned above. In one example, regulatory regions within the mouse γ-interferon gene locus (on chromosome 10) have been shown to interact with the locus on chromosome 11 encoding the T-helper-cell-2 cytokine genes whose expression is known to be linked to that of the interferon gene in these cells.

Even more striking perhaps is activation of odor receptor genes (*OR* genes) in mouse sensory neurons. There are 1300 receptor genes in each neuron, but each neuron expresses just one of these genes. In addition, whichever gene is chosen, expression is activated by the same enhancer (*H*).

Using the 3C technique among others, it was recently shown that in individual neurons, the *H* enhancer is always found in proximity to the promoter of the particular gene expressed in that neuron whether or not that gene is on the same chromosome as the enhancer. This observation led to a model whereby the activators bound at the enhancer randomly select one odor receptor gene in a given cell (presumably whichever gene is first encountered). The fact that there is only a single copy of the enhancer ensures that only one gene can be activated in this way. Support for this model includes an experiment in which additional copies of the enhancer are artificially added, producing neurons that can express more than one odor receptor type each.

Earlier cases of interchromosomal *trans*-activation had been reported in *Drosophila*. In most of these cases (called **transvection**), one allele of a gene was activated by the enhancer associated with the other allele (on the sister chromosome). Thus, it was thought that the two elements were brought together through homologous pairing of the chromosomes. In a case such as the odor receptors, it is not clear how the enhancer and the target gene find each other. The fact that there are 1300 possible genes to choose from reduces the inefficiency of finding one, but still the mechanics remain a mystery. Each of the receptor genes probably has its own regulatory sequences nearby that work together with the *H* enhancer. Perhaps they help recruit the *H* enhancer to the gene.

BOX 17-2 *(Continued)*

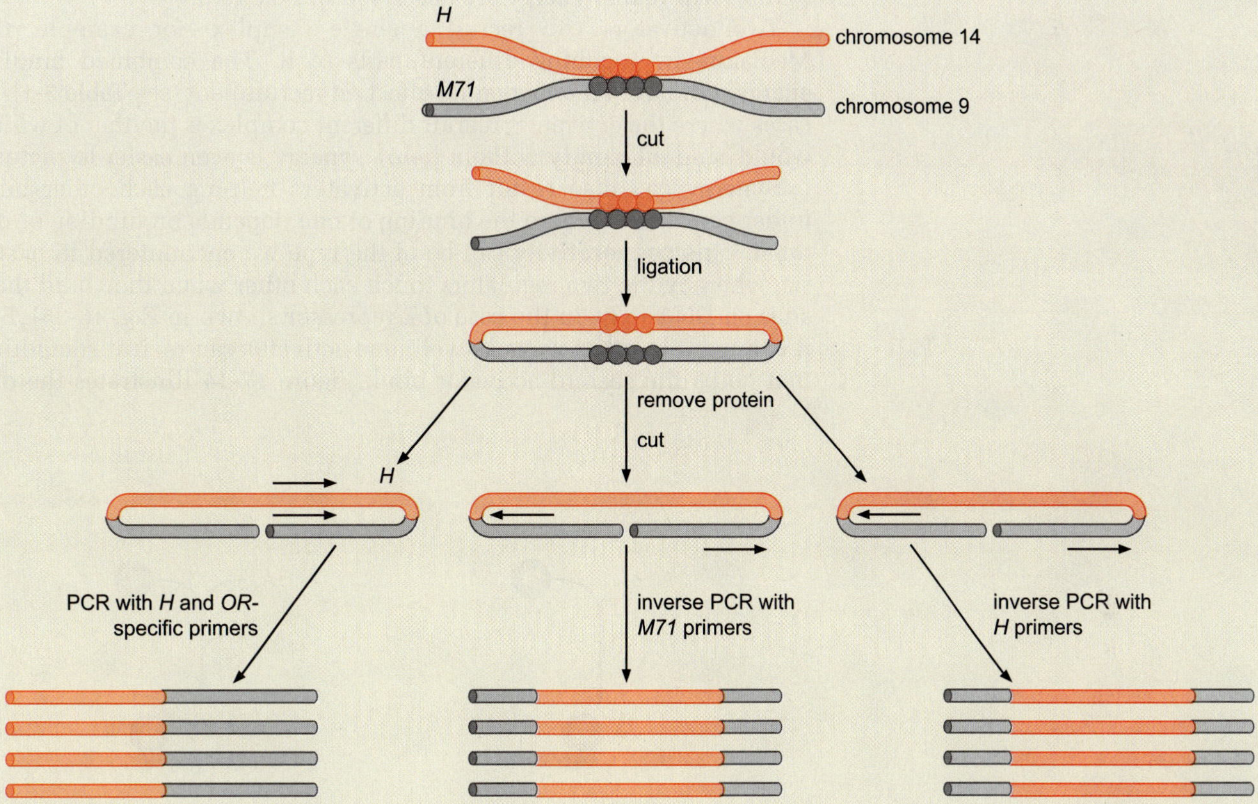

BOX 17-2 FIGURE 1 Chromosome conformation capture reveals the association of the *H* enhancer with olfactory receptor genes. The scheme shows three different 3C strategies to reveal the association of the *H* enhancer element with OR promoters. On the left, a polymerase chain reaction (PCR; see Chapter 21) primer specific to *H* is used with the second primer specific to *OR* promoters. In the center, a pair of inverse primers specific to *M71*, which encodes a mouse odorant receptor, is used to isolate sequences associated with the *M71* gene. On the right, a pair of inverse PCR primers specific to H is used to detect DNA sequences that colocalize with the H element. (Adapted, with permission, from Lomvardas et al. 2006. *Cell* 126: 403–413, Fig. 1A. © Elsevier.)

SIGNAL INTEGRATION AND COMBINATORIAL CONTROL

Activators Work Synergistically to Integrate Signals

In bacteria, we saw examples of signal integration in gene regulation. Recall, for example, that the *lac* genes of *E. coli* are efficiently expressed only when both lactose is present and glucose is absent. The two signals are communicated to the gene through separate regulators: one an activator and the other a repressor. In multicellular organisms, signal integration is used extensively. In some cases, numerous signals are required to switch a gene on. But just like the situation in bacteria, each signal is transmitted to the gene by a separate regulator, so at many genes multiple activators must work together to switch the gene on.

When multiple activators work together, they often do so **synergistically.** That is, the effect of, say, two activators working together is greater (usually much greater) than the sum of each of them working alone. Synergy can result from multiple activators recruiting a single component of the transcriptional machinery, multiple activators each recruiting a different component, or multiple activators helping each other bind to their sites near the gene they control. We briefly consider all three strategies before giving examples. Additional examples of transcriptional synergy are described in Box 19-5.

Two activators can recruit a single complex—for example, the Mediator—by touching different parts of it. The combined binding energy will have an exponential effect on recruitment (see Table 3-1). In cases where the activators recruit different complexes (neither of which would bind efficiently without help), synergy is even easier to picture.

Synergy can also result from activators helping each other bind under conditions where the binding of one depends on binding of the other. This **cooperativity** can be of the type we encountered in bacteria, whereby the two regulators touch each other when they bind their sites on DNA (e.g., in the case of λ repressor shown in Fig. 16-25). But it can work in other ways as well: one activator can recruit something that helps the second activator bind. Figure 17-14 illustrates the dif-

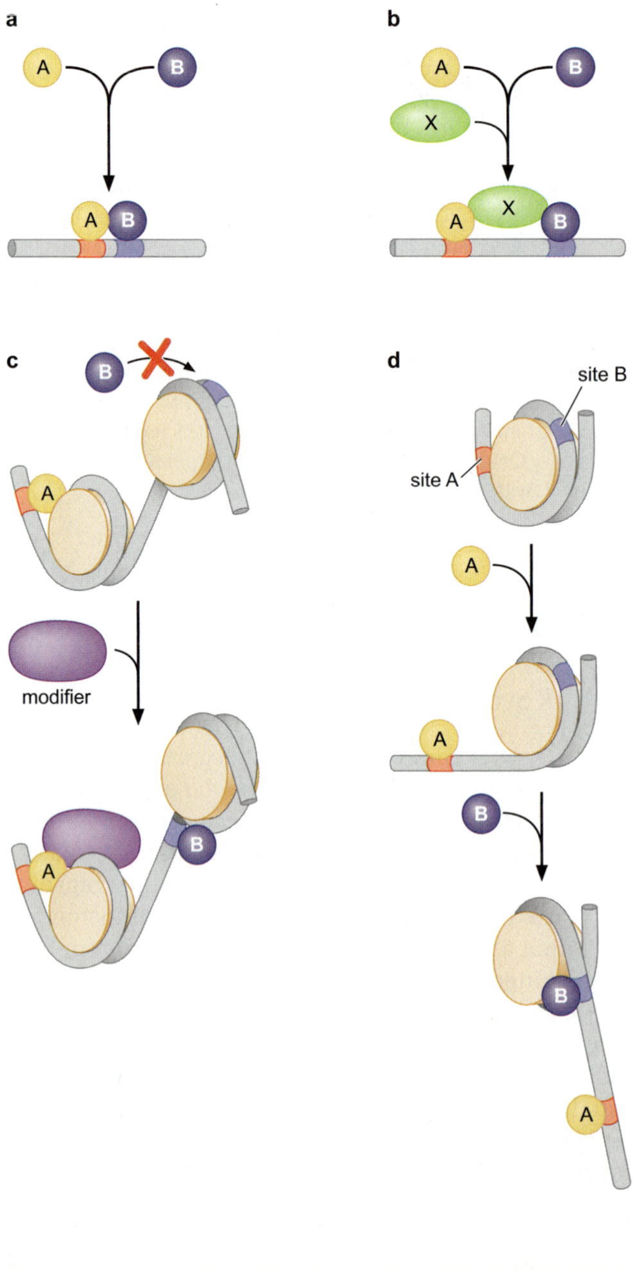

FIGURE 17-14 **Cooperative binding of activators.** Four ways that the binding of one protein to a site on DNA can help the binding of another to a nearby site. (a) Cooperative binding through direct interaction between the two proteins is shown, as we saw for λ repressor in Chapter 16 and shall see between many regulators in eukaryotes as well. (b) A similar effect is achieved by both proteins interacting with a common third protein. (c,d) Indirect effects in which binding of one protein to its site on DNA within nucleosomes helps binding of a second protein are shown. (c) The first protein recruits a nucleosome remodeler whose action reveals a binding site for a second protein. (d) The binding of the first protein to its site on the DNA just where it exits the nucleosome is shown. By binding there, it unwinds the DNA from the nucleosome a little, revealing the binding site for the second protein. Each of these mechanisms can explain how one regulator can help others bind or, indeed, how an activator can help the transcriptional machinery bind to a promoter.

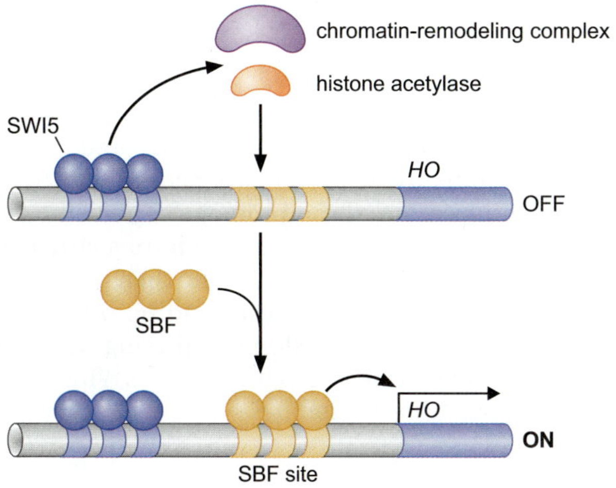

FIGURE 17-15 Control of the HO gene. SWI5 can bind its sites within chromatin unaided, but SBF cannot. Remodelers and histone acetylases recruited by SWI5 alter nucleosomes over the SBF sites, allowing that activator to bind near the promoter and activate the gene. In the figure, for simplicity, the nucleosomes are not drawn. (Adapted, with permission, from Ptashne M. and Gann A. 2002. *Genes & signals*, p. 95, Fig. 2.18. © Cold Spring Harbor Laboratory Press.)

ferent ways activators help each other bind DNA, including "classical" cooperative binding, recruitment of a modifier by one activator to help a second bind, and binding of one activator to nucleosomal DNA uncovering the binding site for another.

Synergy is critical for signal integration by activators. Consider a gene whose product is only needed when two signals are received. Each signal is communicated to the gene by a separate activator. The gene must be efficiently expressed when both activators are present but be relatively impervious to the action of either activator alone.

Signal Integration: The *HO* Gene Is Controlled by Two Regulators—One Recruits Nucleosome Modifiers and the Other Recruits Mediator

The yeast *S. cerevisiae* divides by budding; that is, instead of dividing to produce two identical daughter cells, the so-called mother cell buds to produce a daughter cell. We focus here on the expression of a gene called *HO*. (We need not concern ourselves here with the function of this gene, which is described in Chaper 11.) The *HO* gene is expressed only in mother cells and only at a certain point in the cell cycle (see Interactive Animation 17-1 and Chapter 19). These two conditions are communicated to the gene through two activators: SWI5 and SBF. SWI5 binds to multiple sites some distance from the gene, the nearest being more than 1 kb from the promoter (Fig. 17-15). SBF also binds multiple sites, but these are located closer to the promoter. Why does expression of the gene depend on both activators?

WEB ANIMATION

SBF (which is active only at the correct stage of the cell cycle) cannot bind its sites unaided; their disposition within chromatin prohibits it. SWI5 (which acts only in the mother cell) can bind to its sites unaided but cannot, from that distance, activate the *HO* gene (remember that in yeast, activators typically do not work over long distances). SWI5 can, however, recruit nucleosome modifiers (a histone acetyltransferase followed by the remodeling enzyme SWI/SNF). These act on nucleosomes over the SBF sites. Thus, if both activators are present and active, the action of SWI5 enables SBF to bind, and that activator, in turn, recruits the transcriptional machinery (by directly binding Mediator) and activates expression of the gene.

Signal Integration: Cooperative Binding of Activators at the Human β-Interferon Gene

The human β-interferon gene is activated in cells upon viral infection. Infection triggers three activators: NF-κB, IRF, and Jun/ATF. These proteins bind cooperatively to sites tightly packed within an enhancer located about 1 kb upstream of the promoter. The activators bind the enhancer in a highly cooperative manner to form a structure called an **enhanceosome** (Fig. 17-16). The activators then recruit a so-called coactivator, a protein called CBP (CREB-binding protein) or its close relative p300. This protein has histone-modifying activities and can recruit nucleosome-remodeling activities (e.g., SWI/SNF), as well as the transcriptional machinery itself.

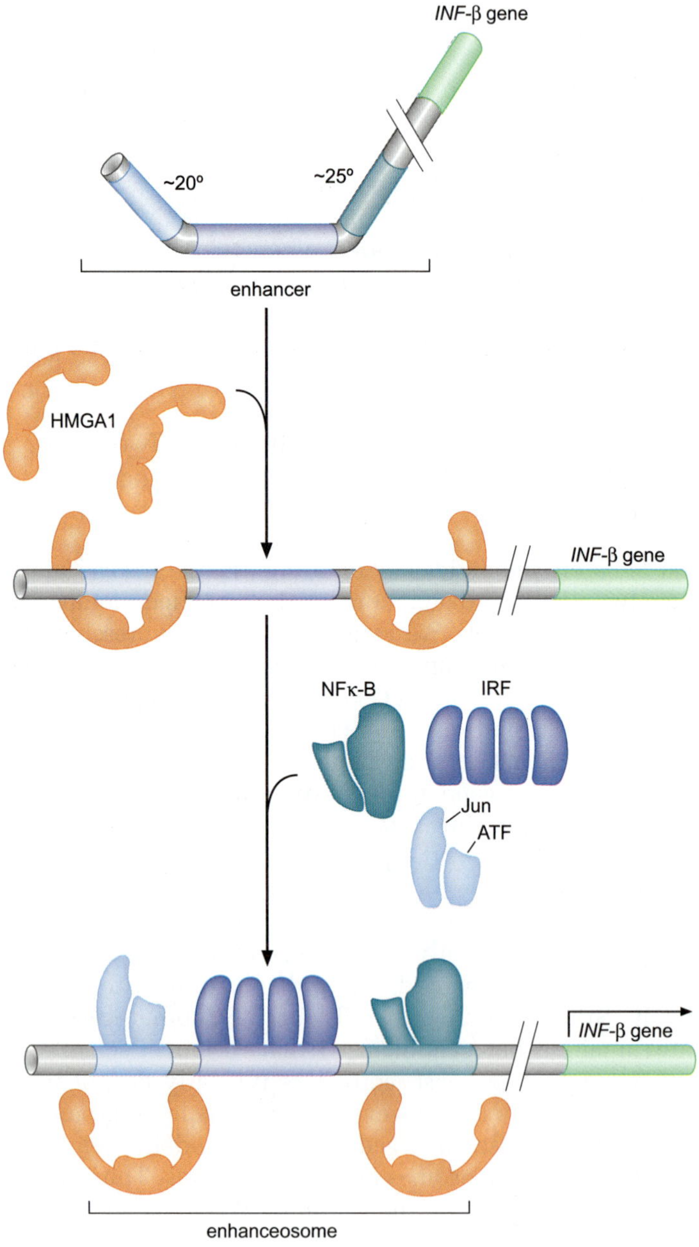

FIGURE 17-16 **The human β-interferon enhanceosome.** Cooperative binding of the three activators, together with the architectural protein HMGA1, activates the β-interferon gene.

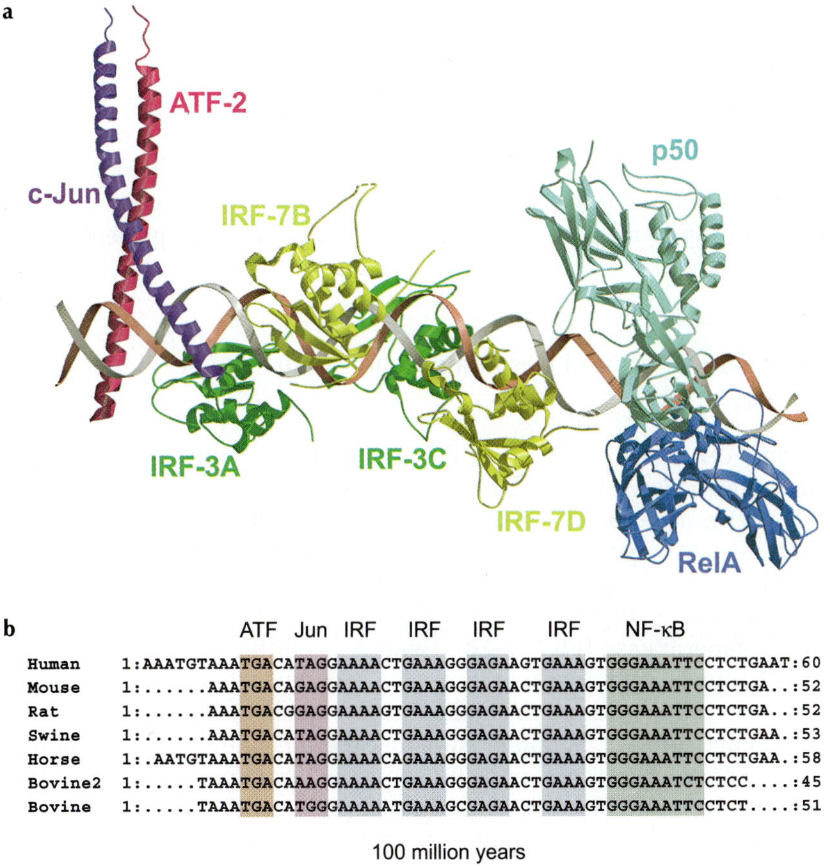

FIGURE 17-17 The enhanceosome structure and sequence. (a) The crystal structure of the enhanceosome, revealing the DNA-binding domains of the activators bound to the enhancer DNA. (b) The conservation of the interferon-β enhancer DNA sequences across species separated by 100 million years. Also indicated are the sequences within the enhancer recognized by each activator. (a, Panne D. et al. 2007. *Cell* 129: 1111. PDB Codes: 2O61, 2O6G.) This image is a combination of structures assembled as a model. Image prepared with MolScript, BobScript, and Raster 3D.

In addition to the activators listed above, another protein binds the enhancer—HMGA1. This protein binds in the minor groove on the opposite face of the DNA and helps in the assembly of the enhanceosome, although it is probably not part of the final structure. Indeed, it seems that it would be impossible for it to remain bound once all the activators are present; the DNA is simply too crowded.

Figure 17-16 shows how the enhanceosome is believed to assemble, and Figure 17-17a shows the crystal structure of the DNA-binding domains of all the activators bound to the enhancer DNA. As shown in Figure 17-16, the enhancer DNA is bent, but once the activators are bound it is straight; HMGA1 straightens the DNA and so helps the final structure form.

A striking feature of the structure is that essentially every base pair of DNA within the enhancer is involved in activator binding, which is why there is thought not to be room for HMGA1 in the final structure. This exhaustive use of the sequence information in the enhancer also likely explains why this enhancer is so highly conserved across organisms as diverse as human, mouse, and horse; indeed, it is even more highly conserved than the coding sequence of the gene itself (see Fig. 17-17b).

As we have noted, the activators bind—and the enhanceosome forms—in a highly cooperative manner, ensuring that all three activators must be present. The following are three ways the regulators might be binding cooperatively: (1) through direct protein–protein interactions between them; (2) by changes in the DNA caused by binding of one protein helping binding of another; and (3) by the fact the activators all interact simultaneously with the coactivator, CBP. All three might operate in this case, although it is hard to know the extent of protein–protein interactions between the activators; not much evidence of direct interactions is apparent in the structure, but then again, only the DNA-binding domain of the activators is in most cases present in the structure.

Combinatorial Control Lies at the Heart of the Complexity and Diversity of Eukaryotes

We encountered simple cases of **combinatorial control** in bacteria. For example, CAP is involved in regulating many genes, in collaboration with other regulators. At the *lac* genes, it works with the Lac repressor; at the *gal* genes, it works with the Gal repressor (see Chapter 16).

There is extensive combinatorial control in eukaryotes. We first consider a generic case (Fig. 17-18). Gene *A* is controlled by four signals (1, 2, 3, and 4), each working through a separate activator (activators 1, 2, 3, and 4). Gene *B* is controlled by three signals (3, 5, and 6), working through activators 3, 5, and 6. Note that there is one signal in common between these two cases, and the activator through which that signal works is the same at both genes. In complex multicellular organisms, such as *Drosophila* and humans, combinatorial control involves many more regulators and genes than shown in this kind of example, and, of course, repressors as well as activators can be involved. How is it that the regulators can intermix so promiscuously?

As we discussed above, multiple activators work synergistically. In fact, even multiple copies of a single activator work synergistically, suggesting that a given activator can interact with multiple targets. This provides an explanation for why different regulators can work together in so many combinations: because each can use any of an array of targets, the combinations that work together are unrestricted.

Both of the examples of signal integration we considered above—the *HO* gene in yeast and the human β-interferon gene—involve activators that also regulate other genes in examples of combinatorial control. Thus, from the yeast example, SWI5 is involved in regulating several other genes. In the mammalian case, NF-κB regulates not only the β-interferon gene, but numerous other genes including the

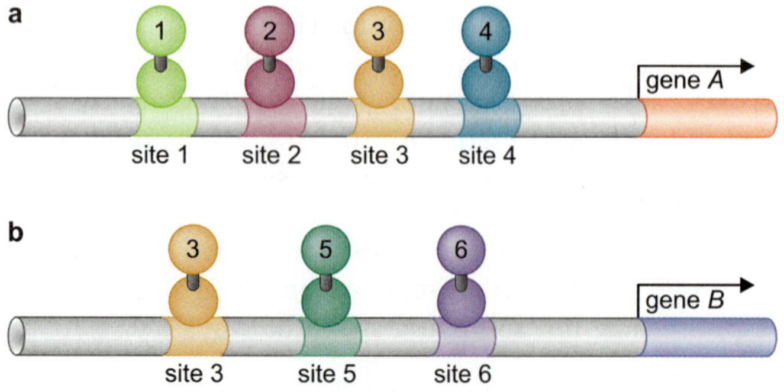

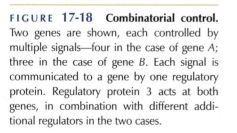

FIGURE 17-18 Combinatorial control. Two genes are shown, each controlled by multiple signals—four in the case of gene *A*; three in the case of gene *B*. Each signal is communicated to a gene by one regulatory protein. Regulatory protein 3 acts at both genes, in combination with different additional regulators in the two cases.

immunoglobulin κ light-chain gene in B cells. Jun/ATF likewise works with other regulators to control other genes. We described earlier that some DNA-binding proteins bind as heterodimers with alternative partners. This offers another level of combinatorial control.

Combinatorial Control of the Mating-Type Genes from *S. cerevisiae*

The yeast *S. cerevisiae* exists in three forms: two haploid cells of different mating types—**a** and α—and the diploid formed when an **a** and an α cell mate and fuse. Cells of the two mating types differ because they express different sets of genes: **a**-specific genes and α-specific genes. These genes are controlled by activators and repressors in various combinations, as we now briefly describe.

The **a** cell and the α cell each encodes cell-type-specific regulators: **a** cells make the regulatory protein **a**1, and α cells make the proteins α1 and α2. A fourth regulatory protein, called Mcm1, is also involved in regulating the mating-type-specific genes (and many other genes) and is present in both cell types. How do these various regulators work together to ensure that in **a** cells, **a**-specific genes are switched on and α-specific genes are off; vice versa in α cells; and in diploid cells, both sets are kept off?

The arrangement of regulators at the promoters of **a**-specific genes and α-specific genes is shown in Figure 17-19.

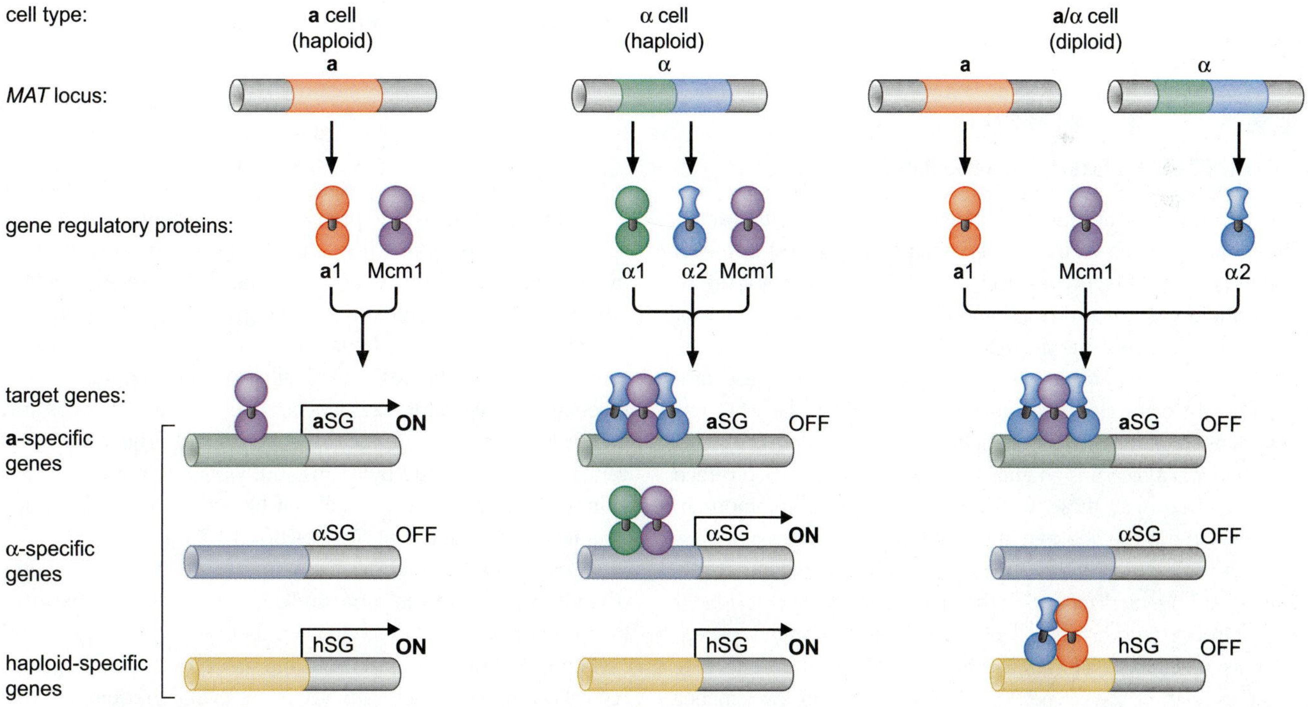

FIGURE 17-19 Control of cell-type-specific genes in yeast. As described in detail in the text, the three cell types of the yeast *S. cerevisiae* (the haploid **a** and α cells, and the **a**/α diploid) are defined by the sets of genes they express. One ubiquitous regulator (Mcm1) and three cell-type-specific regulators (**a**1, α1, and α2) together regulate three classes of target genes. The *MAT* locus is the region of the genome that encodes the mating-type regulators (Chapter 11).

- In **a** cells, the α-specific genes are off because no activators are bound there, whereas the **a**-specific genes are on because Mcm1 is bound and activates those genes.
- In α cells, the α-specific genes are on because Mcm1 is bound upstream and activates them. At these genes, Mcm1 binds to a weak site and does so only when it binds cooperatively with a monomer of the protein α1. This ensures that Mcm1 activates these genes only in α cells. The **a**-specific genes are kept off in α cells by the repressor α2. This repressor binds, as a dimer, cooperatively with Mcm1 at these genes. Two properties of α2 ensure **a**-specific genes are not expressed here: it covers the activating region of Mcm1, preventing that protein from activating; it also actively represses the genes. The mechanism by which α2 acts as a repressor is described in the next section.
- In diploid cells, both **a**- and α-specific genes are off. This is done as follows: the **a**-specific genes bind Mcm1 and α2, just as they do in α cells. This keeps those genes off. The α-specific genes are off because, as in **a** cells, no activators bind there.
- Both the haploid cell types (**a** and α) express another class of genes called haploid-specific genes. These are switched off in the diploid cell by α2, which binds upstream of them as a heterodimer with the **a**1 protein. Only in diploid cells are both of these regulators present.

The molecular details of mating-type gene regulation are now known for other yeast. In Box 17-3, Evolvability of a Regulatory Circuit, we compare how **a**- and α-specific genes are regulated in *S. cerevisiae* and *Candida albicans*. The comparison reveals how a gene regulatory circuit can evolve, a topic we return to in later chapters.

KEY EXPERIMENTS

Box 17-3 Evolvability of a Regulatory Circuit

As described in the text (and shown in Fig. 17-19), the different mating types of the yeast *S. cerevisiae* express some of their genes in a cell-type-specific way. Thus, in α cells, the α-specific genes are expressed and the **a**-specific genes are not, whereas in **a** cells, the **a**-specific genes are expressed and the α-specific genes are not. We know in detail how these programs are controlled by regulators encoded by the *MAT* loci working together with the ubiquitous regulator, Mcm1.

Candida albicans is another species of yeast, widely diverged from *S. cerevisiae*; these two species last shared a common ancestor somewhere between 300 and 900 million years ago. If expressed in terms of the divergence of conserved proteins, these two yeast are as divergent as fish and mammals. *S. cerevisiae* is used in beer and bread making, as well as in laboratory experiments, whereas *C. albicans* is a human pathogen. Nevertheless, just as with *S. cerevisiae*, *C. albicans* comes in two mating types—**a** and α—each characterized by the expression of distinct sets of genes (**a**-specific genes in **a** cells and α-specific genes in α cells). Again, these patterns of expression are controlled by Mcm1 working in conjunction with regulators encoded in the two *MAT* loci.

So the logic of the systems is the same, but the molecular interactions that control expression are not. As we have seen, in **a** cells of *S. cerevisiae*, the **a**-specific genes are switched on by the ubiquitous activator Mcm1 acting alone through a high-affinity binding site upstream of those genes (Fig. 17-19). Thus, no cell-type-specific regulator is involved. In contrast, expression of the **a**-specific genes in *C. albicans* does require a cell-type-specific regulator—**a**2 (encoded by the *MAT***a** locus in that species, but not found in *S. cerevisiae*). The **a**-specific genes are still activated by Mcm1, but working through a low-affinity site to which it only binds if helped by **a**2-binding cooperatively at an adjacent site (see Box 17-3 Fig. 1).

In α cells of *S. cerevisiae*, the **a**-specific genes are switched off by the action of the repressor α2, whereas in *C. albicans*, the absence of **a**2 is enough to keep them off.

The evolutionary transition in the mode of **a**-specific gene regulation from *C. albicans* to *S. cerevisiae* required two changes: (1) **a**-specific gene expression became independent of **a**2, and (2) **a**-specific genes came under the negative control of α2. By sequence comparisons between a number of yeast species, it is possible to describe the likely order of changes that took place in DNA-binding sites, the loss of a transcriptional regulator, and evolution of protein–protein in-

Box 17-3 ***(Continued)***

teractions between regulators. The first of these classes is the so-called *cis*-regulatory changes and the last two are *trans*-regulatory changes (in the sense Jacob and Monod originally defined these terms in the analysis of gene regulation; see Box 16-2, Jacob, Monod, and the Ideas behind Gene Regulation).

Of the several species examined, one (*Kluyveromyces lactis*) is illustrated in the figure (Box 17-3 Fig. 1). This organism falls between *S. cerevisiae* and *C. albicans* on the evolutionary tree and, consistent with that relationship, shows a mode of **a**-specific gene regulation between that seen in those other two cases. Thus, in *K. lactis*, **a**2 is present, and binds with Mcm1 to activate **a**-specific genes in **a** cells. But α2 is also involved in switching off those genes in α cells.

The following reconstruction has been proposed for the evolutionary transition in the regulation of **a**-specific genes. In the ancestral state, **a**2 and Mcm1 activated **a**-specific genes. Subsequently, the interaction between α2 and Mcm1 evolved, along with the evolution of an α2 DNA-binding site, and a strengthening of the Mcm1 DNA-binding site, in the **a**-specific gene promoter. After the divergence of *K. lactis*, the α2–Mcm1-binding specificity increased, and **a**2 was lost.

It is worth noting that the proposed intermediate in this transition between two forms of regulation (positive, in the case of *C. albicans*, negative for *S. cerevisiae*) has both forms. Thus, all the intermediates in the proposed evolutionary pathway retain proper regulation, because a new mode of regulation is added before an older one is dropped. This raises another possibility: different versions of a given gene circuit, such as these we have described for the mating-type regulation, are not strongly selected for or against. Either works equally well, and modern yeast will have whichever they end up with, and may even drift between them.

The case we have described here involves changes over time in all three features that can alter patterns of gene expression: changes in *cis*-regulatory sequences, changes in which regulatory genes are present, and changes in interactions between regulators. These are the very same kinds of changes needed to explain evolution of the phage λ switch (as we discussed in Box 16-4). They are also the same kinds of change underpinning the regulatory changes that account for much of animal diversity, as we see in Chapter 19. The role of *cis*-regulatory changes in particular is discussed in Box 19-7, *cis*-Regulatory Sequences in Animal Development and Evolution.

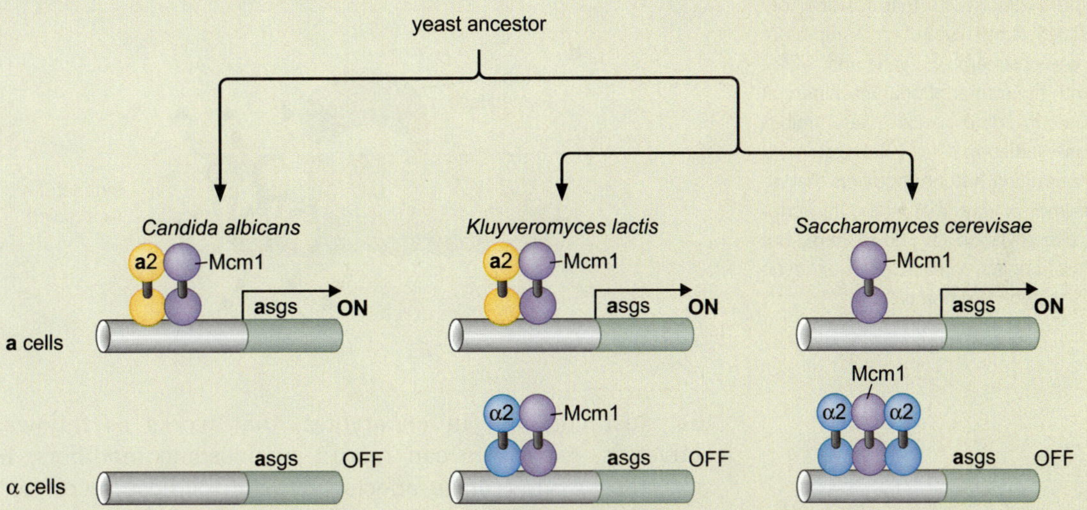

BOX 17-3 FIGURE 1 Regulation of a-specific genes in three species of yeast. (Adapted, with permission, from Rokas A. 2006. *Nature* 443: 401–402, Fig. 1. © Macmillan.)

TRANSCRIPTIONAL REPRESSORS

In bacteria, we saw that many repressors work by binding to sites that overlap the promoter and thus block binding of RNA polymerase. But we also saw other ways they can work: they can bind to sites adjacent to promoters and, by interacting with polymerase bound there, inhibit the enzyme from initiating transcription. They can also interfere with the action of activators.

In eukaryotes, we see all of these except the first (ironically, the most common in bacteria). We also see another form of repression, perhaps

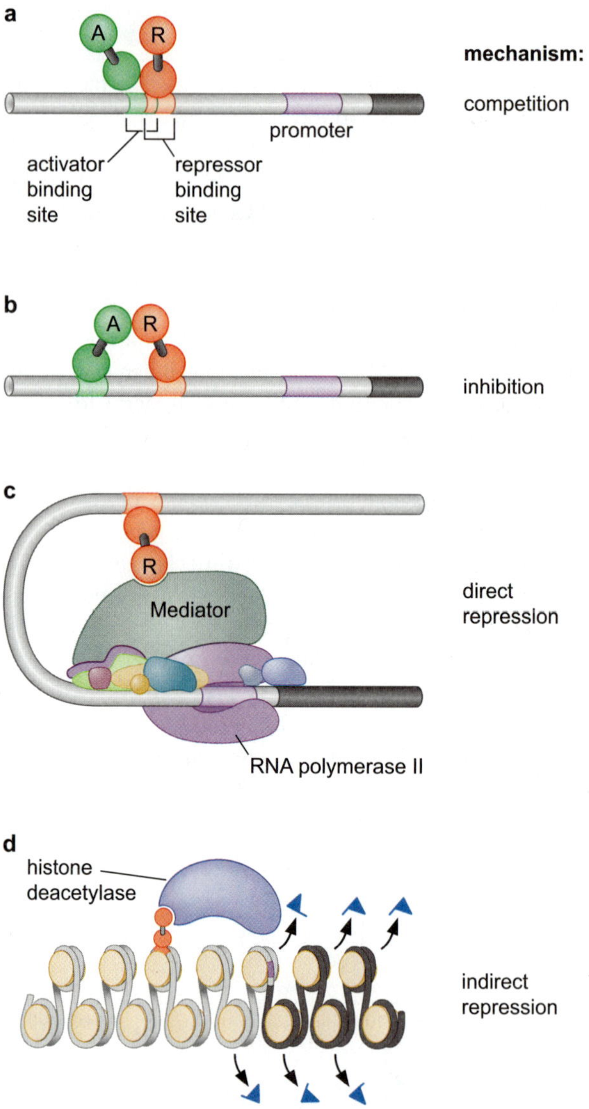

FIGURE 17-20 Ways in which eukaryotic repressors work. Transcription of eukaryotic genes can be repressed in various ways. These include the four mechanisms shown in the figure. (a) By binding to a site on DNA that overlaps the binding site of an activator, a repressor can inhibit binding of the activator to a gene and thus block activation of that gene. In a variation on this theme, a repressor can be a derivative of the same protein as the activator but lack the activating region. In another variation, an activator that binds to DNA as a dimer can be inhibited from doing so by a derivative that retains the region of the protein required for dimerization but lacks the DNA-binding domain. Such a derivative forms inactive heterodimers with the activator. (b) A repressor binds to a site on DNA beside an activator and interacts with that activator, occluding its activating region. (c) A repressor binds to a site upstream of a gene and, by interacting with the transcriptional machinery at the promoter in some specific way, inhibits transcription initiation. (d) Repression is caused by recruiting histone modifiers that alter nucleosomes in ways that inhibit transcription (e.g., deacetylation, as shown here, but also methylation in some cases, or even remodeling at some promoters).

the most common in eukaryotes, that works as follows. As with activators, repressors can recruit nucleosome modifiers, but in this case, the enzymes have effects opposite to those recruited by activators—they compact the chromatin or remove groups recognized by the transcriptional machinery. So, for example, **histone deacetylases** repress transcription by removing actetyl groups from the tails of histones; as we have already seen, the presence of acetyl groups helps transcription. Other enzymes add methyl groups to histone tails, and this frequently represses transcription, although in some cases, it is associated with an actively transcribed gene (see Chapter 7). Histone (and DNA) modifications also form the basis of a type of repression called "silencing," which we consider in some detail later in this chapter.

These various examples of repression are shown schematically in Figure 17-20. Here, we consider just one specific example, the repressor called Mig1 that, like Gal4, is involved in controlling the *GAL* genes of the yeast *S. cerevisiae.*

Figure 17-21 shows the *GAL* genes as we saw them earlier (see Fig. 17-3), but with the addition of a site between the Gal4-binding sites and the promoter: this is where, in the presence of glucose, Mig1

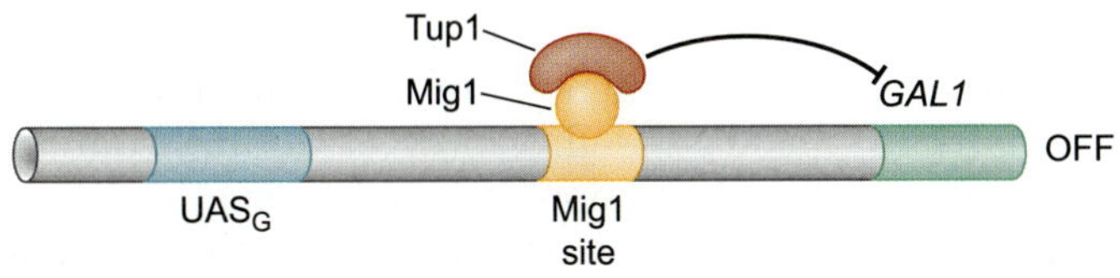

FIGURE 17-21 Repression of the *GAL1* gene in yeast. In the presence of glucose, Mig1 binds a site between the UAS$_G$ and the *GAL1* promoter. By recruiting the Tup1 repressing complex, Mig1 represses expression of *GAL1*. Repression is likely the result of deacetylation of local nucleosomes (Tup1 recruits a deacetylase) and also perhaps of directly contacting and inhibiting the transcriptional machinery. In an experiment not shown, if Tup1 is fused to a DNA-binding domain and a site for that domain is placed upstream of a gene, then expression of the gene is repressed.

binds and switches off the *GAL* genes. Thus, just as in *E. coli*, the cell only makes the enzymes needed to metabolize galactose if the preferred energy source, glucose, is not present. How does Mig1 repress the *GAL* genes?

Mig1 recruits a "repressing complex" containing the Tup1 protein. This complex is recruited by many yeast DNA-binding proteins that repress transcription, including the α2 protein involved in controlling the mating-type-specific genes described above. Tup1 also has counterparts in mammalian cells. Two mechanisms have been proposed to explain the repressing effect of Tup1. First, Tup1 recruits histone deacetylases, which deacetylate nearby nucleosomes. Second, Tup1 interacts directly with the transcriptional machinery at the promoter and inhibits initiation.

SIGNAL TRANSDUCTION AND THE CONTROL OF TRANSCRIPTIONAL REGULATORS

Signals Are Often Communicated to Transcriptional Regulators through Signal Transduction Pathways

As we have seen, whether or not a given gene is expressed very often depends on environmental signals. Signals come in many forms: they can, as was typically the case in bacteria, be small molecules such as sugars, but they can also be proteins released by one cell and received by another. This is particularly common during the development of multicellular organisms (Chapter 19).

There are various ways that signals are detected by a cell and communicated to a gene. In bacteria, we saw that signals control the activities of regulators by inducing allosteric changes in those regulators. Often, this effect is direct: a small molecular signal, such as a sugar, enters the cell and binds the transcriptional regulator directly. But we saw one example where the effect of the signal is indirect (control of the activator NtrC). In that case, the signal (low ammonia levels) induces a kinase that phosphorylates NtrC. This type of indirect signaling is an example of a **signal transduction pathway.**

The term "signal" refers to the initiating ligand itself—the sugar or protein, for example. This is how we have defined it previously. It can also refer to the "information" as it passes from detection of that ligand to the regulators that directly control the genes—that is, as it passes along a signal transduction pathway. In the simplest of bacterial cases, there was no distinction of course, but once a signal transduction pathway is involved, there is. In addition, in eukaryotes, we see—particularly in Chapter 19—that most signals are communicated to genes through signal transduction pathways, sometimes very elaborate ones. In this section, we first look at a few cases of signals being passed along signal transduction pathways in eukaryotes. We then consider more generally how signals, emerging from such pathways, control the transcriptional regulators themselves.

In a signal transduction pathway, the initiating ligand is typically detected by a specific **cell surface receptor:** The ligand binds to an extracellular domain of the receptor, and this binding is communicated to the intracellular domain. From there, the signal is relayed to the relevant transcriptional regulator, often through a cascade of kinases. How is the binding of ligand to the extracellular domain communicated to the intracellular domain? This can be through an allosteric change in the receptor, whereby binding of ligand alters the shape (and thus activity) of the intracellular domain. Alternatively, the ligand can act simply to bring together two or more receptor chains, allowing interactions between the intracellular domains of those receptors to activate each other.

Figure 17-22 shows two examples of signal transduction pathways. The first is a relatively simple case, the **STAT** (signal transducer and activator of transcription) pathway (Fig. 17-22a). In this example, a kinase is bound to the intracellular domain of a receptor. When the receptor is activated by its ligand (a cytokine), it brings together two receptor chains and triggers the kinase in each chain to phosphorylate a particular sequence in the intracellular domain of the opposing receptor. This phosphorylated site is then recognized by a particular STAT protein

FIGURE 17-22 **Two signal transduction pathways from mammalian cells.** Shown are the STAT and Ras pathways. (a) A cytokine is shown binding its receptor, bringing together two receptor chains. Each chain has a kinase called a JAK attached to its intracellular domain. Bringing the chains together (probably accompanied by a conformational change triggered by cytokine binding) leads to phosphorylation of the receptor chains by the JAK kinases (which also phosphorylate each other, stimulating their kinase activity). The sites phosphorylated in the receptor chain are then recognized by cytoplasmic proteins called STATs. Each STAT has a so-called SH2 domain. These domains are found in many proteins involved in signal transduction. They recognize phosphorylated Tyr residues in certain sequence contexts, and this is the basis of specificity in this pathway. That is, the particular STAT recruited to a given receptor determines which genes will subsequently be activated. Once recruited to the receptor, that STAT itself gets phosphorylated by the JAK kinase. This allows two STAT proteins to form a dimer (the SH2 domain on each STAT recognizing the phosphorylated site on the other). The dimer moves to the nucleus where it binds specific sites on DNA (different for different STATs) and activates transcription of nearby genes. (b) The Ras pathway leading into the downstream MAPK pathway. A growth factor (such as epidermal growth factor) binds its receptor, bringing together the chains that, as in the STAT case, then phosphorylate each other. This phosphorylation recruits an adaptor protein called Grb2: this protein has an SH2 domain that recognizes a phosphorylated tyrosine residue in the activated receptor. The other end of Grb2 binds SOS, a guanine nucleotide exchange factor (Ras GEF). SOS in turn binds the Ras protein, which is attached to the inside face of the cell membrane. Ras is a small GTPase, a protein that adopts one conformation when bound to GTP and another when bound to GDP; interaction with SOS triggers Ras to exchange its bound GDP for a GTP and hence undergo a conformational change. In this new conformation, Ras activates a kinase at the top of the so-called MAPK cascade. The first kinase in this pathway is called a MAPK kinase kinase (MAPKKK) (Raf); once activated by Ras, this phosphorylates serine and threonine residues in the next kinase (a MAPK kinase [MAPKK], called Mek). This activates Mek, which in turn phosphorylates and activates the MAPK (Erk). This MAPK then phosphorylates a number of substrates, including transcriptional activators (e.g., Jun) that regulate a number of specific genes, including interferon-β (Fig. 17-16).

that, once bound, gets phosphorylated itself. Once phosphorylated, the STAT dimerizes, moves to the nucleus, and binds DNA.

The other example is more elaborate (Fig. 17-22b): the mitogen-activated protein kinase (**MAPK**) pathway that controls activators such as Jun, one of the activators that works at the interferon-β enhancer we described earlier (Fig. 17-16). In this case, the activated receptor induces a cascade of signaling events, ending in activation of a MAPK that phosphorylates Jun (and other transcriptional regulators). The most common way in which information is passed through signal transduction pathways is via phosphorylation, but proteolysis, dephosphorylation, and other modifications are also used.

Signals Control the Activities of Eukaryotic Transcriptional Regulators in a Variety of Ways

Once a signal has been communicated, directly or indirectly, to a transcriptional regulator, how does it control the activity of that regulator? In bacteria, we saw that the allosteric changes that control transcriptional regulators very often affect the ability of the reg-

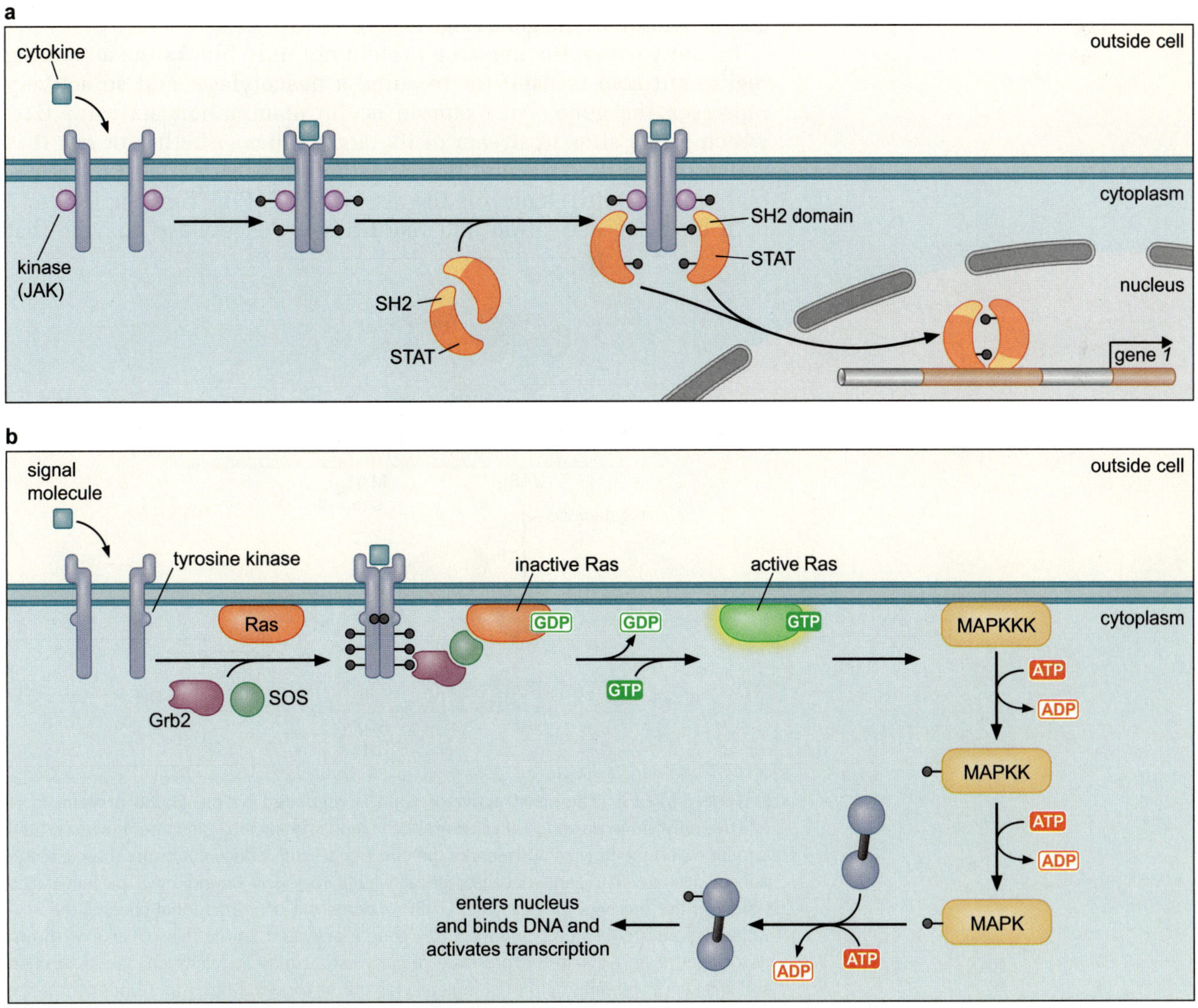

FIGURE 17-22 *(See facing page for legend.)*

ulator to bind DNA. This is true in cases where the signaling ligand itself acts directly on the transcriptional regulator and in cases where the presence of the signaling ligand is communicated to the regulator through a signal transduction pathway. Thus, Lac repressor binds DNA only when free of allolactose, and phosphorylation of NtrC triggers an allosteric change controlling DNA binding by that activator.

In eukaryotes, transcriptional regulators are not typically controlled at the level of DNA binding (although there are exceptions). Regulators are instead usually controlled in one of following two basic ways.

Unmasking an Activating Region This is done either by a conformational change in the DNA-bound activator, revealing a previously buried activating region, or by release of a masking protein that previously interacted with, and eclipsed, an activating region. The conformational changes required in each case can be triggered either by binding ligand directly or through a ligand-dependent phosphorylation.

Gal4 is controlled by a masking protein. In the absence of galactose, Gal4 is bound to its sites upstream of the *GAL1* gene, but it does not activate that gene because another protein, Gal80, binds to Gal4 and occludes its activating region. Galactose triggers the release of Gal80 and activation of the gene (Fig. 17-23).

In many cases, the masking protein not only blocks the activating region but also is itself (or recruits) a deacetylase, and so actively represses the gene. An example is the mammalian activator E2F, which binds sites upstream of its target genes, whether or not it is activating them. A second protein—the repressor called Rb (retinoblastoma protein)—controls the activity of E2F by binding to it and both blocking activation and recruiting a deacetylase enzyme that

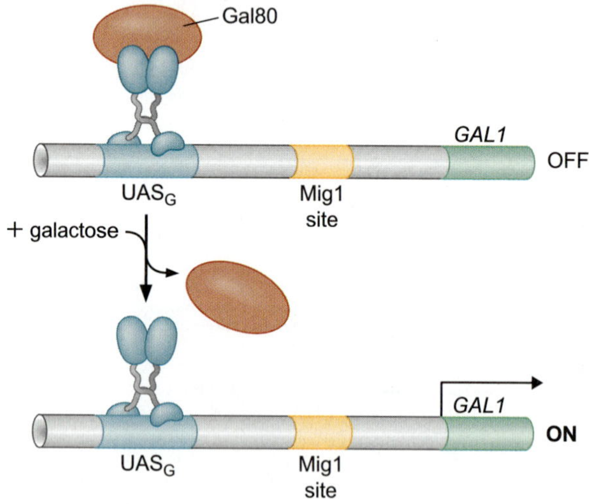

FIGURE 17-23 The yeast activator Gal4 is regulated by the Gal80 protein. Gal4 is active only in the presence of galactose. Even in the absence of galactose, however, Gal4 is found bound to its sites upstream of the *GAL1* gene. But it does not under these circumstances activate that gene because the activating region is bound by a protein called Gal80. In the presence of galactose, Gal80 undergoes a conformational change, the activating regions are revealed, and the *GAL1* gene is activated. In the figure, Gal80 is shown dissociating from Gal4 in the presence of galactose. It may in reality change its position and weaken its binding but not completely fall off. As shown, Mig1 is not bound at its site because there is no glucose present (see Fig. 17-21).

represses the target genes. Phosphorylation of Rb causes release of that protein from E2F and thus activation of the genes. E2F controls genes required to take a mammalian cell through the S phase of the cell cycle (Chapter 7). Phosphorylation of Rb thus controls proliferation in these cells. Mutations affecting this pathway are often associated with uncontrolled cell proliferation and cancer.

Transport into and out of the Nucleus When not active, many activators and repressors are held in the cytoplasm. The signaling ligand causes them to move to the nucleus where they act. There are many variations on this theme. Thus, the regulator can be held in the cytoplasm through interaction with an inhibitory protein or with the cell membrane or it can be in a conformation in which a signal sequence required for its nuclear import is concealed.

Release and transport into the nucleus in response to a signal can be mediated through proteolysis of an inhibitor or tethering region or by allosteric changes. We see an example of this in Chapter 19 when we consider the formation of the dorsoventral axis of the *Drosophila* embryo. There, Cactus is an inhibitory protein that binds the transcriptional regulator Dorsal in the cytoplasm. In response to a specific signal, Cactus is phosphorylated and destroyed, allowing Dorsal to enter the nucleus and act (Fig. 19-13).

Activators and Repressors Sometimes Come in Pieces

We have, on the whole, considered activators and repressors in their simplest forms, although we have alluded to some additional complexities. For example, the activator can come in pieces: the DNA-binding domain and activating region can be on separate polypeptides, which come together on DNA to form the activator. In addition, in considering the regulation of regulators by their signals, we again see examples of protein complexes forming on DNA, and the nature of the complex can determine whether the DNA-binding protein activates or represses nearby genes. For example, we just saw a case (E2F/Rb) where an activator can bind a protein and become a repressor. There are even more elaborate cases, such as the **glucocorticoid receptor (GR).** This mammalian protein can either activate or repress transcription depending on the nature and arrangement of its DNA-binding sites at a given gene.

In the absence of its ligand, GR is held in the cytoplasm through interaction with a protein called Hsp90. Upon ligand binding, the receptor is released and moves to the nucleus. (Thus, GR is another example of a regulator whose activity is controlled by nuclear localization.) Once in the nucleus, the GR binds sites called GREs. These sites come in two types. When bound to one site, it activates transcription; when bound to the other site, it represses, as we now describe.

When bound to the second of these sites, the receptor adopts a conformation that allows it to bind a histone deacetylase. When bound to the first site, the conformation of the receptor is such that it does not bind the histone deacetylase but rather binds the coactivator CBP—the same coactivator bound by the interferon-β enhanceosome (Fig. 17-16). Binding of CBP leads to activation of the nearby gene, partly because CBP is itself a histone acetylase but also because it can recruit components of the transcriptional machinery just as we described for the interferon-β case. Indeed, CBP is recruited to many mammalian genes during activation.

The terms "corepressor" and "coactivator" are often applied to any auxiliary protein that is neither part of the transcriptional machinery nor itself a DNA-binding regulator, but that is nevertheless involved in transcriptional regulation. CBP is an example. The term is also often applied to other nucleosome-modifying complexes.

GENE "SILENCING" BY MODIFICATION OF HISTONES AND DNA

We have thus far considered regulation by activators and repressors that bind near a gene and switch it on or off. The effects are local, and the actions of the regulators are often controlled by specific extracellular signals. We now turn to the mechanisms of **transcriptional silencing.** Silencing, in this context (we see the term applied to a rather different situation in Chapter 18), is a position effect: a gene is silenced because of where it is located, not in response to a specific environmental signal. In addition, silencing can "spread" over large stretches of DNA, switching off multiple genes, even those quite distant from the initiating event. Despite these differences, understanding silencing does not require entirely new principles, just extensions of those we have already encountered in this chapter.

The most common form of silencing is associated with a dense form of chromatin called **heterochromatin.** Heterochromatin was named for its appearance under the light microscope: it appears dense compared to the other chromatin, the **euchromatin**. Heterochromatin is frequently associated with particular regions of the chromosome, notably the telomeres—the structures found at the ends of chromosomes—and the centromeres. As discussed in Chapter 7, telomeres and centromeres are typically composed of repetitive sequences and contain few, if any, protein-coding genes. If a gene is experimentally moved into these regions, that gene is typically switched off. In fact, there are other regions of the chromosome that are also in a heterochromatic state, and in which genes are found, such as in the silent mating-type locus in yeast. And in mammalian cells, about 50% of the genome is estimated to be in some form of heterochromatin.

We have already seen that chromatin can be altered by enzymes that chemically modify the tails of histones. Such modifications affect accessibility of the DNA and therefore affect processes such as replication, recombination, and transcription.

As we have described, both activation and repression of transcription are often associated with modification of nucleosomes to alter the accessibility of a gene to the transcriptional machinery and other regulatory proteins. We have also encountered proteins that recognize modified nucleosomes and bind specifically to them. Heterochromatic silencing can be understood as an extension of these same principles and mechanism, as we describe momentarily.

Transcription can also be silenced by methylation of DNA by enzymes called **DNA methylases.** This kind of silencing is not found in yeast but is common in mammalian cells. Methylation of DNA sequences can inhibit binding of proteins, including the transcriptional machinery, and thereby block gene expression. But methylation can also inhibit expression in another way: some DNA sequences are recognized only when methylated by specific repressors that then switch off nearby genes, often by recruiting histone deacetylase.

Silencing in Yeast Is Mediated by Deacetylation and Methylation of Histones

The telomeres, the silent mating-type locus, and the rDNA genes are all "silent" regions in *S. cerevisiae*. We consider the telomere as an example.

The final 1–5 kb of each chromosome is found in a folded, dense structure, as shown in Figure 17-24. Genes taken from other chromosomal locations and moved to this region are often silenced, particularly if they are only weakly expressed in their usual location. The chromatin at the telomere is less acetylated than that found in most of the rest of the genome, where genes are more readily expressed.

Mutations have been isolated in which silencing is relieved—that is, in which a gene placed at the telomere is expressed at higher levels. These studies implicate three genes encoding regulators of silencing, *SIR2*, *SIR3*, and *SIR4* (*SIR* stands for silent information regulator). The three proteins encoded by these genes form a complex that associates with silent chromatin, and one of them—Sir2—is a histone deacetylase.

The silencing complex is recruited to the telomere by a DNA-binding protein that recognizes the telomere's repeated sequences. At the silent mating-type locus, recruitment is also initiated by a specific DNA-binding protein. In both cases, recruitment of Sirs triggers local deacetylation of histone tails. The deacetylated histones are, in turn, recognized directly by the silencing complex, and so the local deacetylation readily spreads along the chromatin in a self-perpetuating manner, producing an extended region of dense heterochromatin.

Unlike the case of repression by Tup1 where the mechanism is still uncertain, here silencing is clearly caused by the deacetylation of histone tails: loss of Sir2 completely alleviates silencing, and acetylation of the histone tail has a similar effect. The entire heterochromatic structure depends on the continued presence of the DNA-binding protein (Rap1) to remain intact. Thus, despite the reinforcing and spreading of deacetylation by Sir's recognition of deacetylated histones, the DNA-binding protein continues to play a critical part. In addition, of course, it is the

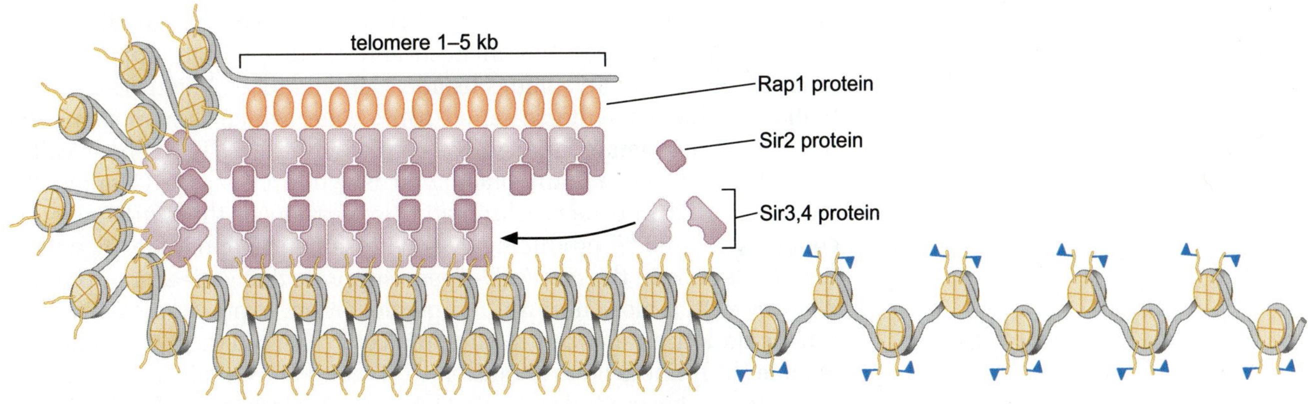

FIGURE 17-24 Silencing at the yeast telomere. Rap1 recruits Sir complex to the telomere. Sir2, a component of that complex, deacetylates nearby nucleosomes. The unacetylated tails themselves then bind Sir3 and Sir4, recruiting more Sir complex, allowing the Sir2 within it to act on nucleosomes further away, and so on. This explains the spreading of the silencing effect produced by deacetylation. (Adapted, with permission, from Grunstein M. et al. 1998. *Cell* 93: 325–328. © Elsevier.)

DNA-binding protein that gives specificity to the whole process—that is, defines where the silencing complex forms. In some cases of silencing, RNA molecules, rather than proteins, provide this critical specificity—we discuss (in Chapter 18) such a case, where the RNAi machinery of another yeast (*Schizosaccharomyces pombe*) is required for silencing at the mating-type loci and centromeres of that organism.

How is the spreading of silenced regions contained—that is, how is it limited to appropriate regions (i.e., the telomere, silenced *MAT* locus, etc.)? We earlier mentioned that insulator elements can block the spread of histone modifications (Fig. 17-12). In addition, other kinds of histone modifications block binding of the Sir2 proteins, and thereby stop spreading. Methylation of the tail of histone H3 is believed to do this.

Histone methyltransferases attach methyl groups to histone tails. As we saw in Chapter 7, these enzymes add methyl groups to specific lysine residues in the tails of histones H3 and H4. Histone methyltransferases have recently been described in *S. cerevisiae*, where they are believed to help repression of some genes and, as just noted, block spreading of Sir2-mediated silencing in others. But histone methylases have been better characterized in higher eukaryotes and in the yeast *S. pombe*. In these organisms, silencing is typically associated with chromatin containing histones that are not only deacetylated, but methylated as well. Thus, methylation of Lys-9 in the H3 tail is a modification associated with silenced heterochromatin in these organisms (Table 7-7). In contrast, other sites of methylation (e.g., Lys-4 on that same tail) are associated with increased transcription.

In *Drosophila,* HP1 Recognizes Methylated Histones and Condenses Chromatin

Just as acetylated residues within histones are recognized by proteins bearing bromodomains, methylated residues bind proteins with chromodomains (see Fig. 7-40). One such protein is the *Drosophila* protein HP1, a component of silent heterochromatin in that organism.

The HP1 protein interacts with modified chromatin containing methylated histone H3. This particular modification is produced by an enzyme encoded by Su(Var)3-9, a suppressor of so-called **variegation**. Variegation is seen in some cases when a gene is moved into a region of heterochromatin. Instead of being silenced in all cells all the time, that gene switches between the silenced and expressed state apparently at random, being "on" in some cells and "off" in others. Variegation is particularly evident for the so-called *white* gene, which is responsible for the normal red pigmentation of the eyes of adult flies. The gene is called *white* because the mutant phenotype is white eyes (see Chapter 22). When inserted into heterochromatin, expression of the *white* gene becomes "variegated," producing eyes with salt and pepper red pigmentation. Mutations in the *Su(Var)3-9* gene suppress this variegation, producing eyes with a more uniform red pigmentation; expression of the *white* gene is no longer silenced in so many cells. The Su(Var)3-9 protein is a histone H3 lysine-9 methyltransferase. Through mechanisms that are not presently understood, the Su(Var)3-9 protein is recruited to heterochromatin, where it attaches methyl groups to histone H3 tails. This modification is essential for the binding of the HP1 protein, which in turn participates in the compaction of the heterochromatin. It is thought that Su(Var)3-9 can also be recruited to specific euchromatic genes by sequence-specific DNA-binding proteins, thereby leading to gene-specific histone methylation and transcriptional repression by HP1.

We have seen how individual types of modification can be involved in gene regulation. But what when multiple forms of modification occur at the same gene? How do their influences interact? It has been proposed that complicated patterns of modification operate as a "histone code." The interactions between histone modifications and the idea of a histone code are described in Box 17-4, Is There a Histone Code?

ADVANCED CONCEPTS

Box 17-4 Is There a Histone Code?

It has been proposed that a **histone code** exists. According to this idea, different patterns of modifications on histone tails at a given gene could be "read" to mean different things (Fig. 7-39). The "meaning" would result from the particular pattern of modifications in each case recruiting a distinct set of proteins; the particular set depends on the number, type, and disposition of recognition domains carried by those proteins.

We have already encountered proteins that recognize specific acetylation or methylation "marks" on histones (e.g., TFIID and HP1). There are also proteins that phosphorylate serine residues in H3 and H4 tails, and proteins that bind those modifications. Thus, multiple modifications at several positions in the histone tails are possible (Fig. 7-39). Add to this the observation that many of the proteins that carry modification-recognizing domains are themselves enzymes that modify histones further, and we start to see how a process of recognizing and maintaining patterns of modification could in principle be achieved.

Consider one simple case—Lys-9 on the tail of histone H3 (see Fig. 7-39). Different modification states of this residue could be interpreted to have different meanings. Thus, acetylation of this residue is associated with actively transcribed genes. This residue is recognized by various histone acetylases bearing bromodomains, and these stimulate acetylation of nearby nucleosomes. When Lys-9 is unmodified, it is associated with silenced regions (as we saw in *S. cerevisiae* above). Unacetylated histones often recruit deacetylating enzymes, reinforcing and maintaining the deacetylated state (as we saw in the spreading of silenced regions in *S. cerevisiae*). Finally, that same lysine can in some organisms be methylated: in that case, the modified residue binds proteins (e.g., HP1) that establish and maintain a heterochromatic state.

But can *combinations* of modifications have distinct meanings? One example of how a histone modification can apparently influence a second modification present nearby is again illustrated by the *Drosophila* HP1 protein (Box 17-4 Fig. 1). During metaphase, the HP1 protein is temporarily lost from mitotic chromosomes, even though they retain the essential "mark"—namely, histone H3 methylation of lysine 9. Loss of HP1 binding is associated with phosphorylation of the neighboring serine residue at position 10 of H3. This phosphorylation is mediated by a cell cycle kinase called Aurora B. That kinase becomes active only during M phase of the cell cycle, thereby causing the release of HP1 from the heterochromatin of metaphase chromosomes.

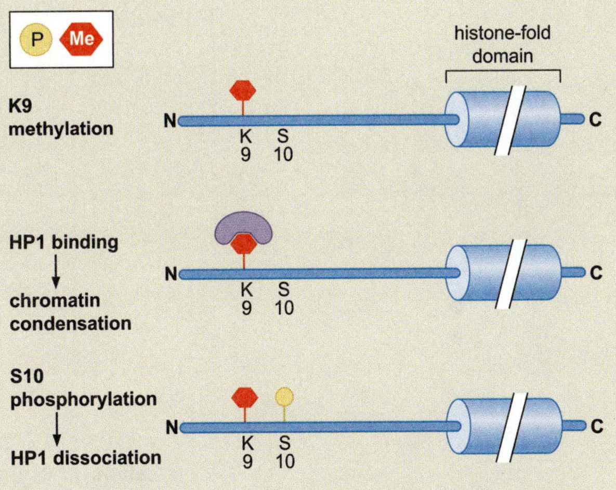

BOX 17-4 FIGURE 1 Influence of one chromatin modification on another. Modifications are shown on the tail of histone H3. Methylation of Lys-9 (K9) recruits HP1, which then effects chromatin condensation. Phosphorylation of the adjacent Ser residue (S10) displaces HP1 from methylated Lys-9, without removing the methyl group.

The dissociation of HP1 by the Aurora B kinase seems to be required for the attachment of the mitotic spindles to the centrosomes and the subsequent separation of sister chromatids during cytokinesis. When this process is complete, the phosphorylation of serine 10 is lost because of diminished Aurora B kinase activity, and HP1 reassociates with the chromosome to maintain the heterochromatin. Consistent with this model, mutations that eliminate Aurora B kinase lead to aberrant segregation.

Despite these observations, it remains highly controversial that a specific code exists, with complex patterns of histone modifications at a given locus generating a highly specific readout. Many of the modifications seen at a gene are likely just to be part of the process by which a gene is activated or repressed, rather than being the initiating signal; that is, they are a consequence of the gene being "on" or "off," not the cause. Site-specific DNA-binding proteins (or, in some cases, as we shall see in Chapter 18, small RNA molecules) remain the strongest provider of specificity determining when a given gene is expressed.

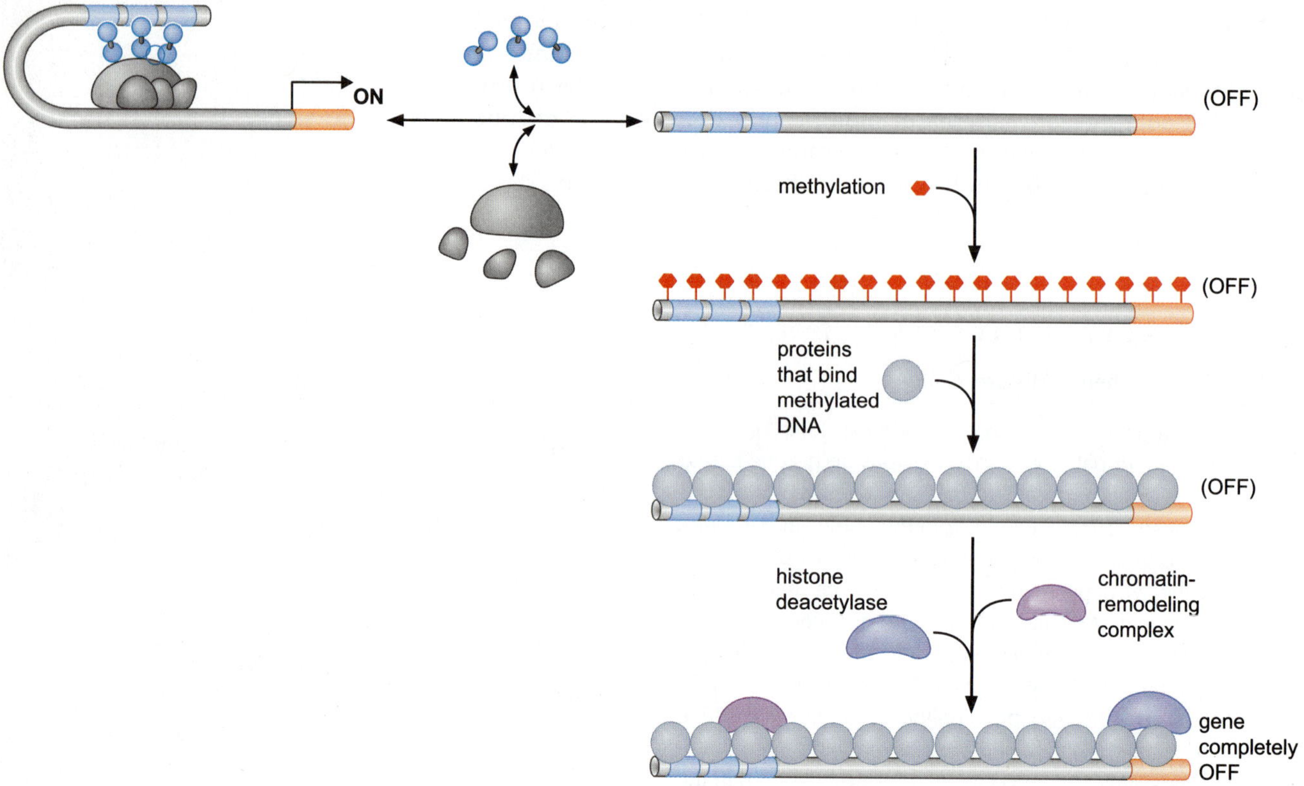

FIGURE 17-25 Switching a gene off through DNA methylation and histone modification. In its unmodified state, the mammalian gene shown can readily switch between being expressed or not expressed in the presence of activators and the transcription machinery, as shown in the top line. In this situation, expression is never firmly shut off—it is leaky. Often that is not good enough; sometimes, a gene must be completely shut off, on occasion permanently. This is achieved through methylation of the DNA and modification of the local nucleosomes. Thus, when the gene is not being expressed, a DNA methyltransferase (a methylase) can gain access and methylate cytosines within the promoter sequence, the gene itself, and the upsteam activator binding sites. The methyl group is added to the 5′ position in the cytosine ring, generating 5-methylcytosine (see Chapter 6). This modification alone can disrupt binding of the transcription machinery and activators in some cases. But it can also increase binding of other proteins (e.g., MeCP2) that recognize DNA sequences containing methylcytosine. These proteins, in turn, recruit complexes that remodel and modify local nucleosomes, switching off expression of the gene completely.

DNA Methylation Is Associated with Silenced Genes in Mammalian Cells

Some mammalian genes are kept silent by methylation of nearby DNA sequences (we are now talking about *DNA* methylation, not *histone* methylation). In fact, large regions of the mammalian genome are marked by methylation of DNA sequences, and often DNA methylation is seen in regions that are also heterochromatic. This is because methylated sequences are often recognized by DNA-binding proteins (such as MeCP2) that recruit histone deacetylases and histone methylases, which then modify nearby chromatin. Thus, methylation of DNA can mark sites where heterochromatin subsequently forms (Fig. 17-25).

DNA methylation lies at the heart of a phenomenon called **imprinting**, as we now describe. In a diploid cell, there are two copies of most genes: one copy on a chromosome inherited from the father and the other copy on the equivalent chromosome from the mother. In most cases, the two alleles are expressed at comparable levels. This is hardly surprising—they carry the same regulatory sequences and are in the presence of the same regulators; they are also located in an equivalent region of two very similar chromosomes. But there are a few cases where one copy of a gene is expressed while the other is silent.

Two well-studied examples are the human *H19* and insulin-like growth factor 2 (*Igf2*) genes (Fig. 17-26). These are located close to each other on human chromosome 11. In a given cell, one copy of *H19* (that on the maternal chromosome) is expressed, whereas the other copy (on the paternal chromosome) is switched off; for *Igf2* the reverse is true—the paternal copy is on and the maternal copy is off.

Two regulatory sequences are critical for the differential expression of these genes: an enhancer (downstream of the *H19* gene) and an insulator (called the imprinting control region [ICR], located between the *H19* and *Igf2* genes). The enhancer (when bound by activators) can, in principle, activate either of the two genes. So why does it activate only *H19* on the maternal chromosome and *Igf2* on the paternal chromosome? The answer lies in the role of the ICR and its methylation state. Thus, the enhancer cannot activate the *Igf2* gene on the maternal chromosome because on that chromosome, the ICR binds a protein, CTCF, that blocks activators at the enhancer from activating the *Igf2* gene. On the paternal chromosome, in contrast, the ICR element and the *H19* promoter are methylated. In that state, the transcriptional machinery cannot bind the *H19* promoter, and CTCF cannot bind the ICR. As a result, the enhancer now activates the *Igf2* gene. The *H19* gene is further repressed on the paternal chromosome by the binding of MeCP2 to the methylated ICR. This, as we have seen, recruits deacetylases, and these repress the *H19* promoter.

Box 17-5, Transcriptional Repression and Human Disease, describes two cases where loss of repression causes human diseases: one involves MeCP2 and the other involves defects in imprinting.

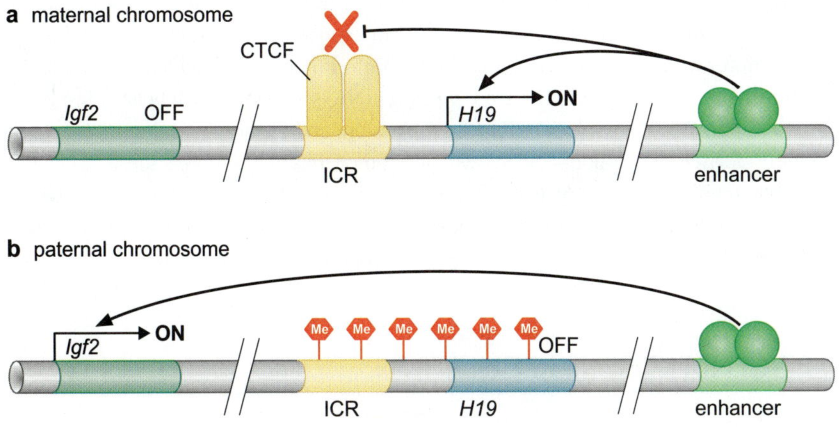

FIGURE 17-26 Imprinting. Shown are two examples of genes controlled by imprinting—the mammalian *Igf2* and *H19* genes. As described in the text, in a given cell, the *H19* gene is expressed only from the maternal chromosome, whereas *Igf2* is expressed from the paternal chromosome. The methylation state of the insulator element determines whether or not the ICR binding protein (CTCF) can bind and block activation of the *H19* gene from the downstream enhancer.

MEDICAL CONNECTIONS

Box 17-5 Transcriptional Repression and Human Disease

Several human diseases are caused by the derepression of specific genes. Here, we consider two conditions, each caused by the loss of repression of a gene whose mechanism of repression was described in the text. First, Rett syndrome, which is caused by the loss of the repressor protein MeCP2, and second, Beckwith–Weidemann syndrome, which is caused by loss of binding sites for CTCF in the ICR of the *Igf2* gene.

Rett syndrome (RTT) is a severe autism spectrum disorder found in 1 in 10,000 girls. This condition is characterized by loss of language and motor skills in early childhood, microcephaly, seizures, stereotypical behaviors (such as repetitive handwringing), and intermitted hyperventilation. It is a common cause of sporadic mental retardation. RTT is caused by a mutation in the X-linked gene encoding the repressor protein MeCP2. We encountered this transcriptional regulator earlier in the text; it recognizes methylated DNA sequences and silences transcription of nearby genes through recruitment of histone deacetylases (Fig. 17-26). Mice carrying a disrupted MeCP2 gene have symptoms similar to those of RTT patients, and this is preserved in mice in which MeCP2 loss is restricted to the brain.

Because the MeCP2 gene is found on the X chromosome, girls with a defective copy (the RTT patients) are a mosaic: in those cells in which the X chromosome carrying the mutant copy (allele) of the gene is inactivated, wild-type MeCP2 is made; but in those cells in which the wild-type copy of the gene is on the inactive copy of the X chromosome, no MeCP2 is made. Boys carrying the mutant gene on their (single) X chromosome lack MeCP2 in all cells and usually die from respiratory failure within a year or two.

RTT is thought to be a neurodevelopmental condition rather than a neurodegenerative disorder, because patients—and the knockout mice—show abnormal neuronal morphology but not neuronal death. Even so, it was thought likely that the MeCP2 deficiency was critical at a particular point during development, after which even restoring its function would not reverse the phenotype. But recently a mouse was constructed in which MeCP2 expression could be manipulated, allowing the mouse to be grown to adulthood without MeCP2 expression, before switching on expression of that regulator. Remarkably, making MeCP2 in adulthood was sufficient to reverse the effects of its absence throughout earlier development. This exciting finding makes therapeutic intervention in humans more feasible, if still difficult.

The link between MeCP2 and the symptoms of RTT is not fully understood. As we have seen, MeCP2 is a repressor of gene expression, and one of its target genes encodes brain-derived neurotrophic factor (BDNF). This protein, a growth factor, has roles in brain development and in synaptic changes associated with learning and memory. Recently, it has been found that neural activity leads to phosphorylation of MeCP2, a modification that causes the repressor to dissociate from DNA, presumably allowing expression of its target genes. Disruption of these activities—inappropriate expression of BDNF, for example—have obvious appeal as an explanation for at least some of the cognitive symptoms of RTT.

There is an ongoing search for the links between MeCP2 and BDNF, and between these proteins and the disease, as there is for other possibly relevant genes regulated by MeCP2. The broad array of symptoms—from cognitive impairment to unusual gait—suggests that there are probably a number of genes whose misexpression is required for the full disease.

Beckwith–Wiedemann syndrome (BWS) is a developmental disorder affecting 1 in 15,000. The condition is characterized by overgrowth (children with this condition are born prematurely and are larger than normal) and increased susceptibility to a variety of childhood cancers (including Wilms' tumor). The syndrome is also associated with disrupted expression of imprinted genes on chromosome 11p15.5, including the insulin-like growth factor 2 gene (*Igf2*) discussed in the text (Fig. 17-26).

As we have described, the *Igf2* gene is usually expressed monoallelically; that is, only one of the two alleles (in this case the paternal allele) is expressed as a result of imprinting of the other. At the same time, the *H19* gene is expressed only from the maternal allele. Many cases of BWS are associated with biallelic expression of *Igf2* and no expression of *H19*, the result of methylation of ICR on both chromosomes. IGF2 is a fetal growth factor, and H19 is a regulatory RNA (see Chapter 18) believed to be involved in tumor suppression, and so the phenotype of the condition—overgrowth and tumor sensitivity—makes sense.

As with RTT, the symptoms of BWS are mimicked in suitably manipulated mice: overexpression of *Igf2* produces general overgrowth of mice and the appearance of specific tumors.

EPIGENETIC GENE REGULATION

Patterns of gene expression must sometimes be inherited. A signal released by one cell during development causes neighboring cells to switch on specific genes. These genes may have to remain switched on in those cells for many cell generations, even if the signal that induced them is present only fleetingly. The inheritance of gene expression patterns, in the absence of both mutation and the initiating signal, is called **epigenetic** regulation.

Contrast this with some of the examples of gene regulation we have discussed. If a gene is controlled by an activator, and that activator is only active in the presence of a given signal, then the gene will remain on only as long as the signal is present. Indeed, under normal conditions, the *lac* genes of *E. coli* will only be expressed while lactose is present and glucose absent. Likewise, the *GAL* genes of yeast are expressed only as long as glucose is absent and galactose present, and human β-interferon is made only while cells are stimulated by viral infection.

Some States of Gene Expression Are Inherited through Cell Division Even When the Initiating Signal Is No Longer Present

We have already encountered examples of gene regulation that can be inherited epigenetically. Consider the maintenance of a bacteriophage λ lysogen (Chapter 16). In a lysogen, the phage is in a dormant state within the bacterial host cell. This state is associated with a specific pattern of gene expression and in particular with sustained expression of the λ repressor protein (see Fig. 16-26).

Lysogenic gene expression is established in an infected cell in response to poor growth conditions. Once established, however, the lysogenic state is maintained stably despite improvements in growth conditions: moving a lysogen into rich growth medium does not lead to induction. Indeed, induction essentially never occurs until a suitable inducing signal (such as UV light) is received.

Maintenance of the lysogenic state through cell division is thus an example of epigenetic regulation. This epigenetic control results from a two-step strategy for repressor synthesis. In the first, synthesis is initially established through activation of the repressor (*c*I) gene by the activator CII (which is sensitive to growth conditions). In the second step, repressor synthesis is maintained by autoregulation: repressor activates expression of its own gene (see Fig. 16-30). In this way, when the lysogenic cell divides, each daughter cell inherits a copy of the dormant phage genome and some repressor protein. This repressor is sufficient to stimulate further repressor synthesis from the phage genome in each cell (Fig. 17-27). Much of gene regulation during the development of muticellular organisms works in just this way. We shall see examples in Chapter 19.

Another known mechanism of epigenetic regulation is provided by DNA methylation, an example of which we saw in our description of imprinting. DNA methylation is reliably inherited through cell division, as shown in Figure 17-28. Thus, certain DNA methylases can methylate, at low frequency, previously unmodified DNA; but far more efficiently, so-called **maintenance methylases** modify hemimethylated DNA, the very substrate provided by replication of fully methylated DNA. In mammalian cells, DNA methylation may be the primary marker of regions of the genome that are silenced. After DNA replication, hemimethylated sites in both daughter cells are remethylated. These can then be recognized by the repressor MeCP2, which in turn recruits histone deacetylases and methylases, reestablishing silencing (Fig. 17-25).

Nucleosome modifications could in principle provide the basis for epigenetic inheritance, although no examples of this have yet been found. Consider a gene switched off by a stretch of methylated histones. When that region of the chromosome is replicated during cell

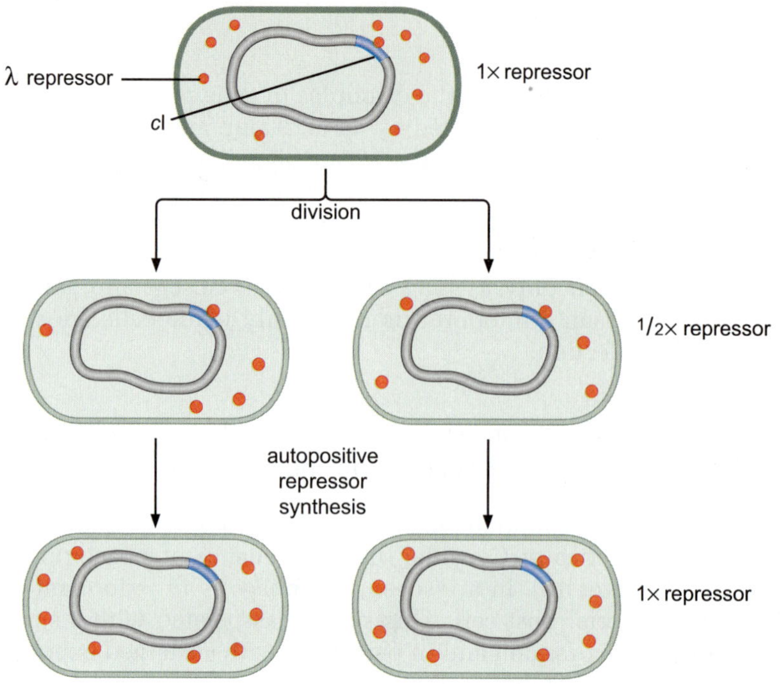

FIGURE 17-27 Epigenetic control of the maintenance of the lysogenic state.

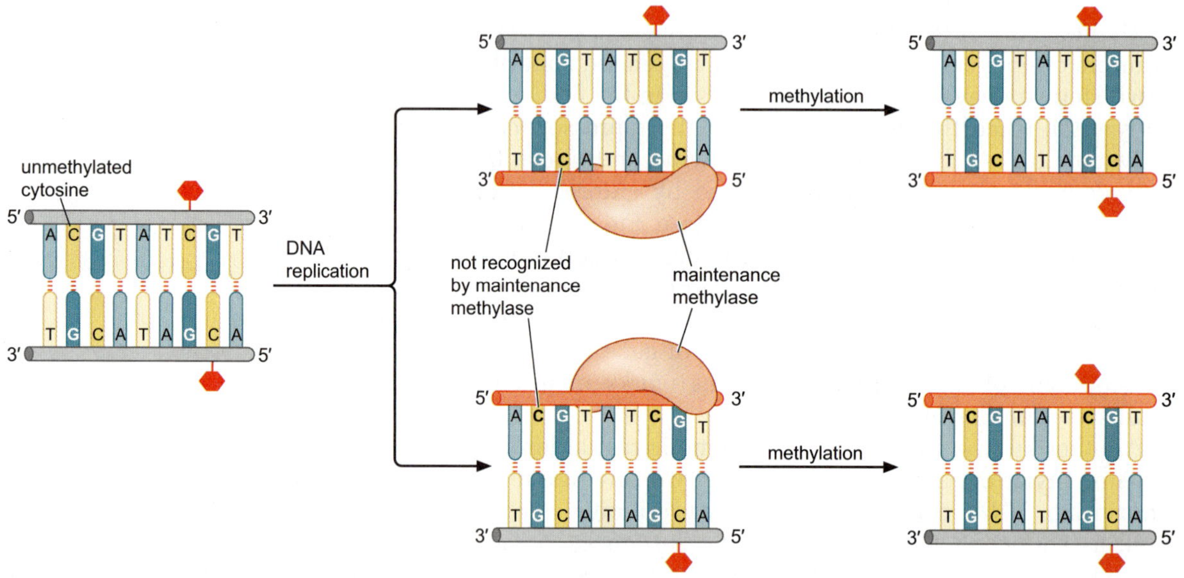

FIGURE 17-28 Patterns of DNA methylation can be maintained through cell division. As we saw in Fig. 17-25, DNA involved in expression of a vertebrate gene can become methylated and expression of that gene switched off. This initial methylation is performed by a de novo methylase. For the shutdown state to keep a gene off permanently, the methylation state must be inherited through cell division. This figure shows how that is achieved. A DNA sequence is shown in which two cytosines are present on each strand—one methylated, the other not. This pattern is maintained through cell division, because, upon DNA replication, a maintenance methylase recognizes the hemimethylated DNA, and adds a methyl group to the unmethylated cytosine within it. The completely unmethylated sequence is not recognized by this enzyme and so remains unmethylated. Thus, both daughter DNA duplexes end up with the same pattern of methylation as the parent. (Adapted, with permission, from Alberts B. et al. 2002. *Molecular biology of the cell*, 4th ed., p. 431, Fig 7-81. © Garland Science/Taylor & Francis LLC.)

division, the methylated histones from the parental DNA molecule end up distributed equally between the two daughter duplexes (see Fig. 7-42). Thus, each of the daughter molecules carries some methylated and some unmethylated nucleosomes. The methylated nucleosomes could recruit proteins bearing chromodomains, including the histone methylase itself, which could then methylate the adjacent unmodified nucleosomes. In this way, the state of chromatin modification could be maintained through generations using the same strategy employed to achieve spreading. Although this is an appealing model, it is yet to be seen to operate in the absence of DNA methylation, DNA-binding proteins, or regulatory RNAs (Chapter 18).

In a final demonstration of the power of transcription factors to determine patterns of gene expression, in Box 17-6, Using Transcription Factors to Reprogram Somatic Cells into Embryonic Stem Cells, we describe recent experiments in which four transcription factors are shown to drive a somatic cell into a pluripotent state.

MEDICAL CONNECTIONS

Box 17-6 Using Transcription Factors to Reprogram Somatic Cells into Embryonic Stem Cells

A major hurdle in the treatment of certain human aliments, such as Parkinson's disease, spinal cord injury, and diabetes, by tissue replacement is rejection of the transplant by the patient's immune system. One way to circumvent this rejection problem would be to use the patient's own somatic cells to generate embryonic stem cells. These embryonic stems cells would have the capacity to differentiate into somatic cells and thereby provide a source of customized adult cells for the transplantation therapy (see Box 19-6). Up until recently, the only way to convert a somatic cell into an embryonic stem cell was by somatic cell nuclear transfer, a procedure in which the nuclear contents of the somatic cell are introduced into an enucleated oocyte. However, somatic cell nuclear transfer has proven to be technically difficult in humans and has raised ethical objections. Recently, an exciting new approach to the reprogramming of somatic cells has been developed that promises to bypass the need for oocytes. This new approach uses the engineered synthesis of four key transcription factors to reverse differentiation and convert somatic cells into embryonic-like stem cells.

Previous research on embryonic stem cells identified approximately 24 transcription factors that are required for, or associated with, the pluripotent state but are absent in somatic cells. (Pluripotency refers to the potential to differentiate into many of the cell types of the adult animal.) These findings led to the hypothesis that artificially causing the synthesis of a subset of these factors in an adult cell might trigger reprogramming into a pluripotent state. As a test of this hypothesis, fibroblast cells from the skin of a mouse were engineered to express various combinations of the genes (using a retrovirus to deliver cDNA copies of the genes into the cells) for the 24 candidate transcription factors. (Fibroblasts are a somatic cell type that is responsible for the production of a matrix that provides structural support for tissues.) Remarkably, 4 of the 24 genes were sufficient to do the trick! These four were the genes for transcription factors called Oct4, Sox2, c-Myc, and Klf4. Fibroblast cells engineered to express the four genes seemed to have entered an embryonic-like stem cell state as judged by morphology, capacity for proliferation, and expression of certain genes (e.g., a gene called *Nanog*) that are known to be *on* in embryonic stem cells but *off* in differentiated cells.

Do these Oct4-, Sox2-, c-Myc-, and Klf4-expressing cells exhibit other features of embryonic stem cells? A striking characteristic of pluripotent cells is that the chromatin at noncoding control elements for key regulatory genes involved in development exhibits a distinctive modification pattern known as bivalent domains. Bivalent domains are characterized by large stretches of methylation at lysine 4 of histone H3 and smaller stretches of methylation at lysine 27 (see Chapter 7, in particular Table 7-7). Lysine 4 methylation is known to *promote* transcription by recruiting remodeling enzymes and histone acetylases, whereas methylation of H3 at lysine 27 inhibits transcription by causing chromatin compaction. (Evidently, these bivalent domains hold developmental regulatory genes silent while at the same time keeping them poised to be switched on.) When fibroblast cells were examined after reprogramming by Oct4, Sox2, c-Myc, and Klf4, chromatin at control elements that had principally exhibited only lysine 4 or lysine 27 methylation in unadulterated fibroblasts was found to exhibit the bivalent pattern characteristic of embryonic stem cells.

These striking results notwithstanding, true pluripotency requires a demonstration that the reprogrammed cell is capable of developing into an adult animal. This acid test has not yet

Box 17-6 *(Continued)*

been met, but it has been possible to generate chimeric mice in which the reprogrammed cells had contributed to all adult cell types and late-stage mouse embryos that were entirely derived from engineered fibroblast cells (see Box 17-6 Fig. 1).

The discovery that the expression of four particular genes suffices to reprogram an adult cell into a pleuripotent cell is no doubt a breakthrough in stem cell research. Nonetheless, the use of a retrovirus and the engineered expression of a known oncogene (Myc) rules out the application of this technology in human therapeutics. In the future, however, it should be possible to elucidate the chain of events leading from Oct4, Sox2, c-Myc, and Klf4 to pluripotency. If so, then based on this knowledge it might be possible to identify pharmaceutical agents that trigger reprogramming in a benign manner that does not depend on the engineered expression of potentially harmful genes.

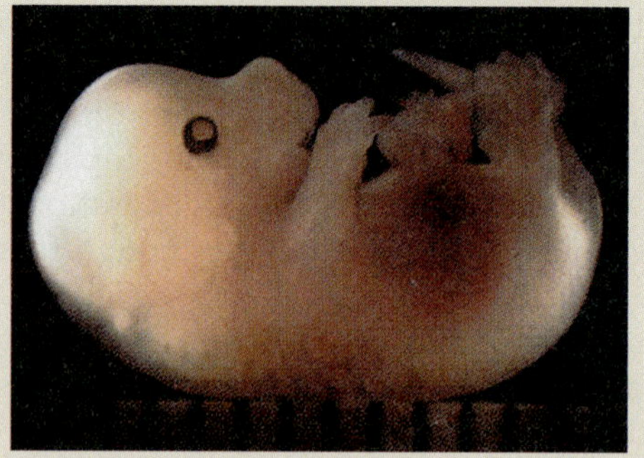

Box 17-6 Figure 1 Late-stage mouse embryo entirely derived from a fibroblast cells expressing the embryonic stem cell transcription factors Oct4, Sox2, c-Myc, and Klf4. (Reprinted, with permission, from Wernig M. et al. 2004. *Nature* 448: 318–324, Fig. 5e. © Macmillan.)

SUMMARY

As in bacteria, transcription initiation is the most frequently regulated step in gene expression in eukaryotes, despite the additional steps that can be regulated in these organisms. Also as in bacteria, transcription initiation is typically regulated by proteins that bind to specific sequences on DNA near a gene and either switch that gene on (activators) or switch it off (repressors). This conservation of regulatory mechanism holds in the face of several complexities in the organization and transcription of eukaryotic genes not found in bacteria, as we now summarize.

Nucleosomes and their modification. The DNA in a eukaryotic cell is wrapped in histones to form nucleosomes. Thus, the DNA sequences to which the transcriptional machinery and the regulatory proteins bind are in many cases occluded. Enzymes that modify histones, by adding (or removing) small chemical groups, alter the histones in two possible ways: changing how tightly the nucleosomes are packed (and thus how accessible the DNA within them is) and forming (or removing) binding sites for other proteins involved in transcribing the gene. Other enzymes "remodel" the nucleosomes: they use the energy from ATP hydrolysis to move the nucleosomes around, influencing which sequences are available.

Many regulators and larger distances. Genes of multicellular eukaryotes are typically controlled by more regulatory proteins than their bacterial counterparts, some bound far from the gene. This reflects the larger number of physiological signals that control a typical gene in multicellular organisms.

The elaborate transcriptional machinery. The enzyme RNA polymerase is largely conserved between bacteria and eukaryotes (Chapter 12). But the eukaryotic enzyme contains more subunits, and there are approximately 50 or so additional proteins that bind at the typical eukaryotic promoter along with polymerase. Many of these proteins come to the promoter as large protein complexes.

In eukaryotes, just as we saw in bacteria, activators predominantly work by recruitment. In these organisms, however, the activators do not recruit polymerase directly, or alone. Thus, they recruit the other protein complexes required to initiate transcription of a given gene. RNA polymerase itself is brought in along with these other complexes. The activator can recruit histone-modifying enzymes as well, and the effects of those modifications may help the transcriptional machinery bind the promoter or initiate efficient transcription.

The activators can interact with one or more of many different components of the transcriptional machinery or the nucleosome modifiers. Gal4, for example, recruits Mediator, SAGA, and TFIID to promoters as required. In other cases, factors required for efficient initiation or elongation might be needed after the polymerase has bound—these too can be recruited by activators. This explains how activators can so readily work together in large numbers and various combinations and accounts for the widespread use of signal integration and combinatorial control we see, particularly in multicellular organisms.

Some activators work from sites far from the gene, requiring that the DNA between their binding sites and the promoter loops out. How loops can form over the very large distances called for in some cases is not clear, but it might involve changes in the chromatin structure between the activator-binding site and the promoter, bringing those two elements closer together. DNA sequences called insulators bind proteins that interfere with the interaction between activators bound at distant enhancers and their promoters. These could work by inhibiting mechanisms that facilitate looping (such as changes in chromatin structure). Insulators help ensure that activators work only on the correct genes.

Eukaryotic repressors work in various ways, just as they do in bacteria. However, the simplest and most common mechanism seen in bacteria is for the repressor to bind to a site overlapping the promoter, thus blocking binding of RNA polymerase. That mechanism is not typically seen in eukaryotes. Most commonly, eukaryotic repressors work by recruiting histone modifiers that reduce transcription. For example, whereas a histone acetylase is typically associated with activation, a histone deacetylase—that is, an enzyme that removes acetyl groups—acts to repress a gene.

In some cases, long stretches of nucleosomal DNA can be kept in a relatively inert state by appropriate nucleosome modification, most notably deacetylation and methylation. In this way, groups of genes can be kept in a "silent" state without the need for specific repressors bound at each individual gene. Once established, this condition can be extended because the modification enzymes themselves are often preferentially recruited to nucleosomes that are in that state. Thus, the modification state recruits the enzymes that produce that particular pattern of modification. This means that once initiated, the silent state can be extended and inherited rather easily.

In some eukaryotic organisms, such as mammals, silent genes are also associated with methylated DNA. Methylated sequences can either block the binding of the transcription machinery and activators or specifically bind a class of repressors that recruit histone-modifying enzymes that repress nearby genes. DNA methylation can be maintained through cell division, and thus patterns of gene expression controlled by that methylation can be as well.

If expression of a gene is maintained in some state through cell division—in the absence of either a mutation or the signal that initiated that pattern—it is said to be inherited epigenetically. Various mechanisms allow this to happen. Inheritance of a λ lysogen is one example (and relies on expression of repressor being initiated from one promoter but then becoming self-sustaining from another). DNA methylation can affect gene expression and readily be inherited, as we have seen. Epigenetic inheritance through the use of histone modifications is often discussed as a possibility, but the extent to which they fulfill this role in the absence of input from DNA-binding proteins, regulatory RNAs, or DNA methylation remains unclear.

BIBLIOGRAPHY

Books

Allis C.D., Jenuwein T., Reinberg D., and Caparros M.-L., eds. 2007. *Epigenetics.* Cold Spring Harbor Laboratory Press, Cold Spring Harbor, New York

Carey M. and Smale S.T. 2000. *Transcriptional regulation in eukaryotes: Concepts, strategies, and techniques.* Cold Spring Harbor Laboratory Press, Cold Spring Harbor, New York.

Ptashne M. and Gann A. 2002. *Genes and signals.* Cold Spring Harbor Laboratory Press, Cold Spring Harbor, New York.

DNA Recognition

Garvie C.W. and Wolberger C. 2001. Recognition of specific DNA sequences. *Mol. Cell.* **8:** 937–946.

Harrison S.C. 1991. A structural taxonomy of DNA-binding domains. *Nature* **353:** 715–719.

Activation

Bjorklund S. and Gustafsson C.M. 2005. The yeast Mediator complex and its regulation. *Trends Biochem. Sci.* **30:** 240–244.

Bulger M. and Groudine M. 2002. TRAPping enhancer function. *Nat. Genet.* **32:** 555–556.

Fry C.J. and Peterson C.L. 2001. Chromatin remodeling enzymes: Who's on first? *Curr. Biol.* **11:** R185–R197.

Jones K.A. and Kadonaga J.T. 2000. Exploring the transcription-chromatin interface. *Genes Dev.* **14:** 1992–1996.

Kim Y.J. and Lis J.T. 2005. Interactions between subunits of *Drosophila* Mediator and activator proteins. *Trends Biochem. Sci.* **30:** 245–249.

Kornberg R.D. 2005. Mediator and the mechanism of transcriptional activation. *Trends Biochem. Sci.* **30:** 235–239.

Lefstin J.A. and Yamamoto K.R. 1998. Allosteric effects of DNA on transcriptional regulators. *Nature* **392:** 885–888.

Malik S. and Roeder R.G. 2005. Dynamic regulation of pol II transcription by the mammalian Mediator complex. *Trends Biochem. Sci.* **30:** 256–263.

Myers L.C. and Kornberg R.D. 2000. Mediator of transcriptional regulation. *Annu. Rev. Biochem.* **69:** 729–749.

Naar A.M., Lemon B.D., and Tjian R. 2001. Transcriptional coactivator complexes. *Annu. Rev. Biochem.* **70:** 475–501.

Ptashne M. and Gann A. 1997. Transcriptional activation by recruitment. *Nature* **386:** 569–577.

Struhl K. 1999. Fundamentally different logic of gene regulation in eukaryotes and prokaryotes. *Cell* **98:** 1–4.

Repression

Maldonado E., Hampsey M., and Reinberg D. 1999. Repression: Targeting the heart of the matter. *Cell* **99:** 455–458.

Smith R.L. and Johnson A.D. 2000. Turning genes off by Ssn6-Tup1: A conserved system of transcriptional repression in eukaryotes. *Trends Biochem. Sci.* **25:** 325–330.

Nucleosome Modifiers and Transcriptional Regulation

Berger S.L. 2002. Histone modifications in transcriptional regulation. *Curr. Opin. Genet. Dev.* **12:** 142–148.

Flaus A. and Owen-Hughes T. 2001. Mechanisms for ATP-dependent chromatin remodeling. *Curr. Opin. Genet. Dev.* **11:** 148–154.

Jenuwein T and Allis C.D. 2001. Translating the histone code. *Science* **293:** 1074–1080.

Marmorstein R. and Roth S.Y. 2001. Histone acetyltransferases: Function, structure, and catalysis. *Curr. Opin. Genet. Dev.* **11:** 155–161.

Narlikar G.J., Fan H.Y., and Kingston R.E. 2002. Cooperation between complexes that regulate chromatin structure and transcription. *Cell* **108:** 475–487.

Peterson C.L. and Workman J.L. 2000. Promoter targeting and chromatin remodeling by the SWI/SNF complex. *Curr. Opin. Genet. Dev.* **10:** 187–192.

Silencing, Imprinting, and Epigenetics

Bird A.P. and Wolffe A.P. 1999. Methylation-induced repression—Belts, braces, and chromatin. *Cell* **99:** 451–454.

Gartenberg M.R. 2000. The Sir proteins of *Saccharomyces cerevisiae:* Mediators of transcriptional silencing and much more. *Curr. Opin. Microbiol.* **3:** 132–137.

Goldberg A.D., Allis C.D., and Bernstein E. 2007. Epigenetics: A landscape takes shape. *Cell* **128:** 635–638.

Gottschling D.E. 2000. Gene silencing: Two faces of SIR2. *Curr. Biol.* **10:** R708–R711.

———. 2004 Summary: Epigenetics—From phenomenon to field. *Cold Spring Harbor Symp. Quant. Biol.* **69:** 507–519.

Grunstein M. 1998. Yeast heterochromatin: Regulation of its assembly and inheritance by histones. *Cell* **93:** 325–328.

Klose R.J. and Bird A.P. 2005. Genomic DNA methylation: The mark and its mediators. *Trends Biochem. Sci.* **31:** 89–97.

Ptashne M. 2007. On the use of the word "epigenetic." *Curr. Biol.* **17:** R233–R236.

Richards E.J. and Elgin S.C. 2002. Epigenetic codes for heterochromatin formation and silencing: Rounding up the usual suspects. *Cell* **108:** 489–500.

Tilghman S.M. 1999. The sins of the fathers and mothers: Genomic imprinting in mammalian development. *Cell* **96:** 185–193.

Wolffe A.P. 2000. Transcriptional control: Imprinting insulation. *Curr. Biol.* **10:** R463–R465.

Wood A.J. and Oakey R.J. 2006. Genomic imprinting in mammals: Emerging themes and established theories. *PloS Genet.* **2:** 1677–1685.

Combinatorial Control and Synergy

Arnosti D.N. and Kulkarin M.M. 2005. Transcriptional enhancers: Intelligent enhanceosomes or flexible billboards? *J. Cell Biol.* **94:** 890–898.

Carey M. 1998. The enhanceosome and transcriptional synergy. *Cell* **92:** 5–8.

Johnson A.D. 1995. Molecular mechanisms of cell type determination in budding yeast. *Curr. Opin. Genet. Dev.* **5:** 552–558.

Maniatis T., Falvo J.V., Kim T.H., Kim T.K., Lin C.H., Parekh B.S., and Wathelet M.G. 1998. Structure and function of the interferon-β enhanceosome. *Cold Spring Harb. Symp. Quant. Biol.* **63:** 609–620.

Merika M. and Thanos D. 2001. Enhanceosomes. *Curr. Opin. Genet. Dev.* **11:** 205–208.

Rokas A. 2006. Evolution: Different paths to the same end. *Nature* **443:** 401–402.

Long-Range Interactions

Dean A. 2006. On a chromosome far, far away: LCRs and gene expression. *Trends Genet.* **22:** 38–45.

Dorsett D. 1999. Distant liaisons: Long-range enhancer–promoter interactions in *Drosophila. Curr. Opin. Genet. Dev.* **9:** 505–514.

Gaszner M.and Felsenfeld G. 2006. Insulators: Exploiting transcriptional and epigenetic mechanisms. *Nat. Rev. Genet.* **7:** 703–713.

Krueger C. and Osborne C.S. 2006. Raising the curtains on interchromosomal interactions. *Trends Genet.* **22:** 637–639.

Li Q., Barkess G., and Qian H. 2006. Chromatin looping and the probability of transcription. *Trends Genet.* **22:** 197–202.

Signals and Signal Transduction

Bromberg J.F. 2001. Activation of STAT proteins and growth control. *Bioessays* **23:** 161–169.

Brown M.S., Ye J., Rawson R.B., and Goldstein J.L. 2000. Regulated intramembrane proteolysis: A control mechanism conserved from bacteria to humans. *Cell* **100:** 391–398.

Darnell J.E., Jr. 1997. STATs and gene regulation. *Science* **277:** 1633–1635.

Hill C.S. and Treisman R. 1995. Transcriptional regulation by extracellular signals: Mechanism and specificity. *Cell* **80:** 199–211.

Hunter T. 2000. Signaling—2000 and beyond. *Cell* **100:** 113–127.

Pawson T. and Nash P. 2000. Protein–protein interactions define specificity in signal transduction. *Genes Dev.* **14:** 1027–1047.

Repression and Disease

Gabellini D., Green M.R., and Tupler R. 2004. When enough is enough: Genetic diseases associated with transcriptional derepression. *Curr. Opin. Genet. Dev.* **14:** 301–307.

Gabellini D., Tupler R., and Green M.R. 2003. Transcriptional derepression as a cause of genetic diseases. *Curr. Opin. Genet. Dev.* **13:** 239–244.

Kriaucianis S. and Bird A. 2003. DNA methylation and Rett syndrome. *Hum. Mol. Genet.* **12:** R221–R227.

Miller G. 2007. Medicine. Rett symptoms reversed in mice. *Science* **315:** 749.

CHAPTER

18 Regulatory RNAs

OUTLINE

We discussed in the previous two chapters how transcription is regulated in prokaryotes and eukaryotes. We learned that this control is achieved using regulatory proteins—typically, sequence-specific DNA-binding proteins that either activate or repress transcription of nearby genes. The mechanistic details of gene regulation have been studied since François Jacob and Jacques Monod proposed their model of repression almost 50 years ago (Box 16-2). At that time, they could not say whether the *trans* factors (repressors) were proteins or RNA. It transpired that in the cases they studied (and indeed most other cases), the regulators were proteins that worked by binding the operator sites on DNA. But in their original paper they suggested that the regulators could just as easily be RNA molecules—indeed, they favored that possibility.

The idea that RNA molecules might be regulators was largely forgotten as more and more protein regulators were found in both prokaryotes and eukaryotes. But in recent years, there has been an explosion in the study of RNA regulators, particularly in eukaryotes, that operate at the level of transcription and especially translation. This new field emerged from two sources: the discovery of microRNAs, first reported in the early 1990s, and then the discovery of the phenomenon known as RNA interference in the late 1990s. Before we describe these forms of regulation—how they work and the applications they afford researchers—we consider cases of RNA-mediated gene regulation first described in bacteria.

REGULATION BY RNAs IN BACTERIA

Small RNAs have been recognized in prokaryotes for many years. Some are involved in regulating the replication of plasmids, and others are involved in regulating gene expression (see the discussion of Tn*10* in Chapter 11). Of the latter group, some of these RNAs control transcription—the 6S RNA of *Escherichia coli,* for example. This RNA binds to the σ^{70} subunit of RNA polymerase and down-regulates transcription from many σ^{70} promoters. The 6S RNA accumulates at high levels in stationary phase (the growth phase bacteria enter as nutrients become depleted and the cells stop dividing; see Chapter 22). In stationary phase, an alternative σ factor, σ^S, is made. This σ competes with σ^{70} for core polymerase and directs the enzyme to promoters expressing genes for the multiple stress responses needed to survive stationary phase. By down-regulating transcription from σ^{70} promoters, 6S RNA helps this shift in expression to σ^{S} promoters.

In recent years, attention has focused on another group of small RNA molecules in bacteria that regulate translation and mRNA degradation. Interest in these small RNAs has been heightened by their similarity to RNAs that regulate gene expression in eukaryotes—the small interfering and microRNAs we discuss in the second half of this chapter. The bacterial small RNAs (called **sRNAs**) are larger (80–110 nucleotides) than those regulatory RNAs from eukaryotes (which range from 21 to 30 nucleotides). And they are not generally formed by processing of larger dsRNA precursors (as those eukaryotic RNA regulators are); instead, they are encoded in their final form by small genes. Many of these genes have been identified by bioinformatics, with close to 100 sRNAs being uncovered in *E. coli.* Of these, about a dozen have been characterized. Most sRNAs work by base pairing with complementary sequences within target mRNAs and directing destruction of the mRNA, inhibiting its translation or even in some cases *stimulating* translation.

Binding of an sRNA to its target mRNA is in most cases aided by the bacterial protein Hfq. This RNA chaperone is needed because the complementarity between the the sRNAs and their target mRNAs is typically imperfect and short, and so their interaction is weak. Hfq facilitiates base pairing; also, by binding the sRNAs even before they are paired with their targets, Hfq increases the stability of these regulators.

A well-studied sRNA from *E. coli* is the 81-nucleotide RybB RNA. This sRNA binds several target mRNAs and triggers their destruction because the double-strand stretch of heteroduplex formed upon pairing is recognized as a substrate by the nuclease RNase E. Most of the mRNAs targeted by RybB encode iron storage proteins. Free iron is required by the cell under certain circumstances, but high levels are toxic. RybB regulates the levels of free iron by controlling the levels of iron storage proteins. RybB is expressed from a promoter recognized by a special σ factor called σ^{E} (like σ^{S}, a stress response σ factor). Expression of the gene encoding σ^{E} is itself regulated by RybB, and so this sRNA is part of an autonegative regulatory loop for σ^{E}.

The stationary-phase σ factor σ^{S}, mentioned above, is encoded by the *rpoS* gene of *E. coli.* Translation of *rpoS* mRNA is stimulated by two sRNAs: DsrA and RprA. Activation is achieved by a switch in alternative RNA base pairing: the small RNAs bind to a region of the mRNA that otherwise would pair with the ribosome-binding site, inhibiting translation. The *rpoS* gene is also acted on negatively by another small RNA, OxyS. Figure 18-1 shows these two mechanisms.

We shall return to *trans*-acting regulatory RNAs in the second half of this chapter where we consider their role in regulating gene expression in eukaryotes. But before turning to that topic, we consider other examples in bacteria of gene regulation mediated through alternative RNA pairing that operate in *cis*—that is, RNA regulatory elements that control expression of the genes within whose mRNAs they reside. The most striking examples are the so-called **riboswitches** that control metabolic operons and **attenuation** in biosynthetic operons. The *trp* genes of *E. coli* are the classic example of the latter mechanism and are where RNA-mediated regulation was discovered (we shall describe this case in detail in Box 18-1).

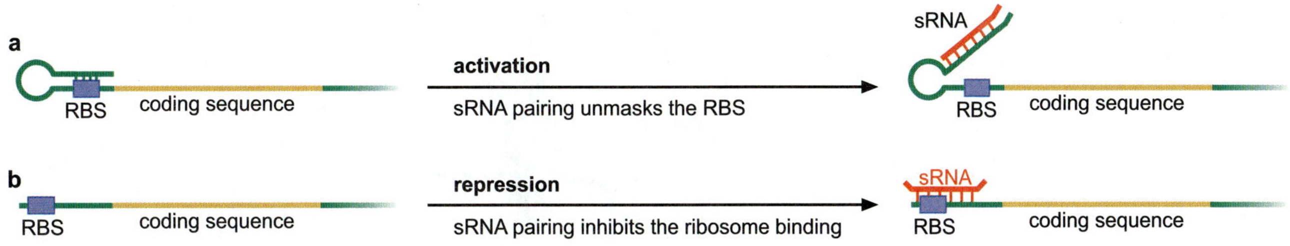

FIGURE 18-1 Activation and repression of translation by sRNAs. RBS indicates the ribosome-binding site. When this is occluded by base pairing with another RNA molecule—as in part b—or another region of the same RNA molecule—as in part a—translation is inhibited. (Adapted, with permission, from Gottesman S. et al. 2006. *Cold Spring Harbor Symp. Quant. Biol.* 71: 1–11, Fig. 1. © Cold Spring Harbor Laboratory Press.)

Riboswitches Reside within the Transcripts of Genes Whose Expression They Control through Changes in Secondary Structure

Riboswitches control gene expression in response to changes in the concentrations of small molecules. These regulatory elements are typically found within the 5′-untranslated regions of the genes they control, and they can regulate expression at the level of transcription or translation. They do this through changes in RNA secondary structure, as we shall see.

Each riboswitch is made up of two components: the **aptamer** and the **expression platform** (Fig. 18-2). The aptamer binds the small-molecule ligand and, in response, undergoes a conformational change, which in turn causes a change in the secondary structure of the adjoining expression platform. These conformational changes alter expression of the associated gene by either terminating transcription or inhibiting the initiation of translation. Both mechanisms are illustrated in the example shown in Figure 18-3, and which we now describe.

Riboswitches are, not surprisingly, typically found upstream of genes involved in the synthesis of the metabolite ligand recognized by the riboswitch in question. For example, in *Bacillus subtilis*, many genes involved in the use of the amino acid methionine have a 200-nucleotide-long untranslated leader RNA that acts as a SAM (*S*-adenosylmethionine)-sensing riboswitch. RNA polymerase initiates transcription at the promoter and transcribes through this leader region before entering the coding sequence of the downstream genes. Once transcribed into RNA, the leader region can adopt alternative structures through alternative patterns of intramolecular base pairing

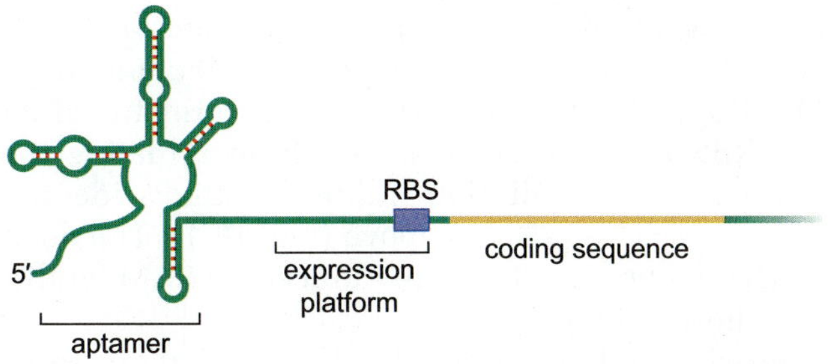

FIGURE 18-2 Organization of riboswitch RNAs. As described in the text, the aptamer binds the controlling metabolite, causing changes in the structure of the adjoining expression platform. The aptamers identified to date vary in size from 70 to 200 nucleotides; the expression platforms vary more in both size and character.

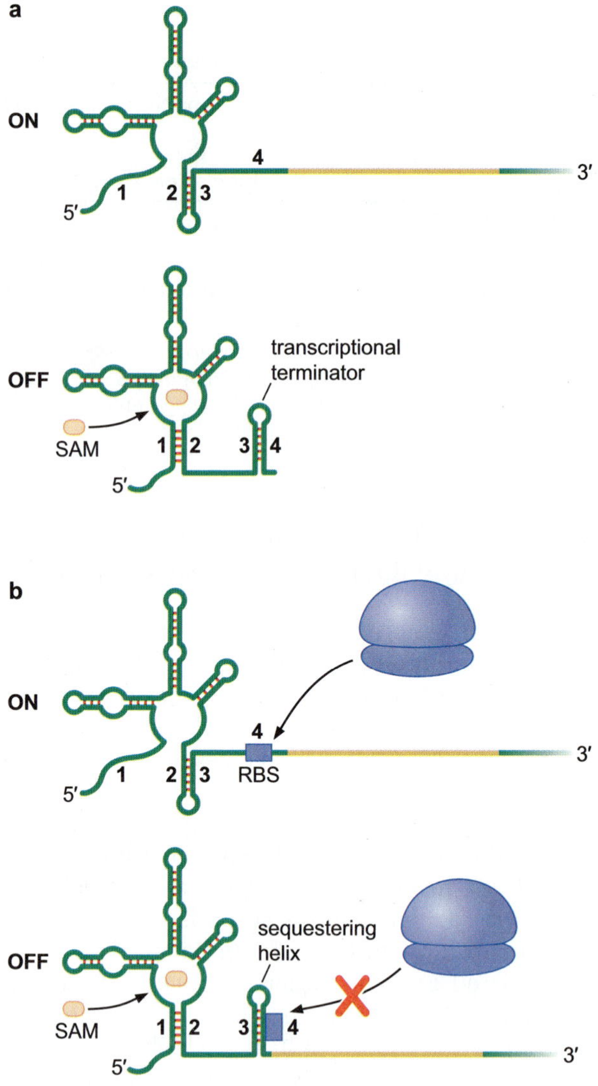

FIGURE 18-3 Riboswitches regulate transcription termination or translation initiation. Shown are two examples of a SAM-sensing riboswitch, in one case (a) regulating transcription termination, in the other (b) translation initiation. Numbers 1–4 indicate different sequence elements within the RNA upstream of the coding region (which is shown in yellow). In the absence of SAM, regions 2 and 3 form a stem loop; in the presence of SAM, regions 1 and 2 form a stem loop, and regions 3 and 4 do likewise. The consequence of that change in secondary structure controls transcription or translation as shown. (a) A stem loop of regions 3 and 4 produces a transcriptional terminator, which triggers RNA polymerase to terminate transcription immediately after transcribing those regions and before entering the downstream coding region. The stem loop in this case is followed by a stretch of Us in the mRNA, another feature of transcriptional terminator (Fig. 12-12). (b) The stem loop formed by regions 3 and 4 inhibits translation initiation by sequestering the ribosome binding site, as shown.

(Fig. 18-3a). One arrangement includes a stem-loop transcriptional terminator (see Chapter 12). SAM—the ligand for this riboswitch—binds to the aptamer and stabilizes the secondary structure that includes this transcriptional terminator (as shown in the bottom part of Fig. 18-3a). Under these circumstances, transcription is terminated before polymerase has a chance to transcribe the downstream protein-coding segment of the gene. This form of transcriptional regulation is also called **attenuation**. In another case—at another gene—a SAM-sensing riboswitch can work by regulating translation. In that case, as shown in Figure 18-3b, the alternative secondary structure stabilized by SAM binding to the aptamer includes a stem loop that, although not a transcriptional terminator, does include the ribosome-binding site (RBS; within region 4). This conformational change sequesters the RBS and blocks ribosomes from initiating translation. This form of translation inhibition is thus essentially identical to that described for *trans*-acting sRNAs above (Fig. 18-1). The details of the changes in RNA secondary structure induced by SAM binding to a riboswitch are shown in Figure 18-4.

Many riboswitches have been identified and these respond to a range of different metabolites. Ligands include lysine and other amino

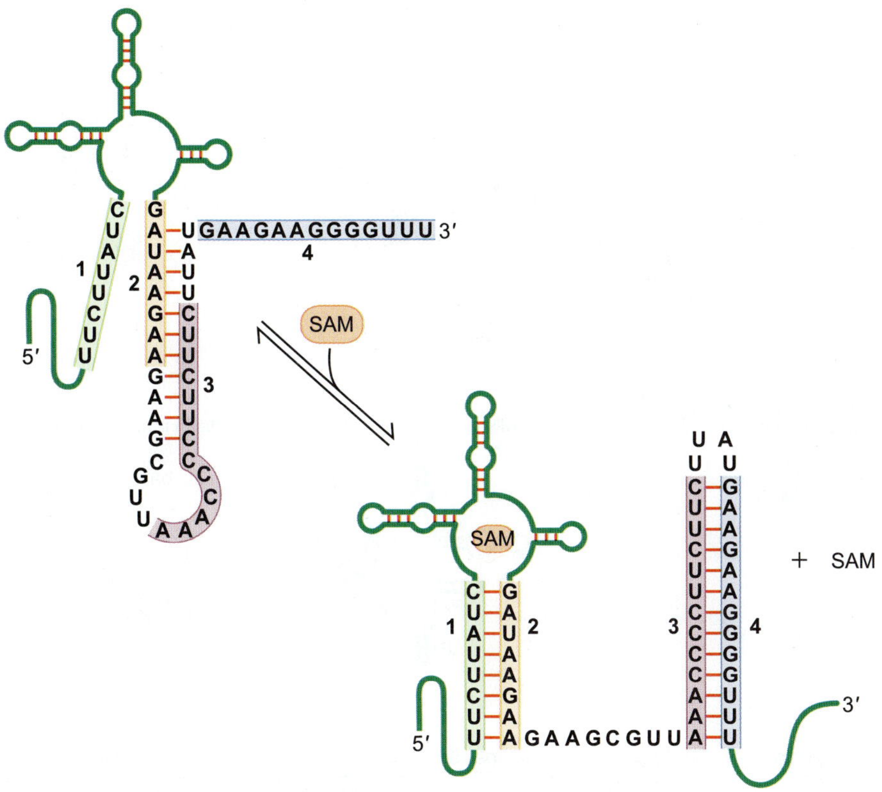

FIGURE 18-4 Changes in secondary structure of a SAM-sensing riboswitch. The sequences of regions 1–4 (described in Fig. 18-3) are here shown in detail and color-coded. The base pairing found in the two alternative secondary structures—that is, with and without SAM bound—are shown. (Adapted from Winkler W.C. et al. 2003. *Nat. Struct. Biol.* 10: 701–707, Fig. 5b. © Macmillan.)

acids, vitamin B12, coenzyme thiamine pyrophosphate (TPP), flavin mononucleotide (FMN), and guanine (Fig. 18-5).

Another kind of riboswitch responds to uncharged tRNAs, rather than to small-molecule ligands. Thus, certain genes, notably genes for aminoacyl-tRNA synthetases (see Chapter 14), are controlled by attenuation mediated by a 200–300-nucleotide-long, untranslated, leader RNA that directly and specifically interacts with the cognate, uncharged tRNA for the synthetase; the charged form of the tRNA does not fit in the binding pocket provided by the RNA secondary structure. Binding of uncharged tRNA stabilizes the leader RNA in its antitermination structure so that transcription into the adjacent synthetase gene can proceed. Specificity is achieved in part by a "codon–anticodon" interaction between the tRNA and the leader RNA. Because uncharged (but not charged) tRNA can bind to the leader, transcriptional readthrough is only stimulated when the cognate amino acid is in short supply and the level of uncharged tRNA in the cell rises.

Although most prevalent in bacteria, riboswitches are found in other organisms as well, including archaea, fungi, and plants. In some cases in these higher organisms, riboswitches are even involved in controlling alternative splicing (Chapter 13). Thus, for example, in a recent case described in the fungus *Neurospora crassa*, three TPP aptamers were identified, two of which inhibited, and the third stimulated, expression of genes through regulation of RNA splicing.

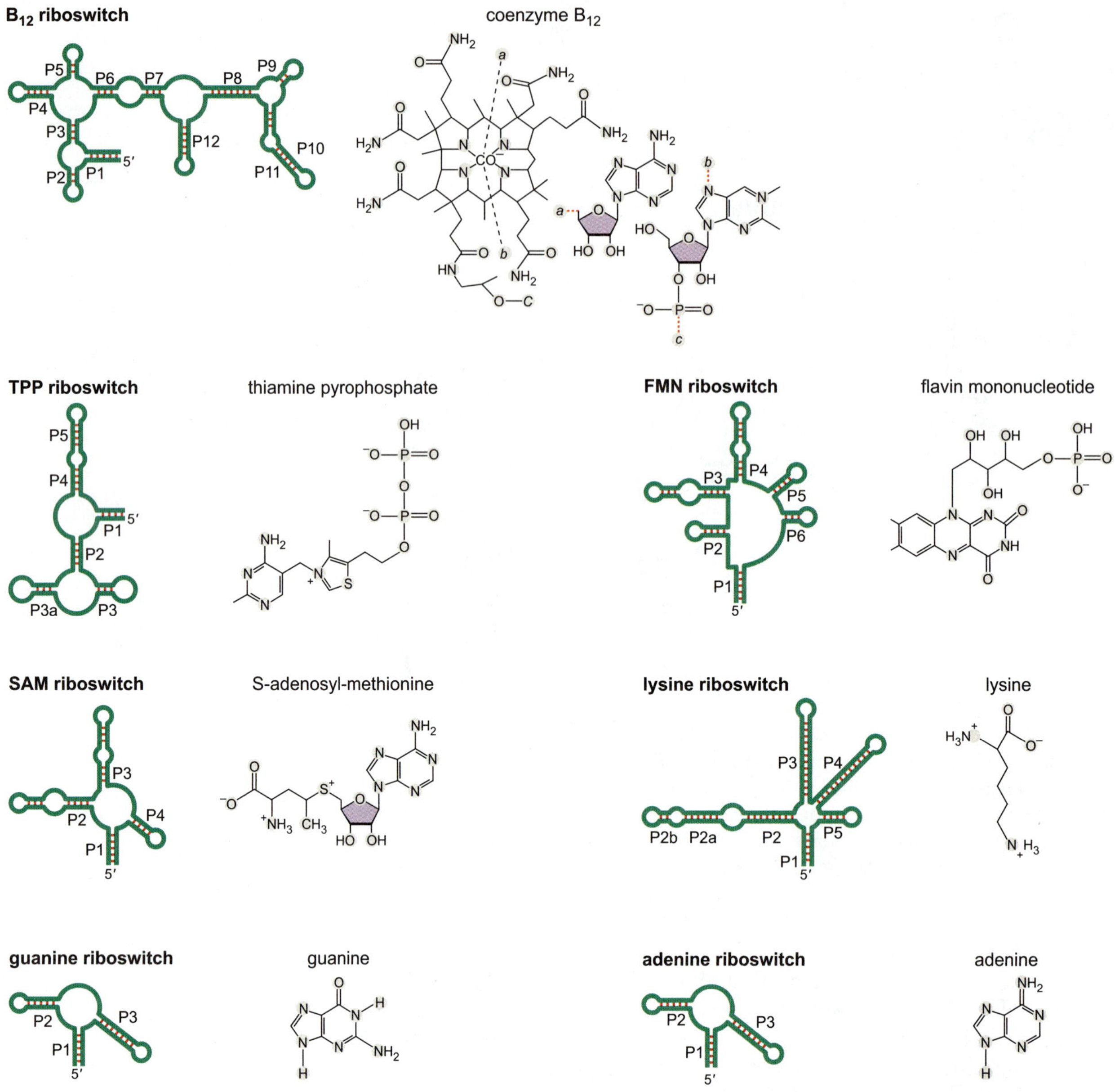

FIGURE 18-5 **Riboswitches respond to a range of metabolites.** The secondary structure of seven riboswitches and the metabolites they sense are shown here. (Adapted, with permission, from Mandal M. et al. 2003. *Cell* 113: 577–586, Fig. 7A. © Elsevier.)

Attenuation mediated by alternative RNA secondary structures was first discovered in studies of the tryptophan operon of *Escherichia coli*, as described in Box 18-1, Amino Acid Biosynthetic Operons Are Controlled by Attenuation. The *trp* operon contains genes responsible for the biosynthesis of tryptophan. Their expression is controlled in response to the level of tryptophan in the cell as measured by the availability of charged $tRNA^{trp}$. This example turned out to be one of several similar cases—attenuation linked to amino acid availability is a common control mechanism for amino acid biosynthetic operons. Understanding how the *trp* operon was controlled was the first time that alternative RNA secondary structures were shown to regulate gene expression in any organism.

ADVANCED CONCEPTS

Box 18-1 Amino Acid Biosynthetic Operons Are Controlled by Attenuation

In *E. coli*, the five contiguous *trp* genes encode enzymes that synthesize the amino acid tryptophan. These genes are expressed efficiently only when tryptophan is limiting (Box 18-1 Fig. 1). The genes are controlled by a repressor, just as the *lac* genes are, although in this case, it is the *absence* of its ligand (tryptophan) that relieves repression.

Even after RNA polymerase has initiated a *trp* mRNA molecule, however, it does not always complete the full transcript. As with riboswitches, the decision to make a complete transcript is controlled by attenuation; in this case, most transcripts are terminated prematurely, before they include even the first *trp* gene (*trpE*). But attenuation is overcome if tryptophan levels are low in the cell; when tryptophan is limiting, polymerase does not terminate and instead transcribes all of the *trp* genes. Whether or not attenuation occurs depends on the ability of RNAs to form alternative secondary structures, just as it did with the riboswitches. In this case, however, the choice between alternative structures formed by the leader RNA is not controlled by binding of ligand directly to that RNA; instead, the choice of alternatives relies on the coupling of transcription and translation in bacteria.

The sequence of the 5′ end of *trp* operon mRNA includes a 161-nucleotide leader sequence upstream of the first codon of *trpE* (Box 18-1 Fig. 2). Near the end of this leader sequence, and before *trpE*, is a transcription terminator, composed of a characteristic hairpin loop in the RNA (made from sequences in regions 3 and 4 of Box 18-1 Fig. 2), followed by eight uridine residues (see Fig. 12-12). Transcription usually stops after this terminator (and, we might have thought, should always stop), yielding a leader RNA 139 nucleotides long. This is the RNA product seen in the presence of high levels of tryptophan.

Three features of the leader sequence allow the terminator to be passed by RNA polymerase when the cellular concentration of tryptophan is low. First, there is a second hairpin (besides the terminator hairpin) that can form between regions 1 and 2 of the leader (see Box 18-1 Fig. 2). Second, region 2 also is complementary to region 3; thus, yet another hairpin consisting of regions 2 and 3 can form, and when it does, it prevents the terminator hairpin (3, 4) from forming. Third, the leader RNA contains an open reading frame encoding a short "leader" peptide of 14 amino acids, and this open reading frame is preceded by a strong ribosome-binding site (see Box 18-1 Fig. 2).

The sequence encoding the leader peptide has a striking feature: two tryptophan codons in a row. When tryptophan is scarce, there is very little charged tryptophan tRNA available, and the ribosome stalls when it reaches the two tryptophan codons. Under these circumstances, RNA around the tryptophan codons is within the ribosome and cannot be part of a hairpin loop. The consequence of this is shown in Box 18-1 Figure 3.

A ribosome caught at the tryptophan codons (part b) masks region 1, leaving region 2 free to pair with region 3; thus, the terminator hairpin (formed by regions 3 and 4) cannot be made, and transcription is not attenuated. If, on the other hand, there is enough tryptophan (and therefore enough charged Trp tRNA) for the ribosome to proceed through the tryptophan codons, the ribosome blocks sequence 2 by the time RNA containing regions 3 and 4 has been made. Thus, the terminator forms, attenuating transcription, and the *trp* genes are not transcribed.

The *trp* operon is controlled by repression and attenuation, providing a two-stage response to progressively more stringent tryptophan starvation. But attenuation alone can provide robust regulation: other amino acid operons such as *leu* and *his* rely entirely on attenuation for their control. In the case of the leucine operon, its leader peptide has four adjacent leucine codons, and the histidine operon leader peptide has seven histidine codons in a row.

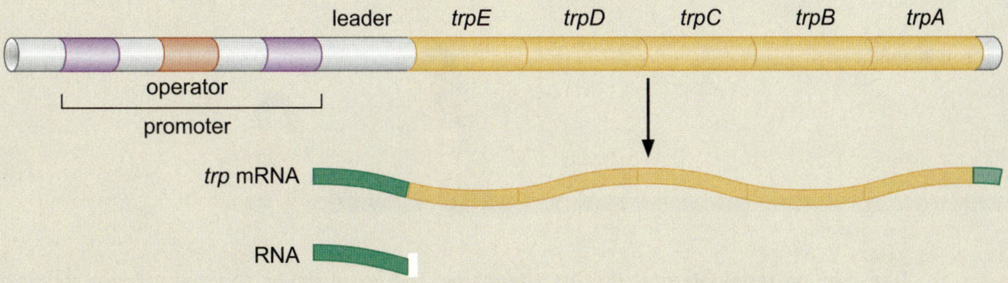

Box 18-1 Figure 1 The *trp* operon. The tryptophan operon of *E. coli,* showing the relationship of the leader (see text) to the structural genes that code for the Trp enzymes. The gene products are anthranilate synthetase (product of *trpE*), phosphoribosyl anthranilate transferase (*trpD*), phosphoribosyl anthranilate isomerase-indole glycerol phosphate synthetase (*trpC*), tryptophan synthetase β (*trpB*), and tryptophan synthetase α (*trpA*).

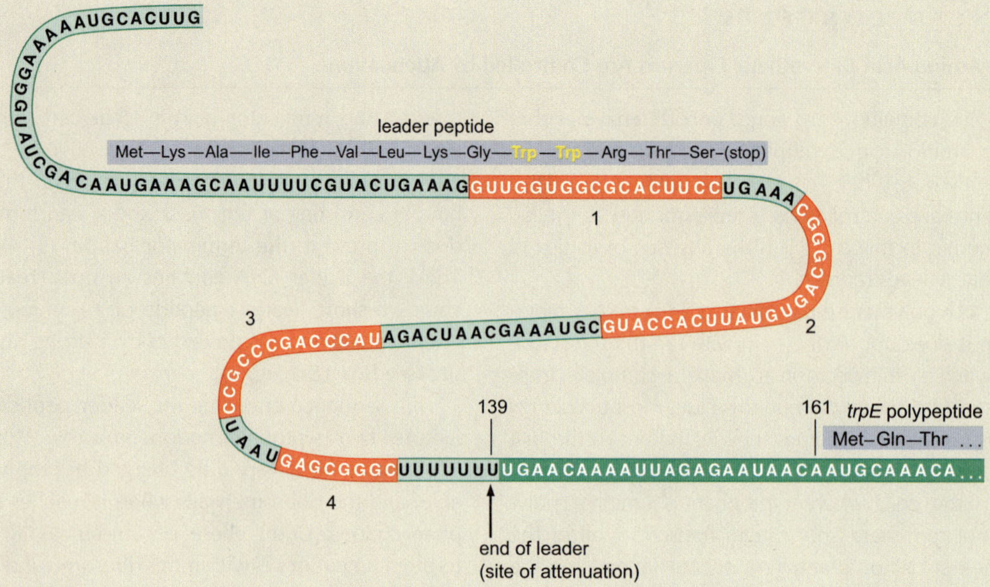

Box 18-1 Figure 2 ***trp* operator leader RNA.** Features of the nucleotide sequence of the *trp* operon leader RNA.

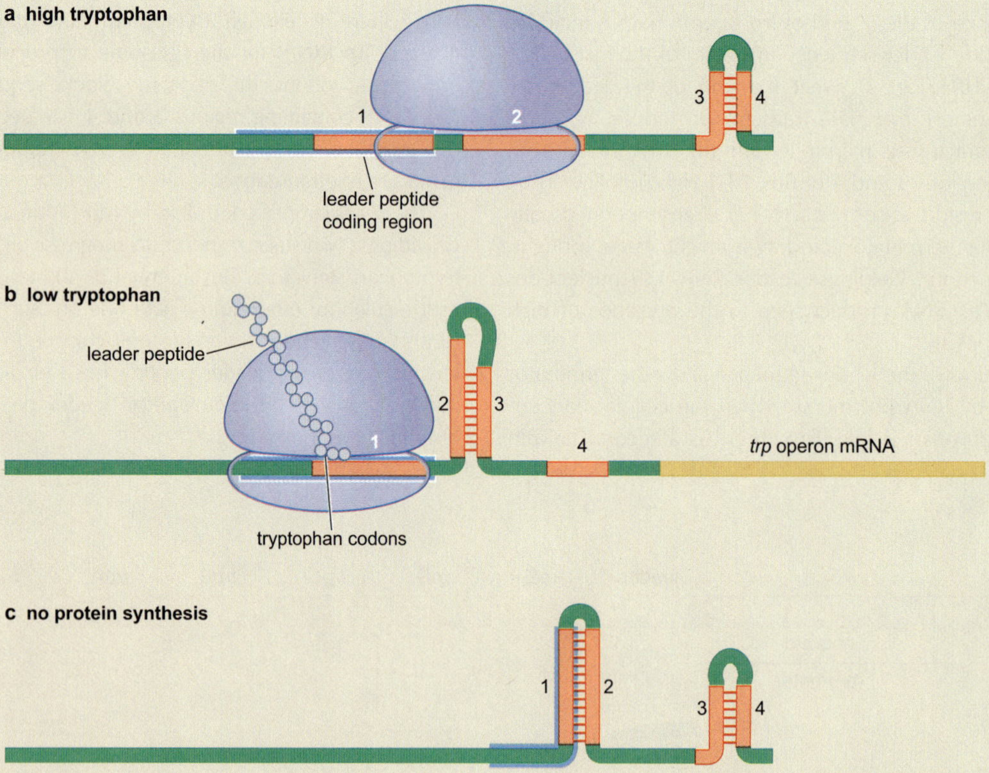

Box 18-1 Figure 3 Transcription termination at the *trp* attenuator. The figure shows how transcription termination at the *trp* operon attenuator is controlled by the availability of tryptophan. The blue box shows the leader peptide coding region. (a) Conditions of high tryptophan: sequence 3 can pair with sequence 4 to form the transcription termination hairpin. (b) Conditions of low tryptophan: the ribosome stalls at adjacent tryptophan codons, leaving sequence 2 free to pair with sequence 3, thereby preventing formation of the 3-4 termination hairpin. (c) No protein synthesis: if no ribosome begins translation of the leader peptide AUG, the hairpin forms by pairing of sequences 1 and 2, preventing formation of the 2-3 hairpin, and allowing formation of the hairpin at sequences 3-4. The Trp enzymes are not expressed.

RNA INTERFERENCE IS A MAJOR REGULATORY MECHANISM IN EUKARYOTES

We have seen how RNA molecules can regulate expression of genes in prokaryotes. We have (in Chapter 17) already seen examples of regulatory RNA elements in the transcripts of some eukaryotic genes, but these function through binding regulatory proteins (the HIV TAT protein was one such case). It is, however, now apparent that RNAs have a more widespread and unanticipated role in gene regulation in eukaryotes. These new modes of regulation are described in this section.

Several types of very short RNAs repress—or silence—expression of genes with homology to those short RNAs. This silencing, called **RNA interference (RNAi)**, manifests in different ways: sometimes by inhibiting translation of the mRNA, in other cases through destruction of the mRNA, and in yet others by transcriptional silencing of the promoter that directs expression of that mRNA. As we describe below, these short RNAs are generated by special enzymes from longer double-stranded RNAs (dsRNAs) of various origins.

It is still unclear how widespread the action of regulatory RNAs will turn out to be, and the detailed mechanism used to silence the target genes in any given case is still emerging. But it is already clear that the roles of these RNAs range from developmental regulation (well studied in the worm *Caenorhabditis elegans* and the plant *Arabidopsis*; see Chapter 22) to mechanisms that protect organisms against viral infection. Furthermore, RNAi has been adapted for use in several organisms as a powerful experimental tool, providing an easy means of "turning off" expression of any specific gene.

Short RNAs That Silence Genes Are Produced from a Variety of Sources and Direct the Silencing of Genes in Three Different Ways

Before describing aspects of the production and function of these short silencing RNAs in more detail, we first provide an overview of how this type of silencing works (illustrated in Fig. 18-6).

The small RNAs have different names depending on their origin. Those made artificially or produced in vivo from dsRNA precursors are typically called **small interfering RNAs (siRNAs)**. Another group of regulatory RNAs are the **microRNAs (miRNAs)**. The latter are derived from precursor RNAs which are encoded by genes expressed in cells where those miRNAs have specific regulatory functions.

Both siRNAs and miRNAs are generated from longer RNA molecules by the enzyme **Dicer**, an RNase-III-like enzyme (Fig. 5-24) that recognizes and digests longer dsRNA or the stem-loop structures formed by miRNA precursors (see below). The siRNA and miRNA products are typically 21–23 nucleotides long; their production is shown as the first step in Figure 18-6.

These small RNAs inhibit expression of homologous target genes in three ways: they trigger destruction of the mRNA encoded by the target gene, they inhibit translation of the mRNA, or they induce chromatin modifications within the target gene and thereby silence its transcription. Remarkably, whichever route is used in any given case, much of the same machinery is required. This machinery includes a complex called the **RNA-induced silencing complex (RISC)**. A RISC

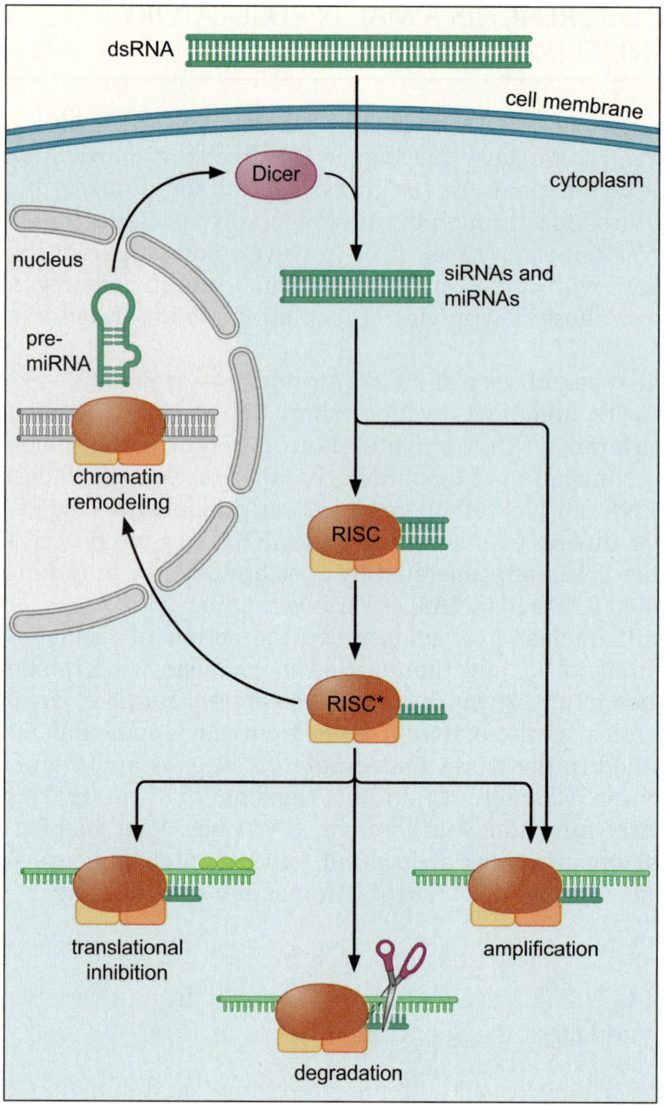

FIGURE 18-6 **RNAi silencing.** RNAi switches off the expression of a gene when dsRNA molecules that have homology to that gene are introduced, or made, in the cell. This effect involves processing of the dsRNA to make siRNAs and miRNAs by the enzyme Dicer. Another enzyme involved only in the case of miRNAs—Drosha—is not shown here, but is described later. The siRNAs and miRNAs direct a complex called RISC (RNA-induced silencing complex) to repress genes in three ways. It attacks and digests mRNA that has homology with the siRNA; it interferes with translation of those mRNAs; or it directs chromatin-modifying enzymes to the promoters that direct expression of those mRNAs. (Adapted, with permission, from Hannon G.J. 2002. *Nature* 418: 244–251, Fig. 5. © Macmillan.)

contains, in addition to the siRNA or miRNA, various proteins including a member of the **Argonaute** family.

The siRNA or miRNA must be denatured to give a **guide RNA**—the strand that gives the RISC specificity, as we shall see—and a **passenger RNA**, which usually gets discarded. The resulting complex, the mature RISC, is then directed to target RNAs containing sequences complementary to the guide RNA. These target RNAs are degraded or their translation is inhibited. Typically, the choice depends in part on how closely the guide RNA matches the target mRNA: if the sequences are highly complementary, the target is degraded; if the match is not as good (i.e., if there are several base-pairing mismatches), the response is more often inhibition of translation. In those cases where the target RNA is degraded, Argonaute is the catalytic subunit that carries out the initial mRNA cleavage; for this reason Argonaute is often called "**Slicer**" and mRNA cleavage is called "slicing."

A RISC can also be directed into the nucleus where it recruits other proteins that modify the chromatin around the promoter of the gene complementary to the guide RNA (shown on the left of Fig. 18-6). This modification leads to silencing of transcription (Chapter 17). Establishing silencing in the centromeric regions of the yeast *Schizosaccharomyces pombe*, for example, requires the RNAi machinery. In this case, it is believed that regions of the centromere (see Chapter 7) are transcribed to produce RNAs that hybridize with other RNAs from the same region. The resulting dsRNAs are recognized by Dicer and cleaved to produce the siRNAs responsible for directing the RNAi machinery to the centromeres.

Another feature of RNAi silencing is worth noting—its extreme efficiency. Thus, very small amounts of dsRNA are often enough to induce a near complete shutdown of target gene expression. A factor adding to efficiency could be the action of an **RNA-dependent RNA polymerase** (RdRP), an additional enzyme required in many cases of RNAi including centromeric silencing in fission yeast. This polymerase can amplify the inhibitory signal: the RdRP generates dsRNA after recruitment to the mRNA by the original siRNA (shown on the right of Fig. 18-6). This feedback process generates large amounts of siRNA. RdRP has not yet been identified in mammalian cells, but high efficiency likely still results from the fact that slicing is catalytic—that is, each RISC can cleave several mRNAs.

Thus, although in the first section of this chapter we saw examples of small RNAs regulating gene expression in *trans* in bacteria, the mechanism of both production and action of such RNAs in eukaryotes is very different.

SYNTHESIS AND FUNCTION OF miRNA MOLECULES

miRNAs Have a Characteristic Structure That Assists in Identifying Them and Their Target Genes

As we have seen, miRNAs are one class of regulatory RNA that silences gene expression through the RNAi pathway. Indeed, in terms of their production and function, miRNAs are perhaps the best understood class. As mentioned above, miRNAs are encoded in the genome as segments of longer transcripts. Their characteristic structure helps identify them and predict the target genes they might regulate.

The functional form of an miRNA is typically about 21 or 22 nucleotides (it can vary from 19 to 25 nucleotides). These short RNAs are generated by two RNA cleavage reactions from a longer RNA transcript (called a pri-miRNA, for *pri*mary) that carries a hairpin-shaped secondary structure. The first cleavage liberates the stem loop, called the pre-miRNA; the second generates the mature miRNA from the pre-miRNA. One of the first identified, and best-characterized, miRNAs is *let-7*, which regulates development at the larval-to-adult transition in the worm *C. elegans* (see Chapter 22). The structure of the pre-miRNAs for *let-7* and some other naturally occurring miRNAs are shown in Figure 18-7.

It was thought initially that one "arm" of pre-miRNA stem-loop structure would be the regulatory miRNA. Recently, however, numerous examples have been identified where both "arms" of the structure give rise to functional miRNAs, each with its own set of target genes (in these cases the two miRNAs are labeled red and blue in Fig. 18-7). It now appears that having miRNAs produced from both arms is common. The pre-miRNAs can be encoded by any part of a transcript: that is, they

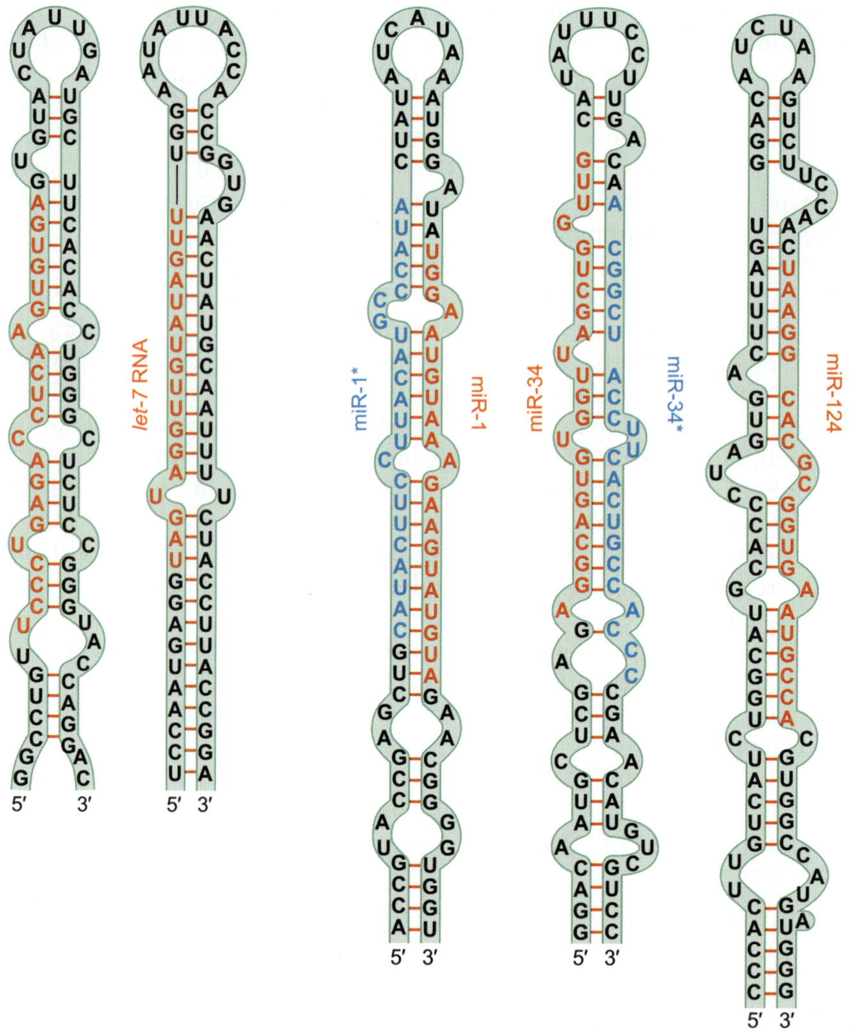

FIGURE 18-7 Structure of some pre-miRNAs prior to processing to generate the mature miRNAs. The sequences in red are miRNAs. In some cases, both "arms" of a stem loop can generate a functional miRNA. In such cases, the second miRNA is shown in blue—for example, miR-1 (red) and mi-R1* (blue), as well as with miR-34 (red) and miR34* (blue). The miRNAs shown are all from the worm. *lin-4* and *let-7* were identified genetically; those called miR are found by bioinformatics. (Modified, with permission, from Lim L.P. et al. 2003. *Genes Dev.* 17: 991, Fig. 6. © Cold Spring Harbor Laboratory Press.)

might fall within coding regions, within leader regions, or within introns (Fig. 18-8).

The distinctive secondary structure of a primary transcript carrying an miRNA (pri-miRNA) has made it possible to predict their presence based on the calculated secondary-structure fold of the RNA sequence. Furthermore, in many cases, candidates for the regulated target genes can also be predicted, because silencing depends on sequence complementarity between the target and the mature miRNA. The base pairing between miRNA and target RNA is initiated by interactions of so-called "seed residues"—typically the sequence between bases 2 and 9 of the 22-nucleotide miRNA. This is the region of highest complementarity, and so it is the region most useful in identifying candidate target genes. Of course, establishing that an miRNA really exists requires that its

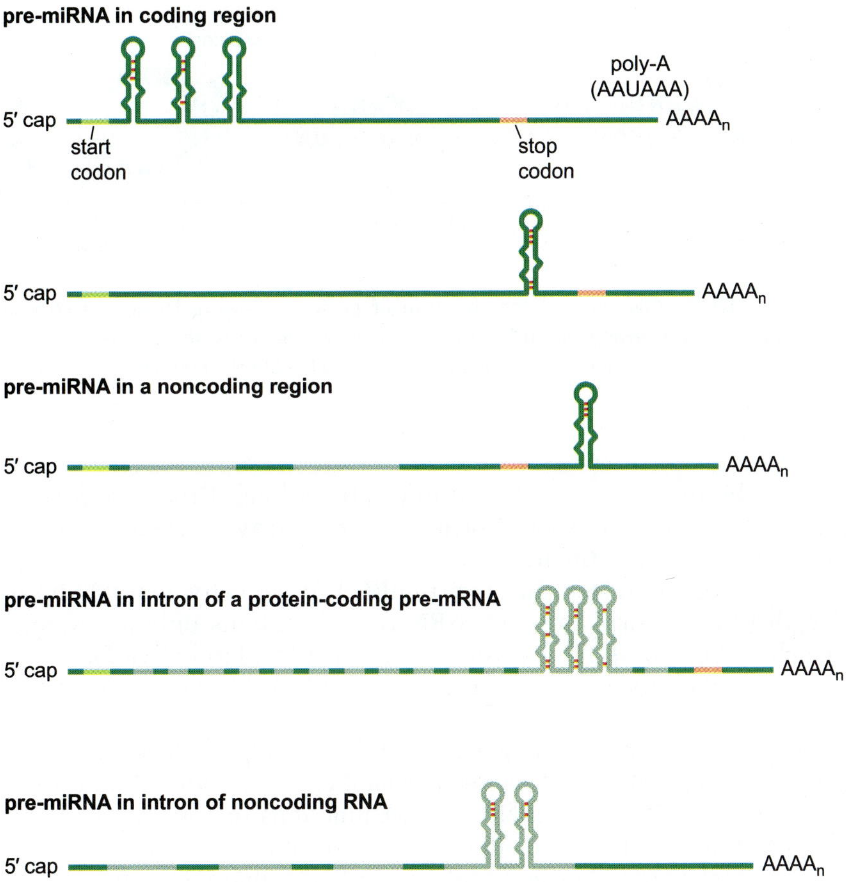

FIGURE 18-8 **miRNAs are coded in both introns and exons in RNA.** Intronic sequences are shown in light green. Start and stop codons are indicated by lime green and pink, respectively.

presence be detected in cells (e.g., by northern blotting) and that gene expression from target mRNA be affected by its presence.

The two cleavage reactions required to generate the miRNA from these primary transcripts are mediated by two distinct RNases. One is Dicer, which we have already introduced, and is required for almost all cases of RNAi. The other, specifically required for miRNA processing, is **Drosha**. A characteristic of both these enzymes is that they recognize and cleave RNAs on the basis of the structure of their substrates rather than their specific sequence. We now turn our attention to how these enzymes work.

An Active miRNA Is Generated through a Two-Step Nucleolytic Processing

Two specialized RNA-cleaving enzymes are required to process the initial pri-miRNA transcript, containing the stem-loop structure, into the mature miRNA. The first enzyme is Drosha, a member of the RNase III family of enzymes. Drosha makes two cleavages that cut the stem-loop region of the RNA (pre-miRNA) out of the primary transcript RNA (pri-miRNA). This enzyme works together with an essential specificity subunit protein (called Pasha in some organisms and DGCR8 in others), and together these two proteins form an active **microprocessor complex**. The pre-miRNA generated by Drosha

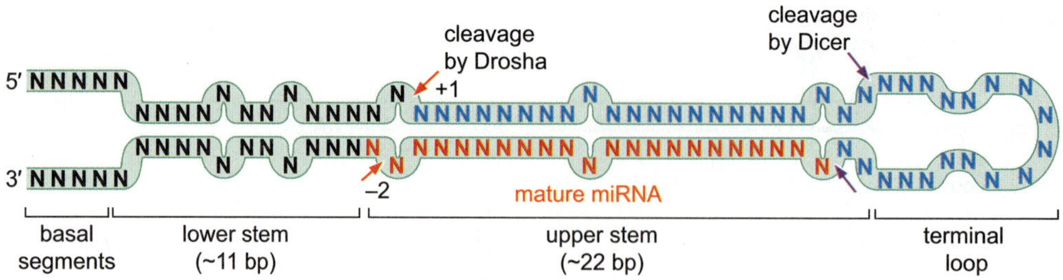

FIGURE 18-9 Overview of the structure of pri-RNA showing Dicer and Drosha cleavage sites. The region in red becomes the mature miRNA. Note that the basal segments must be single-stranded for proper recognition by the Drosha complex.

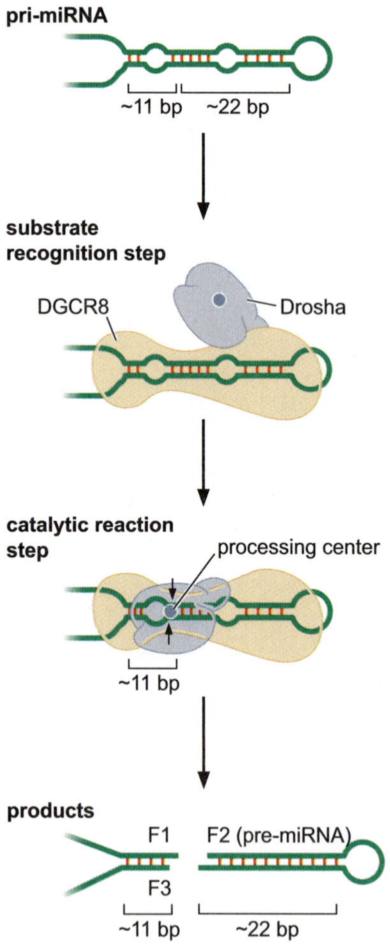

FIGURE 18-10 Recognition and cleavage of pri-miRNA by the microprocessor complex. Three fragments are generated by cleavage, labeled F1, F2 (the pre-miRNA), and F3.

is usually approximately 65–70 nucleotides long. Drosha resides in the cell's nucleus, and the Drosha-catalyzed cleavage event occurs in this cellular compartment.

The base-paired stem in the pri-miRNA is typically about 33 bp in length (three helical turns of dsRNA) and contains only a few mismatches (Fig. 18-9). At the "top" of the stem is a loop of variable size (usually relativity large, about 10 nucleotides); the sequence of this loop region is not critical for the processing reactions. Importantly, for processing by Drosha, single-stranded RNA (ssRNA), lacking significant secondary structure, is needed flanking each side (5′ and 3′) of the stem-loop. It is the ssRNA–dsRNA junctions that are in large part responsible for determining the cleavage specificity of Drosha.

The stem region can be divided into two functional segments: an approximately 11-bp lower stem and an approximately 22-bp upper stem (Fig. 18-9). Drosha cleaves 11 bp away from the dsRNA–ssRNA junctions—that is, between the lower and upper stems in the pri-miRNA (Fig. 18-10). The two cleavages thus generate the approximately 65-nucleotide pre-miRNA composed of the 22 bp (two helical turns) of dsRNA and the top loop. The RNase III family enzymes are specific for dsRNA and cleave it in a manner that leaves a 2 nucleotide overhang on the 3′ ends of the dsRNA product. This 3′ overhang is important for recognition of that RNA molecule by the next enzyme in the pathway, Dicer.

Dicer Is the Second RNA-Cleaving Enzyme Involved in miRNA Production

The pre-miRNA liberated by Drosha is exported to the cytoplasm where the second RNA cleavage reaction, carried out by Dicer, takes place. As with Drosha, Dicer selects its cleavage sites using a measuring, rather than sequence-specific, mechanism. A high-resolution structure of Dicer provides insight into how this likely occurs.

Dicer is constructed of three modules (Fig. 5-23): two RNase III domains and a dsRNA-binding domain called the PAZ domain (named for three proteins that contain this domain: Piwi, Argonaute, and Zwille).

Figure 18-11a shows in cartoon form the organization of the Dicer protein and how it is believed to interact with a dsRNA molecule. In the right panel, Figure 18-11b, is the structure of Dicer, modeled with a substrate RNA. The overall structure of the protein is shaped like a hatchet. The PAZ domain is at the bottom of the handle, where it forms a binding pocket for the 3′ end of the dsRNA substrate. The

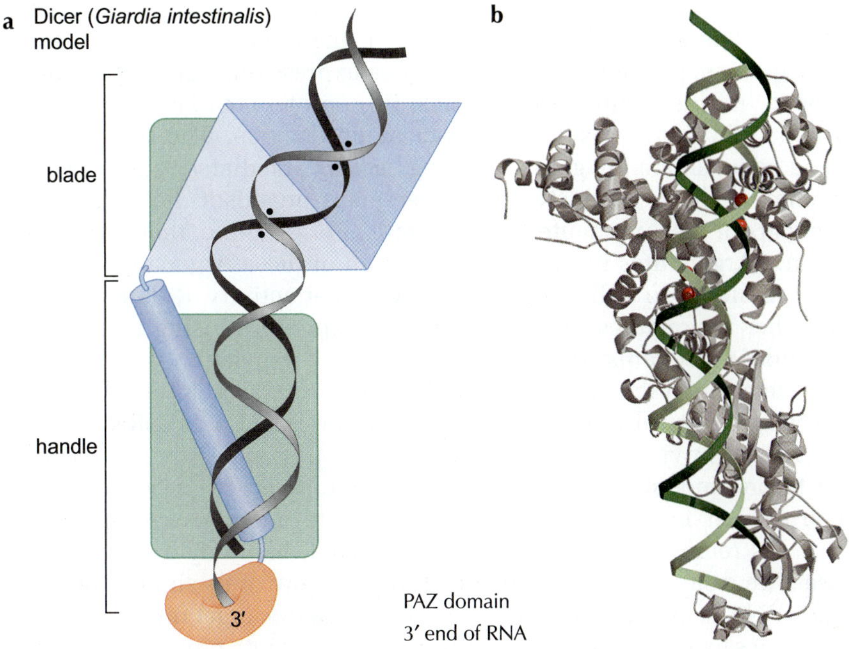

FIGURE 18-11 Dicer structure and organization. (a) The scheme shows Dicer organization. (b) Dicer structure modeled with dsRNA reveals how length is measured. The protein is shown in gray, with nuclease active sites indicated by the red spheres (and as black dots in part a). The RNA is in green. The structure shown contains only the RNase III and PAZ domains. The Dicer protein also contains ATPase and other domains. (b, MacRae I.J. et al. 2006. *Science* 311: 195–198. PDB Code: 2FFL; note that the RNA was modeled into the structure and was not part of the crystal structure.) Image prepared with MolScript, BobScript, and Raster 3-D.

handle of the hatchet is formed by a linker domain and contains a positively charged binding surface for the RNA molecule. The top "blade" region comprises the two RNase domains, arranged in a symmetrical dimer. Each RNase domain carries an active site and is responsible for cleaving one of the two strands of the substrate RNA. Thus, Dicer will act on any dsRNA, regardless of sequence, and will cleave this molecule 22 nucleotides from its end. The PAZ domain anchors the 3′ terminus of the substrate RNA to position the active sites of the enzyme approximately 22 nucleotides away in a ruler-like fashion (see Fig. 18-11). Indeed, the occurrence of differently sized PAZ domains correlates with the different sizes of Dicer products found in different organisms.

Incorporation of a Guide Strand RNA into RISC Makes the Mature Complex That Is Ready to Silence Gene Expression

The actions of Drosha and then Dicer generate the 21–25-nucleotide RNA molecule that will guide regulation of gene expression. The active form of the regulatory miRNA is the single-strand form—at this stage called the "guide" RNA—incorporated into a RISC protein complex. Within this complex, the RNA guide strand recruits RISC to a target RNA. It has been argued that the length of approximately 22 nucleotides is just long enough to uniquely specify a single target gene in the large genomes of complex eukaryotes using RNA–RNA base pairing.

RISC is a multicomponent complex that includes the guide miRNA. The guide RNA base-pairs with the target mRNA, and, as a consequence, promotes silencing of gene expression. The central component of this regulatory complex is a protein called Argonaute, which is in many cases an RNA-cleaving enzyme. The best understood mechanism of gene silencing is RISC-mediated cleavage—or slicing—of the target mRNA. However, many organisms have multiple members of the Argonaute protein family. For example, there are eight distinct Argonautes in humans, but not all of these Argonautes, when incorporated into a RISC complex, have slicer activity. RISCs containing other Argonautes must silence gene expression using non-slicer-dependent mechanisms, such as repression of translation.

Generation of the active RISC and slicing occur as follows. The short dsRNA generated by Dicer is incorporated into RISC. The dsRNA is denatured to provide the guide strand and the passenger strand. According to one model, the passenger strand is cleaved by Argonaute and then ejected from the complex; according to another, the passenger strand is removed without first being cleaved, likely by an RNA helicase. The resulting RISC—called mature RISC—with its single-stranded guide RNA is now ready to recognize and slice the target mRNA.

WEB STRUCTURAL TUTORIAL

As we saw with Dicer, the structure of an Argonaute protein provides a framework for understanding the mechanism of target RNA recognition and cleavage by RISC (Fig. 18-12; see also Structural Tutorial 18-1). Like Dicer, Argonaute has both a PAZ domain and an RNase domain. The PAZ domain specifically recognizes the 3′ end of the guide RNA. The bound guide RNA is base-paired to the target RNA, and the architecture of the complex is such that this binding positions the active site of the RNase

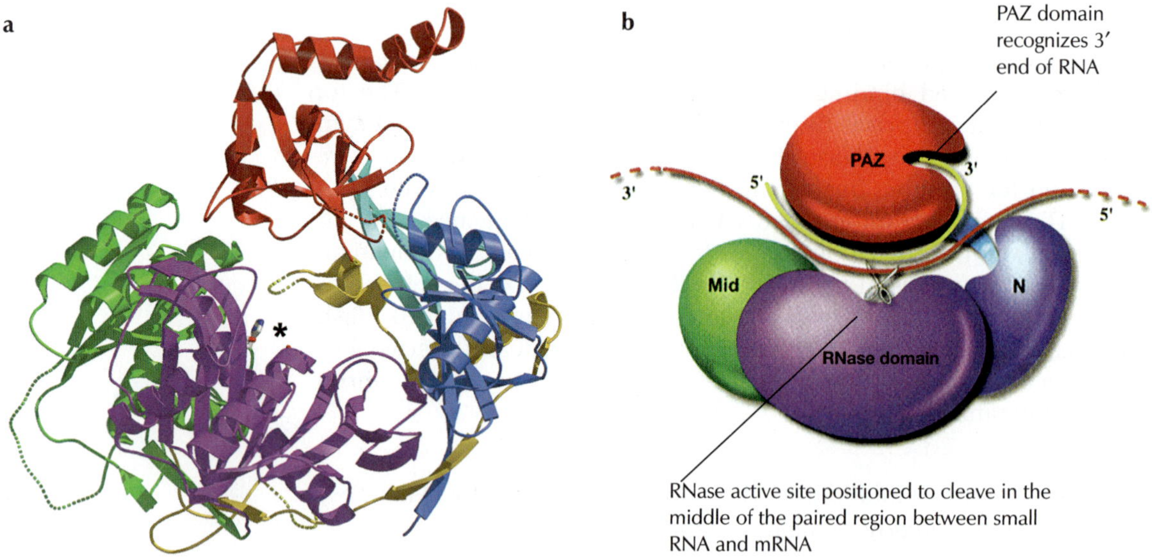

FIGURE 18-12 Argonaute structure, showing RNA-binding regions and an RNase H–like nuclease domain. (a) Crystal structure of Argonaute. The domains are colored as in part b, with the blue domain being the amino-terminal part of the protein, and the green domain in the middle. (b) Cartoon of the Argonaute domains. The arrow shows the RNase active site positioned to cleave in the middle of the paired region between small RNA and miRNA. (a, Song J.J. et al. 2004. *Science* 305: 1434–1437, Fig. 4C. PDB Code: Iu04. Image prepared with MolScript, BobScript, and Raster 3-D. b, Adapted, with permission, from Song J.J. et al. 2004. *Science* 305: 1434–1437, Fig. 4C. © AAAS.)

domain appropriately to cleave the target RNA strand. Cleavage occurs nearly in the middle of the guide RNA–target RNA duplex, between the 10th and 11th nucleotide from the 5′ end of the guide RNA.

As we have already mentioned, in some cases, mature RISC can inhibit translation rather than slicing mRNA. The mechanism of this translational repression is still under scrutiny, but it appears that miRNAs lead, in some cases, to the sequestration of mRNA in so-called processing bodies (P-bodies) within the cytoplasm where translation is repressed. Binding of miRNA can also destabilize the poly-A tail (see Chapter 12) of the target mRNA, disrupting translation initiation (Chapter 14).

siRNAs Are Regulatory RNAs Generated from Long Double-Stranded RNAs

As we have seen, the critical steps in synthesis of an active miRNA are Drosha cleavage, followed by Dicer cleavage, followed by incorporation of the guide RNA into the RISC complex. But only miRNAs are made from large hairpin precursors. In contrast, the precursor RNA for the siRNA pathway is a longer dsRNA. As a consequence of this different initial substrate, Drosha is not needed for the generation of siRNAs. Cleavage by Dicer is still required, however, and again generates a suitable 21–22-nucleotide RNA for incorporation into RISC. In plants, even miRNAs are generated by Dicer alone; it is not clear how they manage to forgo prior action of Drosha.

A fascinating series of observations and experiments led to our current understanding of small regulatory RNAs in eukaryotes. These began in the late 1980s with seemingly mystifying results of attempts to overexpress pigment genes in petunias (to make them a deeper purple, but ending up with the flowers turning white). Next was the surprising discovery of regulatory genes from worms whose products turned out to be miRNAs, and then experiments showing that introducing dsRNAs into worms silenced complementary gene expression. This story is described in Box 18-2, History of miRNAs and RNAi.

Small RNAs Can Transcriptionally Silence Genes by Directing Chromatin Modification

We have now seen how miRNAs and siRNAs can silence genes by inhibiting the translation of target mRNAs or directing their destruction. Regulatory RNAs can also act at the level of transcription, switching off expression of target genes by directing histone modification of the promoter. This mechanism has been most extensively studied in centromeric silencing in the fission yeast *S. pombe*.

We noted in Chapter 17 that in yeast, genes placed in certain regions of the genome are typically silenced. In the case we described in detail in that chapter, genes placed near the telomeres in *Saccharomyces cerevisiae* were silenced. Genes in the mating-type locus of that yeast and of *S. pombe* are also silenced. In *S. pombe*, the centromeres are another silenced region of the genome. In both organisms, silencing involves histone modifications. But unlike cases of silencing in *S. cerevisiae*, which lacks the RNAi machinery, centromeric silencing in *S. pombe* requires that pathway.

The centromeres of *S. pombe* have a sequence organization more like that of higher eukaryotes (e.g., flies and humans) than that of *S. cerevisiae* (see Fig. 7-8). Each centromere has a central region, of

Box 18-2 History of miRNAs and RNAi

In 1989, Richard Jorgensen, working at the biotech company Advanced Genetic Sciences in Oakland, California, was trying to make petunia plants with more deeply purple flowers than existing strains. The strategy seemed straightfoward: he would introduce into the plants an additional copy of the pigment gene (encoding chalcone synthase) under the control of a strong promoter. These plants would make more chalcone synthase and the flowers would be more purple. What he actually got were plants with varying degress of paler flowers, many that were sectored—with purple and white regions—and even some that were completely white (Box 18-2 Fig. 1).

BOX 18-2 FIGURE 1 **Petunia flower.** An example of the effects of overexpressing the pigment gene chalcone synthase in what would otherwise be a completely purple petunia flower. (Courtesy of Richard A. Jorgensen, University of Arizona.)

Although disappointing, these results were intriguing. In attempting to understand what was going on, Jorgensen uncovered various features of the phenomenon, called **cosupression** (because expression of both the transgene and the endogenous gene are repressed). The greater the expression of the transgene, the lower the level of chalcone synthase; this was true whether increased expression resulted from multiple copies of the transgene or from use of stronger promoters driving the transgene. It was also noted that some plants had variegated patterns of pigmentation, and that different variegation patterns could be found in different flowers on the same plant. These patterns were sometimes inherited, but on other occasions apparently altered at random. These observations suggested to Jorgensen and others (particularly Marjori Matzke, who also was investigating this phenomenon) that they were dealing with an epigenetic phenomenon.

Other investigators were trying to make plants resistant to viral infection. One approach was to overexpress in plants a dominant-negative derivative of a common viral replication factor: this protein was expected to block replication of any infecting virus that used this common replicative mechanism. Although the dominant-negative viral product blocked replication of the potato virus from which it was derived, its specificity of action was surprisingly tightly restricted to that virus. It was also shown that the protein itself was not even needed—just the RNA.

Meanwhile other researchers were using antisense RNA to knock down expression of the *par-1* gene in worms. Their intention was to prove that this gene was responsible for a particular developmental phenotype. Antisense RNA produced in vitro and injected into the developing worm induced the phenotype predicted for the loss of *par-1* expression. But it was found that *sense* RNA had the same effect. This was only included in the experiment as a negative control of course; it was not expected to have any effect on expression. RNAs unrelated to the *par-1* gene had no effect.

The explanation for this RNA-dependent gene repression was provided by Andrew Fire and Craig Mello in experiments that earned them the 2006 Nobel Prize in Physiology or Medicine. They demonstrated that it was in fact neither sense nor antisense RNA that silenced the gene—it was the dsRNA produced by a mix of the two. It turned out that the RNA preparations of sense or antisense were both contaminated with small amounts of the opposite strand, and it was the resulting double-strand population that caused silencing. When dsRNAs were deliberately prepared, they were shown to be very potent in eliminating expression of the target gene. Hence, the phenomenon of RNAi had been discovered, a finding published in 1998.

Mechanistic insights came thick and fast from several labs. First, dsRNAs were shown to trigger degradation of homologous mRNAs in extracts from *Drosophila* cells, an assay that led to the identification of RISC. The identification of siRNAs—the species that directs RISC to the target genes—was reported in plants in 1999. Dicer, the nuclease that creates them, was described in 2001. And the final major component of the pathway, Slicer, was identified in 2005 when the crystal structure of Argonaute revealed the protein to be an RNase.

In addition to being needed to generate the siRNAs, Dicer was shown also to be required for miRNAs to function during development. The first miRNA and its target had been described in 1993, by Victor Ambros and Gary Ruvkun, respectively. At the time, this observation was seen as a neat but eccentric oddity; the *lin-4* gene encoded a small RNA that acted on a target gene, *lin-14*, by virtue of sequence complementarity between the miRNA and regions of the 3′ UTR of the target genes (Box 18-2 Fig. 2). Later, other miRNAs were found in worms, some with homology to similar genes in animals and plants, suggesting that this mechanism of regulation was more widespread. Everything was tied together when it became clear that siRNAs and miRNAs work through the same pathways. Thus, the picture emerged of a world of tiny RNAs involved in gene regulation—some exogenously supplied, others inbuilt as part of gene regulatory programs employed during development. The field developed very rapidly, as the dates in this account reveal, moving from obscure phenomenology to a Nobel Prize and demanding of its own chapter in textbooks, in just 15 years. The accelerated progress was perhaps largely a consequence of the range of species (yeast, plant, and worm) studied and approaches (genetics, biochemistry, structural studies, and bioinformatics) employed.

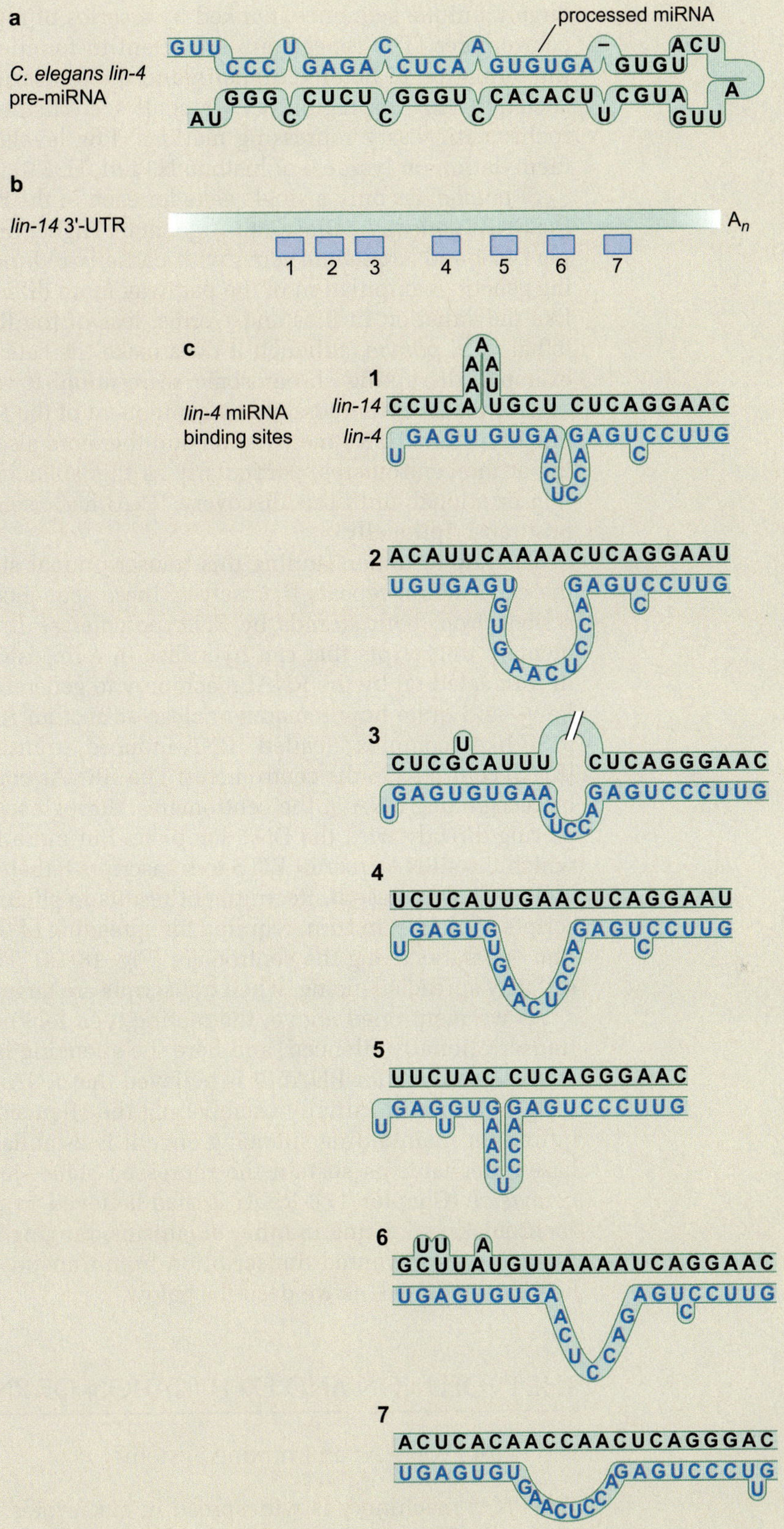

BOX 18-2 FIGURE 2 microRNA *lin-4* binds within the 3′-untranslated region (UTR) of its target gene *lin-14*. (a) The *lin-4* pre-miRNA before processing by Dicer. The sequence of the miRNA is shown in blue. (b) The seven sequences within the *lin-4* 3′-UTR that can base-pair with the *lin-4* miRNA to various extents, as shown in part c. (Modified from Ha I. 1996. *Genes Dev.* 10: 3041–3050, Fig. 1. © Cold Spring Harbor Laboratory Press.)

largely unique sequence, flanked by a series of repeats common to all centromeres. The repeats are important to function and contribute to the formation of heterochromatin and the transcriptional silencing associated with the region, as we shall see. Histones within the heterochromatin carry repressing markers: low levels of acetylation, and methylation on lysine 9 of histone H3 tail (H3K9) (see Table 7-7).

S. pombe has only a single gene for each of the major components of the RNAi pathway—Dicer and Argonaute. Higher organisms have multiple Dicer and Argonaute genes with partially redundant functions, making genetic manipulation of the pathway more difficult. In addition, unlike the situation in flies and worms, loss of the RNAi pathway is not lethal to *S. pombe*, although it does make the cells grow poorly by, for example, disrupting chromosome segregation. It was a surprise, however, to discover that loss of any component of the RNAi pathway led to loss of histone H3K9 methylation and therefore also loss of gene silencing at the centromeres, particularly as this silencing was known to be transcriptional; until this discovery, RNAi had been thought to act only posttranscriptionally.

The key to understanding this transcriptional silencing seems to be the centromeric repeats themselves; these sequence elements are transcribed from both strands by RNA polymerase II, producing complementary transcripts that can hybridize to form dsRNAs. The RNAs are in turn acted on by the RNAi machinery to generate siRNAs that somehow—and quite how remains unclear—direct an Argonaute-containing RISC-like complex (called RNA-induced transcriptional silencing [RITS] complex) to the centromeres. The siRNAs could in theory do this by recognizing DNA at the centromeres, through sequence-specific base pairing directly with the DNA template. But more likely are models in which the siRNAs recruit RITS to transcripts tethered to the centromere by RNA polymerase II. Recruitment results in slicing of centromic transcripts, which is, in turn, required for spreading of the histone modification apparatus along the centromere (Fig. 18-13). Thus transcription itself may spread silencing, when transcripts are targeted by RNAi.

As we mentioned above, the mating-type loci of *S. pombe* are also transcriptionally silenced, and here the silencing is not lost in mutant strains defective for RNAi. It is believed that RNAi acts in this case as well, but only in initially establishing the silenced state—it is not required for maintaining silencing once it is established; other, protein-based mechanisms sustain the repressed state—just as they do in *S. cerevisiae* (Chapter 17). RNAi is also believed to play a part in heterochromatin silencing in other organisms, ranging from flies to plants. Silencing of unwanted transcription from transposons also appears to be RNA-mediated, as we describe below.

THE EVOLUTION AND EXPLOITATION OF RNAi

Did RNAi Evolve As an Immune System?

The RNAi machinery is widespread in eukaryotes, although not ubiquitous. It does not occur in *S. cerevisiae*, for example, as we just noted. It is believed, however, that at least the basic system existed in the most recent common ancestor to all eukaryotes, but was subsequently lost in some lineages.

But what does RNAi do, biologically? There are miRNAs of course—and the RNAi machinery is required to produce and use those regula-

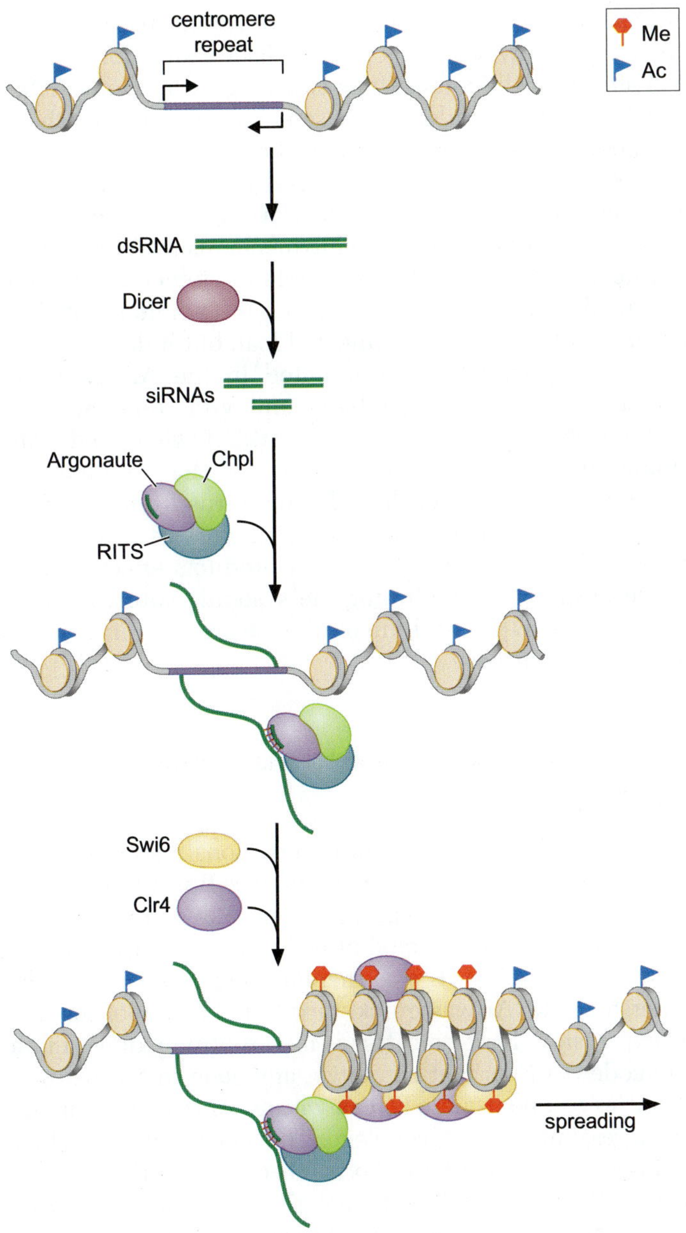

FIGURE 18-13 A model for RITS recruitment and the silencing of centromeres. Shown at the top are nucleosomes around the repeat sequences at a centromere in *S. pombe*. The repeat sequences are transcribed from both strands by RNA polymerase II, generating dsRNA that is a substrate for Dicer. The siRNAs thereby produced are loaded into the Argonaute-containing complex RITS. As shown in the middle, the siRNA-containing RITS is then recruited to the PolII-tethered transcripts being generated by continued transcription of the centromeric repeats, through complementarity between the siRNA and the transcript. This complex then recruits factors (Clr4 and Swi6) that locally modify nucleosomes by adding the H3K9 silencing markers. Another subunit of RITS, Chp1, contains a chromodomain (Fig. 7-40), which, by binding to the methylated nucleosomes, likely stabilizes the binding of RITS. Although not shown in the figure, "slicing" of the transcripts by Argonaute (within RITS) generates substrate RNAs for the RdRP, which synthesizes a complementary strand and thus generates further substrate for Dicer. This process is required for the nucleosome modification—and thus the region silenced—to spread. (Redrawn, with permission, from Martienssen R. and Moazed D. 2007. *Epigenetics* [ed. D. Allis et al.], p. 157, Fig. 4. © Cold Spring Harbor Laboratory Press.)

tors—but some organisms have the RNAi machinery and no miRNAs (including *S. pombe*). It is, in fact, believed that miRNAs evolved to take advantage of the existence of the RNAi machinery rather than being the reason that machinery exists. One ancient function the RNAi machinery might have served (and still serves) is protecting organisms from transposons and viruses.

Transposons are found in all eukaryotes and, in some cases, make up a substantial amount of a genome. In humans, for example, about 45% of our genome is made of sequences that were once transposons. Transposons are often transcriptionally silent and packaged into heterochromatin. In some RNAi mutants, however, the histone modifications associated with transposon silencing are lost. In addition, in plants and worms, several siRNAs have been identified that correspond to transposons. And in some cases in both these organisms, the loss of RNAi reactivates transposons, causes them to jump, and leads to high levels of spontaneous mutagenesis. Not as many transposons are reactivated as are known to generate siRNAs, however. This might reflect a situation similar to that described above for the mating-type silencing in *S. pombe*: RNAi might be essential for initiating silencing of some transposons, but the silencing then becomes self-sustaining without further need for the siRNAs.

The protective effect of RNAi on viral infection of plants has been widely observed. Indeed, the effects were recognized long before RNAi was known to be an underlying mechanism. When one leaf on a plant is infected by a virus, a factor able to silence replication of the virus is spread systematically throughout the whole plant. This factor does not protect that originally infected leaf, but it does stop the infection from spreading. In plants mutated in the Argonaute or Dicer genes, infection spreads unhindered and viral replication is much higher. The protecting signal comprises siRNAs generated from the viral genome itself. Viruses have retaliated: they often carry genes whose function is to protect the infecting virus from host RNAi. One example is HcPro from potato virus Y, which acts to reduce production or stability of siRNAs. Other viral products effect other steps in the defense mechanism, including the systemic spreading of siRNAs.

Links between RNAi and human disease are described in Box 18-3, RNAi and Human Disease.

RNAi Has Become a Powerful Tool for Manipulating Gene Expression

The discovery of RNAi arose from observations made by investigators attempting to manipulate gene expression (see Box 18-2). In the cases of both cosuppression in plants and antisense RNA in worms, it was attempts to understand unexpected blips in those manipulations that led to the discovery of RNAi. It was therefore perhaps not surprising that once understood, RNAi was quickly exploited as a tool for manipulating gene expression. In worms, this is now routinely done. Libraries exist that encode dsRNAs that can target any gene in the worm genome. These libraries can be used to screen worms for the consequences of inhibiting expression of any given gene. The general way in which this is done is shown in Figure 18-14. Worms feed on bacteria. In the lab, they are fed *E. coli*, and it turns out that the quickest route to a worm's genome is through its stomach: any desired dsRNA can be expressed in the *E. coli* on which the worms feed, and this delivers enough substrate for the RNAi response to be triggered in cells of the worm, switching off genes homologous to the original dsRNA.

It would, of course, be of great benefit to screen for genes in this way in mammalian cells, where traditional genetic screens are not feasible. It was established that artificially synthesized siRNAs, made in vitro and introduced into mammalian cells in culture, trigger an RNAi response and down-regulate appropriate target genes, but the efficiency of transfection (getting the RNAs into the cells) was low. Longer dsRNA molecules are also problematic because they trigger a response that shuts down all translation in the cell, a response evolved to block viral replication; many viruses have RNA genomes. Despite these drawbacks this approach can still be fruitfully employed.

Being able to express the RNAs in cells (rather than having to introduce them in high concentrations) would get around the problems mentioned above and have a further benefit: the RNAs would be pro-

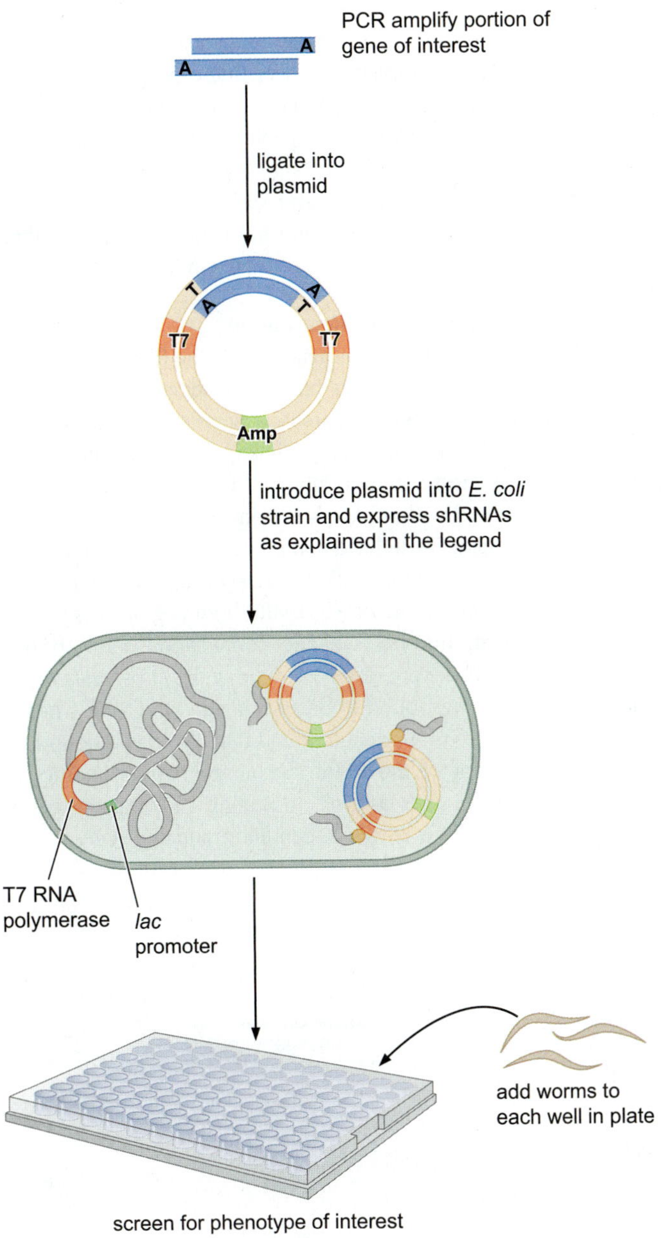
PCR amplify portion of
gene of interest
A
A
ligate into
plasmid
T
A
A
T
T7
T7
Amp
introduce plasmid into E. coli
strain and express shRNAs
as explained in the legend
T7 RNA
polymerase
lac
promoter
add worms to
each well in plate
screen for phenotype of interest

FIGURE 18-14 RNA interference can be induced in worms by feeding bacteria expressing dsRNAs. See Chapter 21 for details of the molecular manipulations required in the first steps of this scheme. The shRNA expression from the plasmid is under the control of a promoter recognized not by *E. coli* RNA polymerase, but rather by a single-subunit RNA polymerase from a phage called T7. The gene for that polymerase is expressed artificially in the cells used in this scheme, under control of the *lac* promoter (Chapter 16). Thus, production of the shRNAs can be controlled using an inducer of the *lac* promoter.

duced in a sustained manner rather than being present transiently as they are with transfection. But this approach presents its own formidable hurdles: how can one ensure that both complementary strands are expressed in the same cell and find each other efficiently to form the active dsRNA substrate for RNAi?

Instead of trying to deliver dsRNA into cells, investigators found it more fruitful to mimic miRNAs. To this end, libraries have been generated in which short genes are synthesized as oligonucleotides and cloned in plasmids. Each short gene is designed to give a transcript that will fold into a stem loop. These are processed by dicer in the cell to generate an siRNA that will direct silencing of its target genes. These short synthetic genes are called **short hairpin RNA** genes (**shRNAs**). By using an appropriately designed shRNA, any individual gene in the genome can be targeted. Or, with a suitable library, a ge-

Box 18-3 RNAi and Human Disease

Cancer

A general decrease in levels of many miRNAs is often seen in cancers. This decrease has been taken to indicate that those miRNAs usually have a tumor-suppressing effect. Despite this general trend, other specific miRNAs are up-regulated in some cancers. Analogous to protein-coding genes implicated in cancer, the miRNAs in question are described as being tumor suppressors (if their absence increases cancer) or oncogenic (if their increased expression leads to cancer). Their targets tend to be genes involved in cell cycle progression (proliferation) or apoptosis.

Of the 300 miRNAs identified in humans, more than one-half are located in regions of the genome regularly disrupted in cancers. Thus, in many cancers, the genes for these miRNAs are deleted or amplified, depending on the nature of the chromosomal rearrangement. So, for example, the miRNAs *miR-15* and *miR-16* induce apoptosis of cells by down-regulating the *BCL2* gene (see Box 18-3 Fig. 1). The most common form of adult leukemia in the western hemisphere is chronic lymphocytic leukemia (CLL), a disease associated with deletions in a region of chromosome 13 (13q14). This region of the genome contains the miRNA *miR-15a* and *miR-16a* genes; indeed, these are the only two genes included in the smallest deletions associated with CLL. Thus, when these genes are deleted, apoptosis is down-regulated and tumors can more readily arise and develop.

In another region of chromosome 13 (13q31) is found *miR-17-92*, an oncogenic miRNA. Compared to normal tissue, expression of this gene is significantly increased in many cancers, including lung cancer, and especially in its most aggressive forms (e.g., small-cell lung cancer). In addition, overexpression of this miRNA in transgenic mice drives tumorigenesis. Among the many predicted targets of *miR-17-92* are two tumor suppressor genes, *PTEN* and *RB2*. One definite target is the cell cycle progression regulator E2F1. Both these and other examples of miRNAs in cancer are shown in Box 18-3 Figure 1.

Fragile X Mental Retardation

Through biochemical analysis of the RISC complex, a number of associated proteins have been identified. One of these is the Fragile X mental retardation protein (FMRP). The gene encoding this protein (*FMR1*) is X-linked and its mutation is the cause of the most common inherited form of mental retardation. FMRP is an RNA-binding protein involved in gene regulation. Patients lacking FMRP have a range of developmental defects, as well as the mental retardation, due to disrupted gene expression.

Drosophila has an FMRP homolog. In flies deficient for this gene, unusual synaptic connections between neurons and muscles were observed. One of the *Drosophila* Argonaute proteins was found to be associated with FMRP, whereas separate studies of Argonaute found that FMRP was bound to that component of the RNAi machinery. Similar findings followed in human cells as well, indicating an intriguing connection between the Fragile X condition and the RNAi gene regulatory pathway.

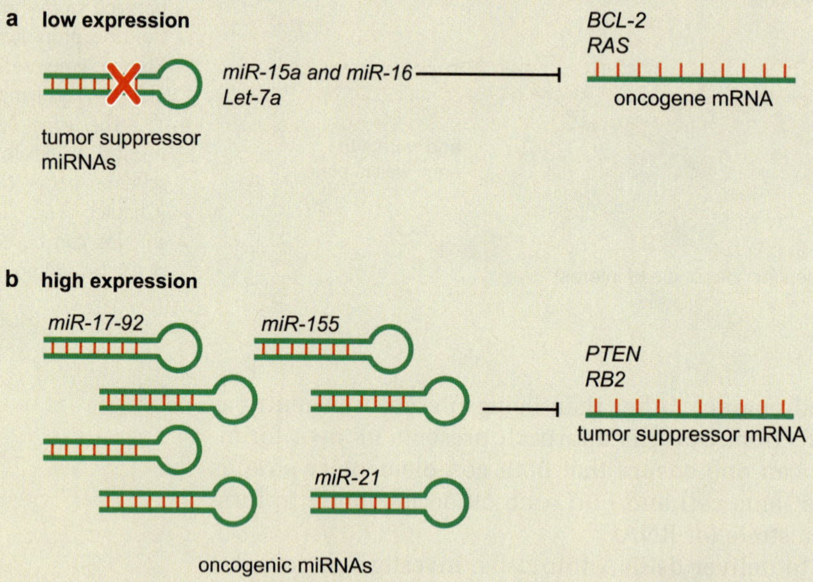

Box 18-3 Figure 1 miRNAs as tumor suppressors of oncogenes. (a) In this model, an miRNA that normally down-regulates an oncogene can function as a tumor suppressor gene. The loss of function of the miRNA by mutation or deletion, for example, might result in an abnormal expression of the target oncogene, which would then contribute to tumor formation. (b) Here, the amplification or overexpression of an miRNA that down-regulates a tumor suppressor or other important genes involved in differentiation may contribute to tumor formation by stimulating proliferation, angiogenesis, and invasion. (Redrawn, with permission, from Garzon R. et al. 2006. *Trends Mol. Med.* 12: 580–587, Fig. 2. © Elsevier.)

netic screen can be carried out. In such a library, for example, each plasmid would encode a shRNA directed against a different gene. The whole library is transfected into cells such that each cell receives a different shRNA. Cells with a particular phenotype are chosen, and the gene whose repression led to that phenotype can be identified.

REGULATORY RNAs AND X-INACTIVATION

X-inactivation Creates Mosaic Individuals

Female mammals have two X chromosomes, whereas males have only one X and a Y chromosome. Although this is the basis of sex determination—what enables males to differ from females—it also poses a problem: any gene encoded by the X chromosome would, if left unchecked, be expressed at twice the level in females as in males. This imbalance would potentially cause disruption to metabolic and other cellular processes. Avoiding such problems requires what is called **dosage compensation**. In mammals this is achieved by females **inactivating** one of their two X chromosomes. This action results in none of the genes on that copy of the chromosome being expressed. In placental mammals, inactivation occurs at the 32- to 64-cell stage, and the choice of which X chromosome to inactivate—the maternal or paternal copy—is apparently made at random in each cell at that time. Once selected in each cell, the same copy remains inactivated in all the descendants of that cell.

A consequence of inactivation being random in each cell is that females are mosaics—some of their cells express the paternal and others the maternal X chromosome. This is usually of little consequence, although it can influence the severity of symptoms of X-linked diseases, depending on the proportion of cells in which the mutated gene is expressed or silenced. A more familiar example is the calico (or tortoiseshell) cat (Fig 18-15). In cats, one gene on the X chromosome influences whether fur is orange or black—one allele of that gene gives rise to orange fur, another allele gives black. In cats heterozygous for this gene, the different patches of black and orange fur reveal regions made up of cells in which one or the other X chromosome was inactivated. This observation also explains why all calico cats are female. The white comes from effects of an autosomal gene.

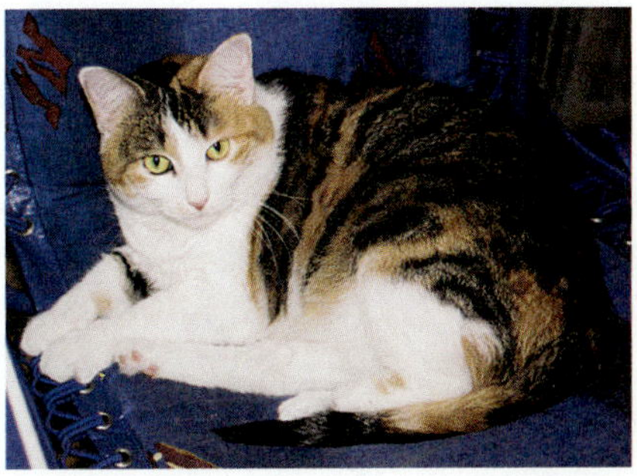

FIGURE 18-15 Visualizing X-inactivation: the calico cat. The patches of orange and black fur provide an indirect visualization of X-chromosome inactivation, as described in the text. (Courtesy VG.)

Xist Is an RNA Regulator That Inactivates a Single X Chromosome in Female Mammals

How is an X chromosome inactivated, and how is inactivation inherited through the remainder of development? The initiating regulator is an RNA molecule called *Xist*. This RNA is encoded within the locus known to be vital for X-inactivation, the *Xic* (X-inactivation center) on the X chromosome. *Xist* RNA coats the X chromosome from which it is expressed. This is shown in the in situ in Figure 18-16a. It is not clear what causes this coating nor how it is restricted to one X chromosome (i.e., why it acts only in *cis*). It is, however, known that the action of *Xist* is central to inactivation and does not require other X-chromosomal sequences beyond *Xic*: when expressed ectopically from an autosomal location (i.e., from a non-sex chromosome), *Xist* can, to varying extents, silence genes along that chromosome. That is, it "inactivates" the autosome from which it is expressed.

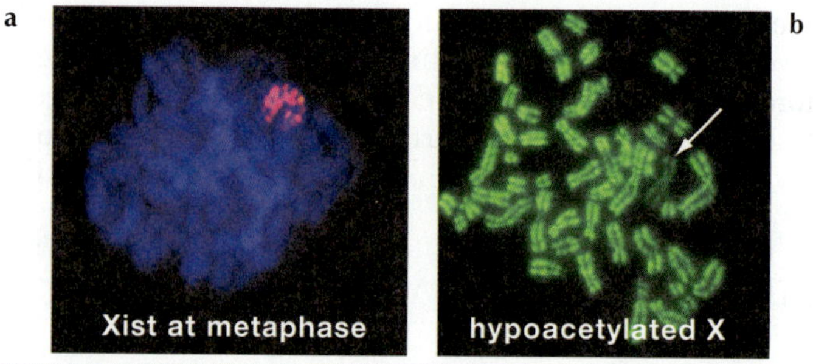

FIGURE 18-16 **Visualizing X-inactivation: molecular markers.** (a) Localization of *Xist* RNA along the inactive X-chromosome is shown by in situ hybridization in metaphase cells. (b) Chromosomes are stained for acetylation on histone H4. The arrow points to the inactivated X chromosome, which has much lower levels of acetylation than the other chromosomes. (Reprinted, with permission, from Brockdorff N. and Turner B.M. 2007. *Epigenetics* [ed. Allis et al.], p. 327. © Cold Spring Harbor Laboratory Press.)

Xist RNA itself does not cause silencing, but it recruits other factors that modify and condense chromatin and perhaps methylate DNA as well (just as we saw in other examples of silencing in Chapter 17). It is these modifications that cause silencing and ensure it is inherited; once firmly established, *Xist* itself is no longer required. One difference in histone modification of the inactivated X chromosome is shown in Figure 18-16b. There, the single inactivated X chromosome is much less acetylated than is the rest of the genome. As we saw in earlier chapters (7 and 17), deacetylated histones are associated with regions of the genome that are not transcribed.

How does a cell choose which X chromosome to inactivate? The answer is not yet known. But another RNA regulator might be key. This other RNA is also encoded by the *Xic* locus but on the opposite strand and overlapping the *Xist* gene. It is called *Tsix* (*Xist* spelled backward) and acts as a negative regulator of *Xist*. Indeed, if *Tsix* is mutated on a given X chromosome, it is that chromosome that will be chosen for inactivation. Thus, a balance between the production and stability of the *Xist* and *Tsix* RNAs may tilt the outcome one way or the other in each cell.

Dosage compensation is necessary in all animals—for example, worms and flies—just as it is in mammals. But in each case the mechanisms for achieving compensation are different. For example, in *Drosophila*, it is achieved by *increasing* expression of *X*-linked genes in the male (rather than decreasing them in the female). But there, too, the mechanism involves noncoding regulatory RNAs. In this case, the RNAs (called *roX1* and *roX2*) are involved in recruiting chromatin-modifying complexes to genes on the *X* chromosome in males, where they help activate transcription.

Noncoding RNAs are also found within some clusters of imprinted genes in the mouse. We discussed imprinting in gene expression in Chapter 17. In addition to the mechanisms described there, it is now believed that in some cases regulatory RNAs play a part, working in a manner analogous to *Xist* and *Tsix* in X-inactivation. The noncoding RNAs in question are regulated by the imprinting control region (ICR; see Fig. 17-26) found at each cluster, and they silence nearby genes on the same chromosome. Both DNA and histone methylation have been implicated in the silencing mechanism.

SUMMARY

Despite it being proposed as long ago as 1961 that RNA molecules were likely agents of gene regulation, it is only in the last few years that their widespread occurrence and significance in that role have come to light. Before that, attenuation of the *trp* operon in *E. coli* was a rare case where RNA sequences in the 5′ region of an mRNA were known to control expression of the downstream genes. In that case, alternative patterns of intramolecular base pairing within that region of the RNA give rise to alternative secondary structures that communicate different outcomes to the genes. In one conformation, transcription is terminated before it enters the coding region of the downstream genes, whereas in another conformation, it allows that transcription to continue, and the genes are expressed. Those genes encode enzymes involved in the synthesis of tryptophan, and the decision of whether or not to express them is based on the availability of uncharged $tRNA^{trp}$: if the uncharged species is rare in the cell, more Trp is needed and the genes are expressed; if there is ample charged $tRNA^{trp}$ in the cell, transcription of the *trp* genes is terminated.

Riboswitches control genes in a similar way: alternative secondary structures in the 5′-untranslated region of genes determine whether transcription of those genes continues (or, in other cases, whether translation is initiated). With riboswitches, the choice of alternative secondary structure depends on the direct binding to the RNA of the ligands that control the given gene. These are small metabolites (vitamin B12, SAM, amino acids) and the genes regulated by each one encode enzymes that use the metabolite in question.

E. coli also encodes small RNAs (called sRNAs) that act in *trans* to regulate genes. Thus, small genes encode short RNAs that base-pair with mRNAs bearing complementary sequences. This situation either inhibits translation of those target mRNAs, triggers their destruction, or even, in some cases, *stimulates* their translation. The actions of bacterial small RNAs are similar in many regards to those of small RNAs that regulate genes in eukaryotic cells, although the machinery used to produce these eukaryotic RNA regulators and the machinery used in carrying out their effects on target genes are quite different.

Eukaryotes use dsRNA as an agent of gene silencing, in a process called RNAi. Special enzymes (Drosha and, most generally, Dicer) recognize dsRNA and generate from that short (21–22-nucleotide) RNAs that are the active species for gene silencing. The dsRNA substrates from which these are generated can arise from two complementary strands base-pairing or a single molecule folding into a secondary structure with a characteristic double-strand region in an appropriate context. The former are acted on by Dicer, and the short products are called siRNAs; the latter require the action of Drosha followed by Dicer, and the products are called miRNAs. Both Dicer and Drosha have RNase domains and cut the substrate RNA on the basis of size and structure, rather than specific sequence.

Once produced, siRNAs and miRNAs act in essentially the same way. They are incorporated into a machine called RISC where one of the RNA strands is selected as the so-called guide RNA and directs the mature RISC complex to target RNAs with complementarity to that guide RNA. Once there, RISC either "slices" the RNA (through its catalytic subunit Argonaut, which includes an RNaseH-related domain) or inhibits the translation of the mRNA. Which route to silencing is chosen depends largely on how good the base-pairing match is between the guide RNA and the target—the higher the match, the more likely it is to trigger slicing. The guide RNA can also direct RISC with associated histone-modifying complexes to promoter regions where it silences genes transcriptionally by modifying their promoters. It is likely that even in these cases, recruitment to the promoter is through base pairing between the guide RNA and an mRNA, but in this instance, one still being made and thus still associated with RNA polymerase II at the gene.

miRNAs are encoded by genes within organisms where they typically act as regulators of genes involved in development—those from worms and plants are well-studied examples. miRNAs have also been associated with cancer, with some miRNAs being classified as tumor suppressors and others as oncogenes. The dsRNAs that give rise to siRNAs can arise from various sources ranging from infecting viruses, to transcribed repeat regions (centromeres or transposons), to dsRNA introduced into a cell deliberately by an experimenter who wants to down-regulate expression of a specific gene. This latter use of RNAi has become a regular tool and is particularly useful in systems where traditional genetics is not feasible.

Finally, we saw another case where noncoding RNAs regulate an important biological process. Female animals have two X chromosomes, whereas males have just one (and a Y chromosome). To ensure both sexes express comparable amounts of X-chromosome gene products, a mechanism of dosage compensation must correct for this unequal chromosome number. Mammals do this by inactivating one of the X chromosomes in females. An RNA molecule (*Xist*) encoded on the X chromosome regulates this process.

BIBLIOGRAPHY

Books

Hannon G.J. 2003. *RNAi: A guide to gene silencing.* Cold Spring Harbor Laboratory Press, Cold Spring Harbor, New York.

Regulatory RNAs. 2006. *Cold Spring Harbor Symp. Quant. Biol.*, vol. 71. Cold Spring Harbor Laboratory Press, Cold Spring Harbor, New York.

Bacterial Small RNAs

Coppins R.L., Hall K.B., and Groisman E.A. 2007. The intricate world of riboswitches. *Curr. Opin. Microbiol.* **10:** 176–181.

Gottesman S. 2005. Micros for microbes: Non-coding regulatory RNAs in bacteria. *Trends Genet.* 21: 399–404. Review.

Storz G. and Haas D. 2007. A guide to small RNAs in microorganisms. *Curr. Opin. Microbiol.* **10:** 93–95.

Riboswitches and Attenuation

Gilbert S.D. and Batey R.T. 2006. Riboswitches: Fold and function. *Chem. Biol.* **13:** 805–807.

Gollnick P., Babitzke P., Antson A., and Yanofsky C. 2005. Complexity in regulation of tryptophan biosynthesis in *Bacillus subtilis*. *Annu. Rev. Genet.* **39:** 47–68.

Winkler W.C. 2005. Riboswitches and the role of noncoding RNAs in bacterial metabolic control. *Curr. Opin. Chem. Biol.* **9:** 594–602.

Winkler W.C. and Breaker R.R. 2005. Regulation of bacterial gene expression by riboswitches. *Annu. Rev. Microbiol.* **59:** 487–517.

Yanofsy C. 2000. Transcription attenuation: Once viewed as a novel regulatory strategy. *J. Bacteriol.* **182:** 1–8.

Mechanisms of RNAi

Baulcombe D. 2005. RNA silencing. *Trends Biochem. Sci.* **30:** 290–293.

Cook A. and Conti E. 2006. Dicer measures up. *Nat. Struct. Mol. Biol.* **13:** 190–192.

Hannon G.J. and Rossi J.J. 2004. Unlocking the potential of the human genome with RNA interference. *Nature* **431:** 371–378.

Pei Y. and Tuschl T. 2006. On the art of identifying effective and specific siRNAs. *Nat. Methods* **3:** 670–676.

Peters L. and Meister G. 2007. Argonaute proteins: Mediators of RNA silencing. *Mol. Cell* **26:** 611–623.

Rana T.M. 2007. Illuminating the silence: Understanding the structure and function of small RNAs. *Nat. Rev. Mol. Cell Biol.* **8:** 23–36.

Tang G. 2005. siRNA and miRNA: An insight into RISCs. *Trends Biochem. Sci.* **30:** 106–114.

Tolia N.H. and Joshua-Tor L. 2007. Slicer and the Argonautes. *Nat. Chem. Biol.* **3:** 36–43.

Valencia-Sanchez M.A., Liu J., Hannon G.J., and Parker R. 2006. Control of translation and mRNA degradation by miRNAs and siRNAs. *Genes Dev.* **20:** 515–524.

Zaratiegui M., Irvine D.V., and Martienssen R.A. 2007. Noncoding RNAs and gene silencing. *Cell* **128:** 763–776.

MicroRNAs

Ambros V. 2004. The functions of animal microRNAs. *Nature* **431:** 350–355.

Bushati N. and Cohen S.M. 2007. microRNA functions. *Annu. Rev. Cell Dev. Biol.* **23:** 175–205.

Hannon G.J., Rivas F.V., Murchison E.P., and Steitz J.A. 2006. The expanding universe of noncoding RNAs. *Cold Spring Harbor Symp. Quant. Biol.* **71:** 551–564.

History of miRNAs and RNAi

Baulcombe D. 1994. Replicase-mediated resistance: A novel type of virus resistance in transgenic plants? *Trends Microbiol.* **2:** 60–63.

Lee R., Feinbaum R., and Ambros V. 2004. A short history of a short RNA. *Cell* (suppl. 2) **116:** S89–S92, 1 p. following S96.

Ruvkun G., Wightman B., and Ha I. 2004. The 20 years it took to recognize the importance of tiny RNAs. *Cell* **116**(suppl. 2)**:** S93-S96, 2 p. following S96.

MicroRNAs and Disease

Calin G.A. and Croce C.M. 2006. MicroRNA signatures in human cancers. *Nat. Rev. Cancer* **6:** 857–866.

Dalmay T. and Edwards D.R. 2006. MicroRNAs and the hallmarks of cancer. *Oncogene* **25:** 6170–6175.

Garzon R., Fabbri M., Cimmino A., Calin G.A., and Croce C.M. 2006. MicroRNA expression and function in cancer. *Trends Mol. Med.* **12:** 580-587.

Zhang B., Pan X., Cobb G.P., and Anderson T.A. 2007. microRNAs as oncogenes and tumor suppressors. *Dev. Biol.* **302:** 1–12.

Xist RNA and X-inactivation

Alexander M.K. and Panning B. 2005. Counting chromosomes: Not as easy as 1, 2, 3. *Curr. Biol.* **15:** R834–R836.

Deng X and Meller V.K. 2006. Non-coding RNA in fly dosage compensation. *Trends Biochem. Sci.* **31:** 526–532.

Heard E. 2005. Delving into the diversity of facultative heterochromatin: The epigenetics of the inactive X chromosome. *Curr. Opin. Genet. Dev.* **15:** 482–489.

Ng K., Pullirsch D., Leeb M., and Wutz A. 2006. *Xist* and the order of silencing. *EMBO Rep.* **8:** 34–39.

Pauler F.M., Koerner M.V., and Barlow D.P. 2007. Silencing by imprinted noncoding RNAs: Is transcription the answer? *Trends Genet.* **23:** 284–292.

CHAPTER

19 Gene Regulation in Development and Evolution

THERE ARE MORE THAN 200 DIFFERENT CELL types in a human being, all of which arise from a single cell, the fertilized egg. These genetically identical cells come to differ from one another by expressing distinct sets of genes during development. For example, developing muscle cells express specialized forms of actin, myosin, and tropomyosin that are absent in other organs such as the liver or kidney. To appreciate the extent of differential gene expression, consider the following. A typical invertebrate, such as a fruit fly or nematode worm, contains approximately 15,000–20,000 genes, whereas vertebrates contain almost double this number, between 25,000 and 30,000 genes. Whole-genome microarray methods make it possible to identify which genes are expressed in a given tissue. As an example, approximately 7% or 8% (~1500 genes) of all genes in the genome of the nematode worm *Caenorhabditis elegans* are expressed in the muscles (Fig. 19-1). Different cell types—for example, a muscle cell and a neuron—express somewhat different, but overlapping, subsets of genes. Typically, less than one-half of the genes expressed in one cell type are also expressed in another given cell type, and a specific cell may be defined by the expression of about 100 "signature" genes that are responsible for its unique characteristics (see Box 19-1, Microarray Assays: Theory and Practice).

How do cells that are derived from the same fertilized egg establish different programs of gene expression? Most differential gene expression is regulated at the level of transcription initiation, and we described the basic mechanisms of this regulation in the preceding two chapters. In the first half of this chapter, we describe how cells communicate with each other during development to ensure that each expresses a particular set of genes required for their proper development. Simple examples of each of these strategies are then described. In the next part, we describe how these strategies are used in combination with the transcriptional regulatory mechanisms described in Chapter 17 to control the development of an entire organism—in this case, the fruit fly. In the final part of the chapter, we discuss how changes in gene regulation can cause diversity of animal morphology during evolution. A particularly important class of developmental control genes, the homeotic genes, is described.

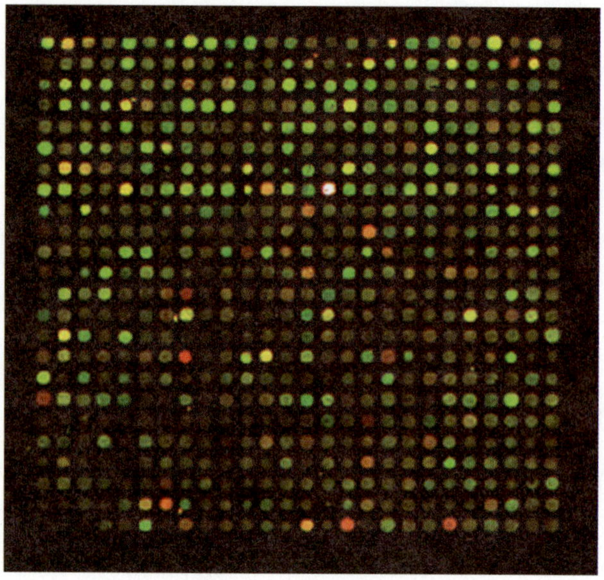

FIGURE 19-1 Microarray grid comparing expression patterns in two tissues (muscles and neurons) in *Caenorhabditis elegans*. Each circle in the grid contains a short DNA segment from the coding region of a single gene in the *C. elegans* genome. RNA was extracted from muscles and neurons, and labeled with fluorescent dyes (red and green, respectively). Thus, the red circles indicate genes expressed in muscle, whereas the green reflect genes expressed in neurons. The yellow circles indicate genes expressed in both cell types. It is clear that the two samples express distinct sets of genes. (Courtesy of Stuart Kim, Stanford University.)

TECHNIQUES

Box 19-1 Microarray Assays: Theory and Practice

Microarray assays permit the genome-wide analysis of gene expression profiles. The microarray, typically encompassing thousands to tens of thousands of known sequences immobilized on a microscope slide, can be subjected to a series of hybridization experiments performed in parallel. To generate the arrayed material for the microarray, protein-coding sequences are prepared using the polymerase chain reaction (PCR; see Chapter 20). The most common amplification method involves the use of short oligonucleotide sequences (typically on the order of 20 nucleotides in length) that bracket an exon for a particular protein-coding gene in the genome. Paired oligonucleotides, each pair representing an exon for every protein-coding gene, are then hybridized to genomic DNA and amplified by PCR. The resulting amplified genomic DNA fragments are then attached to glass slides in a series of spots. Each spot on the slide therefore contains a discrete amplified DNA fragment representing a unique protein-coding gene. Slides the size of a typical microscope slide can carry as many as 250,000 PCR fragments. The capacity increases every year and the most recent microarrays contain millions of DNA fragments or oligonucleotides. This collection represents the entire protein-coding capacity of the human genome on a single slide. In fact, as we discuss in Chapter 20, it is now possible to replicate entire genomes on single slides. These whole-genome tiling arrays permit the analysis of both coding and noncoding sequences.

To investigate whole-genome patterns of gene expression, the slide is hybridized with differentially labeled fluorescent RNA probes. Consider the case shown in Figure 19-1, which compares gene activity in the muscles and neurons of the nematode worm, *C. elegans*. Total mRNA was isolated from each tissue and labeled with different dyes. It is possible to label the muscle mRNAs red and the neuronal mRNAs green. These two samples of labeled mRNAs are then simultaneously hybridized on the same glass slide containing PCR fragments representing each of the nearly 20,000 genes in the *C. elegans* genome. When both samples hybridize to a particular spot, or gene fragment, a yellow color is emitted. This hybridization result indicates that the particular gene is significantly expressed in both tissues. Spots that strongly stain red correspond to genes that are mainly expressed in the muscles but not in neurons. Conversely, those spots that stain green represent genes that are expressed in neurons but not in muscles.

The basic method can be used to compare the gene expression profiles of any two samples. For example, there have been extensive studies that compare mRNA profiles in normal tissues and tumors. It is also possible to isolate RNA from normal yeast cells, or *Drosophila* embryos, and compare these with mutant yeast cells, or mutant fly embryos.

THREE STRATEGIES BY WHICH CELLS ARE INSTRUCTED TO EXPRESS SPECIFIC SETS OF GENES DURING DEVELOPMENT

We have already seen how gene expression can be controlled by "signals" received by a cell from its environment. For example, the sugar lactose activates the transcription of the *lac* operon in *Escherichia coli*, whereas viral infection activates the expression of the β-interferon gene in mammals. In this chapter, we focus on the strategies that are used to instruct genetically identical cells to express distinct sets of genes and thereby differentiate into diverse cell types. The three major strategies are **mRNA localization**, **cell-to-cell contact**, and **signaling through the diffusion of a secreted signaling molecule** (Fig. 19-2). Each of these strategies is introduced briefly in the following sections.

Some mRNAs Become Localized within Eggs and Embryos because of an Intrinsic Polarity in the Cytoskeleton

One strategy to establish differences between two genetically identical cells is to distribute a critical regulatory molecule asymmetrically during cell division, thereby ensuring that the daughter cells inherit different amounts of that regulator and thus follow different pathways of development. Typically, the asymmetrically distributed molecule is an mRNA. These mRNAs can encode RNA-binding proteins or cell-signaling molecules, but most often they encode transcriptional activators or repressors. Despite this diversity in the function of their protein products, a common mechanism exists for localizing mRNAs. Typically, they are transported along elements of the cytoskeleton, actin filaments, or microtubules. The asymmetry in this process is provided by the intrinsic asymmetry of these elements.

Actin filaments and microtubules undergo directed growth at the + ends (Fig. 19-3). An mRNA molecule can be transported from one end of a cell to the other end by means of an "adaptor" protein, which binds to a specific sequence within the noncoding **3′ untranslated trailer region (3′ UTR)** of an mRNA. Adaptor proteins contain two domains. One recognizes the 3′ UTR of the mRNA, whereas the other associates with a specific component of the cytoskeleton, such as myosin. Depending on the specific adaptor used, the mRNA-adaptor complex either "crawls" along an actin filament or directly moves with the + end of a growing microtubule. We will see how this basic process is used to localize mRNA determinants within the egg or to restrict a determinant to a single daughter cell after mitosis.

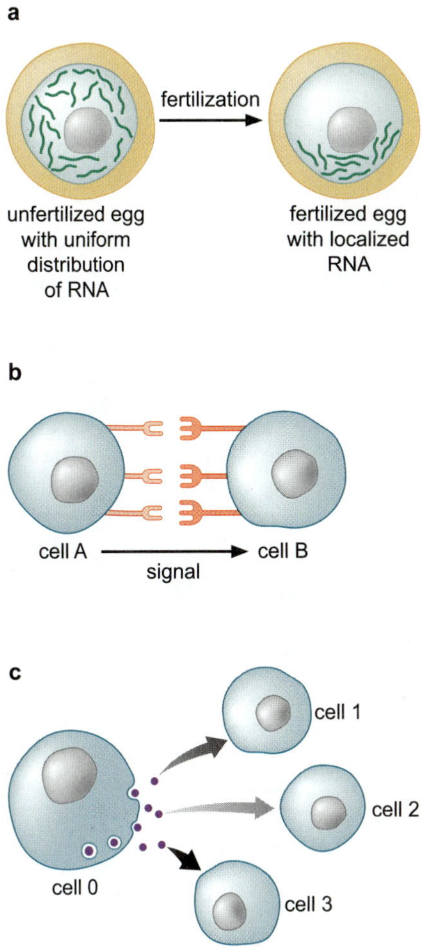

FIGURE 19-2 The three strategies for initiating differential gene activity during development. (a) In some animals, certain "maternal" RNAs present in the egg become localized either before or after fertilization. In this example, a specific mRNA (green squiggles) becomes localized to vegetal (bottom) regions after fertilization. (b) Cell A must physically interact with cell B to stimulate the receptor present on the surface of cell B. This is because the "ligand" produced by cell A is tethered to the plasma membrane. (c) In this example of long-range cell signaling, cell 0 secretes a signaling molecule that diffuses through the extracellular matrix. Different cells (1, 2, 3) receive the signal and ultimately undergo changes in gene activity.

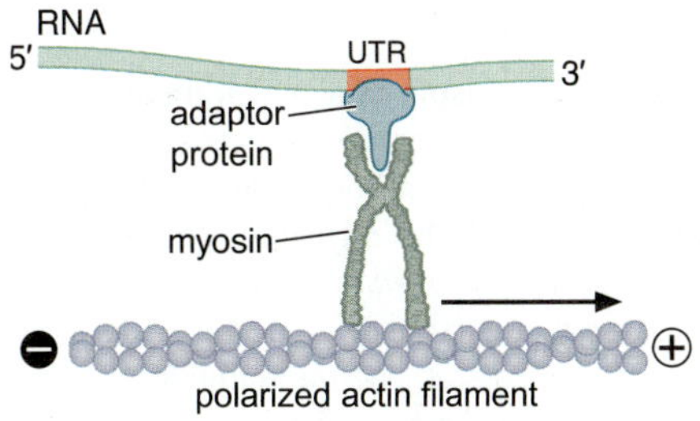

FIGURE 19-3 An adaptor protein binds to specific sequences within the 3′ UTR of the mRNA. The adaptor also binds to myosin, which "crawls" along the actin filament in a directed fashion, from the "–" end to the growing "+" end of the filament.

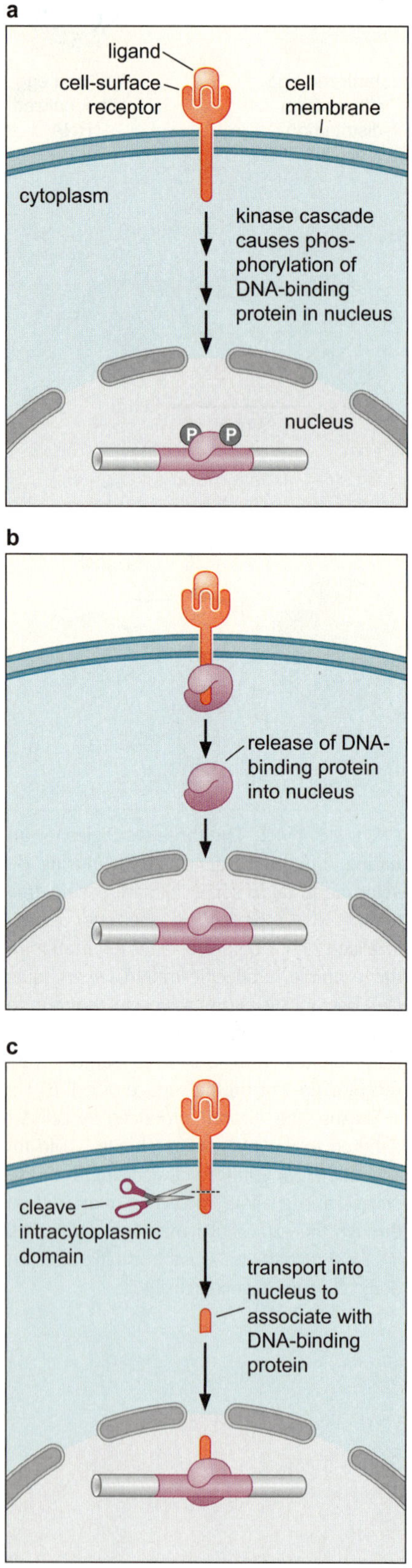

Cell-to-Cell Contact and Secreted Cell-Signaling Molecules Both Elicit Changes in Gene Expression in Neighboring Cells

A cell can influence which genes are expressed in neighboring cells by producing extracellular signaling proteins. These proteins are synthesized in the first cell and then either deposited in the plasma membrane of that cell or secreted into the extracellular matrix. These two approaches have features in common, so we consider them together here. We then see how secreted signals can be used in other ways.

A given signal (of either sort) is generally recognized by a specific receptor on the surface of recipient cells. When that receptor binds to the signaling molecule, it triggers changes in gene expression in the recipient cell. This communication from the cell-surface receptor to the nucleus often involves **signal transduction pathways** of the sort we considered in Chapter 17. Here, we summarize a few basic features of these pathways.

Sometimes, ligand–receptor interactions induce an enzymatic cascade that ultimately modifies regulatory proteins already present in the nucleus (Fig. 19-4a). In other cases, activated receptors cause the release of DNA-binding proteins from the cell surface or cytoplasm into the nucleus (Fig. 19-4b). These regulatory proteins bind to specific DNA-recognition sequences and either activate or repress gene expression. Ligand binding can also cause proteolytic cleavage of the receptor. Upon cleavage, the intracytoplasmic domain of the receptor is released from the cell surface and enters the nucleus, where it associates with DNA-binding proteins and influences how those proteins regulate transcription of the associated genes (Fig. 19-4c). For example, the transported protein might convert what was a transcriptional repressor into an activator. In this case, target genes that were formerly repressed prior to signaling are now induced. We consider examples of each of these variations in cell signaling in this chapter.

Signaling molecules that remain on the surface control gene expression only in those cells that are in direct, physical contact with the signaling cell. We refer to this process as **cell-to-cell contact**. In contrast, signaling molecules that are secreted into the extracellular matrix can work over greater distances. Some travel over a distance of just 1 or 2 cell diameters, whereas others can act over a range of 50 cells or more. Long-range signaling molecules are sometimes responsible for positional information, which is discussed in the next section.

FIGURE 19-4 **Different mechanisms of signal transduction.** A ligand (or "signaling molecule") binds to a cell-surface receptor. (a) The activated receptor induces latent cellular kinases that ultimately cause the phosphorylation of DNA-binding proteins within the nucleus. This phosphorylation causes the regulatory protein to activate (or repress) the transcription of specific genes. (b) The activated receptor releases a dormant DNA-binding protein from the cytoplasm so that it can now enter the nucleus. Once in the nucleus, the regulatory protein activates (or represses) the transcription of specific genes. (c) The activated receptor is cleaved by cellular proteases that cause a carboxy-terminal portion of the receptor to enter the nucleus and interact with specific DNA-binding proteins. The resulting protein complex activates the transcription of specific genes.

Gradients of Secreted Signaling Molecules Can Instruct Cells to Follow Different Pathways of Development Based on Their Location

A recurring theme in development is the importance of a cell's position within a developing embryo or organ in determining what it will become. Cells located at the front of a fruit fly embryo (i.e., in **anterior** regions) will form portions of the adult head such as the antenna or brain but will not develop into **posterior** structures such as the abdomen or genitalia. Cells located on the top, or **dorsal**, surface of a frog embryo can develop into portions of the backbone in the tadpole or adult but do not form **ventral**, or "belly," tissues such as the gut. These examples illustrate the fact that the fate of a cell—what it will become in the adult—is constrained by its location in the developing embryo. The influence of location on development is called **positional information**.

The most common way of establishing positional information involves a simple extension of one of the strategies we have already encountered in Chapter 17—the use of secreted signaling molecules (Fig. 19-5). A small group of cells synthesize and secrete a signaling molecule that becomes distributed in an **extracellular gradient** (Fig. 19-5a). Cells located near the "source" receive high concentrations of the secreted protein and develop into a particular cell type. Those

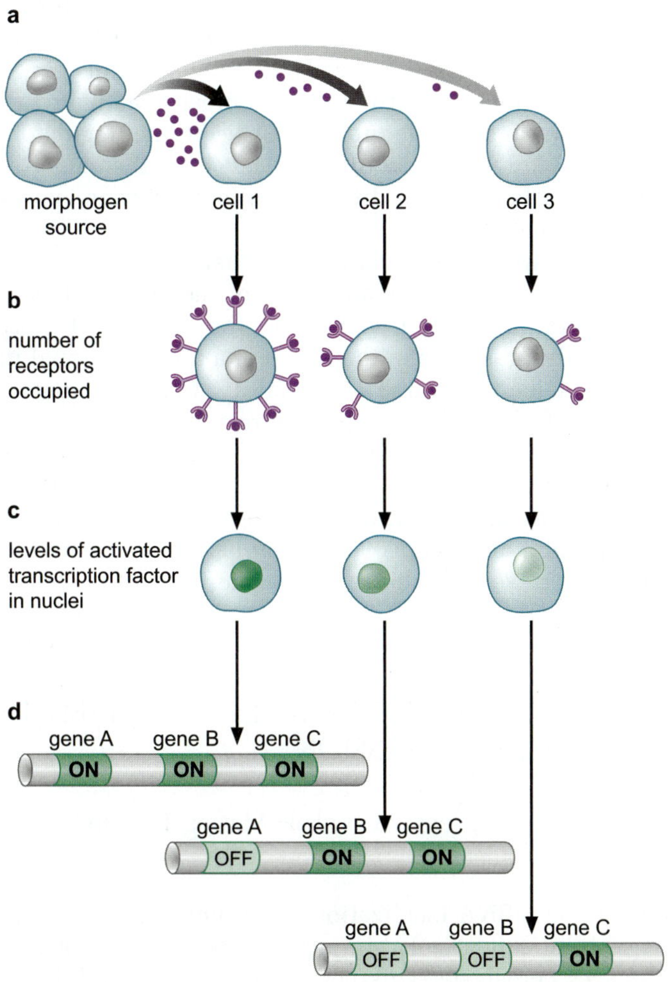

FIGURE 19-5 **A cluster of cells produces a signaling molecule, or morphogen, that diffuses through the extracellular matrix.** (a) Cells 1, 2, and 3 receive progressively lower amounts of the signaling molecule because they are located progressively farther from the source. (b) Cells 1, 2, and 3 contain progressively lower numbers of activated surface receptors. (c) The three cells contain different levels of one or more regulatory proteins. In the simplest scenario, there is a linear correlation between the number of activated cell-surface receptors and the amount of a regulatory factor that enters the nuclei. (d) The different levels of the regulatory factor lead to the expression of different sets of genes. Cell 1 expresses genes A, B, and C because it contains the highest levels of the regulatory factor. Cell 2 expresses genes B and C, but not A, because it contains intermediate levels of the regulatory factor. These levels are not sufficient to activate gene A. Finally, cell 3 contains the lowest levels of the regulatory factor and expresses only gene C because expression of genes A and B requires higher levels.

cells located at progressively farther distances follow different pathways of development as a result of receiving lower concentrations of the signaling molecule. Signaling molecules that control position information are sometimes called **morphogens**.

Cells located near the source of the morphogen receive high concentrations of the signaling molecule and therefore experience peak activation of the specific cell-surface receptors that bind it. In contrast, cells located far from the source receive low levels of the signal and, consequently, only a small fraction of their cell-surface receptors are activated. Consider a row of three cells adjacent to a source of a secreted morphogen. Something like 1000 receptors are activated in the first cell, whereas only 500 receptors are activated in the next cell, and just 200 in the next (Fig. 19-5b). These different levels of receptor occupancy are directly responsible for differential gene expression in the responding cells.

As we have seen, binding of signaling molecules to cell-surface receptors leads (in one way or another) to an increase in the concentration of specific transcriptional regulators, in an active form, in the nucleus of the cell. Each receptor controls a specific transcriptional regulator (or regulators), and this controls expression of particular genes. The number of cell-surface receptors that are activated by the binding of a morphogen determines how many molecules of the particular regulatory protein appear in the nucleus. The cell closest to the morphogen source—containing 1000 activated receptors—will possess high concentrations of the transcriptional activator in its nucleus (Fig. 19-5c). In contrast, the cells located farther from the source contain intermediate and low levels of the activator, respectively. Thus, there is a correlation between the number of activated receptors on the cell surface and the amount of transcriptional regulator present in the nucleus. How are these different levels of the same transcriptional regulator able to trigger different patterns of gene expression in these different cells?

In Chapter 16, we learned that a small change in the levels of the λ repressor determines whether an infected bacterial cell is lysed or lysogenized. Similarly, small changes in the amount of morphogen, and hence small differences in the levels of a transcriptional regulator within the nucleus, determine cell identity. Cells that contain high concentrations of a given transcriptional regulator express a variety of target genes that are inactive in cells containing intermediate or low levels of the regulator (Fig. 19-5d). The differential regulation of gene expression by different concentrations of a regulatory protein is one of the most important and pervasive mechanisms encountered in developmental biology. We consider several examples in the course of this chapter.

EXAMPLES OF THE THREE STRATEGIES FOR ESTABLISHING DIFFERENTIAL GENE EXPRESSION

The Localized Ash1 Repressor Controls Mating Type in Yeast by Silencing the *HO* Gene

Before describing mRNA localization in animal embryos, we first consider a case from a relatively simple single-cell eukaryote, the yeast *Saccharomyces cerevisiae*. This yeast can grow as haploid cells that

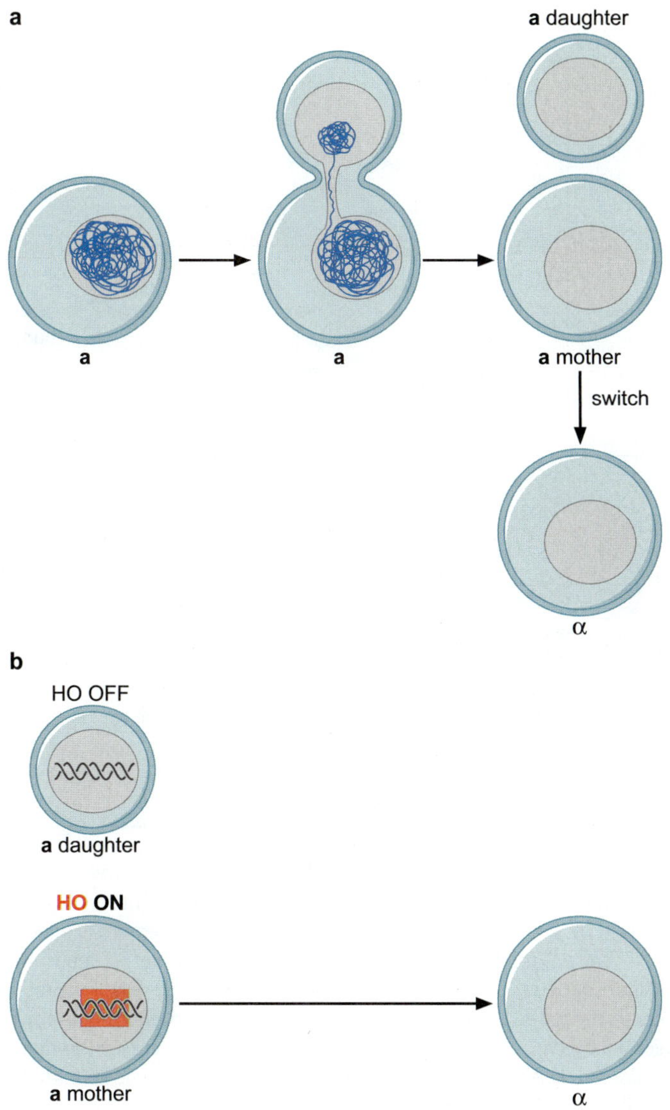

FIGURE 19-6 A haploid yeast cell of mating type a undergoes budding to produce a mother cell and smaller daughter cell. (a) Initially, both cells are mating type **a**, but sometimes the mother cell can undergo switching to the α type. (b) Because of the localized Ash1 transcriptional repressor, the daughter cell is unable to express the *HO* gene and thus cannot undergo switching. In contrast, the mother cell can switch because it lacks Ash1 and is able to express *HO*.

divide by budding (Fig. 19-6). Replicated chromosomes are distributed between two asymmetric cells—the larger progenitor cell, or mother cell, and a smaller bud, or daughter cell (Fig. 19-6a). These cells can exist as either of two mating types, called **a** and α, as discussed in Chapters 10 and 17.

A mother cell and its daughter cell can exhibit different mating types. This difference arises by a process called **mating-type switching**. After budding to produce a daughter, a mother cell can "switch" mating type with, for example, an **a** cell giving rise to an **a** daughter, but subsequently switching to the α mating type (Fig. 19-6b).

Switching is controlled by the product of the *HO* gene. We saw in Chapter 10 that the HO protein is a sequence-specific endonuclease. HO triggers gene conversion within the mating-type locus by creating a double-strand break at one of the two silent mating-type cassettes. We also saw in Chapter 17 how HO is activated in the mother cell. It is kept silent in the daughter cell because of the selective expression of a repressor called Ash1 (Fig. 19-7), and this is why the daughter

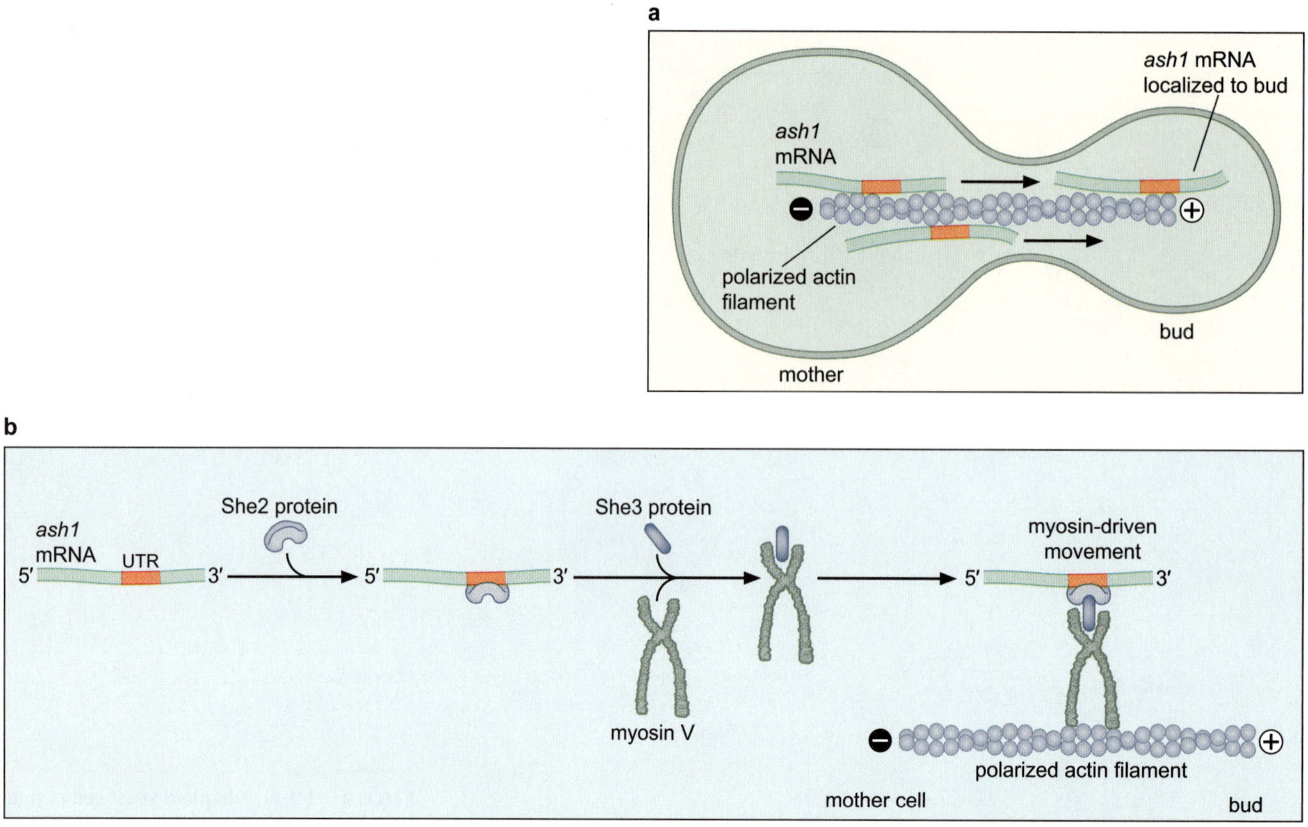

FIGURE 19-7 Localization of *ash1* mRNA during budding. (a) The *ash1* gene is transcribed in the mother cell during budding. The encoded mRNA moves from the mother cell into the bud by sliding along polarized actin filaments. Movement is directed and begins at the "–" ends of the filament and extends with the growing "+" ends. (b) The *ash1* mRNA transport depends on the binding of the She2 and She3 adaptor proteins to specific sequences contained within the 3′ UTR. These adaptor proteins bind myosin, which "crawls" along the actin filament and brings the *ash1* mRNA along for the ride. (Adapted, with permission, from Alberts B. et al. 2002. *Molecular biology of the cell,* 4th ed., p. 971, Fig. 16-84a. © Garland Science/Taylor & Francis LLC.)

cell does not switch mating type. The *ash1* gene is transcribed in the mother cell prior to budding, but the encoded RNA becomes localized within the daughter cell through the following process. During budding, the *ash1* mRNA attaches to the growing ends of microtubules. Several proteins function as "adaptors" that bind the 3′ UTR of the *ash1* mRNA and also to the microtubules. The microtubules extend from the nucleus of the mother cell to the site of budding, and in this way, the *ash1* mRNA is transported to the daughter cell. Once localized within the daughter cell, the *ash1* mRNA is translated into a repressor protein that binds to, and inhibits the transcription of, the *HO* gene. This silencing of *HO* expression in the daughter cell prevents that cell from undergoing mating-type switching.

In the second half of this chapter, we will see the localization of mRNAs used in the development of the *Drosophila* embryo. Once again, this localization is mediated by adaptor proteins that bind to the mRNAs, specifically, to sequences found in their 3′ UTRs (see Box 19-2, Review of Cytoskeleton: Asymmetry and Growth).

A second general principle that emerges from studies on yeast mating-type switching is seen again when we consider *Drosophila* development: the interplay between broadly distributed activators and localized repressors to establish precise patterns of gene expression within individual cells. In yeast, the SWI5 protein is responsible for activating expression of the *HO* gene (see Chapter 17). This activator is present in both the mother cell and the daughter cell during budding, but its ability to turn on *HO* is restricted to the mother cell because of the presence of the Ash1 repressor in the daughter cell. In other words, Ash1 keeps the *HO* gene off in the daughter cell despite the presence of SWI5.

ADVANCED CONCEPTS

Box 19-2 Review of Cytoskeleton: Asymmetry and Growth

The cytoskeleton is composed of three types of filaments: intermediate filaments, actin filaments, and microtubules. Actin filaments and microtubules are used to localize specific mRNAs in a variety of different cell types, including budding yeast and *Drosophila* oocytes. Actin filaments are composed of polymers of actin. The actin polymers are organized as two parallel helices that form a complete twist every 37 nm. Each actin monomer is located in the same orientation within the polymer, and as a result, actin filaments contain a clear polarity. The plus (+) end grows more rapidly than the minus (–) end, and consequently, mRNAs slated for localization move along with the growing end (Box 19-2 Fig. 1).

Microtubules are composed of polymers of a protein called tubulin, which is a heterodimer composed of related α and β chains. Tubulin heterodimers form extended, asymmetric protofilaments. Each tubulin heterodimer is located in the same orientation within the protofilament. Thirteen different protofilaments associate to form a cylindrical microtubule, and all of the protofilaments are aligned in parallel. Thus, as seen for actin filaments, there is an intrinsic polarity in microtubules, with a rapidly growing "+" end and more stable "–" end (Box 19-2 Fig. 2).

Both actin and tubulin function as enzymes. Actin catalyzes the hydrolysis of ATP to ADP, whereas tubulin hydrolyzes GTP to GDP. These enzymatic activities are responsible for the dynamic growth, or "treadmilling," seen for actin

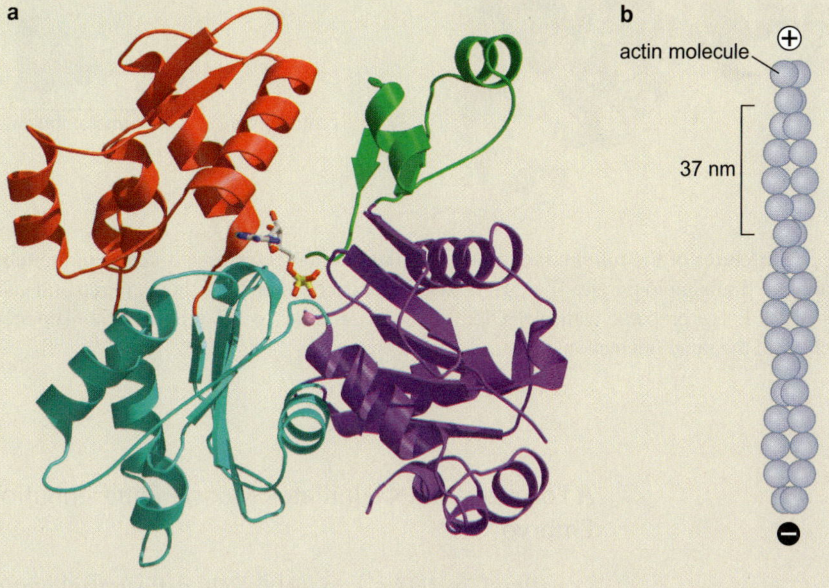

Box 19-2 Figure 1 Structures of the actin monomer and filament. Crystal structure of the actin monomer. (a) The four domains of the monomer are shown, in different colors, with ATP (in red and yellow) in the center. The "–" end of the monomer is at the top and the "+" end is at the bottom. (Otterbein L.R. et al. 2001. *Science* 293: 708–711.) Image prepared with MolScript, BobScript, and Raster 3D. (b) The monomers are assembled, as a single helix, into a filament.

Box 19-2 ***(Continued)***

filaments and microtubules. Typically, it is the actin or tubulin subunits at the "–" end of the filament that mediate the hydrolysis of ATP or GTP, and as a result, these subunits are somewhat unstable and lost from the "–" end. In contrast, newly added subunits at the "+" end have not hydrolyzed ATP or GTP, and this causes them to be more stable components of the filament.

Directed growth of actin filaments or microtubules at the "+" ends depends on a variety of proteins that associate with the cytoskeleton. One such protein is called profilin, which interacts with actin monomers and augments their incorporation into the "+" ends of growing actin filaments. Other proteins have been shown to enhance the growth of tubulin protofilaments at the "+" ends of microtubules.

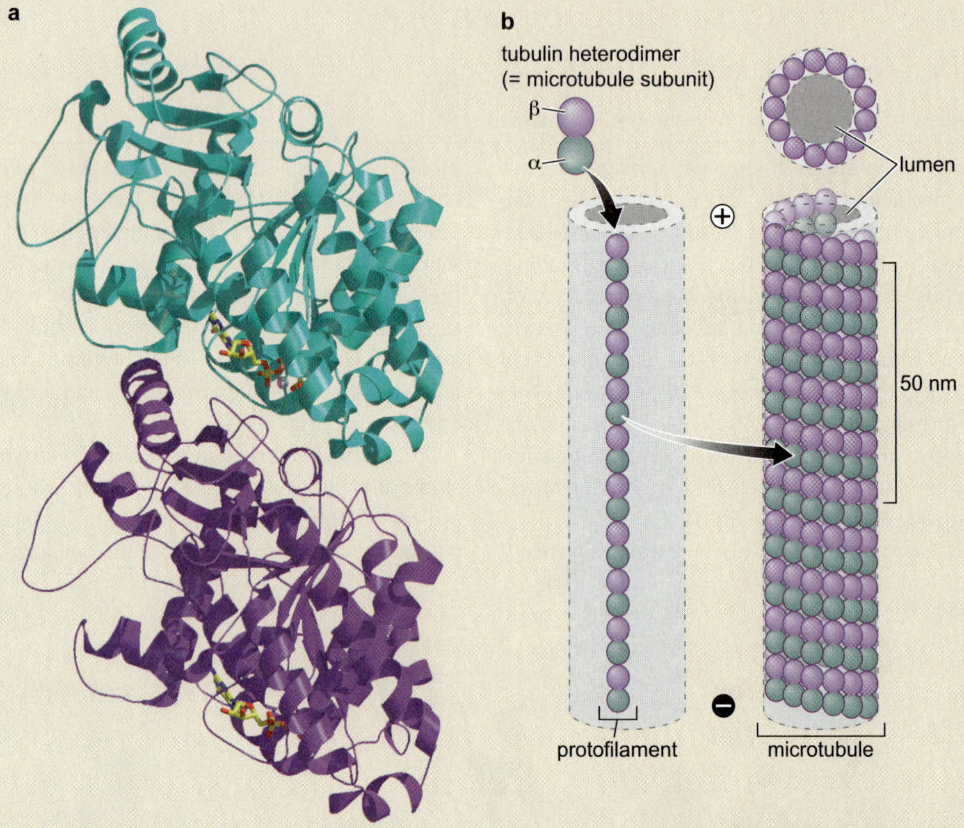

BOX 19-2 FIGURE 2 Structures of the tubulin monomer and filament. (a) The crystal structure of the tubulin monomer shows the α subunit in turquoise and the β subunit in purple. The GTP molecules in each subunit are shown in red and yellow. (Lowe J. et al. 2001. *J. Mol. Biol.* **313:** 1045–1057.) Image prepared with MolScript, BobScript, and Raster 3D. (b) The protofilament of tubulin consists of adjacent monomers assembled in the same orientation.

A Localized mRNA Initiates Muscle Differentiation in the Sea Squirt Embryo

Localized mRNAs can establish differential gene expression among the genetically identical cells of a developing embryo. Just as the fate of the daughter cell is constrained by its inheritance of the *ash1* mRNA in yeast, the cells in a developing embryo can be instructed to follow specific pathways of development through the inheritance of localized mRNAs (see Box 19-3, Overview of *Ciona* Development).

ADVANCED CONCEPTS

Box 19-3 Overview of *Ciona* Development

Adult sea squirts are immobile filter feeders that live in shallow ocean waters (Box 19-3 Fig. 1). They are hermaphrodites and possess both sperm and eggs. They can self-fertilize but prefer not to do so. Instead, sperm from one animal typically fertilizes eggs from another. The resulting embryos are transparent and composed of relatively few cells (hundreds, rather than the tens of thousands seen in vertebrate embryos). These embryos develop rapidly into swimming tadpoles just 18–24 hours after fertilization. Complete cell lineages are known for each of the major tissues. This makes it possible to visualize the sequence of cell divisions from fertilization to the formation of specialized tissues in the tadpole. For example, the tadpole tail contains 36–40 muscle cells (depending on the species), and the lineage that forms these cells can be traced back to the fertilized egg.

The tail muscles represent the first cell lineage that was visualized in any animal embryo, about 100 years ago. This visualization was made possible by a yellow pigment that is present in the unfertilized eggs of certain ascidians. The pigment is initially distributed throughout the egg but becomes localized to vegetal (bottom) regions shortly after fertilization (Box 19-3 Fig. 2). The localized pigment is inherited by just two of the cells, or blastomeres, in eight-cell embryos. These two cells give rise to most of the tail muscles in the tadpole. The yellow pigment is not the actual muscle "determinant"; that is, it is not responsible for programming the cells to form muscle. Rather, the pigment is merely a visible marker that is associated with the determinant.

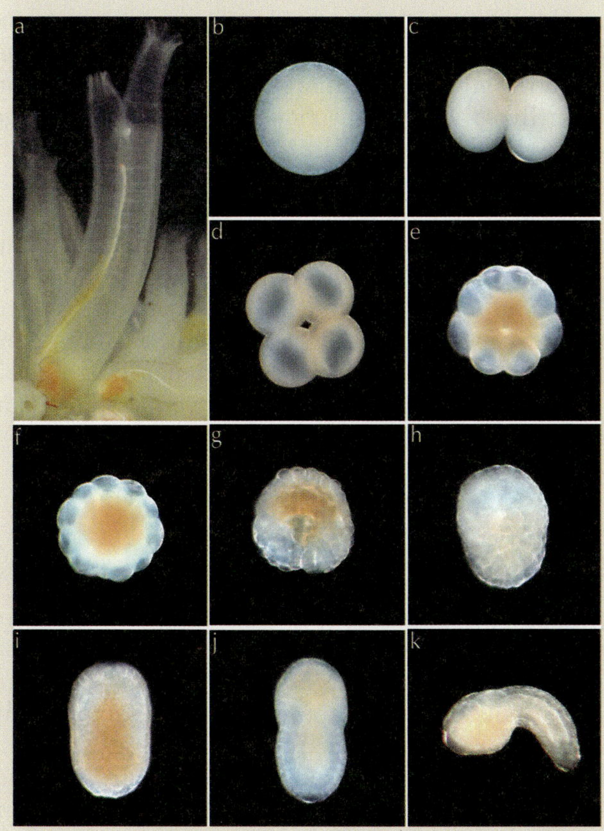

BOX 19-3 FIGURE 1 ***Ciona* life cycle.** (a) The adult sea squirt is shown in the upper left panel. The orange material corresponds to developing eggs and the white duct is the sperm duct. (b–k) Progressively older embryos are shown. A young tadpole is seen in the lower right panel. This stage is reached 12–14 hours after fertilization. (Reproduced, with permission, from Satoh N. et al. 2003. *Trends Genet.* 19: 376–381, Fig. 2. © Elsevier.)

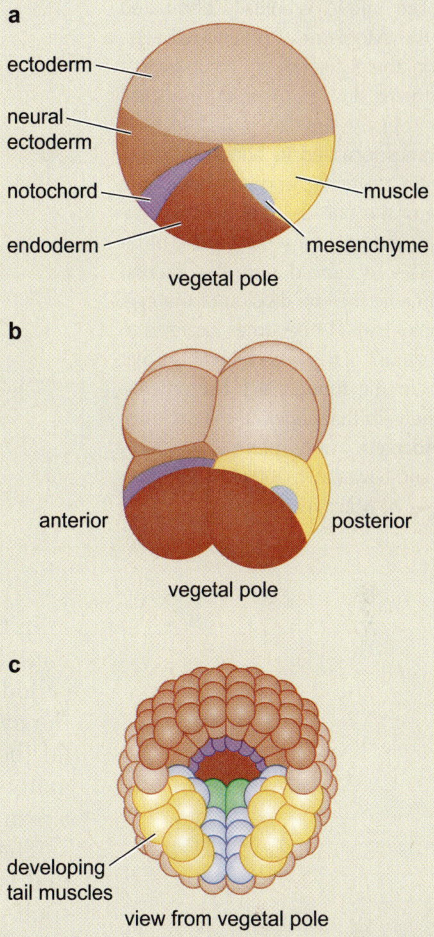

BOX 19-3 FIGURE 2 **Early cleavages in ascidians.** The fertilized, one-cell ascidian embryo contains a number of localized "determinants" that control the development of different tissues. For example, the yellow determinant is inherited by cells that form the tail muscles. The red determinant is inherited by cells that form the endoderm, or gut. (Redrawn, with permission, from Gilbert S.E. 1997. *Developmental biology*, 5th ed., p. 179, Fig. 5.17. © Sinauer.)

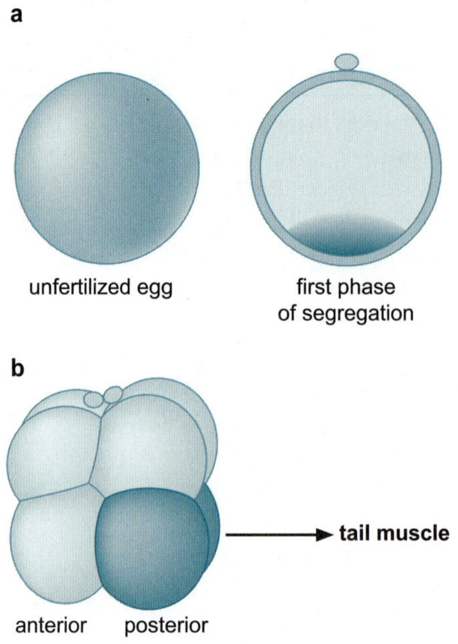

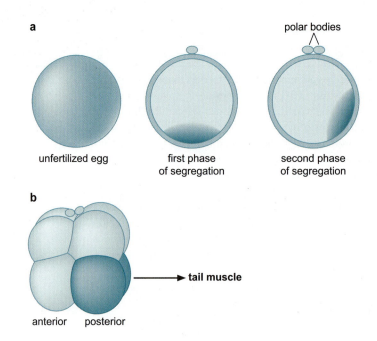

FIGURE 19-8 The Macho-1 mRNA becomes localized in the fertilized egg of a sea squirt. (a) The mRNA is initially distributed throughout the cytoplasm of unfertilized eggs. At fertilization, the egg is induced to undergo a highly asymmetric division to produce a small polar body (top). At this time, the Macho-1 mRNA becomes localized to bottom (vegetal) regions. Shortly thereafter, and well before the first division of the one-cell embryo, the Macho-1 mRNA undergoes a second wave of localization. This occurs during the second highly asymmetric meiotic division of the egg. (b) The Macho-1 mRNA becomes localized to a specific quadrant of the one-cell embryo that corresponds to the future B4.1 blastomeres. These are the cells that generate the tail muscles. (a, Adapted, with permission, from Nishida H. and Sawada K. 2001. *Nature* 409: 725, Fig. 1c–e. © Macmillan.)

In the case of muscle differentiation in sea squirts, a major determinant for programming cells to form muscle is a regulatory protein called Macho-1. Macho-1 mRNA is initially distributed throughout the cytoplasm of unfertilized eggs but becomes restricted to the vegetal (bottom) cytoplasm shortly after fertilization (Fig. 19-8). It is ultimately inherited by just two of the cells in eight-cell embryos, and as a result these two cells go on to form the tail muscles.

The Macho-1 mRNA encodes a zinc finger DNA-binding protein that is believed to activate the transcription of muscle-specific genes, such as actin and myosin. Thus, these genes are expressed only in muscles because Macho-1 is made only in those cells. In the second part of this chapter, we see how regulatory proteins synthesized from localized mRNAs in the *Drosophila* embryo activate and repress gene expression and control the formation of different cell types.

Cell-to-Cell Contact Elicits Differential Gene Expression in the Sporulating Bacterium, *Bacillus subtilis*

The second major strategy for establishing differential gene expression is cell-to-cell contact. Again, we begin our discussion with a relatively simple case, this one from the bacterium *Bacillus subtilis.* Under adverse conditions, *B. subtilis* can form spores. The first step in this process is the formation of a septum at an asymmetric location within the sporangium, the progenitor of the spore. The septum produces two cells of differing sizes that remain attached through abutting membranes. The smaller cell is called the **forespore**; it ultimately forms the spore. The larger cell is called the mother cell; it aids the development of the spore (Fig. 19-9). The forespore influences the expression of genes in the neighboring mother cell, as described below.

The forespore contains an active form of a specific σ factor, σ^F, that is inactive in the mother cell. In Chapter 16, we saw how σ factors associate with RNA polymerase and select specific target promoters for expression. σ^F activates the *spoIIR* gene that encodes a secreted signaling protein. SpoIIR is secreted into the space between the abutting

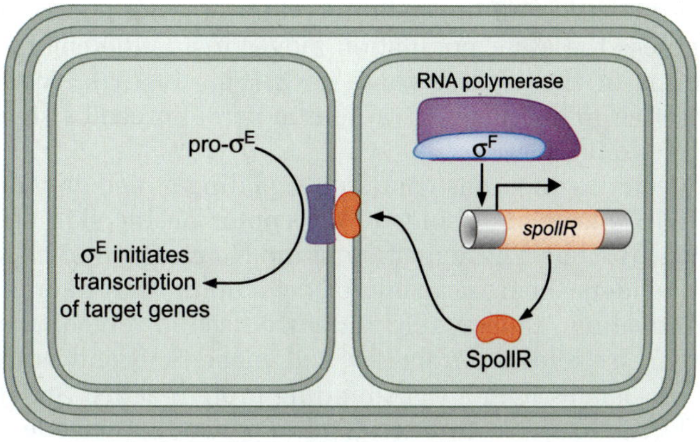

FIGURE 19-9 Asymmetric gene activity in the mother cell and forespore of *Bacillus subtilis* depends on the activation of different classes of σ factors. The *spoIIR* gene is activated by σ^F in the forespore. The encoded SpoIIR protein becomes associated with the septum separating the mother cell (on the left) and forespore (on the right). It triggers the proteolytic processing of an inactive form of σ^E (pro-σ^E) in the mother cell. The activated σ^E protein leads to the recruitment of RNA polymerase and the activation of specific genes in the mother cell. (Redrawn, with permission, from Stragier P. and Losick R. 1996. *Annu. Rev. Genet.* 30: 297–341, Fig. 3a. © Annual Reviews.)

membranes of the mother cell and the forespore where it triggers the proteolytic processing of pro-σ^E in the mother cell. Pro-σ^E is an inactive precursor of the σ^E factor. The pro-σ^E protein contains an amino-terminal inhibitory domain that blocks σ^E activity and tethers the protein to the membrane of the mother cell (Fig. 19-9). SpoIIR induces the proteolytic cleavage of the amino-terminal peptide and the release of the mature and active form of σ^E from the membrane. σ^E activates a set of genes in the mother cell that is distinct from those expressed in the forespore. In this example, SpoIIR functions as a signaling molecule that acts at the interface between the forespore and the mother cell and elicits differential gene expression in the abutting mother cell through the processing of σ^E. Induction requires cell-to-cell contact because the forespore produces small quantities of SpoIIR that can interact with the abutting mother cell but that are insufficient to elicit the processing of σ^E in the other cells of the population.

A Skin–Nerve Regulatory Switch Is Controlled by Notch Signaling in the Insect Central Nervous System

We now turn to an example of cell-to-cell contact in an animal embryo that is surprisingly similar to the one just described in *B. subtilis*. In that earlier example, SpoIIR causes the proteolytic activation of σ^E, which, in its active state, directs RNA polymerase to the promoter sequences of specific genes. In the following example, a cell-surface receptor is cleaved, and the intracytoplasmic domain moves to the nucleus where it binds a sequence-specific DNA-binding protein that activates the transcription of selected genes.

For this example, we must first briefly describe the development of the ventral nerve cord in insect embryos (Fig. 19-10). This nerve cord functions in a manner that is roughly comparable to the spinal cord of humans. It arises from a sheet of cells called the *neurogenic ectoderm*. This tissue is subdivided into two cell populations: one group re-

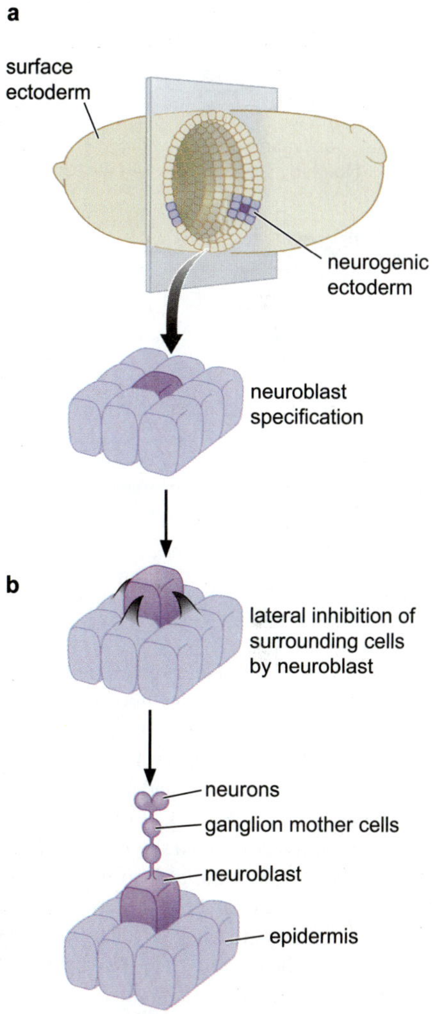

FIGURE 19-10 The neurogenic ectoderm forms two major cell types: Neurons and skin cells (or epidermis). (a) Cells in the early neurogenic ectoderm can form either type of cell. However, once one of the cells begins to form a neuron or "neuroblast" (dark cell in the center of the grid of cells), it inhibits all of the neighboring cells that it directly touches. (b) This inhibition causes most of the cells to remain on the surface of the embryo and form skin cells. In contrast, the developing neuron moves into the embryo cavity and forms neurons.

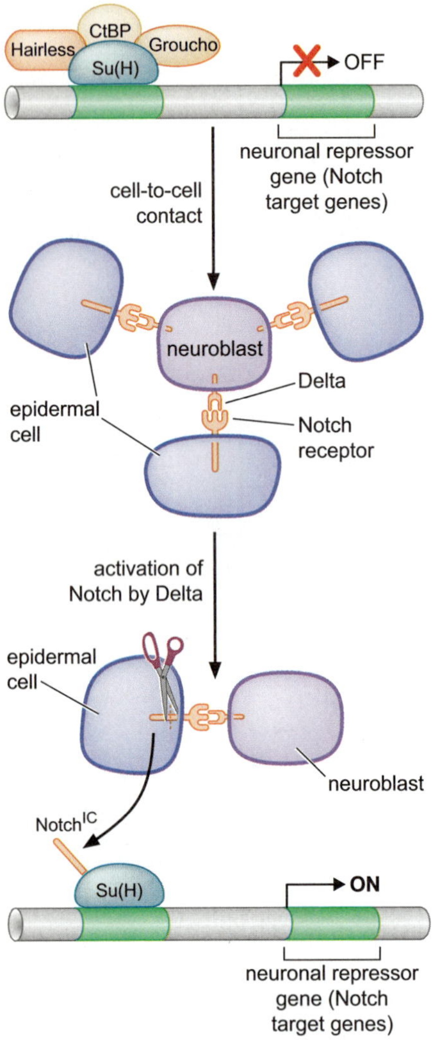

FIGURE 19-11 Notch-Su(H) regulatory switch. The developing neuron (neuroblast) does not express neuronal repressor genes (top). These genes are kept off by a DNA-binding protein called Su(H) and associated corepressor proteins (Hairless, CtBP, Groucho). The neuroblast expresses a signaling molecule, called Delta, that is tethered to the cell surface. Delta binds to the Notch receptor in neighboring cells that are in direct physical contact with the neuron. Delta–Notch interactions cause the Notch receptor to be activated in the neighboring cells, which differentiate into epidermis. The activated Notch receptor is cleaved by cellular proteases (scissors), and the intracytoplasmic region of the receptor is released into the nucleus. This piece of the Notch protein causes the Su(H) regulatory protein to function as an activator rather than a repressor. As a result, the neuronal repressor genes are activated in the epidermal cells so that they cannot develop into neurons.

mains on the surface of the embryo and forms ventral skin (or epidermis), whereas the other population moves inside the embryo to form the neurons of the ventral nerve cord (Fig. 19-10a). This decision about whether to become skin or neuron is reinforced by signaling between the two populations.

The developing neurons contain a signaling molecule on their surface called **Delta**, which binds to a receptor on the skin cells called **Notch** (Fig. 19-10b). The activation of the Notch receptor on skin cells by Delta renders them incapable of developing into neurons, as follows. Activation causes the intracytoplasmic domain of Notch (Notch^{IC}) to be released from the cell membrane and enter nuclei, where it associates with a DNA-binding protein called Su(H). The resulting Su(H)-Notch^{IC} complex activates genes that encode transcriptional repressors which block the development of neurons.

Notch signaling does not cause a simple induction of the Su(H) activator protein but instead triggers an on/off regulatory switch. In the absence of signaling, Su(H) is associated with several corepressor proteins, including Hairless, CtBP, and Groucho (Fig. 19-11). Su(H) complexed with any of these proteins actively represses Notch target genes. When Notch^{IC} enters the nucleus, it displaces the repressor proteins in complex with Su(H), turning that protein into an activator instead. Thus, Su(H) now activates the very same genes that it formerly repressed.

Delta–Notch signaling depends on cell-to-cell contact. The cells that present the Delta ligand (neuronal precursors) must be in direct physical contact with the cells that contain the Notch receptor (epidermis) in order to activate Notch signaling and inhibit neuronal differentiation. In the next section, we see an example of a secreted signaling molecule that influences gene expression in cells located far from those that send the signal.

A Gradient of the Sonic Hedgehog Morphogen Controls the Formation of Different Neurons in the Vertebrate Neural Tube

We now turn to an example of a long-range signaling molecule, a morphogen, that imposes positional information on a developing organ. For this example, we continue our discussion of neuronal differentiation, but this time, we consider the neural tube of vertebrates. In all vertebrate embryos, there is a stage when cells located along the future back—the dorsal ectoderm—move in a coordinated fashion toward internal regions of the embryo and form the neural tube, the forerunner of the adult spinal cord.

Cells located in the ventralmost region of the neural tube form a specialized structure called the **floorplate** (Fig. 19-12). The floorplate is the site of expression of a secreted cell-signaling molecule called Sonic hedgehog (Shh), which functions as a gradient morphogen.

Shh is secreted from the floorplate and forms an extracellular gradient in the ventral half of the neural tube (Fig. 19-12a). Neurons develop within the neural tube into different cell types based on the amount of Shh protein they receive. This is determined by their location relative to the floorplate; cells located near the floorplate receive the highest concentrations of Shh, and those located farther away receive lower levels. The extracellular Shh gradient leads to different degrees of activation of Shh receptors in different cells in the neural tube. The Shh gradient specifies at least four different types of neurons (Fig. 19-12b).

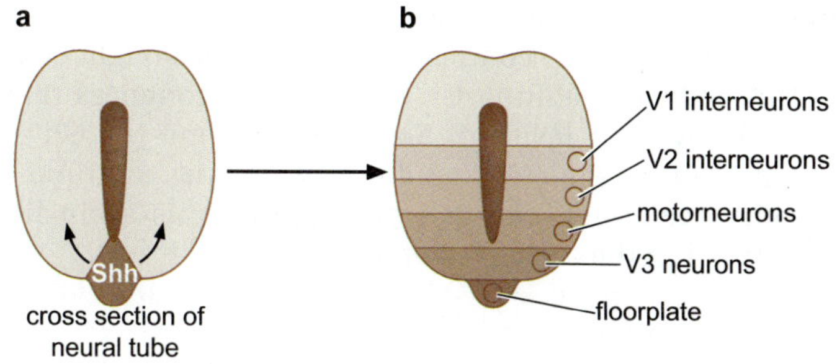

FIGURE 19-12 Formation of different neurons in the vertebrate neural tube. (a) The secreted signaling molecule Sonic hedgehog (Shh) is expressed in the floorplate of the developing neural tube (see the brown circle at the bottom of the diagram). The Shh protein diffuses through the extracellular matrix of the neural tube. The highest levels are present in ventral (bottom) regions and progressively lower levels in more lateral regions (arrows). (b) The graded distribution of the Shh protein leads to the formation of distinct neuronal cell types in the ventral half of the neural tube. High and intermediate levels lead to the development of the V3 neurons and motor neurons, respectively. Low and lowest levels lead to the development of the V2 and V1 interneurons. (Adapted, with permission, from Jessell T. 2000. *Nat. Rev. Genet.* 1: 20–29, Fig. 2c. © Macmillan.)

Cells located near the floorplate—those that receive the highest concentrations of Shh—have a high number of Shh receptors activated on their surface. This instructs those cells to form a neuronal cell type called **V3**, which is distinct from the other neurons that arise from the Shh gradient. Cells located in more lateral regions of the neural tube (farther from the floorplate) receive progressively lower levels of the Shh protein. This results in fewer Shh receptors being activated in those cells, which therefore become motor neurons. Yet lower levels of Shh direct the formation of the V2 and V1 interneurons, respectively (Fig. 19-12b).

How does this differential activation of Shh receptors produce different cell types? The activation of the Shh receptor causes a transcriptional activator called **Gli** to activate the expression of specific "target" genes. The induction of the Gli activator is controlled, in part, by its regulated transport into the nucleus. Binding of Shh to its receptor on the cell surface allows a previously inactive form of Gli to enter the nucleus of that cell in an active form. The extracellular Shh gradient present in the neural tube thus leads to the formation of a corresponding Gli activator gradient. That is, the amount of active Gli in the nucleus of any given cell depends on how far that cell is from the floorplate—the closer it is, the higher the concentration of Gli.

Once in the nucleus, Gli activates gene expression in a concentration-dependent fashion. Peak concentrations of Gli, present in cells immediately adjacent to the floorplate, activate target genes needed for the differentiation of the V3 neurons. Slightly lower levels of Gli activate target genes that specify the formation of motor neurons, whereas intermediate and low levels of Gli induce the formation of the V2 and V1 interneurons, respectively. We see, in the next section, that the different binding affinities of Gli recognition sequences within the regulatory DNAs of the various target genes likely have an important role in this differential regulation of Shh-Gli target genes. Thus, V1 genes can be activated by low levels of Gli because they have high-affinity recognition sequences for that activator in their nearby regulatory DNA. In contrast, V3 target genes might contain regulatory DNA with low-affinity Gli recognition sequences that can be activated only by peak levels of Shh signaling and the Gli activator. This principle of a regulatory gradient producing multiple "thresholds" of gene expression and cell differentiation is again illustrated particularly well in the early *Drosophila* embryo.

THE MOLECULAR BIOLOGY OF *DROSOPHILA* EMBRYOGENESIS

WEB ANIMATION

In this section, we focus on the early embryonic development of the fruit fly, *Drosophila melanogaster* (see Interactive Animation 19-1). The molecular details of how development is regulated are better understood in this system than in any other animal embryo. The various mechanisms of cell communication discussed in the first half of this chapter and those of gene regulation discussed in the previous chapters are brought together in this example.

Localized determinants and cell-signaling pathways are both used to establish positional information that results in gradients of regulatory proteins that pattern the anteroposterior (head–tail) and dorsoventral (back–belly) body axes. These regulatory proteins—activators and repressors—control the expression of genes whose products define different regions of the embryo. A recurring theme is the use of complex regulatory DNAs—particularly complex enhancers—to bring transcriptional activators and repressors to genes where they function in a combinatorial manner to produce sharp on/off patterns of gene expression.

An Overview of *Drosophila* Embryogenesis

Life begins for the fruit fly as it does for humans: adult males inseminate females. A single sperm cell enters a mature egg, and the haploid sperm and egg nuclei fuse to form a diploid, "zygotic" nucleus. This nucleus undergoes a series of nearly synchronous divisions within the central regions of the egg. Because there are no plasma membranes separating the nuclei, the embryo now becomes what is called a **syncytium**—that is, a single cell with multiple nuclei. With the next series of divisions, the nuclei begin to migrate toward the cortex or periphery of the egg. Once located in the cortex, the nuclei undergo another three divisions leading to the formation of a monolayer of approximately 6000 nuclei surrounding the central yolk. During a 1-hour period, from 2 to 3 hours after fertilization, cell membranes form between adjacent nuclei.

Before the formation of cell membranes, the nuclei are **totipotent** or uncommitted: they have not yet taken on an identity and can still give rise to any cell type. Just after cellularization, however, nuclei have become irreversibly "**determined**" to differentiate into specific tissues in the adult fly. This process is described in Box 19-4, Overview of *Drosophila* development. The molecular mechanisms responsible for this dramatic process of determination are described in the following sections of this chapter.

Box 19-4 Overview of *Drosophila* Development

After the sperm and egg haploid nuclei fuse, the diploid, zygotic nucleus undergoes a series of ten rapid and nearly synchronous cleavages within the central yolky regions of the egg. Large microtubule arrays emanating from the centrioles of the dividing nuclei help direct the nuclei from central regions toward the periphery of the egg (Box 19-4 Fig. 1). After eight cleavages, the 256 zygotic nuclei begin to migrate to the periphery. During this migration, they undergo two more cleavages (Box 19-4 Fig. 1, nuclear cleavage cycle 9). Most, but not all, of the resulting approximately 1000 nuclei enter the cortical regions of the egg (Box 19-4 Fig. 1, nuclear cleavage cycle 10). The others ("vitellophages") remain in central regions where they have a somewhat obscure role in development.

Once the majority of the nuclei reach the cortex at about 90 minutes following fertilization, they first acquire competence to transcribe Pol II genes. Thus, as in many other organisms such as *Xenopus*, there seems to be a "midblastula transition," whereby early blastomeres (or nuclei) are transcriptionally silent during rapid periods of mitosis. Although causality is unclear, it does seem that DNA undergoing intense bursts of replication cannot simultaneously sustain transcription. These and other observations have led to the suggestion that there is competition between the large macromolecular complexes promoting replication and transcription. Because transcriptional competence is only achieved when the nuclei reach the cortex, it has been suggested that peripheral regions contain localized determinants. However, recent gene-expression studies have stripped much of the mystery from the cortex. For example, the segmentation gene, *hunchback*, is uniformly transcribed in all of the nuclei present in the anterior half of the early embryo. This expression encompasses both the peripheral nuclei that have entered cortical regions and the vitellophages that remain in the yolk.

After the nuclei reach the cortex, they undergo another three rounds of cleavage (for a total of 13 divisions after fertilization), leading to the dense packing of about 6000 columnar-shaped nuclei enclosing the central yolk (Box 19-4 Fig. 1, nuclear cleavage cycle 14). Technically, the embryo is still a syncytium, although histochemical staining of early embryos with antibodies against cytoskeletal proteins indicates a highly structured meshwork surrounding each nucleus. During a 1-hour period, from 2 to 3 hours after fertilization, the embryo undergoes a dramatic cellularization process, whereby cell membranes are formed between adjacent nuclei (Box 19-4 Fig. 1, nuclear cleavage cycle 14). By 3 hours after fertilization, the embryo has been transformed into a cellular blastoderm, comparable to the "hollow ball of cells" that characterize the blastulae of most other embryos.

One of the most compelling aspects of classical embryology is the intrinsic beauty of the material. The early embryos of most marine organisms, such as ascidians, are visually stunning. Unfortunately, the *Drosophila* embryo is rather ugly; its salvation has been the unprecedented visualization of gene-expression patterns. The differential gene activity that has been so graphi-

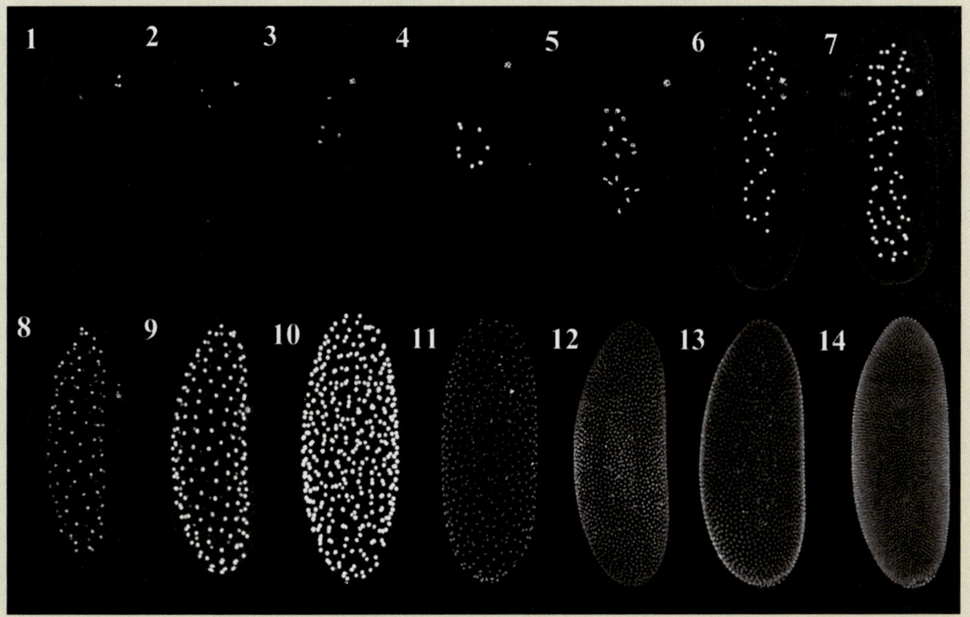

BOX 19-4 FIGURE 1 ***Drosophila* embryogenesis.** *Drosophilia* embryos are oriented with the future head pointed up. The numbers refer to the number of nuclear cleavages. Nuclei are stained white within the embryos. For example, stage 1 contains the single zygotic nucleus resulting from the fusion of the sperm and egg pronuclei. The stained material in the upper right areas of stages 1–7 are polar bodies. The zygotic nucleus of stage 1 and the nuclei of stages 2, 3, etc., are in central regions of the embryo. Stage 2 contains two nuclei arising from the first division of the zygotic neucleus. At stage 10, there are approximately 500 nuclei and most are arranged in a single layer at the cortex (periphery of the embryo). At nuclear cleavage cycle 14, there are more than 6000 nuclei densely packed in a monolayer in the cortex. Cellularization occurs during this stage. (Courtesy of W. Baker and G. Shubiger.)

cally visualized in the early embryo using a variety of molecular and histochemical tools is not simply a manifestation of cell-fate specification. Rather, some of the first genes to be visualized encode regulatory proteins that actually dictate cell fate. Thus, molecular studies have literally illuminated the mysterious process of cell-fate specification and determination.

When the nuclei enter the cortex of the egg, they are totipotent and can form any adult cell type. The location of each nucleus, however, now determines its fate. The 30 or so nuclei that migrate into posterior regions of the cortex encounter localized protein determinants, such as Oskar, which program these naïve nuclei to form the germ cells (Box 19-4 Fig. 2). Among the putative determinants contained in the polar plasm are large nucleoprotein complexes, called polar granules. The posterior nuclei bud off from the main body of the embryo along with the polar granules, and the resulting pole cells differentiate into either sperm or eggs, depending on the sex of the embryo. The microinjection of polar plasm into abnormal locations, such as central and anterior regions, results in the differentiation of supernumerary pole cells.

Cortical nuclei that do not enter the polar plasm are destined to form the somatic tissues. Again, these nuclei are totipotent and can form any adult cell type. However, within a very brief period (no more than an hour), each nucleus is rapidly programmed (or specified) to follow a particular pathway of differentiation. This specification process occurs during the period of cellularization, although there is no reason to believe that the deposition of cell membranes between neighboring nuclei is critical for determining cell fate. Different nuclei exhibit distinct patterns of gene transcription prior to the completion of cell formation. By 3 hours after fertilization, each cell possesses a fixed positional identity, so that cells located in anterior regions of the embryo will form head structures in the adult fly, whereas cells located in posterior regions will form abdominal structures.

Systematic genetic screens by Eric Wieschaus and Christiane Nüsslein-Volhard identified approximately 30 "segmentation genes" that control the early patterning of the *Drosophila* embryo. This involved the examination of thousands of dead embryos. At the midpoint of embryogenesis, the ventral skin, or epidermis, secretes a cuticle that contains many fine hairs, or denticles. Each body segment of the embryo contains a characteristic pattern of denticles. Three different classes of segmentation genes were identified on the basis of causing specific disruptions in the denticle patterns of dead embryos. Mutations in the so-called "gap" genes cause the deletion of several adjacent segments (Box 19-4 Fig. 3). For example, mutations in the gap gene *knirps* cause the loss of the second through seventh abdominal segments (normal embryos possess eight such segments). Mutations in the "pair-rule" genes cause the loss of alternating segments.

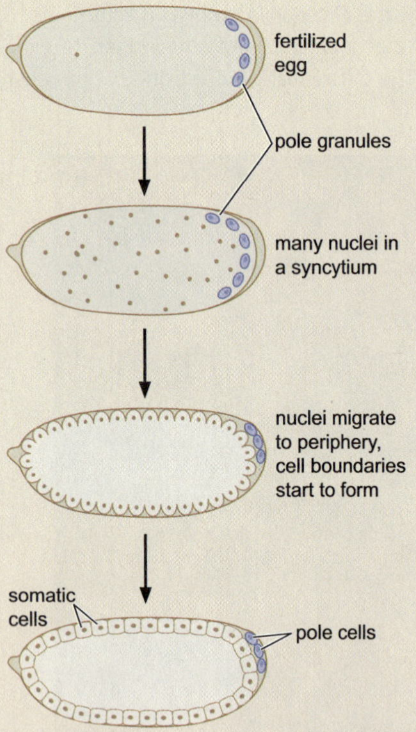

Box 19-4 Figure 2 Development of germ cells. Polar granules located in the posterior cytoplasm of the unfertilized egg contain germ cell determinants and the Nanos mRNA, which is important for the development of the abdominal segments. Nuclei (central dots) begin to migrate to the periphery. Those that enter posterior regions sequester the polar granules and form the pole cells, which form the germ cells. The remaining cells (somatic cells) form all of the other tissues in the adult fly. (Adapted, with permission, from Schneiderman H.A. 1976. *Symp. R. Entomol. Soc. Lond.* 8: 3–34. © Royal Entomological Society.)

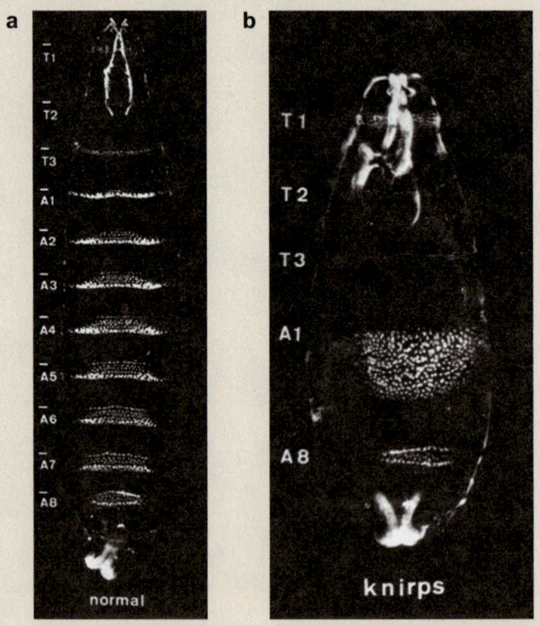

Box 19-4 Figure 3 Dark-field images of normal and mutant cuticles. (a) The pattern of denticle hairs in this normal embryo are slightly different among the different body segments (labeled T1 through A8 in the image). (b) The Knirps mutant (having a mutation in the gap gene *knirps*), shown here, lacks the second through seventh abdominal segments. (Reprinted, with permission, from Nüsslein-Volhard C. and Wieschaus E. 1980. *Nature* 287: 795–801. © Macmillan. Images courtesy of Eric Wieschaus, Princeton University.)

A Morphogen Gradient Controls Dorsoventral Patterning of the *Drosophila* Embryo

The dorsoventral patterning of the early *Drosophila* embryo is controlled by a regulatory protein called Dorsal, which is initially distributed throughout the cytoplasm of the unfertilized egg. After fertilization, and after the nuclei reach the cortex of the embryo, the Dorsal protein enters nuclei in the ventral and lateral regions but remains in the cytoplasm in dorsal regions (Fig. 19-13). The formation of this Dorsal gradient in nuclei across the embryo is very similar, in principle, to the formation of the Gli activator gradient within ventral cells of the vertebrate neural tube.

Regulated nuclear transport of the Dorsal protein is controlled by a cell-signaling molecule called **Spätzle**. This signal is distributed in a ventral-to-dorsal gradient within the extracellular matrix present between the plasma membrane of the unfertilized egg and the outer egg shell. After fertilization, Spätzle binds to the cell-surface Toll receptor. Depending on the concentration of Spätzle, and thus the degree of receptor occupancy in a given region of the syncytial embryo, Toll is activated to a greater or lesser extent. There is peak activation of Toll receptors in ventral regions—where the Spätzle concentration is highest—and progressively lower activation in more lateral regions. Toll signaling causes the degradation of a cytoplasmic inhibitor, Cactus,

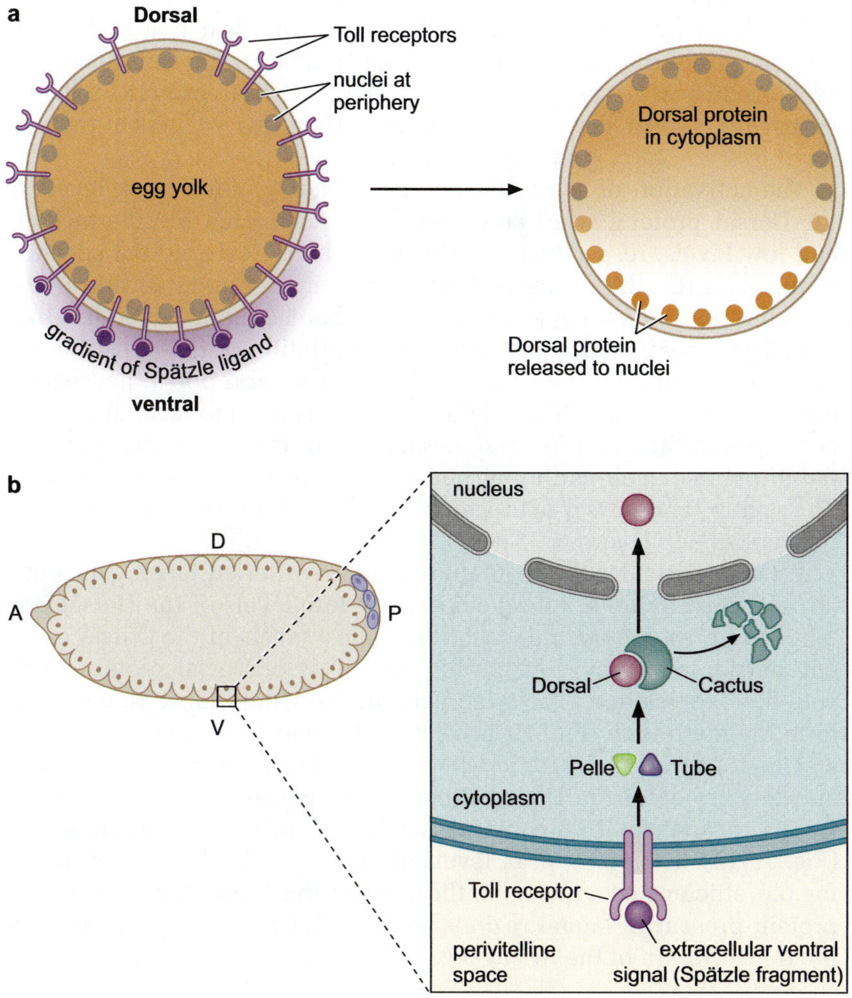

FIGURE 19-13 Spätzle-Toll and Dorsal gradient. (a) The circles represent cross sections through early *Drosophila* embryos. The Toll receptor is uniformly distributed throughout the plasma membrane of the precellular embryo. The Spätzle signaling molecule is distributed in a gradient with peak levels in the ventralmost regions. As a result, more Toll receptors are activated in ventral regions than in lateral and dorsal regions. This gradient in Toll signaling creates a broad Dorsal nuclear gradient. (b) Side view of the embryo with anterior to the left and dorsal surface up; details of the Toll signaling cascade to the right. Activation of the Toll receptor leads to the activation of the Pelle kinase in the cytoplasm. Pelle either directly or indirectly phosphorylates the Cactus protein, which binds and inhibits the Dorsal protein. Phosphorylation of Cactus causes its degradation, so that Dorsal is released from the cytoplasm into nuclei.

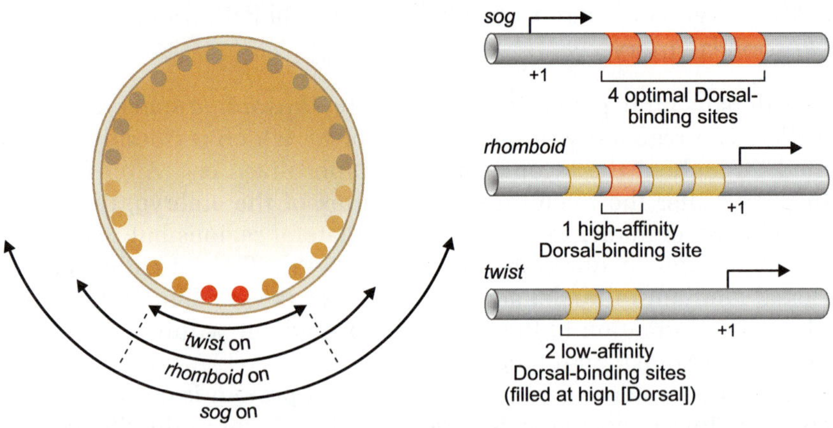

FIGURE 19-14 Three thresholds and three types of regulatory DNAs. The *twist* 5′ regulatory DNA contains two low-affinity Dorsal-binding sites that are occupied only by peak levels of the Dorsal gradient. As a result, *twist* expression is restricted to ventral nuclei. The *rhomboid* 5′ enhancer contains a cluster of Dorsal-binding sites. Only one of these sites represents an optimal, high-affinity Dorsal recognition sequence. This mixture of high- and low-affinity sites allows both high and intermediate levels of the Dorsal gradient to activate *rhomboid* expression in ventrolateral regions. Finally, the *sog* intronic enhancer contains four evenly spaced optimal Dorsal-binding sites. These allow high, intermediate, and low levels of the Dorsal gradient to activate *sog* expression throughout ventral and lateral regions.

and the release of Dorsal from the cytoplasm into nuclei. This leads to the formation of a corresponding Dorsal nuclear gradient in the ventral half of the early embryo. Nuclei located in the ventral regions of the embryo contain peak levels of the Dorsal protein, whereas those nuclei located in lateral regions contain lower levels of the protein.

The activation of some Dorsal target genes requires peak levels of the Dorsal protein, whereas others can be activated by intermediate and low levels, respectively. In this way, the Dorsal gradient specifies three major thresholds of gene expression across the dorsoventral axis of embryos undergoing cellularization about 2 hours after fertilization. These thresholds initiate the differentiation of three distinct tissues: mesoderm, ventral neurogenic ectoderm, and dorsal neurogenic ectoderm (Fig. 19-14). Each of these tissues goes on to form distinctive cell types in the adult fly. The mesoderm forms flight muscles and internal organs, such as the fat body, which is analogous to our liver. The ventral and dorsal neurogenic ectoderm form distinct neurons in the ventral nerve cord.

We now consider the regulation of three different target genes that are activated by high, intermediate, and low levels of the Dorsal protein—*twist*, *rhomboid*, and *sog*. The highest levels of the Dorsal gradient-that is, in nuclei with the highest levels of Dorsal protein—activate the expression of the twist gene in the ventralmost 18 cells that form the mesoderm (Fig. 19-14). The twist gene is not activated in lateral regions, the neurogenic ectoderm, where there are intermediate and low levels of the Dorsal protein. The reason for this is that the *twist* 5′ regulatory DNA contains two low-affinity Dorsal-binding sites (Fig. 19-14). Therefore, peak levels of the Dorsal gradient are required for the efficient occupancy of these sites; the lower levels of Dorsal protein present in lateral regions are insufficient to bind and activate the transcription of the *twist* gene.

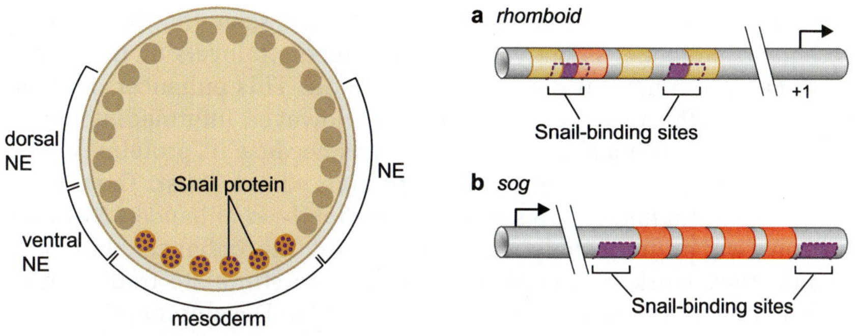

FIGURE 19-15 **Regulatory DNAs.** (a) The *rhomboid* enhancer contains binding sites for both Dorsal and the Snail repressor. Since the Snail protein is only present in ventral regions (the mesoderm), *rhomboid* is kept off in the mesoderm and restricted to ventral regions of the neurogenic ectoderm (ventral NE). (b) The intronic *sog* enhancer also contains Snail repressor sites. These keep *sog* expression off in the mesoderm and restricted to broad lateral stripes that encompass both ventral and dorsal regions of the neurogenic ectoderm (NE).

The *rhomboid* gene is activated by intermediate levels of the Dorsal protein in the ventral neurogenic ectoderm. The rhomboid 5′-flanking region contains a 300-bp enhancer located about 1.5 kb 5′ of the transcription start site (Fig. 19-15a). This enhancer contains a cluster of Dorsal-binding sites, mostly low-affinity sites as seen in the *twist* 5′ regulatory region. At least one of the sites, however, is an optimal, high-affinity site that permits the binding of intermediate levels of Dorsal protein—the amount present in lateral regions. In principle, the *rhomboid* enhancer can be activated by both the high levels of Dorsal protein present in the mesoderm and the intermediate levels present in the ventral neurogenic ectoderm, but it is kept off in the mesoderm by a transcriptional repressor called **Snail**. The Snail repressor is only expressed in the mesoderm; it is not present in the neurogenic ectoderm. The 300-bp *rhomboid* enhancer contains binding sites for the Snail repressor, in addition to the binding sites for the Dorsal activator. This interplay between the broadly distributed Dorsal gradient and the localized Snail repressor leads to the restricted expression of the *rhomboid* gene in the ventral neurogenic ectoderm. We have already seen how the localized Ash1 repressor blocks the action of the SWI5 activator in the daughter cell of budding yeast, and further along in this chapter, we see the extensive use of this principle in other aspects of *Drosophila* development.

The lowest levels of the Dorsal protein, present in lateral regions of the early embryo, are sufficient to activate the *sog* gene in broad lateral stripes that encompass both the ventral and dorsal neurogenic ectoderm. Expression of *sog* is regulated by a 400-bp enhancer located within the first intron of the gene (Fig. 19-15b). This enhancer contains a series of four evenly spaced high-affinity Dorsal-binding sites that can therefore be occupied even by the lowest levels of the Dorsal protein. As seen for *rhomboid*, the presence of the Snail repressor precludes activation of *sog* expression in the mesoderm despite the high levels of Dorsal protein found there. Thus, the differential regulation of gene expression by different thresholds of the Dorsal gradient depends on the combination of the Snail repressor and the affinities of the Dorsal-binding sites.

The occupancy of Dorsal-binding sites not only is determined by the intrinsic affinities of the sites but also depends on protein–protein interactions between Dorsal and other regulatory proteins bound to

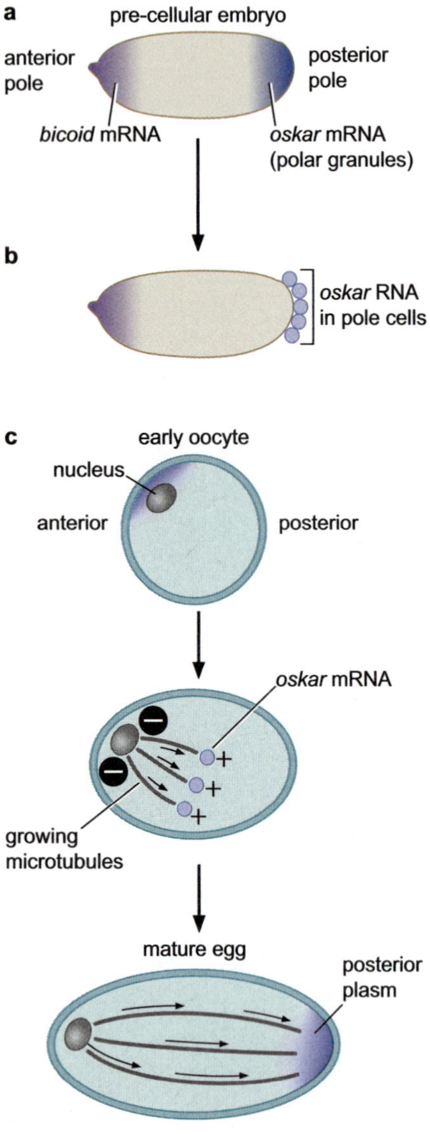

FIGURE 19-16 Localization of maternal mRNAs in the *Drosophila* egg and embryo. (a) The unfertilized *Drosophila* egg contains two localized mRNAs: *bicoid* in anterior regions and *oskar* in posterior regions. (b) The Oskar protein helps coordinate the assembly of the polar granules in the posterior cytoplasm. Nuclei that enter this region bud off the posterior end of the embryo and form the pole cells. During the formation of the *Drosophila* egg, polarized microtubules are formed that extend from the oocyte nucleus and grow toward the posterior plasm. The *oskar* mRNA binds adaptor proteins that interact with the microtubules, and thereby transport the RNA to the posterior plasm. The "–" and "+" symbols indicate the direction of the growing strands of the microtubules.

the target enhancers. For example, we have seen that the 300-bp *rhomboid* enhancer is activated by intermediate levels of the Dorsal gradient in the ventral neurogenic ectoderm. This enhancer contains mostly low-affinity Dorsal-binding sites. However, intermediate levels of Dorsal are sufficient to bind these sites because of protein–protein interactions with another activator protein called **Twist**. Dorsal and Twist bind to adjacent sites within the *rhomboid* enhancer. Not only do the two proteins help each other bind the enhancer, but once bound, they work in a synergistic fashion to stimulate transcription (see Box 19-5, The Role of Activator Synergy in Development).

Segmentation Is Initiated by Localized RNAs at the Anterior and Posterior Poles of the Unfertilized Egg

At the time of fertilization, the *Drosophila* egg contains two localized mRNAs. One, the *bicoid* mRNA, is located at the anterior pole, and the other, the *oskar* mRNA, is located at the posterior pole (Fig. 19-16a). The oskar mRNA encodes an RNA-binding protein that is responsible for the assembly of **polar granules**. These are large macromolecular complexes composed of a variety of different proteins and RNAs. The polar granules control the development of tissues that arise from posterior regions of the early embryo, including the abdomen and the pole cells, which are the precursors of the germ cells (Fig. 19-16b).

The *oskar* mRNA is synthesized within the ovary of the mother fly. It is first deposited at the anterior end of the immature egg, or **oocyte**, by "helper" cells called **nurse cells**. Both the oocyte and associated nurse cells arise from specialized stem cells within the ovary (see Box 19-6, Stem Cells). As the oocyte enlarges to form the mature egg, the *oskar* mRNA is transported from anterior to posterior regions. This localization process depends on specific sequences within the 3′ UTR of the *oskar* mRNA (Fig. 19-17). We have already seen how the 3′ UTR of the *ash1* mRNA mediates its localization to the daughter cell of budding yeast by interacting with the growing ends of microtubules. A remarkably similar process controls the localization of the *oskar* mRNA in the *Drosophila* oocyte.

The *Drosophila* oocyte is highly polarized. The nucleus is located in anterior regions; growing microtubules extend from the nucleus into the posterior cytoplasm. The *oskar* mRNA interacts with adaptor proteins that are associated with the growing + ends of the microtubules and are thereby transported away from anterior regions of the egg, where the nucleus resides, into the posterior plasm. After fertilization, the cells that inherit the localized *oskar* mRNA (and polar granules) form the pole cells.

The localization of the *bicoid* mRNA in anterior regions of the unfertilized egg also depends on sequences contained within its 3′ UTR. The nucleotide sequences of the oskar and *bicoid* mRNAs are distinct. As a result, they interact with different adaptor proteins and become localized to different regions of the egg. The importance of the 3′ UTRs in determining where each mRNA becomes localized is revealed by the following experiment. If the 3′ UTR from the *oskar* mRNA is replaced with that from *bicoid*, the hybrid *oskar* mRNA is located to anterior regions (just as *bicoid* normally is). This mislocalization is sufficient to induce the formation of pole cells at abnormal locations in the early embryo (see Fig. 19-17). In addition,

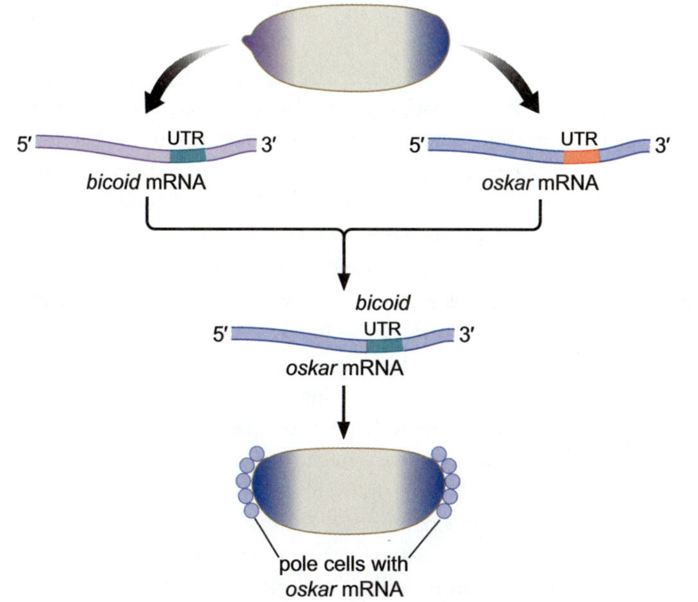

FIGURE 19-17 The *bicoid* and *oskar* mRNAs contain different UTR sequences. The *bicoid* UTR causes it to be localized to the anterior pole, and the distinct *oskar* UTR sequence causes localization in the posterior plasm. An engineered *oskar* mRNA that contains the *bicoid* UTR is localized to the anterior pole, just like the normal *bicoid* mRNA. This mislocalization of *oskar* causes the formation of pole cells in anterior regions. Pole cells also form from the posterior pole because of localization of the normal *oskar* mRNA in the posterior plasm.

the mislocalized polar granules suppress the expression of genes required for the differentiation of head tissues. As a result, embryonic cells that normally form head tissues are transformed into germ cells.

Bicoid and Nanos Regulate *hunchback*

The Bicoid regulatory protein is synthesized prior to the completion of cellularization. As a result, it diffuses away from its source of synthesis at the anterior pole and becomes distributed in a broad concentration gradient along the length of the early embryo. Both high and intermediate concentrations of Bicoid are sufficient to activate *hunchback*, which is essential for the subdivision of the embryo into a series of segments (Fig. 19-18). The *hunchback* gene is actually transcribed from two promoters: one is activated by the Bicoid gradient, and the other controls expression in the developing oocyte. The latter, "maternal" promoter leads to the synthesis of a *hunchback* mRNA that is evenly distributed throughout the cytoplasm of unfertilized eggs. The translation of this maternal transcript is blocked in posterior regions by an RNA-binding protein called **Nanos** (Fig. 19-18). Nanos is found only in posterior regions because its mRNA is, in turn, selectively localized there through interactions between its 3′ UTR and the polar granules we encountered earlier.

Nanos protein binds specific RNA sequences, NREs (Nanos response elements), located in the 3′ UTR of the maternal *hunchback* mRNAs, and this binding causes a reduction in the *hunchback*

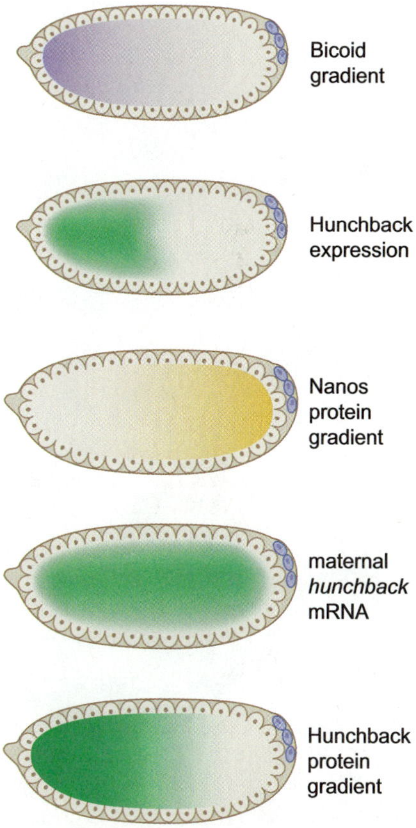

FIGURE 19-18 Hunchback protein gradient and translation inhibition by Nanos. The broad anteroposterior Bicoid protein gradient produces a sharp threshold of *hunchback* gene expression, as *hunchback* is activated by both high and intermediate levels of the Bicoid gradient. The *Nanos* mRNA is associated with polar granules; after its translation, the protein diffuses from posterior regions to form a gradient. The maternal *hunchback* mRNA is distributed throughout the early embryo, but its translation is arrested by the Nanos protein, which binds to specific sequences in the *hunchback* 3′ UTR. The Nanos gradient thereby leads to the formation of a reciprocal Hunchback gradient in anterior regions.

KEY EXPERIMENTS

Box 19-5 The Role of Activator Synergy in Development

Perhaps as little as a twofold difference in the levels of the Dorsal protein determine whether a naïve embryonic cell forms a muscle cell or neuron. This regulatory switch in cell identity depends on the sharp lateral limits of the Snail expression pattern, which demarcate the boundary between the presumptive mesoderm and neurogenic ectoderm (Box 19-5 Fig. 1). Cells that express Snail invaginate to form mesoderm, whereas cells located in more lateral regions (and lack Snail expression) form derivatives of the neurogenic ectoderm.

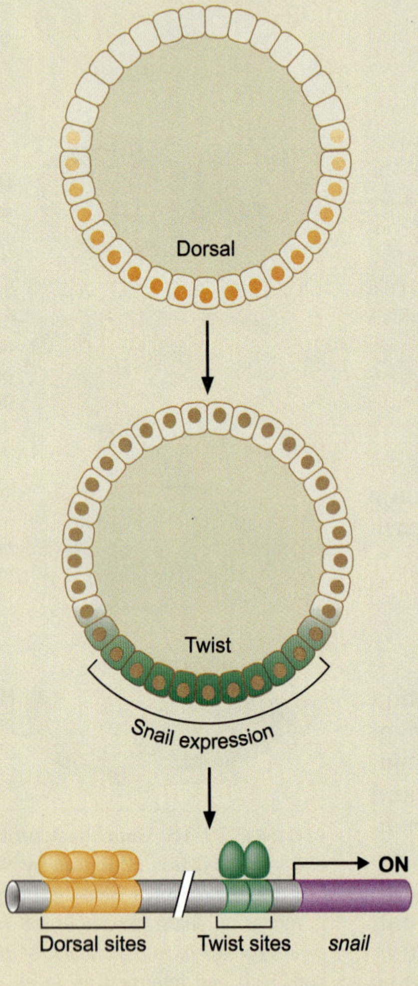

Box 19-5 Figure 1 Model for Dorsal–Twist synergy. The broad Dorsal nuclear gradient activates the twist gene in ventral regions. The Dorsal and Twist proteins work synergistically to activate a variety of genes in ventral and ventrolateral regions. It has been suggested thar Dorsal recruits chromatin-modifying complexes, whereas Twist stimulates transcription by interacting with Mediator or TFIID complexes. (Redrawn from Stathopoulos A. and Levine M. 2002. *Dev. Biol.* 246: 57–67, Fig. 2, p. 59. © Elsevier.)

The formation of the sharp Snail borders depends, in part, on the multiplication of the Dorsal and Twist gradients. The idea is that the broad Dorsal gradient triggers a slightly steeper Twist pattern, and then the Dorsal and Twist proteins function synergistically within the limits of the *snail* 5′ regulatory DNA to activate expression (Box 19-5 Fig. 1).

There is a cluster of low-affinity Dorsal sites located about 1 kb upstream of the transcription start site of the *snail* gene and two Twist-binding sites near the *snail* promoter. Because of the distance separating these sites, it is unlikely that Dorsal and Twist physically interact to facilitate cooperative binding to DNA. Instead, they might make separate contacts with different rate-limiting transcription complexes ("promiscuous synergy," see Chapter 17). For example, Dorsal might render the *snail* 5′ regulatory region in an "open" conformation by recruiting an enzymatic complex that modifies chromatin, such as SWI/SNF or HAT. This opening of the *snail* 5′ regulatory region might facilitate the binding of Twist, which subsequently recruits the TFIID–Pol II complex to the core promoter (see Chapter 17). We see later in this chapter that Bicoid and Hunchback function in a synergistic fashion to activate *eve* stripe 2. A similar principle is used to specify the dorsal mesoderm in a vertebrate embryo, as we now discuss.

The dorsal mesoderm of the *Xenopus* embryo is the source of important signaling molecules that control the development of the CNS during gastrulation. The formation of the dorsal mesoderm depends on localized mRNAs in the unfertilized egg, including *VegT*. The *VegT* gene encodes a sequence-specific transcription factor that leads to the activation of the *Xnr* gene throughout the presumptive mesoderm. *Xnr* encodes a transforming growth factor-β (TGF-β) signaling molecule that is necessary but not sufficient to activate gene expression within the dorsal mesoderm. Instead, activation depends on Xnr and Wnt signaling.

After fertilization, a process called cortical rotation occurs, during which the internal cytoplasm of the egg rotates relative to the plasma membrane (Box 19-5 Fig. 2a). Cortical rotation leads to the stabilization of β-catenin along one side of the early embryo, which corresponds to the future dorsal surface. A cell-surface protein, β-catenin, is normally released into nuclei upon activation of Frizzled receptors by secreted, extracellular signaling proteins called Wnts. However, cortical rotation may circumvent the need for Wnts and directly induces Frizzled receptors to release β-catenin. Once in the nucleus, β-catenin interacts with a sequence-specific transcription factor, called Tcf or Pangolin.

The Tcf/β-catenin complex activates a target gene called *siamois*, which encodes a homeodomain regulatory protein. Siamois expression is distributed throughout dorsal regions, where there are high levels of β-catenin. This Siamois expression profile intersects with the Xnr signaling molecules distributed throughout the mesoderm (Box 19-5 Fig. 2b). The point of

Box 19-5 (*Continued*)

intersection corresponds to the dorsal mesoderm; Siamois functions synergistically with Xnr to activate target genes in the dorsal mesoderm. One of the first genes to be activated is called *goosecoid*, which encodes a homeodomain regulatory protein.

The 5′ regulatory DNA of the *goosecoid* gene contains binding sites for Siamois as well as for "Smad" proteins. Smads are transcription factors that are induced by the activation of TGF-β cell surface receptors (Box 5 Fig. 2b). In the absence of signaling, Smads are inactive because of their association with the intracytoplasmic domains of the TGF-β receptors at the cell surface. Upon signaling, however, the Smads are released into nuclei. This results in the binding of Smads to the *goosecoid* 5′ regulatory DNA. Smads and Siamois now function synergistically to activate *goosecoid* expression within the dorsal mesoderm. The site of expression corresponds to the one region of the embryo where there are high levels of both activators.

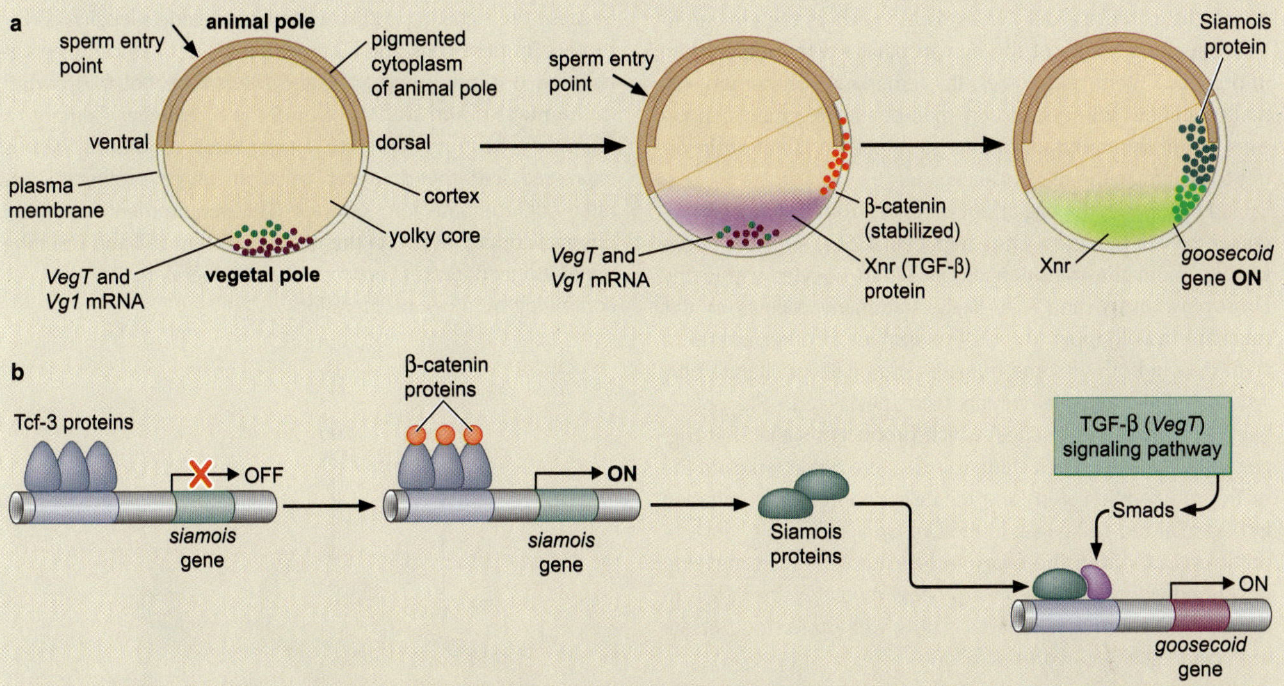

Box 19-5 Figure 2 Specification of the dorsal mesoderm in the *Xenopus* embryo. (a) The *Xenopus* egg contains a number of localized mRNAs including *VegT* and *Vg1*. *VegT* encodes a T-box DNA-binding protein, whereas *Vg1* encodes an activin/TGF-β signaling molecule. They lead to the expression of Xnr in vegetal regions. Cortical rotation occurs after fertilization and leads to the stabilization of β-catenin along the future dorsal surface. The point of intersection between the Xnr and β-catenin domains defines the dorsal mesoderm and leads to the activation of a number of genes such as *goosecoid*. (b) β-catenin in dorsal regions leads to the activation of the *siamois* gene, which encodes a homeobox regulatory protein. The Xnr signaling molecule leads to the activation of another class of regulatory proteins, Smads. Both regulatory proteins, Smads and Siamois, are located only in the dorsal mesoderm. In this region, they work synergistically to activate the *goosecoid* gene. (a, Adapted, with permission, from Alberts B. et al. 2002. *Molecular biology of the cell*, 4th ed., p. 1211, Fig. 21.66. © Garland Science/Taylor & Francis Books LLC. b, Adapted from Gilbert S.E. 2000. *Developmental biology*, 6th ed, p. 322, Fig. 1025. © Sinauer; and, with permission, from Moon R. and Kimelman D. 1998. *BioEssays* 20: 542, Fig. 3. © Wiley Liss.)

poly-A tail, which in turn destabilizes the RNA and inhibits its translation (see Chapter 14). Thus, we see that the Bicoid gradient activates the zygotic *hunchback* promoter in the anterior half of the embryo, whereas Nanos inhibits the translation of the maternal *hunchback* mRNA in posterior regions (see Fig. 19-18). This dual regulation of *hunchback* expression produces a steep Hunchback protein gradient, with the highest concentrations located in the anterior half of the embryo and sharply diminishing levels in the posterior half.

MEDICAL CONNECTIONS

Box 19-6 Stem Cells

Stem cells are often "pluripotent," that is, they have the capability to form many different specialized cell types. Indeed, embryonic stem cells in mammalian embryos, cells from the inner cell mass (ICM), give rise to all the complex tissues seen in adults, including diverse blood cells and the different cell types comprising skin, muscles, and inner organs. The considerable interest in and attention now given to stem cells is due in large part to their potential for therapeutic uses in human medicine (see Box 17-6). One day, it may be possible to use stem cells to replace our "worn parts," such as the critical insulin-secreting β-cells of the human pancreas that deteriorate in diabetics. Stem cells possess the remarkable property of virtually limitless self-renewal. In therapeutic treatments, however, it will be important to limit stem cell proliferation in order to minimize the risk of cancer.

In *Drosophila*, the egg or oocyte arises from a stem cell precursor called the germ-line stem cell (GSC). Quite a lot is known about the transition of GSCs into oocytes within the *Drosophila* ovary, and it is likely that many aspects of this mechanism will apply to the development of other classes of stem cells in both flies and humans. Stem cells proliferate only when in direct physical contact with specialized cells, collectively known as the "niche," which produce a signal that triggers proliferation. When stem cells become detached from the niche, proliferation stops and the cells undergo differentiation into specialized cell types. In the *Drosophila* example, detachment of GSCs from the ovary niche causes them to develop into nondividing oocytes, in a process mediated by signal-induced repression. This process is now well understood at the molecular level and works as follows.

Niche cells within the *Drosophila* ovary, called Cap cells, secrete a diffusible signaling molecule called Dpp. Activation of the Dpp receptor within the associated GSCs results in silencing of a critical regulatory gene called *bam*: When transcription of *bam* is blocked, GSCs proliferate. This silencing of *bam* expression depends on direct physical contact between Cap cells and GSCs, similar to the process that results in activation of Notch signaling during formation of the insect nervous system. As GSCs proliferate, some of the daughter cells become detached from the Cap cells and so are no longer targets of Dpp signaling. In the absence of signaling, *bam* transcription is activated and the cell stops proliferating; instead it differentiates into an oocyte (Box 19-6 Fig. 1).

The basic choice between stem cell proliferation and oocyte differentiation therefore depends on the ON/OFF regulation of *bam* expression. And this regulation is now known to be mediated by a silencer element in the 5′ regulatory region of *bam*, having the sequence GRCGNC(N)$_5$GT CTG (Box 19-6 Fig. 2). Dpp signaling triggers nuclear transport of two Smad regulatory proteins, called Mad and Medea. These proteins bind the two half-sites in the silencer element and, in turn, recruit a transcriptional repressor, called ZF6-6 or Schnurri, that prevents transcription of *bam*. This recruitment of Schnurri and consequent repression of *bam* occurs only in GSCs that remain in contact with the Cap cells. As a result, these cells divide to produce more stem cells. In contrast, in GSC daughter cells that detach from the Cap cells, *bam* is actively transcribed because the signaling pathway leading to gene silencing is disrupted. In these cells, the Dpp receptor is not activated (as signaling is disrupted) and Mad and Medea are not transported to the nucleus and so do not bind the 5′ silencer element or recruit the Schnurri repressor. Under these conditions, *bam* is expressed and the daughter cells no longer proliferate but rather differentiate into oocytes. This requirement for direct physical contact between the niche and stem cell and resulting signal-induced repression may be a general mechanism for continuing stem cell proliferation.

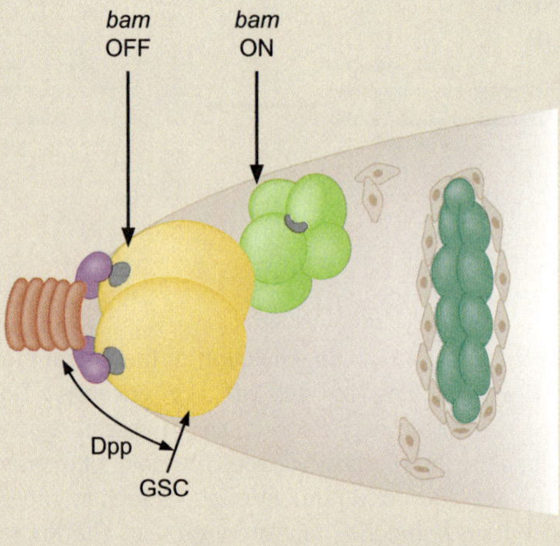

BOX 19-6 FIGURE 1 *bam* expression in developing oocytes. The scheme represents the patterns of expression and distribution of *bam* mRNA and protein. Cap cells (shown in purple) secrete Dpp, which activates its receptor on germ-line stem cells (GSCs shown in yellow), resulting in a signaling process that ultimately represses *bam* expression. As GSCs detach from the Cap cells, Dpp signaling is lost and *bam* mRNA is expressed, leading to production of high levels of its protein in the cytoplasm. In the presence of Bam protein, the detached daughter cells develop into oocyte progenitor (green) and further into eight-cell cysts (dark green). (Adapted, with permission, from Chen D. and McKearin D.M. 2003. *Development* 130: 1159–1170, Fig. 1. © Company of Biologists.)

Box 19-6 *(Continued)*

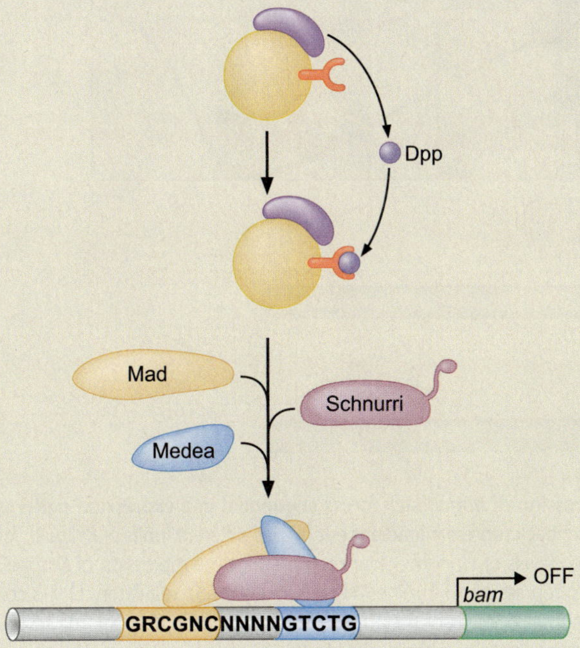

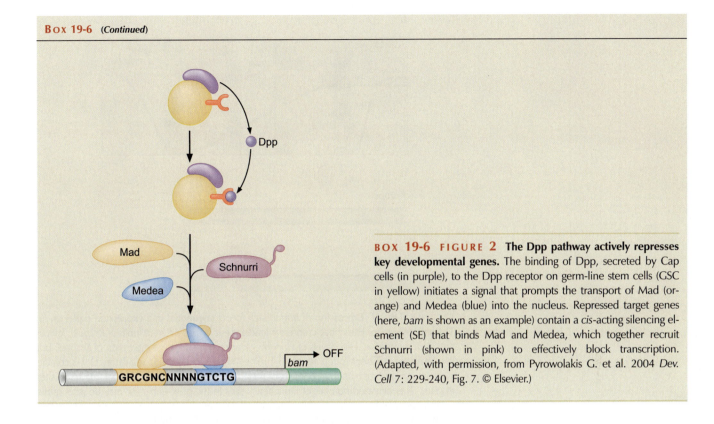

Box 19-6 Figure 2 The Dpp pathway actively represses key developmental genes. The binding of Dpp, secreted by Cap cells (in purple), to the Dpp receptor on germ-line stem cells (GSC in yellow) initiates a signal that prompts the transport of Mad (orange) and Medea (blue) into the nucleus. Repressed target genes (here, *bam* is shown as an example) contain a *cis*-acting silencing element (SE) that binds Mad and Medea, which together recruit Schnurri (shown in pink) to effectively block transcription. (Adapted, with permission, from Pyrowolakis G. et al. 2004 *Dev. Cell* 7: 229-240, Fig. 7. © Elsevier.)

The Gradient of Hunchback Repressor Establishes Different Limits of Gap Gene Expression

Hunchback functions as a transcriptional repressor to establish different limits of expression of the so-called "gap" genes, *Krüppel*, *knirps*, and *giant* (discussed in Box 19-4). We will see that Hunchback also works in concert with the proteins encoded by these gap genes to produce segmentation stripes of gene expression, the first step in subdividing the embryo into a repeating series of body segments.

The Hunchback protein is distributed in a steep gradient that extends through the presumptive thorax and into the abdomen. High levels of the Hunchback protein repress the transcription of *Krüppel*, whereas intermediate and low levels of the protein repress the expression of *knirps* and *giant*, respectively (Fig. 19-19a). We have seen that the binding affinities of the Dorsal activator is responsible for producing different thresholds of gene expression. The Hunchback repressor gradient might not work in the same way. Instead, the *number* of Hunchback repressor sites may be a more critical determinant for distinct patterns of *Krüppel*, *knirps*, and *giant* expression (Fig. 19-19b). The *Krüppel* enhancer contains only three Hunchback-binding sites and is repressed by high levels of the Hunchback gradient. In contrast, the *giant* enhancer contains seven Hunchback sites and is repressed by low levels of the Hunchback gradient. The underlying mechanism here is unknown. Perhaps different thresholds of repression are produced by the additive effects of the individual Hunchback repression domains.

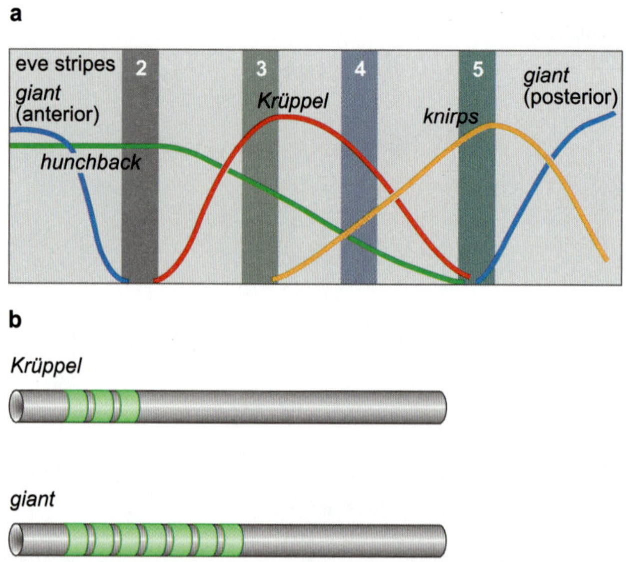

FIGURE 19-19 Expression of *hunchback* forms sequential gap expression patterns. (a) The anteroposterior Hunchback repressor gradient establishes different limits of *Krüppel, knirps,* and *giant* expression. High levels of Hunchback are required for the repression of *Krüppel,* but low levels are sufficient to repress *giant.* (b) The *Krüppel* and *giant* 5´ regulatory DNAs contain different numbers of Hunchback repressor sites. There are three sites in *Krüppel,* but seven sites in *giant.* The increased number of Hunchback sites in the *giant* enhancer may be responsible for its repression by low levels of the Hunchback gradient. (a, Redrawn, with permission, from Gilbert S.E. 1997. *Developmental biology,* 5th ed., p. 565, Fig. 14-23. © Sinauer.)

Hunchback and Gap Proteins Produce Segmentation Stripes of Gene Expression

A culminating event in the regulatory cascade that begins with the localized *bicoid* and *oskar* mRNAs is the expression of a "**pair-rule**" gene called *even-skipped*, or simply *eve*. The *eve* gene is expressed in a series of seven alternating, or "pair-rule," stripes that extend along the length of the embryo (Fig. 19-20). Each eve stripe encompasses four cells, and neighboring stripes are separated by "interstripe" regions—also four cells wide—that express little or no *eve*. These stripes foreshadow the subdivision of the embryo into a repeating series of body segments.

The *eve* protein-coding sequence is rather small, less than 2 kb in length. In contrast, the flanking regulatory DNAs that control *eve* expression encompass more than 12 kb of genomic DNA: about 4 kb located 5´ of the eve transcription start site, and about 8 kb in the 3´-flanking region (see Fig. 19-20). The 5´ regulatory region is responsible for initiating stripes 2, 3, and 7, and the 3´ region regulates stripes 1, 4, 5, and 6. The 12 kb of regulatory DNA contains five separate enhancers that together produce the seven different stripes of *eve* expression seen in the early embryo. Each enhancer initiates the expression of just one or two stripes. (In Box 19-7, *cis*-regulatory Sequences in Animal Development and Evolution, we discuss further aspects and examples of the modular organization of regulatory elements within animal genomes.) We now consider the regulation of the enhancer that controls the expression of *eve* stripe 2.

The stripe-2 enhancer is 500 bp in length and located 1 kb upstream of the eve transcription start site. It contains binding sites for four differ-

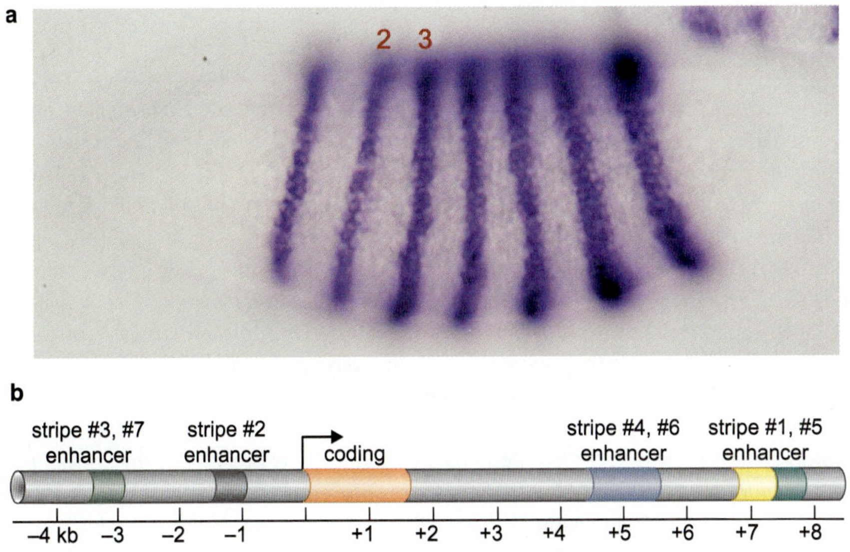

FIGURE 19-20 Expression of the *eve* gene in the developing embryo. (a) *eve* expression pattern in the early embryo. (b) The *eve* locus contains more than 12 kb of regulatory DNA. The 5´ regulatory region contains two enhancers, which control the expression of stripes 2, 3, and 7. Each enhancer is 500 bp in length. The 3´ regulatory region contains three enhancers which control the expression of stripes 4 and 6, stripe 1, and stripe 5, respectively. The five enhancers produce seven stripes of *eve* expression in the early embryo. (a, Image courtesy of Michael Levine.)

ent regulatory proteins: Bicoid, Hunchback, Giant, and Krüppel (Fig. 19-21). We have seen how Hunchback functions as a repressor when controlling the expression of the gap genes; in the context of the *eve* stripe-2 enhancer, it works as an activator. In principle, Bicoid and Hunchback can activate the stripe-2 enhancer in the entire anterior half of the embryo because both proteins are present there, but Giant and Krüppel function as repressors that establish the edges of the stripe-2 pattern—the anterior and posterior borders, respectively (see Fig. 19-21).

Gap Repressor Gradients Produce Many Stripes of Gene Expression

eve stripe-2 is formed by the interplay of broadly distributed activators (Bicoid and Hunchback) and localized repressors (Giant and Krüppel). The same basic mechanism applies to the regulation of the other *eve*

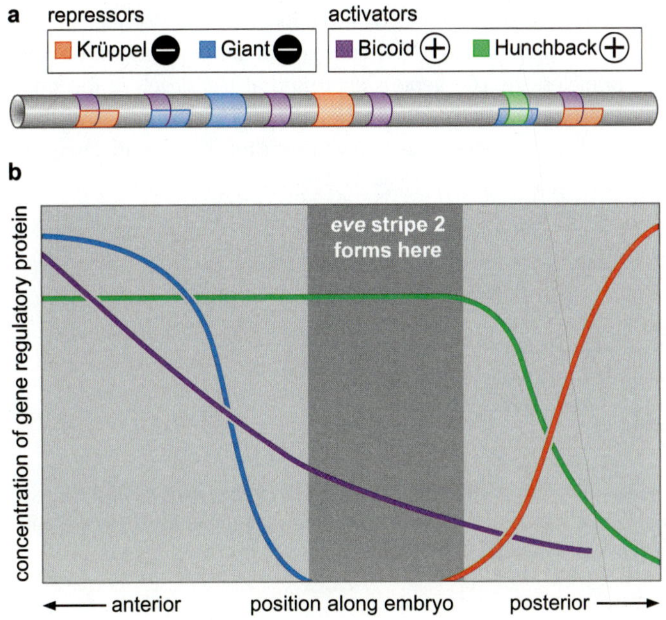

FIGURE 19-21 Regulation of *eve* stripe 2. (a) The 500-bp enhancer contains a total of 12 binding sites for the Bicoid, Hunchback, Krüppel, and Giant proteins. The distributions of these regulatory proteins in the early *Drosophila* embryo is summarized in the diagram shown in b. There are high levels of the Bicoid and Hunchback proteins in the cells that express *eve* stripe 2. The borders of the stripes are formed by the Giant and Krüppel repressors. (Giant is expressed in anterior and posterior regions. Only the anterior pattern is shown; the posterior pattern, which is regulated by Hunchback, is not shown.) (Adapted, with permission, from Alberts B. et al. 2002. *Molecular biology of the cell,* 4th ed.: a, p. 409, Fig. 7-55; b, p. 410, Fig. 7-56. © Garland Science/Taylor & Francis LLC.)

KEY EXPERIMENTS

Box 19-7 *cis*-Regulatory Sequences in Animal Development and Evolution

cis-regulatory sequences are organized in a modular fashion within animal genomes. In general, there are separate enhancers for the individual components of a complex expression pattern. Consider a gene that is expressed in multiple tissues and organs within a developing mouse embryo, such as the liver, pancreas, and pituitary gland. Odds are that the gene contains separate enhancers for each of these sites of expression. We have seen that the eve locus contains five separate enhancers located in the 5´- and 3´-flanking regions (see Fig. 19-20). Each enhancer directs the expression of just one or two of the seven eve stripes in the early *Drosophila* embryo. This type of modular organization facilitates morphological diversity via evolution of *cis*-regulatory sequences, as we discuss below.

Modular Organization Circumvents Pleiotropy

How do patterns of gene expression change during evolution? There is emerging evidence that nucleotide changes within critical activator binding sites eliminate gene expression within a specific tissue or cell type during evolution. Consider the example of pelvic fins in stickleback fish. There are natural variants of sticklebacks that lack pelvic fins. When mated with individuals containing fins, it was possible to identify a major genetic locus responsible for reduced fins. It maps within the 5´-flanking region of the *Pitx1* gene. *Pitx1* is a developmental control gene that is essential for the development of several different tissues in mice, including the thymus, olfactory pit, and hindlimbs. In sticklebacks, it would appear that reduced fins result from point mutations in critical activator sites within the pelvic fin ("hindlimb") enhancer (Box 19-7 Fig. 1). These mutations disrupt expression in the developing pelvic fins, but they do not interfere with the activities of the other enhancers required for regulating *Pitx1* in the thymus, olfactory pit, and other tissues where the *Pitx1* gene is active.

Specific alterations within a modular, *cis*-regulatory region are also responsible for the evolution of distinct pigmentation patterns in different species of *Drosophila*. The classical *yellow* (*y*) locus is critical for pigmentation, and simple mutations in the gene result in flies with a yellow body color that lack localized foci of melanin. The *y* gene is regulated by separate enhancers for expression in the bristles, wings, and abdomen, as we now describe.

D. melanogaster adults (particularly males) contain intense pigmentation in the posterior abdominal segments. This pigmentation is due to the direct activation of the *y* abdominal enhancer by the Hox protein, Abd-B. Drosophilids lacking abdominal segmentation, such as *Drosophila kikkawai*, contain point mutations in a critical Abd-B activator site. This causes a loss of *y* expression in the abdomen and the observed loss of pigmentation.

A separate enhancer controls *y* expression in the wings. In some *Drosophila* species, this enhancer directs a spot of pigmentation in a specific quadrant of the adult male wing (Box 19-7 Fig. 2). This spot is a critical component of the courtship ritual. Species lacking the mating spot contain point mutations in the wing enhancer, causing the restricted loss of *y* gene activity without compromising its function in other tissues such as the bristles and abdominal cuticle.

Changes in Repressor Sites Can Produce Big Changes in Gene Expression

The simple loss of critical activator sites within discrete enhancer modules can explain the localized loss of *Pitx1* and *y* gene activities. New patterns of gene expression might arise through the loss of repressor elements.

Most or all of the enhancers active in the early *Drosophila* embryo have repressor binding sites that are responsible for creating sharp boundaries of gene expression. For example, the eve stripe-2 enhancer contains binding sites for the Giant and Krüppel repressors, which produce sharp anterior and posterior borders of gene expression (see Fig. 19-21). Mutations in these sites cause a dramatic expansion in the normal expression pattern: a broad band of expression rather than a tight stripe.

A possible example of evolution via repressor elements is seen for the lactase (*LCT*) gene in human populations. In most primates, the *LCT* gene is expressed at high levels in the small intestines of infants, during the time they obtain milk from their mothers. However, the *LCT* gene is shut off after adolescence.

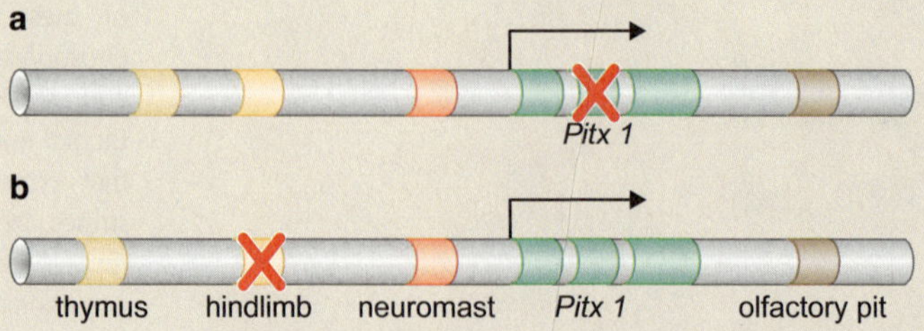

BOX 19-7 FIGURE 1 **The developmental control gene *Pitx1*.** (a) The panel shows the structure of the *Pitx1* gene with 5´ upstream sequences. Shown here is a lethal null mutation (of a laboratory mouse) within the coding region (second exon) of the gene. (b) In the wild stickleback, a viable regulatory mutation within the 5´ upstream sequence results in reduced pelvic fin size.

Box 19-7 *(Continued)*

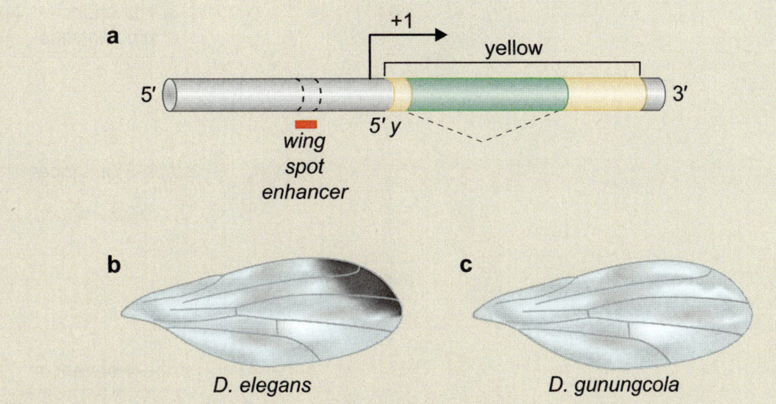

Box 19-7 Figure 2 The *yellow* (*y*) locus of *Drosophila*. (a) The panel shows the structure and upstream regulatory sequences (enhancer sequences) of the *yellow* gene. (b) The normal pigmentation (the "mating spot") of the adult male wing in one species of *Drosophila*. (c) The wing of a species lacking pigmentation; these organisms carry a mutation in the 5′ spot enhancer.

Certain populations of humans are unusual in retaining *LCT* gene expression as adults. This persistence correlates with pastoral societies that use dietary milk long after weaning. Individual populations with persistent *LCT* expression contain nucleotide substitutions in an intronic sequence within the *MCM6* gene, located immediately 5′ of *LCT* (Box 19-7 Fig. 3).

These nucleotide changes might damage repressor elements that normally bind a silencer protein responsible for repressing *LCT* expression in the small intestines of adolescents and adults. Such a loss of critical *cis*-regulatory elements would be comparable to the inactivation of the hindlimb/pelvic fin enhancer in the *Pitx1* gene in sticklebacks or the inactivation of the abdominal and wing enhancers in the *y* gene of *Drosophila*. But in the case of the lactase gene, a novel pattern of gene expression is evolved, temporal persistence of *LCT* activity, because of the loss of repression elements.

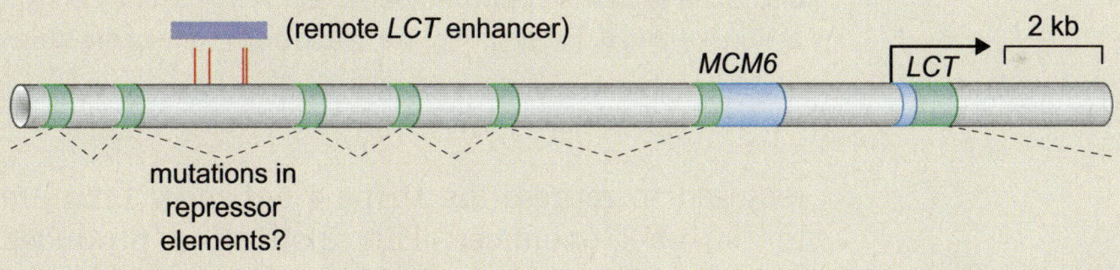

Box 19-7 Figure 3 Structure of the *LCT* gene and its 5′ upstream regulatory region.

enhancers as well. For example, the enhancer that directs the expression of *eve* stripe 3 can be activated throughout the early embryo by ubiquitous transcriptional activators. The stripe borders are defined by localized gap repressors: Hunchback establishes the anterior border, whereas Knirps specifies the posterior border (Fig. 19-22).

The enhancer that controls the expression of *eve* stripe 4 is also repressed by Hunchback and Knirps. However, different concentrations of these repressors are required in each case. Low levels of the Hunchback gradient that are insufficient to repress the *eve* stripe-3 enhancer are sufficient to repress the *eve* stripe-4 enhancer (Fig. 19-22). This differential regulation of the two enhancers by the Hunchback repressor gradient produces distinct anterior borders for the stripe-3 and stripe-4 expression patterns. The Knirps protein is also distributed in a gradient in the precellular embryo. Higher levels of this gradient are

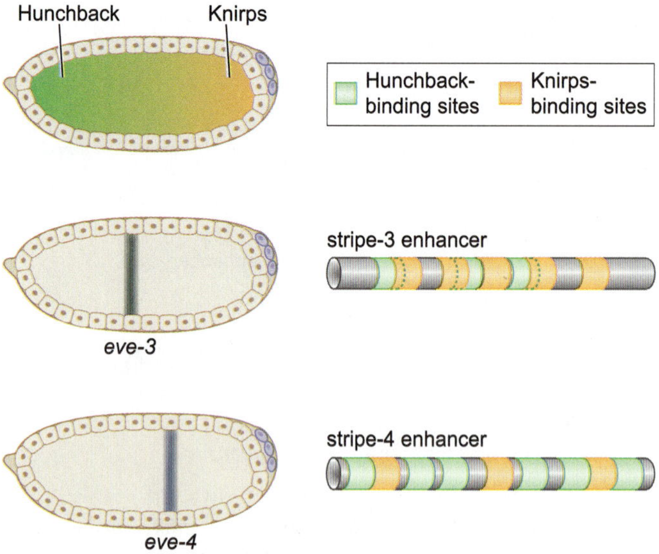

FIGURE 19-22 Differential regulation of the stripe-3 and stripe-4 enhancers by opposing gradients of the Hunchback and Knirps repressors. The two stripes are positioned in different regions of the embryo. The eve stripe-3 enhancer is repressed by high levels of the Hunchback gradient but low levels of the Knirps gradient. Conversely, the stripe-4 enhancer is repressed by low levels of the Hunchback gradient but high levels of Knirps. The stripe-3 enhancer contains just a few Hunchback binding sites, and as a result, high levels of the Hunchback gradient are required for its repression. The stripe-3 enhancer contains many Knirps-binding sites, and consequently, low levels of Knirps are sufficent for repression. The stripe-4 enhancer has the opposite organization of repressor-binding sites. There are many Hunchback sites, and these allow low levels of the Hunchback gradient to repress stripe-4 expression. The stripe-4 enhancer contains just a few Knirps sites, so that high levels of the Knirps gradient are required for repression. Note that the stripe-3 enhancer actually directs the expression of two stripes, 3 and 7. The stripe-4 enhancer directs the expression of stripes 4 and 6. For simplicity, we consider only one of the stripes from each enhancer.

required to repress the stripe-4 enhancer than are needed to repress the stripe-3 enhancer. This distinction produces discrete posterior borders of the stripe-3 and stripe-4 expression patterns.

We have seen that the Hunchback repressor gradient produces different patterns of Krüppel, Knirps, and Giant expression. This differential regulation might be due to the increasing number of Hunchback-binding sites in the Krüppel, Knirps, and Giant enhancers. A similar principle applies to the differential regulation of the stripe-3 and stripe-4 enhancers by the Hunchback and Knirps gradients. The *eve* stripe-3 enhancer contains relatively few Hunchback binding sites but many Knirps sites, whereas the *eve* stripe-4 enhancer contains many Hunchback sites but relatively few Knirps sites (see Fig. 19-22). Similar principles are likely to govern the regulation of the remaining stripe enhancers that control the *eve* expression pattern (as well as the expression of other pair-rule genes).

Short-Range Transcriptional Repressors Permit Different Enhancers to Work Independently of One Another within the Complex *eve* Regulatory Region

We have seen that *eve* expression is regulated in the early embryo by five separate enhancers. In fact, there are additional enhancers that control *eve* expression in the heart and central nervous system (CNS) of older embryos. This type of complex regulation is not a peculiarity of *eve*. There are genetic loci that contain even more enhancers distributed over even larger distances. For example, several genes are known to be regulated by as many as ten different enhancers, perhaps more, that are scattered over distances approaching 100 kb (as we shall discuss below). Thus, genes engaged in important developmental processes are often regulated by multiple enhancers. How do these enhancers work independently of one another to produce additive patterns of gene expression? In the case of *eve*, five seperate enhancers produce seven different stripes.

Short-range transcriptional repression is one mechanism for ensuring enhancer autonomy—the independent action of multiple enhancers to generate additive patterns of gene expression. This means that repressors bound to one enhancer do not interfere with the activators bound to another enhancer within the regulatory region of the same gene. For example, we have seen that the Krüppel repressor binds to the *eve* stripe-2 enhancer and establishes the posterior border of the stripe-2 pattern. The Krüppel repressor works only within the limits of the 500-bp stripe-2 enhancer. It does not repress the core promoter or the activators contained within the stripe-3 enhancer, both of which map more than 1 kb away from the Krüppel repressor sites within the stripe-2 enhancer (Fig. 19-23). If Krüppel were able to function over long distances, then it would interfere with the expression of *eve* stripe 3, because high levels of the Krüppel repressor are present in that region of the embryo where the *eve* stripe-3 enhancer is active.

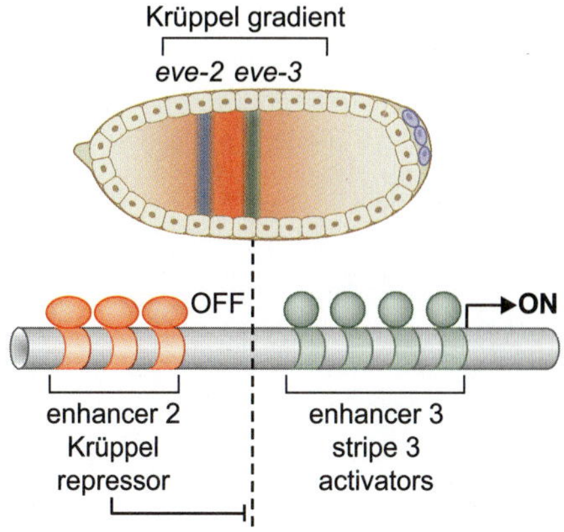

FIGURE 19-23 Short-range repression and enhancer autonomy. Different enhancers work independently of one another in the eve regulatory region because of short-range transcriptional repression. Repressors bound to one enhancer do not interfere with activators in the neighboring enhancers. For example, the Krüppel repressor binds to the stripe-2 enhancer and keeps stripe-2 expression off in central regions of the embryo. The eve stripe-3 enhancer is expressed in these regions. It is not repressed by Krüppel because it lacks the specific DNA sequences that are recognized by the Krüppel protein. In addition, Krüppel repressors bound to the stripe-2 enhancer do not interfere with the stripe-3 activators because they map too far away. Krüppel must bind no more than 100 bp from upstream activators to block their ability to stimulate transcription. The stripe-2 and stripe-3 enhancers are separated by a 1.5-kb spacer sequence.

HOMEOTIC GENES: AN IMPORTANT CLASS OF DEVELOPMENTAL REGULATORS

The genetic analysis of *Drosophila* development led to the discovery of an important class of regulatory genes, the homeotic genes, which cause the morphological diversification of the different body segments. Some homeotic genes control the development of mouth parts and antennae from head segments, whereas others contol the formation of wings and halteres from thoracic segments. The two best-studied homeotic genes are *Antp* and *Ubx*, responsible for suppressing the development of antennae and wings, respectively.

Antp (*Antennapedia*) controls the development of the middle segment of the thorax, the mesothorax. The mesothorax produces a pair of legs that are morphologically distinct from the forelegs and hindlegs. *Antp* encodes a homeodomain regulatory protein that is normally expressed in the mesothorax of the developing embryo. The gene is not expressed, for example, in the developing head tissues. But, a dominant *Antp* mutation, caused by a chromosome inversion, brings the Antp protein-coding sequence under the control of a "foreign" regulatory DNA that mediates gene expression in head tissues, including the antennae (see Fig. 19-24). When misexpressed in the head, *Antp* causes a striking change in morphology: legs develop instead of antennae.

FIGURE 19-24 A dominant mutation in the *Antp* gene results in the homeotic transformation of antennae into legs. The fly on the right is normal. Note the rudimentary set of antennae at the front end of the head. The fly on the left is heterozygous for a dominant *Antp* mutation (*AntpD*/+). It is fully viable and mainly normal in appearance except for the remarkable set of legs emanating from the head in place of antennae. (Courtesy of Matthew Scott.)

Ubx (*Ultrabithorax*) encodes a homeodomain regulatory protein that controls the development of the third thoracic segment, the metathorax. *Ubx* specifically represses the expression of genes that are required for the development of the second thoracic segment, or mesothorax. Indeed, *Antp* is one of the genes that it regulates: Ubx represses *Antp* expression in the metathorax and restricts its expression to the mesothorax of de-

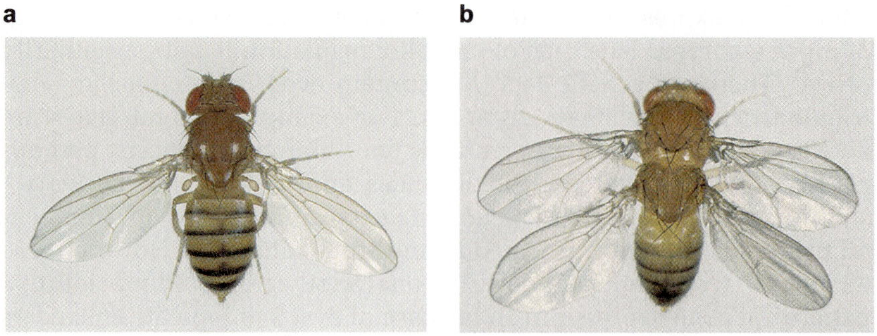

FIGURE 19-25 ***Ubx* mutants cause the transformation of the metathorax into a duplicated mesothorax.** (a) A normal fly is shown that contains a pair of prominent wings and a smaller set of halteres just behind the wings. (b) A mutant that is homozygous for a weak mutation in the *Ubx* gene is shown. The metathorax is transfomred into a duplicated mesothorax. As a result, the fly has two pairs of wings rather than one set of wings and one set of halteres. (Courtesy of E.B. Lewis.)

veloping embryos. Mutants that lack the Ubx repressor exhibit an abnormal pattern of *Antp* expression. The gene is not only expressed within its normal site of action in the developing mesothorax but also misexpressed in the developing metathorax. This misexpression of *Antp* causes a transformation of the metathorax into a duplicated mesothorax.

In adult flies, the mesothorax contains a pair of legs and wings, whereas the metathorax contains a pair of legs and halteres (see Fig. 19-25). The halteres are considerably smaller than the wings and function as balancing structures during flight. *Ubx* mutants exhibit a spectacular phenotype: they have four fully developed wings, because of the transformation of the halteres into wings. This mutant phenotype stems, in part, from the misexpression of *Antp*.

The expression of *Ubx* in the different tissues of the metathorax depends on the regulatory sequences that encompass more than 80 kb of genomic DNA. A mutation called *Cbx* (*Contrabithorax*) disrupts this Ubx regulatory DNA without changing the *Ubx* protein-coding region. The *Cbx* mutation causes *Ubx* to be misexpressed in the mesothorax, in addition to its normal site of expression in the metathorax (Fig. 19-26). Ubx now represses the expression of *Antp*, as well as the other genes needed for the normal development of the mesothorax. As a result, the mesothorax is transformed into a duplicated copy of the normal metathorax. This is a striking phenotype: the wings are transformed into halteres, and the resulting *Cbx* mutant flies look like wingless ants.

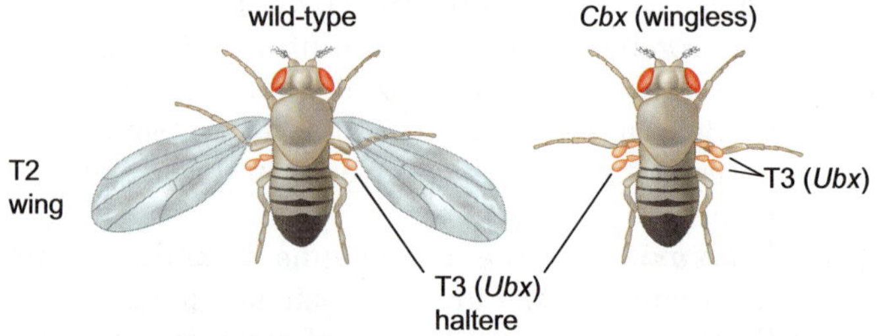

FIGURE 19-26 **Misexpression of *Ubx* in the mesothorax results in the loss of wings.** The *Cbx* mutation disrupts the regulatory region of *Ubx*, causing its misexpression in the mesothorax and results in its transformation into the metathorax.

Changes in Homeotic Gene Expression Are Responsible for Arthropod Diversity

The emerging field known as "evo–devo" lies at the cusp of two traditionally isolated areas of research: evolutionary biology and developmental biology. The impetus for evo–devo research is that genetic analysis of development in flies, nematode worms, and other model organisms has identified the key genes responsible for evolutionary diversity. The homeotic genes represent premiere examples of such genes.

The *Drosophila* genome contains a total of eight homeotic genes organized in two gene clusters or complexes: the Antennapedia complex and the Bithorax complex (see Box 19-8, The Homeotic Genes of *Drosophila* Are Organized in Special Chromosome Clusters). A typical intertebrate genome contains eight to ten homeotic genes, usually located within just one complex. Vertebrates have duplicated the ancestral Hox complex and contain four clusters. Changes in the expression and function of individual homeotic genes are responsible for altering limb morphology in arthropods and the axial skeletons of vertebrates. We conclude this chapter with examples of how changes in Ubx activity have produced evolutionary modifications in insects and other arthropods.

Arthropods Are Remarkably Diverse

Arthropods embrace five groups: trilobites (sadly extinct), hexapods (such as insects), crustaceans (shrimp, lobsters, crabs, and so on), myriapods (centipedes and millipedes), and chelicerates (horseshoe crabs, spiders, and scorpions). The success of the arthropods derives, in part, from their modular architecture. These organisms are composed of a series of repeating body segments that can be modified in seemingly limitless ways. Some segments carry wings, whereas others have antennae, legs, jaws, or specialized mating devices. We know more about the evolutionary processes responsible for the diversification of arthropods than for any other group of animals.

Changes in *Ubx* Expression Explain Modifications in Limbs among the Crustaceans

Crustaceans include most, but not all, of the arthropods that swim. Some live in the ocean, whereas others prefer fresh water. They include some of our favorite culinary dishes, such as shrimp, crab, and lobster. One of the most popular groups of crustaceans for study is *Artemia*, also known as "sea monkeys." Their embryos arrest as tough spores that can be purchased at toy stores. The spores quickly resume development upon addition of salt water.

The heads of these shrimp contain feeding appendages. The thoracic segment nearest the head, T1, contains swimming appendages that look like those further back on the thorax (the second through 11th thoracic segments, T2–T11). *Artemia* belongs to an order of crustaceans known as **branchiopods**. Consider a different order of crustaceans, called **isopods**. Isopods contain swimming limbs on the second through eighth thoracic segments, just like the branchiopods. But, the limbs on the first thoracic segment of isopods have been modified. They are smaller than the others and function as feeding

ADVANCED CONCEPTS

Box 19-8 Homeotic Genes of *Drosphila* Are Organized in Special Chromosome Clusters

Antp and *Ubx* represent only two of the eight homeotic genes in the *Drosophila* genome. The eight homeotic genes of *Drosophila* are located in two clusters, or gene complexes. Five of the eight genes are located within the Antennapedia complex, and the remaining three genes are located within the Bithorax complex (see Box 19-8 Fig. 1). Do not confuse the names of the complex with the individual genes within the complex. For example, the Antennapedia complex is named in honor of the *Antennapedia* (*Antp*) gene, which was the first homeotic gene identified within the complex. There are four other homeotic genes in the Antennapedia complex: *labial* (*lab*), *proboscipedia* (*pb*), *Deformed* (*Dfd*), and *Sex combs reduced* (*Scr*). Similarly, the Bithorax complex is named in honor of the *Ultrabithorax* (*Ubx*) gene, but there are two others in this complex: *abdominal-A* (*abd-A*) and *Abdominal-B* (*Abd-B*). Another insect, the flour beetle, contains a single complex of homeotic genes that includes homologs of all eight homeotic genes contained in the Drosophila Antennapedia and Bithorax complexes. The two complexes probably arose from a chromosomal rearrangement within a single ancestral complex.

There is a colinear correspondence between the order of the homeotic genes along the chromosome and their patterns of expression across the anteroposterior axis in developing embryos (see Box 19-8 Fig. 1). For example, the *lab* gene, located in the 3´-most position of the Antennapedia complex, is expressed in the anteriormost head regions of the developing *Drosophila* embryo. In contrast, the *Abd-B* gene, which is located in the 5´-most position of the Bithorax complex, is expressed in the posteriormost regions (see Box 19-8 Fig. 1). The significance of this colinearity has not been established, but it must be important because it is preserved in each of the major groups of arthropods (including flour beetles), as well as all vertebrates that have been studied, including mice and humans.

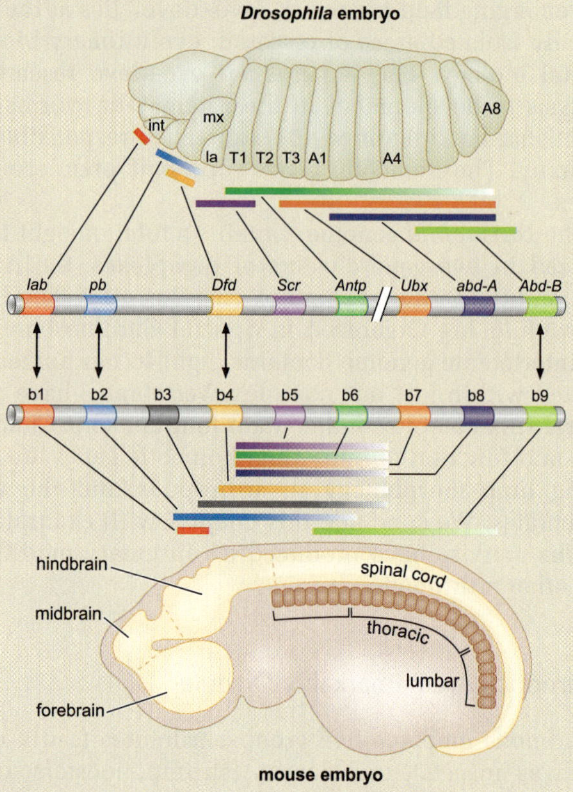

BOX 19-8 FIGURE 1 Organization and expression of *Hox* genes in *Drosophila* and in the mouse. The figure compares the colinear sequences and transcription patterns of the *Hox* genes in *Drosophila* and in the mouse. (Adapted, with permission, from McGinnis W. and Krumlauf R. 1992. *Cell* 68: 283–302, Fig. 2. © Elsevier.)

Mammalian Hox *Gene Complexes Control Anteroposterior Patterning*

Mice contain 38 *Hox* genes arranged within four clusters (Hoxa, Hoxb, Hoxc, Hoxd). Each cluster or complex contains nine or ten *Hox* genes and corresponds to the single homeotic gene cluster in insects that formed the Antennapedia and Bithorax complexes in *Drosophila* (Box 19-8 Fig. 2). For example, the *Hoxa-1* and *Hoxb-1* genes are most closely related to the *lab* gene in *Drosophila*, whereas *Hoxa-9* and *Hoxb-9*—located at the other end of their respective complexes—are similar to the *Abd-B* gene.

In addition to this "serial" homology between mouse and fly *Hox* genes, each mouse Hox complex exhibits the same type of colinearity as that seen in *Drosophila*. For example, *Hox* genes located at the 3´ end of each complex, such as the *Hoxa-1* and *Hoxb-1*, are expressed in the anteriormost regions of developing mouse embryos (future hindbrain). In contrast, *Hox* genes located near the 5´ end of each complex, such as *Hoxa-9* and *Hoxb-9*, are expressed in posterior regions of the embryo (thoracic and lumbar regions of the developing spinal cord). The *Hoxd* complex exhibits sequential expression across the anteroposterior axis of the developing limbs. A comparable pattern is not observed in insect limbs, suggesting that the *Hoxd* genes have acquired "novel" regulatory DNAs during vertebrate evolution. Indeed, we have already seen in Chapter 17 that a specialized "global control region" (GCR) coordinates the expression of the individual *Hoxd* genes in developing limbs.

Altered Patterns of Hox *Expression Create Morphological Diversity in Vertebrates*

Mutations in mammalian *Hox* genes cause disruptions in the axial skeleton, which consists of the spinal cord and the different vertebrae of the backbone. These alterations are evocative of some of the changes in morphology we have seen for the *Antp* and *Ubx* mutants in *Drosophila*.

Consider the *Hoxc-8* gene in mice, which is most closely related to the *abd-A* gene of the *Drosophila* Bithorax complex. It is normally expressed near the boundary between the developing rib cage and lumbar region of the backbone, the ante-

BOX 19-8 (Continued)

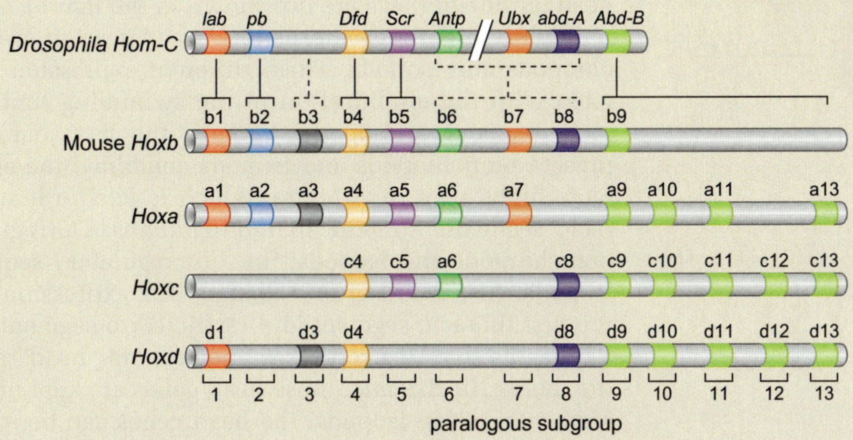

BOX 19-8 FIGURE 2 Conservation of organization and expression of the homeotic gene complexes in *Drosophila* and in the mouse. (Adapted, with permission, from Gilbert S.E. 2000. *Developmental biology*, 6th ed., Fig. 11.36a. © Sinauer.)

rior "tail." (The *abd-A* gene is expressed in the anterior abdomen of the *Drosophila* embryo.) The first lumbar vertebra normally lacks ribs. However, mutant embryos that are homozygous for a knockout mutation in the *Hoxc-8* gene exhibit a dramatic mutant phenotype. The first lumbar vertebra develops an extra pair of vestigial ribs. This type of developmental abnormality is sometimes called a "homeotic" transformation, one in which the proper structure develops in the wrong place. In this case, a vertebra that is typical of the posterior thoracic region develops within the anterior lumbar region.

Maintenance of* Hox *Gene Expression Patterns

Localized patterns of *Hox* gene expression are established in early fly and mouse embryos by combinations of sequence-specific transcriptional activators and repressors. Some of these regulatory proteins are modulated by cell signaling pathways, such as the FGF and Wnt pathways. In *Drosophila*, many of the same gap repressors that establish localized stripes of *eve* expression also control the initial patterns of *Hox* gene expression. These patterns are maintained throughout the life cycle long after the gap repressors are lost.

Consider as an example, the *Abd-B Hox* gene in *Drosophila*. It is specifically expressed in the posterior abdomen, including the primordia of the fifth through eighth abdominal segments. *Abd-B* expression is initially repressed by the Hb, Kr, and Kni gap repressors in the head, thorax, and anterior abdomen of the early *Drosophila* embryo. These are the same repressors that establish localized stripes of eve expression (see Figs. 19-19 and 19-20). These repressors restrict *Abd-B* expression to the posterior abdominal segments.

The maintenance of *Abd-B* expression, as well as the expression of most other *Hox* genes in flies and mammals, depends on a large protein complex, called the polycomb repression complex (PRC). The PRC binds to *Abd-B* regulatory sequences in cells that fail to activate the gene in the early embryo: the progenitors of the head, thorax, and anterior abdomen. In all of these cells, the PRC causes methylation of lysine 27 on histone H3, and this methylation correlates with the repression of the associated *Abd-B* transcription unit. Conversely, a ubiquitous activator complex, the trithorax complex (TRC), binds to *Abd-B* regulatory sequences in cells that express the gene in the early embryo (i.e., the posterior abdominal segments). The binding of the TRC leads to the methylation of lysine 4 on histone H3, and this correlates with active transcription of *Abd-B*.

Thus, the PRC and TRC maintain on/off states of *Hox* gene expression depending on the initial expression patterns of these genes in the early embryo. If a given *Hox* gene is repressed in a particular cell, PRC binds and keeps the gene off in all the descendants of that cell. Conversely, if a given *Hox* gene is activated in a particular cell, then TRC will bind and ensure stable expression of the gene in all of its descendants. TRC and PRC serve to maintain a regulatory "memory" of *Hox* gene expression patterns.

MicroRNAs Modulate* Hox *Activity

Many *Hox* gene complexes contain microRNA (miRNA or miR) genes. For example, the fly *ANT-C* contains *miR-10*, and *BX-C* contains *miR-iab4*. The encoded miRNAs are thought to inhibit or attenuate the synthesis of different Hox proteins. the *iab4* miRNA inhibits Ubx protein synthesis in abdominal tissues. Vertebrate *Hox* complexes also contain miR genes, including *miR-10*. The *miR-196* gene is located in 5′ regions of several vertebrate *Hox* complexes. The encoded miRNA is thought to inhibit the synthesis of the Hoxb8 protein in posterior regions of mouse embryos. See Chapter 18 for more details about how miRNAs block or attenuate protein synthesis.

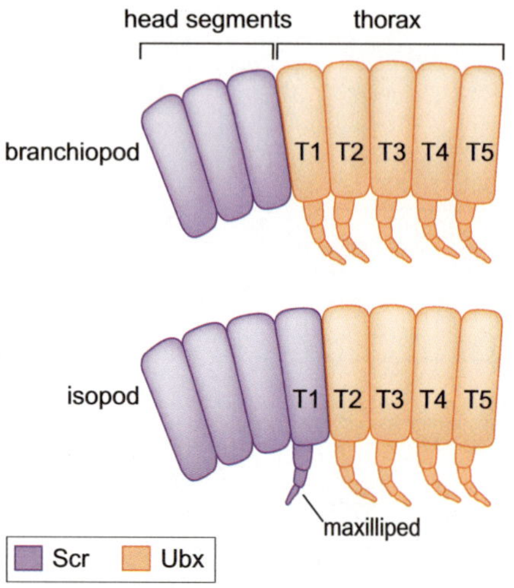

FIGURE 19-27 Changing morphologies in two different groups of crustaceans. In branchiopods, *Scr* expression is restricted to head regions where it helps promote the development of feeding appendages, whereas *Ubx* is expressed in the thorax where it controls the development of swimming limbs. In isopods, *Scr* expression is detected in both the head and the first thoracic segment (T1), and as a result, the swimming limb in T1 is transformed into a feeding appendage (the maxilliped). This posterior expansion of Scr was made possible by the loss of *Ubx* expression in T1 because Ubx normally represses *Scr* expression. (Adapted from Levine M. 2002. *Nature* 415: 848–849, Fig. 2. © Macmillan.)

limbs (Fig. 19-27). These modified limbs are called **maxillipeds** (otherwise known as jaw feet), and look like appendages found on the head (although these are not shown in the figure).

Slightly different patterns of *Ubx* expression are observed in branchiopods and isopods. These different expression patterns are correlated with the modification of the swimming limbs on the first thoracic segment of isopods. Perhaps the last shared ancestor of the present branchiopods and isopods contained the arrangement of thoracic limbs seen in *Artemia* (which is itself a branchiopod): all thoracic segments contain swimming limbs. During the divergence of branchiopods and isopods, the *Ubx* regulatory sequences changed in isopods. As a result of this change, *Ubx* expression was eliminated in the first thoracic segment and restricted to segments T2–T8. It is easy to imagine that Ubx represses one or more "head" patterning genes in the thorax. In *Artemia*, these head genes are kept off in all 11 thoracic segments, but in isopods, the head genes can be expressed in the T1 segment because of the loss of the *Ubx* repressor. Indeed, expression of the *Scr* gene is restricted to head regions of branchiopods, but it is expressed in T1 of isopods. The expression of *Scr* in T1 causes maxillipeds to develop in place of normal swimming limbs (see Fig. 19-27).

What is the basis for the different patterns of *Ubx* expression in isopods and branchiopods? There are several possible explanations, but the most likely one is that the *Ubx* regulatory DNA of isopods acquired mutations. By this model, the *Ubx* enhancer no longer mediates expression in the first thoracic segment. In fact, there is a tight correlation between the absence of *Ubx* expression in the thorax and the development of feeding appendages in different crustaceans. For example, lobster embryos lack *Ubx* expression in the first two thoracic segments and contain two pairs of maxillipeds. Cleaner shrimp lack *Ubx* expression in the first three thoracic segments and contain three pairs of maxillipeds.

Why Insects Lack Abdominal Limbs

All insects have six legs, two on each of the three thoracic segments; this applies to every one of the more than 1 million species of insects. In contrast, other arthropods, such as crustaceans, have a variable number of limbs. Some crustaceans have limbs on every segment in both the thorax and abdomen. This evolutionary change in morphology, the loss of limbs on the abdomen of insects, is not due to altered expression of pattern-determining genes, as seen in the case of maxilliped formation in isopods. Rather, the loss of abdominal limbs in insects is due to functional changes in the Ubx regulatory protein.

In insects, *Ubx* and *abd-A* repress the expression of a critical gene that is required for the development of limbs, called *Distal-less* (*Dll*). In developing *Drosophila* embryos, *Ubx* is expressed at high levels in the metathorax and anterior abdominal segments; *abd-A* expression extends into more posterior abdominal segments. Together, *Ubx* and *abd-A* keep *Dll* off in the first seven abdominal segments. Although *Ubx* is expressed in the metathorax, it does not interfere with the expression of *Dll* in that segment, because *Ubx* is not expressed in the developing T3 legs until after the time when *Dll* is activated. As a result, *Ubx* does not interfere with limb development in T3.

In crustaceans, such as the branchiopod *Artemia* already mentioned, there are high levels of both *Ubx* and *Dll* in all 11 thoracic segments. The expression of *Dll* promotes the development of swimming

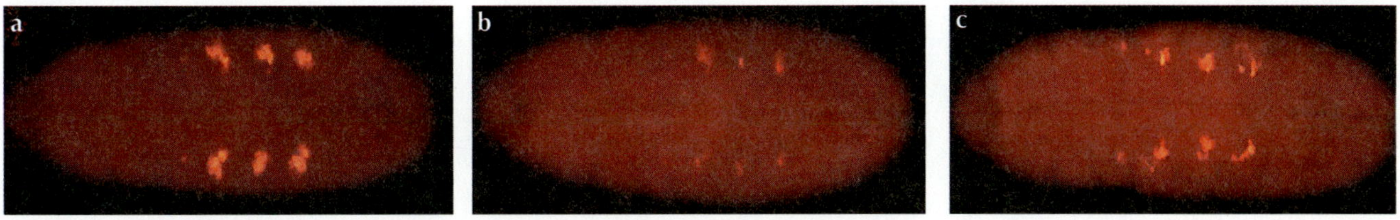

FIGURE 19-28 Evolutionary changes in Ubx protein function. (a) The *Dll* enhancer (*Dll*304) is normally activated in three pairs of "spots" in *Drosophila* embryos. These spots go on to form the three pairs of legs in the adult fly. (b) The misexpression of the *Drosophila* Ubx protein (DmUbxHA) strongly suppresses expression from the *Dll* enhancer. (c) In contrast, the misexpression of the Ubx protein from the brine shrimp *Artemia* (AfUbxHA) causes only a slight suppression of the *Dll* enhancer. (Adapted, with permission, from Ronshaugen M. et al. 2002. *Nature* 415: 914-917, Fig. 2c. © Macmillan. Images courtesy of William McGinnis and Matt Ronshaugen.)

limbs. Why does Ubx repress *Dll* expression in the abdominal segments of insects but not crustaceans? The answer is that the Ubx protein has diverged between insects and crustaceans. This was demonstrated in the following experiment.

The misexpression of *Ubx* throughout all of the tissues of the presumptive thorax in transgenic *Drosophila* embryos suppresses limb development because of the repression of *Dll* (Fig. 19-28). In contrast, the misexpression of the crustacean Ubx protein in transgenic flies does not interfere with *Dll* gene expression and the formation of thoracic limbs. These observations indicate that the *Drosophila* Ubx protein is functionally distinct from Ubx in crustaceans. The fly protein represses *Dll* gene expression, whereas the crustacean Ubx protein does not.

What is the basis for this functional difference between the two Ubx proteins? (They share only 32% overall amino acid identity, but their homeodomains are virtually identical—59/60 matches.) It turns out that the crustacean protein has a short motif containing 29 amino acid residues that blocks repression activity. When this sequence is deleted, the crustacean Ubx protein is just as effective as the fly protein at repressing *Dll* gene expression (Fig. 19-29).

Both the crustacean and fly Ubx proteins contain multiple repression domains. As discussed in Chapter 17, it is likely that these domains interact with one or more transcriptional repression complexes. The "antirepression" peptide present in the crustacean Ubx protein might interfere with the ability of the repression domains to recruit these complexes. When this peptide is attached to the fly protein, the hybrid protein behaves like the crustacean Ubx protein and no longer represses *Dll*.

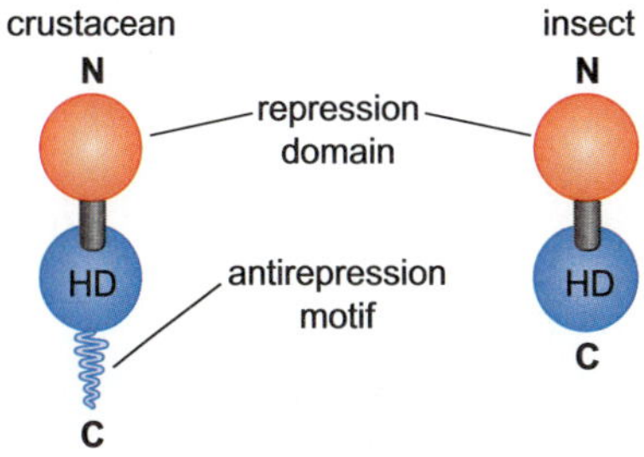

FIGURE 19-29 Comparison of Ubx in crustaceans and in insects. (Left) Ubx in crustaceans. The carboxy-terminal antirepression peptide blocks the activity of the amino-terminal repression domain. (Right) Ubx in insects. The carboxy-terminal antirepression peptide was lost through mutation. (Adapted, with permission, from Ronshaugen M. et al. 2002. *Nature* 415: 914–917, Fig. 4b. © Macmillan.)

Modification of Flight Limbs Might Arise from the Evolution of Regulatory DNA Sequences

Ubx has dominated our discussion of morphological change in arthropods. Changes in the Ubx expression pattern appear to be responsible for the transformation of swimming limbs into maxillipeds in crustaceans. Moreover, the loss of the antirepression motif in the Ubx protein likely accounts for the suppression of abdominal limbs in insects. In this final section on that theme, we review evidence that changes in the regulatory sequences in *Ubx* target genes might explain the different wing morphologies found in fruit flies and butterflies.

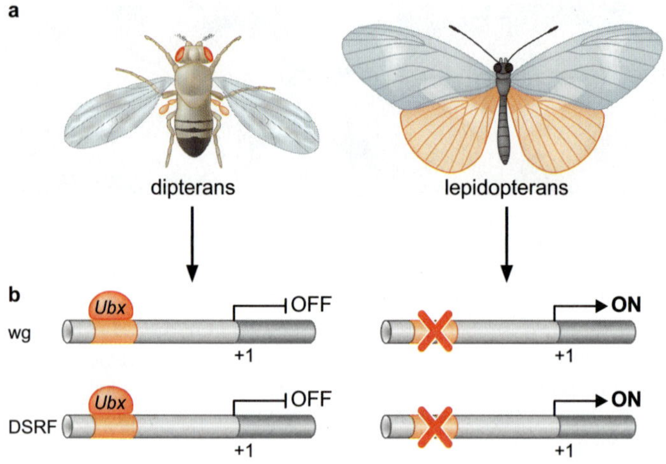

FIGURE 19-30 Changes in the regulatory DNA of *Ubx* target genes. (a) The Ubx repressor is expressed in the halteres of dipterans and hindwings of lepidopterans (orange). (b) Different target genes contain Ubx repressor sites in dipterans. These have been lost in lepidopterans.

In *Drosophila*, *Ubx* is expressed in the developing halteres where it functions as a repressor of wing development. Approximately five to ten target genes are repressed by Ubx. These genes encode proteins that are crucial for the growth and patterning of the wings (Fig. 19-30) and all are expressed in the developing wing. In *Ubx* mutants, these genes are no longer repressed in the halteres, and as a result, the halteres develop into a second set of wings.

Fruit flies are dipterans, and all of the members of this order contain a single pair of wings and a set of halteres. It is likely that Ubx functions as a repressor of wing development in all dipterans. Butterflies belong to a different order of insects, the lepidopterans. All of the members of this order (which also includes moths) contain two pairs of wings rather than a single pair of wings and a set of halteres. What is the basis for these different wing morphologies in dipterans and lepidopterans?

The two orders diverged from a common ancestor more than 250 million years ago. This is about the time of divergence that separates humans and nonmamalian vertebrates such as frogs. It would seem to be a sufficient period of time to alter *Ubx* gene function through any or all of the three strategies that we have discussed. The simplest mechanism would be to change the *Ubx* expression pattern so that it is lost in the progenitors of the hindwings in lepidoptera. Such a loss would permit the developing hindwings to express all of the genes that are normally repressed by Ubx. The transformation of swimming limbs into maxillipeds in isopods provides a clear precedent for such a mechanism. However, there is no obvious change in the *Ubx* expression pattern in flies and butterflies; *Ubx* is expressed at high levels throughout the developing hindwings of butterflies.

That leaves us with two possibilities. First, the Ubx protein is functionally distinct in flies and butterflies. The second is that each of the approximately five to ten target genes that are repressed by Ubx in *Drosophila* have evolved changes in their regulatory DNAs so that they are no longer repressed by Ubx in butterflies (see Fig. 19-30). It seems easier to modify repression activity than to change the regulatory sequences of five to ten different *Ubx* target genes.

Surprisingly, it appears that the less likely explanation—changes in the regulatory sequences of several *Ubx* target genes—accounts for the different wing morphologies. The Ubx protein appears to function in the same way in fruit flies and butterflies. For example, in butterflies, the loss of Ubx in patches of cells in the hindwing causes them to be transformed into forewing structures (see Fig. 19-30a for the difference between forewings and hindwings). This observation suggests that the butterfly Ubx protein functions as a repressor that suppresses the development of forewings. Although not proven, it is possible that the regulatory DNAs of the wing-patterning genes have lost the Ubx-binding sites (Fig. 19-30b). As a result, they are no longer repressed by Ubx in the developing hindwing.

SUMMARY

The cells of a developing embryo follow divergent pathways of development by expressing different sets of genes. Most differential gene expression is regulated at the level of transcription initiation. There are three major strategies: mRNA localization, cell-to-cell contact, and the diffusion of secreted signaling molecules.

mRNA localization is achieved by the attachment of specific 3′ UTR sequences to the growing ends of microtubules. This mechanism is used to localize the *ash1* mRNA to the daughter cells of budding yeast. It is also used to localize the *oskar* mRNA to the posterior plasm of the unfertilized egg in *Drosophila.*

In cell-to-cell contact, a membrane-bound signaling molecule alters gene expression in neighboring cells by activating a cell signaling pathway. In some cases, a dormant transcriptional activator, or coactivator protein, is released from the cell surface into the nucleus. In other cases, a quiescent transcription factor (or transcriptional repressor) already present in the nucleus is modified so that it can activate gene expression. Cell-to-cell contact is used by *B. subtilis* to establish different programs of gene expression in the mother cell and forespore. A remarkably similar mechanism is used to prevent skin cells from becoming neurons during the development of the insect central nervous system.

Extracellular gradients of secreted cell-signaling molecules can establish multiple cell types during the development of a complex tissue or organ. These gradients produce intracellular gradients of activated transcription factors, which, in turn, control gene expression in a concentration-dependent fashion. An extracellular Sonic Hedgehog gradient leads to a Gli activator gradient in the ventral half of the vertebrate neural tube. Different levels of Gli regulate distinct sets of target genes and thereby produce different neuronal cell types. Similarly, the Dorsal gradient in the early *Drosophila* embryo elicits different patterns of gene expression across the dorsoventral axis. This differential regulation depends on the binding affinities of Dorsal-binding sites in the target enhancers.

The segmentation of the *Drosophila* embryo depends on a combination of localized mRNAs and gradients of regulatory factors. Localized *bicoid* and *oskar* mRNAs, at the anterior and posterior poles, respectively, lead to the formation of a steep Hunchback repressor gradient across the anteroposterior axis. This gradient establishes sequential patterns of Krüppel, Knirps, and Giant in the presumptive thorax and abdomen. These four proteins are collectively called gap proteins; they function as transcriptional repressors that establish localized stripes of pair-rule gene expression. Individual stripes are regulated by separate enhancers located in the regulatory regions of pair-rule genes such as eve. Each enhancer contains multiple binding sites for both activators and gap repressors. It is the interplay of broadly distributed activators, such as Bicoid, and localized gap repressors that establish the anterior and posterior borders of individual pair-rule stripes. Separate stripe enhancers work independently of one another to produce composite, seven-stripe patterns of pair-rule expression. This enhancer autonomy is due, in part, to short-range transcriptional repression. A gap repressor bound to one enhancer does not interfere with the activities of a neighboring stripe enhancer located in the same gene.

Homeotic genes encode regulatory proteins responsible for making the individual body segments distinct from one another. The two best-studied homeotic genes, *Antp* and *Ubx*, control the development of the second and third thoracic segments, respectively, of the fruit fly. The misexpression of *Ubx* in the developing wings causes the development of wingless flies, whereas the misexpression of *Antp* in the head causes a transformation of antennae into legs.

In terms of sheer numbers and diversity, the arthropods can be considered the most successful of all animal phyla. More is known about the molecular basis of arthropod diversity than any other group of animals. For example, changes in the expression profile of the *Ubx* gene are correlated with the conversion of swimming limbs into maxillipeds in different groups of crustaceans. Functional changes in the Ubx protein might account for the repression of abdominal limbs in insects. Finally, changes in Ubx target enhancers might explain the different morphologies of the halteres in dipterans and the hindwings of lepidopterans.

BIBLIOGRAPHY

Books

Carroll S.B., Grenier J.K., and Weatherbee S.D. 2004. *From DNA to diversity: Molecular genetics and the evolution of animal design*, 2nd ed. Blackwell, Malden, Massachusetts.

Davidson E.H. 2006. *The regulatory genome: Gene regulatory networks in development and evolution.* Academic Press, San Diego.

Gilbert S.F. 2006. *Developmental biology*, 8th ed. Sinauer Associates, Sunderland, Massachusetts.

Wolpert L., Smith J., Jessell T., Lawrence P., Robertson E., and Meyerowitz E. 2007. *Principles of development*, 3rd ed. Oxford University Press, Oxford.

mRNA Localization

King M.L., Messitt T.J., and Mowry K.L. 2005. Putting RNAs in the right place at the right time: RNA localization in the frog oocyte. *Biol. Cell* **97:** 19–33.

Rongo C. and Lehmann R. 1996. Regulated synthesis, transport and assembly of the *Drosophila* germ plasm. *Trends Genet.* **12:** 102–109.

Weil T.T., Forrest K.M., and Gavis E.R. 2006. Localization of bicoid mRNA in late oocytes is maintained by continual active transport. *Dev. Cell* **11:** 251–262.

Cell-to-Cell Contact

Greenwald I. 1998. LIN-12/Notch signaling: Lessons from worms and flies. *Genes Dev.* **12:** 1751–1762.

Kimble J. and Simpson P. 1997. The LIN-12/Notch signaling pathway and its regulation. *Annu. Rev. Cell Dev. Biol.* **13:** 333–361.

Schweisguth F. 2004. Notch signaling activity. *Curr. Biol.* **14:** R129–R138.

Shapiro L., McAdams H.H., and Losick R. 2002. Generating and exploiting polarity in bacteria. *Science* **298:** 1942–1946.

Morphogen Gradients

Ashe H.L. and Briscoe J. 2006. The interpretation of morphogen gradients. *Development* **133:** 385–394.

Ephrussi A. and St Johnston D. 2004. Seeing is believing: The bicoid morphogen gradient matures. *Cell* **116:** 143–152.

Moussian B. and Roth S. 2005. Dorsoventral axis formation in the *Drosophila* embryo—Shaping and transducing a morphogen gradient. *Curr. Biol.* **15:** R887–R899.

Segmentation

Clyde D.E., Corado M.S., Wu X., Pare A., Papatsenko D., and Small S. 2003. A self-organizing system of repressor gradients establishes segmental complexity in *Drosophila. Nature* **426:** 849–853.

Fujioka M., Emi-Sarker Y., Yusibova G.L., Goto T., and Jaynes J.B. 1999. Analysis of an *even-skipped* rescue transgene reveals both composite and discrete neuronal and early blastoderm enhancers, and multi-stripe positioning by gap gene repressor gradients. *Development* **126:** 2527–2538.

Jaeger J. and Reinitz J. 2006. On the dynamic nature of positional information. *BioEssays* **28:** 1102–1111.

Homeotic Genes

Browne W.E. and Patel N.H. 2000. Molecular genetics of crustacean feeding appendage development and diversification. *Semin. Cell Dev. Biol.* **11:** 427–435.

Krumlauf R. 1994. *Hox* genes in vertebrate development. *Cell* **78:** 191–201.

Lemons D. and McGinnis W. 2006. Genomic evolution of *Hox* gene clusters. *Science* **313:** 1918–1922.

Lewis E.B. 1978. A gene complex controlling segmentation in *Drosophila. Nature* **276:** 565–570.

Maeda R.K. and Karch F. 2006. The ABC of the BX-C: The bithorax complex explained. *Development* **133:** 1413–1422.

Popadić A., Abzhanov A., Rusch D., and Kaufman T.C. 1998. Understanding the genetic basis of morphological evolution: The role of homeotic genes in the diversification of the arthropod bauplan. *Int. J. Dev. Biol.* **42:** 453–461.

Weatherbee S.D., Nijhout H.F., Grunert L.W., Halder G., Galant R., Selegue J., and Carroll S. 1999. Ultrabithorax function in butterfly wings and the evolution of insect wing patterns. *Curr. Biol.* **9:** 109–115.

CHAPTER

20 Genome Analysis and Systems Biology

OUTLINE

TECHNOLOGICAL ADVANCES HAVE TRANSFORMED THE nature of molecular biology. It is now possible to identify every component—every gene and protein—engaged in a complex cellular process such as the differentiation of a naïve stem cell into heart muscles. Prior to the advent of large-scale DNA-sequencing technologies and proteomics methods, molecular biologists sought to obtain general principles from the systematic dissection of just a subset of the total components—those believed to be the key rate-limiting regulatory agents of the process under study. The ability to identify and characterize every component of a process provides the opportunity for a new line of inquiry: what are the underlying design principles? In this chapter, we discuss the emerging disciplines of genomics and systems biology, which have arisen from the marriage of traditional experimental molecular biology and computational analysis.

GENOMICS OVERVIEW

Prior to the advent of whole-genome sequencing, investigators were severely limited in the scope of DNA sequence comparisons. At best, they could look at the DNA sequences of just a few individual genes among a small set of organisms. With the advent of powerful, automated DNA sequencing machines, it is now possible to obtain complete information about the organization and genetic composition of entire genomes. In fact, as of this writing, more than 1000 different genomes have been sequenced and assembled. It is therefore possible to compare the complete genetic composition of many different microbes, plants, and animals. First, we consider the basic methods that are used for the annotation of genomes—that is, the use of both experimental and computational methods for the identification of every gene (including intron–exon structure) and associated regulatory sequences within a complex genome. We then discuss some of the insights obtained from the systematic analysis of individual genomes and comparisons of diverse genomes.

Bioinformatics Tools Facilitate the Genome-wide Identification of Protein-Coding Genes

Genome sequence assemblies correspond to contiguous blocks of millions of sequential As, Gs, Cs, and Ts encompassing every chromosome of the organism in question (see Chapter 21 for a detailed account of how genome assemblies are created from smaller sequence "scaffolds"). They are large, tedious, and uninformative unless "annotated." As described in the next few pages, **annotation** is the systematic identification of every stretch of genomic DNA that contains protein-coding information or noncoding sequences that specify regulatory RNAs such as microRNAs (miRNAs) (see Chapter 18). The detailed intron–exon structure of every transcription unit is identified, and in cases where the genome in question corresponds to a model organism (e.g., yeast and fruit flies), it is possible to assign potential or known functions to most of the genes in the genome. Only when this information is available is it possible to catalog the complete coding capacity of the genome and compare its contents with those of other genomes.

For the genomes of bacteria and simple eukaryotes, genome annotation is relatively straightforward, amounting essentially to the identification of open reading frames (ORFs). Although not all ORFs—especially small ones—are real protein-coding genes, this process is fairly effective, and the key challenge is in correctly assigning the functions of these genes.

For animal genomes with complex intron–exon structures, the challenge is far greater. In this case, a variety of bioinformatics tools are required to identify genes and determine the genetic composition of complex genomes. Computer programs have been developed that identify potential protein-coding genes through a variety of sequence criteria (Fig. 20-1), including the occurrence of extended ORFs that are flanked by appropriate 5′ and 3′ splice sites. As discussed in Chapter 13, splice donor and acceptor sites are short and somewhat degenerate sequences, but they nevertheless help identify exon–intron boundaries when considered in the context of additional information, such as expressed sequence tag (EST) sequence data, which we shall consider below. Nonetheless, computational methods have not yet been refined to the point of complete accuracy. Something like three-fourths of all genes can be identified in this way, but many are missed, and even among the predicted genes that are identified, small exons—particularly noncoding exons—are often overlooked.

Whole-Genome Tiling Arrays Are Used to Visualize the Transcriptome

Once a whole-genome sequence is assembled for an organism, it can be used to comprehensively reveal all protein-coding and noncoding (e.g., introns and miRNA genes) sequences that are expressed in specific cells or tissues. This genome-wide representation of transcription

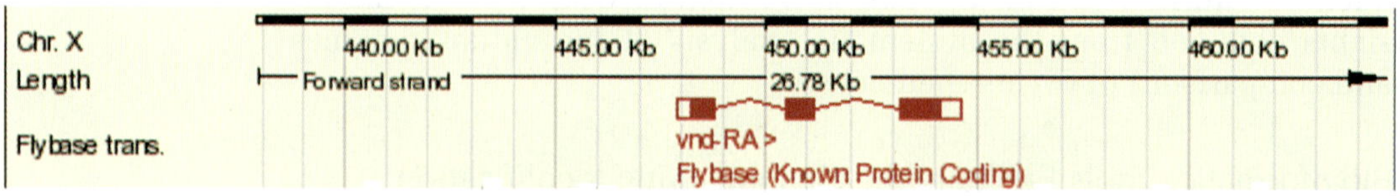

FIGURE 20-1 Structure of the *vnd* locus in *Drosophila*. An ~25-kb interval on the *X* chromosome that contains the *vnd* gene. The *vnd* transcription unit contains three exons and two introns. The unfilled portions of the 5′ (left) and 3′ (right) exons indicate noncoding sequences that do not contribute to the final protein product. FlyBase is the standardized database that is used to analyze the *Drosophila* genome.

is known as the **transcriptome**. Synthetic, single-stranded DNAs of 50 nucleotides in length are spotted on a glass or silicon slide. Typically, one oligonucleotide is produced for every 100–150 bp of DNA sequence in a sequential manner across the genome, resulting in a "tiling array." The technology for genome-wide tiling is advancing rapidly, and it is now feasible to produce complete arrays on a single glass slide or silicon chip that is typically just 1 cm^2 in size. For example, 1 million 50-mers encompass the entire *Dosophila* genome, and all of these oligonucleotides can be spotted on a single chip. Each spot on the chip (i.e., each oligonucleotide sequence) is so small that hybridization signals are detected by microsensors attached to a microscope, as we describe below.

To visualize the transcriptome, the tiling arrays are hybridized with fluorescently labeled RNA (or cDNA) probes. These probes might be derived from a specific cell type, such as the tail muscles of the sea squirt tadpole or yeast cells grown in a particular medium. The end result is a series of hybridization signals superimposed on all of the predicted protein-coding sequences across the genome (Fig. 20-2).

Whole-genome tiling arrays provide immediate information about the intron–exon structure of individual transcription units (Fig. 20-2). This is due to the unstable nature of intronic transcripts. Although total RNA is typically used for these experiments, the exonic sequences are more stable than the introns, which decay rapidly after their removal from primary transcripts (see Chapter 13). After labeling and hybridization to the tiling array chip, exonic sequences display more intense signals than introns.

Another useful feature of whole-genome tiling arrays is that they detect noncoding genes, such as those specifying miRNAs. These RNAs are usually processed from larger precursor RNAs (pri-RNAs) derived from transcription units that are 1–10 kb in length (see Chapter 18). The pri-RNA transcription units are easily detected by hybridization to tiling arrays. In some cases, miRNA genes contain introns that must be processed prior to the final production of the mature miRNA. Other types of noncoding transcripts are also detected, including "antisense" RNAs within the introns of protein-coding genes. It is possible that such RNAs function in a regulatory capacity to control the expression or function of protein-coding genes.

Tiling arrays have led to a rather startling observation: about one-third of a typical genome is transcribed, even though just a fraction of

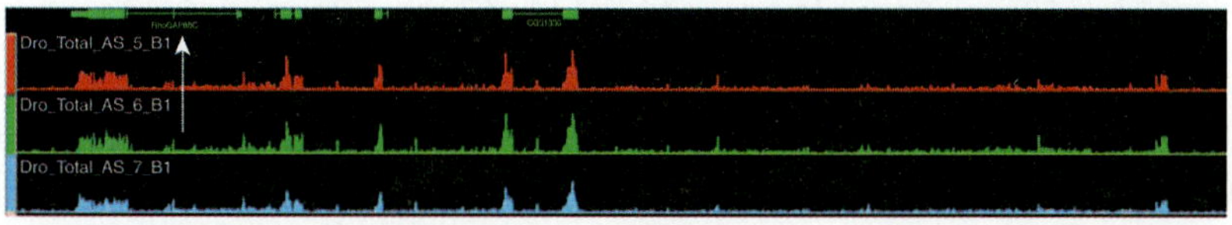

FIGURE 20-2 Whole-genome tiling array. A 50-kb interval on *Drosophila* chromosome 3 that contains four different genes. The intron–exon structure of each transcription unit is shown at the top of the figure. The white arrow indicates the large intronic region that might contain a small ("micro-") exon. Total RNA was extracted from progressively older embryos (red, young; green, older; and blue, still older embryos) and hybridized to the tiling array, which contains 25-nucleotide sequences every 35 bp throughout the entire genome. Strong hybridization signals coincide with the exons, whereas there are weaker signals in the intronic regions. (Reprinted, with permission, from Manak et al. 2006. *Nat. Genet.* 38: 1151–1158, Fig. 5. © Macmillan.)

this transcription corresponds to protein-coding sequences. For example, just 5% of the human genome corresponds to protein-coding information. It appears that most of the additional transcription is due to vast tracts of intronic DNA sequences. Many genes have remote, 5′-noncoding exons that reside far (sometimes a megabase or more) from the main body of the coding sequence. In some cases, these intronic regions produce miRNAs and additional types of noncoding RNAs.

Regulatory DNA Sequences Can Be Identified by Using Specialized Alignment Tools

Genome technologies are effective at identifying genes and determining the structures of their transcription units. Once identified, a host of bioinformatics methods permit the determination of potential protein structure and function, for example, whether the protein contains any known domains or motifs or shares other features with known proteins. In particular, the Basic Local Alignment Search Tool or BLAST algorithm provides a powerful approach for searching, comparing, and aligning either protein or nucleic acid sequences. BLAST searches permit the rapid comparison of a given exon sequence with a vast database of protein-coding information (Fig. 20-3). Significant sequence alignments with protein-coding sequences of known function (e.g., DNA-binding protein, replication factor, or membrane receptor) provide immediate insights into the potential activities of the gene and its putative protein products. Simple BLAST searches can also reveal the identities of noncoding transcripts that produce miRNAs (see Chapter 18).

In contrast to protein-coding sequences, the identification and characterization of regulatory sequences, those stretches of DNA controlling where and when the associated genes are ON and OFF in an organism, are extremely challenging as we saw in Chapter 17. In fact, some refer to the regulatory sequences as the **dark matter** of the genome. Genome-wide methods are only now becoming available for the identification of this important class of DNA sequence information.

A subset of vertebrate regulatory sequences can be identified using

```
Size=6 Zinc finger protein SLUG (Neural crest
transcription factor Slug) (Snail homolog 2)
Length = 268

Score = 217 bits (553), Expect = 1e   -56
Identities = 98/144 (68%), Positives = 113/144 (78%), Gaps = 4/144
(2%), Frame =  -1

Query: 264 LPCKCPICGKAFSRPWLLQGHIRTHTGEKPFQCPDCPRSFADRSNLRAHQQTHVDVKKYA
           LPC C ICGKAFSRPWLLQGHIRTHTGEKPF CP C R+FADRSNLRAH QTH DVKKY
Sbjct: 183 LPCVCKICGKAFSRPWLLQGHIRTHTGEKPFSCPHCNRAFADRSNLRAHLQTHSDVKKYQ

Query: 84  CQVCHKSFSRMSLLNKHSSSNCTI 13
           C+ C K+FSRMSLL+KH  S C +
Sbjct: 243 CKNCSKTFSRMSLLHKHEESGCCV 266
```

FIGURE 20-3 BLAST analysis of an exon encoding a regulatory protein. The `Query` sequence corresponds to the unknown genomic interval that is being analyzed. It is used to survey a huge database of known protein-coding sequences. The `Sbjct` (subject) is the identified sequence sharing significant similarity with the `Query` sequence. In this case, the `Sbjct` corresponds to a regulatory protein called SLUG.

variations in the BLAST searches developed for characterizing protein-coding sequences. Cell-specific enhancers contain clustered binding sites for one or more sequence-specific DNA-binding proteins (see Chapter 17). In some cases, this clustering is sufficient for the identification of short stretches of DNA sequence alignment. A computer program called VISTA aligns the sequences contained in different genomes over short windows, on the order of 10–20 bp, and thereby identifies imperfectly conserved noncoding sequences over stretches of just 50–75 bp (Fig. 20-4). Pufferfish and mice share approximately 10,000 short noncoding sequences. It is conceivable that many of these correspond to tissue-specific enhancers. However, it is likely that both animals, particularly mice, have at least 100,000 enhancers. Thus, these simple sequence alignments fail to capture the vast majority of regulatory sequences.

Tissue-specific enhancers can also be identified by scanning genomic DNA sequences for potential binding sites of known regulatory proteins. Consider the case of the α-catenin gene, which encodes a cell adhesion molecule. The gene is expressed in several different tissues, but it exhibits particularly strong expression in heart precursor cells called cardiomyocytes. It was possible to identify a heart-specific enhancer by surveying the flanking and intronic sequences of α-catenin for matches to the binding sites of known heart cell regulatory proteins, including MEF2C, GATA-4, and E47/HAND (Fig. 20-5). Each of these proteins recognizes a spectrum of short sequence motifs of 6–10 bp. The spectrum of binding sites for each factor is described by a position-weighted matrix (PWM), which can be determined using a variety of computational and experimental methods such as SELEX (in vitro selection) assays (covered in detail in Chapter 21). When these PWMs were used to survey the α-catenin locus, a single cluster of putative MEF2C-, GATA-4-, and E47/HAND-binding sites was identified. Experimental studies confirmed that this cluster of binding sites, located in the 5′-flanking region of the gene, function as a bona fide enhancer.

In principle, it is possible to identify additional heart-specific enhancers by surveying the entire mouse or human genome for clusters of MEF2C, E47/HAND-binding sites. This type of approach has been used to identify enhancers; however, the efficiency is poor (see Box

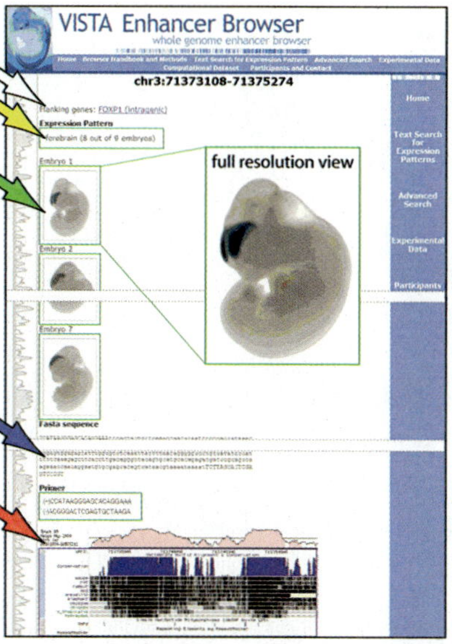

FIGURE 20-4 VISTA alignment of regulatory sequences. The red arrow on the bottom shows sequence conservation in the noncoding region between two highly conserved protein-coding genes in the mouse genome. The noncoding sequence was attached to a *lacZ* reporter gene and expressed in transgenic mouse embryos. The green arrow and box shows the expression pattern generated by the conserved noncoding sequence. This sequence functions as an authentic enhancer to direct *lacZ* reporter gene expression in the developing forebrain of the mouse embryo. (Reprinted, with permission, from Visel A. et al. 2007. *Nucleic Acids Res.* 35: D88–D92, Fig. 1. © Oxford University Press.)

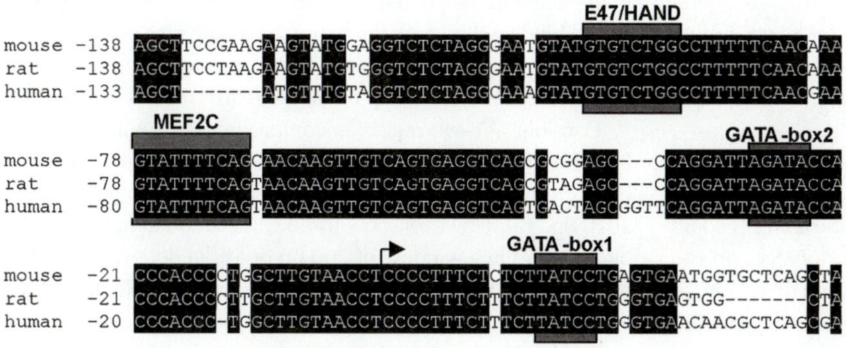

FIGURE 20-5 In silico identification of a heart enhancer. An ~140-bp sequence in the 5′-flanking region of the α-catenin gene is conserved in the mouse, rat, and human genomes. The conserved sequence contains binding sites for three critical regulators of heart differentiation: E47/HAND, MEF2C, and GATA. The mouse sequence has been shown to function as an authentic heart-specific enhancer. In principle, it could be identified by either VISTA alignments (see Fig. 20-4) or the clustering of heart regulatory proteins. (Portion reprinted, with permission, from Vanpoucke G. et al. 2004. *Nucleic Acids Res.* 32: 4155–4165, Fig. 1.)

20-1, Bioinformatics Methods for the Identification of Complex Enhancers). At best, only 10–20% of all enhancers regulated by a given combination of factors are identified, and many of the binding clusters correspond to false positives that do not function as enhancers. Better success is achieved when examining specific candidate genes, such as α-catenin, as discussed above.

The ChIP-Chip Assay Is the Best Method for Identifying Enhancers

As we discussed earlier, tiling arrays represent the single most effective method for determining the detailed structure of individual genetic loci on a whole-genome scale. An adaptation of this method—the ChIP-chip assay—is also the best current technique for the whole-genome identification of regulatory sequences.

This method connects tiling array technology with a method known as chromatin immunoprecipitation (ChIP), which is described in Chapter 21. In the ChIP procedure, cells, tissues, organs, or even whole embryos are treated with formaldehyde to cross-link DNA-binding proteins and their associated DNA sequences with other associated proteins. The cross-linked chromatin is sheared to small fragments of about 200 bp. An antibody against the DNA-binding protein of interest is used to isolate the fragments bound by the protein. In conventional ChIP, the cross-linking is reversed and immunoprecipitated DNA is used as a template for amplification by the polymerase

TECHNIQUES

Box 20-1 Bioinformatics Methods for the Identification of Complex Enhancers

A variety of computer programs have been developed to identify regulatory DNAs within genomes that have been completely sequenced, known as "whole-genome assemblies." These programs take advantage of the fact that regulatory DNAs contain dense clusters of DNA-binding sites. For example, the *eve* stripe-2 enhancer is 500 bp and contains 12 separate binding sites for four different regulatory proteins: Bicoid, Hunchback, Krüppel, and Knirps (see Fig. 19-21). Thus, there is more than one binding site per 50 bp over the length of the enhancer. This density of binding sites is typical of enhancers that direct localized patterns of gene expression in the early *Drosophila* embryo.

As discussed in this chapter, a number of regulatory proteins have been implicated in the regulation of pair-rule stripes of gene expression in the *Drosophila* embryo, including Bicoid, Hunchback, Krüppel, Giant, and Knirps. Unfortunately, an insufficient number of Giant-binding sites have been identified to determine the range of sequences that this protein is likely to recognize. In contrast, there is extensive DNA-binding information for the other four regulatory proteins, as well as for a homedomain protein called Caudal, which is expressed in a broad gradient in the posterior half of the embryo where it functions as a transcriptional activator.

Bicoid, Caudal, Hunchback, Krüppel, and Knirps each bind DNA as a monomer and recognize relatively simple sequences that are present in extremely high copy number in the *Drosophila* genome. Bicoid, for example, recognizes a simple sequence that contains an ATTA-core motif with a few flanking G/C residues. On average, there is a potential Bicoid-binding site every 1 kb in the *Drosophila* genome. Therefore, the use of Bicoid-binding sites for identifying segmentation enhancers would be futile because there are more than 100,000 such sites in the genome (nearly ten sites per gene). However, clusters of Bicoid-binding sites, together with the binding sites of regulatory proteins that work together with Bicoid, provides a powerful filter for eliminating fortuitous binding sites (or "noise").

Consider a 1-Mb region encompassing the *eve* locus (Box 20-1 Fig. 1). There are thousands of Bicoid-, Caudal-, Hunchback-, Krüppel-, and Knirps-binding sites in this interval (Box 20-1 Fig. 1a). However, only three clusters contain at least 13 binding sites in a window of 700 bp or less (a density of nearly one binding site per 50 bp [Box 20-1 Fig. 1b]). Remarkably, these three clusters coincide with the 5′ and 3′ regulatory regions of the *eve* gene. One cluster corresponds to the *eve* stripe-3/7 enhancer, another cluster coincides with the *eve* stripe-2 enhancer, and the third cluster is located in the 3′ regulatory region and coincides with the *eve* stripe-4/6 enhancer (Box 20-1 Fig. 1).

Clustering of DNA-binding sites has proven to be a valuable tool for identifying enhancers in the *Drosophila* genome. How-

Box 20-1 *(Continued)*

ever, the current computer programs are not 100% accurate. In the best cases, only approximately one-third of the identified clusters correspond to actual enhancers. It is conceivable that a higher hit rate will be obtained by placing spatial constraints on the arrangement of binding sites rather than relying solely on simple clustering of sites. We saw in Chapter 17, for example, that the interferon enhanceosome contains binding sites with fixed spacing, including helical phasing between neighboring sites.

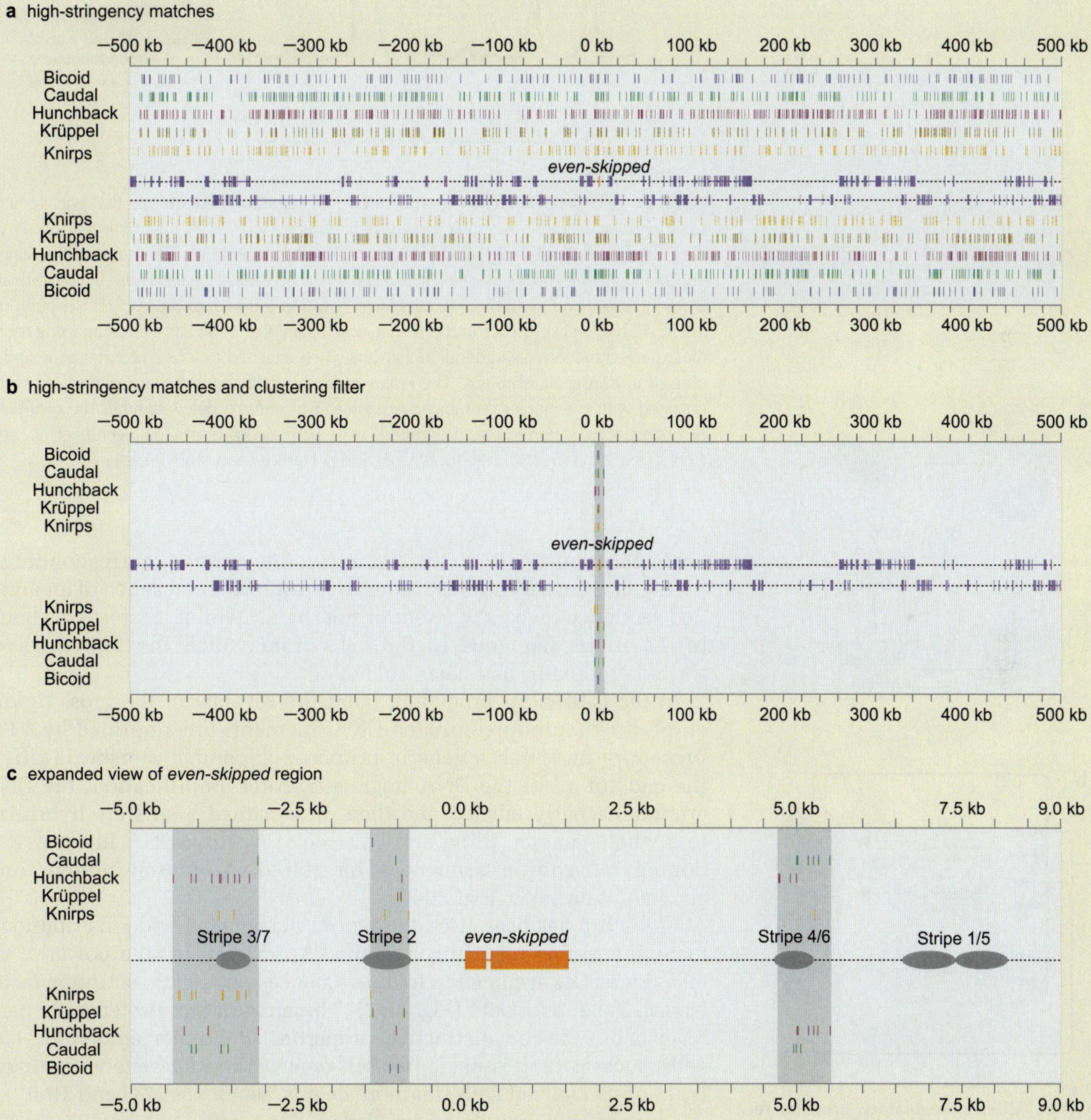

Box 20-1 Figure 1 Clusters of binding sites identify *eve* stripe enhancers. (a) Individual Bicoid-, Caudal-, Hunchback-, Krüppel-, and Knirps-binding sites in a 1-Mb region that contains the *even-skipped* locus (in center along with other intron–exon structures of neighboring genes). (b) High-density clustering of binding sites is uniquely detected near eve and not elsewhere in the 1-Mb region. (c) There are three high-density clusters of binding sites associated with eve. These coincide with the stripe-3/7, stripe-2, and stripe-4/6 enhancers. (Redrawn, with permission, from Berman P. et al. 2002. *Proc. Natl. Acad. Sci.* 99: 757–762, Fig. 1, p. 759. © National Academy of Sciences.)

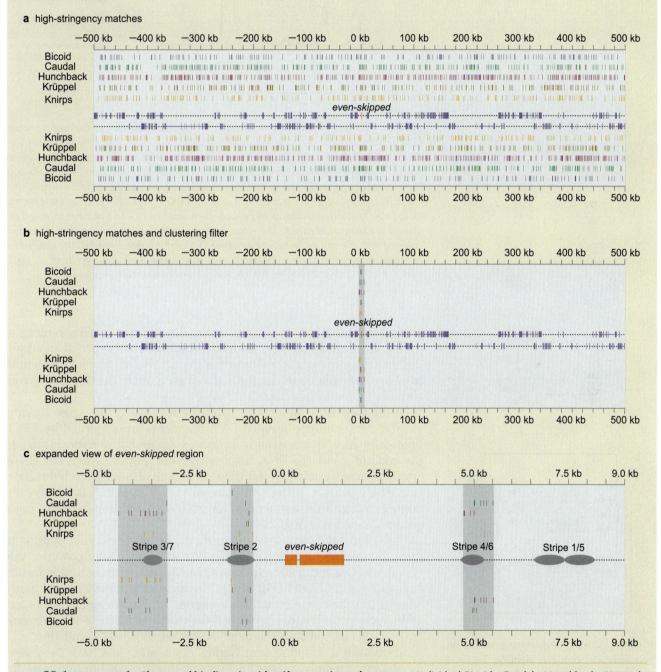

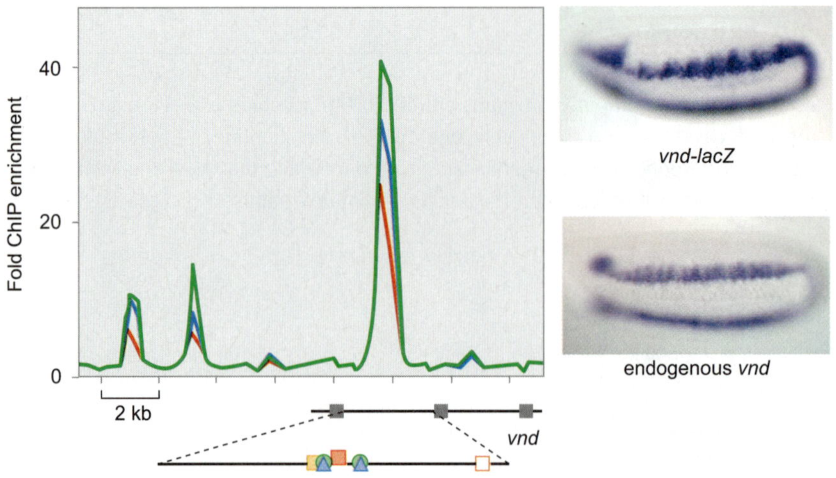

FIGURE 20-6 ChIP-chip identification of regulatory DNAs at the *vnd* locus. A *Drosophila* whole-genome tiling array was hybridized with DNA fragments associated with the Dorsal (red), Twist (blue), and Snail (green) regulatory proteins. The region of the tiling array containing the *vnd* locus (see Fig. 20-1) is shown in the left panel. There is a large peak of Dorsal-, Twist-, and Snail-binding sites located within intron 1 of the *vnd* transcription unit. There are two weaker peaks in the 5´-flanking region. Genomic DNA encompassing each of these peaks functions as authentic enhancers when attached to a *lacZ* reporter gene and expressed in transgenic embryos. The embryos on the right show the *lacZ* expression pattern obtained with the intronic enhancer sequence. The embryo below exhibits the distribution of endogenous *vnd* mRNAs. (Reprinted, with permission, from Zeitlinger J. et al. 2007. *Genes Dev.* 21: 385–390, Fig. 3b. © Cold Spring Harbor Laboratory Press.)

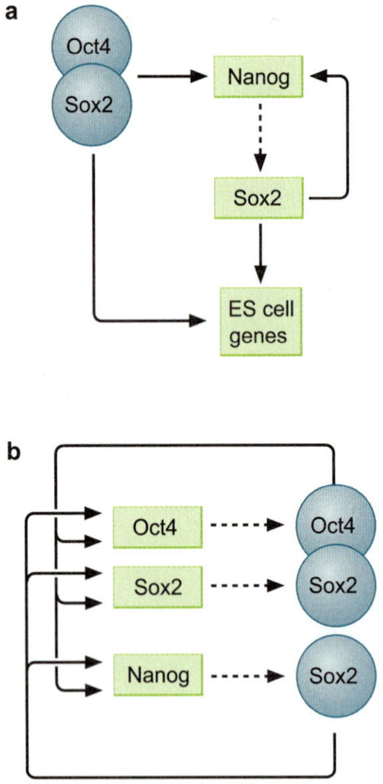

FIGURE 20-7 Regulatory proteins responsible for stem cell identity. The Oct4 and Sox2 regulatory proteins form a complex that activates the expression of Nanog. Nanog is also regulated by Sox2 alone. All three proteins, Oct4, Sox2, and Nanog, are important for maintaining the "stemness" of embryonic stem cells.

chain reaction (PCR) with oligonucleotide primers corresponding to particular genes of interest. Thus, the presence or absence of an amplified sequence reveals whether or not the protein of interest was bound to that DNA sequence in the cells from which the formaldehyde-treated chromatin had been isolated.

In the ChIP-chip procedure, after the reversal of the cross-linking, all of the immunoprecipitated DNA fragments are amplified by a PCR procedure in which a generic primer is appended nonspecifically to the ends of all of the DNA fragments. After amplification, the DNAs are fluorescently labeled and then, in a critical final step, hybridized to a whole-genome tiling array, thereby revealing DNA intervals containing recognition sequences for the DNA-binding protein on a genome-wide basis (Fig. 20-6).

ChIP-chip has been used to identify the genome-wide distribution of many interesting regulatory proteins. An example of the power of this technique is its application to the sequence-specific transcription factors Nanog, Sox2, and Oct4 (Fig. 20-7). These regulatory proteins are partly responsible for the distinctive properties of human embryonic stem cells, such as their capacity for self-renewal and for generating diverse types of specialized cells (for more details, see Boxes 17-6 and 19-6).

Antibodies directed against Nanog, Sox2, and Oct4 have been used for the comprehensive identification of the in vivo binding sites for these proteins in stem cells (see Fig. 20-8). More than 100 potential target enhancers have been identified that are jointly regulated by all three proteins. Some of these enhancers are associated with genes that are known to be important regulators of development, such as *Hoxb1*, which is related to the homeotic genes of *Drosophila* (see Chapter 19).

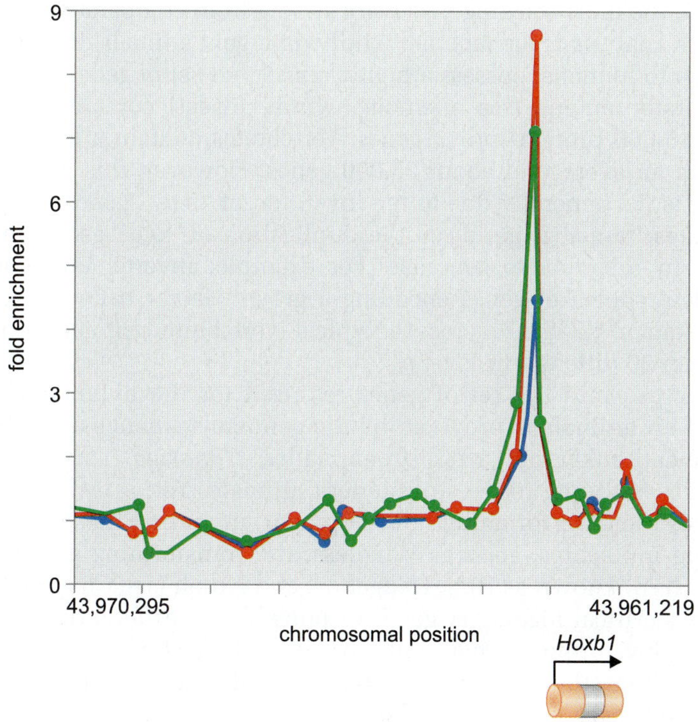

FIGURE 20-8 ChIP-chip identification of enhancers regulated by stem cell factors. A human whole-genome tiling array was hybridized with DNA fragments associated with Nanog (green), Sox2 (red), and Oct4 (blue). All three proteins bind to a 5′-flanking sequence associated with the *Hoxb1* gene. (Adapted, with permission, from Boyer L.A. et al. 2005. *Cell* 122: 947–956, Fig. 2b. © Elsevier.)

The combination of tiling arrays and ChIP-chip assays represents a powerful one–two punch for the identification of all regulatory DNA sequences engaged in a specific process. Let us reconsider the specific example of the α-catenin gene, as well as other genes activated in developing heart cells. Total RNA can be extracted from the heart cells and hybridized to tiling arrays. This identifies all of the genes, both coding and noncoding, that are significantly expressed in the heart precursor cells at specific stages of development. ChIP-chip assays are then used to identify the in vivo binding sites for MEF2C, E47/HAND, and GATA-4, several of the critical regulators of heart-specific gene expression we considered earlier. Significant ChIP signals associated with the expressed genes in the tiling arrays should identify most or all heart cell–specific enhancers.

Diverse Animals Contain Remarkably Similar Sets of Genes

We now consider specific examples of comparative genome analysis, with a particular focus on the comparison of animal genomes. About 100 different animal genomes have been fully sequenced and assembled, but the majority of these sequences correspond to just a few animal groups, centered around the human genome, as well as those of key model organisms such as the fruit fly, *Drosophila melanogaster*, and the nematode worm, *Caenorhabditis elegans*. Thus, several primate genomes (chimpanzees, rhesus monkey, etc.) have been determined to help identify the distinctive features of the human genome (see below). Twelve different species of *Drosophila* have been sequenced to help understand the diversification of distinct species of fruitflies. Currently, just one-third of all animal phyla are represented by a member species with a complete genome sequence assembly.

By far, the most startling discovery arising from comparative genome sequence analysis is the fact that wildly divergent animals, from the sea anemone to humans, possess a highly conserved set of genes. A typical invertebrate genome (sea anemone, worm, insect) contains approximately 15,000 protein-coding genes. Vertebrates contain a larger number, with an average of about 25,000 genes. However, this larger gene number is not generally due to the invention of "new" genes unique to vertebrates; rather, it is due to the duplication of "old" genes already present in invertebrate genomes. For example, invertebrates contain just a few copies of genes encoding a growth factor called fibroblast growth factor (FGF), whereas a typical vertebrate genome contains more than 20 different FGF genes.

A glimpse into the set of genes required for the distinctive attributes of all animals is provided by the genome sequence assembly of a single-cell eukaryote, a protozoan, called *Monosiga*. This organism is the closest living relative of modern animals. Yet, it lacks many of the genes required for animal development, including those encoding signaling molecules, such as Wingless, the transforming growth factor-β (TGF-β) known as Dpp, Hedgehog, and Notch (see Chapter 19). It also lacks critical regulatory genes responsible for differential gene activity in developing animal embryos, including *Hox* genes and *Hox* clusters. Thus, the evolutionary transition of simple eukaryotes into modern animals required the creation of a large number of novel genes not seen among the simple organisms that lived in the ancient oceans more than 1 billion years ago.

Many Animals Contain Anomalous Genes

Despite a constant set, or "tool kit," of basic genes required for the development of all animals, every genome contains its own distinctive—and sometimes surprising—attribute. Consider the case of the sea squirt. It contains a gene encoding cellulose synthase (Fig. 20-9). This enzyme is used by plants to produce cellulose, the major biopolymer

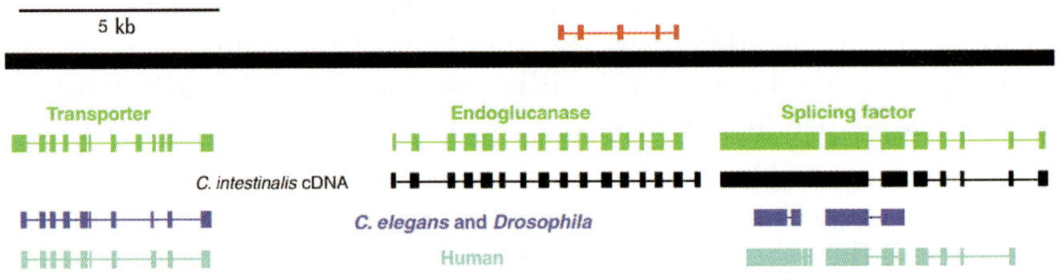

FIGURE 20-9 A plant gene in the *Ciona* genome compared with sequences from other animals. A 20-kb region of one of the *Ciona* scaffolds is shown. This sequence contains an endoglucanase gene, which encodes an enzyme that is required for the degradation and synthesis of cellulose, a major component of plant cell walls. The red rectangles on top represent the *Kerrigan-1* gene of *Arabidopsis*. The gene finder program identified 15 putative exons in the *Ciona* gene, indicated as green rectangles. In reality, there is a 5′ exon present in the cDNA (black rectangles below) that was missed by the computer program. Similarly, a flanking gene, which encodes an RNA splicing factor, is predicted to contain a small intron in a large coding region, whereas the cDNA sequence suggests that there is no intron. There is also a discrepancy in the size of the 5′-most exon. The flanking genes are conserved in worms, flies, and humans, whereas the endoglucanase gene is unique to *Ciona*, which contains a cellulose sheath. Note differences in the detailed intron–exon structures of the flanking genes among the different animal genomes. (Reprinted, with permission, from Dehal et al. 2002. *Science* 298: 2157–2167, Fig. 8. © AAAS.)

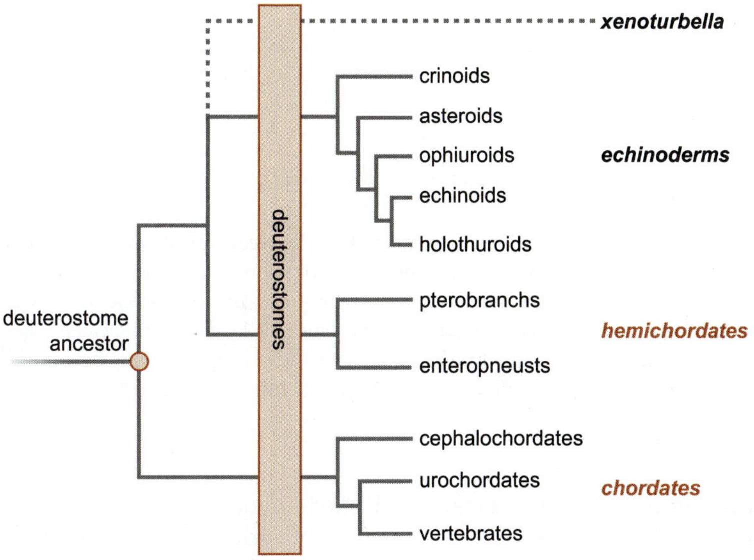

FIGURE 20-10 Deuterostome phylogeny. The deuterostomes include four animal phyla: Xenoturbellida, Echinodermata, Hemichordata, and Chordata. There are five classes of organisms within the echinoderms, two classes of hemichordates, and three classes of chordates. Note that the closest living relatives of the vertebrates are the urochordates, which include the sea squirts (see Box 19-3). (Adapted, with permission, from Gerhart J. 2006. *J. Cell Physiol.* 209: 677–685. © Wiley-Liss, Inc.)

of wood. It is absent in virtually all animals, so what is it doing in the sea squirt? The adult is immobile and sits in tide pools where it filters seawater (see Chapter 19). It contains a rubbery protective sheath composed of tunicin, a biopolymer related to plant cellulose. However, prior to the genome assembly, it was unclear whether the sea squirt contained its own endogenous cellulose synthase gene or employed a symbiotic organism for producing the tunicin sheath. Indeed, there are numerous examples of animals using simple symbionts for unusual genetic functions. For example, termites and wood-eating cockroaches contain symbiotic bacteria in their hindguts that contain the necessary genes required for digesting wood.

Another surprise came from the analysis of the sea urchin genome; it contains two genes, *RAG1* and *RAG2*, required for the rearrangement of immunoglobin genes in humans and other vertebrates (see Chapter 11). One of the distinctive attributes of vertebrates is the ability to mount an adaptive immune response upon infection or injury. This includes the production of specific antibodies that recognize foreign antigens with great specificity and precision. Invertebrates possess a general innate immunity, but they lack the capacity to produce an adaptive immune response. Prior to the sea urchin genome assembly, it was thought that an ancestor of the modern vertebrates acquired a virus or transposon containing the *RAG1* and *RAG2* genes. However, the identification of these genes in sea urchins suggests that this is not true. Instead, the *RAG* genes were acquired by a much more distant ancestor, a progenitor of the so-called Deuterostomes, which diverged into modern echinoderms (e.g., sea urchins) and chordates (e.g., vertebrates) (see Fig. 20-10). It would appear that several descendants of this hypothetical ancestor, such as sea squirts, lost the *RAG* genes.

Synteny Is Evolutionarily Ancient

One of the striking findings of comparative genome analysis is the high degree of **synteny**, conservation in genetic linkage, between distantly related animals. There is extensive synteny between mice

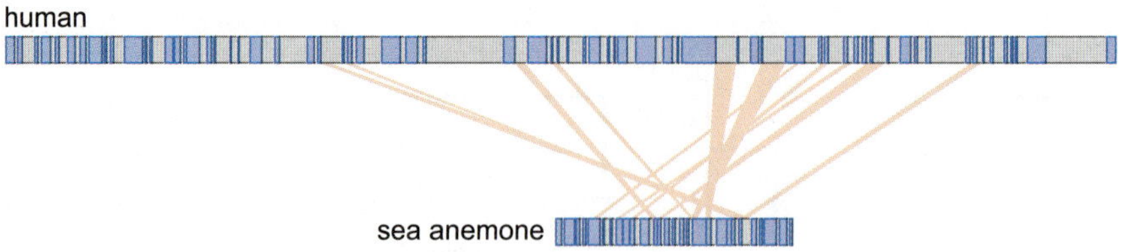

FIGURE 20-11 **Conservation of genetic linkage between sea anemones and humans.** The top diagram shows a 4-Mb region of human chromosome 10 (the q24 region). The lines show alignments between 11 different genes in this interval and corresponding sequences within a 1-Mb region of a sea anemone chromosome. All 11 genes are located together in both chromosomes, but the exact order of the genes has changed during the course of the ~700 million years since humans and sea anemones last shared a common ancestor.

and humans. In many cases, this linkage even extends to the pufferfish, which last shared a common ancestor with mammals more than 400 million years ago. What is even more remarkable is that some of the linkage relationships are conserved between humans and simple invertebrates, such as sea anemones, which last shared a common ancestor more than 700 million years ago, well before the Cambrian radiation that produced most of the modern animal phyla (Fig. 20-11).

Genetic linkage is essential in prokaryotes, where linked genes are coregulated within a common operon (Chapter 16). Such linkage is generally absent in metazoan genomes. In other words, neighboring genes are no more likely to be coexpressed (e.g., in blood cells) than unlinked genes. Early comparative genome analyses appeared to confirm that genetic linkage bore no impact on gene regulation. For example, there is no obvious synteny in the arrangement of related genes in mammalian genomes (e.g., mouse and human) and invertebrate genomes such as *C. elegans* and *Drosophila*. However, there is emerging evidence that the genomes of nematode worms and fruit flies are highly "derived." That is, they have undergone distinctive rearrangements and changes not seen in other genomes. Evidence for this view stems from the analysis of the genome of *Nematostella*, a simple sea anemone.

Sea anemones are ancient creatures. They appear in pre-Cambrian fossils, before the first appearance of Arthropods (e.g., trilobites) and annelids. Despite their simplicity and ancient history, they contain several genes that have been lost in flies and worms. What is even more remarkable is that about half of the genetic linkages seen in the human genome are retained, albeit in a somewhat scrambled order, in the *Nematostella* genome (Fig. 20-11). Consider the q24 region of human chromosome 10. This region contains 11 genes within a 4-Mb interval, including the gene for actin and *SLK*, which encodes a kinase required for cell division. In the smaller *Nematostella* genome, these 11 genes are not only present but also linked within a 1-Mb interval. The conservation of this local synteny raises the possibility that linkage might influence gene function in some subtle manner, which we are currently unable to explain. By sequencing additional animal genomes, particularly those representing ancient creatures such as sponges and flatworms, it might be possible to reconstruct the ancestral karyotype—the exact chromosome complement and genetic linkages of the metazoan ancestor that generated all the modern animal phyla seen today.

Deep Sequencing Is Being Used to Explore Human Origins

The ability to sequence large quantities of DNA quickly and inexpensively has created an opportunity to perform experiments that were impossible to imagine even a year ago. One recent example concerns the analysis of the Neanderthal genome.

Modern humans appeared approximately 100,000 years ago and last shared a common ancestor with Neanderthals about 500,000 years ago. There is evidence that modern humans and Neanderthals coexisted in certain locations prior to the disappearance of the Neanderthals about 30,000 years ago. It has been suggested that the two groups mated, resulting in the occurrence of at least some "Neanderthal genes" in the modern human genome. To test this possibility, scientists are attempting to sequence the complete Neanderthal genome.

Neanderthal DNA samples have been obtained from well-preserved fossils. However, the DNA is heavily contaminated with bacteria and fungi. Nonetheless, the ability to generate hundreds of thousands of short DNA sequence "reads" (see Chapter 21) permits the identification of authentic Neanderthal DNA among the mixture of contaminating DNAs. In fact, just 2–3% of the total DNA obtained from a well-preserved Neanderthal fossil corresponds to authentic Neanderthal DNA that matches chimpanzee and human reference genome sequences. A total of 1 Mb of Neanderthal genomic DNA sequence was identified, and these sequences are 99.5% identical to the modern human genome sequences. The detailed comparison of these Neanderthal sequences with the chimpanzee and human genomes suggests that there was no comingling of Neanderthals and modern humans. Scientists are now sequencing the entire Neanderthal genome. It is amazing to think that the genomes of extinct organisms can be "resurrected."

We have seen how genomic studies have revealed a great deal about various features of coding sequences as well as noncoding regulatory sequences across a wide spectrum of genomes. We now turn our attention to the newly emerging field of systems biology. With this approach, we hope to understand more about the nature of the biological circuits that control complex regulatory processes such as animal development.

SYSTEMS BIOLOGY

Molecular biology owes its success to tackling relatively simple systems, making it possible to investigate underlying mechanisms in great detail. This traditional approach has begun to give way, however, to more ambitious, holistic strategies in which higher and more complex levels of biological organization are examined by a combination of quantitative and high-throughput measurements, modeling, reconstruction, and theory. This interdisciplinary line of investigation has come to define the emerging field of systems biology. Systems biology draws on mathematics, engineering, physics, and computer science, as well as molecular and cellular biology. The objective is to describe the emergent properties of the web of interactions that govern the workings of living things and to do so in a manner that is quantitative and predictive. This approach can apply to biological systems operating at many levels, such as information transfer, signal transduction, cell division, and cytoskeleton dynamics. Here, and as appropriate for a text on the molecular biology of the gene, we focus on the systems biology of gene regulatory circuits. The hope is that this approach will reveal principles of gene control that cannot be understood from the study of individual components in isolation.

Systems biology is closely allied with another new field, that of synthetic biology. Like systems biology, synthetic biology seeks to elucidate design principles of biological circuits. However, synthetic biology attempts to do so by the creation of artificial networks that mimic the features of natural pathways of gene control. This approach enables us to test our models for how regulatory systems work. By virtue of their relative simplicity, such artificial networks can be analyzed in a more quantitative manner than the generally more complex regulatory circuits found in natural systems.

Systems biology is of high interest not only in what it tells us about the logic of gene control, but also in the context of evolution. The principal driving force in the evolution of higher organisms is, as we saw in Chapter 19, in the changes to networks that govern gene expression, not in the genes themselves. For example, animals have similar sets of genes but express these genes in different places and at different times (sometimes radically different). In other words, regulatory networks are relatively plastic in evolution, whereas the genes they control are relatively static.

Here, we present a brief introduction to systems biology and synthetic biology with particular emphasis on natural and reconstructed circuits of gene control. We concentrate on the principles of the design of genetic circuits and on an intuitive understanding of the behavior of alternative wiring diagrams for gene control, but not on the detailed mathematics underlying much of this field. Systems biology is a new field, but, as we shall see, some of its principles derive from classic studies of gene control, particularly those presented in Chapter 16. In this chapter, however, we introduce the formalized language that helps extend these simple examples to a range of biologically diverse regulatory systems.

Transcription Circuits Consist of Nodes and Edges

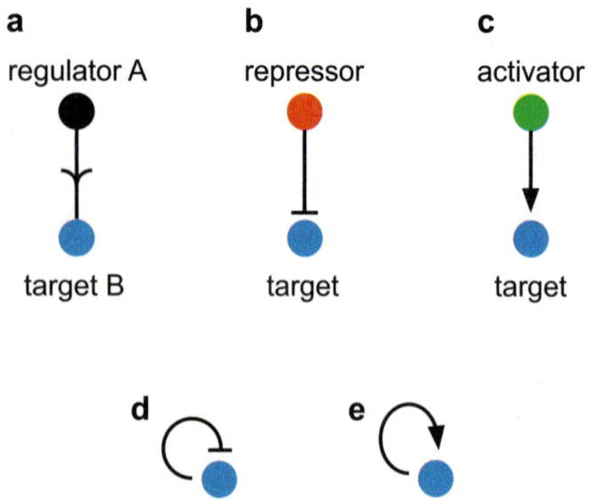

FIGURE 20-12 Simple networks consisting of nodes and edges. (a) A simple switch. Two versions of the switch are shown with negative (b) and positive (c) signs. (d) Negative autoregulation; (e) positive autoregulation.

Regulatory circuits can be described as simple networks consisting of **nodes** and **edges**. Nodes are the genes and are represented by dots; edges represent the regulation of one gene by the product of another and are shown by lines (Fig. 20-12a). Edges can convey directionality to indicate whether A regulates B or visa versa. Edges can also have signs to indicate whether the regulation is negative or positive. Thus, a line ending with a "⊥," extending from gene A to gene B, indicates that the gene A product is a negative regulator of gene B (Fig. 20-12b). Conversely, a line with an arrowhead extending from gene A to gene B indicates that the gene A product acts positively on the expression of gene B (Fig. 20-12c).

Let us begin with a simple two-node switch in which the product of gene A controls the expression of gene B (Fig. 20-12a). Thus, in response to a signal, the regulatory protein encoded by gene A triggers the expression of gene B. The regulatory protein can be a repressor; in which case, transcription is triggered by the presence of an inducer, which inactivates the repressor (Fig. 20-12b). Alternatively, the regulator can be an activator whose ability to trigger transcription occurs in response to a signaling molecule (Fig. 20-12c).

The lactose operon (described in Fig. 16-6) is governed by two regulatory proteins and provides interwoven examples of both kinds of regulation: transcription is triggered by the presence of inducer, which inactivates the Lac repressor, and by a rise in the concentration of cAMP, which promotes the binding of the CAP (catabolite activator protein) activator to DNA. Thus, the lactose operon is not a simple switch: its expression requires *both* the absence of repressor and the presence of CAP bound by its ligand cAMP. This is the logic of an "AND Gate," a term from electrical engineering that denotes that two input conditions must be met in order for there to be an output. Here, the conditions are relief of repression and positive activation. AND gates are depicted by the symbol shown in Fig. 20-13.

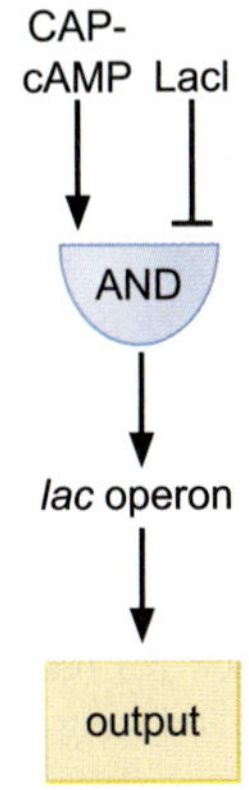

FIGURE 20-13 The AND gate. The lactose operon is subject to the logic of an AND gate in which the output (transcription of the operon) requires both the presence of CAP-cAMP AND the absence of LacI repressor.

Negative Autoregulation Dampens Noise and Allows a Rapid Response Time

Frequently, regulatory genes control their own transcription as well as the transcription of other target genes. This control is known as **autoregulation**, and its sign can be either negative or positive, each with its own characteristic properties (Fig. 20-12d,e). We consider negative autoregulation first; in this case, the gene for a repressor is negatively controlled by its own product. A classic example of negative autoregulation is the *c*I gene of bacteriophage λ, which we discussed in Chapter 16. (Recall that *c*I is also an example of positive autoregulation, as we discuss below.) Thus, the binding of the CI repressor to operator site O_{R3} blocks transcription of its own gene (Fig. 16-27).

What is the biological significance of negative autoregulation, and why has it been repeatedly selected in evolution? One explanation was presented in Chapter 16: negative autoregulation is a homeostasis mechanism ensuring that the level of the regulatory protein is held at a constant level. Thus, should the level of CI fall sufficiently low to relieve repression of *c*I and other target genes, the resulting increase in transcription would raise the cellular concentration of repressor and restore repression. Conversely, should expression of the gene overshoot and produce more repressor than needed, then negative autoregulation will ensure that the gene is kept silent while the level of repressor falls through dilution during cell growth and division or by proteolytic degradation or both.

Negative autoregulation provides another, perhaps less obvious, benefit: rapid response time. Consider the opposing imperatives of producing repressor as rapidly as possible but not wastefully producing excess repressor. The use of a strong promoter would ensure rapid production but would, in steady state, lead to overaccumulation; on the other hand, the use of a comparatively weak promoter could achieve the proper level of repressor but would take a long time to do so. Negative autoregulation allows the best of both worlds: a relatively strong promoter can be used to drive rapid accumulation of the regulatory protein, whereas self-inhibition of transcription shuts off excess accumulation when the appropriate level of repressor is reached. Both mathematical modeling and experiments have confirmed that negative autoregulation allows a more rapid response for the same level of protein accumulation than simple regulation (Fig. 20-14).

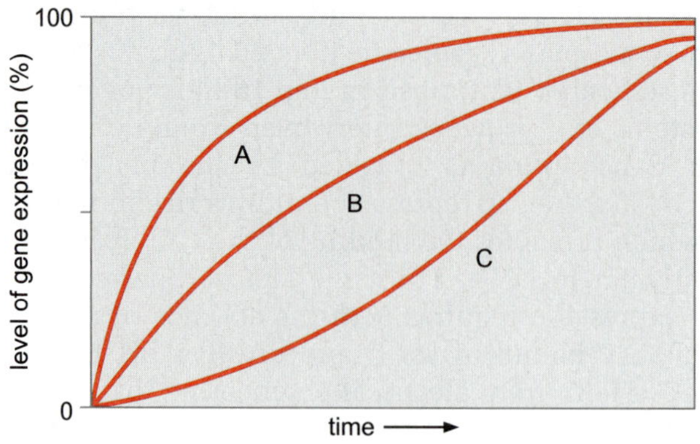

FIGURE 20-14 Kinetics in response to an inducer. A simple switch (B), a negative autoregulatory switch (A), and a positive autoregulatory switch (C) respond with different kinetics to an inducing signal.

Gene Expression Is Noisy

Implicit in our discussion of the role of negative autoregulation in homeostasis is the concept of noise in gene expression. Until recently, it was assumed that the level of expression of a gene in a homogeneous population of cells is relatively constant from cell to cell. Now, however, we appreciate that the levels of gene expression vary substantially among individuals in a population and even between two copies of the same gene in the same cell. We therefore define **noise** as the variation in gene expression under seemingly uniform conditions. The existence of noise indicates that stochasticity influences the level of expression of individual genes. **Stochasticity** indicates that a process is characterized to some degree by randomness. As we shall see, some regulatory motifs are designed to cope with noise and other motifs are designed to exploit it.

Noise in gene expression comes from two sources—intrinsic and extrinsic; both lead to differences in gene expression in a population. **Intrinsic noise** refers to variation in the level of expression of individual genes within a cell and is due to stochastic events within the machinery for gene expression. A classic experiment that demonstrates intrinsic noise employs *Escherichia coli* cells harboring two copies of the same gene. One copy of the gene is joined to a reporter encoding a red fluorescent protein and the other to a reporter encoding a green fluorescent protein. Absent intrinsic noise, both gene copies should produce equal amounts of the red and green fluorescent proteins and hence the cells should be yellow. What is observed instead is that many cells are conspicuously red and others conspicuously green (Fig. 20-15a). Thus, the level of expression of each gene is not identical within any given cell. That is, in some cells, one copy of the gene (e.g., the one tagged with the green fluorescent protein) is more actively expressed than the other copy (the one tagged with the red fluorescent protein); in other cells, the opposite is true.

Extrinsic noise refers to differences in gene expression between cells in a seemingly homogeneous population or to changes in gene expression in the same cell over time. This noise is likely caused by microheterogeneity in the environment of individual cells or by fluctuations in the capacity of cells to carry out transcription or protein synthesis over time. An example of extrinsic noise is illustrated in Figure 20-15b in which the level of expression of both genes is seen to vary over time. In this case, the level of expression of both genes in an individual cell varies in unison, rising for a period and then falling. This means that the overall capacity of individual cells to support the

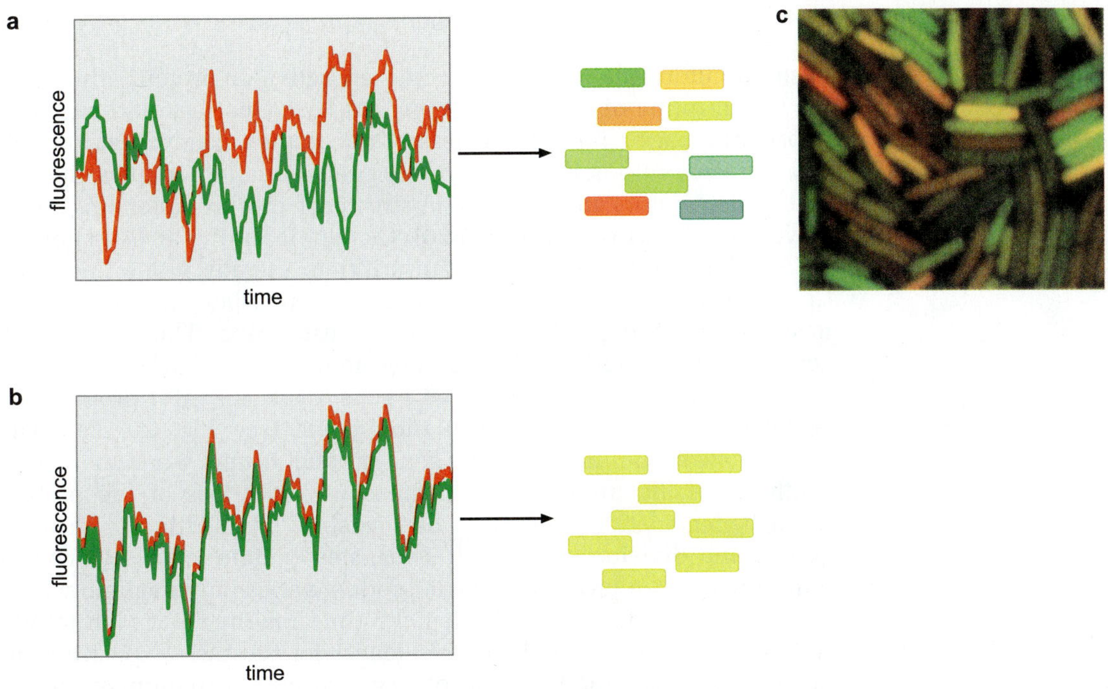

FIGURE 20-15 **Intrinsic versus extrinsic noise.** Cells of *Escherichia coli* harboring two copies of the same gene—in one case fused to a reporter generating a red fluorescent protein and in the other to a reporter generating a green fluorescent protein. (a) The predicted results if both genes varied in expression both over time and among individuals within the same cell (intrinsic noise). (b) The predicted results if the level of expression of both genes varied in synchrony over time because of extrinsic noise. (c) A fluorescent micrograph documents intrinsic noise from the observation that in addition to yellow cells, some cells are red and others green. (Reprinted, with permission, from Elowitz M.B. et al. 2002. *Science* 297: 1184–1186. © AAAS.)

expression of some or all genes fluctuates with time. In the fluorescent micrograph shown in Figure 20-15c, we observe that some cells are yellow, whereas others are red or green, as a result of intrinsic noise.

Returning to negative autoregulation, we see that this regulatory motif helps cells cope with noise by allowing cells to compensate for variations in the level of expression of the autoregulated gene. The negative autoregulatory circuit governing bacterophage λ CI synthesis is therefore said to be robust. **Robustness** indicates that the output of a regulatory circuit is insensitive to a particular parameter. Thus, the ability of the CI autoregulatory circuit to achieve a steady-state level of repressor is robust with respect to noise in the expression of the *c*I gene. As we shall see, other regulatory motifs also help cells meet the challenge of coping with various sources of stochasticity, such as the fluctuations in signals that trigger gene expression.

Stochasticity is not a peculiarity of *E. coli*. Indeed, it is likely to be widespread among living things. For example, the coat pattern of a cat cloned by transfer of a somatic nucleus into an enucleated embryonic stem cell is not identical to that of the cat from which its genome was derived. Because both the clone and the cat from which it was derived are genetically identical, one might have expected that the cats would have identical coat patterns. That they do not suggests that the cascade of genetic events governing coat pattern is not wholly hardwired and must involve stochastic processes. As a second example, the fingerprints of identical twins are not identical.

Positive Autoregulation Delays Gene Expression

Positive autoregulation occurs when an activator protein stimulates the transcription of its own gene (Fig. 20-12e). Once again, the *c*I gene of bacteriophage λ provides a classic but complex example: at low cellular concentrations the CI repressor preferentially occupies the O_{R2} and O_{R1} operators that lie just upstream of the promoter (P_{RM}) that drives *c*I transcription (Fig. 16-26). CI protein sitting at O_{R2} contacts RNA polymerase to stimulate transcription, thereby promoting more CI synthesis. Of course, and as we have seen, when CI reaches high levels, it also occupies O_{R3} to repress transcription. Therefore, the *c*I gene is subject to both positive and negative autoregulation.

Now let us consider the case of a gene for an activator protein that is subject to positive autoregulation alone (see Fig. 20-12e). The steady-state accumulation of the gene product occurs when the rate of synthesis of the protein is in balance with the loss of the protein through degradation (should it be unstable) or its dilution through growth and division of the cell. Thus, **steady state** refers to a condition in which the level of the gene product varies only negligibly over time. The important point is that the time required to reach steady state after a gene is switched on is longer for the case of positive autoregulation than for the case of negative autoregulation or for no feedback at all (Fig. 20-14). Or, to be more precise, the time at which half-maximal accumulation occurs is longer for positive autoregulation than for the alternative regulatory switches. This is because the rate of production, which increases over time, depends on the accumulation of the activator in the first place.

Positive autoregulation can be useful in biological processes that unfold slowly, such as development, which can benefit from the slow accumulation of proteins involved in morphogenesis. For example, in the ancient (or primordial) developmental process of sporulation in the bacterium *Bacillus subtilis* (to which we will return), the principal regulatory proteins that govern late events in spore formation (the alternative RNA polymerase σ factors σ^G and σ^K) stimulate transcription of their own structural genes as well as the genes for morphogenetic proteins. Thus, the σ factors as well as the products of the genes they control accumulate slowly because the production of σ^G and σ^K depends on their own synthesis.

Positive autoregulation has an additional benefit. It is the basis for an extreme type of regulatory switch known as a bistable switch as we explain below.

Some Regulatory Circuits Lock in Alternative Stable States

All of the regulatory circuits we have considered so far are reversible in the sense that once the signal that turned a gene or genes ON is removed, the circuit switches back to the OFF state. In some cases, however, when the gene(s) is switched ON, it remains locked ON for relatively long periods of time. This is known as a bistable switch. A well-studied example of a bistable switch is the circuit that governs whether *B. subtilis* becomes genetically competent. Competence is a specialized state in which the bacterium has stopped growing and has acquired the capacity to take up naked DNA from its environment and incorporate homologous sequences into its genome by genetic recombination. The master regulator for competence is the DNA-binding protein ComK, an activator of approximately 100 genes, including its

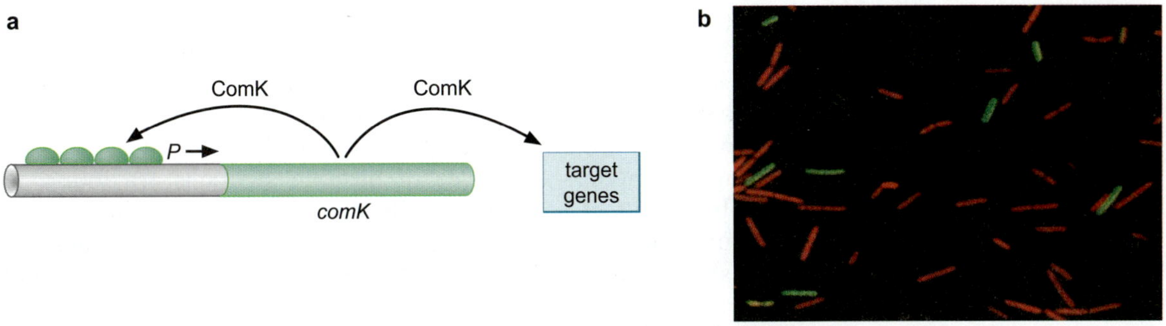

FIGURE 20-16 Bistability. (a) Bistability is governed by a positive autoregulatory switch in which the regulatory protein ComK stimulates transcription from *comK* gene itself as well as from target genes. Cooperatively among ComK molecules bound at the promoter create a nonlinear response, which is hypersensitive to small, stochastic changes in the level of ComK. The switch is poised at a knife edge between flipping into an ON or a OFF state. (b) A classic example of bistablity in which a population of *Bacillus subtilis* cells (red) are either ON (green) or OFF for the expression of a reporter gene under the control of a promoter activated by the competence regulator ComK. The reporter encodes a protein that fluoresces green. (b, Reprinted, with permission, from Dubnau D. and Losick R. 2006. *Mol. Microbiol.* 61: 564–572. © Blackwell Science.)

own (Fig. 20-16). What renders the switch stable is cooperativity in the binding of multiple ComK molecules to the promoter region for *comK*. As we saw in Box 16-3 in the case of the λ repressor (which is itself responsible for a classic example of bistability to which we return below), cooperativity of this sort imparts nonlinearity on the output of the switch as a function of the concentration of the activator. In other words, the output is highly sensitive to changes in the level of ComK (the opposite of robustness).

Whether or not cells have the potential to turn on *comK* is governed by a regulatory pathway operating at the level of the proteolytic stability of the ComK protein. Nevertheless, the ultimate decision to activate *comK* is stochastic. That is, under conditions when ComK is not subject to degradation, only some of the cells in the population become competent. This can be vividly seen using cells harboring a fluorescent reporter (the gene for the green fluorescent protein) for ComK-directed gene activity. Figure 20-16 shows that cells bifurcate into a subpopulation in which *comK* is ON and a subpopulation in which it is OFF. This is because the positive feedback loop is poised on a knife edge between having insufficient ComK to switch *comK* ON and just enough (a threshold amount) to trigger the positive autoregulatory loop necessary to turn on ComK-controlled genes (see Box 20-2, Bistability and Hysteresis). Thus, noise in the expression of the *comK* gene resulting in small variations in the levels of ComK between cells enables the activator to reach a threshold concentration in some cells and not others. This example of positive autoregulation illustrates how noise in gene expression can be exploited to drive cells into alternative states.

Positive autoregulation is not the only basis for bistability. A switch that exists stably in two alternative states is also achieved by the use of mutually repressing repressors, that is, two repressors that negatively control each others' transcription. As mentioned above, bacteriophage λ provides a classic example of a bistable switch but one based on a double-negative regulatory circuit rather than positive autoregulation; the mutually antagonistic actions of the CI and Cro repressors together with cooperativity lock in the alternative lysogenic and lytic states of the virus (Chapter 16). Returning to the language of systems biology, we would say that bacteriophage λ has a two-node switch linked in both directions by negative edges.

Although numerous examples of bistable switches are found in bacteria, bistability is by no means limited to microbes. Thus, for example, during embryogenesis, the nematode *C. elegans* generates bilaterally symmetric gustatory neurons called ASE left and ASE right that express genes for alternative taste receptors. A double-negative feedback loop that can be stably maintained in one state or another dictates whether a common precursor cell will express one set of receptors or the other. In this case the switch is not thrown stochastically. Rather, upstream signals dictate which direction the switch is thrown while the double-negative feedback loop subsequently locks the switch in its predetermined state.

Feed-Forward Loops Are Three-Node Networks with Beneficial Properties

An important contribution from the field of systems biology is the finding that among the myriad kinds of simple regulatory circuits that are theoretically possible, only a small number are commonly found

KEY EXPERIMENTS

Box 20-2 Bistability and Hysteresis

An experiment that demonstrates that positive autoregulation is the basis for the bistablity of the *comK* switch is based on the use of a modified copy of *comK* that has been brought under the control of a promoter whose activity can be modulated up or down in response to an inducer (Box 20-2 Fig. 1a). In cells harboring the modified gene alone, no bistability is observed and the level of ComK-directed gene expression increases in a more or less uniform manner in response to increasing levels of inducer, exhibiting a unimodal distribution of expression levels among cells in the population at any give concentration of inducer (Box 20-2 Fig. 1b). However, in cells harboring both the modified gene and the normal autoregulated gene, increasing concentrations of inducer cause the cells to bifurcate into a subpopulation exhibiting a low level of ComK activity and a subpopulation exhibiting a high level of ComK activity (Box 20-2 Fig. 1c). In other words, production of ComK from the modified gene "primes the pump" for the autoregulated gene, causing the switch to be thrown ON in more and more cells as the level of ComK is increased.

Strictly speaking, the use of the term bistability requires that a switch exhibit a property called **hysteresis**. Hysteresis is a kind of memory that implies that a switch that has been thrown ON under a particular set of conditions does not immediately switch OFF when those conditions are removed or reversed. Consider, for example, the hysteretic properties of ferromagnetic material. When exposed to a magnetic field, the material becomes magnetized and, importantly, remains so even when the external magnetic field is removed. Now let us return to our example of cells harboring both ComK and a modified copy of ComK that responds to inducer. As we saw, adding more and more inducer causes the level of ComK to rise until it exceeds the threshold, causing the positive autoregulatory switch to be thrown ON. Now consider what happens when we lower the level of inducer such that less and less ComK is produced from the engineered copy of the gene. We observe that as the level of inducer is lowered, ComK remains ON even at concentrations of inducer that were insufficient to throw the switch when inducer was increasing. In other words, ComK remembers that it is in the ON state even when the original conditions that switched it ON are reversed.

The switch governing the decision between the lysogenic and lytic modes of propagation of bacteriophage λ is also hysteretic. When the prophage is induced in response to a brief exposure of lysogenic cells to a DNA-damaging agent, the phage irreversibly enters the lytic mode of growth. That is, the phage does not reenter the lysogenic state (resume synthesizing CI repressor) even after the inducing signal (the DNA-damaging agent) is removed. As a counterexample, when the lactose operon is switched ON by the presence of lactose, the operon returns to its OFF state when the inducer is removed from the medium.

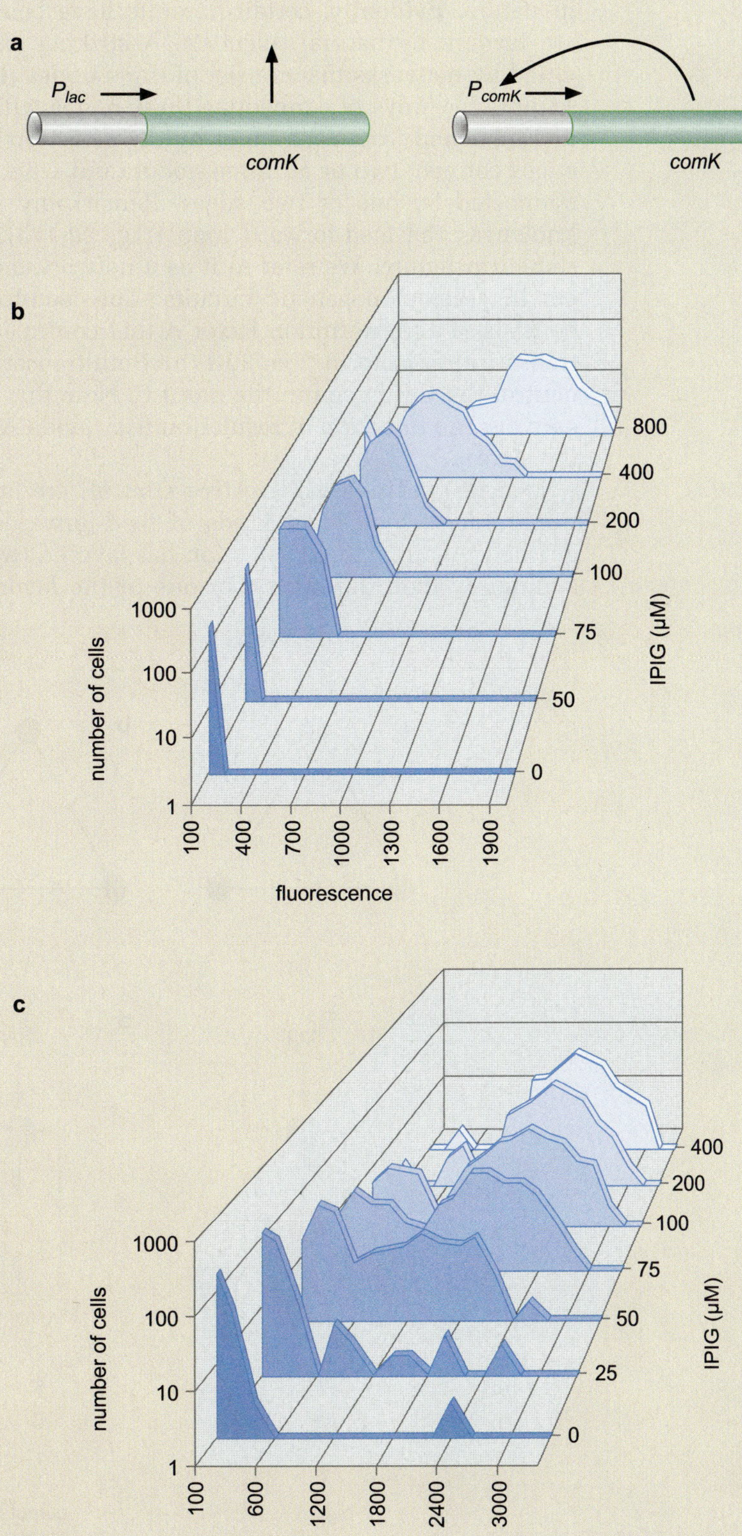

BOX 20-2 FIGURE 1 Bistability of the *comK* switch. (a) The experiment shows that positive autoregulation causes bistability. The panel shows a modified *comK* gene in which the *comK* promoter is replaced by a promoter that responds to the inducer IPTG (that of the *lac* operon). (b) A graded response occurs when *comK* is under the control of an IPTG-inducible promoter in cells harboring only the modified *comK* gene. (c) A bimodal distribution is seen when the positive autoregulation is left intact and the system is primed with a lactose-inducible copy of *comK*. Note that the cells in a and b harbored a fusion of the gene for the green fluorescent protein to a promoter under the control of ComK. (b,c, Reprinted, with permission, from Maamar H. and Dubnau D. 2005. *Mol. Microbiol.* 56: 615–624, Fig. 4E,J. © Blackwell Science.)

in nature. Evidently, certain circuits have beneficial properties that are favored by natural selection. A striking example of this is provided by networks that consist of three nodes (Fig. 20-17a). There are 13 possible ways of connecting three nodes with edges. These can be distinguished from each other by the direction of the edges, whether edges connect two or all three nodes, and whether pairs of nodes are connected by one or two edges. Remarkably, 1 of the 13 patterns, known as the feed-forward loop (Fig. 20-17b), is greatly overrepresented in nature. We refer to it as a network motif because it is a recurring theme in genetic circuits. The feed-forward network motif consists of a transcription factor A that controls the gene for a second transcription factor B (Fig. 20-17b). Both transcription factors, in turn, control the third gene in the motif C. Note that Figure 20-17b simply conveys the *direction* of regulation (e.g., node A controls node B), not the sign.

If signs (positive vs. negative control) are attributed to the directional edges, then eight kinds of feed-forward loops can be distinguished. Again, natural selection has favored two that are found more commonly than the others. In one of the favored feed-forward loop

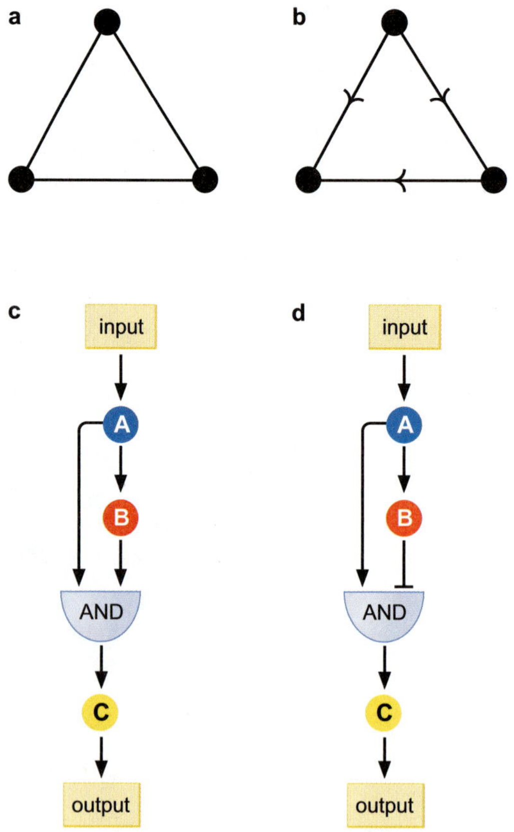

FIGURE 20-17 **Types of networks.** (a) A "three-node" network where each node is a gene and the genes are joined to each other by edges. (b) The feed-forward loop, the most common three-node network found in nature. (c) A "coherent" form of the feed-forward loop in which both the direct and indirect edges leading to the target gene have a positive sign. (d) An "incoherent" form of the loop in which the direct edge has a positive sign and the indirect edge has a negative sign.

motifs (known as a coherent motif), both the direct and the indirect pathways leading to the target gene, representing the output, have the same sign (i.e., both A and B are activators) (Fig. 20-17c). In the other favored motif (known as an incoherent motif), the two pathways have different signs, with the target gene C being subject to positive control by A in the direct pathway and negative control by B in the indirect pathway (Fig. 20-17d). In both cases, expression of the target gene is subject to the logic of an AND Gate; that is, transcription of C requires both A "AND" B in the former and A "AND NOT" B in the later.

Since both motifs are favored among all other feed-forward loops and indeed among all possible three-node networks, it is reasonable to expect that they have favorable properties that have been the basis for their selection in evolution. Indeed, computational modeling and experiment reveal that each motif has characteristics that make them useful in regulatory circuits. For example, the coherent feed-forward loop has the property of requiring a sustained input in order for the target gene C to be transcribed (Fig. 20-17c). In other words, this kind of feed-forward loop is a persistence detector that only responds to a signal that is long-lived or persistent. This property derives from the fact that turning on the target gene depends on both the primary activator A and sufficient accumulation of the secondary activator B. Thus, the input signal must persist long enough for the secondary activator B to reach the threshold concentration needed to turn on the target gene C. In other words, by imposing a delay in the response to an input, the coherent feed-forward loop helps the cell distinguish a true, sustained signal from a stochastic fluctuation (noise) in signal intensity.

The incoherent feed-forward motif has its own beneficial property (Fig. 20-17d). It is a pulse generator that causes gene expression to switch ON and then OFF. Thus, activator A turns on target gene C, but over time the accumulation of repressor B causes the target gene to turn OFF. Thus, the incoherent feed-forward loop is useful when gene expression is required for only a brief period of time.

Feed-Forward Loops Are Used in Development

These insights reveal simplifying design principles in otherwise complex pathways of gene control. In some cases, a combination of coherent and incoherent feed-forward loops is used to produce elaborate patterns of gene activity. A dramatic example comes from the process of sporulation referred to above whose regulatory circuit is a linked series of coherent and incoherent feed-forward loops (Fig. 20-18). The coherent loops ensure that the input to the circuit is persistent and hence that development is not triggered at the wrong time or at the wrong place. Likewise, the incoherent loops are used to generate successive pulses of gene expression over the course of morphogenesis.

Yet another example is seen in the mechanisms that govern dorsoventral patterning in the *Drosophila* embryo. As discussed in Chapter 19, this process is initiated by the maternal regulatory protein, Dorsal, which becomes distributed in a broad gradient. A direct target of Dorsal is the *twist* gene, which is activated at intermediate-high to high levels of the regulatory protein. Twist too is a regulatory protein, and it works in concert with Dorsal to activate a variety of target genes, such as *snail*. This regulatory motif is thus a clear example of a coherent feed-forward loop. In addition, however, *snail* en-

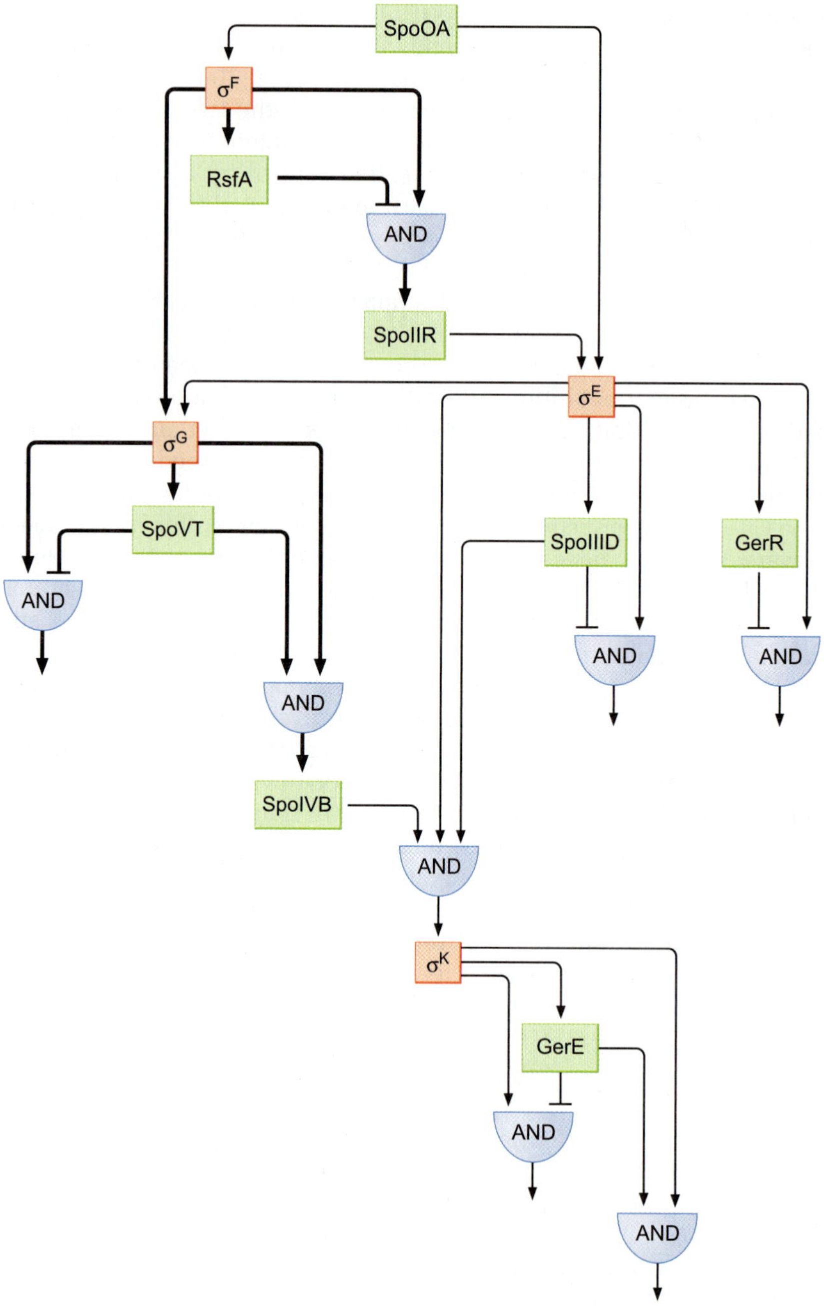

FIGURE 20-18 The circuitry governing spore formation is a linked series of feed-forward loops. The names refer to regulatory or signaling proteins. Not shown for simplicity is that the σ^G and σ^K factors are subject to positive autoregulation. (Redrawn, with permission, from Wang S.T. et al. 2006. *J. Mol. Biol.* 358: 16–37, Fig. 5. © Elsevier.)

codes a transcriptional repressor, and many target genes of Dorsal and Twist are also repressed by Snail. Such target genes are thereby regulated by an incoherent feed-forward loop. Thus, the network of *dorsal*, *twist*, *snail*, and downstream genes consists, as in the case of bacterial sporulation, of linked coherent and incoherent feed-forward loops. In the case of *Drosophila* embryogenesis, the feed-forward loops are used to govern dorsoventral patterning. Thus, in the mesoderm where the levels of Dorsal (and Twist) and hence Snail are high, targets of Snail-mediated repression are OFF, whereas in the neurogenic ectoderm where the levels of Dorsal and hence Snail are low, these genes are ON.

Some Circuits Generate Oscillating Patterns of Gene Expression

We generally think of regulation in terms of switching genes ON or OFF or adjusting their levels of expression. However, another kind of gene control of wide importance in biology is oscillation in which the expression of large numbers of genes is periodically UP-regulated and then DOWN-regulated at regular intervals over time. Elucidating the circuitry that governs this oscillatory behavior, and doing so in a quantitative manner, is one of the premier challenges of systems biology.

A relatively simple example of an oscillating regulatory circuit is the cell cycle of the bacterium *Caulobacter crescentus* (Fig. 20-19). Here, the master regulators, CtrA and GcrA, rise and fall in abundance out of phase with each other in a periodic manner. Their alternating presence drives gene expression in an oscillatory pattern over the course of the cell cycle.

A well-known example of oscillatory behavior is the clock that drives the periodic expression of large numbers of genes at different times during the cycle of day and night. In flies and mammals, this circadian rhythm is governed in part by a negative-feedback loop involving the activator proteins, Clock and Cycle, and the autorepressor Per (Period). The Clock and Cycle proteins bind to the regulatory region for, and stimulate the transcription of, the *per* gene. When the Per protein accumulates to a critical level, it is able to counteract the action of Clock and Cycle and shut off its own synthesis. Once *per* is switched OFF in this manner, Per protein, which is proteolytically unstable, is depleted from the cell. This leads to a subthreshold level

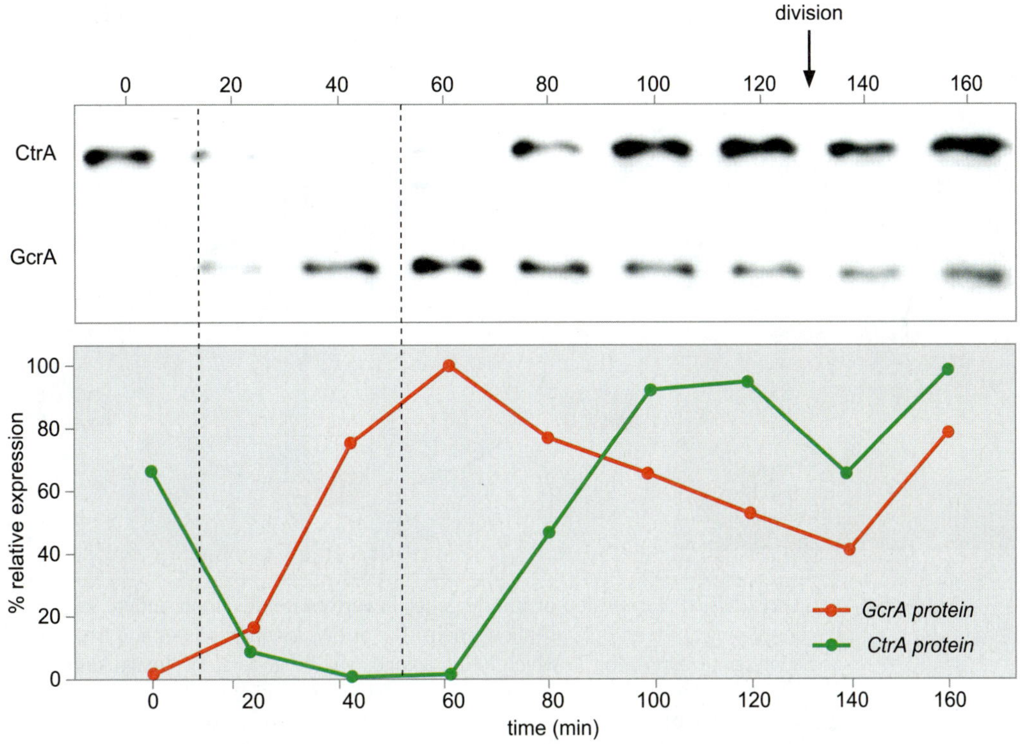

FIGURE 20-19 The regulators CtrA and GcrA rise and fall in abundance out of phase with each other during the *Caulobacter* cell cycle. (Adapted, with permission, from Holtzendorff J. et al. 2004. *Science* 304: 983–987, Fig. 3B,C. © AAAS.)

of the autorepressor, which is insufficient to block activation by Clock and Cycle. The *per* gene is thereby turned back ON. This ON/OFF cycle of *per* expression helps define the 24-hour cycling of gene activity. It is critically dependent on the timing of Per protein synthesis and degradation. Changes in Per protein stability can change the frequency of oscillations to produce aberrant ON/OFF cycles once every 22 hours or 26 hours in place of the normal 24-hour cycle. Nevertheless, just how the circadian clock maintains its 24-hour cycle and does so in a robust manner is not fully understood and undoubtedly involves additional, yet to be elucidated mechanisms.

Interestingly, negative autoregulation also seems to be involved in another, unrelated example of periodic gene expression: the formation of somites in vertebrate embryos. Somites are condensed blocks of mesoderm cells that form the repeating muscle segments and vertebrae of the spinal column (Fig. 20-20a). They form in a head-to-tail manner and—in zebrafish at least—depend on the ON/OFF oscillating activities of the regulatory genes, *her1* and *her7*. These genes are expressed cyclically in the future somite cells, up to the time when these cells are ready to differentiate and form a physical somite. As each new batch of cells matures, it halts its oscillation, in such a way that some cells become arrested at the peak of their oscillation cycle and others at the trough, in a regular spatial order that marks out the pattern of the forming somite. Just behind each newly established somite is the next group of future somite cells that go through the same process, involving another ON/OFF cycle of gene activity, thereby producing a new somite.

The oscillating ON/OFF expression of *her1* and *her7* in zebrafish has been subject to mathematical modeling and computer simulations. Her1 and Her7 are autorepressor proteins that are thought to bind to the regulatory regions of the *her1* and *her7* genes, shutting off transcription. But this repressed state of affairs lasts only a little while: once the Her1 and

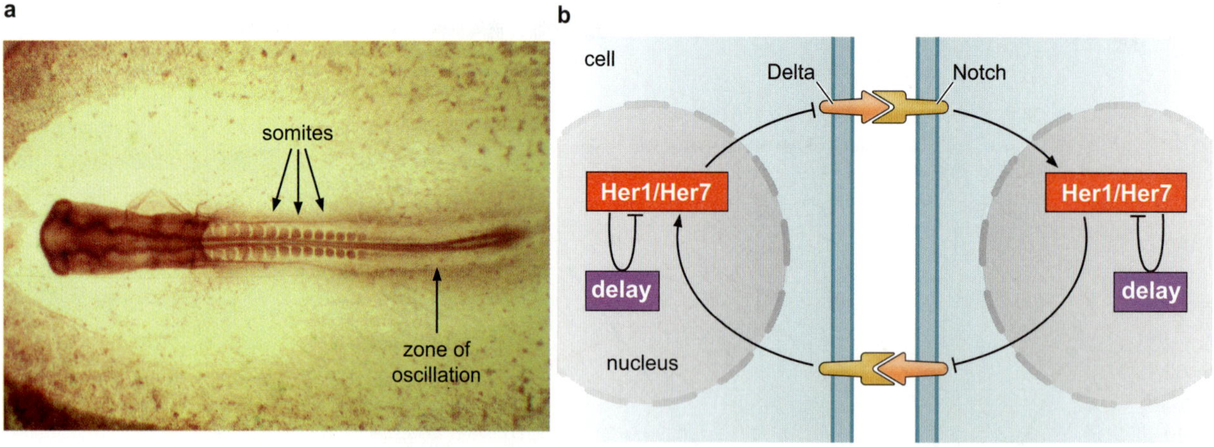

FIGURE 20-20 Expression of somite genes in vertebrate development. (a) Somites in the developing chick embryo are shown here. The arrows identify somites and the zone of oscillatory gene expression in which future somites will be generated. (Image kindly provided by Julian Lewis.) (b) Shown here is the model for generating and synchronizing oscillatory expression of somite genes in zebrafish. Oscillatory gene expression is governed by a negative-feedback loop involving autorepressors Her1 and Her7. Synchronization of the oscillations between cells is achieved by Delta/Notch signaling. Her1/Her7 inhibit production of the Delta ligand. Conversely, Notch signaling stimulates production of the Her proteins. (Adapted from a figure kindly provided by Julian Lewis.)

Her7 repressor proteins diminish below a critical threshold because of their depletion by proteolysis, the block to transcription is relieved, and a new cycle of protein synthesis begins, restoring the repressed state, and so on, in repeated cycles. The oscillating levels of the repressor gene products regulate expression of other genes so as to define the pattern of each new somite. In zebrafish, a new somite forms about every 30 minutes. The key feature of the model that explains the timing of somite formation is the delay between when the *her1* and *her7* genes are switched ON and the accumulation of the autorepressors to a concentration sufficient to shut OFF their own synthesis (Fig. 20-20b).

The Her1/Her7 autoregulatory loop nicely explains how somite genes are expressed in an ON/OFF cycle in individual cells, but how is oscillatory gene expression in one cell kept in synchrony with that of other nearby cells in the prospective somite? Synchronization is achieved by an intercellular pathway of cell–cell signaling. In addition to repressing their own genes, Her1 and Her7 inhibit the expression of a gene coding for a cell-surface protein called Delta. Delta binds to the receptor protein Notch on neighboring cells (see Chapter 19). When activated by the Delta ligand, the Notch signaling system, in turn, stimulates the expression of the genes for Her1 and Her7. When Her protein levels are high in the ON/OFF cycle (and hence *her* gene expression low), production of the Delta ligand is low (Fig. 20-20b). As a consequence, Notch signaling and hence expression of *her* genes in adjacent cells is also low. Conversely, when the levels of Her1 and Her7 are low (and hence *her* gene expression high), Delta levels rise and thereby stimulate *her* gene expression in adjacent cells. In each case, the signal delivered from the neighbor via Notch collaborates with the cell's internal Her1 and Her7 to keep the cell and its neighbor oscillating in synchrony. Thus, interlocking cycles of negative autoregulation and intercellular signaling generate and coordinate oscillatory behavior among the cells that give rise to the somite.

Synthetic Circuits Mimic Some of the Features of Natural Regulatory Networks

A complementary approach to understanding the design principles that govern regulatory networks is to construct relatively simple circuits that mimic the features of natural systems, the goal of the field of synthetic biology. A dramatic example of successful circuit design that extends our discussion of oscillation is the "repressilator." The repressilator is a three-node network that was created in *E. coli* and that consists of three regulatory proteins linked to each other in a circular fashion in which the sign of all three edges is negative. The repressilator consists of the genes for the bacterial repressors λCI, LacI, and TetR such that CI represses the gene for LacI, which, in turn, represses the gene for TetR, which, to complete the network, represses the gene for CI. One might have anticipated that such a three-node circuit would result in a low steady-state level of transcription of all three genes. Instead, however, the repressilator exhibits a striking oscillatory pattern of transcription with a periodicity of about 2 hours. Presumably, fluctuations in the levels of the three repressors due to noise in the expression of their genes prevent the system from achieving steady state and results instead in an oscillatory pattern of expression. Still, the oscillatory behavior of the repressilator is far less robust than that of the natural systems considered above, which highlights the fact that the synthetic circuit is inadequate in mimicking the more intricate (but not yet fully elucidated) circuitry of natural oscillators.

Several other networks have been created synthetically that exhibit diverse stereotyped patterns of behavior. One example is a library of artificial circuits created from multiple transcription factors and multiple promoters in a variety of combinations. Members of this circuit library respond differentially to different combinations of input signals. Another example comes from the construction of "sender" and "responder" strains that create banded patterns of gene expression on agar plates. The sender strain is in the center of the plate and produces a signaling molecule that diffuses out from the center to create a gradient. Each of two responder strains, which are present throughout the plate, respond differentially to high and low concentrations of the signaling molecule by producing distinguishable, chromogenic reporter proteins. As a result, one responder strain produces coloring in a halo pattern that is close to the sender cells and the other produces a halo that is further away from the sender cells.

Prospects

The methods used in systems biology permit the systematic identification of every component engaged in a complex cellular process. The ability to obtain such information is prompting a paradigmatic shift in the way biologists analyze data. Instead of asking *how* a process works, it is now possible to ask *why* it is organized in a particular fashion. Looking ahead, the insights gained from systems biology in combination with the increasing sophistication of synthetic biology may some day make it possible to create artificial cells with the minimal circuitry for self-propagation. If so, then the future holds the prospect of artificial cells with tailor-made features, such as the capacity to efficiently metabolize pollutants, recycle waste materials, convert sunlight into fuel, or combat human disease.

SUMMARY

Large-scale DNA sequencing projects have led to the determination of complete genome assemblies for a variety of animals, including flies, sea squirts, fish, mice, and humans. A variety of experimental and computational methods permit the systematic identification of protein-coding genes, noncoding genes (such as those that specify microRNAs), and regulatory DNA sequences. Whole-genome tiling arrays have been particularly effective for the identification of transcribed sequences and the genome-wide identification of introns and exons. ChIP-chip assays represent a highly effective means for identifying regulatory DNA sequences, such as enhancers, which direct tissue-specific patterns of gene expression.

By far the most surprising result arising from comparative genome analyses is the fact that highly diverse animal genomes contain a similar set of genes. The lowly sea anemone and human genomes encode similar genetic "tool kits." These are the genes that guide the development of diverse cell types, and include cell signaling pathways such as FGF, Notch, EGF, Wnts, Hedgehog, and TGF-β. Regulatory genes are also highly conserved, such as those encoded by *Hox* complexes, and other key transcription factors containing the full complement of DNA binding domains, including HLH, bZIP, HMG, and zinc fingers. The recent comparison of sea anemones and humans reveals extensive synteny. Not only do the two genomes contain similar genes, but at least half the genes are contained within conserved linkage groups.

The development of powerful new sequencing technologies has opened the door to vast vistas of discovery. It is now possible to reconstruct the genomes of extinct organisms. DNA preparations obtained from Neanderthal fossils have been used to obtain genome sequence information. It should be possible to assemble a complete Neanderthal genome within the next few years.

Systems biology is a newly emerging field that seeks to describe complex levels of biological organization by using a combination of quantitative and high-throughput measurements, modeling, reconstruction, and theory. When applied to regulatory circuits, systems biology attempts to reveal principles of gene control that cannot be understood from the study of individual components in isolation. The complementary field of synthetic biology also seeks to elucidate design principles, but it attempts to do so by the creation of artificial regulatory networks that mimic features of natural circuits.

Transcription networks consist of nodes, which represent genes, and edges, which represent the regulation of one gene by another. In a simple, two-node regulatory motif, one gene controls the expression of another, and this regulation can be either negative or positive. Another simple motif is autoregulation, in which a gene regulates its own expression. Negative autoregulation, in which a gene represses its own expression, has the property of dampening noise, which is the variation in gene expression under seemingly uniform conditions. Positive autoregulation has the property of allowing steady-state expression to be reached slowly. An extreme form of positive autoregulation is the bistable switch in which a gene can be either OFF or ON for long periods of time.

Another common motif in regulatory networks is the feed-forward loop. A feed-forward loop is a three-node motif in which a regulatory gene (gene A) governs both the expression of a target gene and the expression of a second regulatory gene (gene B). This second regulatory gene also controls the expression of the target gene. Thus, in a feed-forward loop gene A controls the expression of the target gene both directly and indirectly via gene B. The expression of the target is subject to an AND gate in that expression is subject to two conditions: in one case, the presence of both activators and in the other, the presence of the activator and the absence of the repressor.

Some regulatory circuits in nature generate oscillating cycles of gene expression as observed in the cell cycle, development, and circadian rhythms. The design of these circuits is such that the appearance of one regulatory protein leads to its own disappearance and the appearance of a second regulatory protein. The second regulatory protein, in turn, causes its own disappearance and the reappearance of the first regulatory protein, thereby generating a continuing ON/OFF cycle of gene expression. A synthetic network consisting of three repressors linked in tandem in a circular circuit mimics natural oscillators in that it generates a cyclic pattern of gene expression but not with the robustness of natural oscillators.

BIBLIOGRAPHY

Books

Alon U. 2006. *An introduction to systems biology: Design principles of biological circuits.* Chapman & Hall/CRC, Boca Raton, Florida.

Sussman H.E. and Smith M.A., eds. 2006. *Genomes.* Cold Spring Harbor Laboratory Press, Cold Spring Harbor, New York.

Genomics

Arnosti D.N. 2003. Analysis and function of transcriptional regulatory elements: Insights from *Drosophila. Annu. Rev. Entomol.* **48:** 579–602.

Berman B.P., Nibu Y., Pfeiffer B.D., Tomancak P., Celniker S.E., Levine M., Rubin G.M., and Eisen M.B. 2002. Exploiting transcription factor binding site clustering to identify *cis*-regulatory modules involved in pattern formation in the *Drosophila* genome. *Proc. Natl. Acad. Sci.* **99:** 757–762.

Frazer K.A., Pachter L., Poliakov A., Rubin E.M., and Dubchak I. 2004. VISTA: Computational tools for comparative genomics. *Nucleic Acids Res.* **32:** W273–W279.

Green R.E., Krause J., Ptak S.E., Briggs A.W., Ronan M.T., Simons J.F., Du L., Egholm M., Rothberg J.M., Paunovic M., and Pääbo S. Analysis of one million base pairs of Neanderthal DNA. *Nature* **444:** 330–336.

Osoegawa K., Mammoser A.G., Wu C., Frengen E., Zeng C., Catanese J.J., and de Jong P.J. 2001. A bacterial artificial chromosome library for sequencing the complete human genome. *Genome Res.* **11:** 483–496.

Putnam N.H., Srivastava M., Hellsten U., Dirks B., Chapman J., Salamov A., Terry A., Shapiro H., Lindquist E., Kapitonov V.V., Jurka J., Genikhovich G., Grigoriev I.V., Lucas S.M., Steele R.E., Finnerty J.R., Technau U., Martindale M.Q., and Rokhsar D.S. 2007. Sea anemone genome reveals ancestral eumetazoan gene repertoire and genomic organization. *Science* **317:** 86–94.

Systems Biology

Alon U. 2007. Network motifs: Theory and experimental approaches. *Nat. Rev. Genet.* **8:** 450–461.

Bintu L., Buchler N.E., Garcia H.G., Gerland U., Hwa T., Kondev J., and Phillips R. 2005. Transcriptional regulation by the numbers: Models. *Curr. Opin. Genet. Dev.* **15:** 116–124.

Bintu L., Buchler N.E., Garcia H.G., Gerland U., Hwa T., Kondev J., Kuhlman T., and Phillips R. 2005. Transcriptional regulation by the numbers: Applications. *Curr. Opin. Genet. Dev.* **15:** 125–135.

Crosson S., McAdams H., and Shapiro L. 2004. A genetic oscillator and the regulation of cell cycle progression in *Caulobacter crescentus. Cell Cycle* **3:** 1252–1254.

Dubnau D. and Losick R. 2006. Bistability in bacteria. *Mol. Microbiol.* **61:** 564–572.

Endy D. 2005. Foundations for engineering biology. *Nature* **438:** 449–453.

McAdams H.H., Srinivasan B., and Arkin A.P. 2004. The evolution of genetic regulatory systems in bacteria. *Nat. Rev. Genet.* **5:** 169–178.

McGrath P.T., Viollier P., and McAdams H.H. 2004. Setting the pace: Mechanisms tying *Caulobacter* cell-cycle progression to macroscopic cellular events. *Curr. Opin. Microbiol.* **7:** 192–197.

Raser J.M. and O'Shea E.K. 2005. Noise in gene expression: Origins, consequences, and control. *Science* **309:** 2010–2013.

Sprinzak D. and Elowitz M.B. 2005. Reconstruction of genetic circuits. *Nature* **438:** 443–448.

Vilar J.M., Guet C.C., and Leibler S. 2003. Modeling network dynamics: The *lac* operon, a case study. *J. Cell Biol.* **161:** 471-476.

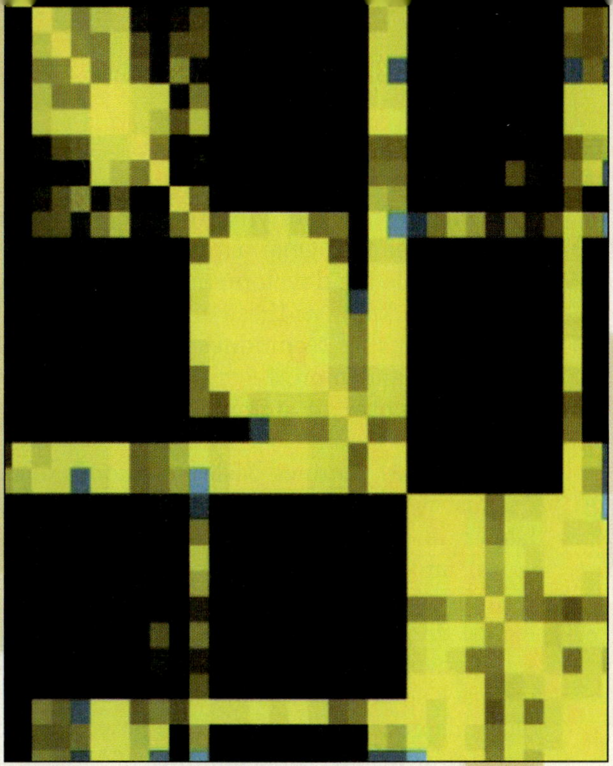

PART

5

METHODS

In Parts 2 and 3, we outlined our understanding of the molecular mechanisms underlying the central dogma; Part 4 focused on the mechanisms of gene expression and how differential gene expression controls the development and evolution of diverse animals. Most of what we know in these areas stems from the study of a few model organisms using techniques of genetics, molecular biology and biochemistry, and more recently genome analysis. The last part of this book is devoted to summarizing some of these methods and organisms.

Chapter 21 outlines basic techniques of molecular biology and biochemistry, which allow molecules (DNA, RNA, and proteins) to be isolated from cells (isolated, that is, from complex mixtures of such molecules) and studied in pure form in vitro. Chapter 22 outlines key features of a few model organisms whose study underpins modern biological thinking: bacteriophage and bacteria, yeast, the plant *Arabidopsis*, the worm *Caenorhabditis elegans*, the *Drosophila* fruit fly, and the mouse. Genetic analysis of these organisms has enabled the study of biological processes in vivo. The power of molecular biology—and the revolution in our understanding of biology gained from it during the last 50 years—stems from using a combination of in vivo genetic approaches and in vitro biochemical approaches.

A golden era of molecular biology was launched once it became possible to isolate specific DNA segments representing individual genes. In earlier times, it was possible to obtain bulk DNA from an organism, but only during the mid-1970s were methods developed that permitted the isolation of specific genes. The use of restriction enzymes and gel electrophoresis to isolate specific DNA fragments is described early in Chapter 21, and this is followed by a consideration of how such fragments can be amplified and expressed in vivo.

Next, we turn to techniques associated with in vitro amplification by polymerase chain reaction (PCR) and DNA sequencing, both of which require the chemical synthesis of DNA fragments for use as primers. This technique is briefly described. PCR permits the purification of virtually unlimited quantities of any given DNA segment—even when starting with just a single DNA molecule. PCR amplification has revolutionized many scientific disciplines, including forensics, medicine, ecology, and, of course, molecular biology.

In the mid- to late-1970s and 1980s, methods for DNA sequencing were still manual and somewhat laborious. During the 1990s, stimulated by the ambitions of the Human Genome Project, DNA sequencing became highly mechanized and has now developed to the point where it is possible to determine the exact nucleotide sequence of entire genomes in just days or weeks.

The first part of Chapter 21 describes the basic strategy for sequencing using the dideoxy chain termination method and its recent application to high-throughput sequencing. This part concludes with a description of exciting new approaches for whole-genome and very-high-throughput nanotechnology sequencing. In the second part we deal with methods of protein purification and analysis and close with an outline of the new field of proteomics. The third part of Chapter 21 presents methods for studying interactions between proteins and nucleic acids. Featured here are descriptions of DNA footprinting, gel mobility shift assays, and chromatin immunoprecipitation.

Chapter 22, in which we describe a handful of model organisms, stresses the principle that researchers employ the simplest organism in which the problem of interest can be studied. The simplest organisms of all—in terms of genome complexity and rapidity of the life cycle—are bacterial viruses, or bacteriophage. The study of bacteria and bacteriophage determined many of the basic features of DNA function, including the induction of gene expression, DNA replication, recombination, and repair. *Escherichia coli* and its phage were the key organisms of study in elucidating the genetic code during the early 1960s.

In the 1970s, molecular biologists were getting restless. Many felt that prokaryotes such as bacteria and their viruses had been conquered and that to answer the next round of biological questions demanded experiments on eukaryotes. Most accessible of these is the yeast, *Saccharomyces cerevisiae.* It has a very rapid life cycle, like bacteria, but nonetheless exhibits many of the properties of more elaborate eukaryotic cells. Yeast has been used for a variety of studies, including DNA replication, the cell cycle, and transcription regulation: these studies proved to be most valuable because in each case, it was found that yeast contain many of the molecular machines used in higher eukaryotes as well.

Chapter 22 ends with the three most popular animal models, the nematode worm, *C. elegans*, the fruit fly, *Drosophila melanogaster*, and the house mouse, *Mus musculus*, and the most widely used plant model, *Arabidopsis thaliana.* One of the big surprises in the past 20 years is the realization that many genetic processes are highly conserved among a broad spectrum of organisms, from simple metazoans to humans. Exhaustive genetic screens in the fruit fly, for example, have identified many of the signaling pathways and regulatory genes that control basic developmental processes common to higher animals as well. The development of highly sophisticated gene manipulation methods in transgenic mice have permitted researchers to determine what processes are controlled by the genetic pathways found in fruit flies. Genetically altered mice also provide models for testing ideas about, and treatments for, many human disorders, including Alzheimer's disease, Parkinson's disease, and rheumatoid arthritis.

PHOTOS FROM THE COLD SPRING HARBOR LABORATORY ARCHIVES

Seymour Benzer, 1975 Symposium on the Synapse. Using phage genetics, Benzer defined the smallest unit of mutation, which turned out later to be a single nucleotide (Chapter 22). This same work also provided an experimental definition of the gene—which he called a cistron—using functional complementation tests. Later, his studies focused on behavior, using the fruit fly as a model.

Werner Arber and Daniel Nathans, 1978 Symposium on DNA: Replication and Recombination. These two shared, with Hamilton O. Smith, the 1978 Nobel Prize in Physiology or Medicine for the characterization of type II restriction enzymes and their application to the molecular analysis of DNA (Chapter 21). This was one of the key discoveries in the development of recombinant DNA technology in the early 1970s.

Dale Kaiser, 1985 Symposium on Molecular Biology of Development. Kaiser contributed much to the early studies of bacteriophage λ propagation (Chapter 16). One aspect of this work led him to recognize that DNA molecules with complementary single-strand ends can readily be joined together, a finding critical to the development of recombinant DNA technologies.

Paul Berg, 1963 Symposium on Synthesis and Structure of Macromolecules. Berg was a pioneer in the construction of recombinant DNA molecules in vitro, work reflected in his share of the 1980 Nobel Prize in Chemistry.

Walter Gilbert and David Botstein, 1986 Symposium on Molecular Biology of *Homo sapiens*. Gilbert, who invented a chemical method for sequencing DNA, is shown here with Botstein during the historic debate about whether it was feasible and sensible to attempt to sequence the human genome. Botstein, after working with phage for many years, contributed much to the development of the yeast *Saccharomyces cerevisiae* as a model eukaryote for molecular biologists; he was also an early figure in the emerging field of genomics (Chapters 20 and 22).

Albert Keston, Sidney Udenfriend, and Frederick Sanger, 1949 Symposium on Amino Acids and Proteins. Keston—inventor of the test tape for detecting glucose—and Udenfriend—developer of screens for, and tests of, antimalarial drugs—are here shown with Sanger, the only person to win two Nobel Prizes in Chemistry. The first, in 1958, was for developing a method to determine the amino acid sequence of a protein; the second, 22 years later, was for developing the method for sequencing DNA that is now used almost exclusively, including in the automated machines used to sequence whole genomes (Chapters 20 and 21). Beyond the obvious technological achievement, determining that a protein had a defined sequence revealed for the first time that it likely had a defined structure as well.

Francis S. Collins and Maynard V. Olson, 1992 Genome Mapping and Sequencing Meeting. Collins was one of the early "gene hunters," finding first the much sought after cystic fibrosis gene in 1989. In 1993 he took over from James Watson as Director of the National Center for Human Genome Research and remains in the comparable position today as Director of the National Human Genome Research Institute. He is here seen listening to Olson in front of a poster about yeast artificial chromosomes (YACs), the vectors Olson had created a few years earlier and which allowed a tenfold jump in the size of DNA fragments that could at the time be cloned (Chapters 20 and 21).

Eric S. Lander (speaking), 1986 Symposium on Molecular Biology of *Homo sapiens*. Lander was to become a leading figure in the public Human Genome Project and first author on the paper it produced reporting that sequence in 2001 (Chapter 20). As in the photo on the facing page of Gilbert and Botstein, Lander is here giving forth his views at the 1986 debate on whether it was worth trying to sequence the human genome. Beside him, David Page, whose work has focused on the structure, function, and evolution of the Y chromosome, appears thoughtful; in the foreground, Nancy Hopkins (a developmental biologist and an author on the fourth edition of this book)—and, in the background, James Watson—seem more amused.

Leroy Hood and J. Craig Venter, 1990 Genome Mappping and Sequencing Meeting. Hood invented automated sequencing, building his first machine in the mid-1980s. It was Venter who later took first and greatest advantage of automated sequencing: by marrying the raw sequencing power such machines offered with a shotgun strategy, he greatly accelerated the sequencing of whole genomes including that of the human (Chapters 20 and 21).

CHAPTER

21 Techniques of Molecular Biology

OUTLINE

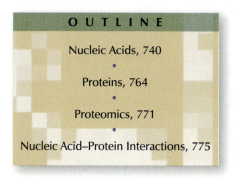

THE LIVING CELL, AS WE HAVE SEEN, IS AN extraordinarily complicated entity, producing thousands of different macromolecules and harboring a genome that ranges in size from a million to billions of base pairs. Understanding how the genetic processes of the cell work requires a variety of challenging experimental approaches, including the use of suitable model organisms in which the tools of genetic analysis are available, as discussed in Chapter 22. These also include, as discussed here, methods for separating individual macromolecules from the myriad mixtures found in the cell and for dissecting the genome into manageably sized segments for manipulation and analysis of specific DNA sequences. The successful development of such methods has been one of the major driving forces in the field of molecular biology during the last several decades, as well as one of its greatest triumphs.

Recently, it has become possible to apply molecular approaches to the large-scale analysis of the full complement of RNAs and proteins in the cell and to determine the nucleotide sequence of entire genomes. These genomic and proteomic approaches, and the rapidly increasing number of genome sequences becoming available, make it possible to undertake large-scale comparisons of the genomes of different organisms or to identify all of the phosphorylated proteins in a particular cell type.

In this chapter, we provide a brief introduction to these molecular, genomic, and proteomic methods and to the principles upon which they are based. As we shall see, the methods of molecular biology depend on, and were developed from, an understanding of the properties of biological macromolecules themselves. For example, an understanding of the structure and base-pairing characteristics of DNA and RNA gave rise to the development of techniques of hybridization and sequencing that allow for the rapid and detailed analysis of gene structure and gene expression. Insight into the activities of DNA polymerases, restriction endonucleases, and DNA ligases gave birth to the techniques of DNA cloning and the polymerase chain reaction (PCR), which allow scientists to isolate essentially any DNA segment—even some from extinct life-forms—in unlimited quantities.

The diversity of protein structure and function makes their analysis more difficult because each protein is in many ways its own unique puzzle. Nevertheless, scientists have developed tools to purify individual proteins and investigate their function. More recently the application of mass spectrometry has greatly enhanced the ability to identify the presence of very small amounts of a protein in a complex mixture or even the particular amino acids on a protein that are modified (e.g., by phosphate). These techniques are allowing researchers to investigate protein populations in a global manner rather than one at a time.

This chapter is divided into four parts. The first part is devoted to techniques for the manipulation and characterization of nucleic acids, from the isolation of RNAs and DNAs to the sequencing of entire genomes and comparative genomics. The next two parts are concerned with the isolation and analysis of proteins: the purification of individual proteins and proteomic methods for analyzing the full array of proteins in a cell or tissue. Although these categories of techniques are dissimilar in detail, many of the procedures for isolating and manipulating nucleic acids and proteins are, as we shall see, based on common underlying principles. The fourth part of this chapter describes the analysis of nucleic acid–protein interactions, approaches that help us to explore how these separate components come together and interact to facilitate the inner workings of the cell.

Finally, a note: it is important to appreciate that when we talk about isolating and purifying a given macromolecule in the ensuing discussion, we rarely (if ever) mean that a single molecule is isolated. Rather, the goal of these procedures is to isolate a large population of identical molecules away from all of the other kinds of molecules in the cell.

NUCLEIC ACIDS

Electrophoresis through a Gel Separates DNA and RNA Molecules according to Size

We begin by discussing the separation of DNA and RNA molecules by the technique of **gel electrophoresis**. Linear DNA molecules separate according to size when subject to an electric field through a **gel matrix**, an inert, jelly-like porous material. Because DNA is negatively charged, when subject to an electrical field in this way, it migrates through the gel toward the positive pole (Fig. 21-1). DNA molecules

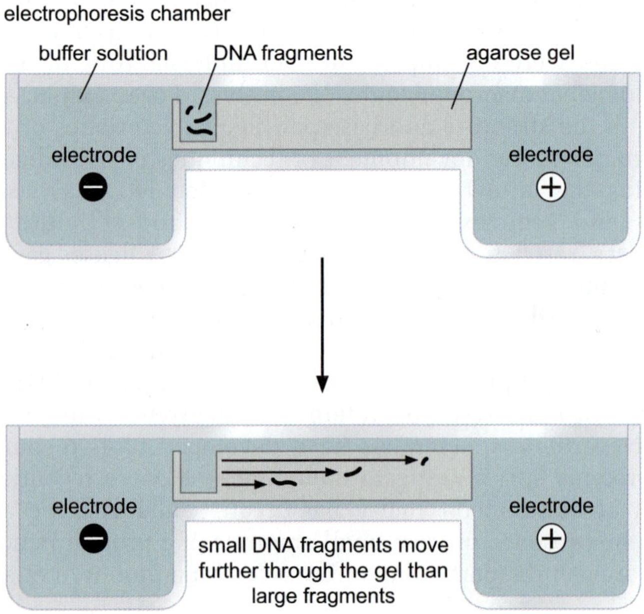

FIGURE 21-1 **DNA separation by gel electrophoresis.** The figure shows a gel from the side in cross section. Thus, the "well" into which the DNA mixture is loaded onto the gel is indicated at the left, at the head of the gel. This is also the end at which the cathode of the electric field is located, the anode being at the foot of the gel. As a result, the DNA fragments, which are negatively charged, move through the gel from the head to the foot. The distance they travel is inversely related to the size of the DNA fragment, as shown. (Adapted, with permission, from Micklos D.A. and Freyer G.A. 2003. *DNA science: A first course*, 2nd ed., p. 114. © Cold Spring Harbor Laboratory Press.)

are flexible and occupy an effective volume. The gel matrix acts as a sieve through which DNA molecules pass; large molecules (with a larger effective volume) have more difficulty passing through the pores of the gel and so migrate through the gel more slowly than do smaller DNAs. This means that once the gels have been electrophoresced or "run" for a given time, molecules of different sizes are separated because they have moved different distances through the gel.

After electrophoresis is complete, the DNA molecules can be visualized by staining the gel with fluorescent dyes, such as **ethidium**, which binds to DNA and intercalates between the stacked bases (see Fig. 6-28). Each band reveals the presence of a population of DNA molecules of a specific size.

Two alternative kinds of gel matrices are used: **polyacrylamide** and **agarose**. Polyacrylamide has high resolving capability but can separate DNAs over only a narrow size range. Thus, electrophoresis through polyacrylamide can resolve DNAs that differ from each other in size by as little as a single base pair but only with molecules of up to several hundred (just under 1000) base pairs. Agarose has less resolving power than polyacrylamide but can separate DNA molecules of up to tens, and even hundreds, of kilobases.

Very long DNAs are unable to penetrate the pores even in agarose. Instead, they snake their way through the matrix with one end leading the way and the other end trailing from behind. As a consequence, DNA molecules above a certain size (30–50 kb) migrate to a similar extent and cannot be resolved. These very long DNAs can be resolved from one another if the electric field is applied in pulses that are oriented orthogonally to each other. This technique is known as **pulsed-field** gel electrophoresis (Fig. 21-2). Each time the orientation of the electric field changes, the DNA molecule, which is snaking its way through the gel, must reorient to the direction of the new field. The larger the DNA, the longer it takes to reorient. Pulsed-field gel electrophoresis can be used to determine the size of entire bacterial chromosomes and chromosomes of lower eukaryotes, such as fungi, that is, molecules of up to several megabases in length.

Electrophoresis separates DNA molecules not only according to their molecular weight, but also according to their shape and topological properties (Fig. 6-26). A circular DNA molecule that is relaxed or nicked migrates more slowly than does a linear molecule of equal mass. In addition, as we have seen, supercoiled DNAs, which are compact and have a small effective volume, migrate more rapidly during electrophoresis than do less supercoiled or relaxed circular DNAs of equal mass.

Electrophoresis is used to separate RNAs as well. Linear double-stranded DNAs have a uniform secondary structure, and their rate of migration during electrophoresis is proportional to their molecular weight. Like DNAs, RNAs have a uniform negative charge. But RNA molecules are usually single-stranded and have, as we have seen (Chapter 6), extensive secondary and tertiary structures, which influences their electrophoretic mobility. To deal with this, RNAs can be treated with reagents, such as glyoxal, that react with the RNA in such a way as to prevent the formation of base pairs (glyoxal forms adducts with amino groups in the bases, thereby preventing base pairing). Glyoxylated RNAs are unable to form secondary or tertiary structures and hence migrate with a mobility that is approximately proportional to molecular weight. As we see in a later section, electrophoresis is used in a similar way to separate proteins on the basis of their size.

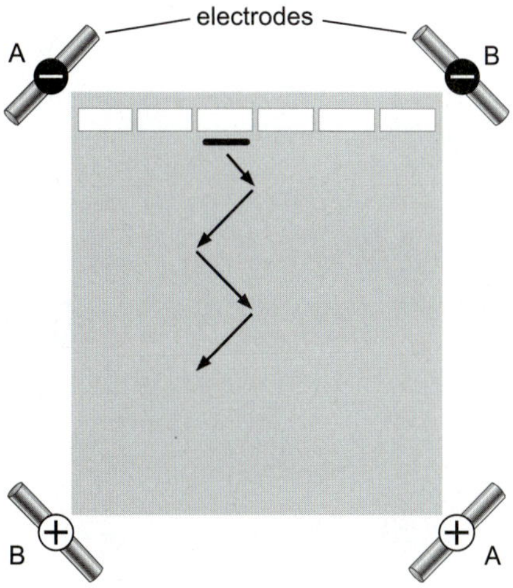

FIGURE 21-2 **Pulsed-field gel electrophoresis.** In this figure, the agarose gel is shown from above with the head of the gel and a series of sample wells, at the top. A and B represent two sets of electrodes. These are switched on and off alternately, as described in the text. When A is on, the DNA is driven toward the bottom right corner of the gel where the anode of that pair is situated. When A is switched off, and B is switched on, the DNA moves toward the bottom left corner. The arrows thus show the path followed by the DNA as electrophoresis proceeds. (Adapted, with permission, from Sambrook J. and Russell D.W. 2001. *Molecular cloning: A laboratory manual*, 3rd ed., Fig. 5-7. © Cold Spring Harbor Laboratory Press.)

Restriction Endonucleases Cleave DNA Molecules at Particular Sites

Most naturally occurring DNA molecules are much larger than can readily be managed, or analyzed, in the lab. For example, chromosomes are extremely long single DNA molecules that can contain thousands of genes and more than 100 Mb of DNA (see Chapter 7). If we are to study individual genes and individual sites on DNA, the large DNA molecules found in cells must be broken into manageable fragments. This is done using **restriction endonucleases**. These are nucleases that cleave DNA at particular sites by the recognizing specific sequences.

Restriction enzymes used in molecular biology typically recognize short (4–8-bp) target sequences, usually palindromic, and cut at a defined position within those sequences. Consider one widely used restriction enzyme, EcoRI, so named because it was found in certain strains of *Escherichia coli*, and was the first (I) such enzyme found in that species. This enzyme recognizes and cleaves the sequence 5′-GAATTC-3′. (Because the two strands of DNA are complementary, we need specify only one strand and its polarity to describe a recognition sequence unambiguously.)

This hexameric sequence (like any other) would be expected to occur once in every 4 kb on average. (This is because there are four possible bases that can occur at any given position within a DNA sequence, and so the chances of finding any given specific 6-bp sequence is 1 in 4^6.) Consider a linear DNA molecule with six copies of the GAATTC sequence: EcoRI would cut it into seven fragments in a range of sizes reflecting the distribution of those sites in the molecule. Subjecting the EcoRI-cut DNA to electrophoresis through a gel would separate the seven fragments from each other on the basis of their different sizes (Fig. 21-3). Thus, in the experiment shown, EcoRI has dissected the DNA into specific fragments, each corresponding to a particular region of the molecule.

If the same DNA molecule had been cleaved with a different restriction enzyme—for example, HindIII, which also recognizes a 6-bp target but of a different sequence (5′-AAGCTT-3′)—the molecule would have been cut at different positions and generated fragments of

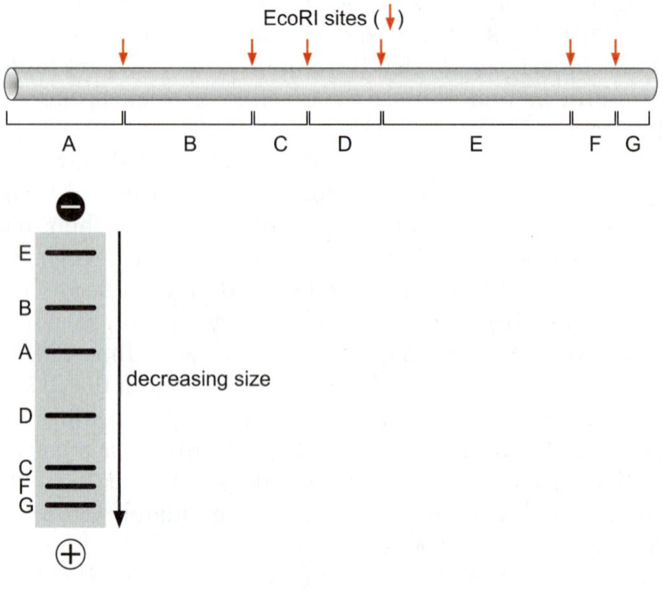

FIGURE 21-3 Digestion of a DNA fragment with endonuclease EcoRI. At the top is shown a DNA molecule and the positions within it at which EcoRI cleaves. When the molecule, digested with that enzyme, is run on an agarose gel, the pattern of bands shown is observed.

TABLE 21-1 Some Restriction Endonucleases and Their Recognition Sequences

Enzyme	Sequence	Cut Frequency[a]
Sau3A1	5´-GATC-3´	0.25 kb
EcoRI	5´-GAATTC-3´	4 kb
NotI	5´-GCGGCCGC-3´	65 kb

[a]Frequency = 1/4*n*, where *n* is the number of base pairs in the recognition sequence.

different sizes. Thus, the use of multiple enzymes allows different regions of a DNA molecule to be isolated. It also allows a given molecule to be identified. Thus, a given molecule will generate a characteristic series of patterns when digested with a set of different enzymes.

Other restriction enzymes such as Sau3A1 (which is found in the bacterium *Staphylococcus aureus*) recognize tetrameric sequences (5´-GATC-3´) and so cut DNA more frequently, approximately once every 250 bp. At the other extreme are enzymes that recognize octomeric sequences such as NotI, which recognizes the octameric sequence 5´-GCGGCCGC-3´ and cuts, on average, only once every 65 kb (Table 21-1).

Of note, some restriction enzymes are sensitive to methylation. That is, methylation of a base (or bases) within a recognition sequence inhibits enzyme activity at that site. Restriction enzymes differ not only in the specificity and length of their recognition sequences but also in the nature of the DNA ends they generate. Thus, some enzymes, such as HpaI, generate flush ends; others, such as EcoRI, HindIII, and PstI, generate staggered ends (Fig. 21-4). For example, EcoRI cleaves covalent (phosphodiester) bonds between G and A at staggered positions on each strand. The hydrogen bonds between the 4 bp between these cut sites are easily broken to generate 5´ protruding ends of four nucleotides in length (Fig. 21-5). Note that these ends are complementary to each other. They are said to be "sticky" because they readily anneal through base pairing to each other or to other DNA molecules cut with the same enzyme. This is a useful property that we consider in the discussion on DNA cloning.

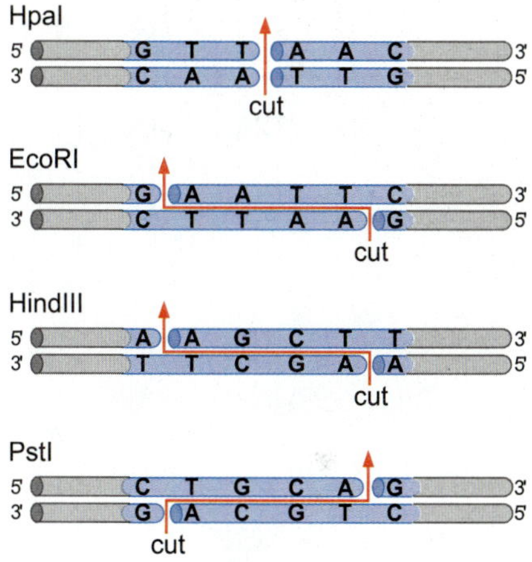

FIGURE 21-4 Recognition sequences and cut sites of various endonucleases. As shown, different endonucleases not only recognize different target sites but also cut at different positions within those sites. Thus, molecules with blunt ends or with 5´ or 3´ overhanging ends can be generated.

DNA Hybridization Can Be Used to Identify Specific DNA Molecules

As we saw in Chapter 6, the capacity of denatured DNA to reanneal (that is, to re-form base pairs between complementary strands) allows for the formation of hybrid molecules when homologous, denatured DNAs from two different sources are mixed with each other under the appropriate conditions of ionic strength and temperature. This process of base pairing between complementary single-stranded polynucleotides is known as **hybridization**.

Many techniques rely on the specificity of hybridization between two DNA molecules of complementary sequence. For example, this property is the basis for detecting specific sequences within complicated mixtures of nucleic acids. In this case, one of the molecules is a **probe** of defined sequence—either a purified fragment or a chemically synthesized DNA molecule. The probe is used to search mixtures of nucleic acids for molecules containing a complementary sequence. The probe DNA must be labeled so that it can be readily located once it has found its target sequence. The mixture being probed is typically either separated by size on a gel or distributed as a library of clones (see below).

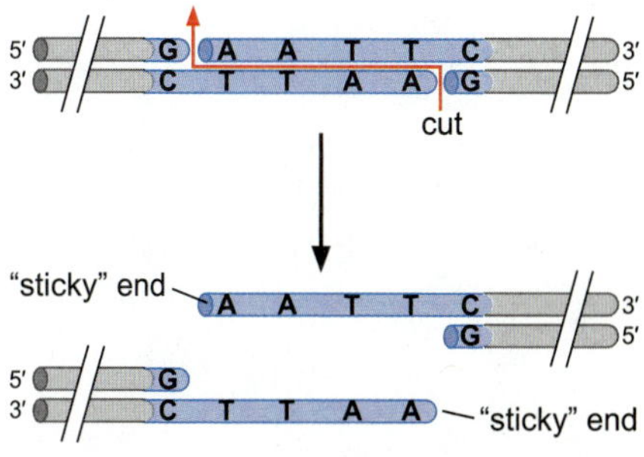

FIGURE 21-5 Cleavage of an EcoRI site. EcoRI cuts the two strands within its recognition site to give 5´ overhanging ends. These are called "sticky" ends—they readily adhere to other molecules cut with the same enzyme because they provide complementary single-strand ends that come together through base pairing.

There are two basic methods for labeling DNA. The first involves synthesizing new DNA in the presence of a labeled precursor, as we describe below. The other involves adding a label to the end of an intact DNA molecule. Thus, for example, the enzyme polynucleotide kinase adds the γ-phosphate from ATP to the 5′-OH group of DNA. If that phosphate is radioactive, this process labels the DNA molecule to which it is transferred.

Labeling by incorporation (the other mechanism) is often carried out by using PCR, which we discuss below, with a labeled precursor, or even by hybridizing short random hexameric oligonucleotides to DNA and allowing a DNA polymerase to extend them. The labeled precursors are most commonly nucleotides modified with either a fluorescent moiety or radioactive atoms. Typically, the fluorescent moiety need only be attached to the base of one of the four nucleotides used as precursors for DNA synthesis (~25% of labeling is generally sufficient for most purposes).

DNA labeled with fluorescent precursors can be detected by illuminating the DNA sample with appropriate wavelength UV light and monitoring the longer wavelength light that is emitted in response. Radioactively labeled precursors typically have radioactive ^{32}P or ^{35}S incorporated into the α-phosphate of one of the four nucleotides. Recall that this phosphate is retained in the product DNA (see Chapter 8). Radioactive DNA can be detected by exposing the sample of interest to X-ray film or by photomultipliers that emit light in response to excitation by the β particles emitted from ^{32}P and ^{35}S.

There are many ways that hybridization is used in the identification of specific DNA or RNA fragments. The two most common are described below.

Hybridization Probes Can Identify Electrophoretically Separated DNAs and RNAs

It is often desirable to monitor the abundance or size of a particular DNA or RNA molecule in a population of many other similar molecules. For example, this can be useful when determining the amount of a specific mRNA that is expressed in two different cell types or the length of a restriction fragment that contains the gene being studied. This type of information can be obtained using blotting methods that localize specific nucleic acids after they have been separated by electrophoresis.

Suppose that the yeast genome has been cleaved with the restriction enzyme EcoRI and the investigator wants to know the size of the fragment that contains the gene of interest. When stained with ethidium bromide, the thousands of DNA fragments generated by cutting the yeast genome are too numerous to resolve into discretely visible bands, and they look like a smear centered around 4 kb. The technique of **Southern blot hybridization** (named after its inventor Edward Southern) will identify within the smear the size of the particular fragment containing the gene of interest.

In this procedure, the cut DNA is separated by gel electrophoresis, and the gel is soaked in alkali to denature the double-stranded DNA fragments. These fragments are then transferred from the gel to a positively charged membrane to which they adhere, creating an imprint, or "blot" of the gel. During the transfer process, the DNA fragments are bound to the membrane in positions that mirror their corresponding positions in the gel after electrophoresis. After DNAs of interest are bound to the membrane, the charged membrane is incubated with a mixture of nonspecific DNA fragments to saturate all the remaining binding sites on the membrane. Because the DNA in this mixture is randomly distributed on the membrane and, if chosen properly, will not contain the sequence of interest, it will not interfere with subsequent detection of a specific gene.

The DNA bound to the membrane is then incubated with probe DNA containing a sequence complementary to a sequence within the gene of interest. Because all of the nonspecific binding sites on the membrane are occupied, the only way that the probe DNA can associate with the membrane is by hybridizing to any complementary DNA present on the membrane. This probing is done under conditions of salt concentration and temperature close to those at which nucleic acids denature and renature. Under these conditions, the probe DNA will hybridize tightly to only its exact complement. Often the probe is in high molar excess compared to its immobilized target on the filter, thereby favoring hybridization rather than the reannealing of the denatured DNA. In addition, the immobilization of the denatured DNA on the filter tends to interfere with renaturation anyway. A variety of films or other media sensitive to the light or electrons emitted by the labeled DNA can detect where on the blot the probe hybridizes. When, for example, an X-ray film is exposed to the filter and then developed, an **autoradiogram** is produced in which the pattern of exposure on the film corresponds to the position of the hybrids on the blot (Fig. 21-6).

A similar procedure called northern blot hybridization (to distinguish it from Southern blot hybridization) can be used to identify a particular mRNA in a population of RNAs. Because mRNAs are relatively short (typically less than 5 kb), there is no need for them to be digested with any enzymes (there are only a limited number of specific RNA-cleaving enzymes anyway). Otherwise, the protocol is fairly similar to that described for Southern blotting. The separated mRNAs are transferred to a positively charged membrane and probed with a radioactive DNA of choice. (In this case, hybrids are formed by base pairing between complementary strands of RNA and DNA.)

An investigator might carry out northern blot hybridization to ascertain the amount of a particular mRNA present in a sample rather than its size. This measure is a reflection of the level of expression of the gene that encodes that mRNA. Thus, for example, one might use northern blot hybridization to ask how much more mRNA of a specific type is present in a cell treated with an inducer of the gene in question compared to an uninduced cell. As another example, northern blot hybridization might be carried out to compare the relative levels of a particular mRNA (and hence the expression level of the gene in question) among different tissues of an organism. Because an excess of DNA probe is used in these assays, the amount of hybridization is related to the amount of mRNA present in the original sample, allowing the relative amounts of mRNA to be determined.

The principles of Southern and northern blot hybridization also underlie gene microarray analysis, which we consider in detail in Chapter 20. The availability of vast amounts of sequence information has enabled development of this "reverse hybridization" experiment. A microarray is constructed by using several hundred to several thousand known DNA sequences, each corresponding to a different gene in the organism under study. Each sequence is individually fixed in an array on the microarray chip (see Fig. 19-1). Note that the use of terms here is the reverse of their use in Southern or northern analysis. In microarray analysis, the fixed, unlabeled sequences are called the

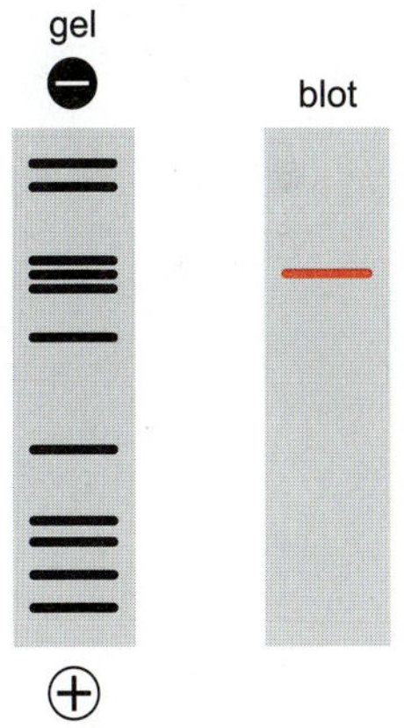

FIGURE 21-6 **A Southern blot.** DNA fragments, generated by digestion of a DNA molecule by a restriction enzyme, are run out on an agarose gel. Once stained, a pattern of fragments is seen. When transferred to a filter and probed with a DNA fragment homologous to just one sequence in the digested molecule, a single band is seen, corresponding to the position on the gel of the fragment containing that sequence.

probes, as these are known, whereas the target is composed of amplified, labeled cDNA generated from total RNA from a cell or tissue. When target sequences are hybridized to the array of probe DNAs, the intensity of the hybridization signal to each DNA species in the array is a measure of the level of expression of the gene in question.

Isolation of Specific Segments of DNA

Much of the molecular analysis of genes and their function requires the separation of specific segments of DNA from much larger DNA molecules and their selective amplification. Isolating a large amount of a single pure DNA molecule facilitates the analysis of the information encoded in that particular DNA molecule. Thus, the DNA can be sequenced and analyzed, or it can be cloned and expressed to allow the study of its protein product.

The ability to purify specific DNA molecules in significant quantities allows them to be manipulated in various other ways as well. For example, recombinant DNA molecules can be created and used to alter the expression of a particular gene (e.g., by fusing its coding sequence to a heterologous promoter). Alternatively, purified DNA sequences can be recombined to generate DNAs that encode so-called fusion proteins—that is, hybrid proteins made up of parts derived from different proteins. The techniques of DNA cloning and amplification by PCR have become essential tools in asking questions about the control of gene expression, maintenance of the genome, and protein function.

DNA Cloning

The ability to construct recombinant DNA molecules and maintain them in cells is called **DNA cloning**. This process typically involves a vector that provides the information necessary to propagate the cloned DNA in the cell and an **insert DNA** that is inserted within the vector and includes the DNA of interest. Key to creating recombinant DNA molecules are the restriction enzymes that cut DNA at specific sequences and other enzymes that join the cut DNAs to one another. By creating recombinant DNA molecules that can be propagated in a host organism, a particular DNA fragment can be both purified from other DNAs and amplified to produce large quantities.

In the remainder of this section, we describe how DNA molecules are cut, recombined, and propagated. We then discuss how large collections of such hybrid molecules, called libraries, can be created. In a library, a common vector carries many alternative inserts. We describe how libraries are made and how specific DNA segments can be identified and isolated from them.

Cloning DNA in Plasmid Vectors

Once DNA is cleaved into fragments by a restriction enzyme, it typically needs to be inserted into a vector for propagation. That is, the DNA fragment must be inserted into a second DNA molecule (the vector) to be replicated in a host organism. The most common host used to propagate DNA is the bacterium *E. coli*. Vector DNAs typically have three characteristics.

1. They contain an origin of replication that allows them to replicate independently of the chromosome of the host. (Note that yeast vectors also require a centromere.)
2. They contain a selectable marker that allows cells that contain the vector (and any attached DNA) to be readily identified.
3. They have unique sites for one or more restriction enzymes. This allows DNA fragments to be inserted at a defined point within the vector such that the insertion does not interfere with the first two functions.

The most common vectors are small (~3 kb) circular DNA molecules called **plasmids**. These molecules were originally derived from circular DNA molecules that are found naturally in many bacteria and single-cell eukaryotes (Chapter 22). In many cases (although not in yeast), these DNAs carry genes encoding resistance to antibiotics. Thus, naturally occurring plasmids already have two of the characteristics desirable for a vector: they can propagate independently in the host and they carry a selectable marker. A further benefit is that these plasmids are sometimes present in multiple copies per cell. This increases the amount of DNA that can be isolated from a population of cells.

In some cases, these plasmids also have useful unique restriction sites. However, since their discovery, plasmids have been simplified and modified such that a typical plasmid vector now has greater than 20 unique restriction sites within a small region, known as a multiple site. This modification allows a much more diverse array of restriction enzymes to be used to cut the target DNA.

Inserting a fragment of DNA into a vector is a relatively simple process (Fig. 21-7). Suppose that a plasmid vector has a unique recognition site for EcoRI. The vector is prepared by digesting it with EcoRI, which linearizes the plasmid. Because EcoRI generates protruding 5′ ends that are complementary to each other (see Fig. 21-5), the sticky ends are capable of reannealing to re-form a circle with two nicks. Treatment of the circle with the enzyme **DNA ligase** and ATP would seal the nicks to re-form a covalently closed circle. The target DNA is prepared by cleaving it with a restriction enzyme, in this case with EcoRI, to generate potential insert DNAs. Vector DNA is mixed with an excess of insert DNAs cleaved by EcoRI under conditions that allow sticky ends to hybridize. DNA ligase is then used to link the compatible ends of the two DNAs. Adding an excess of the insert DNA relative to the plasmid DNA ensures that the majority of vectors will reseal with insert DNA incorporated (Fig. 21-7).

Some vectors not only allow the isolation and purification of a particular DNA but also drive the expression of genes within the insert DNA. These plasmids are called **expression vectors** and have transcriptional promoters, derived from the host cell, immediately adjacent to the site of insertion. If the coding region of a gene (without its promoter) is placed at the site of insertion in the proper orientation, then the inserted gene will be transcribed into mRNA and translated into protein by the host cell. Expression vectors are frequently used to express heterologous or mutant genes to assess their function. They can also be used to produce large amounts of a protein for purification. In addition, the promoter in the expression vector can be chosen such that expression of the insert is regulated by the addition of a simple compound to the growth media—for example, a sugar or an amino acid (see Chapter 16 for a discussion of transcriptional regulation in prokaryotes). This ability to control when the gene will be expressed is particularly useful if the gene product is toxic.

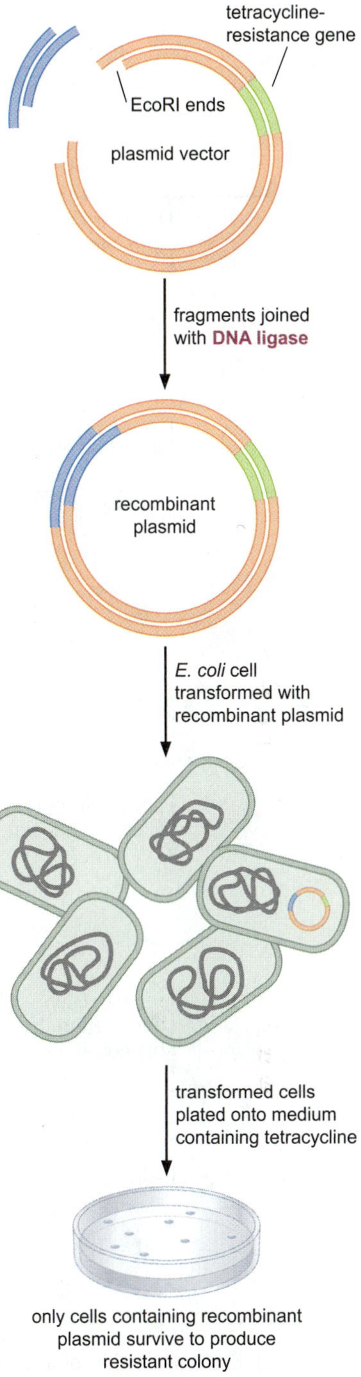

FIGURE 21-7 Cloning in a plasmid vector. A fragment of DNA, generated by cleavage with EcoRI, is inserted into the plasmid vector linearized by that same enzyme. Once ligated (see text), the recombinant plasmid is introduced into bacteria, by transformation (see text). Cells containing the plasmid can be selected by growth on the antibiotic to which the plasmid confers resistance. (Adapted, with permission, from Micklos D.A. and Freyer G.A. 2003. *DNA science: A first course*, 2nd ed., p. 129. © Cold Spring Harbor Laboratory Press.)

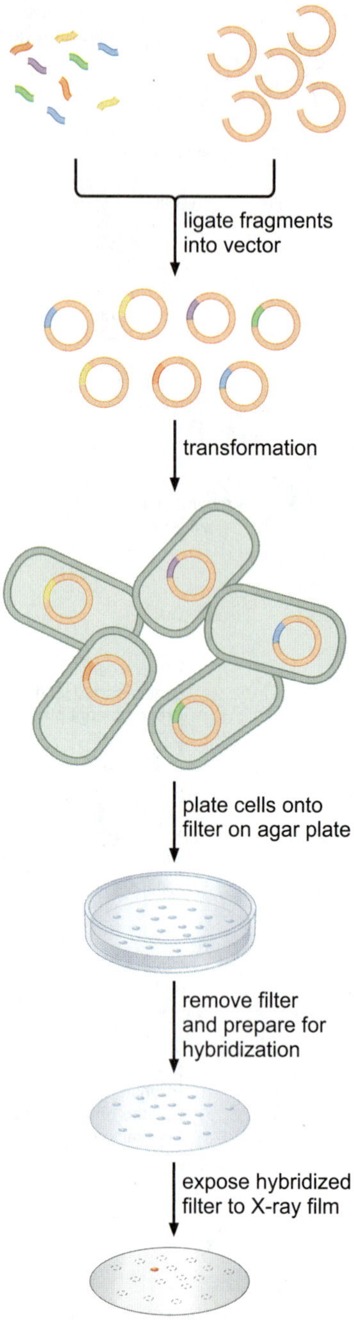

FIGURE 21-8 Construction of a DNA library. To construct the library, genomic DNA and vector DNA, digested with the same restriction enzyme, are incubated together with ligase. The resulting pool or library of hybrid vectors (each vector carrying a different insert of genomic DNA, represented in a different color) is then introduced into *E. coli*, and the cells are plated onto a filter placed over agar medium. Once colonies have grown, the filter is removed from the plate and prepared for hybridization: cells are lysed, the DNA is denatured, and the filter is incubated with a labeled probe. The clone of interest is identified by autoradiography.

Vector DNA Can Be Introduced into Host Organisms by Transformation

Propagation of the vector with its insert DNA is achieved by introducing the recombinant DNA into a host cell by transformation. As we discussed in Chapter 2, **transformation** is the process by which a host organism can take up DNA from its environment. Some bacteria, but not *E. coli*, can do this naturally and are said to have **genetic competence**. *E. coli* can be rendered competent to take up DNA, however, by treatment with calcium ions. Although the exact mechanism for DNA uptake is not known, it is likely that the Ca^{2+} ions shield the negative charge on the DNA, allowing it to pass through the cell membrane. Thus, calcium-treated cells are said to be competent to be transformed. An antibiotic to which the plasmid imparts resistance is then included in the medium to select for the growth of cells that have taken up the plasmid DNA—these cells are called transformants. Cells harboring the plasmid will be able to grow in the presence of the antibiotic, whereas those lacking it will not.

Transformation is a relatively inefficient process. Only a small percentage of the DNA-treated cells take up the plasmid. It is this low efficiency of transformation that makes it necessary to use selection with the antibiotic. The inefficiency of transformation also ensures that, in most cases, each cell receives only a single molecule of DNA. This property makes each transformed cell and its progeny a carrier of a unique DNA molecule. Thus, transformation effectively purifies and amplifies one DNA molecule away from all other DNAs in the transforming mixture.

Libraries of DNA Molecules Can Be Created by Cloning

It is trivial to generate a specific clone if the starting donor DNA is simple. Thus, if the starting DNA is small (derived from a small virus, for example, with a genome of perhaps only 10 kb), then this can be accomplished simply by separating the DNA fragments after digestion with restriction enzymes and gel electrophoresis. Once separated, DNAs of different sizes can be excised from the gel and purified prior to insertion into a vector.

This is harder to do if the starting DNA is more complex (e.g., the human genome). In this case, simple electrophoretic separation of DNA treated with a restriction enzyme will result in very many fragments distributed in a broad range of sizes around the average distance between cut sites. Thus, it is easier under these circumstances to clone the whole population of fragments and separate the individual clones afterward.

A DNA **library** is a population of identical vectors that each contains a different DNA insert (Fig. 21-8). To construct a DNA library, the target DNA (e.g., human genomic DNA) is digested with a restriction enzyme that gives a desired average insert size. The insert size can be of any size ranging from less than 100 bp to more than a megabase (for very large insert sizes, the DNA is typically incompletely cut with a restriction enzyme). The cleaved DNA is then mixed with the appropriate vector cut with the same restriction enzyme in the presence of ligase. This creates a large collection of vectors with different DNA inserts.

Different kinds of libraries are made using insert DNA from different sources. The simplest are derived from total genomic DNA cleaved with a restriction enzyme; these are called **genomic libraries.** This type of library is most useful when generating DNA for sequencing a genome. If, on the other hand, the objective is to clone a DNA fragment encoding a particular gene, then a genomic library can be used efficiently only when the organism in question has relatively little noncoding DNA. For an organism with a more complex genome, this type of library is not suitable for this task because many of the DNA inserts will not contain coding DNA sequences.

To enrich for coding sequences in the library, a **cDNA library** is created. This is made as shown in Figure 21-9. Instead of starting with genomic DNA, mRNA is converted into DNA sequence. The process that allows this is called **reverse transcription** and is performed by a special DNA polymerase (reverse transcriptase) that can make DNA from an RNA template (see Chapter 11). When treated with reverse transcriptase, mRNA sequences can be converted into double-stranded DNA copies called **cDNAs** (for **copy DNAs**). From this point on, construction of the library follows the same strategy as does construction of a genomic library—the cDNA products and vector are treated with the same restriction enzyme and the resulting fragments are then ligated into the vector.

To isolate individual inserts from a library, *E. coli* cells are transformed with the entire library. Each transformed cell typically contains only a single vector with its associated insert DNA. Thus, each cell that propagates after transformation will contain multiple copies of just one of the possible clones from the library. The colony produced from cells carrying any cloned sequence of interest can be identified and the DNA retrieved. There are various ways to identify the clone. For example, as we describe below, hybridization with a unique DNA or RNA probe can identify a population of cells that include a particular insert DNA.

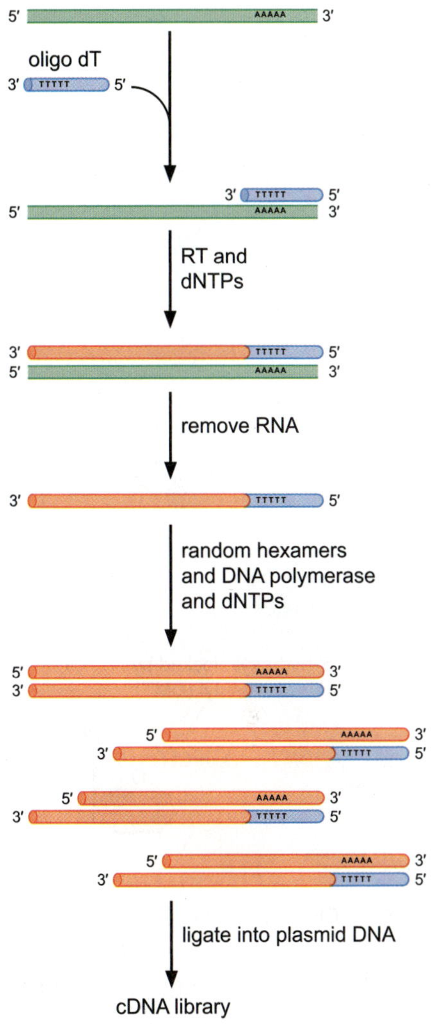

FIGURE 21-9 Construction of a cDNA library. The RNA-dependent DNA polymerase reverse transcriptase (RT) transcribes RNA into DNA (copy or cDNA). In the first step (first strand synthesis), oligos of poly-T sequence serve as primers by hybridizing to the poly-A tails of the mRNAs. Reverse transcriptase extends the dT primer to complete a DNA copy of the mRNA template. The product is a duplex composed of one strand of mRNA and its complementary strand of DNA. The RNA strand is removed by treatment with base (NaOH), and the remaining single-stranded DNA now serves as template for the second step (second strand synthesis). Short random sequences of DNA usually ~6 bp long (called random hexamers) serve as primers by hybridizing to various sequences along the copy DNA template. These primers are then extended by DNA polymerase to create double-stranded DNA products that can be cloned into a plasmid vector (see Fig. 21-8) to create a cDNA library.

Hybridization Can Be Used to Identify a Specific Clone in a DNA Library

When attempting to clone a gene, a common step is to identify fragments of that gene among clones in a library. This can be achieved using a DNA probe whose sequence matches part of the gene of interest. Such a probe can be used to identify colonies of cells harboring clones containing that region of the gene, as we now describe.

The process by which a labeled DNA probe is used to screen a library is called colony hybridization. A typical cDNA library will have thousands of different inserts, each contained within a common vector (see above). After transformation of a suitable bacterial host strain with the library, the cells are plated out on petri dishes containing solid growth medium (usually agar; see Chapter 22). Each cell grows into an isolated colony of cells, and each cell within a given colony contains the same vector and insert from the library (there are typically a few hundred colonies per dish).

The same type of positively charged membrane filter used in the Southern and northern blotting techniques is used here to secure small amounts of DNA for probing. In this case, pieces of the membrane are pressed on top of the dish of colonies, and imprints of cells (including some DNA) from each colony are lifted onto the filter (note that some cells from each colony remain on the plate). Thus, the filter retains a sample of each DNA clone positioned on the filter in a pattern that matches the pattern of colonies on the plate. This ensures that once the desired clone has been identified by probing the filter, the colony of cells carrying that clone can be readily identified on the plate and the plasmid containing the appropriate insert DNA can be purified.

Probing of the filters is carried out as follows. They are treated under conditions that cause the cells on the membrane to break open and the DNA to leak out and bind to the filter at the same location as the cells the DNA was derived from. The filters can then be incubated with the labeled probe under the same conditions that were used in the northern and Southern blotting experiments.

As we mentioned earlier and discuss in Chapter 22, bacteriophage (particularly λ) have also been modified for use as vectors. When libraries are made using a phage vector, they can be screened in much the same way as just described for the screening of plasmid libraries. The difference is that the plaques formed by growth of the phage on bacterial lawns are screened rather than colonies (see Chapter 22).

Chemically Synthesized Oligonucleotides

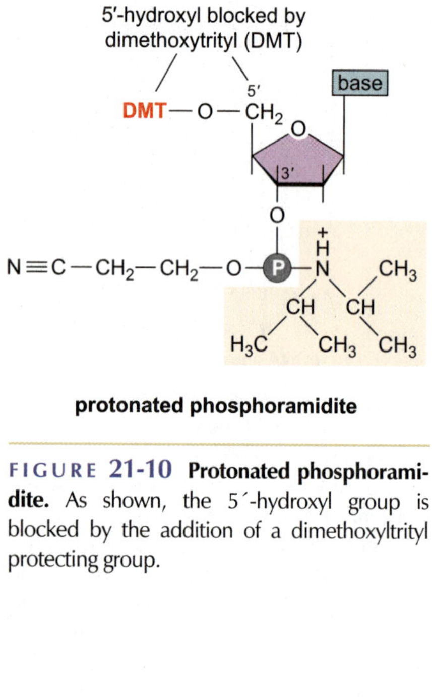

FIGURE 21-10 Protonated phosphoramidite. As shown, the 5′-hydroxyl group is blocked by the addition of a dimethoxyltrityl protecting group.

Short, custom-designed segments of DNA known as **oligonucleotides** are critical for several techniques we describe in this chapter. Although DNA polymerases are the most efficient machines for synthesizing DNA molecules, DNA can also be synthesized chemically. The most common methods of chemical synthesis are performed on solid supports using machines that automate the process. The precursors used for nucleotide addition are chemically protected molecules called **phosphoamidines** (Fig. 21-10). In contrast to the direction of chain growth used by DNA polymerases, growth of the DNA chain is by addition to the 5′ end of the molecule.

Chemical synthesis of DNA molecules up to 60–100 bases long is efficient and accurate and takes only a few hours. It is a routine procedure: an investigator can simply program a DNA synthesizer to make any desired sequence by typing the base sequence into a computer controlling the machine. But as the synthetic molecules get longer, the final product is less uniform because of the inherent failures that occur during any cycle of the process. Thus, molecules more than 100 nucleotides or so are difficult to synthesize in the quantity and with the accuracy desirable for most molecular analysis.

The rather short DNA molecules that can readily be made, however, are well suited for many purposes. For example, a custom-designed oligonucleotide harboring a mismatch to a segment of cloned DNA can be used to create a directed mutation in that cloned DNA. This method, called **site-directed mutagenesis**, is performed as follows. The oligonucleotide is hybridized to the cloned fragment and used to prime DNA synthesis with the cloned DNA as template. In this way, a double-strand molecule with one mismatch is made. The two strands are then separated and that with the desired mismatch is amplified further.

Custom-designed oligonucleotides can be used in this manner to introduce restriction sites into cloned DNAs, which are then used to create fusions between a coding sequence and another coding sequence or a promoter or ribosome-binding site. As another example, synthetic oligonucleotides that have been labeled fluorescently or radioactively can be used as probes in hybridization experiments. Moreover, custom-designed oligonucleotides are critical in PCR, which we describe next, and are an indispensable feature of the DNA-sequencing strategies that we describe below. Therefore, a common feature in designing experiments to construct new molecular clones of genes to detect specific DNAs, to amplify DNAs, and to sequence DNAs is to design and have synthesized a short synthetic DNA oligonucleotide of desired sequence.

The Polymerase Chain Reaction Amplifies DNAs by Repeated Rounds of DNA Replication in Vitro

A powerful method for amplifying particular segments of DNA, distinct from cloning and propagation within a host cell, is the **polymerase chain reaction (PCR)**. This procedure is carried out entirely biochemically, that is, in vitro. PCR uses the enzyme DNA polymerase that directs the synthesis of DNA from deoxynucleotide substrates on a single-stranded DNA template. As seen in Chapter 8, DNA polymerase adds nucleotides to the 3′ end of a custom-designed oligonucleotide when it is annealed to a longer template DNA. Thus, if a synthetic oligonucleotide is annealed to a single-strand template that contains a region complementary to the oligonucleotide, DNA polymerase can use the oligonucleotide as a primer and elongate its 3′ end to generate an extended region of double-stranded DNA.

How is this enzyme and reaction exploited to amplify specific DNA sequences? Two synthetic, single-strand oligonucleotides are synthesized. One is complementary in sequence to the 5′ end of one strand of the DNA to be amplified, and the other is complementary to the 5′ end of the other strand (Fig. 21-11). The DNA to be amplified is then denatured and the oligonucleotides are annealed to their target sequences. At this point, DNA polymerase and deoxynucleotide substrates are added to the reaction and the enzyme extends the two primers. This reaction generates double-stranded DNA over the region of interest on *both* of the strands of DNA. Thus, two double-strand copies of the starting fragment of DNA are produced in this, the first, cycle of the PCR.

Next, the DNA is subject to another round of denaturation and DNA synthesis using the same primers (note that only the sequence between the primers is in fact precisely amplified). This process generates four copies of the fragment of interest. In this way, additional repeated cycles of denaturation and primer-directed DNA synthesis amplify the region between the two primers in a geometric manner (2, 4, 8, 16, 32, 64, and so forth). Thus, a fragment of DNA that was originally present in vanishingly small amounts is amplified into a large quantity of a double-stranded DNA (see Fig. 21-11).

In a sense, DNA cloning and the PCR rely on the same concept: repeated rounds of DNA duplication—whether carried out by cycles of cell division or cycles of DNA synthesis in vitro—amplify tiny samples of DNA into large quantities. In cloning, however, we often rely on a selective reagent or other device to locate the amplified sequence in an already existing library of clones, whereas in PCR, the selective reagent, the pair of oligonucleotides, limits the amplification process to the particular DNA sequence of interest from the beginning (see Box 21-1, Forensics and the Polymerase Chain Reaction).

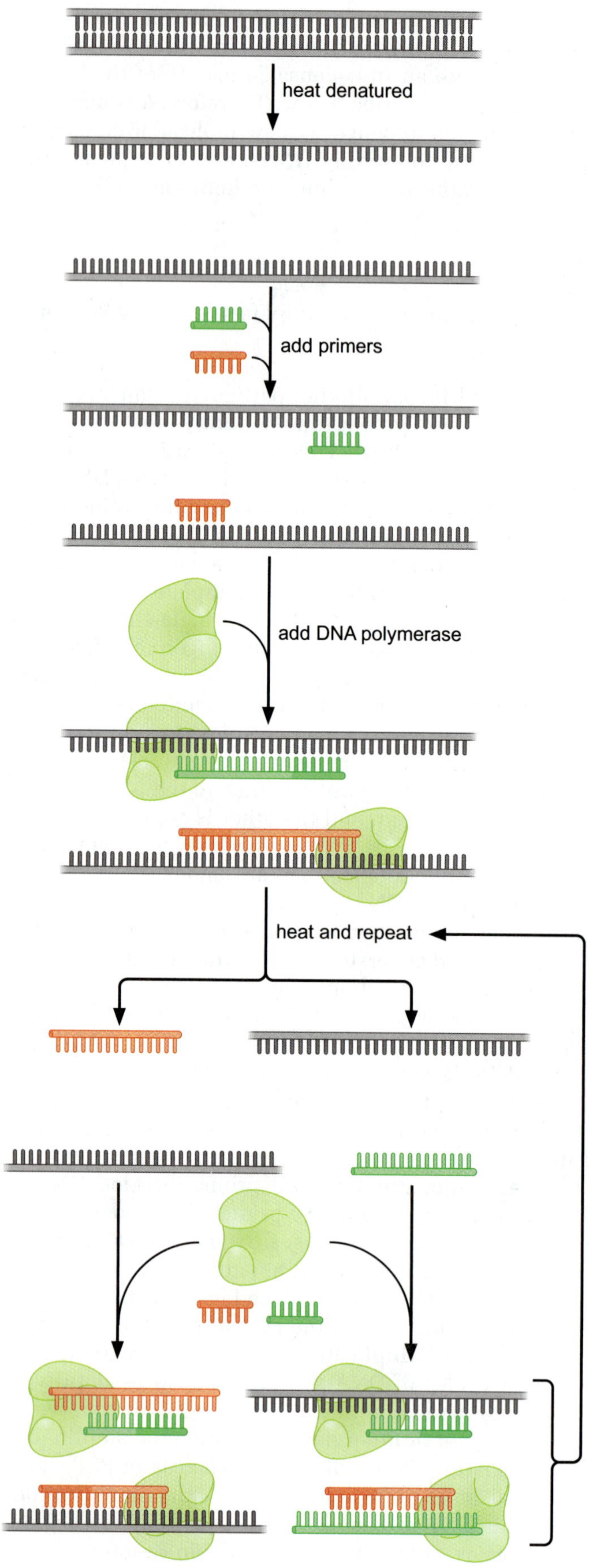

FIGURE 21-11 **Polymerase chain reaction (PCR).** In the first step of the PCR, the DNA template is denatured by heating and annealed with synthetic oligonucleotide primers (dark orange and dark green) corresponding to the boundaries of the DNA sequence to be amplified. DNA polymerase is then used to copy the single-stranded template by extension from the primers (light orange and light green). In the next step, DNA is once again denatured, annealed with primers, and used as a template for a fresh round of DNA synthesis. Note that in this second cycle, the primers can prime synthesis from the newly synthesized DNAs as well as from the original template DNA. When DNA polymerase extends the green-labeled primer that had annealed to newly synthesized (orange-labeled) template from the previous round of DNA synthesis (or orange-labeled primer from green-labeled template), the polymerase proceeds all the way to the end of the template and then falls off (in the figure [bottom], the polymerases have not yet reached the end of the templates). Thus, in this second cycle, DNA will have been synthesized that precisely spans the DNA sequence to be amplified. Thereafter, further rounds of denaturation, priming, and DNA synthesis (not shown) will generate DNAs that correspond to the sequence interval set by the two primers. This DNA will increase in abundance geometrically with each subsequent cycle of the chain reaction.

TECHNIQUES

Box 21-1 Forensics and the Polymerase Chain Reaction

Imagine being in a forensic laboratory and having a DNA sample from a suspected criminal. We want to determine whether the suspect's DNA contains a polymorphism that is present in DNA found at the scene of the crime. Polymorphisms are alternative DNA sequences (alleles) found in a population of organisms at a common, homologous region of the chromosome, such as a gene. A polymorphism can be as simple as alternative, single-base-pair differences at the same site in the chromosome among different members of the population or differences in the length of a simple nucleotide repeat sequence such as CA (see Chapter 9). What we want to do is amplify DNA surrounding and including the site of the polymorphism so that we can subject it to nucleotide sequencing (below) and determine if there is a match to the sequence found in the crime scene sample. The nucleotide sequence of the amplified DNA helps to determine (along with checks for additional polymorphisms) whether the two DNA samples match. This approach to defining the DNA sequence is called DNA profiling or DNA fingerprinting, intended as an analogy between identification using DNA and identification using conventional fingerprinting techniques. DNA profiling was first used in 1985 (the U.S. Federal Bureau of Investigation [FBI] began using the technique in 1988) and since that time has become widely used in the analysis of crime scene evidence, both to convict and to exonerate suspected individuals.

Nested Sets of DNA Fragments Reveal Nucleotide Sequences

We next consider how nucleotide sequences are determined. In a sense, nucleotide sequencing represents the ultimate in probing a genome with high selectivity. It is now possible to determine the entire sequence of nucleotides for a genome, as has now been done for organisms ranging in complexity from bacteria to *Homo sapiens*. This allows us to find any specific sequence with great rapidity and accuracy through the use of a computer and appropriate algorithms. In other words, our "selective reagent" when dealing with nucleotide sequences is a string of bases that we feed into a computer. The increasing availability of large numbers of genome sequences makes it possible to search with high precision for copies of related sequences both within and between organisms in silico. Obviously, nucleotide sequencing generates extraordinarily powerful databases as we describe below.

The underlying principle of DNA sequencing is based on the separation, by size, of nested sets of DNA molecules. Each of the DNA molecules starts at a common 5′ end, and terminates at one of several alternative 3′ endpoints. Members of any given set have a particular type of base at their 3′ ends. Thus, for one set, the molecules all end with a G, for another a C, for a third an A, and for the final set a T. Molecules within a given set (e.g., the G set) vary in length depending on where the particular G at their 3′ end lies in the sequence. Each fragment from this set therefore indicates where there is a G in the DNA molecule from which they were generated. How these fragments are generated is discussed below (and is shown later in Fig. 21-14).

The different lengths of these fragments can be determined by electrophoresis through a polyacrylamide gel. Running the G set on a gel in this way gives a ladder of fragments, with each rung corresponding to a fragment whose length reveals the position of a G in the DNA sequence. The four nested sets can be run out on the gel side-by-side, generating four ladders and revealing where there are Gs, Cs, As, and Ts within the sequence. Comparing the positions of the rungs in these four ladders reveals the entire sequence of the starting DNA molecule. Alternatively, the four nested sets can be differentially labeled with distinct fluorophores, allowing them to be subjected to electrophoresis as a single mixture and distinguished later using fluorometry.

How are nested sets of DNA molecules created? Two methods were invented for doing this. In one, DNA molecules are radioactively labeled at their 5´ termini and then subjected to four different regimens of chemical treatment that cause them to break preferentially at Gs, Cs, Ts, or As. This chemical procedure is no longer in wide use, and we shall not consider it further. The other procedure, which employs **chain-terminating nucleotides** and in vitro DNA synthesis, continues to be used to this day and is the technology upon which modern, automatic sequencing machines called **Sequenators** are based.

In the chain-termination method, DNA is copied by DNA polymerase from a DNA template starting from a fixed point specified by hybridization of an oligonucleotide primer. As seen in Chapter 8, DNA polymerase uses 2´-deoxynucleoside triphosphates as substrates for DNA synthesis, and DNA synthesis occurs by extending the 3´ end. (The chain-termination method relies on the principles of enzymatic synthesis of DNA, which we discussed in Chapter 8.) The chain-termination method employs special, modified substrates called 2´-,3´-dideoxynucleotides (ddNTPs), which lack the 3´-hydroxyl group on their sugar moiety as well as the 2´-hydroxyl (Fig. 21-12). DNA polymerase will incorporate a 2´-,3´-dideoxynucleotide at the 3´ end of a growing polynucleotide chain, but once incorporated, the lack of a 3´-hydroxyl group prevents the addition of further nucleotides causing elongation to terminate (Fig. 21-13).

Now suppose that we "spike" (or add) a cocktail of the nucleotide substrates with the modified substrate 2´-,3´-dideoxyguanosine triphosphate (ddGTP) at a ratio of one ddGTP molecule to 100 2´-deoxy-GTP molecules (dGTP). This will cause DNA synthesis to abort at a frequency of 1 in 100 times the DNA polymerase encounters a C on the template strand (Fig. 21-14a). Because all of the DNA chains commence growth from the same point, the chain-terminating nucleotides

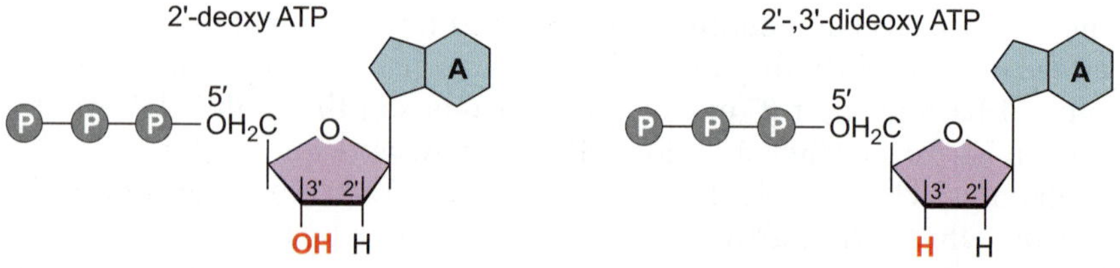

FIGURE 21-12 Dideoxynucleotides used in DNA sequencing. On the left is 2´-deoxy ATP. This can be incorporated into a growing DNA chain and allow another nucleotide to be incorporated directly after it. On the right is 2´-,3´-dideoxy ATP. This can be incorporated into a growing DNA chain, but once in place it blocks further nucleotides being added to the same chain.

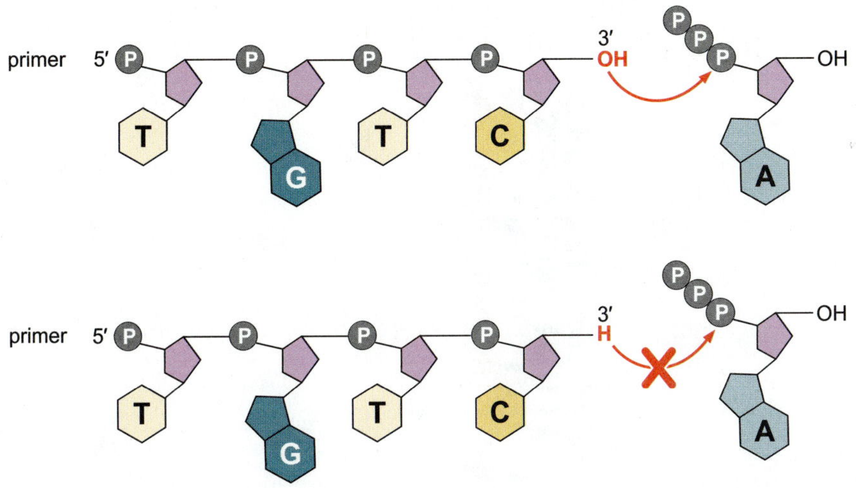

FIGURE 21-13 **Chain termination in the presence of dideoxynucleotides.** In the top line is a DNA chain being extended at the 3′ end with addition of an adenine nucleotide onto the previously incorporated cytosine. The presence of dideoxycytosine in the growing chain (shown at the bottom) blocks further addition of incoming nucleotides as described in the text.

will generate a nested set of polynucleotide fragments, all sharing the same 5′ end but differing in their lengths and hence their 3′ ends. The length of the fragments therefore specifies the position of Cs in the template strand. The fragments can be labeled at their 5′ end either by the use of a radioactively labeled primer or a primer that has been tagged with a fluorescent adduct, or at their 3′ end with fluorescently labeled derivatives of ddGTP. Upon electrophoresis through a polyacrylamide gel, the nested set of fragments yield a ladder of fragments, each rung of the ladder representing a C on the template strand (Fig. 21-14b). If we similarly spike DNA synthesis reactions with

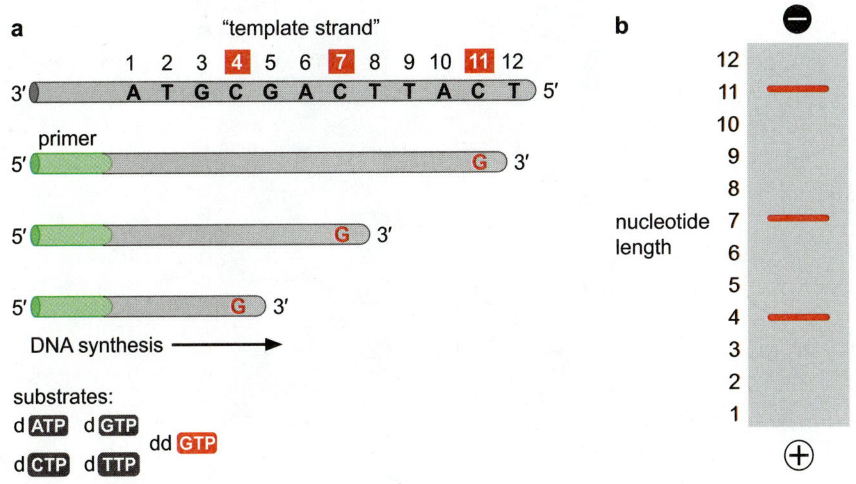

FIGURE 21-14 **DNA sequencing by the chain-termination method.** As described in the text, chains of different length are synthesized in the presence of dideoxynucleotides. The length of the chains produced depend on the sequence of the DNA template and which dideoxynucleotide is included in the reaction. In the figure, the sequence of the template is shown at the top of a. In this reaction, all bases are present as deoxynucleotides, but G is present in the dideoxy form as well. Thus, when the elongating chain reaches a C in the template, it will, in some fraction of the molecules, add the ddGTP instead of dGTP. In those cases, chains terminate at that point. (b) Fragments separated on a polyacrylamide gel. The lengths of fragments seen on the gel reveal the positions of cytosines in the template DNA being sequenced in the reaction described.

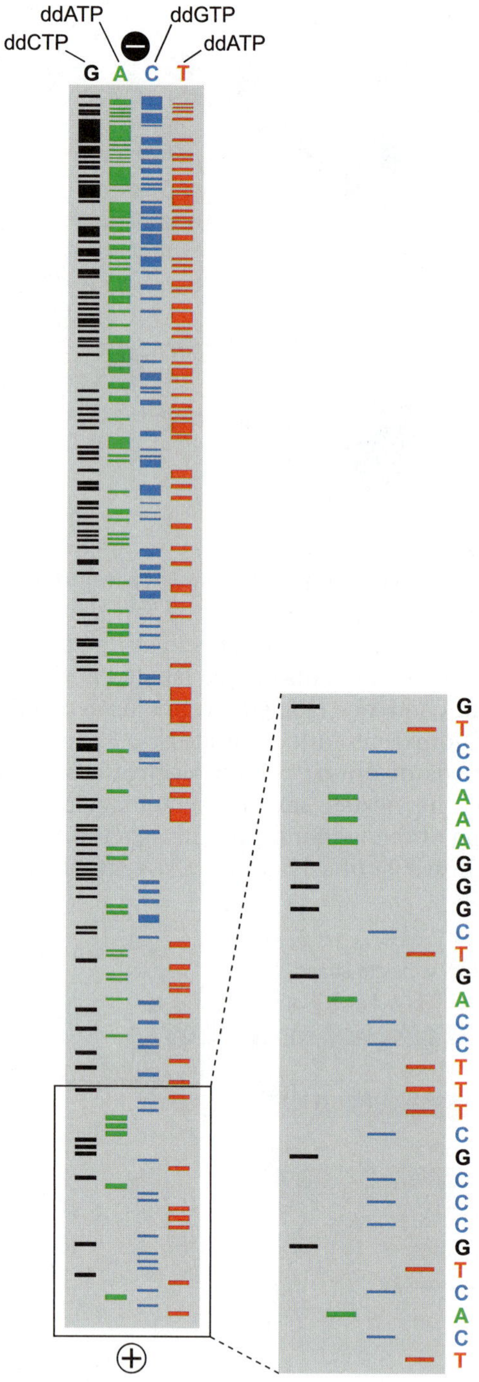

FIGURE 21-15 DNA-sequencing gel. The lengths of DNA chains, terminated with the dideoxynucleotide indicated at the top of each lane, are determined by resolving on a polyacrylamide gel, as shown. Reading the gel from bottom to top gives the 5′ to 3′ sequence.

ddCTP, ddATP, and ddTTP, then in toto we will generate four nested sets of fragments, which together provide the full nucleotide sequence of the DNA. To read that sequence, the fragments generated in each of the four reactions were resolved on a polyacrylamide gel (Fig. 21-15).

As we shall see below, this conceptually simple approach, developed initially to sequence short, defined DNA fragments, has undergone a series of technical adaptations and improvements that allow the analysis of whole genomes (see Box 21-2, Sequenators Are Used for High-Throughput Sequencing).

KEY EXPERIMENTS

Box 21-2 Sequenators Are Used for High-Throughput Sequencing

When the sequencing of the human genome was first envisioned, it seemed like a daunting, virtually hopeless enterprise. After all, the complete human genome consists of a staggering 3 billion (3×10^9) base pairs, and the early methods for determining the nucleotide sequence of even short DNA fragments were quite tedious. In the 1980s and early 1990s, an individual researcher could produce only a few hundred base pairs, perhaps 500 bp, of DNA sequence in a day or two of concentrated effort. Several technical innovations have greatly accelerated the speed and reliability of DNA sequencing.

As we described in the preceding section, the chain-termination method produces nested sets of DNAs that differ in size by just a single nucleotide. Initially, large polyacrylamide gels were used to fractionate these nested DNAs (see Fig. 21-15). However, in recent years, cumbersome gels have been replaced by short columns, which permit the resolution of nested DNAs in just 2–3 hours. These short reusable columns permit the fractionation of DNA fragments ranging from 700 to as many as 800 bp, similar to the capacity of the far more cumbersome polyacrylamide gels that they have replaced.

A major technical advance in DNA sequencing came from the use of **fluorescent chain-terminating nucleotides**. In principle, it is possible to label each of the nested DNAs from a fragment with a single "color." The color of each nested DNA depends on the identification of the last nucleotide. For example, DNAs ending with a T residue at position 50 in the template DNA might be labeled red, whereas those nested DNAs ending with a G residue at position 51 might be labeled black. Thus, each nested DNA has a unique size and color. As they are fractionated on the sequencing columns based on size, fluorescent sensors detect the color of each nested DNA (Box 21-2 Fig. 1). In this way, a single column produces 600–800 bp of DNA sequence after less than 3 hours of size separation.

Automated sequencing machines—**Sequenators**—have been developed that have 384 separate fractionation columns. In principle, these machines can generate more than 200,000 nucleotides (200 kb) of raw DNA sequence in just a few hours. In a 9-hour day, each machine can produce three sequencing "runs" and more than one-half a megabase (500 kb) of sequence information. A cluster of 100 such machines could generate the equivalent of one human genome, 3×10^9 bp, in just 2 months. There are currently five major sequencing centers in the United States and the United Kingdom. Each contains large clusters of automated DNA-sequencing machines. Together, these five centers produce a staggering 60×10^9 bp of raw DNA sequence information per year. This corresponds to the equivalent of 20 human genomes per year!

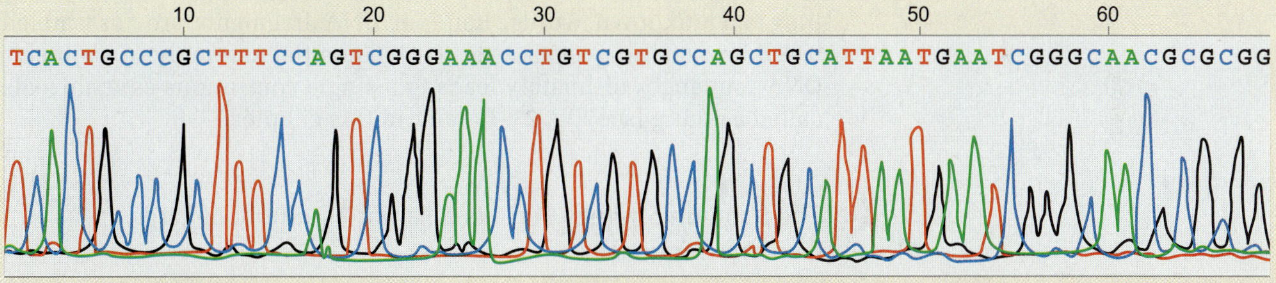

BOX 21-2 FIGURE 1 **DNA sequence readout.** In this reaction, as described in the text, fluorescently end-labeled dideoxynucleotides are used and the chains are separated by column chromatography. The profile of positions of As is represented in green, Ts in red, Gs in black, and Cs in blue.

Shotgun Sequencing a Bacterial Genome

The bacterium *Hemophilus influenzae* was the first free-living organism to have a complete genome sequence and assembly. It was a logical choice because it has a small, compact genome that is composed of just 1.8 million base pairs (Mb) of DNA. The *H. influenzae* genome was sheared into many random fragments with an average size of 1 kb. These pieces of genomic DNA were cloned into a plasmid DNA vector to create a library. DNA was prepared from individual recombinant DNA colonies and separately sequenced on Sequenators using the dideoxy method discussed above. This method is called "shotgun" sequencing. Random recombinant DNA colonies are picked, processed, and sequenced. To ensure that every single nucleotide in the genome was captured in the final genome assembly, 30,000–40,000 separate recombinant clones were sequenced. A total of about 20 Mb of raw genome sequence was produced (600 bp of sequence is produced in an average reaction, and 600 bp x 33,000 different colonies = 20 Mb of total DNA sequence). This is called **10x sequence coverage**. In principle, every nucleotide in the genome should have been sequenced ten times.

This method might seem tedious, but it is considerably faster and less expensive than the techniques that were originally envisioned. One early strategy called for systematically sequencing every defined restriction DNA fragment on the physical map of the bacterial chromosome. A drawback of this procedure is that most of the known restriction fragments are larger than the amount of DNA sequence information generated in a single reaction. Consequently, additional rounds of digestion, mapping, and sequencing would be required to obtain a complete sequence for any given defined region of the genome. These additional steps of cloning and restriction mapping are considerably more time-consuming than the repetitive automated sequencing of random DNA fragments. In other words, the computer is much faster at assembling random DNA sequences than the time required to clone and sequence a complete set of restriction fragments spanning a bacterial genome.

The approximately 30,000 sequencing reads derived from random genomic DNA fragments are directly entered into the computer, and programs are used to assemble overlapping DNA sequences. This process is conceptually similar to the assembly of a giant dense crossword puzzle in which the determined words give clues to the overlapping but unknown words. Random DNA fragments are "assembled" based on matching sequences. The sequential assembly of such short DNA sequences ultimately leads to a single continuous assembly, also called a contig (see Fig. 21-17 later in this chapter).

The Shotgun Strategy Permits a Partial Assembly of Large Genome Sequences

From our preceding discussion, it is obvious that sequencing short 600-bp DNA fragments is incredibly fast and efficient. In fact, the automated sequencing machines are so efficient that they far surpass our ability to assemble and annotate the raw DNA sequence information. In other words, the rate-limiting step in determining the complete DNA sequence of complex genomes, such as the human genome, is the analysis of the data, rather than the production of the data per se. This problem is rapidly becoming even more severe as the methods for sequencing are increasingly faster and more powerful. It is now possible to generate 1 billion base pairs (gigabase pairs) of DNA sequence information in one "run" on an automated machine (see the section entitled "The $1000 Human Genome Is within Reach" below). We now consider how the shotgun-sequencing method used to determine the complete sequence of the H. influenzae genome was adapted for much larger and complicated animal genomes.

The average human chromosome is composed of 150 Mb. Thus, the 600 bp of DNA sequence provided by a typical sequencing reaction

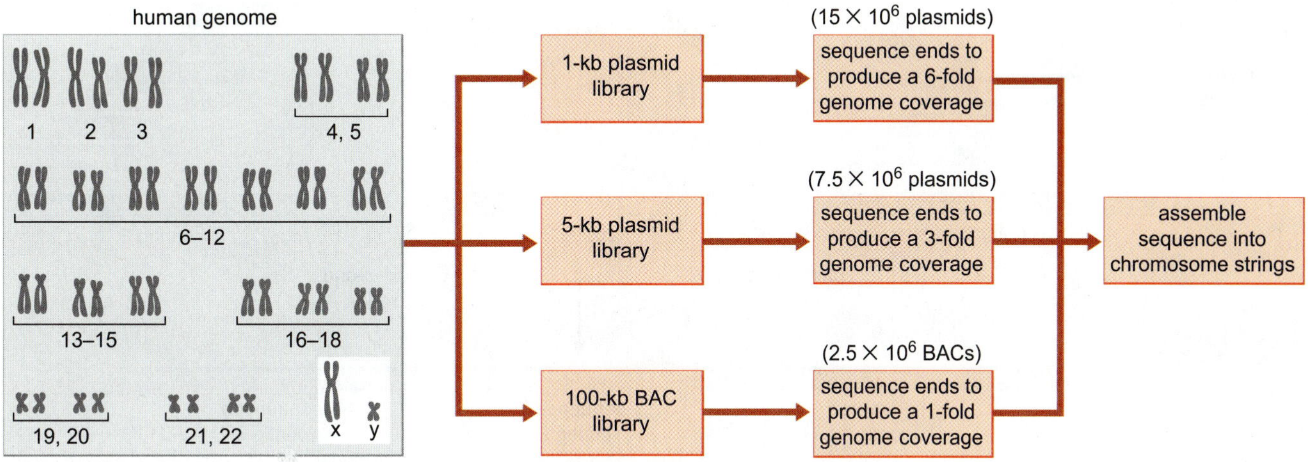

FIGURE 21-16 **Strategy for construction and sequencing of whole-genome libraries.** Contiguous sequences are determined for the shotgun sequencing of the short genomic DNA fragments. Contigs are extended by the use of end sequences derived from the larger fragments carried in the 5-kb and 100-kb insert clones as described in the text. (Adapted, with permission, from Hartwell L. et al. 2003. *Genetics: From genes to genomes*, 2nd ed., Fig. 10-13. © McGraw-Hill.)

represents only 0.0004% of a typical chromosome. Consequently, to determine the complete sequence of the chromosome, it is necessary to generate a large number of sequencing reads from many short DNA fragments (Fig. 21-16). To achieve this goal, DNA is prepared from each of the 23 chromosomes that constitute the human genome and then sheared into small fragments by passage through small-gauge pressurized needles. The collection of small fragements, each derived from individual chromosomes, is then reduced into pools. Typically, two or three pools are constructed for fragments of differing (increasing) sizes—for example, fragments of 1, 5, or 100 kb in length. These fragments are then randomly cloned into bacterial plasmids as we described earlier to make libraries.

Recombinant DNA, containing a random portion of a human chromosome, can be rapidly isolated from bacterial plasmids and then quickly sequenced using automated sequencing machines. To ensure that every sequence is sampled in the complete chromosome, an average of 2 million random DNA fragments are processed. With an average of 600 bp of DNA sequence per fragment, this procedure produces more than 1 billion base pairs of sequence data, or nearly ten times the amount of DNA in a typical chromosome. As discussed earlier for the sequencing of the bacterial chromosome, by sampling about ten times the amount of sequence in a chromosome, we can be confident that every portion of the chromosome will be captured.

The process of producing "shotgun" recombinant libraries and huge excesses of random DNA-sequencing reads seems very wasteful. However, a cluster of 100 384-column automated sequencing machines can generate tenfold coverage of a human chromosome in just a few weeks. This approach is considerably faster than the methods involving the isolation of known regions within the chromosome and sequentially sequencing a known set of staggered DNA fragments. Thus, the key technological insight that facilitated the sequencing of the human genome was the reliance on automated **shotgun sequenc-**

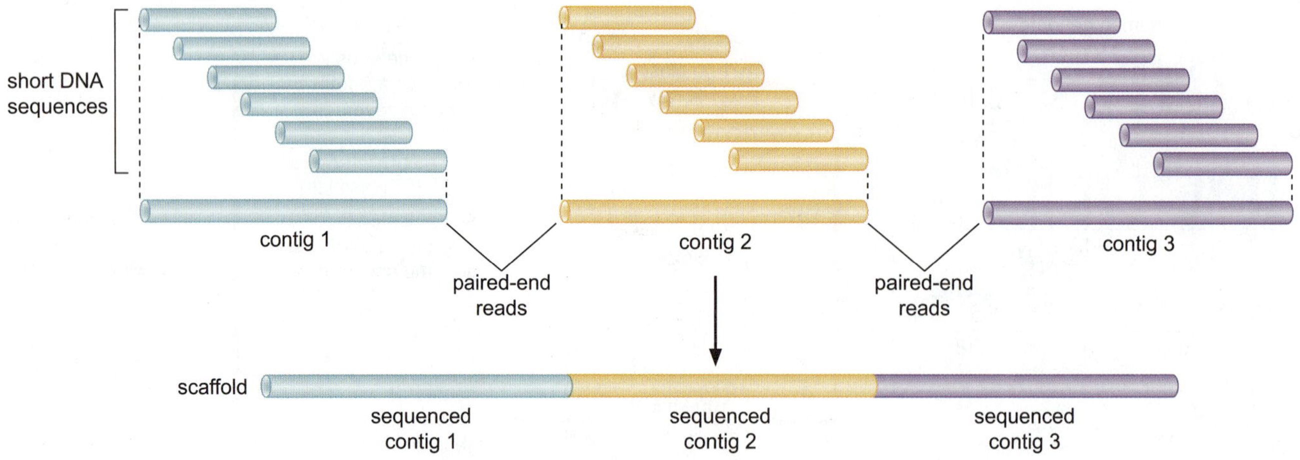

FIGURE 21-17 **Contigs are linked by sequencing the ends of large DNA fragments.** For example, one end of a random 100-kb genomic DNA fragment might contain sequence matches within contig 1, whereas the other end matches sequences in contig 2. This places the two contigs on a common scaffold. (Adapted, with permission, from Griffiths A.J.F. et al. 2002. *Modern genetics*, 2nd ed., Fig. 9-29b. © W.H. Freeman.)

ing and the subsequent use of the computer to assemble the different pieces. The combination of automated sequencing machines and computers proved to be a potent one–two punch that led to the completion of the human genome sequence years earlier than originally planned.

Sophisticated computer programs have been developed that assemble the short sequences from random shotgun DNAs into larger contiguous sequences called **contigs**. Sequences or "reads" that contain identical sequences are assumed to overlap and are joined to form larger contigs (Fig. 21-17). The sizes of these contigs depend on the amount of sequence obtained—the more sequence, the larger the contigs and the fewer gaps in the sequence.

Individual contigs are typically composed of 50,000–200,000 bp. This is still far short of a typical human chromosome. However, such contigs are useful for analyzing compact genomes. For example, the *Drosophila* genome contains an average of one gene every 10 kb, so a typical contig has several linked genes. Unfortunately, more complex genomes often contain considerably lower gene densities. The human genome contains an average of one gene every 100 kb, so a typical contig is often insufficient to capture an entire gene, let alone a series of linked genes. We now consider how relatively short contigs are assembled into larger **scaffolds** that are typically 1–2 Mb in length.

The Paired-End Strategy Permits the Assembly of Large-Genome Scaffolds

A major limitation to producing larger contigs is the occurrence of repetitive DNAs. Such sequences complicate the assembly process because random DNA fragments from unlinked regions of a chromosome or genome might appear to overlap because of the presence of the same repetitive DNA sequence. One method that is used to overcome this difficulty is called **paired-end sequencing**. This is a simple technique that has produced powerful results (see Fig. 21-18).

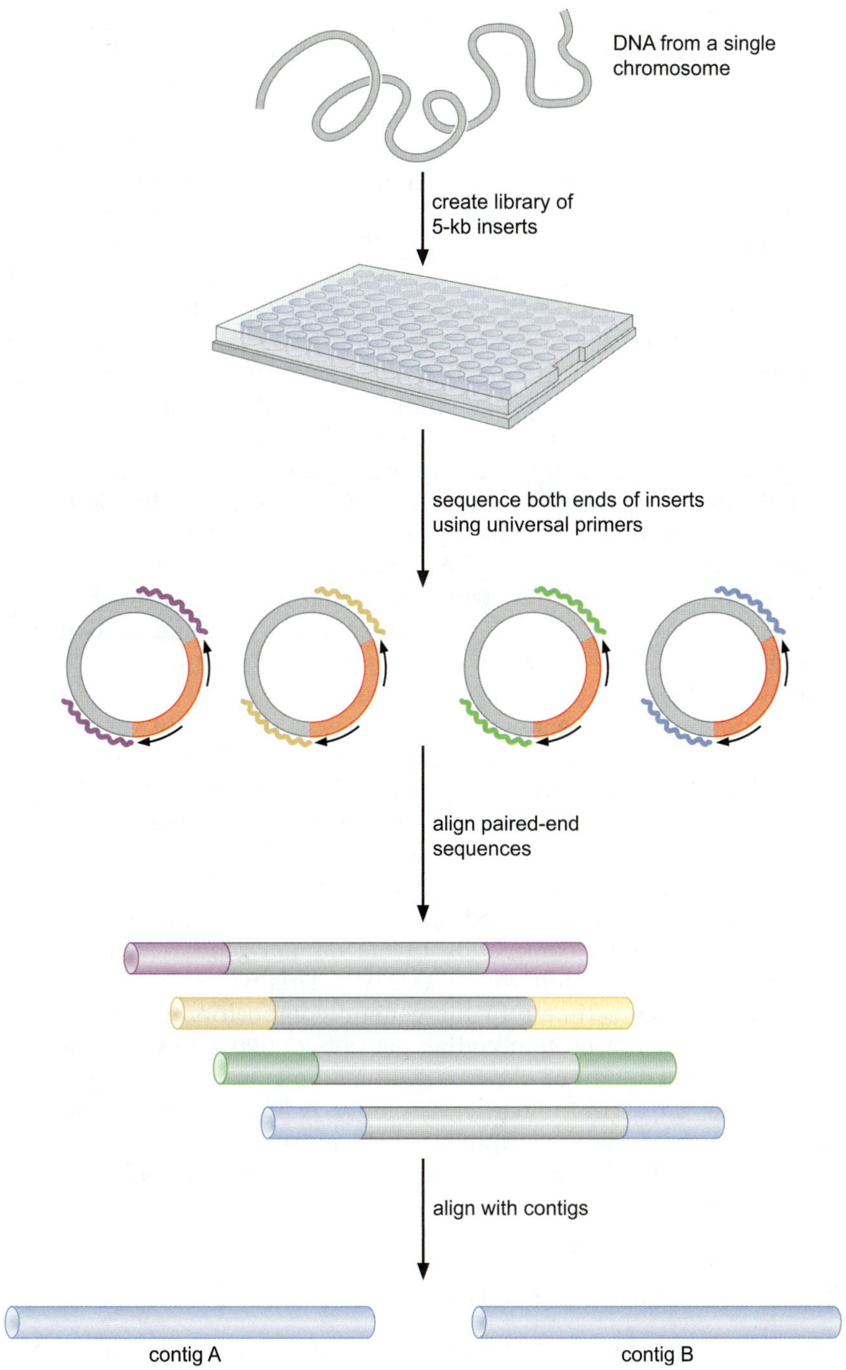

FIGURE 21-18 A "shotgun" library containing random genomic DNA inserts of 5 kb in length. Each well on the plate contains a different insert. Sequences 600 bp in length are determined for both ends of each genomic DNA (color coded). These paired-end sequences are used to align differenty contigs. In this example, the 5-kb genomic DNA fragment with the blue sequences contains matching sequences with contig A and contig B.

In addition to producing shotgun DNA libraries composed of short DNA fragments, the same genomic DNA is also used to produce recombinant libraries composed of larger fragments, typically between 3 and 100 kb in length. Consider a DNA sample from a single human chromosome. Some of the DNA is used to produce 1-kb fragments, whereas another aliquot of the same sample is used to produce 5-kb fragments. The end result is the construction of two libraries, one with small inserts and a second with larger inserts (see Fig. 21-16).

Universal primers are made that anneal at the junction between the plasmid and both sides of the large inserted DNA fragment. Individual runs will produce about 600 bp of sequence information at each end of the random insert. A record is kept of what end sequences are derived from the same inserted fragment. One end might align with sequences contained within contig A, whereas the other end aligns with a different contig, contig B. Contigs A and B are now assumed to derive from the same region of the chromosome since they share sequences with a common 5-kb fragment. Most repetitive DNA sequences are less than 2 or 3 kb in length, so the "paired-end" sequences from the 5-kb insert are sufficient to span contigs interrupted by repetitive DNAs.

The preceding results usually produce contigs that are less than 500 kb in length. To obtain long-range sequence data, on the order of several megabases or more, it is necessary to obtain paired-end sequence data from large DNA fragments that are at least 100 kb in length. These can be obtained using a special cloning vector called a **BAC (bacterial artificial chromosome)** that can accommodate very large inserts, up to hundreds of kilobases of DNA. The principle of how these are used to produce long-range sequence information is the same as that described for the 5-kb inserts. Primers are used to obtain 600-bp sequencing reads from both ends of the BAC insert. These sequences are then aligned to different contigs, which can then be assigned to the same scaffold by virtue of sharing sequences from a common BAC insert. The use of BACs often permits the assignment of multiple contigs into a single scaffold of several megabases (see Fig. 21-17).

The quality of the genome assembly is a measure of the average scaffold size. Those that exceed an average of 1 Mb or more are considered to be high-quality assemblies. For example, the pufferfish genome is 800 Mb in length and the complete assembled sequence is positioned on about 500 different scaffolds, each with an average size of 1.6 Mb. This assembly is sufficient for most analyses, such as the identification of all protein-coding genes. When Bill Clinton and Tony Blair announced the completion of the human genome sequence in 2000, the average scaffold size was 2 Mb. This was sufficient to produce an accurate estimate of the genetic composition of the human genome in terms of protein-coding genes (~25,000 genes). During the past seven years, tedious experimental methods have been used to sequence the "gaps" in the earlier assembly. A finished human genome sequence is essentially complete, which means that there is a single sequence scaffold for each of the 23 chromosomes comprising the human genome.

The $1000 Human Genome Is within Reach

The sequencing of the first two human genomes (one from the National Institutes of Health and the other from a private company) cost more than $300 million. There is now a campaign to use nanotechnology to produce rapid and inexpensive genome sequencing. The goal is to make the technology sufficiently rapid, simple, and inexpensive to permit the sequencing of individual genomes for clinical diagnosis. The first generation of high-throughput, nanotechnology sequencing machines is now available.

The 454 Life Sciences sequencing machine generates up to 400 Mb of sequence information in a 4-hour "run." The basic principle is very clever. Small fragments of DNA (genomic, cDNA, etc.) are mixed with small beads. The mixture is sufficiently dilute so that a single DNA molecule binds to a single bead. Next, the DNA-containing beads are dispersed on a silicon plate consisting of 400,000 regularly spaced picoliter wells. The small size of the wells ensures that each one cap-

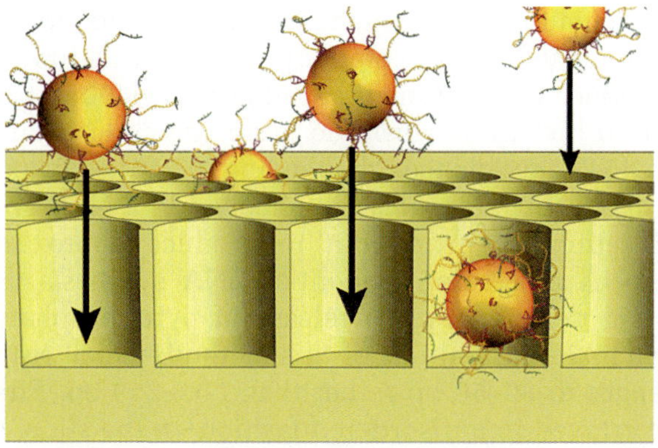

FIGURE 21-19 Cartoon of individual pores in the 454 sequencing apparatus. Each pore contains a small bead with an amplified DNA sequence. Sequential rounds of sequencing are detected by the release of pyrophosphate and light. Further description of the method is given in the text. (Reprinted, with permission, from Margulies M. et al. 2005. *Nature* 437: 376–380, Fig. 1a. © Macmillan.)

tures no more than a single bead. PCR is performed directly on the bead-tethered DNAs to amplify each DNA molecule. Thus, a homogeneous population of DNA molecules is created in each well, which is then used as a template for an additional round of DNA synthesis in a reaction that includes bioluminescent proteins as well as DNA polymerase (Fig. 21-19). This second round of DNA synthesis is carried out in stepwise fashion with the plate being separately exposed to dATP, dGTP, dCTP, and dTTP sequentially, with a washing cycle between each pulse of deoxynucleotide substrate. The incorporation of a deoxynucleotide depends on the presence of the complementary base in the template and results in the liberation of pyrophosphate. This release promotes an enzymatic reaction that produces pulses of light, which are detected by a microprocessor attached to a computer. The light pulses indicate which nucleotide is incorporated in each well at each round of synthesis, thereby producing the sequence of the DNA contained in all 400,000 wells. Sequential addition of each nucleotide is continued until approximately 200–250 bases of sequence have been determined from each DNA fragment.

454 sequencing has produced the complete genome of the lead author of this textbook (for some reason, the company seems less interested in the genomes of the other authors). At 100 Mb of genome sequence per "run," complete 1x coverage of Watson's genome required just 30 runs (2–3 weeks on one machine). If started now (at the time of this writing), the total cost would be about $100,000, a small fraction of the cost of the first human genome sequence. Severalfold coverage is required to ensure that every gene is captured in this sequencing effort, raising the cost to more like half a million dollars. Moreover, the sequence information is not necessarily sufficient to produce a de novo genome assembly. Rather, the finished human genome sequence produced by the National Institutes of Health is used as a template for comparison. Each of the 200–250-bp sequence reads produced by 454 sequencing will be identified on the finished genome until Watson's variants of every gene are identified. Thus, the meaning of sequencing a human genome has shifted. Because we have a finished whole-genome sequence assembly in hand, it might be sufficient to generate large numbers of short sequencing reads to obtain a comprehensive atlas of an individual's unique genetic composition.

The next generation of sequencing machines is approaching the goal of the $1000 genome. A company called Solexa (now part of Illumina) has produced a machine that can generate 400 million sequencing reads of 30 bp in one run. The basic principle is similar to that seen for the 454 Life Sciences sequencing machine. The difference is that individual DNA molecules are attached to a glass slide. Limited PCR amplification is carried out to produce approximately 1000 copies per DNA molecule. Sequential DNA synthesis reactions are performed and detected by the release of pyrophosphate. However, there are two major limitations of this method, which restricts the sequencing length to 30–50 bp rather than 200–250 bp. First, the DNA samples are arrayed in a disordered fashion on the slide, so the emitted light during synthesis is more difficult to detect. Second, the beads used for 454 sequencing contain at least tenfold more amplified copies of the DNA molecule than those used for Solexa sequencing. Consequently, compared with 454 sequencing, Solexa sequencing has a lower signal-to-noise ratio of emitted light.

Despite these limitations the Solexa machine is amazing: it generates more than 1 Gbp of DNA sequence in a single run. Thus, only three runs are required to obtain 1x coverage of a human genome, and each run costs just over $3000. Thus, 1x coverage can be achieved for just $10,000. But the problem is that the reads are very short and might not produce unique matches to the reference genome assembly. This problem can be overcome, in part, by producing deeper coverage (more reads) and, more importantly, by improving the detection of the randomized DNA samples to obtain longer reads. It is thought that 50 bp is easily within the realm of technical feasibility, and this achievement would greatly increase the ability to compare a test sequence with the assembled standard.

PROTEINS

Specific Proteins Can Be Purified from Cell Extracts

The purification of individual proteins is critical to understanding their function. Although in some instances, the function of a protein can be studied in a complex mixture, these studies can often lead to ambiguities. For example, if you are studying the activity of one specific DNA polymerase in a crude mixture of proteins (such as a cell lysate), other DNA polymerases and accessory proteins may be partly or completely responsible for any DNA synthesis activity that you observe. For this reason, the purification of proteins is a major part of understanding their function.

Each protein has unique properties that make its purification somewhat different. This is in contrast to different DNAs, which all share the same helical structure and are only distinguished by their precise sequence. The purification of a protein is designed to exploit its unique characteristics, including size, charge, shape, and, in many instances, function.

Purification of a Protein Requires a Specific Assay

To purify a protein requires an assay that is unique to that protein. For the purification of a DNA, the same assay is almost always used, hybridization to its complement. As we shall see in the discussion of immunoblotting, an antibody can be used to detect specific proteins in the same way. In many instances, it is more convenient to use a more direct measure for the function of the protein. For example, a specific DNA-binding protein can be assayed by determining its interaction with the appropriate DNA (e.g., using an electrophoretic mobility shift assay, described in the section Nucleic Acid–Protein Interactions, at the end of this chapter). Similarly, a DNA or RNA polymerase can be detected by incorporation assays by adding the appropriate template and radioactive nucleotide precursor to a crude extract in a manner similar to the methods used to label DNA described above. As discussed in Box 8-1, incorporation assays are useful for monitoring the purification and function of many different enzymes catalyzing the synthesis of polymers such as DNA, RNA, or proteins.

Preparation of Cell Extracts Containing Active Proteins

The starting material for almost all protein purifications are extracts derived from cells. Unlike DNA, which is very resilient to temperature, even moderate temperatures readily denature proteins once they are released from a cell. For this reason, most extract preparation and protein purification is performed at 4°C. Cell extracts are prepared in a number of different ways. Cells can be lysed by detergent, shearing forces, treatment with low ionic salt (which causes cells to osmotically absorb water and pop easily), or rapid changes in pressure. In each case, the goal is to weaken and break the membrane surrounding the cell to allow proteins to escape. In some instances, this process of treating the membrane is performed at very low temperatures by freezing the cells prior to applying shearing forces (typically, using a coffee grinder or blender similar to the one in many kitchens).

Proteins Can Be Separated from One Another Using Column Chromatography

The most common method for protein purification is **column chromatography**. In this approach to protein purification, protein fractions are passed through glass columns filled with appropriately modified small acrylamide or agarose beads. There are various ways columns can be used to separate proteins. Each separation technique varies on the basis of different properties of the proteins. Three basic approaches are described here. The first two, in this section, separate proteins on the basis of their charge or size, respectively. These methods are summarized in Figure 21-20.

Ion-Exchange Chromatography

In this technique, the proteins are separated by their surface ionic charge using beads that are modified with either positively charged or negatively charged chemical groups. Proteins that interact weakly with the beads (such as a weak positively charged protein passed over beads modified with a negatively charged group) are released from the beads (or eluted) in a low-salt buffer. Proteins that interact more strongly require more salt to be eluted. In either case, the salt masks the charged regions allowing the protein to be released from the beads. Because each protein has a different charge on its surface, they will each be eluted from the column

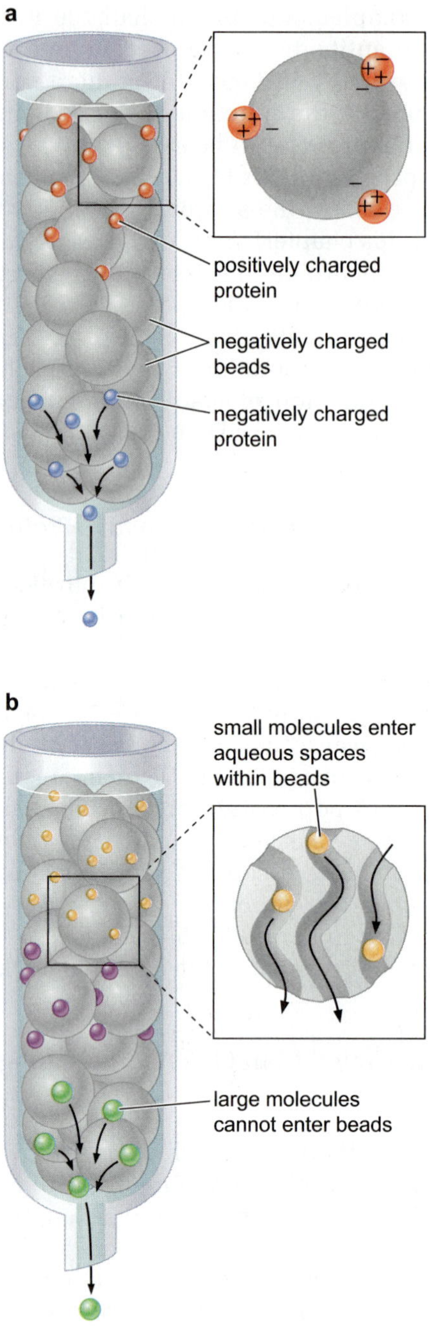

FIGURE 21-20 Ion-exchange and gel-filtration chromatography. As described in the text, these two commonly used forms of chromatography separate proteins on the basis of their charge and size, respectively. Thus, in each case, a glass tube is packed with beads, and the protein mixture is passed through this matrix. The nature of the beads dictates the basis of protein separation. (a) Ion-exchange chromatography. In this example, the beads are negatively charged. Thus, positively charged proteins bind to them and are retained on the column, whereas negatively charged proteins pass through. Increasing the concentration of salt in the surrounding buffer can elute bound proteins by competing for the negative charges on the column. (b) Gel-filtration chromatography. The beads contain aqueous spaces into which small proteins can pass, slowing down their progress through the column. Larger proteins cannot enter the beads, allowing them to pass more rapidly through the column.

at a characteristic salt concentration. By gradually increasing the concentration of salt in the eluting buffer, even proteins with similar charge characteristics can be separated into different fractions as they elute from the column.

Gel-Filtration Chromatography This technique separates proteins on the basis of size and shape. The beads used for this type of chromatography do not have charged chemical groups attached. Instead, each bead has a variety of different-sized pores penetrating their surface (similar to the pores that DNA passes through in agarose or acrylamide gels). Small proteins can enter all the pores and therefore can access more of the column and take longer to elute (in other words, they have more space to explore). Large proteins can access less of the column and elute more rapidly.

For each type of column, chromatography fractions are collected at different salt concentrations or elution times and assayed for the protein of interest. The fractions with the most activity are pooled and subjected to additional purification.

By passing proteins through a number of different columns, they are increasingly purified. Although it is rare that an individual column will purify a protein to homogeneity by repeatedly separating fractions that contain the protein of interest (as determined by the assay for the protein), a series of chromatographic steps can result in a fraction that contains many molecules of a specific protein and few molecules of any other protein. For example, although there are many proteins that elute in high salt from a positively charged column (indicating a high negative charge) or slowly from a gel-filtration column (indicating a relatively small size), there will be far fewer that satisfy both of these criteria.

Affinity Chromatography Can Facilitate More Rapid Protein Purification

Specific knowledge of a protein can frequently be exploited to purify that protein more rapidly. For example, if a protein binds ATP during its function, the protein can be applied to a column of beads that are coupled to ATP. Only proteins that bind to ATP will bind to the column, allowing the large majority of proteins that do not bind ATP to pass through the column. The ATP-binding proteins can be further separated by sequentially adding solutions with increasing concentrations of ATP, which will elute proteins according to their affinity for ATP (the more ATP required to elute, the higher the affinity). This approach to purification is called **affinity chromatography**. Other reagents can be attached to columns to allow the rapid purification of proteins; these include specific DNA sequences (to purify DNA-binding proteins) or even specific proteins that are suspected to interact with the protein to be purified. Thus, before beginning a purification, it is important to think about what information is known about the target protein and to try to exploit this knowledge.

One very common form of protein affinity chromatography is **immunoaffinity chromatography**. In this approach, an antibody that is specific for the target protein is attached to beads. Ideally, this antibody will interact only with the intended target protein and allow all other proteins to pass through the beads. The bound protein can then be eluted from the column using salt, a pH gradient, or, in some cases, mild detergent. The primary difficulty with this approach is that frequently the antibody binds the target protein so tightly that the protein must be denatured before it can be eluted. Because protein denaturation is often irreversible, the target protein obtained in this manner may be inactive and therefore less useful.

Proteins can be modified to facilitate their purification. This modification usually involves adding short additional amino acid sequences to the beginning (amino terminus) or the end (carboxyl terminus) of a target protein. These additions, or "tags," can be generated using molecular cloning methods. The peptide tags add known properties to the modified proteins that assist in their purification. For example, adding six histidine residues in a row to the beginning or end of a protein will make the modified protein bind tightly to a column with Ni^{+2} ions attached to beads—a property that is uncommon among proteins in general. In addition, specific **epitopes** (a sequence of 7–10 amino acids recognized by an antibody) have been defined that can be attached to any protein. This procedure allows the modified protein to be purified using immunoaffinity purification and a heterologous antibody that is specific for the added epitope. Importantly, such antibodies and epitopes can be chosen such that they bind with high affinity under one condition (e.g., in the presence of Ca^{+2}) but readily elute under a second condition (e.g., in the absence of Ca^{+2}). This avoids the need to use denaturing conditions for elution.

Immunoaffinity chromatography can also be used to rapidly precipitate a specific protein (and any proteins tightly associated with it) from a crude extract. In this case, precipitation is achieved by attaching the antibody to the same type of bead used in column chromatography. Because these beads are relatively large, they rapidly sink to the bottom of a test tube along with the antibody and any proteins bound to the antibody. This process, called **immunoprecipitation**, is used to rapidly purify proteins or protein complexes from crude extracts. Although the protein is rarely completely pure at this point, this is often a useful method to determine what proteins or other molecules (e.g., DNA, see the section on chromatin immunoprecipitation later in this chapter) are associated with the target protein.

Separation of Proteins on Polyacrylamide Gels

Proteins have neither a uniform negative charge nor a uniform structure. Rather, they are constructed from 20 distinct amino acids, some of which are uncharged, some are positively charged, and still others are negatively charged (Fig. 5-4). Also, as we discussed in Chapter 5, proteins have extensive secondary and tertiary structures and are often in multimeric complexes (quarternary structure). If, however, a protein is treated with the strong ionic detergent **sodium dodecyl sulfate (SDS)** and a reducing agent, such as mercaptoethanol, the secondary, tertiary, and quarternary structure is usually eliminated. Once coated with SDS, the protein behaves as an unstructured polymer. SDS ions coat the polypeptide chain, giving it a uniform negative charge. Mercaptoethanol reduces disulfide bonds, disrupting intramolecular and intermolecular disulfide bridges formed between cysteine residues. Under these conditions, as is the case with mixtures DNA and RNA, electrophoresis can be used to resolve mixtures of proteins according to the length of individual polypeptide chains (Fig. 21-21). After electrophoresis, the proteins can be visualized with a stain, such as **Coomassie brilliant blue**, that binds to protein nonspecifically. When

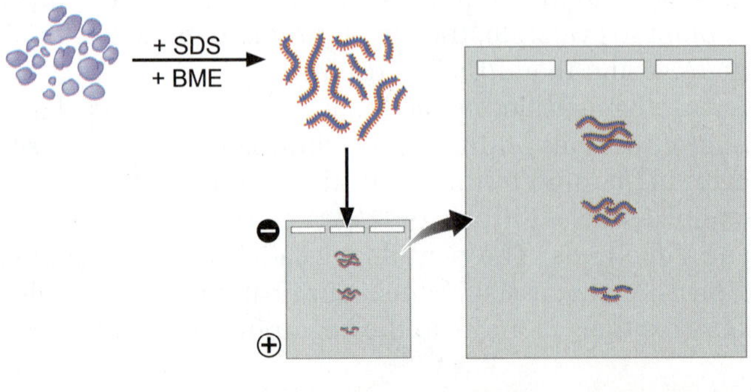

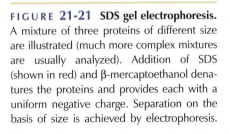

FIGURE 21-21 SDS gel electrophoresis. A mixture of three proteins of different size are illustrated (much more complex mixtures are usually analyzed). Addition of SDS (shown in red) and β-mercaptoethanol denatures the proteins and provides each with a uniform negative charge. Separation on the basis of size is achieved by electrophoresis.

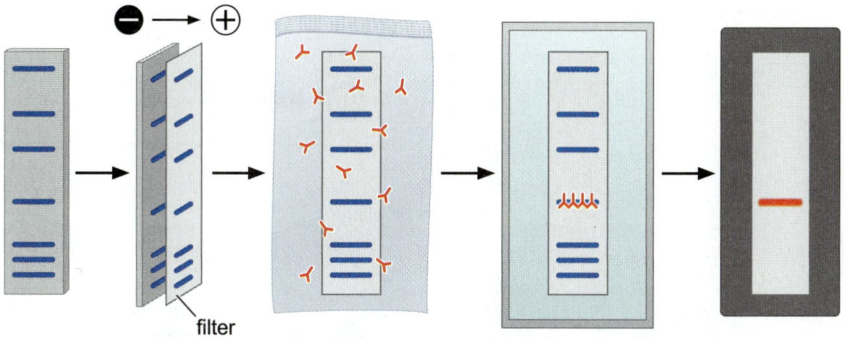

FIGURE 21-22 **Immunoblotting.** After proteins are separated by electrophoresis, they are transferred to filter paper (again using an electric field) in a manner that retains the same relative position of the proteins. After blocking nonspecific protein-binding sites, antibody to the protein of interest is added to the filter paper. The site of antibody binding is then detected using an attached enzyme that creates light when it acts on its substrate.

the SDS is omitted, electrophoresis can be used to separate proteins according to properties other than molecular weight, such as net charge and isoelectric point (see below).

Antibodies Are Used to Visualize Electrophoretically Separated Proteins

Proteins are, of course, quite different from DNA and RNA, but the procedure known as **immunoblotting**, by which an individual protein is visualized amid thousands of other proteins, is analogous in concept to Southern and northern blot hybridization (Fig. 21-22). Indeed, another name for immunoblotting is "western blotting" in homage to its similarity to these earlier techniques. In immunoblotting, electrophoretically separated proteins are transferred to a filter that nonspecifically binds proteins. As for Southern blotting, proteins are transferred to the membrane such that their position on the membrane mirrors their position in the original gel. As in the other blotting techniques we have discussed, once the proteins are attached to the membrane, all the remaining nonspecific binding sites are blocked by incubating with a solution of proteins unrelated to those being studied (often this is powdered milk, which primarily contains albumin proteins). The filter is then incubated in a solution of an antibody that specifically recognizes the protein of interest. The antibody can only bind to the filter it it finds its target protein on the filter. Finally, a chromogenic enzyme that is artificially attached to the antibody (or to a second antibody that binds the first antibody) is used to visualize the filter-bound antibody. Southern, northern, and immunoblotting have in common the use of **selective reagents** to visualize particular molecules in complex mixtures.

Protein Molecules Can Be Directly Sequenced

Although more complex than the sequencing of nucleic acids, protein molecules can also be sequenced: that is, the linear order of amino acids in a protein chain can be directly determined. Two widely used methods for determining protein sequence are Edman degradation using an automated protein sequencer and tandem mass spectrometry. The ability to determine a protein's sequence is very valuable for protein identification. Furthermore, because of the vast resource of complete or nearly complete genome sequences, the determination of even a small stretch of protein sequence is often sufficient to identify the gene which encoded that protein by finding a matching open reading frame.

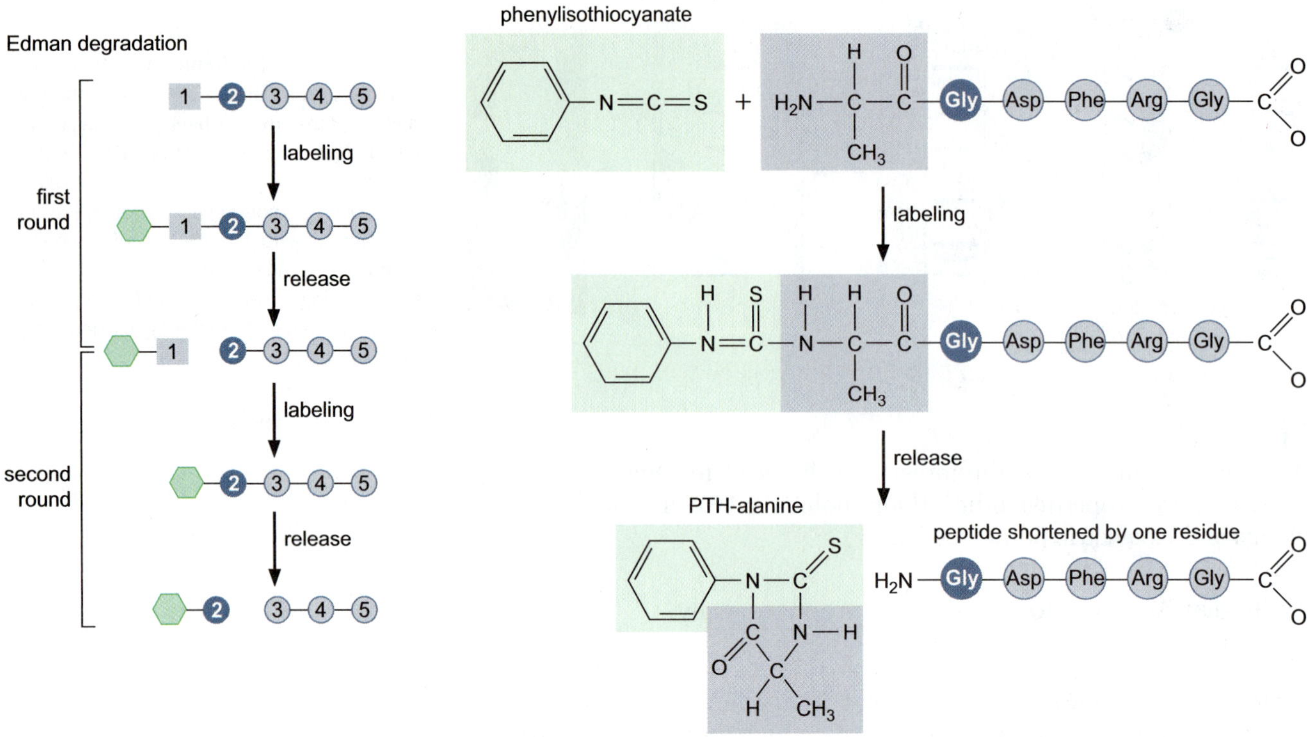

FIGURE 21-23 Protein sequencing by Edman degradation. The amino-terminal residue is labeled and can be removed without hydrolyzing the rest of the peptide. Thus, in each round, one residue is identified, and that residue represents the next one in the sequence of the peptide.

Edman degradation is a chemical reaction in which the amino acid's residues are sequentially released from the amino terminus of a polypeptide chain (Fig. 21-23). One key feature of this method is that the amino-terminal-most amino acid in a chain can be specifically modified by a chemical reagent called **phenylisothiocyanate (PITC)**, which modifies the free α-amino group. This derivatized amino acid is then cleaved off the polypeptide by treatment with acid under conditions that do not destroy the remaining protein. The identity of the released amino acid derivative can be easily determined by its elution profile using a column chromatography method called high-performance liquid chromatography (HPLC) (each of the amino acids has a characteristic retention time). Each round of peptide cleavage regenerates a normal amino terminus with a free α-amino group. Thus, Edman degradation can be repeated for numerous cycles, and thereby reveal the sequence of the amino-terminal segment of the protein. In practice, 8–15 cycles of degradation are commonly performed for protein identification. This number of cycles is nearly always sufficient to uniquely identify an individual protein.

Amino-terminal sequencing by automated Edman degradation is a widespread and robust technique. Problems arise, however, when the amino terminus of a protein is chemically modified (e.g., by formyl or acetyl groups). Such blockage may occur in vivo or during the process of protein isolation. When a protein is amino-terminally blocked, it can usually be sequenced after digestion with a protease to reveal an internal region for sequencing.

Tandem mass spectrometry (**MS/MS**) can also be used to determine protein sequence. Mass spectrometry is a method in which the mass of very small samples of a material can be determined with great accuracy. Very briefly, the principle is that material travels through the instrument (in a vacuum) in a manner that is sensitive to its mass/charge ratio. For small biological macromolecules such as peptides and small proteins, the mass of a molecule can be determined with the accuracy of a single dalton.

To use MS/MS to determine protein sequence, the protein of interest is usually digested into short peptides (often less than 20 amino acids) by digestion with a specific protease such as trypsin. This mixture of peptides is subjected to mass spectrometry and each individual peptide will be separated from the others in the mixture by its mass/charge ratio. The individual peptides are then captured and fragmented into all the component peptides, and the mass of each of these component fragments is then determined (Fig. 21-24). Deconvolution of these data reveals an unambiguous sequence of the initial peptide. As with Edman degradation, sequence of a single, approximately 15-amino-acid peptide from a protein is nearly always sufficient to identify the protein by comparison of the sequence of that predicted from DNA sequences.

MS/MS has revolutionized protein sequencing and identification. Only very small amounts of material are needed, and complex mixtures of proteins can be simultaneously analyzed.

PROTEOMICS

Determining the global levels of gene expression provides a rapid snapshot of the activity of a cell; however, there are important additional levels of regulation that cannot be monitored in this manner. Indeed, the level of transcription of a gene gives only a rough estimate of the level of expression of the encoded protein. If the mRNA is short-lived or poorly translated, then even an abundant mRNA will produce relatively little protein. In addition, many proteins are post-translationally modified in ways that profoundly affect their activities, and transcription profiling gives no data regarding this level of regulation.

The availability of whole-genome sequences in combination with high-throughput analytic methods for protein separation and identification has ushered in the field of proteomics. The goal of proteomics is the identification of the full set of proteins produced by a cell or tissue under a particular set of conditions (called a proteome), their relative abundance, their modifications, and their interacting partner proteins. Whereas microarray analysis makes it possible to profile gene expression or DNA content on a genome-wide basis, the tools of proteomics seek to capture a similar snapshot of the cell's entire repertoire of proteins.

Combining Liquid Chromatography with Mass Spectrometry Identifies Individual Proteins within a Complex Extract

A powerful method to identify all the proteins in a complex mixture such as a crude cell extract uses a combination of liquid chromatography and mass spectrometry (described in the preceding section of this chapter). Although ideally one would simply analyze all the proteins

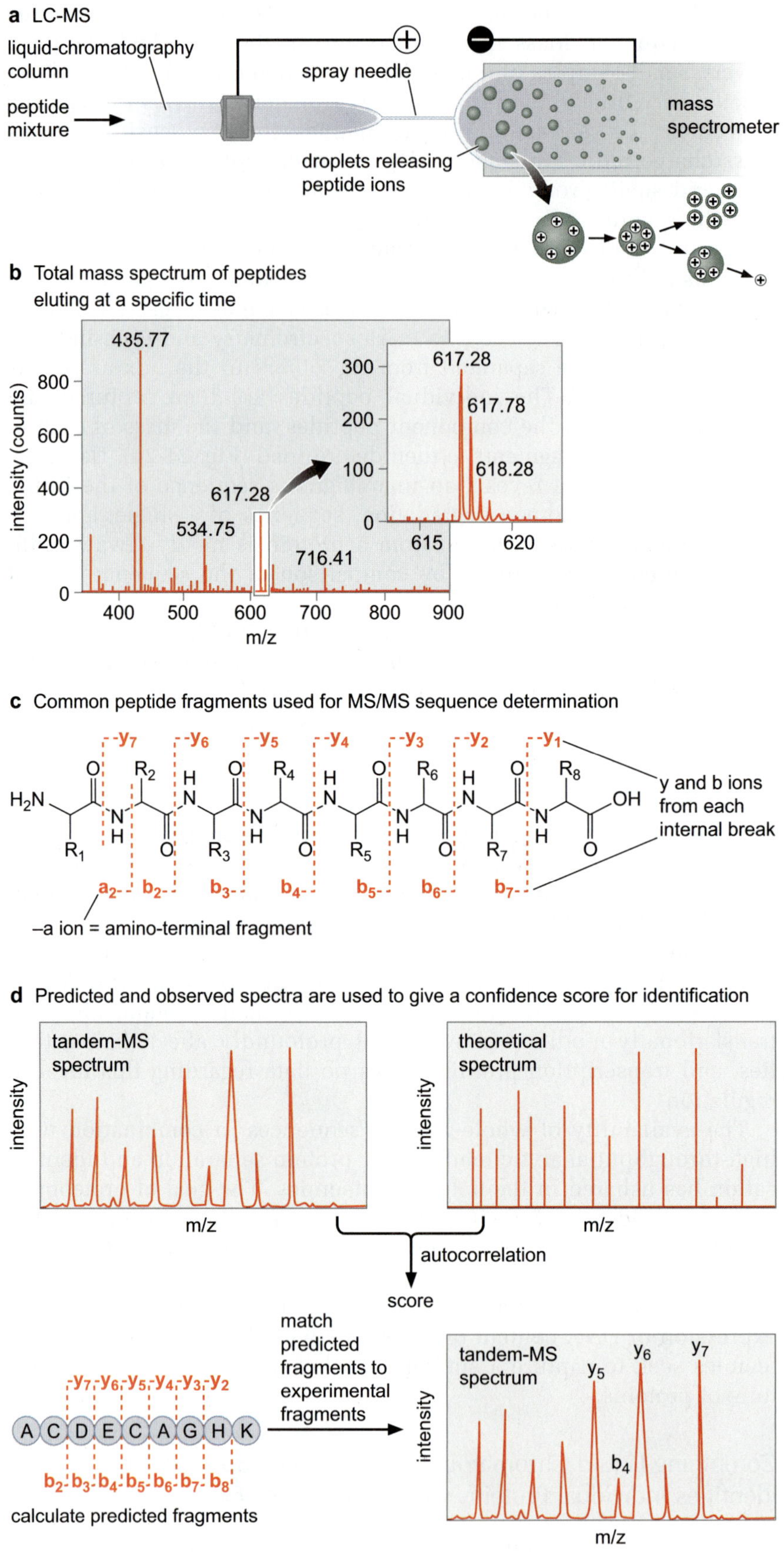

FIGURE 21-24 Using liquid chromatography–MS/MS to analyze the content of a protein mixture. (a) A peptide mixture is subjected to liquid chromatography followed by mass spectrometry. (b) As sets of peptides elute from the chromatography column, they are separated by mass and the results are displayed according to their mass/charge ratio (m/z). Selected sets of related peptides (the differences between these closely related peaks are due to the presence of different atomic isotopes in the peptide) are fragmented and the resulting peptide fragments are analyzed in a second round of mass spectroscopy. (c) Fragmentation of the peptide commonly breaks the peptide in the sites shown in the figure. The possible subpeptides that are generated are called b peptides (amino-terminal fragments), y peptides (carboxy-terminal fragments), and the a_2 peptide (the shortest amino-terminal fragment). (d) The observed spectra are compared to all the possible theoretical spectra that are generated from the amino acid sequences of the proteins encoded by the organism from which the proteins were isolated. Typically, only a subset of peptides can be unambiguously identified. For example, Ile and Leu have identical masses. Nevertheless, clear identification of as few as three or four peptide fragments from a parental peptide is usually sufficient to identify the protein.

in a cell extract directly by mass spectrometry, in practice the very high number of proteins present in such a mixture results in more peptides than can be resolved. Instead researchers have developed powerful methods in which peptides are separated by two types of liquid chromatography before mass spectrometric analysis (LC-MS: Fig. 21-25). In this approach, a crude cell extract is first digested with a sequence-specific protease (e.g., trypsin, which cleaves proteins after Arg and Lys residues) to generate peptides. The resulting mixture of peptides is fractionated by ion exchange chromatography (peptides are separated based on ionic interactions with the charged column material) and reverse phase chromatography (peptides are separated based on hydrophobic interactions with the column material). This procedure separates the highly complex, initial collection of peptides into many lower complexity mixtures of peptides that can be distinguished from one another and sequenced more readily. Each subset of peptides is subjected to tandem mass spectrometry (MS/MS, discussed above) to sequence as many peptides in the population as possible. Finally, given a complete genome sequence for the organism under study and the peptide sequences from the mass spectrometric analysis, the tools of bioinformatics make it possible to assign each peptide to a particular protein-coding sequence (gene) in the genome.

In practice this method detects only a subset of the proteins in a complex mixture of proteins such as that derived from an entire cell. A typical analysis will only detect approximately 1000 different proteins. Nevertheless, additional fractionation methods and enhanced sensitivity of mass spectrometry will likely increase the completeness of these protein profiles in the future. Although LC-MS analysis is very good at identifying which proteins are present in a cell extract, currently it is more difficult to determine the relative abundance of proteins by this approach. To address this weakness, new technologies that quantify the abundance are being developed and have been used in some cases.

Proteome Comparisons Identify Important Differences beween Cells

Although knowing the full complement of proteins in a cell has intrinsic value, in most cases it is the differences between two cell types or between cells exposed to two different growth conditions that are most valuable. By determining the proteome in each situation, the differences in the proteins present can be determined. In turn, this analysis can identify proteins that are likely to be responsible for cellular differences and, therefore, represent good candidates for further study.

The value of comparative proteomics can be seen in an analysis of different cancer cells. It is frequently found that different individuals with apparently the same type of cancer respond very differently to the same chemotherapeutic treatment. By comparing the proteomes of different tumor samples, the apparently similar cells are found to have important differences in the proteins that they express. These differences can become valuable markers to distinguish between the different tumor types. More importantly, these markers can be used to select the most effective chemotherapies for each patient.

Mass Spectrometry Can Also Monitor Protein Modification States

Because the modification state of a protein can profoundly affect its function, efforts are also underway to comprehensively identify the modifica-

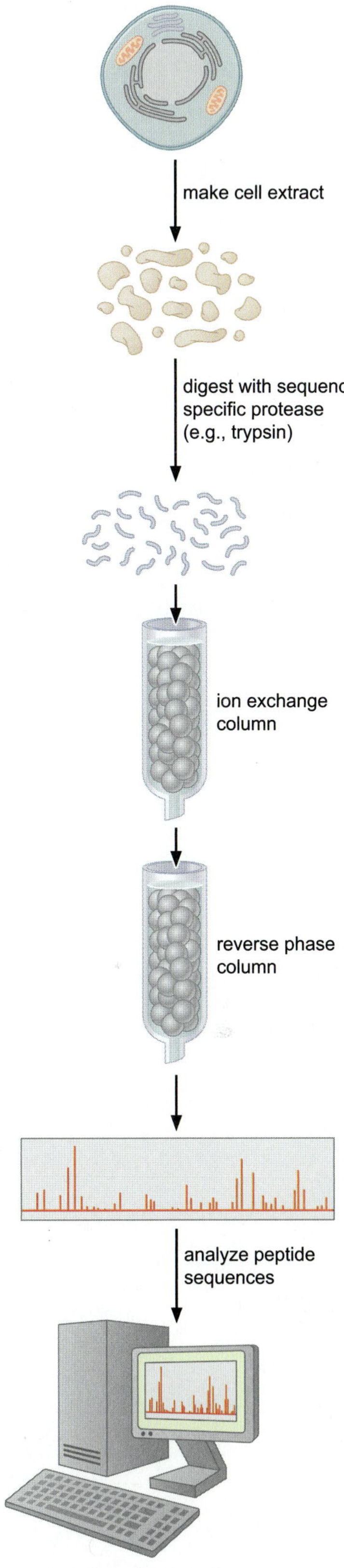

FIGURE 21-25 Separation of proteins by liquid chromatography followed by mass spectrometric analysis. The steps of the method are illustrated in the figure and described in the text.

tion state of proteins in the cell. Specific modifications are commonly used to alter the actvity or stability of a protein. For example, phosphorylation of proteins is used extensively to control their activity. Phosphorylation can cause a protein to alter its conformation in a functionally important manner (e.g., many protein kinases are only active after they are phosphorylated). Alternatively, the attachment of a phosphate can create a new binding site for another protein on the surface of the protein, leading to the assembly of new protein complexes. Other protein modifications include methylation, acetylation, and ubiquitylation. The last of these involves the attachment of the 76-amino-acid protein ubiquitin to a lysine residue via a pseudopeptide bond. Modification of a protein with multiple ubiquitin typically targets the protein for degradation.

Each type of modification causes a discrete change in the molecular mass of the protein. This can be monitored by mass spectrometry, and methods have been developed to identify proteomes that include only those protens with a particular modification. For example, the complete set of phosphorylated proteins in the cell is called the "phospho-proteome." Methods to identify the subset of proteins that include a particular modification have been developed and generally exploit affinity resins that will specifically bind the modification of interest. For example, resins that include immobilized Fe+3 (also called immobilized metal affinity chromatography [IMAC]) specifically bind phosphorylated peptides. Mixtures of peptides derived from crude cell extracts can be incubated with such a resin and the small proportion of peptides that bind are enriched for phophopeptides. These peptides can then be analyzed using LC-MS to identify the proteins that are modified and the sites of modification. This information is a valuable tool to identify the kinase that modified the protein and to test the importance of the modification by generating mutant proteins that cannot be modified.

Protein–Protein Interactions Can Yield Information about Protein Function

Proteomics is also concerned with identifying all the proteins that associate with another protein in a cell to generate what are called interactomes. A complete interactome for a cell would indicate all interactions between proteins in the cell. In what can be considered guilt by association, such interactions can be used to evaluate what processes a protein may be involved in. Proteins that are part of the same protein complex will almost always be involved in the same cellular process.

One method for determining protein–protein interactions is the yeast two-hybrid assay (see Box 17-1), whereby the protein of interest serves as "bait" and a library of proteins can be tested as potential "prey." A second approach is to use affinity resins or immunoprecipitation to rapidly purify a protein of interest along with any associated proteins. The resulting mixture of proteins can then be analyzed by LC-MS to identify the associated proteins. By repeating this procedure with all the proteins in a cell, it is possible to obtain a comprehensive interaction diagram of protein–protein interactions within a cell.

The latter approach has been applied to the yeast *Saccharomyces cerevisiae*. Over 6000 *S. cerevisiae* proteins were purified by affinity chromatography (the gene for each protein was genetically modified or "tagged" to append a short carboxy-terminal extension that is known to bind two affinity resins) and mass spectrometry was used to

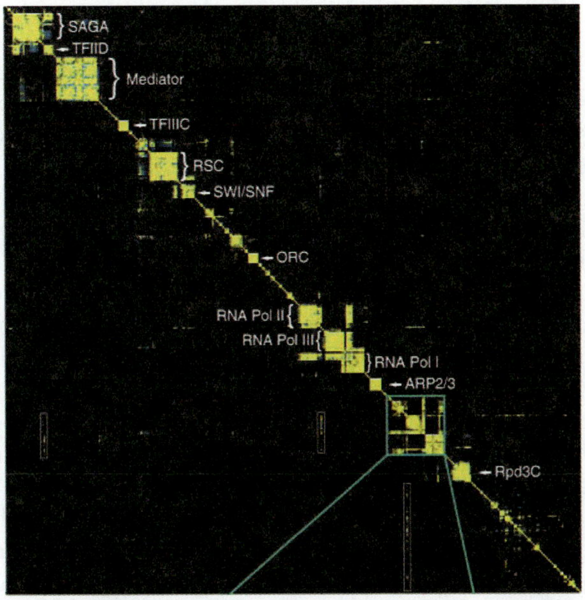

FIGURE 21-26 The physical interactome map of *S. cerevisiae*. Shown here are the results of affinity purification/mass spectrometry studies of all the proteins in *S. cerevisiae*. The figure is actually composed of a series of columns of boxes indicating which proteins co-precipitated with a given protein. If a protein is co-precipitated with the "tagged" protein, the box is yellow. If not, the box is black. In this view, proteins that are in the same complex have been clustered together on both the vertical and horizontal axis; thus, complexes are observed on the diagonal. A subset of all the complexes (many of which have been discussed elsewhere in the text) are labeled and shown in the image presented here. The boxed area is expanded and represented as the part opener image on the first page of Part 5. (Reprinted, with permission, from Collins S.R. et al. 2007. *Mol. Cell. Proteom.* 6: 439–450, Fig. 3b. © American Society for Biochemistry and Molecular Biology.)

identify any additional proteins that copurified with the tagged protein. Comparison of these data identified hundreds of protein complexes present in the cell—many of which were already known, but some of which were novel. The effectiveness of this study can be seen by the detection of a large number of well-documented protein complexes (e.g., RNA polymerase II; Fig. 21-26).

NUCLEIC ACID–PROTEIN INTERACTIONS

We now turn our attention to the various methods that can be used to detect the interactions between nucleic acids and proteins. These interactions are critical to determining the specificity and precision of the events described in this book. Be it transcription, recombination, DNA replication, DNA repair, mRNA splicing, or translation, the proteins that mediate these events must recognize particular nucleic acid structures or sequences to ensure these events occur at the right place and time in the cell.

Consistent with the importance of understanding nucleic acid–protein intereaction, there are a number of robust assays that can be used to measure these events both in vivo and in vitro. In the following sections we will consider several of these assays, comparing their strengths and weaknesses.

The Electrophoretic Mobility of DNA Is Altered by Protein Binding

Just as electrophoretic mobility can be used to determine the relative sizes of DNA, RNA, or protein molecules, it can also be used to detect protein–DNA interactions. If a given DNA molecule has a protein bound to it, migration of that DNA–protein complex through the gel is retarded compared to migration of the unbound DNA molecule. This forms the basis of an assay to detect specific DNA-binding activities. The general approach is as follows. A short DNA fragment containing the binding site of interest is radioactively labeled so it can be detected in small quantities by polyacrylamide gel electrophoresis and autoradiography. A fluorescent label can also be used, but it is important that the fluorophor (see Box 8-1 Fig. 1b) is not in a position that interferes with DNA binding. The resulting DNA "probe" is then mixed with the protein of interest and the mixture is separated on a nondenaturing gel. If the protein binds to the probe DNA, the protein–DNA complex migrates more slowly resulting in a shift in the location of the labeled DNA (Fig. 21-27). For this reason this assay is referred to as a **electrophoretic mobility-shift assay** (EMSA) or, more colloquially, a band or gel shift assay. We have described this assay for a dsDNA fragment; however, it can also be used to detect binding to ssDNA or to RNA.

EMSA can also be used to monitor the association of multiple proteins with the same DNA. These interactions can each be due to sequence-specific DNA binding. Alternatively, after an initial sequence-specific interaction, the subsequent protein can bind to the first DNA-bound protein. In either case, as an additional protein binds, it will further reduce the mobility of the DNA fragment. Using the EMSA in this way can be a very powerful method to identify how a series of proteins interact with DNA interdependently. Different proteins binding to the same DNA probe also can be distinguished because proteins of different size will affect the mobility of the DNA to different extents—the larger the protein, the slower the migration. If two proteins cause a shift to the same extent, then a second method can be used to distinguish which one is bound. The addition of an antibody directed against a protein will cause a "supershift" if that protein is associated with the DNA. Thus, by adding an antibody to a po-

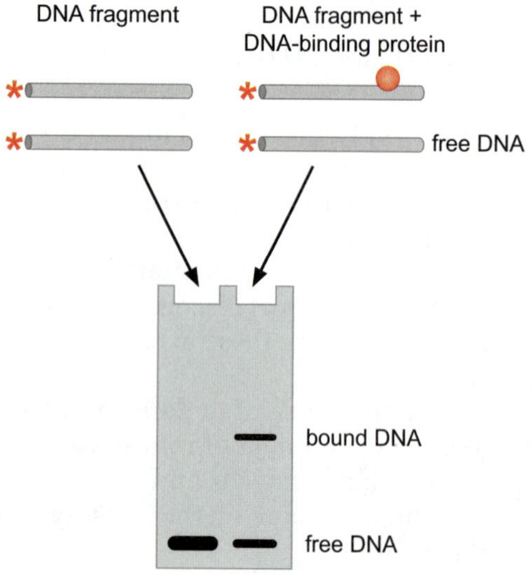

FIGURE 21-27 **Electrophoretic mobility-shift assay.** The principle of the mobility-shift assay is shown schematically. A protein is mixed with radiolabeled probe DNA containing a binding site for that protein. The mixture is resolved by acrylamide gel electrophoresis and visualized using autoradiography. DNA not mixed with protein runs as a single band corresponding to the size of the DNA fragment (left lane). In the mixture with the protein, a proportion of the DNA molecules (but not all of them at the concentrations used) binds the DNA molecule. Thus, in the right-hand lane, there is a band corresponding to free DNA and another corresponding to the DNA fragment in complex with the protein.

tential binding protein, the presence of the protein in the protein–DNA complex can be assessed.

One weakness of EMSAs is that they do not reveal intrinsically what sequence in the DNA the protein binds. Two types of additional experiments can be performed to identify the protein-binding site within the DNA probe. One approach is to add short dsDNA oligomers to the protein prior to incubation with the DNA probe. If the protein-binding site is contained within the oligomer, then the protein will fail to bind the DNA probe. Alternatively, mutations can be made in the DNA probe to assess their effect on protein binding. Although these approaches can be taken without knowledge of potential binding sites, in most instances prior experiments or the conservation of certain DNA sequences within the DNA probe help to simplify the choice of sequences to test.

DNA-Bound Protein Protects the DNA from Nucleases and Chemical Modification

How can a protein-binding site in DNA, such as an operator, be identified more readily? A series of powerful approaches allows identification of the DNA site bound by the protein and of the chemical groups in the DNA (methyl, amino, or phosphate) the protein contacts. The basic principle that underlies these methods is as follows. If a DNA fragment is labeled with a radioactive atom only at one end of one strand, then the location of any break in this strand can be deduced from the size of the labeled fragment that results. The size, in turn, can be determined by high-resolution denaturing electrophoresis in a polyacrylamide gel followed by detection of the labeled ssDNA fragments. For reasons that will be become clear these methods are generally called DNA footprinting.

The most common of these approaches is **nuclease protection footprinting**. After incubating the DNA and the end-labeled DNA together, the resulting complexes are briefly exposed to a DNA nuclease (most often DNase I, which cuts one strand of the target dsDNA). DNA sites bound by protein are protected from nuclease cleavage, creating a region of the DNA without cut sites (Fig. 21-28). The resulting "footprint" is revealed by the absence of bands of particular sizes. The related **chemical protection footprinting** relies on the ability of a bound protein to protect bases in the binding site from base-specific chemical reagents that (after a further reaction) give rise to backbone cuts. In both methods, it is important that the number of nuclease cut sites or chemical modification is titrated to be approximately one per DNA probe. This is because only the cut site that is nearest the labeled DNA end will be detected after gel electrophoresis and labeled DNA detection.

By changing the order of the first two steps, a third method, **chemical interference footprinting**, determines which features of the DNA structure are *necessary* for the protein to bind. Before protein is added to the DNA, an average of one chemical change per DNA is made. The modified DNA is incubated with the DNA-binding protein, and protein–DNA complexes are isolated. One popular method to separate the protein-bound DNA from unbound DNA is to use the EMSA. After detecting the labeled DNA in the EMSA gel, the shifted (protein-bound) and unshifted (unbound) DNA can easily be separated. If a modification at a particular site does not prevent binding of the protein, DNA isolated from the complex will contain that modification.

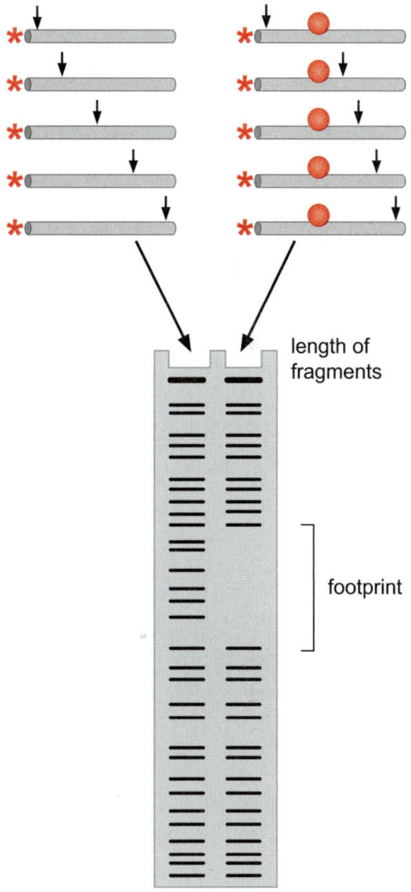

FIGURE 21-28 Nuclease protection footprinting. The stars represent the radioactive labels at the ends of the DNA fragments, arrows indicate sites where DNase cuts, and red circles represent Lac repressor bound to operator. On the left, DNA molecules cut at random by DNase are separated by size using gel electrophoresis. On the right, DNA molecules are first bound to repressor and then subjected to DNase treatment. The "footprint" is indicated on the right. This corresponds to the collection of fragments generated by DNase cutting at sites in free DNA but not in DNA with repressor bound to it. In the latter case, these sites are inaccessible because they are within the operator sequence and hence covered by repressor.

If, on the other hand, a modification prevents the protein from recognizing the DNA, then no DNA modified at the site will be found in the protein-bound DNA sample. As with the chemical protection assay, the sites of chemical modification are detected by treating the DNA with reagents that cleave the DNA at sites of chemical modification.

The reagents used for chemical modification can probe very specific aspects of the DNA. For example, the chemical ethylnitrosourea (ENU) specifically modifies the phosphate residues in the backbone of DNA. Other chemicals specifically modify certain bases in the major or the minor groove. Using a variety of chemicals can provide a precise understanding of the contacts a particular protein makes with the bases and with the phosphates in the sugar–phosphate backbone of DNA.

Footprinting is a powerful approach that immediately identifies the site on the DNA that a protein binds to; however, as a group, these methods require more robust DNA binding by a protein. In general, for a DNA footprinting assay to be effective, greater than 90% of the DNA probe must be bound by protein. This level of binding is required because the footprinting assay detects the lack of a signal (due to protection from cleavage or modification of the DNA) rather than the appearance of a new band. This situation is in contrast to the more sensitive EMSA in which protein binding to the labeled DNA results in the formation of a new band in a region of the gel that, in the absence of protein binding, lacks any DNA molecules.

Chromatin Immunoprecipitation Can Detect Protein Association with DNA in the Cell

Although in vitro assays for protein–DNA binding can be informative, it is often important to determine whether a protein binds a particular DNA site in a living cell. For any particular DNA-binding protein, there are many potential binding sites in the entire genome of a cell. Despite this, in many instances only a subset of these sites will be occupied. In some instances, binding of other proteins may inhibit association of the protein with a potential DNA-binding site. In other instances, binding of adjacent proteins may be required for robust binding. In either case, knowing whether a protein (e.g., a transcriptional regulator) is bound to a particular site (e.g., at a particular promoter region) in the cell can be a powerful piece of evidence that it acts to regulate an event occurring at that site (e.g., transcriptional activation).

Chromatin immunoprecipitation, often just called ChIP, is a powerful technique to monitor protein–nucleic acid interactions in the cell. In outline, the technique is performed as follows: formaldehyde is added to living cells, cross-linking DNA to any bound proteins and proteins bound tightly to other proteins. The cross-linked cells are lysed and the DNA is broken into small fragments (200–300 bp each). Using an antibody specific for the protein of interest (e.g., a transcription regulator), the fragments of DNA attached to that protein can be separated from the majority of the DNA in the cell by immunoprecipitation (or IP). Once the immunoprecipitation is complete, the cross-linking between protein and DNA is reversed, allowing analysis of the DNA sequences that are present in the IP.

The most important step of a ChIP experiment is to determine whether a particular region of DNA is bound by the protein and therefore present in the IP. This can be accomplished by one of two basic approaches. To determine if a particular region of DNA (e.g., a pro-

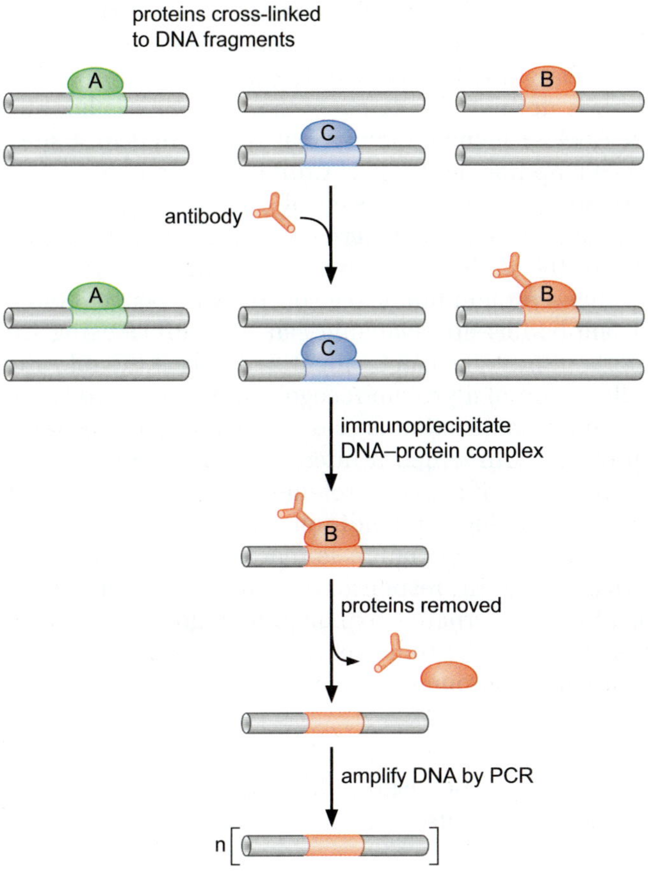

FIGURE 21-29 Chromatin immunoprecipitation (ChIP).

moter) is bound by the protein of interest, PCR can be performed using primers that are targeted to that region. If the protein was bound to that DNA at the time of cross-linking, the sequence will be present in the IP and will be amplified. There are two important controls that are generally included in this assay. First, PCR primers targeting another region of DNA (one to which the protein is known or expected not to bind) are used; in that case, no DNA should be amplified (Fig. 21-29). Second, before performing the IP, a small amount of the total DNA is set aside and both the test and the control primers are used to amplify the DNA in this unfractionated sample. If the PCR primers amplify with the same efficiency, both sequences should be amplified equally from this starting population of DNA. This control ensures that any differences in the extent of PCR amplification of the ChIP DNA using the two different primer sets are due to a difference in abundance and not different efficiency of PCR.

A second approach to identify the DNA sequences associated with a particular protein in the cell is to use tiling DNA microarrays. In this approach the DNA that is cross-linked to the protein and the total DNA isolated from the cell are labeled with two different fluorophores (for this example, we will refer to them as red and green, respectively). The two populations are mixed together and hybridized to the microarray. Regions with a high red:green ratio will represent binding sites of the protein. Those with a low red:green ratio are regions that are not bound. This approach is particularly powerful as whole genomes can be examined simultaneously and no prior knowledge of the potential binding site is required.

Although this microarray technique is very powerful and therefore routinely used, it does have limitations of which the investigator needs to be aware. First, similar to the EMSA, the resolution of ChIP is limited. It is not possible to show that a protein is bound to a specific short DNA sequence, merely that it is bound to a site within a given 200–300 bp fragment. Thus, ChIP is adequate to show that a regulatory protein is bound upstream of one rather than another gene, but it does not show exactly where upstream of the gene the protein is bound. As for the EMSA, mutations in the DNA would be necessary to test whether a protein binds to a specific site. Second, only proteins for which antibodies are available can be studied using ChIP. Even more important, proteins can be identified only if the relevant epitope (the specific region of a protein recognized by an antibody) is exposed when the protein in question is cross-linked to the DNA (and perhaps to other proteins with which it interacts at the gene). In an extension of this complication, if a given protein is not detected under one environmental or physiological condition, but then is detected under another, the obvious interpretation is that the protein binds to that region of DNA only in response to the change in environmental conditions. But an alternative explanation might be that the protein in question is bound all the time, yet its epitope is concealed by another protein, present under one set of conditions but not the other.

In Vitro Selection Can Be Used to Identify a Protein's DNA- or RNA-Binding Site

As more and more DNA-binding proteins are identified and understood, the amino acid motifs associated with sequence-specific DNA binding have become relatively easy to identify (e.g., helix-turn-helix motifs). Despite these findings, our understanding of these nucleic acid–binding protein domains has not evolved to the point where the primary amino acid sequence of a protein is sufficient to reveal the DNA sequence to which it binds. And yet, this information is often very important for identifying potential regulatory regions that can be targeted for subsequent analysis.

How can the DNA sequence recognized by a particular protein be identified? One powerful approach, called **in vitro selection** or **SELEX** (for systematic evolution of ligands by exponential enrichment), involves the use of the sequence specificity of the protein to probe a diverse library of oligonucleotides. By characterizing the enriched DNA, the sequences that bind tightly to the protein can be identified.

The first step in this method is to produce a large library of ssDNA oligonucleotides using chemical DNA synthesis (which we describe in the first part of this chapter). Importantly, the middle 10–12 bases of these oligonucleotides are randomized (by adding a mixture of all four nucleotide precursors to these steps in oligonucleotide synthesis). The randomized region of each nucleotide is flanked on either side by defined sequences. After the oligonucleotide library is synthesized, a short primer is annealed to the defined 3′ end of the oligonucleotides and extended to convert the randomized ssDNA library to a randomized dsDNA library.

Enriching for oligonucleotides that bind the protein of interest can be accomplished using methods similar to those we have already discussed. After incubation of the protein with the library of

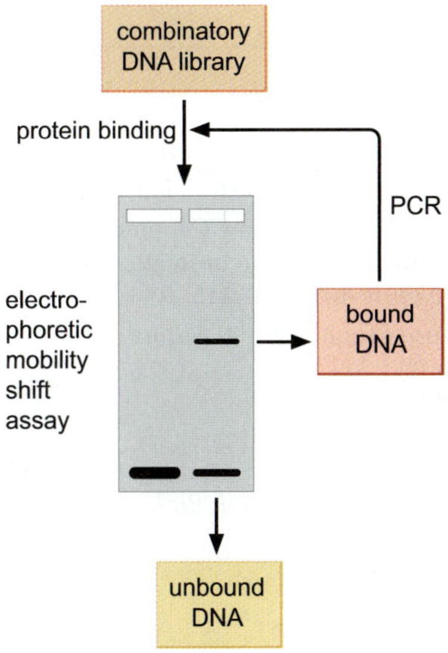

FIGURE 21-30 In vitro selection scheme. A combinatorial DNA library in which the middle 10–12 bases are randomized is bound to the protein of interest. Protein-bound DNA is separated from unbound DNA using an EMSA. Bound DNA is eluted from the gel and subjected to PCR using primers directed against constant regions flanking the random regions of the DNA. These sequences are subjected to two to five more cycles of binding and enrichment to identify the highest affinity.

oligonucleotides, the entire reaction can be separated in a EMSA assay. The DNA in the shifted complex will be strongly enriched for DNA sequences that are tightly bound by the protein. Alternatively, if an antibody is available that recognizes the protein of interest, an immunoprecipitation, similar to the ChIP assay, can be used to separate protein and bound DNA from unbound DNA. Regardless of the mechanism of enrichment, PCR is then used to amplify the bound DNA (using short oligonucleotides that hybridize to the nonrandomized end regions of the oligonucleotides). This amplification step is necessary because only a small percentage of the starting oligonucleotides will bind to the protein. Repeating the binding, enrichment, and amplification steps will greatly enrich for the sequences that are most tightly bound by the protein of interest (Fig. 21-30). Typically, three to five rounds of enrichment are performed to identify the DNA sequences that are most tightly associated with the protein of interest.

The DNA sequence specificity of the protein can be determined by sequencing a subset of the enriched DNAs. Typically only a subset of the sequences within the randomized region will be conserved, as most DNA-binding proteins do not recognize more than six or seven nucleotides. Computational analysis is generally used to assist in identifying the most conserved sequences. The final sequence of bases can be represented by a sequence logo, in which the size of the G, A, T, or C chararacters represents the frequency of appearance of each nucleotide in the library of enriched oligonucleotides (Fig. 21-31).

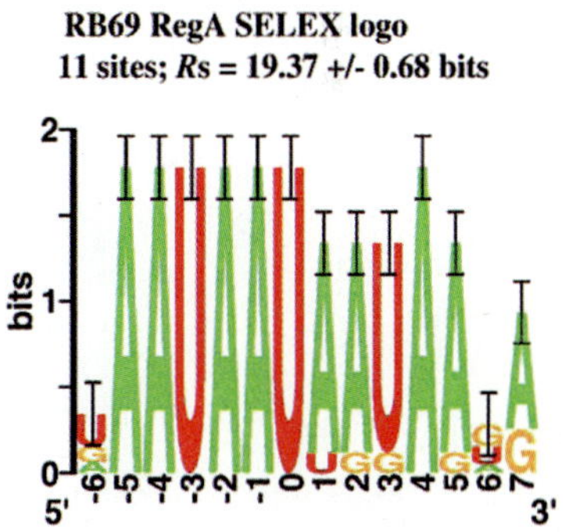

FIGURE 21-31 SELEX sequence logo. In vitro selection was used to isolate RNAs that bind the translational repressor protein RB69 RegA. The image shows the logo of selected sequences. The letter height is proportional to the frequency of each base at that position, with the most frequently occurring base at the top. (Reprinted, with permission, from Dean T.R. et al. 2005. *Virology* 336: 26–36, Fig. 4a. © Elsevier.)

BIBLIOGRAPHY

Books

Griffiths A.J.F., Gelbart W.M., Lewontin R.C., and Miller J.H. 2002. *Modern genetic analysis*, 2nd ed. W.H. Freeman, New York.

Hartwell L., Hood L., Goldberg M.L., Reynolds A.E., Silver L.M., and Veres R.C. 2003. *Genetics: From genes to genomes*, 2nd ed. McGraw-Hill, New York.

Sambrook J. and Russell D.W. 2001. *Molecular cloning: A laboratory manual*, 3rd ed. Cold Spring Harbor Laboratory Press, Cold Spring Harbor, New York.

Snustad D.P. and Simmons M.J. 2002. *Principles of genetics*, 3rd ed. Wiley, New York.

Genomic Analysis

Human genome. 2001. *Nature* **409:** 813–960.

Human genome. 2001. *Science* **291:** 1145–1434.

International Human Genome Sequencing Consortium. 2004. Finishing the euchromatic sequence of the human genome. *Nature* **431:** 931–945.

Mouse genome. 2002. *Nature* **420:** 509–590.

Proteomic Analysis

Yates J.R., 3rd, Gilchrist A., Howell K.E., and Bergeron J.J. 2005. Proteomics of organelles and large cellular structures. *Nat. Rev. Mol. Cell. Biol.* **6:** 702–714.

CHAPTER

22 Model Organisms

A WELL-KNOWN ADAGE IN MOLECULAR BIOLOGY is that fundamental problems are most easily solved in the simplest and most accessible system in which the problem can be addressed. For this reason, over the years molecular biologists have focused their attention on a relatively small number of so-called model organisms. Among the most important of these in order of increasing complexity are *Escherichia coli* and its phage, the T phage and phage λ; baker's yeast *Saccharomyces cerevisiae;* the mustard-like weed, *Arabidopsis thaliana*; the nematode *Caenorhabditis elegans;* the fruit fly *Drosophila melanogaster;* and the house mouse *Mus musculus*.

What is it that model systems have in common? An important feature of all model systems is the availability of powerful tools of traditional and molecular genetics, making it possible to manipulate and study the organism genetically. A second common feature is that the study of each model system attracted a critical mass of investigators. This meant that ideas, methods, tools, and strains could be shared among scientists investigating the same organism, facilitating rapid progress.

For example, beginning in the 1940s a circle of scientists gathered around Max Delbrück, Salvador Luria, and Alfred D. Hershey, spending the summers at the Cold Spring Harbor Laboratories in New York studying the multiplication of the T phage of *E. coli*. These scientists, called the Phage Group, were among those who were important in establishing the field of molecular biology. Many of the members of the Phage Group were physicists attracted to phage, not only because of their relative simplicity, but because the large numbers of phage that could be studied in each experiment generated results that were quantitative and statistically significant. By the late 1950s Cold Spring Harbor offered an annual phage course, where ever-growing numbers of investigators came to learn the new system. This was a case where focusing on the same model organism guaranteed faster progress than would have been made if these individuals had studied many different organisms.

The choice of a model organism depends on what question is being asked. When studying fundamental issues of molecular biology, it is often convenient to study simpler unicellular organisms or viruses. These organisms can be grown rapidly and in large quantities and typically allow genetic and biochemical approaches to be combined. Other questions—for example, those concerning development—can often only be addressed by using more complicated model organisms.

OUTLINE

Thus, the T phage (and its best-known member, T4, in particular) proved to be an ideal system for tackling fundamental aspects of the nature of the gene and information transfer. Meanwhile, yeast, with its powerful mating system for genetic analysis, became the premier system for elucidating fundamental aspects of the eukaryotic cell. Evolutionary conservation from fungi to higher cells has meant that discoveries made in yeast frequently hold true for humans. The nematode and the fruit fly also offer well-developed genetic systems for tackling problems that cannot be effectively addressed in lower organisms, such as development and behavior. Finally, the mouse, though less facile to study than nematodes and fruit flies, is a mammal and hence the best model system for gaining insights into human biology and human disease.

In this chapter, we describe some of the most commonly studied experimental organisms and present the principal features and advantages of each as a model system. We also consider the kind of experimental tools that are available for studying each organism and some of the biological problems that have been studied in each case. This chapter is not intended as a comprehensive presentation of all the model organisms that have had an important impact in molecular biology.

BACTERIOPHAGE

Bacteriophage (and viruses in general) offer the simplest system to examine the basic processes of life. Their genomes, typically small, are replicated—and the genes they encode expressed—only after being injected into a host cell (in the case of phage, a bacterial cell). The genome can also undergo recombination during these infections.

Because of the relative simplicity of the system, phage were used extensively in the early days of molecular biology—indeed, they were vital to the development of that field. Even today they remain a system of choice when studying the basic mechanisms of DNA replication, gene expression, and recombination. In addition, they have been important as vectors in recombinant DNA technology (Chapter 21) and are used in assays for assessing the mutagenic activity of various compounds.

Phage typically consist of a genome (DNA or RNA, most commonly the former) packaged in a coat of protein subunits, some of which form a head structure (in which the genome is stored) and some a tail structure. The tail attaches the phage particle to the outside of a bacterial host cell, allowing the genome of the phage to be passed into that cell. There is specificity here: each phage attaches to a specific cell surface molecule (usually a protein) and so only cells bearing that "receptor" can be infected by a given phage.

Phage come in two basic types—**lytic** and **temperate.** The former, examples of which include the T phage, grow only lytically. That is, as shown in Figure 22-1, when the phage infects a bacterial cell, its DNA is replicated to produce multiple copies of its genome (up to several hundred copies) and expresses genes that encode new coat proteins. These events are highly coordinated to ensure new phage particles are constructed before the host cell is lysed to release them. The progeny phage are then free to infect further host cells.

Temperate phage (such as phage λ) can also replicate lytically. But they can adopt an alternative developmental pathway called **lysogeny** (Fig. 22-2). In lysogeny, instead of being replicated, the phage genome

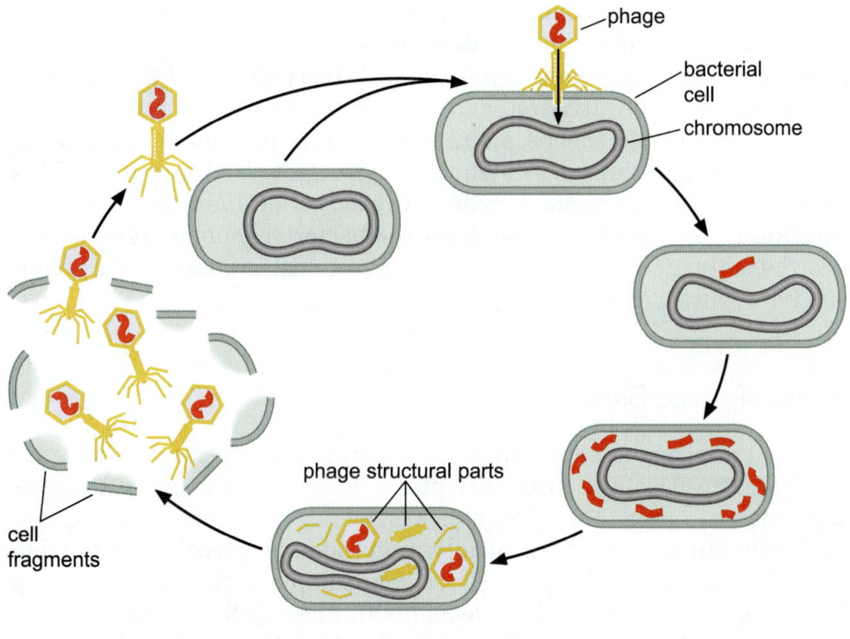

FIGURE 22-1 The lytic growth cycle of a bacteriophage. The phage particle sticks to the outer surface of a suitable bacterial host cell (one bearing the appropriate receptor) and injects its genome, usually a DNA molecule. That DNA is replicated, and the genes expressed to produce many new phage. Once the progeny phage are assembled into mature particles, the bacterial cell is lysed, and the progeny is released to infect another host cell.

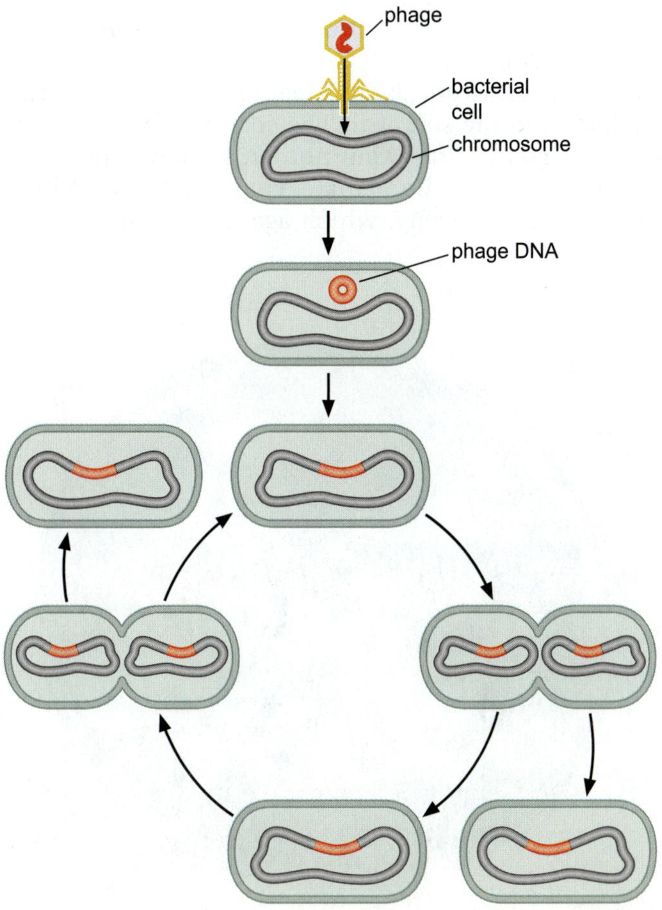

FIGURE 22-2 The lysogenic cycle of a bacteriophage. The initial steps of infection are the same as seen in the lytic case (see Fig. 22-1). But once the DNA has entered the cell, it is integrated into the bacterial chromosome where it is passively replicated as part of that genome. Also, the genes encoding the coat proteins are kept switched off. The integrated phage is called a prophage. The lysogen can be stably maintained for many generations, but it can also switch to the lytic cycle efficiently under appropriate circumstances. See Chapter 16 for a fuller description of these matters.

is integrated into the bacterial genome, and the coat protein genes are not expressed. In this integrated, repressed state the phage is called a **prophage.** The prophage is replicated passively as part of the bacterial chromosome at cell division, and so both daughter cells are lysogens. The lysogenic state can be maintained in this way for many generations but is also poised to switch to lytic growth at any time. This switch from the lysogenic to lytic pathway, called **induction,** involves excision of the prophage DNA from the bacterial genome, replication, and the activation of genes needed to make coat proteins and to regulate lytic growth (shown in Fig. 16-24).

Assays of Phage Growth

For bacteriophage to be useful as an experimental system, methods are needed to propagate and quantify phage. Propagation is needed to generate material (high titer phage stocks for use in experiments) or for DNA extraction. Phage are typically propagated by growth on a suitable bacterial host in liquid culture. Thus, for example, a vigorously growing flask of bacterial cells can be infected with phage. After a suitable time, the cells lyse, leaving a clear liquid suspension of phage particles.

To quantify the numbers of phage particles in a solution, a plaque assay is used (Fig. 22-3). This is done as follows: phage are mixed with, and adsorb to, bacterial cells into which they inject their DNA. The mix is then diluted, and those dilutions are added to "soft agar," which contains many more (and uninfected) bacterial cells. These mixtures are poured onto a hard agar base in a petri dish, where the soft agar sets to form a jelly-like top layer in which the bacterial cells are suspended; some are infected, but most are not. The plates are then incubated for several hours to allow bacterial growth and phage infection to take their course.

Each infected cell (from the original mix) will lyse during subsequent incubation in the soft agar. The consistency of the agar allows the progeny phage to diffuse, but not far, so they infect only bacterial cells growing in the immediate vicinity. Those cells, in turn, lyse, releasing more progeny, which again infect local cells, and so

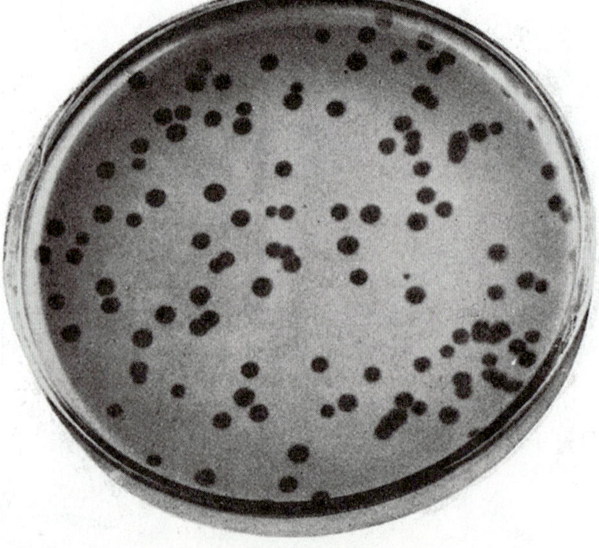

FIGURE 22-3 Plaques formed by phage infection of a lawn of bacterial cells. In the case shown, the plaques are produced by a lytic T phage. (Reprinted, with permission, from Stent G.S. 1963. *Molecular biology of bacterial viruses,* Fig. 1. © W.H. Freeman.)

on. The result of multiple rounds of infection is formation of a **plaque,** a circular clearing in the otherwise opaque lawn of densely grown uninfected bacterial cells. This is because the uninfected bacterial cells grow into a dense population within the soft agar, whereas those bacterial cells located in areas around each initial infection are killed off, leaving a clear patch. Knowing the number of plaques on a given plate, and the extent to which the original stock was diluted before plating, makes it trivial to calculate the number of phage in that original stock.

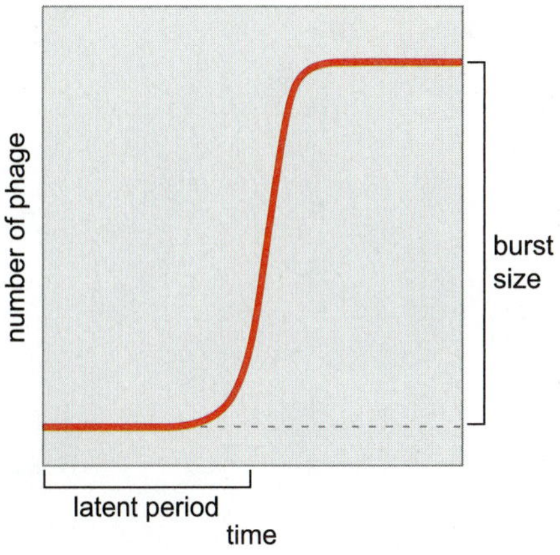

FIGURE 22-4 The single-step growth curve. As described in the text, the single-step growth curve reveals the length of time it takes a phage to undergo one round of lytic growth and also the number of progeny phage produced per infected cell. These are the latent period and burst size, respectively.

The Single-Step Growth Curve

This classic experiment revealed the life cycle of a typical lytic phage and paved the way for many subsequent experiments that examined that life cycle in detail. The essential feature of this procedure is the synchronous infection of a population of bacteria and the elimination of any reinfection by the progeny. This allows the progress of a single round of infection to be followed (Fig. 22-4).

Phage were mixed with bacterial cells for 10 minutes. This is long enough for phage to adsorb to bacterial cells, but it is too short for infection to progress much further. This mixture is then diluted (with fresh growth media) by a factor of 10,000. This dilution ensures that only those cells that bound phage in the initial incubation will contribute to the infected population; also, it ensures that progeny phage produced from those infections will not find host cells to infect.

The diluted population of infected cells is then incubated to allow infection to proceed. At intervals, a sample can be removed from the mixture and the number of free phage counted using a plaque assay. Initially that number is very low (comprising just the phage from the initial infection that did not infect a cell before being diluted).

Once sufficient time has elapsed for infected cells to lyse and release their progeny, a big increase in the number of free phage is detected. (This takes about 30 minutes for the lytic phage T4.) The time lapse between infection and release of progeny is called the **latent period,** and the number of phage released is called the **burst size.**

Phage Crosses and Complementation Tests

Being able to count the number of phage within a population allows researchers to measure whether a given phage derivative can grow on a given bacterial host cell (and the efficiency with which it does so—e..g., the burst size). Also, the plate assay allows certain types of phage derivatives to be distinguished because of the different plaque morphologies they produce. Differences in host range and plaque morphologies were very often the result of genetic differences between otherwise identical phage. In the early days of molecular biology, this provided genetic markers in a system in which they could be analyzed, enabling researchers to ask how genetic information is encoded and functions.

The ability to perform mixed infections—in which a single cell is infected with two phage particles at once—makes genetic analysis possible in two ways. First, it allows one to perform phage crosses. Thus, if two different mutants of the same phage (and thus harboring homologous chromosomes) coinfect a cell, recombination—and thus genetic exchange—can occur between the genomes. The frequency of this genetic exchange can be used to order genes on the genome. A high recombination frequency indicates that the mutations are relatively far apart, whereas a low frequency indicates that the mutations are located close to each other. The large numbers of phage particles that can be used in such experiments ensure that even very rare events will occur (recombination between two very closely positioned mutations) as long as there is a way to screen for—or better still, select for—the rare event. Second, coinfection also allows one to assign mutations to complementation groups; that is, one can identify when two or more mutations are in the same or in different genes. Thus, if two different mutant phage are used to coinfect the same cell and as a result each provides the function that the other was lacking, the two mutations must be in different genes (complementation groups). If, on the other hand, the two mutants fail to complement each other, then that can be taken as evidence that the two mutations are likely located in the same gene.

Transduction and Recombinant DNA

Phage crosses and complementation tests allow the genetics of the phage themselves to be analyzed. These same vehicles and techniques can, however, also be used to investigate the genetics of other systems. Initially these observations were restricted to bacterial genes inadvertently picked up during an infection (as we describe below). With the advent of recombinant DNA techniques in the 1970s, however, these studies were extended to DNA from any organism.

During infection, a phage might occasionally (and accidentally) pick up a piece of bacterial DNA. The most common way in which a phage picks up a section of the host DNA is when a prophage excises from the bacterial chromosome during induction of a lysogen. That process involves a site-specific recombination event (see Chapter 11), and if that event occurs at slightly the wrong position, phage DNA is lost and bacterial DNA included. As long as that exchange does not eliminate part of the phage genome required for propagation, the resulting recombinant phage can still grow and can be used to transfer the bacterial DNA from one bacterial host to another. This process is known as **specialized transduction.** The bacterial DNA included in the specialized transducing phage is amenable to the same kind of genetic analysis as is possible for the phage itself.

Because of its ability to promote specialized transduction, it was natural that phage λ was chosen as one of the original cloning vectors (Chapter 21). Thus, by eliminating many of the sites for a particular restriction enzyme and leaving only one (insertion vector) or two (replacement vector) in a region of the phage not essential for lytic growth, λ can be made to accept the insertion (in vitro) of DNA from any source. That DNA can be propagated and analyzed much more easily than it could in its organism of origin. The restriction endonuclease sites in λ were eliminated by repeatedly selecting phage that plated with higher and higher efficiencies on strains expressing the restriction system in question. By enriching for resistance to endonuclease in this way, and then, in vitro, mapping which sites were lost and which retained, the desired derivative was identified.

Many different λ vectors were developed, all differing in the restriction sites used and in how recombinant phage could be identified. One selection system worked as follows: a λ derivative was derived in which a solitary restriction site was retained within the *c*I gene, the gene that encodes the repressor (see Chapter 16). In the parent vector, therefore, this gene is intact and the phage can, if it chooses, form a lysogen; the phage, therefore, forms turbid plaques. When a piece of DNA is inserted at this site, however, the resulting recombinant phage has a disrupted *c*I gene, cannot form lysogens, and so forms only clear plaques.

This change in plaque morphology provides an easy way of distinguishing recombinant from nonrecombinant phage. Moreover, this approach can be made into a selection (rather than a screen) if the bacterial strain used is an *hfl* strain (see Box 16-5). On that strain, any phage that can form a lysogen invariably does so. Thus, only recombinant phage produce plaques on the *hfl* strain.

BACTERIA

The attraction of bacteria such as *E. coli* or *Bacillus subtilis* as experimental systems is that they are relatively simple cells and can be grown and manipulated with comparative ease. Bacteria are single-celled organisms in which all of the machinery for DNA, RNA, and protein synthesis is contained in the same cellular compartment (bacteria have no nucleus).

Bacteria usually have a single chromosome—typically much smaller than the genome of higher organisms. Also, bacteria have a short generation time (the cell cycle can be as short as 20 minutes) and a genetically homogenous population of cells (a clone) can easily be generated from a single cell. Finally, bacteria are convenient to study genetically because, on the one hand, they are haploid (which means that the phenotypes of mutations, even recessive mutations, manifest readily), and, on the other hand, because genetic material can be conveniently exchanged between bacteria.

Molecular biology owes its origin to experiments with bacterial and phage model systems. Up until the famous fluctuation analysis experiments of Luria and Delbrück in 1943, the study of bacteria (bacteriology) had remained largely outside the realm of traditional genetics. Taking a statistical approach, Luria and Delbrück demonstrated that bacteria can undergo a change in which they become resistant to infection by a particular phage. Critically, they showed that this change arises spontaneously, rather than as a response (adaptation) to the phage. Thus, like other organisms, bacteria can inherit traits (e.g., sensitivity or resistance to a phage), and occasionally this inheritance can undergo a spontaneous change (mutation) to an alternative inheritable state. The experiments of Luria and Delbrück showed that, like other organisms, bacteria exhibit genetically determined characteristics. But because of their simplicity, bacteria would be ideal experimental systems in which to elucidate the nature of the genetic material and the trait-determining factors (genes) of Gregor Mendel.

Assays of Bacterial Growth

Bacteria can be grown in liquid or on solid (agar) medium. Bacterial cells are large enough (~2 μm in length) to scatter light, allowing the growth of a bacterial culture to be monitored conveniently in liquid culture by the increase in optical density. Actively growing bacteria that are dividing with a constant generation time increase in numbers exponentially. They are said to be in the **exponential phase of growth.** As the population increases to high numbers of cells, the growth rate slows and bacteria enter the **stationary phase** (Fig. 22-5).

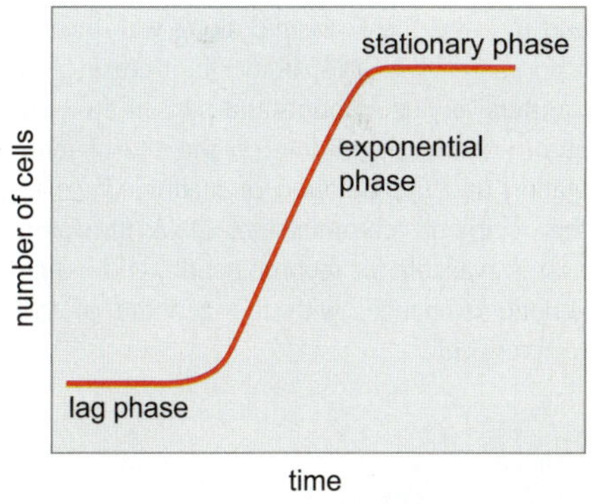

FIGURE 22-5 **Bacterial growth curve.** As described in the text, bacterial cells, such as *E. coli*, can grow very rapidly when not overcrowded and when propagated in well oxygenated rich medium. This phase of growth is called the exponential phase because the cells are replicating exponentially. Once the number of cells gets too high, and the culture becomes very dense, growth tails off into the so-called stationary phase. Cells taken from the stationary phase and diluted to low density in fresh medium will again enter exponential phase growth, but only after a lag phase. The rate of cell number increases in each of these phases is shown.

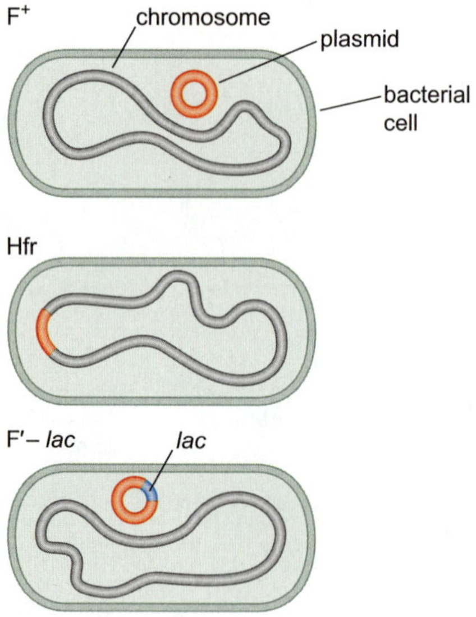

FIGURE 22-6 The three forms of F-plasmid-carrying cells. F^+ cells harbor a single copy of the F-plasmid which replicates as an independent minichromosome. In an Hfr strain, the F-plasmid is integrated into the bacterial chromosome and is replicated as part of that larger molecule. In an F′-strain, an F-plasmid that had previously been integrated into the host chromosome excises, bringing with it a region of adjacent host DNA. All three cell types can be transferred to a recipient F^- cell. If the donor cell is an F^+ strain, it copies and transfers just the F-plasmid; if an F′, it copies and transfers the F-plasmid along with the incorporated host DNA; if an Hfr, it copies and transfers varying amounts and parts of the host chromosome, depending on the site of integration and the duration of mating. Once in the recipient, chromosomal DNA from the host is available for recombination, and hence genetic exchange, with the genome of the recipient cell.

The number of bacteria can be determined by diluting the culture and plating the cells on solid (agar) medium in a petri dish. Single cells grow into macroscopic colonies consisting of millions of cells within a relatively brief period of time. Knowing how many colonies are on the plate and how much the culture was diluted makes it possible to calculate the concentration of cells in the original culture.

Bacteria Exchange DNA by Sexual Conjugation, Phage-Mediated Transduction, and DNA-Mediated Transformation

A principal advantage of bacteria as a model system in molecular biology is the availability of facile systems for genetic change. Genetic exchange makes it possible to map mutations, to construct strains with multiple mutations, and to build partially diploid strains for distinguishing recessive from dominant mutations and for carrying out *cis–trans* analyses.

Bacteria often harbor autonomously replicating DNA elements known as **plasmids** (Fig. 22-6). Some of these plasmids, such as the fertility plasmid of *E. coli* (known as the **F-factor**) are capable of transferring themselves from one cell to another. Thus, a cell harboring an F-factor (which is said to be F^+) can transfer the plasmid to an F^- cell. F-factor-mediated conjugation is a replicative process. Thus, the F^+ cell transfers a copy of the F-factor, while still retaining a copy, such that the products of conjugation are two F^+ cells. Sometimes the F-factor integrates into the chromosome and as a consequence mobilizes conjugative transfer of the host chromosome to an F^- cell. A strain harboring such an integrated F-factor is said to be an **Hfr** (for high frequency recombinant) **strain** and is enormously useful for carrying out genetic exchange.

Precisely which parts of the host chromosome are transferred during any given example of this exchange varies for two reasons. First, different Hfr strains have the F-plasmid integrated at different locations within the host chromosome. Transfer of the host chromosome into the recipient cell takes place linearly, starting with that region of the chromosome closest to one end of the integrated F-plasmid. Thus, where the plasmid is integrated determines which part of the chromosome is transferred first. Also, it is rare that the entire chromosome gets transferred before mating is broken off. Thus, genes far from the transfer start point are transferred with low frequency, and distant genes may never get transferred in a given mating. Note that a complete copy of the integrated F-factor is transferred last, if at all.

A third and extremely important form of the F-factor is the F′-plasmid. The F′ is a fertility plasmid that contains a small segment of chromosomal DNA, which is transferred along with the plasmid from cell to cell with high frequency For example, one such F′ of historic importance is F′-*lac*, an F-factor that contains the lactose operon. F′-factors can be used to create partially diploid strains that have two copies of a particular region of the chromosome. This was precisely how François Jacob and Jacques Monod created partially diploid strains for carrying out their *cis–trans* analyses of mutations in the lactose operon repressor gene and the operator site at which the repressor binds (see Box 16-3).

The F-factor can undergo conjugation only with other *E. coli* strains; however, certain other conjugative plasmids are promiscuous and can transfer DNA to a wide variety of unrelated strains—even to yeast. Such promiscuous conjugative plasmids provide a convenient means for introducing DNA, including DNA that has been modified by recombinant DNA technology, into bacterial strains that are otherwise lacking in their own systems of genetic exchange.

Yet another powerful tool for genetic exchange is phage-mediated transduction (Fig. 22-7). **Generalized transduction** is mediated by phage that occasionally package a fragment of chromosomal DNA during maturation of the virus rather than viral DNA. When such a phage particle infects a cell, it introduces the segment of chromosomal DNA from its previous host in place of infectious viral DNA. The injected chromosomal DNA can recombine with the chromosome of the infected host cell, effecting the permanent transfer of genetic information from one cell to another. This kind of transduction is called generalized transduction because any segment of host chromosomal DNA can be transferred from one cell to another. Depending on the size of the virion, some generalized transducing phages transduce only a few kilobases of chromosomal DNA, whereas others transduce well over 100 kb of DNA.

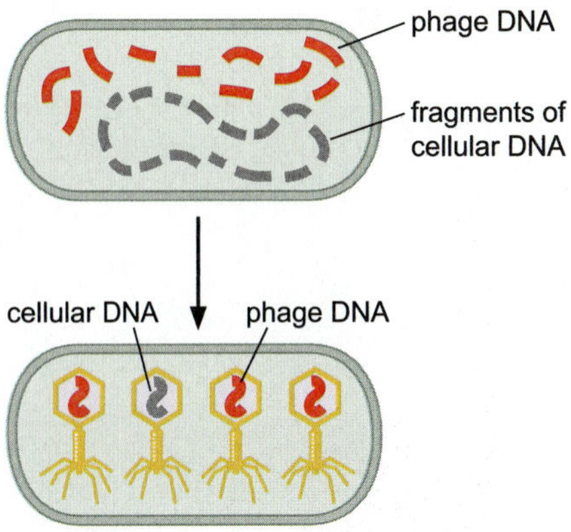

FIGURE 22-7 Phage-mediated generalized transduction. As described in the text, during some phage infections, the host chromosome is fragmented, and segments of that DNA can be packaged in the phage particles instead of the replicated phage DNA. This host DNA is thereby delivered to another cell in the same way as the phage genome ordinarily would. Once in the new host, the DNA can be recombined with the chromosome found there, promoting genetic exchange.

Another kind of phage-mediated transduction is called **specialized transduction,** as already mentioned. This process involves a lysogenic phage such as λ that has incorporated a segment of chromosomal DNA in place of a segment of phage DNA. Such a specialized transducing phage can, upon infection, transfer this bacterial DNA to a new bacterial host cell.

Finally, we come to the case of DNA-mediated transformation, which we described in Chapter 21. Certain experimentally important bacterial species (for example, *B. subtilis* but not *E. coli*) possess a natural system of genetic exchange that enables them to take up and incorporate linear, naked DNA (released or obtained from their siblings) into their own chromosome by recombination. Often the cells must be in a specialized state known as "genetic competence" to take up and incorporate DNA from their environment. Genetic competence is especially useful as it is possible to use recombinant DNA technology to modify a cloned segment of chromosomal DNA and then have it taken up and incorporated into the chromosomes of competent recipient cells.

Bacterial Plasmids Can Be Used as Cloning Vectors

As we have seen, bacteria frequently harbor circular DNA elements known as plasmids that can replicate autonomously. Such plasmids can serve as convenient vectors for bacterial DNA as well as foreign DNA. Indeed, the initial (and successful) attempts to clone recombinant DNA involved a plasmid (pSC101) of *E. coli* that contains a unique restriction site for *Eco*RI into which DNA could be inserted without impairing the capacity of the plasmid to replicate (Chapter 21).

Transposons Can Be Used to Generate Insertional Mutations and Gene and Operon Fusions

As we discussed in Chapter 11, **transposons** are not only fascinating genetic elements in their own right but are enormously useful tools for carrying out molecular genetic manipulations in bacteria. For example, transposons that integrate into the chromosome with low-sequence specificity (i.e., with a high degree of randomness), such as Tn*5* and Mu, can be used to generate a library of insertional mutations on a genome-wide basis (Fig. 22-8).

Such mutations have two important advantages over traditional mutations induced by chemical mutagenesis. One advantage is that the insertion of a transposon into a gene is more likely to result in complete inactivation (a null mutation) of the gene (when such is desired) than a

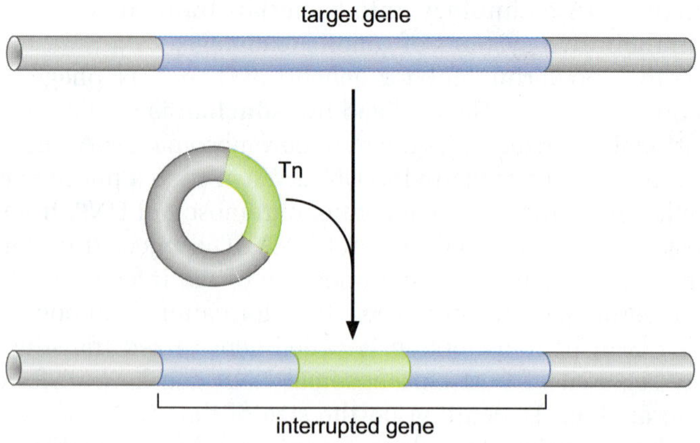

FIGURE 22-8 Transposon-generated insertional mutagenesis. The transposon, carried into a cell on a plasmid, can then transpose from that vehicle into the host genome. Because of the high density of coding regions (genes) on a typical bacterial chromosome, the transposon will very often insert into a gene. A marker carried on the transposon (such as antibiotic resistance) allows cells harboring insertions to be isolated. Knowing the sequence at the ends of the transposon, and of the genome into which it has inserted, makes identifying its location straightforward.

simple nucleotide substitution created by a mutagen. The second advantage is that, having inactivated the gene, the presence of the inserted DNA makes it easy to isolate and clone that gene. Even more simply, with the appropriate DNA primers, the identity of the inactivated gene can be determined by DNA sequence analysis from chromosomal DNA harboring the transposon insertion.

Transposons can also be used to create gene and operon fusions on a genome-wide basis. Modified transposons have been created that harbor a reporter gene such as a promoter-less *lacZ* (e.g., Tn*5lac*). When this transposon inserts into the chromosome (in the appropriate orientation), transcription of the reporter is brought under the control of the disrupted target gene. Such a fusion is known as an operon or transcriptional fusion (Fig. 22-9).

Other fusion-generating transposons have been created that harbor a reporter gene lacking both a promoter and sequences for the initiation of translation. In these cases, expression of the reporter requires both that it be brought under the transcriptional control of the target gene and that it be introduced into the reading frame of the target gene so that it can be translated properly. A fusion in which the reporter is joined both transcriptionally and translationally to the target gene is known as a gene fusion.

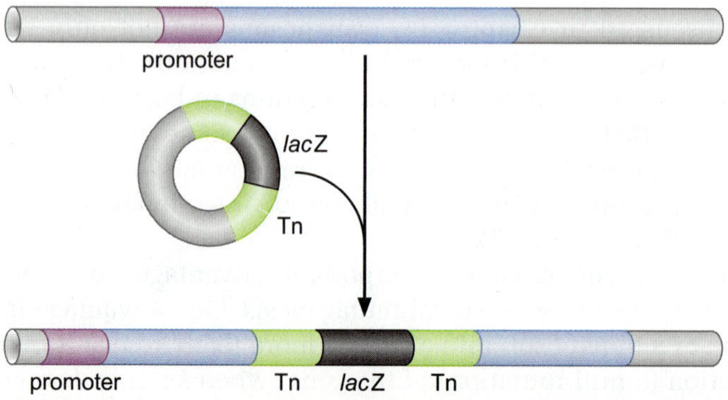

FIGURE 22-9 Transposon-generated *lacZ* fusions. The method of transposon mutagenesis outlined in Fig. 22-8 can be modified to allow insertion of a reporter gene (e.g., *lacZ*) into any region of the genome. This allows expression of a host gene (the one in which the transposon–*lacZ* fusion is inserted) to be assessed simply by measuring the level of expression of *lacZ* in that strain.

Studies on the Molecular Biology of Bacteria Have Been Enhanced by Recombinant DNA Technology, Whole-Genome Sequencing, and Transcriptional Profiling

With the advent of recombinant DNA technologies, such as DNA cloning, the availability of whole-genome sequences and methods for studying gene transcription on a genome-wide basis have, of course, revolutionized molecular biological studies of higher cells. But these same technologies have had an impact on the study of bacterial model systems as well, especially when used in conjunction with the traditional tools of bacterial genetics. For example, the development of tailor-made derivatives of transposons for creating gene fusions is facilitated by recombinant DNA methodologies. As another example, the use of genetic competence in combination with recombinant methods for creating precise mutations and gene fusions has expanded the kinds and number of molecular genetic manipulations. The availability of microarrays representing all of the genes in a bacterium has made it possible to study gene expression on a genome-wide basis. In combination with the tools described above, the function of genes identified as being expressed under a particular set of conditions can be rapidly and conveniently elucidated. Methods for rapidly identifying proteins that interact with each other (such as two-hybrid analysis; see Box 17-1), which have had a great impact in yeast and other eukaryotic systems, are also powerful tools for elucidating networks of interactions among bacterial proteins. The availability of whole-genome sequences and promiscuous conjugative plasmids has created opportunities for carrying out molecular genetic manipulations in bacterial species that otherwise lack sophisticated, traditional tools of genetics.

Biochemical Analysis Is Especially Powerful in Simple Cells with Well-Developed Tools of Traditional and Molecular Genetics

Since the earliest days of molecular biology, bacteria have occupied center stage for biochemical studies of the machinery for DNA replication, information transfer, and gene regulation, among many other topics. There are several reasons for this. First, large quantities of bacterial cells can be grown in a defined and homogenous physiological state. Second, the tools of traditional and molecular genetics make it possible to purify protein complexes harboring precisely engineered alterations or to overproduce and thereby obtain individual proteins in large quantities. Third, and of great importance, the machinery for carrying out DNA replication, gene transcription, protein synthesis, and so forth is much simpler (having far fewer components) in bacteria than in higher cells, as we have seen repeatedly in this text. Thus, elucidating fundamental mechanisms proceeds more rapidly in bacteria in which fewer proteins need to be isolated and in which mechanisms are generally more streamlined than in higher cells.

Bacteria Are Accessible to Cytological Analysis

Despite their apparent simplicity and the absence of membrane-bound cellular compartments (e.g., a nucleus and a mitochondrion), bacteria are not simply bags of enzymes, as had been thought for many decades. Instead, as we now know, proteins and protein complexes have characteristic locations within the cell. Even the chromosome is highly organized inside bacteria. Despite their small size, bacteria are accessible to the tools of cytology, such as immunofluoresence microscopy for localizing proteins in fixed cells with specific antibodies, fluorescence microscopy with the green fluorescent protein for localizing proteins in living cells, and fluorescence in situ hybridization (FISH) for localizing chromosomal regions and plasmids within cells. The applications of such methods have provided invaluable insights into several of the molecular processes considered in this text. For example, we now know that the replication machinery of the bacterial cell is relatively stationary and is localized to the cell center (Chapter 8). This finding tells us that the DNA template is threaded through a relatively stationary replication "factory" during its duplication as opposed to the traditional view in which the DNA polymerase traveled along the template like a train on a track. As another example, the application of cytological methods have taught us (again contrary to the traditional view) that during replication the two newly duplicated origin regions of the chromosome migrate toward opposite poles of the cell. Cytological methods are an important part of the arsenal for molecular studies on the bacterial cell.

Phage and Bacteria Told Us Most of the Fundamental Things about the Gene

Molecular biology owes its origin to experiments with bacterial and phage model systems. Indeed, as we saw in Chapter 2, groundbreaking work with a pneumococcus bacterium led to the discovery that the genetic material is DNA. Since then, experiments with *E. coli* and its phage have led the way, as we have seen throughout this book. For example, the experiment of Hershey and Martha Chase convinced people that the genetic material of phage is DNA; the experiment of Matthew Meselson and Franklin W. Stahl proved that DNA replicates semiconservatively in *E. coli*; the phage crosses of Francis H. Crick and Sydney Brenner (Chapter 15) revealed that the genetic code is built of triplet codons; the elegent genetic studies carried out by Charles Yanofsky in *E. coli* demonstrated genetic colinearity; and the work of Jacob and Monod (see Box 16-3) uncovered the fundamental strategies of gene regulation. There are countless other examples where, by choosing these simplest of systems, fundamental processes of life were understood.

An important example comes from the classic work of Seymour Benzer, who examined intensely a single genetic locus in phage T4, called *r*II. Wild-type T4 is capable of growing in either of two strains of *E. coli* known as B and K, but *r*II mutants grow only in strain B. This makes it possible to detect wild-type phage (arising, e.g., from recombination between two different *r*II mutants) at frequencies of less than 0.01%. That is, a single wild-type phage can be detected among 10,000 *r*II mutant phage when plated on a lawn of strain K bacteria where only the rare recombinant will form a plaque.

Taking advantage of this seemingly arcane property of *r*II mutations, Benzer carried out recombination experiments between pairs of *r*II mutants and was thereby able to map the order of such mutations at a high level of resolution (approaching or reaching that of the nucleotide base pair). He also devised a "complementation" test (discussed above) for showing that the *r*II locus comprises two adjacent genes. Benzer introduced the term **cistron** to describe the gene (based on the words *cis* and *trans*). As an aside, it is interesting to note that it was this work that enabled this same locus to be exploited by Crick and Brenner in their genetic studies on the genetic code.

BAKER'S YEAST, *SACCHAROMYCES CEREVISIAE*

Unicellular eukaryotes offer many advantages as experimental model systems. They have relatively small genomes compared to other eukaryotes (see Chapter 7) and a similarly smaller number of genes. Like *E. coli*, they can be grown rapidly in the laboratory (~90 minutes per cell division under ideal conditions), allowing cloned populations to be propagated from a single precursor cell. Despite this simplicity, yeast cells have the central characteristics of all eukaryotic cells. They contain a discrete nucleus with multiple linear chromosomes packaged into chromatin, and their cytoplasm includes a full spectrum of intracellular organelles (e.g., mitochondria) and cytoskeletal structures (such as actin filaments).

The best studied unicellular eukaryote is the budding yeast *S. cerevisiae*. Often referred to as brewer's or baker's yeast because of its use as a fermenting agent, *S. cerevisiae* has been intensely studied for more than 100 years. In experiments in the 1860s, Louis Pasteur identified this yeast as the catalyst for fermentation (prior to Pasteur's work, sugar was believed to break down spontaneously into alcohol and carbon dioxide). These studies eventually led to the identification of the first enzymes and the development of biochemistry as a experimental approach. The genetics of *S. cerevisiae* has been studied since the 1930s, resulting in the characterization of many of its genes. Thus, like *E. coli*, *S. cerevisiae* allows investigators to attack fundamental problems of biology using both genetic and biochemical approaches.

The Existence of Haploid and Diploid Cells Facilitate Genetic Analysis of *S. cerevisiae*

S. cerevisiae cells can grow in either a haploid state (one copy of each chromosome) or diploid state (two copies of each chromosome) (Fig. 22-10). Conversion between the haploid and diploid states is

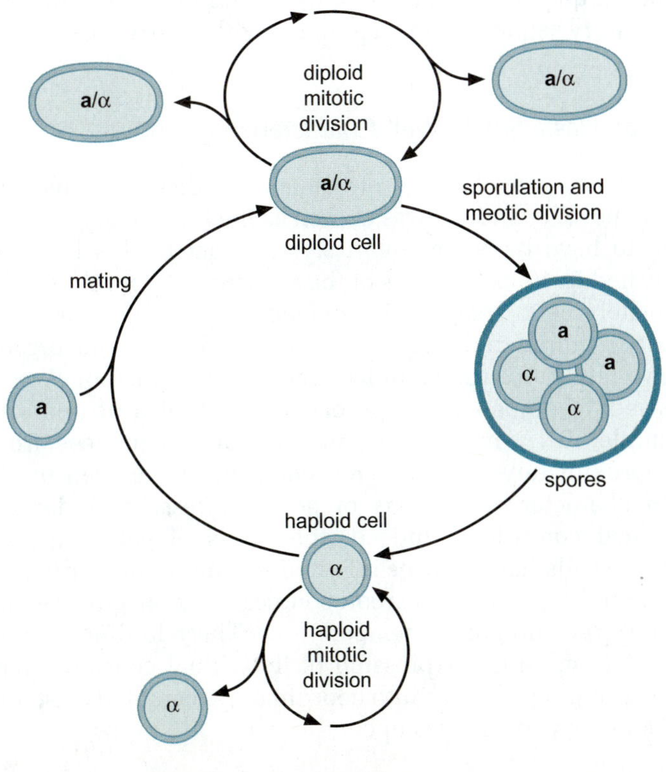

FIGURE 22-10 The life cycle of the budding yeast *S. cerevisiae*. *S. cerevisiae* exists in three forms. Two haploid cell types, **a** and α, and the diploid product of mating between these two. Replication of these different cell types, mating and sporulation, are shown.

mediated by mating (haploid to diploid) and sporulation (diploid to haploid). There are two haploid cell types called **a** and α cells. When grown together, these cells mate to form **a**/α diploid cells. Under conditions of reduced nutrients, **a**/α diploids undergo meiotic division (see Chapter 7) to generate a structure known as the ascus that contains four haploid spores (two **a** spores and two α spores). When growth conditions improve, these spores can germinate and grow as haploid cells or mate to re-form **a**/α diploids.

In the laboratory, these cell types can be manipulated to perform a variety of genetic assays. Genetic complementation can be performed by simply mating two haploid strains, each of which contains one of the two mutations whose complementation is being tested. If the mutations complement each other, the diploid will be a wild type for the mutant phenotype. To test the function of an individual gene, mutations can be made in haploid cells in which there is only a single copy of that gene. For example, to ask if a given gene is essential for cell growth, the gene can be deleted in a haploid. Only deletions of nonessential genes can be tolerated by haploid cells.

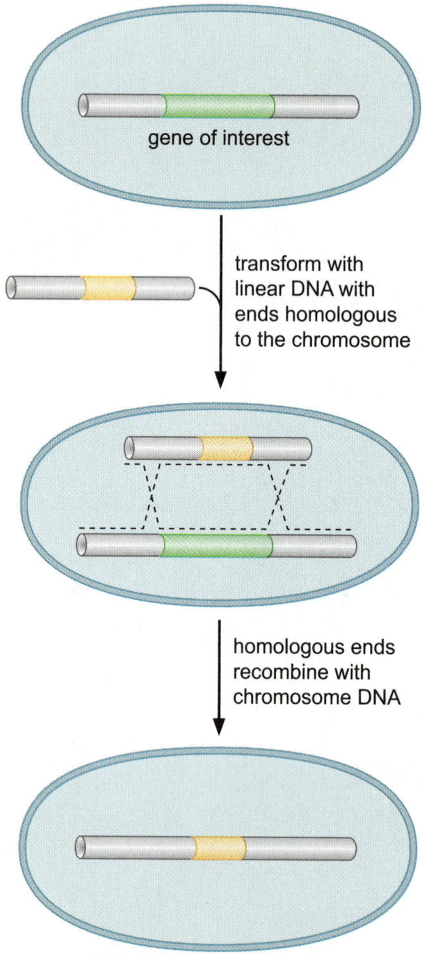

FIGURE 22-11 **Recombinational transformation in yeast.** Any region of the yeast genome can readily be replaced by the sequence of choice. The DNA to be inserted is flanked with DNA sequences homologous to the sequences flanking the region in the chromosome to be replaced. When the donor fragments are introduced to the cell, high levels of homologous recombination in this organism ensure a high frequency of recombination with the chromosome, resulting in the genetic exchange shown. The inserted DNA may differ from the resident sequence by as little as a single base pair, or at the other extreme, it can be very different in length and sequence. Thus, very elaborate genetic modifications can be achieved.

Generating Precise Mutations in Yeast Is Easy

The genetic analysis of *S. cerevisiae* is further enhanced by the availability of techniques used to precisely and rapidly modify individual genes. When linear DNA with ends homologous to any given region of the genome is introduced into *S. cerevisiae* cells, very high rates of homologous recombination are observed resulting in the replacement of chromosomal sequences with DNA used in the transformation (Fig. 22-11). This property can be exploited to make precise changes within the genome. This approach can be used to precisely delete the coding region of an entire gene, change a specific codon in an open reading frame, or even change a specific base pair in a promoter. The ability to make such precise changes in the genome allows very detailed questions concerning the function of particular genes or their regulatory sequences to be pursued with relative ease.

S. cerevisiae Has a Small, Well-Characterized Genome

Because of its rich history of genetic studies and its relatively small genome, *S. cerevisiae* was chosen as the first eukaryotic (nonviral) organism to have its genome entirely sequenced. This landmark was accomplished in 1996. Analysis of the sequence (1.3×10^6 bp) identified approximately 6000 genes and provided the first view of the genetic complexity required to direct the formation of a eukaryotic organism.

The availability of the complete genome sequence of *S. cerevisiae* has allowed "genome-wide" approaches to studies of this organism. For example, DNA microarrays that include sequences from each of the approximately 6000 *S. cerevisiae* genes have been used extensively to characterize patterns of gene expression under different physiological conditions. Indeed, the levels of gene expression in *S. cerevisiae* cells have now been tested for hundreds of different conditions, including different carbon sources (such as glucose vs. galactose), cell types, and growth temperatures. These findings are not only useful to determine the expression of individual genes but have also led to the grouping of genes into coordinately regulated sets, which all respond similarly to changes in conditions.

Other genome-wide resources include a library of 6000 strains, each deleted for only one gene. Greater than 5000 of these strains are viable as haploids, indicating that the majority of yeast genes are nonessential under the ideal growth conditions in the laboratory. This collection of strains has allowed the development of new genetic screens in which every gene in the *S. cerevisiae* genome can be tested individually for its role in a particular process. The use of microarrays has also allowed the genome-wide mapping of binding sites for transcriptional regulators using chromatin immunoprecipitation techniques (see Chapter 21).

S. cerevisiae Cells Change Shape as They Grow

As *S. cerevisiae* cells progress through the cell cycle, they undergo characteristic changes in shape (Fig. 22-12). Immediately after a new cell is released from its mother, the daughter cell appears slightly elliptical in shape. As the cell progresses through the cell cycle, it forms a small "bud" that will eventually become a separate cell. The bud grows until it reaches a size slightly smaller than the "mother" cell from which it arose. At this point the bud is released from the mother and both cells start the process again.

Simple microscopic observation of *S. cerevisiae* cell shape can provide a lot of information about the events occurring inside the cell. A cell that lacks a bud has yet to start replicating its genome. This is because in a wild-type *S. cerevisiae* cell, the emergence of a new bud is linked to the initiation of DNA replication. Similarly, a growing cell with a very large bud is almost always in the process of executing chromosome segregation.

The powerful genetic, biochemical, and genomic tools available to study *S. cerevisiae* have made it a favored eukaryotic organism for the analysis of basic molecular and cell-biological questions. Studies of *S. cerevisiae* have made fundamental contributions to our understanding

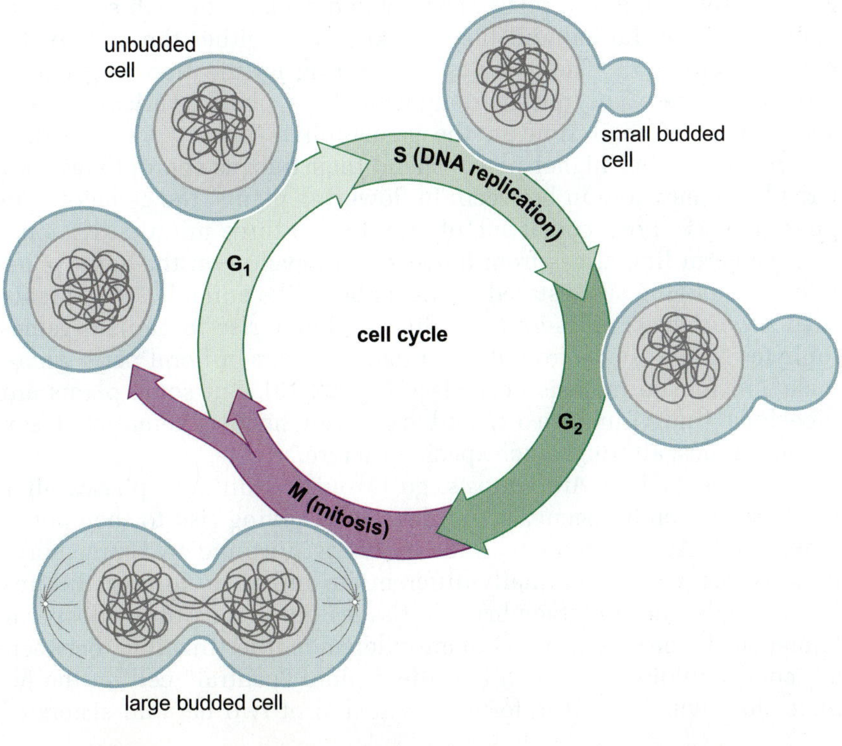

FIGURE 22-12 The mitotic cell cycle in yeast. *S. cerevisiae* divides by budding. The development of a daughter bud through the mitotic cycle is shown and described in the text.

of eukaryotic transcription and gene regulation, DNA replication, recombination, translation, and splicing. Genetic studies in baker's yeast have identified proteins involved in all of these events. Perhaps most importantly, the proteins and genes identified as critical to these fundamental events in *S. cerevisiae* are almost always conserved in other eukaryotes including human. Thus, what is learned with this simple model eukaryote is almost always relevant to the same events in the more complex organisms.

ARABIDOPSIS

Plant science has the longest history of all the life sciences, with its roots in agriculture and botanical medicine: through Mendel, plant science laid the foundations for genetics; and through Charles Darwin, Barbara McClintock, William Bateson, and others, for cytogenetics, development, physiology, and evolution. Plant science continues to make important advances in fundamental areas like RNA interference, while impacting the economy and the environment as it always has. In the last few decades, the humble mustard-like weed, *A. thaliana*, has emerged as a model system on a parallel with *Drosophila*, *C. elegans*, and the mouse. Even more so than its animal counterparts, *Arabidopsis* illustrates most key aspects of plant biology, especially among the angiosperms (flowering seed plants). And just as maize revolutionized plant genetics in the 20th century, *Arabidopsis* promises to revolutionize plant genomics and most aspects of plant biology into the future.

Arabidopsis Has a Fast Life Cycle with Haploid and Diploid Phases

Like yeast, all plants have both haploid and diploid life cycle phases, which are named according to their products—the diploid phase (like yeast) supports meiosis to generate spores and is therefore named the "sporophyte" (spore-bearing plant). These haploid spores germinate to give rise to the haploid phase, from which gametes of each sex differentiate, and so the haploid phase is known as either the male or female gametophyte. The gametes fuse during fertilization to generate diploid zygotes. The relative length of these phases varies—mosses spend most of their time in the gametophyte phase, whereas *Arabidopsis* and other higher plants spend their time as sporophytes, and ferns lie somewhere in between. In flowering plants, the gametophyte phase is very short, consisting of only two to three mitotic divisions, and the germ line arises from flowers that develop on the adult plant rather than being sequestered in the embryo like animals. Most plants (like *Arabidopsis*) are hermaphrodites and give rise to haploid gametophytes of both sexes from differentiated flowers or floral parts where male and female meiosis occur (see Fig. 22-13). But some plants are dioecious (individual sexes) and can even have differentiated sex chromosomes, although such species are rare.

Seed plants, like *Arabidopsis*, go through additional phases after fertilization—embryogenesis and dormancy, giving rise to the eponymous seed. As in mammals, the embryo is nurtured by extraembryonic tissues, which terminally differentiate and go no further. But unlike animals, these extraembryonic tissues (known as the endosperm "inner seed") are the product of an independent fertilization, between a second haploid sperm cell and the diploid "central" cell on the female side, which itself is formed by fusion of two haploid sisters of

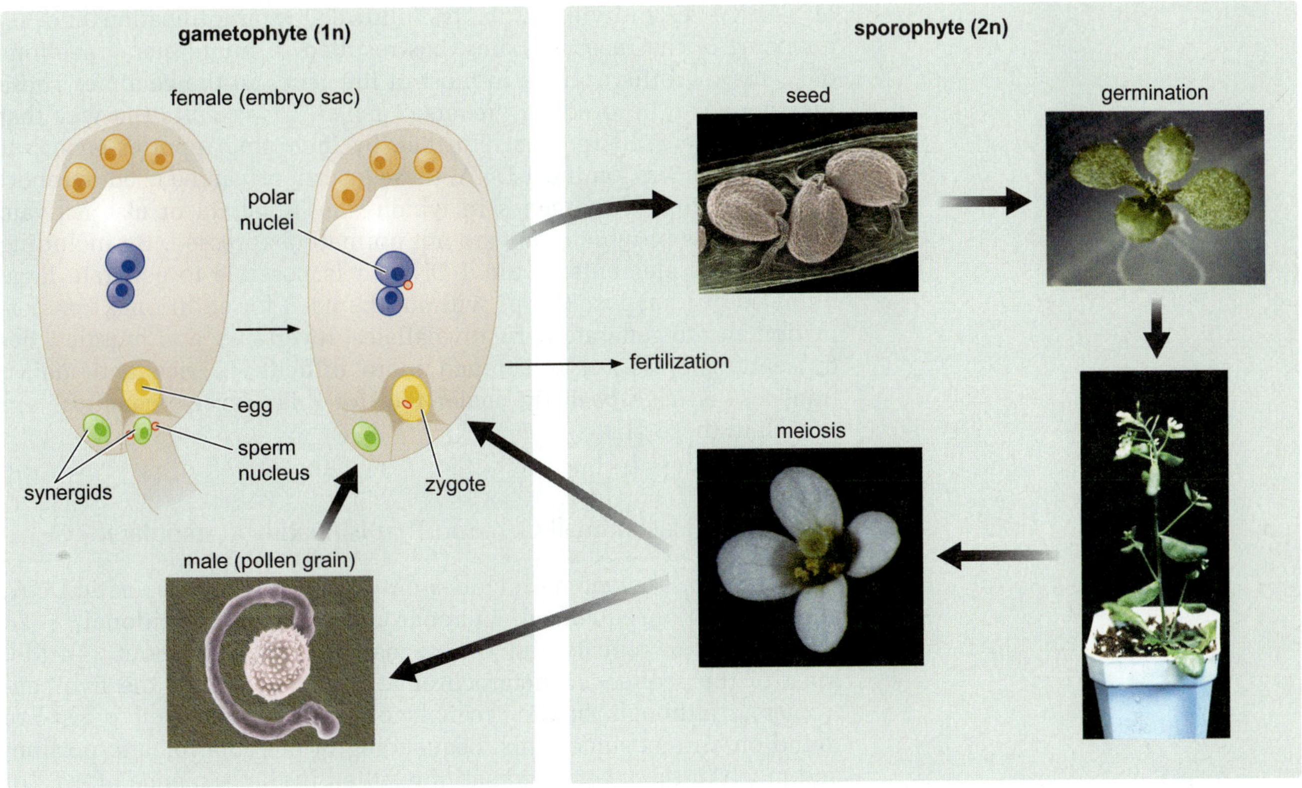

FIGURE 22-13 **The life cycle phases of *Arabidopsis*.** (Courtesy of Rob Martienssen.)

the egg cell. Rapid division of the triploid nucleus is followed by cellularization and starch and protein accumulation, which provides important nutrients to the embryo. The endosperm is ephemeral in many plants (such as *Arabidopsis*), being gobbled up by the embryo as it grows, but can survive until germination in other plants, providing starch in staple crops like wheat and maize.

Arabidopsis Is Easily Transformed for Reverse Genetics

Infection with the soil bacterium, *Agrobacterium tumefaciens*, and its relatives leads to the induction of tumorous growths (galls) because of the transfer of hormone biosynthesis genes from the bacterial Ti (tumor-inducing) plasmid into the chromosomal DNA of the host plant. The tumor-inducing genes are found in the transfer DNA (T-DNA) portion of the plasmid, flanked by directly repeated border sequences required for transfer. By replacing the tumor-inducing genes with genes of interest, it is possible to transform plants. *Arabidopsis* can be transformed by simply spraying the plants with, or dipping them into, a concentrated culture of *Agrobacterium* in a surfactant solution to promote infection. Transient infection occurs almost immediately and is useful for transient expression studies, but stable transformation is thought to occur several days or weeks later, possibly on infection of the female gametophyte, before fertilization. By including a selectable marker gene (for various types of herbicide resistance), it is possible to select transformed plants by germinating seed on media, or soil, containing herbicides.

The efficiency of *Arabidopsis* transformation is so high that it can be used for mutagenesis; random insertion of hundreds of thousands of T-DNAs in individual plants, followed by amplification and sequencing of the insertion sites, has resulted in numerous collections of plants with disruptions in most of the genes in the genome. These insertions can be used for "reverse" genetics in exactly the way that deletions are used in yeast. Further, by including reporter genes, or strong enhancers, on the T-DNA, these insertions can be used to report the expression of the genes in which they integrate or else activate them in cells in which they are not normally expressed. By including transposable elements in the T-DNA, it is possible to generate large numbers of transposon hops without the need for additional transformation and to generate derivative alleles, revertants, and mosaics. Because transformation was much more difficult in rice and maize, transposons have been the major tool for this "reverse genetics" approach in these plants.

Arabidopsis Has a Small Genome That Is Readily Manipulated

The *Arabidopsis* genome includes only 105 Mb of euchromatic DNA, about 15 Mb of sequenced heterochromatin, and an additional 15–25 Mb of satellite repeats and rDNA, making a total of about 140 Mb. Most of the sequenced heterochromatin flanks each of the five centromeres, although smaller regions of heterochromatin (knobs) are found on chromosome arms. Sequencing of the euchromatic portion, and much of the heterochromatin, resulted in the sequence of 99% of the 29,000 *Arabidopsis* genes. Sequencing of many other plant genomes has revealed that several rounds of genome duplication (polyploidy) have occurred in the "eudicots" a major branch of the angiosperm evolutionary tree that includes *Arabidopsis*. The most recent duplication was only a few million years ago, so that about 25% of *Arabidopsis* genes have retained a functional homolog, resulting in substantial genetic redundancy and complicating reverse genetic strategies. On the other hand, forward genetics has been very powerful in *Arabidopsis*, perhaps in part because of this redundancy, which allows heavy doses of mutagens (such as ethylmethane sulfate [EMS], ethylnitrosourea [ENS], or irradiation) to be used without killing the plants, so that relatively small numbers of mutagenized seed can achieve saturation. Seed can be mutagenized directly and recessive mutations recovered by simply allowing the seed to germinate and self-pollinate.

The availability of the genome sequence and several polymorphic strains has made positional cloning of mutations identified by forward genetics extremely straightforward. EMS mutagenesis can even be used in a reverse genetics strategy, known as **tilling**, in which DNA from mutagenized plants is screened for point mutations in genes of interest. With the emergence of very-high-throughput sequencing methods, this strategy is likely to become even more practical and can recover a full spectrum of allelic variation in each gene. The availability of the genome sequence has enabled a host of other genomic technologies, such as tiling microarrays, high-throughput protein localization, and proteomic technologies, to name but a few.

RNA interference via small RNA (19–30 nucleotides) is an important endogenous and exogenous mechanism for regulating genes and was first discovered in plants (see Chapter 18). In *Arabidopsis* at least three classes of small RNA—microRNA (miRNA); *trans*-acting, short interfering RNA (tasiRNA); and siRNA associated with repeats—differ in size and biogenesis, but all can regulate genes by matching their sequence and promoting "slicing" via endonuclease activity, translational arrest, or chromatin and DNA modification. These small RNAs are derived from single-strand "hairpin" structure precursors or from double-stranded RNA that is the product of RNA-dependent RNA polymerase. Genomic methodology using RNA interference in *Arabidopsis* includes VIGS (virally induced gene silencing), cosuppression, hairpin silencing, and artificial miRNA.

Epigenetics

Epigenetic variation is generally defined as "mutations" that are chromosomally inherited but do not involve a change in nucleotide sequence. These "epimutations" are usually reversible at a significantly higher frequency than regular mutations and are associated with chemical modification of DNA and associated proteins (especially histones). Plants have been at the forefront of epigenetics research for several decades, and *Arabidopsis* is no exception. Like mammalian genomes, but unlike those of yeast, worms, and flies, plant genomes are heavily modified by cytosine methylation, which, along with histone modification, has epigenetic consequences for expression of both genes and repetitive elements found in the genome. These modifications are guided by a variety of factors, including RNA interference, resulting in the phenomenon of RNA-dependent DNA methylation, first discovered in plants, and RNA-dependent histone modification, which also occurs in fission yeast and other organisms.

When silencing of a given gene differs between male and female germ lines, imprinting results in expression from (usually) the maternally inherited allele. Imprinted expression is prevalent in the extraembryonic endosperm tissue, reminiscent of imprinting in the mammalian placenta. In these well-studied examples, demethylation of imprinted genes occurs in the central cell, resulting in maternal expression in the endosperm.

Epigenetic effects are often influenced by the environment, and in a dramatic example, plants remember the cold of winter by flowering in the following spring. This memory is induced by cold, retained by clonally propagated cells, but erased by meiosis, resulting in the familiar flowering habit of crops like winter wheat. In *Arabidopsis*, this process (vernalization) is regulated by RNA processing and histone modification, and involves the polycomb complex, also involved in cellular memory in animals.

Plants Respond to the Environment

Unlike animals, plants are rooted to the spot and cannot flee environmental assault, resulting in properties not usually found in animals—such as grazing tolerance. The innate immune system, first molecularly characterized as the "gene-for-gene" response in plants, includes many components conserved in animals, but it is highly diversified and can recognize viruses, microbes, worms, insects, and even other plants. In addition to this "biotic" stress, plants must withstand and respond to "abiotic" stress, including changes in light intensity, circadian rhythm, nutrient, and salt and water stress, to name but a few. Many of these environmental triggers have profound effects on development—for example, by inducing or delaying flowering to optimize seed production.

Light plays a central role in plant biology, because of the photosynthetic chloroplast, which is derived from an ancient symbiotic prokaryote and responsible for most of the organic carbon fixed in the biosphere. Even in photosynthetic research, *Arabidopsis* is replacing classical physiological models—such as tobacco and spinach—because of the ease of genetic and genomic manipulation.

Development and Pattern Formation

Plant development has influenced crop domestication and breeding, and therefore human history, more than any other aspect of plant biology, with dramatic innovations affecting inflorescence architecture, seed shattering, and leaf shape, selected by ancient farmers and sophisticated breeders alike. Cauliflower, popcorn, and kale each differ by only a handful of genes from progenitor species that would only be recognized as weeds to modern-day farmers. Because flowering plants are a recent evolutionary group, many of the genes responsible have since been identified using *Arabidopsis* as a model.

More generally, plants and animals diverged from a common but unicellular ancestor, so that multicellular development evolved independently in each kingdom. Therefore we see that essential general principles, such as the central importance of transcription factors and signaling hierarchies (peptides, hormones, and receptors), are recognized and present in each kingdom, whereas specific molecules are only rarely conserved. Some mechanisms, such as cell cycle and MAP (mitogen-activated protein) kinase cascades, are very familiar, but most are distinct. For example, homeotic and heterochronic identities are specified by transcription factors and miRNA in both lineages, but the molecules are not conserved, involving mostly MADS (MCM1, agamous, deficiens, and serum response) transcription factors in plants and *Hox* genes in animals. Intercellular communication involves hormones in both kingdoms, but these have only general similarities (with the exception, perhaps, of plant and animal steroids). Indeed, the highly connected supracellular vascular system of plants allows macromolecules, such as mRNA, small RNA, and transcription factors themselves, to pass directly between cells, whereas this phenomenon has only rarely been observed in animals.

Common developmental mechanisms, then, are likely to have had a function in unicellular or oligocellular ancestors, and may have been co-opted to serve similar functions independently. Perhaps conserved epigenetic mechanisms, such as the Polycomb system, served functions in genome organization, genome defense, chromosome biology, and cellular differentiation, rather than multicellular transcriptional memory, in the ancestral unicellular eukaryote. *Arabidopsis* is playing a major role in identifying these conserved functions within and between kingdoms.

THE NEMATODE WORM, *CAENORHABDITIS ELEGANS*

Brenner, after making seminal contributions in molecular genetics, identified a small metazoan in which to study the important questions of development and the molecular basis of behavior. Learning from the success of molecular genetic studies in phage and bacteria, he wanted the simplest possible organism that had differentiated cell types, but that was also amenable to microbiological-like genetics. In 1965 he settled on the small nematode worm *C. elegans* because it contained a variety of suitable characteristics. These include a rapid generation time to enable genetic screens, hermaphrodite reproduction producing hundreds of "self-progeny" so that large numbers of animals could be generated, sexual reproduction so that genetic stocks could be constructed by mating, and a small number of transparent cells so that development could be followed directly.

Brenner set two ambitious initial goals that would be essential for the long-term success of this endeavor. One was a complete mapping of all cells by reconstructing serial section electron micrographs (completed by John White in 1986), and the other was the mapping of the cell lineage (completed by John E. Sulston in 1983). Seven years later Brenner established the genetics of the new model organism with the isolation of more than 300 morphological and behavioral mutants. These defined moe than 100 complementation groups mapping to six linkage groups. Nearly 30 years later there are 400 laboratories worldwide that study *C. elegans*. Because of its simplicity and experimental accessibility, it is now one of the most completely understood of all metazoans.

C. elegans Has a Very Rapid Life Cycle

C. elegans is cultured on petri dishes and fed a simple diet of bacteria. They grow well at a range of temperatures, growing twice as fast at 25°C than at 15°C. At 25°C fertilized embryos complete development in 12 hours and hatch into free-living animals capable of complex behaviors. The hatchling worm passes through four juvenile or larval stages (L1–L4) over the course of 40 hours to become a sexually mature adult (Fig. 22-14).

The adult hermaphrodite can produce up to 300 self-progeny over the course of about 4 days or can be mated with rare males to produce up to 1000 hybrid progeny. The adult lives for about 15 days. Under

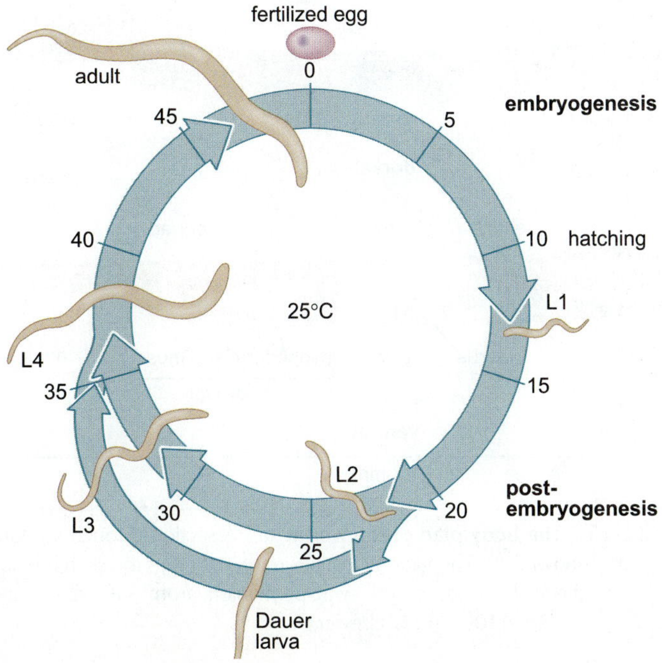

FIGURE 22-14 The life cycle of the worm, *C. elegans*. Shown is the life cycle in hours of development, from first-stage juvenile to adult, as described in the text. The alternative developmental stage—a dauer—is also shown.

stressful conditions (low food, increased temperatures, or high population density), the L1 stage animal can enter an alternative developmental pathway leading to what is called a **dauer.** Dauers are resistant to environmental stresses and can live many months while waiting for environmental conditions to improve. The study of mutants that fail to enter the dauer stage, or that enter it inappropriately, have identified genes expressed in specific neurons that function to sense environmental conditions, genes expressed throughout the animal that control body growth, and genes that control life span. Activation of these latter genes in the adult can dramatically extend the life span of the animal and homologs of these genes have been implicated in life extension in mammals.

C. elegans Is Composed of Relatively Few, Well-Studied Cell Lineages

C. elegans has a simple body plan (Fig. 22-15). The prominent organ in the adult hermaphrodite is the gonad, which contains the proliferating and differentiating germ cells (sperm and oocytes), fertilization chamber (spermatheca), and uterus for temporary storage of young embryos. The embryos pass from the uterus to the outside through the vulva, a structure formed from 22 epidermal cells. Mutations that disrupt the formation of the vulva do not interfere with production of embryos but do prevent the eggs from being laid. Consequently, the embryos develop and hatch inside the uterus. The hatched worms then devour their mother and become trapped inside her skin (cuticle layer) forming a "bag of worms." This readily identified phenotype has allowed the isolation of hundreds of vulva-less mutants identifying scores of genes that function to control the generation, specification, and differentiation of the vulval cells. Among these genes are components of a highly conserved receptor tyrosine kinase signaling pathway that controls cell proliferation.

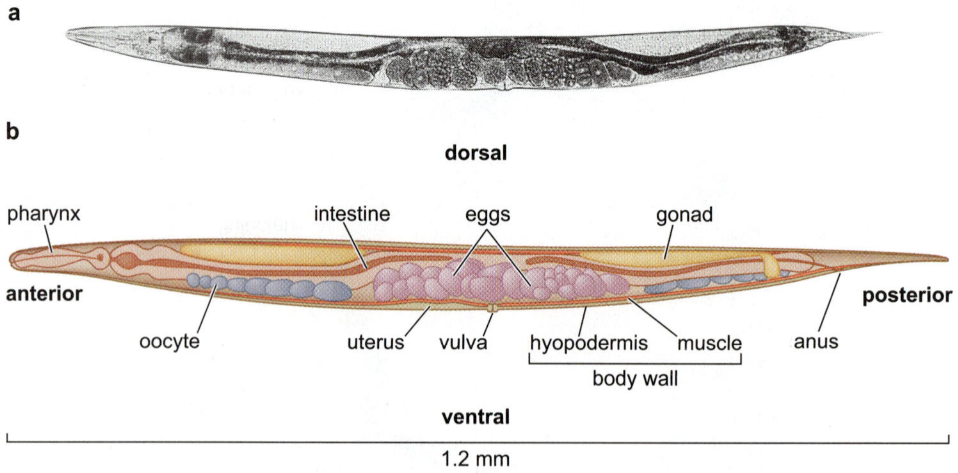

FIGURE 22-15 The body plan of the worm. (a) A section through an adult hermaphrodite worm is shown. (b) The various organs are identified in the sketch below (b) and are described in the text. (a, Reprinted, with permission, from Sulston J.E. and Horvitz H.R. 1977. *Dev. Biol.* 56: 110–156. © Elsevier.)

Many of the mammalian homologs of these genes are oncogenes and tumor-supressor genes that when altered can lead to cancer. In *C. elegans*, mutations that inactivate this pathway eliminate vulva development because the vulval cells are never generated, whereas mutations that activate this pathway cause overproliferation of the vulva precursor cells, resulting in a multiple vulva phenotype. Because the animal is transparent and the vulva is generated from only 22 cells, it is possible to describe the mutant defect with cellular resolution such that the type of mutation can be associated with a specific cellular transformation.

The Cell Death Pathway Was Discovered in *C. elegans*

The most notable achievement to date in *C. elegans* research has been the elucidation of the molecular pathway that regulates apoptosis or cell death. Early analysis of cell lineages noted that the same set of cells died in every animal, suggesting that cell death was under genetic control. The first cell death defective (*ced*) mutants isolated were defective for the consumption of the cell corpse by neighboring cells; thus in the mutants cell corpses persisted for many hours. Using these *ced* mutants, H. Robert Horvitz and his colleagues isolated many additional *ced* mutants that failed to produce persistent cell corpses. These mutants proved to be defective at initiating the cell death program. Analysis of the *ced* mutants showed that, in all but one case, developmentally programmed cell death is cell autonomous—that is, the cell commits suicide. In males, a cell known as the linker cell is killed by its neighbor. The molecular identification of the *ced* genes provided the means to identify proteins in mammals that carry out essentially the identical biochemical reactions to control cell death in all animals; in fact, expressing human homologs in *C. elegans* can substitute for a mutated *ced* gene. Cell death is as important as cell proliferation in development and disease and is the focus of intense research to develop therapeutics for the control of cancer and neurodegenerative diseases.

RNAi Was Discovered in *C. elegans*

In 1998 a remarkable discovery was announced. The introduction of double-stranded RNA (dsRNA) into *C. elegans* silenced the gene homologous to the dsRNA. This unexpected discovery and subsequent analysis of RNA interference (RNAi) is significant in two respects. One is that RNAi appears to be universal because introduction of dsRNA into nearly all animal, fungal, or plant cells leads to homology-directed mRNA degradation. Indeed, much of what we know about RNAi comes from studies in plants (Chapter 18). The second was the rapidity with which experimental investigation of this mysterious process revealed the molecular mechanisms (see Fig. 18-5). These investigations intersected with the analysis of another RNA-mediated gene regulatory process that involves tiny endogenous miRNAs that have been shown to regulate gene expression in plants and animals, coordinate genome rearrangements in ciliates, and regulate chromatin structure in yeast. The first two miRNAs were discovered in genetic screens in *C. elegans*. A fraction of these worm miRNAs is conserved in flies and mammals, where their functions are just beginning to be revealed. Recent studies suggest that the human genome may contain something like 1000 miRNA genes.

THE FRUIT FLY, *DROSOPHILA MELANOGASTER*

We are approaching the 100th anniversary of the fruit fly as a model organism for studies in genetics and developmental biology. In 1908 Thomas Hunt Morgan and his research associates at Columbia University placed rotting fruit on the window ledge of their laboratory in Schermerhorn Hall. Their goal was to isolate a small, quickly reproducing animal that could be cultured in the lab and used to study the inheritance of quantitative traits, such as eye color. Among the menagerie of creatures that were captured, the fruit fly emerged as the animal of choice. Adults produced large numbers of progeny in just 2 weeks. Culturing was done in recycled milk bottles using an inexpensive concoction of yeast and agar.

Drosophila Has a Rapid Life Cycle

The salient features of the *Drosophila* life cycle are a very rapid period of embryogenesis, followed by three periods of larval growth prior to metamorphosis (Fig. 22-16). Embryogenesis is completed within 24 hours after fertilization and culminates in the hatching of a first-instar larva. As we discussed in Chapter 19, the early periods of *Drosophila* embryonic development exhibit the most rapid nuclear cleavages known for any animal. A first-instar larva grows for 24 hours and then molts into a larger, second-instar larva. The process is repeated to yield a third-instar larva that feeds and grows for 2–3 days.

One of the key processes that occurs during larval development is the growth of the imaginal disks, which arise from invaginations of the epidermis in mid-stage embryos (Fig. 22-17). There is a pair of disks for every set of appendages (e.g., a set of foreleg imaginal disks and a set of wing imaginal disks). There are also imaginal disks for eyes, antennae, the mouthparts, and genitalia. Disks are initially small and composed of fewer than 100 cells in the embryo but contain tens of thousands of cells in mature larvae. The development of the wing imaginal disk has become an important model system for understand-

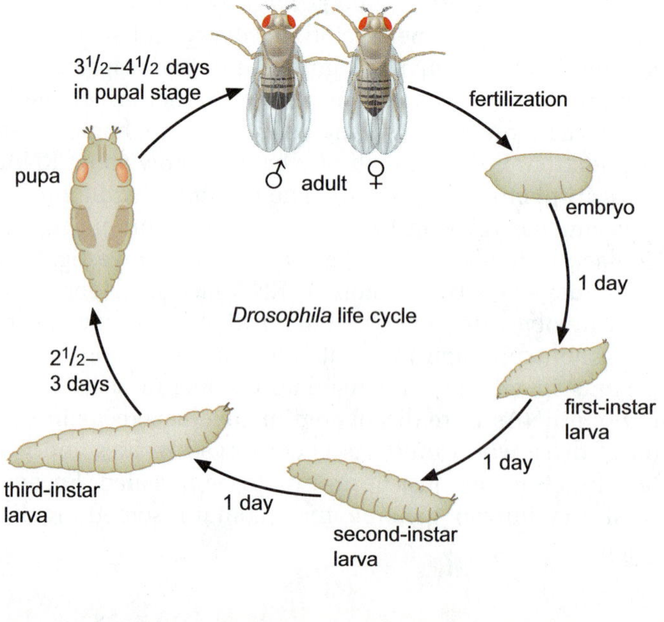

FIGURE 22-16 The *Drosophila* life cycle. The various stages of development of the fly, shown here, are described in the text.

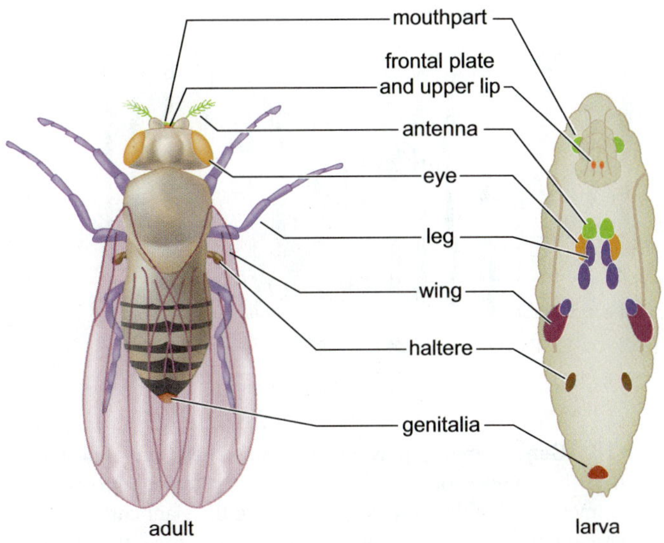

FIGURE 22-17 **Imaginal disks in *Drosophila.*** The position of various imaginal disks in the larva are shown on the right. On the left are shown the limbs and the organs they form in the adult fly. These disks are initially formed as small groups of cells in the embryo but have grown to tens of thousands of cells in the mature larva. These disks develop into their respective adult structures during pupation.

ing how gradients of secreted signaling molecules such as Hedgehog and Dpp (TGF-β) control complex patterning processes. Imaginal disks differentiate into their appropriate adult structures during metamorphosis (or pupation).

The First Genome Maps Were Produced in *Drosophila*

In 1910 the Morgan lab identified a spontaneous mutant male fly that had white eyes rather than the brilliant red seen for normal strains. This single fly launched an incisive series of genetic studies that led to two major discoveries: genes are located on chromosomes, and each gene is composed of two alleles that assort independently during meiosis (see Mendel's first law; Chapter 1). The identification of additional mutations led to the demonstration that genes located on separate chromosomes segregate independently (Mendel's second law), whereas those linked on the same chromosome do not.

An undergraduate at Columbia University, Alfred H. Sturtevant (a member of the Morgan lab), developed a simple mathematical algorithm for mapping the distances between linked genes based on recombination frequencies. By the 1930s, extensive genetic maps were produced that identified the relative positions of numerous genes controlling a variety of physical characteristics of the adult, such as wing size and shape and eye color and shape.

Hermann J. Muller, another scientist trained in the Morgan fly lab, provided the first evidence that environmental factors, such as ionizing radiation, can cause chromosome rearrangements and genetic mutations. Large-scale "genetic screens" are routinely performed by feeding adult males a mutagen, such as EMS, and then mating them with normal females. The F_1 progeny are heterozygous and contain one normal chromosome and one random mutation. A variety of methods are used to study these mutations, as described below.

In addition to its remarkable fecundity (a single female can produce thousands of eggs) and rapid life cycle, the fruit fly was found to possess several very useful features that guaranteed it a sustained and prominent role in experimental research. It contains only four chromosomes: two large autosomes, chromosomes 2 and 3, a smaller

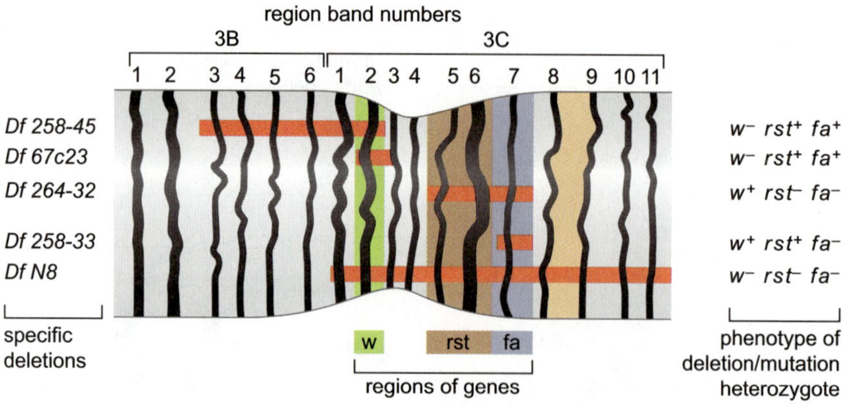

FIGURE 22-18 Genetic maps, polytene chromosomes, and deficiency mapping. Endoreplication in the absence of cytokinesis generates enlarged chromosomes in some tissues of the fly, most notably the salivary glands where the giant chromosomes are composed of a thousand chromatids. It was possible, for the first time, to correlate the occurrence of genes for certain traits with specific physical segments of chromosomes. For example, white eye flies were correlated with deletions in the 3C region of the *X* chromosome. (With permission from Hartwell L. et al. 2004. *Genetics: From genes to genomes,* 2nd ed., p. 816, Fig. D-4. © McGraw–Hill.)

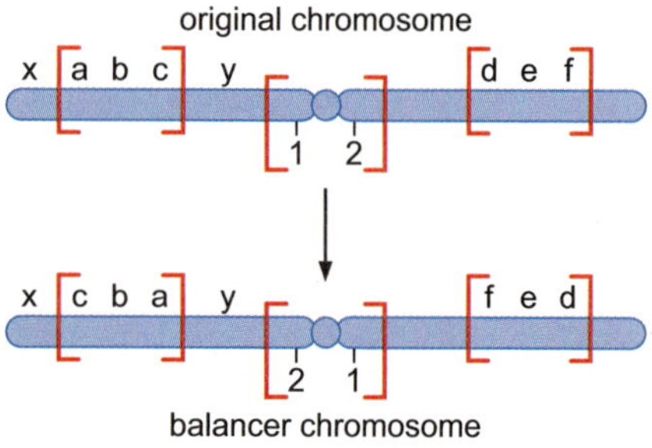

FIGURE 22-19 Balancer chromosome. Balancer chromosomes (bottom panel) contain a series of inversions when compared with the original, parental chromosome (top panel). In this diagram, a hypothetical chromosome has two arms. The left arm of the balancer chromosome has an internal inversion that reverses the order of genes a, b, and c in the original chromosome. Similarly, the arm on the right of the balancer chromosome has an inversion that reverses the order of genes d, e, and f. In addition, there might be an inversion centered around the centromere, in this case reversing the order of genes 1 and 2. The balancer chromosome thus has a significantly different order of genes when compared with the original. As a result, there is a suppression of recombination between the chromosomes in heterozygotes containing one copy of each.

X chromosome (which determines sex), and a very small fourth chromosome. Calvin B. Bridges—yet another of Muller's colleagues—discovered that certain tissues in *Drosophila* larvae undergo extensive endoreplication without cell division. In the salivary gland, this process produces remarkable giant chromosomes composed of approximately 1000 copies of each chromatid. Bridges used these **polytene chromosomes** to determine a physical map of the *Drosophila* genome (the first produced for any organism) (Fig. 22-18).

Bridges identified a total of approximately 5000 "bands" on the four chromosomes and established a correlation between many of these bands and the locations of genetic loci identified in the classical recombination maps. For example, female fruit flies that are heterozygous for the recessive *white* mutation exhibit normal red eyes. However, similar females that contain the *white* mutation and a small deletion in the other *X* chromosome, which removes polytene bands 3C2–3C3, exhibit white eyes. This is because there is no longer a normal, dominant copy of the gene. This type of analysis led to the conclusion that the white gene is located somewhere between polytene bands 3C2 and 3C3 on the *X* chromosome.

A variety of additional genetic methods were created to establish the fruit fly as the premiere model organism for studies in animal inheritance. For example, **balancer chromosomes** were created that contain a series of inversions relative to the organization of the native chromosome (Fig. 22-19). Critically, such balancers fail to undergo recombination with the native chromosome during meiosis. As a result, it is possible to maintain permanent cultures of fruit flies that contain recessive, lethal mutations. Consider a null mutation in the *even-skipped* (*eve*) gene, which we discussed in Chapter 19. Embryos that are homozygous for this mutation die and fail to produce viable larvae and adults. The *eve* locus maps on chromosome 2 (at polytene band 46C). The null mutation can be maintained in a population that is heterozygous for a "normal" chromosome containing the null allele of *eve* and a balancer second chromosome, which contains a normal copy of the gene. Because the *eve* null allele is strictly recessive, these flies are completely viable. However, only heterozygotes are observed among adult progeny in successive generations. Embryos that contain two copies of the balancer chromosome die because some of the inversions produce recessive disruptions in critical genes. In addition, embryos that contain two copies of the normal chromosome die because they are homozgyous for the *eve* null mutation.

Genetic Mosaics Permit the Analysis of Lethal Genes in Adult Flies

Mosaics are animals that contain small patches of mutant tissue in a generally "normal" genetic background. Such small patches do not kill the individual because most of the tissues in the organism are normal. For example, small patches of *engrailed/engrailed* homozygous mutant tissue can be produced by inducing mitotic recombination in developing larvae using X-rays. When such patches are created in posterior regions of the developing wings, the resulting flies exhibit abnormal wings that have duplicated anterior structures in place of the normal posterior structures. The analysis of genetic mosaics provided the first evidence that Engrailed is required for subdividing the appendages and segments of flies into anterior and posterior compartments.

The most spectacular genetic mosaics are gynandromorphs (Fig. 22-20). These are flies that are literally half male and half female. Sexual identity in flies is determined by the number of *X* chromosomes. Individuals with two *X* chromosomes are females, whereas those with just one *X* are males. (The *Y* chromosome does not define sexual identity in flies as it does in mice and humans: in flies, *Y* is only needed for the production of sperm.) Rarely, one of the two *X* chromosomes is lost at the first mitotic division following the fusion of the sperm and egg pronuclei in a newly fertilized *XX* embryo.

This *X* instability occurs only at the first division. In all subsequent divisions, nuclei containing two *X* chromosomes give rise to daughter nuclei with two *X* chromosomes, whereas nuclei with just one *X* chromosome give rise to daughters containing a single *X*. As we discussed in Chapter 19, these nuclei undergo rapid cleavages without cell membranes and then migrate to the periphery of the egg. This migration is coherent and there is little or no intermixing of nuclei containing one *X* chromosome with nuclei containing two *X* chromosomes. Thus, half the embryo is male and half is female, although the "line" separating the male and female tissues is random. Its exact position depends on the orientation of the two daughter nuclei after the first cleavage. The line sometimes bisects the adult into a left half that is female and a right half that is male. Suppose that one of the *X* chromosomes contains the recessive white allele. If the wild-type *X* chromosome is lost at the first division, then the right half of the fly, the male half, has white eyes (the male half has only the mutant *X* chromosome), whereas the left half (the female side) has red eyes. (Remember that the female half has two *X* chromosomes and that one contains the dominant, wild-type allele.)

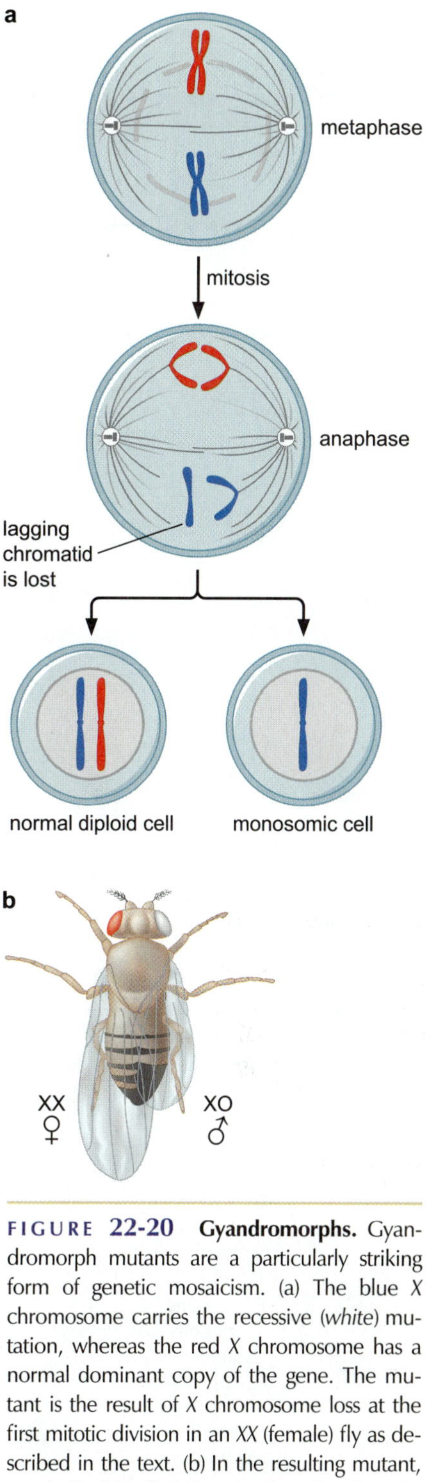

FIGURE 22-20 Gyandromorphs. Gyandromorph mutants are a particularly striking form of genetic mosaicism. (a) The blue *X* chromosome carries the recessive (*white*) mutation, whereas the red *X* chromosome has a normal dominant copy of the gene. The mutant is the result of *X* chromosome loss at the first mitotic division in an *XX* (female) fly as described in the text. (b) In the resulting mutant, one half of the fly is female, the other is male.

The Yeast FLP Recombinase Permits the Efficient Production of Genetic Mosaics

What was not anticipated during the classical era of genetic analysis is the fact that *Drosophila* possesses several favorable attributes for molecular studies and whole-genome analysis. Most notably, the genome is

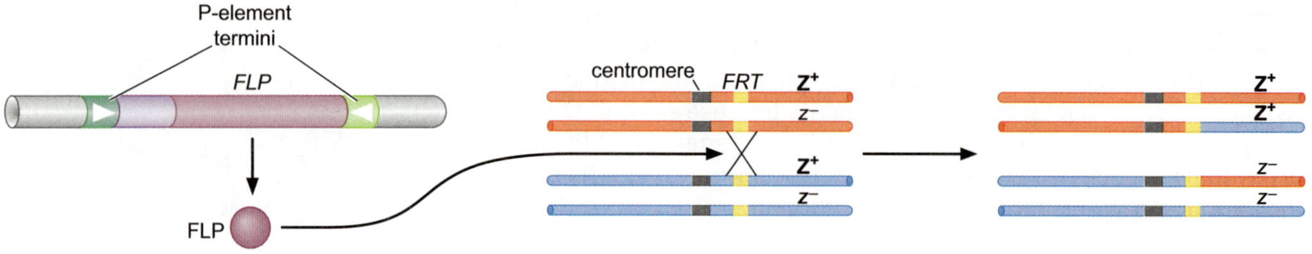

FIGURE 22-21 **FLP-FRT.** The use of this site-specific recombination system from yeast (described in Chapter 11) promotes high levels of mitotic recombination in flies. The recombination is controlled by expressing the recombinase in flies only when required.

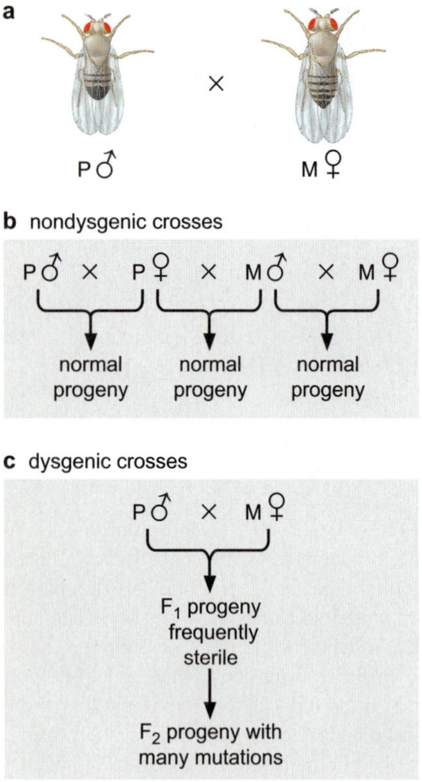

FIGURE 22-22 **Hybrid dysgenesis.** P-element transposons reside passively in P strains because they express a repressor that keeps the transposons silent. When P strains are mated with an M strain lacking such a repressor, the transposons are mobilized within the pole cells and often integrate into genes required for germ cell formation. This explains the high frequency of sterility in the offspring from such crosses.

relatively small. It is composed of only approximately 150 Mb and contains fewer than 14,000 protein coding genes. This represents just 5% of the amount of DNA that makes up the mouse and human genomes. As the fruit fly entered the modern era, several methods were established that improved some of the older techniques of genetic manipulation and also led to completely new experimental methods, such as the production of stable transgenic strains carrying recombinant DNAs.

As we discussed earlier, genetic mosaics are produced by mitotic recombination in somatic tissues. Initially, X-rays were used to induce recombination, although this method is inefficient and produces small patches of mutant tissue. More recently, the frequency of mitotic recombination was greatly enhanced by the use of the FLP recombinase from yeast (Fig. 22-21). FLP recognizes a simple sequence motif, FRT, and then catalyzes DNA rearrangement (see Chapter 11). FRT sequences were inserted near the centromere of each of the four chromosomes using P-element transformation (see below). Heterozygous flies are then produced that contain a null allele in gene *Z* on one chromosome and a wild-type copy of that gene on the homologous chromosome. Both chromosomes contain the FRT sequences. These flies are stable and viable as there is no endogenous FLP recombinase in *Drosophila*. It is, however, possible to introduce the recombinase in transgenic strains that contain the yeast FLP protein-coding sequence under the control of the heat-inducible hsp70 promoter. Upon heat shock, FLP is synthesized in all cells. FLP binds to the FRT motifs in the two homologs containing gene *Z* and catalyze mitotic recombination (Fig. 22-21). This method is quite efficient. In fact, short pulses of heat shock are often sufficient to produce enough FLP recombinase to produce large patches of z^-/z^- tissue in different regions of an adult fly. FRT recognition sequences have been inserted throughout the *Drosophila* genome via P-transformation. It is now possible to create small deletions for just about any gene by inducing rearrangements between FRT sites flanking the gene of interest using the FLP recombinase.

It Is Easy to Create Transgenic Fruit Flies that Carry Foreign DNA

P-elements are transposable DNA segments that are the causal agent of a genetic phenomenon called **hybrid dysgenesis** (Fig. 22-22; see also Box 19-4). Consider the consequences of mating females from the "M" strain of *D. melanogaster* with males from the "P" strain (same species, but different populations). The F_1 progeny are often sterile. The reason is that the P strain contains numerous copies of the P-element trans-

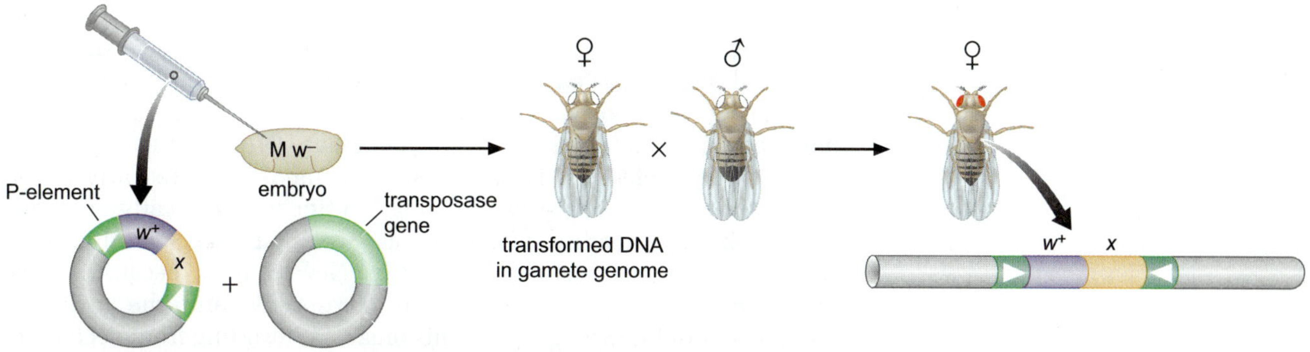

FIGURE 22-23 P-element transformation. P-elements can be used as vectors in the transformation of fly embryos. Thus, as discussed in the text, sequences of choice can be inserted into a modified P-element. A single copy of this recombinant molecule is stably incorporated into a single location of a fly chromosome.

poson that are mobilized in embryos derived from M eggs. These eggs lack a repressor protein that inhibits P-element mobilization. P-element excision and insertion is limited to the pole cells, the progenitors of the gametes (sperm in males and eggs in females). Sometimes the P-elements insert into genes that are essential for the development of these germ cells, and, as a result, the adult flies derived from these matings are sterile.

P-elements are used as transformation vectors to introduce recombinant DNAs into otherwise normal strains of flies (Fig. 22-23). A full-length P-element transposon is 3 kb in length. It contains inverted repeats at the termini that are essential for excision and insertion. The intervening DNA encodes both a repressor of transposition and a transposase that promotes mobilization. The repressor is expressed in the developing eggs of P strains. As a result, there is no movement of P-elements in embryos derived from females of the P strain (these contain P-elements). Movement is seen only in embryos derived from eggs produced by M-strain females, which lack P-elements.

Recombinant DNA is inserted into defective P-elements that lack the internal genes encoding repressor and transposase. This DNA is injected into posterior regions of early, precellular embryos (as we saw in Chapter 19, this is the region that contains the polar granules). The transposase is injected along with the recombinant P-element vector. As the cleavage nuclei enter posterior regions, they acquire both the polar granules and recombinant P-element DNA together with transposase. The pole cells bud off from the polar plasm and the recombinant P-elements insert into random positions in the pole cells. Different pole cells contain different P-element insertion events. The amount of recombinant P-element DNA and transposase is calibrated so that, on average, a given pole cell receives just a single integrated P-element. The embryos are allowed to develop into adults and then mated with appropriate tester strains.

The recombinant P-element contains a "marker" gene such as *white*$^+$ and the strain used for the injections is a *white*$^-$ mutant. The tester strains are also *white*$^-$, so that any F_2 fly that has red eyes must contain a copy of the recombinant P-element. This method of P-element transformation is routinely used to identify regulatory sequences such as those governing *eve* stripe 2 expression (which we discussed in Chapter 19). In addition, this strategy is used to examine protein-coding genes in various genetic backgrounds.

In summary, *Drosophila* offers many of the sophisticated tools of classical and molecular genetics that, as we have seen, are available in microbial model systems. One conspicuous exception has been the absence of methods for precise manipulation of the genome by homologous recombination with recombinant DNA, such as in the creation of gene deletions. However, such methods were recently developed and are now being streamlined for routine use. Ironically, such manipulations are readily available, as we shall see, in the more complicated model system, the mouse. Nevertheless, because of the wealth of genetic tools available in *Drosophila* and the extensive groundwork of knowledge about this organism resulting from decades of investigation, the fruit fly remains one of the premier model systems for studies of development and behavior.

THE HOUSE MOUSE, *MUS MUSCULUS*

By the standards of *C. elegans* and *Drosophila*, the life cycle of the mouse is slow and cumbersome. Embryonic development, or gestation, occurs over a period of 3 weeks and the newborn mouse does not reach puberty for another 5–6 weeks. Thus, the effective life cycle is roughly 8–9 weeks, more than five times longer than that of *Drosophila*. The mouse, however, enjoys a special status because of its exalted position on the evolutionary tree: it is a mammal and, therefore, related to humans. Of course, chimpanzees and other higher primates are closer to humans than mice, but they are not amenable to the various experimental manipulations available in mice.

Thus, the mouse provides the link between the basic principles, discovered in simpler creatures like worms and flies, and human disease. For example, the *patched* gene of *Drosophila* encodes a critical component of the Hedgehog receptor (Chapter 19). Mutant fly embryos that lack the wild-type *patched* gene activity exhibit a variety of patterning defects. The orthologous genes in mice are also important in development. Unexpectedly, however, certain *patched* mutants cause various cancers, such as skin cancer, in both mice and humans. No amount of analysis in the fly would reveal such a function. In addition, methods have been developed that permit the efficient removal of specific genes in otherwise normal mice. This "knockout" technology continues to have an enormous impact on our understanding of the basic mechanisms underlying human development, behavior, and disease. We briefly review the salient features of the mouse as an experimental system.

The chromosome complement of the mouse is similar to that seen in humans: there are 19 autosomomes in mice (22 in humans), as well as X and Y sex chromosomes. There is extensive synteny between mice and humans: extended regions of a given mouse chromosome contain the same set of genes (in the same order) as the "homologous" regions of the corresponding human chromosomes. The mouse genome has been sequenced and assembled. As discussed in Chapter 20, the mouse has virtually the same complement of genes as those present in the human genome: each contains approximately 25,000 genes, and there is a one-to-one correspondence for more than 85% of these genes. Most, if not all, of the differences between the mouse and human genomes stem from the selective duplication of certain gene families in one lineage or the other. Comparative genome analysis confirms what we have known for some time: the mouse is an excellent model for human development and disease.

Mouse Embryonic Development Depends on Stem Cells

Mouse eggs are small and difficult to manipulate. Like human eggs, they are just 100 microns in diameter. Their small size prohibits grafting experiments of the sort done in zebrafish and frogs, but microinjection methods have been developed for introducing recombinant DNA into the mouse germ line so as to create transgenic strains, as discussed below. In addition, it is possible to harvest enough mouse embryos, even at the earliest stages, for in situ hybridization assays and the visualization of specific gene expression patterns. Such visualization methods can be applied to both normal embryos and mutants carrying disruptions in defined genetic loci.

Figure 22-24 shows an overview of mouse embryogenesis. The initial divisions of the early mouse embryo are very slow and occur with an average frequency of just once every 12–24 hours. The first obvious diversification of cell types is seen at the 16-cell stage, called the **morula** (Fig. 22-24, panel 6). The cells located in outer regions form tissues that do not contribute to the embryo but instead develop into the placenta. Cells located in internal regions generate the inner cell mass (ICM). At the 64-cell stage, there are only 13 ICM cells, but these form all of the tissues of the adult mouse. The ICM is the prime source of embryonic stem cells, which can be cultured and induced to form any adult cell type upon addition of the appropriate growth factors. Human stem cells have become the subject of considerable social controversy, but offer the promise of providing a renewable source of tissues that can be used to replace defective cells in a variety of degenerative diseases such as diabetes and Alzheimer's.

At the 64-cell stage (about 3–4 days after fertilization) the mouse embryo, now called a **blastocyst,** is finally ready for implantation. Interactions between the blastocyst and uterine wall lead to the formation of the placenta, a characteristic of all mammals except the primitive egg-laying platypus. After formation of the placenta, the embryo enters gastrulation, whereby the ICM forms all three germ layers: endoderm, mesoderm, and ectoderm. Shortly thereafter, a fetus emerges that contains a brain, a spinal cord, and internal organs such as the heart and liver.

The first stage in mouse gastrulation is the subdivision of the ICM into two cell layers: an inner hypoblast and an outer epiblast, which form the endoderm and ectoderm, respectively. A groove called the **primitive streak** forms along the length of the epiblast and the cells that migrate into the groove form the internal mesoderm. The anterior end of the primitive streak is called the **node;** it is the source of a variety of signaling molecules that are used to pattern the anterior–posterior axis of the embryo, including two secreted inhibitors of TGF-β signaling, Chordin and Noggin. Double mutant mouse embryos that lack both genes develop into fetuses that lack head structures such as the forebrain and nose.

It Is Easy to Introduce Foreign DNA into the Mouse Embryo

Microinjection methods have been developed for the efficient expression of recombinant DNA in transgenic strains of mice. DNA is injected into the egg pronucleus, and the embryos are placed into the oviduct of a female mouse and allowed to implant and develop. The

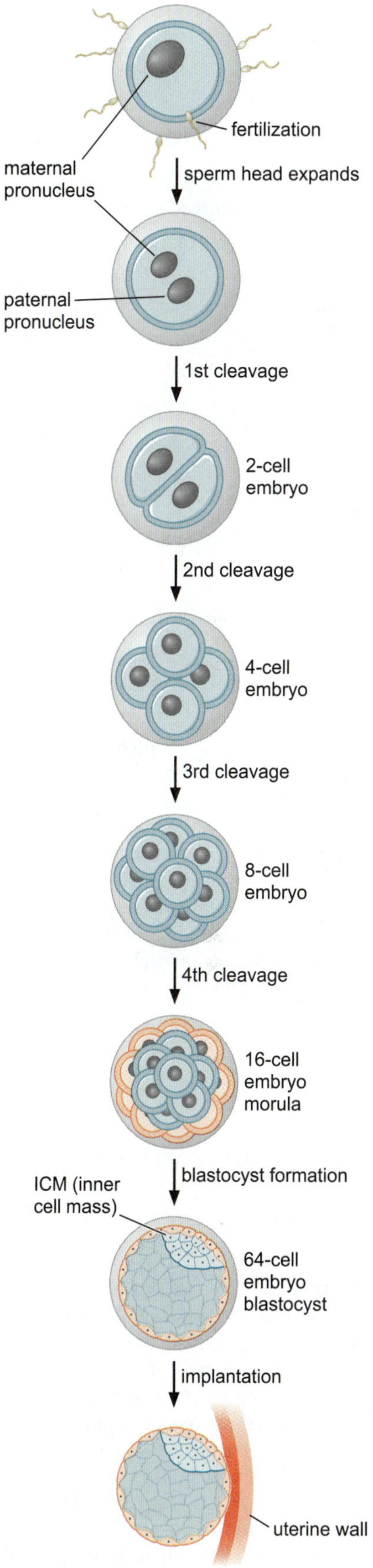

FIGURE 22-24 Overview of mouse embryogenesis.

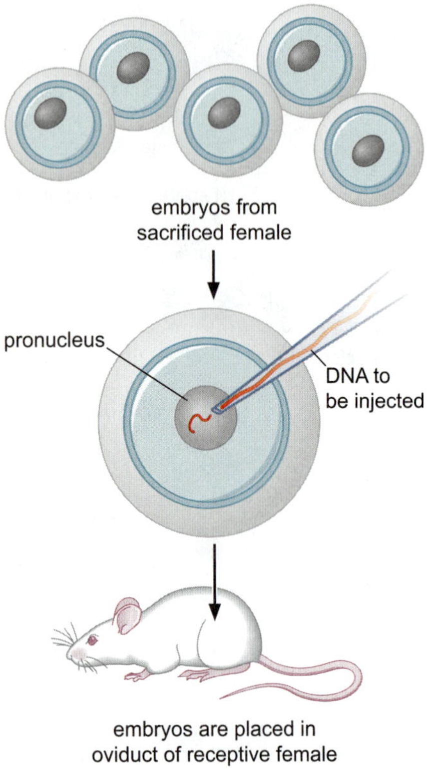

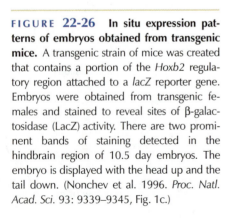

FIGURE 22-25 Creation of transgenic mice by microinjection of DNA into the egg pronucleus. One-cell embryos are obtained from a newly mated female mouse. Recombinant DNA is injected into the nucleus, and the embryo is then implanted into the oviduct of a surrogate. After several days, the embryo implants and ultimately forms a fetus that contains integrated copies of the recombinant DNA.

injected DNA integrates at random positions in the genome (Fig. 22-25). The efficiency of integration is quite high and usually occurs during early stages of development, often in one-cell embryos. As a result, the fusion gene inserts into most or all of the cells in the embryo, including the ICM cells that form the somatic tissues and germline of the adult mouse. Approximately 50% of the transgenic mice that are produced using this simple method of microinjection exhibit **germline transformation;** that is, their offspring also contain the foreign recombinant DNA.

Consider as an example a fusion gene containing the enhancer from the *Hoxb2* gene attached to a *lacZ* reporter gene. Embryos and fetuses can be harvested from transgenic strains carrying this reporter and stained to reveal the pattern of *lacZ* expression. In this case, staining is observed in the hindbrain (Fig. 22-26). Transgenic mice have been used to characterize several regulatory sequences, including those that regulate the β-globin genes and *HoxD* genes. Both complex loci contain long-range regulatory elements (the LCR and GCR, respectively) that coordinate the expression of the different genes over distances of several hundred kilobases (see Chapter 17).

Homologous Recombination Permits the Selective Ablation of Individual Genes

The single most powerful method of mouse transgenesis is the ability to disrupt, or "knock out," single genetic loci. This permits the creation of mouse models for human disease. For example, the *p53* gene encodes a regulatory protein that activates the expression of genes required for DNA repair. It has been implicated in a variety of human cancers. When *p53* function is lost, cancer cells become highly invasive because of rapid accumulation of DNA mutations. A strain of mice has been established that is completely normal except for the removal of the *p53*

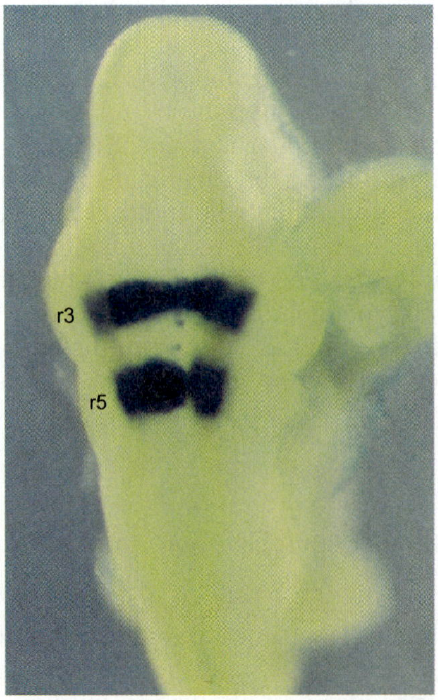

FIGURE 22-26 In situ expression patterns of embryos obtained from transgenic mice. A transgenic strain of mice was created that contains a portion of the *Hoxb2* regulatory region attached to a *lacZ* reporter gene. Embryos were obtained from transgenic females and stained to reveal sites of β-galactosidase (LacZ) activity. There are two prominent bands of staining detected in the hindbrain region of 10.5 day embryos. The embryo is displayed with the head up and the tail down. (Nonchev et al. 1996. *Proc. Natl. Acad. Sci.* 93: 9339–9345, Fig. 1c.)

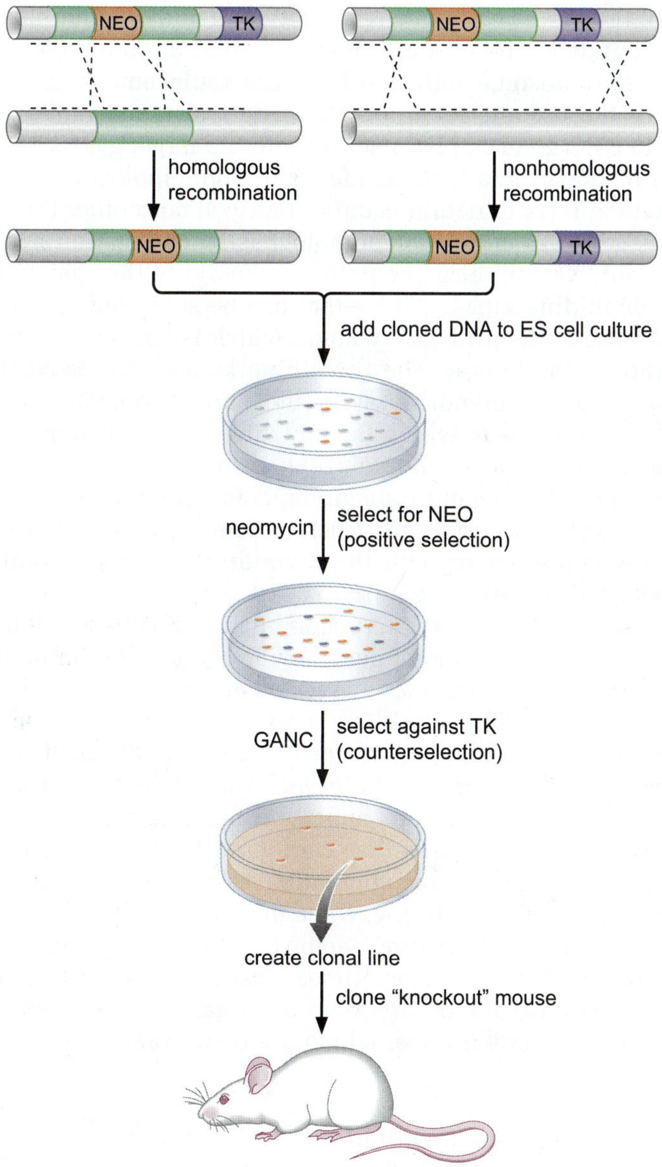

FIGURE 22-27 Gene knockout via homologous recombination. The figure outlines the method used to create a cell line lacking any given gene. Homologous recombination that occurs within a target gene (shown in green) results in the incorporation of NEO and disruption of that gene. Nonhomologous, or random, recombination can result in the incorporation of the disrupted gene containing NEO, and the gene encoding thymidine kinase (TK). Clones carrying both constructs survive exposure to neomycin, but the clones also carrying TK are subsequently counterselected by growth in gancyclovir (GANC). Clones containing the NEO insertion via homologous recombination are the only survivors. Once produced, these cells can be cloned and used to generate a complete mouse lacking that same gene (see Fig. 22-25).

gene. These mice, which are highly susceptible to cancer, die young. There is the hope that these mice can be used to test potential drugs and anticancer agents for use in humans. Although *Drosophila* contains a *p53* gene, and mutants have been isolated, it does not provide the same opportunity for drug discovery as does the mouse model.

Gene disruption experiments are done with embryonic stem (ES) cells (Fig. 22-27). These cells are obtained by culturing mouse blastocysts so that ICM cells proliferate without differentiating. A recombinant DNA is created that contains a mutant form of the gene of interest. For example, the protein coding region of a given target gene is modified by deleting a small region near the beginning of the gene that removes codons for essential amino acids from the encoded protein and causes a frameshift in the remaining coding sequence. The modified form of the target gene is linked to a drug-resistance gene, such as NEO, which confers resistance to neomycin. Only those ES cells that contain the transgene are able to grow in medium containing the antibiotic. The NEO gene is placed downstream of the modified target gene, but upstream of a flanking region of homology with the chromosome such that double recombination with the chromosome will result in the replacement of the target gene with the mutant gene and the drug resistance gene. (Alternatively, the NEO gene can be inserted into the target gene.)

There is, however, a high incidence of nonhomologous recombination in which recombination occurs illicitly at sites other than the endogenous gene. To enrich for homologous recombination events, the recombinant vector also contains a marker—the gene for the enzyme thymidine kinase (TK)—that can be subjected to counter selection by use of the drug gancyclovir, which is converted into a toxic compound by the kinase. The thymidine kinase gene is carried outside the region of homology with the chromosome in the vector. Hence, transformants in which the mutant gene has been incorporated into the chromosome by homologous recombination will shed the thymidine kinase gene, but transformants in which incorporation into the chromosome occurred by illicit recombination will frequently contain the entire vector with the thymidine kinase gene and hence can be selected against.

As a result of this procedure, recombinant ES cells are obtained in which one copy of the target gene corresponds to the mutant allele. These recombinant ES cells are harvested and injected into the ICM of normal blastocysts. The hybrid embryos are inserted into the oviduct of a host mouse and allowed to develop to term. Some of the adults that arise from the hybrid embryos possess a transformed germ line and therefore produce haploid gametes containing the mutant form of the target gene. The ES cells that were used for the original transformation and homologous recombination assays give rise to both somatic tissues and the germ line. Once mice are produced that contain transformed germ cells, matings among siblings are performed to obtain homozygous mutants. Sometimes these mutants must be analyzed as embryos because of lethality. With other genes, the mutant embryos develop into full-grown mice, which are then examined using a variety of techniques.

Mice Exhibit Epigenetic Inheritance

Studies on manipulated mouse embryos led to the discovery of a very peculiar mechanism of non-Mendelian, or epigenetic, inheritance. This phenomenon is known as **parental imprinting** (Fig. 22-28). The basic idea is that only one of the two alleles for certain genes is active. This is because the other copy is selectively inactivated either in the developing sperm cell or the developing egg. Consider the case of the *Igf2* gene. It encodes an insulin-like growth factor that is expressed in the gut and liver of developing fetuses. Only the *Igf2* allele inherited from the father is actively expressed in the embryo. The other copy, although perfectly normal in sequence, is inactive. The differential activities of the maternal and paternal copies of the *Igf2* gene arise from the methylation of an associated silencer DNA that represses *Igf2* expression. During spermiogenesis, the DNA is methylated, and as a result, the *Igf2* gene can be activated in the developing fetus. The methylation inactivates the silencer. In contrast, the silencer DNA is not methylated in the developing oocyte. Hence, the *Igf2* allele inherited from the female is silent. In other words, the paternal copy of the gene is "imprinted"—in this case, methylated—for future expression in the embryo. This specific example is discussed in greater detail in Chapter 17.

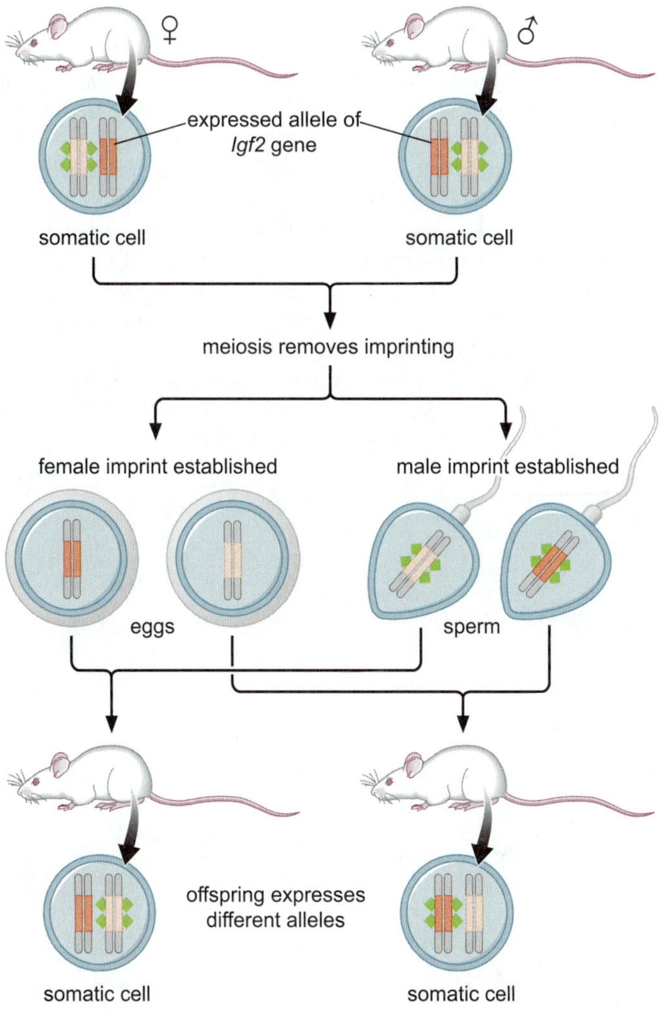

FIGURE 22-28 Imprinting in the mouse. The permanent silencing of one allele of a given gene in a mouse. As outlined in the text, and described in detail in Chapter 17, imprinting ensures that only one copy of the mouse *Igf2* gene is expressed in each cell. It is always the copy carried on the paternal chromosome that is expressed.

There are approximately 30 imprinted genes in mice and humans. Many of the genes, including the preceding example of *Igf2*, control the growth of the developing fetus. It has been suggested that imprinting has evolved to protect the mother from her own fetus. The Igf2 protein promotes the growth of the fetus. The mother attempts to limit this growth by inactivating the maternal copy of the gene.

We have considered how every organism must maintain and duplicate its DNA to survive, adapt, and propagate. The overall strategies for achieving these basic biological goals are similar in the vast majority of organisms and, therefore, may be examined rather successfully using simple organisms. It is, however, clear that the more intricate processes found in higher organisms, such as differentiation and development, require more complicated systems for regulating gene expression and that these can be studied only in more complex organisms. We have seen that a wide range of powerful experimental techniques can be used with success to manipulate the mouse and to explore various complex biological problems. As a result, the mouse has served as an excellent model system for studying developmental, genetic, and biochemical processes that are likely to occur in more highly evolved mammals. The recent publication and annotation of the mouse genome has underscored the importance of the mouse as a model for further exploring and understanding problems in human development and disease.

BIBLIOGRAPHY

Burke D., Dawson D., and Stearns T. 2000. *Methods in yeast genetics.* Cold Spring Harbor Laboratory Press, Cold Spring Harbor, New York.

Hartwell L.H., Hood L., Goldberg M.L., Reynolds A.E., Silver L.S., and Veres R.C. 2004. *Genetics: From genes to genomes*, 2nd ed. McGraw–Hill, New York.

Miller J.H. 1972. *Experiments in molecular genetics.* Cold Spring Harbor Laboratory Press, Cold Spring Harbor, New York.

Nagy A., Gertsenstein M., Vintersten K., and Behringer R. 2003. *Manipulating the mouse embryo*, 3rd ed. Cold Spring Harbor Laboratory Press, Cold Spring Harbor, New York.

Sambrook J. and Russell D.W. 2001. *Molecular cloning: A laboratory manual*, 3rd ed. Cold Spring Harbor Laboratory Press, Cold Spring Harbor, New York.

Snustad D.P. and Simmons M.J. 2002. *Principles of genetics*, 3rd ed. John Wiley and Sons, New York.

Stent G.S. and Calendar R. 1978. *Molecular genetics: An introductory narrative.* W.H. Freeman and Co., San Francisco.

Sullivan W., Ashburner M., and Hawley R.S. 2000. Drosophila *protocols.* Cold Spring Harbor Laboratory Press, Cold Spring Harbor, New York.

Wolpert L., Beddington R., Lawrence P., Meyerowitz E., Smith J., and Jessell T.M. 2002. *Principles of development*, 2nd ed. Oxford University Press, Oxford.

Index

A

Page references followed by f denote figures; those followed by t denote tables.

B

C

D

F

G

H

I

J

K

L

N

Q

R

U

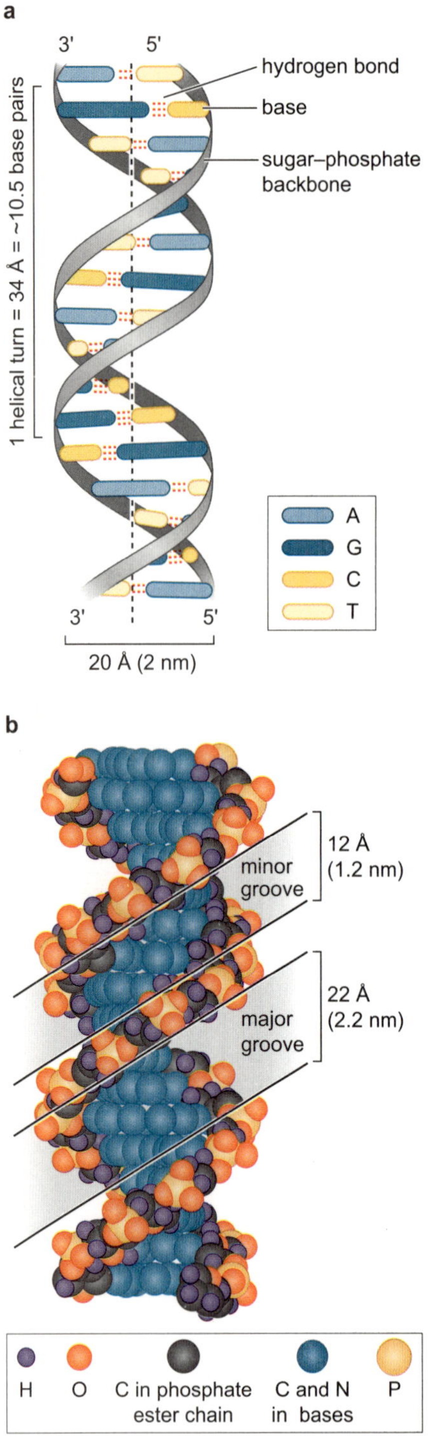

(a) Schematic model of the double helix.
(b) Space-filling model of the double helix.